# NATO ADVANCED STUDY INSTITUTES SERIES

*Proceedings of the Advanced Study Institute Programme, which aims at the dissemination of advanced knowledge and the formation of contacts among scientists from different countries*

The series is published by an international board of publishers in conjunction with NATO Scientific Affairs Division

| | | |
|---|---|---|
| A | Life Sciences | Plenum Publishing Corporation |
| B | Physics | London and New York |
| C | Mathematical and Physical Sciences | D. Reidel Publishing Company Dordrecht, Boston and London |
| D | Behavioural and Social Sciences | Sijthoff & Noordhoff International Publishers |
| E | Applied Sciences | Alphen aan den Rijn and Germantown U.S.A. |

Series C – Mathematical and Physical Sciences

*Volume 83 - Ordered Sets*

Ordered Sets

# Ordered Sets

*Proceedings of the NATO Advanced Study Institute held at Banff, Canada, August 28 to September 12, 1981*

edited by

IVAN RIVAL

*Department of Mathematics and Statistics, The University of Calgary, Canada*

D. Reidel Publishing Company

Dordrecht : Holland / Boston : U.S.A. / London : England

Published in cooperation with NATO Scientific Affairs Division

**Library of Congress Cataloging in Publication Data**

NATO Advanced Study Institute (1981: Banff, Alta.)
Ordered sets.

(NATO Advanced Study Institutes Series. Series C, mathematical and physical sciences, v. 83)
Bibliography: p.
1. Ordered sets–Congresses. I. Rival, Ivan, 1947–
II. Title. III. Series.
QA171.48.N27 1981 511.3′2 82-544
ISBN 90-277-1396-0 AACR2

---



Published by D. Reidel Publishing Company
P.O. Box 17, 3300 AA Dordrecht, Holland

Sold and distributed in the U.S.A. and Canada
by Kluwer Boston Inc.,
190 Old Derby Street, Hingham, MA 02043, U.S.A.

In all other countries, sold and distributed
by Kluwer Academic Publishers Group,
P.O. Box 322, 3300 AH Dordrecht, Holland

D. Reidel Publishing Company is a member of the Kluwer Group

Printed in The Netherlands

# CONTENTS

TOP ROW (LEFT TO RIGHT) M. Erné, Mrs. D. Ray-Chaudhuri, D. K. Ray-Chaudhuri, F. East, J. Koslowski, Mrs. L. Lea, J. Lea, F. Farmer, B. Jónsson, S. Devitt, W. R. Pulleyblank, R. K. Guy, J. K. Lenstra, R. J. Nowakowski, R. W. Quackenbush, N. Sauer, R. Jamison-Waldner, O. Pretzel, Q. Stout, K. Leeb, D. Scott, S. Goberstein, Master R. C. Rival, D. Duffus, B. Sands, G. McNulty, R. Freese, J. E. Baumgartner, D. J. Miller, R. E. Woodrow, T. J. Schneider, W. Poguntke, W. T. Trotter, K. Bogart, R. P. Stanley, H. A. Kierstead, J. Constantin, M. Aigner, K. Głazek, K. M. Koh, Cl. Flament, R. Fraïssé, Mlle J. Fraïssé, W. Luxemburg, E. Harzheim, E. Graczynska-Wronska, G. Steiner, S. Todorčevic.

SECOND ROW (LEFT TO RIGHT) Mrs. E. Crapo, H. Crapo, Mrs. K. Greene, C. Greene, G. Kalmbach, J. P. Barthélemy, P. Edelman, I. Broere, Mrs. E. Hoffman, A. Hoffman, M. Jambu-Giraudet, A. Szendrei, I. Rival, Master D. E. Rival, Mrs. H. Rival, J. B. Miller, R. Wille, G. N. Robertson, M. K. Bennett, Z. Marcusz, G. Birkoff, J. Larson, P. Erdös, P. M. Winkler, E. C. Milner, J. Borwein, D. B. West, Mrs. M. Dilworth, R. P. Dilworth, H. Bauer, F. Yaqub, G. Bruns, M. Robinson, Mme E. Corominas, D. Schweigert, I. Rosenberg, Mme C. Fraïssé, Mme L. Benzaken, Mrs. J. Kelly, D. Kelly, Mrs. T. Luxemburg.

FRONT ROW, KNEELING (LEFT TO RIGHT) A. Stralka, K. Keimel, H. Priestley, J. Walker, H. J. Bandelt, B. A. Davey, K. A. Baker, Mrs. G. Baker, R. L. Graham, Miss N. Borwein, Mrs. J. Borwein, B. Monjardet, A. Garsia, A. Björner, R. Bonnet, M. Pouzet, D. Daykin, Cl. Benzaken, O. Cogis, E. Corominas.

# PREFACE

This volume contains all twenty-three of the principal survey papers presented at the Symposium on Ordered Sets held at Banff, Canada from August 28 to September 12, 1981. The Symposium was supported by grants from the NATO Advanced Study Institute programme, the Natural Sciences and Engineering Research Council of Canada, the Canadian Mathematical Society Summer Research Institute programme, and the University of Calgary. We are very grateful to these Organizations for their considerable interest and support.

Over forty years ago on April 15, 1938 the first Symposium on Lattice Theory was held in Charlottesville, U.S.A. in conjunction with a meeting of the American Mathematical Society. The principal addresses on that occasion were *Lattices and their applications* by G. Birkhoff, *On the application of structure theory to groups* by O. Ore, and *The representation of Boolean algebras* by M. H. Stone. The texts of these addresses and three others by R. Baer, H. M. MacNeille, and K. Menger appear in the Bulletin of the American Mathematical Society, Volume 44, 1938.

In those days the theory of ordered sets, and especially lattice theory was described as a "vigorous and promising younger brother of group theory." Some early workers hoped that lattice-theoretic methods would lead to solutions of important problems in group theory.

Some twenty years later in April of 1959 the Symposium on Partially Ordered Sets and Lattice Theory was held in Monterey, U.S.A. in conjunction with a meeting of the American Mathematical Society. The papers at that meeting were divided into four sections: *lattice structure theory* (C. C. Chang, R. A. Dean, R. P. Dilworth, J. Hartmanis, A. Horn, P. M. Whitman); *complemented modular lattices* (K. D. Fryer, I. Halperin, B. Jónsson, J. E. McLaughlin); *Boolean algebras* (C. C. Chang, P. Dwinger, P. R. Halmos, L. Henkin, R. S. Pierce, A. Tarski); *applications of lattice theory* (F. W. Anderson, L. W. Anderson, G. Birkhoff, M. Hall, Jr.). The full texts appear in the Proceedings of Symposia in Pure Mathematics, Volume II, American Mathematical Society, 1961.



*I. Rival (ed.), Ordered Sets, ix–x.*

The hopes for lattice theory had changed in the twenty years since that first Symposium in 1938. It was now to answer questions concerned with lattices and ordered sets themselves. Many of the problems identified by the early 1960s as the leading problems have now been completely solved. For instance, the embedding of finite lattices into finite partition lattices and the word problem for free modular lattices.

More than twenty years have now elapsed since that Monterey meeting. The emphasis today has shifted considerably. The leading edge problems today are both more purely order-theoretic and more concrete in character than ever before. The subject has a wider scope due to the important links that have been forged especially with modern combinatorial theory. With the escape from the algebraic influence traditional affinities, for example, with set theory have also come to the surface. This association of new and old friends is the basis for the current vitality of the theory of ordered sets.

It was the aim of this Symposium and of this volume to document this vitality through expository lectures and articles. The collective record of these twenty-three articles provides an introduction to all current topics in the theory of ordered sets and its applications. Researchers will welcome the accounts of the ten problem sessions held on different evenings of the Symposium -- almost two hundred open and unsolved problems in all, neatly classified and briefly motivated. And the volume ends with an edited reference list of the theory of ordered sets.

A meeting of this size and of this duration must owe much to many. The scientific programme committee consisted of R. P. Dilworth, R. Wille, and myself. My discussions with the Director of the Scientific Affairs Division of NATO, M. di Lullo were very helpful in designing the format of the scientific sessions. Many of the participants themselves played a special role in the preparations for the Symposium -- refereeing manuscripts, designing the scientific and social programmes, even transporting participants to the Symposium site. I am very grateful to D. Duffus, D. Kelly, and R. E. Woodrow. One of my colleagues, Bill Sands, worked tirelessly to make this Symposium a success. His influence and contributions have been felt at all stages of this production. From refereeing to proofreading, very little has escaped his devoted attention. I am especially grateful to my wife Hetje who has allowed the science and the society of the mathematical community to be a part of our family life.

Calgary, Canada, December 20, 1981 Ivan Rival

## PARTICIPANTS

M. Aigner (Germany)
K. Baker (U.S.A.)
H.J. Bandelt (Germany)
J.P. Barthélemy (France)
H. Bauer (Germany)
J.E. Baumgartner (U.S.A.)
M.K. Bennett (U.S.A.)
C. Benzaken (France)
G. Birkhoff (U.S.A.)
A. Björner (Sweden)
K.P. Bogart (U.S.A.)
R. Bonnet (France)
J.M. Borwein (U.S.A.)
I. Broere (South Africa)
G. Bruns (Canada)
O. Cogis (France)
J. Constantin (Canada)
E. Corominas (France)
H. Crapo (Canada)
B.A. Davey (Australia)
D.E. Daykin (England)
J.S. Devitt (Canada)
R.P. Dilworth (U.S.A.)
D. Duffus (U.S.A.)
F.R. East (England)
P.H. Edelman (U.S.A.)
P. Erdös (Hungary)
M. Erné (Germany)
F.D. Farmer (U.S.A.)
Cl. Flament (France)
R.J. Fraïssé (France)
R.S. Freese (U.S.A.)
E. Fried (Hungary)
A. Garsia (U.S.A.)
J.N. Ginsburg (Canada)
K. Głazek (Poland)
H.H. Glover (U.S.A.)
S.M. Goberstein (U.S.A.)
C. Godsil (Austria)
E.W. Graczyńska-Wrońska
R.L. Graham (U.S.A.)
C. Greene (U.S.A.)
R.K. Guy (Canada)
P.L. Hammer (Canada)
L.H. Harper (U.S.A.)
E. Harzheim (Germany)
A.J. Hoffman (U.S.A.)
M. Jambu-Giraudet (France)
R.E. Jamison-Waldner (U.S.A.)
B. Jónsson (U.S.A.)
G. Kalmbach (Germany)
K. Keimel (Germany)
D. Kelly (Canada)
H.A. Kierstead (U.S.A.)
M. Klamkin (Canada)
K.M. Koh (Singapore)

J. Koslowski (Germany)
J.A. Larson (U.S.A.)
E.L. Lawler (U.S.A.)
J. Lawson (U.S.A.)
J.W. Lea (U.S.A.)
J.K. Lenstra (Netherlands)
H.F.J. Lowig (Canada)
W.A.J. Luxemburg (U.S.A.)
G.F. McNulty (U.S.A.)
D.J. Miller (Canada)
J.B. Miller (Australia)
E.C. Milner (Canada)
M.W. Mislove (U.S.A.)
B. Monjardet (France)
R.J. Nowakowski (Canada)
W. Poguntke (Germany)
M. Pouzet (France)
O. Pretzel (England)
H.A. Priestley (England)
W.R. Pulleyblank (Canada)
R.W. Quackenbush (Canada)
D.K. Ray-Chaudhuri (U.S.A.)
I. Rival (Canada)
G.N. Robertson (U.S.A.)
I.G. Rosenberg (Canada)
J.G. Rosenstein (U.S.A.)
B. Sands (Canada)
N.W. Sauer (Canada)
T.J. Schneider (U.S.A.)
D. Schweigert (Germany)
D.S. Scott (England)
R.P. Stanley (U.S.A.)
G. Steiner (Canada)
Q.F. Stout (U.S.A.)
A.R. Stralka (U.S.A.)
A. Szendrei (Hungary)
S. Todorčevic (Yugoslavia)
W.T. Trotter, Jr. (U.S.A.)
J.W. Walker (U.S.A.)
W. Weiss (Canada)
D.B. West (U.S.A.)
R. Wille (Germany)
P.M. Winkler (U.S.A.)
R.E. Woodrow (Canada)
F.M. Yaqub (Lebanon)

# SCIENTIFIC PROGRAMME

Saturday, August 29, 1981.

Chair: E. Corominas

J.E. Baumgartner, *Order types of real numbers and other uncountable orderings. (I)*

E.C. Milner, *The cofinality of partially ordered sets.*

Chair: P. Erdös

A.J. Hoffman, *Ordered sets and linear programming. (I)*

D.B. West, *Extremal problems in partially ordered sets.*

Sunday, August 30, 1981.

Chair: W.A.J. Luxemburg

J.E. Baumgartner, *Order types of real numbers and other uncountable orderings. (II)*

A.J. Hoffman, *Ordered sets and linear programming. (II)*

Problem Session 1. Chair: E.C. Milner, *Order types.*

Monday, August 31, 1981.

Chair: Cl. Benzaken

D.S. Scott, *Some ordered sets in computer science. (I)*

C. Greene, *The Möbius function of a partially ordered set.*

Problem Session 2. Chair: P.L. Hammer, *Combinatorics.*

Tuesday, September 1, 1981.

Chair: D.K. Ray-Chaudhuri

D.S. Scott, *Some ordered sets in computer science. (II)*

W.T. Trotter, Jr., *Order dimension (I): introduction, history and applications.*

D. Kelly, *Order dimension (II): recent results.*

Chair: R.K. Guy

R.L. Graham, *Linear extensions of partial orders and the FKG inequality.*

Problem Session 3. Chair: K.A. Baker, *Linear extensions of finite ordered sets.*

Wednesday, September 2, 1981.

Chair: M. Aigner

A. Björner, *Cohen-Macaulay ordered sets (I): lexicographic shellability and order homology.*

J.K. Lenstra, *Machine scheduling with precedence constraints. (I)*

Special Session: *Scheduling.*

Chair: J.K. Lenstra

E.L. Lawler, *Series-parallel sequencing.*

Special Session: *Medians and convexity in graphs.*

Chair: R.E. Jamison-Waldner

J.P. Barthélemy, *Medians and generalized medians in graphs.*

H.J. Bandelt, *Intervals in graphs and median graphs.*

R.E. Jamison-Waldner, *Antimatroids and chordal graphs.*

P.H. Edelman, *Antimatroids and meet-distributive lattices.*

R.J. Nowakowski, *Longest path convexity in graphs.*

Problem Session 4. Chair: R.L. Graham, *Scheduling and sorting.*

Thursday, September 3, 1981.

Chair: I.G. Rosenberg

A. Garsia, *Cohen-Macaulay ordered sets (II): the poset ring and Reisner's ring.*

J.K. Lenstra, *Machine scheduling with precedence constraints. (II)*

R.W. Quackenbush, *Enumeration in classes of ordered structures.*

Contributed Lecture Session (I).

Chair: N. Sauer

D.K. Ray-Chaudhuri, *Some recent results in geometries.*

G.N. Robertson, *The Wagner graph as an excluded minor.*

H. Crapo, *Concurrence geometries.*

Contributed Lecture Session (II).

Chair: E. Harzheim

M. Jambu-Giraudet, *Groups and lattices of automorphisms of 2-homogeneous chains.*

Q.F. Stout, *Searching countably infinite totally ordered sets.*

Problem Session 5. Chair: B. Sands, *Graphs and enumeration.*

<u>Friday, September 4, 1981</u>.

Chair: G. Bruns

R.P. Dilworth, *The role of order in lattice theory.*

R. Freese, *Some order-theoretic questions about free lattices and free modular lattices.*

Special Session: *Algebraic and topological aspects of ordered sets.*

Chair: L. Harper

J. Walker, *Ortholattices and Kneser's conjecture.*

F. Farmer, *Homotopy spheres and formal languages.*

P. Edelman, *The order polynomial of a partially ordered set.*

G. Kalmbach, *Ordered sets and homology.*

A talk of general interest by G. Birkhoff on *Perspectives on partially ordered sets.*

<u>Sunday, September 6, 1981</u>.

Chair: M.S. Klamkin

R. Wille, *Restructuring lattice theory: an approach based on hierarchies of concepts.*

Cl. Flament, *Uses of ordered sets in the social sciences.*

K. Bogart, *Distance and the amalgamation of preference.*

Problem Session 6. Chair: J.K. Lenstra, *Social science and operations research.*

Monday, September 7, 1981.

Chair: R. Fraïssé

G. Birkhoff, *Ordered sets in geometry.*

G.F. McNulty, *Infinite ordered sets: a recursive perspective.*

Special Session: *Recursion.*

Chair: G.F. McNulty

H. Kierstead, *Covering ordered sets with recursive chains and the theory of recursive dimension.*

J. Rosenstein, *Recursive linear orders.*

Special Session: *Well quasi-orders.*

Chair: G.N. Robertson

E. Corominas, *Different kinds of better quasi-orders.*

R. Fraïssé, *Classical results about barriers and better quasi-ordering relations.*

Problem Session 7. Chair: R.E. Woodrow, *Recursion and game theory.*

Tuesday, September 8, 1981.

Chair: K. Keimel

I. Rival, *The retract construction.*

M. Pouzet, *Linear extensions of ordered sets.*

Problem Session 8. Chair: I. Rival, *Order-preserving maps.*

Wednesday, September 9, 1981.

Chair: A.R. Stralka

R. Stanley, *Cohen-Macaulay ordered sets (III): applications to algebra, combinatorics and geometry.*

B.A. Davey, *Exponentiation and duality (I): the interplay between ordered sets and algebras.*

B. Jónsson, *Arithmetic of ordered sets. (I)*

Special Session: *Linear extensions.*

Chair: P. Winkler

R.P. Stanley, *Mixed volumes and linear extensions.*

P. Winkler, *More on the XYZ conjecture.*

Special Session: *Social sciences,*

Chair: K. Bogart

B. Monjardet, *Metrics on ordered sets and 'social distances'.*

J.P. Barthélemy, *Aggregation of preferences: axiomatic approaches.*

Special Session: *Order dimension.*

Chair: K. Baker

J.K. Lenstra, *The acyclic subgraph problem.*

B. Sands, *Finitely generated planar lattices of finite height.*

O. Pretzel, *Dimension and interval orders.*

O. Cogis, *The generalization of dimension to arbitrary relations.*

Problem Session 9. Chair: R. Wille, *Lattices.*

Thursday, September 10, 1981.

Chair: J.D. Lawson

D. Duffus, *Exponentiation and duality (II): the logarithmic property.*

M.W. Mislove, *An introduction to the theory of continuous lattices.*

B. Jónsson, *Arithmetic of ordered sets. (II).*

Special Session: *Continuous lattices and topologies on ordered sets.*

Chair: K. Keimel

J.D. Lawson, *Continuous posets and a lemma of M.E. Rudin.*

H.-J. Bandelt, *Complemented continuous lattices.*

A.R. Stralka, *The Frink ideal topology on distributive lattices.*

H.A. Preistley, *Representable posets and intrinsic topologies.*

M. Erné, *Relativisation of intrinsic topologies on posets to subsets.*

Problem Session 10. Chair: M. Pouzet, *Miscellaneous.*

Friday, September 11, 1981.

Contributed Lecture Session (III).

Chair: G. Kalmbach

M.K. Bennett, *Lattice-theoretical characterization of affine n-space.*

R. Jamison-Waldner, *Convexity invariants of ordered sets.*

P. Winkler, *Spanning trees determined by edge orderings.*

Contributed Lecture Session (IV).

Chair: R. Bonnet

M. Aigner, *Production of posets.*

K. Leeb, *The dual of Harzheim-Rado's theorem on chains.*

J.B. Miller, *Coups and alignments.*

Contributed Lecture Session (V).

Chair. F. Farmer

R. Nowakowski, *A game of cop and robber on graphs.*

K. Leeb, *Jungles, a generalization of forests yielding fast growing functions and large ordinals.*

K. Głazek, *Some characterization of distributive lattices in the class of semirings.*

H. Bauer, *On cancellation, refinement and automorphisms of exponents.*

Contributed Lecture Session (VI).

Chair: D. Daykin

I. Broere, *Dense subsets of partially ordered sets.*

D. Schweigert, *Algebraic description of ordered sets.*

H. Glover, *A partial order for generalized Kuratowski graphs.*

M. Erné, *How to extend lattice properties to partially ordered sets.*

# PART I

# STRUCTURE AND ARITHMETIC OF ORDERED SETS

# ARITHMETIC OF ORDERED SETS

Bjarni Jónsson*
Department of Mathematics
Vanderbilt University
Nashville, Tennessee 37235

ABSTRACT

The fundamental operations of ordinal and cardinal arithmetic have been extended to arbitrary binary relations and their isomorphism types. (Whitehead and Russell [1912], pp. 291 ff.) When applied to (partially) ordered sets, these operations yield new ordered sets. A unified treatment of these operations was initiated in Birkhoff [1937] and [1942], where it was shown that most of the basic laws of arithmetic apply in this general setting. See also Birkhoff [1967], pp. 55 ff.

Many of the deeper properties of ordered addition were listed without proofs in Lindenbaum and Tarski [1926], and these properties were developed axiomatically in Tarski [1956]. More specifically, the operations treated there were ordinal addition, as applied to pairs of types and to $\omega$-sequences of types, and conversion. The axioms for ordinal algebras hold for the isomorphism types of arbitrary binary relations, and most of the known properties of these operations are consequences of the axioms. However, some facts about ordinal addition cannot even be formulated within this framework. A notable example is Aronszajn's [1952] characterization of commuting pairs of isomorphism types.

The lexicographic product has an important role in the development of the arithmetic of ordinal numbers, and in the study of total order (Hausdorff [1914]). For finite reflexive relations (digraphs) this operation was investigated in Dörfler and Imrich [1972].

Every relation has a unique representation as the cardinal sum (disjoint union) of connected relations, and the study of cardinal addition therefore largely reduces to the study of cardinal numbers. The most important fact about direct (or



I. Rival (ed.), Ordered Sets, 3–41.

cardinal, or Cartesian) products is Hashimoto's [1951] result that any two representations of a connected ordered set as a direct product have isomorphic refinements. From this various cancellation and unique factorization results follow. Without the assumption of connectedness, the unique factorization property fails, even for finite ordered sets, but the cancellation property holds for arbitrary finite relations with a reflexive element (Lovász [1967]).

The (cardinal) power $A^B$, where $A$ and $B$ are ordered sets, is defined (Birkhoff [1937]), to be the set of all isotone maps from $B$ to $A$, with the order inherited from the direct power $A^{|B|}$. As Birkhoff noted, this operation satisfies the usual laws for exponents. More difficult problems arise when we consider conditions under which the cancellation laws for the bases and for exponents hold:

$$A^C \simeq A^D \text{ implies } C \simeq D \ ,$$

$$A^C \simeq B^C \text{ implies } A \simeq B \ ,$$

and conditions which insure that two representations

$$P \simeq A^C \text{ and } P \simeq B^D$$

have common refinements

$$A \simeq E^X,\ B \simeq E^Y,\ C \simeq YZ,\ D \simeq XZ \ .$$

These problems were investigated in Bergman, McKenzie and Nagý [a], Duffus [1978], Duffus, Jónsson and Rival [1978], Duffus and Rival [1978], Duffus and Wille [1979], Fuchs [1965], Jonsson and McKenzie [a], Novotný [1960], and Wille [1980].

## 1. INTRODUCTION

Although this survey is primarily concerned with (partially) ordered sets, most of the basic concepts can be applied to much more general structures, and many of the results are valid in a quite general setting. In order to keep the notation simple, the discussion will be confined to binary structures, or digraphs, i.e., to systems $A = (X,R)$ consisting of a set $X$ and a binary relation $R$ on $X$. For brevity we henceforth write simply "structure" for "binary structure".

Except in Section 2, where the basic operations are defined, we usually omit all mention of the universe (vertex set) $X$ and of the relation (edge set) $R$ of a structure $A = (X,R)$, referring to the members of $X$ as elements of $A$, and writing $x \to y$ for $xRy$ (or $(x,y) \in R$). When speaking specifically of ordered sets, we denote the relation by the customary symbol $\leq$, and the symbols $\geq$, $<$ and $>$ are then to be interpreted in the usual manner. A natural number $n$, which is of course assumed to coincide with the set $\{0,1,\ldots,n-1\}$, will be identified with the antichain $(n,=)$, while $\underline{n}$ will denote the chain $(n,\leq)$. The structures $1 = \underline{1}$ and $0 = \underline{0}$ (and their isomorphic copies) will be referred to, respectively, as the unit structure and the zero structure, or the null structure.

## 2. THE BASIC OPERATIONS

The cardinal sum and the ordinal sum of two structures

$$A = (X,R), \qquad B = (Y,S)$$

are defined to be, respectively, the structures

$$A + B = (X \cup Y, R \cup S), \qquad A \oplus B = (X \cup Y, R \cup S \cup (X \times Y)) .$$

Here it is assumed that the sets $X$ and $Y$ are disjoint. If this is not the case, we simply replace $A$ and $B$ by isomorphic copies whose universes are disjoint. In particular, it is in this sense that the sums $A + A$ and $A \oplus A$ must be interpreted. The direct product and the lexicographic product of $A$ and $B$ are defined to be, respectively, the structures

$$A \cdot B = (X \times Y, T), \qquad A \circ B = (X \times Y, U),$$

where, for $x,x' \in X$ and $y,y' \in Y$,

$$(x,y)T(x',y') \quad \text{iff} \quad xRx' \text{ and } ySy',$$

$$(x,y)U(x',y') \quad \text{iff} \quad \begin{array}{l} \text{either } x \neq x' \text{ and } xRx' \quad , \\ \text{or else } x = x' \text{ and } ySy' \quad . \end{array}$$

The operation of exponentiation is defined by letting

$$A^B = (Z,V) \quad ,$$

where $Z$ is the set of all maps $f$ from $Y$ to $X$ with the property that

$$f(y)Rf(y') \quad \text{whenever } ySy' \quad ,$$

and for all $f,g \in Z$,

$$fVg \quad \text{iff} \quad f(y)Rg(y) \quad \text{for all } y \in Y \quad .$$

The converse of $A$ is the structure $A^{\vee} = (X,R^{\vee})$, where $xR^{\vee}y$ iff $yRx$.

When applied to ordered sets, these operations obviously yield new ordered sets. In $A+B$, the order relation on each of the sets $X$ and $Y$ is preserved, while an element of $X$ is never comparable with an element of $Y$. In $A \oplus B$, the order relation on each of the sets $X$ and $Y$ is also preserved, but in this case each element of $X$ is less than every element of $Y$. Pictorially, the diagram for $A+B$ is obtained by placing the diagrams for $A$ and $B$ beside each other, but the diagram for $A \oplus B$ is obtained by placing the diagram for $B$ above the diagram for $A$ and connecting each maximal element of $A$ to each minimal element of $B$. The diagram for $A \circ B$ is obtained by replacing each point in the diagram for $A$ by a copy of the diagram for $B$, but the diagram for $A \cdot B$ is less easy to visualize. The power $A^B$ consists of all isotone maps from $B$ to $A$, with their pointwise ordering. The diagram for $A^{\vee}$ is simply the diagram for $A$ turned upside down. The six operations are illustrated in Fig. 1.

The cardinal sum and the direct product of arbitrarily many structures $A_k = (X_k,R_k)$, $k \in K$,

$$\sum_{k \in K} A_k, \qquad \prod_{k \in K} A_k,$$

are defined in an obvious manner. If all the structures $A_k$ are equal to the same structure $A$, then the cardinal sum and the direct product can be written as a scalar product and a direct power, $K \cdot A$ and $A^K$, respectively. To define the ordinal sum of the structures $A_k$ we require the index set $K$ to be totally ordered,

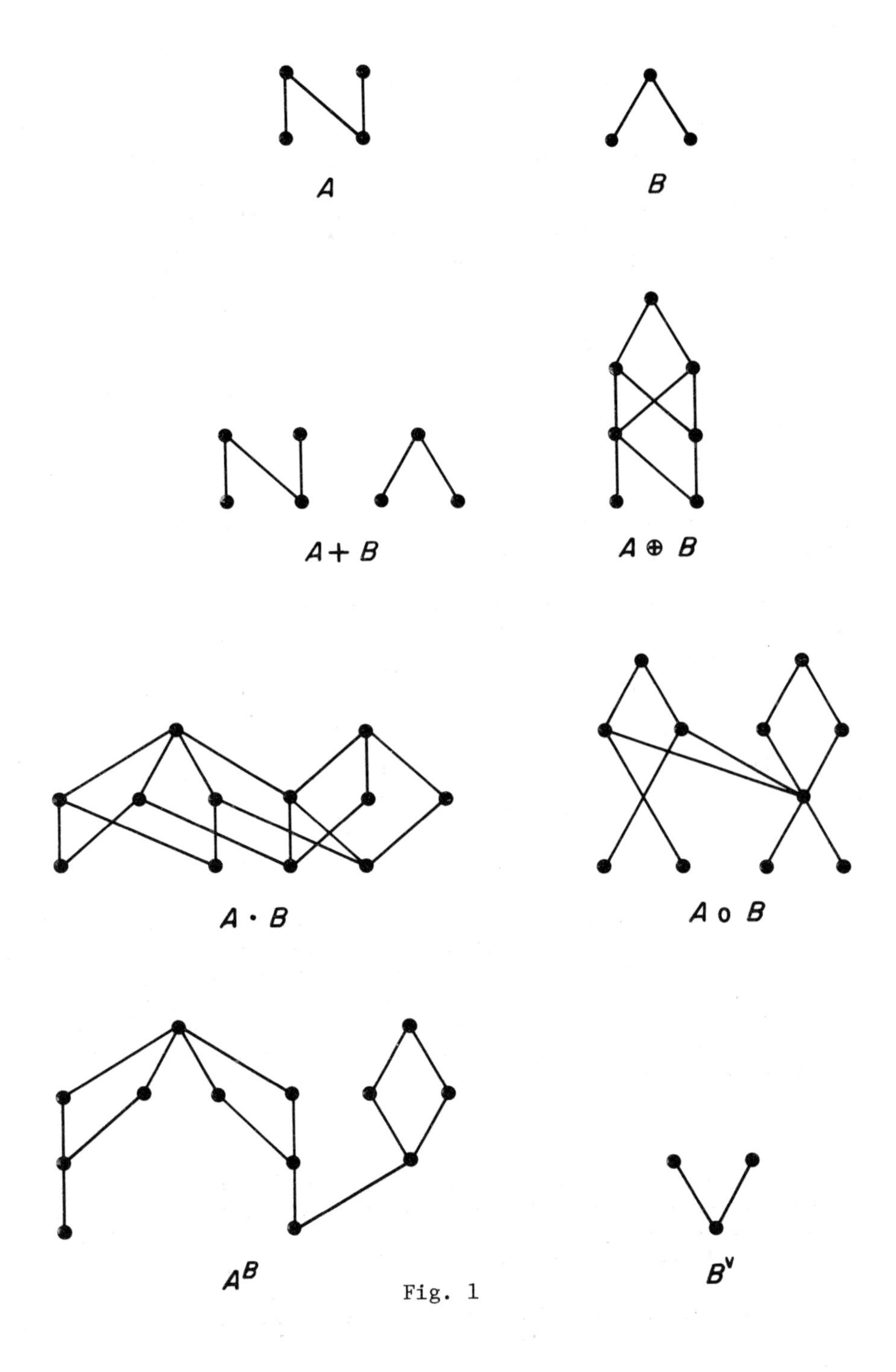


Fig. 1

and we then let

$$\bigoplus_{k \in K} A_k = (Y,S)$$

where $Y$ is the union of the sets $X_k$ and $S$ is the union of the relations $R_k$ and of the Cartesian products $X_k \times X_m$ with $k < m$. Lexicographic products with infinitely many factors will not be needed here. For finite sequences of structures $A_0, A_1, \ldots$, the lexicographic product is defined recursively by the formulas

$$\mathop{\circ}_{k<0} A_k = 1, \qquad \mathop{\circ}_{k<n+1} A_k = \left(\mathop{\circ}_{k<n} A_k\right) \circ A_n \quad .$$

In particular, the lexicographic powers of a structure $A$ are defined by the formulas

$${}^0A = 1, \qquad {}^{(n+1)}A = {}^nA \circ A \quad .$$

The exponent is placed in front of the base in order to distinguish this operation from the direct powers discussed earlier.

The cardinal sum and the ordinal sum have a common generalization: the sum over an arbitrary structure. Given a structure $A = (X,R)$ and structures $B_x = (Y_x, S_x)$, $x \in X$, we let

$$\sum_{x,A} B_x = (Z,T)$$

where $Z$ is the union of the sets $Y_x$ and, for $y \in Y_x$ and $y' \in Y_{x'}$,

$$yTy' \quad \text{iff either } x \neq x' \text{ and } xRx'$$
$$\text{or else } x = x' \text{ and } yS_xy' \quad .$$

It will be assumed that with each structure $A$ there has been associated an entity $\tau(A)$, called the isomorphism type of $A$, in such a way that $\tau(A) = \tau(B)$ iff $A \simeq B$. Since all the operations discussed here are invariants under isomorphism, they induce operations on the isomorphism types. These operations will be denoted by the same symbols as the corresponding operations on the structures. Thus

$$\tau(A) + \tau(B) = \tau(A+B) \quad ,$$

and similarly for the other operations. Many of the results discussed here can be formulated as theorems about these operations on isomorphism types, and we shall use that mode of expression whenever it is more convenient.

The operations of ordinal addition and lexicographic

multiplication were considered for ordered sets in Hausdorff [1906-1907] and [1914], and in Whitehead and Russell [1912] these operations were defined for arbitrary binary relations. In Birkhoff [1937] and [1942] a unified treatment of the various operations was initiated. The sum over an arbitrary structure was investigated in Jónsson [1956]. It is possible to define lexicographic products indexed by an arbitrary structure and lexicographic powers with an arbitrary structure as exponent. See, for example, Birkhoff's two papers and Schmidt [1955]. However, these concepts will not be needed here.

## 3. ELEMENTARY ARITHMETIC PROPERTIES

The cardinal addition is obviously both commutative and associative, and it has the null structure as a neutral element. Furthermore, every structure is the sum of its (connected) components. From this it follows that the cancellation law,

$$A+B \simeq A+C \text{ implies } B \simeq C \quad ,$$

holds whenever $A$ is finite. In fact, this holds for every structure $A$ that does not have an infinite set of pairwise isomorphic components. The cardinal multiplication is also commutative and associative (up to isomorphism), and it is distributive over cardinal addition,

$$A(B+C) \simeq AB + AC \quad ,$$

and similarly for infinite sums. This operation also has a neutral element, the unit structure $\underline{1}$. An infinite structure need not have a representation as a direct product of directly indecomposable structures. Example: Any infinite, atomless Boolean algebra. For a finite structure such a representation obviously exists, but it need not be unique, even for ordered sets. A simple counterexample, due to Hashimoto and Nakayama [1950], is the ordered set,

$$P = \underline{1} + \underline{2} + \underline{2}^2 + \underline{2}^3 + \underline{2}^4 + \underline{2}^5 \quad .$$

Here we are using the notation explained in the introduction, e.g., $\underline{2}$ is the Boolean algebra of order 2, and $\underline{2}^n$ the Boolean algebra of order $2^n$. Since every direct factor of a Boolean algebra is a Boolean algebra, every direct factor of $P$ must be a polynomial in $\underline{2}$. From this it follows that, apart from $P$ and $\underline{1}$, the only direct factors of $P$ are

$$A = \underline{1} + \underline{2}, \qquad B = \underline{1} + \underline{2}^2 + \underline{2}^4,$$

$$C = \underline{1} + \underline{2}^3, \qquad D = \underline{1} + \underline{2} + \underline{2}^2.$$

These four factors are therefore directly indecomposable, and since $P \simeq AB \simeq CD$, this violates the unique factorization property. As will be clear later, the cause of this misbehavior is the lack of connectedness, but we drop this topic for now and proceed to the next operation.

The ordinal addition is not commutative, but it is associative. Every structure is the ordinal sum of ordinally indecomposable structures, and this representation is unique. However, this result is not quite as trivial as the corresponding fact about cardinal addition, and we shall therefore look at it more closely later. The lexicographic multiplication is also non-commutative, but associative. It is right distributive over both cardinal and ordinal addition,

$$(A+B)\circ C = (A\circ C) + (B\circ C), \qquad (A\oplus B)\circ C = (A\circ C) \oplus (B\circ C) \quad ,$$

and similarly for infinite sums. However, the left distributive laws fail. The unique factorization property fails, even for finite ordered sets (Chang [1955] and [1961a], Dörfler and Imrich [1972]). E.g., let $A = (2\circ\underline{2}) + 1$, $B = C = 2$, $D = (\underline{2}\circ 2) + 1$. Then $A\circ B \simeq C\circ D$. (Fig. 2).

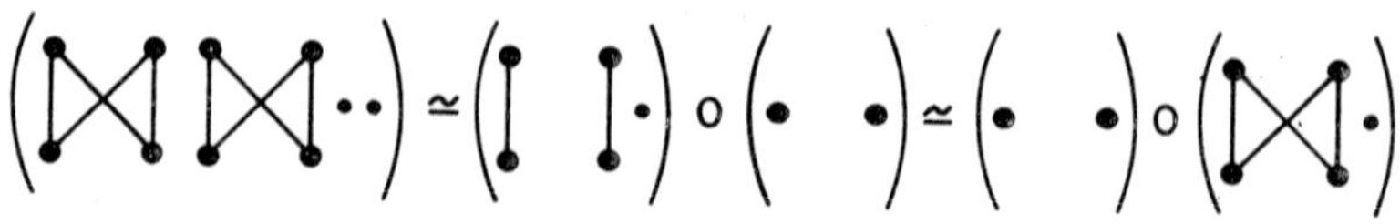

Fig. 2

The cardinal exponentiation satisfies the usual laws for exponents,

$$A^{B+C} \simeq A^B A^C, \qquad (A^B)^C \simeq A^{BC}, \qquad (AB)^C \simeq A^C B^C \quad ,$$

and of course $A^{\underline{1}} \simeq A$ and $\underline{1}^A \simeq \underline{1}$. It is also useful to observe that if $C$ is connected, then

$$(A+B)^C = A^C + B^C \quad .$$

The ordinal exponentiation also satisfies the expected laws,

$${}^{(m+n)}A \simeq {}^{m}A\circ {}^{n}A, \qquad {}^{mn}A \simeq {}^{m}({}^{n}A) \quad ,$$

and ${}^{\underline{1}}A \simeq A$. Finally, the conversion satisfies the obvious identities,

$$(A+B)^{\vee} = A^{\vee} + B^{\vee}, \qquad (AB)^{\vee} = A^{\vee}B^{\vee} \quad ,$$

$$(A \oplus B)^{\vee} = B^{\vee} \oplus A^{\vee}, \qquad (A \circ B)^{\vee} = A^{\vee} \circ B^{\vee},$$

$$(A^{C})^{\vee} = (A^{\vee})^{C^{\vee}} \quad , \qquad A^{\vee\vee} = A \quad .$$

## 4. DIRECT DECOMPOSITIONS: REFINEMENT PROPERTIES

Two direct decompositions

$$\varphi : P \simeq \prod_{i \in I} A_i, \qquad \psi : P \simeq \prod_{j \in J} B_j, \tag{1}$$

are said to be isomorphic if there exist a bijection $\lambda : I \to J$ and isomorphisms $\alpha_i : A_i \simeq B_{\lambda(i)}$, $i \in I$. If $\lambda$ and $\alpha_i$ can be so chosen that the diagram

$$\begin{array}{ccc} & P & \\ \varphi \swarrow & & \searrow \psi \\ \prod_{i \in I} A_i & \xrightarrow{\alpha} & \prod_{j \in J} B_j \end{array}$$

commutes, where $\alpha$ is the direct product of the isomorphisms $\alpha_i$, then $\varphi$ and $\psi$ are said to be equivalent. We say that $\psi$ is a refinement of $\varphi$ if there exist isomorphisms

$$\alpha_i : A_i \simeq \prod_{j \in J_i} B_j,$$

where the sets $J_i$ constitute a partitioning of $J$ such that the diagram

$$\begin{array}{ccc} P & \xrightarrow{\psi} & \prod_{j \in J} B_j \\ \varphi \downarrow & & \downarrow \beta \\ \prod_{i \in I} A_i & \xrightarrow{\alpha} & \prod_{i \in I} \prod_{j \in J_i} B_i \end{array}$$

commutes. Here $\alpha$ is the direct product of the isomorphisms $\alpha_i$, and $\beta$ is the natural isomorphism.

A structure $P$ is said to have the unique factorization property if $P$ can be represented as a direct product of directly indecomposable structures and if any two such representations $\varphi$ and $\psi$ are isomorphic. If $\varphi$ and $\psi$ are always equivalent, then $P$ is said to have the strict unique factorization property. There are many familiar structures that have the unique factorization property but do not have the strict unique factorization property, for example finite Abelian groups that are not cyclic, and finite antichains whose order is not prime.

A structure $P$ is said to have the refinement property if any two direct decompositions $\varphi$ and $\psi$ of $P$ have isomorphic refinements. If $\varphi$ and $\psi$ always have a common refinement, then $P$ is said to have the strict refinement property. It is easy to see that $P$ has the refinement property iff, for any two direct decompositions $\varphi$ and $\psi$, as in (1), there exist direct decompositions

$$\alpha_i : A_i \simeq \prod_{j \in J} C_{i,j}, \qquad \beta_j : B_j \simeq \prod_{i \in I} C_{i,j},$$

and that $P$ has the strict refinement property iff $\alpha_i$ and $\beta_j$ can always be so chosen that the diagram

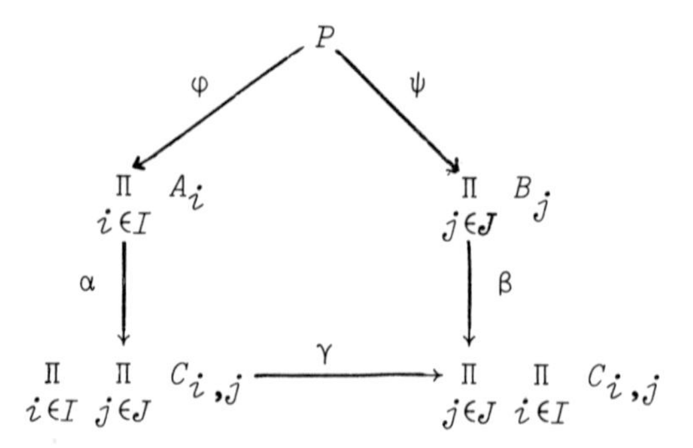


commutes. Here $\alpha$ and $\beta$ are the obvious direct products, and $\gamma$ is the natural isomorphism.

It was proved in Hashimoto [1951] that every connected ordered set has the refinement property. Special cases of this result had been obtained earlier in Birkhoff [1933], [1934] and [1940], Hashimoto [1948], and Nakayama [1948]. Although Hashimoto does not consider the strict refinement property, his construction yields refinements that can be shown to be equivalent, and not just isomorphic. The distinction between the two properties is important; e.g., the stronger version plays an essential role in the proofs of several of the results on exponentiation that will be discussed later. The strict refinement property was explicitly formulated and was investigated in detail in Chang, Jónsson and Tarski [1964]. These investigations were continued in McKenzie [1971], and the strongest results (as far as binary structures are concerned) can be found in that paper.

To establish the strict refinement property for a particular structure, it suffices to consider pairs of direct decompositions with just two factors each.

THEOREM 4.1. (Chang, Jónsson and Tarski [1964]) *A structure P has the strict refinement property iff, for any two direct decompositions,*

$$\phi:AB \simeq P \quad \textit{and} \quad \psi:CD \simeq P \quad ,$$

*there exist direct decompositions*

$$\alpha:WX \simeq A, \quad \beta:YZ \simeq B, \quad \gamma:WY \simeq C, \quad \delta:XZ \simeq D$$

*such that, for all $w \in W$, $x \in X$, $y \in Y$ and $z \in Z$,*

$$\phi(\alpha(w,x),\beta(y,z)) = \psi(\gamma(w,y),\delta(x,z)) \quad .$$

Any bijection $\phi:A\times B \to P$ induces a binary operation on $P$, corresponding to the rectangle multiplication

$$(a,b)(a',b') = (a,b')$$

on $A\times B$. An operation induced by a direct product decomposition $\phi:AB \simeq P$ of a structure $P$ is called a decomposition function of $P$. It is easy to see that a second decomposition, $\phi':A'B' \simeq P$ induces the same decomposition function just in case there exist isomorphisms $\alpha:A \simeq A'$ and $\beta:B \simeq B'$ such that the diagram

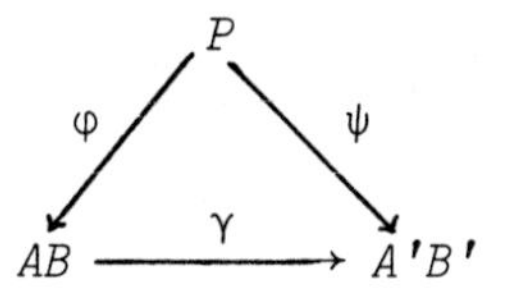


commutes, where $\gamma$ is the direct product of $\alpha$ and $\beta$. For this reason, decomposition functions are a suitable tool for investigating direct decompositions of a structure, and especially for studying the strict refinement property. Decomposition functions have a simple intrinsic characterization. They must of course satisfy the familiar axioms for rectangular bands, and the only additional condition is that they be homomorphisms from $PP$ to $P$.

THEOREM 4.2. (Chang, Jónsson and Tarski [1964]) *A binary operation $\kappa$ on $P$ is a decomposition function of the structure $P$ iff it satisfies the identities*

$$p\kappa p = p \quad \textit{and} \quad (p\kappa q)\kappa r = p\kappa r = p\kappa(q\kappa r)$$

*and, for all $p,q,p',q' \in P$,*

$$p \to p' \quad \textit{and} \quad q \to q' \quad \textit{implies} \quad (p\kappa q) \to (p'\kappa q') \quad .$$

The next step is to translate the strict refinement property into a condition on the decomposition functions.

THEOREM 4.3. (Chang, Jónsson and Tarski [1964]) *A structure $P$ has the strict refinement property iff, for any decomposition functions $\kappa$ and $\lambda$ of $P$, and for any $p, q \in P$,*

$$(p\kappa q)\lambda p = p\kappa(q\lambda p) \quad .$$

The strategy now is to find out as much as possible about the relationship between the elements $(p\kappa q)\lambda p$ and $p\kappa(q\lambda p)$ in an arbitrary structure, in the hope of being able to show that in certain types of structures these two elements must be equal. The strongest result of this kind was proved in McKenzie [1971]. Two elements $p$ and $q$ in a structure $P$ are said to be indistinguishable, -- in symbols, $p \approx q$, -- if for every $r \in P$ the conditions

$$p \to r \quad \text{and} \quad q \to r$$

either both hold or both fail, and likewise the conditions

$$r \to p \quad \text{and} \quad r \to q \quad .$$

THEOREM 4.4. (McKenzie [1971]) *If $\kappa$ and $\lambda$ are decomposition functions of a reflexive structure $P$, then*

$$(p\kappa q)\lambda p \approx p\kappa(q\lambda p)$$

*whenever $p$ and $q$ belong to the same component of $P$.*

In a reflexive structure, $p \approx q$ implies that $p \to q$ and $q \to p$. In an ordered set, $\approx$ is therefore the identity relation. Hence Hashimoto's theorem follows from Theorem 4.4.

THEOREM 4.5. (Hashimoto [1951]) *Every connected ordered set has the strict refinement property.*

Incidentally, McKenzie was able to use this theorem to prove the refinement property for a large class of structures that do not have the strict refinement property. This was done by applying the strict refinement property to the quotient structure $P/\approx$, and by then "lifting" the decomposition to $P$. The last step works, in particular, if $P$ is finite.

THEOREM 4.6. (McKenzie [1971]) *Every reflexive, connected, finite structure has the refinement property.*

## 5. DIRECT PRODUCTS: UNIQUE FACTORIZATION AND CANCELLATION PROPERTIES

As an immediate consequence of Hashimoto's theorem we obtain the following unique factorization result.

THEOREM 5.1. *A connected ordered set has, up to equivalence, at most one representation as a direct product of directly indecomposable ordered sets.*

A structure will be said to be (finitely) factorable if it is isomorphic to a direct product of (finitely many) directly indecomposable structures. There are many kinds of finiteness conditions that insure that an ordered set $P$ is finitely factorable. Let $\dim(P)$ be the Dushnik, Miller dimension of $P$, that is, the smallest cardinal $m$ such that the order relation on $P$ is the intersection of $m$ total order relations. Also let $\mathrm{width}(P)$ be the supremum of the cardinalities of the antichains of $P$.

THEOREM 5.2. *Every connected ordered set that satisfies one of the conditions (i), (ii) and (iii) below is finitely factorable:*

(*i*) $\dim(P)$ *is finite.*
(*ii*) $\mathrm{width}(P)$ *is finite.*
(*iii*) *Some maximal chain in $P$ is finite.*

THEOREM 5.3. *If $A$ is a non-empty, finitely factorable, connected ordered set, then for any ordered sets $B$ and $C$,*

$$AB \simeq AC \quad \text{implies} \quad B \simeq C \quad .$$

If $B$ and $C$ are also connected, this easily follows from Hashimoto's theorem. The general case is reduced to this special case by observing that if $B_i$ is a component of $B$, then $AB_i$ is a component of $AB$, and is therefore mapped onto $AC_i$, where $C_i$ is a component of $C$. Thus $AB_i \simeq AC_i$, hence $B_i \simeq C_i$.

The conditions imposed on $A$ can be relaxed considerably if we also place some restrictions on $B$ and $C$.

THEOREM 5.4. *Suppose $A$, $B$ and $C$ are ordered sets whose components are finitely factorable, with $A$ non-empty, and suppose none of them has an infinite set of pairwise isomorphic components. Then*

$$AB \simeq AC \quad \text{implies} \quad B \simeq C \quad .$$

The proof employs an idea first used in Hashimoto [1948]. Let $P_i$, $i \in I$, be up to isomorphism a complete list of the directly indecomposable ordered sets that occur as factors in components of $A$, $B$ or $C$. Let $F$ be the set of all finitely non-zero maps from $I$ to $\omega$, and for $f \in F$ let

$$Q_f = \prod_{i \in I} P_i^{f(i)} \quad .$$

Then every component of $A$, $B$ and $C$ is isomorphic to some $Q_f$, and hence

$$A \simeq \sum_{f \in F} m(f) Q_f, \qquad B \simeq \sum_{f \in F} n(f) Q_f, \qquad C \simeq \sum_{f \in F} p(f) Q_f,$$

where $m(f)$, $n(f)$ and $p(f)$ are natural numbers. We can think of $A$, $B$ and $C$ as members of the ring of power series in the indeterminates $P_i$ over the integers. The conclusion of the theorem therefore follows from the fact that a ring of power series, in any number of indeterminates, over an integral domain is always an integral domain. An easy argument shows that for the same class of ordered sets considered in the preceding theorem, the $n$-th root property holds:

$$A^n \simeq B^n \quad \text{implies} \quad A \simeq B \quad .$$

If we enlarge the class slightly by allowing infinitely many isomorphic components, then both properties fail. E.g., if

$$A = \underline{1} + \underline{2} + \underline{2}^2 + \ldots, \qquad B = \aleph_0 \cdot \underline{1}, \qquad C = \aleph_0 \underline{1} + \underline{2},$$

then $AB \simeq AC \simeq \aleph_0 A$, but $B \not\simeq C$. More interesting is the fact that even for connected sets these and most other arithmetic properties that one might think of fail without some kind of finiteness condition; in fact, they even fail for Boolean algebras. Numerous counterexamples can be found in Kinoshita [1953], Hanf [1957], and Adámak, Koubek and Trnková [1975]. However, these are eclipsed by a recent result due to Ketonen.

THEOREM 5.5. (Ketonen [1978]) *Every countable commutative semigroup can be embedded in the semigroup of all isomorphism types of countable Boolean algebras (with the operation of direct multiplication).*

The proof of this result is quite difficult; it is accomplished by developing a complete set of invariants for countable Boolean algebras.

Ketonen's theorem settles in the negative many difficult questions concerning the arithmetic of the isomorphism types of countable Boolean algebras. Many of these had been settled previously, by various ingenious constructions, for non-denumerable Boolean algebras in the papers just mentioned, but for countable Boolean algebras, and in many cases for countable structures in general, no counterexamples had been obtained. Now explicit constructions of such examples are no longer needed. It is of course trivial that the cancellation law fails, for if $A$ is the countably infinite atomless Boolean algebra, then $AA \simeq A1$. On the other hand, we did not know previously whether the Cantor, Bernstein property,

$$ABC \simeq A \quad \text{implies} \quad AB \simeq A \quad ,$$

fails. It does, and quite badly, for if $n$ is any positive integer, then there exist countable Boolean algebras $A$ and $B$ such that $A \simeq AB^m$ iff $m$ is a multiple of $n$. The $n$-th root property also fails for every integer $n > 1$. Indeed, there exists a countable Boolean algebra $A$ such that there are infinitely many non-isomorphic Boolean algebras $B$ with $A \simeq B^n$. These horror stories could be continued ad infinitum. Indeed, Murphy's Law applies here: Everything that could go wrong will go wrong.

Almost everything. There is a special case of the Cantor, Bernstein property that holds for countable, reflexive structures.

THEOREM 5.6. (McKenzie [1971]) *If $A$ is a countable, reflexive structure and $B$ and $C$ are finite structures, then*

$$ABC \simeq A \quad \textit{implies} \quad AB \simeq A \quad .$$

For Boolean algebras this result is rather trivial; it follows from the fact that if a compact, separable Hausdorff space has infinitely many isolated points, then the subspace obtained by removing a single isolated point is homeomorphic to the original space. However, even for Boolean algebras it is essential that $A$ be countable.

THEOREM 5.7. (Hanf [1957]) *There exists a Boolean algebra $H$ such that*

$$H\cdot\underline{2}\cdot\underline{2} \simeq H \quad \textit{and} \quad H\cdot\underline{2} \not\simeq H \quad .$$

The proof is based on a simple but ingenious construction. Choose an infinite set $U$ and an involution $\alpha$ of $U$ that leaves no point fixed. Let $A$ consist of all subsets $X$ of $U$ that do not separate $x$ and $\alpha(x)$, except in finitely many cases. I.e. a subset $X$ of $U$ belongs to $A$ iff the set

$$\{x \in X : \alpha(x) \notin X\}$$

is finite. The point of the proof is that if we adjoin a single point to $U$, then any involution of the new set must split up infinitely many of the pairs $x$, $\alpha(x)$, but if we adjoin two points, then these can be matched against each other, and all the old pairs left intact.

As C.C. Chang observed (cf. Hanf [1957]), Hanf's result provides us with a situation where even a finite factor cannot be cancelled, namely

$$(\underline{1}+\underline{2})\cdot(H\cdot\underline{2}) \simeq (\underline{1}+\underline{2})\cdot H \quad .$$

There is a reasonably comprehensive class of ordered sets that behave quite nicely under multiplication, namely the class of all countably complete lattices.

THEOREM 5.8. (Tarski [1949]) *The class of all isomorphism types of countably complete lattices is a cardinal algebra under direct multiplication.*

We should actually speak of suitable sets of isomorphism types rather than the class of all isomorphism types, which is a proper class. However, there is no danger of encountering the set-theoretic paradoxes here, and in any case this is a technical point that can be handled routinely, so we shall ignore it, here and in the future.

The numerous theorems about cardinal algebras that were proved by Tarski can now be applied, in particular, to the isomorphism types of countably complete lattices. Since this subject is, from our present point of view, rather specialized, we only recall here three such theorems, refering the reader to Tarski's book for further details.

An isomorphism type $a$ is said to divide an isomorphism type $b$, -- in symbols $a|b$, -- if $ax = b$ for some $x$.

THEOREM 5.9. (Tarski [1956], Corollary 2.34 and Theorems 1.31 and 2.28) *If $a$, $b$, $c$, $a_i$, $b_j$, $(i,j \in \omega)$ are isomorphism types of countably complete lattices, and if $n$ is a positive integer then the following statements hold:*

(*i*) *If $a^n = b^n$, then $a = b$.*
(*ii*) *If $a|b$ and $b|a$, then $a = b$.*
(*iii*) *If $a_i|b_j$ for all $i,j \in \omega$, then there exists $c$ such that $a_i|c$ and $c|b_j$ for all $i,j \in \omega$.*

The first two statements are obviously the $n$-th root property and the Cantor, Bernstein property. The third one is the so-called interpolation property.

In Lovász [1967], radically different methods are used to study direct decompositions of finite structures, and even for ordered sets this approach yields some new results. The principal technique is to count the number of homomorphisms from various structures $D$ into the structure $A$ under consideration, i.e., the number of elements in the power $A^D$.

LEMMA 5.10. *For any finite structures $A$ and $B$, if for every finite, connected structure $D$, $A^D$ and $B^D$ have the same number of elements, then $A \simeq B$.*

To state his fundamental theorem, Lovász uses polynomials with fixed structures as coefficients,

$$f(X) = C_0 + C_1X + C_2X^2 + \ldots + C_nX^n \quad .$$

THEOREM 5.11. *If all the coefficients $C_i$ are finite, and if at least one $C_i$, $i > 0$ has a reflexive element, then for any finite structures $A$ and $B$,*

$$f(A) \simeq f(B) \quad \textit{implies} \quad A \simeq B \quad .$$

COROLLARY 5.12. *For any finite structures $A$ and $B$, and any positive integer $n$,*

$$A^n \simeq B^n \quad \textit{implies} \quad A \simeq B \quad .$$

COROLLARY 5.13. *For any finite structures $A$, $B$ and $C$, if $A$ has a reflexive element, then*

$$AB \simeq AC \quad \textit{implies} \quad B \simeq C \quad .$$

## 6. EXPONENTIATION: SOME EXAMPLES

It has already been noted that the (cardinal) exponentiation satisfies the standards laws for exponents. Before describing the deeper properties of this operation, we shall look at some examples where the operation yields familiar or interesting ordered sets. Many of these examples were treated in Birkhoff [1937a] and [1942].

If $A$ is a lattice and $C$ is any ordered set, then $A^C$ is a lattice. In fact, it is a sublattice of a direct power of $A$, and therefore satisfies every identity that holds in $A$. In particular, if $A$ is distributive, then so is $A^C$. The members of the lattice $\underline{2}^C$ are the characteristic functions $\chi_F$ of the order filters $F$ on $C$ (including the empty filter). Indeed, the correspondence $F \to \chi_F$ is an isomorphism from the lattice of all filters on $C$ to $\underline{2}^C$. This observation can be used to show that every finite distributive lattice $L$ is determined up to isomorphism by the set $J(L)$ of all join irreducible elements of $L$.

THEOREM 6.1. *For every finite ordered set $C$, $J(\underline{2}^C) \simeq C^{\vee}$. Conversely, for any finite distributive lattice $L$, $\bar{L} \simeq \underline{2}^{J(L)^{\vee}}$.*

This is a simple consequence of the fact that in the lattice of all order filters of $C$ the join irreducible members are precisely the principal order filters.

COROLLARY 6.2. *If $A$ and $B$ are finite distributive lattices, then the three lattices*

$$A^{J(B)^{\vee}}, \quad B^{J(A)^{\vee}}, \quad \underline{2}^{J(A)^{\vee}\cdot J(B)^{\vee}}$$

*are isomorphic to each other; in fact, they are all isomorphic to $A * B$, the free distributive product of $A$ and $B$.*

This follows from the fact that $J(A*B) \simeq J(A)J(B)$.

COROLLARY 6.3. *The bounded distributive lattice freely generated by a finite ordered set $P$ is isomorphic to $\underline{2}^{\underline{2}^P}$.*

COROLLARY 6.4. *For any positive integers $m$ and $n$, and the corresponding ordinals $\underline{m}$ and $\underline{n}$:*

(*i*) $\underline{2}^n$ *is a Boolean algebra of order* $2^n$.
(*ii*) $\underline{2}^{\underline{n}}$ *is a chain of length $n$, i.e.,* $\underline{n} \oplus 1$.
(*iii*) $\underline{2}^{2^n}$ *is the free Boolean algebra on $n$ generators.*
(*iv*) $\underline{2}^{\underline{2}^n}$ *is the free bounded distributive lattice on $n$ generators.*
(*v*) $(\underline{m+1})^{\underline{n}} \simeq \underline{2}^{\underline{m}\cdot\underline{n}}$ *is the bounded distributive lattice freely generated by the union of two chains of lengths $m-2$ and $n-2$.*

In the next two sections we will discuss three properties that hold for large classes of ordered sets. To set the stage, we list here some examples that show that none of these properties holds in general. The three properties are:

I. The cancellation law for bases. $A^C \simeq A^D$ implies $C \simeq D$.
II. The cancellation law for exponents. $A^C \simeq B^C$ implies $A \simeq B$.
III. The refinement property for powers. $A^C \simeq B^D$ implies that, for some $E$, $X$, $Y$, $Z$, $A \simeq E^X$, $B \simeq E^Y$, $C \simeq YZ$, $D \simeq XZ$.

The first example below appeared in Fuchs [1965], the remaining ones in Jónsson, McKenzie [a].

EXAMPLE 1. (Failure of I). If $A$ is an antichain, then $A^C$ is an antichain whose members are the maps from $C$ to $A$ that are constant on each component of $C$. Hence $A^C$ and $A^D$ are isomorphic whenever $C$ and $D$ have the same number of components.

EXAMPLE 2. (Failure of I and III). Let $H$ be Hanf's Boolean algebra with $H\cdot\underline{2} \not\simeq H$ and $H\cdot\underline{2}\cdot\underline{2} \simeq H$ (Theorem 5.7), and recall Chang's observation that $(\underline{1}+\underline{2})\cdot H \simeq (\underline{1}+\underline{2})\cdot H\cdot\underline{2}$. Let $C = H$, $D = H\cdot\underline{2}$, and $A = \underline{2}^{\underline{1}+\underline{2}}$ ($\simeq \underline{2}\cdot\underline{3}$). Then $A^C \simeq A^D$, and I is violated. Property III is also violated if we take $B = A$.

The next example does not have direct bearing on the three properties under consideration, but it will be used in connection with Example 4, and it is of independent interest.

EXAMPLE 3. A power of a connected ordered set need not be connected. Let $A = C = 2 \oplus 2$. Then the four isomorphisms from $C$ to $A$ are isolated points of $A^C$.

EXAMPLE 4. (Failure of II and III). Let $C = 2 \oplus 2$, let $A$ be a sum of infinitely many copies of $C$, and let $B = A+\underline{1}$. Then $B^C \simeq A^C+\underline{1} \simeq A^C$, because $A^C$ is the sum of infinitely many copies of $C^C$, and therefore has infinitely many isolated points. Thus II is violated, and so is III if we take $D = C$.

EXAMPLE 5. (Failure of II and III). Let $A = \underline{1} \oplus (\underline{1}+\underline{1})$ and $B = \underline{1} \oplus (\underline{1}+\underline{2})$ (Fig. 3), and let $C$ be the direct product of

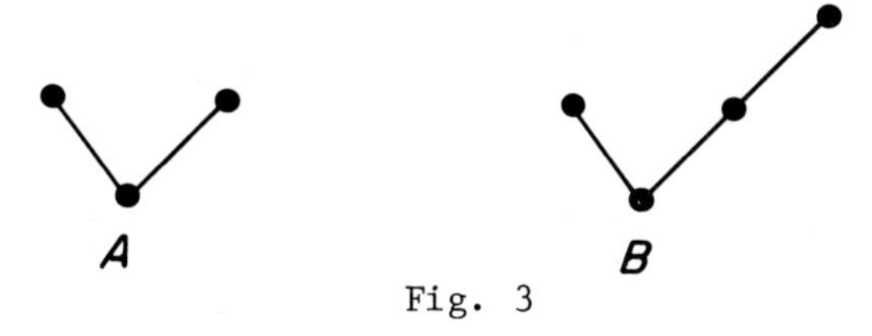


Fig. 3

infinitely many copies of $\underline{2}$. Each member of $A^C$ must map the whole set $C$ into one of the two arms of $A$, because the range of the map must have a largest element. Thus $A^C$ is obtained from two copies of $\underline{2}^C$ by identifying their bottom elements, and $B^C$ is obtained in the same manner from a copy of $\underline{2}^C$ and a copy of $\underline{3}^C$. However, $\underline{3}^C \simeq \underline{2}^{\underline{2}\cdot C} \simeq \underline{2}^C$. Hence $A^C \simeq B^C$. Again II is violated, and so is III if we take $D = C$.

## 7. EXPONENTS: THE CANCELLATION LAW FOR BASES

Having shown that the properties I, II and III do not hold in general, we now present some positive results. From Theorem 6.1 it follows that if $C$ and $D$ are finite ordered sets, then $\underline{2}^C \simeq \underline{2}^D$ implies $C \simeq D$. This was generalized in Novotný [1960].

THEOREM 7.1. *If $A$ is a non-trivial chain, and if $C$ and $D$ are any ordered sets, then*

$$A^C \simeq A^D \quad \textit{implies} \quad C \simeq D \quad .$$

The first cancellation result for bases other than chains appeared in Fuchs [1965].

THEOREM 7.2. *If $A$, $B$ and $C$ are finite ordered sets, and if $A$ is finite and has a smallest element, then*

$$A^C \simeq A^D \quad \textit{implies} \quad C \simeq D \quad .$$

No further work was done on powers of ordered sets for well over a decade, until the subject was taken up by D. Duffus and I. Rival. The key idea introduced by them, which has played a major role in all later investigations, was their concept of a logarithmic property. The inspiration for this came from Birkhoff's theorem, 6.1. It readily follows from this theorem that if $L$ is a finite distributive lattice, and if $C$ is a finite ordered set, then

$$J(L^C) \simeq J(L)\cdot C^{\vee} \quad . \tag{1}$$

Hence, if $L^C \simeq L^D$, where $D$ is also an ordered set, then $J(L)\cdot C^{\vee} \simeq J(L)\cdot D^{\vee}$, and disregarding the uninteresting case in which $L$ consists of just one element, we infer by Theorem 5.4 that $C \simeq D$.

The formula (1) was called by Duffus and Rival the logarithmic property. In order to extend this technique to other bases, they needed a suitable generalization of the notion of a join irreducible element. There are several candidates, but they chose to consider the set $V(P)$ of all elements $a \in P$ with the property that $a$ has a unique lower cover in $P$. The first task was to show that $V$ does in fact have the logarithmic property.

THEOREM 7.3. (Duffus and Rival [1978]) *For any finite ordered sets $A$ and $C$, if $A$ has a smallest element, then*

$$V(A^C) \simeq V(A)\cdot C^{\vee} \quad .$$

THEOREM 7.4. (Duffus and Rival [1978]) *If $A$, $C$ and $D$ are finite ordered sets, and if $A$ is not an antichain, then*

$$A^C \simeq A^D \quad \textit{implies} \quad C \simeq D \quad .$$

If $A$ has a smallest element, then this follows readily from the preceding theorem. This is Fuchs's theorem. In the general case the argument is more complicated and involves applications of the logarithmic property to certain maximal principal filters in $A$.

There are other conditions that are known to imply the cancellation law for bases. We state these results here, although their proofs depend on results that will be discussed in the next two sections.

THEOREM 7.5. (Jónsson, McKenzie [a]) *If $A$ is a lower bounded ordered set of finite length, and if $C$ and $D$ are finite ordered sets, then*

$$A^C \simeq A^D \quad \textit{implies} \quad C \simeq D \quad .$$

The next theorem involves a modified version of the "logarithm". For an ordered set $P$, let $J(P)$ be the set of all $a \in P$ with the property that there exists $a' \in P$ such that $a \nleq a'$ but $x \leq a'$ whenever $x < a$. This is perhaps the most natural notion of a strictly join irreducible element, but it does not lead to a logarithmic property, even for finite ordered sets with a zero. We therefore introduce another related concept. Let $J'(P)$ be the set of all $a \in P$ with the property that there exist $a',b \in P$ with $a \leq b$, $a' \leq b$, and $a \nleq a'$, but $x \leq a'$ whenever $x < a$. Observe that $J'(P) \subseteq J(P)$, and that equality holds whenever $P$ is updirected. In particular, equality holds if $P$ is a lattice, or is upper bounded.

We need one other notion. For $X \subseteq P$ let $\nabla X$ be the set of all $a \in P$ with the property that there do not exist $a',b \in P$ with $a \leq b$, $a' \leq b$ and $a \nleq a'$ such that $x \leq a'$ whenever $x \in X$ and $x < a$. The members of $\nabla X$ may be thought of as least upper bounds of subsets of $X$. The condition

$$P = \nabla J'(P) \quad ,$$

which will occur in the next theorem and in some theorems in the next section, may therefore be thought of as saying that every member of $P$ is the join of strictly join irreducible members. It is not hard to see that this condition holds whenever $P$ satisfies the minimal condition. Hint: Show that, in an arbitrary ordered set $P$, $P - J'(P)$ has no minimal element.

THEOREM 7.6. (Jónsson, McKenzie [a]) *Suppose $A$ is a lower bounded ordered set with more than one element, $J'(A)$ is connected and $\nabla J'(A) = A$, and suppose that either $A$ has finite length, or else it is exponentially indecomposable. Then, for any ordered sets $C$ and $D$,*

$$A^C \simeq A^D \quad \textit{implies} \quad C \simeq D \quad .$$

The assumption that $A$ be exponentially indecomposable means that $A \simeq X^Y$ implies $Y \simeq \underline{1}$. This theorem provides us with many ordered sets $A$, other than chains, that cancel as bases regardless of what the exponents are. E.g., if $A_0$ is a bounded ordered set of finite length, then $A = A_0 \oplus \underline{1}$ has this property. Recall that, by Example 2 in the preceding section, this property fails for $A = \underline{2}\cdot\underline{3}$.

We mention here two more results whose proofs make essential use of the logarithmic property.

THEOREM 7.7. (Duffus and Wille [1979]) *If $A$ and $B$ are finite connected ordered sets with more than one element, then each of the conditions $A^A \simeq B^B$ and $A^B \simeq B^A$ implies that $A \simeq B$.*

## 8. EXPONENTIATION: THE CANCELLATION LAW FOR EXPONENTS, AND THE REFINEMENT PROPERTY

The logarithmic property is not as directly applicable to the refinement problem and to the cancellation problem for exponents as it is to the cancellation problem for bases. An isomorphism

$$\varphi : A^C \simeq B^D$$

induces, under suitable conditions, an isomorphism

$$\psi : J'(A) \cdot C^{\vee} \simeq J'(B) \cdot D^{\vee} \quad .$$

If $A \simeq B$, then $J'(A) \simeq J'(B)$, and cancellation yields $C \simeq D$. On the other hand, if $C \simeq D$, then cancellation of the right hand factor yields only $J'(A) \simeq J'(B)$, not $A \simeq B$. Also, it is not clear why a refinement for $\psi$ should yield a refinement for $\varphi$. The trouble is that an ordered set cannot be reconstructed from its logarithm. In Duffus and Rival [1978], the case when the base is a finite lattice was treated by using the operation $P(L) = J(L) \cup M(L)$, where $M(L)$ is the set of all meet irreducible elements of $L$. If $L$ is a lattice of finite height, then $P(L)$ does determine $L$; in fact, $L$ is isomorphic to the MacNeille completion of $P(L)$. (Banaschewski [1956], Schmidt [1956]). Of course $P$ does not have the logarithmic property, but Duffus and Rival were nonetheless able to prove that

$$K^C \simeq L^C \quad \text{implies} \quad K \simeq L \quad ,$$

under the assumption that $K$ and $L$ are finite lattices and that $C$ is a finite, bounded ordered set. In Wille [1980], a different kind of "logarithm" was considered, again only for lattices. Wille's scaffolding, $G(L)$, of a lattice $L$ is a certain subset of $L$ with the order relation induced by $L$, and with a partial binary operation obtained by restricting the join operation in $L$. For lattices $L$ of finite length, $G(L)$ determines $L$, and if $C$ is a finite ordered set, then $G(L^C) \simeq G(L)C^{\vee}$. (On the right hand side, $C^{\vee}$ is regarded as a scaffolding with a trivial partial operation.) From this the cancellation law is obtained under the assumptions that $K$ and $L$ are lattices of finite length, and $C$ is a finite ordered set.

The results proved in Jónsson, McKenzie [a] use quite different methods, although the logarithm does play a role there too. Since the proofs are quite long, the best we can do is to illustrate the technique by treating a very special case. We shall prove that

$$\underline{A}^2 \simeq \underline{B}^2 \quad \text{implies} \quad A \simeq B \quad .$$

This was first proved under the assumption that $A$ was atomic. The present proof, without that restriction, is based on Bergman, McKenzie and Nagý [a].

We are given an isomorphism

$$\phi : \underline{A}^2 \simeq \underline{B}^2 \quad .$$

The members of $\underline{A}^2$ are ordered pairs of elements $x,y \in A$ with $x \leq y$; we shall write simply $xy$ for $(x,y)$. We want to associate with each member $a$ of $A$ a member $b$ of $B$. The image of the diagonal element $aa$ is an ordered pair $b_0b_1$, but unfortunately it need not be a diagonal element. This is seen e.g. by considering the automorphism of $\underline{3}^{\underline{2}}$ that interchanges 11 and 02. (Fig. 4) We claim, however, that the counterimages of the two elements $b_0b_0$ and $b_1b_1$ are diagonal elements. Consider, e.g., the counterimage $xy$ of $b_1b_1$. Let the image of $yy$ be $pq$, and let the counterimage of $b_0q$ be $uv$. We now have the situation in Fig. 5. From the fact that $b_0 \leq b_1 \leq q$, we infer that the meet of $b_1b_1$ and $b_0q$ exists and is equal to $b_0b_1$. Hence $xy \wedge uv = aa$. But $av$ is a lower bound for $xy$ and $uv$, whence $av \leq aa$, $v \leq a$, and thus $a = u = v$, $b_0q = b_0b_1$, $b_1 = p = q$, $xy = yy$ and, finally, $x = y$.

Thus there exist elements $a_0, a_1 \in A$ with $\phi(a_0a_0) = b_0b_0$ and $\phi(a_1a_1) = b_1b_1$. Of course $a_0 \leq a_1$, so that $a_0a_1 \in \underline{A}^2$. Our next goal is to show that the image of $a_0a_1$ is a diagonal element.

Let $\phi(a_0a_1) = bc$. The element $bb$ lies between the elements $b_0b_0$ and $bc$, whose counterimages are $a_0a_0$ and $a_0a_1$. Hence the counterimage of $bb$ is of the form $a_0y$ with $a_0 \leq y \leq a_1$. We are going to show that $y = a_1$. Letting $\phi(yy) = pq$, we have the situation in Fig. 6. From the fact that $b_0 \leq b \leq c \leq b_1$ we infer that $b_0b_1 \vee bb = bb_1 = b_0b_1 \vee bc$, hence $aa \vee a_0y = aa \vee a_0a_1 = aa_1$ and, consequently, $a \vee y = a_1$. From this we see that $aa \vee yy = a_1a_1$, hence $b_0b_1 \vee pq = b_1b_1$. But $b_0 \leq p \leq q \leq b_1$, so that $b_0b_1 \vee pq = pb_1$. Thus $p = b_1$, and we successively infer that $q = b_1$, $y = a_1$ and, finally, $b = c$, which was our claim.

We have shown that the situation in Fig. 7 prevails. In particular, we have associated with each element $a$ of $A$ an element $b$ of $B$. It is obvious that this map is isotone. Furthermore, the same construction with the roles of $A$ and $B$ interchanged would lead from $b$ to $a$. Consequently, our map is one-to-one and onto, and its inverse is also isotone. In other words, the map is an isomorphism from $A$ to $B$.

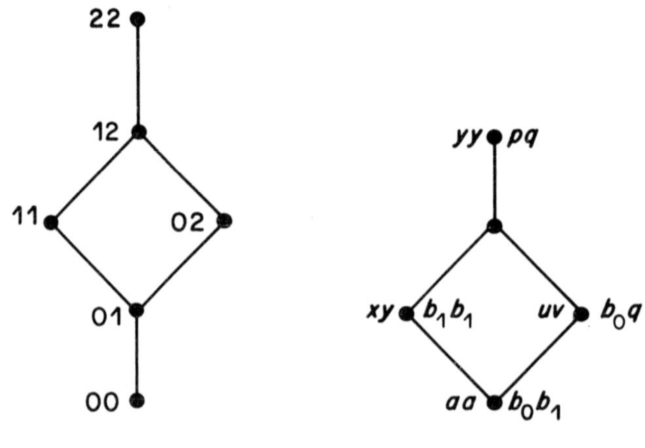


Fig. 4 Fig. 5

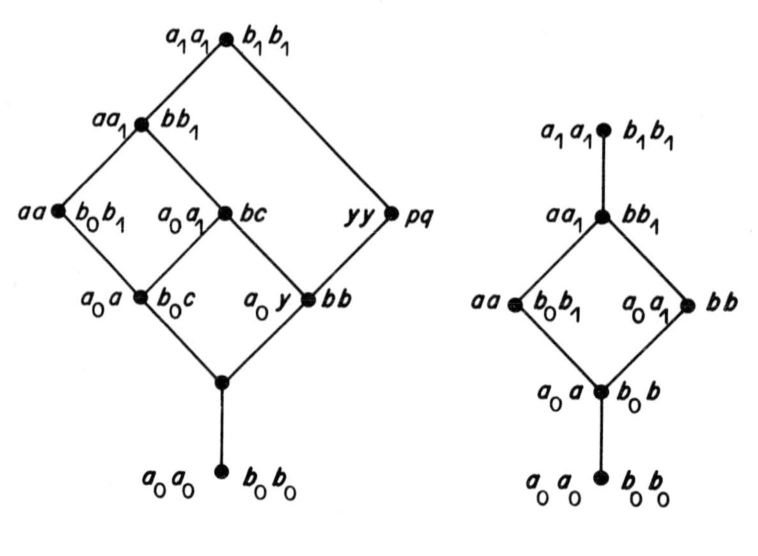


Fig. 6 Fig. 7

This argument suggests an approach that can be tried for other exponents. Suppose

$$\phi : A^C \simeq B^C \quad ,$$

and for $a \in A$ and $b \in B$ let $\langle a \rangle$ and $\langle b \rangle$ be the constant maps from $C$ whose sole values are $a$ and $b$, respectively. Consider an element $a \in A$, and let $f = \phi(\langle a \rangle)$. For $c \in C$, try to prove that the counterimage of the constant function $\langle f(c) \rangle$ is a constant function $\langle a_c \rangle$. Finally, observe that the function $g$ with $g(c) = a_c$ is a member of $A^C$, and try to prove that $\phi(g)$ is a constant function, $\langle b \rangle$. This will give us the required map from $A$ to $B$.

There is certainly a great deal of wishful thinking in this "outline", but the remarkable fact is that it works under a variety of rather general conditions. Part (i) of the next theorem comes from Bergman, McKenzie, Nagý [a], the other parts, as well as Theorem 8.2, come from Jónsson, McKenzie [a].

THEOREM 8.1. *For any ordered sets $A$, $B$ and $C$, the cancellation law for exponents,*

$$A^C \simeq B^C \quad \textit{implies} \quad A \simeq B$$

*holds whenever one of the following conditions is satisfied:*

(*i*) *$C$ is a chain.*
(*ii*) *$A^C$ is atomic, and $C$ is finitely factorable and upper bounded.*
(*iii*) *$A$ is a meet semilattice, and $C$ is finite and upper bounded.*
(*iv*) *$A$ is lower bounded, $\nabla J'(A) = A$, and $C$ is finitely factorable and connected.*

THEOREM 8.2. *For any ordered sets $A$, $B$, $C$ and $D$, the refinement property for powers,*

$$A^C \simeq B^D \textit{ implies that for some } E,\ X,\ Y,\ Z,$$

$$A \simeq E^X, \quad B \simeq E^Y, \quad C \simeq YZ, \quad D \simeq XZ,$$

*holds whenever one of the following conditions is satisfied:*

(*i*) *$A^C$ is atomic, and $C$ and $D$ are finitely factorable and upper bounded.*
(*ii*) *$A$ and $B$ are meet semilattices, and $C$ and $D$ are finite and upper bounded.*
(*iii*) *$A$ is lower bounded, $\nabla J'(A) = A$, $C$ is finitely factorable and $C$ and $D$ are connected.*

## 9. DIRECT DECOMPOSITIONS OF A POWER

Any representation of a structure $A$ as a direct product induces a direct decomposition of the power $A^B$, and so does every representation of $B$ as a cardinal sum. Under some very general conditions these are, up to equivalence, the only direct decompositions of $A^B$. It clearly suffices to consider the case when $B$ is connected.

THEOREM 9.1. (Jónsson, McKenzie [a]) *For any ordered sets $A$ and $B$, if $B$ and $A^B$ are connected, then every direct decomposition of $A^B$ is equivalent to one induced by a direct decomposition of $A$.*

The proof can be divided into three steps. First, one shows that it is sufficient to consider direct decompositions with two factors,

$$\psi : A^B \simeq C_0 C_1 \quad .$$

Next, we construct the required direct decomposition $\phi$ of $A$. There is an obvious injection from $A$ to $A^B$, taking $a \in A$ into the constant map $\langle a \rangle$, and setting $\phi(a) = \psi(\langle a \rangle)$, we clearly have a representation of $A$ as a subdirect product of subsets $A_0$ and $A_1$ of $C_0$ and $C_1$. One must show that $\phi$ does in fact map $A$ onto the full direct product $A_0 A_1$.

The isomorphism $\phi$ induces an isomorphism

$$\phi^B : A^B \simeq A_0^B A_1^B \quad ,$$

and the final step is to show that $\psi$ and $\phi^B$ are equivalent. Here the strict refinement property plays the crucial role. There exist direct decompositions

$$\alpha_0 : WX \simeq A_0^B, \qquad \alpha_1 : YZ \simeq A_1^B,$$

$$\gamma_0 : WY \simeq C_0, \qquad \gamma_0 : XZ \simeq C_1,$$

such that for all $w \in W$, $x \in X$, $y \in Y$ and $z \in Z$,

$$(\phi^B)^{-1}(\alpha_0(w,x),\alpha_1(y,z)) = \psi^{-1}(\gamma_0(w,y),\gamma_1(x,z)) \quad .$$

The crucial step, not a difficult one, is to show that $X$ and $Y$ must be one-element sets. This yields the required isomorphisms between $A_i^B$ and $C_i$, $i = 0,1$.

We shall now indicate how this theorem, together with Theorems 8.1 and 8.2, can be used to prove Theorems 7.5 and 7.6. In 7.5 we can represent $A$ as a direct product of finitely many directly indecomposable ordered sets, and $C$ and $D$ as the sums of finitely many connected ordered sets,

$$A \simeq \prod_{i \in I} A_i, \qquad C = \sum_{j \in J} C_j, \qquad D = \sum_{k \in K} D_k \quad .$$

Thus

$$\prod_{i \in I} \prod_{j \in J} A_i^{C_j} \simeq \prod_{i \in I} \prod_{k \in K} A_i^{D_k} \quad .$$

By Theorem 9.1, all the factors on both sides are directly indecomposable. Consequently

$$A_i^{C_j} \simeq (A_{\lambda(i,j)})^{D_{\mu(i,j)}} \quad ,$$

where $(i,j) \to (\lambda(i,j), \mu(i,j))$ is a bijection from $I \times J$ to $I \times K$. We can write $A_i \simeq E_i^{U_i}$, with $E_i$ exponentially indecomposable. From the fact that $A_i$ is directly indecomposable it follows that $U_i$ is connected. Hence, by 8.1(iv), $U_i C_j \simeq U_{\lambda(i,j)} D_{\mu(i,j)}$. Summing over $i$ and $j$, we obtain $UC \simeq UD$, where $U$ is the sum of the sets $U_i$. Since $U$, $C$ and $D$ are sums of finitely many finitely factorable sets, we conclude that $C \simeq D$.

The proof of 7.6 is similar, except that it uses 8.2 in place of 8.1.

The paper Davey and Duffus [a] includes a discussion of some of the same properties of the exponentiation as have been considered here. In particular, that paper treats in greater detail the role of the logarithmic property. In Bauer [a], important new techniques are introduced that yield improvements of some of the previously known cancellation and refinement properties.

## 10. A GENERAL REFINEMENT PROPERTY FOR ADDITION

In Section 3 we mentioned the fact that every structure has a unique representation as the ordinal sum of ordinally indecomposable structures. This result and the corresponding refinement property are special cases of two theorems about sums over arbitrary reflexive structures, Theorems B6 and B5 in Jónsson [1956]. We shall describe these results briefly.

Two representations

$$P = \textstyle\sum_{x,A} B_x, \qquad P = \textstyle\sum_{y,C} D_y$$

are said to be equivalent if there exists an isomorphism $\lambda : A \simeq C$ such that $B_x = D_{\lambda(x)}$ for all $x$ in $A$. We say that the second representation is a refinement of the first if there exists a representation

$$C = \textstyle\sum_{x,A} E_x$$

such that

$$B_x = \sum_{y, E_x} D_y \qquad \text{for all } x \text{ in } A \quad .$$

The notion of an indecomposable structure is defined relative to a class $K$ of index structures. We say that a non-empty structure $P$ is $K$-indecomposable if it does not have a representation

$$P = \sum_{x,A} B_x$$

with $A \in K$, $|A| > 1$, and each $B_x$ non-empty.

THEOREM 10.1. (Jónsson [1956], Theorem B5) *Suppose $K$ is a universal class of reflexive structures, with the property that*

$$\sum_{x,A} E_x \in K$$

*whenever $A$ and all the structures $E_x$ belong to $K$. Then any system of representations*

$$P = \sum_{x,A_i} B_{i,x}, \qquad i \in I \quad ,$$

*with $A_i \in K$ for all $i \in I$, have a common refinement*

$$P = \sum_{f,F} B_f$$

*with $F \in K$. In fact, let $G$ be the direct product of the structures $A_i$, for $f$ in $G$ let $B_f$ be the intersection of all the structures $B_{i,f(i)}$ with $i \in I$. We can then take $F$ to be the substructure of $G$ consisting of all those functions $f$ for which $B_f$ is non-empty.*

The proof of this theorem is not overly difficult. The crucial step is to show that $F$ belongs to $K$ (although $G$ obviously need not be a member of $K$).

COROLLARY 10.2. (Jónsson [1956], Theorem B6) *If $K$ is as in Theorem 10.1, then every non-empty structure $P$ has, up to equivalence, a unique representation*

$$P = \sum_{x,A} B_x$$

*with $A \in K$ and each $B_x$ $K$-indecomposable.*

The special case of these results that is primarily of interest here is the one in which $K$ consists of all totally ordered sets.

THEOREM 10.3. *Every system of representations of a structure as an ordinal sum of substructures has a common refinement.*

COROLLARY 10.4. *Every structure can be represented as the ordinal sum of ordinally indecomposable structures. This representation is unique up to equivalence.*

COROLLARY 10.5. *For any structures $A$, $B$, $C$ and $D$, if*

$$A \oplus B = C \oplus D$$

*then one and only one of the following statements holds:*

(*i*) $A = C$ *and* $B = D$.
(*ii*) $A \cap D \neq \phi$, $A = C \oplus (A \cap D)$ *and* $D = (A \cap D) \oplus B$.
(*iii*) $B \cap C \neq \phi$, $C = A \oplus (B \cap C)$ *and* $B = (B \cap C) \oplus D$.

## 11. ORDINAL ADDITION AND DIRECTED REFINEMENT ALGEBRAS

A long list of properties of ordinal addition was published, without proofs, in Lindenbaum and Tarski [1926], and in Tarski [1956] these properties were developed axiomatically. For this purpose, Tarski introduced two kinds of algebraic structures, ordinal algebras and directed refinement algebras.

Reversing Tarski's order, we consider the latter kind of structure first. We also modify Tarski's definition by postulating the existence of a neutral element. Thus we define a directed refinement algebra to be a monoid $(M,\oplus,0)$ with the following two properties:

For any $a,b,c \in M$, if $b = a \oplus b \oplus c$, then $b = a \oplus b$.
For any $a,b,c,d \in M$, if $a \oplus b = c \oplus d$, then for some $x \in M$,

either $a = c \oplus x$ and $d = x \oplus b$
or $c = a \oplus x$ and $b = x \oplus d$ .

THEOREM 11.1. (Tarski [1956]) *The class of all isomorphism types of (binary) structures is a directed refinement algebra under the operation of ordinal addition.*

The proof of the theorem is easy. The second axiom is an immediate consequence of Corollary 10.5. To verify the first axiom we have to show that, given an isomorphism

$$\phi : A \oplus B \oplus C \simeq B \quad ,$$

there exist an isomorphism

$$\psi : A \oplus B \simeq B \quad .$$

Actually, the standard proof of the Cantor, Bernstein Theorem provides us with such an isomorphism: Letting

$$B_0 = \phi(A) \cup \phi^2(A) \cup \dots \quad ,$$

we set $\psi(u) = \phi(u)$ for $u \in A \cup B_0$, and $\psi(u) = u$ for $u \in B-B_0$.

It follows from this theorem that every arithmetic property that holds in arbitrary directed refinement algebras is true of ordinal addition. We recall here some of the more interesting results proved in this manner by Tarski. Following his notation, we write $a \le b$ if $b = a \oplus x$ for some $x$, and $a \le^* b$ if $b = x \oplus a$ for some $x$. Observe that, in the algebra of isomorphism types, scalar multiples corresponding to ordinal addition should be written $\underline{n}\circ a$ rather than $na$. E.g. $a \oplus a = \underline{2}\circ a$.

THEOREM 11.2. (Tarski [1956]) *For any isomorphism types $a$, $b$, and $c$, and any positive integers $m$ and $n$, the following statements hold*:

(*i*) *If $a \le b$ and $b \le^* a$, then $a = b$.*
(*ii*) *If $\underline{n}\circ a \le \underline{n}\circ b$, then $a \le b$.*
(*iii*) *If $\underline{n}\circ a \le^* \underline{n}\circ b$, then $a \le^* b$.*
(*iv*) *If $\underline{n}\circ a = \underline{n}\circ b$, then $a = b$.*
(*v*) *If $\underline{m}\circ a \le \underline{n}\circ a$ and $n < m$, then $a \oplus a = a$.*
(*vi*) *If $\underline{m}\circ a = \underline{n}\circ b$, then $a \oplus b = b \oplus a$.*
(*vii*) *If $\underline{m}\circ a = \underline{n}\circ b$, and $m$ and $n$ are relatively prime, then for some $c$, $a = \underline{n}\circ c$ and $b = \underline{m}\circ c$.*
(*viii*) *If $\underline{m}\circ a = \underline{n}\circ b$, then for some $c$, and some positive integers $p$ and $q$, $a = \underline{p}\circ c$ and $b = \underline{q}\circ c$.*

## 12. ORDINAL ADDITION AND ORDINAL ALGEBRAS

There are arithmetic properties that are true for isomorphism types but do not hold in directed refinement algebras in general. Example: $(\underline{2}\circ a) \oplus b = b \oplus (\underline{2}\circ a)$ implies $a \oplus b = b \oplus a$. There are also many properties that involve infinite sums, and therefore cannot be formulated within the framework of directed refinement algebras. Ordinal algebras are defined to be structures $(M, \bigoplus, \oplus, *, 0)$ such that $(M, \oplus, 0)$ is a monoid, $*$ is an operation of rank one, and $\bigoplus$ is an operation on $\omega$-sequences of elements. The axioms involving the unary operation are simple:

$$a^{**} = a \quad \text{and } (a \oplus b)^* = b^* \oplus a^* \quad .$$

There are three axioms involving the operation $\bigoplus$. The first one is the associative law:

$$a_0 \oplus \bigoplus_{k \in \omega} a_{1+k} = \bigoplus_{k \in \omega} a_k \quad .$$

The second is the directed refinement postulate: If

$$\bigoplus_{k\in\omega} a_k = b \oplus c, \qquad c \neq 0 \quad ,$$

then for some $n \in \omega$ and $x,y \in M$,

$$b = \bigoplus_{k<n} a_k \oplus x, \qquad c = y \oplus \bigoplus_{k\in\omega} a_{n+k+1}, \qquad a_n = x \oplus y \quad .$$

Here $\bigoplus_{k<n} a_k$ is defined recursively, using the binary operation $\oplus$.

The final axiom is the remainder postulate: If

$$a_k = b_k \oplus a_{k+1} \qquad \text{for } k = 0,1,2,\ldots,$$

then for some $c \in M$,

$$a_n = \bigoplus_{k\in\omega} b_{n+k} \oplus c \qquad \text{for } n = 0,1,2,\ldots \quad .$$

THEOREM 12.1. *The isomorphism types of structures form an ordinal algebra under the operations of ordinal addition of* $\omega$*-sequences and of ordered pairs of types, and the operation of conversion.*

The proof is routine. To verify the remainder postulate, we start with a structure whose isomorphism type is $a_0$, and then successively for $n = 0,1,2,\ldots$ choose structures $B_n$ and $A_{n+1}$ with isomorphism types $b_n$ and $a_{n+1}$ such that

$$A_n = B_n \oplus A_{n+1} \quad .$$

For $c$ we then take the isomorphism type of the intersection of the structures $A_n$.

THEOREM 12.2. *If* $(M,\bigoplus,\oplus,*,0)$ *is an ordinal algebra, then* $(M,\oplus,0)$ *is a directed refinement algebra.*

The proof is not difficult. The old version of the directed refinement property readily follows from the new one once it has been verified that the sum of a sequence of zeroes is zero, and that $a \oplus b$ can therefore be written as the sum of the sequence $a,b,0,0,\ldots$ . The proof that $a \oplus b \oplus c = b$ implies $a \oplus b = b$ is somewhat more involved.

Some of the theorems proved by Tarski for ordinal algebras are listed below. In addition to the notation introduced in the preceding section we define

$$\omega\circ a = \bigoplus_{k\in\omega} a_k \quad ,$$

where $a_k = a$ for $k = 0,1,\ldots,$ and

$$\omega \circ * a = (\omega \circ a^*)*$$

THEOREM 12.3. (Tarski [1956]) *For any isomorphism types of structures $a$, $b$, $c$, and for any positive integers $m$ and $n$, the following statements hold:*

(*i*) *If* $\omega \circ a = b \oplus c$ *and* $c \neq 0$, *then for some* $x$, $y$, *and some* $k \in \omega$, $b = (\underline{k} \circ a) \oplus x$, $c = y \oplus (\omega \circ a)$, *and* $x \oplus y = a$.

(*ii*) $a \oplus b = b$ *iff* $\omega \circ a \leq b$.

(*iii*) $a \leq a^*$ *iff* $a = a^*$.

(*iv*) *If* $b \leq \omega \circ a$, $b \leq^* \omega \circ a$, *and* $b \neq 0$, *then* $b = \omega \circ a$.

(*v*) *If* $a \leq b \oplus c$, *then either* $a \leq b$ *or* $a = b \oplus x$, *for some* $x \leq c$.

(*vi*) *The following conditions are equivalent:*
- ($vi_1$) $a \oplus b \oplus c = b$.
- ($vi_2$) $\omega \circ a \leq b$ *and* $\omega \circ * c \leq^* b$.
- ($vi_3$) $b = (\omega \circ a) \oplus x \oplus (\omega \circ * c)$ *for some* $x$.

(*vii*) *If* $a \oplus b \oplus c = c$ *and* $\omega \circ (a \oplus b) = \omega \circ (b \oplus a)$, *then* $a \oplus c = b \oplus c = c$.

(*viii*) *If* $(\underline{n} \circ a) \oplus b = b$, *then* $a \oplus b = b$.

(*ix*) *If* $\omega \circ a \leq \underline{n} \circ b$, *then* $\omega \circ a \leq b$.

(*x*) *If* $a \oplus (\underline{n} \circ c) = b \oplus (\underline{n} \circ c)$, *then* $a \oplus c = b \oplus c$.

(*xi*) $\omega \circ a \leq b$ *iff* $\underline{k} \circ a \leq b$ *for all* $k \in \omega$.

(*xii*) $\omega \circ (\underline{n} \circ a) = \omega \circ a$.

(*xiii*) *If* $(\underline{m} \circ a) \oplus b = b \oplus (n \circ a)$, *then* $a \oplus b = b \oplus a$.

All the properties listed above hold in every ordinal algebra, as is shown in Tarski [1956]. As is also shown there, certain other properties which can be formulated for arbitrary ordinal algebras are not valid in general, although they hold for isomorphism types. Among these is the general associative law;

$$\bigoplus_{k \in \omega} \bigoplus_{p < q_k} a_{n_k + p} = \bigoplus_{p \in \omega} a_p \quad ,$$

where $0 = n_0 < n_1 < n_2 < \cdots$ and $q_k = n_{k+1} - n_k$, and even the special case

$$\omega \circ (a \oplus b) = a \oplus (\omega \circ (b \oplus a)) \quad .$$

There are still other properties that cannot be formulated in the language of ordinal algebras, e.g. properties that involve cardinalities of the structures, or involve addition with infinite index sets other than $\omega$. A particularly interesting example of this arises in connection with the problem of characterizing those pairs of isomorphism types that commute with each other. The history of this problem is related in Tarski [1956], p. 80. Obviously $a \oplus b = b \oplus a$ holds if $a$ and $b$ are multiples of the same type,

$$a = \underline{m} \circ c \quad \text{and} \quad b = \underline{n} \circ c \quad ,$$

and also if one of them absorbs the other, both on the left and on the right, i.e., if either

$$a = (\omega \circ b) \oplus c \oplus (\omega \circ * b)$$

or

$$b = (\omega \circ a) \oplus c \oplus (\omega \circ * a) \quad .$$

Tarski proved that if either $a$ or $b$ is countable, and if $a \oplus b = b \oplus a$, then one of these three conditions must be satisfied. On the other hand, Lindenbaum constructed uncountable totally ordered sets for which this is not the case. Neither result has been published, but in Aronszajn [1952] the commuting pairs were completely characterized, and the earlier results derived as corollaries. This characterization is rather involved, and will not be described here. It does involve summations over non-denumerable sets of real numbers.

## 13. LEXICOGRAPHIC PRODUCTS

Lexicographic products of reflexive structures were investigated in Chang [1955], [1961] and [1961a], and in Chang, Morel [1960]. Somewhat later, several graph-theorists became interested in this subject and, unaware of the earlier work, rediscovered many of the basic results.

The paper Chang, Morel [1960] contains two fundamental results.

THEOREM 13.1. *For any non-empty, reflexive structures $A$, $B$, $C$ and $D$, if $A \circ B \simeq C \circ D$ and $|B| = |D| < \aleph_0$, then $A \simeq C$.*

THEOREM 13.2. *For any non-empty, reflexive structures $A$, $B$, $C$ and $D$, if $A \circ B \simeq C \circ D$ and $A$ and $C$ are finite, then one of the structures $B$ and $D$ is isomorphic to a substructure of the other.*

The proof of the first theorem is not hard. Let $\varphi : A \circ B \simeq C \circ D$ be the given isomorphism, and let $n = |B| = |D|$. For $a \in A$ define $aB = \{(a,x) : x \in B\}$, and for $c \in C$ define $cD$ similarly. The set $C \times D$ is partitioned in two ways into $n$-element subsets, on the one hand the sets $cD$, on the other hand the sets $\varphi(aB)$. By deBruijn [1943], the two partitions have a common transveral $T$. It follows that there exists a bijection $\psi$ from $A$ to $C$ such that $\psi(a) = c$ iff $\varphi(aB)$ and $cD$ have the same representative in $T$. The proof is then completed by showing that $\psi : A \simeq C$.

The proof of the second theorem is more difficult, and we shall not attempt to describe it.

These two results have a number of interesting corollaries.

COROLLARY 13.3. *For any reflexive structures* $A$, $B$ *and* $C$, *if* $A \circ C \simeq B \circ C$, *and if* $C$ *is non-empty and finite, then* $A \simeq B$.

COROLLARY 13.4. *For any finite, reflexive structures* $A$ *and* $B$, *and any positive integer* $n$, *if* ${}^nA \simeq {}^nB$, *then* $A \simeq B$.

COROLLARY 13.5. *For any non-empty, finite, reflexive structures* $A$, $B$, $C$ *and* $D$, *if* $A \circ B \simeq C \circ D$, *and if* $|B| \simeq |D|$, *then* $A \simeq C$ *and* $B \simeq D$.

It is obvious that the factorization of a finite, reflexive structure into a lexicographic product of lexicographically indecomposable structures is not in general unique in the strict sense, for non-isomorphic, lexicographically indecomposable structures sometimes commute. E.g., any two finite chains commute, and so do any two antichains, and any two complete graphs. The factorization is not even unique up to the order of the factors. Counterexamples can be obtained from the observation that if $m$ and $n$ are natural numbers and $X$ is any reflexive structure, then

$$m \circ ((X \circ m) + n) \simeq ((m \circ X) + n) \circ m \quad . \tag{1}$$

If $m$ is a prime, then $m$ is lexicographically indecomposable, and $(X \circ m) + n$ and $(m \circ X) + n$ will also be indecomposable for suitable choices of $X$ and $n$. The case $m = 2$, $n = 1$, and $X = \underline{2}$ is illustrated in Fig. 2.

To describe another related counterexample we need the notions of complementation and of coaddition. The complement $A^-$ of a reflexive structure $A$ is defined to be the reflexive structure with the same elements as $A$ such that, for any distinct members $p$ and $q$ of $A$, the condition $p \to q$ holds in $A^-$ just in case it fails in $A$. The coaddition, $+^-$, is defined by the formula

$$A +^- B = (A^- + B^-)^- \quad .$$

Pictorially, the graph of $A +^- B$ is obtained by drawing the graphs of $A$ and $B$ and joining each vertex of $A$ to each vertex of $B$ by a double arrow. This operation was considered in the 1950's by A. Tarski, who called it square addition, and it was used in Chang [1955] and [1961a], and in Jónsson [1956]. The above definition using complementation can be found in Imrich [1972], where the terms "Kosumme" and "Zykovsumme" are used.

The coaddition has exactly the same arithmetic properties as cardinal addition, for the correspondence $A \to A^-$ is an isomorphism between the two operations. It is also easy to check that

$$(A \circ B)^- = A^- \circ B^- \quad .$$

With the aid of these observations we infer from (1) that

$$m^{-}\circ((X\circ m^{-})+^{-}n^{-}) \simeq ((m^{-}\circ X)+^{-}n^{-})\circ m^{-}$$

whenever $m$ and $n$ are natural numbers and $X$ is a reflexive graph. (Observe that $m^{-}$ and $n^{-}$, the complements of the antichains $m$ and $n$, are complete graphs.) It is also clear that $(X\circ m^{-}) +^{-} n^{-}$ and $(m^{-}\circ X) +^{-} n^{-}$ are lexicographically indecomposable iff $(X^{-}\circ m) + n$ and $(m\circ X^{-}) + n$ are.

The principal result from lexicographic products, Theorem 13.9 below, shows that the lack of uniqueness of the factorization is completely accounted for by the situations described above. This result appeared in Chang [1955] and [1961a], and a different proof can be found in Dörfler and Imrich [1972]. The formulation used here is essentially the one given by Dörfler and Imrich.

LEMMA 13.6. *For any non-empty, finite, reflexive structures* $A$ *and* $B$, $A\circ B \simeq B\circ A$ *iff either* $A$ *and* $B$ *are both chains, or both antichains, or both complete graphs, or else* $A \simeq {}^{m}E$ *and* $B \simeq {}^{n}E$ *for some structure* $E$ *and natural numbers* $m$ *and* $n$.

LEMMA 13.7. *Suppose* $A$, $B$, $C$ *and* $D$ *are non-empty, finite, reflexive, lexicographically indecomposable structures with* $A\circ B \simeq C\circ D$ *and* $|C| \leq |A|$. *Then one of the following conditions holds:*

(*i*) $A \simeq C$ *and* $B \simeq D$.
(*ii*) $A \simeq D$ *and* $B \simeq C$, *and all four factors are chains, or all are antichains, or all are complete graphs.*
(*iii*) *For some non-empty, reflexive structure* $X$ *and positive integers* $m$ *and* $n$, *either* $B \simeq C \simeq m$ *and*

$$A \simeq (m\circ X) + n, \qquad D \simeq (X\circ m) + n,$$

*or else* $B \simeq C \simeq m^{-}$ *and*

$$A \simeq (m^{-}\circ X) +^{-} n^{-}, \qquad D \simeq (X\circ m^{-}) +^{-} n^{-} \quad .$$

LEMMA 13.8. *For any finite, reflexive structure* $X$ *and natural numbers* $m$ *and* $n$, *if one of the structures*

$$(m\circ X) + n \qquad \text{and} \qquad (X\circ m) + n$$

*is lexicographically indecomposable, then both are. The same is true of the structures*

$$(m^{-}\circ X) +^{-} n^{-} \qquad \text{and} \qquad (X\circ m^{-}) +^{-} n \quad .$$

By an elementary transformation of a sequence $(A_0,A_1,\ldots,A_n)$

of lexicographically indecomposable structures we shall mean a transformation that replaces two successive terms $A_i$ and $A_{i+1}$ by lexicographically indecomposable structures $A_i'$ and $A_{i+1}'$ with $A_i \circ A_{i+1} \simeq A_i' \circ A_{i+1}'$, leaving all the remaining terms unchanged. The preceding two lemmas completely describe all the elementary transformations.

With this terminology, the principal result can be stated very simply.

THEOREM 13.9. *For any two representations of a non-empty, finite, reflexive structure* $P$ *as a lexicographic product of lexicographically indecomposable structures,*

$$P \simeq A_0 \circ A_1 \circ \ldots \circ A_m \simeq B_0 \circ B_1 \circ \ldots \circ B_n \quad ,$$

*we have* $m = n$, *and the sequence* $(B_0, B_1, \ldots, B_n)$ *can be obtained from* $(A_0, A_1, \ldots, A_n)$ *by a finite number of elementary transformations.*

This result has rather interesting consequences concerning greatest common divisors and least common multiples.

THEOREM 13.10. *Any two non-empty, finite, reflexive structures* $P$ *and* $Q$ *have, relative to the lexicographic multiplications, a greatest common left divisor, and if they have any common left multiple, then they have a least common left multiple. The statement obtained by replacing "left" by "right" is also true.*

*This work was supported by NSF Grant MCS 7901735.

REFERENCES

N. Aronszajn (1952) Characterization of types of order satisfying $\alpha_0+\alpha_1 = \alpha_1+\alpha_0$, *Fund. Math.* 31, 65-96.

B. Banaschewski (1956) Hüllensysteme und Erweiterungen von Quasi-Ordnungen, *Z. Math. Logik, Grundlagen Math.* 2, 117-130.

H. Bauer (a) Garben und Automorphismen geordneter Mengen, preprint.

C. Bergman, R. McKenzie and Sz. Nagý (a), How to cancel exponents, to appear.

G. Birkhoff (1937), Extended arithmetic, *Duke Math. J.* 3, 311-316.

G. Birkhoff (1933) On the combination of subalgebras, *Proc. Camb. Phil. Soc.* 29, 441-464.

G. Birkhoff (1934) On the lattice theory of ideals, *Bull. Amer. Math. Soc.* 40, 613-619.

G. Birkhoff (1937a) Rings of sets, *Duke Math. J.* 3, 443-454.

G. Birkhoff (1940) Lattice theory, *Amer. Math. Soc. Colloq. Publ.* 35.

G. Birkhoff (1942) Generalized arithmetic, *Duke Math. J.* 9, 283-302.

G. Birkhoff (1948), Lattice theory, *Amer. Math. Soc. Colloq. Publ.* 25, 2nd ed.; 3rd ed. (1967).

N.G. de Bruijn (1943) Gemeenschappelijke Representantensystemen van twee Klassenindeelingen van een Versameling, *Niew Archief voor Wiskunde*, 22, 48-52.

C.C. Chang (1955), Cardinal and ordinal factorization of relation types, Ph.D. Thesis, U.C. Berkeley.

C.C. Chang (1961) Cardinal and ordinal multiplication of relational types, *Proc. Symposia in Pure Math.*, *Amer. Math. Soc.*, vol. 2, pp. 123-128.

C.C. Chang (1961a) Ordinal factorization of finite relations, *Trans. Amer. Math. Soc.* 101, 259-293.

C.C. Chang and Anne C. Morel (1960) Some cancellation properties for ordinal products of relations, *Duke Math. J.* 27, 171-182.

C.C. Chang, B. Jónsson and A. Tarski (1964) Refinement properties for relational structures, *Fund. Math.* 55, 249-281.

B.A. Davey and D. Duffus (a) Exponentiation and duality, this volume.

M.M. Day (1945) Arithmetic of ordered systems, *Trans. Amer. Math. Soc.* 58, 1-43.

W. Dörfler and W. Imrich (1972), Das lexicographische Produkt gerichteter Graphen, *Monatsheft für Math.* 76, 21-30.

D. Duffus (1978) Toward a theory of partially ordered sets, Ph.D. Thesis, University of Calgary.

D. Duffus, B. Jónsson and I. Rival (1978), Structure results for function lattices, *Canad. J. Math.* 30, 392-400.

D. Duffus and I. Rival (1978) A logarithmic property for exponents of partially ordered sets, *Canad. J. Math.* 30, 797-807.

D. Duffus and R. Wille (1979) A theorem on partially ordered sets of order preserving maps, *Proc. Amer. Math. Soc.* 76, 14-16.

E. Fuchs (1965) Isomorphismus der Kardinalpotenzen, *Arch. Math.* (Brno) 1, 83-93.

W. Hanf (1957) On some fundamental problems concerning isomorphisms of Boolean algebras, *Math. Scand.* 5, 205-217.

J. Hashimoto (1948) On the product decomposition of partially ordered sets, *Math. Japan* 1, 120-123.

J. Hashimoto (1951) On direct product decomposition of partially ordered sets, *Ann. of Math.* 54, 315-318.

J. Hashimoto and T. Nakayama (1950) On a problem of G. Birkhoff, *Proc. Amer. Math. Soc.* 1, 141-142.

F. Hausdorff (1906-1907) Unterzuchungen über Ordnungstypen, I-III, Ber. Math. - *Phys. Kl. Kg. Sächs Ges. Wiss.* 58, 106-169; IV-V, ibid 59, 84-159.

F. Hausdorff (1914) Grundzüge der Mengenlehre, Leipzig.

W. Imrich (1972) Assoziative Produkte von Graphen, *Österreich. Akad Wiss. Math.-Natur. Kl. S.-B.* II 180, 203-239.

B. Jónsson (1956) A unique decomposition theorem for relational addition, an appendix to A. Tarski (1956).

B. Jónsson and R. McKenzie (a) Powers of partially ordered sets: cancellation and refinement properties, to appear in *Math. Scand.*

J. Ketonen (1978) The structure of countable Boolean algebras, *Ann. of Math.* 108, 41-89.

S. Kinoshita (1953) A solution of a problem of R. Sikorski, *Fund. Math.* 40, 39-41.

A. Lindenbaum and A. Tarski (1926) Communication sur les recherches de la théorie des ensembles, *Comptes Rendus des Séances de la Societé des Sciences et des Lettres de Varsovie*, Cl. III, 19, 299-330.

L. Lovász (1967) Operations with structures, *Acta Math. Acad. Sci. Hungar.* 18, 321-328.

R. McKenzie (1971) Cardinal multiplication of structures with a reflexive relation, *Fund. Math.* 70, 59-101.

T. Nakayama (1948) Remark on direct product decompositions of a partially ordered system, *Math. Japan* 1, 49-50.

M. Novotný (1960) Über gewisse Eigenschaften von Kardinaloperationen, Spisy Přírod Fac. Univ. Brno 1960, 465-484.

J. Schmidt (1955) Lexicographische Operationen, *Zeitschr. Math. Logik, Grundlagen Math.* 1, 127-171.

J. Schmidt (1956) Zur Kennzeichnung der Dedekind-McNeilleschen Hülle einer geordneten Menge, *Arch. Math.* 7, 241-249.

A. Tarski (1949) Cardinal Algebras, *Oxford University Press.*

A. Tarski (1956) Ordinal algebras, *Studies in Logic and the Foundations of Mathematics.*

A.N. Whitehead and B. Russell (1912) *Principia Mathematica*, vol. 2.

R. Wille (1980) Cancellation and refinement results for function lattices, *Houston J. of Math.* 6, 431-437.

# EXPONENTIATION AND DUALITY

B.A. Davey
Department of Pure Mathematics
La Trobe University
Victoria, Australia 3083

D. Duffus
Department of Mathematics
Emory University
Atlanta, Georgia 30322

ABSTRACT

In a unified treatment of cardinal and ordinal arithmetic G. Birkhoff [18], [20] defined (cardinal) exponentiation of ordered sets: for ordered sets $X$ and $Y$, $X^Y$ (the power) is the set of all order-preserving maps of $Y$ (the exponent) to $X$ (the base) ordered componentwise. Our aim is to review a significant body of results concerning powers of ordered sets and to present some central open problems arising in recent work. Roughly, we have two topics: first, an analysis of the structure of powers and their symmetries; second, a study of duality results for lattice-ordered algebras.

Birkhoff [18] observed that exponentiation of ordered sets satisfies the usual laws for exponents and raised questions concerning cancellation laws for bases and exponents [20]: *Does either* $X^Y \cong X^Z$ *or* $Y^X \cong Z^X$ *imply* $Y \cong Z$ ? Early results appear in M. Novotný [59] and E. Fuchs [40]. A *logarithmic property* is used by D. Duffus and I. Rival [37] to obtain cancellation of bases for finite ordered sets. Using extensions of this method and the relation between symmetries of $X^Y$ and both product and exponential decompositions of $X^Y$, B. Jónsson and R. McKenzie [52] obtained many cancellation and refinement theorems. For instance, under conditions on ordered sets $A$, $B$, $C$, $D$ (too varied and technical to state here) they show that $A^C \cong B^C$ implies $A \cong B$, and that $A^C \cong B^D$ yields $E$, $X$, $Y$, $Z$ and isomorphisms $A \cong E^X$,



*I. Rival (ed.), Ordered Sets, 43–95.*

$B \cong E^Y$, $C \cong Y \cdot Z$, $D \cong X \cdot Z$. (See [16], [36], [39], [87] for related results.)

It remains to be seen if, in the finite case, cancellation of exponents is always possible. Other open problems concern simplification of conditions insuring refinement properties and unification of existing results (cf. [52]).

Concerning symmetries of powers, it is easy to see that for ordered sets $X$ and $Y$, $\mathrm{Aut}(X) \times \mathrm{Aut}(Y)$ is embeddable in $\mathrm{Aut}(X^Y)$. In [52] two questions are raised: under what conditions is the embedding full, and, if not full, can $\mathrm{Aut}(X^Y)$ still be described in terms of $\mathrm{Aut}(X)$ and $\mathrm{Aut}(Y)$? Jónsson and McKenzie [52] obtained some results: $\mathrm{Aut}(X^Y) \cong \mathrm{Aut}(X) \times \mathrm{Aut}(Y)$ if $X$ is a subdirectly irreducible lattice and $Y$ is finite and connected. Duffus and Wille [38] provided answers to both questions when $X$ is a lattice of finite length. Most recently, Jónsson [51], building on techniques of [52], has obtained much stronger results.

Birkhoff [19] proved that every finite distributive lattice $D$ is isomorphic to $\underset{\sim}{2}^{J(D)^d}$ where $\underset{\sim}{2}$ is the two-element chain and $J(D)^d$ is the dual of the ordered subset of join-irreducible elements of $D$. Implicit in this is the fact that the category $\underset{\sim}{D}_F$ of finite bounded distributive lattices is dually equivalent to the category $\underset{\sim}{P}_F$ of finite ordered sets. This duality was extended to infinite bounded distributive lattices by H.A. Priestley [61], [62]: the category $\underset{\sim}{D}$ of all bounded distributive lattices is dually equivalent to the category $\underset{\sim}{P}$ of all compact Hausdorff ordered spaces which are embeddable in a power of $\underset{\sim}{2}$.

Further extension of this duality is possible. A finite lattice $L$ is *order-polynomial-complete* if for every positive integer $n$, every order-preserving map of $L^n$ to $L$ is an $n$-ary polynomial (alias, algebraic) function [70], [86], [87]. For instance, Wille [85] showed that every finite simple lattice whose greatest element is the join of atoms is order-polynomial-complete. A proof of Priestley's duality result utilizing the order-polynomial-completeness of $\underset{\sim}{2}$ is given in B.A. Davey [28]. The implicit possibility that a similar duality holds for each order-polynomial-complete lattice was exploited in B.A. Davey, D. Duffus, R.W. Quackenbush and I. Rival [30] and further refined in B.A. Davey and I. Rival [32].

The dual equivalence of $\underset{\sim}{D}$ (respectively, $\underset{\sim}{D}_F$) and $\underset{\sim}{P}$ (respectively, $\underset{\sim}{P}_F$) allows representation of both the objects and the morphisms of $D$ in terms of those of $\underset{\sim}{P}$ . Moreover the objects and morphisms of $\underset{\sim}{P}$ , especially of $\underset{\sim}{P}_F$ , are so pictorial that this duality is a very powerful tool for the study of particular bounded distributive lattices and the class they comprise, and for generating examples and counterexamples. Further, this duality reveals interplay between ordered sets and distributive-lattice-ordered algebras -- algebraic structures often arising in logic such as Boolean algebras, Heyting algebras, and pseudo-complemented distributive lattices. Provided that one can describe those ordered spaces in $\underset{\sim}{P}$ which arise as duals of these enriched structures and the continuous order-preserving maps in $\underset{\sim}{P}$ which are dual to their homomorphisms, the Birkhoff-Priestley result immediately yields a duality for the class of enriched structures. In this setting, natural algebraic questions can lead to order-theoretic considerations which are interesting in their own right. (B.A. Davey [28] and B.A. Davey and H. Werner [33] provide a general approach to duality theory.)

## 1. INTRODUCTION

Both topics of the title are rooted in work of G. Birkhoff begun in the 1930's. Developing an "extended arithmetic" of (partially) ordered sets he defined (cardinal) exponentiation [18] : for ordered sets $X$ and $Y$, $X^Y$ (the power) is the collection of all order-preserving maps of $Y$ (the exponent) to $X$ (the base) ordered by $f \leq g$ if $f(y) \leq g(y)$ for all $y \in Y$. He obtained results concerning the arithmetic of this operation [18], [20] and posed several problems. At the same time, Birkhoff [17] obtained his representation theorem for distributive lattices (cf. [74]): every distributive lattice is isomorphic to a ring of sets. Exponentiation appears in the finite version [19] : every finite distributive lattice $D$ is isomorphic to $\underset{\sim}{2}^P$ where $\underset{\sim}{2}$ denotes the two-element chain and $P$ is isomorphic to $J(D)^d$, the dual of the ordered subset of join irreducible elements of $D$. Implicit in this correspondence is a dual category equivalence between finite bounded distributive lattices and finite ordered sets.

It is our aim to review two areas of investigation arising from Birkhoff's work.

First, as an operation on ordered sets, exponentiation invites questions of an arithmetical nature. For instance, for which ordered sets $A$, $B$, $C$, and $D$ does

$A^B \cong A^C$ imply $B \cong C$ (cancellation of bases),
$B^A \cong C^A$ imply $B \cong C$ (cancellation of exponents),
$A^C \cong B^D$ imply that there exist ordered sets $E$, $X$, $Y$, and $Z$ such that $A \cong E^X$, $B \cong E^Y$, $C \cong YZ$, and $D \cong XZ$ (refinement of powers)?

Recently, several approaches have given results on these questions. One makes use of a "logarithmic" property of certain isomorphism-invariant subsets of an ordered set. Roughly, if for an ordered set $A$, a subset $S(A)$ satisfies $S(A^B) \cong S(A) \times B$, we have a technique applicable to cancellation of bases, provided we have cancellation for products. These approaches are discussed in section 3. In section 4 we outline methods for obtaining an isomorphism of $B$ onto $C$ from one of $B^A$ onto $C^A$. Such a link between isomorphisms of powers and isomorphisms of bases leads one to ask about the relation between the automorphism group of a power $X^Y$ and those of $X$ and $Y$. In section 5 we state some recent results describing this relation (results based on the methods of sections 3 and 4).

Second, we discuss extensions of Birkhoff's representation theorem for distributive lattices and attendant duality theorems. Section 6 contains results concerning order-polynomial-complete lattices, one of which shows that for every such finite lattice $L$, regarded as a suitable lattice-ordered algebra, every finite member of the variety it generates is of the form $L^P$ for some finite ordered set $P$. (This is an extension of Birkhoff's representation theorem.) Moreover, a dual category equivalence holds. The dual category equivalences obtained between categories of distributive-lattice-ordered algebras and categories of ordered topological spaces afford interplay between these objects. In section 7 we display some of this interplay and both its application to the study of distributive-lattice-ordered algebras and how natural algebraic questions give rise to interesting order-theoretic considerations.

It is not our object to be exhaustive. Rather, we attempt to highlight some central ideas and some successful strategies recently employed in studies and applications of exponentiation. Where appropriate we outline the proof of a special case rather than the general result. Enough detail is provided, we hope, to impart the flavour of these investigations.

## 2. PRELIMINARIES

For ordered sets $X_i$ $(i \in I)$, $X_i + X_j$ (or, $\sum_{i\in I} X_i$) denotes the (*disjoint*) *sum* of $X_i$ and $X_j$ (or, of $X_i$ $(i \in I)$). An ordered set $X$ is connected if $X = Y + Z$ implies $Y = \emptyset$ or $Z = \emptyset$; $X = \sum_{i\in I} X_i$ is a sum representation of $X$ by its connected components if each $X_i$ is connected. Let $X_iX_j$, or $X_i \times X_j$ (or, $\prod_{i\in I} X_i$) denote the usual (*direct*) *product* of $X_i$ and $X_j$ (or, of $X_i$ $(i \in I)$). We make use of the various properties of sum and product (see [50]). Let us mention only two. J. Hashimoto [46] showed that for connected ordered sets $A$, $B$, $C$, and $D$, an isomorphism $AB \cong CD$ has a common refinement. That is, there exist ordered sets $R$, $S$, $T$, and $U$ and isomorphisms $A \cong RS$, $B \cong TU$, $C \cong RT$, and $D \cong SU$. (In fact, connected ordered sets have the strict refinement property - see [50]). Note that in the presence of some finiteness condition the common refinement property yields unique factorization into product nondecomposable ordered sets (that is, admitting only trivial product decompositions) and, of course, cancellation for products. Concerning the latter, L. Lovász [57] showed that for all finite ordered sets $A$, $B$, and $C$, $AB \cong AC$ implies $B \cong C$.

As Birkhoff [18], [20] observed, exponentiation satisfies the usual laws of exponents for numbers. For ordered sets $A$, $B$, and $C$,

$$(AB)^C \cong A^C B^C, \qquad A^{BC} \cong (A^B)^C, \qquad A^{B+C} \cong A^B A^C.$$

Moreover, if $C$ is connected then $(A+B)^C \cong A^C + B^C$. Using $n$ to denote an antichain (totally unordered set) of cardinality $n$, $1^A \cong 1$, $nA$ is isomorphic to the $n$-fold sum, and $A^n$, to the $n$-fold product.

Let us also note that if $A$ is a lattice then $A^B$ is a lattice, in fact, $A^B$ satisfies precisely the lattice identities satisfied by $A$ [20]. For instance, letting $\underset{\sim}{n}$ denote an $n$-element chain (totally ordered set) then $\underset{\sim}{n}^B$ is a distributive lattice, actually $\underset{\sim}{n}^B \cong \underset{\sim}{2}^{\underset{\sim}{n-1} \times B}$. Still concerning a lattice $A$, $J(A)$ (respectively, $M(A)$) denotes the ordered set of nonzero join irreducible (respectively, nonunit meet irreducible) elements of $A$, while $0_A$ (respectively, $1_A$) denotes the zero or minimum element (respectively, unit or maximum element) of $A$, if it exists. For arbitrary ordered sets $A$ and $B$, the ordered subset of constant maps in $A^B$ is called the *diagonal* $\Delta(A^B)$ of $A^B$. Obviously, $\Delta(A^B) \cong A$ (cf. Figure 1).

Recall that $a$ *covers* $a'$ in $A$, denoted by $a' \prec a$, if $a' < a$ and $a' \le b < a$ implies $a' = b$.

LEMMA 2.1 (cf. [21],[37]). *For ordered sets $A$ and $B$, $g \prec f$ in $A^B$ if and only if there exists $b \in B$ with $g(b) \prec f(b)$ in $A$ and $g|B-\{b\} = f|B-\{b\}$.* □

The *length* $\ell(A)$ of an ordered set $A$ is the supremum of the lengths of its chains, where the length of a chain $C$ is $|C| - 1$.

LEMMA 2.2 (cf. [21],[37]). *For finite ordered sets $A$ and $B$, $\ell(A^B) = \ell(A) \cdot |B|$.*

Proof. By Lemma 2.1, $\ell(A^B) \leq \ell(A)\cdot|B|$. For the opposite inequality let $C \subseteq A$ be a chain with $\ell(A) = \ell(C) = n$. Then $C^B \subseteq A^B$ and $C^B \cong \underset{\sim}{2}^{\underset{\sim}{n}\times B}$; the length of a distributive lattice $D$ is $|J(D)|$ [19] (cf. Theorem 7.10), so $\ell(C^B) = \ell(C)\cdot|B| = \ell(A)\cdot|B|$. □

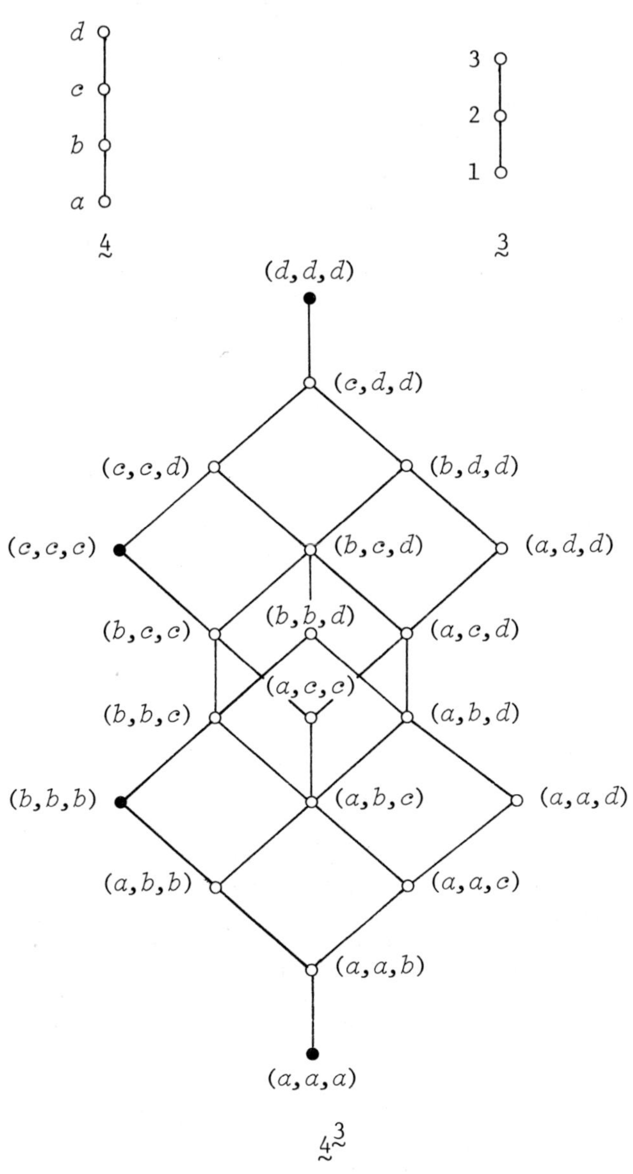


Figure 1. The diagonal elements of $\underset{\sim}{4}^{\underset{\sim}{3}}$

Birkhoff [20] conjectured that for all finite ordered sets $A$, $B$, and $C$, $B^A \cong C^A$ implies $B \cong C$, and he observed that $A^B \cong A^C$ does not imply that $B \cong C$. For the latter, note that if $A$ is an antichain then $A^B$ is the set of all maps that are constant on each connected component of $B$, so $A^B \cong A^C$ provided only that $B$ and $C$ have the same number of connected components. B. Jónsson and R. McKenzie [52] provide another example based on a Boolean algebra $H$ satisfying $\mathbf{2}H \ncong H$ and $\mathbf{2}^2 H \cong H$. Let $B$ be the sum of countably many copies of $H$, and let $C = B + \mathbf{2}H$. Then $(\mathbf{1}+\mathbf{2})B \cong (\mathbf{1}+\mathbf{2})C$, so with $A = \mathbf{2}^{\mathbf{1}+\mathbf{2}}$ we have $A^B \cong A^C$ and $B \ncong C$. Another example of Jónsson and McKenzie [52] shows that neither cancellation of exponents nor refinement of powers holds without restriction. Let $A = \mathbf{2}^{\omega}$, the lattice of subsets of a countable set, and let $B$ and $C$ be the ordered sets of Figure 2. Then $B^A \cong C^A$ — both ordered sets are two copies of $\mathbf{2}^{\mathbf{2}^{\omega}}$ with minimum elements identified. (See [52] for further examples.)

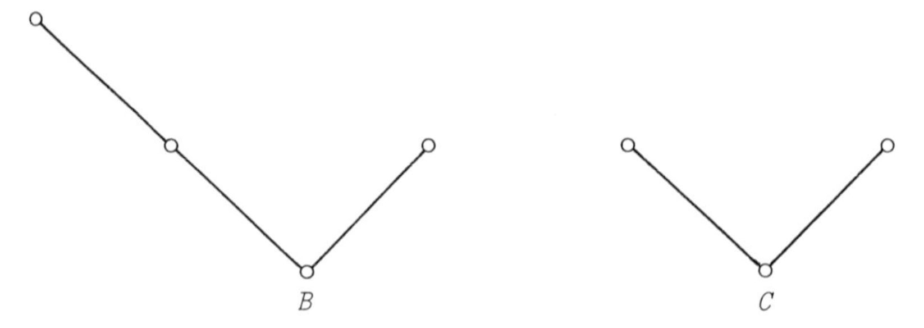


Figure 2. With $A = \mathbf{2}^{\omega}$, $B^A \cong C^A$.

We end this section with a list of results bearing on cancellation and refinement problems. (The list is not complete; because of technical conditions some of the results of [52] are not included or are deferred to section 3.) Note that the following theorems are not independent, rather, the listing is roughly chronological. Also, a few of the more easily stated open problems are given.

A. Cancellation of bases : $A^B \cong A^C$ *implies* $B \cong C$ *if*

*(1)* $A$ *is a nontrivial chain* (Novotny [59]),

*(2)* $A$, $B$, *and* $C$ *are finite ordered sets,* $A$ *is nontrivial with a minimum element* (Fuchs [40]),

*(3)* $A$, $B$, *and* $C$ *are finite,* $A$ *is nontrivial and not an antichain* (Duffus and Rival [37], cf. Theorem 3.2),

*(4) $A$ is a lattice of finite length, $B$ and $C$ are finite* (Wille [87], cf. section 3),

*(5) $A$ is nontrivial of finite length and has a minimum element, $B$ and $C$ are finite* (Jónsson and McKenzie [52]).

Statements (1), (3), and (5) are independent, and (3) settles the finite case. See also Theorem 3.6 where extra conditions are imposed on $A$ and no assumptions are made about $B$ and $C$. Some extension of (5) is possible, for instance, dropping the condition that $A$ has a minimum element.

B. Cancellation of exponents : $B^A \cong C^A$ *implies* $B \cong C$ *if*

*(1) $A$, $B$, and $C$ are finite and $A$ is an antichain* (Lovász [57]),

*(2) $B$ and $C$ are finite lattices, $A$ is finite and bounded* (Duffus and Rival [37], cf. Theorem 4.1),

*(3) $B$ and $C$ are lattices of finite length, $A$ is finite* (Wille [87], cf. section 3),

*(4) $A$ is a chain* (Bergman, McKenzie, and Nagý [16]),

*(5) $B^A$ is atomic, $A$ is a product of finitely many product nondecomposable ordered sets and has a maximum element* (Jónsson and McKenzie [52]),

*(6) $B$ and $C$ are meet semilattices, $A$ is finite with a maximum element* (Jónsson and McKenzie [52]).

Statements (1), (3), (4), (5), and (6) are independent. See also Theorem 3.5(a). It is not known whether cancellation of exponents holds for all finite $A$, $B$, and $C$.

C. Refinement of powers : $A^C \cong B^D$ *implies there exist ordered sets* $E$, $X$, $Y$, *and* $Z$ *such that* $A \cong E^X$, $B \cong E^Y$, $C \cong YZ$, *and* $D \cong XZ$ *if*

*(1) $A$ and $B$ are product nondecomposable lattices of finite length, $C$ and $D$ are finite* (Wille [87]),

*(2) $A$ and $B$ are lattices of finite length, $C$ and $D$ are finite and connected* (Wille [87], cf. Theorem 3.7),

*(3) $A^C$ is atomic, $C$ and $D$ are products of finitely many product nondecomposable ordered sets and have maximum elements* (Jónsson and McKenzie [52]),

*(4) $A$ and $B$ are meet semilattices, $C$ and $D$ are finite with maximum elements* (Jónsson and McKenzie [52]).

Statement (2) is contained in Theorem 3.5(d). Refinement of powers does not hold for all finite ordered sets — see example 2.4 of [52]. However, it is an open question whether refinement holds if $A$, $B$, $C$, and $D$ are finite and connected, or, more strongly, if $C$, $D$, and $A^C$ are finite and connected.

Jónsson and McKenzie [52] also suggest that refinement can be considered if $A^C$ is a product nondecomposable lattice, or if $C$ and $D$ are bounded and have no infinite chains.

## 3. LOGARITHMIC PROPERTIES

Let $\phi$ be an isomorphism of the power $B^A$ onto $C^A$. If the constant maps in $B^A$ were mapped by $\phi$ onto constants of $C^A$ then cancellation of exponents would follow. This is far from the case. (See Figure 1 : there is an automorphism $\phi$ of $\underset{\sim}{4}^{\underset{\sim}{3}}$ such that $\phi(\Delta(\underset{\sim}{4}^{\underset{\sim}{3}})) \neq \Delta(\underset{\sim}{4}^{\underset{\sim}{3}})$.) We shall see in the next section that something can be salvaged from the diagonal. For now, let us look elsewhere for isomorphism-invariant subsets of powers.

In some special cases an ordered set is determined by a particular subset. As already noted, for any finite distributive lattice $D$, $D \cong \underset{\sim}{2}^{J(D)^d}$ and conversely, $J(\underset{\sim}{2}^A) \cong A^d$ for any finite ordered set $A$. If $A$ and $B$ are finite ordered sets and $D^A \cong D^B$ then $\underset{\sim}{2}^{J(D)^d \times A} \cong \underset{\sim}{2}^{J(D)^d \times B}$. Thus, $J(D)^d \times A \cong J(D)^d \times B$. Our question concerning cancellation of bases becomes : *does* $J(D)^d \times A \cong J(D)^d \times B$ *imply* $A \cong B$? This follows from cancellation for products of finite ordered sets [57]. For more general results we require a subset $S(A)$ of an ordered set $A$ such that $S(A^B) \cong S(A) \times B$ — a "logarithm".

According to E. Fuchs [40] M. Novotny obtained the result, $A^B \cong A^C$ implies $B \cong C$, where $A$ is a join semilattice with minimum element, by relating "irreducible" elements of $A^B$ to those of $A$. Fuchs [40] showed cancellation of bases holds if $A$, $B$, and $C$ are finite and $A$ has a minimum element; his proof also rests on description of the "irreducible" elements of a power.

Let us present a logarithmic property obtained by D. Duffus and I. Rival [37]. For an ordered set $A$, let $V(A)$ denote the ordered subset of those elements of $A$ each with a unique lower cover in $A$ (cf. Figure 3).

LEMMA 3.1 ([37]). *Let* $A$ *and* $B$ *be ordered sets each with no infinite chains. If* $A$ *has a minimum element then* $V(A^B) \cong V(A) \times B^d$.

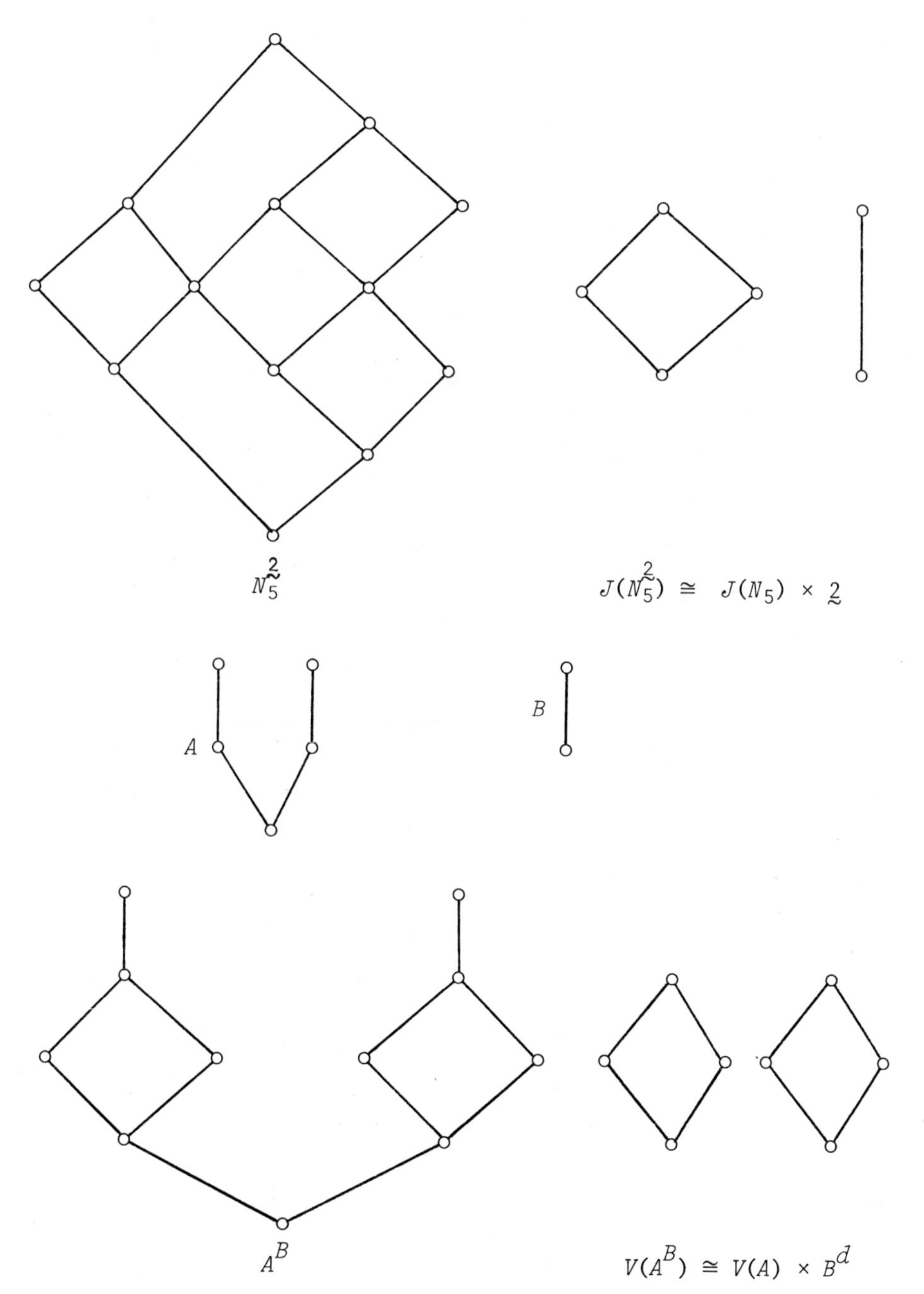


Figure 3. Illustrations of the logarithmic property

Proof. Let $a \in V(A)$ have unique lower cover $a_*$ and let $b \in B$. Define $f_{a,b} \in A^B$ by $f_{a,b}(x) = a$ if $x \geq b$ and $f_{a,b}(x) = 0_A$ if $x \ngeq b$. It is clear that $f_{a,b}$ has a unique lower cover in $A^B$, namely, the map $g \in A^B$ with $g(x) = a$ if $x > b$, $g(b) = a_*$, and $g(x) = 0_A$ if $x \ngeq b$. Also, the set of all such $f_{a,b} \in A^B$ is isomorphic to $V(A) \times B^d$ via the map assigning $(a,b)$ to $f_{a,b}$. The result follows by showing that for each $f \in V(A^B)$, $f = f_{a,b}$ for some $(a,b) \in V(A) \times B^d$. This is done by showing first that $f(B)$ contains a unique nonzero element $a$ of $A$, second, that $a \in V(A)$, and third, that $f^{-1}(a)$ has a minimum element $b \in B$. □

If $A$ is a finite lattice then $V(A) = J(A)$, so $J(A^B) \cong J(A) \times B^d$, extending the observations concerning distributive lattices. (See Figure 3, where $N_5$ is the nonmodular five-element lattice.)

The logarithm gives Fuchs' [40] result : if $A, B$, and $C$ are finite and $A$ has a minimum element and $A^B \cong A^C$ then $V(A^B) \cong V(A^C)$ so $V(A) \times B^d \cong V(A) \times C^d$ giving $B \cong C$. More generally, there is

THEOREM 3.2 ([37]). *Let $A$ be a finite ordered set that is not an antichain and let $B$ and $C$ be finite ordered sets. Then $A^B \cong A^C$ implies $B \cong C$.*

Sketch of proof. The logarithmic property cannot be applied directly — this requires a minimum element of $A$. Yet we can consider a suitable subset of $A^B$ with minimum.

For an element $x$ of an ordered set $X$, let $[x)$ denote the order filter of all $y \in X$ such that $y \geq x$, and let the *depth* $\delta(x)$ of $x$ be length $\ell([x))$ of $[x)$. Note that $x$ has maximum depth if and only if $\delta(x) = \ell(X)$.

Choose $f \in A^B$ such that $\delta(f) = \ell(A^B)$. By Lemmas 2.1 and 2.2, $\ell(A) \cdot |B| = \delta(f) \leq \sum_{b \in B} \delta(f(b))$, so each $f(b)$ has maximum depth in $A$. To simplify the discussion suppose $A$ has a unique element $a$ of maximum depth. Then $f(B) = \{a\}$ and $[f) \subseteq A^B$ satisfies $[f) \cong [a)^B$. Also, $V([f)) \cong V([a)^B) \cong V([a)) \times B^d$ by Lemma 3.1. Note that $V([a)) \neq \emptyset$ since $\delta(a) = \ell(A) \geq 1$.

Let $\phi$ be an isomorphism of $A^B$ onto $A^C$ and let

$\phi(f) = g$. Of course, $g$ has maximum depth in $A^C$ and $[f) \cong [g)$. It follows that

$$V([a)) \times B^d \cong V([f)) \cong V([g)) \cong V([a)) \times C^d$$

so $B \cong C$, by Lemma 3.1.

In general $A$ has more than one element of maximum depth and $f$ is identified with an $n$-tuple $(a_1,a_2,\ldots,a_n)$ where $\delta(a_i) = \ell(A)$, $n$ is the number of connected components of $B$, and $a_1,a_2,\ldots,a_n$ are the images of $B$ under $f$. One must argue from this and the correspondence $\phi$ provides between the set of all functions of maximum depth in $A^B$ and that set in $A^C$. □

Let us apply the logarithmic property to an isomorphism $A^A \cong B^B$.

L.M. Gluskin [43] showed that an ordered set $A$ is determined up to duality by $E(A)$, the set of all order-preserving maps of $A$ to $A$, considered as a semigroup. The constant maps are readily identified in $E(A)$ and the order of $A$ can be recovered from the two-valued maps. Does $A^A$, the same set considered as an ordered set, determine $A$?

THEOREM 3.3 ([39]). *For finite, connected ordered sets* $A$ *and* $B$, $A^A \cong B^B$ *implies* $A \cong B$, *and* $A^B \cong B^A$ *implies* $A \cong B$.

Sketch of proof. Let $\phi$ be an isomorphism of $A^A$ onto $B^B$. Let $f \in A^A$ satisfy $\delta(f) = \ell(A^A)$. Since $A$ is connected, $f(A)$, consisting only of elements of maximum depth in $A$, is connected and, thus, is a singleton. Say $f(A) = \{a\}$. Let $\phi(f) = g$ where $g(B) = \{b\}$ for some $b \in B$. Then $[a)^A \cong [f) \cong [g) \cong [b)^B$, so Lemma 3.1 yields

$$V([a)) \times A^d \cong V([b)) \times B^d. \tag{1}$$

That $A$ and $B$ are connected allows application of the common refinement theorem [46] to (1); $V([a))$ and $V([b))$ are not connected but this is no difficulty (cf. [35], [50]) so there are ordered sets $R$, $S$, $T$, $U$, such that

$$V([a)) \cong RS^d, \quad A \cong TU, \quad V([b)) \cong RT^d, \quad B \cong SU.$$

By considering $a$ and $b$ as elements of $TU$ and $SU$, respectively, and using the "small" size of $V([a))$ and $V([b))$

relative to $|A|$ and $|B|$, one can show that $|S| = |T| = 1$. Thus, $A \cong U \cong B$.

Essentially the same argument shows that $A^B \cong B^A$ implies $A \cong B$. □

It is not known whether both finiteness and connectedness are required for the conclusion of Theorem 3.3. (For antichains $A$ and $B$ of some infinite cardinalities, $A \cong B$ need not follow from $A^A \cong B^B$.) One might think that at least for finite ordered sets $A$ and $B$ we could write $A \cong \sum A_i$, $B \cong \sum B_j$, use the various laws of exponents, and examine the isomorphisms present between connected components of $(\sum A_i)^{\sum A_i}$ and $(\sum B_j)^{\sum B_j}$. One problem : $A_i$ being connected does not imply $A_i^{A_i}$ is connected (cf. [35], [52]).

There are choices for the logarithm of $A$ aside from $V(A)$ which are more useful if $A$ is infinite. Jónsson and McKenzie [52] employ a modified version of strict join irreducibility. Let $J'(A)$ be the set of all $a \in A$ such that there exist $a', b \in A$ with $a < b$, $a' \leq b$, $a \nleq a'$ and $x \leq a'$ whenever $x < a$. The element $b$ is introduced in order to obtain the following description : if $A$ has a minimum element then $f \in J'(A^B)$ if and only if there exist $a \in J'(A)$ and $b \in B$ such that $f(x) = a$ if $x \geq b$ and $f(x) = 0_A$ if $x \ngeq b$ [52]. In other words, $J'(A^B) \cong J'(A) \times B^d$.

Let us briefly and, because of the detail and length of argument, loosely describe Jónsson and McKenzie's use of the logarithm $J'$ [52].

Let $\phi$ be an isomorphism of $A^B$ onto $C^D$ where $A^B$, $A$, and $C$ have minimum elements. Of course, $\phi$ induces an isomorphism $\psi$ of $J'(A) \times B^d$ onto $J'(C) \times D^d$. The strict refinement property for product decompositions is applied to each of the isomorphisms induced by $\psi$ of a connected component of $J'(A) \times B^d$ to a component of $J'(C) \times D^d$ (see [52], section 7). (Very loosely, if $f \in J'(A^B)$ corresponds to $(a,b) \in J'(A) \times B^d$ then comparabilities of $f$ with $g \in A^B$ can be linked to comparabilities between $a$ and elements of the range of $g$ and these followed to components of $J'(C) \times D^d$ via the isomorphism given by strict refinement.)

Now impose a condition on $A$ ensuring that each element $x$ of $A$ is the supremum of $\{y \in J'(A) \mid y \leq x\}$, the condition denoted by $\nabla J'(A) = A$. Then $\nabla J'(A^B) = A^B$. Jónsson and McKenzie then show that $\phi$ satisfies certain general conditions which give cancellation or refinement (see section 3 of [52] and section 4 below). A selection of their results follows.

THEOREM 3.5 ([52]). *Let $A$, $B$, $C$, and $D$ be ordered sets such that $A$ has a minimum element, $\nabla J'(A) = A$, and $C$ and $D$ are connected.*

(a) *If $C$ has a factorization into finitely many product nondecomposable ordered sets then $A^C \cong B^D$ implies there exist ordered sets $E$, $X$, $Y$, and $Z$ with $A \cong E^X, B \cong E^Y$, $C \cong YZ$, and $D \cong XZ$. In particular, if $C \cong D$ then $A \cong B$.*

(b) *If $J'(A)$ is connected then $A^C \cong B^D$ implies that there exist ordered sets $E$, $X$, $Y$, and $Z$ with $A \cong E^X$, $B \cong E^Y$, $C \cong YZ$, and $D \cong XZ$.* □

Concerning cancellation of bases, the logarithmic property of $J'$ is used to obtain

THEOREM 3.6 ([52]). *Let $A$ be a nontrivial ordered set of finite length, with a minimum element, and such that $J'(A)$ is connected and $\nabla J'(A) = A$. Then for any ordered sets $B$ and $C$, $A^B \cong A^C$ implies $B \cong C$.* □

Let us present one other variant of the logarithm useful for lattices of finite length.

Throughout the rest of this section $L$, $M$, and $N$ denote lattices of finite length, and $P$ and $Q$ denote finite ordered sets. R. Wille [84] defined the *scaffolding* $G(L)$ of $L$ to be the set of subirreducible elements of $L$. An element $x$ of $L$ is *subirreducible* if $x \neq 0_L$ and there is a homomorphism $\phi$ of $L$ onto a subdirectly irreducible lattice such that $x$ is the infimum of $\phi^{-1}(\phi(x))$. Regard $G(L)$ as a $(2,3)$-relational structure : the binary relation is the ordering induced by $L$ and the ternary relation $\nu$ is given by $\nu(x,y,z)$ if $x = y = z$ or $x \vee y = z$ and $x \nleq y$, $y \nleq x$. (Observe that if $L$ is distributive then $G(L) = J(L)$ and $\nu$ is the diagonal relation.) A subset $X$ of $G(L)$ is an *ideal* if $z \leq x \in X$ and $x \in G(L)$ imply $z \in X$, and if $x,y \in X$, $z \in G(L)$, and $\nu(x,y,z)$ imply $z \in X$.

A lattice of finite length is determined by its scaffolding.

The map that associates $x \in L$ with the ideal $\{y \in G(L) \mid y \leq x\}$ is an isomorphism of $L$ onto the lattice $I(G(L))$ of all ideals of $G(L)$ [84]. Moreover, Wille [87] obtains a logarithmic property : $G(L^P) \cong G(L) \times P^d$ where $P$ has the diagonal ternary relation and the isomorphism is a (2,3)-isomorphism. Thus

(1) $L^P \cong L^Q$ implies $G(L) \times P^d \cong G(L) \times Q^d$,

(2) $L^P \cong M^P$ implies $G(L) \times P^d \cong G(M) \times P^d$,

where the latter isomorphisms in (1) and (2) are (2,3)-isomorphisms. Some work is required to extend the usual common refinement of connected ordered sets to obtain cancellation (cf. [87]). Once accomplished (1) gives $P \cong Q$ and (2) yields $G(L) \cong G(M)$, from which $L \cong M$ follows [87].

Wille [87] obtains several other theorems with the scaffolding. We give one example of a refinement result.

THEOREM 3.7 ([87]). *Let $L^P \cong M^Q$ and let $P$ and $Q$ be connected. Then there exists a lattice $K$ and ordered sets $R$, $S$, and $T$ such that $L \cong K^R$, $M \cong K^S$, $P \cong ST$, and $Q \cong RT$.* □

The proof rests on two facts. First, if $G(L)$ is (2,3)-isomorphic to $GP^d$ and $\nu$ is the diagonal relation on $P$ then there is a lattice $K$ such that $G(K) \cong G$ and $L \cong K^P$ [87]. Second, $L^P \cong M^Q$ implies $G(L) \times P^d \cong G(M) \times Q^d$. Applying strict refinement for product decompositions to the (2,3)-isomorphism $G(L) \times P^d \cong G(M) \times Q^d$ we obtain order-isomorphisms $G(L) \cong GR^d$, $P \cong ST^d$, $G(M) \cong GS^d$, and $Q \cong RT^d$. One can show that the ternary relations on $R$, $S$, and $T$ projected by the appropriate isomorphisms from the ternary relations on $G(L)$ and $G(M)$ are each diagonal relations. Now apply the first fact to $G(L) \cong GR^d$ and $G(M) \cong GS^d$.

## 4. ANOTHER APPROACH.

Usually a logarithm $S(A)$ of an ordered set $A$ does not itself give cancellation of exponents. From $B^A \cong C^A$, one has $S(B) \times A \cong S(C) \times A$, from which $S(B) \cong S(C)$ may follow. It is necessary that, through some construction, $S(B) \cong S(C)$ implies $B \cong C$. If this fails, another approach is required.

In this section we outline two examples of alternative methods. The first, due to Duffus and Rival [37], is, at least in its strategy, straightforward and will be described in some detail.

THEOREM 4.1 ([37]). *Let* $L$ *and* $K$ *be finite lattices and let* $Q$ *be a finite bounded ordered set. If* $L^Q \cong K^Q$ *then* $L \cong K$. □

Lemma 3.1 shows that $J(L) \cong J(K)$ and, by its dual, $M(L) \cong M(K)$. Neither isomorphism alone shows $L \cong K$. However, if $P(L) = J(L) \cup M(L)$, the ordered subset of all irreducible elements, satisfies $P(L) \cong P(K)$, we can conclude that $L \cong K$. The conclusion is obtained from this fact : if $N(A)$ denotes the normal completion of an ordered set $A$ then $N(P(L)) \cong L$ (cf. [14], [58], [69]). Duffus and Rival proved that a fixed isomorphism $\phi$ of $L^Q$ onto $K^Q$ induces an isomorphism $\psi$ of $P(\Delta(L^Q))$ onto $P(\Delta(K^Q))$, and so conclude that $P(L) \cong P(K)$ and $L \cong K$.

Let $\phi$ be an isomorphism of $L^Q$ onto $K^Q$. Let $0_Q = 0$ and let $1_Q = 1$. Also, for $a \in L$ and $x \in Q$ let $a^x$ and ${}^x a$ denote elements of $L^Q$ defined by

$$a^x(y) = \begin{cases} a & \text{if } y \geq x, \\ 0 & \text{if } y \not\geq x \end{cases} \qquad {}^x a(y) = \begin{cases} a & \text{if } y \leq x, \\ 1 & \text{if } y \not\leq x. \end{cases}$$

Lemma 3.1 and its dual show that

$$J(L^Q) = \{a^x \mid a \in J(L),\ x \in Q\},\ M(L^Q) = \{{}^x a \mid a \in M(L),\ x \in Q\}.$$

Of course, $\Delta(L^Q) = \{a^0 \mid a \in L\} = \{{}^1 a \mid a \in L\}$.

Let $a \in J(L)$. Then $a^0 \in J(\Delta(L^Q))$ and while $\phi(a^0)$ need not be an element of $J(\Delta(K^Q))$ it can be associated by the isomorphism $\phi$ to an element of $J(\Delta(K^Q))$ in a natural way. Assume for the moment that $Q$ is product nondecomposable. We claim that there exist $b \in J(L)$, $p, q \in J(K)$ and a sequence of nonzero elements $x_1, x_2, \ldots, x_n$ of $Q$ satisfying

(i) $\phi(a^0) = p^{x_1}$, $\phi(b^0) = p^0$;

(ii) $\phi(b^{x_1}) = p^{x_2}$, $\phi(b^{x_2}) = p^{x_3}, \ldots, \phi(b^{x_{n-1}}) = p^{x_n}$;

(iii) $\phi(b^{x_n}) = q^0$.

Since $\phi(a^0) \in J(K^Q)$ there is $p \in J(K)$, $x \in Q$ such that

$\phi(a^0) = p^x$. If $x = 0$ take $a = b$, $p = q$, and $x_i = x$. Otherwise $x > 0$ in $Q$ and $p^x < p^0$. Then $\phi^{-1}(p^0) \in J(L^Q)$ and $\phi^{-1}(p^0) > a^0$ so there is some $b \in J(L)$ such that $\phi^{-1}(p^0) = b^0$. We show (ii) and (iii) hold by proving that for any $y \in Q$, if $\phi(b^y) = q^z$ then $p = q$ or $z = 0$.

Suppose that $\phi(b^y) = q^z$ and $p \neq q$ in $J(K)$, $z > 0$ in $Q$. Since $b^0 = \phi^{-1}(p^0) > \phi^{-1}(p^x) = a^0$, we have $b > a$ and $p > q$ (see Figure 4). Thus, $a^0 \wedge b^y = a^y$, an element of $J(L^Q)$, so $\phi(a^y) = p^x \wedge q^z$ belongs to $J(K^Q)$. It follows that $x \vee z$ exists in $Q$, say $t = x \vee z$, and $\phi(a^y) = q^t$. Let $u \in Q$ satisfy $u \leq x$ and $u \leq z$. Then $a^0 \vee b^y \leq \phi^{-1}(p^u) \leq b^0$. But $\phi^{-1}(p^u) \in J(L^Q)$, so $\phi^{-1}(p^u) = b^0$ and $u = 0$. Thus $x \wedge z = 0$ in $Q$.

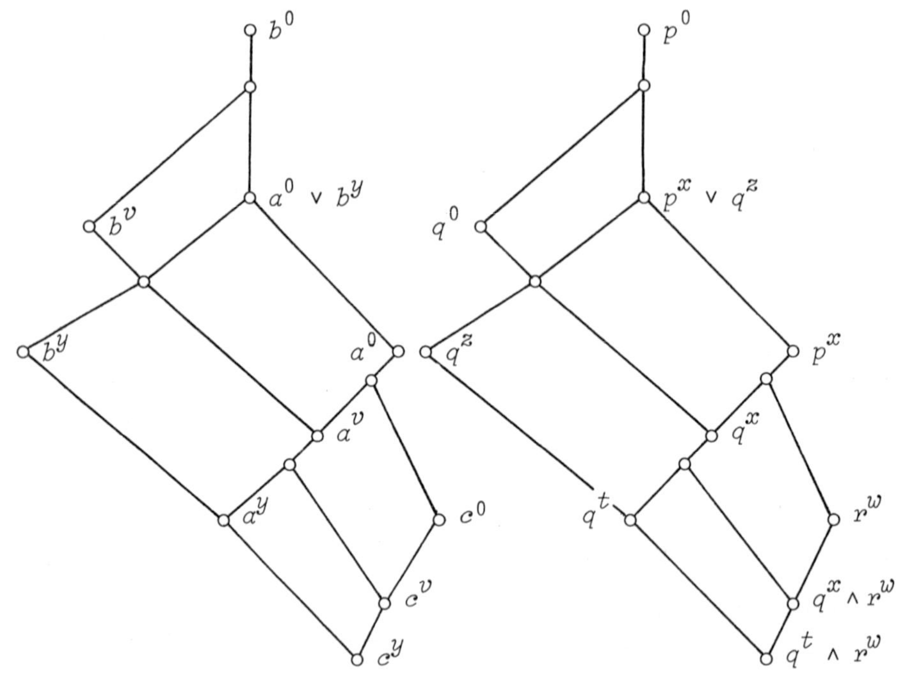


Figure 4. Construction in the proof of Theorem 4.1.

Now consider $q^0 \in J(K^Q)$. Since $q^z < q^0 < p^0$, $b^y < \phi^{-1}(q^0) < b^0$, so $\phi^{-1}(q^0) = b^v$ for some $v \in Q$ satisfying $0 < v < y$. Also, $a^v = a^0 \wedge b^v$ and $q^x = p^x \wedge q^0$.

Choose $c \in L$ with $a \geq c \succ 0_L$ and let $\phi(c^0) = r^w$ for some $r \in J(K)$ and $w \in Q$. It follows that

$$[0_L,c]^Q \cong [0_L{}^0,c^0] \cong [0_K{}^0,r^w] \cong [0_K,r]^{[w,1]}.$$

Apply Lemma 3.1 to $[0_L,c]^Q \cong [0_K,r]^{[w,1]}$ and obtain $J([0_L,c]) \times Q^d \cong J([0_K,r]) \times [w,1]^d$. Since $0_L \prec c$, $|J([0_L,c])| = 1$ and the product nondecomposable $Q^d$ is isomorphic to $J([0_K,r]) \times [w,1]^d$. Because $r^w < p^x$, $0 < w$, so $Q^d \not\cong [w,1]^d$. Thus $Q^d \cong J([0_K,r])$ and $w = 1$.

Finally, $a^v \wedge c^0 = c^v > c^y = a^y \wedge c^0$ and both belong to $J(L^Q)$. On the other hand

$$\phi(c^v) = q^x \wedge r^1 = (q \wedge r)^1 = \phi(c^y),$$

a contradiction. Thus, $p = q$ or $z = 0$.

The proof of Theorem 4.1 is accomplished as follows. Proceed by induction on $|Q|$, which allows the assumption that $Q$ is product nondecomposable. Define a map $\psi$ of $P(\Delta(L^Q))$ to $P(\Delta(K^Q))$ by $\psi(a^0) = q^0$ for $a \in J(L)$, where $q \in J(K)$ is obtained as above, and $\psi({}^1a) = {}^1q$ for $a \in M(L)$, where $q \in M(K)$ is given by a dual argument to that preceding. By symmetry, it is enough to prove that $\psi$ is an order-embedding; this can be accomplished with (i), (ii), (iii), and a detailed study of cases.

In [52] Jónsson and McKenzie describe a general approach to the cancellation and refinement problems which depends upon showing that an isomorphism $\phi$ of $B^A$ onto $C^A$ can be used to obtain an isomorphism $\psi$ of $B$ onto $C$. The approach is a generalization of an argument successful if $A \cong \underset{\sim}{2}$ (cf. [16], [50], [52]). As they describe the basic strategy, let $b \in B$, let $\langle b \rangle \in B^A$ denote the constant map with value $b$, and let $g = \phi(\langle b \rangle)$. For each $a \in A$, let $\langle g(a) \rangle \in C^A$ denote the constant map with value $g(a)$ and *show that there is some* $b_a \in B$ *such that* $\phi(\langle b_a \rangle) = \langle g(a) \rangle$. Now define a map $f$ of $A$ to $B$ by $f(a) = b_a$ for all $a \in A$ and *prove that* $\phi(f) = \langle c \rangle$. The mapping $\psi$ of $B$ to $C$ given by $\psi(b) = c$ is then an isomorphism. They are able to accomplish this in a wide variety of circumstances [52] (cf. Theorem 3.5 and the end of section 2) with arguments far too involved to present here. However, let us make more formal the above reasoning by introducing properties of the isomorphism $\phi$ which, when satisfied, yield $\psi$.

Let $A$, $B$, and $C$ be ordered sets and let $\phi$ be an isomorphism of $B^A$ onto $C^A$. For each $b \in B$ let $\langle b \rangle$ be the constant map of $A$ to $B$ with value $b$. Let

$$\Delta(\phi) = \{f \in B^A \mid \phi(f) \text{ is constant}\},$$
$$R(\phi) = \{b \in B \mid \langle b \rangle \in \Delta(\phi)\}.$$

Define a binary relation $\leq_\phi$ on $R(\phi)$ by $b \leq_\phi b'$ if $b \leq b'$ in $R(\phi)$, and every $f \in B^A$ with $f(A) \subseteq \{b, b'\}$ belongs to $\Delta(\phi)$. Also, let $\bar{\phi}$ be the map of $R(\phi)$ to $R(\phi^{-1})$ given by $\phi(\langle b \rangle) = \langle \bar{\phi}(b) \rangle$ for $b \in R(\phi)$. Some of the properties of interest for $\phi$ are [16], [52] :

$(\phi,1)$ $\leq_\phi$ is an ordering of $R(\phi)$;

$(\phi,2)$ $\bar{\phi}$ is an isomorphism of $(R(\phi), \leq_\phi)$ onto $(R(\phi^{-1}), \leq_{\phi^{-1}})$;

$(\phi,3)$ $\Delta(\phi)$ is the set of all order-preserving maps of $A$ to $(R(\phi), \leq_\phi)$.

One can see, with some effort, how these conditions correspond to the strategy described above.

THEOREM 4.2 ([52]). *Let $\phi$ be an isomorphism of $B^A$ onto $C^A$ and let $(\phi,1)$, $(\phi,2)$, $(\phi,3)$ and $(\phi^{-1},3)$ hold. Then there is an isomorphism $\psi$ of $B$ onto $C$ such that*

$$\langle \psi(b) \rangle = \phi(\bar{\phi}^{-1} \circ \phi(\langle b \rangle))$$

*for all $b \in B$.* □

The proof of Theorem 4.2 is straightforward. The noteworthy accomplishment lies in the formulation of these conditions and the demonstration, in diverse circumstances, of their satisfaction. (See C. Bergman, R. McKenzie, and S. Nagý [16] for the same approach in case $A$ is a chain.)

## 5. AUTOMORPHISM GROUPS

A natural question arises concerning automorphisms and operations on ordered sets : for instance, what are the relationships between the automorphism groups of $A + B$, $AB$, $A^B$, and the automorphism groups of $A$ and $B$. Also, upon decomposition of an ordered set via these operations, one can ask for a description of the automorphism group of the ordered set in terms of the automorphism groups of its components.

For an ordered set $A$, let $\mathrm{Aut}(A)$ denote the group of automorphisms of $A$. As usual, let $G \times H$ denote the direct product of groups $G$ and $H$.

It is easy to see that for any connected ordered sets $A$ and $B$, $\mathrm{Aut}(A+B) \cong \mathrm{Aut}(A) \times \mathrm{Aut}(B)$ if $A \not\cong B$, and $\mathrm{Aut}(A+B)$ is the wreath product of $\mathrm{Aut}(A)$ and the symmetric group on two elements if $A \cong B$. The automorphism group of an ordered set $A$ can be given as a direct product of wreath products of the automorphism groups of its connected components. Let $G \wr S_n$ denote the wreath product of the group $G$ and $S_n$ the symmetric group on $n$ elements. So if $A = \sum_{i \in I} n_i A_i$ where $A_i \not\cong A_j$ then $\mathrm{Aut}(A) \cong \prod_{i \in I} (\mathrm{Aut}(A_i) \wr S_{n_i})$.

The strict refinement property for products of connected ordered sets allows a similar description of $\mathrm{Aut}(A)$ in terms of the automorphism groups of its nondecomposable factors (cf. [35], [51], [66]). If $A \cong A_1^{n_1} A_2^{n_2} \ldots A_k^{n_k}$, each $A_i$ is product nondecomposable, and $A_i \not\cong A_j$, then

$$\mathrm{Aut}(A) = \prod_{i \in 1}^{k} (\mathrm{Aut}(A_i) \wr S_{n_i}).$$

Following the decomposition of an ordered set into its connected components, and these into their product nondecomposable factors, we are led to exponential decompositions of connected product nondecomposable ordered sets and to ask about the relation of $\mathrm{Aut}(A^B)$ to $\mathrm{Aut}(A)$ and $\mathrm{Aut}(B)$. (See B. Jónsson [51] for a more detailed description of this decomposition procedure.)

For any ordered sets $A$ and $B$ there is an obvious embedding $\Phi$ of $\mathrm{Aut}(A) \times \mathrm{Aut}(B)$ in $\mathrm{Aut}(A^B)$ : let $\Phi(\alpha,\beta)(f) = \alpha \circ f \circ \beta^{-1}$ for all $(\alpha,\beta) \in \mathrm{Aut}(A) \times \mathrm{Aut}(B)$, $f \in A^B$. One might think that the success of Jónsson and McKenzie [52] in producing from an automorphism $\phi$ of $A^B$ an automorphism $\psi$ of $A$ (see §4) indicates that the embedding $\Phi$ is onto when their conditions on $\phi$ hold (cf. $(\phi,1)$, $(\phi,2)$, $(\phi,3)$ in section 4). As they point out ([52], section 11) this is not so. They give the following example. Let $A$ be an ordered set composed of $\underset{\sim}{2}$ and $\underset{\sim}{3}$ with minimum elements identified and let $B \cong \underset{\sim}{2}$. Then $|\mathrm{Aut}(A)| = |\mathrm{Aut}(B)| = 1$ while $\mathrm{Aut}(A^B)$ has order 2. Moreover, $A$ and $B$ are both connected, product nondecomposable, and exponentially nondecomposable (that is,

$A \cong C^D$ implies that $|D| = 1$).

Let us first state results which provide conditions ensuring that the embedding is onto.

THEOREM 5.1 ([52]). *Let* $A$ *be a subdirectly irreducible lattice and let* $B$ *be a finite, connected ordered set. Then* $\mathrm{Aut}(A^B) \cong \mathrm{Aut}(A) \times \mathrm{Aut}(B)$. □

Without the condition of subdirect irreducibility it is easy to construct a counterexample. Take ordered sets $B$ and $C$ such that $\mathrm{Aut}(CB) \ncong \mathrm{Aut}(C) \times \mathrm{Aut}(B)$. Let $A = \underset{\sim}{2}^C$. Then $\mathrm{Aut}(A^B) \cong \mathrm{Aut}(\underset{\sim}{2}^{CB}) \cong \mathrm{Aut}(CB)$ while $\mathrm{Aut}(A) \times \mathrm{Aut}(B) \cong \mathrm{Aut}(C) \times \mathrm{Aut}(B)$. It is also easy to see that connectedness of the exponent $B$ is required.

In [52] it is asked whether $\mathrm{Aut}(A^B) \cong \mathrm{Aut}(A) \times \mathrm{Aut}(B)$ if $A$ is a product nondecomposable, exponentially nondecomposable lattice and $B$ is a (finite) connected ordered set.

THEOREM 5.2 ([38]). *Let* $A$ *be a nontrivial lattice of finite length that is product and exponentially nondecomposable and let* $B$ *be a finite, connected ordered set. Then* $\mathrm{Aut}(A^B) \cong \mathrm{Aut}(A) \times \mathrm{Aut}(B)$. □

Duffus and Wille [38] use the scaffolding (see section 3) of a lattice of finite length : $G(A^B) \cong G(A) \times B^d$ and $\mathrm{Aut}(A^B) \cong \mathrm{Aut}(G(A^B))$. Under the conditions of Theorem 5.2, $\mathrm{Aut}(G(A) \times B^d) \cong \mathrm{Aut}(G(A)) \times \mathrm{Aut}(B)$. They [38] also give a complete description of $\mathrm{Aut}(A^B)$ when $A$ is a nontrivial lattice of finite length and $B$ is a finite, connected ordered set.

Recently, Jónsson [51] obtained a description of $\mathrm{Aut}(A^B)$ under very weak conditions on $A^B$. Extending the techniques in [52], using the analysis of an isomorphism $\phi$ of powers $A^B$ and $C^D$ (cf. section 4) and the logarithmic property of the set of join irreducibles (cf. section 3), he has obtained

THEOREM 5.3 ([51]). *Let* $A^B$ *be a bounded, product nondecomposable ordered set satisfying the descending chain condition, and let* $A$ *be exponentially nondecomposable. Then*

$$\mathrm{Aut}(A^B) \cong \mathrm{Aut}(A) \times \mathrm{Aut}(B). \quad \square$$

Concerning a description of $\mathrm{Aut}(A^B)$ when the natural embedding of $\mathrm{Aut}(A) \times \mathrm{Aut}(B)$ is not onto, Jónsson and McKenzie [52] have two results (see [52], section 11). We state one of these.

THEOREM 5.4 ([52]). *Let $A$ be an ordered set with a minimum element, let $A = \nabla J'(A)$, and let $\mathrm{Aut}(A)$ be trivial. Let $B$ be a connected, product nondecomposable ordered set. If $J'(A)$ has $n$ connected components then $\mathrm{Aut}(A^B)$ is isomorphic to an extension of a subgroup of $(\mathrm{Aut}(B))^n$ by an elementary Abelian 2-group.* □

Again, both properties $(\phi,1)$, $(\phi,2)$, $(\phi,3)$ of an automorphism $\phi$ of $A^B$ and the nature of elements of $J'(A^B)$ are employed.

As Jónsson [51] remarks, there is at this point little known about the relationship, in general, of the automorphism groups $\mathrm{Aut}(A)$, $\mathrm{Aut}(B)$, and $\mathrm{Aut}(A^B)$.

## 6. ORDER-POLYNOMIAL-COMPLETENESS AND DUALITY

G. Birkhoff [19] proved that every finite distributive lattice $L$ is determined up to isomorphism by its ordered set $J(L)$ of nonzero join irreducible elements; in fact, $L$ is isomorphic to the power $\underset{\sim}{2}^P$ where $P = J(L)^d$. It is implicit in Birkhoff's work that this correspondence between lattices and ordered sets is in fact a dual category equivalence between the category $\underset{\sim}{D}_F$ of finite bounded distributive lattices and the category $\underset{\sim}{P}_F$ of finite ordered sets.

It was not until the late 1960's that this duality was extended by H.A. Priestley to infinite distributive lattices. An ordered topological space $X$ is called *totally order-disconnected* if for all $x,y \in X$ with $x \nleq y$ there exists a clopen order ideal (that is, hereditary subset) $A \subset X$ with $x \in A$ and $y \notin A$; note that $X$ is a compact totally order-disconnected space precisely when it is homeomorphic and order-isomorphic to a closed subset of a power of the (discretely topologized) ordered set $\underset{\sim}{2}$.

THEOREM 6.1 ([61], [62]). *The category $\underset{\sim}{D}$ of bounded distributive lattices and $(0,1)$-homomorphisms is dually equivalent to the category $\underset{\sim}{P}$ of compact totally order-disconnected spaces and continuous order-preserving maps.* □

The lattice of $n$-ary polynomial (alias, algebraic) functions on a lattice $L$ is, by definition, the sublattice of the direct power $L^{|L^n|}$ generated by the $n$ projections and the constant maps, and consequently is a sublattice of the power $L^{(L^n)}$. If for every positive integer $n$, each order-preserving map of $L^n$ into $L$ is an $n$-ary polynomial function, we say that $L$ is *order-polynomial-complete*. This concept was first studied by D. Schweigert [70] and R. Wille [85], [86]; see also [53] and [71]. That $\underset{\sim}{2}$ is order-polynomial-complete was proved by Birkhoff [19]. Indeed, let $\phi : \underset{\sim}{2}^n \to \underset{\sim}{2}$ be order-preserving and consider the order filter $F = \phi^{-1}(1)$: if $F$ is empty, then $\phi = 0$; if $F = \underset{\sim}{2}^n$, then $\phi = 1$; and in general

$$\phi(x_1,\ldots,x_n) = \bigvee(\bigwedge(x_i \mid a_i = 1) \mid a \in F);$$

where the join and meet of the empty set denote 0 and 1 respectively. More generally, extending a result of Schweigert [70], Wille proved the following result, thereby providing a multitude of examples.

PROPOSITION 6.2 ([85]). *Every finite simple lattice whose unit is a join of atoms is order-polynomial-complete.* □

An important idea in the early investigations was the following reduction result.

PROPOSITION 6.3 ([70]). *A finite lattice $L$ is order-polynomial-complete if and only if every element of the power $L^L$ is a unary polynomial function.* □

Note that in 6.2 and 6.3 it is assumed that $L$ is finite. It is conjectured that an order-polynomial-complete lattice is necessarily finite.

A proof of Theorem 6.1 utilizing the order-polynomial-completeness of $\underset{\sim}{2}$ was given in B.A. Davey [28]. The implied possibility of obtaining a similar duality for each order-polynomial-complete lattice $L$ was exploited in B.A. Davey, D. Duffus, R.W. Quackenbush, and I. Rival [30] and further refined in B.A. Davey and I. Rival [32]. Let $L_c$ denote the algebra obtained from the lattice $L$ by making each element the value of a nullary operation, and let $\underset{\sim}{L}_c$ denote the variety (of lattices with constants) generated by $L_c$; for example, if $L$ is the two-element chain $\underset{\sim}{2}$, then $\underset{\sim}{L}_c$ is the variety $\underset{\sim}{D}$

of bounded distributive lattices. For finite $L$ we define the contravariant hom-functor

$$\mathrm{D}(-) := \underset{\sim}{L}_c(-, L_c) : \underset{\sim}{L}_c \to \underset{\sim}{P}$$

so that $\mathrm{D}(A) = \underset{\sim}{L}_c(A, L_c) \subseteq L^{|A|}$ inherits its order and topology from the power $L^{|A|}$. Since join and meet are order-preserving there is a natural contravariant hom-functor

$$\mathrm{E}(-) := \underset{\sim}{P}(-, L_c) : \underset{\sim}{P} \to \underset{\sim}{L}_c$$

for which the operations on $\mathrm{E}(X) = \underset{\sim}{P}(X, L_c)$ are defined pointwise. We regard $L_c$ as an object of $\underset{\sim}{P}$ by giving it the discrete topology; hence, if $P$ is a finite ordered set, then $\mathrm{E}(P)$ is the power $L^P$. For each $A \in \underset{\sim}{L}_c$ and each $X \in \underset{\sim}{P}$ there are natural evaluation maps $\eta_A : A \to \mathrm{ED}(A)$, defined by $\eta_A(a)(g) = g(a)$ for all $g \in \mathrm{D}(A)$, and $\varepsilon_X : X \to \mathrm{DE}(X)$, defined by $\varepsilon_X(x)(\phi) = \phi(x)$ for all $\phi \in \mathrm{E}(X)$.

The equivalence of (1), (4), and (5) in the following result was first established by R. Wille [85], [86]; the equivalence of (1), (2), and (3) is due to Davey, Duffus, Quackenbush, and Rival [30]; the proof sketched below comes from Davey and Rival [32].

THEOREM 6.4. *Let $L$ be a finite non-trivial lattice. The following are equivalent :*

*(1)* *$L$ is order-polynomial-complete;*

*(2)* *the functors D and E are mutually inverse dual category equivalences : for each $A \in \underset{\sim}{L}_c$ the map $\eta_A : A \to \mathrm{ED}(A)$ is an isomorphism, and for each $X \in \underset{\sim}{P}$ the map $\varepsilon_X : X \to \mathrm{DE}(X)$ is an order-isomorphism and a homeomorphism;*

*(3)* *every finite member of $\underset{\sim}{L}_c$ is isomorphic to $L^P$ for some finite ordered set $P$;*

*(4)* *$L^2$ has precisely four sublattices which contain the diagonal $\Delta = \Delta(L^2)$, namely $\Delta$, $\leq$, $\geq$, and $L^2$;*

*(5)* *the identity map and the constant map onto $\{0\}$ are the only maps $\delta$ of $L$ to itself which identically satisfy $\delta(x) \leq x$ and $\delta(x \vee y) = \delta(x) \vee \delta(y)$.*

Of course, Theorem 6.1 is the special case of this theorem

which arises when $L$ is $\underset{\sim}{2}$. For any lattice $L$ and ordered set $P$, $L^P$ is a diagonal sublattice of $L^{|P|}$; that is $L^P$ is a sublattice of $L^{|P|}$ which contains the constant maps. Hence (1) ⇒ (3) tells us that if $L$ is a finite order-polynomial-complete lattice then every lattice which could possibly be a power of $L$ is one. The equivalence of (1) and (2) has a surprising corollary : *if $L$ is a finite non-trivial order-polynomial-complete lattice, then $\underset{\sim}{L}_c$ is equivalent as a category to the category $\underset{\sim}{D}$ of bounded distributive lattices.* It is proved in [32] that this theorem extends at once to lattice-ordered algebras provided we assume that, as for lattices, all fundamental operations are order-preserving : in (5) we must replace the condition that $\delta$ is join-preserving by $\delta(f(x_1,\ldots,x_n)) \leq f(\delta(x_1),\ldots,\delta(x_n))$ for all fundamental operations on $L$.

We now sketch the proof of 6.4. By the main result of B.A. Davey and H. Werner [33] (see also B.A. Davey [28]), to prove that D,E is a dual category equivalence it suffices to show that

(A) *every algebra in $\underset{\sim}{L}_c$ can be embedded into a direct power of $L_c$;*

(B) *$L$ is injective in $\underset{\sim}{P}_F$;*

(C) *for all positive integers $n$ each $\phi \in \underset{\sim}{P}(L^n,L)$ is an $n$-ary term function on $L_c$;*

(D) *if $X,Y \in \underset{\sim}{P}_F$ with $X \subsetneq Y$ then there exist $\phi,\psi \in \underset{\sim}{P}(Y,L)$ with $\phi \neq \psi$ but $\phi|X = \psi|X$.*

Since lattices have distributive congruence lattices, Jónsson's lemma [49] implies that every algebra in $\underset{\sim}{L}_c$ can be embedded in a product of homomorphic images of subalgebras of $L_c$. But $L_c$ has no proper subalgebras, and it is easily seen that an order-polynomial-complete lattice is simple and so has no proper homomorphic images; whence (*A*) follows. It is straightforward to see that $\underset{\sim}{2}$ is injective in $\underset{\sim}{P}_F$ and that any finite lattice is a retract of a power of $\underset{\sim}{2}$; since retractions and products preserve injectivity we conclude that $L$ is injective in $\underset{\sim}{P}_F$. Since the $n$-ary term functions on the algebra $L_c$ are precisely the $n$-ary polynomial functions on the lattice $L$, (*C*) is just the definition of order-polynomial-completeness. Given $X, Y \in \underset{\sim}{P}_F$ with $X \subsetneq Y$ it is easy to find maps $\phi, \psi \in \underset{\sim}{P}(Y, \underset{\sim}{2})$ with $\phi \neq \psi$ and $\phi | X = \psi | X$; since $\underset{\sim}{2}$ can be embedded into $L$, (D) follows. Thus (1) implies (2). For a finite member $A$ of $\underset{\sim}{L}_c$, $P = \mathrm{D}(A) = \underset{\sim}{L}_c(A, L_c)$ is finite and so discretely topologized, giving $A \cong \mathrm{ED}(A) = \underset{\sim}{P}(\mathrm{D}(A), L_c) = L^P$, whence (2) implies (3). That (3) implies (4) is a simple exercise : note that $L^P$ is isomorphic to $\Delta$ when $P$ has one element, is isomorphic to $\leq$ and to $\geq$ when $P$ is a two-element chain, and is isomorphic to $L^2$ when $P$ is a two-element antichain. If $\delta$ is a map satisfying the conditions of (5), then $K := \{(x,y) \mid \delta(y) \leq x \leq y\}$ is a subalgebra of $L_c^2$, and given (4) it follows that $K = \Delta$ and so $\delta = \mathrm{id}_L$, or $K = \leq$ so $\delta$ is the constant map onto $\{0\}$; hence (5) follows from (4). If $K$ is a diagonal sublattice of $L^2$, then the maps $\delta$ and $\delta'$ of $L$ to itself defined by

$$\delta(y) := \wedge(x \mid (x,y) \in K) \quad \text{and} \quad \delta'(y) := \wedge(y \mid (x,y) \in K)$$

satisfy the conditions of (5). Hence if (5) holds then $\delta$ and $\delta'$ are either the identity on $L$ or the constant map onto $\{0\}$; the four combinations yield $K \in \{\Delta, \leq, \geq, L^2\}$. Consequently (5) implies (4). To prove that (4) implies (1) we call on the beautiful result of K.A. Baker and A.F. Pixley [7] which says that for a finite lattice $L$ a map $f : L^n \to L$ is a polynomial function if and only if every diagonal sublattice of $L^2$ is closed under $f$. Since $\Delta$, $\leq$, $\geq$, and $L^2$ are closed under $f$ if and only if $f$ is order-preserving, (1) follows from (4).

We close this section with a brief discussion of some other dualities which are closely related to the Birkhoff-Priestley duality for bounded distributive lattices. M.H. Stone developed a representation theory for arbitrary distributive lattices in 1937 [75] as an extension of his representation for Boolean algebras [74], [76]. Compared with the Birkhoff-Priestley duality, Stone's duality has the disadvantage (from the standpoint of this Symposium) that it represents a bounded distributive lattice in terms of a purely topological object with no ordering (even in the finite case), and moreover although this object is compact it is no longer Hausdorff. In fact it is relatively straightforward to show that the Birkhoff-Priestley dual and the Stone dual of a bounded distributive lattice are equivalent in that each has a natural definition in terms of the other; see W.H. Cornish [22] and T.P. Speed [72]. There have been several attempts to generalize the Birkhoff-Priestley result to arbitrary lattices : in A. Urquhart [79] arbitrary lattices are represented in terms of compact Hausdorff spaces endowed with two quasi-orders, while R. Wille [84] and G. Gierz and K. Keimel [42] give an alternative generalization whereby arbitrary lattices are represented via compact Hausdorff topological partial semilattices (in the finite-length case, this is the scaffolding discussed in section 3). Both of these results have the defect that although they are representation theorems they do not yield categorical dualities; they give no information about non-onto lattice homomorphisms. It is natural to ask for an extension of the duality between $\underset{\sim}{D}_F$ and $\underset{\sim}{P}_F$ where, instead of extending $\underset{\sim}{D}_F$ to the category $\underset{\sim}{D}$ of all bounded distributive lattices we extend $\underset{\sim}{P}_F$ to the category $\underset{\sim}{Q}$ of all ordered sets. It is proved in K.H. Hofmann, M. Mislove, and A. Stralka [48] (see also [41]) that $\underset{\sim}{Q}$ is dually equivalent to the category $\underset{\sim}{DB}$ of bounded distributive topological lattices, whose topology is Boolean, and continuous (0,1)-homomorphisms; or, if one wishes to dispense with the topology, $\underset{\sim}{Q}$ is dually equivalent to the category $\underset{\sim}{DBA}$ of distributive bi-algebraic lattices and complete (0,1)-homomorphisms : the dual of $X \in \underset{\sim}{Q}$ is the power $\underset{\sim}{2}^X$ and the dual of an order-preserving map $f : X \to Y$ is the natural induced map $\hat{f} : \underset{\sim}{2}^Y \to \underset{\sim}{2}^X$; the dual of $A \in \underset{\sim}{DBA}$ is the ordered set $J_c(A)$ of completely join irreducible elements of $A$, and for a morphism $\phi : A \to B$ the dual $\hat{\phi} : J_c(B) \to J_c(A)$ is defined by $\hat{\phi}(x) = \wedge\phi^{-1}([x))$.

## 7. ORDERED SETS AND DISTRIBUTIVE-LATTICE-ORDERED ALGEBRAS.

Let $\mathcal{O}(P)$ denote the lattice of clopen order ideals of an object $P \in \underset{\sim}{P}$; clearly $\mathcal{O}(P)^d \cong \underset{\sim}{P}(P,\underset{\sim}{2})$, and since the finite members of $\underset{\sim}{P}$ are just ordered sets with the discrete topology, in the finite case $\mathcal{O}(P)^d \cong \underset{\sim}{2}^P$. Since $\mathcal{O}(P)$ is usually easier to visualize than $\underset{\sim}{2}^P$, the most pictorial presentation of the correspondence between finite distributive lattices $L$ and finite ordered sets $P$ is given by

$$L \cong \mathcal{O}(J(L)) \quad \text{and} \quad P \cong J(\mathcal{O}(P)).$$

These isomorphisms remain valid for arbitrary $L \in \underset{\sim}{D}$ and $P \in \underset{\sim}{P}$ provided we replace $J(L)$ by the set of prime ideals of $L$ ordered by set inclusion (and endowed with an appropriate topology). Note that for finite $L \in \underset{\sim}{D}$, the sets $J(L)$ of join irreducibles, $M(L)$ of meet irreducibles, and $PI(L)$ of prime ideals are order - isomorphic, and for arbitrary $L \in \underset{\sim}{D}$, $PI(L)$ and $\underset{\sim}{D}(L,\underset{\sim}{2})$ are dually order-isomorphic. Thus in practice we do not work with the hom-sets $\mathrm{D}(L) = \underset{\sim}{D}(L,\underset{\sim}{2})$ and $\mathrm{E}(P) = \underset{\sim}{P}(P,\underset{\sim}{2})$ but with more easily visualized (dually) order-isomorphic copies. The fact that we can not only represent the objects but also the morphisms of $\underset{\sim}{D}$ (respectively, $\underset{\sim}{D}_F$) in terms of objects and morphisms of $\underset{\sim}{P}$ (respectively, $\underset{\sim}{P}_F$) and moreover the fact that there are many objects in $\underset{\sim}{D}$ which cannot easily be pictured but have duals in $\underset{\sim}{P}$ which can be, makes this a very powerful tool for

(A) *the study of particular bounded distributive lattices,*

(B) *the study of bounded distributive lattices as a class, and*

(C) *for generating examples and counterexamples.*

Many algebraic structures which arise in logic are bounded distributive lattices with some extra structure; the most notable examples being Boolean algebras, Post algebras, Heyting algebras, pseudocomplemented and doubly pseudocomplemented distributive lattices, de Morgan algebras, the Łukasiewicz algebras; see, for example, R. Balbes and Ph. Dwinger [11]. Provided one is able to

(1) describe those ordered spaces in $\underset{\sim}{P}$ which arise as duals of these enriched structures, and

(2) describe those continuous order-preserving maps in $\underset{\sim}{P}$ which are dual to the homomorphisms of these enriched structures,

then the Birkhoff-Priestley duality immediately yields a duality for the class of enriched structures. Since 1970 there have been many papers which utilize these ideas to study distributive lattices and distributive-lattice-ordered algebras. We now present a selection of results which demonstrate the techniques involved. To maximize the order-theoretic and to minimize the topological considerations we shall concentrate on the finite case.

## 7.1 Distributive lattices.

Our first lemma summarizes the basic properties of the transfer from $\underset{\sim}{P}$ to $\underset{\sim}{D}$.

LEMMA 7.1. *Let $P$ and $Q$ belong to $\underset{\sim}{P}$.*

(a) *$O(P)$ is isomorphic to a $(0,1)$-sublattice of $O(Q)$ if and only if there is a continuous order-preserving map from $Q$ onto $P$.*

(b) *$L$ is a homomorphic image of $O(P)$ if and only if there is a closed subset $Q$ of $P$ such that $L$ is isomorphic to $O(Q)$.*

(c) *the congruence lattice* $\mathrm{Con}(\mathcal{O}(P))$ *of* $\mathcal{O}(P)$ *is dually isomorphic to the lattice* $C\ell(P)$ *of closed subsets of* $P$. *For each closed subset* $Q$ *of* $P$ *the equivalence relation* $\Theta(Q)$ *defined by*

$$A \equiv B(\Theta(Q)) \Leftrightarrow A \cap Q = B \cap Q, \quad \text{for } A,B \in \mathcal{O}(P)$$

*is a congruence on* $\mathcal{O}(P)$ *and* $\mathcal{O}(P)/\Theta(Q) \cong \mathcal{O}(Q)$.

(d) $\mathcal{O}(P + Q) \cong \mathcal{O}(P) \times \mathcal{O}(Q)$.

(e) $\mathcal{O}(P \times Q) \cong \mathcal{O}(P) * \mathcal{O}(Q)$ *(where* $*$ *is the coproduct in* $\underset{\sim}{D}$*).* □

These results are straightforward : (a) says that the duals of the embeddings in $\underset{\sim}{D}$ are the onto maps in $\underset{\sim}{P}$, which follows from the fact that $\underset{\sim}{2}$ is injective in $\underset{\sim}{D}$; (b) says that the duals of onto maps in $\underset{\sim}{D}$ are the embeddings in $\underset{\sim}{P}$, which in turn follows from the fact that $\underset{\sim}{2}$ is injective in $\underset{\sim}{P}$ (see [28]); (c) is a consequence of (b); (d) and (e) follow from simple category-theoretic considerations. The finite case of Lemma 7.1, translated from $\underset{\sim}{P}$ to $\underset{\sim}{D}$, warrants repeating.

LEMMA 7.2. *Let* $L$ *and* $K$ *be finite distributive lattices.*

(a) $L$ *is isomorphic to a* (0,1)*-sublattice of* $K$ *if and only if there is an order-preserving map of* $J(K)$ *onto* $J(L)$.

(b) $L$ *is a homomorphic image of* $K$ *if and only if there is an order-embedding of* $J(L)$ *into* $J(K)$.

(c) $\mathrm{Con}(L)$ *is isomorphic to* $\underset{\sim}{2}^n$ *where* $n = |J(L)|$.

(d) $J(L \times K) \cong J(L) + J(K)$.

(e) $J(L * K) \cong J(L) \times J(K)$.

(f) $L * K \cong L^{J(K)^d} \cong K^{J(L)^d}$.

(g) $L$ *is directly indecomposable if and only if* $J(L)$ *is connected.* □

Parts (f) and (g) are easy corollaries of (e) and (d) respectively.

We note in passing that 7.1(c) and 7.2(c) have been variously generalized : if $L$ is a finite simple lattice and $P \in \underset{\sim}{P}$, then $\mathrm{Con}(\underset{\sim}{P}(P,L))$ is dually isomorphic to $C\ell(P)$ [32], and when $P$ is finite this yields $\mathrm{Con}(L^P) \cong \underset{\sim}{2}^n$ where $n = |P|$ [30], [36]; if $L$ is an arbitrary lattice and $P$ is finite, then $\mathrm{Con}(L^P) \cong (\mathrm{Con}(L))^n$ [32], [36]; if $L$ is a finite lattice, $P \in \underset{\sim}{P}$ and

$K = \mathcal{O}(P)$, then $\mathrm{Con}(\underset{\sim}{P}(P,L)) \cong \mathrm{Con}(L) * \mathrm{Con}(K)$ [68] (recall that $\mathrm{Con}(L)$ is a distributive lattice), which, when $P$ is finite, reduces to the previous result by an application of 7.2(f).

We now present a collection of results which illustrate some of the combinations of algebraic, topological, order-theoretic, and category-theoretic techniques at our disposal. To this end we shall, wherever possible, work within the dual category $\underset{\sim}{P}$, even when a direct algebraic proof exists.

Since each object in $\underset{\sim}{P}$ is order-isomorphic and homeomorphic to a closed subspace of $\underset{\sim}{2}^I$ for some set $I$, the following trivial result is quite useful. Given $A \subseteq B \subseteq \underset{\sim}{2}^I$ we say that a map $\psi \in B$ is *locally in* $A$ if for each finite subset $J$ of $I$ there exists $\phi$ in $A$ such that $\psi$ and $\phi$ agree on $J$.

LEMMA 7.3. *Let* $B$ *be a subspace of* $\underset{\sim}{2}^I$ *and let* $A \subseteq B$. *Then* $A$ *is a closed subset of* $B$ *if and only if* $A$ *contains every member of* $B$ *which is locally in* $A$. □

An ordered set P is called *representable* if it is order-isomorphic to the set $PI(L)$ of some (not necessarily bounded) distributive lattice $L$; we then say that $P$ is *representable by* $L$. We may adjoin universal bounds to $L$ and this has the effect of adjoining universal bounds to $PI(L)$. Hence $P$ is representable if and only if there exists $X \in \underset{\sim}{P}$ such that $X$ has universal bounds $0,1$ and $P$ is order-isomorphic to $X-\{0,1\}$. Thus 7.3 yields the following characterization of representable ordered sets. The constant maps into $\underset{\sim}{2}$ will be denoted by $\hat{0}$ and $\hat{1}$.

THEOREM 7.4 ([26]). *An ordered set* $P$ *is representable by a distributive lattice* [*with zero, with unit, with zero and unit*] *if and only if for some set* $I$, $P$ *is order-isomorphic to a subset* $Q$ *of* $\underset{\sim}{2}^I-\{\hat{0},\hat{1}\}$ *such that if an element* $\psi$ *of* $\underset{\sim}{2}^I$ *is locally in* $Q$, *then* $\psi$ *is in* $Q \cup \{\hat{0},\hat{1}\}$ [*in* $Q \cup \{\hat{1}\}$, *in* $Q \cup \{\hat{0}\}$, *in* $Q$]. □

In Figure 5 we give five representable ordered sets. The elements of $P_1$ are the following maps from $\omega$ into $\underset{\sim}{2}$ : $x_n$ takes the value 1 at $n$ and 0 otherwise, and $y_n$ takes the value 1 on $\{n,n+1\}$ and 0 otherwise. Any map locally in $P_1$ is in $P_1 \cup \{\hat{0}\}$ and so $P_1$ is representable by a distributive lattice with unit. We obtain $P_2$ from $P_1$ by extending the domain of the $x_n$ and $y_n$ from the natural numbers

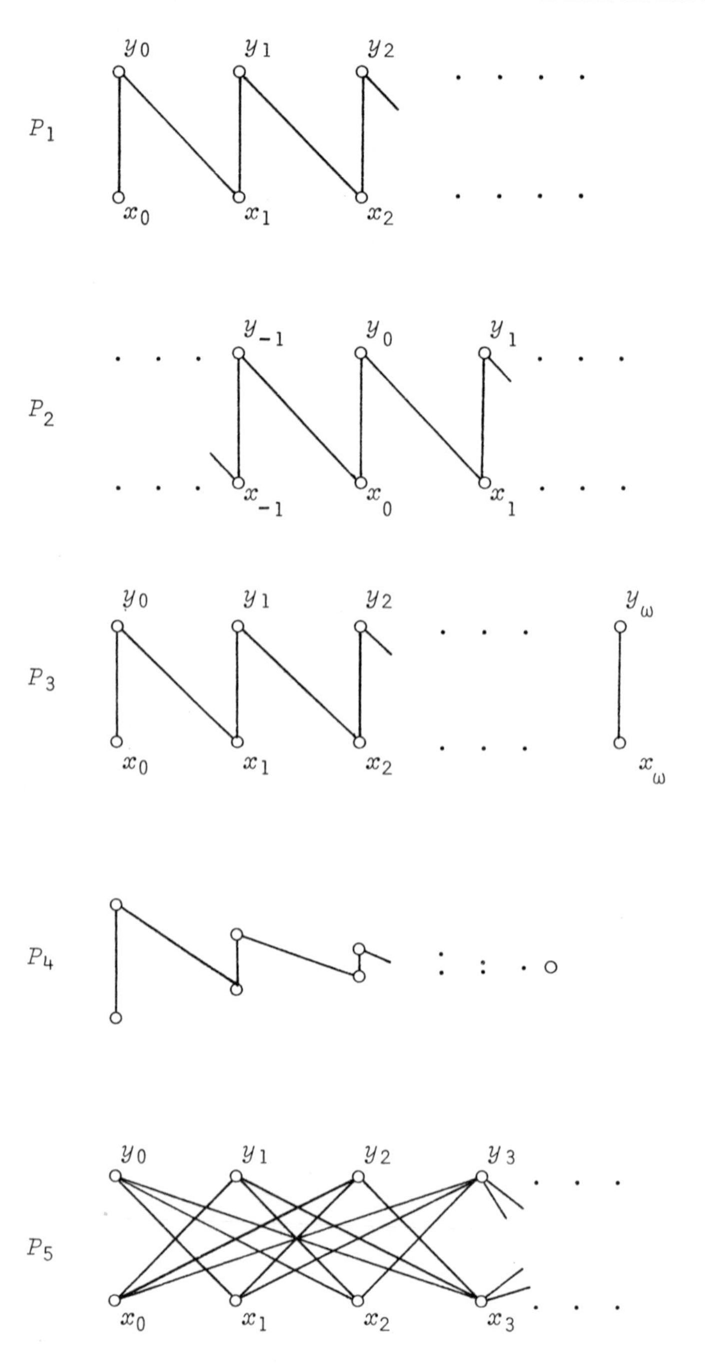


Figure 5. Some representable ordered sets

to the integers; $P_2$ is also representable by a distributive lattice with unit. The easiest way to see that $P_3$ is representable is to define a topology on it by declaring $\{x_n \mid n \leq \omega\}$ to be the one-point compactification of $\{x_n \mid n < \omega\}$ and similarly for $\{y_n \mid n \leq \omega\}$ (as the diagram suggests). Checking a few cases shows that $P_3$ is now totally order-disconnected and so is an object of $\underset{\sim}{P}$; consequently $P_3$ is representable by a bounded distributive lattice. Similarly $P_4$ is represented by a bounded distributive lattice. (The reader may like, as an exercise, to find simple descriptions of $P_3$ and $P_4$ as closed subsets of $\underset{\sim}{2}^{\omega}-\{\hat{0},\hat{1}\}$.) In $P_5$ we have $y_n > x_m$ exactly when $n \neq m$ : as maps from $\omega$ to $\underset{\sim}{2}$ we define $x_n$ to be 1 at $n$ and 0 otherwise, and $y_n$ is 0 at $n$ and 1 otherwise. T. Tan observed that although $P_5$ is representable it is not representable by a distributive lattice with either a zero or a unit. Representable ordered sets are considered in [3], [9], [73].

Let $F\underset{\sim}{D}(P)$ denote the free bounded distributive lattice generated by the ordered set $P$; i.e. there is an order-embedding $e$ of $P$ into $F\underset{\sim}{D}(P)$ and for all $K \in \underset{\sim}{D}$ if $\alpha$ is an order-preserving map of $P$ into $K$ then there is a unique (0,1)-homomorphism $\bar{\alpha}$ of $F\underset{\sim}{D}(P)$ into $K$ such that $\bar{\alpha} \circ e = \alpha$. The usual arguments show that the image of $P$ under $e$ generates $F\underset{\sim}{D}(P)$ and that $F\underset{\sim}{D}(P)$ is unique up to isomorphism. The definition of $F\underset{\sim}{D}(P)$ yields a bijection between $\mathrm{D}(F\underset{\sim}{D}(P)) = \underset{\sim}{D}(F\underset{\sim}{D}(P),\underset{\sim}{2})$ and the power $\underset{\sim}{2}^P$ which is clearly an order-isomorphism and becomes a homeomorphism once we endow $\underset{\sim}{2}^P$ with the subspace topology from the direct power $\underset{\sim}{2}^{|P|}$. Since

$$F\underset{\sim}{D}(P) \cong \mathrm{ED}(F\underset{\sim}{D}(P)) \cong \mathrm{E}(\underset{\sim}{2}^P) \cong \mathcal{O}(\underset{\sim}{2}^P)^d,$$

we have established the following :

THEOREM 7.5 ([32]). *(a)* $\mathcal{O}(\underset{\sim}{2}^P)^d$ *is a free bounded distributive lattice generated by the ordered set* $P$*, where the power* $\underset{\sim}{2}^P$ *is topologized as a subspace of* $\underset{\sim}{2}^{|P|}$. *The order-embedding* $e$ *of* $P$ *into* $\mathcal{O}(\underset{\sim}{2}^P)^d$ *is given by* $e(x) = \{f \in \underset{\sim}{2}^P \mid f(x) = 0\}$.

*(b)* $\mathcal{O}(\underset{\sim}{2}^{\kappa})$ *is a free* $\kappa$*-generated bounded distributive lattice.* □

Let $L \in \underset{\sim}{D}$ and consider the free bounded distributive lattice $F\underset{\sim}{D}(L) = \mathcal{O}(\underset{\sim}{2}^L)^d$ generated by the ordered set $L$.

Clearly any map in $\underset{\sim}{2}^L$ which is locally a (0,1)-homomorphism is a (0,1)-homomorphism and hence $D(L) = \underset{\sim}{D}(L,\underset{\sim}{2})$ is a closed subset of $\underset{\sim}{2}^L$ by 7.3. It is natural to ask to which congruence on $F\underset{\sim}{D}(L)$ this closed set corresponds. Let us say that a congruence $\Theta$ on $F\underset{\sim}{D}(L)$ *separates* $L$ if $\theta \circ e : L \to F\underset{\sim}{D}(L)/\Theta$ is an order-embedding where $\theta : F\underset{\sim}{D}(L) \to F\underset{\sim}{D}(L)/\Theta$ is the natural map. By 7.1(c) the congruence $\Theta_L$ on $F\underset{\sim}{D}(L) = \mathcal{O}(\underset{\sim}{2}^L)^d$ corresponding to $D(L)$ is defined by

$$A \equiv B\ (\Theta_L) \Leftrightarrow A \cap D(L) = B \cap D(L), \quad \text{for } A,B \in \mathcal{O}(\underset{\sim}{2}^L)^d.$$

If $a \nleq b$ in $L$, then by the prime-ideal theorem, there exists a homomorphism $\psi \in D(L) = \underset{\sim}{D}(L,\underset{\sim}{2})$ with $\psi(a) = 0$ and $\psi(b) = 1$. Since $\psi \in e(a) \cap D(L)$ and $\psi \notin e(b)$, it follows that $e(a) \cap D(L) \nsubseteq e(b) \cap D(L)$, whence $\Theta_L$ separates $L$. Now let $X$ be any closed subset of $\underset{\sim}{2}^L$ such that the corresponding congruence $\Theta(X)$ on $\mathcal{O}(\underset{\sim}{2}^L)^d$ separates $L$. We claim that $D(L) \subseteq X$. Let $\psi \in D(L)$; to prove that $\psi \in X$ it suffices to show that $\psi$ is locally in $X$. Let $J$ be a finite subset of $L$; we must find $\phi \in X$ such that $\psi|J = \phi|J$. Defining

$$a = \bigvee\{x \in J \mid \psi(x) = 0\} \quad \text{and} \quad b = \bigwedge\{x \in J \mid \psi(x) = 1\}$$

it suffices to find $\phi \in X$ with $\phi(a) = 0$ and $\phi(b) = 1$. Since $\psi(a) = 0$ and $\psi(b) = 1$ we have $a \nleq b$ and thus, since $\Theta(X)$ separates $L$, $e(a) \cap X \nsubseteq e(b) \cap X$ by the definition of $\Theta(X)$ (see 7.1(c)). Let $\phi \in (e(a) \cap X)-(e(b) \cap X)$; then $\phi \in X$ and $\phi(a) = 0$, $\phi(b) = 1$ as required.

Hence $D(L)$ is the smallest closed subset of $\underset{\sim}{2}^L$ for which the corresponding congruence separates $L$ or equivalently, the congruence corresponding to $D(L)$ is the largest congruence on $F\underset{\sim}{D}(L)$ which separates $L$. Let $\bar{1}_L : F\underset{\sim}{D}(L) \to L$ extend the identity map $1_L : L \to L$. Since the kernel of $\bar{1}_L$ is clearly a maximal congruence on $F\underset{\sim}{D}(L)$ which separates $L$ it now follows that the kernel of $\bar{1}_L$ is the largest congruence on $F\underset{\sim}{D}(L)$ which separates and that it corresponds to the closed set $D(L)$. The following result of I. Rival and B. Sands [65] is an easy corollary of this discussion.

LEMMA 7.6 ([65]). *Let $L$ and $K$ be distributive lattices. If there is an order-embedding of $L$ into $K$, then $L$ is a homomorphic image of a sublattice of $K$. If $L$ is finite, the converse also holds.*

Proof. We may assume that $L,K \in \underset{\sim}{D}$ for otherwise we adjoin universal bounds and proceed. Let $\alpha : L \to K$ be an order-embedding and let $\bar{\alpha} : F\underset{\sim}{D}(L) \to K$ be the induced homomorphism.

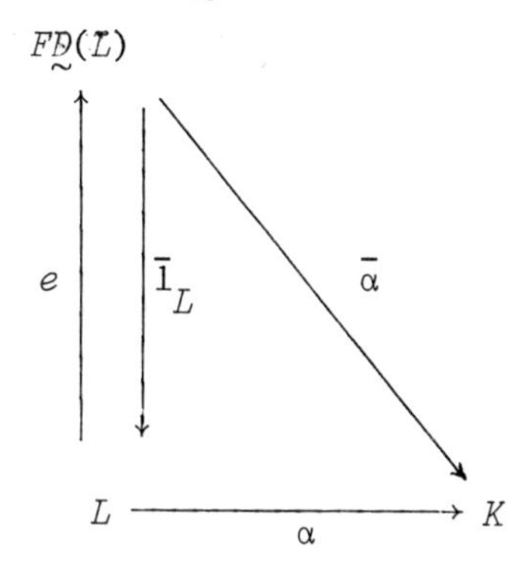


Since the kernel of $\bar{\alpha}$ separates $L$ we have $\ker(\bar{\alpha}) \leq \ker(\bar{1}_L)$ and consequently $L$ is a homomorphic image of $\mathrm{Im}(\bar{\alpha})$ which is a sublattice of $K$. For finite $L$ the converse is clear. □

Note that even if $K$ is bounded, if $L$ has no universal bounds then the sublattice of $K$ referred to in 7.6 need not be a (0,1)-sublattice.

Before presenting the main results of I. Rival and B. Sands [65] we require a description of the finite projective members of $\underset{\sim}{D}$. As is often the case by considering $\underset{\sim}{D}$ we also obtain information about the category of (not necessarily bounded) distributive lattices.

THEOREM 7.7. *(a) A finite distributive lattice $L$ is projective in the category of bounded distributive lattices if and only if $J(L)$ is a lattice, in which case $J(L)$ is a meet-subsemilattice of $L$.*

*(b) A finite distributive lattice $L$ is projective in the category of distributive lattices if and only if $J(L) \cup \{0,1\}$ is a lattice, or equivalently, if $J(L) \cup \{0\}$ is a meet-subsemilattice of $L$.* [8].

Proof. Recall that $L$ is projective in $\underset{\sim}{D}$ if and only if $L$ is a retract of $F\underset{\sim}{D}(\kappa) = O(\underset{\sim}{2}^{\kappa})$ for some $\kappa$, and if $L$ is finite we may choose $\kappa = |L|$. But since functors preserve retractions, $L$ is a retract of $O(\underset{\sim}{2}^{n})$ if and only if $J(L)$ is a retract of $JO(\underset{\sim}{2}^{n}) \cong \underset{\sim}{2}^{n}$. It is easy to show than an ordered set $P$ is a retract of $\underset{\sim}{2}^{n}$ for some $n$ if and only if $P$ is a finite lattice, and that if $J(L)$ is a finite lattice then the meet in $L$ of two join irreducibles is again join irreducible, which completes the proof of (a).

It is a simple exercise to show that $L$ is projective in the category of distributive lattices if and only if $L$ with universal bounds adjoined is projective in $\underset{\sim}{D}$, and consequently (b) follows from (a). □

Since an ordered set is a retract of a power of $\underset{\sim}{2}$ if and only if it is a complete lattice, the argument given above shows that if $L$ is projective in $\underset{\sim}{D}$, then $D(L) = PI(L)$ is a complete lattice. It would be of interest to know what additional conditions on $PI(L)$ are needed to characterize the projectivity of $L$ in $\underset{\sim}{D}$. (R. Balbes and A. Horn [12] have given an algebraic characterization which is somewhat unsatisfactory.) This is equivalent to describing the injectives in $\underset{\sim}{P}$. Note that the injectives in the category $\underset{\sim}{O}$ of ordered sets are precisely the complete lattices [15].

For (not necessarily bounded) distributive lattices $L$ and $K$ consider the following property :

> $E(L,K)$ : *if there is an order-embedding of* $L$ *into* $K$, *then there is an embedding of* $L$ *into* $K$.

Even when $L$ and $K$ are bounded, the embedding referred to in $E(L,K)$ need not be a (0,1)-embedding. Let $L$ be projective in the category of distributive lattices. If there is an order-embedding of $L$ into $K$, then by 7.6, $L$ is a homomorphic image of a sublattice of $K$ and consequently, since $L$ is projective, there is an embedding of $L$ into $K$ : hence $E(L,K)$ holds for all $K$. The class of distributive lattices $L$ such that $E(L,K)$ holds for all distributive lattices $K$ is much wider than the class of projective distributive lattices as the next result shows. If there is an order-embedding of $P$ into $Q$ we write $P < Q$.

THEOREM 7.8 ([65]). *Let* $L$ *be a finite distributive lattice and let* $Q$ *be a finite lattice with* $J(L) < Q$. *Then* $L$ *satisfies* $E(L,K)$ *for all distributive lattices* $K$ *if and only if there is an embedding of* $L$ *into* $O(P)$ *for every ordered set* $P$ *with* $J(L) < P < Q$.

Proof. Assume $E(L,K)$ holds for all distributive lattices $K$. If $J(L) < P < Q$, then by 7.2(b), $L$ is a homomorphic image of $O(P)$ and consequently $L < O(P)$. Since $E(L,O(P))$ holds there is an embedding of $L$ into $O(P)$.

Conversely assume that the condition holds and suppose that $K$ is a distributive lattice with $L < K$. We may suppose that the image of $L$ in $K$ generates $K$ and hence that $K$ is finite. By 7.6 there is a homomorphism $\gamma$ from $K$ onto $L$. On the other hand, invoking 7.2(b), there is a homomorphism $\beta$ from $\mathcal{O}(Q)$ onto $L$ since $J(L) < Q$. But by 7.7, $\mathcal{O}(Q)$ is projective in $\underset{\sim}{D}$ and since all our maps are onto and therefore preserve bounds, there is a (0,1)-homomorphism $\alpha$ from $\mathcal{O}(Q)$ into $K$ such that $\alpha \circ \gamma = \beta$. Let $K' = \mathrm{Im}(\alpha)$; then the restriction of $\gamma$ to $K'$ is onto and hence $J(L) < J(K') < J(\mathcal{O}(Q)) \cong Q$ by 7.2(b). By our assumption there is an embedding of $L$ into $\mathcal{O}(J(K')) \cong K'$ and so there is an embedding of $L$ into $K$. □

This theorem yields another proof that if $L$ is a finite projective distributive lattice, then $E(L,K)$ holds for all distributive lattices $L$ : by 7.7 we may choose $Q = J(L) \cup \{0,1\}$; whence if $J(L) < P < Q$ it follows that $\mathcal{O}(P)$ is, to within isomorphism, either $L$ or is $L$ with one or two new universal bounds adjoined, and so there is an embedding of $L$ into $\mathcal{O}(P)$.

We turn now to the characterization of those distributive lattices $K$ which satisfy $E(L,K)$ for all distributive lattices $L$. We shall need the fact that distributive lattices have the *congruence extension property* : if $L$ is a sublattice of $K$, then every congruence on $L$ is the restriction to $L$ of a congruence on $K$. A simple argument shows that the congruence extension property for distributive lattices follows from the congruence extension property for bounded distributive lattices and (0,1)-sublattices which in turn follows, via 7.1(a)(c), from the trivial fact that if $P,Q \in \underset{\sim}{P}$, $\phi$ is a continuous order-preserving map from $P$ onto $Q$, and $X \in \mathcal{C}\ell(Q)$; then $\phi^{-1}(X) \in \mathcal{C}\ell(P)$.

THEOREM 7.9 ([65]). *Let $K$ be a distributive lattice. If every homomorphic image of $K$ is embeddable in $K$, then $E(L,K)$ holds for all distributive lattices $L$. If $K$ is finite, then the converse also holds.*

Proof. If there is an order-embedding of $L$ into $K$ then by 7.6, $L$ is a homomorphic image of a sublattice of $K$. By the congruence extension property, $L$ is embeddable into a homomorphic image of $K$ and hence, by hypothesis, $L$ is embeddable into $K$. If $K$ is finite and $L$ is a homomorphic image of $K$, then there is an order-embedding of $L$ into $K$. Thus if $E(L,K)$ holds, $L$ is embeddable into $K$. □

One would naturally expect the arithmetic of a finite distributive lattice $L$ to be strongly reflected in $J(L)$. Recall that the *dimension* of $(P;\leq)$ is the minimum number of linear orderings of $P$ whose intersection is the given ordering, $\leq$. Our proofs of the following results of G. Birkhoff, R. Wille, and R.P. Dilworth are instructive illustrations of the machinery now available to us.

THEOREM 7.10. *Let* $L$ *be a finite distributive lattice.*

*(a) The length of* $L$ *equals* $|J(L)|$ [19].

*(b)* $L$ *is generated by* $n$ *chains if and only if the dimension of* $J(L)$ *is at most* $n$ [83].

*(c) The dimension of* $L$ *equals the width of* $J(L)$. [34].

Proof. (a) Let $J(L) = \{a_1,\ldots,a_m\}$ where $a_1 < a_2 < \ldots < a_m$ is a linear extension of the ordering on $J(L)$. Then

$$\emptyset \subset \{a_1\} \subset \{a_1,a_2\} \subset \ldots \subset \{a_1,\ldots,a_m\} = J(L)$$

is a maximal chain in $\mathcal{O}(J(L))$, from which the result follows.

(b) The dimension of an ordered set $P$ is just the minimum number $n$ such that $P$ can be order-embedded into a product of $n$ chains. Since $L$ is generated by the chains $C_1,\ldots,C_n \subseteq L$ if and only if it is a homomorphic image of the coproduct $C_1 * \ldots * C_n$, a simple application of 7.2(b)(e) yields (b).

(c) By 7.2(d) if $K = C_1 \times \ldots \times C_n$ is a finite product of finite chains, then $J(K)$ is a disjoint union of finite chains. Consequently for any subset $P$ of $J(K)$ there is an order-preserving map of $J(K)$ onto $P$. Hence by 7.2(a)(b) every homomorphic image of $C_1 \times \ldots \times C_n$ can be embedded into $C_1 \times \ldots \times C_n$. Thus by 7.9, if a finite distributive lattice $L$ can be order-embedded into $C_1 \times \ldots \times C_n$, there is an embedding $\gamma$ of $L$ into $C_1 \times \ldots \times C_n$. Let $\pi_i : C_1 \times \ldots \times C_n \to C_i$ be the $i$-th projection and define $C_i' = \mathrm{Im}(\pi_i \circ \gamma)$; then there is a (0,1)-embedding of $L$ into $C_1' \times \ldots \times C_n'$. Hence the dimension of $L$ is the smallest $n$ such that there is a (0,1)-embedding of $L$ into a product of $n$ finite chains. Transferring from $\underset{\sim}{D}$ to $\underset{\sim}{P}$ via 7.2, the dimension of $L$ is the smallest $n$ such that there is an order-preserving map from the disjoint union of $n$ finite chains onto $J(L)$, and thus, finally, the dimension of $L$ is the smallest $n$ such that $J(L)$ is a union of $n$ chains. This, by a result of Dilworth [34], is exactly the width of $L$. □

For further applications of the Birkhoff-Priestley duality to finite distributive lattices we refer the reader to [5], [60], and for applications to arbitrary distributive lattices see [1], [2], [10], [23], [27], [67].

## 7.2 Distributive-lattice-ordered algebras.

As we indicated earlier any class $\underset{\sim}{K}$ of bounded-distributive-lattice ordered algebras may be viewed as a subcategory of $\underset{\sim}{D}$ and hence we immediately obtain a duality between $\underset{\sim}{K}$ and the corresponding subcategory of $\underset{\sim}{P}$. We shall concentrate our attention on pseudocomplemented distributive lattices and a natural generalization of de Morgan algebras called Ockham algebras.

An algebra $\langle L;\vee,\wedge,*,0,1\rangle$ is a *p-algebra* (i.e. a pseudocomplemented lattice) if $\langle L;\vee,\wedge,0,1\rangle$ is a bounded lattice and $*$ is a unary operation defined by

$$x \wedge a = 0 \Leftrightarrow x \leq a^*,$$

and an algebra $\langle A;\vee,\wedge,*,+,0,1\rangle$ is a *double p-algebra* if the deletion of the operation $+$ yields a $p$-algebra and the deletion of the operation $*$ yields a dual $p$-algebra. The category $\underset{\sim}{B}_\omega$ of distributive $p$-algebras and the category $\underset{\sim}{B}{}^{\omega}_{\omega}$ of distributive double $p$-algebras are both varieties. It is easily seen that every finite distributive lattice is a double $p$-algebra.

H.A. Priestley described the dual category of $\underset{\sim}{B}{}^{\omega}_{\omega}$ in [63],[64].

THEOREM 7.11 ([63]). *Let $P$ be an object of $\underset{\sim}{P}$.*

*(i) $O(P)$ is a p-algebra iff for every clopen order-ideal $A$ of $P$, $[A)$ is open in $P$; pseudocomplements in $O(P)$ are given by $A^* = P-[A)$.*

*(ii) $O(P)$ is a dual p-algebra if and only if for every clopen dual order-ideal $B$ of $P$, $(B]$ is open in $P$; dual pseudocomplements in $O(P)$ are given by $A^+ = (P-A]$.* □

Objects of $\underset{\sim}{P}$ which satisfy the condition of 7.11(i) will be called *p-spaces*, those which satisfy the condition of 7.11(ii) will be called *dual p-spaces*, and those which satisfy both conditions will be called *double p-spaces*. Clearly every finite ordered set $P$ is a double $p$-space since as an object of $\underset{\sim}{P}$ the topology on $P$ is discrete; an easy check shows that $P_3$ and $P_4$ of Figure 5 are double $p$-spaces.

For each object $P$ in $\underset{\sim}{P}$ let $\mathrm{Min}(P)$ and $\mathrm{Max}(P)$ denote respectively the set of minimal elements and the set of maximal elements of $P$, and for all $x \in P$ define

$$\mathrm{Min}(x) = \mathrm{Min}(P) \cap (x] \quad \text{and} \quad \mathrm{Max}(x) = \mathrm{Max}(P) \cap [x).$$

THEOREM 7.12 ([4],[64]). *Let $P$ and $Q$ be p-spaces and let $\phi : P \to Q$ be continuous and order-preserving; then*

*$\phi^{-1} : \mathcal{O}(Q) \to \mathcal{O}(P)$ is a p-algebra homomorphism if and only if $\phi(\mathrm{Min}(x)) = \mathrm{Min}(\phi(x))$ for all $x \in P$. If $P$ and $Q$ are double p-spaces, then $\phi^{-1} : \mathcal{O}(Q) \to \mathcal{O}(P)$ is a double p-algebra homomorphism if and only if it also satisfies $\phi(\mathrm{Max}(x)) = \mathrm{Max}(\phi(x))$ for all $x \in P$.* □

Theorems 7.11 and 7.12 describe the objects and morphisms of categories dual to $\underset{\sim}{B}_\omega$ and to $\underset{\sim}{B}^\omega_\omega$. Having described the dual category we can immediately determine the (double) $p$-algebra congruences since a closed subset $X$ of a (double) $p$-space $P$ will induce a (double) $p$-space congruence on $\mathcal{O}(P)$ if and only if the inclusion map of $X$ into $P$ is a (double) $p$-space morphism.

LEMMA 7.13 ([29]). *Let $P$ be a double p-space and let $X$ be a closed subset of $P$. Then $\Theta(X)$ is a (double) p-algebra congruence on $\mathcal{O}(P)$ if and only if $\mathrm{Min}(x) \subseteq X$ (and $\mathrm{Max}(x) \subseteq X$) for all $x \in X$.* □

Call a subset $X$ of $P$ a **-subset* if it satisfies $\mathrm{Min}(x) \subseteq X$ for all $x \in X$, a *+-subset* if it satisfies $\mathrm{Max}(x) \subseteq X$ for all $x \in X$, and a *c-subset* if it satisfies both conditions.

THEOREM 7.14 ([29]). *Let $P$ be a (double) p-space; then the lattice of congruences on the (double) p-algebra $\mathcal{O}(P)$ is dually isomorphic to the lattice of closed (c-subsets) *-subsets of $P$.* □

In any variety an important role is played by the subdirectly irreducible algebras (that is, those algebras which have a smallest nontrivial congruence); Theorem 7.14 is clearly a tool for their description. Let $P$ be a $p$-space. Hence the $p$-algebra $\mathcal{O}(P)$ is subdirectly irreducible if and only if $P$ has a largest proper closed *-subset. It is easily seen that $\mathrm{Min}(P)$ is a closed subset of $P$; indeed $\mathrm{Min}(P)$ corresponds to the congruence $\Phi$ on $\mathcal{O}(P)$ defined by $A \equiv B(\Phi)$ if and only if $A^* = B^*$. Consequently $\mathrm{Min}(P) \cup \{x\}$ is a closed *-subset for all $x \in P$ and hence $\mathcal{O}(P)$ is subdirectly irreducible if and only if $|P| = 1$ or $P = \mathrm{Min}(P) \cup \{x\}$ for some non-minimal element $x$. Transferring back to $\underset{\sim}{B}_\omega$ we obtain the result of K.B. Lee [56] and H. Lakser [55] that a distributive $p$-algebra is subdirectly irreducible if and only if it is a Boolean algebra with a new unit adjoined. Via a similar line of reasoning it is shown in B.A. Davey [29] that if $\mathcal{O}(P)$ is a subdirectly irreducible double $p$-algebra then $P$ contains at most one element which is neither minimal nor maximal. In the finite case we can conclude that $L$ is subdirectly irreducible if and only if $\mathcal{J}(L)$ is connected and contains at most one element which is neither minimal nor maximal. $\mathcal{O}(P_3)$ and $\mathcal{O}(P_4)$ are subdirectly irreducible distributive double $p$-algebras.

The subvarieties of $\underset{\sim}{B}_\omega$ are very neatly described in terms of dual spaces. Let $\underset{\sim}{B}_{-1}$ be the class of trivial algebras, $\underset{\sim}{B}_0$ be the class of Boolean algebras, and for $n \geq 1$ let $\underset{\sim}{B}_n$ be all algebras $L$ in $\underset{\sim}{B}_\omega$ such that $|\mathrm{Min}(x)| \leq n$ for all $x$ in the dual space of $L$. Then the lattice of subvarieties of $\underset{\sim}{B}_\omega$ is

$$\underset{\sim}{B}_{-1} \subset \underset{\sim}{B}_0 \subset \underset{\sim}{B}_1 \subset \underset{\sim}{B}_2 \subset \dots \subset \underset{\sim}{B}_\omega \qquad [56].$$

The algebras in $\underset{\sim}{B}_1$ are called *Stone algebras*. $\mathcal{O}(P_3)$ and $\mathcal{O}(P_4)$ are in $\underset{\sim}{B}_2$ but not in $\underset{\sim}{B}_1$. The lattice of subvarieties of $\underset{\sim}{B}_\omega^\omega$ is much less tractable : using the duality and a neat graph-theoretic construction, A. Urquhart [82] has shown that the lattice of subvarieties is uncountable and that there is an equational class which is finitely based but not generated by its finite members.

One of the uses to which a duality is usually put is the description of free algebras and coproducts. This has been done for all the subvarieties of $\underset{\sim}{B}_\omega$ and for certain subvarieties of $\underset{\sim}{B}_\omega^\omega$ in A. Urquhart [77], [78], and B.A. Davey and M. Goldberg [31] : we consider briefly the case of $\underset{\sim}{B}_1^1$, the variety of double Stone algebras.

The description of coproducts in any subvariety of $\underset{\sim}{B}_\omega^\omega$ amounts to describing the products in the corresponding dual category. Since these products are in general not just the usual direct product, some ingenuity is required. The dual category of $\underset{\sim}{B}_1^1$ is the class $\underset{\sim}{P}_1^1$ of all double $p$-spaces $P$ such that $x \in P$ dominates a unique minimal and is dominated by a unique maximal. It is easily seen that a product in $\underset{\sim}{P}_1^1$ is just the direct product and so agrees with the product in $\underset{\sim}{P}$; it follows that the $\underset{\sim}{B}_1^1$-coproduct of $L, K \in \underset{\sim}{B}_1^1$ is their $\underset{\sim}{D}$-coproduct $L * K$. Note that for all $P, Q, R \in \underset{\sim}{P}$ we have

$$P \times (Q + R) \cong (P \times Q) + (P \times R) \;:$$

translating this into $\underset{\sim}{B}_1^1$ via $\underset{\sim}{D}$ gives

$$L * (M \times N) \cong (L * M) \times (L * N)$$

for all $L, M, N \in \underset{\sim}{B}_1^1$. From our earlier discussion of the dual spaces of subdirectly irreducibles in $\underset{\sim}{B}_{\omega}^{\omega}$ it follows that the only subdirectly irreducible algebras in $\underset{\sim}{B}_1^1$ are the two-, three-, and four-element chains, whence $\underset{\sim}{B}_1^1$ is generated as a variety by $\underset{\sim}{4}$. Thus the free $\underset{\sim}{B}_1^1$-algebra on one generator, $F\underset{\sim}{B}_1^1(1)$, is the subalgebra of $\underset{\sim}{4}^4$ generated by the identity map, and pencil and paper yields

$$F\underset{\sim}{B}_1^1(1) \cong \underset{\sim}{2} \times \underset{\sim}{2} \times \underset{\sim}{3} \cong (F\underset{\sim}{D}(0))^2 \times F\underset{\sim}{D}(1). \tag{1}$$

Since coproducts distribute over products in $\underset{\sim}{B}_1^1$, for $m < \omega$ we have

$$\begin{aligned} F\underset{\sim}{B}_1^1(m+1) &\cong F\underset{\sim}{B}_1^1(m) * F\underset{\sim}{B}_1^1(1) \cong F\underset{\sim}{B}_1^1(m) * (F\underset{\sim}{D}(0)^2 \times F\underset{\sim}{D}(1)) \\ &\cong F\underset{\sim}{B}_1^1(m)^2 \times (F\underset{\sim}{B}_1^1(m) * F\underset{\sim}{D}(1)). \end{aligned} \tag{2}$$

Combining (1) and (2) we conclude

$$F\underset{\sim}{B}_1^1(m) \cong F\underset{\sim}{D}(0)^{a_{m,0}} \times \dots \times F\underset{\sim}{D}(m)^{a_{m,m}}$$

where the natural numbers $a_{m,k}$ satisfy

(a) $a_{m,k} = 0$, when $m < k$,

(b) $a_{m,m} = 1$,

(c) $a_{m,k} = 2\, a_{m-1,k} + a_{m-1,k-1}$ when $m \geq k$.

Solving these recurrence relations yields

$$F\underset{\sim}{B}_1^1(m) \cong \prod_{k=0}^{m} \left[ F\underset{\sim}{D}(k) \right]^{\binom{m}{k} 2^{m-k}} \qquad [31], [47].$$

In exactly the same way we can derive the description of $F\underset{\sim}{B}_1(m)$ given by R. Balbes and A. Horn [13].

Further applications of this duality may be found in [4], [6], [54], [81].

An algebra $\langle L;\vee,\wedge,\sim,0,1\rangle$ is an *Ockham algebra* if $\langle L;\vee,\wedge,0,1\rangle$ is a bounded lattice and $\sim$ is a dual $(0,1)$-endomorphism of $L$; that is, satisfies

$$\sim(x \wedge y) = \sim x \vee \sim y, \ \sim(x \vee y) = \sim x \wedge \sim y, \ \sim 0 = 1, \ \sim 1 = 0.$$

Denote the variety of distributive Ockham algebras by $\underset{\sim}{H}$. The subvariety given by the identity $\sim\sim x = x$ is the variety $\underset{\sim}{M}$ of *de Morgan algebras* and the subvariety of $\underset{\sim}{M}$ determined by the extra identity $x \wedge \sim x \leq y \vee \sim y$ is the variety of *Kleene algebras*. The dual category $\underset{\sim}{S}$ of $\underset{\sim}{H}$ is very easily described : since the objects of $\underset{\sim}{H}$ are bounded distributive lattices equipped with a dual endomorphism, an object of $\underset{\sim}{S}$ (which we refer to as an *Ockham space*) will be an object $\underset{\sim}{P}$ of $\underset{\sim}{P}$ equipped with an order-reversing continuous map $g : P \to P$, and the morphisms of $\underset{\sim}{S}$ will be continuous order-preserving maps which preserve $g$ [80]. Given $P \in \underset{\sim}{S}$ the operation $\sim$ is defined on $\mathcal{O}(P)$ by $\sim A = P - g^{-1}(A)$ for all $A \in \mathcal{O}(P)$.

The techniques illustrated above for distributive (double) $p$-algebras may now be applied to $\underset{\sim}{H}$ and $\underset{\sim}{S}$. Call a subset $X$ of an Ockham space $P$ a *$g$-subset* if $X$ is closed under $g$; that is $x \in X$ implies $g(x) \in X$.

THEOREM 7.15 ([80]). *Let $P$ be an Ockham space.*

(a) *If $X$ is a closed subset of $P$, then $\Theta(X)$ is a congruence on the Ockham algebra $\mathcal{O}(P)$ if and only if $X$ is a $g$-subset of $P$.*

(b) *The lattice of congruences on $\mathcal{O}(P)$ is dually isomorphic to the lattice of closed $g$-subsets of $P$.* □

Thus we conclude that $\mathcal{O}(P)$ is subdirectly irreducible if and only if $P$ contains a largest proper closed $g$-subset. For $x \in P$ define $g^{\omega}(x) = \{g^{n}(x) \mid n \geq 0\}$. If $P$ is a finite Ockham space and $\mathcal{O}(P)$ is subdirectly irreducible, then there exists $x \in P$ such that $P = g^{\omega}(x)$ : simply choose an $x$ not contained in the largest proper $g$-subset. Conversely if there exists $x \in P$ such that $P = g^{\omega}(x)$ then either $g^{\omega}(g(x))$ is the largest proper $g$-subset, whence $\mathcal{O}(P)$ is subdirectly irreducible, or $g^{\omega}(g(x)) = P$ in which case the empty set is the largest proper $g$-subset, whence $\mathcal{O}(P)$ is simple and therefore subdirectly irreducible. We have established the finite case of

THEOREM 7.16 ([80]). *Let $P$ be an Ockham space.*

(a) *$\mathcal{O}(P)$ is subdirectly irreducible if and only if there is an open set $U$ in $P$ such that $g^{\omega}(x)$ is dense in $P$ for all $x \in U$.*

(b) *If $P$ is finite, then $\mathcal{O}(P)$ is subdirectly irreducible if and only if $g^{\omega}(x) = P$ for some $x \in P$.* □

The remainder of this section is devoted to a selection of results from M.S. Goldberg [45].

For $m > n$ and $n \geq 0$, let $m_n$ be the Ockham space defined on the set $\mathbb{Z}_m = \{0,\ldots,m-1\}$ by imposing the discrete order and defining $g$ to be the map $\gamma_n$ where $\gamma_n(k) = k + 1$ for $0 \leq k < m - 1$, and $\gamma_n(m-1) = n$. (See Figure 6).

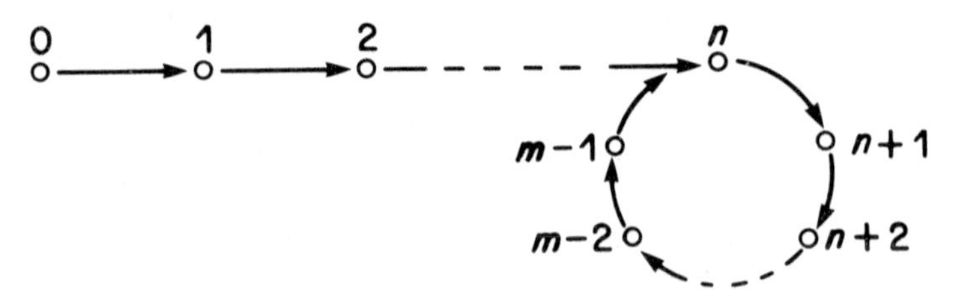


Figure 6. The Ockham space $m_n$

Any order on $\mathbb{Z}_m$ with respect to which $\gamma_n$ is order-reversing produces the dual space $P$ of a subdirectly irreducible Ockham algebra. Clearly all dual spaces of finite subdirectly irreducible Ockham algebras arise in this way. We refer to $T = \{0,1,\ldots,n-1\}$ as the *tail* of $P$; to $L = \{n,n+1,\ldots,m-1\}$ as the *loop* of $P$, and to any element $e$ such that $P = g^{\omega}(e)$ as an *end* of $P$. If $P$ has a non-empty tail, then it has a unique end (namely 0) and $\mathcal{O}(P)$ is subdirectly irreducible but not simple; if $P$ is just a loop, then every element of $P$ is an end and $\mathcal{O}(P)$ is simple. Since the identity map on $\mathbb{Z}_m$ is an Ockham-space morphism of $m_n$ onto $P$, it follows that a finite distributive Ockham algebra is subdirectly irreducible if and only if it is isomorphic to a subalgebra of $L_{m,n} = \mathcal{O}(m_n)$ for some $m$ and $n$.

For the remainder of this section we assume that $P$ is a finite Ockham space obtained by imposing an order on the space $m_n$. By Jónsson's Lemma [49], every subdirectly irreducible algebra $A$ in the variety $\underset{\sim}{V}$ generated by $\mathcal{O}(P)$ is a homomorphic image of a subalgebra of $\mathcal{O}(P)$ and thus $J(A)$ is a $g$-subset of an $\underset{\sim}{S}$-morphic image of $P$. But it is obvious that every $g$-subset of an $\underset{\sim}{S}$-morphic image of $P$ is the dual space of a subdirectly irreducible and hence the subdirectly irreducibles in $V$ are precisely the homomorphic images of subalgebras of $\mathcal{O}(P)$; in symbols $\mathrm{Si}(\underset{\sim}{V}) = \mathrm{HS}(\mathcal{O}(P))$. For $m_n$ we have more : the subdirectly irreducibles in the variety $\underset{\sim}{P}_{m,n}$ generated by $L_{m,n} = \mathcal{O}(m_n)$ are precisely the subalgebras of $L_{m,n}$. It is natural to ask : *Under what conditions on* $P$ *is it true that the subdirectly irreducibles in the variety* $\underset{\sim}{V}$ *generated by* $\mathcal{O}(P)$ *are precisely the subalgebras of* $\widetilde{\mathcal{O}}(P)$?

A subdirectly irreducible algebra $L$ in $\underset{\sim}{V}$ is called *maximal* if it has no proper extension in $\mathrm{Si}(\underset{\sim}{V})$. If $\underset{\sim}{X}$ is a class of Ockham spaces, then $\mathrm{G}(\underset{\sim}{X})$ and $\mathrm{M}(\underset{\sim}{X})$ denote respectively the classes of closed $g$-subsets and $\underset{\sim}{S}$-morphic images of spaces in $\underset{\sim}{X}$. Remembering that $\mathrm{Si}(\underset{\sim}{V}) = \mathrm{HS}(\widetilde{\mathcal{O}}(P))$ we say that $Z$ in $\mathrm{GM}(P)$ is *maximal* if every onto $\underset{\sim}{S}$-morphism $\phi : Y \to Z$ with $Y \in \mathrm{GM}(P)$ is an isomorphism. Thus we have $\mathrm{Si}(\underset{\sim}{V}) = \mathrm{S}(\mathcal{O}(P))$ precisely when $P$ is the only maximal in $\mathrm{GM}(P)$.

LEMMA 7.17 ([45]). *The only retracts of* $P$ *are* $P$ *and its loop* $L$.

Sketch of proof. Let $R \subseteq P$ and let $\phi : P \to R$ be a retraction. Let $e$ be an end of $P$ and $e_R$ an end of $R$; then $e_R = g^k(e)$ for some $k \in \mathbb{Z}_m$, and since $\phi$ is onto, $\phi(e) = e_R$. Hence

$$e_R = \phi(e_R) = \phi g^k(e) = g^k\phi(e) = g^k(e_R);$$

thus either $k = 0$ in which case $e_R = e$ and $R = P$, or $k > 0$ in which case $R$ is the loop.

For the converse, let $e$ be an end of $P$ and let $\phi : P \to L$ be determined by $\phi(e) = g^k(e)$ where $k$ is the unique natural number such that $n \le k < m$ and $k \equiv 0 \bmod (m-n)$; since $(P;g)$ is generated by $e$, any $\underset{\sim}{S}$-morphism out of $P$ is uniquely determined by its value at $e$. The map $\phi$ wraps the tail of $P$ backwards around the loop, and a careful check shows that it is an $\underset{\sim}{S}$-morphism. □

Let $e$ be a fixed end of $P$ and for $k \le n$ define $P_k = g^\omega(g^k(e))$. Let $Z$ be a maximal in $\mathrm{GM}(P)$. Since $\mathrm{MG} = \mathrm{GM}$, $Z$ is an $\underset{\sim}{S}$-morphic image of a $g$-subset of $P$; and since $Z$ is maximal it is (isomorphic to) a $g$-subset of $P$. Thus $Z \cong P_k$ for some $k \le n$. If $n = 0$, then $P$ is a loop and $Z \cong P_0 = P$. If $n = 1$, then $P_1$ is the loop of $P$ and by 7.17 is an $\underset{\sim}{S}$-morphic image of $P$, whence $Z \cong P$. Finally, if $n \ge 2$, an easy argument shows that for $2 \le k \le n - 2$, the map $g^2$ is an $\underset{\sim}{S}$-morphism of $P_{k-2}$ onto $P_k$. Hence up to isomorphism the only maximals in $\mathrm{GM}(P)$ are $P$ and $P_1$.

Now if $m = 1$, then $P$ is the only object in $\mathrm{GM}(P)$; and if $m = 2$, either $P$ or $P_1$ is the loop : see Figure 7.

Thus for $m \leq 2$, the only maximal in $\mathrm{GM}(P)$ is $P$. Furthermore if $P$ is a 3-element chain, then $P_1$ is either the loop, in which case $P_1$ is not maximal in $\mathrm{GM}(P)$, or the middle element

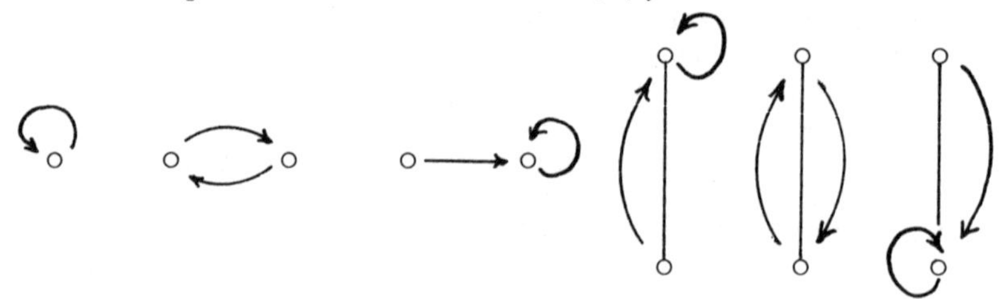

Figure 7. Dual space of subdirectly irreducibles with $m \leq 2$.

of $P$ is a $g$-fixed point, in which case there can be no $\underset{\sim}{S}$-morphism from $P$ onto $P_1$, whence $P_1$ is maximal in $\mathrm{GM}(P)$ : see Figure 8. It is not hard to see that if $m \geq 4$, and $P$ is a chain, then $P_1$ is maximal in $\mathrm{GM}(P)$.

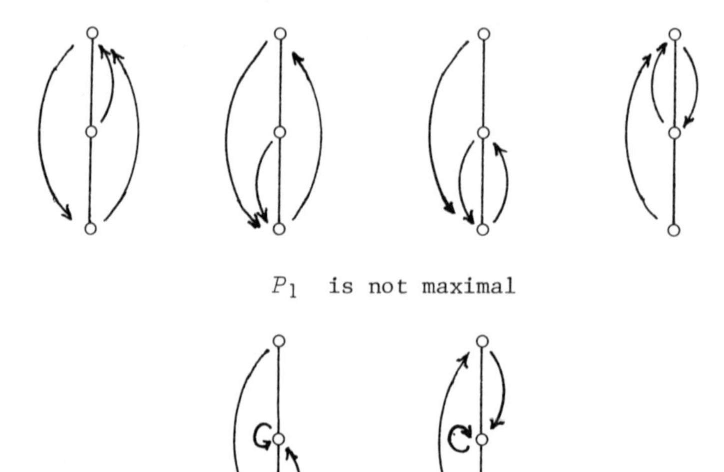

$P_1$ is not maximal

$P_1$ is maximal

Figure 8 : the case where $P$ is a 3-element chain

In general for $m \geq 3$, the situation is quite complicated, but a careful combinatorial analysis yields the following most satisfactory answer to our question.

THEOREM 7.18 ([45]). *$P$ is the only maximal in* $\mathrm{GM}(P)$ *and hence the subdirectly irreducibles in the variety generated by* $O(P)$ *are precisely the subalgebras of* $O(P)$ *if and only if* *one of the following holds :*

(a) $P$ *is isomorphic to one of the spaces in Figure 7;*

(b) $P$ *is isomorphic to* $m_n$;

(c) $P_1$ *is the loop of* $P$ *(regardless of the order on* $P$*);*

(d) $P$ *is isomorphic to one of the spaces in Figure 9, or to its order-theoretic dual.* □

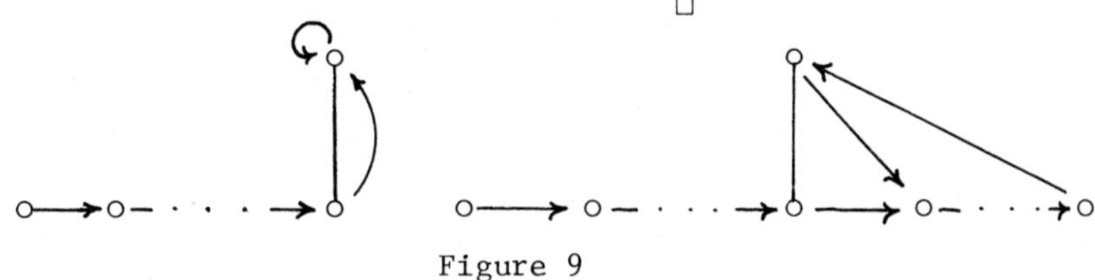

Figure 9

A continuation of the combinatorial analysis required to prove Theorem 7.18 leads to a description of the injective and weak injective algebras in the variety $\underset{\sim}{V}$ generated by $\mathcal{O}(P)$. The proof amounts to a careful analysis of projectivity in GM($P$). It turns out that the only subdirectly irreducibles which can be (weak) injectives in $\underset{\sim}{V}$ are $\mathcal{O}(P)$, $\mathcal{O}(P_1)$, and $\mathcal{O}(L)$ : for example, if $P_1$ is not maximal in GM($P$), then $\mathcal{O}(P)$ and $\mathcal{O}(L)$ are injective while $\mathcal{O}(P_1)$ is not even weak injective, and if $P_1$ is maximal in GM($P$), then $\mathcal{O}(P)$, $\mathcal{O}(P_1)$, and $\mathcal{O}(L)$ are weak injective in $\underset{\sim}{V}$ but may fail to be injective [45].

The duality can also be used to describe free algebras and coproducts in various subvarieties of $\underset{\sim}{H}$ [24], [25], [45], [80], and to yield information on the lattice of subvarieties [44].

BIBLIOGRAPHY

[1] M.E. Adams, The Frattini sublattice of a distributive lattice, *Algebra Univ.* 3(1973), 216-228.

[2] M.E. Adams, A problem of A. Monteiro concerning relative complementation of lattices, *Coll. Math.* 30(1974), 61-67.

[3] M.E. Adams, The poset of prime ideals of a distributive lattice, *Algebra Univ.* 5(1975), 141-142.

[4] M.E. Adams, Implicational classes of pseudocomplemented distributive lattices, *J. London Math. Soc.* (2) 13(1976), 381-384.

[5] M.E. Adams and D. Kelly, Homomorphic images of free distributive lattices that are not sublattices, *Algebra Univ.* 5(1975), 143-144.

[6] M.E. Adams and J. Sichler, Endomorphism monoids of distributive double $p$-algebras, *Glasgow Math. J.* 20(1979), 81-86.

[7] K.A. Baker and A.F. Pixley, Polynomial interpolation and the Chinese remainder theorem for algebraic systems, *Math. Z.* 143(1975), 165-174.

[8] R. Balbes, Projective and injective distributive lattices, *Pacific J. Math.* 21(1967), 405-420.

[9] R. Balbes, On the partially ordered set of prime ideals of a distributive lattice, *Canad. J. Math.* 23(1971), 866-874.

[10] R. Balbes, Solution of a problem concerning the intersection of maximal filters and maximal ideals in a distributive lattice, *Algebra Univ.* 2(1972), 389-392.

[11] R. Balbes and P. Dwinger, *Distributive Lattices*, University of Missouri Press, Columbia, Missouri (1974).

[12] R. Balbes and A. Horn, Projective distributive lattices, *Pacific J. Math.* 33(1970), 273-279.

[13] R. Balbes and A. Horn, Stone lattices, *Duke Math. J.* 37(1970), 537-546.

[14] B. Banaschewski, Hüllensysteme und Erweiterungen von Quasi-Ordnungen, *Z. Math. Logik Grundlagen Math.* 2(1956), 117-130.

[15] B. Banaschewski and G. Bruns, Categorical characterization of the Dedekind-MacNeille completion, *Arch. Math.* 18(1967), 369-377.

[16] C. Bergman, R. McKenzie, and S. Nagý, How to cancel a linearly ordered exponent, preprint (1979).

[17] G. Birkhoff, On the combination of subalgebras, *Proc. Camb. Phil. Soc.* 29(1933), 441-464.

[18] G. Birkhoff, Extended arithmetic, *Duke Math. J.* 3(1937), 311-316.

[19] G. Birkhoff, Rings of sets, *Duke Math. J.* 3(1937), 443-454.

[20] G. Birkhoff, Generalized arithmetic, *Duke Math. J.* 9(1942), 283-302.

[21] G. Birkhoff, *Lattice Theory*, American Mathematical Society, Providence, Rhode Island (1967).

[22] W.H. Cornish, On H. Priestley's dual of the category of bounded distributive lattices, *Mat. Vesnik* 12(27)(1975), 329-332.

[23] W.H. Cornish, Ordered topological spaces and the coproduct of bounded distributive lattices, *Coll. Math.* 36(1976), 27-35.

[24] W.H. Cornish and P.R. Fowler, Coproducts of de Morgan algebras, *Bull. Austral. Math. Soc.* 16(1977), 1-13.

[25] W.H. Cornish and P.R. Fowler, Coproducts of Kleene algebras, *J. Austral. Math. Soc.* 27(1979), 209-220.

[26] B.A. Davey, A note on representable posets, *Algebra Univ.* 3(1973), 345-347.

[27] B.A. Davey, Free products of bounded distributive lattices, *Algebra Univ.* 4(1974), 106-107.

[28] B.A. Davey, Topological duality for prevarieties of universal algebras, in G.-C. Rota, ed., *Studies in Foundations and Combinatorics*, Advances in Mathematics Supplementary Studies, Volume 1, Academic Press (1978), 61-99.

[29] B.A. Davey, Subdirectly irreducible distributive double $p$-algebras, *Algebra Univ.* 8(1978), 73-88.

[30] B.A. Davey, D. Duffus, R.W. Quackenbush, and I. Rival, Exponents of finite simple lattices, *J. London Math. Soc.* (2) 17(1978), 203-211.

[31] B.A. Davey and M.S. Goldberg, The free $p$-algebra generated by a distributive lattice, *Algebra Univ.* 11(1980), 90-100.

[32] B.A. Davey and I. Rival, Exponents of lattice-ordered algebras, *Algebra Univ.* (to appear).

[33] B.A. Davey and H. Werner, Dualities and equivalences for varieties of algebras, *Coll. Math. Soc. János Bolyai* (to appear).

[34] R.P. Dilworth, A decomposition theorem for partially ordered sets, *Ann. Math.* 51(1950), 161-166.

[35] D. Duffus, *Toward a Theory of Finite Partially Ordered Sets*, PhD thesis, University of Calgary, Calgary, Alberta (1978).

[36] D. Duffus, B. Jónsson, and I. Rival, Structure results for function lattices, *Canad. J. Math.* 30(1978), 392-400.

[37] D. Duffus and I. Rival, A logarithmic property for exponents of partially ordered sets, *Canad. J. Math.* 30(1978), 797-807.

[38] D. Duffus and R. Wille, Automorphism groups of function lattices, *Proceedings of a Meeting on Universal Algebra, Estergom* (1978).

[39] D. Duffus and R. Wille, A theorem on partially ordered sets of order-preserving mappings, *Proc. Amer. Math. Soc.* 76(1979), 14-16.

[40] E. Fuchs, Isomorphismus der Kardinalpotenzen, *Arch. Math. (Brno)* 1(1965), 83-93.

[41] L. Geissinger and W. Graves, The category of complete algebraic lattices, *J. Comb. Theory* 13(1972), 332-338.

[42] G. Gierz and K. Keimel, Topologische Darstellung von Verbänden, *Math. Z.* 150(1976), 83-99.

[43] L.M. Gluskin, Semigroups of isotone transformations, *Uspehi Mat. Nauk* 16(1961), 157-162.

[44] M.S. Goldberg, *Distributive p-Algebras and Ockham Algebras : a Topological Approach*, PhD thesis, La Trobe University, Bundoora, Australia (1979).

[45] M.S. Goldberg, Distributive Ockham algebras : free algebras and injectivity, *Bull. Austral. Math. Soc.* 24(1981), 161-203.

[46] J. Hashimoto, On direct product decompositions of partially ordered sets, *Ann. of Math.* (2) 54(1951), 315-318.

[47] T. Hecht and T. Katriňák, Free double-Stone algebras, in *Lattice Theory*, Coll. Math. Soc. János Bolyai 14(1974), pp.77-95.

[48] K.H. Hofmann, M. Mislove, and A. Stralka, *The Pontryagin Duality of Compact 0-Dimensional Semilattices and Its Applications*, Lect. Notes in Math. 396, Springer, Berlin-Heidelberg-New York (1974).

[49] B. Jónsson, Algebras whose congruences are distributive, *Math. Scand.* 21(1967), 110-121.

[50] B. Jónsson, The arithmetic of ordered sets, this volume, 3-41.

[51] B. Jónsson, Powers of partially ordered sets : the automorphism group, preprint (1980).

[52] B. Jónsson and R. McKenzie, Powers of partially ordered sets : cancellation and refinement properties, *Math. Scand.* (to appear).

[53] M. Kamara and D. Schweigert, Eine Charakterisierung polynomvollständiger Polaritätsverbände, *Arch. Math.* 30(1978), 661-664.

[54] T. Katriňák, The free {0,1}-distributive product of distributive $p$-algebras, *Bull. Soc. Roy. Sci. Liège* 47(1978), 323-328.

[55] H. Lakser, The structure of pseudocomplemented distributive lattices. I. Subdirect decomposition, *Trans. Amer. Math. Soc.* 156(1971), 334-342.

[56] K.B. Lee, Equational classes of distributive pseudo-complemented lattices, *Canad. J. Math.* 22(1970), 881-891.

[57] L. Lovász, Operations with structures, *Acta Math. Acad. Sci. Hungar.* 18(1967), 321-328.

[58] H.M. MacNeille, Partially ordered sets, *Trans. Amer. Math. Soc.* 42(1937), 416-460.

[59] M. Novotný, Über gewisse Eigenschaften von Kardinal operationen, *Spisy Prirod Fac. Univ. Brno* (1960), 465-484.

[60] R. Nowakowski and I. Rival, Distributive cover-preserving sublattices of modular lattices, *Nanta Math.* 11(1979), 110-123.

[61] H.A. Priestley, Representation of distributive lattices by means of ordered Stone spaces, *Bull. London Math. Soc.* 2(1970), 186-190.

[62] H.A. Priestley, Ordered topological spaces and the representation of distributive lattices, *Proc. London Math. Soc.* (3) 24(1972), 507-530.

[63] H.A. Priestley, Stone lattices : a topological approach, *Fund. Math.* 84(1974), 127-143.

[64] H.A. Priestley, The construction of spaces dual to pseudocomplemented distributive lattices, *Quart. J. Math. Oxford* Ser.(2) 26(1975), 215-228.

[65] I. Rival and B. Sands, Weak embeddings and embeddings of finite distributive lattices, *Archiv. Math.* 26(1975), 346-352.

[66] G. Sabidussi, Graph multiplication, *Math. Z.* 72(1960), 446-457.

[67] J. Schmid, Tensor products of distributive lattices and their Priestley duals, *Acta Math. Acad. Sci. Hungar.* 35(1980),387-392.

[68] E.T. Schmidt, Remark on generalized function lattices, *Acta Math. Acad. Sci. Hungar.* 34(1979), 337-339.

[69] J. Schmidt, Zur Kennzeichnung der Dedekind-MacNeilleschen Hülle einer geordneten Menge, *Arch. Math.* 7(1956), 241-249.

[70] D. Schweigert, Über endliche, ordnungspolynomvollständiger Verbände, *Monatsch. Math.* 78(1974), 68-76.

[71] D. Schweigert, Affine complete ortholattices, *Proc. Amer. Math. Soc.* 67(1977), 198-200.

[72] T.P. Speed, Profinite posets, *Bull. Austral. Math. Soc.* 6(1972), 177-183.

[73] T.P. Speed, On the order of primes, *Algebra Univ.* 2(1972), 85-87.

[74] M.H. Stone, The theory of representations of Boolean algebras, *Trans. Amer. Math. Soc.* 40(1936), 37-111.

[75] M.H. Stone, Topological representation of distributive lattices and Brouwerian logics, *Casopis Pest. Mat. Fys.* 67(1937), 1-25.

[75] M.H. Stone, Applications of the theory of Boolean rings to general topology, *Trans. Amer. Math. Soc.* 41(1937), 375-481.

[77] A. Urquhart, Free distributive pseudo-complemented lattices, *Algebra Univ.* 3(1973), 13-15.

[78] A. Urquhart, Finite free products of pseudocomplemented distributive lattices (unpublished manuscript, 1973).

[79] A. Urquhart, A topological representation theory for lattices, *Algebra Univ.* 8(1978), 45-58.

[80] A. Urquhart, Distributive lattices with a dual homomorphic operation, *Studia Logica* 38(1979), 201-209.

[81] A. Urquhart, Projective distributive *p*-algebras, *Bull. Austral. Math. Soc.* 24(1981), 269-275.

[82] A. Urquhart, Equational classes of distributive double *p*-algebras, *Algebra Univ.* (to appear).

[83] R. Wille, Note on the order dimension of partially ordered sets, *Algebra Univ.* 5(1975), 443-444.

[84] R. Wille, Subdirekte Produkte vollständiger Verbände, *J. Reine Angew. Math.* 283/284(1976), 53-70.

[85] R. Wille, Eine Charakterisierung endlicher ordnungspolynomvollständiger Verbände, *Arch. Math. (Basel)* 28(1977), 557-560.

[86] R. Wille, A note on algebraic operations and algebraic functions on finite lattices, *Houston J. Math.* 3(1977), 593-597.

[87] R. Wille, Cancellation and refinement properties for function lattices, *Houston J. Math.* 6(1980), 431-437.

# THE RETRACT CONSTRUCTION

Ivan Rival
Department of Mathematics and Statistics
The University of Calgary
Calgary, Canada T2N 1N4

ABSTRACT

The retract theme is a common one in the setting of ordered algebras. For instance, retracts play a considerable role in the algebraical theory of projective lattices ($\equiv$ retracts of free lattices). [9], [13] The study of retracts for general ordered sets is much less familiar.

A subset $Q$ of an ordered set $P$ is a *retract* of $P$ if there is an order-preserving map $g$ of $P$ to $Q$ which is the identity map on $Q$. The term is used up to isomorphism: an ordered set $Q$ is a *retract* of an ordered set $P$ if there are order-preserving maps $f$ of $Q$ to $P$ and $g$ of $P$ to $Q$ such that $g \circ f$ is the identity map on $Q$. In either case $g$ is a *retraction* map.

For instance, if $P$ contains a subset $Q \cong \underset{\sim}{2}$ (the two-element chain) then $Q$ is a retract; in fact, if $a < b$ in $P$ then $\{a,b\} \cong \underset{\sim}{2}$ and

$$g(x) = \begin{cases} a & \text{if } x \leq a \\ b & \text{if } x \not\leq a \end{cases}$$

is a retraction. On the other hand, if $P$ contains $Q \cong 2$ (the two-element antichain) then $Q$ is a retract of $P$ if and only if $P$ is disconnected. If $A \subseteq P$ is a (connected) component of $P$, $B = P - A \neq \emptyset$, $a \in A$, and $b \in B$, then

$$g(x) = \begin{cases} a & \text{if } x \in A \\ b & \text{if } x \in B \end{cases}$$

is a retraction; if $P$ is connected then each retract must be connected.



*I. Rival (ed.), Ordered Sets, 97–122.*

Retracts as Subobjects. It is fairly common to treat any subset of an ordered set with the inherited order as a 'subobject'. The trouble with this convention is that very little of an ordered set is really inherited by an arbitrary subset. For instance, neither the completeness nor the connectivity of an ordered set is inherited by every subset. These and many other syntactical properties of an ordered set are, however, preserved by their retracts. One of the most interesting of these properties is the fixed point property. [5] The theme of 'retract as subobject' has recently been rather carefully exploited in a series of papers aimed at fashioning a *structure* theory for ordered sets. [7], [16], [20]

Retractions as Choice Functions. It is usually quite difficult to decide just which subsets of an ordered set are its retracts. As a rule we learn a great deal structurally about an ordered set when we can classify some or all of its retracts.

The retraction maps themselves are at times surprisingly simple. Among the earliest results are these two for lattices.

The *supremum* map. Let $P$ be a complete ordered set. Then $Q \subseteq P$ is a retract if and only if $Q$ is complete. In this case,

$$g(x) = \sup_Q\{y \in Q \mid y \leq x\}$$

is the retraction map. (Note that $\sup_Q \emptyset = 0_Q$, the least element of $Q$.) [1]

The *weaving* map. Let $\varphi$ be a homomorphism of a lattice $P$ onto a countable lattice $Q$. Then $Q$ is a retract of $P$. Let $Q = \{q_1, q_2, q_3, \ldots\}$ and let $p_i \in \varphi^{-1}(q_i)$. Define the map $f$ of $Q$ to $P$ by $f(q_1) = p_1$ and

$$f(q_i) = \left(p_i \vee \bigvee_{j<i} (f(q_j) \mid q_j < q_i)\right) \wedge \bigwedge_{j<i} (f(q_i) \mid q_j > q_i) .$$

Then $g = \varphi$ is a retraction ($g \circ f = \mathrm{id}_Q$). [4]

Some retraction maps of more recent vintage are these.

The *folding* map. Every crown of minimum order in an ordered set of length one is a retract. [14], [18]

The *well order* map. Let $C$ be a maximal chain in an ordered set $P$. Then $C$ is a retract of $P$. Let $\underset{\sim}{\alpha}$ be a well order of $C$ and for each $x \in P$ let $N(x)$ consist of all those $y \in C$ such that either $y = x$ or $y$ is noncomparable with $x$. Then

$$g(x) = \inf_{\underset{\sim}{\alpha}} N(x)$$

is a retraction map. [8]

Some retraction maps are quite cunning.

An example of a result from the structure theory is this. *Every countable lattice is a retract of a direct product of chains each isomorphic to the chain of rational numbers.* [16]

INTRODUCTION

A subset $Q$ of an ordered set $P$ is a *retract* of $P$ if there is an order-preserving map $g$ of $P$ to $Q$ which is the identity map on $Q$. This term is used up to isomorphism: an ordered set $Q$ is a *retract* of an ordered set $P$ if there are order-preserving maps $f$ of $Q$ to $P$ and $g$ of $P$ to $Q$ such that $g \circ f = id_Q$ [$id_Q$ is the identity map on $Q$]. A common mnemonic is this "commutative diagram":

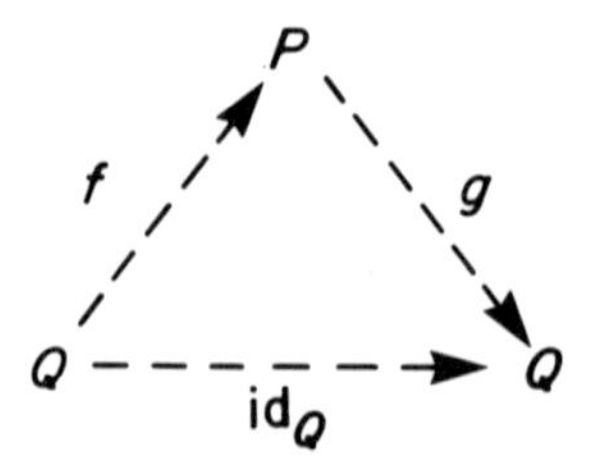


$g$ is called a *retraction* and $f$ is called a *section*.

Loosely speaking there are two general themes that arise in the study of retracts of ordered sets. The first is a "syntactical" theme; it concerns the study of those properties that are preserved by, or that characterize, the retracts of an ordered set. Insofar as many of these same properties are not usually preserved by the subsets of an ordered set it may eventually be that the retracts of an ordered set will be the preferred subobject.

The second theme is an "algorithmic" one. It turns on these twin questions.

*Which subsets of P are retracts? Which quotients [images of order-preserving maps] of P are retracts?*

The construction of the corresponding retraction or section can be intricate.

In outline this paper features several types of retractions. Three of them, the *supremum* map, the *well order* map, and the *greedy* map are applied primarily to infinite ordered sets. The last two, the *dismantling* map and the *folding* map are combinatorial in character and are applied mainly to finite ordered sets.

I am grateful to Bill Sands for suggestions concerning the representation of the rationals, and for examples and counter-examples concerning the greedy map for countable lattices.

THE SUPREMUM MAP

An ordered set $P$ is *complete* if every subset $S$ of $P$ has both a supremum $\sup_P S$ and an infimum $\inf_P S$. It does not at this time seem possible to describe the formal character of those properties that are inherited by the retracts of an ordered set. For instance, we know that "completeness" -- a property that is inherited by the retracts -- cannot be described by any set of first order sentences.

Neither every subset nor every quotient of a complete ordered set need be complete.

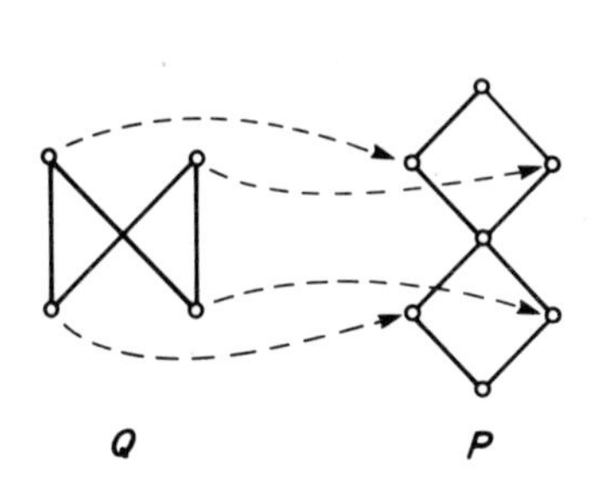


**$Q$ is a noncomplete subset of the complete ordered set $P$.**

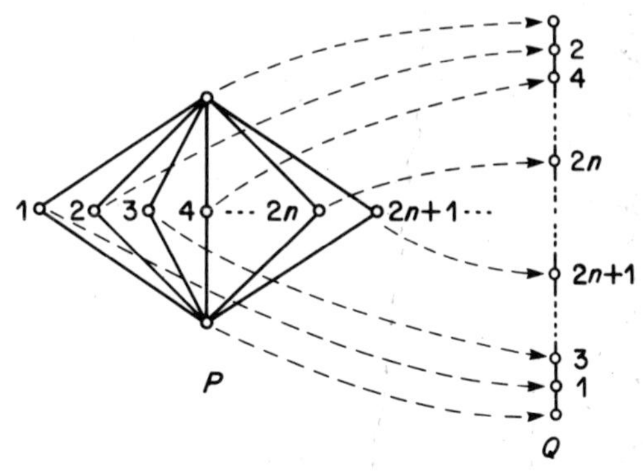


**$Q$ is a noncomplete quotient of the complete ordered set $P$.**

*Every retract of a complete ordered set is complete.*

If $g$ is a retraction of the complete ordered set $P$ onto the subset $Q$ of $P$ then

$$\sup_Q S = g(\sup_P S)$$

for, if $x \in S$ is an upper bound of $S$ [$x \geq s$ for each $s \in S$] then $x \geq \sup_P S$ and, as $g$ is order-preserving

$$g(x) \geq g(\sup_P S) .$$

*Every subset of a complete ordered set which is itself complete is a retract.*

Actually, more is true. A complete ordered set is a retract of *any* ordered set of which it is a subset. For instance, if an ordered set $P$ contains a subset $\{a,b\}$ with $a < b$ then

$$g(x) = \begin{cases} a & \text{if } x \leq a \\ b & \text{if } x \nleq a \end{cases}$$

is a retraction. In general, if $Q$ is a complete ordered set and $Q \subseteq P$ then the *supremum map*

$$g(x) = \sup_Q\{y \in Q | y \leq x\}$$

is a retraction. [Note that $\sup_Q \emptyset = 0_Q$ , the least element of $Q$.]

One consequence is that

*every complete ordered set $P$ is a retract of the power $\underset{\sim}{2}^{|P|}$.*

$\underset{\sim}{2}^{|P|}$ is the $|P|$-fold direct product of ordered sets each isomorphic to the two-element chain $\underset{\sim}{2}$ . For each $x \in P$ let the map $f_x$ of $P$ to $\underset{\sim}{2}$ be defined by

$$f_x(y) = \begin{cases} 0 & \text{if } y \leq x \\ 1 & \text{if } y \nleq x \,. \end{cases}$$

Then the map $f$, whose $x$th projection is $\pi_x \circ f = f_x$ embeds $P$ into $\underset{\sim}{2}^{|P|}$: $f$ is a section.

The "completeness" theme is a common one in the theory of ordered sets -- even when it is not required that *every* subset have both a supremum and an infimum. One instance is this: An ordered set is *chain-complete* if every maximal chain is a complete ordered set.

*Every retract of a chain-complete ordered set $P$ is itself chain-complete.*

If $g$ is a retraction of $P$ to $Q$ and $C$ is a noncomplete maximal chain in $Q$ then $C$ contains an increasing subset $A$ and a decreasing subset $B$ such that, for each $a \in A, b \in B$, $a < b$ and there is *no* $c \in Q$ satisfying $a < c < b$ [$A$ or $B$ may be empty]. As $P$ is chain-complete there is $x \in P$ such that, for each $a \in A, b \in B$

$$a < x < b$$

so

$$a = g(a) \leq g(x) \leq g(b) = b$$

and $g(x) \in Q$, which is impossible.

The "chain-completeness" condition in turn is a useful one in the study of "fixed points". An ordered set $P$ has the *fixed point property* if every order-preserving map of $P$ to itself has a fixed point; in symbols,

$$\forall\ \sigma \in P^P\ \exists\ a \in P\ (\sigma(a) = a)\ ,$$

(which is a first order sentence). Otherwise, $P$ is *fixed point free.* Neither subsets nor quotients need preserve this property either.

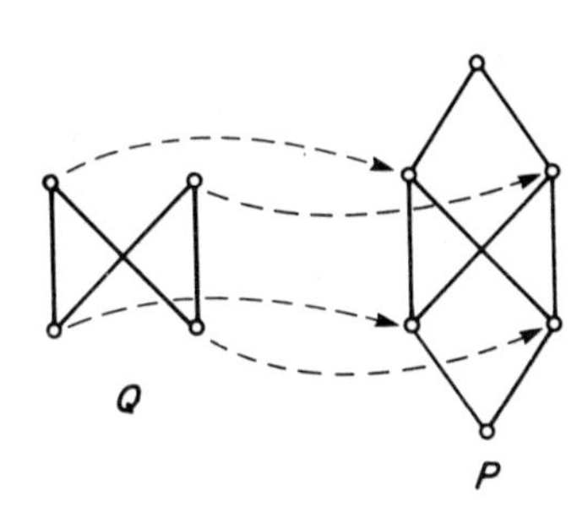


**The subset $Q$ of $P$ is fixed point free but $P$ has the fixed point property.**

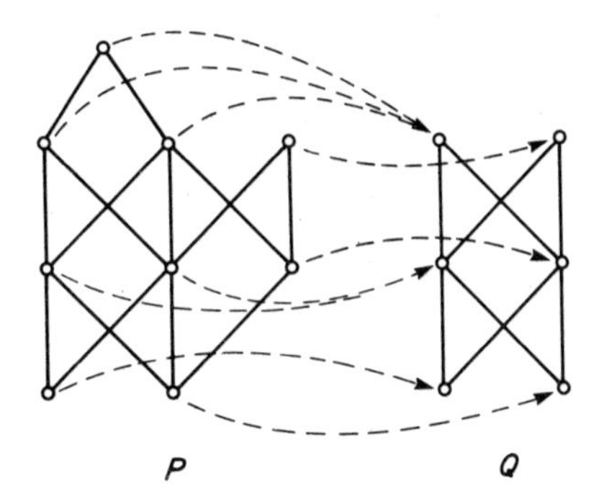


**The quotient $Q$ of $P$ is fixed point free but $P$ has the fixed point property.**

However,

> *every retract of an ordered set with the fixed point property has the fixed point property.*

Let $P$ be an ordered set with the fixed point property and let $g$ be a retraction of $P$ onto the subset $Q$. Let $\sigma \in Q^Q$. Then $\sigma \circ g \in Q^P$ and, in particular $\sigma \circ g \in P^P$, so there is $a \in P$ satisfying $\sigma \circ g(a) = a$. But $\sigma(g(a)) \in Q$ so $a \in Q$ and this means that $g(a) = a$. Therefore, $\sigma(g(a)) = g(a)$, which is to say, $\sigma$ has a fixed point.

> *If $P$ has the fixed point property then $P$ is chain-complete* [12], [8].

Otherwise, $P$ contains a noncomplete maximal chain $C$ which, in turn, must contain a subset $D = A \cup B$ in which $A = \{a_i | i \in I\}$ is an increasing, well ordered subset without supremum in $C$, $B = \{b_j | j \in J\}$ is a decreasing, dually well ordered subset without infimum in $C$, and for each $i \in I$ and each $j \in J, a_i < b_j$ and there is *no* $c \in P$ satisfying $a_i \leq c \leq b_j$. Now, the map $g$ of $P$ onto $D$ -- a variation of the "supremum map" -- prescribed by

$$g(x) = \begin{cases} \sup_A\{a_i \mid a_i \leq x\} & \text{unless } x \geq a_i \text{ for each } i \in I \\ \inf_B\{b_j \mid b_j \geq x\} & \text{if } x \geq a_i \text{ for each } i \in I \end{cases}$$

is a retraction. Furthermore, the map $\sigma$ of $D$ to $D$ defined by $\sigma(a_i) = a_{i+1}$ and $\sigma(b_j) = b_{j+1}$ is order-preserving and has no fixed points. Then the map $\sigma \circ g$ is an order-preserving map of $P$ to $P$ with no fixed points.

This fact is the substance of the classical result that

*a lattice with the fixed point property is complete* [3].

THE WELL ORDER MAP

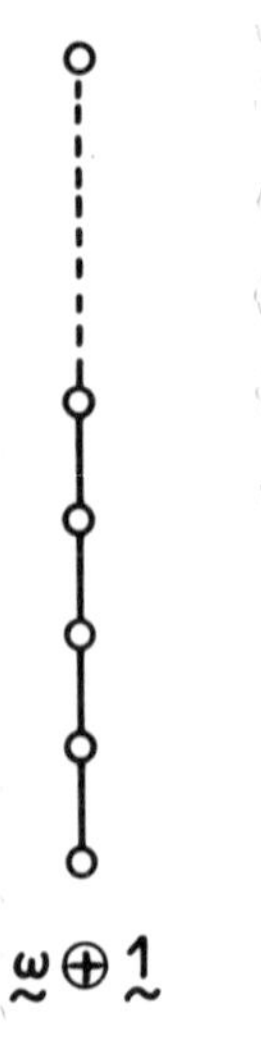

**$\underset{\sim}{\omega}$ is not a retract of $\underset{\sim}{\omega} \oplus \underset{\sim}{1}$.**

The supremum map shows that

*every complete chain in an ordered set is a retract.*

In particular, every finite chain is a retract. However, not *every* chain is a retract; for example, $\underset{\sim}{\omega}$ is *not* a retract of $\underset{\sim}{\omega} \oplus \underset{\sim}{1}$ [the linear sum of $\underset{\sim}{\omega} = \{0 < 1 < 2 < \ldots\}$ and then the singleton chain $\underset{\sim}{1} = \{0\}$].

It is an important fact in the theory of retracts that

*every maximal chain of an ordered set is a retract* [8].

Let $C$ be a maximal chain in an ordered set $P$. For each $x \in C$ put

$$N(x) = \{c \in C \mid x \text{ is noncomparable with } c \text{ or } x = c\}.$$

Then $N(x) = \{x\}$ if and only if $x \in C$. Let $\underset{\sim}{\alpha}$ be a well order of $C$. We use $\underset{\sim}{\alpha}$ to construct a retraction -- the *well order* map -- of $P$ onto $C$. Define this map $g$ of $P$ to $C$ by choosing $g(x)$ as the *least* element of $N(x)$ *with respect to the well order* $\underset{\sim}{\alpha}$; in symbols, the well order map is

$$g(x) = \inf_{\underset{\sim}{\alpha}} N(x).$$

Here is an example of the use of the supremum map together with the well order map.

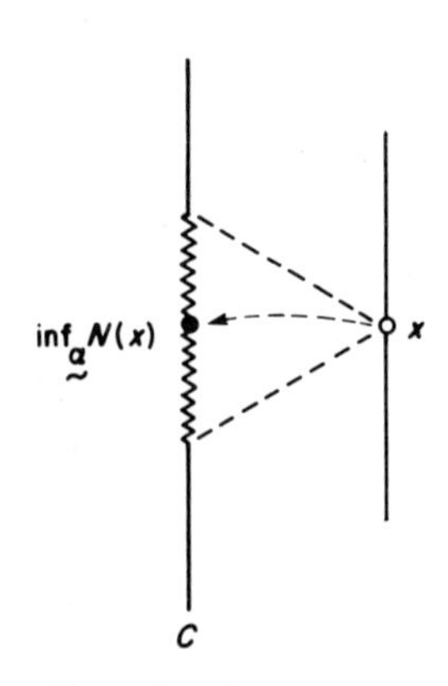

The well order map.

*The chain* $\mathbb{Q}$ *of the rational numbers is a retract of a direct product of chains each isomorphic to* $\underset{\sim}{\omega}$, $\underset{\sim}{\omega}^d$ *or* $\underset{\sim}{\omega} \oplus \underset{\sim}{\omega}^d$ [7].

We shall prove this result by examining in turn how the chains $\underset{\sim}{\omega} \oplus \underset{\sim}{\omega}^d$, $\underset{\sim}{\omega}$ and $\underset{\sim}{\omega}^d$ are used. The first step in the proof is this.

A. *The chain* $\overline{\mathbb{Q}}$ *of the rationals in the unit interval* $[0,1]$ *of* $\mathbb{R}$ *is a retract of the direct product of (uncountably many) chains, each isomorphic to* $\underset{\sim}{\omega} \oplus \underset{\sim}{\omega}^d$. For each irrational $\lambda \in [0,1]$ let $\lambda_0 < \lambda_1 < \lambda_2 < \ldots$ be an ascending sequence in $\overline{\mathbb{Q}}$ converging to $\lambda$ and let $\lambda^0 > \lambda^1 > \lambda^2 > \ldots$ be a descending sequence in $\overline{\mathbb{Q}}$ converging to $\lambda$. Then

$$\sup_{\mathbb{R}}\{\lambda_i \mid i = 0,1,2,\ldots\} = \lambda = \inf_{\mathbb{R}}\{\lambda^i \mid i = 0,1,2,\ldots\} .$$

Let $C_\lambda$ be the union of the two sequences: then

$$C_\lambda \cong \underset{\sim}{\omega} \oplus \underset{\sim}{\omega}^d .$$

In fact, each $C_\lambda$ is a retract of $\overline{\mathbb{Q}}$. For each $\lambda$ define $g_\lambda$ of $\overline{\mathbb{Q}}$ to $C_\lambda$ by the supremum map

$$g_\lambda(q) = \begin{cases} \sup_{C_\lambda}\{\lambda_i \mid \lambda_i \leq q\} & \text{if } q < \lambda \\ \inf_{C_\lambda}\{\lambda^i \mid \lambda^i \geq q\} & \text{if } q > \lambda . \end{cases}$$

We use this map in turn to construct the retraction of $K = \Pi_\lambda C_\lambda$ to $\overline{\mathbb{Q}}$.

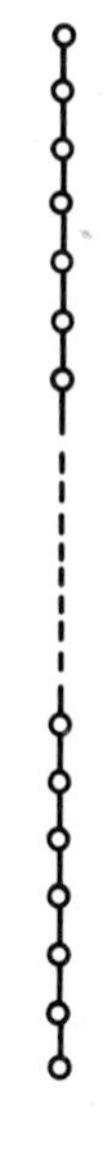
$\underset{\sim}{\omega} \oplus \underset{\sim}{\omega}^d$

At first, we define the "embedding" $f$ of $\overline{\mathbb{Q}}$ to $K$ such that its $\lambda$th projection $\pi_\lambda \circ f = g_\lambda$. Evidently $q \leq q'$ implies $f(q) \leq f(q')$. If $q \nleq q'$ then there is an irrational $\lambda$ such that $q' < \lambda < q$; then $g_\lambda(q') = \lambda_i < \lambda^j = g_\lambda(q)$ for some $i,j$ so $f(q) \nleq f(q')$. This shows that $f(\overline{\mathbb{Q}}) \cong \overline{\mathbb{Q}}$.

Now we can define the retraction of $K$ onto $f(\overline{\mathbb{Q}})$. Let $C$ be a maximal chain of $K$ containing $f(\overline{\mathbb{Q}})$. According to the well order map, $C$ itself is a retract of $K$. It is enough to show that $f(\overline{\mathbb{Q}})$ is a retract of $C$. Again a variation of the supremum map is used. For all $x \in C$ let $U(x) = \{y \in f(\overline{\mathbb{Q}}) \mid y \geq x\}$ and let $D(x) = \{y \in f(\overline{\mathbb{Q}}) \mid y \leq x\}$. In fact, the map $g$ defined by

$$g(x) = \begin{cases} \sup_{f(\overline{\mathbb{Q}})} D(x) & \text{if it exists} \\ \inf_{f(\overline{\mathbb{Q}})} U(x) & \text{otherwise} \end{cases}$$

is the retraction. The reason why this map is well-defined is interesting. Suppose that for some $x \in C$ neither $\inf_{f(\overline{\mathbb{Q}})} U(x)$ nor $\sup_{f(\overline{\mathbb{Q}})} D(x)$ exists. Then neither $\inf_{\overline{\mathbb{Q}}} f^{-1}(U(x))$ nor $\sup_{\overline{\mathbb{Q}}} f^{-1}(D(x))$ exists; hence, there is an irrational number $\lambda$ such that $d < \lambda < u$ for all $d \in f^{-1}(D(x))$ and for all $u \in f^{-1}(U(x))$. Consider $\pi_\lambda(x) = x_\lambda$ in $C_\lambda$. If $x_\lambda = \lambda^i$, say, then choose $q \in \overline{\mathbb{Q}}$ such that $g_\lambda(q) = \lambda^{i+1}$. Then $q > \lambda$ and $f(q) \in U(x)$. However, $g_\lambda(q) = \lambda^{i+1} < \lambda^i = x_\lambda$ which is impossible.

The next step in the proof is this simple observation.

B. *The chain* $\mathbb{Z} \cong \underset{\sim}{\omega}^d \oplus \underset{\sim}{\omega}$ *of integers is a retract of the direct product* $\underset{\sim}{\omega} \times \underset{\sim}{\omega}^d$.

The required retraction is illustrated.

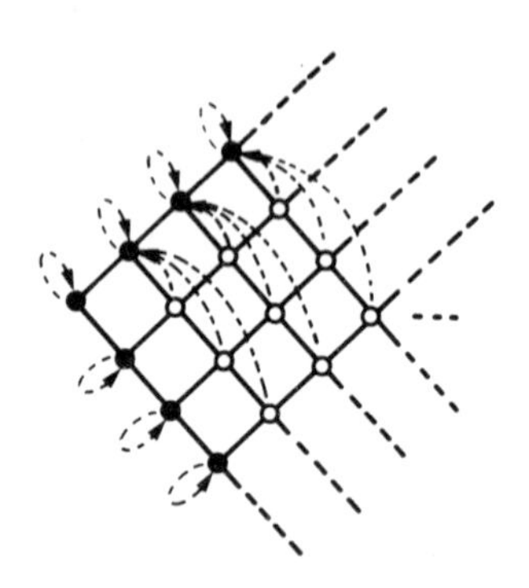

The last step in the proof is this.

C. $\mathbb{Q}$ *is a retract of* $\mathbb{Z} \times \overline{\mathbb{Q}}$.

It helps to represent $\mathbb{Q}$ as a linear sum of intervals of rational numbers indexed by $\mathbb{Z}$. To do this, let

$$\cdots < q_{-2} < q_{-1} < q_0 = \frac{1}{2} =$$

$$= q_0 < q_1 < q_2 < \cdots$$

be sequences in $\mathbb{Q} \cap (0,1)$ such that

$$\inf_{\mathbb{Q}} \{q_m \mid m = 0,-1,-2,\ldots\} = 0$$

and

$$\sup_{\mathbb{Q}} \{q_n \mid n = 0,1,2,\ldots\} = 1 .$$

Then

$$\mathbb{Q} \cong \bigoplus_{m \leq -1} (q_{m+1}, q_m] \oplus (q_{-1}, q_1) \oplus \bigoplus_{n \geq 1} [q_n, q_{n+1})$$

and, in $\mathbb{Z} \times \overline{\mathbb{Q}}$,

$$\mathbb{Q} \cong \left(\bigcup_{m\leq -1} \{m\}\times(q_{m-1},q_m]\right) \oplus \left(\{0\}\times(q_{-1},q_1)\right) \oplus \left(\bigcup_{n\geq 1} \{n\}\times[q_n,q_{n+1})\right) .$$

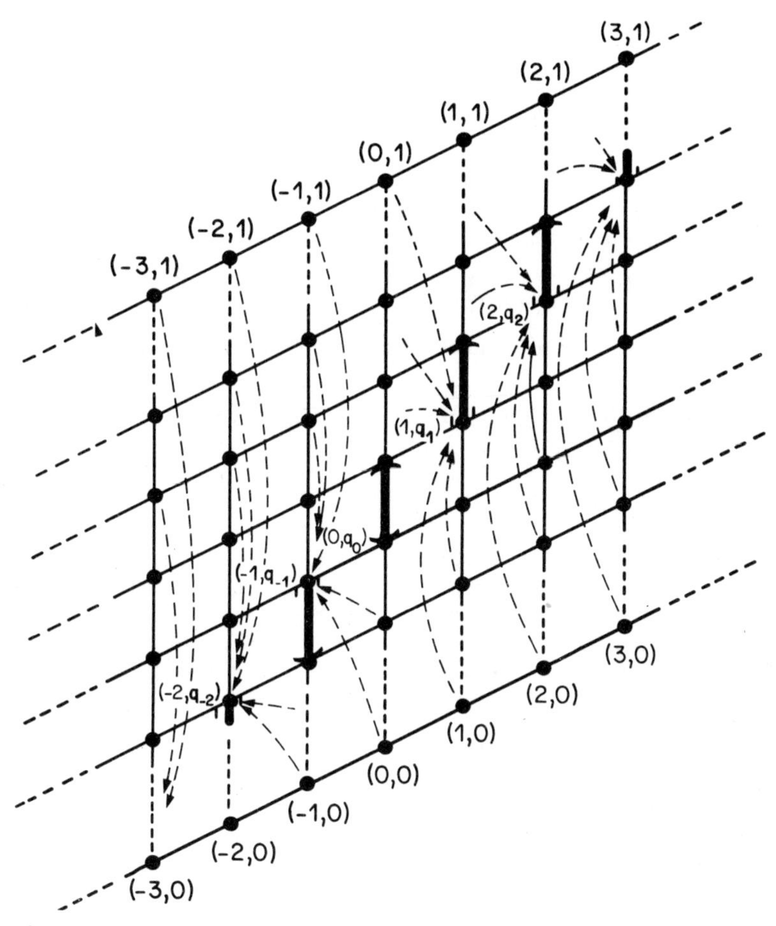


The map $g$ of $\mathbb{Z} \times \overline{\mathbb{Q}}$ to $\mathbb{Q}$ defined below is a retraction: if $m \leq -1$,

$$g(m,q) = \begin{cases} (m,q_m) & \text{if } q_m < q \\ (m,q) & \text{if } q_{m-1} < q \leq q_m \\ (m-1,q_{m-1}) & \text{if } q \leq q_{m-1} \end{cases} ;$$

if $n \geq 1$

$$g(n,q) = \begin{cases} (n,q_n) & \text{if } q < q_n \\ (n,q) & \text{if } q_n \le q < q_{n+1} \\ (n+1,q_{n+1}) & \text{if } q_{n+1} \le q \end{cases} ;$$

$$g(0,q) = \begin{cases} (-1,q_{-1}) & \text{if } q \le q_{-1} \\ (0,q) & \text{if } q_{-1} < q < q_1 \\ (1,q_1) & \text{if } q_1 \le q \end{cases} .$$

These ideas are at the heart of several other results.

*Every lattice of finite width is a retract of a direct product of chains* [20].

More intricate is the construction which proves that

*every countable lattice is a retract of a direct product of chains each isomorphic to* $\mathbb{Q}$ [16].

## THE GREEDY MAPS

The construction of a retract $Q$ of an ordered set $P$ usually consists of two separate steps. The first step is to identify $Q$ with a subset of $P$ isomorphic to $Q$; the second step is to construct a retraction map $g$ of $P$ onto $Q$. Whether, for a given subset $Q$ of $P$, there *is* such a map $g$, is, at times, difficult to answer. Even when such a map exists finding it may require some cunning.

There are certain simple conditions that the embedding of $Q$ into $P$ must satisfy if there is to be a retraction. We describe one of the most obvious of these now. First, for $S \subseteq P$ let $S^* = \{x \in P | x \ge s$ for each $s \in S\}$ and let $S_* = \{x \in P | x \le s$ for each $s \in S\}$. For $x \in P$, set $U_Q(x) = \{x\}^* \cap Q$ and $D_Q(x) = \{x\}_* \cap Q$. Now put

$$S_Q(x) = D_Q(x)^* \cap U_Q(x)_* \cap Q .$$

For there to be a retraction $g$ of $P$ to $Q$ it must be that $g(x) \le g(y) = y$ for each $y \in U_Q(x)$ and $g(x) \ge g(y) \ge y$ for each $y \in D_Q(x)$; in particular,

*if $Q$ is a retract of $P$ then $S_Q(x) \neq \emptyset$ for every $x \in P$.*

[Note that if $x \in Q$ then $S_Q(x) = \{x\}$.]

Another way to express the same idea is this. We call the pair $(D,U)$ of subsets of $P$ a *gap* of $P$ if $D \subseteq U_*$, $U \subseteq D^*$, and $D^* \cap U_* = \emptyset$. For instance, $(\{x \in \mathbb{Q} \mid x < \sqrt{2}\}, \{x \in \mathbb{Q} \mid x > \sqrt{2}\})$ is a gap of $\mathbb{Q}$. If there is a retraction of $P$ to $Q$ then $(D_Q(x), U_Q(x))$ cannot be a gap of $Q$ for any $x \in P$. Equivalently,

*if $Q$ is a retract of $P$ then every gap of $Q$ is a gap of $P$.*

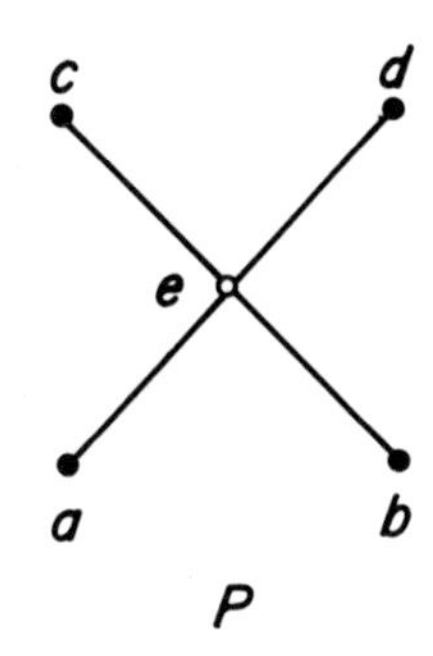


**$Q = \{a,b,c,d\} \subseteq P$ and $(\{a,b\}, \{c,d\})$ is a gap of $Q$ but not a gap of $P$ so $Q$ is not a retract of $P$.**

There are instances when this necessary condition is also a sufficient one. For example,

*a chain $C \subseteq P$ is a retract of $P$ if and only if every gap of $C$ is a gap of $P$.*

However, an arbitrary choice for $g(x) \in S_C(x)$ will not do. In fact, some variation of the well order map is essential for $g$. [Well order $C$ by $\underset{\sim}{\alpha}$, say, and define $g(x) = \inf_{\underset{\sim}{\alpha}} S_C(x)$.]

The well order map for chains is surprisingly simple -- its effectiveness is, however, confined to chains. In general, the choice of $g(x) \in S_Q(x)$ must explicitly account for the requirement that $g$ be order-preserving.

One idea is to define $g$ inductively so that it is order-preserving at every stage and each $g(x) \in S_Q(x)$. To do this let $\underset{\sim}{\alpha}$ be a well order of $P$. Then the *greedy* map

$$g(x) = \inf_{\underset{\sim}{\alpha}}(S_Q(x) \cap \{g(y) \mid y \in P,\ \underset{\sim}{\alpha}(y) < \underset{\sim}{\alpha}(x),\ y < x\}^* \cap$$

$$\{g(z) \mid z \in P,\ \underset{\sim}{\alpha}(z) < \underset{\sim}{\alpha}(x),\ z > x\}_*)$$

is a retraction of $P$ onto $Q$ provided only that $g$ is well-defined (or, equivalently provided only that the intersection above is nonempty for each $x \in P$). This greedy map is in an obvious sense a generalization of the well order map (as well as the supremum map).

For a particular well order $\underset{\sim}{\alpha}$ this greedy map may work while for other well orders it can fail. For instance, the subset consisting of the shaded elements in this ordered set is a retract, yet the greedy map with the indicated well order is not a retraction. Even for lattices the choice of the well order is important. For example, the well order $\underset{\sim}{\alpha} \cong \underset{\sim}{\omega} \oplus \underset{\sim}{\omega}$ as applied to the lattice in the illustration does not admit $Q$ [the shaded elements] as a retract of $P$ by the greedy map. Note, however, that each gap of $Q$ is a gap of $P$.

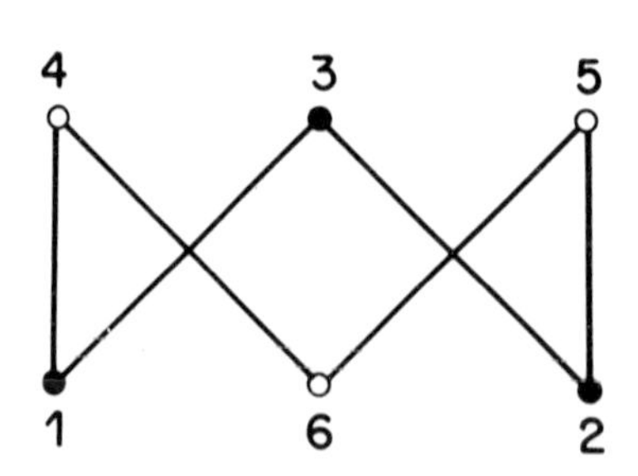


In fact, *if $Q,P$ are lattices, $Q \subseteq P$ and $P$ countable then $Q$ is a retract of $P$ if each gap of $Q$ is a gap of $P$.* Indeed, the greedy map works provided that an $\underset{\sim}{\omega}$ order is taken for $P$. The proof of this works in the obvious way by induction on $\underset{\sim}{\omega}$. We have by induction that, for each $x \in P$,

$$g(y) \leq g(z)$$

for each $y,z \in P$ satisfying $\underset{\sim}{\alpha}(y) < \underset{\sim}{\alpha}(x)$, $\underset{\sim}{\alpha}(z) < \underset{\sim}{\alpha}(x)$ and $y < x < z$. Finally, the sets

$$\{g(y)\,|\,y \in P,\ \underset{\sim}{\alpha}(y) < \underset{\sim}{\alpha}(x),\ y < x\}$$

and

$$\{g(z)\,|\,z \in P,\ \underset{\sim}{\alpha}(z) < \underset{\sim}{\alpha}(x),\ z > x\}$$

are finite. From this it follows directly that $g$ is well-defined.

Actually, in [16] arguments quite a bit more subtle than this greedy map are used to show this.

> *Let $Q$ and $P$ be lattices, $Q$ countable and $Q \subseteq P$. Then $Q$ is a retract of $P$ if and only if each gap of $Q$ is a gap of $P$.*

The other general approach to the construction of a retract $Q$ of $P$ consists of these two steps. The first is to construct an order-preserving map $g$ of $P$ onto $Q$; the second step is to construct a section $f$ of $Q$ to $P$. In effect, once the quotient $Q$ of $P$ is

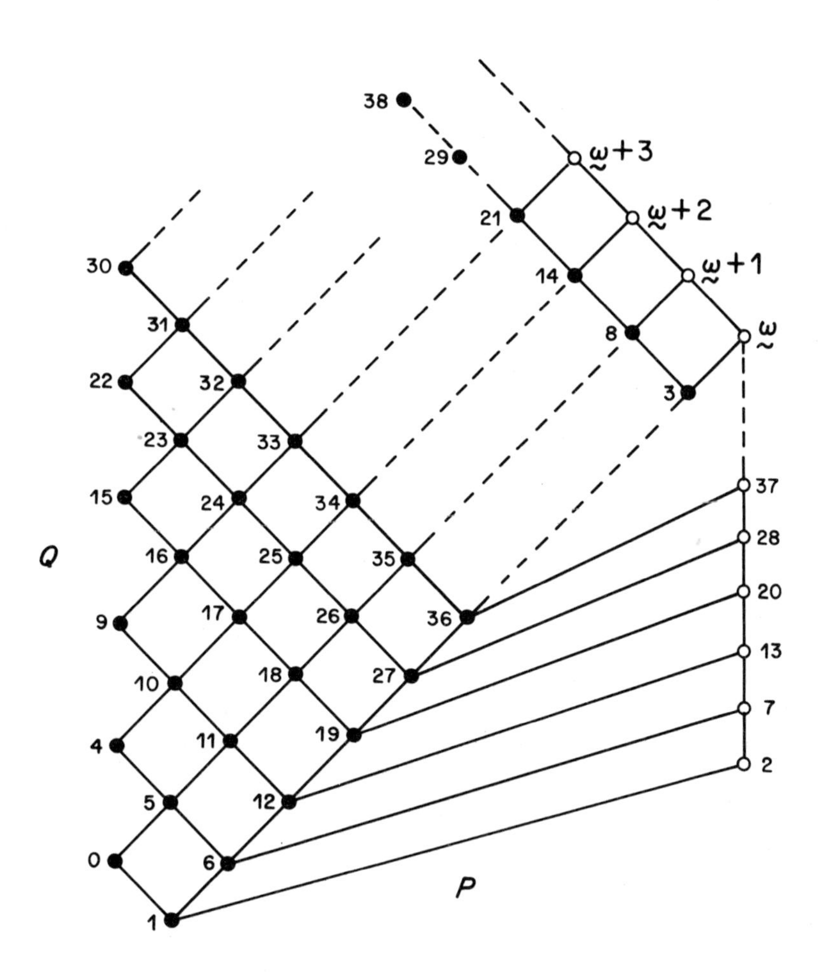


prescribed, the problem is to select $f(x) \in g^{-1}(\{x\})$ for each $x \in Q$ such that $f$ is order-preserving. There is a classical example of this approach taken from lattice theory. Its origin is somewhat unclear although there is a reference to it in [4] as the *weaving map*. It can also be considered as a "greedy" map.

Let $P,Q$ be lattices, let $Q$ be countable and let an order-preserving map $g$ of $P$ onto $Q$ be given satisfying in addition, that $g(x \vee y) = g(x) \vee g(y)$ and $g(x \wedge y) = g(x) \wedge g(y)$ [that is, $g$ is a *lattice homomorphism* of $P$ onto $Q$]. Let $Q = \{q_1, q_2, q_3, \ldots\}$ be an enumeration of $Q$ and for each $i = 1,2\ldots$ choose $p_i \in g^{-1}(\{q_i\})$ Simple greed will not do: the map $f(q_i) = p_i$ is not necessarily order-preserving. The idea of the greedy map is to ensure the order-preserving character of the choice for $f$ at each step: this approach does work. We define $f(q_1) = p_1$ and

$$f(q_i) = \left[p_i \vee \bigvee_{j<i} (f(q_j) \mid q_j < q_i)\right] \wedge \bigwedge_{j<i} (f(q_j) \mid q_j > q_i) .$$

This map exists because $P$ is a lattice and the joins and meets are all finite. Furthermore, $f$ is order-preserving because the lattice operations are.

The countable condition on $Q$ is necessary. There is an interesting reason for this. Every lattice is a homomorphic image of a free lattice. For instance, $Q = \underset{\sim}{\omega}_1$ is a homomorphic image of $FL(\aleph_1)$. As *every chain in a free lattice is countable* [10], $FL(\aleph_1)$ cannot even contain $\underset{\sim}{\omega}_1$ as a subset!

This "weaving" map is, of course, not new. One modern use of it yields the curious result:

> *every homomorphic image of a complete lattice of finite width is itself a complete lattice* [17].

THE DISMANTLING MAP

Let $P$ be an ordered set and let $a,b \in P$. The element $a$ is an *upper cover* of $b$ (or $b$ is a *lower cover* of $a$) if, whenever $c \in P$ satisfies $a \geq c > b$ then $a = c$. $a$ is *irreducible* if either $a$ has precisely one upper cover or $a$ has precisely one lower cover. Let $I(P)$ denote the set of irreducible elements of $P$. Call $P$ *dismantlable* if its elements can be labelled $P = \{a_1, a_2, \ldots, a_n\}$ such that $a_i \in I(P - \{a_1, a_2, \ldots, a_{i-1}\})$, for each $i = 1,2,\ldots,n-1$. Examples of dismantlable ordered sets abound: every finite lattice, every finite ordered set with a greatest (or least) element, every fence, etc.

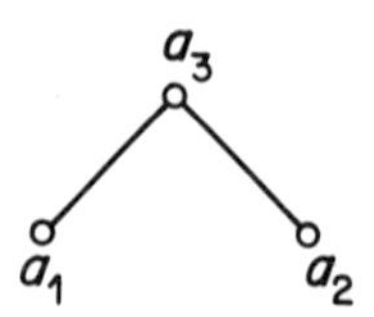


**A dismantlable ordered set.**

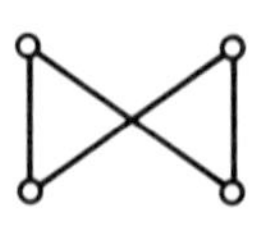

**A nondismantlable ordered set.**

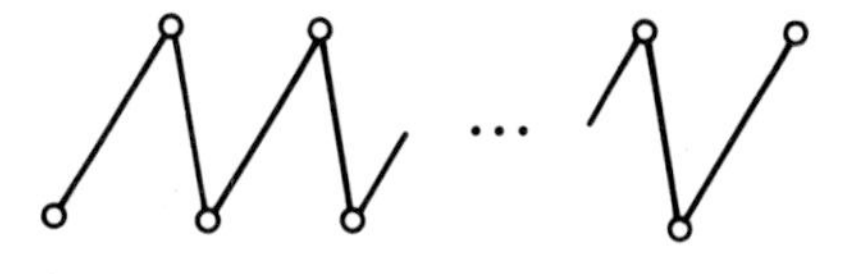

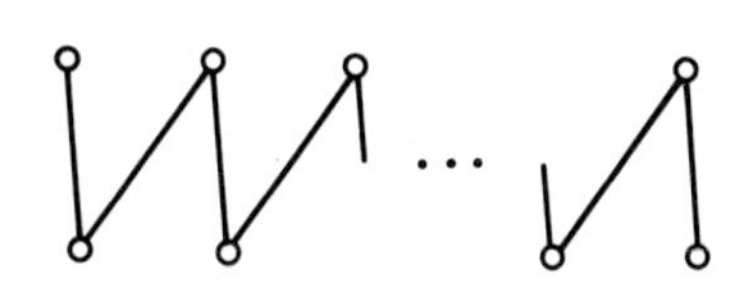

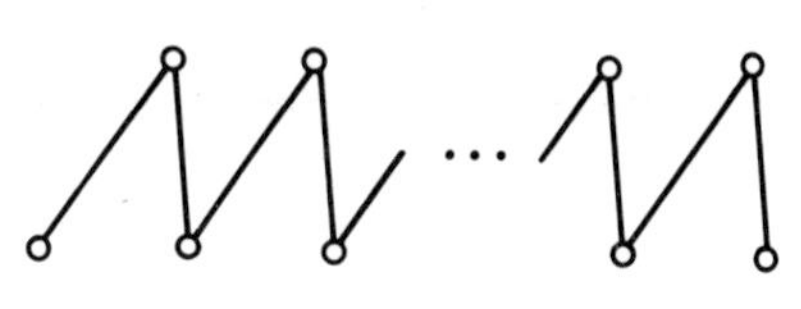

Fences

It is fairly obvious that dismantlability is not inherited by all of the subsets or all of the quotients of a dismantlable ordered set. In contrast,

*every retract of a dismantlable ordered set is dismantlable* [5].

We can prove this by induction on the order $|P|$ of a dismantlable ordered set $P$. Let $a \in I(P)$ such that $P - \{a\}$ is dismantlable and let $g$ be a retraction of $P$ to $Q$. If $a \notin Q$ then $Q$ is a retract of $P - \{a\}$ whence, by the induction hypothesis $Q$ is dismantlable. If $a \in I(Q)$ let $a^*$ be the unique upper cover, say, of $a$ *in* $Q$ and define $g'$ of $P - \{a\}$ to $Q - \{a\}$ by

$$g'(z) = \begin{cases} g(z) & \text{if } g(z) \neq a \\ a^* & \text{if } g(z) = a . \end{cases}$$

Then $g'$ is a retraction of $P - \{a\}$ onto $Q - \{a\}$ and the induction hypothesis implies that $Q - \{a\}$ (so $Q$ too) is dismantlable. Let $a \in Q - I(Q)$ and let $a^*$ be the unique upper cover of $a$ *in* $P$. Since $a \notin I(Q)$, $a^* \notin Q$; since $a^* \geq a$, $g(a^*) \geq g(a) = a$. If $g(a^*) > a$ then $a$ is not maximal in $Q$ so there are distinct $b,c$ in $Q$ each an upper cover of $a$ in $Q$; hence $b \geq a^*$ and $c \geq a^*$ in $P$, so $b = g(b) \geq g(a^*)$ and $c = g(c) \geq g(a^*)$. Therefore, $g(a^*) = a$. Finally,

$$Q' = (Q - \{a\}) \cup \{a^*\} \cong Q .$$

Let $x \in Q - \{a\}$. Then $a < x$ implies $a^* < x$; $x < a$ implies $x < a^*$; $a^* < x$ implies $a < x$. Let $x < a^*$. Then $x = g(x) \leq g(a^*) = a$. Hence, $Q \cong Q'$. As $Q' \subseteq P - \{a\}$ we need only observe that $Q'$ is a retract of $P - \{a\}$ to conclude that $Q'$ (and so $Q$) is dismantlable. Define $g'$ of $P - \{a\}$ to $Q'$, just as above, by $g'(z) = g(z)$ if $g(z) \neq a$ and $g'(z) = a^*$ if $g(z) = a$. Then $g'$ is a retraction.

How are the retracts of a dismantlable ordered set constructed? The simple-minded map will usually do.

Let $P = \{a_1, a_2, \ldots, a_n\}$ be a dismantlable ordered set with $a_i \in I(P - \{a_1, a_2, \ldots, a_{i-1}\})$ for each $i = 1,2,\ldots,n-1$. Then for each $0 \leq m \leq n$, $Q = \{a_{m+1}, a_{m+2}, \ldots, a_n\}$ is a retract of $P$. Indeed, for each $0 \leq i \leq m$ let $g_i$ be the map of $P - \{a_1, a_2, \ldots, a_{i-1}\}$ to $P - \{a_1, a_2, \ldots, a_i\}$ that is the identity map on $P - \{a_1, a_2, \ldots, a_i\}$ and that maps $a_i$ to $b_i$ where $b_i$ is either the unique upper cover or the unique lower cover of $a$ in $P - \{a_1, a_2, \ldots, a_{i-1}\}$. Then

$$g = g_m \circ g_{m-1} \circ \cdots \circ g_1$$

is a retraction of $P$ onto $Q$. We call each of these maps, each $g_i$ and the composition $g$ itself, *dismantling* maps.

**A retraction of $g$ constructed by a sequence of dismantling maps: $g = g_8 \circ g_7 \circ \cdots \circ g_2 \circ g_1$.**

There are instances in which *every* retract can be constructed using dismantling maps alone.

> *Every retract of a finite ordered set which contains no crowns is constructed using only dismantling maps* [6].

Still this seems far from the general situation. The subset $Q = \{a_1, a_3, a_5, a_6, a_7\}$ is a retract of the dismantlable ordered set illustrated, yet as $I(P) \subseteq Q$, $Q$ cannot be determined by a dismantling map or a sequence of dismantling maps.

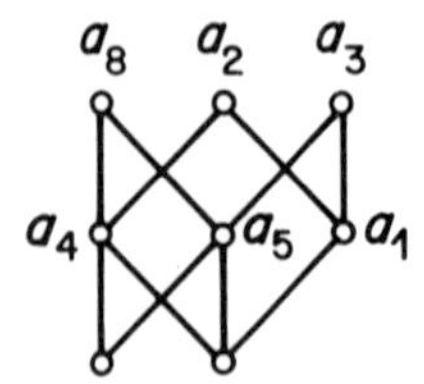


Dismantlable ordered sets all have the fixed point property [18]. In fact, an ordered set $P$ has the fixed point property if there is a subset $Q$ of $P$ with the fixed point property and a sequence of dismantling maps of $P$ onto $Q$. This dismantling map together with the supremum map can be combined to establish an interesting fixed point result. Let $\overline{L}$ stand for the ordered set $L - \{0,1\}$.

> *For a finite semimodular lattice $L$, $\overline{L}$ has the fixed point property if and only if $\overline{\underset{\sim}{2}^n}$ is not a retract of $\overline{L}$, where $n$ is the length of $L$* [2].

The heart of the proof is this. Suppose

$$\sup_P\{a|a \text{ is an atom of } L\} = 1_L$$

in a semimodular lattice of length $n \geq 2$. There is then an independent set $A = \{a_1, a_2, \ldots, a_n\}$ of atoms of $L$ and

$$S = \{\sup_L B | B \subseteq A\}$$

is a cover-preserving sublattice of $L$ isomorphic to $\underset{\sim}{2}^n$. Let $c_1, c_2, \ldots, c_n$ be the $n$ coatoms of $S$ labelled so that $a_i \not\leq c_i$ for each $i = 1, 2, \ldots, n$. We define a map $g$ of $L$ to $S$ as follows. If $a$ is an atom of $L$ then

$$g(a) = a_m$$

where

$$m = \min\{i | a \not\leq c_i\}$$

and, in general, for $x \in L$

$$g(x) = \sup_L\{g(a) | a \text{ is an atom of } L \text{ and } a \leq x\} .$$

In fact, $g|\overline{L}$ is a retraction of $\overline{L}$ onto $\overline{S} \cong \overline{2^n}$.

If $\overline{\underset{\sim}{2}^n}$ is not a retract of $\overline{L}$ then

$$S = \sup_L\{a|a \text{ is an atom of } L\} < 1_L .$$

The final step is to show that there is a sequence of dismantling maps of $\overline{L}$ onto the subset

$$\{x \in \overline{L} | x \leq s\} .$$

Since this subset of $\overline{L}$ is finite and has a greatest element it has the fixed point property. It follows then that $\overline{L}$ must also have the fixed point property. To construct the dismantling maps by induction choose a minimal element $b$ from the set

$$\{x \in \overline{L} | x \not\leq s\} .$$

Then $b$ has a unique lower cover, that is, $b$ is irreducible.

An ordered set $P$ is *connected* if, for any partition of $P$ into nonempty subsets $A$ and $B$ there is $a \in A$ and $b \in B$ such that $a > b$ or $a < b$. Connectivity in turn is a property inherited by all retracts of a connected ordered set. (Actually, every quotient of a connected ordered set is connected.) Connectivity itself can be neatly characterized in this way. *$P$ is connected if and only if* $\underset{\sim}{2}$ *is not a retract* [$\underset{\sim}{2}$ is the two-element antichain]. A surprising characterization of dismantlability is this one [7]:

> *for a finite ordered set $P$, $P$ is dismantlable if and only if $P^P$ is connected.*

## THE FOLDING MAP

An alternate description of connectivity is this: *$P$ is connected if and only if for each $a, b \in P$ there is a fence in $P$ containing $a$ and $b$.* It is common to define the *distance function* $d_P$ of $P \times P$ to $\mathbb{N} \cup \{\infty\}$ by

$$d_P(x,y) = \inf\{|F|-1 \mid x,y \in F \text{ and } F \text{ is a fence in } P\} .$$

Of course, the distance function of a subset $Q$ of $P$ need not coincide with the restriction of $d_P$ to $Q \times Q$. If $Q$ is a quotient of $P$, say $g$ is an order-preserving map of $P$ to $Q$, then

$$d_Q(g(x),g(y)) \leq d_P(x,y) .$$

If $f$ is a section of $Q$ to $P$

$$d_P(f(x),f(y)) = d_Q(x,y) .$$

We call a subset $Q$ of $P$ *isometric* if, for each $x,y \in Q$,

$$d_Q(x,y) = d_P(x,y) .$$

*Every retract is isometric.*

Let $\max P = \{x \in P \mid x \text{ maximal in } P\}$, let $\min P = \{x \in P \mid x \text{ minimal in } P\}$ and let $\text{ext } P = \min P \cup \max P$. A subset $Q$, with $|Q| > 1$, of a finite ordered set $P$ is *spanning* if $\max Q \subseteq \max P$ and $\min Q \subseteq \min P$. Let $f$ be a section of $Q$ to the finite ordered set $P$. For each $x \in \max Q$ ($x \in \min Q$) choose $h(x) \in \max P$ ($h(x) \in \min P$) such that $f(x) \leq h(x)$ ($h(x) \leq f(x)$). For all $x \in Q - \text{ext } Q$, let $h(x) = f(x)$. Then $h$ is a section of $Q$ to $P$ and $h(Q)$ is a spanning subset of $P$. This shows that

> *every retract $Q$, with $|Q| > 1$, of a finite ordered set $P$ is isomorphic to a spanning subset of $P$ which is also a retract* [5].

Particular spanning subsets are always retracts.

> *Every isometric spanning fence of a finite ordered set is a retract.* [8].

Let $F = \{x_0 < x_1, x_1 > x_2, x_2 < x_3, \ldots, x_{n-2} > x_{n-1}, x_{n-1} < x_n\}$, say, be an isometric spanning subset of $P$. For each $i = 0,1,2,\ldots,|F| - 1$, set

$$A_i = \{x \in \text{ext } P \mid d_P(x_0,x) = i\}$$

and

$$A_\infty = \text{ext } P - \bigcup_{i=0}^{|F|-1} A_i .$$

We define the *folding* map $g$ of ext $P$ to $F$ by

$$g(x) = \begin{cases} x_i & \text{if } x \in A_i \\ x_{n-1} & \text{if } x \in \min A_\infty \\ x_n & \text{if } x \in \max A_\infty \, . \end{cases}$$

(For convenience, let min $P$ contain all of the isolated points of $P$.) Then $g$ is a retraction and $g(\min P) \subseteq \min F, g\ (\max P) \subseteq \max F$.

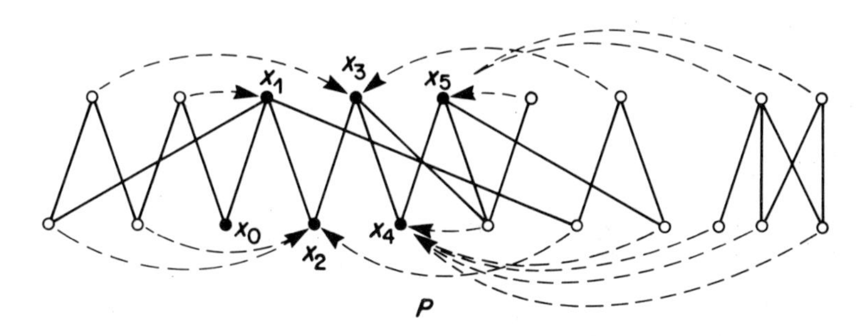


The folding map of $P$ to $\{x_0, x_1, x_2, x_3, x_4, x_5\}$.

We can extend $g$ to a retraction $g'$ of $P$ to $F$ as follows. Observe that for each $x \in P$

$$|g(\{y \in P | y \geq x\} \cap \max P)| = 1$$

or

$$|g(\{y \in P | y \leq x\} \cap \min P)| = 1 \, .$$

Then $g' | \text{ext } P = g$ and

$$g'(x) = g(\{y \in P | y \geq x\} \cap \max P)$$

if *dexter* has order one and

$$g'(x) = g(\{y \in P | y \leq x\} \cap \min P)$$

otherwise.

A particularly illustrative instance of the folding map is used to prove this fact [14], [18].

*Every crown of minimum order in an ordered set of length one is a retract.*

A *crown* in an ordered set $P$ is a subset $C_{2n} = \{x_1, y_1, x_2, y_2, \ldots, x_n, y_n\}$ whose elements satisfy precisely these comparabilities; $x_1 < y_1, y_1 > x_2, x_2 < y_2, y_2 > x_3, x_3 < y_3, \ldots, x_{n-1} < y_{n-1}, y_{n-1} > x_n, x_n < y_n, y_n > x_1$, and in the case $n = 2$, there is *no* $z \in P$ such that $x_1, x_2 \leq z \leq y_1, y_2$. The proof of this result is most transparent when it is cast in graph theoretical terms. The comparability graph $G$ of a length one ordered set $P$ is identical to the undirected graph associated with its diagram. Let $c_0, c_1, c_2, \ldots, c_{n-1}$ be vertices of a "cycle" in $G$ which corresponds to a crown of least order in $P$. Let us suppose that $G$ is an undirected graph except that the edges of the cycle have this orientation: $c_0 \to c_{n-1} \to c_{n-2} \to \ldots \to c_2 \to c_1 \to c_0$.

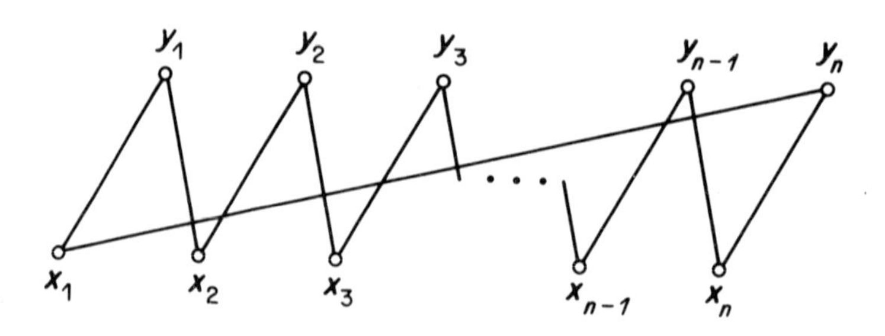


A crown $C_{2n}$.

The *distance* $d_G(a,b)$ between vertices $a$ and $b$ in $G$ is the least length ($\equiv$ number of edges) -- if it exists -- of a "path" containing both $a$ and $b$. (A *path* of $G$ is a sequence $a = a_0, a_1, a_2, \ldots, a_n = b$ of distinct vertices of $G$ such that $a_{i+1}$ is adjacent to $a_i$ and $a_{i+1} \to a_i$ if there is an arrow on the edge. The "arrows", whenever they exist, are all in the same "direction".)

With this definition in hand we partition the vertices of $G$ into the subsets $A_0, A_1, A_2, \ldots$ and $B$ as follows: a vertex $a$ belongs to $A_i$ just if $d_G(c_0, a) = i$ while $a$ belongs to $B$ if there is no path in $G$ from $a$ to $c_0$ (that is, $a$ and $c_0$ are in different connected components of $G$). We define this folding map

$$g(a) = \begin{cases} c_i & \text{if } a \in A_j \text{ where} \\ & j \equiv i \pmod{n} \text{ and } i = 0,1,2,\ldots,n-1 \\ c_0 & \text{if } a \in B \,. \end{cases}$$

In fact, $g$ corresponds to a retraction of $P$ to the crown $\{c_0, c_1, c_2, \ldots, c_{n-1}\}$.

These ideas are exploited further in [14] and [18] to establish this result.

> *An ordered set P of length one has the fixed point property if and only if (i) P is connected, (ii) P contains no crowns, and (iii) P contains no infinite paths.*

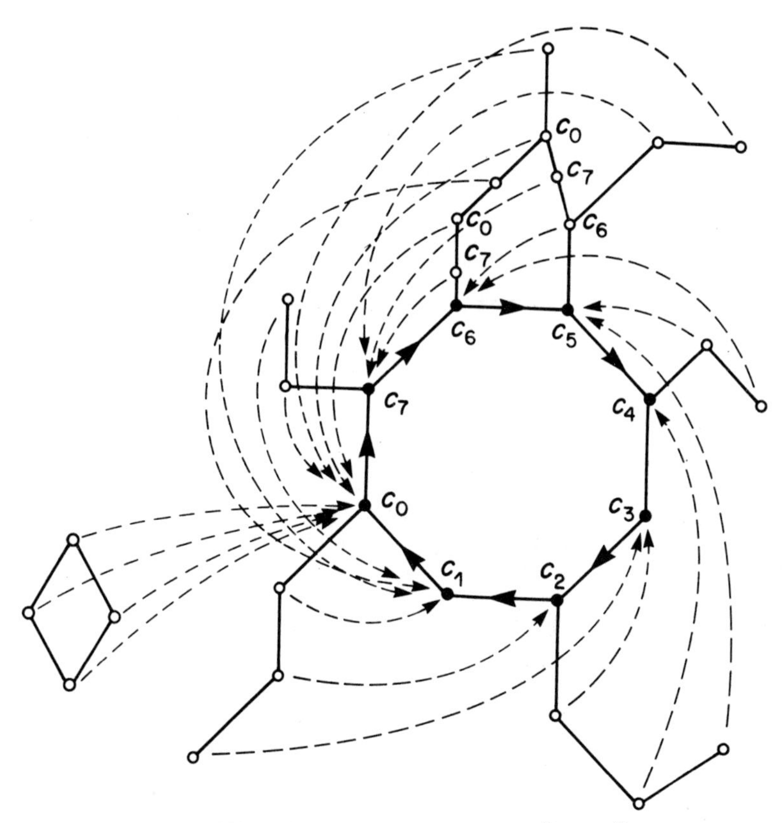


The folding map onto a "cycle".

The failure of each of these conditions amounts to a retraction: (i) to a retraction of $P$ to 2; (ii) and (iii) to appropriate folding maps [19].

Variations on the folding map are also central to proofs of these facts [7].

> *Let $P$ be a finite, connected ordered set. (i) If there is an order-preserving map of $P$ onto a finite fence $F$ then $F$ is a retract of $P$. (ii) If $d_P(x,y) \leq n$ for all $x, y \in P$, and there is an order-preserving map of $P$ onto a crown $C_{2n}$ then $C_{2n}$ is a retract of $P$.*

In [15] the folding map and the greedy map (for an arbitrary well order) are combined to establish a general retraction theorem for (reflexive) graphs.

## A STRUCTURE THEORY

The retract concept is the cornerstone of a structure theory for ordered sets [7]. This structure theory is much akin to the familiar subdirect representation theory of general algebra. It has these three fundamental ingredients -- all turning on the class $\underset{\sim}{R}(P)$ of (all isomorphism types of) the retracts of an ordered set $P$.

I. *Representation.* A family $(P_i | i \in I)$ of ordered sets *represents* $P$ if $P_i \in \underset{\sim}{R}(P)$ for each $i \in I$, and $P \in \underset{\sim}{R}\left(\prod_{i \in I} P_i\right)$.

II. *Irreducibility.* An ordered set $P$ is *irreducible* if, for each representation $(P_i | i \in I)$ of $P$, $P \in \underset{\sim}{R}(P_i)$, for some $i \in I$.

III. *Order variety.* An *order variety* is a class $K$ of ordered sets which is closed under the formation of all direct products of nonempty families of members of $K$ and under the formation of all retracts of members of $K$.

One result from this theory is this.

> *Every chain has a unique representation by irreducible chains, each isomorphic to one of $\underset{\sim}{1}$, $\underset{\sim}{2}$, $\underset{\sim}{\alpha}$, $\underset{\sim}{\beta}^d$, or $\underset{\sim}{\alpha} \oplus \underset{\sim}{\beta}^d$ where $\underset{\sim}{\alpha}$ and $\underset{\sim}{\beta}$ are regular ordinals* [7].

REFERENCES

[1] G. Birkhoff (1937) Generalized arithmetic, *Duke Math. J.*, 3, 283-302.

[2] A. Björner and I. Rival (1980) A note on fixed points in semimodular lattices, *Discrete Math.* 29, 245-250.

[3] A.C. Davis (1955) A characterization of complete lattices, *Pacific J. Math.* 5, 311-319.

[4] R.A. Dean (1961) Sublattices of free lattices, in: *Proceedings of Symposia in Pure Mathematics* (R.P. Dilworth, ed.) American Math. Soc., Providence.

[5] D. Duffus, W. Poguntke and I. Rival (1980) Retracts and the fixed point problem for finite partially ordered sets, *Canad. Math. Bull.* 23, 231-236.

[6] D. Duffus and I. Rival (1979) Retracts of partially ordered sets, *J. Australian Math. Soc. (Ser. A)* 27, 495-506.

[7] D. Duffus and I. Rival (1981) A structure theory for ordered sets, *Discrete Math.* 35, 53-118.

[8] D. Duffus, I. Rival and M. Simonovits (1980) Spanning retracts of a partially ordered set, *Discrete Math.* 32, 1-7.

[9] R. Freese and J.B. Nation (1978) Projective lattices, *Pacific J. Math.* 75, 93-106.

[10] F. Galvin and B. Jónsson (1961) Distributive sublattices of a free lattice, *Canad. J. Math.* 13, 265-272.

[11] J. Keisler (1965) Some applications of infinitely long formulas, *J. Symbolic Logic* 30, 339-349.

[12] S.R. Kogalovskii (1964) On linearly complete ordered sets [Russian], *Uspehi Mat. Nauk* 19 no. 2 (116), 147-150.

[13] R. McKenzie (1972) Equational bases and nonmodular lattice varieties, *Trans. Amer. Math. Soc.* 174, 1-43.

[14] R.J. Nowakowski and I. Rival (1979) Fixed-edge theorem for graphs with loops, *J. Graph Theory* 3, 339-350.

[15] R.J. Nowakowski and I. Rival, The smallest graph variety containing all paths, (to appear).

[16] M. Pouzet and I. Rival, Every countable lattice is a retract of a direct product of chains, (to appear).

[17] M. Pouzet and I. Rival, Quotients of complete ordered sets, (to appear).

[18] I. Rival (1976) A fixed point theorem for finite partially ordered sets, *J. Combinatorial Theory (A)* 21, 309-318.

[19] I. Rival (1980) The problem of fixed points in ordered sets, *Annals of Discrete Math.* 8, 283-292.

[20] I. Rival and R. Wille (1981) The smallest order variety containing all chains, *Discrete Math.* 35, 203-212.

NOTE. Quite recently I have learned of these further and interesting works on retracts of (irreflexive) graphs.

P. Hell (1974) Retracts in graphs, *Springer-Verlag Lecture Notes in Math.* #406, 291-301.

P. Hell (1974) Absolute planar retracts and the four colour conjecture, *J. Combinatorial Theory (B)* 17, 5-10.

P. Hell (1976) Graph retractions, *Atti dei convegni lincei* 17, Vol. II, 263-268.

# PART II

# LINEAR EXTENSIONS

# LINEAR EXTENSIONS OF ORDERED SETS

R. Bonnet and M. Pouzet

Département de Mathématiques
Université Claude Bernard - Lyon 1
43 Boulevard du 11 Novembre 1918
69622 Villeurbanne Cédex
France

ABSTRACT

The best known connection between partial orders and linear orders is the Szpilrajn theorem:

> *Any partial order on a set can be extended to a linear order on the same set.*

From this, it follows that any partial order is the intersection of its linear extensions; equivalently, every ordered set can be represented as some subset of a Cartesian product of chains.

A natural question to ask is whether these results work with orders having some prescribed property $(P)$. For instance, *does an order with the property* $(P)$ *have a linear extension with this same property* $(P)$? As an example, the answer is affirmative if $(P)$ is the property that the order is well-founded (i.e. its chains are well-ordered). In this case there is a well-ordered linear extension. Another question is this: *does every linear extension of an order with the property* $(P)$ *also have this property* $(P)$? For example, if a well-founded ordered set has no infinite antichain, then all of its linear extensions are well-ordered. These kinds of questions have been investigated by many authors (Baumgartner, Galvin, McKenzie, and others). The purpose of these lectures is to survey some of the most relevant results.

I. We can characterize those countable order types $\alpha$ such that any ordered set, with a subchain of type $\alpha$, has a linear extension with this property. As well, we can characterize those countable order types $\alpha$ such that any ordered set, without



*I. Rival (ed.), Ordered Sets, 125–170.*

any infinite antichain and without any subchain of type $\alpha$, will have all of its linear extensions with this property.

Typical examples of such $\alpha$ are the rational chain $\eta$ [9], the chain $\omega$ of the natural numbers and its dual $\omega^*$ (which correspond to the example above). A powerful counterexample is $\omega + 1$ [33].

II. We consider the problem [44] of describing the ordered sets which have a complete linear extension. The answer is positive for chain-complete ordered sets (ordered sets for which every maximal chain is complete) which are either countable or without any infinite antichain. An example is given of an uncountable complete lattice which has no complete linear extension.

III. We consider ordered structures generated by chains by a uniform procedure (for instance, Ehrenfeucht-Mostowski models or Boolean algebras). We discuss the result in [12] that a partial order can be extended to a linear order in such a way that the new structure is uniformly generated too.

These lectures will include general facts about linear extensions and some elements of representation theory. For this purpose we consider the properties of the lattice $I(P)$ consisting of the initial segments of an ordering $P$. This lattice reflects the properties of this ordering (by a categorical duality). For instance the linear extensions of $P$ correspond exactly to the maximal chains of $I(P)$, and the sets of the linear extensions of $P$, the intersection of which is the order relation itself, correspond to the sets of generators (in the lattice sense) of $I(P)$. The characterization of such sets, and applications to the computation of the partial order dimension [10] will be given.

## INTRODUCTION

The most popular connection between partial orders and linear orders is the SZPILRAJN theorem (1930): *every partial order can be extended to a linear order*. As a consequence every partial order is the intersection of its linear extensions, and thus the corresponding ordered set can be represented as a subset of a cartesian product of chains.

A natural question in this context is to ask whether every order with a given property (P) has a linear extension with the same property (P). We can also ask for more, namely whether *all* linear extensions of an order have the property (P) provided the order itself satisfies (P). For a typical example consider well-founded ordered sets (every non-empty subset has a minimal element). These are known to have a well-founded linear extension, i.e. well-ordered extension. Moreover if they also have no infinite antichain (i.e. partially well ordered) then *all* their linear extensions are well-ordered.

Many authors have investigated problems of this type. We survey here the most important results, especially those related to the above example:

1. Characterization of the countable order types $\alpha$ such that:

(i) every ordered set without any chain of type $\alpha$ has a linear extension with this property, or
(ii) every ordered set with no chain of type $\alpha$ and no infinite antichain has all its linear extensions with the same property.

Typical examples of such $\alpha$ are (a) the rational chain $\eta$ [9], (b) the chain $\omega$ of the natural numbers and its dual $\omega^*$ (this is the 'well-founded' example given above). An example of a type not satisfying (i) or (ii) is $\omega$+1 [33].

2. For complete linear extensions we have the following results from [44]: every *chain-complete* ordered set (i.e. every maximal chain is complete) which is either countable or does not contain an infinite antichain has a complete linear extension. There are complete lattices of every cardinality greater than $\aleph_1$ having no complete linear extension. But it is not known if there is such a lattice of cardinality $\aleph_1$.

3. More generally for ordered structures uniformly generated by chains (e.g. EHRENFEUCHT-MOSTOWSKI models or countable Boolean algebras) we show that the partial order can be extended to a linear order which is also uniformly generated. This result is useful comparing certain ordered structures.

We need general properties of linear extensions and some elements of representation theory. We use the lattice $\underset{\sim}{I}(\underset{\sim}{P})$ of the initial segments of an ordered set $\underset{\sim}{P}$, which, by a categorical duality, nicely reflects the property of $\underset{\sim}{P}$, e.g. the linear extensions of $\underset{\sim}{P}$ correspond to the maximal chains of $\underset{\sim}{I}(\underset{\sim}{P})$. Moreover the sets of those linear extensions of $\underset{\sim}{P}$ whose intersection is $\underset{\sim}{P}$ are in a 1-1 correspondence with the sets of generators of the lattice $\underset{\sim}{I}(\underset{\sim}{P})$. The characterization of such sets (BOUCHET's theorem) and applications to the computation of the DUSHNIK-MILLER order dimension are given.

## I. ORDERS AND THEIR LINEAR EXTENSIONS

A binary relation on a set $P$ is a subset $P$ of the cartesian product $P\times P$. An order is a binary relation $P$ that is *reflexive, antisymmetric and transitive*. The set $P$ equipped with the order relation is said to be *partially ordered* and is denoted by $\underset{\sim}{P}$. We denote by $P^*$ the dual (or converse) order (i.e. $(x,y) \in P^*$ iff $(y,x) \in P$) and $\underset{\sim}{P}^*$ the corresponding ordered set. For a subset $A$ of $P$ the order induced by $P$ on $A$ is denoted by $P_{|A}$ and the corresponding ordered set by $\underset{\sim}{P}_{|A}$.

We write $x \leq y(P)$ or $x \leq y(\underset{\sim}{P})$ or simply $x \leq y$ instead of $(x,y) \in P$ and we say that $x$ is smaller than $y$. Two elements $x$, $y$ such that $x \leq y$ or $y \leq x$ are comparable, otherwise they are *incomparable*. If all elements of $\underset{\sim}{P}$ are pairwise comparable (incomparable) the order is a *total* or a *linear* (*discrete*) order and $\underset{\sim}{P}$ is a *chain* (an *antichain*).

### I-1. The SZPILRAJN Theorem.

An order $P'$ extends an order $P$ if $P$ is a subset of $P'$, that is to say $x \leq y(\underset{\sim}{P})$ implies $x \leq y(\underset{\sim}{P}')$. The following striking fact is the classical theorem of SZPILRAJN [51].

I-1.1. THEOREM. *Every partial order extends to a linear order.*

The basic idea of the proof is extremely simple: if the order $P$ has two incomparable elements $a$ and $b$, then $P$ extends to a new order $P'$ for which $a \leq b$. Now, if this new order has two incomparable elements $a'$ and $b'$, extend $P'$ into a new order in which $a' \leq b'$, and continue this process.

For a finite set $P$ this process clearly leads to a linear order in a finite number of steps. However for an infinite $P$ we need some set-theoretical machinery to ensure this process is *well defined and terminating*.

As to the first point, we make the following observation: if a binary relation $P$ is reflexive and acyclic (i.e. we never have $x_0 < x_1 \dots x_n < x_0$) then the transitive hull of $P$ (obtained by adjoining to $P$ the ordered pairs $(x,y)$ such that $x = x_0 < \dots < x_n = y$ for some finite sequence of elements $x_1,\dots,x_{n-1} \in P$) is a partial order, in fact it is the least order containing $P$. From this we obtain the simple but useful fact:

LOCAL EXTENSION LEMMA. *Let* $P$ *be a partial order on a set* $P$. *Let* $A$ *be a subset of* $P$ *and* $P'_A$ *an order on* $A$ *which extends* $P_{|A}$. *Then there is an order* $\tilde{P}$ *extending* $P$ *such that* $\tilde{P}_{|A} = P'_A$.

(Proof. Observe that $P \cup P'_A$ has no cycle).

Here we need just the special case when $|A| = 2$: *if* $a$ *and* $b$ *are incomparable for* $P$ *then there is an extension* $P'$ *of* $P$ *for which* $a \leq b$ [and to get the minimal one it is enough to add to $P$ all the pairs $(x,y)$ with $x \leq a$ and $b \leq y(P)$].

We now argue as follows; let $Op$ be the set of all extensions of $P$. This set, ordered by inclusion is closed under the union of chains, and so is inductive. Thus by ZORN's lemma, it has a maximal element. By the extension property, every maximal element of $Op$ is a linear order and the theorem is proved.

Comments.

We have used ZORN's lemma (equivalent to the axiom of choice or the well ordering principle). In fact, the result can be deduced from weaker assumptions such as the ultrafilter axiom or any form of the compactness theorem. The reader unfamiliar with such methods may prefer the following proof:

For any finite subset $A$ of $P$ let $V_A$ denote the set of binary relations $P$ on $P$ such that $P \cap A^2$ is a linear extension of $P_{|A}$. Each set $V_A$ is non-empty (by SZPILRAJN's theorem for finite sets) and closed in the power set $2^{P\times P}$ equipped with the product topology. The intersection of a finite number of them is non-empty (observe that $V_A \cap V_{A'} \supseteq V_{A \cup A'}$). By the compactness theorem, $2^{P\times P}$ is compact, so the intersection of all the $V_A$ is non-empty. Obviously every element $\overline{P}$ of this intersection is a linear order extending $P$.

To conclude the set theoretical aspects of the SZPILRAJN theorem note that this theorem considered as a new axiom is weaker than the ultrafilter axiom (FELGNER) but stronger than the axiom that any set can be totally ordered (MATHIAS, see JECH [32]). However, in this paper we are not concerned so much with independence results; we assume throughout the axiom of choice (and sometimes even stronger assumptions like the generalized continuum hypothesis (G.C.H.)).

A consequence of the SZPILRAJN theorem is:

I-1.2. THEOREM. *Every order is the intersection of its linear extensions.*

Proof. The intersection of all the linear extensions of an order $P$ is obviously an order, say $\breve{P}$, containing $P$. If $\breve{P} \neq P$ then there is $(a,b) \in \breve{P} \setminus P$ such that $a$ and $b$ are incomparable in $P$. Hence there is an extension $P'$ of $P$ with $b \leq a$ and by the SZPILRAJN theorem, a linear extension $\overline{P}'$ of $P'$. This extension is also a linear extension of $P$ containing $(b,a)$, contradicting the fact that $(a,b) \in \breve{P}$.

In other words we have $x \leq y(P)$ iff we have $x \leq y(\overline{P})$ for each linear extension $\overline{P}$ of $P$. This fact, expressed in terms of products of sets, says that $\underset{\sim}{P}$ is order isomorphic to the diagonal of the cartesian product of all the chains $\underset{\sim}{\overline{P}} = (P,\overline{P})$ equipped with the product order.

In particular we get:

I-1.3. THEOREM. *Every ordered set is order isomorphic to a subset of a cartesian product of chains.*

A result of this kind is somewhat different in spirit from the previous one. It permits the representation of a structure as a substructure of a product of some simpler structures.

We can get it directly from the following.

I-1.4. PROPOSITION. *Any ordered set $\underset{\sim}{P}$ is order isomorphic to a field of subsets of $P$ ordered by inclusion.*

Proof. The map $i$ from $P$ into $\mathfrak{P}(P)$ defined by $i(x) = \{y / y \leq x\}$ is an order isomorphism.

Now to conclude it is enough to observe that for any set $P$ the set of subsets of $P$ ordered by inclusion is order-isomorphic to the power set $\underset{\sim}{2}^P$ consisting of 0-1 sequences, ordered as a cartesian product of the two element chain $\underset{\sim}{2} = \{0,1\}$.

In fact using well ordering we can get a new proof of the SZPILRAJN theorem from the theorem I-1.3 or proposition I-1.4. To see that, we recall the notion of the *lexicographic product of ordered sets*: let $(\underset{\sim}{P}_i)_{i\in I}$ be a collection of ordered sets indexed by a well-founded ordered set $\underset{\sim}{I}$ (each non empty subset of $I$ has a minimal element). The *lexicographic product* of the $\underset{\sim}{P}_i$ is the cartesian product of the $P_i$ equipped with the *lexicographic order* defined as follows: $(x_i)_{i\in I} \leq (y_i)_{i\in I}$ iff $x_j \leq y_j$ for each minimal element $j$ of the set $\{i \in I : x_i \neq y_i\}$. If $\underset{\sim}{I}$ is an antichain then the lexicographic product is just the usual cartesian order product. If $I$ is a well ordering then the lexicographic product is the usual ordinal product. Obviously the order defined by the ordinal product extends the cartesian order, and any ordinal product of chains is again a chain.

So, assuming an ordered set $\underset{\sim}{P}$ is isomorphic to a subset of a cartesian product of chains $\underset{\sim}{P}_i$, $i \in I$, then, for any well ordering $I$ on $I$, the ordinal product of the $\underset{\sim}{P}_i$ gives a linear ordering which induces a linear extension of $\underset{\sim}{P}$. For example, with a prescribed well ordering on $P$, the ordinal product of the two element chain gives a linear extension of the power set $2^P$ and then again, a linear extension of $\underset{\sim}{P}$. It is a natural question to ask whether the cartesian order defined by a family of orders $P_i$, $i \in I$ can be obtained as the intersection of all the possible lexicographical orders of such a family. The answer is positive. And, in fact, for $\kappa = |I|$ no more than $\kappa$ lexicographical orders suffices.

To see this let $J$ be a fixed well ordering on $I$ and let $0$ be its least element. For each $i \in I$ let $J_i$ be the well ordering obtained from $J$ after exchanging $0$ and $i$ (i.e. $i$ is the least element for $J_i$). The ordinal product of the $\underset{\sim}{P}_i$ indexed by $\underset{\sim}{J}_i$ gives an extension of the cartesian product and if $(x_j)_{j\in I} \leq (y_j)_{j\in I}$ for this order then necessarily $x_i \leq y_i(P_i)$. So the intersection of all such orders is the cartesian order. This, together with theorem I-1.3 gives the theorem of ORE [41] (see also [30], [39]).

I-1.5. THEOREM. *An order $P$ on the set $P$ is the intersection of a family of $\kappa$ linear orders iff the ordered set $\underset{\sim}{P}$ is order-isomorphic to a subset of a cartesian product of $\kappa$ chains.*

The smallest $\kappa$ is the *order dimension* of $P$ as defined by B. DUSHNIK and E.W. MILLER [18]. Some of the properties of the dimension will be given in the next section.

I-2. Initial Segments of an Ordered Set and BOUCHET's Theorem.

Let $P$ be an ordered set. A subset $I$ of $P$ *is an initial segment* of $P$ if for $x,y \in P$, $x \in I$ whenever $x \leq y$ and $y \in I$. For example, for any subset $A$ of $P$, the set $\{x \in P | x \leq y$ for some $y \in A\}$ is an initial segment (in fact the smallest initial segment containing $A$). If $|A| = 1$ we call it a *principal initial segment*.

We denote by $I(P)$ the collection of all the initial segments of $P$ ordered by inclusion. For example if $P$ is a finite $n$-element chain then $I(P)$ is an $n+1$-element chain. On the other hand if $P$ is an antichain then any subset of $P$ is an initial segment and $I(P)$ is the power set $\mathfrak{P}(P)$.

Because the union and the intersection of any family of initial segments of our ordered set $P$ are again initial segments of $P$, *the ordered set $I(P)$ is a complete sublattice of $\mathfrak{P}(P)$.* It can be completely characterized by lattice properties. For this purpose let us recall that for a lattice $T$, an element $x \in T$ is completely-join-irreducible if for any subset $X$ of $T$, $x \in X$ whenever $x = \vee X$. Let $T^{\vee}$ be the set of such elements and $\nabla T^{\vee} = \{x \in T / x = \vee X$ for some $X \subseteq T^{\vee}\}$. If $\nabla T^{\vee} = T$ we say that $T$ is *join-generated* by $T^{\vee}$. The (complete) lattice $T$ is *completely-join-distributive* if the equation $(\vee X) \wedge x = \vee\{y \wedge x / y \in X\}$ holds for any $X \subseteq T$, $x \in T$. (In this case, for completely-join-irreducible $x$, if $x \leq \vee X$ then $x \leq y$ for some $y \in X$.)

I-2.1. PROPOSITION. *Let $P$ be an ordered set; then $I(P)$ is:*

(1) *a complete, completely-join-distributive lattice;*
(2) *join-generated by the completely-join-irreducible elements, and $P$ is isomorphic to the set $I(P)^{\vee}$ by the map $i(x) = \{y / y \leq x\}$.*

*Conversely, an ordered set $T$ satisfying* (1) *and* (2) *is order isomorphic to $I(T^{\vee})$ by the map $\varphi(x) = \{y / y \in T^{\vee}, y \leq x\}$.*

Proof. $I(P)$ satisfies (1) as a complete sublattice of $\mathfrak{P}(P)$. (In fact it is completely distributive). Every principal initial segment has a greatest element, so it is completely-join-irreducible. Every initial segment $I$ is the union of the principal initial segments contained in $I$ ($I = \bigcup_{x \in I} \{y / y \leq x\}$), so the completely-join-irreducible initial segments are principal and they generate $I(P)$.

Now let $T$ satisfy (1). The map $\varphi : T \to I(T^{\vee})$ defined by $\varphi(x) = \{y / y \in T^{\vee}, y \leq x\}$ is a lattice homomorphism (the meets are obviously preserved, for the joins it follows from complete-join-distributivity). Let $\psi : I(T^{\vee}) \to T$ be defined by $\psi(I) = \vee I$. Then $\varphi \circ \psi$ is the identity map and consequently $\varphi$ induces an isomorphism from $\nabla T^{\vee}$ onto $I(T^{\vee})$.

As a corollary we get the converse of our first observation about initial segments.

COROLLARY I-2.2. *Any complete sublattice of a power set is isomorphic to the lattice of initial segments of some ordered set.*

[Let $T$ be a complete sublattice of $\mathfrak{P}(P)$. To verify (2) let $A \in T$ and for any $x \in A$, let $A_x = \cap \{A'/x \in A' \in T, A' \subseteq A\}$. Observe that every $A_x$ is completely-join-irreducible and $A = \cup A_x$].

Any lattice $T$ satisfying the descending chain condition (in other words $T$ is well founded) satisfies condition (2). [If $T - \nabla T^{\vee} \neq \emptyset$ pick a minimal element $x$. If $x = \vee\{y/y < x\}$ then $x \in \nabla T^{\vee}$ and, on the other hand $x \in T^{\vee}$. Contradiction]. From this remark, (or by I-2.2) follows the classical result:

I-2.3. PROPOSITION [5]. *Any finite distributive lattice $T$ is isomorphic to the lattice $I(T^{\vee})$ where $T^{\vee}$ is the set of its join-irreducible elements.*

Now it is natural to ask how the knowledge of the lattice properties of $I(P)$ can be used to study the linear extensions of $P$. For this we will study the connections between the lattices $I(P)$ and $I(P')$, when $P'$ is a linear extension of $P$.

From the previous lattice characterization of such sets it follows that *$P$ is isomorphic to $P'$ iff $I(P)$ is isomorphic to $I(P')$*. It is a particular case of the following result:

I-2.4. PROPOSITION. *Let $P$, $P'$ be two ordered sets.*

1. *If $f: P \to P'$ is increasing then the map $If : I(P') \to I(P)$ defined by $If(I') = f^{-1}(I')$ is a complete lattice homomorphism.*

2. *Conversely if $g: I(P') \to I(P)$ is a complete lattice homomorphism then there is exactly one increasing map $f: P \to P'$ such that $If = g$.*

Proof. (1). If $f: P \to P'$ is increasing then for any initial segment $I'$ of $P'$, its inverse image $f^{-1}(I')$ is an initial segment of $P$. The result follows.

(2) For $x \in P$ let us consider the set $K' = \cap \{I'/I' \in I(P')$ and $x \in g(I')\}$. This set is a principal initial segment: because $g$ preserves meets $x \in g(K')$, so $K'$ is the smallest initial segment such that $x \in g(K')$; if $K' = \cup I_i$ for some family $I_i$ then, because $g$ preserves joins, we get $g(K') = \bigcup_i g(I_i)$, so $x \in g(I_i)$ for some

$I_i$, and then $K = I_i$. Thus $K'$ is join-irreducible, and hence principal.

We can define $f(x)$ as the greatest element of $K'$. It is then routine to verify that $f$ is increasing, $\underset{\sim}{I}f = g$ (it suffices to do this for the principal initial segments) and that $f$ is unique.

N.B. The same method of proof gives slightly more: if $\underset{\sim}{T}$, $\underset{\sim}{T}'$ are two complete, completely-join-distributive lattices, then for every complete lattice homomorphism $g:\underset{\sim}{T}' \to \underset{\sim}{T}$ there is exactly one increasing map $f:\underset{\sim}{T}^{\vee} \to \underset{\sim}{T}'^{\vee}$ such that the following diagram commutes:

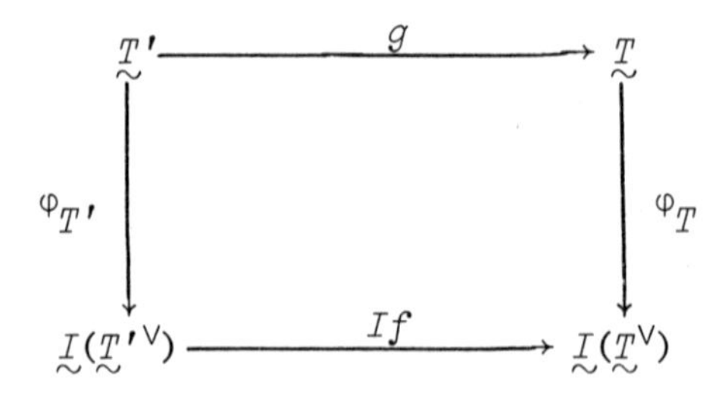


The proposition I-2.4 expresses a categorical duality between ordered sets and lattices of initial segments. It is the straightforward generalization of the duality between finite ordered sets and finite distributive lattices. One of the most striking consequences is:

I-2.5. THEOREM [9]. *Let $\underset{\sim}{P}$ be an ordered set and $\underset{\sim}{I}(\underset{\sim}{P})$ the lattice of initial segments of $\underset{\sim}{P}$.*

1. *If $P'$ is a linear extension of $P$ then $\underset{\sim}{I}(\underset{\sim}{P}')$ is a maximal chain of $\underset{\sim}{I}(\underset{\sim}{P})$.*

2. *If $C$ is a maximal chain of $\underset{\sim}{I}(P)$ then there is a unique linear extension $P'$ of $P$ such that $\underset{\sim}{I}(\underset{\sim}{P}') = \underset{\sim}{C}$.*

Proof. This can be deduced from I-2.2 and I-2.4 but a direct proof is easy:

(1) Any initial segment of $\underset{\sim}{P}'$ is an initial segment of $\underset{\sim}{P}$ so $\underset{\sim}{I}(P') \subseteq \underset{\sim}{I}(\underset{\sim}{P})$. Because $\underset{\sim}{P}'$ is a chain, $\underset{\sim}{I}(\underset{\sim}{P}')$ is also a chain. It is a maximal chain because if $I'$, $J'$ are two consecutive elements of $\underset{\sim}{I}(P')$ then $|J'-I'| = 1$.

(2) Because $\underset{\sim}{C}$ is maximal, it is complete, so $\underset{\sim}{C}$ is isomorphic to some $\underset{\sim}{I}(\underset{\sim}{Q})$ and by this way we can get the result. But we can also define the order $P'$ as follows: $x \leq y(P')$ iff for any

$I' \in \underset{\sim}{C}$, $x \in I'$ whenever $y \in I'$.

This result was obtained independently by A. BOUCHET and M. POUZET around 1968, we don't know if it appears before our paper [9].

I-2.6. COROLLARY. *For any set P there is a one-to-one correspondence between linear orders on P and maximal chains of* $\mathfrak{P}(P)$. *And in particular (for set theorists) P can be totally ordered iff* $\mathfrak{P}(P)$ *contains a maximal chain.*

For brevity say that a subset $G$ of a complete lattice $\underset{\sim}{T}$ generates $\underset{\sim}{T}$ if $\underset{\sim}{T}$ is the smallest complete sublattice containing $G$. Another instance of the correspondence between linear orders and maximal chains is the following:

I-2.7. LEMMA. *Let* $(P_i)_i$ *be a family of linear orders on the same set P, and let P be* $\cap P_i$. *Then the union of* $\underset{\sim}{I}(\underset{\sim}{P_i})$ *generates* $\underset{\sim}{I}(\underset{\sim}{P})$.

Proof. We have $x \leq y(P)$ iff $x \leq y(P_i)$ for every $P_i$. Writing this in terms of sets we get $\leftarrow x]_P = \cap \leftarrow x]_{P_i}$ (here $\leftarrow x]_P = \{y / y \leq x(P)\}$). But any $I \in \underset{\sim}{I}(\underset{\sim}{P})$ can be written $I = \bigcup_{x \in I} \leftarrow x]_P$. So the result follows. Of course, the sets $\leftarrow x]_{P_i} \in \underset{\sim}{I}(\underset{\sim}{P})$.

This fact added to the previous theorem expresses that *the order dimension of an ordered set* $\underset{\sim}{P}$ *is the smallest number of maximal chains needed to generate* $\underset{\sim}{I}(\underset{\sim}{P})$.

In fact, we can say a little more. For this we denote by dim $\underset{\sim}{P}$ the order dimension of $\underset{\sim}{P}$, by $S^0(\underset{\sim}{P})$ the smallest number of chains we need to cover $\underset{\sim}{P}$ ($S^0\underset{\sim}{P} = \text{Min}\{\kappa / P = \bigcup_{i \in I} C_i$, $\underset{\sim}{P}|C_i$ is a chain, $|I| = \kappa\}$.

I-2.8. THEOREM [10]. *For every ordered set* $\underset{\sim}{P}$, *dim* $\underset{\sim}{P} = \text{Min}\{S^0(\underset{\sim}{G}) / \underset{\sim}{G}$ *generates* $\underset{\sim}{I}(\underset{\sim}{P})\}$.

Proof. Extend each chain in a covering of $G$ into a maximal chain of $\underset{\sim}{I}(\underset{\sim}{P})$. Here presumably $G \subseteq \mathfrak{P}(P)$ and ordered by $\subseteq$.

This result is useful only in so far as generators are known. Such sets have also been characterized by A. BOUCHET.

I-2.9. THEOREM [10]. *Let G be a subset of* $\underset{\sim}{I}(\underset{\sim}{P})$. *G generates* $\underset{\sim}{I}(\underset{\sim}{P})$ *iff for each* $x, y \in P$, *if* $x \not\leq y(P)$, *there is some* $I \in G$ *such that* $x \notin I$, $y \in I$.

Proof. Observe first that $G$ generates a complete sublattice $\underset{\sim}{T}$ of $(P)$ iff each element of $T$ can be written as $\bigcup_{j\in L} \bigcap_{i\in K_j} I_i$ where each $I_i \in G$. So assume $G$ generates $\underset{\sim}{I}(P)$, and $x,y \in P$ with $x \not\leq y$. Because $\leftarrow y]_P$ is join-irreducible, $\leftarrow y]_P = \bigcap_i I_i$ for some collection of $I_i \in G$. Then for some $i$, we have $x \notin I_i$, and $y \in I_i$. Conversely if $G$ satisfies the above condition then for $x,y \in P$ with $x \not\leq y$ the set $L_{x,y} = \bigcap\{I/I \in G,\ x \notin I,\ y \in I\}$ belongs to the lattice $\tilde{G}$ generated by $G$. The same holds for $\bigcap_{x/x\not\leq y} L_{x,y}$. But it is just $\leftarrow y]_P$. Because it contains all principal initial segments, $\tilde{G} = \underset{\sim}{I}(\underset{\sim}{P})$.

Many results about order dimension can be deduced from the two previous theorems. We give a few examples:

I-2.10. THEOREM [42]. *For any ordered set* $\underset{\sim}{P}$, *dim* $\underset{\sim}{I}(\underset{\sim}{P}) = S^0(\underset{\sim}{P})$.

Proof. Let $\ell : \underset{\sim}{P} \to \underset{\sim}{I}^2(\underset{\sim}{P})$ defined by $\ell(x) = \{I/I \in I(P) \text{ and } x \notin I\}$. This map is an order-isomorphism from $\underset{\sim}{P}$ onto its image $\underset{\sim}{P}'$. On the other hand $P'$ generates $\underset{\sim}{I}^2(\underset{\sim}{P})$ and is included in any generator of $\underset{\sim}{I}^2(\underset{\sim}{P})$ [If $I \not\subseteq J$ then for any $x \in I-J$ we get $I \notin \ell(x)$, $J \in \ell(x)$ and so $\underset{\sim}{P}'$ generates $\underset{\sim}{I}^2(\underset{\sim}{P})$. Now for any $x$, we have $\leftarrow x] \not\subseteq P\backslash[x\rightarrow$, and we have just one element of $P'$, namely $\ell(x)$, which separates them. So $P'$ is included in any generator]. By BOUCHET's theorem I.2.8 the result follows.

The first corollary is KOMM's theorem [34]:

dim $\mathcal{P}(P) = |P|$ *for any set* $P$. This, added to the theorem of ORE I.1.5 gives a theorem of HIRAGUCHI [31]:

dim $\prod_{i\in I} \underset{\sim}{P}_i = |I|$ *for any family of chains* $\underset{\sim}{P}_i$ *such that each* $|P_i| \geq 2$.

Now let us recall that an ordered set $\underset{\sim}{P}$ *has finite width* if its antichains are finite and have bounded size. The maximum size of these antichains is the width of $\underset{\sim}{P}$, $\omega(\underset{\sim}{P})$. The well known theorem of R.P. DILWORTH states $\omega(\underset{\sim}{P}) = S^0(\underset{\sim}{P})$. Thus a second corollary of I.2.10 is the well known:

I-2.11. THEOREM. *Let* $\underset{\sim}{T}$ *be a finite distributive lattice then dim* $\underset{\sim}{T} = \omega(\underset{\sim}{T}^\vee)$.

Proof. $\underset{\sim}{T} \cong \underset{\sim}{I}(\underset{\sim}{T}^\vee)$ so dim $\underset{\sim}{T} = S^0\underset{\sim}{T}^\vee = \omega(\underset{\sim}{T}^\vee)$.

This result is no longer true for lattices with finite antichains of unbounded size. There are lattices with arbitrarily large dimension but having only finite antichains [42]. Numerous examples can be obtained as follows: let $\kappa$ a cardinal, $\alpha$ a well ordered set with cardinality $\kappa$ and $P$ the direct product $\alpha \times \alpha$. It is an easy exercise to verify that $I(P)$ has no infinite antichain, but $\dim I(P) = S^0(P) = \kappa$.

REMARK

A direct way to prove I-2.10 is the following: for an ordered set $P$ let $D(P) = \{\leftarrow x] / x \in P\} \cup \{P - [x \rightarrow / x \in P\}$ ordered by inclusion. If $P$ is an antichain then $D(P)$ is the classical example already used by B. DUSHNIK, E.W. MILLER. In the general case it is straightforward to prove that $\dim D(P) = S^0 P$. Now because the elements of $D(P)$ are just the completely join and meet irreducibles of $I(P)$, then by a classical result [3], [47] the normal completion (MACNEILLE [38]) $N\, D(P)$ is isomorphic to $I(P)$. Now it is well known that normal completion preserves the dimension [2] so: $\dim I(P) = \dim D(P) = S^0 P$.

## I-3. Algebraic Dimension and Order Dimension of Lattices.

The duality between ordered sets and lattices of initial segments obtained in Proposition I.2.4 gives the following:

I.3.1. *An ordered set $P$ is the surjective image of a direct sum $\coprod_{i \in I} C_i$ of a family $(C_i)_{i \in I}$ of chains iff $I(P)$ can be embedded as a complete sublattice into the direct product $\prod_{i \in I} I(C_i)$.*

Proof. In the category of ordered sets the epimorphisms are the surjective and increasing maps. By duality they are transformed into monomorphisms of complete lattices. So any surjective and increasing map $f: \coprod_{i \in I} C_i \rightarrow P$ gives a complete lattice homomorphism (injective) $If: I(P) \rightarrow I(\coprod_{i \in I} C_i)$.

By duality again we get $I(\coprod_{i \in I} C_i) \cong \prod_{i \in I} I(C_i)$ and the result follows.

This, added to the Theorem I.2.10, gives:

I.3.2. THEOREM. *The smallest cardinal number $\kappa$ such that $I(P)$ can be order embedded into a product of $\kappa$ chains and the smallest cardinal number $\kappa'$ such that $I(P)$ can be embedded, as a complete sublattice, into a product of $\kappa'$ chains are both equal to $S^0(P)$.*

This fact suggests we consider other notions of dimension than the order dimension. So, for a distributive lattice $T$, we define the (algebraic) *lattice dimension* of $T$, $\dim_{DL} T$, as the smallest $\kappa$ such that $T$ can be embedded, as a sublattice, into a product of $\kappa$ chains. Because any distributive lattice $\underset{\sim}{T}$ is a sublattice of a field of sets this concept is well defined. In the same way, for a complete, completely distributive, lattice $\underset{\sim}{T}$ we define the *complete lattice dimension* of $\underset{\sim}{T}$, $\dim_{CDL} \underset{\sim}{T}$. The fact that this concept is well defined is a non trivial result due to G.N. RANEY [46]. A natural question is whether these notions of dimension coincide with the usual order dimension. Of course, they do for $\underset{\sim}{I}(\underset{\sim}{P})$, but in the general case this question is unsolved. We can only note that there is a characterization of lattice dimension similar to BOUCHET's theorem and also that distributive lattices with finite order dimension have the same lattice dimension -- it is in fact the width of their spectrum.

## II - EXTENDABLE AND UNIVERSALLY EXTENDABLE TYPES

### II.1 Extendable Types.

One of the simplest examples of an order property which may be preserved in a linear extension is the "descending chain condition". A partially ordered set $P$ has the descending chain property if there is no infinite strictly decreasing chain, or equivalently, assuming the axiom of choice, every non-empty subset $A$ of $P$ has a minimal element. Thus: *any partially ordered set* $\underset{\sim}{P}$ *satisfying the descending chain condition can be extended to a chain with the same property, that is to say, a well-ordered chain.*

A simple proof is as follows: Let $P_0$ be the set of minimal elements of $\underset{\sim}{P}$. More generally, if $P_\mu$ has been defined for every ordinal $\mu < \delta$, let $P_\delta$ be the set of minimal elements of $P \setminus \bigcup\{P_\mu : \mu < \delta\}$. For some ordinal $\delta_P$ this process terminates with the empty set. Each $P_\mu$ is an antichain of $\underset{\sim}{P}$ and so has a well-ordered linear extension $\underset{\sim}{L}(P_\mu)$. Now the lexicographical sum $\sum_{\mu<\delta_P} \underset{\sim}{L}(P_\mu)$ is a well-ordered linear extension of $\underset{\sim}{P}$. (Another proof of this is given in II-2).

The result may be restated by saying that every partially ordered set $P$ which does not embed $\omega^*$, the order type of the negative integers, extends to a chain with the same property. With this formulation it is natural to ask what types $\alpha$, apart from $\omega^*$, have a similar property. We call such types *extendable types*, and the purpose of this section is to give a detailed proof of the characterization of the countable extendable types obtained by R. BONNET [7].

This problem was studied by several people in France in the late 60's including E. Corominas, R. Fraïssé, P. Jullien and the present authors. Also at about the same time, and quite independently F. Galvin, A. Kostinsky and R. McKenzie worked on the problem. Although their results were not published there was considerable overlap between their results and ours which we shall mention later. They also obtained some results for non-denumerable types but we learned of these too late to include them in this survey, but a brief account will be given in section II-2.

First some definitions and notation. If $\underset{\sim}{P}$ and $\underset{\sim}{Q}$ are two ordered sets we write $P \leq Q$ if $P$ is isomorphic to a subset of $Q$ with the induced order, and the negation of this is written $P \not\leq \underset{\sim}{Q}$. The notation $P \equiv Q$ means that $P \leq Q$ and $Q \leq P$, and $P < Q$ means $P \leq Q$ and $Q \not\leq P$. In the sequel *order type* always refers to the order type of a chain, and we denote these by greek letters $\alpha$, $\alpha'$, $\beta$, $\beta'$, etc. We use the standard notation for the familiar order types, thus $\omega$ is the order type of the chain of the natural numbers ordered by magnitude and $\eta$ is the order type of the rational chain. The partially ordered set $\underset{\sim}{P}$ is said to be $\alpha$-*ordered* if $\alpha \not\leq \underset{\sim}{P}$, i.e. $\underset{\sim}{P}$ does not contain a chain of type $\alpha$. $\underset{\sim}{P}$ is *well-founded* if it is $\omega^*$-ordered and it is *scattered* if it is $\eta$-ordered. In this context, an order type $\alpha$ is *extendable* if every $\alpha$-ordered partially ordered set $\underset{\sim}{P}$ can be extended to a chain which is also $\alpha$-ordered. It is easy to see that if $\alpha$ is extendable and $\beta \equiv \alpha$, then $\beta$ and $\beta^*$ are also extendable.

II-1.1. A Description of the Countable Extendable Types.

As we already observed $\omega$ and $\omega^*$ are extendable. The next important case is $\eta$, the order type of the rationals. This was proved by the authors [9] and independently by F. GALVIN and R. McKENZIE.

II-1.1.1 PROPOSITION. $\eta$ *is an extendable type, in other words any scattered partial order has a scattered linear extension.*

Proof. We shall employ the useful notion of a *kernel* due to E. Corominas.

A partially ordered set $\underset{\sim}{P}$ is called an $\eta$-*kernel* if no non-empty initial or final segment of $\underset{\sim}{P}$ (equipped with the induced order) can be extended to a scattered chain, i.e. has no $\eta$-ordered linear extension.

Claim 1. *If $\underset{\sim}{P}$ has no scattered extension, then it contains an $\eta$-kernel.*

Indeed, let $G(\underset{\sim}{P})$ be the set of initial segments $I$ of $\underset{\sim}{P}$ which do have a scattered linear extension, $\underset{\sim}{L}(I)$. Clearly, if $A \subseteq I \in G(\underset{\sim}{P})$, then $A$ also has a scattered linear extension $\underset{\sim}{L}(A)$. Also, if $I \in G(\underset{\sim}{P})$ and $J \in G(\underset{\sim}{P}\setminus I)$, then $I \cup J$ is an initial segment of $\underset{\sim}{P}$ and the linear sum $\underset{\sim}{L}(I) + \underset{\sim}{L}(J)$ is a scattered extension of $I \cup J$. Thus $G(\underset{\sim}{P})$ is closed under finite unions. Further, $G(\underset{\sim}{P})$ is closed under the union of chains, that is to say, if $(I_\mu)_{\mu<\delta}$ is a well-ordered increasing sequence of elements of $G(\underset{\sim}{P})$, then $I = \cup\{I_\mu : \mu < \delta\} \in G(\underset{\sim}{P})$. For put $J_\mu = I_\mu \cup\{I_\rho : \rho < \mu\}$ and let $\underset{\sim}{L}(J_\mu)$ be a scattered linear extension of $\underset{\sim}{J_\mu}$ (this exists since $I_\mu \in G(\underset{\sim}{P})$. Then the lexicographic sum $\sum_{\mu<\delta} \underset{\sim}{L}(J_\mu)$ is a scattered chain extending the induced ordered set $I$.

From the above it follows that $G(\underset{\sim}{P})$ is closed under arbitrary unions and so has a largest member $\overset{o}{I}$ $(=\cup G(\underset{\sim}{P}))$. Similarly, there is a largest final segment $\overset{o}{F}$ of $\underset{\sim}{P}$ which has a scattered linear extension. If $N = P\setminus(\overset{o}{I}\cup\overset{o}{F})$, then the induced partially ordered set $\underset{\sim}{N}$ is an $\eta$-kernel.

Claim 2. *If $\underset{\sim}{P}$ contains an $\eta$-kernel, then it is not $\eta$-ordered.*

Let $N_{1/2}$ be an $\eta$-kernel of $\underset{\sim}{P} = P_{1/2}$ and choose $x_{1/2} \in N_{1/2}$. Define $P_{1/4} = \{x \in N_{1/2} : x \leq x_{1/2}\}$, $P_{3/4} = \{x \in N_{1/2} : x \geq x_{1/2}\}$.

Since $N_{1/2}$ is an $\eta$-kernel, $\underset{\sim}{P}_{1/4}$ and $\underset{\sim}{P}_{3/4}$ have no scattered linear extensions. Hence they respectively contain (non-empty) $\eta$-kernels $N_{1/4}$ and $N_{3/4}$, from which we choose elements $x_{1/4}$, $x_{3/4}$. Continuing this process (at the next stage choose elements $x_{1/8}$, $x_{3/8}$, $x_{5/8}$ and $x_{7/8}$), we define a sequence $(x_d)_{d \in D}$ of elements of $\underset{\sim}{P}$ indexed by the set of dyadic fractions, $D$. These elements are ordered in the same way as the dyadic fractions and so $\underset{\sim}{P} \geq \eta$, i.e. $\underset{\sim}{P}$ is not $\eta$-ordered.

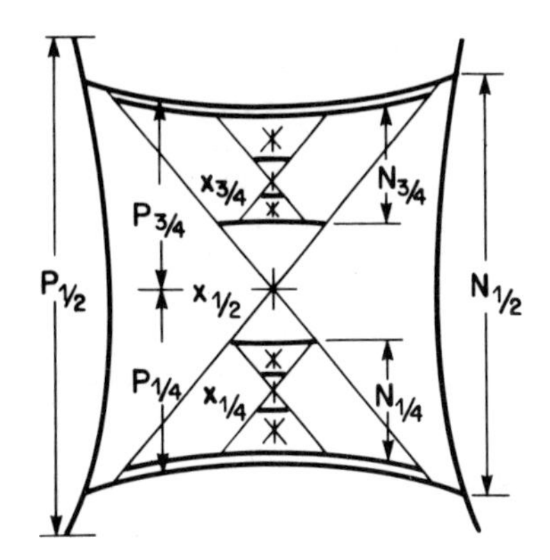


The next two results were also obtained by P. Jullien [33], F. Galvin and R. McKenzie.

II-1.1.2. *The order type of the integers $\omega + \omega^*$ is extendable. More generally, if $\alpha+1$ and $1+\beta$ are extendable, then so also is $\alpha+1+\beta$.*

Proof. Let $\underset{\sim}{P}$ be a $\gamma$-ordered set, where $\gamma = \alpha+1+\beta$. Define $I = \{x \in P \mid \alpha+1 \not\leq \underset{\sim}{P}(x)\}$, $F = \{x \in P \mid 1+\beta \not\leq \underset{\sim}{P}^*(x)\}$, where $\underset{\sim}{P}(x) = \{y \in P \mid y \leq x\}$ and $\underset{\sim}{P}^*(x) = \{y \in P \mid y \geq x\}$. Then $I \cup F = P$. Clearly $\alpha+1 \not\leq \underset{\sim}{I}$, $1+\beta \not\leq \underset{\sim}{F}$ and so there are linear extensions $\underset{\sim}{L}(I)$, $\underset{\sim}{L}(F)$ such that $\alpha+1 \not\leq \underset{\sim}{L}(I)$, $1+\beta \not\leq \underset{\sim}{L}(F)$. The lexicographic sum $\underset{\sim}{L}(I) + \underset{\sim}{L}(P\backslash I)$ is a linear extension of $\underset{\sim}{P}$ which is $\gamma$-ordered (where could the image of the middle element of $\gamma$ be?)

II-1.1.3. *Let $(\alpha_n)_{n<\omega}$ be an increasing sequence of order types and suppose that $\alpha_n+1$ is extendable $(n < \omega)$. Then $\alpha = \sum_{n<\omega} \alpha_n$ is also extendable.* (Thus $\omega^*\omega = (\omega^*+1)\omega$ is extendable).

Proof. Let $\underset{\sim}{P}$ be an $\alpha$-ordered set. Let $G(\underset{\sim}{P})$ be the set of all final segments $F$ of $\underset{\sim}{P}$ such that $\underset{\sim}{F}$ has an $\alpha$-ordered linear extension. As in the proof of II-1.1.1, we see that $G(\underset{\sim}{P})$ is closed under arbitrary unions (since $\alpha \equiv \sum_{n \geq m} \alpha_n$ for any $m < \omega$), and hence $\overset{\circ}{F} = \bigcup G(\underset{\sim}{P}) \in G(\underset{\sim}{P})$. Hence, if $F'$ is any non-empty final segment of $\underset{\sim}{P}'$, where $P' = P \setminus \overset{\circ}{F}$, then $\underset{\sim}{F}'$ does not have an $\alpha$-ordered linear extension. It follows from this that if $P'$ is non-empty then there is an increasing sequence of elements $(x_n)_{n<\omega}$ in $\underset{\sim}{P}'$ so that each interval $[x_n, x_{n+1}] = \{y \in P' : x_n \leq y \leq x_{n+1}\}$ contains a chain isomorphic to $\alpha_n + 1$. Indeed, let $x_0 \in P'$ be arbitrary, and let $P'^*(x_0) = \{y \in P' : y \geq x_0\}$. Then $\underset{\sim}{P}'^*(x_0)$ is not $(\alpha_0+1)$-ordered (any linear extension of $\underset{\sim}{P}'^*(x_0)$ contains a subchain of type $\alpha$ and hence a subchain of type $\alpha_0+1$). So there is $x_1 \in P'^*(x_0)$ so that $[x_0, x_1] \geq \alpha_0+1$. Similarly, since $\alpha_1+1$ is extendable, there is $x_2 \in P'^*(x_1)$ so that $[x_1, x_2] \geq \alpha_1+1$. This process may be continued and leads to the contradiction that $\underset{\sim}{P} \geq \alpha$. Hence $P'$ is empty and $P \in G(\underset{\sim}{P})$.

II-1.1.4. It follows from the above that $\pi_1 = \omega$, $\pi_2 = \omega^*\omega$, $\pi_3 = \omega\omega^*\omega$ are extendable. More generally, define $\pi_{\alpha+1} = \pi^*_\alpha \omega$, and $\pi_\alpha = \sum_{\beta<\alpha} \pi^*_\beta$ for limit $\alpha$. Then by the above we see that $\pi_\alpha$ is extendable for $\alpha < \omega_1$ (since $\pi_{\alpha+1} \equiv (\pi^*_\alpha+1)\omega$ and, for limit $\alpha$, $\pi_\alpha \equiv \sum_{n<\omega} (\pi^*_{\alpha_n}+1)$ for any cofinal sequence $(\alpha_n)_{n<\omega}$ of $\alpha$). Also from II-1.1.2 it follows that $\pi^*_\alpha + \pi_\alpha$ is extendable for $\alpha < \omega_1$.

We shall prove that the countable extendable order types are essentially those we have already described. In other words we shall prove the following theorem of R. BONNET [ 7 ].

II-1.1.5 THEOREM. *The countable order type* $\alpha$ *is extendable iff* $\alpha \equiv \eta$ *or* $\alpha \equiv \pi_\nu$ *or* $\alpha \equiv \pi^*_\nu$ *or* $\alpha \equiv \pi^*_\nu + \pi_\nu$ *for some* $\nu < \omega_1$.

II-1.2. Non-Extendable Types.

The main idea is contained in the following result of P. JULLIEN [33]:

II-1.2.1 PROPOSITION. $\omega+1$ *is not extendable. More generally, if* $\alpha+1 \not\equiv \alpha$, *then* $\alpha+1$ *is not extendable.*

Proof. We illustrate the proof for the case $\alpha = \omega$. Let $Y$ be a set of cardinality $\aleph_1$ and let $\Phi$ be the set of all injective maps $f : \omega \to Y$. Consider the partially ordered set $P = Y \cup (\Phi \times \omega)$ ordered in the following way: $u \leq v$ in $\underset{\sim}{P}$ iff EITHER (i) $u, v \in Y$ and $u = v$, OR (ii) $u, v \in \Phi \times \omega$, $u = (f, m)$, $v = (f', m')$, $f = f'$ and

$m \le m'$, OR (iii) $u \in \Phi\times\omega$, $v \in Y$, $u = (f,m)$ and $v = f(m')$ for some $m' \ge m$. It is easy to see that $\omega+1 \not\le P$.

Now let $\preccurlyeq$ be any linear extension of $\le$. Since $Y$ is an uncountable chain in this linear ordering, there is some $y \in Y$ such that $T = \{t \in Y : t \preccurlyeq y\}$ is infinite. Let $F$ be any injective map $f:\omega \to T$. Since $(f,n) \le f(n) \preccurlyeq y$, it follows that $C = \{(f,n) : n < \omega\} \cup \{y\}$ is a chain of type $\omega+1$ in the ordering $\preccurlyeq$.

Remark. The ordered set $P$ constructed above has cardinality $2^{\aleph_0}$. F. GALVIN constructed an example of cardinality $\aleph_1$. However (as observed by COROMINAS, JULLIEN, GALVIN and McKENZIE) there is no countable example. This leads to the notion of a *weakly extendable type* $\alpha$ (i.e. any $\alpha$-ordered partially ordered set of cardinality $|\alpha|$ has an $\alpha$-ordered linear extension). For a characterization of the countable weakly extendable types see [33].

In order to show that extendable types have a very special form, we need some important facts about scattered types. Recall that an order type $\alpha$ is *indecomposable* iff $\alpha = \alpha_1+\alpha_2$ implies $\alpha \le \alpha_1$ or $\alpha \le \alpha_2$. It is *right-indecomposable* if $\alpha = \alpha_1+\alpha_2$ and $\alpha_2 \ne 0$ implies $\alpha \le \alpha_2$ and $\alpha \not\le \alpha_1$, similarly for left indecomposability. These notions are clearly preserved by $\equiv$. Note that an indecomposable scattered type $\alpha \ne 1$ cannot be both left- and right-indecomposable (otherwise $\alpha+1+\alpha \le \alpha$ and hence $\eta \le \alpha$).

II-1.2.2. PROPOSITION. *If $\alpha$ is an extendable scattered type (of any cardinality) then either $\alpha$ is indecomposable or $\alpha = \alpha_1+\alpha_2$, where $\alpha_1$ and $\alpha_2$ are infinite extendable types, $\alpha_1$ being left-indecomposable and $\alpha_2$ being right-indecomposable.*

In order to prove this we need the following two lemmas and the important theorem of R. LAVER [36]. If the type $\alpha$ can be written as a finite sum of indecomposable types $\alpha = \alpha_1+\alpha_2+\ldots+\alpha_n$, we call this a canonical form for $\alpha$ if $n$ is minimal. Observe that, if $\beta \equiv \alpha$, then $\beta$ also has a canonical form $\beta = \beta_1+\ldots+\beta_n$ with $\beta_i \equiv \alpha_i$ $(1 \le i \le n)$.

LEMMA 1. *If $\alpha$ is extendable and has canonical form $\alpha = \alpha_1+\ldots+\alpha_n$, then every sum of the form $\alpha_p+\alpha_{p+1}+\ldots+\alpha_q$ is extendable where $1 \le p \le q \le n$.*

Indeed, it is enough to verify that $\beta = \alpha_1+\alpha_2+\ldots+\alpha_q$ is extendable. Suppose on the contrary that $\beta$ is not extendable and that $B$ is a $\beta$-ordered set having no $\beta$-ordered linear extension. Let $C$ be a chain of type $\gamma = \alpha_{q+1}+\ldots+\alpha_n$. Then the lexicographic sum $\tilde{B} + C$ is $\alpha$-ordered and has no $\alpha$-ordered linear extension.

LEMMA 2. *Let $\alpha$, $\beta$ be two order types such that $\alpha \ne 1$ is right-indecomposable, $\beta \ne 0$ and $\alpha+\beta \not\le \beta$. Then there is an $(\alpha+\beta)$-ordered* *set which has no* $(\alpha+1+\beta)$*-ordered linear extension.*

The proof is a refinement of JULLIEN's construction to prove II-1.2.1. We replace $\omega$ by a chain of type $\alpha$. Let $Y$ be a set with cardinality $|Y| > |\alpha|$ and let $\Phi$ be the set of all injective maps $f:\alpha \to Y$ and order $Q = Y \,\dot\cup\, (\Phi\times\alpha)$ in the same manner as previously. This gives an $(\alpha+1)$-ordered set with no $(\alpha+1)$-ordered linear extension. Now for each element $y \in Y$ adjoin a chain $\underset{\sim}{C}_y$ of order type $\beta$ and extend the ordering so that $y < x$ for each $x \in C_y$ (elements of different $C_y$ are incomparable). This defines the $(\alpha+\beta)$-ordered set $\underset{\sim}{P}$ (if $\underset{\sim}{P}$ contains a chain $K$ of type $\alpha+\beta$, then $K \cap Q$ contains a chain of type $\alpha+1$). If $\underset{\sim}{L}(P)$ is a linear extension then $\underset{\sim}{L}(Q)$ contains a chain $X$ of type $\alpha+1$ with last element $x$, say. Now there is $y \in Y$ so that $x$ precedes $y$ in $\underset{\sim}{L}(Q)$ and hence $X \cup C_y$ is a chain of type $\alpha+1+\beta$ in $\underset{\sim}{L}(P)$.

In order to conclude the proof of II-1.2.2 we need the following deep result of R. LAVER [36]

THEOREM. *The class* $\mathcal{D}$ *of scattered order types is well-quasi-ordered under embeddability, i.e. if* $(\alpha_n)_{n<\omega}$ *is any* $\omega$*-sequence of scattered types then there are* $n_1 < n_2 < \ldots$ *so that* $\alpha_{n_1} \le \alpha_{n_2} \le \ldots$ .

It follows easily from this, as noted by LAVER, that *every* $\alpha \in \mathcal{D}$ *is a finite sum of indecomposable types.* (If not, consider a minimal type not having this property).

We now conclude the proof of II-1.2.2. Suppose $\alpha$ is a scattered extendable type. By the above $\alpha = \alpha_1+\alpha_2+\ldots+\alpha_n$, where $n$ is finite and the $\alpha_i$ are indecomposable. Suppose $\alpha_i$ is right-indecomposable. If $i < n$ then by Lemma 2 $\alpha_i+\alpha_{i+1}$ is not extendable and this contradicts Lemma 1. Hence $i = n$. Similarly, if $\alpha_i$ is left-indecomposable, then by duality, $i = 1$. From this we see that $n \le 2$ and the result follows.

For the remainder of this section we *assume that all types are scattered and countable.* To show that a right-indecomposable extendable type $\alpha$ is some $\pi_\nu$ we introduce the *box-product* of a sequence of ordered sets, and the notion of *weight* of a scattered type.

II-1.2.4. The Box-Product.

Let $(\underset{\sim}{P}_n)_{n<\omega}$ be a sequence of ordered sets. The box-product, $BP(\underset{\sim}{P}_n)_{n<\omega}$, of this sequence is the set of all finite sequences of the form $\langle x_0, x_1, \ldots, x_{p-1}\rangle$ with $x_i \in P_i$ $(i < p)$ ordered so that $\langle x_i\rangle_{i<p} \le \langle y_j\rangle_{j<q}$ iff $p \le q$ and $x_i \le y_i$ in $\underset{\sim}{P}_i$ for each $i < p$.

We need two fundamental properties of the box-product.

BP1: *Suppose that, for each* $n < \omega$, $\underset{\sim}{P}_n$ *has the property that*

$$\alpha \leq \underset{\sim}{P}_n \quad \textit{implies} \quad \alpha \leq \pi_\nu \quad ,$$

*then the box-product* $BP(\underset{\sim}{P}_n)_{n<\omega}$ *has the same property.*

Proof. Since the projection from $\prod_{n<q} \underset{\sim}{P}_n$ onto $\underset{\sim}{P}_n$ is monotone, we can assume without loss of generality that each $\underset{\sim}{P}_n$ is the type $\pi_\nu$. In the following we identify a chain with its order type, so that $\gamma\cdot\delta$ denotes the lexicographic product (i.e. the chain $\sum_{i\in\delta} \gamma_i$ where $\gamma_i = \gamma$ for all $i$) and $\gamma\times\delta$ denotes the ordered cartesian product of $\gamma$ and $\delta$.

We first prove

$$c(\nu,p,q): \quad \textit{if} \quad \beta \leq (\pi_\nu\cdot p)\times(\pi_\nu\cdot q), \quad \textit{then} \quad \beta \leq \pi_\nu\cdot(p\cdot q) \quad .$$

Clearly $c(0,1,1)$ is true and $c(\nu,1,1)$ implies $c(\nu,p,q)$. We will prove that $c(\mu,1,1)$ holds if $c(\nu,p,q)$ holds for all $\nu < \mu$ and $p,q < \omega$. We may write $\pi_\mu = \sum_{n<\omega} \pi^*_{\mu_n}$ and so a chain $\beta$ in $\pi_\mu\times\pi_\mu$ is a lexicographic sum of the $\beta_{m,n} = \beta \cap (\pi^*_{\mu_m}\times\pi^*_{\mu_n})$. Obviously, the set $C$ of pairs $(m,n) \in \omega\times\omega$ for which $\beta_{m,n} \neq 0$ has type at most $\omega$. Also, by the induction hypothesis $\beta_{m,n} \leq \pi^*_{\mu_p}\cdot p^2$, where $p = \max(m,n)$, and so $\beta \leq \pi_\mu$.

Let $\delta \leq \delta+1 \leq \pi_\nu \times\ldots\times\pi_\nu$ ($r$-times, $r < \omega$). So $\delta \leq (\pi^*_{\nu_1}\cdot p_1)\times\ldots\times(\pi^*_{\nu_r}\cdot p_r)$, in which $\nu_j < \nu$ and $p_j < \omega$. According to $c(\mu,p,q)$, we have $\delta \leq \pi^*_{\nu(r)}p(r)^r$, where $\nu(r) = \mathrm{Max}(\nu_1,\ldots,\nu_r)$ and $p(r) = \mathrm{Max}(p_1,\ldots,p_r)$.

Now, let $\gamma \leq BP(\underset{\sim}{P}_n, n<\omega)$. If the length of elements of $\gamma$ is bounded, then $\gamma \leq \pi_\nu\times\ldots\times\pi_\nu$ ($p$-times). So $\gamma \leq \pi_\nu$. Otherwise $\gamma = \sum_{q<\omega} \gamma_q$ where $\gamma_q \leq \gamma_q+1 \leq \pi_\nu\times\ldots\times\pi_\nu$ ($q$-times). Then $\gamma_q \leq \pi^*_{\nu(q)}\cdot l(q)$ and so $\gamma \leq \pi_\nu$.

BP2. *Let* $(\alpha_n)_{n<\omega}$ *be a sequence of order types with* $\omega$*-sum* $\alpha$. *If* $\alpha_n+1$ *is embeddable in every linear extension of* $\underset{\sim}{P}_n$, *then* $\alpha$ *is embeddable in every linear extension of* $\underset{\sim}{P} = BP(\underset{\sim}{P}_n)_{n<\omega}$.

Proof. Let $\overline{\underset{\sim}{P}}$ be any linear extension of $\underset{\sim}{P}$. The embedding $\varphi_0$ of $\underset{\sim}{P}_0$ into $\underset{\sim}{P}$ defined by $\varphi_0(x) = \langle x\rangle$ induces a linear extension $\overline{\underset{\sim}{P}}_0$ of $\underset{\sim}{P}_0$ which is not $(\alpha_0+1)$-ordered. Since $\overline{\underset{\sim}{P}}_0$ and its image $\overline{\underset{\sim}{Q}}_0$

under $\varphi_0$ are isomorphic there is a chain $B_0 = A_0 + \{a_0\}$ of type $\alpha_0+1$ in $\overline{\underset{\sim}{Q}}_0$ with final element $a_0$. Let $a_0 = \langle t_0 \rangle$. Now the embedding $\varphi_1$ of $\underset{\sim}{P}_1$ into $\underset{\sim}{P}$ defined by $\varphi(x) = \langle t_0, x\rangle$ induces a linear extension $\overline{\underset{\sim}{P}}_1$ of $\underset{\sim}{P}_1$ which is not $(\alpha_1+1)$-ordered. Thus we find a chain $B_1 = A_1 + \{a_1\}$ of type $\alpha_1+1$ in the image of $\overline{\underset{\sim}{P}}_1$ under $\varphi_1$. Since each member of $B_1$ has the form $\langle t_0, x\rangle$, $a_0 = \langle t_0\rangle$ precedes all elements of $B_1$ in the linear ordering $\overline{\underset{\sim}{P}}$. Continuing in this way we see that $\alpha \leq \overline{\underset{\sim}{P}}$.

II-1.2.5. The Weight Function.

Let $\alpha$ be a right-indecomposable countable scattered type, and let $\mu$ be the supremum of all ordinals $\nu$ satisfying $\pi_\nu \leq \alpha$ and $\pi_\nu^* \leq \alpha$. Then $\mu$ is countable. Suppose $\mu$ is a limit ordinal and that $(\mu_n)_{n<\omega}$ is an increasing sequence with limit $\mu$. Since $\pi^*_{\mu_n} \leq \pi_{\mu_n+1} \leq \pi_{\mu_{n+1}} \leq \alpha$, and since $\alpha$ is right-indecomposable, it follows that $\pi_\mu \equiv \sum_{n<\omega} \pi^*_{\mu_n} \leq \alpha$, which is a contradiction. Therefore $\mu$ is a successor ordinal.

In view of the above it follows that for any countable scattered order type $\alpha$ we can define the *weight*, $\pi(\alpha)$, of $\alpha$ to be the largest ordinal $\nu$ such that $\pi_\nu \leq \alpha$ or $\pi_\nu^* \leq \alpha$. (For example, $\pi(\alpha) = 1$ if $\alpha$ is an infinite ordinal). It is easy to see that if $\alpha = \alpha_1+\ldots+\alpha_n$ is scattered, then $\pi(\alpha) = \max(\pi(\alpha_1), \pi(\alpha_2), \ldots, \pi(\alpha_n))$. Also, for a right-indecomposable type $\alpha$, we have $\pi(\alpha) = \nu$ iff $\pi_\nu \leq \alpha$, $\pi_\nu^* \not\leq \alpha$ and $\pi_{\nu+1} \not\leq \alpha$.

II-1.2.6. *If $\alpha$ is right-indecomposable and $\pi(\alpha) = \nu$, then there is a partially ordered set $\underset{\sim}{P}$ such that*

(i) $\gamma \leq \underset{\sim}{P}$ *implies* $\gamma \leq \pi_\nu$,

(ii) $\alpha+1 \leq \overline{\underset{\sim}{P}}$ *for every linear extension* $\overline{\underset{\sim}{P}}$ *of* $\underset{\sim}{P}$.

We call such a $P$ satisfying (i) and (ii) an $\alpha$-example. Thus, if $\alpha$ is an ordinal $\pi(\alpha) = 1$ and (i) asserts that any subchain of $\underset{\sim}{P}$ has type $\leq \omega$ (II-1.2.1 gives an $\omega$-example). In the general case the subchains of $\underset{\sim}{P}$ have type $\leq \pi_\nu$ and $\pi_\nu$ is attained (since $\pi_\nu$ is extendable and $\pi_\nu \leq \alpha < \alpha+1$).

Proof. We can assume that $\alpha > \omega$ and that the result (and its dual) is true for every $\beta$ with $\pi(\beta) < \pi(\alpha)$.

Since $\alpha$ is scattered $\alpha+\alpha \not\leq \alpha$ and so, by Laver's theorem, $\alpha \equiv \sum_{n<\omega} \alpha_n$ where $(\alpha_n)_{n<\omega}$ is a sequence of infinite indecomposable types such that $\alpha_n < \alpha$ for every $n$.

First Step. For $n < \omega$, let $\pi(\alpha_n) = \nu_n$. Then $\nu_n \leq \nu$. Also, if $\alpha_n$ is left indecomposable, then $\nu_n < \nu$ (since $\pi^*_{\nu_n} \leq \alpha_n < \alpha$). For each $n < \omega$, let $\underset{\sim}{P}_n$ be an $\alpha_n$-example. Thus EITHER (a) $\alpha_n$ is right-indecomposable and $\pi_{\nu_n} \leq \alpha_n$, $\delta \leq \pi_{\nu_n} \leq \pi_\nu$ for every $\delta \leq \underset{\sim}{P}_n$, and $\alpha_n+1 \leq \overline{\underset{\sim}{P}}_n$ for any linear extension of $\underset{\sim}{P}_n$; OR (b) $\alpha_n$ is left-indecomposable and $\pi^*_{\nu_n} \leq \alpha_n$, $\delta \leq \pi^*_{\nu_n} \leq \pi_\nu$ for every $\delta \leq \underset{\sim}{P}_n$ and $\alpha_n+1 \leq \alpha_n \leq 1+\alpha_n \leq \overline{\underset{\sim}{P}}_n$ for any linear extension $\overline{\underset{\sim}{P}}_n$.

Now let $\underset{\sim}{P} = BP(\underset{\sim}{P}_n)_{n<\omega}$ be the box product of the $\underset{\sim}{P}_n$. Since the chains of any $\underset{\sim}{P}_n$ are embeddable in $\pi_\nu$, by BP1 the chains of $\underset{\sim}{P}$ are also embeddable in $\pi_\nu$. Further, since $\alpha_n+1 \leq \overline{\underset{\sim}{P}}_n$ for any linear extension $\overline{\underset{\sim}{P}}_n$ of $\underset{\sim}{P}_n$, it follows by BP2 that $\alpha \equiv \sum \alpha_n$ is embeddable in every linear extension $\overline{\underset{\sim}{P}}$ of $\underset{\sim}{P}$.

Second Step. If $\alpha \equiv \pi_\nu$, then Jullien's construction given in II-1.2.1 (replacing $\omega$ by $\pi_\nu$) gives an $\alpha$-example. More generally, suppose $\pi_\nu < \alpha$. Then we replace $\omega$ by the partially ordered set $\underset{\sim}{P}$ constructed above in the Jullien construction to obtain an $\alpha$-example.

Conclusion of Proof of II-1.1.5.

If $\alpha$ is right-indecomposable scattered, countable type and $\alpha < \pi_\nu$, where $\nu = \pi(\alpha)$, then the $\alpha$-example constructed in II-1.2.6 shows that $\alpha$ is not extendable. Hence, if $\alpha$ is scattered, right-indecomposable and extendable, then $\alpha \equiv \pi_\nu$ $(\nu = \pi(\alpha))$. Now II-1.1.5 follows from II-1.2.2 and the dual of the above remark.

II-2. Uncountable Extendable Types.

It was first observed by A. KOSTINSKY (1966, unpublished [25]) that: *if $\kappa$ is a regular cardinal, then $\kappa$ and $\kappa^*$ are extendable.* This can be proved using the following *iterative extension* constructions. Let $\underset{\sim}{P} = \langle P,P\rangle$ be an ordered set and let $(x_\alpha)_{\alpha<\lambda}$ be any enumeration of the elements of $P$. Construct a sequence of extensions $P_\alpha$ $(\alpha \leq \lambda)$ of $P$ as follows. Put $P_0 = P$. Let $\alpha \leq \lambda$ and suppose $P_\beta$ has been defined for $\beta < \alpha$. If $\alpha$ is a limit put $P_\alpha = \bigcup\{P_\beta : \beta < \alpha\}$. If $\alpha = \beta + 1$ is a successor, define $P_\alpha = \langle P,P_\alpha\rangle$ to be the lexicographic sum $\underset{\sim}{I}_\beta + \underset{\sim}{F}_\beta$, which are suborders of $P_\beta$ induced by the sets $I_\beta = \{x \in P : x \leq x_\beta \text{ in } P_\beta\}$ and $F_\beta = P\backslash I_\beta$. We leave the reader to verify that, if $\underset{\sim}{P}$ is a $\kappa^*$-type, then so also is the linear extension $\underset{\sim}{P}_\lambda$. (HINT: if $x \leq y$ in $\underset{\sim}{P}_\lambda$ and $\alpha$ is the smallest index such that $x \leq x_\alpha \leq y$ in $\underset{\sim}{P}_\lambda$, then $x \leq x_\alpha$ in $\underset{\sim}{P}$.)

F. GALVIN and R. MCKENZIE (between 1966-69, unpublished) extended Kostinsky's result and obtained a complete characterization of extendable ordinals $\alpha$ (and their duals $\alpha^*$). We do not give the proof.

THEOREM. *The class of extendable ordinals is the smallest class of ordinals $U$ such that*

*(i)* $\omega \in U$,

*(ii) if* $\alpha \in U$, *then* $\alpha \cdot (\mathrm{cf}(\alpha))^+ \in U$,

*(iii)* $U$ *is closed, i.e. if* $\alpha_\xi \in U$ *for* $\xi < \theta$ *then* $\sup_{\xi<\theta} \alpha_\xi \in U$.

Thus, for example, we see that $\omega, \omega_1, \omega_2, \ldots, \omega_\omega, \omega_{\omega_1}, \omega_{\omega_2}, \ldots, \omega^2_\omega, \ldots, \omega_{\omega_1}, \ldots$ are extendable types.

II-3. Universally Extendable Types.

If $\underset{\sim}{P}$ is an $\omega^*$-type, i.e. $\omega^* \not\leq \underset{\sim}{P}$, then, as we already remarked, $\underset{\sim}{P}$ has some linear extension which is a well order. Of course, we cannot expect that *every* linear extension of $\underset{\sim}{P}$ should be well ordered. For example, if $\underset{\sim}{P} = \aleph_0$ is an antichain of size $\aleph_0$ it can be ordered as an $\omega^*$-chain. However, if we exclude this trivial case and consider partial orders having only finite antichains ($\aleph_0 \not\leq \underset{\sim}{P}$) then it is true that every linear extension is well ordered. This is a well known result of G. HIGMAN [28] (and frequently rediscovered, e.g. WOLK [52]).

In fact we have the following result from [28]:

THEOREM. *If $\underset{\sim}{P}$ is a partial order, then the following statements are equivalent:*

*(i)* $\omega^* \not\leq \underset{\sim}{P}$, *and* $\aleph_0 \not\leq \underset{\sim}{P}$;

*(ii) for every $\omega$-sequence $(x_n)_{n<\omega}$ in $P$, there are $i, j$ such that $i < j < \omega$ and $x_i \leq x_j$;*

*(iii)* $\omega^* \not\leq \overline{\underset{\sim}{P}}$ *for every linear extension* $\overline{\underset{\sim}{P}}$ *of* $\underset{\sim}{P}$;

*(iv)* $\omega^* \not\leq \underset{\sim}{I}(\underset{\sim}{P})$.

In the present context it is natural to ask what other order types enjoy a similar property, and we call an order type $\alpha$ *universally extendable* if whenever $\underset{\sim}{P}$ is a partially ordered set such that $\aleph_0 \not\leq \underset{\sim}{P}$ and $\alpha \not\leq \underset{\sim}{P}$, then $\alpha \not\leq \overline{\underset{\sim}{P}}$ for every linear extension $\overline{\underset{\sim}{P}}$ of $\underset{\sim}{P}$.

The countable universally extendable types are described in the following result of BONNET [7]:

THEOREM. *$\alpha$ is a universally extendable type iff $\alpha$ is $\eta$ or $\pi_\nu$ or $\pi_\nu^*$ (for some $\nu < \omega_1$).*

It is interesting that all the equivalences in the above theorem of HIGMAN also carry over to arbitrary countable universally extendable types.

THEOREM. *Let $\alpha$ be a countable universally extendable type and let $\underset{\sim}{P}$ be any partially ordered set. Then the following statements are equivalent:*

*(i)* $\alpha \not\leq \underset{\sim}{P}$, $\aleph_0 \not\leq \underset{\sim}{P}$;

*(ii)* *If $f:\alpha \to P$, then there are $a_1, a_2 \in \alpha$ such that $a_1 < a_2$ (in $\alpha$) and $f(a_2) \leq f(a_1)$ (in $\underset{\sim}{P}$);*

*(iii)* $\alpha \not\leq \overline{\underset{\sim}{P}}$ *for every linear extension* $\overline{\underset{\sim}{P}}$ *of* $\underset{\sim}{P}$;

*(iv)* $\alpha \not\leq \underset{\sim}{I}(\underset{\sim}{P})$.

Proof. Using I-2.5 and the fact that $2\alpha \leq \alpha$, we see that (iii) and (iv) are equivalent. Clearly (i) and (iii) are equivalent, and so also are (ii) and (iii).

## III. COMPLETE LINEAR EXTENSIONS OF AN ORDERED SET

### III-1. Introduction

Another property for which we can ask whether it can be preserved by linear extensions is completeness. Thus the first question is: *does every complete lattice have a complete linear extension?*

This question is due to I. RIVAL and was studied by M. POUZET and I. RIVAL [44]. They show that the answer is positive for a rather wide class of lattices including the class of complete, completely distributive lattices but that it is, in general, false. Here we survey their results and give a proof of one of their main results.

THEOREM. 1. *Every countable complete lattice has a complete linear extension, and the order is the intersection of its complete linear extensions.*

2. *For every cardinal* $\kappa$, $\kappa > \omega_1$ *there is a complete, distributive lattice which has no complete linear extension.*

The answer for lattices with cardinality $\omega_1$ is unknown. It seems a plausible conjecture that there are complete lattices of cardinality $2^{\aleph_0}$ with no complete linear extension.

### III-2. Example

The direct products of complete chains provide the simplest examples of complete lattices. The complete linear extensions are easy to manufacture: if $\underset{\sim}{P}$ is the direct product $\prod_{i\in I} \underset{\sim}{C_i}$ of a family $(\underset{\sim}{C_i})_{i\in I}$ of complete chains, then for every well ordering $J$ on $I$, the lexicographical product of this family gives a complete linear extension. Of course, the intersection of the complete linear extensions obtained for all the possible $J$ gives the order of the direct product. Such complete lattices are completely distributive. Actually *every completely distributive complete lattice has a complete linear extension. Moreover, the order is the intersection of its complete linear extensions.* The proof involves three steps.

(i) Every completely distributive, complete lattice is a complete sublattice of the direct product $\underset{\sim}{P} = \prod_{i\in I} \underset{\sim}{C_i}$ of a family $(\underset{\sim}{C_i}, i \in I)$ of complete chains. (This is a well known, and non-trivial, result of G.N. RANEY [46]).

(ii) Every complete sublattice of the direct product $\underset{\sim}{P} = \prod_{i \in I} \underset{\sim}{C_i}$ is closed in the product topology of $\underset{\sim}{P}$ (where each $\underset{\sim}{C_i}$ is equipped with the interval topology).

For this observe that the topology can be described abstractly in terms of the lattice operations: a generalized sequence $(x_\alpha)$ of elements of $\underset{\sim}{P}$ converges to $x \in \underset{\sim}{P}$ (in the product topology) if and only if $x = \lim\sup(x_\alpha) = \lim\inf(x_\alpha)$, where

$$\lim\sup(x_\alpha) = \inf_\alpha\{\sup x_\beta \mid \beta \geq \alpha\}$$

and the $\lim\inf(x_\alpha)$ is defined dually.

(iii) Let $A$ be a subset of $\underset{\sim}{P} = \prod_{i \in I} \underset{\sim}{C_i}$. If $A$ is closed (in the product topology of $\underset{\sim}{P}$) then for any well ordering $I$ on $I$, the ordinal product of the $\underset{\sim}{C_i}$ induces a complete linear extension of $A$.

This follows from the compactness property of $\underset{\sim}{P}$. For details see [46].

III-3. Chain-Complete and Locally Chain-Complete Ordered Sets.

The construction we need now to get complete linear extensions of some particular lattices destroys the lattice structure, but preserves some property of completeness. Thus, even to study lattices, it is more convenient to introduce the following definitions. We call an ordered set $\underset{\sim}{P}$ *chain-complete* if every maximal chain is complete (if $\underset{\sim}{P}$ is a lattice this means $P$ is complete). We call a chain *locally complete* if any bounded subset has a supremum and an infimum, and we call an ordered set *locally-chain-complete* if every maximal chain is locally complete (however an ordered set $\underset{\sim}{P}$ is locally-chain-complete iff the ordered set $\underset{\sim}{P}' = \{0\} + \underset{\sim}{P} + \{1\}$, obtained by addition of one smallest and one greatest element, is chain-complete).

The problem of constructing complete linear extensions for complete lattices is a special case of the problem of constructing (locally) complete linear extensions for (locally) chain-complete ordered sets. Of course these problems for chain-complete ordered sets and for locally-chain-complete ordered sets are equivalent. Still, it is not at all clear that this is equivalent to the problem of constructing complete linear extensions for lattices. For instance does a chain-complete ordered set have a complete linear extension if its normal completion has a complete linear extension?

There is another reason to consider chain-complete (or equivalently locally chain-complete) ordered sets instead of complete lattices. Let $\underset{\sim}{P}$ be *a chain-complete ordered set. Then for any pair $x,y$ of noncomparable elements the minimal order* $P'$ *which contains* $P \cup \{(x,y)\}$ *gives a chain-complete ordered set. Thus if any chain-complete extension of* $P$ *has a complete linear extension, then* $P$ *is the intersection of its complete linear extensions.* It is just the situation we have for the ordered sets we consider. But for lattices, such an argument does not work, so we do not know *whether a complete lattice having at least one complete linear extension is the intersection of its complete linear extensions.*

III-4. Direct Sums and Direct Products.

Direct sums and direct products of (locally) complete chains provide the simplest examples of (locally) chain complete ordered sets.

Every disjoint sum of a family of complete chains has a complete linear extension (choose a complete linear ordering on the index set and take the lexicographical sum of the components). The corresponding result holds for locally complete chains but needs some care. Let $\underset{\sim}{P}$ be the disjoint sum of a well ordered family $(\underset{\sim}{C_i})_{i \in \underset{\sim}{I}}$ of locally complete chains. Associate with each $\underset{\sim}{C_i}$ a triple $(X_i, Y_i, Z_i)$ of subsets of $\underset{\sim}{C_i}$ as follows: if $\sup \underset{\sim}{C_i} \in \underset{\sim}{C_i}$ set $X_i = \underset{\sim}{C_i}$, $Y_i = \emptyset = Z_i$; if $\sup \underset{\sim}{C_i} \notin \underset{\sim}{C_i}$ but $\inf \underset{\sim}{C_i} \in \underset{\sim}{C_i}$ set $X_i = \emptyset = Z_i$, $Y_i = \underset{\sim}{C_i}$; otherwise, consider a subsequence $(z_{i_n})$ of $\underset{\sim}{C_i}$ and set $X_i = \leftarrow \inf(z_{i_n})]$, $Y_i = [\sup(z_{i_n}) \rightarrow$, and $Z_i = \underset{\sim}{C_i} - \bigcup_{n \in N} [z_{i_n}, z_{i_{n+1}}[$, where $\leftarrow a]$, $[b \rightarrow$, and $[a,b[$ have the obvious meaning. Finally, the ordering induced by the partition

$$\sum_{i \in \underset{\sim}{I}^*} X_i + \left( \sum_{n \in N} \sum_{i \in \underset{\sim}{I}} [z_{i_n}, z_{i_{n+1}}[ \right) + \sum_{i \in \underset{\sim}{I}} Y_i \quad ,$$

where $\underset{\sim}{I}^*$ is the dual of $\underset{\sim}{I}$, determines a complete linear extension of $P$. From this follows that a *disjoint sum of ordered sets, each of which has a locally complete linear extension, does itself have a locally complete linear extension.*

For direct product the situation is completely different: as we have already seen, every direct product of complete chains has a complete linear extension *but the corresponding result for locally complete chains is false.*

III.4.1. PROPOSITION. *Let* $\kappa$ *and* $\lambda$ *be non-isomorphic, uncountable, regular ordinals. Then* $\kappa^*\times\lambda$ *is locally chain-complete and has no locally complete linear extension. Moreover, the ordered set obtained from* $\kappa^*\times\lambda$ *by adjoining a least element and a greatest element is a complete distributive lattice and has no complete linear extension.*

III.4.2. COROLLARY. *For any cardinal* $\kappa$, $\kappa \geq \omega_2$ *there is a complete distributive lattice without any complete linear extension.*

Proof. Add a least element to $\omega_1^*\times\omega_2$ and above add a complete chain of cardinality $\kappa$. A complete linear extension would give a locally complete extension of $\omega_1^*\times\omega_2$.

Sketch of Proof of Proposition III.4.1.

Let $\underset{\sim}{P} = \kappa^*\times\lambda$ and let $\underset{\sim}{L}$ be a linear extension of $\underset{\sim}{P}$. To prove that $\underset{\sim}{L}$ is not locally complete we will construct a strictly increasing sequence $(I_n)_n$ of initial segments of $\underset{\sim}{L}$ such that $\underset{\sim}{L}-\bigcup_n I_n$ has no minimal element. For this purpose we call an initial segment $I$ of $\underset{\sim}{P}$ *bounded* if there is some principal initial segment $\leftarrow x]_{\underset{\sim}{P}}$ which contains $I$. We define a bounded final segment dually.

The crucial properties of such bounded sets are these.

1) An initial segment $I$ of $\underset{\sim}{P}$ is bounded iff the final segment $P\setminus I$ is unbounded. (Observe that $\kappa$ and $\lambda$ have distinct cofinalities.)

2) The set of bounded initial (resp. final) segments is closed under countable union, and closed by inclusion.

In other words, the set of bounded initial segments is an $\omega_1$-prime-ideal in the lattice $\underset{\sim}{I}(\underset{\sim}{P})$.

Now for each $a \in P$ let $\underset{\sim}{L}(a) = \{x \in P/x \leq a \text{ in } \underset{\sim}{L}\}$ and dually let $\underset{\sim}{L}^*(a) = \{x\in P/a \leq x \text{ in } \underset{\sim}{L}\}$. Let $\underset{\sim}{P}_{\underset{\sim}{L}} = \{a\in P|\underset{\sim}{L}(a) \text{ is bounded}\}$ and $\underset{\sim}{P}_{\underset{\sim}{L}^*} = \{a\in P/L^*(a) \text{ is bounded}\}$.

From the properties listed above follows that $\underset{\sim}{P}_{\underset{\sim}{L}}$ and $\underset{\sim}{P}_{\underset{\sim}{L}^*}$ constitutes a partition of $\underset{\sim}{P}$ into an initial and a final segment. Thus one of them is unbounded. We may suppose that $\underset{\sim}{P}_{\underset{\sim}{L}}$ is unbounded (and therefore $\underset{\sim}{P}_{\underset{\sim}{L}^*}$ is bounded). Then for each $i \in \lambda$ let $f(i)$ be the least member of $\kappa$ such that $(f(i),i) \in \underset{\sim}{P}_{\underset{\sim}{L}}$. We define a countable sequence $(I_n)$ of bounded initial segments of $\underset{\sim}{L}$ inductively as follows: $I_0 = \underset{\sim}{L}(f(0),0)$ and $I_{n+1} = I_n \cup \bigcup_{0\leq j\leq i} \underset{\sim}{L}(f(j),j)$

where $i$ is the least member of $\lambda$ such that there is no $x$ such that $(x,i) \in I_n$. Because this sequence is strictly increasing there is no element $a$ such that $\underset{\sim}{L}(a) = \bigcup_n I_n$. To see that $\underset{\sim}{L} - \bigcup_n I_n$ has no minimal element, observe that each $I_n \subseteq \underset{\sim}{P}_{\underset{\sim}{L}}$ and then $\bigcup_n I_n \subseteq \underset{\sim}{P}_{\underset{\sim}{L}}$. Thus if $\underset{\sim}{L} - \bigcup_n I_n$ has a minimal element $(k,i)$, then $(k,i) \in \underset{\sim}{P}_{\underset{\sim}{L}}$ and there is $k < \ell < \underset{\sim}{\kappa}$ such that $(\ell,i) \in \bigcup_n I_n$, so $(\ell,i) \in I_n$ for some $n$. It follows that $(f(i),i) \in I_{n+1}$, so $(k,i) \in I_{n+1}$ which is a contradiction.

In some respects this result is best possible: for any regular ordinal $\kappa$, both $\kappa^* \times \kappa$ and $\kappa^* \times \omega$ have a locally complete linear extension.

## III-5. LOCALLY CHAIN-COMPLETE ORDERED SETS WITHOUT INFINITE ANTICHAINS.

THEOREM. *Let $\underset{\sim}{P}$ be a locally chain-complete ordered set without infinite antichains. Then $\underset{\sim}{P}$ has a locally complete linear extension and the order is the intersection of these locally complete linear extensions.*

Sketch of Proof.

Loosely speaking the proof consists in showing that the ordering induced on the blocks of a partition of $\underset{\sim}{P}$ into antichains is itself locally complete.

For now let $P$ and $Q$ be two ordered sets. We call a map $f$ from $\underset{\sim}{P}$ to $\underset{\sim}{Q}$ a *quotient map* if for any $a,b \in \underset{\sim}{Q}$ we have $a \leq b$ iff there is an integer $n$ and sequences $x_1,y_1,x_2,y_2,\ldots,x_n,y_n$ of elements of $\underset{\sim}{P}$ satisfying:

$$a=f(x_1)=f(y_1),y_1 \leq x_2,f(x_2)=f(y_2),\ldots,y_{n-1} \leq x_n,f(x_n)=f(y_n)=b \quad .$$

We call $\underset{\sim}{Q}$ an *order quotient* of $\underset{\sim}{P}$.

We call an equivalence relation $\sim$ on an ordered set $\underset{\sim}{P}$ *orderable*, if there is some quotient map from $P$ into $Q$ such that $a \sim b$ iff $f(a) = f(b)$. Now the proof follows from these two lemmas.

LEMMA 1. (Cf. R. BONNET and M. POUZET [9]). *On every ordered set $\underset{\sim}{P}$ there is an orderable equivalence relation $\sim$ whose equivalence classes are antichains of $\underset{\sim}{P}$ and such that the quotient $\underset{\sim}{Q}$*

*is a chain. Moreover, if* $\underset{\sim}{Q}$ *is complete, then* $\underset{\sim}{P}$ *has a complete linear extension.*

Hint. Consider the set $E(\underset{\sim}{P})$ of all orderable equivalence relations of $\underset{\sim}{P}$ each of whose equivalence classes is an antichain of $\underset{\sim}{P}$. Order this set by inclusion (e.g. $\sim$ is contained in $\sim'$ if $a \sim b$ implies $a \sim' b$) and take a maximal member.

LEMMA 2. (M. POUZET and I. RIVAL, [1981]). *Let* $\underset{\sim}{P}$ *be an ordered set and* $\underset{\sim}{Q}$ *be an order quotient of* $\underset{\sim}{P}$. *If* $\underset{\sim}{P}$ *is locally chain complete and has no infinite antichain then* $\underset{\sim}{Q}$ *is locally chain complete and has no infinite antichain.*

Hint. Evidently $\underset{\sim}{Q}$ has no infinite antichain. To prove that $\underset{\sim}{Q}$ is locally chain complete, let $\underset{\sim}{C}$ be a maximal chain of $\underset{\sim}{Q}$. By a well-known result of DUFFUS, RIVAL, SIMONOVITS [16], $\underset{\sim}{C}$ is a retract of $\underset{\sim}{Q}$. That is, there is some increasing map $g:\underset{\sim}{Q} \to \underset{\sim}{C}$ which is the identity onto $\underset{\sim}{C}$. Such retraction being a quotient map it follows that $\underset{\sim}{C}$ is an order quotient of $\underset{\sim}{P}$. Now assume $\underset{\sim}{C}$ is not locally complete, then there are disjoint subsets $A$, $B$ of $\underset{\sim}{C}$ such that $a < b$ for each $a \in A$, $b \in B$, and neither sup $A$ nor inf $B$ exists in $\underset{\sim}{C}$. There are then regular ordinals $\alpha$, $\beta$ and sequences $(a_i | i < \alpha)$, $(b_j | j < \beta)$ such that $(a_i | i < \alpha)$ is cofinal in $A$ and $(b_j | j < \beta)$ is coinitial in $B$. Since, for each $i < \alpha$ and $j < \beta$, $a_i < b_j$, there exist elements $x_{ij} < y_{ij}$ in $P$ satisfying $a_{i'} \leq f(x_{ij}) \leq a_i$ for some $i' < \alpha$, and $b_{j'} \leq f(y_{ij}) \leq b_j$ for some $j' < \beta$.

Now to conclude it suffices to build with the $x_{ij}$ an increasing sequence $(u_{i'})_{i'<\alpha}$ and with the $y_{ij}$ a decreasing sequence $(v_{j'})_{j'<\beta}$ such that $u_{i'} \leq v_{j'}$ for each $i'$, $j'$ and such that their images are respectively cofinal into $A$ and coinitial into $B$. For this the main tool is a partition theorem of B. DUSHNIK, E.W. MILLER [18]: every ordered set $\underset{\sim}{P}$ of cardinal $\kappa$ contains either a chain of cardinal $\kappa$ or an infinite antichain. For details when the $f^{-1}(x)$ are antichains see [44].

## III-5. THE COUNTABLE CASE.

THEOREM. *Every countable, locally chain-complete ordered set has a locally complete linear extension. Moreover the order is the intersection of the locally complete linear extensions.*

Proof. By our previous remark (see III-3) it suffices to show that there is at least a locally complete linear extension. For this we use the same idea of kernel which occurs in the proof of the extendability of the rational chain. We prove that every countable locally chain-complete ordered set with no locally complete linear extension has a nonempty kernel (this being adequately defined), we deduce then such ordered sets would contain an isomorphic copy of the rational chain which contradicts the fact that the ordered set is countable and locally chain-complete.

Let $\mathbf{P}$ be a locally chain-complete ordered set. Let $\mathbf{I}_P$ denote the set of initial segments $I$ of $\mathbf{P}$ so that each convex subset $S$ of $I$ has a locally complete linear extension $L(S)$. (Recall, a subset $A$ of an ordered set $\mathbf{P}$ is convex if for every $x,y,z \in P$, $x \le z \le y$ and $x,y \in A$ implies $z \in A$. Note that $\mathbf{A}$ is locally chain-complete whenever $\mathbf{P}$ is locally chain complete.) Define $\mathbf{F}_P$ the set of final segments dually.

LEMMA 1. *Let $\mathbf{P}$ be a locally chain-complete ordered set.*

(*i*) *If $I \in \mathbf{I}_P$ and $J \in \mathbf{I}_{P\setminus I}$ then $I \cup J \in \mathbf{I}_P$; moreover if $I \in \mathbf{I}_P$ and $\mathbf{P}\setminus I \in \mathbf{F}_P$ then $\mathbf{P} \in \mathbf{I}_P$ and then $\mathbf{P}$ has a locally complete linear extension.*

(*ii*) *Every countable union of members of $\mathbf{I}_P$ is a member of $\mathbf{I}_P$. Especially $\mathbf{I}_P$ is an $\omega_1$-ideal of initial segments.*

Proof. (i) Let $I \in \mathbf{I}_P$ and $J \in \mathbf{I}_{P\setminus I}$. We show first that $I \cup J$ has a locally complete linear extension. If $I$ has a maximal element $a$, say, then $I\setminus\{a\}$ is a convex subset of $I$ and $L(I\setminus\{a\}) + \{a\} + L(J)$ induces a locally complete linear extension of $I \cup J$. If $J$ has a minimal element $b$, say, then again $J\setminus\{b\}$ is a convex subset of $J$ and $L(I) + \{b\} + L(J\setminus\{b\})$ induces a locally complete linear extension of $I \cup J$. If neither $I$ has a maximal element nor $J$ has a minimal element then, as $\mathbf{P}$ is locally chain complete, there is $a \in I$ and $b \in J$ such that $a \nleq b$ and $b \nleq a$. (Otherwise for each $a \in I$ and $b \in J$, $a \le b$ and, as $I \cup J$ is locally chain-complete, a maximal chain of $I \cup J$ would contain either a maximal element of $I$ or a minimal element of $J$). Let $A = \{x \in I \mid x > a\}$ and $B = \{y \in J \mid y < b\}$. Now, $A$ is convex in $I$, $B$ is convex in $J$ and, for each $x \in A$, $y \in B$, $x$ is non-comparable to $y$. We can then construct a locally complete linear extension $L(A\cup B)$ of $A \cup B$ (cf. III-4 or III-5). Finally, $L(I\setminus(A\cup\{a\})) + \{a\} + L(A\cup B) + \{b\} + L(J\ (B\cup\{b\}))$ induces a locally complete linear extension of $I \cup J$. In essentially the same way we can construct a locally complete linear extension for each convex subset of $I \cup J$. It follows that $I \cup J \in \mathbf{I}_P$.

(ii) We now show that for any countable increasing sequence $I_0 \subseteq I_1 \subseteq I_2 \subseteq \ldots$ of members of $I_P$, the union $I = \bigcup_n I_n$ is a member of $\mathbf{I}_P$. Indeed, for each $0 < n < \omega$, choose $a_n \in I_n\setminus I_{n-1}$ and set $A_n = \{x \in P \mid x < a_n\}$, $J_n = (I_n \cup A_{n+1})\setminus(I_{n-1} \cup A_n)$ (and

$J_0 = I_0 \cup A_1$). Evidently, $I = \bigcup_n J_n$ and, for each $0 < n < \omega$, $a_n$ is a minimal element of $J_n$. It follows that $J_n \setminus \{a_n\}$ is a convex subset of $I_{n+1}$ so $\{a_n\}+L(J_n\setminus\{a_n\})$ is a locally complete extension of $J_n$. Then $L(J_0)+\{a_1\}+L(J_1\setminus\{a_1\})+ \ldots$ $\ldots +\{a_n\}+L(J_n\setminus\{a_n\})+ \ldots$ is a locally complete extension of $I$. In a similar fashion we can verify that each convex subset of $I$ also has a locally complete linear extension.

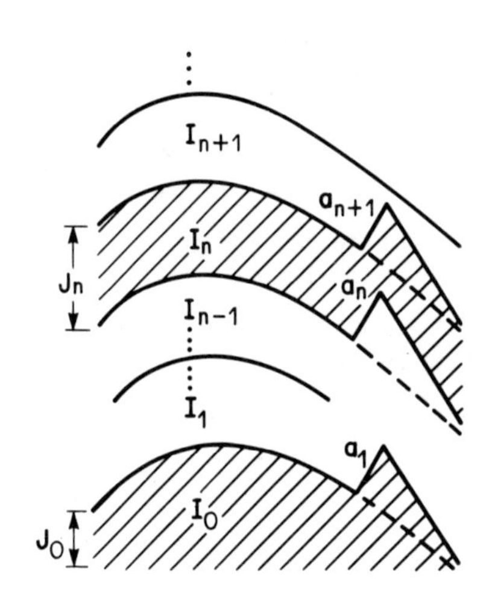


By duality $\underset{\sim}{F}_P$, too, satisfies the claims of this lemma. In particular, if there is $I \in \underset{\sim}{I}_P$ and $F \in \underset{\sim}{F}_P$ such that $I \cup F = P$ then $F\setminus I \in \underset{\sim}{I}_{P\setminus I}$ and so $\underset{\sim}{P}$ has a locally complete linear extension.

A subset $\underset{\sim}{N}$ of $\underset{\sim}{P}$ is a *kernel* whenever $N$ is convex and $\underset{\sim}{I}_N = \underset{\sim}{F}_N = \{\emptyset\}$; i.e. any non-empty initial (or final) segment of $\underset{\sim}{N}$ contains some convex subset without any locally complete linear extension.

LEMMA 2. *Let $\underset{\sim}{P}$ be a countable locally chain complete ordered set. If $\underset{\sim}{P}$ has no locally complete linear extension then it would contain a nonempty kernel $\underset{\sim}{N}$.*

Proof. As $\underset{\sim}{P}$ is countable and both $\underset{\sim}{I}_P$ and $\underset{\sim}{F}_P$ are closed with respect to set inclusion, both $I = \sup \underset{\sim}{I}_P$ and $F = \sup \underset{\sim}{F}_P$ exist and $P\setminus I \cup F$ is nonempty. By (i) this set is a kernel.

To conclude it suffices to show the following.

LEMMA 3. *Let $\underset{\sim}{P}$ be a locally chain complete ordered set. If $\underset{\sim}{P}$ has no locally complete linear extension and any convex subset of $\underset{\sim}{P}$ without locally complete linear extensions has a nonempty kernel then $\underset{\sim}{P}$ contains an isomorphic copy of the chain $\mathbb{R}$ of the real numbers, so $\underset{\sim}{P}$ is uncountable.*

Proof. Because $\underset{\sim}{P}$ is locally chain-complete, it suffices to construct a countable dense chain. For this let $P_{1/2} = \underset{\sim}{P}$ and $N_{1/2}$ be the kernel of $P_{1/2}$. Now let $x_{1/2} \in N_{1/2}$, $P_{1/4} = \{y \in N_{1/2} / y \leq x_{1/2}\}$, $P_{3/4} = \{y \in N_{1/2} / y \geq x_{1/2}\}$. According to

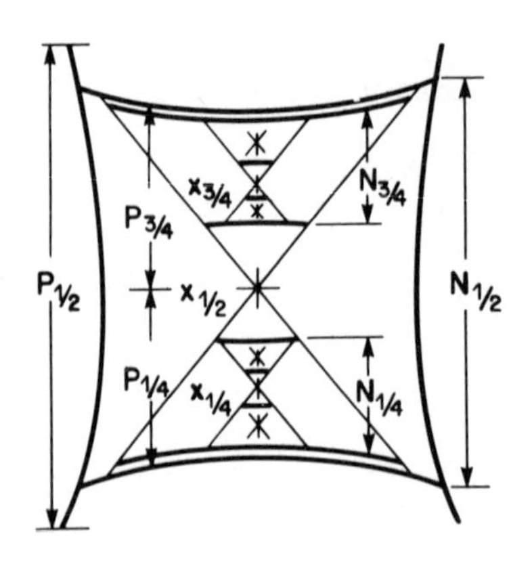


our hypothesis $P_{1/4}$ and $P_{3/4}$ have respectively nonempty kernels $N_{1/4}$ and $N_{3/4}$. We can then choose $x_{1/4} \in N_{1/4}$, $x_{3/4}$ and again define $P_{1/8}$, $P_{3/8}$, $P_{5/8}$, $P_{7/8}$ and their corresponding kernels $N_{1/8}$, $N_{3/8}$, $N_{5/8}$, $N_{7/8}$. In this way we obtain a countable dense chain $\underset{\sim}{C}$ consisting of the elements $\boldsymbol{x}_q$ in $P$, where $q$ is a dyadic fraction in the unit interval.

## IV. LINEAR EXTENSIONS OF ORDERS DEFINED BY SOME UNIFORM PROCEDURE.

### IV-1. Introduction.

There are many constructions of ordered sets where the ordered set $\underset{\sim}{P}$ is built from a family $\underset{\sim}{P}_1,\ldots,\underset{\sim}{P}_n$ of ordered sets by some uniform procedure. As an example consider the direct product $\underset{\sim}{P}$ of chains $\underset{\sim}{P}_1,\ldots,\underset{\sim}{P}_n$. The procedure is uniform in the sense that the order between two elements $\vec{x} = (x_1\ldots x_n)$ and $\vec{y} = (y_1\ldots y_n)$ depends on the order between $x_1$ and $y_1$, $x_2$ and $y_2,\ldots,x_n$ and $y_n$, but not the particular values of $x_1,\ldots,x_n,y_1,\ldots,y_n$. In this case the ordinal products of the $\underset{\sim}{P}_1,\ldots,\underset{\sim}{P}_n$, obtained for all the linear orders on $\{1,\ldots,n\}$, give linear extensions of $\underset{\sim}{P}$ with the same property. Thus it is a natural question to ask whether an ordered set $\underset{\sim}{P}$ defined by some uniform procedure, as above, has a linear extension uniformly defined too.

A positive answer was obtained by C. CHARRETTON and M. POUZET [12] for a particular type of ordered set, the *chain-invariant* ordered sets. The purpose of this section is to expose this result.

### IV-2. Chain-Invariant Ordered Set.

Let $\underset{\sim}{C}$ be a chain. For an integer $m$, we denote by $[\underset{\sim}{C}]^m$ the set of all the $m$-element subsets of $\underset{\sim}{C}$. We identify every $m$-element subset with an increasing $m$-tuple $(c_1,c_2,\ldots,c_m)$, where $c_1 < c_2 < \ldots < c_m$, and we denote it by $\vec{c}$. Call a map $g$ a *local automorphism* of $\underset{\sim}{C}$ if $g$ is an order isomorphism of $\underset{\sim}{C}'$ to $\underset{\sim}{C}''$ where $\underset{\sim}{C}'$ and $\underset{\sim}{C}''$ are subchains of $\underset{\sim}{C}$. Let dom $g$ denote the domain of $g$. For every $\vec{c} = (c_1,\ldots,c_m) \in [C]^m$ with components $c_1,\ldots,c_m \in$ dom $g$, let $\overline{g}(\vec{c})$ denote the $m$-tuple $(g(c_1),\ldots,g(c_m)) \in [\underset{\sim}{C}]^m$. Let $\underset{\sim}{C}$ be a chain, $\underset{\sim}{P}$ be an ordered set and $\Phi$ be a set of maps, where each $\varphi \in \Phi$ is a map from $[\underset{\sim}{C}]^{m(\varphi)}$ to $\underset{\sim}{P}$ (the exponent $m(\varphi)$ is an integer called the *arity* of $\varphi$).

The ordered set $\underset{\sim}{P}$ is $\underset{\sim}{C}$*-invariant modulo* $\Phi$ if for all $\varphi,\psi \in \Phi$, for all $\vec{a} \in [C]^{m(\varphi)}$, $\vec{b} \in [C]^{m(\psi)}$ and for all local automorphisms $g$ of $\underset{\sim}{C}$ such that $\vec{a} \cup \vec{b} \subseteq$ dom $g$, we have $\varphi(\vec{a}) \leq \psi(\vec{b})$ in $\underset{\sim}{P}$ iff $\varphi(g(\vec{a})) \leq \psi(g(\vec{b}))$ in $\underset{\sim}{P}$. The condition of invariance means, roughly, that the order between $\varphi(\vec{a})$ and $\psi(\vec{b})$ depends not on the particular choice of $\vec{a}$ and $\vec{b}$ but on their "relative positions" as subsets of $\underset{\sim}{C}$.

The ordered set $\underset{\sim}{P}$ is $\underset{\sim}{C}$*-generated mod* $\Phi$ if any element $x$

belongs to the range of some $\varphi \in \Phi$. We call an ordered set $\underset{\sim}{P}$ *chain-invariant* if there is some chain $\underset{\sim}{C}$, some set $\Phi$ of maps such that $\underset{\sim}{P}$ is $\underset{\sim}{C}$-invariant and $\underset{\sim}{C}$-generated mod $\Phi$.

Example. The simplest case is when $\underset{\sim}{P} = [\underset{\sim}{C}]^m$ for some $m$ and $\Phi$ contains only the identity map on $[\underset{\sim}{C}]^m$. In this case we say briefly that $[\underset{\sim}{C}]^m$ is *$\underset{\sim}{C}$-invariant*. These conditions mean that the order between $\vec{a}$ and $\vec{b}$ does not depend upon the choice of $\vec{a}$ and $\vec{b}$ but only on the relative position of their components. Precisely these conditions are equivalent to the definability of the order on $[\underset{\sim}{C}]^m$ from the order on $\underset{\sim}{C}$, by way of a free formula $F(\vec{x},\vec{y})$. As typical examples we have the Cartesian order and the lexicographical order and also the interval order. We shall show below that all examples of a $\underset{\sim}{C}$-invariant *linear* order on $[\underset{\sim}{C}]^m$ are lexicographically ordered. For another one, let $B(\underset{\sim}{C})$ be the Boolean algebra generated by $\underset{\sim}{C}$, that is to say the Boolean algebra which consists of the finite unions of intervals $[a,b[$, $a \leq b$ of the chain $\overline{\underset{\sim}{C}} = \{-\infty\} + \underset{\sim}{C} + \{+\infty\}$ (where $-\infty$, $+\infty$ are two new elements added to $\underset{\sim}{C}$). For each integer $m$ let $\varphi_m : [\overline{\underset{\sim}{C}}]^{2m} \to B(\underset{\sim}{C})$ defined by

$$\varphi_m(a_1,b_1,a_2,b_2,\ldots,a_n,b_n) = [a_1,b_2[\cup[a_2,b_2\ldots\cup[a_n,b_n[$$

and let $\Phi$ be the set of such maps. Then $B(\underset{\sim}{C})$ is $\overline{\underset{\sim}{C}}$ invariant and $\overline{\underset{\sim}{C}}$ generated mod $\Phi$. Now to get a $\underset{\sim}{C}$-invariant, mod $\Phi$, linear extension, we can identify each element $X$ of $B(\underset{\sim}{C})$ with the sequence $(a_1,b_1,\ldots,a_n,b_n)$ of minimal length such that $X = [a_1,b_1[\cup[a_2,b_2[\ldots\cup[a_n,b_n[$. Then we can order the set of the strictly increasing sequences of any length by a lexicographical ordering defined in the following way:

$$\vec{u} = (u_1,u_2,\ldots,u_{2n-1},u_{2n}) \leq \vec{v} = (v_1,v_2,\ldots,v_{2m-1},v_{2m})$$

iff either $u_i = v_i$ for every $i \leq n$

or for the smallest $i$ such that $u_i \neq v_i$ we have $u_i > v_i$ if $i$ is odd and $v_i > u_i$ if $i$ is even. The induced ordering on $B(\underset{\sim}{C})$ is a $\overline{\underset{\sim}{C}}$ invariant linear extension of the order on $B(\underset{\sim}{C})$.

Other examples are the Ehrenfeucht-Mostowski-models [22].

## IV-3. Chain-Invariant Ordered Sets and Ramsey's Theorem.

The theorem of F.P. RAMSEY [45] asserts that *for any finite colouring of the m-subsets of an infinite set E there is an infinite subset E' of E all of whose m-subsets are assigned the same colour.*

The following version of this result is central in our investigations.

IV-3.1. THEOREM [12]. *Let $\underset{\sim}{C}$ be an infinite chain, let $\underset{\sim}{P}$ be an ordered set and let $\Phi$ be a finite set of maps $\varphi$ of $[\underset{\sim}{C}]^{n(\varphi)}$ to $\underset{\sim}{P}$. Then $\underset{\sim}{C}$ contains an infinite subchain $\underset{\sim}{C}'$ such that $\underset{\sim}{P}$ is $\underset{\sim}{C}'$ invariant mod $\Phi_{|\underset{\sim}{C}'}$.*

Proof. Let $m = \max\{n(\varphi) \mid \varphi \in \Phi\}$. For each $2m$-element chain $\underset{\sim}{D}$ in $\underset{\sim}{C}$ define a subset $\underset{\sim}{D}^{\Phi}$ of $\underset{\sim}{P}$ by $\underset{\sim}{D}^{\Phi} = \{\varphi(\vec{a}) \mid \varphi \in \Phi\}$, $\vec{a} \in [\underset{\sim}{D}]^{n(\varphi)}$. Now define an equivalence relation $\sim$ on the collection of all $2m$-subsets of $\underset{\sim}{C}$ by $\underset{\sim}{D} \sim \underset{\sim}{D}'$ if there is an order isomorphism $\tilde{h}$ of $\underset{\sim}{D}^{\Phi}$ onto $\underset{\sim}{D}'^{\Phi}$ such that $\tilde{h}(\varphi(\vec{a})) = \varphi(\overline{h}(\vec{a}))$ for all $\varphi \in \Phi$, $\vec{a} \in [\underset{\sim}{D}]^{n(\varphi)}$, where $h$ is the unique isomorphism of $\underset{\sim}{D}$ onto $\underset{\sim}{D}'$. This relation has only finitely many classes, so RAMSEY's theorem can be applied to this partition of the $2m$-subsets of $\underset{\sim}{C}$ to obtain an infinite subchain $\underset{\sim}{C}'$ of $\underset{\sim}{C}$ such that all elements of $[C']^{2m}$ are equivalent. It follows that $\underset{\sim}{P}$ is $\underset{\sim}{C}'$-invariant mod $\Phi$.

In fact this theorem is the same as RAMSEY's theorem, and moreover it contains the ERDÖS-RADO generalization of RAMSEY's theorem to infinite colourings [20]. This is a consequence of the next proposition.

IV-3.2. PROPOSITION. *Let $\underset{\sim}{C}$ be a chain, let $\underset{\sim}{P}$ be an ordered set and $\varphi$ a map from $[\underset{\sim}{C}]^{n(\varphi)}$ to $\underset{\sim}{P}$. If $\underset{\sim}{P}$ is $\underset{\sim}{C}$-invariant mod $\varphi$, and $|\underset{\sim}{C}| \geq 3 \cdot n(\varphi)$, then there is an integer $k$, a $k$ element subset $K$ of $\{1,\ldots m(\varphi)\}$ and an injective map $\psi$ from $[\underset{\sim}{C}]^{k}$ into an ordered set $\underset{\sim}{P}'$ containing $\underset{\sim}{P}$ such that*

(1) $\varphi(\vec{a}) = \psi(\vec{a}_{|K})$ *for every* $\vec{a} \in [\underset{\sim}{C}]^{m(\varphi)}$ [here $\vec{a}_{|K}$ consists of the components of $\vec{a}$ with indices in $K$],

(2) *$\underset{\sim}{P}'$ is $\underset{\sim}{C}$-invariant mod $\psi$, and*

(3) *if $\underset{\sim}{C}$ is dense without endpoints then $\underset{\sim}{P}' = \underset{\sim}{P}$.*

While this looks a little technical, it is really meaningful. For a proof see [12]. The trick is to observe that if $\varphi$ is not injective then there is some index $i$ such that $\varphi(\vec{a})$ does not depend on $i$. After, the proof works by induction on $n(\varphi)$.

Let us return to partition theorems. Consider a colouring of the $m$-subsets of an infinite set $E$, that is, a map $\varphi$ from $[E]^{m}$ into a set $X$ of colours. Order $E$ as a chain and $X$ as an antichain. According to our theorem there is an infinte sub-chain $E'$ of $E$ such that $X$ is invariant mod $\varphi_{|E'}$. By the previous proposition there is a subset $K$ of $\{1,\dots,m\}$ such that $\varphi(\vec{a}) = \varphi(\vec{b})$, e.g. $\vec{a}$ and $\vec{b}$ have the same colour, iff $\vec{a}_{|K} = \vec{b}_{|K}$. If $X$ is finite then $K$ is necessarily empty and we get RAMSEY's theorem. In the general case, $X$ is possibly infinite, and we get the ERDÖS-RADO theorem.

While this machinery looks heavy it does in many instances the work of the RAMSEY theorem quickly. For another application see D. DUFFUS, M. POUZET, I. RIVAL [17].

IV-4. Linear Extensions of a Chain-Invariant Ordered Set.

IV-4.1. THEOREM [12]. *Let $\underset{\sim}{C}$ be an infinite chain, let $\underset{\sim}{P}$ be an ordered set and let $\Phi$ be a set of maps $\varphi$ of $[\underset{\sim}{C}]^{n(\varphi)}$ to $\underset{\sim}{P}$. If $\underset{\sim}{P}$ is $\underset{\sim}{C}$-invariant mod $\Phi$, then the order on $\underset{\sim}{P}$ can be extended to a linear order in such a way that the chain $\overline{\underset{\sim}{P}}$ is $\underset{\sim}{C}$-invariant mod $\Phi$.*

Proof. The proof is in two steps.

(1) *For every finite subchain $D$ of $C$ and for every finite subset $\Phi'$ of $\Phi$, the order on $\underset{\sim}{D}^{\Phi'}$ can be extended to a $\underset{\sim}{D}$-invariant, mod $\Phi'$, linear order.*

Let $P'$ be an arbitrary linear extension of the order $P$, and $\underset{\sim}{P}'$ the corresponding ordered set. By our version of RAMSEY's theorem there is some infinite chain $\underset{\sim}{C}'$ of $\underset{\sim}{C}$ such that $\underset{\sim}{P}'$ is $\underset{\sim}{C}'$-invariant mod $\Phi'_{|C'}$. Now for every finite subchain $\underset{\sim}{D}$ of $\underset{\sim}{C}$ choose an isomorphic subchain $\underset{\sim}{D}'$ of $\underset{\sim}{C}'$ and the isomorphism $g$ from $\underset{\sim}{D}$ onto $\underset{\sim}{D}'$. Because $\underset{\sim}{P}$ is $\underset{\sim}{C}$-invariant, $g$ extends to $\tilde{g}$ from $\underset{\sim}{D}^{\Phi'}$ to $\underset{\sim}{D}'^{\Phi'}$. As order on $\underset{\sim}{D}^{\Phi'}$ take the isomorphic image under $\tilde{g}^{-1}$, of the linear ordering on $\underset{\sim}{D}'^{\Phi'}$ induced by $P'$.

(2) We apply a compactness argument. For any finite sub-chain $\underset{\sim}{D}$ of $\underset{\sim}{C}$, and for any finite $\Phi' \subseteq \Phi$ let $U_{\underset{\sim}{D},\Phi'}$ be the set of linear extensions of $P$ such that $P_{|\underset{\sim}{D}^{\Phi'}}$ is a $\underset{\sim}{D}$-invariant, mod $\Phi'$, linear order. By our previous step these $U_{\underset{\sim}{D},\Phi'}$ are non-empty. They are closed in the set $2^{P\times P}$ of all the binary relations on $\underset{\sim}{P}$ (equipped with the pointwise convergence topology) and they have the finite intersection property. (Note that

$$U_{\underset{\sim}{D}',\Phi'} \cap U_{\underset{\sim}{D}'',\Phi''} \supseteq U_{\underset{\sim}{D}'\cup\underset{\sim}{D}'',\Phi'\cup\Phi''} \quad .)$$

Thus their intersection is non-empty and any element of this intersection is a linear extension of $P$ with the required property.

Note that if $\Phi$ is finite then the compactness argument is useless: to have the invariant linear extension it suffices to have it on some $\underset{\sim}{D}^{\Phi}$ for some $\underset{\sim}{D}$ with $4\ m(\Phi)$ elements ($m(\Phi) = \mathrm{Max}\{m(\varphi) \mid \varphi \in \Phi\}$). Although $\underset{\sim}{D}$ is of very small size we do not know how to prove without RAMSEY's theorem that $\underset{\sim}{D}^{\Phi}$ has such a linear extension.

Remark.

The same method as for SZPILRAJN's theorem does not work here. Consider this example. Assume $\underset{\sim}{P}$ is generated by two unary functions $\varphi$ and $\psi$ and ordered in such a way that $\varphi(c) < \psi(c')$ for $c \neq c'$ and $\varphi(c)$, $\psi(c)$ are pairwise noncomparable. If we decide to extend the order in such a way that $\psi(c) \leq \varphi(c)$ for some $c$, then by $\underset{\sim}{C}$ invariance this must be true for all the other $c$, and we get the cycle $\psi(c) \leq \varphi(c) < \psi(c') \leq \varphi(c') < \psi(c)$.

The conclusion of this theorem remains true if we have a binary relation without cycles instead of a partial order. (Take an arbitrary linear ordered extension of this binary relation and do the two steps of the previous proof.) So it can be applied to the $\in$ relation (see Theorem IV-6.2).

## IV-5. Subchains of Chain-Invariant Ordered Sets.

By our previous theorem the order types of subchains of a $\underset{\sim}{C}$-invariant ordered set are bounded by the order type of its $\underset{\sim}{C}$-invariant linear extensions. Thus, a natural question arises: what are the possible order types of the $\underset{\sim}{C}$-invariant chains? This question is completely solved now and will appear in a forthcoming paper of C. CHARRETTON and M. POUZET. Here we present a solution to this question for the special case of a $\underset{\sim}{C}$-invariant chain generated by one function.

IV-5.1. THEOREM [12]. *Let $C$ be an infinite chain, $\underset{\sim}{P}$ be another chain and $\varphi$ be a surjective map from $[\underset{\sim}{C}]^{m(\varphi)}$ into $\underset{\sim}{P}$. If $\underset{\sim}{P}$ is $C$-invariant mod $\varphi$ then $\underset{\sim}{P}$ is order isomorphic to a subchain of a finite lexicographical product of copies of $\underset{\sim}{C}$ and $\underset{\sim}{C}^*$.*

This theorem and the next lemma have also been obtained by W. Hodges in his Doctoral Thesis [29].

Proof. We start with the following special case.

LEMMA. *Let $\underset{\sim}{C}$ be a dense chain without endpoints and $n$ be an integer. Every $C$-invariant linear order on $[\underset{\sim}{C}]^n$ is a lexicographical order for some linear order on $[1,\ldots,n]$, the components being ordered as $\underset{\sim}{C}$ or $\underset{\sim}{C}^*$.*

Sketch of Proof. 1.1 For every $i \in [1,\ldots,n]$ define a linear order $\leq_i$ on $C$ as follows: $x \underset{i}{\leq} y$ iff there is some $\vec{a} = (a_1 \ldots a_n)$

such that $(a_1 \dots a_{i-1}, x, a_{i+1} \dots a_n) \leq (a_1 \dots a_{i-1}, y, a_{i+1} \dots a_n)$ $(\mathrm{mod}[\underset{\sim}{C}]^n)$. This orders $C$ as $\underset{\sim}{C}$ or $\underset{\sim}{C}^*$.

1.2. Define a linear order $\trianglelefteq$ on the index set $\{1, \dots, n\}$ as follows: $i \trianglelefteq j$ iff $i = j$ or there is $\vec{a}, \vec{b} \in [C]^n$ such that $a_i <_i b_i$, $b_j <_j a_j$ and $a_k = b_k$ for $k \neq i, j$.

It is not hard (but tedious) to verify that $\trianglelefteq$ is a linear order.

2. Let $\leq_{\mathrm{lex}}$ be the lexicographic order associated with $\trianglelefteq$ and $\leq_i$, $i = 1, \dots, n$ (that is to say $\vec{a} \leq_{\mathrm{lex}} \vec{b}$ iff $\vec{a} = \vec{b}$ or $a_{i_1} <_{i_1} b_{i_1}$ for the smallest index $i_1$ (relative to $\trianglelefteq$) such that $a_{i_1} \neq b_{i_1}$). To show that it is, in fact, the order $\leq$ on $[C]^n$ it suffices to prove that $\vec{a} \leq_{\mathrm{lex}} \vec{b}$ implies $\vec{a} \leq \vec{b}$. This implication can be proved by induction on the number $d(\vec{a}, \vec{b})$ of indices $i$ such that $a_i \neq b_i$.

Remarks.

1. The result is true even when $\underset{\sim}{C}$ is not dense, and even more *when* $\underset{\sim}{C}$ *is finite with at least* $2n+1$ *elements*. To get the result when $\underset{\sim}{C}$ has at least $3n$ elements it is enough to know it when the chain is dense: extend $\underset{\sim}{C}$ to a dense chain $\underset{\sim}{C}_1$ and get an order on $[\underset{\sim}{C}_1]^n$, which extends the order on $[\underset{\sim}{C}]^n$. By our lemma, $[\underset{\sim}{C}_1]^n$ is lexicographically ordered, so $[\underset{\sim}{C}]^n$ is too.

2. Following W. Hodges, this result is a very special form of a theorem of K. ARROW [ 1 ] known as ARROW's paradox. Our knowledge about social sciences allows us to say the following. Consider the elements $i$ of $\{1, \dots, n\}$ as individuals and the elements $\vec{a} \in [\underset{\sim}{C}]^n$ as objects they have to select. Each individual $i$ says he prefers $\vec{a}$ to $\vec{b}$ whenever $a_i < b_i$. Now the people agree on a collective preference which linearly orders $[\underset{\sim}{C}]^n$, subject to the condition that if for two pairs of objects $\vec{a}$, $\vec{b}$ and $\vec{a}'$, $\vec{b}'$ the individual preferences are the same then the collective preference is also the same. Then the lemma says there is an individual $i_1$ such that either $\vec{a}$ is collectively preferred to $\vec{b}$ whenever $i_1$ prefers $\vec{a}$ to $\vec{b}$, and $i_1$ is a *dictator*, or $\vec{a}$ is collectively preferred to $\vec{b}$ whenever $i_1$ prefers $\vec{b}$ to $\vec{a}$ and then $i_1$ is *persecuted*. Now when $i_1$ has no preference between objects, then there is a second individual $i_2$ who plays a similar role, and if $i_1$, $i_2$ have no preference then another individual $i_3$ plays their role, and so on, until all the objects are classified.

Proof of the theorem.

If $C$ is a dense chain then, by the Proposition IV-3.2, we get a bijective map from some $[C]^m$ onto $P$. The previous lemma gives the result.

If $C$ is not dense, extend it to a dense chain $C'$ and extend $\varphi$ to some map $\varphi'$ from $[C']^{m(\varphi)}$ to some $P'$. By the Proposition IV-3.2, take the one-to-one $\psi$ from some $[C']^m$ onto $P'$. By our previous lemma $P'$ is, like $[C']^m$, lexicographically ordered. As $\psi(\vec{a}_{|K}) = \varphi'(\vec{a})$ it follows that $P \subseteq \psi([C]^m)$. [Warning: the equality does not hold in general.] So $P$ is ordered as a subset of $[C]^m$.

The theorem IV-4.1 about linear extensions together with this theorem gives the following.

IV-5.2. THEOREM. *Let $P$ be an ordered set which is $C$-generated and $C$-invariant modulo a countable set $\Phi$. Then every chain $D$ of $P$ with a regular cardinality $\kappa$ contains a subchain $D'$ with the same cardinality $\kappa$ which is order isomorphic to a subchain of $C$ or $C^*$.*

Proof. This is obviously true for $\kappa = \omega$. So assume $\kappa > \omega$. By Theorem IV-4.1 we can assume that $P$ is linearly ordered (otherwise, consider a $C$-invariant linear extension of $P$). Because $\kappa$ is regular there is some $\varphi \in \Phi$ such that its range $P_\varphi$ contains a subchain $D_1$ of $D$ with power $\kappa$. By the Theorem IV-5.1 such a $P_\varphi$ is order-isomorphic to a subchain of a finite lexicographic product of copies of $C$ and $C^*$, and so $D_1$ too. The regularity of $\kappa$ now implies that $D_1$ contains a subchain $D'$ of power $\kappa$ which is order isomorphic to a subchain of $C$ or $C^*$.

## IV-6. APPLICATION TO THE COMPARISON OF ORDERED STRUCTURES

The machinery developed in this chapter can be used to produce large families of ordered structures with few substructures in common. For this purpose let us introduce the following definition. For a cardinal $\kappa$ we say that two ordered sets $P$ and $P'$ are *almost disjoint* for *$\kappa$-betweenness* if there is no chain $D$ of $P$, with power $\kappa$ which is isomorphic or reverse isomorphic to a subchain $D'$ of $P'$. As an example the chain $\mathbb{R}$ of the real numbers and the $\omega_1$ chain are almost disjoint for $\omega_1$-betweenness. Now let us consider a family of ordered sets $\{P_i, i \in I\}$ such that each $P_i$ is $C_i$-generated and $C_i$-invariant modulo a countable set of maps $\Phi_i$. From the theorem above it follows that the $P_i$ are pairwise almost disjoint for $\kappa$-betweenness ($\kappa$ regular) whenever the $C_i$ are. Assuming G.C.H., for every successor cardinal $\kappa$ there are $2^\kappa$ chains of power $\kappa$ which are pairwise almost disjoint for $\kappa$-betweenness. For $\kappa = \omega_1$ such chains can be found in the real line [8] [50]. So, classical constructions (as for example the Ehrenfeucht-Mostowski one) can be used to produce invariant ordered sets. One can find in various classes of ordered (or orderable) structures $2^\kappa$ structures of power $\kappa$ which are pairwise almost disjoint for $\kappa$-betweenness. With extra assumptions on the chains, such structures are in fact almost-disjoint (in the sense that no structure of power $\kappa$ can be embedded simultaneously in both of them).

This applies to the classes of Boolean Algebras [ 8 ], of models of Zermelo-Fraenkel set theory, of models of Peano-Arithmetic, of ordered groups, etc. We give two examples.

IV-6.1. THEOREM. [8]. *For any successor cardinal $\kappa$ there are $2^\kappa$ Boolean algebras of power $\kappa$ such that no Boolean algebra of power $\kappa$ can be embedded in both of them.*

IV-6.2. THEOREM. [12] (G.C.H. and Cons. Z.F.C). *For any successor cardinal $\kappa$ there are $2^\kappa$ models of Z.F.C. of power $\kappa$ such that no model of Z.F.C. of power $\kappa$ can be embedded in both of them.*

That there are so many different mathematical universes is not a very surprising fact, after all.

ACKNOWLEDGEMENTS

This work has been done in the LABORATORY on ORDINAL ALGEBRA directed by Professor E. COROMINAS in LYON. The ideas and methods he has introduced play a prominent role in many parts of this work. We have also profited from the particular taste in ordered sets he has. It is a real pleasure for us to express to Professor COROMINAS our gratitude.

We are grateful to F. GALVIN for informing us about the results about extendable types he obtained jointly with McKENZIE. We are grateful to E.C. MILNER for his comments about the first version of this paper. We are grateful to I. RIVAL who provided us with the opportunity to present these results at the Symposium.

Finally, the second author would like to thank the C.N.R.S. for its financial support enabling him to attend this symposium.

REFERENCES

[1] K. ARROW (1951) Social choice in individual values. Cowles Commission Monograph N. 12. John Wiley and Sons Inc., N.Y. Chapman and Hall Ltd., London, 99 pp.

[2] K.A. BAKER (1961) Dimension, join-independence and breadth in partially ordered sets, (unpublished).

[3] B. BANASCHEWSKI (1956) Hüllensysteme und Erweiterungen von Quasi-Ordnungen, Z. Math. Logik, Grundlagen Math. 2, 117-130.

[4] J.E. BAUMGARTNER (1973) All $\aleph_1$ dense sets of reals can be isomorphic, Fund. Math. 79, 101-106.

[5] G. BIRKHOFF (1937) Rings of sets, Duke Math. J.3, 311-316.

[6] G. BIRKHOFF (1967) Lattice theory, Amer. Math. Soc., Providence, vol. 25, 3rd edition.

[7] R. BONNET (1969) Stratification et extension des genres de chaines dénombrables, C.R.A.S. (Paris), 269, 880-882.

[8] R. BONNET,(1980) Very strongly rigid Boolean Algebras, continuum discrete set condition, countable antichain condition, Algebra Universalis 11, 341-364.

[9] R. BONNET et M. POUZET (1969) Extension et stratification d'ensembles dispersés, C.R.A.S. (Paris), 268, 1512-1515.

[10] A. BOUCHET (1971) Étude combinatoire des ordonnés finis, applications, Thèse (Paris).

[11] C.C. CHANG and H.J. KEISLER, Model theory, North Holland, Amsterdam, Vol. 73, 1973.

[12] C. CHARRETTON et M. POUZET (1981) Les chaines dans les Modèles d'Ehrenfeucht-Mostowski, C.R.A.S. (Paris), 285, 597-599. See also The chains in Ehrenfeucht-Mostowski models (to appear in Fund. Math.).

[13] M.M. DAY (1945) Arithmetic of ordered systems, Trans. Amer. Math. Soc. 58, 1-43.

[14] D.H.J. DE JONGH and R. PARIKH (1977) Well partial orderings and hierarchies, Indig. Math. 39, 195-207.

[15] R.P. DILWORTH (1950) A decomposition theorem for partially ordered sets, Ann. of Math. 51, 161-166.

[16] D. DUFFUS, I. RIVAL and M. SIMONOVITS (1980) Spanning retracts of partially ordered sets. Discrete Math. 32, 1-7.

[17] D. DUFFUS, M. POUZET and I. RIVAL (1981) Complete ordered sets with no infinite antichains. Discrete Math. 35, 39-52.

[18] B. DUSHNIK and E.W. MILLER (1941) Partially ordered sets, Amer. J. Math. 63, 600-610.

[19] P. ERDÖS and R. RADO (1956) A partition calculus in set theory, Bull. Amer. Math. Soc. 62, 427-489.

[20] P. ERDÖS and R. RADO (1950) A combinatorial theorem, J. London Math. Soc. 25, 249-255.

[21] P. ERDÖS and A. HAJNAL (1962) On a classification of denumerable order types and an application to the partition calculus, Fund. Math. 51, 117-129.

[22] EHRENFEUCHT and A. MOSTOWSKI (1956) Models of axiomatic theories admitting automorphisms. Fund. Math. 43, 50-68.

[23] R. FRAÏSSÉ (1953) Sur quelques classifications des systèmes de relations, Thèse Univ. Paris (152 p) (cf also Pub. Sci. Univ. Alger Serie A, (1954) 135-182).

[24] R. FRAÏSSÉ (1971) Abritement entre relations et spécialement entre chaines, Inst. Nazionale Di Alta Mathematica, Indam, Rome, Symposia Math, Vol. 5, 203-251. Academic Press.

[25] F. GALVIN (1972 and 1981) Private communications.

[26] A. GLEYZAL (1940) Order types and structures of orders. Trans. Amer. Math. Soc. 48, 451-466.

[27] A. HAJNAL and E.C. MILNER (1971) Some problems for scattered ordered sets, Periodica Math. Hungarica, 1(2), 81-92.

[28] G. HIGMAN (1952) Ordering by divisibility in abstract algebras. Proc. London Math. Soc. 2, 326-336.

[29] W. HODGES (1969) The Ehrenfeucht-Mostowski method of constructing models, Doctoral Dissertation, Univ. of Oxford, 139 pp.

[30] T. HIRAGUCHI (1955) On the dimension of orders, Sciences Report of the Kanazawa University 4, 1-20.

[31] T. HIRAGUCHI (1951) On the dimension of partially ordered sets, Sciences Report of the Kanazawa University 1, 77-94.

[32] T. JECH (1979) Set Theory, Academic Press.

[33] P. JULLIEN (1969) Contribution à l'étude des types d'ordre dispersés, Thèse, Marseille.

[34] H. KOMM (1948) On the dimension of partially ordered sets, Amer. J. Math. 70, 507-520.

[35] J.B. KRUSKAL (1972) The theory of well-quasi ordering, a frequently discovered concept, J. Comb. Th. (A), 13, 297-305.

[36] R. LAVER (1971) On FRAÏSSÉ's order type conjecture, Annals of Math. 93, 89-111.

[37] R. LAVER (1973) An order type decomposition theorem, Annals of Math. 98, 96-119.

[38] H. M. MacNEILLE (1937) Partially ordered sets, Trans. Amer. Math. Soc. 42, 416-460.

[39] V. NOVAK (1965) On the lexicographic product of ordered sets, Czech. Math. Journ. 15, 270-282.

[40] R. LUCE and H. RAIFFA, Games and Decision, N.Y. Wiley, 1952.

[41] O. ORE (1962) Theory of graphs, Amer. Math. Soc. Colloq. Publ. 38.

[42] M. POUZET (1969) Sections initiales d'un ensemble ordonné, C.R.A.S. Paris, 269, 113-116.

[43] M. POUZET (1969) Généralization d'une construction de BEN-DUSHNIK-E.W. MILLER, C.R.A.S. Paris, 269, 877-879.

[44] M. POUZET and I. RIVAL, Which ordered sets have a complete linear extension? Canad. J. Math. (to appear).

[45] F.P. RAMSEY (1930) On a problem of formal logic, Proc. London Math. Soc. 30, 264-286.

[46] G.N. RANEY (1952) Completely distributive complete lattices, Proc. Amer. Math. Soc. 3, 677-680.

[47] J. SCHMIDT (1956) Zur Kennzeichnung der Dedekind MacNeilleschen Hülle einer geordneten Menge, Arch. Math. 7 241-249.

[48] J.H. SCHMERL (1977) $\aleph_0$-categoricity and comparability of graphs, Logic Colloquium, North-Holland, 221-228.

[49] S. SHELAH, Classification theory and the number of non-isomorphic models, Studies in Logic. Vol. 92, North Holland, Amsterdam. 544.

[50] W. SIERPINSKI (1950) Sur les types d'ordre des ensembles lineaires, Fund. Math. 37, 253-264.

[51] E. SZPILRAJN (1930) Sur l'extension de l'ordre partiel, Fund. Math. 16, 386-389.

[52] E.S. WOLK (1967) Partially well ordered sets and partial ordinals. Fund. Math. 60, 175-186.

# DIMENSION THEORY FOR ORDERED SETS

David Kelly
Department of Mathematics and Astronomy
The University of Manitoba
Winnipeg, Canada R3T 2N2

William T. Trotter, Jr.
Department of Mathematics and Statistics
University of South Carolina
Columbia, South Carolina 29208

ABSTRACT

In 1930, E. Szpilrajn proved that any order relation on a set $X$ can be extended to a linear order on $X$. It also follows that any order relation is the intersection of its linear extensions. B. Dushnik and E.W. Miller later defined the *dimension* of an ordered set $P = \langle X;\leq\rangle$ to be the minimum number of linear extensions whose intersection is the ordering $\leq$ .

For a cardinal $m$, $\underset{\sim}{2}^m$ denotes the subsets of $m$, ordered by inclusion. As the notation indicates, $\underset{\sim}{2}^m$ is a product of 2-element chains (linearly ordered sets). Any poset $\langle X;\leq\rangle$ with $|X| \leq m$ can be embedded in $\underset{\sim}{2}^m$. O. Ore proved that the dimension of a poset $P$ is the least number of chains whose product contains $P$ as a subposet. He also showed that the product of $m$ nontrivial chains has dimension $m$. In particular, $\underset{\sim}{2}^m$ has dimension $m$, a result of H. Komm. Thus, every cardinal is the dimension of some poset.

It is usually very difficult to calculate the dimension of any "standard" poset. However, dimension can be related to other parameters of a poset. For example, the dimension of a finite poset does not exceed the size of any maximal antichain. Also, T. Hiraguchi showed that any poset of dimension $d \geq 3$ has at least $2d$ elements. Moreover, any integer $\geq 2d$ is the size of some poset of dimension $d$.



*I. Rival (ed.), Ordered Sets, 171–211.*

Let $d$ be a positive integer. A poset is *$d$-irreducible* if it has dimension $d$ and removal of any element lowers its dimension. Any poset whose dimension is at least $d$ contains a $d$-irreducible subposet. Although there is only one 2-irreducible poset, there are infinitely many $d$-irreducible posets whenever $d \geq 3$. The set of all 3-irreducible posets was independently determined by D. Kelly and W.T. Trotter, Jr. and J.I. Moore, Jr. There is a 3-irreducible poset of any size $n$ not excluded by Hiraguchi; i.e., for any $n \geq 6$. However, R.J. Kimble, Jr. has shown that a $d$-irreducible poset cannot have size $2d + 1$ when $d \geq 4$. If $d \geq 4$ and $n \geq 2d$ but $n \neq 2d + 1$, then there is a $d$-irreducible poset of size $n$.

A finite poset is *planar* if its diagram can be drawn in the plane without any crossing of lines. Planar posets have arbitrary finite dimension. However, K.A. Baker showed that a finite *lattice* is planar exactly when its dimension does not exceed 2. He also showed that the completion of a poset is a lattice that has the same dimension as the poset. Baker's results and three papers of D. Kelly and I. Rival were used to obtain the list of 3-irreducible posets.

The approach of W.T. Trotter and J.I. Moore, Jr. rested on the observation of Dushnik and Miller that a poset has dimension at most 2 if and only if its incomparability graph is a comparability graph. T. Gallai's characterization of comparability graphs in terms of excluded subgraphs was then applied.

Several other connections between dimension theory for posets and graph theory have been established. For example, posets with the same comparability graph have the same dimension. C.R. Platt reduced the planarity of a finite lattice to the planarity of an *undirected* graph obtained by adding an edge to its diagram.

E. Szpilrajn [1930] showed that any order relation on a set $X$ can be extended to a linear order on $X$. He also proved that any order relation is the intersection of its linear extensions. If $\mathcal{C}$ is a family of linear orders (chains) whose intersection is the order relation $\le$, then $\mathcal{C}$ is a *realizer* of $\le$; we also say that $\mathcal{C}$ *realizes* $\le$. B. Dushnik and E.W. Miller [1941] defined the *dimension* of an ordered set (poset) $\langle X;\le\rangle$ to be the minimum cardinality of a realizer of $\le$.

In this survey, we usually deal with the case of finite dimension. For a positive integer $k$, any poset of dimension at least $k$ contains a finite subposet of dimension $k$. (This compactness property follows from the compactness theorem of first-order logic; see Harzheim [1970] and the review by K.A. Baker.) Moreover, any finite poset has finite dimension. Consequently, most of the posets that we consider will be finite.

## 1. INTRODUCTION

In this section, we shall introduce the many definitions and constructions that are required to study dimension. The 18 posets in the first figure are meant to provide the reader with some realistic examples in order to better appreciate the definitions and constructions we give. We want the reader to get a feeling for the main concepts before we get involved in difficult dimension calculations. No subsequent section requires any background except this introduction, and possibly, Section 3.

If a poset is denoted by $P$, then its underlying set will usually also be denoted by $P$, and its order relation by $\le$ or $\le_P$. An extension of $P$ means an extension of $\le_P$. If $P$ is a subposet of $Q$, then $\dim P \le \dim Q$ since any linear extension of $Q$ restricts to a linear extension of $P$. T. Hiraguchi [1951] showed that the dimension increases by at most one when a single point is added. Equivalently, the "one-point removal theorem" is valid; that is, removing one point from a poset decreases its dimension by at most one. We shall consider this and other removal theorems in Sections 5 and 6.

---

This research was supported by the NSERC of Canada and the National Science Foundation.

The converse of a binary relation $R$ is denoted by $R^d$. Clearly, a poset $P$ and its dual $P^d$ have the same dimension. A poset is *d-irreducible* if it has dimension $d \geq 2$ and the removal of any element lowers its dimension. An *irreducible* poset is $d$-irreducible for some $d \geq 2$. By the compactness property, every irreducible poset is finite. By the one-point removal theorem, a poset is $d$-irreducible when $\dim P \geq d$ and $\dim(P - \{x\}) < d$ for every $x \in P$. The only 2-irreducible poset is the 2-element antichain (two incomparable elements). Figure 1 shows all 3-irreducible posets with at most 8 elements (up to isomorphism and duality). (We are using the notation of Kelly [1977].) Two of these posets have 6 elements, three have 8 elements, and the remainder have 7 elements. In Section 3, we show how to check whether a poset has dimension at most two.

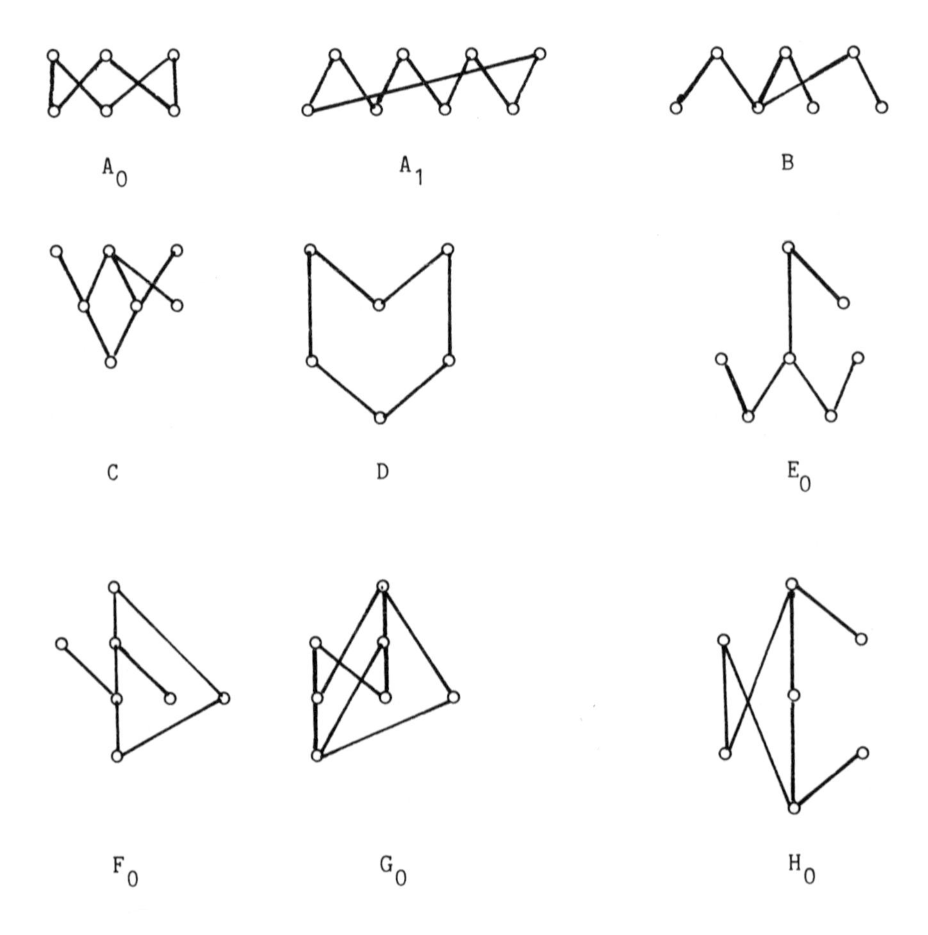


FIGURE 1. The 3-irreducible posets with at most 8 elements.

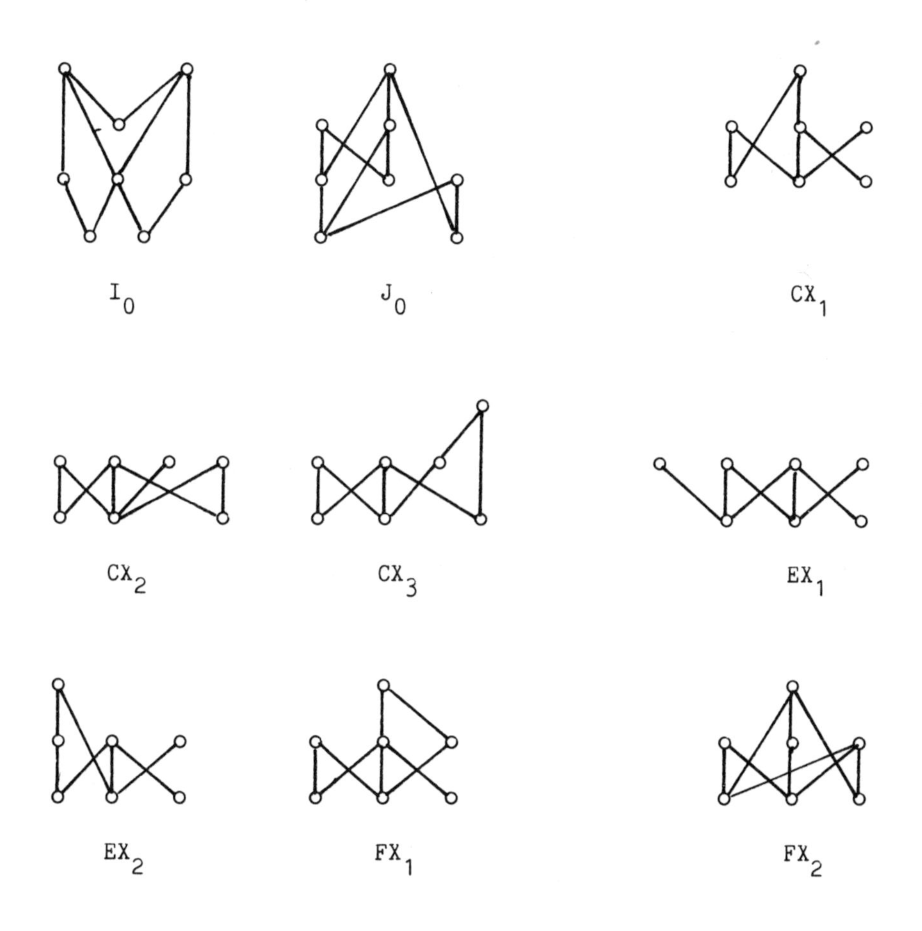


FIGURE 1. (concluded)

By Figure 1, any poset of dimension 3 must have at least 6 elements. In fact, this is a special case of the following inequality of Hiraguchi [1951], valid for $|P| \geq 4$ :

$$\dim P \leq |P|/2.$$

This result is proved in Section 5.

The posets $A_0$ and $A_1$ of Figure 1 are the two smallest members of an infinite family $A_n$ $(n \geq 0)$ of 3-irreducible posets. These posets, called *crowns*, appeared in Harzheim [1970] and Baker, Fishburn and Roberts [1971]. D. Kelly and I.

Rival [1975b] discovered infinitely many more 3-irreducible posets. They presented the four infinite families $E_n$, $F_n$, $G_n$, $H_n$ $(n \geq 0)$, whose smallest members are shown in Figure 1. The list of all 3-irreducible posets was completed independently by D. Kelly [1977], and W.T. Trotter and J.I. Moore [1976a]. Up to duality, this list consists of the seven infinite families $A_n$, $E_n$, $F_n$, $G_n$, $H_n$, $I_n$, $J_n$ $(n \geq 0)$, whose smallest members are shown in Figure 1, together with the remaining posets of that figure. We shall return to the topic of irreducible posets in Section 7.

The following alternative definition is often credited to O. Ore [1962], but appeared earlier in Hiraguchi [1955] : The dimension of a poset $P$ is the minimum size of a family $(C_i \mid i \in I)$ of chains such that $P$ is isomorphic to a subposet of the direct product $\Pi(C_i \mid i \in I)$. If $(C_i \mid i \in I)$ is a family of linear extensions realizing $P$, then the diagonal mapping $\phi$ embeds $P$ into the direct product of these chains ( $\phi(x)$ is the constant function with value $x$ for each $x$ in $P$). To complete the proof of the equivalence of the two definitions, let $P$ be the direct product of the family $(C_i \mid i \in I)$ of chains. For $i \in I$, we define the extension $E_i$ of $P$ by $\langle x, y\rangle \in E_i$ iff $\pi_i(x) < \pi_i(y)$ , where $\pi_i$ denotes the $i$-th projection function. If $E_i'$ is a linear extension of $E_i$ for each $i \in I$, then $(E_i' \mid i \in I)$ realizes $P$. H. Komm [1948] proved that $\dim \mathfrak{2}^{\mathfrak{m}} = \mathfrak{m}$ , where $\mathfrak{2}$ denotes the 2-element chain and $\mathfrak{m}$ denotes a possibly infinite cardinal. It follows that the product of $\mathfrak{m}$ nontrivial chains has dimension $\mathfrak{m}$.

The Ore definition of dimension is often appropriate when no explicit calculation of dimension is required. When the dimension must be calculated, the original Dushnik-Miller definition is used. We shall describe a simple technique which reduces the "bookkeeping" involved in calculating the dimension of a finite poset.

A (linear) extension of a subposet of $P$ will be called a *partial (linear) extension* of $P$. $\mathrm{Tr}(R)$ denotes the transitive closure of a binary relation $R$. For any partial extension $E$ , $\mathrm{Tr}(\leq_P \cup E)$ is an extension of $P$. For a poset $P$, $\mathcal{I}(P) = \{\langle a,b\rangle \mid a \| b \text{ in } P\}$, the set of *incomparable pairs*. Any $u \in \mathcal{I}(P)$ can be considered as a partial linear extension of $P$.

For $u, v \in \mathcal{I}(P)$, $u$ *forces* $v$ ( in symbols, $\langle u,v\rangle \in F$) if $v \in \mathrm{Tr}( \leq_P \cup \{u\} )$. This concept was introduced by I. Rabinovitch and I. Rival [1979], who observed that $F$ is an order relation on $\mathcal{I}(P)$. The incomparable pair $\langle a,b\rangle$ is called a *critical pair* if $x < b$ implies $x < a$, and $x > a$ implies $x > b$. A pair $\langle a,b\rangle \in P^2$ is a *max-min pair* if $a$ is maximal and $b$ is minimal. Observe that every incomparable max-min pair is critical. For a poset $P$, Crit($P$) denotes the set of critical pairs. Crit($P$) is the set of minimal elements of $\langle \mathcal{I}(P);F \rangle$. We denote the set of maximal elements of $\langle \mathcal{I}(P);F \rangle$ by $\mathcal{N}(P)$ ; these pairs have been called *unforced, nonforcing* and *nonforced*. We shall use the last term. Moreover, $\mathcal{N}(P)$ is the converse of Crit($P$). Thus, critical pairs are also called "reversed nonforced pairs". For any finite poset, every incomparable pair is forced by a critical pair.

For $S \subseteq \mathcal{I}(P)$, an *alternating cycle of length* $n$ for $S$ is a sequence of the form $x_1, y_1, x_2, y_2, \ldots, x_n, y_n$ with $\langle x_i,y_i\rangle \in S$, $1 \leq i \leq n$, and $y_i \leq x_{i+1}$ in $P$ (with subscripts taken modulo $n$). $S$ is *cycle-free* if it has no alternating cycle. The above cycle is *minimal* if $y_i \leq x_j$ implies $j = i+1 \pmod n$. If $S$ is not cycle-free, then there is a minimal alternating cycle for $S$. If the above alternating cycle is minimal, then both $\{x_1, x_2, \ldots, x_n\}$ and $\{y_1, y_2, \ldots, y_n\}$ are $n$-element antichains. For example, $x_1 \leq x_i$ would imply that $y_n \leq x_i$. Thus, the length of a minimal alternating cycle for $\mathcal{I}(P)$ cannot exceed the width of the poset $P$.

We say that a family $(E_i \mid i \in I)$ of partial extensions *realizes* $P$ (or $\leq_P$) when $(C_i \mid i \in I)$ realizes $\leq_P$ for any choice of linear extensions $C_i$ of $P$ that extend $E_i$ $(i \in I)$. The dimension of $P$ is the minimum (nonzero) size of such a realizer. Let $P$ be a poset in which every incomparable pair is forced by a critical pair. If $P$ is not a chain, then dim $P$ is specified by each of the following three definitions:

(a) The least number of partial extensions of $P$ whose union contains Crit($P$).

(b) The least number of partial linear extensions of $P$ whose union contains Crit($P$).

(c) The least number of cycle-free sets covering Crit($P$).

The set Crit($P$) is the smallest subset of $\mathcal{I}(P)$ that could be used in the above definitions; thus, Crit($P$) is a "critical" set in dimension calculations. When using definition (a) or (b), we often write a critical pair $\langle a,b \rangle$ as "$a < b$" and call it a *critical inequality*.

Henceforth, any poset is understood to be finite unless the contrary is explicitly stated. $P$ will always denote a poset. A realizer is *irredundant* if no proper subset is a realizer. The dimension is the smallest size of an irredundant realizer consisting of linear extensions, while the *rank* is the largest. Note that realizers used for rank must consist of linear extensions, in contrast to the case for dimension. For example, the realizer consisting of Crit($P$), considered as partial linear extensions, is not permitted. Since any proper extension of $P$ satisfies some nonforced pair, I. Rabinovitch and I. Rival [1979] observed that rank($P$) $\leq |\mathfrak{n}(P)|$. S.B. Maurer, I. Rabinovitch and W.T. Trotter ([1980a], [1980b], [1980c]) describe how to determine the rank from $\mathfrak{n}(P)$, considered as a directed graph. Since $\mathfrak{n}(P)$ and Crit($P$) are converses of each other, any extension satisfying all of $\mathfrak{n}(P)$ will fail every critical inequality. An extension satisfying all of $\mathfrak{n}(P)$ is called a *weak* extension. In Section 5, we show that every irreducible poset, except the 2-element antichain, has a weak extension.

We shall define a subposet $\underset{\sim}{\mathrm{P}}(P)$, the set of *irreducible* elements of $P$, that has the same dimension as $P$. (The term "irreducible" has a completely different meaning when applied to elements than when applied to posets.) $\underset{\sim}{\mathrm{J}}(P)$ is the subposet of *join-irreducible* elements ($a \in \underset{\sim}{\mathrm{J}}(P)$ iff $a = \bigvee S$ implies $a \in S$). If $P$ contains a least element 0, it is join-reducible since we allow $S = \emptyset$. $\underset{\sim}{\mathrm{M}}(P)$, the subposet of meet-irreducible elements, is defined dually. We define

$$\underset{\sim}{\mathrm{P}}(P) = \underset{\sim}{\mathrm{J}}(P) \cup \underset{\sim}{\mathrm{M}}(P),$$

a subposet of $P$.

If $P$ is a lattice and $x \in P$, then $x \in \underset{\sim}{\mathrm{J}}(P)$ iff $x$ has a unique lower cover. In an arbitrary finite poset $P$, it is much harder to recognize the join-irreducible elements. This difficulty is overcome by constructing the finite lattice $L = \underset{\sim}{\mathrm{L}}(P)$ which contains $P$ as a subposet and satisfies $\underset{\sim}{\mathrm{J}}(L) = \underset{\sim}{\mathrm{J}}(P)$, $\underset{\sim}{\mathrm{M}}(L) = \underset{\sim}{\mathrm{M}}(P)$ and $\underset{\sim}{\mathrm{P}}(L) = \underset{\sim}{\mathrm{P}}(P)$. $\underset{\sim}{\mathrm{L}}(P)$, the *completion* of $P$, is called the "completion by cuts" and defined in Birkhoff [1967]; it is also called the "MacNeille completion". The definition of $\underset{\sim}{\mathrm{L}}(P)$ does not require $P$ to be finite. Whenever a complete lattice contains a poset $P$, it contains the completion of $P$ as a subposet. We shall not define the completion. Given a finite poset $P$ and a lattice $L$, one can use the following result to determine whether $L = \underset{\sim}{L}(P)$.

LEMMA 1.1. (Banaschewski [1956] or Schmidt [1956]) *If the poset* $P$ *is a subposet of the finite lattice* $L$, *then* $L = \underset{\sim}{L}(P)$ *iff* $\underset{\sim}{P}(L) \subseteq P$.

In particular, the completion of a finite chain is the chain itself. Using definition (a) or (b) of dimension, the next result shows that $\underset{\sim}{P}(P)$ and $P$ have the same dimension.

LEMMA 1.2. (Kelly [1981]) *For a finite poset* $P$,

$$\mathrm{Crit}(P) \subseteq \underset{\sim}{M}(P) \times \underset{\sim}{J}(P).$$

*Proof*. Let $\langle a,b\rangle \in \mathrm{Crit}(P)$ and suppose that $a = \bigwedge S$ with $a \notin S$. For all $x \in S$, $x > a$, and therefore, $x > b$. Consequently, $a = \bigwedge S \geq b$, a contradiction.

PROBLEM 1.3. Let $P$ be an irreducible poset. Since $\underset{\sim}{P}(P)$ has the same dimension as $P$, $\underset{\sim}{P}(P) = P$. By definition (a), every element of $P$ appears in some critical pair. For $x \in P$, it follows by Lemma 1.2 that $x \in \underset{\sim}{M}(P) - \underset{\sim}{J}(P)$ if and only if $x$ is the first element of some critical pair, but never the second. If $x$ appears in both positions in critical pairs, then $x$ must be doubly irreducible (i.e., $x \in \mathrm{Irr}(P) = \underset{\sim}{J}(P) \cap \underset{\sim}{M}(P)$). We do not know whether the converse holds.

PROBLEM 1.4. From the Ore definition,

$$\dim(P \times Q) \leq \dim P + \dim Q,$$

for any posets $P$ and $Q$. In general, this inequality can be strict. For example, the product of two nontrivial antichains is an antichain and thus has dimension two, although the sum of the dimensions is four. A poset is bounded if it has a least element (zero) and a greatest element (one). K.A. Baker [1961] showed that equality holds above if $P$ and $Q$ are bounded posets. Let us indicate a proof. Suppose the linear extensions $C_1, C_2, \ldots, C_{m+n}$ realize $P \times Q$, $\langle 0,1\rangle < \langle 1,0\rangle$ holds in $C_1, \ldots, C_m$, and $\langle 1,0\rangle < \langle 0,1\rangle$ holds in $C_{m+1}, \ldots, C_{m+n}$. If $\langle a,b\rangle \in \mathcal{J}(P)$, then $\langle a,1\rangle < \langle b,0\rangle$ must hold in $C_i$ for some $1 \leq i \leq m$. Consequently, $m \geq \dim P$ and, similarly, $n \geq \dim Q$. Therefore, $\dim(P \times Q) \geq \dim P + \dim Q$. Thus, dimension is additive for direct products of bounded posets. Let $b(P, Q) \in \{0,1,2\}$ count how many of $P$ and $Q$ are not bounded. We *conjecture* that

$$\dim(P \times Q) \geq \dim P + \dim Q - b(P, Q).$$

In the $b = 2$ case, it would suffice to prove (or disprove) this conjecture for irreducible posets $P$ and $Q$.

K.A. Baker [1961] proved that $\underset{\sim}{L}(P)$ has the same dimension as $P$ for any (possibly infinite) poset $P$. The completion of a chain is a complete chain. A product of complete chains is a complete lattice, and therefore contains, as a subposet, the completion of each of its subposets. Using the Ore definition of dimension, Baker's result now follows easily. (A proof using the Dushnik-Miller definition is given in Kelly [1981].) Recall that $\underset{\sim}{P}(P) = \underset{\sim}{P}(\underset{\sim}{L}(P))$. If $P$ is finite, then

$$\dim P = \dim \underset{\sim}{P}(P) = \dim \underset{\sim}{P}(\underset{\sim}{L}(P)) = \dim \underset{\sim}{L}(P),$$

because the operator $\underset{\sim}{P}$ preserves dimension in the first and last equalities. This proves Baker's result for finite $P$.

For $n \geq 3$, let us define the poset

$$S_n = \underset{\sim}{P}(\underset{\sim}{2}^n).$$

$S_n$ has $2n$ elements and dimension $n$. We call $S_n$ the *standard* poset of dimension $n$. Let $S_n = \{a_1, a_2, \ldots, a_n, b_1, b_2, \ldots, b_n\}$, where $a_i$ is minimal and $a_i \| b_i$ for $1 \leq i \leq n$. The following $n-1$ partial linear extensions realize $S_n - \{b_1\}$ :

$a_3, a_4, \ldots, a_n, a_1, b_2, a_2$
$a_2, a_1, b_3, a_3$
$b_i, a_i \qquad (4 \leq i \leq n).$

Consequently, $S_n$ is $n$-irreducible. Observe that $S_3$ is the poset $A_0$ of Figure 1. For an infinite cardinal $\mathfrak{m}$, the poset $S_{\mathfrak{m}}$, consisting of one-element subsets of $\mathfrak{m}$ and their complements, is the *standard* poset of dimension $\mathfrak{m}$.

We can associate a hypergraph to a finite poset $P$ so that the chromatic number of the hypergraph is the dimension of the poset. The hypergraph $\mathcal{H}(P)$ is defined as follows:

The vertex set of $\mathcal{H}(P)$ is $\mathrm{Crit}(P)$.

An edge of $\mathfrak{H}(P)$ is any set of the form

$$\{\langle x_i, y_i\rangle \mid 1 \le i \le n\}$$

where $x_1, y_1, x_2, y_2, \ldots, x_n, y_n$ is a minimal alternating cycle for $\mathrm{Crit}(P)$.

By definition (c), $P$ has dimension at most $k$ if and only if the hypergraph $\mathfrak{H}(P)$ is (vertex) k-colorable. By Kelly, Schönheim and Woodrow [1981], a graph can be constructed that is k-colorable if and only if $\mathfrak{H}(P)$ is.

Calculating the chromatic number of $\mathfrak{H}(P)$ is not usually the easiest way to determine the dimension of a poset $P$. Even the hypergraph itself may be difficult to calculate. For example, the only posets $P$ of Figure 1 for which $\mathfrak{H}(P)$ is a graph are $A_0$, $A_1$, C, D, and $F_0$. However, $\mathfrak{H}(G_n)$ is a graph whenever $n \ge 1$. The simple structure of this graph will be exploited in Section 7. In Figure 2, we show $G_2$ and its hypergraph; the labelling indicates a weak linear extension (so that $i > j$ as integers whenever $\langle i, j\rangle$ is critical).

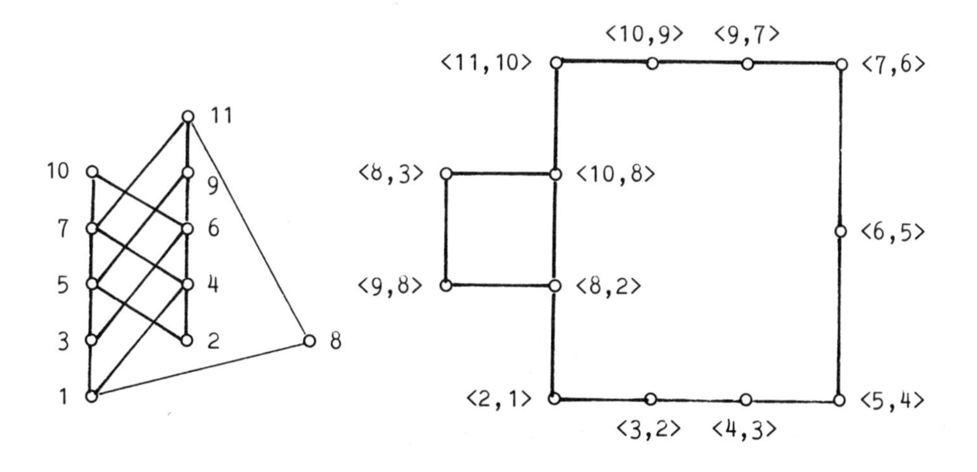


$G_2$ $\mathfrak{H}(G_2)$

FIGURE 2. A poset and its hypergraph

The connection between dimension and graph (or hypergraph) coloring has been exploited to determine the computational complexity of dimension. Section 4 is devoted to this topic.

## 2. TWO INTERPRETATIONS OF DIMENSION

THEOREM 2.1. (Trotter and Moore [1976b]) *Let $G$ be a possibly infinite connected graph. Let $P$ be all connected induced subgraphs of $G$, ordered by inclusion. ($P$ includes the empty set.) The dimension of $P$ is the number of noncut vertices of $G$. (A vertex $v$ of $G$ is a noncut vertex when the removal of $v$ leaves a connected subgraph.)*

*Proof*. A critical pair for $P$ is $\langle A,\{u\}\rangle$ where $A$ is a connected component of $G - \{u\}$. Since every incomparable pair is forced by a critical pair, we can use definition (c) of dimension.

Let $v_i$, $0 \leq i < \mathfrak{m}$, be the noncut vertices of $G$. The subposet

$$\{ \{v_i\} \mid 0 \leq i < \mathfrak{m} \} \cup \{ G - \{v_i\} \mid 0 \leq i < \mathfrak{m} \}$$

is isomorphic to the standard $\mathfrak{m}$-dimensional poset. Thus, dim $P \geq \mathfrak{m}$. Next, we define $\mathfrak{m}$ cycle-free sets covering Crit($P$). For $0 \leq i < \mathfrak{m}$, $S_i$ consists of the critical pairs $\langle A,\{u\}\rangle$ with $d(x,v_i) > d(u,v_i)$ for all $x \in A$. Let $\langle A,\{u\}\rangle$ be a critical pair. If $u = v_i$, then $\langle A,\{u\}\rangle$ occurs in $S_i$. Otherwise, we can assume that $u$ is a cut vertex. Choose a noncut vertex $v_i$ in a connected component of $G - \{u\}$ different from $A$. For $x \in A$, $d(x,v_i) = d(x,u) + d(u,v_i) > d(u,v_i)$. Consequently, $\langle A,\{u\}\rangle$ occurs in $S_i$. Since each $S_i$ is cycle-free, dim $P = \mathfrak{m}$.

THEOREM 2.2. (Wille [1975]) *Let $L$ be a finite distributive lattice and $P = \underset{\sim}{J}(L)$, the subposet of join-irreducible elements.* dim $P$ *is the least number of chains whose union generates $L$.*

*Remark.* Applying the decomposition theorem of Dilworth [1950], the above statement means that dim $P$ is the minimum width of a generating set of $L$.

*Proof.* If $P$ is embedded in the direct product of finite chains $C_1, C_2, \ldots, C_n$, then applying the usual duality theory for finite distributive lattices (see, for example, Birkhoff [1967]), $L$ is a homomorphic image of the free $\{0,1\}$-distributive product of $\{0,1\}$-chains $D_1, D_2, \ldots, D_n$, where $|D_i| = |C_i| + 1$ for $1 \le i \le n$. For $1 \le i \le n$, $D_i'$, the image of $D_i$ in $L$, is a chain. Clearly, $D_1' \cup D_2' \cup \ldots \cup D_n'$ generates $L$.

If $L$ is generated by the $\{0,1\}$-chains $D_i$, $1 \le i \le n$, then $L$ is a homomorphic image of the free $\{0,1\}$-distributive product of $D_1, D_2, \ldots, D_n$. Reversing the previous argument, we conclude that $P$ is embedded in the direct product of $n$ chains. This completes the proof.

## 3. DIMENSION AT MOST TWO

As usual, all posets are finite. For a poset $P$, $C(P)$ will denote its covering digraph; $\langle a, b\rangle$ is in $C(P)$ exactly when $a \prec b$ in $P$. The usual diagram ("Hasse diagram") of $P$ is a *representation* of $C(P)$ in the plane where each edge $\langle a,b\rangle$ is represented by a straight line segment with $a$ lower than $b$. The poset $P$ is *planar* if $C(P)$ can be represented in the plane with no crossing of edges. A bounded planar poset is a lattice. A lattice is planar iff its dimension does not exceed two. If a poset is not a lattice we could take its completion, and determine whether that is planar. In fact, to prove that a poset has dimension at most two, it suffices to embed it in any planar lattice. If a planar poset $P$ has a zero, its dimension does not exceed three (Trotter and Moore [1977]). However, planar posets have arbitrary finite dimension (Kelly [1981]).

Let $\phi\colon L \to C_1 \times C_2$ be a (poset) embedding of the lattice $L$ into the direct product of the two chains $C_1$ and $C_2$. We identify each chain $C_i$ with the subposet $\{1,2,\ldots,|C_i|\}$ of the reals. We shall write $\langle a_1, a_2\rangle$ for $\phi(a)$. As in

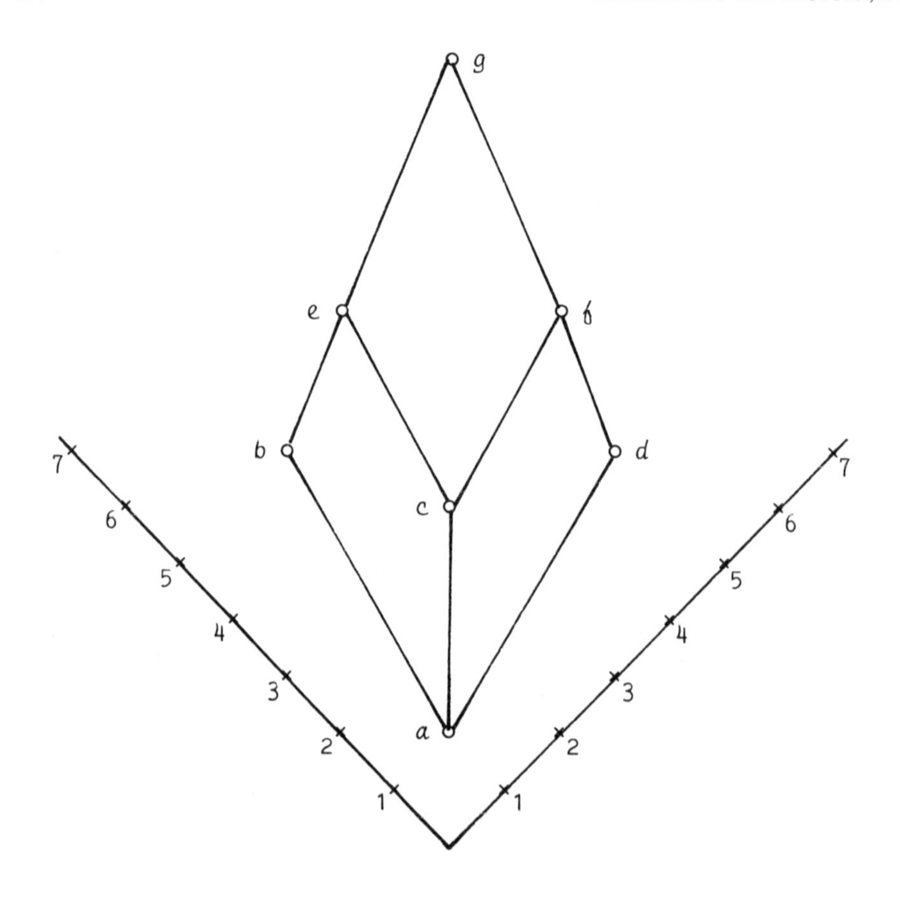


FIGURE 3. A planar representation

Figure 3, we draw axes at $45^{\circ}$ to the horizontal so that the line $x_1 + x_2 = 0$ is horizontal and the line $x_1 - x_2 = 0$ is vertical. Each $a \in L$ is plotted at the point with coordinates $\langle a_1, a_2 \rangle$. Joining covering pairs by straight lines, we obtain a planar representation of $L$. Certainly, $a$ is plotted below $b$ whenever $a < b$ in $L$. Let $a \prec b$ and $c \prec d$ in $L$ with $|\{a, b, c, d\}| = 4$, and suppose that the corresponding edges cross. It would then follow that $\langle a_1, a_2 \rangle$ and $\langle c_1, c_2 \rangle$ are both less than $\langle b_1, b_2 \rangle$ and $\langle d_1, d_2 \rangle$, and therefore, that $a \vee c \leq b \wedge d$ holds in $L$. Since $a \prec b$, we conclude from $a \leq a \vee c \leq b$ that $a \vee c \in \{a, b\}$. If $a \vee c = b$, then $c < b \leq b \wedge d \leq d$ would imply $b = d$, a contradiction. Therefore, $a \vee c = a$. By symmetry, $a \vee c = c$, implying $a = c$, a contradiction.

The planar representation of the lattice $L$ in Figure 3 is derived from the diagonal mapping and the following two linear extensions of $L$:

$$C_1 = a, b, c, e, d, f, g$$

$$C_2 = a, d, c, f, b, e, g$$

Let $G$ be the *incomparability graph* of $P$, the graph with vertex set $P$ and edge set $\mathcal{I}(P)$. In Section 8, we show that $G$ determines the dimension of $P$. This result is usually stated using the complement of $G$, which is called the *comparability graph* of $P$ and denoted by Comp($P$). We shall also write $\mathcal{I}(P)$ for the incomparability graph of $G$. A *transitive orientation* of a graph $G$ is a strict order relation $\lambda$ on the vertex set of $G$ such that $x$ and $y$ are comparable with respect to $\lambda$ iff $xy$ is an edge of $G$. Clearly, a graph having a transitive orientation is the same thing as a comparability graph. A transitive orientation $\lambda$ of $\mathcal{I}(P)$ is called a *conjugate* order for $\leq_P$ (or for $P$). B. Dushnik and E.W. Miller [1941] showed that dim $P \leq 2$ iff $P$ has a conjugate order. K.A. Baker [1961] combined this result with a result of J. Zilber (ex. 7(c), p.32 of Birkhoff [1967]) to show the equivalence for lattices between planarity and dimension at most two.

For a poset $P$ embedded in the product $C_1 \times C_2$ of two chains, the *standard* conjugate order $\lambda$ is defined by:

$$x \lambda y \text{ iff } \pi_1(x) < \pi_1(y) \quad (\text{for } x || y \text{ in } P).$$

If $\lambda$ is a conjugate order for $\leq_P$, then $\leq_P \cup \lambda$ and $\leq_P \cup \lambda^d$ are two linear orders that realize $P$. Consequently, given any conjugate order $\lambda$ for a poset $P$, there is an embedding of $P$ into the product of two chains for which $\lambda$ is standard.

A conjugate order $\lambda$ for a lattice $L$ has a geometric interpretation: for $x||y$ in $L$, $x \lambda y$ iff $x$ is "to the left of" $y$ in some planar representation of $L$. Given a planar representation of a lattice $L$, the set of lower covers of any element has an obvious left-to-right ordering $\lambda$. If $x||y$ in $L$, then let $x'$ and $y'$ be the lower covers of $x \vee y$ that are comparable with $x$ and $y$ respectively. We define $x \lambda y$ if $x' \lambda y'$, and $y \lambda x$ if $y' \lambda x'$; $\lambda$ is the left-to-right ordering of $L$ with respect to the planar representation. For example, in Figure 3, $b$ is to the left of $f$. (For more details, see Kelly and Rival [1975a].)

If the planar representation of a lattice $L$ is obtained from an embedding into the product of chains $C_1$ and $C_2$ by the procedure we gave above, observe that the left-to-right ordering coincides with the standard conjugate order.

Let $\lambda$ be a conjugate order for a poset $P$. We can assume that $P$ is a subposet of a product $C_1 \times C_2$ of chains and that $\lambda$ is standard. Since $C_1 \times C_2$ is a lattice it contains $\underset{\sim}{L}(P)$ as a subposet. Clearly, the standard conjugate order for $\underset{\sim}{L}(P)$ extends $\lambda$. We have shown that any conjugate order for a poset $P$ can be extended to a conjugate order for $\underset{\sim}{L}(P)$. (In fact, the extension is unique.) Thus, any conjugate order $\lambda$ for an arbitrary poset $P$ has the following interpretation: $\lambda$ is the restriction to $P$ of the left-to-right ordering of $\underset{\sim}{L}(P)$ with respect to some planar representation.

M.C. Golumbic ([1977a], [1977b], [1980]) has developed an algorithm to decide whether a graph is a comparability graph. Since this algorithm has complexity $O(n^3)$, where $n$ is the number of vertices, deciding whether $\dim(P) \leq 2$ has complexity $O(|P|^3)$ by the Dushnik-Miller result.

D. Kelly and I. Rival [1975a] determined the minimum list $\mathcal{L}$ of lattices such that every nonplanar lattice contains a lattice in $\mathcal{L}$ as a subposet. In particular, each lattice in $\mathcal{L}$ is nonplanar. In contrast to K. Kuratowski's [1930] characterization of planar graphs, $\mathcal{L}$ is infinite. $\mathcal{P} = \{ \underset{\sim}{P}(L) \mid L \in \mathcal{L} \}$ is an infinite set of 3-irreducible posets because, for each $L \in \mathcal{L}$, the *subposet* $L - \{x\}$ is a planar lattice for each $x \in \underset{\sim}{P}(L)$ (Kelly and Rival [1975b]). $\mathcal{P}$ contains the first nine posets of Figure 1.

Let $G(P)$ be the ordinary graph corresponding to the covering digraph $C(P)$, and let $G_{01}(P)$ be the graph obtained from $G(P)$ by adding an extra edge from 0 to 1. C.R. Platt [1976] showed that a lattice $L$ is planar iff $G_{01}(L)$ is planar. If $P$ is an $n$-element poset with dimension at most 2, then $|\underset{\sim}{L}(P)| \leq n^2$. Clearly, one cannot form the completion in any polynomial algorithm to check whether $\dim(P) \leq 2$. If a subposet $Q$ of $\underset{\sim}{L}(P)$ is not a lattice, one can add a point $x$ to $Q$ so that $Q \cup \{x\}$ is a subposet of $\underset{\sim}{L}(P)$. Starting with $P$, this procedure is repeated until a lattice $L$ is obtained or $Q$ has more that $n^2$ elements. If $L$ is obtained, $G_{01}(L)$ is formed and Platt's result is applied. Thus, we have indicated another proof that dimension at most 2 is polynomial, but with a much less efficient algorithm than Golumbic's.

## 4. COMPUTATIONAL COMPLEXITY OF DIMENSION

M. Yannakakis [1981] has shown that it is NP-complete to determine if the dimension of a finite poset is at most 3. The reader is referred to the book of M.R. Garey and D.S. Johnson [1979] for the precise definitions and background on computational complexity. The intuitive idea (which more than suffices for us) is that any algorithm for deciding an NP-complete problem will use an excessive amount of time, even for small examples. M. Yannakakis associated to each finite graph $G$ a poset $Y(G)$ such that $G$ is 3-colorable iff $\dim(Y(G)) \leq 3$. We shall present his construction of $Y(G)$. Since 3-colorability is NP-complete (Garey, Johnson and Stockmeyer [1976]), the above equivalence proves that the dimension at most 3 problem is also NP-complete. Since the dimension $\leq 2$ problem is polynomial (see Section 3), it follows that the dimension equal to 3 problem is NP-complete. Consequently, dimension equal to $k$ is NP-complete for each fixed $k \geq 3$. By similar methods, M. Yannakakis [1981] also showed that the dimension $\leq 4$ problem for posets of unit length is NP-complete, but he left open the complexity of the dimension $\leq 3$ problem for posets of unit length. Using a reduction to hypergraph 2-colorability, which is NP-complete by Lovász [1973], E.L. Lawler and O. Vornberger [1981] proved the weaker result that the dimension problem is NP-complete.

Let $G$ be a finite graph with its vertices labelled 1, 2, ..., $n$ that has $m$ edges $E_1, E_2, \ldots, E_m$. The underlying set of the poset $Y(G)$ consists of $a_i, b_i, a_i', b_i'$ $(1 \leq i \leq n)$ and $c_k, d_k, c_k', d_k'$ $(1 \leq k \leq m)$. The ordering of $Y(G)$ is indicated in Figure 4. The comparabilities in the left-hand diagram are the same for every graph with $n$ vertices and $m$ edges. For $i \in E_k$, let

$$u_{ik} = \begin{cases} c_k, & \text{if } i = \min E_k ; \\ d_k, & \text{if } i = \max E_k . \end{cases}$$

and

$$v_{ik} = \begin{cases} d_k, & \text{if } i = \min E_k ; \\ c_k, & \text{if } i = \max E_k . \end{cases}$$

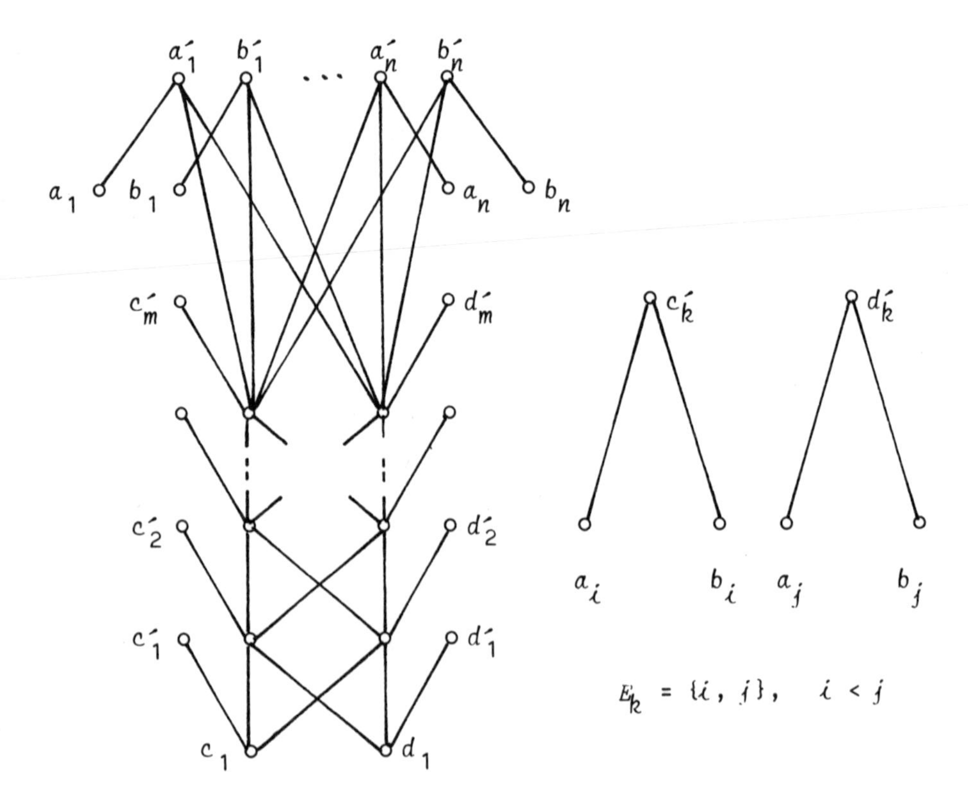


FIGURE 4. The associated poset $Y(G)$

As indicated in Figure 4, the following comparabilities also hold in $Y(G)$, where $i \in E_k$:

$$a_i < u'_{ik} \quad , \qquad b_i < u'_{ik}.$$

Let $1 \leq i, j \leq n$ and $1 \leq k \leq m$. The critical inequalities for $Y(G)$ are:

$$a'_i < b_j \ , \ b'_i < a_j$$

$$a'_i < a_j \ , \ b'_i < b_j \qquad (i \neq j)$$

$$c'_k < d_k \ , \ d'_k < c_k$$

$$a_i < c_1 , \ b_i < c_1 , \ a_i < d_1 , \ b_i < d_1$$

$$v'_{ik} < a_i \ , \ v'_{ik} < b_i \qquad (i \in E_k) .$$

We first assume that $\dim(Y(G)) \leq 3$, and let $L_0$, $L_1$ and $L_2$ be three linear extensions realizing $Y(G)$. Any isolated vertex is colored arbitrarily. If $i \in E_k$, then the following three critical inequalities must hold in three different linear extensions:

$$a_i' < b_i \quad , \quad b_i' < a_i \quad , \quad u_{ik}' < v_{ik} \ .$$

If the last one holds in $L_\alpha$, then we set $\mathrm{col}(i) = \alpha$. If $E_k = \{i, j\}$ and $i$ and $j$ are both assigned the color $\alpha$, then $c_k' < d_k$ and $d_k' < c_k$ would both hold in $L_\alpha$, a contradiction. Thus, $G$ is 3-colorable.

Let $G$ be a graph that is colorable with the three colors 0, 1, 2. We shall define three partial linear extensions that realize $Y(G)$. If $i \notin E_k$, then we define $u_{ik}'$ to be the empty sequence. The range of $i$ is $1, 2, \ldots, n$, in this order when we write $i\uparrow$, and in the opposite order when we write $i\downarrow$. When no arrow appears, any linear order is suitable. Also, $k$ ranges over the set $\{1, 2, \ldots, m\}$. For $\alpha = 0, 1, 2$, $L_\alpha$ is the following linear extension of $Y(G)$:

$$\begin{aligned}
&(a_i\, b_i \mid \mathrm{col}(i) = \alpha) \oplus (u_{ik}'\, c_k\, d_k \mid \mathrm{col}(i) = \alpha,\ k\uparrow) \\
&\oplus (a_i'\, b_i' \mid \mathrm{col}(i) = \alpha) \\
&\oplus (a_i\, a_i'\, b_i\, b_i'\, u_{ik}' \mid \mathrm{col}(i) = \alpha + 1 \ ,\ i\uparrow) \\
&\oplus (b_i\, b_i'\, a_i\, a_i'\, u_{ik}' \mid \mathrm{col}(i) = \alpha + 2 \ ,\ i\downarrow) \ .
\end{aligned}$$

(The ordinal sum, indicated by $\oplus$, is explained in the next section; addition is modulo 3.) Every critical inequality holds in one of these extensions. Therefore, $\dim(Y(G)) \leq 3$.

## 5. DIMENSION, WIDTH AND CARDINALITY

In this section, we shall derive some upper bounds on the dimension of a poset, in terms of parameters such as width and cardinality. Essential use will be made of R.P. Dilworth's decomposition theorem (Dilworth [1950]). All posets in this section are finite.

Let $P$ be a poset and $(Q_x \mid x \in P)$ be a family of posets indexed by $P$. The *ordinal sum* of this family is the poset $\langle R;\leq\rangle$, where $R$ consists of the pairs $\langle x,y\rangle$ with $x \in P$ and $y \in Q_x$, and $\langle x_1,y_1\rangle \leq \langle x_2,y_2\rangle$ iff $x_1 < x_2$ or $x_1 = x_2 = x$ and $y_1 \leq y_2$ in $Q_x$. The ordinal sum is *improper* if the base poset $P$ is trivial or if each $Q_x$ is trivial. If the base poset is the chain $1 < 2 < \ldots < n$, we write the ordinal sum as $Q_1 \oplus Q_2 \oplus \ldots \oplus Q_n$ . A poset is *decomposable* if it can be written as a proper ordinal sum. In particular, an indecomposable poset with more than two elements is *connected*; that is, it cannot be written as a disjoint union $P + Q$ of two posets.

A subset $S$ of a poset $P$ is *partitive* when both the following conditions hold whenever $x, x' \in S$, $y \notin S$ :

$$x < y \text{ iff } x' < y ;$$

$$x > y \text{ iff } x' > y .$$

A partitive subset $S$ is *proper* when $1 < |S| < |P|$. If a poset has a proper partitive set, it is decomposable, and conversely.

THEOREM 5.1. (Hiraguchi [1951]) *If* $\langle R;\leq\rangle$ *is the ordinal sum of the family* $(Q_x \mid x \in P)$ *over* $P$, *then* dim $R$ *is the maximum of the dimensions of* $P$ *and the posets* $Q_x$ , $x \in P$.

To prove Theorem 5.1, a realizer of $R$ is constructed from a realizer of $P$ by replacing each $x \in P$ by a linear exten sion of $Q_x$. This result also appears in Novák [1961]. Theo rem 5.1 implies that any irreducible poset is indecomposable. Recall that a *weak* extension of a poset $P$ satisfies every nonforced pair of $P$. The next result shows that every irre-ducible poset except $\underset{\sim}{1} + \underset{\sim}{1}$, the 2-element antichain, has a weak extension.

LEMMA 5.2. *Each nontrivial poset contains* $\underset{\sim}{1} + \underset{\sim}{1}$ *or has a weak extension.*

*Proof*. We write $X$ for the underlying set of $P$. Observe that $n(P)$ is a transitive relation on $X$. If $n(P)$ is not antisymmetric, then $n(P)$ contains both $\langle x, y\rangle$ and $\langle y, x\rangle$ for some $x$, $y$ in $X$. The antichain $\{x, y\}$ is then a partitive subset of $P$. We can now assume that $n(P)$ is a strict order relation on $X$. Since $\langle x, y\rangle \in n(P)$ and $y < z$ implies that $x < z$, it follows that $\leq_P \cup\, n(P)$ is an order relation on $X$.

For a nonempty subset $S$ of $P$, the *down-set* for $S$, denoted by $D(S)$, is the set of all $y \in P$ such that $y < x$ for some $x \in S$. The *up-set* $U(S)$ is defined dually. The down-set for $\{x\}$ is written as $D(x)$. Observe that the principal ideal $(x] = \{x\} \cup D(x)$ and the principal filter $[x) = \{x\} \cup U(x)$. If $A$ and $B$ are disjoint subsets of a poset $P$, then an extension $E$ of $P$ is said to *put* $A$ *under* $B$ when $a < b$ in $E$ for each incomparable pair $\langle a, b\rangle$ of $P$ with $a \in A$ and $b \in B$.

THEOREM 5.3. (Rabinovitch [1973]) *Let $A$ and $B$ be disjoint subsets of a poset $P$. There does not exist an extension of $P$ putting $A$ under $B$ iff there are $a_1, a_2 \in A$ and $b_1, b_2 \in B$ with $b_1 < a_1$, $b_2 < a_2$, $b_1 \| a_2$ and $b_2 \| a_1$ (so that these four elements form $\underset{\sim}{2} + \underset{\sim}{2}$).*

*Proof.* Let $S = (A \times B) \cap \mathcal{I}(P)$. If $S$ is cycle-free, then clearly the desired extension exists. If, on the other hand, $S$ contains a minimal alternating cycle $a_1, b_1, \ldots, a_n, b_n$, then $a_1, a_2, b_n, b_1$ form the required copy of $\underset{\sim}{2} + \underset{\sim}{2}$.

If $C$ is a chain in $P$, then an extension putting $C$ under $P - C$ is called an *upper extension* of $C$. Lower extensions are defined dually. T. Hiraguchi [1951] first established the following result, which we obtain as an immediate corollary of Theorem 5.3. In fact, if $c_1 < c_2 < \ldots < c_n$ is a chain in $P$, then the least upper extension of this chain is:

$$(c_1] \oplus ((c_2] - (c_1]) \oplus \ldots \oplus ((c_n] - (c_{n-1}]) \oplus (P - (c_n]).$$

COROLLARY 5.4. *Any chain in a poset has an upper extension.*

THEOREM 5.5. (Hiraguchi [1955]) *The dimension of a poset does not exceed its width.*

*Proof.* Let $P$ be a poset of width $n$. By the theorem of Dilworth [1950], $P$ can be covered by chains $C_i$, $1 \le i \le n$. For each $i$, let $E_i$ be an upper extension of $C_i$, whose existence is guaranteed by Corollary 5.4. If $a \in C_i$ and $a \| b$ in $P$, then $a < b$ in $E_i$. Therefore, $(E_i \mid 1 \le i \le n)$ realizes $P$, so that $\dim P \le n$.

Since the standard $n$-dimensional poset has width $n$, the inequality in Theorem 5.5 is the best possible. We give another example of an irreducible poset with equal width and dimension.

EXAMPLE 5.6. For each $n \geq 4$, the poset $P_n$ of Figure 5 has dimension $n$. (For clarity, we have only indicated the incom parable max-min pairs of $S_{n-1}$.)

*Proof.* By Theorem 5.5, $\dim(P_n) \leq n$. Suppose that the linear extensions $C_1, C_2, \ldots, C_{n-1}$ realize $P_n$. We can assume that $c_i < a_i$ holds in $C_i$ for $1 \leq i \leq n-1$. Suppose $b_i < u$ in $C_j$. It then follows that $b_i < a_j$ in $C_j$, from which we conclude that $i = j$. Thus, $b_i < u < c_i < a_i$ holds in $C_i$ for each $i$. In a similar manner, we deduce that $v < a_i$ in $C_i$ for each $i$. Consequently, $v < d$ in each $C_i$ and, therefore, in $P_n$. This contradiction shows that $P_n$ has dimension $n$.

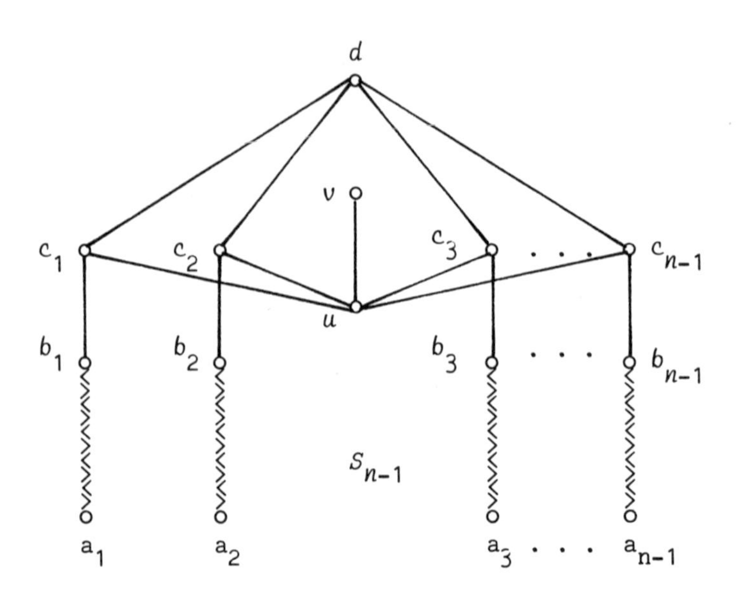


FIGURE 5. The poset $P_n$

For $x \in P$, $I(x)$, or $I_P(x)$, will denote the set of all $y \in P$ with $x \| y$. Observe that the sets $D(x)$, $U(x)$, $I(x)$, $\{x\}$ form a partition of $P$. If $S$ is a subset of $P$ and $E$ is an extension of $P$, then $E(S)$ denotes the restriction of $E$ to $S$. Clearly, $E(S)$ is a partial extension of $P$. We can now state and prove the one-point removal theorem.

THEOREM 5.7. (Hiraguchi [1951]) *Removing one point from a poset decreases its dimension by at most one.*

*Proof*. For a poset $P$, let $E=E_1, E_2, \ldots, E_n$ be extensions realizing $Q = P - \{x\}$. We define two extensions of $P$:

$$M_1 = E(D(x) \cup I(x)) \oplus \{x\} \oplus E(U(x)),$$

$$M_2 = E(D(x)) \oplus \{x\} \oplus E(U(x) \cup I(x)) .$$

Any incomparable pair of $P$ involving $x$ is satisfied in $M_1$ or $M_2$. Let $\langle a,b \rangle \in \mathcal{I}(Q)$, and assume that $b < a$ in $E_2$, $E_3, \ldots, E_n$. It follows that $a < b$ in $E_1$. Suppose that $a \nless b$ in both $M_1$ and $M_2$. This can only happen if $b \in D(x)$ and $a \in U(x)$. This implies $b < a$ in $P$, a contradiction. Therefore, the partial extensions $M_1$, $M_2$, $E_2$, $E_3, \ldots, E_n$ realize $P$, so that $\dim P \le \dim Q + 1$.

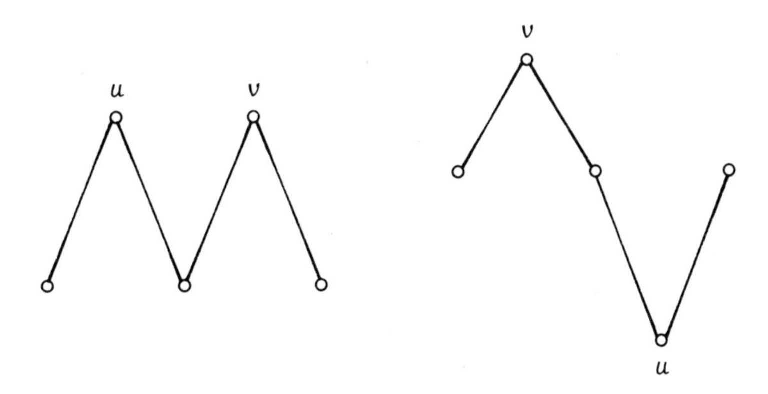


FIGURE 6. A pair of 2-dimensional posets

THEOREM 5.8. (Kimble [1973], Trotter [1975b]) *If* $A$ *is an antichain in a poset* $P$ *, then* $\dim P \leq \max(2, |P-A|)$ .

*Proof.* Let $A$ be an antichain in a poset $P$ with $|P-A| \leq 2$. We shall show that $\dim P \leq 2$. The general result will then follow by Theorem 5.7. By induction on $|P|$ , we can assume that $P$ is connected. We can also assume that $A$ is maximal. Let $P^*$ denote the poset obtained from $P$ by collapsing each partitive antichain to a single point. By Theorem 5.1, $\dim P \leq \max(2, \dim P^*)$. If $|P-A|$ is at most one, then $\dim P$ is at most two because $|P^*| \leq 2$ in this case. Let $P-A = \{u,v\}$. If $u||v$ , then $P^*$ is a subposet of the left poset of Figure 6, or its dual. If $u < v$ , then $P^*$ is a subposet of the right poset of Figure 6. Since both posets of Figure 6 are 2-dimensional, the proof is complete.

We now combine the preceding results to obtain Hiraguchi's inequality.

THEOREM 5.9. (Hiraguchi [1951]) *If* $P$ *is a poset with at least 4 elements, then* $\dim P \leq |P|/2$ .

*Proof.* Let $A$ be an antichain in $P$ of maximum size. If $|A| \leq |P|/2$ , then $\dim P \leq |P|/2$ by Theorem 5.5. Otherwise, $|P-A| \leq |P|/2$ , and the result follows by Theorem 5.8.

We conclude this section with two additional inequalities whose proofs utilize Dilworth's decomposition theorem.

THEOREM 5.10. (Trotter [1975b]) *If* $M$ *is the set of maximal elements of a poset* $P$, *then* $\dim P \leq 1 + \text{width}(P - M)$.

*Proof.* Let $n = \text{width}(P - M)$ and let

$$P - M = C_1 \cup C_2 \cup \dots \cup C_n$$

be a partition into chains. Let $E_i$ be a lower extension of $C_i$ for $1 \leq i \leq n$, with $E_n$ linear. Also, let

$$E_{n+1} = E_n(P - M) \oplus (E_n)^{\mathrm{d}}(M),$$

a linear extension of $P$. Since $(E_1, E_2, \dots, E_n, E_{n+1})$ is a realizer of $P$, the result follows.

THEOREM 5.11. (Trotter [1974b]) *If* $A$ *is an antichain in a poset* $P$ *and* $A \neq P$, *then* $\dim P \leq 1 + 2\,\mathrm{width}(P - A)$.

*Proof.* Let $n = \mathrm{width}(P - A)$ and let

$$P - A = C_1 \cup C_2 \cup \ldots \cup C_n$$

be a partition into chains. As above, let $E_i$ be a lower extension of $C_i$ for each $i$, with $E_n$ linear. Also, let $F_i$ be an upper extension of $C_i$ for each $i$. These $2n$ extensions, together with $(E_n)^d(A)$, realize $P$.

Equality holds in Theorem 5.10 for the poset $P_n$ of Example 5.6. The inequality in Theorem 5.11 is shown to be the best possible in Trotter [1974b]. If $A$ and $B$ are antichains in a poset $P$, then $\langle A, B\rangle$ is a *double antichain* iff $a \leq b$ whenever $a \in A$ and $b \in B$. The *double width* of $P$ is the maximum value of $|A| + |B|$ for a double antichain $\langle A, B\rangle$. O. Pretzel [1977] has proved an analog of Dilworth's decomposition theorem for double width, and has generalized the inequality of Theorem 5.11 by replacing twice the width by the double width.

## 6. MORE REMOVAL THEOREMS

As usual, all posets are finite. Removing an antichain can lower the dimension drastically. If all the maximal elements are removed from a standard $n$-dimensional poset, then one is left with an antichain, a poset of dimension 2. In fact, if all but two maximal elements are removed from $S_n$, the dimension becomes two. For example, if $n=7$, the remaining poset can be embedded in the planar lattice of Figure 7. However, the dimension drops by at most two when a chain is removed from a poset.

THEOREM 6.1. (Hiraguchi [1951]) *If* $C$ *is a chain in a poset* $P$, *then* $\dim P \leq 2 + \dim(P - C)$.

*Proof.* Add a lower and an upper extension of $C$ to a realizer of $P-C$ to obtain a realizer for $P$ (consisting of partial extensions).

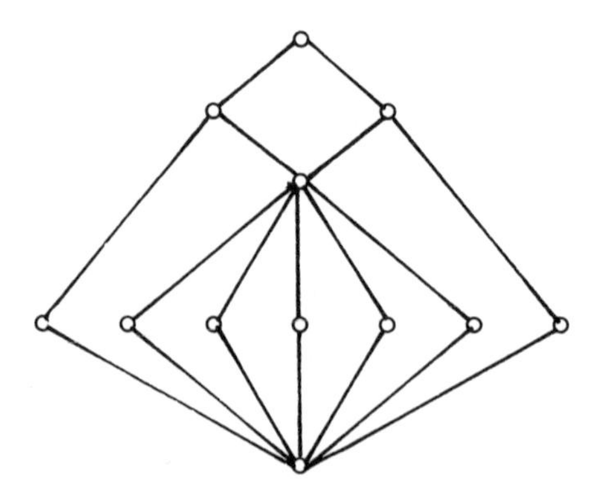

FIGURE 7. A planar lattice

If $n \geq 4$ and a 2-element chain is removed from $S_n$, then the dimension drops by 2. Since the remaining poset has width $n - 1$, this may have been unexpected. Hiraguchi's [1951] original proof of his inequality (Theorem 5.9) depended on many removal theorems, as did the later proof of Bogart [1973]. The maximum decrease in dimension is one for the two-point removal theorems and two for the four-point ones. In both cases, at least one four-point removal theorem was required. Chain removal theorems give conditions under which removal of a chain decreases the dimension by at most one.

THEOREM 6.2. (Bogart [1973]) *If $C$ is a chain in $P$ and each point in $P - C$ is incomparable with at most one point in $C$, then* $\dim P \leq 1 + \dim(P - C)$.

*Proof.* Let $C$ be $c_1 < c_2 < \ldots < c_n$. Let $E = E_1, E_2, \ldots, E_t$ be extensions realizing $P - C$. For $1 \leq i \leq n$, let $D_i = D(c_i) - C$ and $U_i = U(c_i) - C$. Also, let $D_0 = \emptyset = U_{n+1}$ and $D_{n+1} = P = U_0$. The assumptions imply that each $x \in P - C$ is in $(D_{i+1} - D_i) \cap (U_i - U_{i+1})$ for some (unique) value of $i$ in $\{0, 1, \ldots, n\}$. It is then immediate that $P$ is realized by $E_2, E_3, \ldots, E_t$ and the following two extensions of $P$:

$$E(D_1) \oplus c_1 \oplus E(D_2 - D_1) \oplus c_2 \oplus \ldots \oplus c_n \oplus E(P - D_n)$$

$$E(P - U_1) \oplus c_1 \oplus E(U_1 - U_2) \oplus c_2 \oplus \ldots \oplus c_n \oplus E(U_n)$$

This completes the proof.

PROBLEM 6.3. A pair $\langle a, b\rangle$ of distinct elements in a poset $P$ is called *removable* if $\dim(P - \{a,b\}) \geq \dim P - 1$. If a poset with at least three elements is not irreducible, then it has a removable pair. (First, choose an element whose removal does not lower the dimension; the second element can be chosen arbitrarily.) We believe that every d-irreducible poset ($d \geq 3$) contains a removable pair. In fact, we *conjecture* that every critical pair of such a poset is removable.

The *rank* of a cover $a \prec b$ is the number of incomparable pairs $\langle x,y\rangle$ with $a < x$ and $y < b$. T. Hiraguchi [1951] showed that a cover of rank at most one is removable. This result is generalized in the following theorem.

THEOREM 6.4. *If the cover* $a \prec b$ *in* $P$ *has rank at most* $(\dim P) - 3$, *then it is removable.*

*Proof.* Let $L = L_1, L_2, \ldots, L_n$ be linear extensions realizing $P - \{a,b\}$. By the one-point removal theorem, $n \geq (\dim P) - 2$. Thus, there are at most $n-1$ incomparable pairs $\langle x,y\rangle$ such that $a < x$ and $y < b$. If $\langle x,y\rangle$ is any of these pairs, we can assume that $y < x$ holds in $L_i$ for some $2 \leq i \leq n$. $P$ is realized by $L_2, L_3, \ldots, L_n$ and the following two linear extensions:

$$L(\mathrm{D}(a)) \oplus a \oplus L(\mathrm{D}(b) - \mathrm{D}(a)) \oplus b \oplus L(P - \mathrm{D}(b))$$

$$L(P - \mathrm{U}(a) \oplus a \oplus L(\mathrm{U}(a) - \mathrm{U}(b)) \oplus b \oplus L(\mathrm{U}(b))$$

Thus, $\dim P \leq 1 + \dim(P - \{a,b\})$.

Except in the $\dim P = 3$ case, Theorem 6.4 yields Hiraguchi's result that covers $a \prec b$ of rank at most one are removable. However, if $\dim P = 3$, it is clear that $P - \{a,b\}$ is not a chain. (Otherwise, $P$ would have width at most 2.) Thus, $P - \{a,b\}$ has dimension at least 2, and we can reason as in the proof of Theorem 6.4. Note that each chain in the standard

$n$-dimensional poset $(n \geq 4)$ is a cover of rank $n-2$ whose removal decreases the dimension by two.

The *rank* of an incomparable pair $\langle a,b\rangle$ is the number of incomparable pairs $\langle x,y\rangle$ where $x$ is comparable to both $a$ and $b$, and $y$ is incomparable to both $a$ and $b$. An incomparable pair of rank at most one is removable. We shall present a more general result.

THEOREM 6.5. *An incomparable pair* $\langle a,b\rangle$ *of rank at most* $(\dim P) - 3$ *is removable.*

*Proof.* Let $L = L_1, L_2, \ldots, L_n$ be linear extensions realizing $P-\{a,b\}$. There are at most $n-1$ incomparable pairs $\langle x,y\rangle$ for which $x$ is comparable with both $a$ and $b$, and $y$ is incomparable with both $a$ and $b$. For each of these pairs $\langle x,y\rangle$ we can assume that, if $x > a$, then $x < y$ holds in $L_i$ for some $i \geq 2$, and if $x < a$, then $x > y$ holds in $L_i$ for some $i \geq 2$. $P$ is realized by $L_2, L_3, \ldots, L_n$ and the following two linear extensions:

$$L(\mathrm{D}(a)) \oplus a \oplus L(P - \mathrm{D}(a)) - \mathrm{U}(b)) \oplus b \oplus L(\mathrm{U}(b))$$

$$L(\mathrm{D}(b)) \oplus b + L(P - \mathrm{D}(b)) - \mathrm{U}(a)) \oplus a \oplus L(\mathrm{U}(a))$$

This completes the proof.

Since a pair of maximal elements in $S_n$ $(n \geq 4)$ is not removable, Theorem 6.5 is the best possible result. In particular, any incomparable max-min pair is removable. An incomparable pair $\langle a,b\rangle$ is a *DU-pair* if $\mathrm{D}(a) \subseteq \mathrm{D}(b)$ and $\mathrm{U}(a) \subseteq \mathrm{U}(b)$.

THEOREM 6.6. *Any* DU-*pair* $\langle a,b\rangle$ *is removable.*

*Proof.* Let $E = E_1, E_2, \ldots, E_n$ be extensions realizing $P - \{a,b\}$. $P$ is realized by $E_2, E_3, \ldots, E_n$ and the following two extensions:

$$E(\mathrm{D}(a)) \oplus a \oplus E(\mathrm{D}(b) - \mathrm{D}(a)) \oplus b \oplus E(P - \mathrm{D}(b))$$

$$E(P - \mathrm{U}(b)) \oplus b \oplus E(\mathrm{U}(b) - \mathrm{U}(a)) \oplus a \oplus E(\mathrm{U}(a))$$

## 7. IRREDUCIBLE POSETS

In Section 1, we mentioned that the list of all 3-irreducible posets was discovered independently by D. Kelly [1977], and W.T. Trotter and J.I. Moore [1976a]. We shall discuss the two solutions. The different approaches originate from the two characterizations of a poset $P$ of dimension at most 2: the completion of $P$ is planar and $\mathcal{I}(P)$ is a comparability graph.

Using arguments with a geometric flavor, D. Kelly and I. Rival [1975a] characterized planar lattices by the exclusion *as subposets* of a family $\mathcal{L}$ of nonplanar lattices; moreover, $\mathcal{L}$ is the minimum such collection. If $\mathcal{P} = \{\underset{\sim}{P}(L) \mid L \in \mathcal{L}\}$, then a lattice $L$ is planar iff it does not contain any poset in $\mathcal{P}$. As we mentioned in Section 3, every poset in $\mathcal{P}$ is 3-irreducible (Kelly and Rival [1975b]) and $\mathcal{P}$ includes the first nine posets of Figure 1. The determination of $\mathcal{L}$ was based on the characterization of dismantlable lattices in Kelly and Rival [1975a] or M. Ajtai [1973]. A finite lattice is *dismantlable* (Rival [1974]) if every sublattice with at least 3 elements contains a doubly irreducible element. K.A. Baker, P.C. Fishburn and F.S. Roberts [1971] showed that every planar lattice is dismantlable. Let $\mathcal{R}$ be a candidate for the set of all 3-irreducible posets. Certainly, $\mathcal{R}$ must contain $\mathcal{P}$. If a poset $P$ has dimension greater than 2, then $L = \underset{\sim}{L}(P)$ is nonplanar, and therefore, contains a poset in $\mathcal{P}$. It remains to show that $P$ contains a poset in $\mathcal{R}$. The first step is to handle the crowns. We can then assume that $P$ does not contain any crowns. By the characterization of dismantlable lattices, this means that $L$ is dismantlable. This has two consequences: we can assume that $L - \{x\}$ is a planar lattice for a doubly irreducible element $x$ of $L$; any element of $L - P$ is simultaneously the join of two elements of $P$ and the meet of two elements of $P$. The remaining details appear in Kelly [1977].

A poset $P$ is 3-irreducible iff $G = \mathcal{I}(P)$ is not a comparability graph but every proper induced subgraph of $G$ is a comparability graph. T. Gallai [1967] provided a forbidden subgraph characterization of comparability graphs by discovering the minimum collection $\mathbb{C}$ of graphs so that a graph fails to be a comparability graph iff it contains a graph from $\mathbb{C}$ as an induced subgraph. It follows that the incomparability graph of a 3-irreducible poset belongs to $\mathbb{C}$. From this observation, it follows that we need only examine graphs from the family $\mathbb{C}$ and determine those whose complement is a comparability graph. Any transitive orientation yields a 3-irreducible poset. In fact, up to duality, there is no choice of orientation because the comparability graph of an irreducible poset is uniquely orderable (see Section 8). The details are given in Trotter and Moore [1976a].

A problem for a class of posets is to determine a formula for the dimension and to specify the irreducible ones. We shall discuss the family $S_n^k$ of W.T. Trotter [1974a]. Other examples appear in Kelly [1981]. Of course, one may only be able to express the dimension in terms of other combinatorial parameters as in Dushnik [1950] (the parameters defined there were investigated further in Spencer [1971] and Trotter [1976b]). For integers $n$, $k$ with $n \geq 0$, the *general crown* $S_n^k$ is the poset of unit length with $n + k$ maximal elements $a_1, a_2, \ldots, a_{n+k}$ and $n + k$ minimal elements $b_1, b_2, \ldots, b_{n+k}$. The order on $S_n^k$ is defined by $b_i < a_j$ iff $j \in \{i, i+1, \ldots, i+k\}$, where addition is modulo $n+k$. Note that $S_n^k$ has the $(n+k)(k+1)$ nonforced pairs $\langle b_i, a_j \rangle$ where $b_i \| a_j$ in $S_n^k$. Clearly, $S_n^0 = S_n$.

THEOREM 7.1. (Trotter [1974a]). For each $n$, $k$ with $n \geq 3$, $k \geq 0$, the dimension of $S_n^k$ is $\lceil 2(n+k)/(k+2) \rceil$.

*Sketch of Proof.* We have already observed that $S_n^k$ has $(n+k)(k+1)$ nonforced pairs. It is relatively easy to show that a linear extension of $S_n^k$ can reverse at most $(k+1)(k+2)/2$ of these nonforced pairs. Therefore,

$$\dim(S_n^k) \geq \lceil 2(n+k)/(k+2) \rceil .$$

We illustrate the construction of a realizer for $S_n^k$ by providing seven partial linear extensions which realize $S_{18}^5$.

$$b_7, a_6, b_6, a_5, b_5, a_4, b_4, a_3, b_1, a_7, b_2, a_8, b_3$$

$$b_{10}, a_9, b_9, a_8, a_7, b_4, a_{10}, b_5, a_{11}, b_6, a_{12}, b_7$$

$$b_{14}, a_{13}, a_{12}, b_{12}, a_{11}, b_{11}, a_{10}, b_8, a_{14}, b_9, a_{15}, b_{10}$$

$$b_{17}, a_{16}, b_{16}, a_{15}, b_{15}, a_{14}, b_{11}, a_{17}, b_{12}, a_{18}, b_{13}, a_{19}, b_{14}$$

$$b_{21}, a_{20}, b_{20}, a_{19}, b_{19}, a_{18}, b_{18}, a_{17}, b_{15}, a_{21}, b_{16}, a_{22}, b_{17}$$

$$a_{23}, b_{23}, a_{22}, b_{22}, a_{21}, b_{18}, a_1, b_{19}, a_2, b_{20}, a_3, b_{21}$$

$$a_5, a_4, b_3, a_2, b_2, a_1, b_1, a_{23}, b_{22}, a_5, b_{23}$$

THEOREM 7.2. (Trotter [1974a]) *Let* $n \geq 3$, $k \geq 0$.

(*a.*) *If* $n + k = q(k + 2) + 1$, *then* $S_n^k$ *is* $(2q + 2)$-*irreducible.*

(*b.*) *If* $n + k = q(k + 2) + \lfloor (k + 2)/2 \rfloor + 1$, *then* $S_n^k$ *is* $(2q + 2)$-*irreducible iff* $k = 0$ *or* $k$ *is odd.*

The study of general crowns is closely related to a graph coloring problem which involves properties of the hypergraph $\mathcal{H}(S_n^k)$. When $n \geq 3$ and $k \geq 1$, $\mathcal{H}(S_n^k)$ is not an ordinary graph. However, there is a family of graphs for which the determination of chromatic number is equivalent to the computation of dimension for general crowns. For $m \geq 1$, $t \geq 0$, define the *web* $W_m^t$ as the graph whose vertex set is $\{w_1, w_2, w_3, \ldots, w_m\}$ with each $w_i$ adjacent to $w_{i+t+j}$ for $j = 1, 2, \ldots, n - 2t - 1$ (cyclically). Note that $W_m^t$ is a regular graph of degree $n - 2t - 1$; also note that $W_m^0$ is a complete graph on $m$ vertices.

THEOREM 7.3. *For* $m \geq 0$ *and* $t \geq 0$, *the chromatic number of the web* $W_m^t$ *is* $\lceil m/(t + 1) \rceil$.

*Proof.* Let $W$ be the vertex set of $W_m^t$. We first show that any independent subset $W' \subseteq W$ has at most $t + 1$ vertices. For each $j$ with $1 \leq j \leq m$, let $I(w_j)$ be the $(2t + 1)$- element subset of $W$ defined by

$$I(w_j) = \{ w_{j-t}, w_{j-t+1}, \ldots, w_j, \ldots, w_{j+t} \}.$$

Observe that $w \in I(w_j)$ if $w$ is not adjacent to $w_j$. Therefore, $W' \subseteq \bigwedge (I(w) \mid w \in W')$ holds for any independent set of vertices $W'$ in $W_m^t$. Since each set $I(w)$ is a set of $2t + 1$ consecutive elements of $W$, it is straightforward to show that if $W'$ is an independent set in $W_m^t$, then

$$|W'| \leq | \bigwedge(I(w) \mid w \in W')| \leq 2t + 2 - |W'|.$$

Therefore, $|W'| \le t + 1$.

Now suppose that $\chi(W_m^t) = s$ and let $W = W_1 \cup \ldots \cup W_s$ be a partition into independent subsets. Then since $|W_i| \le t+1$ for $i = 1,2,\ldots,s$, $|W| = m \le s(t+1)$; i.e., $s \ge m/(t+1)$. Since any subset of $W$ consisting of $t + 1$ consecutive elements is independent, $W$ can be partitioned into $\lceil m/(t + 1) \rceil$ independent subsets.

When $n \ge 3$, $k \ge 0$ and $k$ is even, say $k = 2\ell$, the hypergraph $\mathcal{H}(S_n^k)$ contains the web $W_{n+k}^{\ell}$ as an induced subhypergraph. Consider the vertices $\langle a_{i+\ell}, b_i \rangle$ for $1 \le i \le k$. Thus, $\dim(S_n^k) \ge \chi(W_{n+k}^{\ell}) = \lceil (n+k)/(\ell+1) \rceil = \lceil 2(n+k)/(k+2) \rceil$. When $n \ge 3$, $k \ge 0$, and $k$ is odd, say $k = 2\ell + 1$, then $\mathcal{H}(S_n^k)$ contains the web $W_{2(n+k)}^{k+1}$ as an induced subhypergraph. This may be seen by considering the pairs $\{\langle a_j, b_i \rangle \mid j = i + \ell \text{ or } j = i + k + 1\}$ for $i = 1, 2, \ldots, n+k$.

Each of the present authors has developed a technique for constructing irreducible posets, both under the name *dimension product*.

If each of $P$ and $Q$ is an irreducible poset satisfying certain extra conditions, or equals $\underset{\sim}{2}$, then the *dimension product* $P \otimes Q$ of $P$ and $Q$ introduced by Kelly [1981] is an irreducible poset of dimension $\dim P + \dim Q$. We shall not give these extra conditions here. We remark that we do not know any irreducible posets which violate any of these conditions. In particular, all 3-irreducible posets satisfy these conditions. As a consequence, for any $n > 3$ every 3-irreducible poset can be embedded in an $n$-irreducible poset.

In general, the definition of $P \otimes Q$ is quite complicated. It is always a subposet of $R = \underset{\sim}{P}(\underset{\sim}{L}(P) \times \underset{\sim}{L}(Q))$. Although the expression for $R$ may look formidable, $R$ is a subposet of $(P \cup \{0,1\}) \times (Q \cup \{0,1\})$. Note that $\dim R = \dim P + \dim Q$ by the results of Section 1. In fact, for the 3-irreducible posets in $\mathcal{P}$ and for the general crowns, $P \otimes Q = R$.

Let $P$ be a poset of unit length, and for simplicity, assume no element of $P$ is both minimal and maximal. The *interval dimension* of $P$, denoted by Idim $P$, is the least size of a family $\mathfrak{F}$ of linear extensions such that every max-min incomparable pair is satisfied in some member of $\mathfrak{F}$. $P$ is *d-interval irreducible* if Idim $P = d \geq 2$ and Idim $(P-\{x\}) < d$ for every $x \in P$. For the family $S_k^n$ of general crowns, interval dimension and dimension coincide. Moreover, being irreducible and being interval irreducible are equivalent for this family. Trotter's *dimension product* of two posets of unit length is a poset of unit length where interval dimension is at least the sum of the interval dimensions of the factors. If $P$ is $m$-interval irreducible and $Q$ is $n$-interval irreducible, then Trotter's dimension product is $(m+n)$-interval irreducible, *and* $(m+n)$-irreducible. Trotter's dimension product of $P$ and $Q$ is the disjoint union of $P$ and $Q$ with every minimal element of $P$ put under every maximal element of $Q$, and every minimal element of $Q$ put under every maximal element of $P$. Let $d \geq 6$ and $n \geq 2d+2$. Starting with the general crowns and a single 3-interval irreducible poset on 9 points, one can construct $d$-irreducible posets of unit length on $n$ points using Trotter's dimension product. (The list of all 3-interval irreducible posets of unit length is given in Trotter [1981].)

Recently, W.T. Trotter and J.A. Ross made some striking discoveries about irreducible posets. We have mentioned that Kelly's dimension product, for any 3-irreducible poset and $n > 3$, gives an embedding of $P$ into an $n$-irreducible poset $Q$. Both $Q$ and the embedding are given explicitly. Whenever $n > d \geq 3$, Trotter and Ross [1981a] showed that every $d$-irreducible poset $P$ can be embedded into an $n$-irreducible poset $Q$. Rather than trying to construct such a poset $Q$ explicitly, they construct a poset $R$ of dimension $n$ containing $P$. We can assume that $n = d + 1$. The poset $R$ is constructed so that any subposet of $R$ that does not contain all of $P$ has dimension less than $n$. The construction uses a weak extension of $P$ in an essential way. There are posets of dimension two (for example, the direct product of a 3-element chain with itself) which are not subposets of any 3-irreducible poset. However, Trotter and Ross [1981b] have shown that every poset of dimension $d \geq 3$ is a subposet of a $(d+1)$-irreducible poset. Their construction uses the dimension product $G_m \otimes S_{d-3}$ for a suitably large value of $m$. They utilize special properties of $\mathcal{H}(G_m \otimes S_{d-3})$. (Recall that the hypergraph for $G_m$ appears in Figure 2.)

## 8. COMPARABILITY GRAPHS

We have mentioned comparability graphs in Sections 3 and 7. We refer the reader to the book of Golumbic [1980] for a detailed account of this topic. Since an irreducible poset is indecomposable, it is uniquely orderable. Using this fact, Trotter, Moore and Sumner [1976] showed that the dimension of a poset is determined by its comparability graph. This latter statement is also true for infinite posets, as shown by Arditti and Jung [1980]. Using Golumbic's algorithm for orienting a comparability graph, R. Stanley has observed that the number of linear extensions of a finite poset depends only on its comparability graph.

One might think that the comparability graph also determines the rank. However, the following examples due to Maurer, Rabinovitch and Trotter [1980b] show that posets with the same comparability graph may differ substantially in rank. For $n \geq 2$, let $V_n$ be the poset consisting of an $n$-element antichain with a zero added. Let $P_n = V_n + V_n$, and $Q_n = V_n + (V_n)^d$. Certainly, these two posets have the same comparability graph. Furthermore, $\dim(P_n) = \dim(Q_n) = 2$, but $\operatorname{rank}(P_n) = 2\lceil (n+1)^2/4 \rceil$ and $\operatorname{rank}(Q_n) = n^2 + 1$.

## 9. FURTHER TOPICS

A poset $P$ is an *interval order* if we can associate a closed interval $I_x$ of the reals to each $x \in P$ so that $x < y$ in $P$ iff $I_x$ lies entirely to the left of $I_y$. P.C. Fishburn [1970] showed that a poset is an interval order iff it does not contain $\underset{\sim}{2} \oplus \underset{\sim}{2}$. A *semiorder* is an interval order that does not contain $\underset{\sim}{3} \oplus \underset{\sim}{1}$. I. Rabinovitch ([1973], ]1978a], [1978b]) showed that an interval order has arbitrary dimension but is bounded by its height plus one, while the dimension of a semiorder cannot exceed three. (An improved upper bound for the dimension of an interval order is given in Bogart, Rabinovitch and Trotter [1976].)

D. Kelly [1980] constucted finite modular dismantlable lattices of each finite dimension. (The dimension of a distributive dismantlable lattice cannot exceed two.) I. Rival [1976] showed that the dimension of a finite modular dismantlable lattice $L$ is bounded by the width of $\operatorname{Irr}(L)$. There is a finite list of modular lattices such that a modular lattice $L$ has dimension at most two iff no member of this list is a sublattice of $L$ (Kelly [1980]). A version of this result with exclusion as subposets was proved previously by R. Wille [1974]; the case of finite $L$ appears in Kelly and Rival [1975a]. The dimension of lattices is also considered in Rival and Sands ([1978],[1979]), Sands [1980], and Babai and Duffus [1981].

The dimension of a poset is the number of order relations of a special type (linear orders) that are needed to represent the ordering of the poset as an intersection. If we replace linear orders by interval orders, then we obtain the *interval dimension* of a poset. (Interval dimension is considered in Trotter and Bogart [1976a] and Trotter and Moore [1976a].) There are analogous dimension concepts for graphs and digraphs (i.e.,binary relations; loops may occur). *Ferrers* relations are defined in Riguet [1951] and *threshhold* graphs are defined in Chvátal and Hammer [1977]. In each of these definitions, a finite family is excluded (as induced relations). The *Ferrers* dimension of a digraph and the *threshhold* dimension of a graph are then defined as indicated above. A. Bouchet [1971] showed that the Ferrers dimension of a (reflexive) order relation is the same as its dimension. He also showed that the Ferrers dimension of a binary relation $R$ equals the dimension of the Galois lattice associated with $R$. O. Cogis [1980] associated to each binary relation $R$ a graph $G$ such that the Ferrers dimension of $R$ equals the threshhold dimension of $G$.

For a poset $P$, let $\mathcal{G}(P)$ be the graph whose edges are the 2-edges of $\mathcal{H}(P)$. O. Cogis [1980] has shown that $\dim P \leq 2$ iff $\mathcal{G}(P)$ is 2-colorable. In fact, he proved the corresponding result for the Ferrers dimension of digraphs. Consequently, deciding whether a digraph has Ferrers dimension at most two has polynomial complexity.

REFERENCES

M. Aigner and G. Prins [1971] Uniquely partially orderable graphs, *J. London Math. Soc.* (2) 3, 260-266. MR43 #1866.

M. Ajtai [1973] On a class of finite lattices, *Period. Math. Hungar.* 4, 217-220. MR49 #8906.

J.C. Arditti [1978] Partially ordered sets and their comparability graphs, their dimension and their adjacency, in *Problèmes combinatoires et théorie des graphes, Orsay* 1976, Colloq. int. CNRS No. 260, 5-8.

J.C. Arditti and H.A. Jung [1980] The dimension of finite and infinite comparability graphs, *J. London Math. Soc.* (2) 21, 31-38.

L. Babai and D. Duffus [1981] Dimension and automorphism groups of lattices, *Algebra Universalis* 12, 279-289.

K.A. Baker [1961] Dimension, join-independence, and breadth in partially ordered sets, (unpublished).

K.A. Baker, P.C. Fishburn and F.S. Roberts [1971] Partial orders of dimension 2, *Networks* 2, 11-28. MR46 #104.

B. Banaschewski [1956] Hüllensysteme und Erweiterungen von Quasi-Ordnungen, *Z. Math. Logik Grundlagen Math.* 2, 117-130.

G. Birkhoff [1967] *Lattice Theory, 3rd. ed.*, Colloquium Publication 25, American Mathematical Society, Providence, R.I.

K.P. Bogart [1973] Maximal dimensional partially ordered sets I. Hiraguchi's Theorem, *Discrete Math.* 5, 21-32. MR47 #6562.

K.P. Bogart and W.T. Trotter [1973] Maximal dimensional partially ordered sets II. Characterization of $2n$-element posets with dimension $n$, *Discrete Math.* 5, 33-45. MR47 #6563.

K.P. Bogart, I. Rabinovitch and W.T. Trotter [1976] A bound on the dimension of interval orders, *J. Combinatorial Theory (Ser. A)* 21 (1976), 319-328. MR54 #5059.

A. Bouchet [1971] *Etude combinatoire des ordonnés finis*, Thèse de Doctorat d'Etat, Université de Grenoble.

V. Chvátal and P.L. Hammer [1977] Aggregation of inequalities in integer programming, *Annals of Discrete Math.* 1, 145-162.

O. Cogis [1980] *La dimension Ferrers des graphes orientés*, Thèse de Doctorat d'Etat, Université Curie, Paris.

R.P. Dilworth [1950] A decomposition theorem for partially ordered sets, Ann. Math. 51, 161-166. MR11, p.309.

A. Ducamp [1967] Sur la dimension d'un ordre partiel, in *Theorie des Graphes, Journees Internationales d'Etudes*, Dunod, Paris; Gordon and Breach, New York, 103-112. MR36 #3684.

B. Dushnik [1950] Concerning a certain set of arrangements, *Proc. Amer. Math. Soc.* 1, 788-796. MR12, p.470.

B. Dushnik and E.W. Miller [1941] Partially ordered sets, *Amer. J. Math.* 63, 600-610. MR3, p.73.

P.C. Fishburn [1970] Intransitive indifference with unequal indifference intervals, *J. Math. Psychol.* 7, 144-149.

T. Gallai [1967] Transitiv orientierbare Graphen, *Acta Math. Acad. Sci. Hungar.* 18, 25-66. MR36 #5026.

M.R. Garey and D.S. Johnson [1979] *Computers and intractability. A guide to the theory of NP-completeness*, Freeman, San Francisco.

M.R. Garey, D.S. Johnson and L. Stockmeyer [1976] Some simplified NP-complete graph problems, *Theoret. Comp. Sci.* 1, 237-267.

A. Ghouila-Houri [1962] Caractérisation des graphes non orientés dont on peut orienter les arêtes de manière à obtenir le graphe d'une relation d'ordre, *C.R. Acad. Sci. Paris* 254, 1370-1371. MR30 #2495.

P.C. Gilmore and A.J. Hoffman [1964] A characterization of comparability graphs and of interval graphs, *Canad. J. Math.* 16, 539-548. MR31 #87.

M.C. Golumbic [1977a] Comparability graphs and a new matroid, *J. Combinatorial Theory (Ser. B)* 22, 68-90. MR55 #12575.

M.C. Golumbic [1977b] The complexity of comparability graph recognition and coloring, *Computing* 18, 199-203.

M.C. Golumbic [1980] *Algorithmic Graph Theory and Perfect Graphs*, Academic Press, New York. (Chapter 5) MR81e:68081.

R. Gysin [1977] Dimension transitiv orientierbarer Graphen, *Acta Math. Acad. Sci. Hungar.* 29, 313-316. MR58 #5393.

E. Harzheim [1970] Ein Endlichkeitssatz über die Dimension teilweise geordneter Mengen, *Math. Nachr.* 46, 183-188. MR43 #113.

T. Hiraguchi [1951] On the dimension of partially ordered sets, *Sci. Rep. Kanazawa Univ.* 1, 77-94. MR17, p.19.

T. Hiraguchi [1955] On the dimension of orders, *Sci. Rep. Kanazawa Univ.* 4, 1-20. MR17, p.1045.

D. Kelly [1977] The 3-irreducible partially ordered sets, *Canad. J. Math.* 29, 367-383. MR55 #205.

D. Kelly [1980] Modular lattices of dimension two, *Algebra Universalis* 11, 101-104.

D. Kelly [1981] On the dimension of partially ordered sets, *Discrete Math.* 35 (1981), 135-156.

D. Kelly and I. Rival [1974] Crowns, fences, and dismantlable lattices, *Canad. J. Math.* 26, 1257-1271. MR54 #5064.

D. Kelly and I. Rival [1975a] Planar lattices, *Canad. J. Math.* 27, 636-665. MR52 #2974.

D. Kelly and I. Rival [1975b] Certain partially ordered sets of dimension three, *J. Combinatorial Theory (Ser. A)* 18, 239-242. MR50 #12828.

D. Kelly, J. Schönheim and R.E. Woodrow [1981] Relating vertex colorability of graphs and hypergraphs, Univ. of Calgary Mathematics Research Paper No. 481.

R.J. Kimble [1973] *Extremal Problems in Dimension Theory for Partially Ordered Sets*, Ph.D. thesis, M.I.T.

H. Komm [1948] On the dimension of partially ordered sets, *Amer. J. Math.* 70, 507-520. MR10, p.22.

K. Kuratowski [1930] Sur le problème des courbes gauches en topologie, *Fund. Math.* 15, 271-283.

E. L. Lawler and O. Vornberger [1981] The partial order dimension problem is NP-complete, preprint (Univ. of Calif. at Berkeley).

B. Leclerc [1976] Arbres et dimension des ordres, *Discrete Math.* 14 (1976), 69-76. MR52 #7979.

L. Lovász [1973] Coverings and colorings of hypergraphs, in *Proc. 4th Southeastern Conference on Combinatorics, Graph Theory and Computing*, Utilitas Mathematica Publishing, Winnipeg, 3-12.

L. Lovász, J. Nešetřil, and A. Pultr [1980] On a product dimension of graphs, *J. Combinatorial Theory (Ser.B)* 29, 47-67.

S.B. Maurer and I. Rabinovitch [1978] Large minimal realizers of a partial order, *Proc. Amer. Math. Soc.* 66, 211-216. MR56 #8441.

S.B. Maurer, I. Rabinovitch and W. T. Trotter [1980a] Large minimal realizers of a partial order II, *Discrete Math.* 31, 297-314.

S.B. Maurer, I. Rabinovitch and W. T. Trotter [1980b] A generalization of Turán's theorem, *Discrete Math.* 32, 167-189.

S.B. Maurer, I. Rabinovitch and W.T. Trotter [1980c] Partially ordered sets with equal rank and dimension, *Congressus Numerantium* 29, 627-637.

V. Novák [1961] The dimension of lexicographic sums of partially ordered sets (Czech., English summary), *Časopis Pěst. Mat.* 86, 385-391. MR24 #1847.

O. Ore [1962] *Theory of Graphs*, Colloquium Publication 38, Amer. Math. Soc., Providence, R. I., (Section 10.4) MR27 #740.

C.R. Platt [1976] Planar lattices and planar graphs, *J. Combinatorial Theory (Ser.B)* 21, 30-39.

O. Pretzel [1977] On the dimension of partially ordered sets, *J. Combinatorial Theory (Ser. A)* 22 (1977), 146-152. MR55 #206.

I. Rabinovitch [1973] *The Dimension Theory of Semiorders and Interval Orders*, Ph.D. thesis, Dartmouth College.

I. Rabinovitch [1978a] The dimension of semiorders, *J. Combinatorial Theory (Ser. A)* 25, 50-61. MR58 #16435.

I. Rabinovitch [1978b] An upper bound on the dimension of interval orders, *J. Combinatorial Theory (Ser. A)* 25, 68-71.

I. Rabinovitch and I. Rival [1979] The rank of a distributive lattice, *Discrete Math.* 25, 275-279.

J. Riguet [1951] Les relations de Ferrers, *C.R. Acad. Sci. Paris* 232, 1729-1730.

I. Rival [1974] Lattices with doubly irreducible elements, *Canad. Math. Bull.* 17, 91-95. MR50 #12837.

I. Rival [1976] Combinatorial inequalities for semimodular lattices of breadth two, *Algebra Universalis* 6, 303-311. MR55 #217.

I. Rival and B. Sands [1978] Planar sublattices of a free lattice I, *Canad. J. Math.* 30, 1256-1283.

I. Rival and B. Sands [1979] Planar sublattices of a free lattice II, *Canad. J. Math.* 31, 17-34.

B. Sands [1980] Generating sets for lattices of dimension two, *Discrete Math.* 29, 287-292.

J. Schmidt [1956] Zur Kennzeichnung der Dedekind-MacNeilleschen Hülle einer geordneten Menge, *Arch. Math.* 7, 241-249.

L.N. Shevrin and N.D. Filippov [1970] Partially ordered sets and their comparability graphs, *Siberian Math. J.* 11, 497-509 (648-667 in Russian original).

J. Spencer [1971] Minimal scrambling sets of simple orders, *Acta Math. Acad. Sci. Hungar.* 22, 349-353.

E. Szpilrajn [1930] Sur l'extension de l'ordre partiel, *Fund. Math.* 16 (1930), 386-389.

W.T. Trotter [1974a] Dimension of the crown $S_n^k$, *Discrete Math.* 8, 85-103. MR49 #158.

W.T. Trotter [1974b] Irreducible posets with large height exist, *J. Combinatorial Theory (Ser. A)* 17 (1974), 337-344. MR50 #6935.

W.T. Trotter [1975a] Embedding finite posets in cubes, *Discrete Math.* 12, 165-172. MR51 #5426.

W.T. Trotter [1975b] Inequalities in dimension theory for posets, *Proc. Amer. Math. Soc.* 47, 311-316. MR51 #5427.

W.T. Trotter [1975c] A note on Dilworth's embedding theorem, *Proc. Amer. Math. Soc.* 52, 33-39. MR51 #10188.

W.T. Trotter [1976a] A forbidden subposet characterization of an order-dimension inequality, *Math. Systems Theory* 10, 91-96. MR55 #7856.

W.T. Trotter [1976b] A generalization of Hiraguchi's inequality for posets, *J. Combinatorial Theory (Ser.A)* 20, 114-123. MR52 #10515.

W.T. Trotter [1978a] Combinatorial problems in dimension theory for partially ordered sets, in *Problèmes combinatoires et theorie des graphes,* Colloq. int. CNRS No. 260 (1978), 403-406. MR80g:06002.

W.T. Trotter [1978b] Some combinatorial problems for permutations, *Congressus Numerantium* 19 (1978), 619-632.

W.T. Trotter [1981] Stacks and splits of partially ordered sets, *Discrete Math.* 35, 229-256.

W.T. Trotter and K.P. Bogart [1976a] Maximal dimensional partially ordered sets III. A characterization of Hiraguchi's inequality for interval dimension, *Discrete Math.* 15, 389-400. MR54 #5061.

W.T. Trotter and K.P. Bogart [1976b] On the complexity of posets, *Discrete Math.* 16, 71-82. MR54 #2553.

W.T. Trotter and J.I. Moore [1976a] Characterization problems for graphs, partially ordered sets, lattices, and families of sets, *Discrete Math.* 4, 361-368. MR56 #8437.

W.T. Trotter and J.I. Moore [1976b] Some theorems on graphs and posets, *Discrete Math.* 15, 79-84. MR54 #5060.

W.T. Trotter and J.I. Moore [1977] The dimension of planar posets, *J. Combinatorial Theory (Ser. B)* 22, 54-67. MR55 #7857.

W.T. Trotter, J.I. Moore and D.P. Sumner [1976] The dimension of a comparability graph, *Proc. Amer. Math. Soc.* 60, 35-38. MR54 #5062.

W.T. Trotter and J.A. Ross [1981a] Every $t$-irreducible partial order is a subposet of a $t$+1-irreducible partial order, preprint (Univ. of South Carolina).

W.T. Trotter and J.A. Ross [1981b] For $t \geq 3$, every $t$-dimensional partial order can be embedded in a $t$+1-irreducible partial order, preprint (Univ. of South Carolina).

R. Wille [1974] On modular lattices of order dimension two, *Proc. Amer. Math. Soc.* 43, 287-292. MR48 #8327.

R. Wille [1975] Note on the order dimension of partially ordered sets, *Algebra Universalis* 5, 443-444. MR52 #13536.

M. Yannakakis [1981] The complexity of the partial order dimension problem, preprint (Bell Labs, Murray Hill).

# LINEAR EXTENSIONS OF PARTIAL ORDERS AND THE FKG INEQUALITY

R.L. Graham
Bell Laboratories
Murray Hill, New Jersey 07974

ABSTRACT

Many algorithms for sorting $n$ numbers $\{a_1, a_2, \ldots, a_n\}$ proceed by using binary comparisons $a_i : a_j$ to construct successively stronger partial orders $P$ on $\{a_i\}$ until a linear order emerges (e.g., see Knuth [Kn]). A fundamental quantity in deciding the expected efficiency of such algorithms is $Pr(a_i < a_j \mid P)$, the probability that the result of $a_i : a_j$ is $a_i < a_j$ when all linear orders consistent with $P$ are equally likely. In this talk we discuss various intuitive but nontrivial properties of $Pr(a_i < a_j \mid P)$ and related quantities. The only known proofs of some of these results require the use of the so-called FKG inequality [FKG, SW]. We will describe this powerful result and show how it is used in problems like this.



*I. Rival (ed.), Ordered Sets, 213–236.*

## INTRODUCTION

Many algorithms for sorting n numbers $X = \{x_1,\ldots,x_n\}$ proceed by using binary comparisons $x_i:x_j$ to build successively stronger partial orders P on X until a linear order can be deduced (e.g., see Knuth [Kn]). A fundamental quantity in determining the expected efficiency of such algorithms is $\Pr\{x_i<x_j|P\}$, the probability that the result of $x_i:x_j$ is $x_i < x_j$ when all linear orders consistent with P are equally likely. In this paper we discuss a number of results of this type and show how a fundamental inequality, called the FKG inequality, can be used to prove them as well as a variety of related results.

We begin our discussion with a motivating example. Assume there are two teams of tennis players, say $A = \{a_1,\ldots,a_m\}$ and $B = \{b_1,\ldots,b_n\}$. Suppose that the players are inherently totally ordered by an unknown *linear* ordering which may be any permutation of the players, say $a_{i_1} < a_{i_2} < b_{j_1} < \ldots$, each having probability $\frac{1}{(m+n)!}$ (thus, all are equally likely). Assume further that in a match between two players, the *higher ordered player always wins*. Thus, as we observe the outcomes of various matches, we can learn more about the underlying total order of the players.

Suppose that at some point in time we have seen various matches between A players and have thereby learned some partial ordering $\mathcal{A} = \{a_{i_1}<a_{i_2},a_{i_3}<a_{i_4},\ldots\}$ of A, and, similarly, we know a partial ordering $\mathcal{B} = \{b_{j_1}<b_{j_2},b_{j_3}<b_{j_4},\ldots\}$ of B.

Let $C,C',\ldots$ be observed outcomes of matches between a's and b's, e.g.,

$$C:\ a_1 < b_5,\ a_1 < b_3,\ a_4 < b_2,\ldots\ .$$

*Suppose* that in all matches between a's and b's so far, a's always lose to b's. It is certainly reasonable to conjecture that the events C and C' are *mutually favorable*, i.e.,

$$\Pr(C'|\mathcal{A} \text{ and } \mathcal{B} \text{ and } C) \geq \Pr(C'|\mathcal{A} \text{ and } \mathcal{B}). \tag{1}$$

Since, by definition,

$$\Pr(C' \mid A \text{ and } B \text{ and } C) = \Pr(A \text{ and } B \text{ and } C \text{ and } C')/\Pr(A \text{ and } B \text{ and } C)$$

then (1) is equivalent to

$$\Pr(C \mid A \text{ and } B \text{ and } C') \geq \Pr(C \mid A \text{ and } B).$$

Indeed, inequality (1), first conjectured in [GYY] was very recently proved by Shepp [Sh]. Shepp's proof employed the so-called FKG inequality, a result which has just begun to be exploited in combinatorics.

The same intuition which leads to (1) also is present if in addition to knowing $A$ and $B$, we also know that $C''$ has occurred, i.e., various a's have previously lost to other b's. The occurrence of $C''$ somehow reinforces the feeling that the a's are generally weaker players than the b's and so, $C$ and $C'$ should still be mutually favorable, i.e.,

$$\Pr(C' \mid A \text{ and } B \text{ and } C \text{ and } C'') \geq \Pr(C' \mid A \text{ and } B \text{ and } C''). \qquad (1')$$

Surprisingly, this is *not* the case, as the following example shows.

*Example.* $m = n = 2$, $A = B = \phi$

$$C = \{a_2 < b_2\}, \quad C' = \{a_1 < b_1\}, \quad C'' = \{a_2 < b_1\}.$$

We show the allowable linear orderings in the table below. Thus,

$$\Pr(C'') = \frac{12}{24} = \frac{1}{2},$$

$$\Pr(C \text{ and } C'') = \frac{8}{24} = \frac{1}{3},$$

$$\Pr(C \text{ and } C' \text{ and } C'') = \frac{5}{24},$$

and so,

$$\Pr(C' \mid C \text{ and } C'') = \frac{5/24}{1/3} = \frac{5}{8} < \frac{2}{3} = \Pr(C' \mid C'').$$

Again, however somewhat unexpectedly, (1') *does* hold if $A$ and $B$ are both *linear* orders. This was first shown by Graham, Yao and Yao [GYY] using basically combinatorial techniques. Subsequently, Shepp [Sh] and Kleitman and Shearer [KS] found much shorter proofs, again using the FKG inequality.

| Ordering | C | C' | C" |
|---|---|---|---|
| $a_1 < a_2 < b_1 < b_2$ | 1 | 1 | 1 |
| $a_1 < a_2 < b_2 < b_1$ | 1 | 1 | 1 |
| $a_1 < b_2 < a_2 < b_1$ | 0 | 1 | 1 |
| $a_2 < a_1 < b_1 < b_2$ | 1 | 1 | 1 |
| $a_2 < a_1 < b_2 < b_1$ | 1 | 1 | 1 |
| $a_2 < b_1 < a_1 < b_2$ | 1 | 0 | 1 |
| $a_2 < b_1 < b_2 < a_1$ | 1 | 0 | 1 |
| $a_2 < b_2 < a_1 < b_1$ | 1 | 1 | 1 |
| $a_2 < b_2 < b_1 < a_1$ | 1 | 0 | 1 |
| $b_2 < a_1 < a_2 < b_1$ | 0 | 1 | 1 |
| $b_2 < a_2 < a_1 < b_1$ | 0 | 1 | 1 |
| $b_2 < a_2 < b_1 < a_1$ | 0 | 0 | 1 |

Table

In the following sections of the paper, we will describe the FKG inequality and its recent numerous generalizations, and illustrate its use in connection with problems involving linear extensions and order preserving maps of partial orders.

## THE FKG INEQUALITY AND ITS GENERALIZATIONS

In some sense the FKG inequality has its roots in the old result of Chebyshev which asserts that if f and g are both increasing (or both decreasing) functions on [0,1] then the average value of the product fg is at least as large as the product of the average values of f and g (where the average is taken with respect to some measure $\mu$ on [0,1]). In symbols,

$$\int_0^1 fg\,d\mu \geq \int_0^1 f\,d\mu \int_0^1 g\,d\mu. \tag{2}$$

In the case that $\mu$ is a discrete measure we can restate (2) as follows: If f(k) and g(k) are both increasing (or both decreasing) and $\mu(k) \geq 0$, $k = 1,2,3,\ldots$ then

$$\frac{\sum_k f(k)g(k)\mu(k)}{\sum_k \mu(k)} \geq \frac{\sum_k f(k)\mu(k)}{\sum_k \mu(k)} \cdot \frac{\sum_k g(k)\mu(k)}{\sum_k \mu(k)}$$

i.e.,

$$\sum_k f(k)g(k)\mu(k) \sum_k \mu(k) \geq \sum_k f(k)\mu(k) \sum_k g(k)\mu(k). \qquad (2')$$

The proof of (2') (and (2)) follows once by expanding the inequality

$$\sum_{i,j} (f(i)-f(j))(g(i)-g(j))\mu(i)\mu(j) \geq 0.$$

The FKG inequality, due to Fortuin, Kasteleyn and Ginibre [FKG], was discovered in connection with the proof of certain natural conjectures arising in statistical mechanics (in particular, dealing with correlations of Ising spin systems). They trace rudimentary forms of the inequality back to Griffiths [G] (who was also interested in these questions) and Harris [Ha] (who was investigating the probabilities of certain events in percolation models). It turns out that special cases were also anticipated by Kleitman [K1] (see [K2]) and Marica-Schönheim [MS], among others.

Basically the FKG inequality is an attempt to extend (2') to the case where the underlying set is only *partially ordered*, as opposed to the *totally* ordered index set of integers occurring in (2'). The setting is as follows. Let $(\Gamma, <)$ be a distributive lattice. That is, $\Gamma$ is a (finite) set, partially ordered by $<$, on which two commutative functions $\wedge$ (greatest lower bound) and $\vee$ (least upper bound) are defined which satisfy the distributive laws:

$$\left.\begin{aligned} x \wedge (y \vee z) &= (x \wedge y) \vee (x \wedge z) \\ x \vee (y \wedge z) &= (x \vee y) \wedge (x \vee z) \end{aligned}\right\} \quad \text{for all } x, y, z \in \Gamma.$$

It is well known that without loss of generality we may assume $\Gamma$ is a sublattice of the lattice of all subsets of some finite set partially ordered under inclusion with $x \wedge y = x \cap y$ and $x \vee y = x \cup y$.

Let $\mu:\Gamma \to \mathbb{R}_0$, the nonnegative reals, satisfy

$$\mu(x)\mu(y) \leq \mu(x\vee y)\mu(x\wedge y) \text{ for all } x, y \in \Gamma. \tag{3}$$

We call a function $f:\Gamma \to \mathbb{R}$ *increasing* if

$$x \leq y \Rightarrow f(x) \leq f(y) \text{ for } x, y \in \Gamma,$$

(with *decreasing* defined similarly).

The FKG inequality: If f and g are both increasing (or both decreasing) functions on a distributive lattice $\Gamma$ and $\mu:\Gamma \to \mathbb{R}_0$ satisfies (3) then

$$\sum_{x\in\Gamma} f(x)g(x)\mu(x) \sum_{x\in\Gamma} \mu(x) \geq \sum_{x\in\Gamma} f(x)\mu(x) \sum_{x\in\Gamma} g(x)\mu(x). \tag{4}$$

The original proof [FKG] of (4) was not so simple. Several years after (4) appeared, Holley [Ho] found the following generalization of FKG:

Suppose $\alpha,\beta:\Gamma \to \mathbb{R}_0$ satisfy

$$\alpha(x)\beta(y) \leq \alpha(x\vee y)\beta(x\wedge y) \text{ for all } x, y \in \Gamma.$$

Then for any increasing function $\theta:\Gamma \to \mathbb{R}_0$

$$\sum_{x} \alpha(x)\theta(x) \geq \sum_{x} \beta(x)\theta(x). \tag{5}$$

Rather recently, Ahlswede and Daykin have given a remarkable strengthening of (4) and (5).

THEOREM [AD2]. *Suppose we are given four functions* $\alpha,\beta,\gamma,\delta:\Gamma \to \mathbb{R}_0$ *which satisfy*

$$\alpha(x)\beta(y) \leq \gamma(x\vee y)\delta(x\wedge y) \text{ for all } x, y \in \Gamma. \tag{6}$$

*Then*

$$\alpha(X)\beta(Y) \leq \gamma(X\vee Y)\delta(X\wedge Y) \text{ for all } X,\ Y \subseteq \Gamma \tag{7}$$

*where*

$$X \vee Y = \{x\vee y: x\in X, y\in Y\},$$

$$X \wedge Y = \{x\wedge y: x\in X, y\in Y\} \text{ and}$$

$$\text{for } Z \subseteq \Gamma,\ f(Z) \equiv \sum_{z\in Z} f(z).$$

The proof of (7) is surprisingly simple. To begin with, it follows from our previous observations that it is enough to prove the following result (where $2^{[N]}$ denotes the family of all subsets of $[N] = \{1,2,\ldots,N\}$): Suppose $\alpha,\beta,\gamma,\delta: 2^{[N]} \to \mathbb{R}_0$ satisfy

$$\alpha(x)\beta(y) \leq \gamma(x\cup y)\delta(x\cap y) \text{ for all } x,\ y \in 2^{[N]}. \tag{6'}$$

Then

$$\alpha(X)\beta(Y) \leq \gamma(X\vee Y)\delta(X\wedge Y) \text{ for all } X,\ Y \subseteq 2^{[N]}. \tag{7'}$$

PROOF: The proof is by induction on N. We first consider the case N = 1. Then $2^{[1]} = \{\phi,\{1\}\}$. Denote $\alpha(\phi)$ by $\alpha_0$ and $\alpha(\{1\})$ by $\alpha_1$, with $\beta_0$, $\beta_1$, etc., defined similarly. The system (6') becomes:

$$\begin{aligned} \alpha_0\beta_0 &\leq \gamma_0\delta_0, \\ \alpha_1\beta_0 &\leq \gamma_1\delta_0, \\ \alpha_0\beta_1 &\leq \gamma_1\delta_0, \\ \alpha_1\beta_1 &\leq \gamma_1\delta_1, \end{aligned} \tag{8}$$

It is easily checked that if either X or Y consists of a single element then (7') is an immediate consequence of (8). The only interesting case is $X = \{\phi,\{1\}\} = Y$. In this case, the inequality (7') we are required to prove is

$$(\alpha_0+\alpha_1)(\beta_0+\beta_1) \leq (\gamma_0+\gamma_1)(\delta_0+\delta_1). \tag{9}$$

Equation (9) would follow instantly from (8) if one of the two occurrences of $\gamma_1\delta_0$ in (8) were $\gamma_0\delta_1$ instead. As it is, we have to work a little (but not much) harder. If any of $\alpha_0$, $\beta_0$, $\gamma_0$ or $\delta_0$ is zero, (9) follows at once. Hence, by a suitable normalization, we can assume

$$\alpha_0 = \beta_0 = \gamma_0 = \delta_0 = 1.$$

The system (8) becomes

$$\alpha_1 \leq \gamma_1, \; \beta_1 \leq \gamma_1, \; \alpha_1\beta_1 \leq \gamma_1\delta_1 \tag{8'}$$

and (9) becomes

$$(1+\alpha_1)(1+\beta_1) \leq (1+\gamma_1)(1+\delta_1). \tag{9'}$$

Again (9') is immediate if $\gamma_1 = 0$ so we may assume $\gamma_1 > 0$. Since (9') becomes harder to satisfy as $\delta_1$ decreases, it is enough to prove (9') when $\delta_1$ is as small as possible, i.e., $\delta_1 = \frac{\alpha_1\beta_1}{\gamma_1}$. In this case we need

$$(1+\alpha_1)(1+\beta_1) \leq (1+\gamma_1)\left(1 + \frac{\alpha_1\beta_1}{\gamma_1}\right),$$

i.e.,

$$\alpha_1 + \beta_1 \leq \gamma_1 + \frac{\alpha_1\beta_1}{\gamma_1}.$$

However, this is an immediate consequence of

$$(\gamma_1-\alpha_1)(\gamma_1-\beta_1) \geq 0$$

which is implied by (8'). This proves the result for N = 1.

Assume now that the assertion holds for $N = n - 1$ for some $n \geq 2$. Let $\alpha, \beta, \gamma, \delta : 2^{[n]} \to \mathbb{R}_0$ satisfy the hypotheses (6') with $N = n$ and let $X, Y \subseteq 2^{[n]}$ be given. We will define new functions $\alpha'$, $\beta'$, $\gamma'$, $\delta'$ mapping $2^{[n-1]} = T'$ into $\mathbb{R}_0$ as follows:

$$\alpha'(x') = \sum_{\substack{x \in X \\ x' = x \setminus \{n\}}} \alpha(x), \quad \beta'(y') = \sum_{\substack{y \in Y \\ y' = y \setminus \{n\}}} \beta(y)$$

$$\gamma'(z') = \sum_{\substack{z \in X \cup Y \\ z' = z \setminus \{n\}}} \gamma(z), \quad \delta'(w') = \sum_{\substack{w \in X \cap Y \\ w' = w \setminus \{n\}}} \delta(w).$$

Thus, for $x' \in T'$

$$\alpha'(x') = \begin{cases} \alpha(x') + \alpha(x' \cup \{n\}) & \text{if } x' \in X,\ x' \cup \{n\} \in X, \\ \alpha(x') & \text{if } x' \in X,\ x' \cup \{n\} \notin X, \\ \alpha(x' \cup \{n\}) & \text{if } x' \notin X,\ x' \cup \{n\} \in X, \\ 0 & \text{otherwise.} \end{cases} \tag{10}$$

Observe that with these definitions

$$\alpha(X) = \sum_{x \in X} \alpha(x) = \sum_{x' \in T'} \alpha'(x') = \alpha'(T')$$

and

$$\beta(Y) = \beta'(T'), \quad \gamma(X \vee Y) = \gamma'(T'),$$

$$\delta(X \wedge Y) = \delta'(T').$$

Thus, if

$$\alpha'(x')\beta'(y') \le \gamma'(x' \cup y')\delta'(x' \cap y') \text{ for all } x',\ y' \in T' \quad (11)$$

holds, then by the induction hypotheses

$$\alpha(X)\beta(Y) = \alpha'(T')\beta'(T') \le \gamma'(T')\delta'(T') = \gamma(X \vee Y)\delta(X \wedge Y)$$

since $T' \vee T' = T'$, $T' \wedge T' = T'$, which is (7').

It remains to prove (11). However, by (10) this is *exactly* the computation performed for the case $N = 1$ with $x' \leftrightarrow \phi$ and $x' \cup \{n\} \leftrightarrow \{1\}$. Since we have already treated this case then the induction step is completed. This proves the theorem of Ahlswede and Daykin. □

To prove the FKG inequality from Ahlswede-Daykin it suffices as usual to restrict ourselves to the case that $\Gamma = 2^{[N]} = T$. Observe that if A and B are upper ideals in T (i.e., $x, y \in A \Rightarrow x \cup y \in A$) then the indicator functions $f = I_A$ and $g = I_B$ (with $I_A(x) = 1$ iff $x \in A$) are increasing. Taking $\alpha = \beta = \gamma = \delta = \mu$ in (6') and $X = A$, $Y = B$ in (7') we have

$$\mu(A)\mu(B) \le \mu(A \vee B)\mu(A \wedge B). \quad (12)$$

But

$$\mu(A) = \sum_{x \in A} \mu(x) = \sum_{x \in T} f(x)\mu(x),$$

$$\mu(B) = \sum_{x \in T} g(x)\mu(x),$$

$$\mu(A \wedge B) = \sum_{z \in A \wedge B} \mu(z) = \sum_{z \in T} f(z)g(z)\mu(z)$$

and

$$\mu(A \vee B) = \sum_{z \in A \vee B} \mu(z) \le \sum_{x \in T} \mu(z).$$

Thus, (12) implies

$$\sum_z f(z)g(z)\mu(z) \sum_z \mu(z) \geq \sum_z f(z)\mu(z) \sum_z g(z)\mu(z)$$

which is just the FKG inequality for this case. The general FKG inequality is proved in just this way by first writing an arbitrary increasing function f on T as $f = \sum_i \lambda_i I_{A_i}$ where $\lambda_i \geq 0$ and $A_i$ are suitable upper ideals in T. That is, for

$$f = \sum_i \lambda_i I_{A_i}, \quad \lambda_i \geq 0,$$

we have

$$\sum_{z\varepsilon T} f(z)\mu(z) = \sum_{z\varepsilon T} \sum_i \lambda_i I_{A_i}(z)\mu(z)$$

$$= \sum_i \lambda_i \sum_{z\varepsilon T} I_{A_i}(z)\mu(z), \text{ etc.,}$$

and now apply the preceding inequality.

Before concluding this section, we make several remarks concerning the Ahlswede-Daykin result. Setting $\alpha = \beta = \gamma = \delta = 1$ we obtain

$$|X||Y| \leq |X \vee Y||X \wedge Y| \text{ for all } X, Y \subseteq 2^{[N]}. \tag{13}$$

This was first proved by Daykin [D] and has as immediate corollaries:

(a) (*Marica-Schönheim* [MS])

$$|A| \le |A \backslash A| \text{ for all } A \subseteq 2^{[N]}.$$

(b) (*Kleitman* [K1]). *For any upper ideal* U *and any lower ideal* L *of* $2^{[N]}$,

$$|U \cap L| \cdot 2^N \le |U||L|.$$

(c) (*Seymour* [Se]). *For any two upper ideals* $U_1$, $U_2$ *of* $2^{[N]}$,

$$|U_1||U_2| \le |U_1 \cap U_2| \cdot 2^N.$$

The special cases (a) and (b) actually appeared in the literature before the FKG inequality was found. We leave as an exercise for the reader to show that the starting inequality (2') also follows from this.

A good summary of these and related results can be found in [AD1], [D1], [D2]. Far ranging generalizations of the Ahlswede-Daykin Theorem were obtained by the same authors and appear in [AD2].

LINEAR EXTENSIONS OF TWO CHAINS

Suppose $(X,<)$ is a partially ordered set in which $X = A \cup B$ is a disjoint union of two chains $A = \{a_1 < \ldots < a_m\}$ and $B = \{b_1 < \ldots < b_n\}$. Of course, relations such as $a_i < b_j$ and $b_k < a_\ell$ are also allowed. Consider the set $\Lambda$ of all $(m+n)!$ 1 - 1 mappings of X onto $[m+n] = \{1,2,\ldots,m+n\}$. Assign a probability $\frac{1}{(m+n)!}$ to each $\lambda \in \Lambda$, i.e., assume they are all equally likely. Let Q denote the set of all *linear extensions* $\lambda$ of X, i.e., $\lambda \in \Lambda$ such that

$$x < y \Rightarrow \lambda(x) < \lambda(y).$$

Let P and P' both be unions of subsets of $\Lambda$ of the form

$$\{\lambda : \lambda(a_{i_1}) < \lambda(b_{j_1}), \lambda(a_{i_2}) < \lambda(b_{j_2}), \ldots\}$$

(i.e., a's always lose to b's).

*THEOREM 1 (Graham, Yao, Yao [GYY]).*

$$\Pr(P \mid Q \cap P') \geq \Pr(P \mid Q). \tag{14}$$

PROOF. We will give a proof of this result (due to Shepp [Sh]) based on the FKG inequality. The original proof used explicit combinatorial pairings of certain types of mappings in $\Lambda$.

Define a lattice $\Gamma$ with elements of the form $\bar{x} = \{x_1, \ldots, x_m\}$ with $1 \leq x_1 < \ldots < x_m \leq m + n$. We say that $\bar{x} \leq \bar{x}'$ if $x_i \leq x_i'$ for $1 \leq i \leq m$. Define

$$\bar{x} \vee \bar{x}' = \{\ldots, \max(x_i, x_i'), \ldots\},$$

$$\bar{x} \wedge \bar{x}' = \{\ldots, \min(x_i, x_i'), \ldots\}.$$

It is easily checked that with these definitions $\Gamma$ is a distributive lattice.

For each $\bar{x} \in \Gamma$ we can associate a unique $\lambda_{\bar{x}} \in \Lambda$ by setting:

$$\lambda_{\bar{x}}(a_i) = x_i,$$

$$\lambda_{\bar{x}}(b_j) = y_j$$

where $[m+n] \setminus \{x_1 < \ldots < x_m\} = \{y_1 < \ldots < y_n\}$. Finally, define

$$\mu(\bar{x}) = \begin{cases} 1 \text{ if } \lambda_{\bar{x}} \in Q, \\ 0 \text{ otherwise,} \end{cases}$$

$$f(\bar{x}) = \begin{cases} 1 \text{ if } \lambda_{\bar{x}} \in P, \\ 0 \text{ otherwise,} \end{cases}$$

$$f'(\bar{x}) = \begin{cases} 1 \text{ if } \lambda_{\bar{x}} \in P', \\ 0 \text{ otherwise.} \end{cases}$$

To apply FKG, we must check:

$$\mu(\bar{x})\mu(\bar{x}') \leq \mu(\bar{x}\vee\bar{x}')\mu(\bar{x}\wedge\bar{x}'). \tag{15}$$

Suppose $\mu(\bar{x})\mu(\bar{x}') = 1$. Then $\lambda_{\bar{x}} \in Q$, $\lambda_{\bar{x}'} \in Q$. If $a_i < a_j$ in X then

$$\lambda_{\bar{x}}(a_i) = x_i < x_j = \lambda_{\bar{x}}(a_j)$$
$$\lambda_{\bar{x}'}(a_i) = x_i' < x_j' = \lambda_{\bar{x}'}(a_j)$$

and so,

$$\lambda_{\bar{x}\vee\bar{x}'}(a_i) = \max(x_i, x_i') < \max(x_j, x_j') = \lambda_{\bar{x}\vee\bar{x}'}(a_j).$$

Similarly, if $b_i < b_j$ in X then

$$\lambda_{\bar{x}\vee\bar{x}'}(b_i) < \lambda_{\bar{x}\vee\bar{x}'}(b_j).$$

On the other hand, if $a_i < b_j$ in X then

$$\lambda_{\bar{x}}(a_i) = x_i < y_j = \lambda_{\bar{x}}(b_j)$$

$$\lambda_{\bar{x}'}(a_i) = x_i' < y_j' = \lambda_{\bar{x}'}(b_j)$$

i.e.,

$$x_i \le i + j - 1, \; x_i' \le i + j - 1.$$

Thus,

$$\lambda_{\bar{x} \vee \bar{x}'}(a_i) = \max(x_i, x_i') \le i + j - 1$$

so that

$$\lambda_{\bar{x} \vee \bar{x}'}(a_i) < \lambda_{\bar{x} \vee \bar{x}'}(b_j).$$

The argument for $b_i < a_j$ is similar. This shows that $\lambda_{\bar{x} \vee \bar{x}'} \in Q$, i.e., $\mu(\bar{x} \vee \bar{x}') = 1$. In this same way it follows that $\mu(\bar{x} \wedge \bar{x}') = 1$. Therefore, we have shown that

$$\mu(\bar{x})\mu(\bar{x}') = 1 \Rightarrow \mu(\bar{x} \vee \bar{x}')\mu(\bar{x} \wedge \bar{x}') = 1$$

and consequently (15) always holds.

The final condition to check before applying FKG is that f and f' are decreasing. To see this, suppose $\bar{x} < \bar{x}'$ and $f(\bar{x}') = 1$. Then by definition, $\lambda_{\bar{x}'} \in P = \bigcup_k P_k$ where

$$P_k = \{\lambda : \lambda(a_{i_1}) < \lambda(b_{j_1}), \ldots\}.$$

Thus, for some k, the elements $\bar{x}_i'$ of $\bar{x}'$ satisfy all the constraints $x_{i_1}' \le i_1 + j_1 - 1, \ldots,$ imposed by $P_k$. But since $\bar{x} \le \bar{x}'$, i.e., $x_i \le x_i'$ for all i, then $x_{i_1} \le x_{i_1}' \le i_1 + j_1 - 1, \ldots,$ as well. This implies that $\lambda_{\bar{x}} \in P_k \subseteq P$ and $f(x) = 1$, i.e., f is decreasing.

The FKG inequality can now be applied to the functions we have defined, yielding:

$$\sum_{\bar{x} \in \Gamma} f(\bar{x}) f'(\bar{x}) \mu(\bar{x}) \sum_{\bar{x} \in \Gamma} \mu(\bar{x}) \ge \sum_{\bar{x} \in \Gamma} f(\bar{x}) \mu(\bar{x}) \sum_{\bar{x} \in \Gamma} f'(\bar{x}) \mu(\bar{x}) \tag{16}$$

Interpreting (15) in terms of Q, P and P', we obtain

$$|P\cap P'\cap Q|\,|Q| \geq |P\cap Q|\,|P'\cap Q| \tag{17}$$

i.e.,

$$\Pr(P\cap P'|Q) \geq \Pr(P|Q)\Pr(P'|Q)$$

which implies (14). □

As the example at the beginning of the paper shows, (14) does not necessarily hold if A and B are not linearly ordered. A somewhat different example [GYY] showing this is the following

*Example:*

$$A = \{a_1 < a_2\}, \quad B = \{b_1 < b_3, b_2 < b_3\}.$$

In addition, in $X = A \cup B$ we have $b_1 < a_2$ and $a_1 < b_2$. Let P denote the event $\{a_1 < b_1\}$ and P' denote the event $\{a_2 < b_3\}$. An easy calculation shows

$$\Pr(P|P'\cap Q) = \frac{3}{5} < \frac{5}{8} = \Pr(P|Q).$$

ORDER PRESERVING MAPS

In this section we show how the FKG inequality theorem can be used to prove a broader class of similar results for the class of order preserving maps on a partially ordered set. We say that a mapping $\rho: X \to X'$ between partially ordered sets $(X,<)$ and $(X',<)$ is *order preserving* if $x, y \in X$ with $x < y \Rightarrow \rho(x) < \rho(y)$. Note that $\rho$ is not required to be 1 - 1. (The use of the same symbol < for both partial orders should cause no confusion). Let $R = R(X,X')$ denote the set of order preserving maps of X into X'. As before, we will assume that all $\rho \in R$ are equally likely.

We need one further definition. For $x \in X$ define *range*$(x)$ to be $\{\rho(x):\rho\in R\}$.

*THEOREM 2. Suppose X is the disjoint union of* $A = \{a_1,\ldots,a_m\}$ *and* $B = \{b_1,\ldots,b_n\}$. *Let P and P' both be unions of subsets of the form* $\{\rho:X\to X':\rho(a_i)<\rho(b_j),\rho(a_k)<\rho(b_\ell),\ldots\}$ *(i.e., certain* a's *are always mapped below certain* b's*).*

*Suppose for all* $a \in A$, $b \in A$ *which are related by* $<$,

$$\text{range}(a) \cap \text{range}(b) = \emptyset. \tag{18}$$

*Then*

$$\Pr(P\cap P'|R) \geq \Pr(P|R)\Pr(P'|R). \tag{19}$$

PROOF: Define a lattice $(\Gamma,<)$ by taking the points of $\Gamma$ to be sequences $\bar{z} = (x_1,\ldots,x_m;y_1,\ldots,y_n)$ where $x_i$, $y_j \in X'$ and $\bar{z} \leq \bar{z}'$ means $x_i \leq x_i'$ and $y_j \geq y_j'$ for all i, j. Further, define

$$\bar{z} \vee \bar{z}' = (\ldots,\max(x_i,x_i'),\ldots;\ldots,\min(y_j,y_j'),\ldots)$$

$$\bar{z} \wedge \bar{z}' = (\ldots,\min(x_i,x_i'),\ldots;\ldots,\max(y_j,y_j'),\ldots).$$

It is easy to check that with these operations, $\Gamma$ is distributive.

To each $\bar{z} \in \Gamma$, associate a map $\rho_{\bar{z}}:X \to X'$ by defining

$$\rho_{\bar{z}}(a_i) = x_i,\ \rho_{\bar{z}}(b_j) = y_j.$$

Finally, define the three functions:

$$\mu(\bar{z}) = \begin{cases} 1 \text{ if } \rho_{\bar{z}} \in R, \\ 0 \text{ otherwise,} \end{cases}$$

$$f(\bar{z}) = \begin{cases} 1 \text{ if } \rho_{\bar{z}} \in P, \\ 0 \text{ otherwise,} \end{cases}$$

$$f'(\bar{z}) = \begin{cases} 1 \text{ if } \rho_{\bar{z}} \in P', \\ 0 \text{ otherwise.} \end{cases}$$

In order to apply the FKG inequality, we must check that f and f' are both decreasing (this is basically the same argument as before) and that $\mu$ satisfies the "log supermodularity" condition

$$\mu(\bar{z})\mu(\bar{z}') \leq \mu(\bar{z} \vee \bar{z}')\mu(\bar{z} \wedge \bar{z}') \text{ for all } \bar{z}, \bar{z}' \in \Gamma.$$

Again the argument is quite close to the previous one with one exception. In showing that $\mu(\bar{z})\mu(\bar{z}') = 1 \Rightarrow \mu(\bar{z} \vee \bar{z}')\mu(\bar{z} \wedge \bar{z}') = 1$ we must know (for example) that

$$\min(x_i, x_i') > \max(y_j, y_j')$$

if $x_i < y_j$ and $x_i' < y_j'$. While this is not ordinarily always the case, with the assumption (18) in force, it does indeed hold. Using this observation several times, the verification that $\mu$, f and f' satisfy the hypotheses of FKG is accomplished. The conclusion of FKG now implies (similar to (16) $\Rightarrow$ (17))

$$|P \cap P' \cap R| \, |R| \geq |P \cap R| \, |P' \cap R|$$

which in turn implies (19). □

Whether or not the "range" condition (18) is needed is not currently known.

LINEAR EXTENSIONS OF PAIRS OF GENERAL PARTIAL ORDERS

It is natural to try and extend Theorem 1 to linear extensions of more general partial orders. However, numerous examples like those previously presented show that *some* restrictions on the partial order will be necessary. One such extension, conjectured by Graham, Yao and Yao [GYY], and proved by Shepp [Sh], is the following.

*THEOREM 3. Suppose* $(X,<)$ *is a partial order where* $X$ *is a disjoint union of two partial orders* $A = \{a_1,\ldots,a_m\}$ *and* $B = \{b_1,\ldots,b_n\}$, *i.e.,* $a_i$ *and* $b_j$ *are all pairwise incomparable. Let* $R$ *denote the set of linear extensions of* $X$ *onto* $[m+n]$ *and let* $P$ *and* $P'$ *be sets of maps* $\lambda$ *of* $X$ *onto* $[m+n]$ *which are unions of sets of the form*

$$\{\lambda:\lambda(a_{i_1})<\lambda(b_{j_1}),\lambda(a_{i_2})<\lambda(b_{j_2}),\ldots\}.$$

*If all elements of* $R$ *are given equal probability then* $P$ *and* $P'$ *are positively correlated, i.e.,*

$$\Pr(P|P'\cap R) \geq \Pr(P|R). \tag{20}$$

PROOF: The proof is similar to that of Theorem 2 but with an extra complication. To begin with, we choose a large fixed integer $N$ and define $\Gamma = \Gamma_N$ to be the set of all sequences $\bar{z} = (x_1,\ldots,x_m;y_1,\ldots,y_n)$ partially ordered by setting $\bar{z} \leq \bar{z}'$ iff $x_i \leq x_i'$ and $y_j \geq y_j'$ for all $i$ and $j$ where $1 \leq x_i, y_j \leq N$. As before define

$$\bar{z} \wedge \bar{z}' = (\ldots,\min(x_i,x_i'),\ldots;\ldots,\max(y_j,y_j'),\ldots)$$

$$\bar{z} \vee \bar{z}' = (\ldots,\max(x_i,x_i'),\ldots;\ldots,\min(y_j,y_j'),\ldots).$$

This makes $\Gamma$ into a distributive lattice. For $\bar{z} \in \Gamma$, define a mapping $\rho_{\bar{z}}:\Gamma \to [N]$ by $\rho_{\bar{z}}(a_i) = x_i$, $\rho_{\bar{z}}(b_j) = y_j$. We will assume all $N^{m+n}$ such mappings are equally likely.

Let $\bar{R}$, $\bar{P}$, $\bar{P}'$ be the subsets of mappings $\rho_{\bar{z}}$, $\bar{z} \in \Gamma$, which preserve the order given by $R$, $P$ and $P'$, respectively, i.e., $\lambda(x) < \lambda(y) \Rightarrow \rho_{\bar{z}}(x) < \rho_{\bar{z}}(y)$, with $\mu$, $f$ and $f'$ defined to be the corresponding characteristic functions.

It is straightforward to check that the hypotheses of the FKG inequality are satisfied. The potential difficulty for $\mu$ which required the range disjointness condition (18) does not occur here since $A$ and $B$ are incomparable by hypothesis.

Thus, by the FKG inequality we obtain

$$|\bar{P}\cap\bar{P}'\cap\bar{R}|\,|\bar{R}| \geq |\bar{P}\cap\bar{R}|\,|\bar{P}'\cap\bar{R}|. \tag{21}$$

Define $\hat{R}$ to be the subset of R in which all components of $\bar{z} \in \bar{R}'$ are *distinct* (with $\hat{P}$ and $\hat{P}'$ defined similarly). Since

$$|\hat{R}| = \frac{N!}{(N-m-n)!}$$

then

$$\lim_{N\to\infty} \frac{|\hat{R}|}{|\bar{R}|} = 1.$$

Each $\rho_{\bar{z}}$, $\bar{z} \in \hat{R}$, determines a *linear extension* $\lambda$ of $X \to [m+n]$ in the obvious way, with the relative ordering of the $\rho_{\bar{z}}(a_i)$ and $\rho_{\bar{z}}(b_j)$ determining their ranks in $[m+n]$. Furthermore, for $N \geq m + n$, each linear extension $\lambda: X \to [m+n]$ is generated exactly the same number of times by the $\rho_{\bar{z}}$, $\bar{z} \in \hat{R}$. Of course, the preceding comments also apply to $\hat{P}$ and $\hat{P}'$. Thus,

$$\Pr(P|R) = \lim_{N\to\infty} \frac{|\hat{P}\cap\hat{R}|}{|\hat{R}|} = \lim_{N\to\infty} \frac{|\bar{P}\cap\bar{R}|}{|\bar{R}|}$$

and

$$\Pr(P\cap P'\cap R) = \lim_{N\to\infty} \frac{|\bar{P}\cap\bar{P}'\cap\bar{R}|}{|\bar{R}|}.$$

The desired conclusion (20) now follows and the theorem is proved. □

## CONCLUDING REMARKS

We close the paper with a discussion of a number of questions dealing with linear extensions of partial orders and potential applications of the FKG inequality.

1. It would be quite interesting to know under exactly what conditions two events (= subsets of linear extensions of a partially ordered set) are mutually favorable. Theorems 1

and 3 give some conditions under which this occurs but they certainly do not tell the whole story. Is it true, for example, that Theorem 2 holds for linear extensions rather than order preserving maps?

2. Recently, Stanley [St2] proved the following result (a strengthening of a conjecture of Rivest [R] which was conjectured by Chung, Fishburn and Graham [CFG]). Let X be a partially ordered set with n elements and let $x \in X$ be a fixed element. Let $N_i$ denote the number of linear extensions $\lambda$ of X onto [n] for which $\lambda(x) = i$.

*THEOREM. The sequence $N_i$ is logarithmically concave.*

This result had been established earlier in [CFG] for the case that X can be covered by two chains (= linearly ordered sets). Stanley's proof required the use of the so-called Aleksandrov-Fenchel inequalities from the theory of mixed volumes (see [B] or [Fe]) similar to those for mixed discriminants which have recently been used by Egoritsjev [E], [vL] to prove the infamous van der Waerden permanent conjecture. It is known that the FKG inequality can be used very naturally to prove the log concavity of various sequences of combinatorial interest (e.g., see [SW]). Can Stanley's result be proved using the FKG inequality (or the Ahlswede-Daykin theorem?) Is the analog of Stanley's theorem for *order preserving* maps true as well?

3. Fishburn [Fi1], [Fi2] has recently studied the following problem. If x and y are two elements of a partially ordered set X on n-elements, let us say that x *dominates* y (written $x \to y$) if more linear extensions $\lambda: X \to [n]$ have $\lambda(x) > \lambda(y)$ than $\lambda(y) > \lambda(x)$. He has shown that there are partial orders X for which the cycle $x \to y \to z \to x$ is possible (the smallest such X known has 31 elements). More generally he has constructed partial orders X for which every vertex of this (directed) domination graph D(X) has outdegree at least one. What is the least X such that D(X) has a cycle? Is it possible to characterize those directed graphs which occur as D(X) for some X? What if we require

$$|\{\lambda: \lambda(x)>\lambda(y)\}| > \alpha|\{\lambda: \lambda(y)>\lambda(x)\}|$$

for some (fixed) $\alpha > 1$? What is the largest $\alpha$ for which the corresponding domination graph can have a cycle?

4. The following nice problem of Stanley (see [St1]) is still open. Let X be a partially ordered set on n elements and let $(a_1,a_2,\ldots)$ be a sequence of nonnegative integers satisfying $\sum_k a_k = n$. Define $f_X(a_1,a_2,a_3,\ldots)$ to be the number of order preserving maps $\rho:X \to \{1,2,3,\ldots\}$ such that

$$|x\varepsilon X:\rho(x)=i| = a_i.$$

Is it possible for two different partially ordered sets X and X' to have

$$f_X(a_1,a_2,a_3,\ldots) = f_{X'}(a_1,a_2,a_3,\ldots)$$

for all $(a_1,a_2,a_3,\ldots)$? Stanley has shown that if such X, X' exist then $n \geq 7$.

REFERENCES

[AD1] R. Ahlswede and D. E. Daykin, Inequalities for a pair of maps $S \times S \to S$ with S a finite set, *Math. Zeit.*, 165 (1979), 267-289.

[AD2] ___________, An inequality for the weights of two families of sets, their unions and intersections, *Z. Wahrscheinlichkeitstheorie verw. Gebeite*, 43 (1978), 183-185.

[B] H. Busemann, *Convex Surfaces*, Interscience, New York, 1958.

[CFG] F. R. K. Chung, P. C. Fishburn and R. L. Graham, On unimodality for linear extensions of partial orders, *SIAM Jour. Alg. Disc. Meth.*, 1 (1980), 405-410.

[D] D. E. Daykin, A lattice is distributive iff $|A||B| \leq |A\vee B||A\wedge B|$, *Nanta Math.*, 10 (1977), 58-60.

[E] G. P. Egoritsjev, Solution of van der Waerden's permanent conjecture, (preprint) 13M of the Kirenski Inst. of Physics, Krasnojarsk (1980), (Russian).

[Fe] W. Fenchel, Inégalités quadratique entre les volumes mixtes des corps convexes, *C. R. Acad. Sci.*, Paris, 203

(1936), 647-650.

[Fi1] P. C. Fishburn, On the family of linear extensions of a partial order, *J. Combinatorial Th.* (B), 17 (1974), 240-243.

[Fi2] ______________, On linear extension majority graphs of partial orders, *J. Combinatorial Th.* (B), 21 (1976), 65-70.

[FKG] C. M. Fortuin, P.W. Kasteleyn and J. Ginibre, Correlation inequalities on some partially ordered sets, *Comm. Math. Phys.*, 22 (1971), 89-103.

[GYY] R. L. Graham, A. C. Yao, F. F. Yao, Some monotonicity properties of partial orders, *SIAM Jour.. Alg. Disc. Meth.*, 1 (1980), 251-258.

[G] R. B. Griffiths, Correlations in Ising ferromagnets. *J. Math. Phys.*, 8 (1967), 478-483.

[Ha] T. E. Harris, A lower bound for the critical probability in a certain percolation process, *Proc. Camb. Phil. Soc.*, 56 (1960), 13-20.

[Ho] R. Holley, Remarks on the FKG inequalities, *Comm. Math. Phys.*, 36 (1974), 227-231.

[K1] D. J. Kleitman, Families of non-disjoint sets, *J. Combinatorial Th.*, 1 (1966), 153-155.

[K2] ______________, Extremal hypergraph problems, Surveys in Combinatorics, B. Bollobás, Ed., *London Math. Soc.*, Lecture Note Series 38, Cambridge Univ. Press, (1979), pp. 44-65.

[KS] D. J. Kleitman and J. B. Shearer, A monotonicity property of partial orders (to appear).

[Kn] D. E. Knuth, The Art of Computer Programming, Vol. 3, Sorting and Searching, *Addison-Wesley*, Reading, Mass., 1973.

[MS] J. Marica and J. Schönheim, Differences of sets and a problem of Graham, *Canad. Math. Bull.*, 12 (1969), 635-637.

[R] R. Rivest (personal communication).

[Se] P. D. Seymour, On incomparable collections of sets, *Mathematika*, 20 (1973), 208-209.

[SW] P. D. Seymour and D. J. A. Welsh, Combinatorial applications of an inequality from statistical mechanics, *Math. Proc. Camb. Phil. Soc.*, 77 (1975), 485-495.

[Sh] L. A. Shepp, The FKG inequality and some monotonicity properties of partial orders, *SIAM Jour. Alg. Disc. Meth.*, 1 (1980), 295-299.

[St1] R. P. Stanley, Comments to solution of problem S20, *Amer. Math. Monthly*, 88 (1981), p. 208.

[St2] ____________, Two combinatorial applications of the Aleksandrov-Fenchel inequalities, (to appear).

[vL] J. H. van Lint, Notes on Egoritsjev's proof of the van der Waerden conjecture, Eindhoven Univ. of Tech., Math. Dept. Memorandum 1981-01 (11 pp.).

[D1] D. E. Daykin, Inequalities among the subsets of a set, *Nanta Math.*, 12 (1980), 137-145.

[D2] ____________, A hierarchy of inequalities, *Studies in Appl. Math.*, 63 (1980), 263-274.

# PART III

# SET THEORY AND RECURSION

# ORDER TYPES OF REAL NUMBERS
# AND OTHER UNCOUNTABLE ORDERINGS

James E. Baumgartner
Department of Mathematics
Dartmouth College
Hanover, New Hampshire 03755

ABSTRACT

If $\varphi$ and $\psi$ are order types, then $\varphi \leq \psi$ means that if $A$ and $B$ are ordered sets of type $\varphi$ and $\psi$, respectively, then $A$ is order-isomorphic to a subset of $B$. We write $\varphi < \psi$ if $\varphi \leq \psi$ but it is not the case that $\psi \leq \varphi$. The converse type of $\varphi$ is denoted by $\varphi^*$. If $C$ is a class of order types, then a *basis* for $C$ is a set $B \subseteq C$ such that $(\forall\ \psi \in C)(\exists\ \varphi \in B)\ \varphi \leq \psi$.

The theory of the class of countable order types is really quite nice. It follows from Ramsey's Theorem that $\{\omega, \omega^*\}$ is a basis. The type $\eta$ of rational numbers is universal (i.e., for every countable type $\varphi$, $\varphi \leq \eta$). Best of all, there is Laver's beautiful theorem [11], which asserts that if $\langle \varphi_n : n \in \omega \rangle$ is a sequence of countable order types, then there are $i, j$ such that $i < j$ and $\varphi_i \leq \varphi_j$ .

For the class of uncountable types, no such nice theory seems possible. Already it is clear that $\{\omega_1, \omega_1^*\}$ is not a basis, since the order type $\lambda$ of the real numbers has the property that $\omega_1$ , $\omega_1^* \not\leq \lambda$. This was used by Sierpiński [13] to disprove the natural generalization of Ramsey's Theorem. Work of Dushnik and Miller [5], and of Sierpiński [14], shows also that nothing like Laver's theorem can hold. Sierpiński, for example, constructs for each $\varphi \leq \lambda$ of cardinality $2^{\aleph_0}$ a type $\psi \leq \varphi$ of cardinality $2^{\aleph_0}$.

In [6] Erdös and Rado asked whether $\omega_1$ , $\omega_1^*$ , and all uncountable $\varphi \leq \lambda$ form a basis for the uncountable order types.



*I. Rival (ed.), Ordered Sets, 239–277.*

This question was settled in the negative by Specker (unpublished, but see [3]). Uncountable types $\varphi$ such that $\omega_1$ , $\omega_1^* \not\leq \varphi$ and $\psi \not\leq \varphi$ for all uncountable $\psi \leq \lambda$ are now called *Specker types*. Every Specker type arises from a lexicographically ordered Aronszajn tree (shown to exist by, of course, Aronszajn; see [8, 9 or 10]), which is an uncountable tree with all levels countable, and with no uncountable branches. Counterexamples to Souslin's hypothesis (see [4]) give rise to particularly nice examples of Specker types.

Using the existence of Specker types, Galvin [7] has extended Sierpiński's counterexample to show $\aleph_1 \not\rightarrow [\aleph_1]^2_4$ . Also, Shelah [12] has constructed (in ZFC) a Specker type $\varphi$ such that $\varphi \times \varphi$ (the partial order type obtained from $\varphi$ by using coordinatewise ordering) is the union of countably many chains.

One of the most interesting open questions is whether it is consistent for the uncountable types to have a finite basis. (This is not possible if the continuum hypothesis is true.) A start on this is in [2], where it is shown consistent that there exists a one-element basis for the uncountable order types of reals. See also Avraham and Shelah [1], where further results are obtained.

Shelah's Specker type $\varphi$ has the property that for no uncountable $\psi$ do we have $\psi \leq \varphi, \varphi^*$, so the Specker types cannot have a one-element basis. Very little beyond this is known for Specker types.

There are other interesting classes of uncountable types, notably the class of all $\varphi$ such that $\varphi \rightarrow (\omega)^1_\omega$ , studied in [3]. There is also the class of uncountable *partial* order types, the study of which has hardly begun.

REFERENCES

[1] U. Avraham and S. Shelah (1981) Martin's axiom does not imply that every two $\aleph_1$-dense sets of reals are isomorphic, *Israel J. Math.* 38, 161-176.

[2] J. Baumgartner (1973) All $\aleph_1$-dense sets of reals can be isomorphic, *Fund. Math.* 79, 101-106.

[3] J. Baumgartner (1976) A new class of order types, *Annals Math. Logic* 9, 187-222.

[4] K.J. Devlin and H. Johnsbråten (1974) The Souslin problem, *Lecture Notes in Mathematics* 405, Springer-Verlag.

[5] B. Dushnik and E.W. Miller (1940) Concerning similarity transformations of linearly ordered sets, *Bull. Amer. Math. Soc.* 46, 322-326.

[6] P. Erdös and R. Rado (1956) A partition calculus in set theory, *Bull. Amer. Math. Soc.* 62, 427-489.

[7] F. Galvin and S. Shelah (1973) Some counterexamples in the partition calculus, *J. Combinatorial Theory* 15, 167-174.

[8] T. Jech (1978) *Set Theory*, Academic Press.

[9] K. Kuratowski and A. Mostowski, *Set Theory*, Studies in Logic and the Foundations of Mathematics, Vol. 86, North Holland.

[10] G. Kurepa (1935) Ensembles ordonnés et ramifiés, *Publ. Math. Univ. Belgrade* 4, 1-138.

[11] R. Laver (1971) On Fraïssé's order type conjecture, *Ann. of Math.* 93, 89-111.

[12] S. Shelah (1976) Decomposing uncountable squares to countably many chains, *J. Combinatorial Theory (A)* 21, 110-114.

[13] W. Sierpiński (1933) Sur un problème de la théorie des relations, *Annali R. Scuola Normale Superiore de Pisa Ser. 2*, Vol. 2, 285-287.

[14] W. Sierpiński (1950) Sur les types d'ordre des ensembles linéaires, *Fund. Math.* 37, 253-264.

## 1. INTRODUCTION

Since the theory of uncountable order types does not seem to have been treated systematically in the literature, we have collected here a number of results, both old and new, which concern one particularly interesting aspect of the theory. If $S$ and $T$ are linearly ordered sets of order type $\varphi$ and $\psi$ respectively, then we write $\varphi \leq \psi$ to mean that $S$ is order-isomorphic to a subset of $T$. If $C$ is a class of order types, then a collection $B \subseteq C$ is a *basis* for $C$ if $\forall \varphi \in C \; \exists \psi \in B \; \psi \leq \varphi$. Much of this paper may be understood as part of the search for a simple basis for the class of uncountable order types under various set-theoretical hypotheses. In particular, we concentrate on real types and Specker types.

It should be emphasized that many aspects of the theory of order types are not treated here. For example, there is no discussion of automorphisms of order types, nor of the existence of universal order types of various kinds, and basis questions for other classes of order types, such as in [Ba4], are not treated.

After a brief discussion of uncountable order types in section 1, we attack real types in section 2. The results in this section are all classical in nature, although most of them were only found in the early 1950's by Sierpiński and Ginsburg. We have tried to state these results in the strongest form possible, and in several cases we have made some slight improvements. There is one theorem (Theorem 2.11) which seems to be new.

In section 3 the tools of post-Cohen set theory are applied to the study of $\aleph_1$-dense real types in the absence of the continuum hypothesis. The Weak Diamond Principle of Devlin and Shelah is used to produce many different examples of $\aleph_1$-dense types, and a forcing argument of Shelah is given to show it is consistent that all $\aleph_1$-dense real types are isomorphic.

Specker types are treated in section 4. It is a remarkable fact that almost none of these results, including Specker's original construction of a Specker type, have appeared in print. We show in ZFC that many non-isomorphic $\aleph_1$-dense Specker types exist, and then progressively stronger versions of $\Diamond$ are used to construct examples of Specker types. The last theorem, the construction under $\Diamond^+$ of a type $\varphi$ such that for all uncountable $\psi$, if $\psi \leq \varphi$ then $\varphi \leq \psi$, seems to be new. Section 4 concludes with several open problems.

We have tried to select results not just for their importance but also in order to illustrate the widest possible range of techniques. This has inevitably meant that some nice results could not be included.

We have also tried to make the paper as self-contained as possible. This was easy for section 2 but quite difficult for sections 3 and 4. It seems to be a simple fact of life that the reader who wishes to make serious progress on the problems mentioned herein must be in command of the tools of modern set theory and, accordingly, we have not hesitated to use those tools where necessary. Jech's book [Je] is an excellent reference for material which is used here but not explained.

If $A$ is a set then let $[A]^\kappa = \{x \subseteq A: |X| = \kappa\}$; $[A]^{<\kappa} = \{x \subseteq A: |x| < \kappa\}$. For cardinals $\kappa$, $\lambda$, $\mu$, and for $n \in \omega$ the notation

$$\kappa \to (\lambda)^n_\mu$$

means that $(\forall f:[\kappa]^n \to \mu)(\exists X \in [\kappa]^\lambda)$ $f$ is constant on $[X]^n$. The notation

$$\kappa \to [\lambda]^n_\mu$$

means that $(\forall f:[\kappa]^n \to \mu)(\exists X \in [\kappa]^\lambda)$ $f([X]^n) \neq \mu$.

If $A$ and $B$ are sets, then ${}^BA$ denotes the collection of all functions $f:B \to A$. If $\kappa$ is a cardinal and $f,g \in {}^\kappa\kappa$, then we say *$g$ eventually dominates $f$* if $\{\alpha: f(\alpha) \geq g(\alpha)\}$ is bounded in $\kappa$.

LEMMA 1.1. *Suppose $\kappa$ is a cardinal. Let $A(\kappa) = \{f \in {}^\kappa\kappa: \forall\beta < \kappa\ f(\beta) < |\beta|^+\}$. If $F \subseteq A(\kappa)$ and $|F| = \kappa$ then there is $g \in A(\kappa)$ such that $g$ eventually dominates every element of $F$.*

PROOF. Let $F = \{f_\alpha:\alpha < \kappa\}$. Choose $g$ so that for all $\alpha < \kappa$, $g(\alpha) > \sup\{f_\beta(\alpha):\beta < \alpha\}$. Clearly $g$ works. □

Two sets $A,B \subseteq \kappa$ are *almost-disjoint* if $|A \cap B| < \kappa$.

THEOREM 1.2. *Let $\kappa$ be a cardinal.*

(*a*) *There is a family $F = \langle A_\alpha:\alpha < \kappa^+\rangle$ such that $A_\alpha \subseteq \kappa$, $|A_\alpha| = \kappa$ and $A_\alpha \cap A_\beta$ is bounded in $\kappa$ whenever $\alpha \neq \beta$. Thus in particular $F$ is an almost-disjoint family.*

(*b*) *There is a sequence $\langle A_\alpha:\alpha < \kappa^+\rangle$ of subsets of $\kappa$ such* *that* $|A_\alpha| = \kappa$ *and if* $\alpha < \beta < \kappa^+$ *then* $|A_\alpha - A_\beta| = \kappa$ *and* $A_\beta - A_\alpha$ *is bounded in* $\kappa$.

PROOF. By Lemma 1.1 and transfinite induction, there is $\langle f_\alpha : \alpha < \kappa^+ \rangle$ such that $f_\alpha \in {}^\kappa\kappa$, $f_\alpha(\beta) < |\beta|^+$ for all $\beta < \kappa$, and $f_\beta$ eventually dominates $f_\alpha$ whenever $\alpha < \beta < \kappa^+$. Now for part (a) let $A_\alpha = f_\alpha$, regarded as a subset of $\kappa \times \kappa$, and for part (b) let $A_\alpha = \{(\beta,\gamma) \in \kappa\times\kappa : f_\alpha(\beta) < \gamma\}$. If $\kappa\times\kappa$ is ordered in Gödel's fashion, so that $(\alpha,\beta) < (\gamma,\delta)$ if $\max(\alpha,\beta) < \max(\gamma,\delta)$ or if $\max(\alpha,\beta) = \max(\gamma,\delta)$ and $(\alpha,\beta)$ lexicographically precedes $(\gamma,\delta)$, then $\kappa\times\kappa$ has order type $\kappa$, and the boundedness assertions are easily verified. □

If $(S,\leq)$ is a linearly ordered set, then $tpS$ or $tp(S,\leq)$ is its order type. If $\varphi$ is an order type then $\varphi^*$ denotes the converse order type. Thus if $tp(S,\leq) = \varphi$ then $tp(S,\geq) = \varphi^*$.

We conclude with a very brief discussion of countable order types.

THEOREM 1.3. *The smallest basis for the class of countable order types is* $\{\omega,\omega^*\}$.

PROOF. Every infinite subset of $\omega$ has order type $\omega$, so every basis for the countable types must contain $\omega$. A similar argument applies to $\omega^*$. To see that $\{\omega,\omega^*\}$ is already a basis, it will suffice to apply Ramsey's Theorem [Ra], which asserts

$$\aleph_0 \to (\aleph_0)^n_k \qquad \text{for all } n,k < \omega \quad .$$

Let $(S,\leq_S)$ be an arbitrary countable linear ordering. Without loss of generality we may assume $S = \omega$. Now suppose $m,n \in \omega$ and $m < n$. Let

$$f(\{m,n\}) = \begin{cases} 0 \text{ if } m <_S n \\ 1 \text{ if } n <_S m \end{cases} \quad .$$

By Ramsey's Theorem, there is $X \in [\omega]^{\aleph_0}$ such that $f$ is constant on $[X]^2$. But clearly $tp(X,\leq_S) = \omega$ or $\omega^*$. □

The structure of the countable order types under the embeddability relation $\leq$ turns out to be astonishingly simple. There is a beautiful theorem of Laver [La], verifying a conjecture of Fraïssé, which asserts that if $\langle \varphi_n : n \in \omega \rangle$ is any sequence of countable order types, then for some $m,n$ we have $m < n$ and $\varphi_m \leq \varphi_n$. This is usually expressed by saying that the countable order types are *well quasi-ordered*. Notice in particular that there can be no descending sequence $\varphi_0 > \varphi_1 > \varphi_2 > \cdots$ (where $\varphi < \psi$ iff $\varphi \leq \psi$ and $\psi \not\leq \varphi$) and no infinite set of order types pairwise incomparable under $\leq$.

## 2. ORDER TYPES OF REAL NUMBERS

Now we turn to the case of uncountable order types. Since any uncountable subset of $\omega_1$ is order-isomorphic to $\omega_1$, it is clear that $\omega_1$ and $\omega_1^*$ must be elements of any basis for the uncountable order types. It is also clear that $\{\omega_1,\omega_1^*\}$ is not a basis, since the order-type $\lambda$ of the real numbers has the property that $\omega_1,\omega_1^* \not\leq \lambda$. This immediately yields the failure of the naive generalization of Ramsey's Theorem to uncountable cardinals.

THEOREM 2.1. (essentially due to Sierpiński [Si1]).
$2^{\aleph_0} \not\rightarrow (\aleph_1)^2_2$.

PROOF. Let $<$ be the usual well-ordering on $2^{\aleph_0}$, and let $<^*$ be an ordering such that $tp(2^{\aleph_0},<^*) \leq \lambda$. Define $f:[2^{\aleph_0}]^2 \rightarrow 2$ so that $f\{\alpha,\beta\} = 0$ iff $<$ and $<^*$ agree on $\{\alpha,\beta\}$. Any homogeneous set for $f$ must be a well-ordered or conversely well-ordered set of reals, hence must be countable. □

Theorem 2.1 is not the only difference between the countable and uncountable case. In [Si2], Sierpiński constructed examples of order types of reals sufficient to show that nothing approaching Laver's results can be true for uncountable types, and Ginsburg greatly extended Sierpiński's results in [Gi1], [Gi2], [Gi3]. We shall derive some strong forms of these results below. The methods we use are slightly different from those of Sierpiński and Ginsburg, but the fundamental ideas are the same. Order types of cardinality $2^{\aleph_0}$ are constructed by means of diagonal arguments which make use of the fact that, in an appropriate sense, there are only $2^{\aleph_0}$ order-preserving functions from sets of reals to sets of reals.

Suppose $A$ and $B$ are sets of reals and $f:A \rightarrow B$ is an order-isomorphism. The *functional closure of* $f$, $\overline{f}$, is the set of all pairs $(x,y)$ such that $\forall\varepsilon > 0\ \exists x_1,x_2 \in A\ 0 \leq x-x_1, x_2-x, y-f(x_1), f(x_2)-y < \varepsilon$. Note that $\overline{f}$ is the largest order-isomorphism whose values are uniquely determined by $f$.

LEMMA 2.2. *If $f$ is any order-isomorphism from a set of reals to a set of reals then there is countable $g \subseteq f$ such that $f \subseteq \overline{g}$.*

PROOF. Let $A \subseteq$ domain $(f)$ be a countable set such that for every $x \in$ domain$(f)$, $x$ is a limit point of both $\{y \in A: x < y\}$ and $\{y \in A: y < x\}$. Let $B$ be a countable subset of range$(f)$ with a similar property. By enlarging $A$ and $B$, if necessary, we may assume that $f$ carries $A$ onto $B$. Now it is easy to verify that $g = f|A$ works. □

Next we construct a specific set $Z = \{z_\alpha : \alpha < 2^{\aleph_0}\}$ of reals, which will play a role in many of the proofs in this section.

Let $\langle f_\alpha : \alpha < 2^{\aleph_0} \rangle$ be an enumeration of all countable order-isomorphisms between sets of reals. Now obtain $z_\alpha$ inductively as follows. Let $Z_\alpha$ be the closure of the set $\{z_\beta ; \beta < \alpha\}$ under all the functions $\overline{f_\beta}$ and $\overline{f_\beta}^{-1}$, $\beta < \alpha$. Then $|Z_\alpha| < 2^{\aleph_0}$, so we may choose $z_\alpha \notin Z_\alpha$ arbitrarily.

Note that if we had been given any set of reals of cardinality $2^{\aleph_0}$, we could have chosen the $z_\alpha$'s from the elements of this set.

If $X \subseteq 2^{\aleph_0}$, then let $Z(X) = \{z_\alpha : \alpha \in X\}$.

LEMMA 2.3. *Let $X, Y \subseteq 2^{\aleph_0}$ and suppose $X-Y$ is cofinal in $2^{\aleph_0}$. Then $Z(X)$ is not embeddable in $Z(Y)$.*

PROOF. Suppose on the contrary that $f: Z(X) \to Z(Y)$ is an embedding. By Lemma 2.2, $f \subseteq \overline{f}_\alpha$ for some $\alpha$. Choose $\beta \in X-Y$, $\beta > \alpha$. Then $z_\beta \in Z(X)$ and $\overline{f(z_\beta)} = \overline{f}_\alpha(z_\beta) \in Z_{\beta+1}$. Say $f(z_\beta) = z_\gamma$. Since $z_\gamma \in Z_{\beta+1}$ and $\gamma \neq \beta$, we must have $\gamma < \beta$. But then by the construction of $Z$ we must have $z_\beta = \overline{f}_\alpha^{-1}(z_\gamma) \in Z_{\max(\gamma,\alpha)+1} \subseteq Z_\beta$, a contradiction since $z_\beta \notin Z_\beta$. □

If $\varphi, \psi \leq \lambda$ are types of cardinality $2^{\aleph_0}$ then we say that $\varphi$ and $\psi$ are *incompatible* if there is no type $\chi$ of cardinality $2^{\aleph_0}$ such that $\chi \leq \varphi, \psi$.

Parts (a) and (b) of the next theorem are essentially due to Sierpinski [Si2], although he stated part (a) only for $2^{\aleph_0}$ in place of $2^{2^{\aleph_0}}$, and he did not attempt to find more than two pairwise incompatible types. Part (c) is due to Ginsburg [Gi3]; he also obtained (a) with $(2^{\aleph_0})^+$ in place of $2^{2^{\aleph_0}}$.

THEOREM 2.4. *Let $\varphi \leq \lambda$ be an order type of cardinality $2^{\aleph_0}$.*

(*a*) *There is a family $F \subseteq \{\psi : \psi \leq \varphi, |\psi| = 2^{\aleph_0}\}$ of pairwise incomparable order types such that $|F| = 2^{2^{\aleph_0}}$.*

(*b*) *There is a family $F \subseteq \{\psi : \psi \leq \varphi, |\psi| = 2^{\aleph_0}\}$ of pairwise incompatible order types such that $|F| = (2^{\aleph_0})^+$.*

(*c*) *Suppose $G$ is a family of subsets of $2^{\aleph_0}$ such that $G$ is linearly ordered by inclusion. Then for each $S \in G$ there is $\varphi_S \leq \varphi$ such that for all $S, T \in G$ if $S \subseteq T$ and $S \neq T$ then $\varphi_S < \varphi_T$.*

PROOF. (a) First let us observe that there is a family $F'$ of subsets of $2^{\aleph_0}$ such that $|F'| = 2^{2^{\aleph_0}}$ and if $S, T \in F'$ are distinct, then $|S-T| = |T-S| = 2^{\aleph_0}$. Such a family may be obtained as follows: If $f: 2^{\aleph_0} \to 2$, then $X_f \subseteq 2^{\aleph_0} \times 2 \times 2^{\aleph_0}$ be defined by $X_f = \{(\alpha, i, \beta) : f(\alpha) = i\}$. Then the set of all $X_f$ has the property desired.

As we observed above, we may assume that $Z \subseteq E$, where $tp(E) = \varphi$.

Now for $S \in F'$ let $\varphi_S = tp(Z(S))$, and let $F = \{\varphi_S : S \in F'\}$. If $S,T \in F'$ and $S \neq T$ then since $|S-T| = |T-S| = 2^{\aleph_0}$, Lemma 2.3 asserts that $\varphi_S$ and $\varphi_T$ are incomparable.

(b) This is quite similar, but one uses a set $F'$ of subsets of $2^{\aleph_0}$ such that each element of $F'$ has cardinality $2^{\aleph_0}$ and if $S,T \subset F'$ are distinct then $\sup(S \cap T) < 2^{\aleph_0}$. Such a family was described in the introduction.

Now define $\varphi_S$ and $F$ as in part (a). Suppose $S,T \in F'$ are distinct and $\psi \leq \varphi_S, \varphi_T$. We may find $E \subseteq Z(S)$ such that $\psi = tp(E)$. If $\{\alpha : z_\alpha \in E\}$ is cofinal in $2^{\aleph_0}$, then by Lemma 2.3 we would have $\psi \not\leq \varphi_T$. Hence $|\psi| < 2^{\aleph_0}$.

(c) We may assume that if $S,T \in G$, $S \subseteq T$ and $S \neq T$, then $|T-S| = 2^{\aleph_0}$, for if this is not the case we may choose a decomposition $\{X_\alpha : \alpha < 2^{\aleph_0}\}$ of $2^{\aleph_0}$ into disjoint sets each of cardinality $2^{\aleph_0}$, and then replace each $S \in G$ by $\bigcup\{X_\alpha : \alpha \in S\}$. The argument is now completed as in (a) and (b). □

For the next theorem, we must recall some facts about Borel sets. If $P$ is an uncountable perfect set of real numbers then it is easy to see that $\lambda \leq tp(P)$. Since any uncountable Borel set $B$ must contain a perfect set, it follows that $\lambda \leq tp(B)$ also. Furthermore, any uncountable Borel set may be decomposed into $2^{\aleph_0}$ pairwise disjoint uncountable Borel sets. We also have

LEMMA 2.5. (Ginsburg [Gi3]). *There is a sequence* $\langle B_\alpha : \alpha < 2^{\aleph_0} \rangle$ *of pairwise disjoint uncountable Borel sets with the property that for any uncountable Borel set* $B$, *there is* $\alpha < 2^{\aleph_0}$ *such that* $B \cap B_\alpha$ *is uncountable.*

PROOF. Let $\langle A_\alpha : \alpha < 2^{\aleph_0} \rangle$ enumerate all uncountable Borel sets, and let $\langle C_\alpha : \alpha < 2^{\aleph_0} \rangle$ be an arbitrary sequence of pairwise disjoint uncountable Borel sets. We construct the $B_\alpha$ by induction so that for all $\alpha$, either

$$(1) \qquad \exists \beta \quad B_\alpha \subseteq C_\beta \qquad \text{or}$$

$$(2) \qquad \forall \beta \quad |B_\alpha \cap C_\beta| \leq \aleph_0 \quad .$$

Given $\langle B_\beta : \beta < \alpha \rangle$, obtain a set $B$ as follows. If $\exists \beta < \alpha \;\; A_\alpha \cap B_\beta$ is uncountable, then let $B = C_\gamma$, where $\gamma$ is large enough so that $|C_\gamma \cap B_\beta| \leq \aleph_0$ for all $\beta < \alpha$. If this fails but $\exists \gamma \;\; A_\alpha \cap C_\gamma$ is uncountable, then let $B = A_\alpha \cap C_\gamma$. If $A_\alpha \cap C_\gamma$ is countable for all $\gamma$, then let $B = A_\alpha$. In every case $B \cap B_\beta$ is countable for all $\beta < \alpha$, and $B$ is uncountable. Now decompose $B$ into $2^{\aleph_0}$

disjoint Borel sets. One of these must be disjoint from all $B_\beta$, $\beta < \alpha$; let $B_\alpha$ be such a set.

It is clear from the construction that $\forall\alpha\ \exists\beta \le \alpha\ \ A_\alpha \cap B_\beta$ is uncountable. □

THEOREM 2.6. (Ginsburg [Gi2],[Gi3]). *Let* $\langle\varphi_\alpha : \alpha < 2^{\aleph_0}\rangle$ *be a collection of uncountable types such that* $\varphi_\alpha < \lambda$ *for all* $\alpha$. *Then:*

(a) *There is* $\varphi < \lambda$ *such that* $\varphi$ *is incomparable with* $\varphi_\alpha$ *for all* $\alpha < 2^{\aleph_0}$ *and*

(b) *There is* $\varphi < \lambda$ *such that for all* $\alpha$, $\varphi_\alpha \le \varphi < \lambda$. *Hence if* $\varphi < \lambda$ *is arbitrary, then there is a well ordered sequence* $\langle\psi_\alpha : \alpha < (2^{\aleph_0})^+\rangle$ *of types such that for all* $\alpha$, $\beta$, *if* $\alpha < \beta$ *then* $\varphi \le \psi_\alpha < \psi_\beta < \lambda$.

PROOF. (a) Choose a set $E_\alpha$ of reals for each $\alpha$ so that $tp(E_\alpha) = \varphi_\alpha$. Let $\langle f_\alpha : \alpha < 2^{\aleph_0}\rangle$ enumerate all countable functions from reals to reals, as before, and let $p: 2^{\aleph_0} \to 2^{\aleph_0} \times 2^{\aleph_0}$ be one-to-one and onto. By induction on $\varphi$ we will obtain sequences $\langle x_\alpha : \alpha < 2^{\aleph_0}\rangle$ and $\langle y_\alpha : \alpha < 2^{\aleph_0}\rangle$ of distinct reals, and at the end we will let $\varphi = tp(\{x_\alpha : \alpha < 2^{\aleph_0}\})$.

Suppose $x_\beta$, $y_\beta$ have been obtained for $\beta < \alpha$. Let $p(\alpha) = (\alpha_0, \alpha_1)$. If $E_{\alpha_1} \subseteq \text{domain}(\overline{f}_{\alpha_0})$, then choose $y_\alpha$ in $\overline{f}_{\alpha_0}(E_{\alpha_1})$ distinct from all earlier $x_\beta$, $y_\beta$; otherwise let $y_\alpha$ be arbitrary (but distinct from earlier choices). Since the $x_\alpha$'s and $y_\alpha$'s are distinct, this choice of $y_\alpha$ guarantees that $\overline{f}_{\alpha_0}$ cannot map $E_{\alpha_1}$ into $\{x_\alpha : \alpha < 2^{\aleph_0}\}$.

We want to choose $x_\alpha$ so that $\overline{f}_{\alpha_0}$ does not map $\{x_\alpha : \alpha < 2^{\aleph_0}\}$ into $E_{\alpha_1}$. The only situation in which such an $x_\alpha$ could not be chosen is if $\overline{f}_{\alpha_0}^{-1}(E_{\alpha_1})$ contains all but $< 2^{\aleph_0}$ real numbers. But then, letting $\langle B_\gamma : \gamma < 2^{\aleph_0}\rangle$ be a decomposition of the reals into disjoint uncountable Borel sets, we would have to have $B_\gamma \subseteq \overline{f}_{\alpha_0}^{-1}(E_{\alpha_1})$ for some $\gamma$, and it would follow that

$$\lambda \le tp(B_\gamma) \le tp(E_{\alpha_1}) = \varphi_{\alpha_1} \quad ,$$

contrary to assumption. Hence $x_\alpha$ can be chosen as desired.

It follows immediately that if $\varphi = tp(\{x_\alpha : \alpha < 2^{\aleph_0}\})$, then $\varphi_{\alpha_1} \not\le \varphi \not\le \varphi_{\alpha_1}$ for all $\alpha_1 < 2^{\aleph_0}$.

(b) Let $\langle B_\alpha : \alpha < 2^{\aleph_0}\rangle$ be a sequence as in Lemma 2.5. Since $\lambda \le tp(B_\alpha)$ for each $\alpha$, there is $E_\alpha \subseteq B_\alpha$ such that $tp(E_\alpha) = \varphi_\alpha$. Let $E = \bigcup\{E_\alpha : \alpha < 2^{\aleph_0}\}$, and let $\varphi = tp(E)$. Clearly, $\varphi_\alpha \le \varphi$ for all $\alpha < 2^{\aleph_0}$; we must check that $\varphi < \lambda$. Suppose, on the contrary, that $f$ is an order-preserving mapping from the reals into $E$. If

$Y$ is the range of $f$, then it is easy to see that the (topological) closure of $Y$ contains at most countably many points not already in $Y$. Hence $Y$ is Borel, so by Lemma 2.5, $Y \cap B_\alpha$ is uncountable for some $\alpha$. But since $Y \cap B_\alpha \subseteq E \cap B_\alpha = E_\alpha$, we would have

$$\lambda \leq tp(Y \cap B_\alpha) \leq tp(E_\alpha) = \varphi_\alpha \quad ,$$

contrary to hypothesis. Hence $\varphi < \lambda$. □

One would naively expect that the situation for ascending and descending sequences of order types ought to be symmetrical. As it turns out, this is not the case. Theorem 2.4(c) assures us of finding, below any given $\varphi \leq \lambda$ of cardinality $2^{\aleph_0}$, a descending sequence of length any ordinal $< (2^{\aleph_0})^+$. The following result of Ginsburg [Gi3, Theorem 13] shows that this is the best we can do.

THEOREM 2.7. *Assume that* $2^\kappa = 2^{\aleph_0}$ *for all* $\kappa < 2^{\aleph_0}$. *Then there exists* $\varphi \leq \lambda$ *such that* $|\varphi| = 2^{\aleph_0}$ *and there is no descending sequence of length* $(2^{\aleph_0})^+$ *consisting of types below* $\varphi$.

PROOF. Let $Z$ be the set constructed at the beginning of this section. It will suffice to show that $\varphi = tp(Z)$ will work, but first we need some more observations about $Z$.

The proof of Lemma 2.3 actually establishes the following:

LEMMA 2.8. *If* $Y \subseteq Z$ *and* $f: Y \to Z$ *is order-preserving, then* $\exists \alpha < 2^{\aleph_0}$ $f$ *is the identity on* $\{z_\beta \in Y : \beta \geq \alpha\}$.

Let us call a set $X$ of real numbers *uniform* (Ginsburg: Property C) if every point of $X$ is a two-sided $2^{\aleph_0}$-limit point of $X$, i.e., if $(\forall x \in X)(\forall \varepsilon > 0)$ both $\{y \in X : x < y < x+\varepsilon\}$ and $\{y \in X : x-\varepsilon < y < x\}$ have cardinality $2^{\aleph_0}$.

LEMMA 2.9. *If* $X$ *is a set of reals and* $|X| = 2^{\aleph_0}$, *then there are* $X_1$, $X_2$ *such that* $X = X_1 \cup X_2$, $X_1 \cap X_2 = 0$, $|X_1| < 2^{\aleph_0}$ *and* $X_2$ *is uniform.*

PROOF. This is quite easy. Just let $X_2$ be the set of two-sided $2^{\aleph_0}$-limit points of $X$. □

LEMMA 2.10. *If* $U \subseteq Z$ *is uniform and* $f: U \to Z$ *is order-preserving, then* $f$ *is the identity.*

PROOF. If $x \in U$, then by Lemma 2.8, $x$ is a two-sided $2^{\aleph_0}$-limit point of points $y$ such that $f(y) = y$; hence $f(x) = x$. □

Now we complete the proof of Theorem 2.7. Suppose $\langle \varphi_\alpha : \alpha < (2^{\aleph_0})^+ \rangle$ is a descending sequence with $\varphi_0 \leq \varphi$. Choose

$Z_\alpha \subseteq Z$ such that $tp(Z_\alpha) = \varphi_\alpha$, and let $Z_\alpha = Z^1_\alpha \cup Z^2_\alpha$ be a decomposition as in Lemma 2.9. Since $2^\kappa = 2^{\aleph_0}$ for all $\kappa < 2^{\aleph_0}$,

$$\sum\{(2^{\aleph_0})^\kappa : \kappa < 2^{\aleph_0}\} \leq \sum\{(2^\kappa)^\kappa : \kappa < 2^{\aleph_0}\} = \sum\{2^{\aleph_0} : \kappa < 2^{\aleph_0}\} = 2^{\aleph_0} \quad .$$

It follows that there are only $2^{\aleph_0}$ possible choices for the sets $Z^1_\alpha$; hence we may assume $Z^1_\alpha = Z^1_\beta$ for all $\alpha$, $\beta$.

Suppose $\alpha > \beta$. Then there is order-preserving $f:Z_\alpha \to Z_\beta$. Since $Z^2_\alpha$ is uniform, $f$ is the identity on $Z^2_\alpha$ by Lemma 2.10. Also, since $Z^2_\alpha \cap Z^1_\alpha = 0$ and $Z^1_\alpha = Z^2_\alpha$, we must have $Z^2_\alpha \subseteq Z^2_\beta$. Hence there is some $\alpha < 2^{\aleph_0}$ such that $Z^2_\gamma = Z^2_\alpha$ for all $\gamma \geq \alpha$. But then $Z_\gamma = Z_\alpha$ for all $\gamma \geq \alpha$, contrary to hypothesis. □

PROBLEM: Can one eliminate from Theorem 2.7 the hypothesis that $2^\kappa = 2^{\aleph_0}$ for all $\kappa < 2^{\aleph_0}$?

Although by Theorem 2.7 we cannot find long descending sequences below *arbitrary* order types, it is still conceivable that long descending sequences always exist. This existence question was posed by Ginsburg [Gi3, p. 349]. The next theorem answers it affirmatively. Since the method of proof is "classical", it seems unlikely that the result is new; however, we have been unable to discover it in the literature.

THEOREM 2.11. *There is a sequence* $\langle\varphi_\alpha : \alpha < (2^{\aleph_0})^+\rangle$ *such that for all* $\alpha$, $\beta$, *if* $\alpha < \beta$ *then* $\varphi_\beta < \varphi_\alpha \leq \lambda$.

PROOF. We begin with an argument similar to the construction of $Z$. Let $\langle f_\alpha : \alpha < 2^{\aleph_0}\rangle$ enumerate all countable order-isomorphisms from sets of reals to sets of reals. By induction we will construct a sequence $\langle t_\alpha : \alpha < 2^{\aleph_0}\rangle$ of translations and a sequence $\langle D_\alpha : \alpha < 2^{\aleph_0}\rangle$ of disjoint sets of reals such that $|D_\alpha| \leq |\alpha| + \aleph_0$ for all $\alpha$.

Suppose $t_\beta$, $D_\beta$ have been obtained for all $\beta < \alpha$. Let $D$ be the closure of $\bigcup\{D_\beta : \beta < \alpha\}$ under all functions $f$ and $f^{-1}$, where $f \in \{f_\beta : \beta < \alpha\} \cup \{t_\beta : \beta < \alpha\}$. Choose a translation $t_\alpha$ so that $t_\alpha(D) \cap D = 0$, and let $D_\alpha$ be the closure of $t_\alpha(D)$ under all $f$ and $f^{-1}$ for $f \in \{f_\beta : \beta < \alpha\} \cup \{t_\beta : \beta < \alpha\}$. This completes the construction.

As was observed in the introduction, there exists a sequence $\langle A_\alpha : \alpha < (2^{\aleph_0})^+\rangle$ of subsets of $2^{\aleph_0}$ such that if $\alpha < \beta$ then $|A_\alpha - A_\beta| = 2^{\aleph_0}$ and $(\exists\gamma < 2^{\aleph_0})\ A_\beta - \gamma \subseteq A_\alpha$. Let $E_\alpha = \bigcup\{D_\beta : \beta \in A_\alpha\}$, and let $\varphi_\alpha = tp(E_\alpha)$.

Now let $\alpha < \beta$. Choose $\gamma$ so that $A_\beta - \gamma \subseteq A_\alpha$, and assume that $\gamma$ is chosen so that $\gamma \in A_\alpha - A_\beta$ as well. We claim $t_\gamma : E_\beta \to E_\alpha$. By the choice of $t_\gamma$, $t_\gamma$ carries $\bigcup\{D_\xi : \xi < \gamma\}$ into $D_\gamma \subseteq E_\alpha$. If $\delta \geq \gamma$

and $\delta \in A_\beta$, then $\delta > \gamma$ and $\delta \in A_\alpha$ also. But then $D_\delta$ is closed under $t_\gamma$, so $t_\gamma : E_\beta \to E_\alpha$ as desired.

To see that $\varphi_\alpha \not\leq \varphi_\beta$, fix $f$ and $\gamma > \xi$ so that $\gamma \in A_\alpha - A_\beta$. Then $D_\gamma \subseteq E_\alpha$ but $D_\gamma \cap E_\beta = 0$. Since $\overline{f}_\xi : D_\gamma \to D_\gamma$, $\overline{f}_\xi$ cannot map $E_\alpha$ into $E_\beta$. □

It may be worth remarking that in Theorems 2.4(b), 2.6 and 2.11, the cardinal $(2^{\aleph_0})^+$ cannot be raised, even if $2^{2^{\aleph_0}} > (2^{\aleph_0})^+$. For Theorem 2.4(b), note that if $tp(E_1)$ and $tp(E_2)$ are incompatible, then $|E_1 \cap E_2| < 2^{\aleph_0}$. Thus the observation above follows from the theorem in [Ba2] that it is consistent with $2^{2^{\aleph_0}} > (2^{\aleph_0})^+$ that there is no almost-disjoint family $F$ of subsets of $2^{\aleph_0}$, each of cardinality $2^{\aleph_0}$, such that $|F| > (2^{\aleph_0})^+$. On the other hand, if $2^\kappa = 2^{\aleph_0}$ for all $\kappa < 2^{\aleph_0}$ (and in certain other situations; see [Ba2]) then there is an almost-disjoint family $F$ as above with $|F| = 2^{2^{\aleph_0}}$, and the proof of Theorem 2.4(b) yields a family $G$ of incompatible types such that $|G| = |F|$.

For Theorems 2.6 and 2.11, the situation is somewhat simpler. Using standard Easton-style forcing, it is easy to find models in which $2^{2^{\aleph_0}} > (2^{\aleph_0})^+$ and there is no sequence $\langle A_\alpha : \alpha < (2^{\aleph_0})^{++} \rangle$ of subsets of $2^{\aleph_0}$ such that for all $\alpha < \beta$, $|A_\alpha - A_\beta| = 2^{\aleph_0}$ and $|A_\beta - A_\alpha| < 2^{\aleph_0}$. The proof uses a simple $\Delta$-system argument, which is left to the interested reader to work out, and it will work even when $2^\kappa = 2^{\aleph_0}$ for all $\kappa < 2^{\aleph_0}$. Moreover, the argument easily extends to show that there is no descending sequence of types of length $(2^{\aleph_0})^{++}$. In this case, the situation for ascending sequences of types is exactly similar.

Of course it is also possible to show by forcing that it is *consistent* that the $(2^{\aleph_0})^+$ in the theorems mentioned above can be raised to $2^{2^{\aleph_0}}$ while $2^{2^{\aleph_0}} > (2^{\aleph_0})^+$. Once again, details are left to the reader.

PROBLEM: If there is a descending sequence of types of length $(2^{\aleph_0})^{++}$, must there be a sequence $\langle A_\alpha : \alpha < (2^{\aleph_0})^{++} \rangle$ of subsets of $2^{\aleph_0}$ as above?

It should also be remarked that whereas the results of Sierpiński and Ginsburg presented here are in the writer's opinion the most interesting and important, they form only a fraction of the material in [Si2], [Gi1], [Gi2] and [Gi3], to which the reader is referred for further information.

Rotman [Ro] has generalized many of the Sierpiński-Ginsburg results to real-like orderings of higher cardinality.

Now we shall end this section as we began it, with a counterexample to a partition relation.

THEOREM 2.12. (Shelah [GS]) *(a)* $2^{\aleph_0} \not\to [2^{\aleph_0}]^2_{\aleph_0}$.

*(b)* $cf(2^{\aleph_0}) \not\to [cf 2^{\aleph_0}]^2_{\aleph_0}$.

PROOF. (a) Let $R$ denote the set of real numbers. For each $n \in \omega$ we will construct a one-to-one function $f_n : R \to R$ so that $f : [R]^2 \to \omega$ will work, where $f$ is any function such that if $x < y$, $f_i(x) < f_i(y)$ for all $i \leq n$ and $f_{n+1}(x) > f_{n+1}(y)$, then $f\{x,y\} = n$. Recall that $n = \{i : i < n\}$.

Let $f_0$ be the identity. Suppose $f_i$ has been obtained for all $i < n$. If $s : n \to 2$ then let $x <_s y$ iff for each $i < n$, if $s(i) = 0$ then $f_i(x) < f_i(y)$ while if $s(i) = 1$ then $f_i(x) > f_i(y)$. Let $\langle r_\alpha : \alpha < 2^{\aleph_0}\rangle$ enumerate $R$, and let $\langle X_\alpha : \alpha < 2^{\aleph_0}\rangle$ enumerate all countable subsets of $R$ in such a way that $X_\alpha \subseteq \{r_\beta : \beta < \alpha\}$. For $\alpha < 2^{\aleph_0}$, $s : n \to 2$ and $r \in R$, let

$$X(\alpha,s,r) = \{x \in X_\alpha : x <_s r\} .$$

Now construct $f_n : R \to R$ so that for all $r_\beta$, $f_n(r_\beta) \neq \sup(X(\alpha,s,r_\beta))$, $\inf(X(\alpha,s,r_\beta))$ for any $\alpha < \beta$ and any $s : n \to 2$.

LEMMA 2.13. *For any $n > 0$, any $s : n \to 2$ and any $W \subseteq R$, if $|W| = 2^{\aleph_0}$ then there are $x, y \in W$ such that $x <_s y$.*

PROOF. This is clear for $n = 1$. Suppose it is true for $n$, and let $s : n+1 \to 2$. Fix $W \subseteq R$, $|W| = 2^{\aleph_0}$.

For each $x \in W$, let $h(x) = (f_0(x), \ldots, f_n(x))$. Since $R^{n+1}$ is separable there is countable $X \subseteq W$ such that $h(X)$ is dense in $h(W)$. Say $X = X_\alpha$, and without loss of generality assume $s(n) = 1$. Now suppose $y \in W$. If

$$f_n(y) < \sup f_n(X(\alpha, s|n, y)) ,$$

then there is $x \in X(\alpha, s|n, y)$ such that $x <_s y$ and we are done. Hence if $y = r_\beta$ and $\beta > \alpha$ we must have

$$f_n(y) > \sup f_n(X(\alpha, s|n, y)) .$$

In particular, there is $\varepsilon = \varepsilon(y) > 0$ such that

$$f_n(y) - \sup f_n(X(\alpha, s|n, y)) > \varepsilon .$$

Now there must be $W' \subseteq W$ and $\varepsilon > 0$ such that $|W'| = 2^{\aleph_0}$ and $\varepsilon(y) \geq \varepsilon$ for all $y \in W'$. Furthermore, we can find an interval $I$ of width $< \varepsilon/2$ such that $|W' \cap I| = 2^{\aleph_0}$.

By inductive hypothesis there are $x, y \in W' \cap I$ such that $x <_{s|n} y$. Since $|y-x| < \varepsilon/2$ there is $x' \in X_\alpha$ such that $x' <_{s|n} y$

and $|x'-y| < \varepsilon$. But then $x' \in X(\alpha,s|n,y)$, and this contradicts the assumption that $\varepsilon(y) \geq \varepsilon$. □

Now, constructing $f:[R]^2 \to \omega$ as indicated, the proof of (a) is immediate.

(b) For the proof of (b), simply note that the proof of Lemma 2.13 goes through under the weaker assumption that $\{\alpha : r_\alpha \in W\}$ is cofinal in $2^{\aleph_0}$. □

## 3. $\aleph_1$-DENSE REAL TYPES

In this section we shall be concerned with order types $\varphi \leq \lambda$ such that $|\varphi| = \aleph_1$. We shall be especially interested in such types which are quite uniform. Let us call a set $D$ of real numbers $\aleph_1$-*dense* if for every non-empty open interval $I$, $|D \cap I| = \aleph_1$ (so in particular $|D| = \aleph_1$). If $D$ is $\aleph_1$-dense then $tp(D)$ is also called $\aleph_1$-dense. Note that if $D$ is a set of reals of cardinality $\aleph_1$, then there is a countable set $X \subseteq D$ such that $tp(D-X)$ is $\aleph_1$-dense. Thus every real type of cardinality $\aleph_1$ is "almost" $\aleph_1$-dense.

If $2^{\aleph_0} = \aleph_1$, then the results of the previous section easily yield a profusion of different $\aleph_1$-dense types. The interesting case, therefore, is when $2^{\aleph_0} > \aleph_1$.

A distinction is often made between "classical" and "modern" techniques in set theory. Classical techniques tend to be elementary, involving ideas like the diagonal arguments of section 2, and they do not require much background in set theory. Modern techniques on the other hand are a bit more sophisticated, and make use of such things as trees and stationary sets, model theory (particularly elementary substructures and collapsing arguments), and the theory of forcing and generic sets. The methods of this section are, unfortunately, modern. Although we have tried to make the material as accessible as possible, this section is necessarily less self-contained than the preceding one.

Conditions considerably weaker than the continuum hypothesis are still sufficient to imply the existence of a variety of $\aleph_1$-dense types. If we know only that $2^{\aleph_0} < 2^{\aleph_1}$ then it is not difficult to see that there are distinct $\aleph_1$-dense types. If $D$ is $\aleph_1$-dense then there are only $2^{\aleph_0}$ order-isomorphisms carrying $D$ into the reals. Since there are $2^{\aleph_1}$ $\aleph_1$-dense sets altogether there must be one which is not isomorphic to $D$.

In fact one can do rather better. Shelah has shown that if $2^{\aleph_0} < 2^{\aleph_1}$ then no $\aleph_1$-dense type is embeddable in every $\aleph_1$-dense type, and there is a family of $2^{\aleph_1}$ mutually incomparable $\aleph_1$-dense types. Shelah's results were announced in [AS1] and appear, vastly generalized, in [Sh2].

The proofs of Shelah's results make use of the following combinatorial principle, which was studied by Devlin and Shelah in [DS].

The Weak Diamond Principle (WDP): Suppose $|A| \leq 2^{\aleph_0}$ and $T = \bigcup\{{}^{\alpha}A : \alpha < \omega_1\}$. Then $(\forall F: T \to 2)\ (\exists g: \omega_1 \to 2)\ (\forall f: \omega_1 \to A)$ $\{\alpha : g(\alpha) = F(f|\alpha)\}$ is stationary.

It is proved in [DS] that $2^{\aleph_0} < 2^{\aleph_1}$ implies WDP (and a moment's thought will show that the converse is trivial), so in fact WDP is simply a peculiar way of restating the hypothesis $2^{\aleph_0} < 2^{\aleph_1}$.

Furthermore, WDP is self-improving to the following extent:

LEMMA 3.1. *Let $A$ and $T$ be as in WDP. Suppose that for all $\alpha < 2^{\aleph_0}$, $F_\alpha: T \to 2$. Then $(\exists g: \omega_1 \to 2)(\forall f: \omega_1 \to A)\ (\forall \alpha < 2^{\aleph_0})$ $\{\beta : g(\beta) = F_\alpha(f|\beta)\}$ is stationary.*

PROOF. Let $A' = A \times 2^{\aleph_0}$. Given $s \in {}^{\alpha}A'$, define $F(s) = F_\xi(t)$, where $t(\beta)$ is the first coordinate of $s(\beta)$ for all $\beta < \alpha$, and $s(0) = (t(0), \xi)$. Now let $g: \omega_1 \to 2$ satisfy WDP for $F$. Suppose $f: \omega_1 \to A$ and $\alpha < 2^{\aleph_0}$. Let $f'(\beta) = (f(\beta), \alpha)$ for all $\beta < \omega_1$. If $g(\beta) = F(f'|\beta)$, then clearly $g(\beta) = F_\alpha(f|\beta)$ by the definition of $F$. Hence $\{\beta : g(\beta) = F_\alpha(f|\beta)\}$ is stationary. □

THEOREM 3.2. (Shelah). *If $2^{\aleph_0} < 2^{\aleph_1}$ then there is no $\aleph_1$-dense type $\varphi$ such that $\varphi \leq \psi$ for all $\aleph_1$-dense types $\psi$.*

PROOF. Let $D$ be an $\aleph_1$-dense set, let $T = \bigcup\{{}^{\alpha}2 : \alpha < \omega_1\}$, and let $R$ denote the real numbers. Fix a one-to-one mapping $x: T \to R$ such that for all $f: \omega_1 \to 2$, $x(f) = \{x(f|\alpha+1) : \alpha < \omega_1\}$ is $\aleph_1$-dense. We assert that for some $f$, $D$ is not embeddable in $x(f)$.

Let $\langle d_\alpha : \alpha < \omega_1 \rangle$ enumerate $D$. If $h$ is an order-isomorphism from $D$ into $R$, then define $F_h: T \to 2$ as follows. Let $s \in {}^{\alpha}2$, and let $t$ be such that $h(d_\alpha) = x(t)$, if such a $t$ exists. Now let $F_h(s) = i \in \{0,1\}$ such that $s \frown \langle i \rangle \nsubseteq t$.

Let $g: \omega_1 \to 2$ be as in Lemma 3.1 for the $F_h$. We assert that $D$ is not embeddable in $x(g)$. Suppose on the contrary that $h: D \to x(g)$ is order-preserving. Then $S = \{\alpha : g(\alpha) = F_h(g|\alpha)\}$ is stationary. If $C = \{\alpha : h^{-1}\{x(g|\beta) : \beta < \alpha\} \subseteq \{d_\beta : \beta < \alpha\}\}$, then $C$ is clearly closed unbounded, so there is $\alpha \in C \cap S$, $\alpha$ a limit ordinal. Now $h(d_\alpha) = x(g|\gamma)$ for some $\gamma$, and since $\alpha \in C$ we have $\gamma > \alpha$ (recall that since $h(d_\alpha) \in x(g)$, $\gamma$ must be a successor ordinal). But since $g(\alpha) = F_h(g|\alpha)$, we must have $g|\alpha{+}1 \not\subseteq g|\gamma$ by the definition of $F_h$, a contradiction. □

In the proof of Theorem 3.2 we may regard $g$ as a successful $\omega_1$-sequence of guesses, the success being measured by the functions $F_h$. If now we want to find two $\aleph_1$-dense sets, neither of which is embeddable in the other, then it is natural to try to produce them via two sequences of guesses, $g_1$ and $g_2$. Unfortunately, the number of possibilities at each stage will now be 4, and in WDP we have explicitly assumed that range $F \subseteq 2$. Moreover, Shelah has shown that if 2 is replaced by 4 then the resulting WDP does not follow from $2^{\aleph_0} < 2^{\aleph_1}$. So we must be more circumspect.

The right way to proceed is to alternate the guesses so that there are only two possibilities at each stage. The trouble now is in verifying that each set of guesses separately satisfies WDP. Accordingly, we make the following definition.

Let $X \subseteq \omega_1$. We say that WDP($X$) holds provided that if $A$ and $T$ are as before, and $F: T \to 2$, then $(\exists g: X \to 2)\ (\forall f: \omega_1 \to A)$ $\{\alpha \in X : g(\alpha) = F(f|\alpha)\}$ is stationary.

In [DS], Devlin and Shelah show that $\{X \subseteq \omega_1 : \text{WDP}(X) \text{ fails}\}$ is a non-trivial normal ideal on $\omega_1$. It follows immediately from an old argument of Ulam [Ul] that there is a disjoint decomposition $\langle X_\alpha : \alpha < \omega_1\rangle$ of $\omega_1$ such that WDP($X_\alpha$) holds for all $\alpha$. Using this fact, we can now prove

THEOREM 3.3. (Shelah): *If $2^{\aleph_0} < 2^{\aleph_1}$ then there is a family of $2^{\aleph_1}$ mutually incomparable $\aleph_1$-dense types.*

PROOF. Let $T$, $x$ and the sets $x(f)$ be as in the proof of Theorem 3.2. Let $T^2 = \bigcup\{{}^{\alpha}2 \times {}^{\alpha}2 : \alpha < \omega_1\}$. Regarding $T^2$ as the same thing as $\bigcup\{{}^{\alpha}(2\times 2) : \alpha < \omega_1\}$, it is clear that WDP applies to $T^2$.

Let $\langle X_{\alpha i} : (\alpha, i) \in \omega_1 \times 2\rangle$ be a disjoint decomposition of $\omega_1$ so that WDP($X_{\alpha i}$) holds for all $(\alpha, i)$. Let $h$ be a functionally closed order-preserving mapping from reals to reals. We must define $F_h : T^2 \to 2$.

First suppose $(s,t) \in {}^{\alpha}2 \times {}^{\alpha}2$, $\alpha \in X_{\beta i}$, $s(\beta) \neq i$ and $t(\beta) = i$ Then let $F_h(s,t)$ be a number $j \in \{0,1\}$ such that if $h(x(s)) = x(u)$, then $t \frown j \not\subseteq u$. Otherwise let $F_h(s,t)$ be arbitrary.

Now for each $(\alpha, i)$ there is $g_{\alpha i} : X_{\alpha i} \to 2$ which satisfies WDP($X_{\alpha i}$) with respect to all $2^{\aleph_0}$ of the $F_h$, as in Lemma 3.1. Let $G$ be the set of all functions $g : \omega_1 \to 2$ such that

(*) $\quad (\forall \alpha < \omega_1)$ if $\alpha \in X_{\beta i}$ and $g(\beta) = i$, then $g(\alpha) = g_{\beta i}(\alpha)$.

Note that if $\alpha \in X_{\beta i}$ and $g(\beta) \neq i$ then (*) imposes no requirement on $g(\alpha)$; hence $|G| = 2^{\aleph_1}$.

If $g \in G$ then let $x(g) = \{x(g|\alpha+1) : \alpha < \omega_1\} \cup \{x(g|\alpha) : \exists(\beta, i)$ $\alpha \in X_{\beta i}$ and $g(\beta) \neq i\}$. Note that this definition differs from the one in Theorem 3.2.

To complete the proof it will suffice to show that if $f, g \in G$ and $f \neq g$, then $x(f)$ is not embeddable in $x(g)$. Suppose otherwise. Then there is a functionally closed $h$ carrying $x(f)$ into $x(g)$. Let $C = \{\alpha : h^{-1}\{x(g|\beta) : \beta < \alpha\} \cap x(f) \subseteq \{x(f|\beta) : \beta < \alpha\}$. As before, $C$ is closed unbounded.

Since $f \neq g$ there is $(\beta, i)$ such that $g(\beta) = i$ and $f(\beta) \neq i$. Since $g_{\beta i}$ satisfies WDP$(X_{\beta i})$ there is $\alpha \in C \cap X_{\beta i}$ such that $\alpha$ is a limit ordinal and $g_{\beta i}(\alpha) = F_h(f|\alpha, g|\alpha)$. Now by the definition of $x(f)$ we have $x(f|\alpha) \in x(f)$. Also, since $\alpha \in C$ we know that if $h(x(f|\alpha)) = x(g|\gamma)$ then $\gamma \geq \alpha$. Since $g(\beta) = i$ we have $x(g|\alpha) \notin x(g)$, so in fact $\gamma > \alpha$. But by hypothesis $g(\alpha) = g_{\beta i}(\alpha) = F_h(f|\alpha, g|\alpha)$, which is $j$ such that $g|\alpha \frown \langle j \rangle \nsubseteq g|\gamma$, a contradiction as before. □

It is natural to ask about incompatible types here also, where to say that $\aleph_1$-dense $\varphi$, $\psi$ are incompatible means that there is no uncountable type (equivalently, no $\aleph_1$-dense type) $\chi \leq \varphi, \psi$. It is quite easy to arrange, via the usual sort of Easton-style forcing, that $2^{\aleph_0} < 2^{\aleph_1}$ and there is a family of cardinality $2^{\aleph_1}$ of mutually incompatible $\aleph_1$-dense types. More interestingly, though, Shelah has shown it consistent with $2^{\aleph_0} < 2^{\aleph_1}$ that all $\aleph_1$-dense types are compatible.

Now we turn to the case when $2^{\aleph_0} = 2^{\aleph_1}$. Once again, it is easy to produce a model in which $2^{\aleph_0} = 2^{\aleph_1}$ and there are many distinct $\aleph_1$-dense types. It is also consistent, as the author showed in [Ba1], that $2^{\aleph_0} = 2^{\aleph_1} = \aleph_2$ and all $\aleph_1$-dense sets are isomorphic. It follows that in such a model there is a basis for the uncountable real types consisting of a single element.

The method of [Ba1] is as follows. Let $\langle x_\alpha : \alpha < \omega_1 \rangle$ and $\langle y_\alpha : \alpha < \omega_1 \rangle$ be enumerations of $\aleph_1$-dense sets $X$ and $Y$ respectively. Assuming CH, it is shown that it is possible to decompose $\omega_1$ into disjoint countable sets $\langle A_\alpha : \alpha < \omega_1 \rangle$ such that, if $P$ is the set of all finite order-preserving functions $p$ from $X$ to $Y$ which satisfy:

$$\text{if} \quad p(x_\alpha) = y_\beta \quad \text{then for some } \gamma,\ \alpha, \beta \in A_\gamma \quad ,$$

then $P$ has the countable chain condition (c.c.c.) The heart of the proof is a purely classical diagonal construction. Any uncountable antichain in $P$ may be regarded as a subset of $R^{2n}$ for some $n$, hence has a countable dense subset. By enumerating all such countable approximations to antichains it is possible to construct $\langle A_\alpha : \alpha < \omega_1 \rangle$ so that each one is avoided (i.e., its closure acquires no new elements) from a certain point on. The model is now obtained by iterating these c.c.c. partial orderings $\aleph_2$ times. Since CH holds at each partial stage of the iteration, one may arrange things so that every pair of $\aleph_1$-dense sets is made isomorphic. Unfortunately the details of the diagonal argument are quite complicated, and we shall not present them here.

It is clear that this iteration can be combined with the usual one for making Martin's Axiom (MA) true, and it was asked in [Bal] whether MA implies that all $\aleph_1$-dense sets of reals are isomorphic. This question was answered negatively by Avraham and Shelah in [AS2]. Shelah has also found another proof of the isomorphism result which is at once more sophisticated and much simpler than the one outlined above, and which has the further advantage of yielding the consistency with $2^{\aleph_0} = 2^{\aleph_1} > \aleph_2$ of the assertion that all $\aleph_1$-dense sets are isomorphic.

The heart of Shelah's argument is the following theorem, which is to appear in a paper by Avraham, Rubin and Shelah. For the rest of this section we must assume that the reader is familiar with two-stage forcing extensions and with the basics of forcing with countably closed and c.c.c. partial orderings.

THEOREM 3.4. *Let $A$ and $B$ be $\aleph_1$-dense sets. Then there are partial orderings $P_1 \in V$, $P_2 \in V^{P_1}$, $P_3 \in V^{P_1 \otimes P_2}$ such that*

(*a*) *$P_1$ has the c.c.c.,*

(*b*) *$P_2$ is countably closed in $V^{P_1}$,*

(*c*) *$P_3$ has the c.c.c. in $V^{P_1 \otimes P_2}$, and*

(*d*) *$A$ and $B$ are order-isomorphic in $V^{P_1 \otimes P_2 \otimes P_3}$*

(Here the product $\otimes$ refers to forcing composition.)

PROOF. Let $P_1$ be the usual (c.c.c.) partial ordering for adding $\aleph_1$ Cohen reals. It will be convenient (but not essential) to assume that $A$ and $B$ are disjoint. If this is not the case, then let $r$ be a Cohen real and translate $B$ by $r$. The resulting sets $A$ and $B+r$ are disjoint, and $V^{P_1}$ is obtained by adding $\aleph_1$ Cohen reals to $V[r]$, so it is immaterial whether or not $r \in V$.

Let $\langle a_\alpha : \alpha < \omega_1 \rangle$ and $\langle b_\alpha : \alpha < \omega_1 \rangle$ be enumerations of $A$ and $B$ such that for all $\alpha$, $\{a_{\alpha+n} : n \in \omega\}$ and $\{b_{\alpha+n} : n \in \omega\}$ are dense.

Let $P_2 \in V^{P_1}$ be the following ordering for adding a closed unbounded subset of $\omega_1$ which is eventually contained in every closed unbounded subset of $\omega_1$ lying in $V^{P_1}$. Elements of $P_2$ are pairs $p = (c_p, C_p)$ such that $c_p$ is a closed countable set of limit ordinals in $\omega_1$ and $C_p \subseteq \omega_1$ is closed unbounded. Let $p \leq q$ iff $c_p$ is an end-extension of $c_q$, $C_p \subseteq C_q$ and $c_p - c_q \subseteq C_q$. It should be clear that $P_2$ is countably closed.

If $G$ is $P_2$-generic, then let $C_G = \bigcup\{c_p : p \in G\}$. Let $\langle c_\alpha : \alpha < \omega_1 \rangle$ enumerate $C_G$ in increasing order, and assume for simplicity that $c_0 = 0$. Let $A_\alpha = \{a_\xi : c_\alpha \leq \xi < c_{\alpha+1}\}$, $B_\alpha = \{b_\xi : c_\alpha \leq \xi < c_{\alpha+1}\}$ and $D_\alpha = A_\alpha \cup B_\alpha$. If $\alpha < \beta$ and $(x,y) \in A_\alpha \times B_\beta$ or $A_\beta \times B_\alpha$, then that element of the pair $(x,y)$ which belongs to $D_\beta$ will be called the *top* element of the pair; the other is the *bottom* element. Two ordinals $\alpha$, $\beta$ are *neighbors* if $\alpha \neq \beta$ but for some $n \in \omega$, $\alpha = \beta + n$ or $\beta = \alpha + n$.

Let $P$ be the set of all finite, order-preserving functions from $A$ into $B$, ordered by letting $s \leq t$ iff $s \supseteq t$. Finally, let $P_3$ be the set of all $s \in P$ such that

(1) if $(x,y) \in s \cap (A_\alpha \times B_\beta)$ then $\alpha$ and $\beta$ are neighbors, and

(2) for each $\alpha$ there is at most one pair $(x,y) \in s$ whose top element belongs to $D_\alpha$.

Conditions (1) and (2) are likely to seem strange at first, but they are crucial for simplifying the argument. Note that $P \in V$ but we know only that $P_3 \in V^{P_1 \otimes P_2}$. Also, since each $A_\alpha$ and $B_\alpha$ is dense it is clear that the function added generically by $P_3$ will map $A$ order-isomorphically onto $B$. Thus the only thing to check is that $P_3$ has the c.c.c. in $V^{P_1 \otimes P_2}$.

We shall work in $V^{P_1}$. Suppose by way of contradiction that $p \in P_2$ and $\dot{X}$ is a term of the forcing language such that

$$p \Vdash \dot{X} \text{ is an uncountable antichain in } P_3.$$

Let $F = \{(p,s) : p \Vdash s \in \dot{X}\}$. Find regular $\mu$ large enough so that $P_2$, $F \in H(\mu)$, the collection of all sets hereditarily of cardinality $< \mu$. Now let $\langle N_\alpha : \alpha < \omega_1 \rangle$ be a sequence of countable elementary substructures of $(H(\mu), \in)$ such that $p, P_2$, $F \in N_0$, $N_\alpha \subseteq N_{\alpha+1}$, $N_\alpha \cap \omega_1 < N_{\alpha+1} \cap \omega_1$, and if $\alpha$ is limit then $N_\alpha = \bigcup\{N_\beta : \beta < \alpha\}$.

Let $N = N_0$ and $\alpha_0 = N \cap \omega_1$. Let $M$ be the transitive collapse of $N$ via the mapping $j : N \to M$. Now $P_1$ adds $\aleph_1$ Cohen reals, and since $M$ is countable and transitive it is added by at most $\aleph_0$ of the Cohen reals. Hence $V^{P_1}$ contains reals Cohen-generic over $V[M]$.

Now $j(P_2)$ is a countable non-trivial partial ordering, so it and the Cohen ordering have isomorphic dense subsets. Thus $V^{P_1}$ contains a set $H$ which is $j(P_2)$-generic over $V[M]$. Moreover we may assume $j(p) \in H$. Since the elements of $H$ are pairwise compatible, so are the elements of $j^{-1}(H)$, and since $P_2$ is countably closed there is $p' \in P_2$ such that $p' \leq q$ for all $q \in H$ and $\max(c_{p'}) = N \cap \omega_1$. Let $p^* = (c_{p'}, C_{p'} \cap C^*)$, where $C^* = \{N_\alpha \cap \omega_1 : \alpha < \omega_1\}$.

The following lemma is now the key to the proof.

LEMMA 3.5. *Suppose* $q \leq p^*$ *and* $q \Vdash s \in \dot{X}$. *Then* $\forall p' \leq p$ *if* $p' \in N$ *then there exist* $p'' \leq p'$ *and* $t \in N$ *such that* $s$ *and* $t$ *are compatible and* $p'' \Vdash t \in \dot{X}$.

Before proving Lemma 3.5, let us establish the following corollary, which will complete the proof of Theorem 3.4.

COROLLARY 3.6. $p^* \Vdash \dot{X} \subseteq N$.

PROOF. First note that the collapsing map $j$ is the identity on real numbers, hence on $P \cap N$. Suppose $q \leq p^*$ and $q \Vdash s \in \dot{X}$. Working in $V[M]$, form $E = \{\overline{p} \in j(P_2) : \exists t \in P\ t$ is compatible with $s$ and $(\overline{p},t) \in j(F)\}$, and let $D = E \cup \{\overline{q} \in j(P_2) : \forall q' \leq \overline{q}\ q' \notin E\}$. Since $D$ is dense in $j(P_2)$, there is $\overline{p} \in D \cap H$.

Case 1. $\overline{p} \in E$. Then $(j^{-1}(\overline{p}),t) \in F$ for some $t$ compatible with $s$ and $p^* \leq (j^{-1}(\overline{p}),t)$, so $p^* \Vdash t \in \dot{X}$. Hence $s = t$, and we are done since $t \in N$.

Case 2. $\overline{p} \notin E$. Now $\overline{p}$ and $j(p)$ are compatible so we may assume $\overline{p} \leq j(p)$ without loss of generality. Let $p' = j^{-1}(\overline{p})$. Note that $q$ satisfies the hypothesis of Lemma 3.5. Hence there exist $p'' \leq p'$ and $t \in N$ such that $s$, $t$ are compatible and $p'' \Vdash t \in \dot{X}$. Moreover, since $N$ is an elementary substructure of $(H(\mu),\in)$ there must be such $p'' \in N$. But then clearly $j(p'') \in E$, contradicting the assumption that $\overline{p} \in D-E$. □

In the proof of Lemma 3.5 we will need the following model-theoretic result.

LEMMA 3.7. *Suppose* $\varphi$ *is a first-order formula,* $y_1,\ldots,y_n \in N_\alpha$, $x \notin N_\alpha$, $x,y_1,\ldots,y_n$ *are real numbers and* $(H(\mu),\in) \models \varphi(x,y_1,\ldots,y_n)$. *Then* $x$ *is a two sided limit point of* $\{x' \in N_\alpha : (H(\mu),\in) \models \varphi(x',y_1,\ldots,y_n)\}$.

PROOF. Note that $Z = \{x : \varphi(x,y_1,\ldots,y_n)\}$ is definable in $H(\mu)$ from $y_1,\ldots,y_n$, hence must lie in $N_\alpha$, since $N_\alpha$ is an elementary substructure of $(H(\mu),\in)$. If $Z$ were countable then $Z \subseteq N_\alpha$, contradicting the fact that $x \notin N_\alpha$. Hence there is a countable set $Y \subseteq Z$ such that every element of $Z-Y$ is a two-sided limit point of $Y$. Since the existence of $Y$ may be asserted in $H(\mu)$, such a $Y$ belongs to $N_\alpha$ as well, and since $Y$ is countable, $Y \subseteq N_\alpha$. □

PROOF OF LEMMA 3.5. Let $S$ = domain($s$) ∪ range($s$). We refer to $S$ as the *field* of $s$. Since $q \leq p^*$ and $q \Vdash s \in P_3$, it follows that $\forall\alpha < \omega_1$

(*) $(N_{\alpha+1}-N_\alpha) \cap S$ contains at most one top element of $s$.

Let $s_1,\ldots,s_n$ be an enumeration of $S$ with the property that if $s_i \in N_\alpha$ and $s_j \notin N_\alpha$ then $i < j$. Let $J_1,\ldots,J_n$ be disjoint intervals such that $s_i \in J_i$. We shall say that two functions $t,u \in P$ are *isomorphic* if $|t| = |u|$ and the (unique) order-isomorphism of their fields lifts to an isomorphism between the ordered pairs in $t$ and the ordered pairs in $u$. Note that there are only countably many isomorphism types.

Suppose $s$ has isomorphism type $\tau$. If $t_i \in J_i$ for $1 \leq i \leq n$ and $t = \{(t_i,t_j):(s_i,s_j) \in s\}$, then $t$ is clearly the unique element of $P$ with type $\tau$ and field $\{t_i:1 \leq i \leq n\}$. Such a $t$ is said to be *weakly compatible with* $s$ if $(t_i,t_j) \in t$ is compatible with $(s_i,s_j) \in s$ whenever $t_i \neq s_i$ and $t_j \neq s_j$. Note that if $t$ is weakly compatible with $s$ and for all $(s_i,s_j) \in s$ either ($s_i = t_i$ and $s_j = t_j$) or ($s_i \neq t_i$ and $s_j \neq t_j$), then $t$ is compatible with $s$.

A function $t$ as above is *j-consistent* with $s$ if

(a) $s_i = t_i$ for all $i \leq j$,
(b) $s_i \neq t_i$ for all $i > j$,
(c) $t$ is weakly compatible with $s$, and
(d) $\exists p'' \leq p'\ \ p'' \Vdash t \in \dot{X}$.

Let $T_j = \{s_i \in S: i > j$ and $s_i$ is the top element of a pair in $s$ whose bottom element is among $s_1,\ldots,s_j\}$. If $t$ is $j$-consistent with $s$ then $t$ determines a function $\varepsilon_t:T_j \to 2$ given by

$$\varepsilon_t(s_i) = \begin{cases} 0 & \text{if } t_i < s_i \\ 1 & \text{iff } t_i > s_i \end{cases}.$$

SUBLEMMA 3.8. *For any $\alpha < \omega_1$, if $N_\alpha \cap S = \{s_1,\ldots,s_j\}$ then for any $\varepsilon:T_j \to 2$ there is $t \in N_\alpha$ such that $t$ is $j$-consistent with $s$ and $\varepsilon_t = \varepsilon$.*

PROOF. If $S \subseteq N_\alpha$ then $T_j = 0$ and the sublemma is trivial. Hence there are at most countably many $\alpha$ which fail to satisfy the sublemma. Also, since $S$ is finite there must be a largest $\alpha$ which fails. Suppose $(N_{\alpha+1}-N_\alpha) \cap S = \{s_{j+1},\dots,s_k\}$. Let $\varepsilon_1,\dots,\varepsilon_l$ enumerate all $\varepsilon:T_k \to 2$, and let $t_i \in N_{\alpha+1}$ be $k$-consistent with $s$ such that $\varepsilon_{t^i} = \varepsilon_i$.

By observation (*) at most one of $s_{j+1},\dots,s_k$ is a top element of $s$; suppose $s_{j+1}$ is such an element. (If there is no top element among $s_{j+1},\dots,s_k$ then the proof is a proper subset of the present one, so we omit this case.)

For each $i \le l$, let $\overline{I^i} = \langle \dot{I}^i_{k+1},\dots,\dot{I}^i_n\rangle$ be a sequence of intervals with rational endpoints such that $\dot{I}^i_m \subseteq J_m$, $t^i_m \in \dot{I}^i_m$ and $s_m \notin \dot{I}^i_m$ for all $m$, $k < m \le n$. Now let $\varphi_i(s_1,\dots,s_k,\overline{I^i},\tau,p')$ be a sentence asserting the following (in $(H(\mu),\in)$):

$(\exists t_{k+1} \in \dot{I}^i_{k+1})\dots(\exists t_n \in \dot{I}^i_n)$ if $t$ is the function of type $\tau$ with field $\{s_1,\dots,s_k,t_{k+1},\dots,t_n\}$ then $\exists p'' \le p' \quad p'' \Vdash t \in \dot{X}$.

Let $\varphi(s_1,\dots,s_j,s_{j+1},\overline{I^1},\dots,\overline{I^l},\tau,p')$ be

$$(**) \quad \exists x_{j+2} \dots \exists x_k \bigwedge_{i=1}^{l} \varphi_i(s_1,\dots,s_j,s_{j+1},x_{j+2},\dots,x_k,\overline{I^i},\tau,p') \quad .$$

Now all the parameters in $\varphi$ except $s_{j+1}$ belong to $N_\alpha$, so by Lemma 3.7 $s_{j+1}$ is a two-sided limit of elements of $N_\alpha$ which also satisfy $\varphi$.

Suppose $\varepsilon:T_j \to 2$, and for concreteness suppose $\varepsilon(s_{j+1}) = 0$ (the other case is symmetric). Choose $t_{j+1} < s_{j+1}$ so that $t_{j+1} \in N_\alpha$ and $t_{j+1}$ satisfies (**). Since $N_\alpha$ is an elementary substructure of $H(\mu)$, there exist $t_{j+2},\dots,t_k \in N_\alpha$ such that

$$\bigwedge_{i=1}^{l} \varphi_i(s_1,\dots,s_j,t_{j+1},t_{j+2},\dots,t_k,\overline{I^i},\tau,p')$$

holds in $H(\mu)$ (equivalently: in $N_\alpha$).

Now define $\varepsilon':T_k \to 2$ as follows. If $s_i \in T_k \cap T_j$ let $\varepsilon'(s_i) = \varepsilon(s_i)$. If $s_i \in T_k - T_j$ then for some $m$, $j+2 \le m \le k$, $s_m$ is the bottom element of a pair for which $s_i$ is the top. In this case let $\varepsilon'(s_i) = 0$ if $t_m < s_m$, and let $\varepsilon(s_i) = 1$ if $t_m > s_m$ (note $t_m \ne s_m$ since $t_m \in N_\alpha$, $s_m \notin N_\alpha$). Say $\varepsilon' = \varepsilon_r$.

Since

$$(N_\alpha,\varepsilon) \models \varphi_r(s_1,\dots,s_j,t_{j+1},\dots,t_k,\overline{I^r},\tau,p') \quad ,$$

there are $t_i \in \dot{I}^r_i \cap N_\alpha$, $k+1 \le i \le n$, such that if $t$ is of type $\tau$ with field $\{s_1,\dots,s_j,t_{j+1},\dots,t_n\}$ then $\exists p'' \le p'$

$$p'' \Vdash t \in \dot{X} \quad .$$

We assert that $t$ is weakly compatible with $s$, from which it will follow that $t$ is $j$-consistent with $s$. Suppose $(t_i, t_m) \in t$, $t_i \neq s_i$, $t_m \neq s_m$. Then $i, m > j$. If $i, m > k$ then we are done by the fact that $t^r$ is weakly compatible with $s$ and $t_i, t_i^r \in I_i^r$, $t_m, t_m^r \in I_m^r$ but $s_i \notin I_i^r$, $s_m \notin I_m^r$. We cannot have $j < i, m \leq k$ since $\{s_{j+1}, \ldots, s_k\}$ contains only one top element and by (1) of the definition of $P_3$ it must be connected to a bottom element in $N_\alpha$. So suppose $i > k$ and $j < m \leq k$. Then $s_i \in T_k - T_j$ and by the definition of $\varepsilon' = \varepsilon_r$ in this case we must have $t_i^r < s_i$ iff $t_m < s_m$. But $t_i < s_i$ iff $t_i^r < s_i$ since $t_i, t_i^r \in I_i^r$, $s_i \notin I_i^r$. Hence $(t_i, t_m)$ is compatible with $(s_i, s_m)$ and $t$ is weakly compatible with $s$.

It is now routine to verify that $\varepsilon_t = \varepsilon$. □

Now let $S \cap N = \{s_1, \ldots, s_j\}$. By the genericity of $H$ it is clear that $\max(c_{p^*}) = \alpha_0$ is a limit point of $c_{p^*}$. Hence by (1) of the definition of $P_3$, no element of $S - N$ can be connected to an element of $S \cap N$ in $s$. Sublemma 3.8 yields a $t \in N$ which is $j$-consistent with $s$. But now it is clear that for all $(s_i, s_m) \in s$, either ($s_i = t_i$ and $s_m = t_m$) or ($s_i \neq t_i$ and $s_m \neq t_m$). Hence $t$ and $s$ are compatible and we are done with Lemma 3.5 and Theorem 3.4. □ □

One may now obtain the consistency of the assertion that all $\aleph_1$-dense sets are isomorphic by iterating the three-stage extensions of Theorem 3.4 in $\aleph_2$ stages with countable support over a model of GCH. Details of such iterations may be found in [Ba3], for example.

To obtain the consistency with $2^{\aleph_0} = \aleph_3$, say, of the assertion that all $\aleph_1$-dense sets are isomorphic, proceed as follows. Begin with a model of GCH and add $\aleph_1$ Cohen reals. Then iterate in $\aleph_3$ stages, adding at each stage a closed unbounded set almost contained in every closed unbounded set produced by the previous stage. Finally, do a finite-support c.c.c. iteration designed to make all $\aleph_1$-dense sets isomorphic. Theorem 3.4 may be applied to see that this is always possible.

Another approach is simply to derive the isomorphism of $\aleph_1$-dense sets from the Proper Forcing Axiom. This axiom asserts that if $P$ is proper (a notion invented and studied by Shelah; see [Sh1]), and if $\langle D_\alpha : \alpha < \omega_1 \rangle$ are dense subsets of $P$, then there is a filter $G \subseteq P$ such that $G \cap D_\alpha \neq 0$ for all $\alpha$.

Every countably closed or c.c.c. ordering is proper, and if $P$ is proper and $Q$ is proper in $V^P$, then $P \otimes Q$ is proper. Hence $P_1 \otimes P_2 \otimes P_3 = P$ of Theorem 3.4 is proper. There is a $P$-term $\dot{f}$ such that

$$\Vdash_P \dot{f}:A \to B \text{ is an order-isomorphism.}$$

For each $a \in A$ let $D_a = \{p \in P: \exists b \in B,\ p \Vdash \dot{f}(a) = b\}$ and for each $b \in B$ let $D_b = \{p \in P: \exists a \in A,\ p \Vdash \dot{f}(a) = b\}$. If $G \subseteq P$ is a filter and $G \cap D_a$, $G \cap D_b \neq 0$ for all $a \in A$, $b \in B$, then clearly $\{(a,b): \exists p \in G,\ p \Vdash \dot{f}(a) = b\}$ is an order-isomorphism of $A$ and $B$. But such a $G$ exists by the Proper Forcing Axiom.

Finally, let us state what seems to be the most interesting open problem about real types:

PROBLEM. Is it consistent that all $\aleph_2$-dense sets of reals are order-isomorphic?

## 4. SPECKER TYPES

We have seen that any basis for the uncountable order types must contain $\omega_1$, $\omega_1^*$ together with some of the real types. In fact the results of section 3 show that it is consistent that there exists a basis for the uncountable real types which consists of a single element. But even if we allow all the uncountable real types together with $\omega_1$ and $\omega_1^*$, do we have a basis for the uncountable types? To put the question the other way round, does there exist an uncountable linear ordering, no uncountable subset of which is well-ordered, conversely well-ordered, or embeddable in the reals? This question was asked by Erdös and Rado in their 1956 paper [ER], and it was answered in unpublished work by E.P. Specker.

Specker showed that such orderings exist. Nowadays these orderings are called *Specker orderings*, and their order types are *Specker types*.

It turns out that Specker types are intimately connected with Aronszajn trees. Recall that a *tree* is a partial ordering $(T,\leq_T)$ with the property that for all $t \in T$, $\{s \in T: s <_T t\}$ is well-ordered. The *level* of $t$, $l(t)$, is $tp\{s \in T: s <_T t\}$. The $\alpha$th *level of* $T$ is $T^\alpha = \{t \in T: l(t) = \alpha\}$, and $T_\alpha = \bigcup\{T^\beta: \beta < \alpha\}$. A *branch* through a tree $(T,\leq_T)$ is a maximal linearly ordered (hence well-ordered) subset of $T$. A tree $(T,\leq_T)$ is an *Aronszajn tree* if $0 < |T^\alpha| \leq \aleph_0$ for all $\alpha < \omega_1$ and $T$ has no uncountable branches.

If $(T,\leq_T)$ is an Aronszajn tree and $\leq_\alpha$ is a linear ordering of $T^\alpha$ for each $\alpha < \omega_1$, then there is a natural linear ordering $\leq_l$ of $T$ which we shall describe as the *lexicographical ordering* (determined by $\leq_T$ and the $\leq_\alpha$), defined as follows. If $s \leq_T t$ then let $s \leq_l t$. Suppose $s$, $t$ are incomparable with respect to $\leq_T$. Let $s'$ be the least element of $\{s' \in T: s' \leq_T s \text{ but } s' \not\leq_T t\}$ and let $t'$ be the least element of $\{t' \in T: t' \leq_T t$ but $t' \not\leq_T s\}$. Then $s' \neq t'$ and $s', t' \in T^\alpha$ for some $\alpha$. Now let $s \leq_l t$ iff $s' \leq_\alpha t'$.

THEOREM 4.1. *If $(T, \leq_T)$ is an Aronszajn tree, $\leq_\alpha$ is a linear ordering of $T^\alpha$ for each $\alpha < \omega_1$, and $\leq_l$ is the lexicographical ordering, then $(T, \leq_l)$ is a Specker ordering.*

PROOF. Suppose $X \subseteq T$ is uncountable and well-ordered in type $\omega_1$ by $\leq_l$. Then for each $\alpha$ there is $t_\alpha \in T^\alpha$ such that $\{t \in X: t_\alpha \leq_T t\}$ is uncountable. Moreover there cannot be more than one such $t_\alpha$ since then in the lexicographical ordering $X$ would have an uncountable proper initial segment. Furthermore if $\alpha < \beta$ then clearly $t_\alpha \leq_T t_\beta$. That is, $\{t_\alpha: \alpha \in \omega_1\}$ forms an uncountable branch through $(T, \leq_T)$, a contradiction. Similarly, $X$ cannot be ordered in type $\omega_1^*$.

Suppose $X$ were embeddable in the reals. Then there would be countable $Y \subseteq X$ such that $Y$ is dense in $X$. Say $Y \subseteq T_\alpha$. But then $\{t \in X: t_\alpha \leq_T t\}$ is an uncountable interval of $X$ which contains no element of $Y$. □

Rather surprisingly, a converse of Theorem 4.1 is true.

THEOREM 4.2. *Every Specker ordering is isomorphic to a lexicographically ordered Aronszajn tree.*

PROOF. Let $(S, \leq_S)$ be a Specker ordering. Note that if $I$ is an interval of $S$ with no left-hand endpoint, then since $\omega_1^*$ is not embeddable in $S$ there is a descending sequence $\langle s_n: n < \omega\rangle$ of elements of $I$ coinitial with $I$.

We begin by building an Aronszajn tree $U$ such that each level $U^\alpha$ of $U$ consists of pairwise disjoint intervals of $S$, each with a left-hand endpoint. We will have $I \leq_U J$ iff $J \subseteq I$. Suppose we have obtained $U_\alpha$.

Case 1: $\alpha$ is a successor ordinal $\beta+1$. Let $I \in U^\beta$ and let $p(I)$ denote the left endpoint of $I$. If $I - \{p(I)\}$ has a left-hand endpoint, then put $I - \{p(I)\} \in U^\alpha$. Otherwise, if $I - \{p(I)\} \neq 0$, choose descending $\langle s_n: n \in \omega\rangle \in I$ coinitial in $I - \{p(I)\}$ and put $\{t \in I: s_0 \leq_S t\}$ and $\{t \in I: s_{n+1} \leq_S t <_S s_n\}$ into $U^\alpha$ for each $n$.

Case 2: $\alpha$ is a limit ordinal. Let $B$ be a branch through $U_\alpha$. If $\bigcap B$ is non-empty then it is an interval. If it has a left-hand endpoint, put $\bigcap B \in U^\alpha$. Otherwise choose descending $\langle s_n: n \in \omega\rangle$ coinitial in $\bigcap B$ and proceed as in Case 1. This is to be done for each branch through $U_\alpha$.

Let us see that $U$ is Aronszajn. If $B$ were an uncountable branch through $U$, then, since clearly $I <_U J$ implies $p(I) <_S p(J)$, $\{p(I): I \in U\}$ would be well-ordered in type $\omega_1$, a contradiction. Now suppose some $U^\alpha$ were uncountable. It is clear from the construction of $U$ that the least $\alpha$ for which this happens must be a limit ordinal. So there are uncountably many branches $B$ through $U_\alpha$ for which $\bigcap B \neq 0$. Choose $s_B \in \bigcap B$ for each such branch. Note that if $B \neq B'$ then there is some $I \in U_\alpha$ with $s_B <_S p(I) <_S s_{B'}$. Hence $\{p(I): I \in U_\alpha\}$ is dense in $\{p(I): I \in U_\alpha\} \cup \{s_B: B$ a branch through $U_\alpha\}$, and the latter set is embeddable in the reals, a contradiction. Hence $U$ is Aronszajn.

Let $T = \{p(I): I \in U\}$, with the inherited tree ordering. If $s \in S-T$ then $\{I \in U: s \in I\}$ would be an uncountable branch, so $S = T$. Moreover, as is easy to check, if $\leq_\alpha$ is just $\leq_S$ restricted to $T^\alpha$, then the lexicographical ordering of $T$ is just $\leq_S$. □

COROLLARY 4.3. *Every Specker ordering has cardinality* $\aleph_1$.

Finally, to see that Specker types exist we must prove the well-known

THEOREM 4.4. (Aronszajn; see [Ku]). *There exists an Aronszajn tree.*

PROOF. We will construct $T$ so that $T^\alpha$ will consist of one-to-one functions $f: \alpha \to \omega$ such that $\omega$-range$(f)$ is infinite. We will have $f \leq_T g$ iff $f \subseteq g$. Thus any uncountable branch would give rise to a one-to-one function mapping $\omega_1$ into $\omega$, which is impossible. Thus we need only concentrate on making each $T^\alpha$ countable.

We shall construct $T^\alpha$ so that the following inductive hypothesis will hold:

(*) if $\gamma < \alpha$, $f \in T^\gamma$ and $x \subseteq \omega$-range$(f)$ is finite, then $(\exists g \in T^\alpha)\ f \subseteq g$ and $x \subseteq \omega$-range$(g)$.

Let $T^0 = \{0\}$. Now suppose countable $T^\alpha$ satisfying (*) has been obtained.

Case 1. $\alpha = \beta+1$. Let $T^\alpha$ consist of all extensions of elements of $T^\beta$. Then (*) holds for $T^\alpha$ and $T^\alpha$ is countable.

Case 2. $\alpha$ is a limit ordinal. Suppose $\gamma < \alpha$, $f \in T^\gamma$ and finite $x \subseteq \omega$-range$(f)$ are specified. We must find $g \in T^\alpha$ as in (*), and if we simply let $T^\alpha$ consist of all (countably many) such $g$, we will be done. Fix an ascending sequence $\langle \alpha_n : n \in \omega \rangle$ cofinal in $\alpha$, with $\alpha_0 = \gamma$. We will construct sequences $\langle f_n : n \in \omega \rangle$ and $\langle x_n : n \in \omega \rangle$ so that

(1) $f_0 = f$ and $x_n = x$,
(2) $f_n \in T^{\alpha_n}$ and $f_n \subseteq f_{n+1}$,
(3) $x_n$ is a proper subset of $x_{n+1}$ and $x_n \subseteq \omega\text{-range}(f_n)$.

But this is easy using the inductive hypothesis (*) for each $\alpha_n$. Now let $g = \bigcup\{f_n : n \in \omega\}$. □

By making use of the existence of Specker types, Galvin [GS] was able to find the following improvement of Sierpiński's Theorem 2.1.

THEOREM 4.5. $\aleph_1 \nrightarrow [\aleph_1]^2_4$.

PROOF. Let $\leq_1, \leq_2, \leq_3$ be linear orderings of $\omega_1$ such that $\leq_1$ is the usual well-ordering, $tp(\omega_1, \leq_2)$ is a real type, and $tp(\omega_1, \leq_3)$ is a Specker type. If $\alpha, \beta \in \omega_1$ and $\alpha <_1 \beta$ then let

$$f(\{\alpha,\beta\}) = \begin{cases} 0 & \text{if} \quad \alpha <_2 \beta \quad \text{and} \quad \alpha <_3 \beta \\ 1 & \text{if} \quad \alpha <_2 \beta \quad \text{and} \quad \alpha >_3 \beta \\ 2 & \text{if} \quad \alpha >_2 \beta \quad \text{and} \quad \alpha <_3 \beta \\ 3 & \text{if} \quad \alpha >_2 \beta \quad \text{and} \quad \alpha >_3 \beta \end{cases} .$$

In order to see that $f$ works, we need the following lemma.

LEMMA 4.6. *If $(S, \leq_S)$ is a Specker ordering and $g$ is a one-to-one mapping of $S$ into the unit interval* (0,1), *then* $\exists s \in S$ $\{t \in S : s <_S t$ *and* $g(s) < g(t)\}$ *is uncountable.*

PROOF. For each $s \in S$ let $A(s) = \{r \in R : \{t \in S : t >_S s$ and $g(t) > r\}$ is countable$\}$, and let $a(s) = \inf(A(s))$. Note that if $s \leq_S t$ then $a(s) \geq a(t)$. If $a$ takes on uncountably many values then $a$ maps some uncountable subset of $S$ into the reals in an order-reversing fashion, and since $\lambda^* = \lambda$ we conclude that $S$ cannot be a Specker ordering. Hence $a$ takes on only countably many values and there is an uncountable set $S' \subseteq S$ on which $a$ is constant; say $a(s) = r$ for all $s \in S'$.

If there is $s \in S'$ with $g(s) < r$ then we are done immediately by the definition of $a(s)$, so assume not. Since $\omega_1^* \nleq tp(S')$ it is easy to see that there must be $s \in S'$ such that $\{t \in S' : s <_S t\}$ is uncountable. But $g(t) \geq r$ for all $t \in S'$ with $s <_S t$, contradicting the fact that $r = a(s)$. □

It follows immediately from Lemma 4.6 that if $X \subseteq \omega_1$ is uncountable then there exists $\alpha \in X$ such that $\{\beta \subset X : \alpha <_3 \beta$ and $\alpha <_2 \beta\}$ is uncountable. Hence there is $\beta \in X$ such that $\alpha <_1 \beta$, $\alpha <_2 \beta$, $\alpha <_3 \beta$, and $f(\{\alpha,\beta\}) = 0$. But $tp(\omega_1, >_3)$ is a Specker type and $tp(\omega_1, >_2)$ is a real type so by symmetry we have $f([X]^2) = 4$. □

It is an open problem at present whether the relation $\aleph_1 \to [\aleph_1]_5^2$ is consistent with ZFC. In unpublished work, Galvin has shown that this relation would settle several other old unsolved problems.

Let us turn now to the question of the existence of non-isomorphic Specker types. By making use of types with countable intervals, it is easy to produce non-isomorphic Specker types, so it is natural to impose an additional condition as we did for real types. Let us say that a Specker order $(S, \leq_S)$ (and its associated type) is $\aleph_1$-*dense* if between any two elements of $S$ there are $\aleph_1$ elements of $S$.

Unlike the situation for real types, it is not difficult to produce many non-isomorphic $\aleph_1$-dense Specker types. We shall begin by reviewing the proof of Theorem 4.2. Let us modify the tree $U$ in that proof to obtain a new tree $\overline{U}$ as follows: Let $\overline{U}^{\alpha+1} = U^\alpha$, and if $\alpha$ is a limit ordinal, let $\overline{U}^\alpha$ consist of all intervals of the form $\bigcap B$, where $B$ is a branch through $U_\alpha$. Note that in the latter case $\overline{U}^\alpha$ may contain intervals with no left-hand endpoint. Of course, if $I \in \overline{U}^\alpha$ has a left-hand endpoint, then $I$ also occurs in $\overline{U}^{\alpha+1}$; we regard these two occurrences as distinct elements of $\overline{U}$. We refer to $\overline{U}$ as an *augmented tree associated with* $(S, \leq_S)$.

THEOREM 4.7. *If $(S, \leq_S)$ is a Specker ordering and $U$ and $W$ are augmented trees associated with $(S, \leq_S)$, then $\{\alpha : U^\alpha = W^\alpha\}$ contains a set closed unbounded in $\omega_1$.*

PROOF. Let $C$ be the set of all limit ordinals $\alpha$ such that the set of left-hand endpoints of intervals in $U_\alpha$ coincides with the set of left-hand endpoints of intervals in $W_\alpha$. It is easy to see that $C$ is closed unbounded, and if $\alpha \in C$ then $U^\alpha = W^\alpha$. □

THEOREM 4.8. *There exist $2^{\aleph_1}$ non-isomorphic $\aleph_1$-dense Specker types.*

PROOF. Consider the Aronszajn tree $T$ constructed in Theorem 4.4, but regard the elements of $T$ as functions into the rationals instead of into $\omega$ (just replace $n$ by the $n$th rational, $q_n$). With the lexicographical ordering, $T$ is an $\aleph_1$-dense Specker order. (It has a first element, arising from the minimal element of $T$, but this element can be deleted if endpoints are not desired.)

Now add new elements $\{c_\alpha : \alpha$ is a limit ordinal $< \omega_1\}$ to $T$ to obtain $\overline{T}$ as follows. For each limit $\alpha < \omega_1$, choose a branch $B$ through $T_\alpha$ such that $(\forall t \in T^\alpha)\ B \neq \{s : s <_T t\}$. There are $2^{\aleph_0}$ such branches so this is always possible. Now let $c_\alpha \in \overline{T}^\alpha$ and let $s <_{\overline{T}} c_\alpha$ iff $s \in B$. Since $c_\alpha$ is not extended at higher levels of $\overline{T}$, $T$ is dense in the lexicographical ordering of $\overline{T}$, so $\overline{T}$ is also $\aleph_1$-dense.

Now $\overline{T}$ is canonically isomorphic to an augmented tree $U$ associated with $\overline{T}$ (lexicographically ordered). Just replace each $t \in \overline{T}$ by $\{s \in \overline{T}: t \leq_{\overline{T}} s\}$. Note that for limit $\alpha$, $U^{\alpha}$ contains exactly one singleton, which arises from $c_{\alpha}$.

If $A \subseteq \omega_1$ is stationary, let $T_A = T \cup \{c_{\alpha}:\alpha \in A\}$. By Theorem 4.7 and the remarks above, if $U$ is any augmented tree for $T_A$ then there is closed unbounded $C \subseteq \omega_1$ such that $\forall\alpha \in C$

$$\alpha \in A \quad \text{iff} \quad U^{\alpha} \text{ contains a singleton.}$$

It follows immediately that if $A$ and $B$ are stationary sets such that $(A-B) \cup (B-A)$ is stationary, then $T_A$ and $T_B$ cannot be order-isomorphic as Specker orderings. But there are $2^{\aleph_1}$ stationary sets which pairwise satisfy the condition above. These may be obtained as follows: First let $\langle A_{\alpha}:\alpha < \omega_1\rangle$ be a sequence of disjoint stationary sets; the existence of such a sequence is a result of Ulam [Ul] (but see also [Je]). Now if $X \subseteq \omega_1$ and $X \neq 0$, let $A_X = \bigcup\{A_{\alpha}:\alpha \in X\}$. Then $\{A_X : X \subseteq \omega_1, X \neq 0\}$ is the desired family. □

Of course, Theorem 4.8 is really quite weak. It is conceivable, for example, that the orderings of Theorem 4.8 are all embeddable in each other. Let us say that Specker types $\varphi_1$, $\varphi_2$ are *incompatible* if there is no Specker type $\psi \leq \varphi_1, \varphi_2$. In particular Theorem 4.8 says nothing about the existence of incompatible Specker types.

Nevertheless, there is a remarkable result of Shelah which implies that for some Specker type $\varphi$, $\varphi$ and $\varphi^*$ are incompatible! If $(S, \leq_S)$ is a Specker ordering, then $S \times S$ may be partially ordered coordinatewise:

$$(s_1,t_1) \leq (s_2,t_2) \quad \text{iff} \quad s_1 \leq_S s_2,\ t_1 \leq_S t_2.$$

Shelah [Sh3] constructs a Specker ordering $S$ such that $S \times S$ is the union of $\aleph_0$ chains. Hence $S \times S$ can have no uncountable pairwise incomparable set, which is precisely what an order anti-isomorphism between uncountable subsets of $S$ would be. The construction of $S$ is rather like the proof of Theorem 4.4, but several orders of magnitude more complicated, so we shall not give it here. Let us note, however, that this can be used to show that distinct Specker types may have isomorphic associated Aronszajn trees, for any augmented trees for $S$ and its converse must be isomorphic (indeed identical) on a closed unbounded set.

If we are willing to make assumptions beyond ZFC then more examples of Specker types may be constructed. If $T$ is a tree and $X \subseteq \omega_1$, let $T(X) = \bigcup\{T^{\alpha}:\alpha \in X\}$. Let $\omega$ say that two Aronszajn trees $T$ and $U$ are *almost-isomorphic* if there is closed unbounded $C \subseteq \omega_1$ such that $T(C)$ and $U(C)$ are isomorphic. Note that isomorphic Specker types always have almost-isomorphic augmented trees. Avraham and Shelah [AS1] have shown it consistent with ZFC + $2^{\aleph_0} = 2^{\aleph_1}$ that all Aronszajn trees are almost-isomorphic, but their results do not seem to apply to Specker types. On the other hand, non-isomorphism and non-embeddability results for Aronszajn trees do apply to Specker types, and it is possible to obtain such results using the Weak Diamond Principle (WDP) discussed in section 3.

Let us say that $T$ is *almost-embeddable* in $U$ if there is a closed unbounded set $C \subseteq \omega_1$ such that $T(C)$ is embeddable in $U$.

THEOREM 4.9. (Avraham-Shelah [AS1]). *Assume* $2^{\aleph_0} < 2^{\aleph_1}$. *If $T$ is an Aronszajn tree then there exists an Aronszajn tree $U$ such that $T$ is not almost-embeddable in $U$.*

PROOF. Note that the construction of the tree $T$ in Theorem 4.4 is highly non-canonical. If $\alpha$ is a limit ordinal then there are many possible choices for $T^\alpha$; in particular there are two disjoint sets, each of which could serve as $T^\alpha$. Thus we could construct a tree $\langle U_s : s \in \bigcup\{{}^\alpha 2 : \alpha < \omega_1\}\rangle$ of trees such that each $U_s$ is a possible realization of $T_\alpha$ (where $s \in {}^\alpha 2$), $U_t$ is an end-extension of $U_s$ when $s < t$ (i.e., if $s \in {}^\alpha 2$ then $U_s$ consists of the first $\alpha$ levels of $U_t$), and for limit $\alpha$ and $s \in {}^\alpha 2$, $U^\alpha_{s\frown 0}$ and $U^\alpha_{s\frown 1}$ are disjoint. Note that for any function $f : \omega_1 \to 2$, $U_f = \bigcup\{U_{f|\alpha} : \alpha \in \omega_1\}$ is an Aronszajn tree.

Given an arbitrary Aronszajn tree $T$, we want to find $f$ as above so that $T$ is not almost-embeddable in $U_f$. Without loss of generality, let us assume that $T^\alpha = \{\xi : \omega\cdot\alpha \leq \xi < \omega\cdot(\alpha+1)\}$ for all $\alpha < \omega_1$.

Now let $W = \{(s,t) : \exists\alpha\ s \in {}^\alpha 2$ and $t$ maps a subset of $T_\alpha$ into $U_s\}$. Let $(s_1,t_1) \leq (s_2,t_2)$ iff $s_1 \subseteq s_2$ and $t_2 \cap (T_\alpha \times U_{s_1}) = t_1$, where $\alpha = \text{domain}(s_1)$. As in the proof of Theorem 3.3, it is easy to see that WDP applies to $W$. Let us define $F : W \to 2$ as follows:

Case 1. $\alpha = \text{domain}(s)$ is a limit ordinal and there is closed unbounded $c \subseteq \alpha$ such that $t$ is an isomorphism of $T_\alpha(c)$ into $U_s$; moreover $t$ is cofinal in the sense that for any $\xi \in T^{\alpha+1}$, $\{l(t(\beta)) : \beta <_T \xi\}$ is cofinal in $\alpha$. (Recall that $l(x)$ is the level of $x$.)

In this case, $t$ determines a mapping carrying each $\xi \in T^\alpha$ to a branch $B(\xi) = \{g \in U_s : \exists\beta <_T \xi\ g \subseteq t(\beta)\}$ through $U_s$ of length $\alpha$. Since $U^\alpha_{s\frown 0}$ and $U^\alpha_{s\frown 1}$ are disjoint, there must be $i \in \{0,1\}$ such that for some $\xi \in T^{\alpha+1}$, there is no element of $U^\alpha_{s\frown i}$ which extends every element of $B(\xi)$. Let $F(s,t) = i$.

Case 2. Otherwise. Let $F(s,t) \in \{0,1\}$ be arbitrary.

By WDP there is a function $g:\omega_1 \to 2$ such that for any uncountable branch $B$ through $W$, $\{\alpha:(s,t) \in B$ and $l(s,t) = \alpha$ in $W$ and $F(s,t) = g(\alpha)\}$ is stationary. Let $U = U_g = \bigcup\{U_{g|\alpha}:\alpha \in \omega_1\}$. Suppose $T$ is almost-embeddable in $U$. Let $C \subseteq \omega_1$ be closed unbounded and let $h:T(C) \to U$ be an isomorphism. Let $B = \{(g|\alpha, h \cap (T_\alpha \times U_{g|\alpha})):\alpha \in \omega_1\}$. Then $B$ is an uncountable branch through $W$. If $D = \{\alpha \in C:(g|\alpha, h \cap (T_\alpha \times U_{g|\alpha}))$ satisfies case 1 above$\}$ then it is easy to see that $D$ is closed unbounded. But now it follows from the definition of $F$ and the choice of $g$ that for some $\alpha \in D$, $h \cap (T_\alpha \times U_{g|\alpha})$ cannot be extended to $T^{\alpha+1}$, contrary to hypothesis. □

We may improve this result considerably by using the full Diamond Principle $\Diamond$, which is stated like WDP except that $F$ (and hence $g$) may have range $2^{\aleph_0}$. It is easy to see that $\Diamond$ implies CH, so $\Diamond$ is stronger than WDP. It is well-known that $\Diamond$ holds in $L$, the universe of constructible sets (see [De], for example). With $\Diamond$, it is no longer necessary to restrict our attention to only two ways, $U_{s\frown 0}$ and $U_{s\frown 1}$, of extending each $U_s$; rather all $2^{\aleph_0}$ possible extensions may be considered. With this additional freedom we can easily prove the following:

THEOREM 4.10. *Assume $\Diamond$. If $T$ is an Aronszajn tree, then there exists an Aronszajn tree $U$ such that no Aronszajn tree is embeddable in both $T$ and $U$.*

Details are left to the reader.

From Theorems 4.9 and 4.10, we obtain immediately

COROLLARY 4.11. (*a*) *If $2^{\aleph_0} < 2^{\aleph_1}$ then for any Specker type $\varphi$ there is a Specker type $\psi$ such that $\varphi \not\leq \psi$, $\varphi \not\leq \psi^*$.*

(*b*) *If $\Diamond$, then for any Specker type $\varphi$ there is a Specker type $\psi$ such that for no Specker type $\chi$ do we have $\chi \leq \varphi,\psi$ or $\chi \leq \varphi,\psi^*$.*

By defining appropriate normal ideals as in section 3, Corollary 4.11 may be extended to assert the existence of a family, of cardinality $2^{\aleph_1}$, of Specker types such that (in part (a)) pairwise, $\varphi \not\leq \psi$, $\varphi \not\leq \psi^*$ or (in part (b)) pairwise, there is no Specker type $\chi$ such that $\chi \leq \varphi,\psi$ or $\chi \leq \varphi,\psi^*$.

REMARK. The usual statement of $\Diamond$ is rather different from the one we have given. It runs as follows: There is a sequence $\langle S_\alpha:\alpha < \omega_1\rangle$ such that $S_\alpha \subseteq \alpha$ and $\forall A \subseteq \omega_1$ $\{\alpha:S_\alpha = A \cap \alpha\}$ is stationary in $\omega_1$. We leave the equivalence of these two assertions as an exercise for the reader.

So far we have seen WDP and $\Diamond$ used as devices to avoid certain kinds of embeddings. We shall conclude this paper with an example, which seems to be new, of a rather different kind. Using the combinatorial principle $\Diamond^+$, we will construct a *minimal* Specker type, i.e., a type $\varphi$ such that for all uncountable $\psi \leq \varphi$ we have $\varphi \leq \psi$. Of course, this is completely different from the situation for real types under CH, which $\Diamond^+$ implies.

The principle $\Diamond^+$ is as follows: There exists a sequence $\langle S_\alpha : \alpha < \omega_1 \rangle$ such that each $S_\alpha$ is a countable collection of subsets of $\alpha$ and $\forall A \subseteq \omega_1 \, \exists C$ $C$ is closed unbounded in $\omega_1$ and $\forall \alpha \in C$ $A \cap \alpha, C \cap \alpha \in S_\alpha$. It is well-known that $\Diamond^+$ holds in $L$ and implies $\Diamond$. See [De].

Before proving the theorem we need some terminology and a couple of lemmas.

The *height* of a tree $T$, height$(T)$, is the smallest $\alpha$ such that $T^\alpha = 0$. Let us call a tree $T$ *regular* if every level of $T$ is countable and

(1) $\forall \alpha \, \forall t \in T^\alpha \, \{t' \in T^\alpha : \{s : s <_T t'\} = \{s : s \leq_T t\}\}$ is countably infinite.

(2) $\forall \alpha, \beta$ if $\alpha < \beta <$ height$(T)$, then $\forall s \in T^\alpha \, \exists t \in T^\beta \; s <_T t$.

If $T$ is a tree, $A \subseteq T$ and $t \in T$, then $A$ is *cofinal above* $t$ *in* $T$ if $\forall s \in T$ if $t \leq_T s$ then $\exists r \in A \; s \leq_T r$. If $t \in T_\alpha$ and $A \subseteq T_\alpha$ then we say $A$ is *dense between* $t$ *and* $T^\alpha$ if $\forall s \in T^\alpha$ if $t <_T s$ then $\exists r \in A \; t \leq_T r <_T s$.

LEMMA 4.12. *Suppose $T$ is a regular tree of height $\alpha$ for some countable limit ordinal $\alpha$. Suppose also that $\langle A_n : n \in \omega \rangle$, $\langle t_n : n \in \omega \rangle$ and $\langle f_n : n \in \omega \rangle$ are such that*

(*a*) $A_n \subseteq \{t \in T : t_n \leq_T t\}$,
(*b*) $A_n$ *is cofinal above* $t_n$ *in* $T$,
(*c*) $f_n : T \to A_n$ *is a tree-isomorphism.*

*Then it is possible to define an $\alpha$th level $T^\alpha$ such that $T \cup T^\alpha$ is regular and $\forall n$ $A_n$ is dense between $t_n$ and $T^\alpha$.*

PROOF. First note that it is only necessary to find countable $T^\alpha$ so that $T \cup T^\alpha$ satisfies (2) of the definition of regularity; each point of $T^\alpha$ may then be duplicated $\aleph_0$ times to make (1) true as well.

For simplicity we assume that $A_0 = T$ and $f_0$ is the identity.

Let $<\alpha_n : n \in \omega>$ be a sequence of ordinals cofinal in $\alpha$, and let $<(m_i, n_i) : i \in \omega>$ enumerate all pairs $(m,n) \in \omega\times\omega$ in such a way that each pair appears infinitely often. Now we choose an increasing sequence $<s_i : i \in \omega>$ of elements of $T$ by induction as follows. Let $s_0$ be arbitrary. If $s_i$ has been obtained then consider $(m_i, n_i)$. If $f_{m_i}(s_i) >_T t_{n_i}$ then find $s'$, $s''$ such that $f_{m_i}(s_i) <_T s' <_T s''$ and $s' \in A_{n_i}$, $s'' \in A_{m_i}$. This is possible since $A_{m_i}$, $A_{n_i}$ must both be cofinal above $\max(t_{m_i}, t_{n_i}) \leq_T f_{m_i}(s_i)$, for $t_{n_i} \leq_T f_{m_i}(s_i)$ by hypothesis and $t_{m_i} \leq_T f_{m_i}(s_i)$. Moreover we can choose $s''$ so large that $l(f_{m_i}^{-1}(s'')) \geq \alpha_i$. Let $s_{i+1} = f_{m_i}^{-1}(s'')$. If the condition above is not satisfied, i.e., if $f_{m_i}(s_i) \not>_T t_{n_i}$, then choose $s_{i+1} >_T s_i$ arbitrarily so that $l(s_{i+1}) \geq \alpha_i$.

Now attempt a first approximation to $T^\alpha$ by defining $T^\alpha = \{x_n : n \in \omega\}$, where $x_n >_T f_n(s_i)$ for all $i$. We assert that each $A_n$ is dense between $t_n$ and $T^\alpha$. Suppose $x_m > t_n$. The pair $(m,n)$ occurs infinitely many times as $(m_i, n_i)$, so there must have been such an $i$ when $f_m(s_i) >_T t_n$. But then the $s'$ constructed above has the property that $s' \in A_n$, $s' <_T s'' = f_m(s_{i+1}) < x_m$.

Thus the only remaining thing to check is that $T \cup T^\alpha$ satisfies (2). If (2) fails for some $s$, then repeat the procedure above with $s_0 = s$, and enlarge $T^\alpha$ accordingly. Since $T^\alpha$ is countable, this process will succeed after countably many steps. □

A regular tree $T$ is *doubly ordered* if it also has a linear ordering of each level with the property that $\forall\alpha\ \forall t \in T^\alpha$ $\{t' \in T^\alpha : \{s' : s' <_T t'\} = \{s : s <_T t\}\}$ has the order-type $\eta$ of the rational numbers. Two doubly ordered trees $T$ and $U$ are *doubly isomorphic* if there is a tree-isomorphism of $T$ and $U$ which preserves the lexicographical orderings induced by the orderings of the levels.

LEMMA 4.13. *Suppose $T$ and $U$ are doubly ordered regular trees of height $\alpha \in \omega_1$. Then $T$ and $U$ are doubly isomorphic.*

PROOF. Since a regular tree of (countable) limit height may be extended to a regular tree of successor height, we need only prove the lemma for $\alpha$ a successor ordinal, say $\alpha = \beta+1$. If $\beta$ is a successor ordinal then the lemma is trivial provided we know it is true for $\beta$. Thus we need only treat the case when $\beta$ is a limit ordinal.

If $t_1, \ldots, t_n \in T^\beta$ then the subtree generated by $t_1, \ldots, t_n$ is $\{s \in T : \exists i \in \{1, \ldots, n\}\ s \leq_T t_i\}$. A similar definition may be made for $U$. Let us say that a finite mapping $f$ from a subset of $T^\beta$ to a subset of $U^\beta$ is *successful* if $f$ may be extended to an isomorphism of the subtree generated by domain$(f)$ with the subtree generated by range($f$) which preserves the lexicographical ordering. Note that if $t \in T^\beta$, $u \in U^\beta$ then $\{(t,u)\}$ is successful.

We will argue that if $f$ is successful and $t \in T^\beta$ then there is successful $g \supseteq f$ with $t \in$ domain($g$). Then symmetry and a simple back-and-forth argument will yield the lemma.

Fix $t \in T^\beta$ and let $s \leq_T t$ be the smallest element of $T$ such that $\forall t' \in$ domain($f$) $s \not\leq_T t'$. Let $A = \{t' \in T\colon s'':s' <_T t'\} = \{s:s <_T t\}\}$. By the choice of $s$ there must be at least one $t' \in A$ such that $t' \neq s$ and $\exists t'' \in$ domain($f$) $t' \leq_T t''$. Let $t'_1,\ldots,t'_k$ enumerate all such $t'$. Now since $f$ is successful it extends to the generated subtrees, so we may look at $u'_i = f(t'_i)$. Since $f$ preserves the lexicographical ordering, and since $A$ and $B = \{u' \in U\colon\{s':s' <_U u'\} = \{s':s' <_U u'_i\}$ for each $i\}$ have order-type $\eta$, there must be some $u' \in B$ such that for $i = 1,\ldots,k$, $u' < u'_i$ iff $s < t'_i$ (in the levelwise ordering). Now let $u \in U^\beta$ be such that $u' \leq_U u$, and let $g = f \cup \{(t,u)\}$. It is easy to see that $g$ is successful. □

Now we may derive immediately

LEMMA 4.14. *Suppose $T'$ and $U'$ are doubly ordered regular trees of height $\alpha \in \omega_1$, and $f\colon T' \to U'$ is a double isomorphism. Suppose $T$ and $U$ are doubly ordered regular trees of height $\beta \in \omega_1$ such that $T$ and $U$ are end-extensions of $T'$ and $U'$ respectively. If $(\forall t \in T')$ $T'$ is dense between $t$ and $T^\alpha$ and $(\forall u \in U')$ $U'$ is dense between $u$ and $U^\alpha$, then $f$ may be extended to a double isomorphism of $T$ and $U$.*

Now we are ready to prove the theorem. The Specker type we construct will be obtained from an Aronszajn tree $T$ with the property that every pairwise incomparable subset of $T$ is countable. Such a tree is called a *Souslin tree*, and the completion of a Specker type arising from a Souslin tree is called a *Souslin line*. See [DJ] for a discussion of Souslin's Hypothesis, and for the construction from $\Diamond$ and $\Diamond^+$ of Souslin lines with various automorphism properties.

THEOREM 4.15. *Assume $\Diamond^+$. Then there is a Specker type $\varphi$ such that for all uncountable $\psi$, if $\psi \leq \varphi$ then $\varphi \leq \psi$.*

PROOF. Let $<S_\alpha:\alpha < \omega_1>$ be a $\Diamond^+$-sequence. We may assume that $A,B \in S_\alpha$ implies $A \cap B \in S_\alpha$ for all $\alpha$. We will construct $T$ so that $T^\alpha = \{\xi\colon \omega\cdot\alpha \leq \xi < \omega\cdot(\alpha+1)\}$. Since $T_\alpha$ will be constructed without reference to $S_\alpha$, we may assume that if $A \in S_\alpha$ and $\beta < \alpha$, then $\{\gamma \in A:\beta <_T \gamma\} \in S_\alpha$.

Now we obtain $T$, a doubly ordered regular tree of height $\omega_1$. We construct $T^\alpha$ by induction on $\alpha$. For successor ordinals the construction is trivial. For limit ordinals $\alpha$, $T^\alpha$ will be constructed satisfying an inductive hypothesis, stated below.

Let us call $A \in S_\alpha$ $\alpha$-*good* if $A$ (with the inherited ordering) is a regular tree of height $\alpha$ and $(\exists\beta \in T_\alpha)$ $A$ is cofinal above $\beta$ in $T_\alpha$. If $A,C \in S_\alpha$ then $(A,C)$ is $\alpha$-good if $A$ is $\alpha$-good, $C$ is a closed set of limit ordinals, and $\forall\beta \in C$ $A \cap \beta$ is $\beta$-good. Note that $(A,0)$ is $\alpha$-good iff $A$ is $\alpha$-good.

The inductive hypothesis for the construction of $T^\alpha$ is as follows: For all limit $\beta < \alpha$ and all $\beta$-good $(A,C)$, there is a double isomorphism $f=f(A,C):T_\beta \to A$. Moreover, if $\beta < \gamma$, $(A,C)$ is $\gamma$-good and $\beta \in C$, then $f(A \cap \beta, C \cap \beta) \subseteq f(A,C)$. Also, if $(A,C)$ is $\beta$-good and $A$ is cofinal above $t$ in $T_\beta$ then $A$ is dense between $t$ and $T^\beta$.

Now suppose $(A,C)$ is $\alpha$-good. If $\sup C = \alpha$ then let $f(A,C) = \bigcup\{f(A \cap \beta, C \cap \beta):\beta \in C\}$, which is well-defined (and a double isomorphism) by inductive hypothesis. If $\sup C = \beta < \alpha$ then by Lemma 4.14 and the inductive hypothesis for $T_\beta$, there is a double isomorphism $f(A,C)$ extending $f(A \cap \beta, C \cap \beta)$. Since there are only countably many $\alpha$-good $(A,C)$ we may apply Lemma 4.12 to construct $T^\alpha$ satisfying the inductive hypothesis. This completes the construction. Of course, $\varphi$ is the order type of the lexicographical ordering on $T$.

Next let us verify that $T$, with the tree ordering, is a Souslin tree. Suppose $A \subseteq T$ is a maximal pairwise incomparable set. Let $\bar{A} = \{\alpha:\exists\beta \in A\ \beta \le_T \alpha\}$. Then $\bar{A}$ is cofinal in $T$, so $C = \{\alpha \in \omega_1:\bar{A} \cap \alpha$ is a regular tree of height $\alpha$ cofinal in $T_\alpha\}$ is closed unbounded. By $\diamondsuit^+$ there is limit $\alpha \in C$ such that $\bar{A} \cap \alpha \in S_\alpha$. We may assume $0 \in S_\alpha$ as well. But now if $\beta < \alpha$ then $A_\beta = \{\gamma \in \bar{A} \cap \alpha:\beta <_T \gamma\}$ is cofinal above $\beta$ in $T_\alpha$, so $(A_\beta,0)$ is $\alpha$-good. Hence $A_\beta$ is dense between $\beta$ and $T^\alpha$. Since this happens for all $\beta < \alpha$, every element of $T^\alpha$, and hence every larger element of $T$, lies above some element of $A \cap T_\alpha$. Hence $A \subseteq T_\alpha$ so $A$ is countable and $T$ is a Souslin tree.

Finally suppose $X \subseteq T$ is uncountable. If there is no $\beta \in T$ such that $X$ is cofinal above $\beta$, then $\{\gamma \in T$:no element of $X$ lies above $\gamma$ and $\gamma$ is minimal with this property$\}$ is a pairwise incomparable uncountable set, contrary to the fact that $T$ is a Souslin tree. Thus without loss of generality we may assume that $X \subseteq \{\alpha:\beta <_T \alpha\}$ and $X$ is cofinal above $\beta$. By $\diamondsuit^+$ there is closed unbounded $C_1 \subseteq \omega_1$ such that $\forall\alpha \in C_1$, $X \cap \alpha$, $C_1 \cap \alpha \in S_\alpha$. Let $C_2 = \{\alpha:X \cap \alpha$ is a regular tree of height $\alpha\}$. Since $X$ is cofinal above $\beta$ and $T$ is Souslin, it follows immediately that $X$ is regular. Also, $\{\alpha:X \cap \alpha$ is the first $\alpha$ levels of $X\}$ is clearly closed unbounded. Thus $C_2$ is closed unbounded and, by $\diamondsuit^+$ again, there is closed unbounded $C_3$ such that

$\forall\alpha \in C_3$, $C_2 \cap \alpha$, $C_3 \cap \alpha \in S_\alpha$. But now if $C = C_1 \cap C_2 \cap C_3$, then $\forall\alpha \in C$ $(X \cap \alpha, C \cap \alpha)$ is $\alpha$-good. Hence by the construction of $T$, $f = \bigcup\{f(X \cap \alpha, C \cap \alpha) : \alpha \in C\}$ is a double isomorphism from $T$ to $X$. Thus $\varphi \leq tp(X)$. □

We do not know whether the type $\varphi$ of Theorem 4.15 can be obtained from anything but a Souslin tree, nor whether the assumption $\Diamond^+$ is necessary. Perhaps $\Diamond$ will suffice.

We conclude now with three open problems of considerable interest.

PROBLEM. Is it relatively consistent with ZFC that $\aleph_1 \rightarrow [\aleph_1]^2_5$?

PROBLEM. Is it relatively consistent with ZFC that for any Specker types $\varphi$, $\psi$, either $\varphi$ and $\psi$ are compatible or $\varphi$ and $\psi^*$ are compatible?

PROBLEM Is it relatively consistent with ZFC that the class of uncountable order types has a finite basis?

[1]The preparation of this paper was partially supported by National Science Foundation grant number MCS 79-03376.

REFERENCES

[AS1] U. Avraham and S. Shelah, Isomorphism types of Aronszajn trees, to appear.

[AS2] U. Avraham and S. Shelah, Martin's Axiom does not imply that every two $\aleph_1$-dense sets of reals are isomorphic, *Israel J. Math.* **38** (1981), 161-176.

[Ba1] J. Baumgartner, All $\aleph_1$-dense sets of reals can be isomorphic, *Fund. Math.* **79** (1973), 101-106.

[Ba2] J. Baumgartner, Almost-disjoint sets, the dense set problem and the partition calculus, *Annals Math. Logic* **10** (1976), 401-439.

[Ba3] J. Baumgartner, Iterated forcing, to appear in Proc. 1978 Cambridge Summer School in Set Theory.

[Ba4] J. Baumgartner, A new class of order types, *Annals Math. Logic* **9** (1976), 187-222.

[De] K.J. Devlin, Aspects of constructibility, *Lecture Notes in Mathematics*, vol. 354, Springer, 1973.

[DJ] K.J. Devlin and H. Johnsbråten, The Souslin problem, *Lecture Notes in Mathematics*, vol. 405, Springer, 1974.

[DS] K.J. Devlin and S. Shelah, A weak version of $\Diamond$ which follows from $2^{\aleph_0} < 2^{\aleph_1}$, *Israel J. Math.* **29** (1978), 239-247.

[ER] P. Erdös and R. Rado, A partition calculus in set theory, *Bull. Amer. Math. Soc.* **62** (1956), 427-489.

[GS] F. Galvin and S. Shelah, Some counterexamples in the partition calculus, *J. Comb. Theory* **15** (1973), 167-174.

[Gi1] S. Ginsburg, Some remarks on order types and decompositions of sets, *Trans. Amer. Math. Soc.* **74** (1953), 514-535.

[Gi2] S. Ginsburg, Further results on order types and decompositions of sets, *Trans. Amer. Math. Soc.* **77** (1954), 122-150.

[Gi3] S. Ginsburg, Order types and similarity transformations, *Trans. Amer. Math. Soc.* **79** (1955), 341-361.

[Je] T. Jech, *Set Theory*, Academic Press, 1978.

[Ku] G. Kurepa, Ensembles ordonnés et ramifiés, *Publ. Math. Univ. Belgrade* **4** (1935), 1-138.

[La] R. Laver, On Fraïssé's order type conjecture, *Ann. of Math.* **93** (1971), 89-111.

[Ra] F.P. Ramsey, On a problem of formal logic, *Proc. London Math. Soc.* (2) **30** (1930), 264-286.

[Ro] B. Rotman, On the comparison of order types, *Acta Math. Acad. Sci. Hung.* **19** (1968), 311-327.

[Sh1] S. Shelah, mimeographed letters to E. Wimmers, 1978.

[Sh2] S. Shelah, Classification theory for the non-elementary case I: the number of uncountable models of $\psi \in L_{\omega_1,\omega}$, to appear.

[Sh3] S. Shelah, Decomposing uncountable squares to countably many chains, *J. Comb. Theory* **21** (1976), 110-114.

[Si1] W. Sierpiński, Sur un problème de la théorie des relations, *Ann. Scuola Norm. Sup. Pisa* (2) **2** (1953), 285-287.

[Si2] W. Sierpiński, Sur les types d'ordre des ensembles linéaires, *Fund. Math.* 37 (1950), 253-264.

[Ul] S. Ulam, Zur Masstheorie in der allgemeinen Mengenlehre, *Fund. Math.* 16 (1930), 140-150.

# ON THE COFINALITY OF PARTIALLY ORDERED SETS

E.C. Milner
Department of Mathematics and Statistics
The University of Calgary
Calgary, Canada T2N 1N4

M. Pouzet
Départment de Mathématiques
Université Claude Bernard - Lyon 1
43 Boulevard du 11 Novembre 1918
69622 Villeurbanne Cédex
France

ABSTRACT

If $(P,\leq)$ is a partially ordered set (poset), then a subset $A \subseteq P$ is *cofinal* in $P$ if every element of $P$ is majorized by some element of $A$, and we define the *cofinality* of $(P,\leq)$, $\mathrm{cf}(P,\leq)$, to be the smallest cardinal number $\kappa$ such that $P$ contains a cofinal subset $A$ of cardinality $|A| = \kappa$. This notion is very familiar and has been well studied in the case of linearly ordered sets. In fact, the cofinality $\mathrm{cf}(\kappa)$ of a cardinal number $\kappa$ (an initial ordinal number) plays a basic role in ordinary cardinal arithmetic. In the case of a linearly ordered set $(P,\leq)$, two obvious features about the cofinality $\mathrm{cf}(P,\leq) = \kappa$ are that (i) every cofinal subset of $(P,\leq)$ contains a cofinal subset order isomorphic to the cardinal number $\kappa$ and (ii) the cardinal $\kappa$ is either 1 or an infinite regular cardinal (we assume $P \neq \emptyset$). For general partially ordered sets these features are lost, although in some important cases it is possible to draw similar conclusions. For example, if $\mathrm{cf}(P,\leq) = \aleph_0$ or if $(P,\leq)$ does not contain an infinite antichain, then there is a poset A such that every cofinal subset of $(P,\leq)$ contains a cofinal isomorphic copy of A. Also, if $(P,\leq)$ has no infinite antichain, the cofinality is again a regular cardinal. Of course, in general we do not have such results, but, for example, if the cofinality of a poset is a singular cardinal we show that this implies something about the structure of the cofinal subsets (they must contain large antichains).



*I. Rival (ed.), Ordered Sets, 279–298.*

The purpose of this paper is to survey some of the results obtained in recent studies of the cofinality of posets, to prove some of these and also to present some new results. We extend the notion of cofinality to binary relations, and in particular to directed graphs in which case the cofinality corresponds to the external stability number which has already been studied in the finite case.

The notion of cofinality is a fairly old one, and there are a number of results obtained from rather different viewpoints, for example from the study of the Moore-Smith convergence, but we do not mention such results here. Instead we refer the reader to the papers [6], [9] and [16] and the references contained in these.

## INTRODUCTION

If $(P,\leq)$ is a partially ordered set (poset), then a subset $A \subseteq P$ is *cofinal* in $P$ if every element of $P$ is majorized by some element of $A$, and we define the *cofinality* of $(P,\leq)$, $\mathrm{cf}(P,\leq)$, to be the smallest cardinal number $\kappa$ such that $P$ contains a cofinal subset $A$ of cardinality $|A| = \kappa$. This notion is very familiar and has been well studied in the case of linearly ordered sets. In fact, the cofinality $\mathrm{cf}(\kappa)$ of a cardinal number $\kappa$ (an initial ordinal number) plays a basic role in ordinary cardinal arithmetic. In the case of a linearly ordered set $(P,\leq)$, two obvious features about the cofinality $\mathrm{cf}(P,\leq) = \kappa$ are that (i) every cofinal subset of $(P,\leq)$ contains a cofinal subset order isomorphic to the cardinal number $\kappa$ and (ii) the cardinal $\kappa$ is either 1 or an infinite regular cardinal (we assume $P \neq \emptyset$). For general partially ordered sets these features are lost, although in some important cases it is possible to draw similar conclusions. For example, if $\mathrm{cf}(P,\leq) = \aleph_0$ or if $(P,\leq)$ does not contain an infinite antichain, then there is a poset $A$ such that every cofinal subset of $(P,\leq)$ contains a cofinal isomorphic copy of $A$. Also, if $(P,\leq)$ has no infinite antichain, the cofinality is again a regular cardinal. Of course, in general we do not have such results, but, for example, if the cofinality of a poset is a singular cardinal we show that this implies something about the structure of the cofinal subsets (they must contain large antichains).

The purpose of this paper is to survey some of the results obtained in recent studies of the cofinality of posets, to prove some of these and also to present some new results. We extend the notion of cofinality to binary relations, and in particular to directed graphs in which case the cofinality corresponds to the external stability number which has already been studied in the finite case.

The notion of cofinality is a fairly old one, and there are a number of results obtained from rather different viewpoints, for example from the study of the Moore-Smith convergence, but we do not mention such results here. Instead we refer the reader to the papers [6], [9] and [16] and the references contained in these.

## 1. NOTATION AND DEFINITIONS

We use standard set-theoretic notation. The cardinality of a set $X$ is denoted by $|X|$. An ordinal number $\alpha$ is the set of all smaller ordinals $\alpha = \{\beta : \beta < \alpha\}$, and we sometimes identify $\alpha$ with the ordered set $(\alpha, \subseteq)$ defined on $\alpha$ by the natural order. Lower case greek letters always denote ordinal numbers unless stated otherwise, and, in particular the letters $\kappa$, $\lambda$, $\mu$ will denote cardinal numbers or initial ordinal numbers. As usual $\omega_\alpha$ or $\aleph_\alpha$ denotes the $\alpha$-th term in the sequence $\omega_0, \omega_1, \omega_2, \ldots$ of infinite cardinal numbers. We usually write $\omega$ instead of $\omega_0$. The smallest cardinal number greater than $\kappa$ is its successor $\kappa^+$.

The elements $a,b$ of a partially ordered set (poset) $(P,\leq)$ are *comparable* if $a \leq b$ or $b \leq a$, they are *compatible* if $a \leq c$ and $b \leq c$ for some element $c \in P$. A *chain* is a linearly (or totally) ordered set, i.e. a poset $(P,\leq)$ in which every pair of elements is comparable, an *antichain* is a totally disordered set, i.e. a poset in which no two elements are comparable. A poset is *directed* if every pair of elements is compatible. A *strong antichain* in $(P,\leq)$ is a subset $A \subseteq P$ such that no two elements of $A$ are compatible in $P$.

The *direct sum* of two disjoint posets $(P_1,\leq_1)$, $(P_2,\leq_2)$ is the poset $(P_1 \cup P_2, \leq_1 \cup \leq_2)$. The *direct product* is $(P_1,\leq_1) \otimes (P_2,\leq) = (P,\leq)$, where $P$ is the cartesian product $P_1 \times P_2$ and $(p_1,p_2) \leq (p_1',p_2')$ iff $p_1 \leq_1 p_1'$ and $p_2 \leq_2 p_2'$. In particular, note that in the direct product $\alpha \otimes \beta$ of two well-ordered chains $\alpha$ and $\beta$, the elements $(\alpha_1,\beta_1)$ and $(\alpha_2,\beta_2)$ are incomparable iff $\alpha_1 < \alpha_2$ and $\beta_1 > \beta_2$ or $\alpha_1 > \alpha_2$ and $\beta_1 < \beta_2$. Since there is no infinite descending sequence of ordinals, it follows that $\alpha \otimes \beta$ does not contain any infinite antichain.

For any subset $A$ of the poset $(P,\leq)$ we write $\hat{A} = \{x \in P : (\exists\, a \in A)(x \leq a)\}$, $\check{A} = \{x \in P : (\exists\, a \in A)(a \leq x)\}$. Also, in the special case when $A = \{a\}$ is a singleton we write $\hat{a}$ (resp. $\check{a}$) for $\hat{A}$ ($\check{A}$). $A$ is an *initial segment* of $(P,\leq)$ if $\hat{A} = A$ and a *final segment* is defined dually. We denote the set of all initial segments of $(P,\leq)$ by $I(P)$. A subset $X$ is *cofinal* in the poset $(P,\leq)$ if every element of $P$ is majorized by some element of $X$, i.e. if $\hat{X} = P$. Dually, $X$ is *coinitial* if $\check{X} = P$. The notion of cofinality is clearly transitive in the sense that if $A$ is cofinal in $(B,\leq)$ and $B$ is cofinal in $(P,\leq)$, then $A$ is also cofinal in $(P,\leq)$. Note that if $A \subseteq P$ we write $(A,\leq)$ to denote the induced suborder on $A$ instead of the more precise notation $(A,\leq')$ where $\leq' = \leq \cap (A \times A)$.

A poset $(P,\leq)$ is *well-founded* if it contains no infinite descending chain $x_0 > x_1 > \ldots$ . In this case there is a naturally defined height function $h : P \to O_n$ , the class of ordinal numbers, given inductively by the formula

$$h(x) = \sup\{h(y) + 1 : y < x\} \quad (\forall\, x \in P) .$$

The height function $h$ partitions $P$ into disjoint *levels* $P_\alpha = \{x \in P : h(x) = \alpha\}$. If $\beta < \alpha$ and $x \in P_\alpha$ , then $x$ is above some element $y \in P_\beta$ (i.e. $y < x$). The *height* of the well-founded poset $(P,\leq)$ is the least ordinal $\alpha$ such that $P_\alpha = \emptyset$. A well-founded chain is a *well-order*. A *tree* is a poset $(T,\leq)$ in which $\hat{x}$ is well-ordered by $\leq$ for each $x \in T$.

An $\omega$-*tree* is a tree $(T,\leq)$ of height $\omega$ with a minimal element (i.e. a rooted tree) and such that every point has $\omega$ successors at the next highest level. (It is order isomorphic to the set of finite sequences in $\omega$ ordered by extension.) An $\omega$-*forest* is the direct sum of $\omega$ disjoint $\omega$-trees.

It is easy to see that a poset $(P,\leq)$ contains a well-founded cofinal subset $A$ having cardinality $|A| = \mathrm{cf}(P,\leq)$. In fact, *$A$ may be chosen so that it has height at most* $\mathrm{cf}(P,\leq)$. The following lemma says a little more than this.

LEMMA 1.1. *If* $\mathrm{cf}(P,\leq) = \kappa$, *then $P$ contains a cofinal subset* $\{x_\alpha : \alpha < \kappa\}$ *such that* $x_\beta \not\leq x_\alpha$ $(\alpha < \beta < \kappa)$.

Proof. Let $X = \{p_\nu : \nu < \kappa\}$ be any cofinal subset of $(P,\leq)$ of cardinality $\kappa$, and let $Y = \{p_\nu : \nu < \kappa \text{ and } (\forall\, \sigma < \nu)(p_\nu \not\leq p_\sigma)\}$. Then $Y$ is cofinal in $P$ and so $|Y| = \kappa$. Thus $Y = \{p_{\nu_\alpha} : \alpha < \kappa\}$ where $\nu_0 < \nu_1 < \ldots$ and the lemma holds with $x_\alpha = p_{\nu_\alpha}$ $(\alpha < \kappa)$. □ (It is easy to see that the set $Y = \{x_\alpha : \alpha < \kappa\}$ has height at most $\kappa$ since, by induction, the height of the element $x_\beta$ in $(Y,\leq)$ is at most $\beta$.)

If $\mu$ is a cardinal number we write $(P,\leq) \in \mathcal{D}(\mu)$ if $(P,\leq)$ does not contain any antichain of cardinality $\mu$, and $(P,\leq) \in \mathcal{D}'(\mu)$ if there is no strong antichain of cardinality $\mu$. Since a strong antichain is an antichain $\mathcal{D}(\mu) \Rightarrow \mathcal{D}'(\mu)$. We also write $(P,\leq) \in C(\mu)$ if $P$ is a union of fewer than $\mu$ (disjoint) chains and $(P,\leq) \in C'(\mu)$ if $(P,\leq)$ is the union of fewer than $\mu$ (disjoint) directed sets. Clearly $C(\mu) \Rightarrow C'(\mu)$. Note that if $P$ is the union of $\lambda$ directed sets $D_\alpha$ $(\alpha < \lambda)$ then it is also the union of

the $\lambda$ pairwise disjoint directed sets $D'_\alpha$ $(\alpha < \lambda)$, where
$D'_\alpha = \hat{D}_\alpha \setminus \bigcup_{\beta<\alpha} \hat{D}_\beta$ .

A poset $(P,\leq)$ is *partially well-ordered* (p.w.o.) if it is well-founded and has no infinite antichain (i.e. $(P,\leq) \in \mathcal{D}(\aleph_0)$). It is well-known [8] that this is equivalent to the following assertion: whenever $x_0,x_1,x_2,\ldots$ is a sequence of elements in $P$, then there are $i < j < \omega$ such that $x_i \leq x_j$ .

## 2. THE MAIN RESULTS

One of the important, basic notions in the theory of linearly ordered sets (or chains) is that of cofinality. We extend this to posets in the most natural way by defining the *cofinality* of the poset $(P,\leq)$ to be the least cardinal $\kappa$ such that $P$ contains a cofinal subset of power $\kappa$. We denote this by $\mathrm{cf}(P,\leq)$. Thus

$$\mathrm{cf}(P,\leq) = \min\{|A| : \hat{A} = P\} .$$

By transitivity it is easy to see that any cofinal subset $X$ of $P$ has the same cofinality as $P$, i.e. $\mathrm{cf}(X,\leq) = \mathrm{cf}(P,\leq)$.

Since we identify the ordinal $\alpha$ with the chain $(\alpha,\subseteq)$ of type $\alpha$, the above definition in this case agrees with the usual meaning of $\mathrm{cf}(\alpha)$. Recall that a cardinal number $\kappa$ is *regular* if $\mathrm{cf}(\kappa) = \kappa$ and *singular* if $\mathrm{cf}(\kappa) < \kappa$ ($\aleph_\omega = \aleph_0 + \aleph_1 + \ldots$ is the smallest singular cardinal). In the case when $(P,\leq)$ is a chain it follows from Lemma 1.1 that any cofinal subset of $P$ contains a (well-ordered) cofinal chain of type $\kappa = \mathrm{cf}(P,\leq)$. An immediate consequence of this is that, in this case, $\mathrm{cf}(\kappa) = \kappa$, i.e. the cofinality is always regular.

For general partial orders the situation is not quite so simple. First of all, it is obvious that the cofinality need not be regular (consider an antichain of size $\kappa$, where $\kappa$ is any singular cardinal). However, as we shall see, if the cofinality of $(P,\leq)$ is singular, then this does imply something about the structure of $P$, namely: *every cofinal subset contains 'large' antichains* (Theorem 5). Secondly, it is no longer true that the cofinal subsets of a poset $(P,\leq)$ contain a cofinal isomorphic copy of some fixed poset.

For posets $P,A$ let $\mathrm{CF}(P,A)$ denote the following assertion: *every cofinal subset of $P$ contains a cofinal isomorphic copy of* $A$. Thus $\mathrm{CF}(P,\kappa)$ holds if $P$ is a chain and $\kappa = \mathrm{cf}(P)$. The next simplest case in which the situation is similar is when $P$ is a directed poset having countable cofinality $cf(P) = \kappa \leq \omega$. It is easy to see that in this case we again have

(2.1) $CF(P,\kappa)$.

For, if $\kappa < \omega$, then $\kappa = 1$ since $P$ is directed and $P$ has a greatest element. Suppose $\kappa = \omega$ and $X$ is any cofinal subset of $P$. Then $X$ contains a denumerable cofinal subset $Y = \{y_n : n < \omega\}$. Put $z_0 = y_0$ and choose $z_n \in Y$ inductively so that $z_{n+1} > y_n, z_{n+1} > z_n$. Then $\{z_n : n < \omega\}$ is a cofinal $\omega$-chain in $Y$ which is also cofinal in $X$.

With slightly more effort we can prove a more general result (see §3) that, for any poset $P$ with countable cofinality, there is a poset $A$ such that $CF(P,A)$ holds.

THEOREM 1. *If* $cf(P,\leq) \leq \omega$, *then* $CF(P,A)$ *holds where* $A$ *is either (i) a countable antichain, or (ii) a direct sum of countably many* $\omega$*-chains or (iii) an* $\omega$*-forest or (iv) the direct sum of two or three posets satisfying (i), (ii) or (iii).*

As we have already remarked, in the general case when $cf(P,\leq) > \aleph_0$ there is no poset $A$ such that $CF(P,A)$ holds. For example ([11], [14]), consider the Sierpinski order $\prec$ defined on the set of real numbers $\mathbb{R}$ as follows. Let $<$ be the natural ordering on $\mathbb{R}$ and let $<<$ be any well-ordering of $\mathbb{R}$ and set $x \prec y$ if $x < y$ and $x << y$. Then $(\mathbb{R},\preccurlyeq)$ is a directed poset having no uncountable antichains and it can be shown that there is a family $F$ of subsets of $\mathbb{R}$ such that: $|F| > 2^{\aleph_0}$, *each number of* $F$ *is cofinal in* $(\mathbb{R},\preccurlyeq)$, *and, moreover, whenever* $F_1, F_2 \in F$ *and* $F_1 \neq F_2$ *then* $(F_1,\preccurlyeq)$ *and* $(F_2,\preccurlyeq)$ *do not contain order isomorphic cofinal subsets.* In other words, $(\mathbb{R},\preccurlyeq)$ contains more than $2^{\aleph_0}$ cofinal subsets which themselves are cofinally very different indeed.

It should be noted that in the above example, although $(\mathbb{R},\preccurlyeq)$ does not contain uncountable antichains, i.e. $(\mathbb{R},\preccurlyeq) \in \mathcal{D}(\aleph_1)$, there are infinite antichains. In the case when a poset has only finite antichains the situation is much more pleasant as shown by the following positive structure theorem of M. Pouzet.

THEOREM 2 [14]. *Let* $(P,\leq) \in \mathcal{D}(\aleph_0)$. *Then there is a poset* $A$ *such that* $CF(P,A)$ *holds. In fact,* $A$ *is a direct sum of a finite number of components where each component is either (i) a single point or (ii) the direct product of a finite number of regular distinct cardinal numbers.*

Theorem 2 can be stated in a more precise way in two parts:

(2.2) *If* $(P,\leq) \in \mathcal{D}(\aleph_0)$, *then* $P$ *is a finite union of directed sets.*

(2.3) *If $(P,\leq) \in \mathcal{D}(\aleph_0)$ is directed, then either $P$ has a greatest element or $P$ contains a cofinal subset which is order isomorphic to a direct product of the form $\kappa_1 \otimes \kappa_2 \otimes \ldots \otimes \kappa_n$ where $n$ is finite and $\kappa_1,\ldots,\kappa_n$ are distinct regular infinite cardinal numbers.*

The best known decomposition theorem for ordered sets is Dilworth's theorem [2]:

(2.4) *If $k < \aleph_0$ and $(P,\leq) \in \mathcal{D}(k)$, then $(P,\leq) \in C(k)$.*

It was observed by Perles [13] (and rediscovered by many others) that the finiteness of $k$ in (2.4) is essential. The theorem fails very badly for infinite $k$ since for any cardinal $\kappa$ there is a poset $(P,\leq)$ (the direct product $\kappa \otimes \kappa$ is an example) such that

(2.5) $(P,\leq) \in \mathcal{D}(\aleph_0)$ *and* $(P,\leq) \notin C(\kappa)$.

An easier result than (2.4) is the corresponding result for partitioning a poset into directed sets:

(2.6) *If $k < \aleph_0$ and $(P,\leq) \in \mathcal{D}'(k)$, then $(P,\leq) \in C'(k)$.*

To see this, simply observe that, if $A$ is a strong antichain of maximum cardinality, then the sets $D_a = \check{a}$ $(a \in A)$ are directed. Therefore, the sets $\hat{D}_a$ are also directed and $P = \bigcup \hat{D}_a$. An old result of Erdös and Tarski [3] is that, if $\delta = \delta(P,\leq)$ is the smallest cardinal number such that $(P,\leq) \in \mathcal{D}'(\delta)$, then $\delta$ is not $\aleph_0$ (they also showed $\delta$ is not a singular cardinal). Consequently, if $(P,\leq) \in \mathcal{D}'(\aleph_0)$, then $(P,\leq) \in \mathcal{D}'(k)$ for some $k < \aleph_0$, and so by (2.6) we have

(2.7) $(P,\leq) \in \mathcal{D}'(\aleph_0) \Rightarrow (P,\leq) \in C'(\aleph_0)$.

This is stronger than (1.2) since $\mathcal{D}(\aleph_0) \Rightarrow \mathcal{D}'(\aleph_0)$.

We remark that the Erdös-Tarski theorem is not really needed to prove (2.2). This is almost an immediate consequence of the following well-known fact [8]:

LEMMA 2.1. *$(P,\leq)$ is p.w.o. iff $(I(P),\subseteq)$ is well-founded.*

Proof. If $(J_n : n < \omega)$ is a strictly decreasing sequence in $(I(P),\subseteq)$, then consider $(x_n : n < \omega)$, where $x_n \in J_n \setminus J_{n+1}$. □

Now, if $(P,\leq) \in \mathcal{D}(\aleph_0)$ and $Q$ is a well-founded, cofinal subset of $P$, then $Q$ is p.w.o. If $Q$ is not a finite union of directed sets, then by the well-foundedness of $\mathcal{I}(Q)$, there is a minimal $I \in \mathcal{I}(Q)$ which is also not a finite union of directed sets. But if $x,y \in I$ are incompatible (in $I$), then $I$ is the union of the two proper initial segments $I \cap \hat{X}$, $I \cap \hat{Y}$, where $X = \check{x}$ and $Y = I \setminus X$, and this contradicts the minimality of $I$. Thus $Q$ is a finite union of directed sets and hence so also is $P$. This proves (2.2). □

Like (2.4) the result (2.7) does not extend to higher cardinals. For any infinite cardinal $\kappa$ there is a poset satisfying

(2.8) $(P,\leq) \in \mathcal{D}'(\aleph_1)$ *and* $(P,\leq) \notin C'(\kappa)$.

An example of such a poset (due to J. Baumgartner) is given in [11], alternatively observe that if $\lambda > 2^{\kappa}$, then the free boolean algebra, $B$, with $\lambda$ generators has no uncountable strong antichain and the number of filters needed to cover $B \setminus \{0\}$ is the density of its Stone space which is greater than $\kappa$.

Although (2.7) fails to generalize to higher cardinals we do have the following generalization of (2.2) due to Milner and Prikry [11].

THEOREM 3. *For any cardinal number* $\kappa$ *there is a cardinal* $\theta(\kappa)\left(\leq\left(\sum_{\nu<\kappa} \kappa^{\nu}\right)^{+}\right)$ *such that*

(2.9) $(P,\leq) \in \mathcal{D}(\kappa) \Rightarrow (P,\leq) \in C'(\theta(\kappa))$.

The best value for $\theta(\kappa)$ is not known, but more precise results can be given in some cases. For example, if $\kappa$ is weakly compact $\theta(\kappa) = \kappa$. Also, if $(P,\leq) \in \mathcal{D}(\kappa)$ is a tree and $\kappa$ is regular, $P$ can be covered by fewer than $\kappa^+$ chains (and this result is best possible if there is a $\kappa$-Souslin tree). Also, if $(P,\leq) \in \mathcal{D}(\kappa)$ is a tree and $\kappa$ is singular, then $(P,\leq) \in \mathcal{D}(\kappa_1)$ for some $\kappa_1 < \kappa$ by the Erdös-Tarski theorem, and so $P$ is the union of fewer than $\kappa$ chains. It would be interesting to know if $\theta(\kappa) = \kappa$ is consistent for every $\kappa$.

It should be remarked in this connection that an immediate corollary of Theorem 2 is that if $(P,\leq) \in \mathcal{D}(\aleph_0)$ and $\mathrm{cf}(P,\leq) = \kappa \geq \aleph_0$, then $P$ contains a cofinal subset which is a union of fewer than $\kappa$ chains. This result had previously been noted by R. Laver and A. Hajnal [10].

In view of Theorem 3 it is of interest to study the cofinal subsets of directed posets. One result, which is in the same style as (2.3), is the following theorem of Milner and Prikry [11]. This is formulated in terms of the partition symbol (2.9) which, by definition, means that the following statement is true: *for any map* $f : [\kappa]^2 = \{\{\alpha,\beta\} : \alpha < \beta < \kappa\} \to 2$, EITHER *there is a subset* $X \subseteq \kappa$ *such that* $|X| = \kappa$ *and* $f(\{\alpha,\beta\}) = 0$ *for all* $\alpha, \beta \in X$ $(\alpha \neq \beta)$, OR *there is a subset* $Y \subseteq \kappa$ *such that* $|Y| = \mu$ *and* $f(\{\alpha,\beta\}) = 1$ *for all* $\alpha, \beta \in Y$ $(\alpha \neq \beta)$. For a full discussion of such relations see [2].

THEOREM 4 [11]. *If* $(P,\leq) \in \mathcal{D}(\mu)$ *is directed and* $\mathrm{cf}(P,\leq) = \kappa$ *and if*

(2.9) $$\kappa \to (\kappa,\mu)^2 ,$$

*then* $P$ *contains a cofinal subset which is order isomorphic to a direct product of the form* $\kappa \otimes A$, *where* $A$ *is a poset having cardinality strictly less than* $\kappa$.

By a well-known theorem of Erdös, Dushnik and Miller [5], the relation (2.9) holds for any $\kappa \geq \aleph_0$ when $\mu = \aleph_0$, i.e. $\kappa \to (\kappa,\aleph_0)^2$ $(\forall\ \kappa \geq \aleph_0)$. Thus a finite number of applications of Theorem 4 yields (2.3). In §4 we outline a proof of Theorem 4 for this special case when $\mu = \aleph_0$.

Another immediate corollary of Pouzet's theorem (Theorem 2) is the following:

(2.10) *if* $(P,\leq) \in \mathcal{D}(\aleph_0)$ *and if* $\mathrm{cf}(P,\leq) = \kappa \geq \aleph_0$, *then* $\kappa$ *is a regular cardinal number.*

The fact that the cofinality should again be a regular cardinal, as in the case of linearly ordered sets, seems to be rather a basic result, and in [12] Milner and Sauer asked for a direct proof of this without using the heavy artillery of Pouzet's structure theorem. Such proofs have since been given by Milner and Prikry [11] and by Hajnal and Sauer [7]. Here we shall give a different proof in §4 of this fact. In fact, we shall prove the following result which is a slight strengthening of the result established in [11].

THEOREM 5. *Let* $\mathrm{cf}(P,\leq) = \kappa$ *be a singular cardinal number. If* $\mu$ *is minimal such that* $\kappa^\mu > \kappa$, *i.e. if*

$$\kappa^\rho \leq \kappa \quad (\forall\ \rho < \mu),$$

*then* $(P,\leq)$ *contains an antichain of cardinality* $\mu$.

Note that, since $\kappa^n = \kappa$ for $n < \omega$, Theorem 5 easily implies (2.10). Of course, $\kappa^{\mathrm{cf}(\kappa)} > \kappa$ and so the number $\mu$ in the theorem is at most $\mathrm{cf}(\kappa)$. In this sense the result is best possible since it is easy to give an example of a poset $(P,\leq)$ having singular cofinality $\kappa$ which does not contain any antichain of size $(\mathrm{cf}(\kappa))^+$. (Consider $\mathrm{cf}(\kappa)$ disconnected chains with increasing cardinality and limit $\kappa$.) If we assume $\kappa^\rho = \kappa$ for $\rho < \mathrm{cf}(\kappa)$, then $\mathrm{cf}(P,\leq) = \kappa > \mathrm{cf}(\kappa)$ implies $(P,\leq) \notin \mathcal{D}(\mathrm{cf}(\kappa))$. However, we do not know if this assumption about cardinal exponentiation is really needed here.

After proving Theorem 5 we realized that, in fact, the proof has very little to do with order, and this led us to extend the notion of cofinality to any relational structure. For simplicity we consider a structure $\langle A,R\rangle$ where $R$ is a binary relation on $A$. Call a subset $X \subseteq A$ *cofinal* in $\langle A,R\rangle$ if

$$(\forall\, a \in A)(\exists\, x \in X)(aRx \text{ or } a = x)\ .$$

Then, as before, we may define the cofinality

$$\mathrm{cf}(\langle A,R\rangle) = \min\{|X| : X \text{ is cofinal in } \langle A,R\rangle\}\ .$$

In the case when $\langle A,R\rangle$ is a directed graph, the cofinality corresponds to the external stability number of a graph, a concept that has already been studied in the finite case (see [1]). The notion that corresponds to an antichain in a poset is that of an *independent* set. A set $I \subseteq A$ is *R-independent* if $aRb$ fails to hold for all pairs of distinct elements $a,b \in I$. The main difference between cofinality for an arbitrary relational structure and the case for posets is that we lose transitivity, i.e. if $X$ is cofinal in $Y$ and $Z$ is cofinal in $X$ it does not follow that $Z$ is cofinal in $Y$. We will prove the following result which easily implies Theorem 5.

THEOREM 6. *Let $R$ be a binary relation on $A$ and assume that $|A| = \mathrm{cf}(\langle A,R\rangle) = \kappa > \mathrm{cf}(\kappa)$. Then the set $S = \{X \subseteq A : X$ is $R$-independent$\}$ has cardinality $|S| > \kappa$. In particular, if $\mu$ is the least cardinal number such that $\kappa^\mu > \kappa$, then $A$ contains more than $\kappa$ independent subsets of cardinality $\geq \mu$.*

To see that Theorem 6 implies Theorem 5, apply it to $(Q,\leq)$, where $Q$ is a cofinal subset of $(P,\leq)$ having cardinality $\kappa$. By Theorem 6, $(Q,\leq)$ contains more than $\kappa$ independent sets (i.e. antichains) and these cannot all have cardinality less than $\mu$ by the hypothesis on $\mu$.

The hypothesis $|A| = \text{cf}(\langle A,R\rangle)$ is essential in Theorem 6. For, if $\kappa,\lambda$ are any infinite cardinal numbers with $\kappa > \lambda$, then, as we show in §6, there is a tournament $T(\kappa,\lambda)$ which has cardinality $\kappa$ and cofinality $\lambda$. Of course, a tournament does not contain an independent set of cardinality 2 and this shows that the hypothesis $|A| = \text{cf}(\langle A,R\rangle)$ is needed for Theorem 6.

## 3. PROOF OF THEOREM 1

The proof is in two parts.

1. Let $U = \{x \in P : \check{x} \text{ is directed}\}$, $V = P \setminus \hat{U}$. Then $U,V$ are unrelated final segments of $P$ and a subset $X \subseteq P$ is cofinal in $P$ iff $X \cap U$ and $X \cap V$ are respectively cofinal in $U$ and $V$. Thus a set $X$ is cofinal in $P$ iff it contains a cofinal subset which is the direct sum of cofinal subsets of $U,V$. Thus in order to prove the theorem we need only verify that (if they are non-empty) $\text{CF}((U,\leq),A_1)$ and $\text{CF}((V,\leq),A_2)$ hold, where $A_1$ is of the form (i) a countable antichain or (ii) a direct sum of a countable number of $\omega$-chains or (iii) a direct sum of (i) and (ii), and where $A_2$ is an $\omega$-forest.

If $S \subseteq U$ is a maximal strong antichain of $(P,\leq)$ then the disjoint union of the sets $\check{x}$ $(x \in S)$ is cofinal in $U$ and, since each $\check{x}$ is directed, it follows from (2.1) that there is a suitable $A_1$ so that $\text{CF}((U,\leq),A_1)$.

2. Suppose $V \neq \emptyset$. Then obviously $V$ is infinite and $\text{cf}(V,\leq) = \omega$. We first show that above any element $v \in V$ there is an infinite strong antichain $S(v)$ so that the union of $\check{x}$ $(x \in S(v))$ is cofinal in $\check{v}$. Indeed, let $v \in V$. Then there is an incompatible pair $u_0,v_0 \in \check{v}$. More generally, if $v_n \in \check{v}$ has been defined, there is an incompatible pair $u_{n+1},v_{n+1} \in \check{v}_n$. This defines $u_n,v_n \in \check{v}$ for all $n < \omega$ so that $v_n < u_{n+1}$, $v_n < v_{n+1}$ and $u_n,v_n$ are incompatible. Then $\{u_n : n < \omega\}$ is a strong antichain above $v$. Now let $S(v)$ be a maximal strong antichain containing $\{u_n : n < \omega\}$. The same applies to any cofinal subset of $V$, i.e. if $X$ is cofinal in $V$, then above any element $x \in X$ there is an infinite strong antichain $S(v,X)$ so that $\check{x} = \bigcup\{\check{y} : y \in S(v,X)\}$. Now it is easy to see that any cofinal subset $X$ of $V$ contains a cofinal $\omega$-forest. Indeed let $X$ be a maximal infinite strong antichain in $X$ and define $X_{n+1} = \bigcup\{S(x,X) : x \in X_n\}$ $(n < \omega)$. Then $\bigcup X_n$ is an $\omega$-forest which is cofinal in $X$. □

## 4. PROOFS OF THEOREM 5 AND (2.3)

We first prove Theorem 5.

By Lemma 1.1 we may assume that $(P,\leq)$ is well-founded and that $|P| = \kappa$ (replace $P$ by a cofinal subset). Suppose that $P$ does not contain an antichain of size $\mu$. We shall obtain a contradiction.

Since an initial segment $I \in \mathcal{I}(P)$ is uniquely determined by the antichain $A(I)$, consisting of all the minimal elements of $P \setminus I$, it follows from the hypothesis on $\mu$, that

$$|\mathcal{I}(P)| \leq \sum_{\nu<\mu} \kappa^{\nu} = \kappa \,. \tag{4.1}$$

Let $\mathcal{I}^* = \{I \in \mathcal{I} : \mathrm{cf}(I,\leq) \leq \mathrm{cf}(\kappa)\}$. By Lemma 1.1 there is a cofinal set $\{I_\rho : \rho < \lambda\}$ in $(\mathcal{I}^*,\subseteq)$ so that $I_\beta \not\subseteq I_\alpha$ for $\alpha < \beta < \lambda$, where $\lambda \leq \kappa$. Clearly $\bigcup_{\rho<\lambda} I_\rho = P$, and since $\mathrm{cf}\left(\bigcup_{\rho<\lambda} I_\rho\right) \leq \lambda \cdot \mathrm{cf}(\kappa)$, it follows that $\lambda = \kappa$.

Let $\kappa_\xi$ $(\xi < \mathrm{cf}(\kappa))$ be an increasing sequence cofinal in $\kappa$, and let $I = \bigcup_{\xi<\mathrm{cf}(\kappa)} I_{\kappa_\xi}$. Then $\mathrm{cf}(I) \leq \mathrm{cf}(\kappa)$ and so $I \in \mathcal{I}^*$. Thus $I \subseteq I_\rho$ for some $\rho < \kappa$ and this implies that $I_{\kappa_\xi} \subseteq I_\rho$ for some $\kappa_\xi > \rho$, which contradicts the fact that the $I_\rho$ $(\rho < \kappa)$ are non-decreasing. □

We now give a proof of (2.3), the structural part of Pouzet's theorem. All the ideas used here are to be found in [11] and [14] and a similar argument may be used to prove the more general result of Theorem 4. We prove this only for the case $\mu = \aleph_0$ since this avoids some slightly technical set theoretic discussion about partition relations. In view of Theorem 5 we may assume that: *if* $(P,\leq) \in \mathcal{D}(\aleph_0)$ *and* $\mathrm{cf}(P,\leq) = \kappa \geq \aleph_0$, *then* $\kappa$ *is a regular cardinal.*

The following simple lemma is needed.

LEMMA 4.1. *If* $\mathrm{cf}(P,\leq) = \kappa \geq \aleph_0$ *and if* $(P,\leq) \in \mathcal{D}(\aleph_0)$, *then* $P$ *contains an unbounded* $\kappa$*-chain.*

Proof. By Lemma 1.1, $(P,\leq)$ contains a cofinal subset $Q$ such that $|Q| = \kappa$, $Q$ is well-founded and $Q$ has height at most $\kappa$. Since the levels of $Q$ are antichains in $(P,\leq)$ it follows that $(Q,\leq)$ has height exactly $\kappa$. Since $\kappa \to (\kappa,\omega)^2$ and there is no infinite antichain, it follows that $(Q,\leq)$ contains a $\kappa$-chain. But any $\kappa$-chain in $Q$ is unbounded in $(Q,\leq)$ and hence unbounded in $(P,\leq)$. □

We now prove (2.3).

Since $(P,\leq)$ is directed and does not have a largest element, we may assume that $\mathrm{cf}(P,\leq) = \kappa \geq \aleph_0$. We may replace $(P,\leq)$ by a cofinal subset and assume that $|P| = \kappa$, $(P,\leq)$ is well-founded and has height $\kappa$, so that any $\kappa$-chain in $(P,\leq)$ is unbounded in $(P,\leq)$.

We establish a few facts about the poset $(K,\subseteq)$, where $K = \{\hat{C} : C \text{ is a } \kappa\text{-chain in } (P,\leq)\}$. For each $J \in K$, let $C(J)$ denote a $\kappa$-chain such that $J = \hat{C}(J)$. The argument leading to (4.1) applies here also (in this case $\mu = \aleph_0$) and so

(4.2) $$|K| \leq |I(P)| \leq \kappa .$$

We now show that

(4.3) $(K,\subseteq)$ *does not contain an unbounded* $\kappa$*-chain.*

Indeed, let $\langle J_\nu : \nu < \kappa\rangle$ be an increasing $\kappa$-chain in $(K,\subseteq)$. Let $C(J_\nu) = \{x_{\nu\rho} : \rho < \kappa\}$, where $x_{\nu 0} < x_{\nu 1} < \ldots$ . Since $\kappa$ is regular and $x_{\rho\sigma} \in J_\nu = \hat{C}(J_\nu)$ $(\rho,\sigma \leq \nu)$, it follows that there is $x_\nu \in C(J_\nu)$ such that $x_{\rho\sigma} \leq x_\nu$ for all $\rho,\sigma \leq \nu$. The set $\{x_\nu : \nu < \kappa\}$ contains a $\kappa$-chain $C$ (since $\kappa \to (\kappa,\omega)^2$). Now for any $\rho,\sigma < \kappa$ there is some $x_\nu \in C$ with $\nu > \rho,\sigma$ and so $x_{\rho\sigma} \in \hat{C}$. Thus $J_\rho \subseteq \hat{C} \in K$ and this proves (4.3).

Next we observe that

(4.4) *if* $\kappa > \aleph_0$, *then* $(K,\subseteq) \in \mathcal{D}(\aleph_0)$.

For suppose on the contrary that $\{J_n : n < \omega\}$ is an infinite antichain in $(K,\subseteq)$. Choose $x_{rs} \in J_r \setminus J_s$ for $r \neq s$. By assumption $\kappa > \aleph_0$ and since $\kappa$ is a regular cardinal, there is $x_r \in C(J_r)$ such that $x_r > x_{rs}$ for all $s < \omega$. Since $(P,\leq)$ has no infinite antichain, there are distinct $r,r' < \omega$ such that $x_r \leq x_{r'}$. But this implies that $x_{rr'} \leq x_{r'}$ and so $x_{rr'} \in J_{r'}$, a contradiction.

We now prove that

(4.5) $$\mathrm{cf}(K,\subseteq) < \kappa .$$

Clearly, the cofinality of $(K,\leq)$ is at most $\kappa$ by (4.2). We will assume that $\mathrm{cf}(K,\subseteq) = \kappa$ and obtain a contradiction. If $\kappa > \aleph_0$, then $(K,\subseteq) \in \mathcal{D}(\aleph_0)$ by (4.4) and so contains an unbounded $\kappa$-chain by Lemma 4.1, and this contradicts (4.3). Now suppose $\kappa = \aleph_0$. In this case, by Lemma 1.1 there is a cofinal subset $K' = \{J_n : n < \omega\}$ of $K$ so that

$$(4.6) \qquad J_m \nsubseteq J_n \quad (n < m < \omega) .$$

Now pick $x_m \in J_m \setminus \bigcup_{n<\omega} J_n (m < \omega)$. By Ramsey's theorem [15], since $(P,\leq)$ contains no infinite antichain, there is a chain $C \subseteq \{x_n : n < \omega\}$. Since $K'$ is cofinal in $K$ we must have $C \subseteq J_n$ for some $n < \omega$ and this is again a contradiction. Therefore, (4.5) holds.

If $A$ is any cofinal subset of $(K,\subseteq)$, then

$$(4.7) \qquad \bigcup A = P .$$

For consider any element $x \in P$ and any $\kappa$-chain $\{y_\rho : \rho < \kappa\}$ in $(P,\leq)$. Since $(P,\leq)$ is directed, there is $z_\rho \in P$ such that $x \leq z_\rho$ and $y_\rho \leq z_\rho$. Since $(P,\leq) \in \mathcal{D}(\aleph_0)$ and $\kappa \to (\kappa,\aleph_0)^2$, the set $\{z_\rho : \rho < \kappa\}$ contains a $\kappa$-chain $C \in K$. Since $A$ is cofinal in $(K,\subseteq)$, there is $J \in A$ such that $x \in \hat{C} \subseteq J$. This proves (4.7).

By Lemma 1.1 there is a cofinal subset $A = \{J_\nu : \nu < \lambda\}$ of $(K,\subseteq)$ such that $|A| = \lambda = \mathrm{cf}(K,\subseteq) < \kappa$ and

$$(4.8) \qquad J_\beta \nsubseteq J_\alpha \quad (\alpha < \beta < \lambda).$$

We will show that there is an order isomorphism

$$f : \kappa \otimes (A,\subseteq) \to (P,\leq)$$

such that the image under $f$ is a cofinal subset of $(P,\leq)$. The result (1.3) follows easily from this since we may apply the same argument to the partially ordered set $(A,\subseteq)$ and, since $|A| < \kappa = \mathrm{cf}(P,\leq)$, the process terminates after a finite number of steps.

We shall actually construct the isomorphism $f$ so that, for each $J \in A$, the set $S_J = \{f(\nu,J) : \nu < \kappa\}$ is a cofinal subset of $C(J)$. This will ensure that $\hat{S}_J = J$ and hence that the image under $f$, the set $S = \bigcup_{J\in A} S_J$, is cofinal in $(P,\leq)$ since

$$\hat{S} = \bigcup \hat{S}_J = \bigcup A = P$$

by (4.7).

Since $|A| = \lambda < \kappa = \mathrm{cf}(\kappa)$, we can choose $x_J \in C(J)$ for $J \in A$ so that $x_J \nleq x'$ whenever $x' \in J' \in A$ and $J \nsubseteq J'$. Let $C_1(J) = \{x \in C(J) : x_J \leq x\}$. Then $C_1(J)$ is a final segment of $C(J)$ and we have

$x \nleq x'$ *whenever* $J, J' \in A$, $J \nsubseteq J'$, $x \in C_1(J)$ *and* $x' \in C_1(J')$ .

We define $f$ by transfinite induction as follows. Let $(\alpha,\beta) \in \kappa \times \lambda$ and suppose $f(\nu, I_\rho) \in C_1(J_\rho)$ has already been defined for all $(\nu,\rho) \in L(\alpha,\beta) = \{(\nu,\rho) : (\nu < \alpha)$ or $(\nu = \alpha$ and $\rho < \beta)\}$. Since $C_1(J_\beta)$ is an unbounded $\kappa$-chain in $(P,\leq)$ and $|L(\alpha,\beta)| < \kappa = \mathrm{cf}(\kappa)$, we can choose $z(\alpha,\beta) \in C_1(J_\beta)$ such that

(4.9) $\qquad z(\alpha,\beta) \nleq f(\nu,J_\rho) \qquad (\forall\ (\nu,\rho) \in L(\alpha,\beta))$ .

Also, we can choose $f(\alpha,J_\beta) \in C_1(J_\beta)$ so that $f(\alpha,J_\beta) \geq z(\alpha,\beta)$ and

(4.10) $\quad f(\alpha,J_\beta) > J(\nu,J_\rho) \quad (\forall(\nu,\rho) \in L(\alpha,\beta)$ *such that* $J_\rho \subseteq J_\beta)$ .

This completes the definition of $f$ and it is clear from (4.10) that $S_J = \{f(\nu,J) : \nu < \kappa\}$ is a cofinal subset of $C_1(J)$ for $J \in A$. All that remains is to verify that $f$ is an isomorphism.

Suppose $(\nu,J_\rho) < (\alpha,J_\beta)$ in $\kappa \otimes A$. Then $\nu \leq \alpha$ and $J_\rho \subseteq J_\beta$ and so $\rho \leq \beta$ by (4.8). Thus $(\nu,\rho) \in L(\alpha,\beta)$ and so $f(\nu,J_\rho) < f(\alpha,J_\beta)$ by (4.10). Conversely, suppose $f(\nu,J_\rho) < f(\alpha,J_\beta)$ for some $(\nu,\rho),(\alpha,\beta) \in \kappa \times \lambda$ . Since $f(\nu,J_\rho) \in C_1(J_\rho)$, it follows that $x_{J_\rho} \leq F(\alpha,J_\beta) \in C_1(J_\beta)$ and hence $x_{J_\rho} \in J_\beta$ . It follows, therefore, from the definition of $x_{J_\rho}$, that $J_\rho \subseteq J_\beta$ . If $\nu > \alpha$ then $(\alpha,\beta) \in L(\nu,\rho)$ and $z(\nu,\rho) \leq f(\nu,J_\rho) < f(\alpha,J_\beta)$ which contradicts (4.9). Therefore $\nu \leq \alpha$ and so $(\nu,J_\rho) < (\alpha,J_\beta)$ in $\kappa \otimes A$. □

## 5. PROOF OF THEOREM 6

Let $|A| = \mathrm{cf}(\langle A,R\rangle) = \kappa$ be a singular cardinal. We may assume that there is no independent subset of cardinal $\kappa$, otherwise there would be at least $2^\kappa > \kappa$ different independent sets.

For any set $Y \subseteq A$ we define $U(Y,R) = \{u \in A : u \in Y$ or $(\forall y \in Y)(\neg yRu)\}$. Let $\mathcal{B} = \{U(Y,R) : Y \subseteq A,\ Y$ is $R$-independent and $\mathrm{cf}(\langle U(Y,R),R\rangle) < \kappa\}$.

CLAIM 1. *If $X \subseteq A$ and $|X| < \kappa$, then $X \subseteq U(Y,R)$ for some $U(Y,R) \in \mathcal{B}$.*

To see this let $Y$ be a maximal $R$-independent subset of $A_1 = A \setminus \{a \in A : a \in X$ or $(\exists x \in X)(aRx)\}$. If $X \not\subseteq U(Y,R)$, then there are $x \in X$ and $y \in Y$ so that $yRx$. But this implies that $y \in A \setminus A_1$, a contradiction. Hence $X \subseteq U(Y,R)$. To show that $U(Y,R) \in \mathcal{B}$ it will be enough to show that $X \cup Y$ is cofinal in $U(Y,R)$ (since $|X| < \kappa$ and $|Y| < \kappa$ since $Y$ is $R$-independent). Now if $z \in U(Y,R)$ is arbitrary and if $z \in A \setminus A_1$, then there is $x \in X$ so that $x = z$ or $zRx$. On the other hand, if $z \in A_1$, then either $z \in Y$ or there is some $y \in Y$ such that $yRz$ or $zRy$ holds. But we cannot have $yRz$ by definition of $U(Y,R)$, and so either $z \in Y$ or there is $y \in Y$ such that $zRy$. This proves the Claim 1.

Let $(\kappa_\alpha)_{\alpha<\mathrm{cf}(\kappa)}$ be an increasing sequence of cardinals with limit $\kappa$ and let $\mathcal{B}_\alpha = \{B \in \mathcal{B} : \mathrm{cf}(B) \leq \kappa_\alpha\}$.

CLAIM 2. *If $|\mathcal{B}_\alpha| \leq \kappa$, then there is some set $X_\alpha$ such that $|X_\alpha| \leq \mathrm{cf}(\kappa)$ and $X_\alpha \not\subseteq B$ for all $B \in \mathcal{B}_\alpha$.*

If $\mathcal{B}_\alpha = \emptyset$ the claim is trivial. So we can assume that $\mathcal{B}_\alpha \neq \emptyset$ and that $(B_\xi)_{\xi<\kappa}$ is a sequence containing each member of $\mathcal{B}_\alpha$ at least once. For $\xi < \kappa$ we have $\mathrm{cf}\left(\bigcup_{\xi'<\xi} B_{\xi'}\right) \leq |\xi| \cdot \kappa_\alpha < \kappa$ and so there is an element $x_\xi \in A \setminus \bigcup_{\xi'<\xi} B_{\xi'}$. The set $X_\alpha = \{x_{\kappa_\nu} : \nu < \mathrm{cf}(\kappa)\}$ has the required properties.

Now assume that $|\mathcal{B}_\alpha| \leq \kappa$ for each $\alpha < \mathrm{cf}(\kappa)$. Then by Claim 2, for each $\alpha < \mathrm{cf}(\kappa)$ there is $X_\alpha$ such that $|X_\alpha| \leq \mathrm{cf}(\kappa)$ and $X_\alpha \not\subseteq B$ for each $B \in \mathcal{B}_\alpha$. Thus $X = \bigcup_{\alpha<\mathrm{cf}(\kappa)} X_\alpha$ has power at most $\mathrm{cf}(\kappa)$ and is not contained in any set $B \in \mathcal{B}$. This is a contradiction and therefore $|\mathcal{B}| > \kappa$.

This completes the proof of Theorem 6 since each member $B \in \mathcal{B}$ has the form $U(Y,R)$ where $Y$ is an independent set, i.e. $A$ contains more than $\kappa$ independent sets. □

## 6. TOURNAMENTS

A *tournament* is a relational structure $T = \langle A,R\rangle$ such that for every pair of distinct elements $a,b \in A$ exactly one of the relations $aRb, bRa$ holds. Thus a tournament does not contain an independent set of size 2 and so, by Theorem 6, if $|T| = \mathrm{cf}(T) = \lambda$, then $\lambda$ must be regular. This fact also follows from the following observation which says a little more:

*Let $T = \langle A,R\rangle$ be a tournament such that $|T| = \mathrm{cf}(T) = \lambda$. Thus $T$ contains a chain $C$ of type $\lambda$ such that any subchain $C'$ of $C$ which is cofinal in $C$ is also cofinal in $T$.*

Proof. Let $\{a_\nu : \nu < \lambda\}$ be any enumeration of the elements of $T$. Define a function $\varphi : \lambda \to \lambda$ as follows. Let $\lambda' < \lambda$ and suppose $\varphi(\lambda'')$ has been defined for $\lambda'' < \lambda'$. Since $\{\varphi(\lambda'') : \lambda'' < \lambda'\}$ is not cofinal in $T$, there is a least index $\varphi(\lambda')$ so that $a_{\varphi(\lambda'')} R a_{\varphi(\lambda')}$ holds for all $\lambda'' < \lambda'$. Then $C = \{a_{\varphi(\lambda')} : \lambda' < \lambda\}$ is a chain with the right property. □

We conclude by proving the following result which shows that the cofinality of a tournament need not be regular and so the hypothesis $|A| = \mathrm{cf}(\langle A,R\rangle)$ is needed for Theorem 6:

*For any infinite cardinals $\kappa,\lambda$ with $\lambda < \kappa$, there is a tournament $T(\kappa,\lambda)$ of cardinality $\kappa$ and cofinality $\lambda$.*

Proof. We first construct $T(\lambda^+,\lambda)$. Let $|A| = \lambda^+$ and consider any partition $A = A_0 \cup A_1$ with $|A_0| = \lambda$, $|A_1| = \lambda^+$. Let $A_0 = \{x_\nu : \nu < \lambda\}$ and let $A_{1\rho} (\rho < \lambda^+)$ be pairwise disjoint subsets of $A_1$ each of cardinality $\lambda^+$ and so that $A_1 = \bigcup_{\rho<\lambda^+} A_{1\rho}$. Consider the tournament $T(\lambda^+,\lambda) = \langle A,R\rangle$ which is such that $\langle A_0,R\rangle$ is a chain of order type $\lambda$, $\langle A_1,R\rangle$ is a chain of order type $\lambda^+$ and so that for $a \in A_0$ and $b \in A_1$ the relation $bRa$ holds if and only if $a = x_\nu$ and $b \in A_{1\nu}$ for some $\nu < \lambda$. Clearly $A_0$ is a cofinal subset of $T(\lambda^+,\lambda)$ and so the cofinality is at most $\lambda$. On the other hand, if $X \subseteq A$ and $|X| < \lambda$ then there are $\nu < \lambda$ and $y \in A_{1\nu}$ so that $x_\nu \notin X$ and $xRy$ holds for all $x \in X \cap A_1$. Since $x_\nu \notin X$ we also have $xRy$ for all $x \in X \cap A_0$. Therefore $X$ is not cofinal in $T(\lambda^+,\lambda)$ and so the cofinality is exactly $\lambda$.

Now suppose $\kappa$ is any cardinal greater than $\lambda^+$. We define $T(\kappa,\lambda)$ as follows. Let $T(\lambda^+,\lambda) = \langle A,R\rangle$ and let $B$ be any set of cardinality $\kappa$ disjoint from $A$. Define any tournament $T(\kappa,\lambda) = \langle A \cup B, R_1 \rangle$ where the relation $R_1$ is such that $R \subseteq R_1$ and $bRa$ holds for all pairs $(a,b) \in A \times B$. Clearly any cofinal subset of $T(\kappa,\lambda)$ has a cofinal subset which is cofinal in $T(\lambda^+,\lambda)$. □

ACKNOWLEDGEMENT

The authors wish to express their thanks to Dr. Norbert Sauer for several discussions about the content of this paper especially in connection with Theorem 1 and §6.

REFERENCES

[1] C. Berge (1973) *Graphs and Hypergraphs*, North Holland.

[2] P. Erdös, A. Hajnal, and R. Rado (1965) Partition relations for cardinal numbers, *Acta Math Acad. Sci. Hung.* 16, 93-196.

[3] P. Erdös and A. Tarski (1943) On families of mutually exclusive sets, *Ann. of Math.* 44, 315-329.

[4] R.P. Dilworth (1950) A decomposition theorem for partially ordered sets, *Ann. of Math.* 51, 161-166.

[5] B. Dushnik and E. Miller (1941) Partially ordered sets, *Amer. J. Math.* 63, 605.

[6] S. Ginsburg and J. Isbell (1965) The category of cofinal types I, *Trans. Amer. Math. Soc.* 116, 386-393.

[7] A. Hajnal and N. Sauer, unpublished private communication.

[8] G. Higman (1952) Ordering by divisibility in abstract algebras, *Proc. London Math. Soc.* (3) 2, 326-336.

[9] J. Isbell (1965) The category of cofinal types II, *Trans. Amer. Math. Soc.* 116, 394-416.

[10] R. Laver (1973) An order type decomposition theorem, *Ann. of Math.* 98, 96-119.

[11] E.C. Milner and K. Prikry (1980) The cofinality of a partially ordered set, U. of Calgary, Research Paper No. 474 (submitted to *London Math. Soc.*).

[12] E.C. Milner and N. Sauer (1981) Remarks on the cofinality of a partially ordered set and a generalization of König's Lemma, *Discrete Math.* 35, 165-171.

[13] M.A. Perles (1963) On Dilworth's theorem in the infinite case, *Israel J. of Math.* 1, 108-109.

[14] M. Pouzet (to appear) Parties cofinales des ordres partiels ne contenant pas d'antichaines infinies, *J. London Math. Soc.*

[15] F.P. Ramsey (1930) On a problem of formal logic, *Proc. London Math. Soc.* 29, 264-286.

[16] J. Schmidt (1955) Kofinaletät, *Z. Math. Logik Grundlagen Math.* 1, 271-303. MR #17-951.

# INFINITE ORDERED SETS, A RECURSIVE PERSPECTIVE

George F. McNulty
Department of Mathematics,
Computer Science, and Statistics
University of South Carolina
Columbia, South Carolina 09208

ABSTRACT

This work surveys the development of the theory of infinite ordered sets along the lines suggested by recursion theory. Since recursion theory is principally concerned with explicit effective constructions, this line of research retains much of the character of the mathematics of finite ordered sets. There are two distinct categories into which the results fall: the first is concerned with individual infinite ordered sets on which the ordering can be mechanically determined, while the second concerns whether the elementary sentences true in a given class of ordered sets can be distinguished by an algorithm. Roughly, an elementary sentence is one that refers to the elements of an ordered set rather than, say, sets of elements. "Width 3" is a notion expressible by an elementary sentence, but "finite width" cannot be expressed even by a set of such sentences. The requisite background in recursion theory and in the logic of elementary sentences can be found in Barwise [1977].

RECURSIVE ORDERED SETS. An ordered set $(A,R)$ is *recursive* provided both $A$ and $R$ are recursive (i.e. algorithmically recognizable) sets. Here we focus on the *recursive dimension* of $(A,R)$ (= minimum number of recursive linear extensions of $R$ with intersection $R$) and its connection with the *recursive chain covering number* (= minimum number of recursive chains into which $(A,R)$ can be partitioned) as well as the width of $(A,R)$. Manaster and Rosenstein [1980] have established a theorem that, like the classical Dushnik-Miller theorem [1941], provides a geometric meaning for dimension. In the recursive setting, the connection between width and dimension is subtle. There is a recursive ordered set of width 2 and recursive chain covering



*I. Rival (ed.), Ordered Sets, 299–330.*

number 5; but every recursive ordered set of width $w$ can be covered by

$$\frac{5^{w} - 1}{4}$$

recursive chains (Kierstead [1981]). Kierstead's theorem plays a role similar to that played by Dilworth's theorem [1950]. Kierstead has also shown that every recursive ordered set of width 2 has a recursive dimension no larger than 6! , (though this number is almost surely much too large.) Kierstead, McNulty, and Trotter have devised a recursive ordered set of width 3 which has no finite recursive dimension. They also showed that every crown-free recursive ordered set with recursive chain covering number $c$ has recursive dimension no more than $c$! and that every recursive interval order with recursive chain covering number $c$ has recursive dimension no more than $2c$. Hopkins has shown that every recursive interval order of width 2 has recursive dimension no more than 3. In each case above lower bounds on the recursive dimension have also been discovered, but several are not yet sharp. Manaster and Remmel [1981] have found a Dushnik-Miller style characterization of two dimensional dense recursive ordered sets and it evidently extends to higher dimensions. They have also constructed a dense two dimensional recursive ordered set which is *recursively rigid* (i.e. the identity map is the only recursive automorphism). The theory of recursive ordered sets is at its beginnings and interesting questions seem to multiply.

ELEMENTARY THEORIES OF ORDERED SETS. Let $K$ be a class of ordered sets. The *elementary theory* of $K$, Th$K$, is the set of all elementary sentences true in every member of $K$. Th$A$ denotes Th$\{A\}$. An elementary theory $T$ is *decidable* provided $T$ is recursive; otherwise $T$ is *undecidable*. $T$ is *finitely axiomatizable* if $T$ has a finite subset $S$ such that each sentence in $T$ is a logical consequence of $S$. $T$ is $\aleph_0$-*categorical* provided $A = B$ whenever $A$ and $B$ are countably infinite ordered sets in which each sentence of $T$ is true. Vaught [1954] showed that every finitely axiomatizable $\aleph_0$-categorical theory is decidable. Schmerl [1980] has established $\aleph_0$-categorical versions of Dilworth's theorem and the theorem of Szpilrajn (Marczewski) [1930] and combined them with Vaught's theorem to obtain the decidability of the elementary theory of certain ordered sets. Let $A$ be a finite ordered set; $K_A$ is the class of all ordered sets which have no suborders isomorphic to $A$. Schmerl [1980] proved that Th$K_A$ is decidable if and only if no suborder of $A$ has a Hasse diagram like

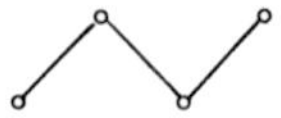

.

The table below displays most known decidability results about ordered sets.

| Theory of | Finitely Axiomatizable? | Decidable? |
|---|---|---|
| Ordered Sets | yes | no, Tarski [1949a] |
| Well Orders | no, Doner, Mostowski & Tarski [1978] | yes, Doner, Mostowski & Tarski [1978] |
| Linear Orders | yes | yes, Ehrenfeucht [1959] |
| Dense Linear Orders | yes | yes, Vaught [1954] |
| Discrete Linear Orders | yes | yes, Robinson & Zakon [1960] |
| Two Dimensional Ordered Sets | no, Baker, Fishburn & Roberts [1971] | no, Manaster & Rosenstein [1980] |
| Dense Two Dimensional Orders | yes, Manaster & Remmel [1981] | no, Manaster & Remmel [1981] |
| Ordered Sets of Width $n$ ($n \geq 2$) | yes | no, Schmerl [1980] |
| Interval Orders | yes, Fishburn [1970] | no, Schmerl [1980] |
| Unit Interval Orders | yes, Fishburn [1970] | no, Schmerl [1980] |
| An $\aleph_0$-categorical Tree | | yes, Schmerl [1980] |
| An $\aleph_0$-categorical Ordered Set of Finite Width | | yes, Schmerl [1980] |
| Lattices | yes | no, Tarski [1949a] |
| Modular Lattices | yes | no, Tarski [1949a] |
| Distributive Lattices | yes | no, Grzegorczyk [1951] |
| Boolean Algebras | yes | yes, Tarski [1949b] |

## §0. INTRODUCTION

An impressive array of explicit constructions is one of the distinguishing features of the mathematics of finite ordered sets. Recursion theory constitutes an elegant mathematical analysis of the notion of explicit (mechanical) construction. The present work surveys the development of the theory of infinite ordered sets along lines suggested by recursion theory. In this development, the explicit constructions, if anything, play a more prominent role and the character of many of the examples and proofs is discrete and combinatorial. Ordered sets are ubiquitous in mathematics so that the connections between the theory of ordered sets and recursion theory or, more generally, mathematical logic are deep and intricate. For the purposes of this survey, a selection of topics is essential. I have chosen, mainly, to expose only two topics: recursive ordered sets and decision problems for ordered sets. These areas seem to me to offer insight into the nature of ordered sets and the first area opens a fresh direction for combinatorial investigation. Other areas such as the analysis of the ordered set of Turing degrees or of the class of models of arithmetic ordered by elementary extension, while they comprehend many difficult, deep, and even surprising theorems, rely on motivation and techniques that have little to do with ordered sets per se. I have omitted topics of this kind. A more regrettable omission is made in connection with questions of computational complexity. In section 4 below there are extremely brief descriptions of a few other interactions between the theory of ordered sets and mathematical logic.

Section 1 is a terse account of those elements of recursion theory and first order logic that play a role in theorems presented in later sections. Rogers [1967] provides an authoritative development of recursion theory. Briefer expositions are available in the books: Cutland [1980] and Davis [1958]. Machtey and Young [1978] is a highly readable account which emphasizes computational complexity. A more concise rendition of the elements of recursion theory can be found in Enderton's article in the Handbook of Mathematical Logic (Barwise [1977]), in Putnam [1973], or in Stillwell [1977]. The papers in which recursion theory originated are gathered in Davis [1965]. Chang and Keisler [1973] is an authoritative guide to first order or elementary logic. Bell and Machover [1977] examine decidability of elementary theories in depth.

Shorter books which provide readable accounts are Enderton [1972] and Manaster [1975]. The expository articles of Barwise, Davis, and Rabin in the Handbook of Mathematical Logic provide a concise rendition of the elements of the subject, as do the articles of Stillwell [1977] and Vaught [1973]. The key papers in the literature are very readable - they are: Tarski, Mostowski, and Robinson [1953], Ershov, Lavrov, Taimanov, and Taitslin [1965] and Rabin [1965].

Section 2 below is devoted to the theory of recursive ordered sets. Roughly, an ordered set (A,R) is *recursive* provided there is an algorithm which, when two points are input, will output the order of the points. The notions of recursive dimension and recursive chain covering are introduced below as natural extensions of the notions for finite ordered sets. The first portion of the section concerns the interplay between width, recursive dimension, and recursive chain coverings. The second portion of the section focusses on recursive linear orders.

Section 3 is an exposition of what is known about the decidability of elementary theories of ordered sets.

## §1 BACKGROUND

### Recursion Theory

Recursion theory has, as its most immediate objective, the illumination of the intuitive notion of an effective or mechanical procedure. The first step, it would seem, is to provide a definition of what is meant by a "computable" function. The literature abounds with such definitions, some of quite dissimilar characters, but all acceptable from the mathematical viewpoint. Any of these definitions would serve our purpose here admirably except at one point: for all their conceptual simplicity, each of these definitions turns out to be intricate in detail. These details don't play an important role in our survey. It is enough to say that a function f of several variables is a *partial recursive function* provided it can be described by a computer program in, say BASIC, in the sense that if an element of the domain of f is input, then upon execution on an ideal machine the program would produce the value of f as the output, while an input outside the domain of f should result in no output. Because the syntax and semantics of BASIC are unambiguous, the collection of partial recursive functions is quite definite. A partial recursive function is, then, a certain kind of function of finitely many variables and these variables range over rational numbers and over strings of symbols taken from a fixed finite alphabet. (This selection of possible values for inputs and outputs is somewhat arbitrary, but convenient for our purposes.)

The central doctrine of recursion theory is the Church-Turing Thesis:

> *A partial function can be calculated by a mechanical procedure if and only if it is a partial recursive function.*

This thesis, dealing as it does with intuitive notions as well as mathematical notions, is not open to proof. Nevertheless, the evidence in its favor is overwhelming. In the sequel we adhere to the Church-Turing Thesis. On the one hand, as soon as a way to compute a given function becomes clear we will claim it to be a partial recursive function - without going to the labor of actually producing a program in BASIC. On the other hand, if we are in some way convinced that no BASIC program describes the partial function f, then we will take a leap of faith to assert that no mechanical procedure of any kind is capable of computing f.

A set is *recursive* provided its characteristic function is a (partial) recursive function. This means that a set X is recursive if there is an algorithm which distinguishes membership in X. Since relations are certain kinds of sets, we come to the notion of a recursive relation. For example the usual order $\leq$ on Q is a recursive relation.

Now the BASIC programs can be very neatly listed: say by length, using a kind of alphabetical order to break ties. $\phi_e$ denotes the partial recursive function described by the $e^{th}$ program on the list. In fact, this list of programs is so nice that there is a partial recursive function u(e,x) such that

$$u(e,x) = \begin{cases} \phi_e(x) & \text{if } e \text{ is a natural number and} \\ & x \text{ is in the domain of } \phi_e. \\ \text{Undefined} & \text{otherwise.} \end{cases}$$

A program for this function u is essentially a compiler for BASIC: it accepts as input (the index e of) a BASIC program p and some other input x and then simulates the behavior of p on input x. Rather inconveniently, the ternary relation $\{(e,x,y) : \phi_e(x)=y\}$ turns out not to be recursive. A proof can be found in Rogers [1967]. The trouble is that the $e^{th}$ program upon input x may indulge in an infinite search and never present any output at all. Fortunately, with the help of the function u, the following 4-ary relation T(e,x,y,n) is recursive. This relation holds provided the $e^{th}$ program upon input x finishes its computation in no more than n steps and outputs y. Usually T(e,x,y,n) is written $\phi_e^n(x)=y$. The recursiveness of this predicate underlies most of the proofs and constructions in the theory of recursive ordered sets.

The Logic of Elementary Sentences

An ordered set is a structure of the form (A,R) where A, is a non-empty set and R is a binary relation on A enjoying certain properties which may be expressed by reference to how R behaves with respect to the *elements* of A. In the course of developing the theory of ordered sets one is led quite naturally to consider more elaborate structures, e.g. $(A,R,C_0,\ldots,C_{w-1})$ where $C_0,\ldots,C_{w-1}$ are subsets of A linearly ordered by R which cover A. Apparently, the language in which properties of (A,R) can be expressed requires a symbol to stand for the relation R, while the language appropriate to $(A,R,C_0,\ldots,C_{w-1})$ should be richer and the language appropriate for rings would be still different. We will describe the *elementary language* $L$ of ordered sets. Languages appropriate for richer structures will be needed below but they have analogous definitions. The language $L$ is provided with two kinds of symbols: logical and non-logical.

The non-logical symbol is R which stands for a binary relation.

The logical symbols are:

| | |
|---|---|
| $v_0,v_1,v_2,\ldots$ | which are variables intended to range over *elements* of a structure |
| $\wedge$ | which stands for "and" |
| $\neg$ | which stands for "not" |
| $=$ | which stands for "equals" |
| $\exists$ | which stands for "there exists" |

and various parentheses and brackets for punctuation.

To avoid subscripts it proves convenient to let x,y,z,w, and u represent variables. Other symbols, like $\rightarrow$ for "implies", $\vee$ for "or", $\forall$ for "for all", can be defined and used with impunity. The symbols can be assembled into strings which are, in the obvious sense, well-formed. Such strings are called *formulas*. For example, let $\phi(x)$ be $\forall y(yRx \vee y=x)$. The formula $\phi(x)$ attributes to x the property of being a "largest" element. The variable x is a free variable in this formula. We cannot determine whether $\phi(x)$ is true unless we fix the value for x. On the other hand the formula $\exists x \forall y(yRx \vee y=x)$ asserts the existence of a largest element and in any structure (A,R) this formula $\exists x\, \phi(x)$ is either true or false. Such formulas with no free variables are called *elementary sentences*.

What properties can be expressed by elementary sentences? To begin with the property of being an ordered set is an elementary property. The property "width k" is also an elementary property for each natural number k, but the property "finite width" cannot be expressed even by a set of elementary sentences. Likewise "linear order" is elementary, but "well-order" is not. Roughly, the reason why we cannot express "well-order" or "finite width" with sets of elementary sentences can be traced to the fact that we only permitted ourselves variables which range over elements, rather than over subsets of a structure. We denote by Th(A,R) the set of all elementary sentences that are true in (A,R). Th(A,R) is called the *elementary theory* of (A,R). Similarly, when K is a class of structures, ThK is the set of sentences that are true in each structure belonging to K. Two structures (A,R) and (A',R') are *elementarily equivalent* provided Th(A,R) = Th(A',R'); $(A,R) \equiv (A',R')$ denotes this equivalence. Though elementary equivalence is a coarser equivalence relation than isomorphism, still many properties are shared by structures which are elementarily equivalent. Sometimes elementary equivalence entails isomorphism: a structure (A,R) which is countably infinite is said to be *ω-categorical* provided $(A,R) \cong (A',R')$ whenever $(A,R) \equiv (A',R')$ and A' is countable. Cantor proved, in essence, that the rationals are ω-categorical.

A structure (A,R) has a *decidable* elementary theory provided Th(A,R) is a recursive set. Thus, Th(A,R) is decidable if and only if there is an algorithm which can recognize those elementary sentences true in (A,R). An elementary theory T is *axiomatizable* if there is a recursive set $\Sigma$ of sentences such that $\Sigma \subseteq T$ and each $\sigma \in T$ is a logical consequence of $\Sigma$. On the basis of a somewhat stronger theorem he had proven, Vaught [1954] noted that if a countably infinite structure is ω-categorical and has an axiomatizable theory, then its elementary theory is decidable. It follows from the work of Cantor and Vaught that the elementary theory of the rationals is decidable. But there are theories of ordered sets which are undecidable and a very combinatorial technique sketched in section 3 can be used to prove such facts.

## §2 RECURSIVE ORDERED SETS

An ordered set (A,R) is *recursive* provided both A and R are recursive. Perhaps the oldest results concerning recursive ordered sets are those dealing with recursive well-orders which can be found in A. Church and S. Kleene [1937], S. Kleene [1938], and A. Church [1938]. These results were added to in W. Markwald [1954] and C. Spector [1955]. It turns out that these recursive ordinals form an initial segment of the ordinals. The theory of recursive ordinals is detailed in Rogers [1967] and has had some significant role in mathematical logic. Nevertheless, the theory of recursive ordered sets is really at its beginnings. All the theorems below arose in the last five years (with one exception) and only a few have found their way into print. Since, from most points of view, any given finite linear order is trivial, it is not surprising that the theory of recursive linear orders is driven by motivations resembling those for arbitrary infinite linear orders. On the other hand, even the finite ordered sets of dimension two embody a wild array of possibilities; the concerns of the theory of recursive ordered sets of dimension at least two bear a strong resemblance to their cousins for the finite ordered sets. So we first take up the theory of recursive dimension, leaving one dimensional recursive ordered sets for the last part of this section.

The Theory of Dimension for Recursive Ordered Sets

The theory of dimension for (finite) ordered sets is surveyed in Kelly and Trotter [1982] to be found elsewhere in this volume. This theory really has its beginning with

THEOREM (Dushnik and Miller [1941])

> *Let* $(A,R)$ *be an ordered set.* $(A,R)$ *can be embedded in a direct product of* $n$ *chains if and only if there exist* $n$ *linear orderings* $L_0,\ldots,L_{n-1}$ *of* $A$ *such that* $R = L_0 \cap L_1 \cap\ldots\cap L_{n-1}$.

Recalling that every countable chain is embeddable in $Q$, we observe that this theorem tells us that $(A,R)$ can be represented in n-dimensional space if and only if $R$ is the intersection of $n$ linear orders of $A$. This provided an order theoretic definition for a geometric notion of dimension. In constructing a notion of recursive dimension we will want to retain the geometric nature of the concept. Fortunately, this is possible.

THEOREM (Manaster and Remmel [1980])

> *Let* $(A,R)$ *be a recursive ordered set.* $(A,R)$ *is recursively isomorphic to a recursive subset of* $Q^n$ *if and only if there are* $n$ *recursive linear orderings* $L_0,L_1,\ldots,L_{n-1}$ *of* $A$ *such that* $R = L_0 \cap L_1 \cap\ldots\cap L_{n-1}$.

Let $(A,R)$ be a recursive ordered set. The *recursive dimension* of $(A,R)$ is the smallest number $n$ such that there are recursive linear orders $L_0,L_1,\ldots,L_{n-1}$ of $A$ with $R = L_0 \cap L_1 \cap\ldots\cap L_{n-1}$. In view of the theorem above, this definition retains a geometric as well as a recursive character.

The next real advance in the theory of dimension came in 1950. The *chain covering number* of $(A,R)$ is the smallest number $c$ such that $A$ can be covered by $c$ subsets linearly ordered by $R$. It is easy to see that the chain covering number is an upper bound on the dimension.

THEOREM (Dilworth [1950])

> *Let* (A,R) *be an ordered set.* (A,R) *has width no more than* w *if and only if* A *can be covered by* w *subsets linearly ordered by* R.

According to Dilworth's Theorem, the width is also an upper bound. The situation among recursive ordered sets is more subtle. For a recursive ordered set (A,R) the *recursive chain covering number* is the smallest number c such that A can be covered by c recursive subsets linearly ordered by R. The connection between width and the recursive chain covering number is considerably more intricate and the formerly easy inequality between the chain covering number and dimension is simply false. An example will be provided below.

THEOREM (Kierstead [1981])

> *There is a recursive ordered set of width* 2 *with recursive chain covering number* 5.

So there is a kind of non-effective nature to the full duality between antichains and chain covers embodied in Dilworth's Theorem. Nevertheless, not everything is lost.

THEOREM (Kierstead [1981])

> *If* (A,R) *is a recursive ordered set of finite width* w, *then the recursive chain covering number of* (A,R) *is no more than* $\frac{5^w-1}{4}$.

The bound $\frac{5^w-1}{4}$ seems unlikely to be sharp. It is unknown whether there is a recursive ordered set of width 2 with recursive chain covering number 6. The most remarkable thing about this theorem is that there is any finite bound at all.

THEOREM (Kierstead)

> *Every recursive ordered set of width* 2 *has recursive dimension no more than* 6.

THEOREM (Kierstead)

> *Every recursive ordered set with recursive chain covering number* 3 *has recursive dimension no more than* 6*; moreover, there is a recursive ordered set with recursive chain covering number* 3 *and recursive dimension* 6.

But the relationship between the recursive chain covering number and the recursive dimension is delicate. We include a proof of the next theorem to give a taste of the arguments at work here.

THEOREM (Kierstead, McNulty, and Trotter)

> *There is a recursive ordered set of width 3 with recursive chain covering number 4 which has no finite recursive dimension.*

PROOF:

We describe the construction of a recursive ordered set $(\omega, R)$ which fulfills this theorem; the points of this set will comprise the set of natural numbers. The construction proceeds through infinitely many stages in such a way that after the $k^{th}$ stage at least all of the comparabilities among $0,1,\ldots,k-1$ have been determined. The construction itself will be an effective process so that the ordered set which is constructed will be recursive: to determine the order relation between two points m and n, it will suffice to go through the construction to stage $\max(m,n)+1$, since by then their comparability will have surely been decided.

For the convenience of our construction, fix some recursive one-to-one list of pairs $(e,n)$ of natural numbers. We will say $(e,n)$ is *higher* than $(e',n')$ if $(e,n)$ occurs later in this list than $(e',n')$. Our recursive ordered set is built in layers labelled with the ordered pairs $(e,n)$ and all the points of layer $(e,n)$ are larger in the recursive order under construction than all the points in layer $(e',n')$ if $(e,n)$ is higher than $(e',n')$.

In the course of the construction we will also build four sets A,B,C, and D which will be linearly ordered in the ordering we are building and shortly after we introduce a new point we will assign it to one of these sets. In this way, we will insure that our ordered set will have recursive chain covering number at most 4.

In order to destroy the possibility of a finite recursive dimension we will build in the layer $(e,n)$ a finite ordered set of width 3 that defeats the assertion (*).

(*) "For each $k < n$, $\phi_e(x,y,k)$ is the characteristic function of a linear order $L_k$ of $\omega$ such that $R = L_0 \cap L_1 \cap \ldots \cap L_{n-1}$."

Suppose, for the moment, that $\phi_e(x,y,k)$ really is the characteristic function of a linear order $L_k$ for each $k < n$. Let $a_0$ and

$b_0$ be (the smallest) natural numbers (not yet used in the construction). Set $a_0$ and $b_0$ incomparable in R. Let $a_1$ and $b_1$ be the next natural numbers and set $a_1$ and $b_1$ incomparable in R and $a_1$ larger than $a_0$ but $b_1$ smaller than $b_0$. Now continue in this way to build two incomparable chains $a_0, a_1, \ldots, a_{i+1}$; $b_0, b_1, \ldots, b_{i+1}$ until the conditions in the next paragraph are met. Depending on the results of this construction, c and d (shown in the diagram) may be added to form R in layer (e,n).

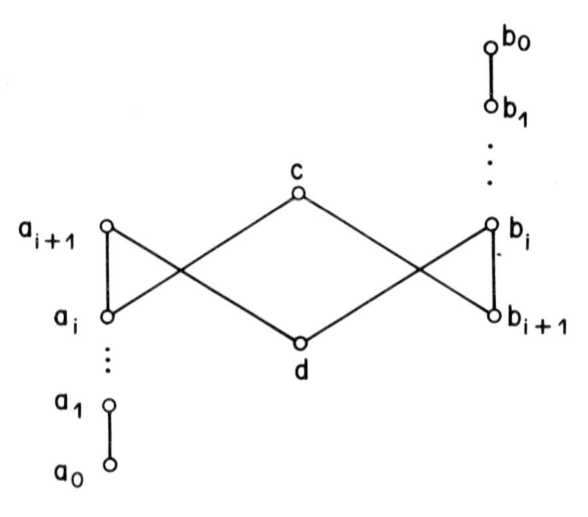

R in some layer (e,n)

After each $a_{i+1}$ and $b_{i+1}$ is adjoined determine, using $\phi_e$, the following:

a) Is R, so far as it is built, the intersection of the linear orders $L_0, L_1, \ldots, L_{n-1}$ ?

b) Given any $k < n$, is it true that in $L_k$ either $a_i$ is larger than $b_i$ or else that $b_{i+1}$ is larger than $a_{i+1}$ ?

If the answer to (a) is "no" we can stop the construction since (*) is defeated. If the answer to (a) is "yes" but the answer to (b) is "no" we add new points to our incomparable chains. The reader should convince himself that this can only go on n steps. If the answer to both questions is "yes" then adjoin the next two natural numbers c and d as indicated. Then c and d will be incomparable in R but any attempt to place them on any of the linear orders $L_0, \ldots, L_{n-1}$ will always put c larger than d. Thus (*) is defeated again.

To insure that $(\omega,R)$ has recursive chain covering number 4 we insert into the construction just given the following provisions: all the $b_i$'s get put into B; all the $a_i$'s get put into A; if and when c and d are adjoined c is put in C and d is put into D.

In the argument just given, we supposed that $\phi_e(x,y,k)$ was a characteristic function for each $k < n$. In particular, this meant that we could always count on getting answers to the questions raised in (a) and (b) above. As it stands, our construction could be stalled forever if $\phi_e$ indulged in an infinite search. To avoid this kind of dead end, we arrange a recursive pattern that visits each layer $(e,n)$ infinitely often. On each visit we perform as much of the construction described above as possible on the basis of allowing $\phi_e$ to execute an additional step in its computation. This will allow all computations to reach conclusions, if they have conclusions. Those $\phi_e$'s that really did indulge in infinite searches cannot fulfill (*) in any case. □

The broad outlines of the proof just given are well known in recursion theory. This kind of argument is called a *diagonal argument*. It is so familiar that once the part carried out in the layer $(e,n)$ is described, the rest of the proof could be regarded as routine. The part dealing with layer $(e,n)$ can be rendered in the following highly suggestive game theoretic setting. The game $G_n$ has two players. The task of Player I is to build an ordered set of width 3 one point at a time. The task of Player II is to generate n linear orders whose intersection is the order that Player I is building. The play of $G_n$ alternates between the players with Player I positioning each new point in the order subject to the constraint of width 3 and with Player II then inserting the new point into an appropriate position on each of his chains so that Player I's order is the intersection of his chains. Player II wins if the game goes through infinitely many plays and Player I wins otherwise. What was demonstrated in the proof above was that for each n, Player I has a winning strategy for game $G_n$ which is recursive and that these strategies for the $G_n$'s can be recursively "woven together".

Some additional conditions must be imposed on a recursive ordered set in order to insure a finite recursive dimension. The construction carried out above produced an ordered set with many (induced) suborders of the form

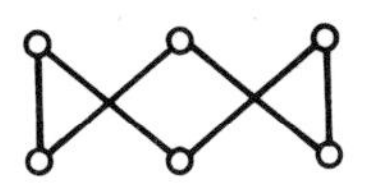

The 3-crown

The ordered set depicted above is called a *3-crown*. More generally, a *crown* is a finite ordered set of height 1 with at least six points and a Hasse diagram of the following kind:

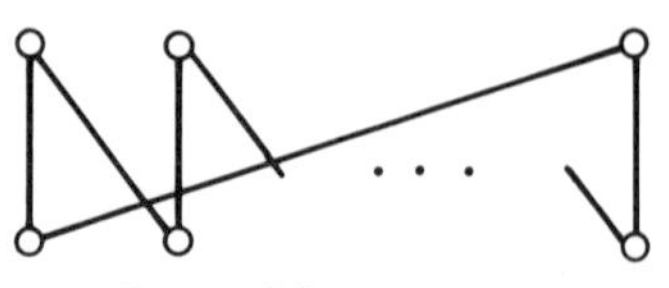

An arbitrary crown

An ordered set is *crown-free* provided none of its suborders is a crown. Crowns enter quite naturally into the investigation of dimension. See, for example, Kelly [1977], Kelly and Trotter [1982], Trotter [1974], and Trotter and Moore [1976], Our construction in the proof above breaks down immediately for crown-free recursive ordered sets. Moreover, we can prove the following theorem:

THEOREM (Kierstead, McNulty, and Trotter)

> *Every crown-free recursive ordered set with recursive chain covering number* $c$ *has recursive dimension no more than* $c!$*; moreover, for each* $c > 3$ *there is a crown-free recursive ordered set with width* $c-1$ *and recursive chain covering number* $c$ *but with recursive dimension at least* $c2^{c-2}$.

An ordered set is an *interval order* provided it has no suborder with a Hasse diagram like

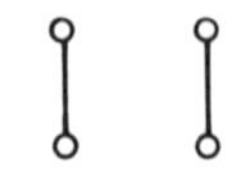

Interval orders have been investigated in Scott and Suppes [1958], Fishburn [1970], Baker, Fishburn, and Roberts [1971], Rabinovitch [1973], and Bogart, Rabinovitch, and Trotter [1976]. Every interval order is crown-free.

THEOREM (Kierstead and Trotter [1982])

> *Every recursive interval order of width* $w$ *has a recursive chain covering number no larger than* $3w-2$ *and for each positive* $w$ *there is a recursive interval order of width* $w$ *with recursive chain covering number* $3w-2$.

THEOREM (Kierstead, McNulty, and Trotter)

*Every recursive interval order with recursive chain covering number* c *has recursive dimension no larger than* 2c*; moreover, for every* $w > 1$ *there is a recursive interval order of width* w *with recursive dimension no less than* $\frac{4}{3}w$.

THEOREM (Hopkins)

*Every recursive interval order of width* 2 *has recursive dimension no more than* 3.

Before we focus on recursive linear orderings there are two more theorems which should be mentioned. The notion of a *dense* two dimensional ordered set is defined in Manaster and Remmel [1981] and it is fairly clear how to extend the definition to higher dimensions. The definition is made in the elementary theory of ordered sets, although it is topological in inspiration.

THEOREM (Manaster and Remmel [1981])

*There is a dense two dimensional recursive ordered set* (D,R) *such that the dense two dimensional recursive ordered sets are exactly those which are recursively isomorphic to recursive suborders of* (D,R).

It is curious that $Q \times Q$ cannot play the role of (D,R).

THEOREM (Manaster and Remmel [1981])

*There is a dense two dimensional recursive ordered set which is recursively rigid (i.e. the identity function is the only recursive automorphism.)*

Recursive Linear Orderings

It is easy to see that if (A,R) is a recursive ordered set then the following are equivalent:

0. (A,R) is a linear order
1. (A,R) has width 1
2. (A,R) has dimension 1
3. (A,R) has recursive dimension 1

Only a little deeper is

4, (A,R) is recursively isomorphic to a recursive suborder of $(Q, <)$.

The theory of recursive linear orders resembles the theory of countable linear orders much more than it resembles the theory of finite linear orders. An excellent reference on linear orders is the book Rosenstein [1981]. As well as exposing the classical theory, Rosenstein includes a chapter in which many of the theorems mentioned below are proved. In fact, it is fair to say that the theory of recursive linear orders is due almost exclusively to Rosenstein, his students, and his collaborators.

For the remainder of this section we will employ the usual notation for order types ( = isomorphism types of linear orders). $\omega$ is the type of the natural numbers, $\omega^*$ is the type of the negative integers, $\eta$ is the type of the rationals. More complicated types arise through the use of ordinal sums and products. For example $\omega^* + \omega$ is the type of the integers and $(\omega^* + \omega)\eta$ looks like the rationals with each point expanded to a copy of the integers. The earliest result concerning recursive linear orders, outside the results mentioned above about recursive ordinals, seems to be an unpublished theorem of S. Tennenbaum originating in the late 1950's. The following theorem is not too troublesome.

THEOREM

*Let* (A,R) *be a recursive linear order.* (A,R) *has a recursive suborder of one of the following types:* $\omega, \omega^*, \omega+\omega^*$ *or* $\omega + (\omega^*+\omega)\eta + \omega^*$.

Tennenbaum showed

THEOREM (Tennenbaum)

*There is a recursive linear order of type* $\omega+\omega^*$ *which has no recursive suborder of either type* $\omega$ *or type* $\omega^*$.

Recently, M. Lerman settled the remaining case

THEOREM (Lerman [1981])

*There is a recursive linear order of type* $\omega+(\omega^*+\omega)\eta + \omega^*$ *which has no recursive suborders of the types:* $\omega,\omega^*$, or $\omega+\omega^*$.

In the same spirit, one might inquire after the possibility of other recursive suborders. The next theorems pursue questions of this sort.

THEOREM (Rosenstein [1981])

*Every recursive linear order of type* $2\eta$ *has a recursive suborder of type* $\eta$.

THEOREM* (Rosenstein [1981])

*There is a recursive linear order of type* $(\omega^* + \omega)\eta$ *that has no recursive suborder of type* $\eta$.

THEOREM* (Lerman and Rosenstein [1981])

*There is a recursive linear order of type* $(\omega^* + \omega)\eta$ *that has no dense recursive subset.*

Actually, the theorem stated above is a cheapened version of a stronger result proven by Lerman and Rosenstein: *There is a recursive linear order of type* $(\omega^* + \omega)\eta$ *that has no dense* $\Sigma_2^0$ *subset.* I have decided not to describe the arithmetical hierarchy here (which would make $\Sigma_2^0$ understandable) and so for the purposes of this survey I sometimes state weakened versions of theorems. These are indicated with (*). The arithmetical hierarchy is described in Rogers [1967] and more briefly in Rosenstein [1981].

An *interval* in a linear order $(A,R)$ is a subset $I$ with the property that if $a,b \in I$ and $c \in A$, with $a\ R\ c\ R\ b$, then $c \in I$. A finite interval $I$ is *maximal* if it is not a proper subset if any finite interval. Let

$$M(A,R) = \{n : (A,R) \text{ has a maximal finite non-empty interval of cardinality } n\}.$$

THEOREM* (Rosenstein [1981] and Lerman [1981])

*For every recursive set* $M$ *of natural numbers with* $0 \notin M$ *there is a recursive linear order* $(A,R)$ *with* $M = M(A,R)$ . *On the other hand, there is a recursive linear order* $(A,R)$ *such that* $M(A,R)$ *is not recursive.*

Aside from refinements involving the arithmetical hierarchy, Rosenstein and Lerman also have been able to impose sharp restrictions on the order types involved. Some of this work stems from Fellner [1976].

The last topic in recursive linear orders that we mention concerns order-morphisms. An ordered set $(A,R)$ is *recursively rigid* provided the identity function on $A$ is the only automorphism of $(A,R)$ that is a partial recursive function. $(A,R)$ is *recursively self-embeddable* if there is an order-morphism from (A,R) into itself which is a partial recursive function different from the identity.

THEOREM (Rosenstein [1981])

*There is a recursive linear order of type* $\omega^* + \omega$ *which is recursively rigid.*

THEOREM (Rosenstein [1981])

*No recursive linear order of type* $\eta$ *is recursively rigid.*

THEOREM (Rosenstein [1981])

*There is a recursive linear order of type* $2\eta$ *which is recursively rigid.*

THEOREM (Rosenstein [1981])

*Every recursive linear order of type* $2\eta$ *is recursively self-embeddable.*

THEOREM (Hay and Rosenstein, in Rosenstein [1982])

*There is a recursive linear order of type* $\omega$ *which is not recursively self-embeddable.*

COROLLARY (Rosenstein [1981])

*There are two recursive linear orders of type* $\omega$ *which are not recursively isomorphic.*

Most of the theorems above that deal with the existence of recursive linear orders with pathological properties require arguments which are similar in spirit to the diagonal argument used to construct an ordered set of width 3 with no finite recursive dimension. On the other hand, some of the results cited just above require an exotic cousin of the diagonal argument, namely the finite injury priority argument. This kind of argument is described in Rosenstein [1981] or in any of our references to recursion theory.

There are several more results on recursive linear orders which fall just beyond the scope of this survey. They are Chen [1976], Feiner [1967], Fellner [1976], Hay, Manaster, and Rosenstein [1982], Watnick [1979] and [1981].

## §3. DECIDABILITY OF ELEMENTARY THEORIES OF ORDERED SETS

A second connection between ordered sets and recursion theory is represented by the decision problem:

> *For a given class* K *of ordered sets, is the elementary theory* Th K *a recursive set of sentences?*

For many classes K the answer to this question is immediate from the next theorem. Let (A,R) be a finite ordered set and denote by $K_{(A,R)}$ the class of all ordered sets which have no suborder isomorphic to (A,R).

THEOREM (Schmerl [1980])

> *Let* (A,R) *be a finite ordered set.* Th $K_{(A,R)}$ *is decidable if and only if* (A,R) *has no suborder with a Hasse diagram like*

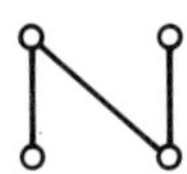

The task of determining which elementary theories of ordered sets are decidable may require stronger techniques than a simple application of this theorem. To establish that a theory is not decidable, the Rabin-Scott technique described below is frequently successful. This technique is, at heart, combinatorial and actually establishes a property stronger than undecidability. To verify that a theory is decidable may require more effort. The reward of this effort is an extremely detailed knowledge of the theory. The method of elimination of quantifiers, described in Rabin [1977], yields this kind of knowledge. Doner, Mostowski, and Tarski [1978] includes a detailed deployment of the method of elimination of quantifiers on the elementary theory of well orders. The theorem of Vaught described at the end of section 1 also offers a technique to establish the decidability of an elementary theory and we will shortly have more to say on this point. Still other techniques are known, but they depend on deeper theorems of model theory.

In the table below, we compile most of the known results that concern the decidability and the finite axiomatizability of elementary theories of ordered sets. While some of the results are more or less representative (like those concerning interval orders which follow from Schmerl's theorem above), the decision problem for an elementary theory not mentioned is usually open.

Results about elementary theories of ordered sets

| Theory of | Finitely Axiomatizable? | Decidable? |
|---|---|---|
| Ordered Sets | yes | no, Tarski [1949a] |
| Well Orders | no, Doner, Mostowski, Tarski [1978] | yes, Doner, Mostowski, Tarski [1978] |
| Linear Orders | yes | yes, Ehrenfeucht [1959] |
| Dense Linear Orders | yes | yes, Vaught [1954] |
| Discrete Linear Orders | yes | yes, Robinson & Zakon [1960] |
| Two Dimensional Ordered Sets | no, Baker, Fishburn & Roberts [1971] | no, Manaster & Rosenstein [1980] |
| Dense Two Dimensional Orders | yes, Manaster & Remmel [1981] | no, Manaster & Remmel [1981] |
| Ordered Sets of Width n ($n \geq 2$) | yes | no, Schmerl 1980 Manaster & Rosenstein [1980] for n=2 |
| Interval Orders | yes, Fishburn [1970] | no, Schmerl [1980] |
| Unit Interval Orders | yes, Fishburn [1970] | no, Schmerl [1980] |
| An $\omega$-Categorical Tree | ? | yes, Schmerl [1976] |
| An $\omega$-Categorical Ordered Set of Finite Width | ? | yes, Schmerl [1980] |
| An $\omega$-Categorical Linear Order | yes, Rosenstein [1969] | yes, Rosenstein [1969] |
| Lattices | yes | no, Tarski [1949a] |
| Modular Lattices | yes | no, Tarski [1949a] |
| Distributive Lattices | yes | no, Grzegorczyk [1951] |
| Boolean Algebras | yes | yes, Tarski [1949b] |

This section concludes with some indication as to how some of these results can be established.

Schmerl [1980] proves that any $\omega$-categorical ordered set of finite width is decidable. This would be a straightforward consequence of Vaught's Theorem mentioned in §1 if every $\omega$-categorical ordered set of finite width were axiomatizable. Schmerl's proof uses the following very nice $\omega$-categorical version of Dilworth's Theorem to expand $\omega$-categorical ordered sets of finite width to finitely axiomatizable, $\omega$-categorical structures to which Vaught's Theorem does apply.

THEOREM (Schmerl [1980])

> *Let* $(A,R)$ *be an* $\omega$*-categorical ordered set of width* $n$. *There are subsets* $L_0, L_1, \ldots, L_{n-1}$ *of* $A$ *linearly ordered by* $R$ *such that* $A = L_0 \cup L_1 \cup \ldots \cup L_{n-1}$ *and* $(A,R,L_0,L_1,\ldots,L_{n-1})$ *is* $\omega$*-categorical and finitely axiomatizable.*

Schmerl also established a restricted $\omega$-categorical version of the well-known theorem of Szpilrajn (Marczewski) [1930].

THEOREM (Schmerl [1980])

> *For any* $\omega$*-categorical ordered set of finite height or finite width, there is a linear ordering* $S$ *of* $A$ *such that* $R \subseteq S$ *and* $(A,R,S)$ *is* $\omega$*-categorical.*

It is therefore tempting to speculate about an $\omega$-categorical version of the Dushnik-Miller theorem. Such speculation is a bit dampened due to the next theorem.

THEOREM (Schmerl [1980])

> *There is a two dimensional* $\omega$*-categorical ordered set whose elementary theory is undecidable.*

To complete this section we will sketch a proof that the elementary theory of interval orders is undecidable. This proof employs the method of Rabin and Scott (see Rabin [1965]) and it was told to me by Ralph McKenzie. The result itself is a corollary to the much more general theorem of Schmerl stated above. The proof is included here because it is so highly combinatorial in character - proceeding almost entirely on the basis of a few simple pictures (this is typical of the Rabin-Scott method).

The next notion, finite inseparability, is stronger than undecidability and it is more technical. The Rabin-Scott method creates new finitely inseparable theories from old ones in a way that avoids the technicalities. It is only necessary to have a place to start - to know some theory which is finitely inseparable. Our purpose here is admirably served by the theory of bipartite graphs.

We call a theory $T$ *finitely inseparable* provided there is a recursive function $f(e,r)$ such that if the domains of $\phi_e$ and $\phi_r$ are disjoint and $T$ is a subset of the domain of $\phi_e$ while every sentence that fails in some finite model of $T$ belongs to the domain of $\phi_r$, then $f(e,r)$ belongs to neither the domain of $\phi_e$ nor the domain of $\phi_r$. Every finitely inseparable theory is undecidable.

The Rabin-Scott Method

We need the notion of an interpretation of a structure. This idea is easy to illustrate. Let $(Z,\leq,0)$ be the integers with their usual order and zero as a distinguished element. Let $(\omega,\leftarrow)$ be the natural numbers with the immediate cover relation (i.e. $n \leftarrow m$, iff $m=n+1$). Then $(\omega,\leftarrow)$ is interpretable in $(Z,\leq,0)$ in view of the fact that

$$\omega = \{z : z \text{ satisfies } 0 \leq z \text{ in } (Z,\leq,0)\} \text{ and}$$
$$\leftarrow = \{(x,y) : x,y \in \omega \text{ and } (x,y) \text{ satisfies } x\leq y \wedge x\neq y \wedge \forall z[x\leq z \wedge x\neq z \rightarrow y\leq z] \text{ in } (Z,\leq,0)\}.$$

In other words $(\omega,\leftarrow)$ is interpretable in $(Z,\leq,0)$ because the universe $\omega$ and the relation $\leftarrow$ were definable in $(Z,\leq,0)$ by elementary formulas.

The Rabin-Scott method depends on finding the right interpretations. Here is the method to show $T$ is finitely inseparable:

*Step 0* Select an elementary theory $T_0$ which is known to be finitely inseparable and which seems "describable" in terms of $T$.

*Step 1* Add finitely many distinguished constants to the language of $T$ and finitely many elementary sentences as new axioms to $T$, obtaining a theory $T_1$ in which $T_0$ is even easier to describe.

*Step 2* Find a formula $\psi(x)$ (in the language of $T_1$) in one free variable and, for each relation symbol $R(x_0,\ldots,x_{n-1})$ of $T_0$ find a formula $\phi_R(x_0,\ldots,x_{n-1})$ in the language of $T_1$ such that

a) If $\mathfrak{B}$ is any model of $T_1$ and $\mathfrak{A}$ is interpretable in $\mathfrak{B}$ by $\psi(x)$ (for the universe of $\mathfrak{A}$) and $\phi_R(x_0,\ldots,x_m)$ (for each relation R in $\mathfrak{A}$), then $\mathfrak{A}$ is a model of $T_0$.

b) If $\mathfrak{A}$ is a finite model of $T_0$ then there is a finite model $\mathfrak{B}$ of $T_1$, into which $\mathfrak{A}$ can be interpreted (by the same formulas).

Once these steps have been carried out we can be secure in the knowledge that T is finitely inseparable. To see this carried out, let T be the elementary theory of interval orders.

*Step 0* We take $T_0$ to be the elementary theory of bipartite graphs. This theory is known to be finitely inseparable. It has two unary relation symbols (one for each part of the graph) and a binary relation to denote the set of edges.

*Step 1* We add one new constant symbol c to the language of interval orders and one new axiom $\exists x\, y\, [x \neq c \wedge y \neq c \wedge xRc \wedge cRy]$. So our theory $T_1$ is the theory of interval orders with a distinguished point that has something above and something below it.

*Step 2* $\psi(x)$ is $cRx \vee xRc$
$\phi_A(x)$ is $cRx$
$\phi_B(x)$ is $xRc$
$\phi_E(x,y)$ is the formula which says

"there is a z incomparable with c which is between x and y and there is nothing properly between z and x nor between z and y."

It should be abundantly clear that in every model of $T_1$ the structure defined by these formulas is a bipartite graph. In fact the reason for the new axiom we added to T to get $T_1$ should be clear: it insures that each part of the graph is nonempty. In this way, condition (a) is fulfilled. So it only remains to interpret an arbitrary finite bipartite graph into some finite interval order by means of the formulas above. The formula by which the notion of "edge" is defined will force us to include many incomparable pairs of points. On the other hand, diverse arrays of points with many incomparabilities are forbidden in interval orders. To satisfy condition (b), we construct a linearly ordered set with an antichain adjoined. The size of the antichain is one more than the number of edges. See the picture on the next page.

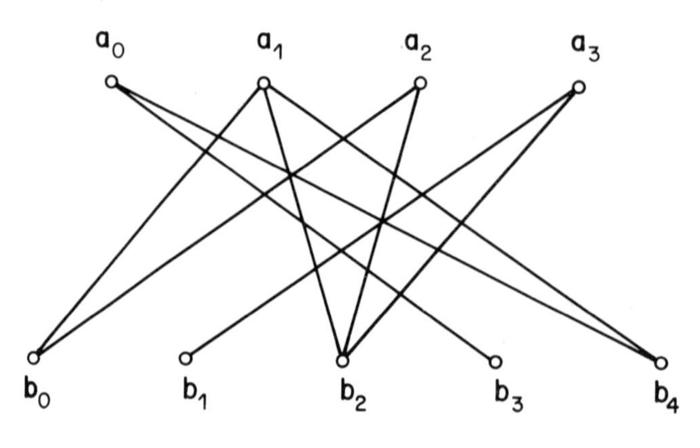


Bipartite graph

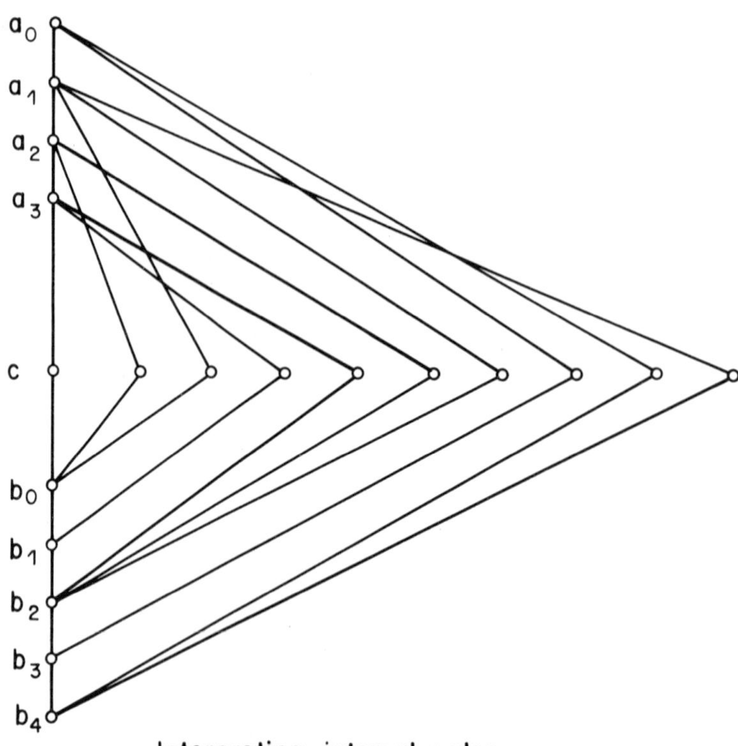


Interpreting interval order

## §4. SOME OTHER CONNECTIONS BETWEEN ORDERED SETS AND LOGIC

### Computational Complexity

We saw above that rational numbers and points of ordered sets and sentences about ordered sets could all be appropriate inputs to an algorithm. In fact, the only reasonable restraint (in the ideal sense) on the inputs and outputs of an algorithm is that they should be hereditarily finite sets. These are just the sets which can be written down with $\emptyset$, {, }, and commas. They are the finite sets which are discrete in every sense [$\{\pi\}$ is a finite set with an indiscrete flavor.] With this in mind there is nothing wrong with using finite ordered sets themselves as inputs and outputs. So it makes sense to ask of a set $K$ of finite ordered sets whether it is recursive. It is also reasonable to ask whether the function, which assigns to a given finite ordered set a parameter like its dimension, is recursive; or whether the set of all pairs of finite ordered sets, such that the first is isomorphic to (embeddable in, a retract of, ...) the second, is recursive. Virtually every question of this kind based on properties encountered in the theory of finite ordered sets is destined to have a positive answer. The interesting question is not the mere existence of an algorithm, but rather the efficiency of the "best" possible algorithm. The classification of recursive sets and functions according to computational complexity is described in Machtey and Young [1978], Hopcroft and Ullman [1979], and Garey and Johnson [1978].

### Elementary Equivalence Among Countable Ordered Sets

One of the most difficult long outstanding open problems in mathematical logic is Vaught's Conjecture [1961] that a countably infinite structure elementarily equivalent to uncountably many pairwise non-isomorphic countable structures must be elementarily equivalent to $2^{\omega}$ pairwise non-isomorphic countable structures. It has been shown by A. Miller [1977] that if Vaught's Conjecture fails, then it fails for some denumerable ordered set. M. Rubin [1974] proved that Vaught's Conjecture is true for linear orders. This result was extended by Steel [1979] who showed that Vaught's Conjecture is true for trees. In any event, this important open problem may be approached by anyone with an expertise in ordered sets and a predilection for model theory. Steel's proof used the theory of admissable sets. C. Wagner [1929] presents a more conventional proof.

### Delving Deeper into Logic

Mathematical logic is rich in methods of describing classes of objects. Above we saw two of these: recursive sets and elementary sentences. We mentioned that recursive sets were actually part of a hierarchy of sets called the arithmetical hierarchy. Most of the results in section 2 have stronger versions in the context of the arithmetical hierarchy. This hierarchy resembles in some ways the Borel and analytic hierarchies of descriptive set theory. (An account of descriptive set theory is in the Handbook of Mathematical Logic, Barwise [1977]; see also Moschovakis [1980]). Descriptive set theory offers another context in which to investigate the theory of ordered sets. In a somewhat different direction, there are formal languages more powerful in their means of expression than the language of elementary sentences described above. For example, one might study the monadic second order theory of ordered sets. In this setting quantification over subsets (but not over relations of higher rank) as well as elements is permitted. Some very interesting results concerning decidability of monadic second order theories of linear orders and trees can be found in Rabin [1969], Shelah [1975], and Gurevich and Shelah [1979]. Still other logics stronger than elementary logic are serving as arenas in which to pursue the theory of ordered sets.

## §5. OPEN PROBLEMS

The problems below are of varying degrees of difficulty, but so far as I know, they are all open.

Problem 0. (Kierstead)

> *Let* $K(w)$ *be the smallest natural number such that any recursive ordered set of width* $w$ *has recursive chain covering number no more than* $K(w)$. *Describe (the asymptotic behavior of) this function.*

Problem 1.

> *For each large enough natural number* $c$, *is there a recursive ordered set with recursive chain covering number* $c$ *which is* $k$*-crown free for all* $k < c$ *and which has no finite recursive dimension?*

Problem 2.

> *A recursive ordered set* $(A,R)$ *is autostable provided whenever* $(A',R')$ *is a recursive ordered set and* $(A,R) \cong (A',R')$ *then* $(A,R)$ *and* $(A',R')$ *are recursively isomorphic. Characterize the autostable ordered sets.* (Cf. Mal'cev [1962] *and* Nurtazin [1974]).

Problem 3 (Schmerl)

*Does every $\omega$-categorical ordered set of finite width have a finitely axiomatizable elementary theory?*

Problem 4 (Vaught [1961] and Miller [1977])

*Is every denumerably infinite ordered set which is elementarily equivalent to uncountably many pairwise non-isomorphic countable ordered sets, also elementarily equivalent to $2^{\omega}$ pairwise non-isomorphic countable ordered sets?*

Problem 5

*For which $\omega$-categorical ordered sets* $(A,R)$ *of dimension* $n$ *is it possible to find* $n$ *linear orderings* $L_0, L_1, \ldots, L_{n-1}$ *of* $A$ *such that*

i) $R = L_0 \cap L_1 \cap \ldots \cap L_{n-1}$, *and*

ii) $(A,R,L_0,L_1,\ldots,L_{n-1})$ *is $\omega$-categorical?*

## §6 REFERENCES

Below * indicates a reference supplying background in the theory of recursive functions, while # indicates a reference concerned with elementary theories. The single most useful reference for background in both areas is the Handbook of Mathematical Logic, Barwise [1977]. The papers Kierstead [1981] and Schmerl [1980] and the book Rosenstein [1981] reveal the techniques used to establish the bulk of the theorems here.


K. Baker, P. Fishburn, and F. Roberts (1971) Partial orders of dimension 2, interval orders, and interval graphs, *Networks 2*, 11-28.

\# J. Barwise (1977a) An introduction to first order logic, pp. 5-46 in Barwise [1977].

* # J. Barwise (1977) *Handbook of Mathematical Logic*, Studies in Logic v. 90, North-Holland Publishing Company, Amsterdam.

* # J. Bell and M. Machover (1977) *A Course in Mathematical Logic*, North-Holland Publishing Company, Amsterdam.

K. Bogart, I. Rabinovich, and W. Trotter (1976) A bound on the dimension of interval orders, *J. Comb. Theory*, 21, 319-328.

\# C. C. Chang and H. H. Keisler (1973) *Model Theory*, North-Holland Publishing Company, Amsterdam.

K. H. Chen (1976) *Recursive Well-Founded Orderings*, Ph.D. Thesis, Duke University.

A. Church (1938) The constructive second class number, *Bull. Amer. Math. Soc.* 44, 229-232.

A. Church and S. Kleene (1937) Formal definitions in the theory of ordinal numbers, *Fundamenta Mathematicae* 28, 11-21.

* N. Cutland (1980) *Computability, an introduction to recursive function theory*, Cambridge University Press, Cambridge.

* M. Davis (1958) *Computability and Unsolvability*, McGraw-Hill Book Company, New York.

* M. Davis (1965) *The Undecidable*, Raven Press, New York.

* M. Davis (1977) Unsolvable problems, pp. 567-594 in Barwise [1977].

R. P. Dilworth (1950) A decomposition theorem for partially ordered sets, *Annals of Math.* 51, 161-166.

J. Doner, A. Mostowski, and A. Tarski (1978) The elementary theory of well-ordering -- a metamathematical study --, in *Logic Colloquium* 77, (MacIntyre, Pacholski and Paris, eds.), North-Holland Publishing Company, Amsterdam, 1-54.

B. Dushnik and E. Miller (1941) Partially ordered sets, *Amer. J. Math.* 63, 600-610.

A. Ehrenfeucht (1959) Decidability of the theory of one linear ordering relation, Notices Amer. Math. Soc. 6, 268-269.

* # H. Enderton (1972) *A Mathematical Introduction to Logic*, Academic Press, New York.

* H. Enderton (1977) Elements of recursion theory, 527-566 in Barwise [1977].

\# Y. Ershov, I. Lavrov, A. Taimanov, and M. Taitslin (1965) Elementary theories (Russian) *Uspehi Matematisheski Nauk* 20. Translation: *Russian Mathematical Surveys* 20, 35-105.

L. Feiner (1967) *Orderings and Boolean Algebras not Isomorphic to Recursive Ones*, Ph.D. Thesis, M.I.T.

S. Fellner (1976) *Recursiveness and Finite Axiomatizability of Linear Orderings*. Ph.D. Thesis, Rutgers University.

P. Fishburn (1970) Intransitive indifference with unequal indifference intervals, *J. Math. Psychol.* 7, 144-149.

M. Garey and D. Johnson (1978) *Computers and Intractability: A Guide to the Theory of NP-Completeness*, W. H. Freeman, San Francisco.

A. Grzegorczyk (1951) Undecidability of some topological theories, *Fund. Math.* 38, 137-152.

Y. Gurevich and S. Shelah (1981) Modest theory of short chains, II, *J. Symbolic Logic* 44, 491-502.

L. Hay, A. Manaster, and J. Rosenstein (1975) Small recursive ordinals, many-one degrees, and the arithmetical difference hierarchy, *Annals of Math. Logic* 8, 297-343.

L. Hay, A. Manaster, and J. Rosenstein (1982) Concerning partial recursive similarity transformations of linearly ordered sets, *Pacific J. Mathematics*.

J. Hopcroft and J. Ullman (1979) *Introduction to Automata Theory, Languages, and Computation*, Addison-Wesley, Reading, Massachusetts.

D. Kelly (1977) The 3-irreducible partially ordered sets, *Canad. J. Math.* 29, 367-383.

D. Kelly and W. T. Trotter (1982) Dimension theory for ordered sets, *this volume*, 171-211.

H. Kierstead (1981) An effective version of Dilworth's theorem, *Trans. Amer. Math. Soc.*

H. Kierstead and W. T. Trotter (1982) An extremal problem in recursive combinatorics.

S. Kleene (1938) On notation for ordinal numbers, *J. Symbolic Logic* 3, 150-155.

M. Lerman (1981) On recursive linear orderings, in: *Logic Year 1979-80*, (Lerman, Schmerl, Soare, eds.), Lecture Notes in Mathematics v. 859, Springer-Verlag, Berlin.

M. Lerman and J. Rosenstein (1981) On recursive linear orderings: II.

* M. Machtey and P. Young (1978) *An Introduction to the General Theory of Algorithms*, North-Holland Publishing Company, Amsterdam.

A. Mal'cev (1962) On recursive Abelian groups, *Soviet Math. Dokl.* 3, 1431-1434.

* # A. Manaster (1975) *Completeness, Compactness, and Undecidability*, Prentice Hall, Englewood Cliffs, New Jersey.

A. Manaster and J. Remmel (1981a) Partial orderings of fixed finite dimension: model companions and density, *J. Symbolic Logic*.

A. Manaster and J. Remmel (1981b) Some recursion theoretic aspects of dense two dimensional partial orders.

A. Manaster and J. Rosenstein (1980a) Two dimensional partial orderings; recursive model theory, *J. Symbolic Logic*, 45, 121-132.

A. Manaster and J. Rosenstein (1980b) Two dimensional partial orderings: undecidability. *J. Symbolic Logic* 45, 133-143.

W. Markwald (1954) Zur Theorie der konstructiven Wohlcrdnungen, *Math. Annalen*, 127, 135-149.

A. Miller (1977) Ph.D. Thesis, University of Wisconsin.

Y. Moschovakis (1981) *Descriptive Set Theory*, Studies in Logic v. 100, North-Holland Publishing Co., Amsterdam.

A. Nurtazin (1974) Strong and weak constructivizations and computable families, *Algebra and Logic* 13, 177-184.

H. Putnam (1973) Recursive functions and hierarchies, in *The Thirteenth Herbert Ellsworth Slaught Memorial Papers, American Mathematical Monthly*, 80 (suppl.), 68-86.

# M. Rabin (1965) A simple method for undecidability proofs and some applications, in *Logic, Methodology, and Philosophy of Science* 11, (Y. Bar-Hille, ed.), North-Holland, Amsterdam, 58-68.

# M. Rabin (1965) Decidable theories, 565-629 in Barwise [1977].

M. Rabin (1969) Decidability of second order theories and automata on infinite trees, *Trans. Amer. Math. Soc.* 141,1-35.

I. Rabinovitch (1973) *The Dimension Theory of Semiorders and Interval Orders*, Ph.D. dissertation, Dartmouth College.

A. Robinson and E. Zakon (1960) Elementary properties of ordered abelian groups, *Trans. Amer. Math. Soc.* 96, 222-235.

* H. Rogers (1967) *Theory of Recursive Functions and Effective Computability*, McGraw-Hill Book Company, New York.

J. Rosenstein (1969) $\aleph_0$ categoricity of linear orderings, *Fundamenta Mathematicae* 44, 1-5.

J. Rosenstein (1981) *Linear Orderings*, Academic Press, New York.

M. Rubin (1974) Theories of linear orderings, *Israel J. of Mathematics* 17, 392-443.

J. Schmerl (1977) On $\aleph_0$ categoricity and the theory of trees, *Fundamenta Mathematicae* 94, 121-128.

J. Schmerl (1980) Decidability and $\aleph_0$ categoricity of theories of partially ordered sets, *J. Symbolic Logic* 45, 585-611.

D. Scott and P. Suppes (1958) Foundational aspects of theories of measurements, *J. Symbolic Logic* 23, 113-128.

S. Shelah (1975) The monadic theory of order, *Annals of Mathematics* 102, 379-419.

C. Spector (1955) Recursive well orderings, *J. Symbolic Logic* 206, 151-163.

J. Steel (1978) On Vaught's Conjecture, in *Cabal Seminar* 76-77, Lecture Notes in Mathematics, Springer-Verlag, Berlin.

J. Stillwell (1977) Concise survey of mathematical logic, *J. Austral. Math. Soc. Ser. A*, 24, 139-161.

E. Szpilrajn (E. Marczewski) (1930) Sur l'extension de l'ordre partial, *Fund. Math.* 16, 386-389.

A. Tarski (1949a) Arithmetical classes and types of Boolean algebras, *Bull. Amer. Math. Soc.* 55, 64 and 1192.

A. Tarski (1949b) Undecidability of the theories of lattices and projective geometries, *J. Symbolic Logic* 14, 77-78.

# A. Tarski with A. Mostowski and R. Robinson (1953) *Undecidable Theories*, Studies in Logic, North-Holland Publishing Company, Amsterdam.

W. T. Trotter (1974) Dimension of the crown $S_n^k$, *Discrete Math.* 8, 85-103.

W. T. Trotter and J. Moore (1976) Characterization problems for graphs, partially ordered sets, lattices, and families of sets, *Discrete Math.* 16, 361-381.

R. Vaught (1954) Applications of the Löwenheim-Skolem-Tarski theorem to problems of completeness and decidability, *Indag. Math.* 16, 467-472.

R. Vaught (1961) Denumerable models of complete theories, in *Infinitistic Methods*, Pergamon, London.

# R. Vaught (1973) Some aspects of the theory of models, in the *Thirteenth Herbert Ellsworth Slaught Memorial Papers, Amer. Math. Monthly* 80 (supp.), 3-37.

C. Wagner (1979) *On Martin's Conjecture*, Ph.D. Thesis, Cornell University.

R. Watnick (1979) *Recursive and Constructive Linear Orderings*, Ph.D. Thesis, Rutgers University.

R. Watnick (1981) Constructive and recursive scattered order types, in *Logic Year (1979-80)*, M. Lerman, J. Schmerl, and R. Soare, eds.), Lecture Notes in Mathematics vol. 859, Springer-Verlag, Berlin.

Footnote:

* This paper was prepared while the author was supported, in part, by NSF Grant MCS-80-01778.

# PART IV

# LATTICE THEORY

THE ROLE OF ORDER IN LATTICE THEORY

R.P. Dilworth
Department of Mathematics
California Institute of Technology
Pasadena, California 91125

ABSTRACT

As an algebraic system, a lattice is a two operation system with equation definable axioms. The usual axiom set is the following:

Idempotent laws: $a \vee a = a, \quad a \wedge a = a$

Commutative laws: $a \vee b = b \vee a, \quad a \wedge b = b \wedge a$

Associative laws: $(a \vee b) \vee c = a \vee (b \vee c),$

$(a \wedge b) \wedge c = a \wedge (b \wedge c)$

Adsorptive laws: $a \vee (a \wedge b) = a, \quad a \wedge (a \vee b) = a$ .

A particularly significant feature of a lattice as an algebraic system is the fact that it has an intrinsic order relation given by

$$a \leq b \text{ if and only if } a \wedge b = a$$

or equivalently

$$a \leq b \text{ if and only if } a \vee b = b .$$

The lattice axioms insure that this relation is a partial order consistent with the lattice operations.

The intrinsic character of the lattice ordering sets lattices apart from most other ordered algebraic systems, such as ordered groups, ordered rings, ordered fields, etc. in which the order relation is imposed on the system and has specified properties. Furthermore, the lattice operations themselves can be defined in terms of the lattice order relation.



*I. Rival (ed.), Ordered Sets, 333–353.*

Least upper bound: $a \vee b \geq a, b;$

$x \geq a, b \Rightarrow x \geq a \vee b$

Greatest lower bound: $a, b \geq a \wedge b;$

$a, b \geq x \Rightarrow a \wedge b \geq x$ .

This special relationship between order properties and algebraic properties for lattices has played an important role in the development of lattice theory. Some of the more significant implications of this relationship will be examined in this paper.

1. A finite partially ordered set is completely determined by the set of covering pairs, namely, those pairs $(a,b)$ for which $a > b$ and there is no $x$ for which $a > x > b$. It follows that a finite lattice is determined by its set of covering pairs. This leads immediately to the lattice diagram of a finite lattice in which the elements of the lattice are represented by points on a sheet of paper so arranged that for a covering pair $(a,b)$ the point corresponding to $a$ lies above the point corresponding to $b$ and is joined to it by a line segment. Infinite lattices with a sufficiently regular structure can also be diagramed in this manner. From the earliest investigations, lattice diagrams have been an important research tool and a useful device for exposition. [5, 21, 25, 31]

2. Associated with any partially ordered set is a lattice (in fact, a complete lattice) obtained as a generalization of Dedekind's completion by cuts. If $A$ is a subset of the partially ordered set and $A^*$ is the set of upper bounds of $A$ and $A_*$ is the set of lower bounds of $A$, then $A \to (A^*)_*$ is a closure operation on the subsets and the closed subsets under set inclusion are a complete lattice -- the normal completion of the partially ordered set. This construction is basic for the study of many embedding problems. When applied to lattices, the completion process preserves the meet and join operations and hence is a lattice embedding. Thus it is not unexpected that the normal completion is useful in the study of purely algebraic problems in lattice theory. [8, 20]

3. Associated with any finite distributive lattice $L$ is the partially ordered set $J$ of its join irreducible elements ($q$ is join irreducible if $q = a \vee b$ implies $q = a$ or $q = b$). The order ideals of $J$ form a lattice isomorphic to $L$. Under this correspondence, many algebraic properties of $L$ are reflected in order-theoretic properties of $J$. Since the order-theoretic analogues of algebraic problems may be easier to attack, this correspondence frequently provides a promising approach to the study of problems concerning distributive lattices. [2, 11]

4. Much of the structure theory of algebraic systems centers on the study of the lattice of congruence relations. In particular, the lattice of congruence relations on a lattice plays a very important role in exhibiting its lattice structure. As a consequence of being an ordered system the lattice of congruence relations on a lattice is distributive. In turn, the congruence distributivity of lattices underlies many of the most interesting structure theorems for lattices. [12, 14, 17]

As an algebraic system, a lattice is a two operation system with equation definable axioms. The usual axiom set is the following:

Idempotent laws: $a \vee a = a \wedge a = a$

Commutative laws: $a \vee b = b \vee a$, $a \wedge b = b \wedge a$

Associative laws: $(a \vee b) \vee c = a \vee (b \vee c)$, $(a \wedge b) \wedge c = a \wedge (b \wedge c)$

Absorptive laws: $a \vee (a \wedge b) = a$, $a \wedge (a \vee b) = a$.

A particularly significant feature of a lattice as an algebraic system is the fact that it has an intrinsic order relation given by

$$a \leq b \text{ if and only if } a \wedge b = a$$

or equivalently

$$a \leq b \text{ if and only if } a \vee b = b.$$

The lattice axioms insure that this relation is a partial order consistent with the lattice operations.

The intrinsic character of the lattice ordering sets lattices apart from most other ordered algebraic systems, such as ordered groups, ordered rings, ordered fields, etc. in which the order relation is imposed on the system with certain specified properties. Furthermore, the lattice operations themselves can be defined in terms of the lattice order relation.

Least upper bound: $a \vee b \geq a, b;\ x \geq a, b \Rightarrow x \geq a \vee b$

Greatest lower bound: $a, b \geq a \wedge b;\ a, b \geq x \Rightarrow a \wedge b \geq x$

This special relationship between order properties and algebraic properties for lattices has played an important role in the development of lattice theory. Some of the more significant implications of this relationship will be examined in this paper.

LATTICE DIAGRAMS. A finite partially ordered set is completely determined by the set of its covering pairs, namely, those pairs $(a,b)$ for which $a > b$ and there is no $x$ such that $a > x > b$. In particular, a finite lattice is determined by its order relation and hence is characterized by its set of covering pairs. This leads immediately to the lattice (or Hasse) diagram for a finite lattice in which the elements of the lattice are represented by points on a sheet of paper so arranged that for a covering pair (a,b), the point corresponding to a lies above the horizontal through the point corresponding to b and the two points are joined by a line segment. As used by Hasse to represent subfield relationships in fields and subgroup relationships in groups, the diagrams proved to be a useful exposition device. But it is in lattice theory that the diagrams have had their greatest impact, not only for purposes of exposition but as a technique for discovering and proving theorems. A few specific examples may help to elucidate the role of lattice diagrams in the development of the subject.

A lattice generated by a single element consists only of the generating element, since the lattice operations are idempotent. The lattices diagramed in Figure 1 are the only lattices with two generators.

Figure 1.

There are many lattices having three generators. Although most of them are infinite, there are many important examples of finite three generated lattices. For example Davey and Rival [7] have shown that every lattice generated by three unordered elements has a sublattice isomorphic to one of twelve specific finite lattices with three unordered generators. However, there are several finite examples of three generated lattices that have been particularly significant in the development of the subject. These examples are diagramed in Figures 2 and 3.

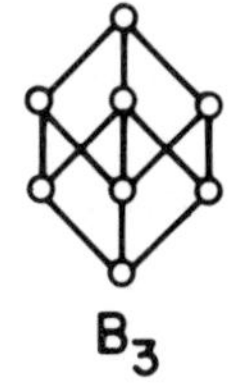


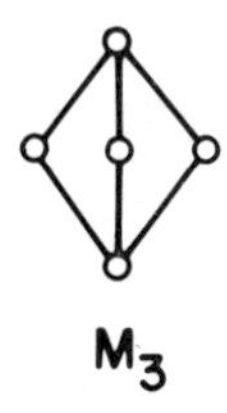


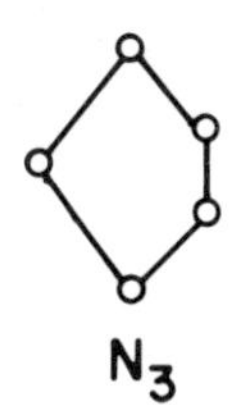


Figure 2.

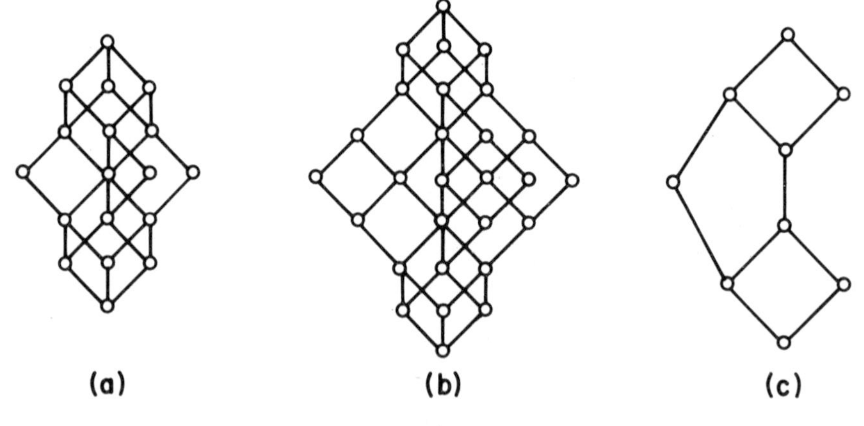


Figure 3.

$B_3$ is the lattice of subsets of a three element set. $M_3$ is the smallest modular, nondistributive lattice. $N_3$ is the smallest nonmodular lattice. Figure 3a is the most general (= free) distributive lattice with three generators. Figure 3b is the most general modular lattice with three generators. Figure 3c is the most general lattice with three generators two of which are a comparable pair.

The free modular lattice on three generators was first determined by Richard Dedekind [8] around the turn of the century. There is no evidence in his collected works that he drew the lattice diagram of Figure 3b, but he did succeed in finding three submodules of a certain module which indeed generated the lattice of Figure 3b.

Note that there is a copy of $M_3$ in the central area of Figure 3b. If these elements are identified, Figure 3b becomes 3a and hence is distributive. Thus every modular nondistributive lattice must have a copy of $M_3$ as a sublattice. Similarly Figure 3c contains a copy of $N_3$. If the comparable pair in this copy of $N_3$ is identified Figure 3c becomes a sublattice of Figure 3a and hence is distributive. Thus every nonmodular lattice must have a copy of $N_3$ as a sublattice. It follows that modular lattices may be characterized as those lattices having no sublattice isomorphic to $N_3$. Similarly distributive lattices may be characterized as those lattices having no sublattice isomorphic to $N_3$ or to $M_3$. The characterization of classes of lattices in terms of the exclusion of a specified set of sublattices occurs frequently in lattice theory and has been particularly useful in the study of varieties of lattices.

In order to illustrate the role of lattice diagrams in theorem proving it will be convenient to examine in some detail a theorem on complemented lattices. Now a lattice is *complemented* if it has a unit (= largest) element 1, a null (= smallest) element 0, and for each x there is a y such that $x \vee y = 1$ and $x \wedge y = 0$. If each quotient sublattice $a/b = \{x \mid a \geq x \geq b\}$ is complemented, the lattice is said to be *relatively complemented*. If the lattice is distributive, complements are unique. For if z is another complement of x, then

$$z = z \wedge 1 = z \wedge (x \vee y) = (z \wedge x) \vee (z \wedge y) = z \wedge y = (x \wedge y) \vee (z \wedge y) = (x \vee z) \wedge y = y.$$

It was conjectured in the late 1930's and early 1940's that the converse also holds, namely, a uniquely complemented lattice is distributive. This turned out not to be the case (Dilworth [10]). Nevertheless, a number of affirmative theorems for restricted classes of lattices were proved. The most noteworthy of these positive results is the elegant theorem of McLaughlin [21] which asserts that *every uniquely complemented atomic lattice is distributive*. A lattice is *atomic* if it has a null element 0 and for each $x > 0$ there exists an element p covering 0 such that $x \geq p$. McLaughlin actually proved the more general result that a *complemented atomic lattice with unique comparable complements is modular*. This was accomplished by showing that the hypotheses imply that the lattice is relatively complemented. For if the lattice is relatively complemented and nonmodular it contains a copy of $N_3$. Let a, b, c with $a \geq b$ be the generating elements of $N_3$ and let x be a relative complement of $b \wedge c$ in $c/0$ and let y be a relative complement $a \vee c$ in $1/x$. Then, as the diagram in Figure 4 shows, a and b are comparable complements of y.

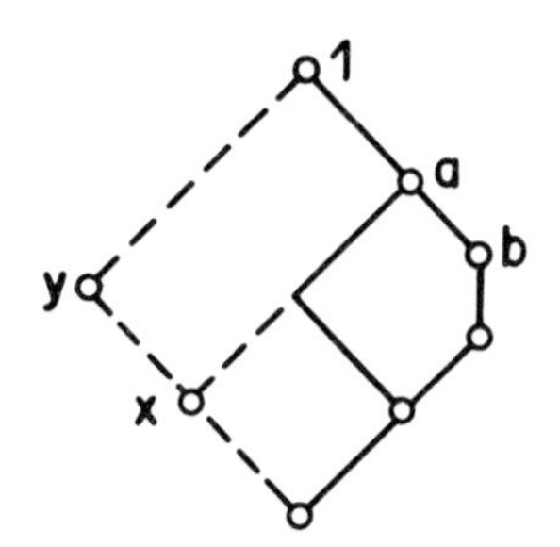


Figure 4.

The McLaughlin proof proceeds in a series of lemmas involving extensive lattice calculations each step of which is transparent but for which the strategy is obscure. The key lemma concerning a uniquely complemented atomic lattice L which gets the proof underway is the following:

A) If $a \not\geq p$, where p is an atom of L, then $a \vee p$ covers a. In order to understand the proof, consider the diagram of Figure 5.

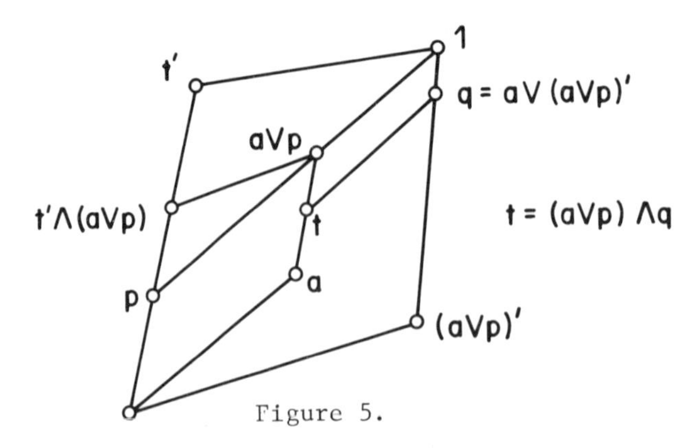


Figure 5.

In Figure 5 we have $a \not\geq p$, $q = a \vee (a \vee p)'$ and $t = (a \vee p) \wedge q$. Now $q \neq 1$ since otherwise a and $(a \vee b)$ would be comparable complements of $(a \vee p)'$. Also 1 covers q since any element between 1 and q would be a complement of p. Now let $a \vee p > x \geq a$. If $x \not\leq q$, then $x \vee (a \vee p)' = x \vee a \vee (a \vee p)' = x \vee q = 1$ and x and $a \vee p$ are comparable complements of $(a \vee p)'$. Hence

$$a \vee p > x \geq a \Rightarrow q \geq x \Rightarrow t \geq x.$$

Thus $a \vee p$ covers t. Now $t' \wedge (a \vee p) \not\leq t$ since otherwise $t' \wedge (a \vee p) = 0$ and $a \vee p$ and t are comparable complements of $t'$. Thus since $a \vee p \geq a \vee (t' \wedge (a \vee p)) \geq a$ we have $a \vee (t' \wedge (a \vee p)) = a \vee p$. But then $a \vee t' = a \vee p \vee t' \geq t \vee t' = 1$ and a and t are comparable complements of $t'$. Hence $a = t$ and $a \vee p$ covers a.

If $a > b$, then $a \wedge b' \neq 0$ and hence $a \wedge b' \geq p$ for some atom p. Thus $a \geq p$ and $b \not\geq p$. From this observation follows the stronger form of the lemma.

*If* a *covers* $a \wedge b$, *then* $a \vee b$ *covers* b.

Now consider the diagram of Figure 6.

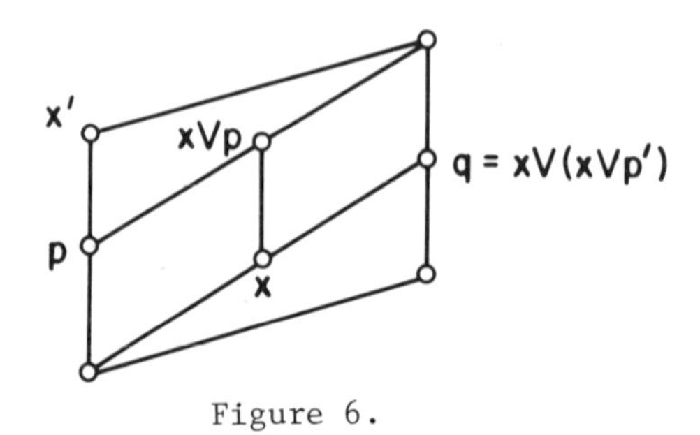


Figure 6.

If $x \neq 1$ and p is an atom in $x'$, then $q = x \vee (x \vee p)'$ is not equal to 1 and is covered by 1. Hence the dual of L is atomic. But then the dual of the above lemma holds and we have

B) a *covers* $a \wedge b$ *if and only if* $a \vee b$ *covers* b.

Finally, the diagram of Figure 7 outlines a proof of relative complementation with respect to 0.

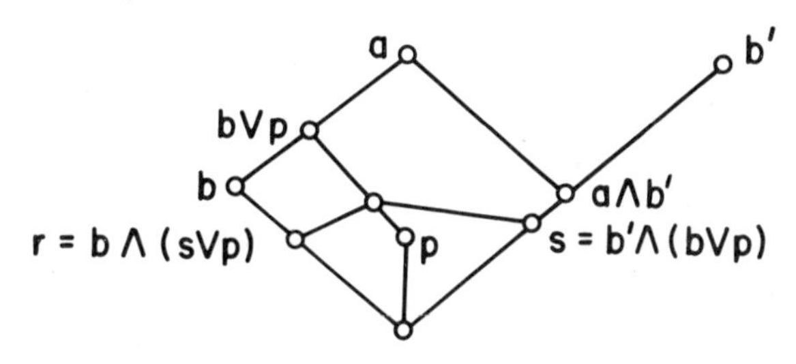


Figure 7.

It will clearly suffice to show that each atom p contained in a is also contained in $b \vee (a \wedge b')$. Also it is clear that we may assume $p \not\leq b$ and $p \not\leq b'$. Hence $s \neq 0$ since otherwise b and $b \vee p$ would be comparable complements of $b'$. Since $b \vee p$ covers b by A), $b \vee s = b \vee p$ covers b and hence s covers $b \wedge s = 0$ by B). But then $s \vee p$ covers p by A) and $b \vee (s \vee p) = b \vee p$ covers b. Again by B) $s \vee p$ covers $b \wedge (s \vee p) = r$ and hence $r \vee s = s \vee p \geq p$. It follows that $p \leq b \vee (a \wedge b')$.

By duality, L is relatively complemented with respect to 1 and hence the proof of the McLaughlin theorem is complete. Furthermore, Figures 4-7 contain all of the essential elements of the proof.

The mere existence of the representation of lattices by diagrams in the plane has been the source of some interesting investigations of the structure of lattices. For example, a finite lattice is *planar* if it has a lattice diagram in which no two of the line segments corresponding to covering pairs intersect. The class of planar lattices can be characterized by the exclusion of an appropriate collection of subpartially ordered sets [19]. Furthermore a finite planar lattice is *dismantlable*, namely, there exists a chain of sublattices

$$L = L_0 \supset \cdots \supset L_n = (0)$$

where each $L_i$ is obtained by removing one element from $L_{i-1}$. Clearly the removed element must be both join and meet irreducible in $L_{i-1}$. In order to see that a planar lattice is dismantlable, consider the covering segments from 1. One of these, say (1,a,) lies furthest to the right. Similarly, there is a covering segment below a, which is furthest to the right. Continuing in this way produces a maximal chain

$$1 > a_1 > a_1 > a_2 > \cdots > a_n = 0$$

which is a right boundary of the lattice. Since $a_1$ is maximal in 1, it is meet irreducible. If it is join irreducible it can be removed and a proper sublattice remains. Hence we may suppose that $a_1$ is reducible. Let $\kappa$ be a maximal such that $a_\kappa$ is reducible. Then there is a covering pair $(a_\kappa, b_{\kappa+1})$ whose segment lies to the left of $(a_\kappa, a_{\kappa+1})$. Let C be a maximal chain from $a_\kappa$ to 0 beginning with $a_\kappa > b_{\kappa+1}$. Now $a_{\kappa+1}$ is join irreducible. If it is meet irreducible it can be removed and a sublattice remains. Hence we assume that it is meet reducible and there is a covering pair $(c_\kappa, a_{\kappa+1})$ whose segment lies to the left of $(a_\kappa, a_{\kappa+1})$. Let D be a maximal chain to 1 beginning with $a_{\kappa+1} < c_\kappa$. Then C and D must intersect and hence a segment of C must intersect a segment of D contrary to planarity (Figure 8).

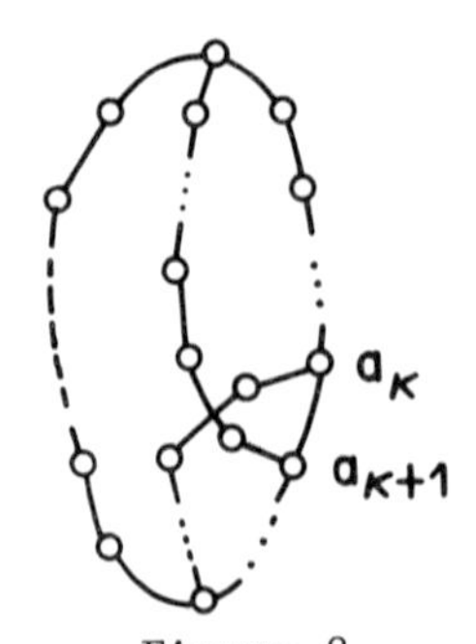


Figure 8.

Although lattice diagrams are particularly appropriate for the study of finite lattices, they can also be useful for the representation of certain infinite lattices. Thus if a finite portion of the lattice determines in a regular manner the structure of the whole, then a diagram may be a satisfactory representation. For example the lattice of Figure 9 is an infinite lattice which is infinitely distributive but whose normal completion is not even modular. The dots indicate that the given configuration continues indefinitely. The construction of this lattice illustrates an important technique of joining two different types of configurations in a way which cannot be accomplished in a finite lattice.

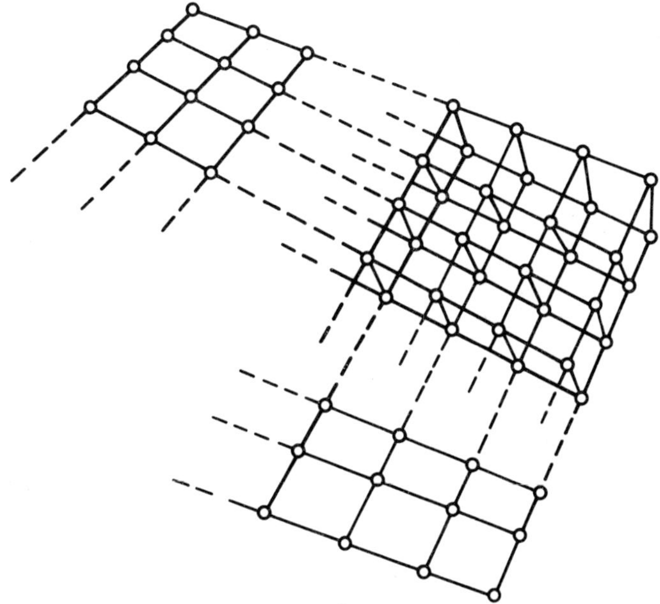

Figure 9.

In recent years, R. Wille [29] has made extensive use of lattice diagrams in the study of various aspects of finite lattices. In particular, he has sketched in detail lattices which are freely generated by partially ordered sets and, in particular, has shown that FL(P) is finite if and only if P contains no copies of $1 + 1 + 1$, $2 + 2$, or $1 + 4$ in which case FL(P) is a subdirect product of $D_2$, $N_3$, and the lattice of Figure 10.

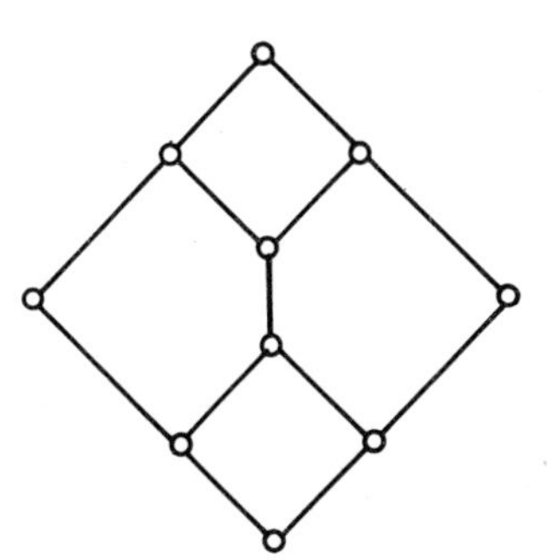

Figure 10.

I. Rival and R. Wille [25] have tackled the general problem of determining when FL(P) is "drawable", i.e. has a reasonable lattice diagram. Their main result states that FL(P) is "drawable" if and only if P contains no copy of $1 + 1 + 1$, $2 + 3$, or $1 + 5$. Moreover, in this case, they show that FL(P) is isomorphic to a sublattice of FL(H) where H is the partially ordered set diagramed in Figure 11.

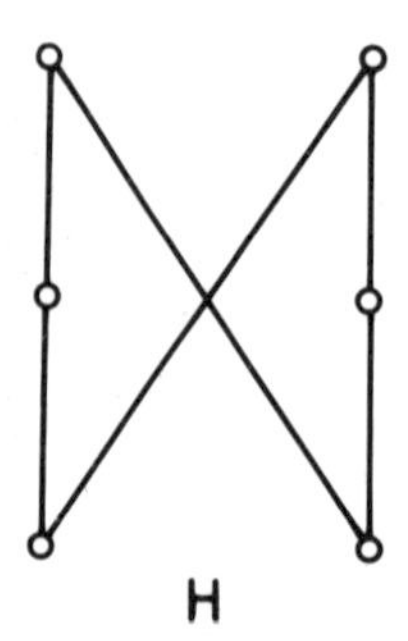


Figure 11.

COMPLETION OF PARTIALLY ORDERED SETS. Among the lattice theoretic studies in which order played an important role, perhaps the most significant was the study of lattice embedding. One of the earliest embedding questions to be investigated was the problem of embedding a partially ordered set in a complete lattice (MacNeille [20]).

Let us suppose that a partially ordered set P is embedded in a complete lattice L. Then there is a natural map of subsets of P into subsets of P given by

$$\varphi(A) = \{x \in P \mid x \leq \vee A\}.$$

The mapping $\varphi$ clearly has the following properties:

1) $\varphi(A) \supseteq A$

2) $A \supseteq B \Rightarrow \varphi(A) \supseteq \varphi(B)$

3) $\varphi(\varphi(A)) = \varphi(A)$.

A mapping $\varphi$ on the subsets of P having properties 1), 2), and 3) is called a *closure operator*. If $x \in P$, $(x]$ will denote $\{y \mid y \leq x\}$. The mapping $\varphi$ defined above has the additional property

4) $\varphi(\{x\}) = (x]$.

A closure operator on the subsets of P having property 4) is called an *embedding* operator.

Now let $\varphi$ be an arbitrary embedding operator on the subsets of P. Let

$$L = \{A \subseteq P \mid \varphi(A) = A\}.$$

L is clearly partially ordered by set inclusion. If $\{A_\alpha\}$ is a collection of subsets belonging to L, then

$$\varphi(\bigcap_\alpha A_\alpha) \subseteq \bigcap_\alpha \varphi(A_\alpha) = \bigcap_\alpha A_\alpha \subseteq \varphi(\bigcap_\alpha A_\alpha).$$

Hence $\varphi(\bigcap_\alpha A_\alpha) = \bigcap_\alpha A_\alpha$ and L is closed under set intersections.

Since $P \in L$, L is a complete lattice. Furthermore the one-to-one mapping $x \to (x]$ is an order preserving embedding of P in L. From these considerations it is clear that embedding operators provide the appropriate tool for embedding a partially ordered set in a complete lattice.

There is a natural partial ordering of the embedding operators on a partially ordered set P; namely

$\varphi \geq \psi$ if and only if $\varphi(A) \geq \psi(A)$ all $A \subseteq P$. Under this partial ordering there is a unique smallest embedding operator $\eta$ given by

$$\eta(A) = \bigcup\{(a] \mid a \in A\}.$$

It is easily verified that the set union of $\eta$-closed subsets is again closed so that $\eta$ embeds P in a distributive complete lattice. The $\eta$-closed subsets of P are called *order ideals*. There is also a unique largest embedding operator $\nu$ given by

$$\nu(A) = \bigcap\{(x] \mid (x] \supseteq A\}.$$

$\nu$ is the most efficient embedding operator in the sense that if P is embedded in a complete lattice L, then there is an order preserving embedding of the lattice of $\nu$-closed subsets into L. It is customary to refer to the lattice of $\nu$-closed subsets as the *normal* completion of P. Furthermore, the embedding of P in the lattice of $\nu$-closed subsets preserves all meets and joins which exist in P.

If P is a lattice, there is a particularly important embedding operator for P, called the *ideal* embedding operator given by

$$\theta(A) = \{x \in P \mid x \leq a_1 \vee \cdots \vee a_n \text{ where } a_i \in A\}.$$

The $\theta$-closed subsets of P are the *ideals* of P. The lattice of ideals of P has the important property that any lattice identity satisfied by P is also satisfied by the lattice of ideals. Hence if P belongs to a variety of lattices, the lattice of ideals of P also belongs to the variety.

In contrast with the ideal operator, the operator $\nu$ need not preserve identities. Indeed, the lattice of Figure 9 is a distributive lattice whose normal closure is even non-modular. However, if the lattice is also complemented, i.e. is a Boolean algebra, then the normal completion is also a Boolean algebra.

The normal completion of a partially ordered set P is an order-theoretic construct. Nevertheless it has proved to be a useful tool for both order-theoretic and algebraic studies of lattices. Two examples will illustrate the use of the normal completion.

Let L be a complete lattice and let S be a subset of L such that every element of L is both a join of elements of S and a meet of elements of S. Now S inherits a natural partial ordering from L. Then it is easily verified that the normal closure of S is isomorphic to L. Thus all of the information concerning the lattice L is contained in the subset S. In particular, if L is a finite lattice and S is the set of elements of L which are either join irreducible or meet irreducible, then L can be obtained from S by means of the normal completion. In the same way, if L is a finite geometric lattice and S is the bi-partite partially ordered set consisting of the atoms and coatoms, then L can be reconstructed from S.

As a second example, consider the following question which arose in the early days of lattice theory: Does a lattice equation which holds in every finite lattice hold in general? An affirmative answer was supplied by the following argument (Dean [8]). Let u, v be lattice polynomials in the variables $x_1, \ldots, x_n$. Furthermore, let u, v represent distinct elements in the free lattice generated by $x_1, \ldots, x_n$. Let P be the partially ordered set of all lattice polynomials in $x_1, \ldots, x_n$ whose length does not exceed the sum of the lengths of u and v. Then P is a finite partially ordered set and its normal completion L is a finite lattice. Since all existing joins and meets are preserved in the embedding, u and v are obtained from $x_1, \ldots, x_n$ in L exactly as in the free lattice and hence u = v fails in L. Thus a purely algebraic question is settled by an order-theoretic construction.

DISTRIBUTIVE LATTICES. An element q of a lattice L is *join irreducible* if

$$q = a \vee b \Rightarrow q = a \text{ or } q = b.$$

An easy inductive argument shows that each element of L is a join of a finite number of join irreducibles. For if $a \in L$ is join irreducible there is nothing to prove. Otherwise $a = b \vee c$ where $a > b$, $a > c$. But then if b and c are joins of a finite number of join irreducibles, a is a join of a finite number of join irreducibles. Let J(L) denote the join irreducibles of L.

Now let L be a finite distributive lattice. If $a \in L$, then $J_a = \{q \in J(L) | q \leq a\}$ is an order ideal ($\eta$-closed subset) of J(L). Clearly $\bigvee J_a = a$ and it is easily verified that the mapping

$$a \to J_a$$

is an isomorphism from L to the lattice of order ideals of L. Hence the lattice structure of L is determined by the order structure of J(L). Thus it is not surprising that algebraic properties of L are closely related to order properties of J(L).

If P is an arbitrary partially ordered set, we have noted that the order ideals of P form a complete distributive lattice L. J(L) consists of the principal order ideals (p) where $p \in P$. Hence J(L) is order isomorphic to P. It is convenient to represent L as $\underset{\sim}{2}^P$ consisting of all order preserving maps of P into the two-element lattice.

As a first example of the relation of algebraic properties of L to order properties of P, let P, Q be disjoint partially ordered sets and let $L = \underset{\sim}{2}^P$, $M = \underset{\sim}{2}^Q$. Let P + Q denote the partially ordered set whose elements consist of the set union $P \cup Q$ and the order relation present in P and in Q. From a diagramatic point of view P + Q are simply P and Q set side by side. Then it is easily verified that

$$L \otimes M \cong \underset{\sim}{2}^{P+Q}$$

since every order ideal $\underset{\sim}{2}^{P+Q}$ is of the form $A \cup B$ where A is an order ideal of P and B is an order ideal of Q. This is just a very beginning example of the extensive theory of the arithmetic of partially ordered sets.

Returning to finite distributive lattices L it has been noted above that

$$L \cong \underset{\sim}{2}^{J(L)}.$$

Let $0 = x_0 < x_1 < x_2 < \cdots < x_n = 1$ be a maximal chain in L. Then there is exactly one join irreducible $q_i$ such that $q_i \leq x_i$, $q_i \not\leq x_{i-1}$. Furthermore $q_1, \ldots, q_n$ are all of the join irreducibles of L. Hence we have

$$\ell(L) = |J(L)|.$$

Next note that the mapping $a \to J_a = \{q \in J(L) | q \leq a\}$ gives a representation of L as a lattice of subsets of J(L). If L is represented as a lattice of subsets of a finite set S, then it is easily verified that

$$|S| \geq |J(L)|.$$

Hence $|J(L)|$ is the minimum size of a representative set for L.

Now let D be another finite distributive lattice. We ask: When is D a homomorphic image of L? If $q \in J(L)$, let $q^*$ denote the unique element of L covered by q. Now suppose that f is a homomorphic mapping of L onto D. Let $Q = \{q \in J(L) | f(q) > f(q^*)\}$. Q is a partially ordered set under the ordering of L. If $q \in Q$, then it can be shown that $f(q) \in J(D)$ and f is an isomorphism of Q onto J(D). Hence J(D) is isomorphic to a subpartially ordered set of J(L). Conversely, if J(D) is isomorphic to a subset Q of J(L), let $\theta$ be the congruence relation on L generated by the pairs $(q,q^*)$ where $q \notin Q$. Then $L/\theta$ is isomorphic to the lattice of order ideals of Q and hence is isomorphic to D. Thus D is a homomorphic image of L if and only if J(D) is isomorphic to a subset of J(L).

We next ask: When is D isomorphic to a sublattice of L? Let us suppose D is isomorphic to a sublattice $L_0$ of L under the mapping f. Let

$$Q_0 = \{q \in J(L) | q \leq f(p) \text{ for some } p \in J(D)\}.$$

Clearly $Q_0$ is an order ideal of J(L). It can be shown that for each $q \in Q_0$, there is a unique $p \in J(D)$ such that $q \leq f(p)$, $q \nleq f(p^*)$. Let $g(q) = p$. Then g is an order preserving map of $Q_0$ onto J(D). Conversely, if g is an order preserving map of an order ideal $Q_0$ of L onto J(D) and if $d \in D$, let

$$f(d) = \{q \in Q_0 | g(q) \leq d\}.$$

Then f(d) is an order ideal of $Q_0$ and hence is an order ideal of J(L). It is easy to see that f preserves meets and joins and is an injection. Since L is isomorphic to the lattice of order ideals of J(L) it follows that D is isomorphic to a sublattice of L. Thus D is isomorphic to a sublattice of L if and only if there is an order preserving map of an ideal of J(L) onto J(D).

We conclude this section with two further examples of the analogues of properties of L in J(L).

L *is a sublattice of a direct product of* $\kappa$ *chains if and only if* J(L) *is a set union of* $\kappa$ *chains.*

*Every element of* L *is covered by* $\kappa$ *or less elements if and only if each anti-chain of* J(L) *has at most* $\kappa$ *elements.*

In summary, it seems to be a general principle that direct products for L are reflected in set unions in J(L), sublattices are reflected in order-preserving maps and homomorphic images are reflected in subsets.

CONGRUENCE LATTICES. If L is a lattice, an equivalence relation $\theta$ is a *congruence relation* if it satisfies

$$a \;\theta\; b \Rightarrow a \wedge c \;\theta\; b \wedge c \text{ and } a \vee c \;\theta\; b \vee c.$$

There is a natural partial ordering of congruence relations given by

$$\theta \leq \varphi \text{ if and only if } a \;\theta\; b \Rightarrow a \;\varphi\; b.$$

The equality relation is the smallest congruence relation, while the largest congruence relation relates all elements of the lattice. The set of congruence relations on L will be denoted by Con L. If $\Theta$ is a subset of Con L, the relation

$$a \bigwedge\Theta\; b \text{ if and only if } a \;\theta\; b \text{ all } \theta \in \Theta$$

is clearly a congruence relation and is the meet of $\Theta$ under the partial ordering of congruence relations. Thus Con L is a complete lattice. It is not hard to show that $a \bigvee\Theta\; b$ holds if and only if there exists a sequence

$$a = x_0, x_1, \ldots, x_n = b$$

of elements of L and congruence relations $\theta_1, \ldots, \theta_n$ in Con $\Theta$ such that $x_{i-1} \;\theta_i\; x_i$ for $i = 1, \ldots, n$.

It is a remarkable fact, first observed by Funayama and Nakayama [14], that Con L is distributive. The key to the proof is the order preserving map

$$f : x \to (a \vee b) \wedge ((a \wedge b) \vee x)$$

which squeezes the lattice L into the quotient lattice $a \vee b / a \wedge b$. Note that $f(a) = a$, $f(b) = b$ and if $a \;\theta\; b$, then $f(x) \;\theta\; a \wedge (a \vee x) = a$ and hence $f(x) \;\theta\; f(y)$ all x and y. Now let

$$a \;\theta \wedge \bigvee\theta_\alpha\; b.$$

Then $a \;\theta\; b$ and $a \bigvee\theta_\alpha\; b$. Hence there exist $x_0, x_1, \ldots, x_n$ and $\alpha_1, \ldots, \alpha_n$ such that

$$a = x_0 \;\theta_{\alpha_1} x_1 \;\theta_{\alpha_2} \cdots \theta_{\alpha_n} x_n = b.$$

But then $a = f(a) = f(x_0) \;\theta_{\alpha_1}\; f(x_1) \;\theta_{\alpha_2} \cdots \theta_{\alpha_n} f(x_n) = f(b) = b$.

Since $f(x_{i-i})\,\theta\, f(x_i)$ we have $f(x_{i-i})\,\theta \wedge \theta_{\alpha_i}\, f(x_i)$ and thus

$$a \bigvee_{\alpha}(\theta \wedge \theta_\alpha)\ b.$$

It follows that $\theta \wedge \vee\theta_\alpha \leq \bigvee_\alpha(\theta \wedge \theta_\alpha)$. However, it is trivial that $\theta \wedge \vee\theta_\alpha \geq \bigvee_\alpha(\theta \wedge \theta_\alpha)$. Hence we have

$$\theta \wedge \vee\theta_\alpha = \vee(\theta \wedge \theta_\alpha).$$

The distributivity of Con L underlies much of the structure theory of lattices. For example, let

$$L = L_1 \otimes L_2 \otimes \cdots \otimes L_n.$$

The kernel of the projection $L \to L_i$ is a congruence relation $\theta_i \in$ Con L. The congruence relations $\theta_i$ have the properties

1) $\theta_1 \wedge \cdots \wedge \theta_n = 0$

2) $\theta_i \vee (\theta_1 \wedge \cdots \wedge \theta_{i-1} \wedge \theta_{i+1} \wedge \cdots \wedge \theta_n) = 1$

3) $\theta_i, \ldots, \theta_n$ pairwise permute.

Two congruence relations $\theta$, $\psi$ *permute* if $a\ \theta\ c\ \psi\ b \Rightarrow a\ \psi\ d\ \theta\ b$ for some d.

Conversely, if $\theta_1, \ldots, \theta_n$ are congruence relations satisfying 1), 2), and 3). Then

$$L \cong L/\theta_1 \otimes \cdots \otimes L/\theta_n.$$

From these considerations it is clear that direct product representations of L are closely related to direct decompositions of the lattice Con L. Since direct decompositions into directly indecomposable elements in a distributive lattice are unique it is not surprising that there are uniqueness theorems for direct product representations of lattices. For example,

*Every finite lattice has a unique representation as a direct product of directly indecomposable factors.*

*A relatively complemented lattice satisfying the ascending chain condition is uniquely a direct product of finitely many simple lattices.*

The distributivity of Con L also provides uniqueness theorems for subdirect representations. For example,

*Let* $L \cong \text{sub} \prod_\alpha M_\alpha$ *and* $L \cong \text{sub} \prod_\beta N_\beta$ *where* $\alpha \in A$, $\beta \in B$, *be two*

*irredundant subdirect representations of a lattice* L *such that each* $M_\alpha$ *and each* $N_\beta$ *is subdirectly irreducible, then there is a one-to-one mapping* $\varphi$ *of* A *onto* B *and for each* $\alpha \in A$, *there is an isomorphism* $h_\alpha$ *of* $M_\alpha$ *onto* $N_{\varphi(\alpha)}$ *such that* $\Pi_{\varphi(\alpha)} = h_\alpha \Pi_\alpha$ *where* $\Pi$ *is the projection operator.*

The distributivity of Con L has important implications for the study of varieties of lattices. If $\mathcal{K}$ is a class of algebras of a given type, $\mathcal{K}^v$, the variety generated by $\mathcal{K}$, is the class of all algebras satisfying the identities satisfied by the algebras of $\mathcal{K}$. Let $H(\mathcal{K})$, $S(\mathcal{K})$, $P(\mathcal{K})$ respectively denote the class of all algebras isomorphic to homomorphic images of algebras of $\mathcal{K}$, subalgebras of algebras of $\mathcal{K}$, direct products of algebras of $\mathcal{K}$. A classical theorem of Birkhoff asserts

$$\mathcal{K}^v = HSP\mathcal{K}.$$

A variety $\mathcal{V}$ is *congruence distributive* if Con A is distributive for all $A \in \mathcal{V}$. If $\mathcal{K}$ is a subclass of a congruence distributive variety, then a beautiful theorem of Jónsson [17] gives the following improvement.

$$\mathcal{K}^v = P_s HSP_u \mathcal{K}$$

where $P_s\mathcal{K}$ is the class of all algebras isomorphic to subdirect products of algebras of $\mathcal{K}$ and $P_u\mathcal{K}$ is the class of all algebras isomorphic to ultra products of algebras of $\mathcal{K}$.

Jónsson's formula applies to the variety generated by a class of lattices since Con L is distributive for each lattice L. The formula is the key to many interesting and surprising results on varieties of lattices. We give one example.

*If* L *and* M *are finite non-isomorphic subdirectly irreducible lattices such that* L *has at least as many elements as* M, *then there is an identity which holds in* M *but not in* L.

REFERENCES

[1] M. Ajtai (1973) On a class of finite lattices, *Periodica Math. Hungar.* 4(2-3), 217-230.

[2] G. Birkhoff (1933) On the combination of subalgebras, *Proc. Comb. Phil. Soc.* 29, 441-464.

[3] G. Birkhoff and M. Ward (1939) A characterization of Boolean algebras, *Annals of Math.* 40, 609-610.

[4] G. Bruns (1961) Distributivität und subdirekte Zerlegbarkeit Vollständiger Verbände, *Arch. Math.* 12, 61-66.

[5] P. Crawley (1961) Decomposition theory of nonsemimodular lattices, *Trans. Amer. Math. Soc.* 99, 246-254.

[6] P. Crawley (1960) Lattices whose congruences form a Boolean algebra, *Pacific J. Math.* 10, 787-795.

[7] B. Davey and I. Rival (1976) Finite sublattices of three generated lattices, *J. Austral. Math. Soc.* 21 (Series A), 171-178.

[8] R.A. Dean (1956) Component subsets of the free lattice on n generators, *Proc. Amer. Math. Soc.* 7, 220-226.

[9] R. Dedekind (1900) Über die von drei Moduln erzeugte Dualgruppe, *Math. Annalen* 53, 371-404.

[10] R.P. Dilworth (1945) Lattices with unique complements, *Trans. Amer. Math. Soc.* 57, 123-154.

[11] R.P. Dilworth (1950) A decomposition theorem for partially ordered sets, *Annals of Math.* 51, 161-166.

[12] R.P. Dilworth (1950) The structure of relatively complemented lattices, *Annals of Math.* 51, 348-359.

[13] D. Duffus and I. Rival (1976) Crowns in dismantlable partially ordered sets, *Coll. Math. Soc. J. Bolyai* 18, 271-292.

[14] N. Funayama and T. Nakayama (1942) On the distributivity of the lattice of congruences, *Proc. Imp. Acad. Tokyo* 18, 553-554.

[15] G. Grätzer and E.T. Schmidt (1962) On congruence lattices of lattices, *Acta Math. Acad. Sci. Hungar.* 14, 179-185.

[16] J. Hashimoto (1957) Direct, subdirect decompositions and congruence relations, *Osaka Math. J.* 9, 87-112.

[17] B. Jónsson (1967) Algebras whose congruence lattices are distributive, *Math. Scand.* 21, 110-112.

[18] D. Kelly and I. Rival (1974) Crowns, fences and dismantlable lattices, *Canad. J. Math* 26, 1257-1271.

[19] D. Kelly and I. Rival (1975) Planar lattices, *Canad. J. Math.*, 636-665.

[20] H. MacNeille (1937) Partially ordered sets, *Trans. Amer. Math. Soc.* 42, 416-460.

[21] J.E. McLaughlin (1956) Atomic lattices with unique comparable complements, *Proc. Amer. Math. Soc.* 7, 864-866.

[22] G.N. Raney (1953) A subdirect-union representation for completely distributive complete lattices, *Proc. Amer. Math. Soc.* 4, 518-522.

[23] I. Rival (1974) Lattices with doubly irreducible elements, *Canad. Math. Bull.* 17, 91-95.

[24] I. Rival and B. Sands, Pictures in lattice theory, to appear in *Annals of Discrete Math.*

[25] I. Rival and R. Wille (1979) Lattices generated by partially ordered sets: which can be "drawn"? *J. Reine Angew Math.* 310, 56-80.

[26] H. Rolf (1958) The free lattice generated by a set of chains, *Pacific J. Math.* 8, 585-595.

[27] E.T. Schmidt (1962) Über der Kongruenzverbände der Verbände, *Publ. Math. Debrecen* 9, 243-256.

[28] A. Tarski (1946) A remark on functionally free algebras, *Annals of Math.* 47, 163-165.

[29] R. Wille (1977) On lattices freely generated by finite partially ordered sets, *Coll. Math. Soc. J. Bolyai* 17, 581-593.

[30] R. Wille (1972) Primitive subsets of lattices, *Alg. Universalis* 2, 95-98.

[31] R. Wille (1976) Aspects of finite lattices, *Proc. of the Conference on higher combinatorics*, Berlin.

# SOME ORDER THEORETIC QUESTIONS ABOUT FREE LATTICES AND FREE MODULAR LATTICES

Ralph Freese
Department of Mathematics
University of Hawaii at Manoa
Honolulu, Hawaii 96822

ABSTRACT

In this paper we look at some of the problems on free lattices and free modular lattices which are of an order theoretic nature. We review some of the known results, give some new results, and present several open problems.

Every countable partially ordered set can be order embedded into a countable free lattice [6]. However, free lattices contain no uncountable chains [25], so the above result does not extend to arbitrary partially ordered sets. The problem of which partially ordered sets can be embedded into a free lattice is open. It is not enough to require that the partially ordered set does not have any uncountable chains. In fact, there are partially ordered sets of height one, which cannot be embedded into any free lattice [23]. The importance of these concepts to projective lattices is discussed.

If $a > b$ and there is no $c$ with $a > c > b$ we say $a$ *covers* $b$ and write $a \succ b$. Covers in free lattices and free modular lattices are important to lattice structure theory. We discuss the connection between covers and structure theory and give some of the more important results about covers. Alan Day has shown that every quotient sublattice (i.e. interval) of a finitely generated free lattice contains a covering [9]. R.A. Dean, on the other hand, has some results on noncovers in free lattices. The analogous problems for free modular lattices are open.

Suppose $w(x_1,\ldots,x_n)$ is a lattice word and $L$ is a lattice. If we replace all but one of the variables with fixed elements from $L$ we obtain a function $f(x) = w(x,a_2,\ldots,a_n)$ from $L$ to $L$.



*I. Rival (ed.), Ordered Sets, 355–377.*

Call functions of this form unary polynomials; they always preserve order. For which lattices does every unary polynomial have a fixed point? We give some results on this new problem without completely solving it.

The word problem for free modular lattices with five or more generators is unsolvable [22]. The solvability of the word problem for FM(4) is open. There is evidence that four generated subdirectly irreducible lattices form a much more restricted class than five generated ones [26]. However, we will give a proof that there are $2^{\aleph_0}$ four generated simple modular lattices. We will also discuss Herrmann's result that the free modular lattice generated by two complemented pairs has a solvable word problem and the prospects for solving the word problem for FM(4). These problems are related to covers in FM(4) and various problems on covers in free modular lattices will be discussed.

We also give Wille's classification of those partially ordered sets $P$ such that the free modular lattice generated by $P$ is finite [50].

In this paper we examine some of the problems on free lattices and free modular lattices which are of an order theoretic nature. Many are combinatorial. We review some of the known results and show their importance to lattice structure theory. We also give some new results. Several open problems, both old and new, are scattered throughout the text.

## §1 PARTIALLY ORDERED SETS IN FREE LATTICES

If $X$ is a set, then the free lattice on $X$ is a lattice, denoted FL($X$), generated by $X$ such that any set map from $X$ to a lattice can be extended to a homomorphism. The free modular and distributive lattices are defined similarly and denoted FM($X$) and FD($X$). Free lattices have an interesting arithmetical and combinatorial structure as can be seen from the papers of Whitman, Dilworth, Dean, and Jónsson. Many of the important questions about free lattices have to do with their order theory.

The definition of free lattices guarantees that every lattice is a homomorphic image of a free lattice. However, not all lattices can be embedded into a free lattice. Free lattices satisfy the following conditions [32]:

(W) $u = \bigwedge u_i \leq \bigvee v_j = v$ implies either $u_i \leq v$ for some $i$ or $u \leq v_j$ for some $j$,

(SD) $u \vee v = u \vee w$ implies $u \vee v = u \vee (v \wedge w)$
$u \wedge v = u \wedge w$ implies $u \wedge v = u \wedge (v \vee w)$.

These are rather restrictive conditions. Very recently, J.B. Nation has shown that *a finite lattice can be embedded into a free lattice if and only if it satisfies* (W) *and* (SD), proving a long-standing conjecture of Bjarni Jónsson. The problem of characterizing infinite sublattices of free lattices remains open. A. Kostinsky does have a characterization of finitely generated sublattices (see [37]).

If $f$: FL($X$) $\rightarrow$ L is an onto homomorphism, then $f^{-1}(a) = \{u \in \text{FL}(X): f(u) = a\}$ is a convex sublattice of FL($X$) for each $a \in L$; i.e., if $u, v \in f^{-1}(a)$ and $u \leq w \leq v$ then $w \in f^{-1}(a)$. $L$ is *projective* if there is a $u_a \in f^{-1}(a)$ such that the $u_a$'s form a sublattice isomorphic to $L$. Call a map $g: L \to \mathrm{FL}(X)$ a *transversal* for $f$ if $g(a) \in f^{-1}(a)$. Then $L$ is projective if $f$ has a transversal which is a homomorphism. In particular, if $L$ is projective, $f$ has an order-preserving transversal. *For which lattices does $f$: $\mathrm{FL}(X) \to L$ have an order-preserving transversal?* This property is important since it together with three other conditions form necessary and sufficient conditions for $L$ to be projective. If $L$ is countable, then $f$: $\mathrm{FL}(X) \to L$ does have an order-preserving transversal (Crawley and Dean [6]). In fact if $L = \{a_0, a_1, a_2, \ldots\}$ then define

$$g(a_n) = (g_0(a_n) \vee \bigvee(g(a_i): a_i < a_n \text{ and } i < n))$$
$$\wedge \bigwedge(g(a_j): a_j > a_n \text{ and } j < n),$$

where $g_0(a_n)$ is any element from $f^{-1}(a_n)$. On the other hand, if $L$ contains an uncountable chain, then $f$ has no order-preserving transversal. This follows from Galvin and Jónsson's theorem that $\mathrm{FL}(X)$ *has no uncountable chains*. Galvin and Jónsson gave the following ingenious proof. Let $G$ be the group of those automorphisms of $\mathrm{FL}(X)$ induced from permutations of $X$ which move only finitely many elements. The orbits of $G$ are antichains since if $u < u\sigma$ for $\sigma \in G$ then $\sigma^n = 1$ for some $n$ and hence $u < u\sigma < u\sigma^2 < \ldots < u\sigma^n = u$. Let $X_0$ be a countable subset of $X$. Each element $w$ of $\mathrm{FL}(X)$ is a word in finitely many letters. Clearly we can find a $\sigma \in G$ so that $w\sigma$ is a word in $X_0$. Since $\mathrm{FL}(X_0)$ is countable, it follows that $G$ has countably many orbits. *Thus* $\mathrm{FL}(X)$ *is a countable union of antichains and thus contains no uncountable chain.* These arguments work for every variety of lattices and also for Boolean algebras.

The problem of when $f$ has an order transversal is solved by the following theorem.

THEOREM 1. *Let $f$ be a homomorphism from* $\mathrm{FL}(X)$ *onto $L$. Then $f$ has an order-preserving transversal if and only if $L$ satisfies*

(*) *for each element $a$ in $L$ there is a finite subset $S(a)$ of $L$ such that if $a \le b$ in $L$ then there is a $c \in S(a) \cap S(b)$ with $a \le c \le b$.*

If $L$ satisfies (*), then $L$ can be arranged into a well-ordered sequence $(a_\alpha: \alpha < \kappa)$ such that each $b \in S(a)$ either comes before $a$ or at worst only finitely many places after $a$. We define $g$ on $a_\alpha$ inductively. First choose any inverse image $u_\alpha$ of $a_\alpha$. Then just as in Crawley's and Dean's proof above, adjust $u_\alpha$

so that its order is correct with respect to those elements of $S(a_\alpha)$ that come before $a_\alpha$ and the (finitely many) elements which directly precede $a_\alpha$ back to the last limit ordinal. It is easy to see that $g$ is an order-preserving transversal.

The proof of the converse is based on the fact that each element of FL($X$) is in a sublattice generated by a finite subset of $X$.

For an application of this theorem consider the ordinal sum of two free lattices FL($X$) and FL($Y$) (this is the lattice on FL($X$) $\cup$ FL($Y$) with the additional order relation $u \leq v$ for all $u \in$ FL($X$) and $v \in$ FL($Y$)). We denote this lattice FL($X$) $\dot{+}$ FL($Y$). It is a little surprising that FL($X$) $\dot{+}$ FL($Y$) is projective if and only if either $X$ or $Y$ is finite, or they are both countable. To see that FL($X$) $\dot{+}$ FL($Y$) is not projective when $Y$ is infinite and $X$ is uncountable suppose it satisfied (*). Let $X_i = \{x \in X\colon |S(x)| \leq i\}$. Then $X = \bigcup_{i<\omega} X_i$ and since $X$ is uncountable there is a $k$ with $X_k$ infinite.

An element in a free lattice can be below at most finitely many generators. Choose $y_1 \in Y$. Since each $a \in S(y_1) \cap$ FL($Y$) is below at most finitely many elements of $Y$, and $S(y_1) \cap$ FL($Y$) is finite, but $Y$ is infinite, there is a $y_2 \in Y$ such that $y_2 \not\leq a$ for all $a \in S(y_1) \cap$ FL($Y$). Continuing in this way we can find $y_1, \ldots, y_{k+1} \in Y$ such that $y_j \not\leq a$ for all $a \in$ FL($Y$) $\cap \bigcup_{i<j} S(y_i)$. For each $x \in X_k$ and for each $i$ there is an element $a_{xy_i} \in S(x) \cap S(y_i)$ with $x \leq a_{xy_i} \leq y_i$. Again, since $S(y_i)$ is finite, the dual of the above remark implies that $a_{xy_i} \in$ FL($X$) for at most finitely many $x \in X_k$. Since $X_k$ is infinite, we can choose $x \in X_k$ such that $a_{xy_i} \in$ FL($Y$) for $i = 1, \ldots, k+1$. But for such an $x$ the $a_{xy_i}$ must be distinct. Hence $|S(x)| \geq k+1$, a contradiction.

A related but unsolved problem is: *which partially ordered sets can be order embedded into a free lattice?* All of the lattices FL($X$) $\dot{+}$ FL($Y$) are sublattices of free lattices since they are sublattices of FL($X$) $\dot{+}$ $\underset{\sim}{1}$ $\dot{+}$ FL($Y$), which are in fact projective [23]. Thus the complete bipartite partially ordered set induced on $X \cup Y$ can be embedded into a free lattice.

Since every partially ordered set can be embedded into a lattice, Crawley's and Dean's result can be used to show that $\mathrm{FL}(X)$, $|X| = \aleph_0$, *is an* $\aleph_0$*-universal partially ordered set.* That is, every countable partially ordered set can be order embedded into $\mathrm{FL}(X)$. The same proof in fact shows that $\mathrm{FD}(X)$ *is an* $\aleph_0$*-universal partially ordered set.* On the other hand the result of Galvin and Jónsson shows that if $P$ contains an uncountable chain then it cannot be embedded into a free lattice. With a more involved argument one can show that *the partially ordered set induced from the atoms and coatoms of the Boolean algebra of subsets of an uncountable set cannot be embedded into a free lattice* [23]. Thus not every bipartite partially ordered set can be embedded into a free lattice. In fact the problem of *which bipartite partially ordered sets can be embedded into a free lattice* is open.

Excluded partially ordered sets play a role in Nation's proof of Jónsson's conjecture cited above. If a lattice $L$ satisfies (W) and the following partially ordered set is embedded

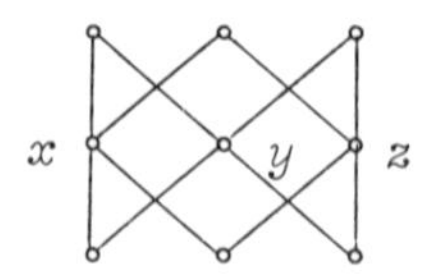


in $L$, then $x \nleq y \vee z$ and cyclically and dually. It follows from [49] that the sublattice generated by $x$, $y$, and $z$ is free and hence infinite. Actually only five of the six relations of the form $x \nleq y \vee z$ are necessary to force $L$ to be infinite. In fact if $L$ contains

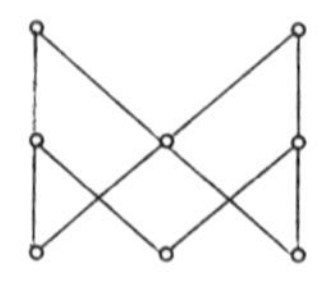

then $L$ must be infinite [35]. Ivan Rival has a similar result for lattices of breadth two which satisfy (W) [43].

## §2 COVERS IN FREE LATTICES

We say that $a$ *covers* $b$ in a lattice if $a > b$ and there is no element $c$ with $a > c > b$. We write $a \succ b$. If $X$ is infinite $\mathrm{FL}(X)$ has no coverings. However, finitely generated free lattices have many coverings. Coverings in free lattices and free modular lattices are particularly important to lattice theory. If $a \succ b$ in a lattice $L$, then by Dilworth's characterization of lattice congruences there is a unique maximal congruence $\psi$ on $L$ with $(a, b) \notin \psi$. If $L$ is a free lattice, then $L/\psi$ is a finite, subdirectly irreducible lattice. Lattices of this form are called *splitting lattices* and were extensively studied by Ralph McKenzie [37]. By Jónsson's celebrated theorem, a subdirectly irreducible lattice cannot be in the join of finitely many lattice varieties without being in one of them. A splitting lattice cannot be in the join of an arbitrary collection of lattice varieties without being in one of them. Because of this property splitting lattices have played a very important role in lattice structure theory. Splitting modular lattices have even stronger structural properties (see §5).

McKenzie has shown that *one can recursively decide for lattice words* $u(x_1,\ldots,x_n)$ *and* $v(x_1,\ldots,x_n)$ *if* $u \succ v$ *in* $\mathrm{FL}(x_1,\ldots,x_n)$ [37]. Perhaps the most important result on coverings is that of Alan Day who shows that the *finitely generated free lattices are weakly atomic*. A lattice is *weakly atomic* if $a < b$ implies there exist $c$ and $d$ with $a \le c \prec d \le b$. Thus there is an abundance of coverings and thus of splitting lattices.

The next theorem contrasts these results by showing that free lattices do not satisfy stronger atomicity properties.

THEOREM 2 (R.A. Dean, unpublished). *The element* $x \vee (y \wedge z)$ *of* $\mathrm{FL}(x, y, z)$ *has no cover.*

To prove this we define the length of a lattice word to be the number of join and meet signs which appear in it and length of an element of $\mathrm{FL}(X)$ as the minimum of the lengths of the words which represent the element.

Now suppose $x \vee (y \wedge z) \prec w$ in $\mathrm{FL}(x, y, z)$. Then $w = x \vee (y \wedge z) \vee w$ and we can choose an element $u$ of minimal length such that $w = x \vee (y \wedge z) \vee u$. Since $x \vee y$ does not cover $x \vee (y \wedge z)$, $u$ cannot be $y$ or $z$ or $x$. By its minimality $u$ must be join-irreducible. Thus $u = \bigwedge u_i$ where each $u_i$ has smaller length than the length of $u$. Notice also that $u \le y$ and $u \le z$ cannot both hold. Thus we may assume $u \not\le y$. Now

$$x \vee (y \wedge z) \le x \vee (y \wedge z) \vee (u \wedge y) \le x \vee (y \wedge z) \vee u = w.$$

Since $x \vee (y \wedge z) \prec w$, we must have one equality. If $x \vee (y \wedge z) \vee (u \wedge y) = x \vee (y \wedge z) \vee u = w$, then $u \le x \vee (y \wedge z) \vee (u \wedge y)$. Applying (W) we get that for some $i, u_i \le x \vee (y \wedge z) \vee (u \wedge y) = w$. But $u \le u_i$ and so

$w = x \vee (y \wedge z) \vee u_i$, violating the minimality of $u$.

If $x \vee (y \wedge z) = x \vee (y \wedge z) \vee (u \wedge y)$, then, since $x$, $y$, and $z$ are meet-prime, a few applications of (W) gives us $u \leq z$. Let $\sigma \in \text{Aut}(\text{FL}(x, y, z))$ be the automorphism with $x\sigma = x$, $y\sigma = z$, and $z\sigma = y$. Since $u \leq z$ and $u \nleq y$, $u\sigma \neq u$ and so $u\sigma$ and $u$ are incomparable. However $(x \vee (y \wedge z))\sigma = x \vee (y \wedge z)$ and thus $x \vee (y \wedge z) \prec w\sigma$. (W) implies that $x \vee (y \wedge z)$ is meet-irreducible and hence can have at most one cover. Thus $w\sigma = w$; that is, $x \vee (y \wedge z) \vee u = x \vee (y \wedge z) \vee (u\sigma)$. This implies $u \leq x \vee (y \wedge z) \vee (u\sigma)$. Since $u \nleq x$ or $y \wedge z$ or $u\sigma$, (W) yields that for some $i, u_i \leq x \vee (y \wedge z) \vee (u\sigma) = w$, which again violates the minimality of $u$.

## §3 LATTICES FREELY GENERATED BY PARTIALLY ORDERED SETS

The lattice freely generated by a partially ordered set $P$, denoted FL($P$), is defined by the property that any order preserving map from $P$ into a lattice $L$ can be uniquely extended to a homomorphism from FL($P$) to $L$. The structure of these lattices was described by Dilworth [16] and Dean [11]. They are projective if and only if $P$ satisfies (*) of Theorem 1. M.E. Adams and D. Kelly [1] have shown that *if P has no chains of cardinality* $\kappa$ *then neither does* FL(P) ($\kappa$ an infinite cardinal). Moreover this result holds in any variety of lattices. In particular it holds for modular and distributive lattices.

R. Wille has shown that *if P is a finite partially ordered set then* FL($P$) *is finite if and only if neither* 1 + 1 + 1, 2 + 2, *nor* 1 + 4 *can be embedded in P*. Here, for example, 2 + 2 is the partially ordered set consisting of two two-element chains.

In a related problem I. Rival and R. Wille [45] characterized those $P$ for which FL($P$) can be "drawn."

THEOREM 3. *Let P be a finite partially ordered set. The following are equivalent*:

(i) FL($x$, $y$, $z$) *is not a sublattice of* FL($P$);
(ii) FL($P$) *has breadth at most* 3;
(iii) *neither* 1 + 1 + 1, 2 + 3, *nor* 1 + 5 *can be order embedded into P*;
(iv) FL($P$) *is a sublattice of* FL($H$).

Here $H$ is the following partially ordered set:

H = 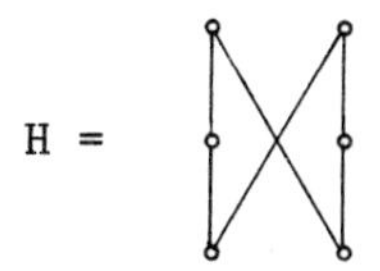

One will readily agree that FL($x$, $y$, $z$) cannot be drawn. For example FL($X$), $|X| = \aleph_0$, is a sublattice of FL($x$, $y$, $z$). To show FL($H$) is "drawable," they draw it.

The lattices FL(2 + 2) and FL(1 + 4) are both infinite but have nice diagrams given by H. Rolf in [46].

Wille has also characterized those partially ordered sets $P$ with FM($P$) finite. (FM($P$) is the free modular lattice generated by $P$.)

THEOREM 4. *For a partially ordered set P the following are equivalent:*

(i) FM($P$) *is finite;*
(ii) FM($P$) *is a subdirect product of copies of the two element lattice and of* $M_3$;
(iii) *neither* 1 + 1 + 1 + 1 *nor* 1 + 2 + 2 *can be embedded in P.*

In a similar vein, I. Rival, W. Ruckelshausen, and B. Sands have shown that a certain infinite partially ordered set they call the "herringbone" can be order embedded into every infinite, finitely generated lattice of finite width [41]. Dwight Duffus and Ivan Rival have shown that if a distributive lattice $D$ can be order embedded into a lattice $L$ then $D$ is a homomorphic image of a sublattice of $L$ [18].

## §4 FIXED POINTS FOR LATTICE POLYNOMIALS

Let $L$ be a lattice and $w(x_1,\ldots,x_n)$ a lattice word (or term). If we replace all but one of the variables in $w$ with fixed elements of $L$, we obtain a function $f(x) = w(x, a_2,\ldots,a_n)$ from $L$ to $L$. Such a function is called a *unary algebraic function* or *unary polynomial* on $L$. Although such functions will not in general be homomorphisms, they will always preserve order. Recently there has been a great deal of interest in order preserving maps on a partially ordered set which have a fixed point; i.e., an element $x$ with $f(x) = x$. In this section we will address a similar, but new problem: *for which lattices does every unary algebraic function have a fixed point?* Let $\mathcal{F}$ denote those lattices which have this property. We will not completely answer this question but will give some partial results which are intended to draw interest to the problem. Recall the result of Anne Morel and A. Tarski [8], [47]: *every order-preserving map from a lattice $L$ to itself has a fixed point if and only if $L$ is complete.* Thus $\mathcal{F}$ *contains all complete lattices.* Not all lattices in $\mathcal{F}$ are complete. For example, $\omega \in \mathcal{F}$ ($\omega$ is the natural numbers under their usual order). In fact every distributive lattice is in $\mathcal{F}$. More generally, *every locally finite lattice is in* $\mathcal{F}$. (A lattice is locally finite if every finitely generated sublattice is finite.) To see this result let $f(x) = w(x, a_2, \ldots, a_n)$ be a unary polynomial on a locally finite lattice $L$. Let $L_1$ be the (finite) sublattice generated by $a_2, \ldots, a_n$. Then $f$ restricted to $L_1$ is a unary polynomial on $L_1$ and has a fixed point since $L_1$ is finite. This will be a fixed point for $L$.

On the other hand we can ask if there are any lattices not in $\mathcal{F}$. The answer is yes. In fact $f(x) = ((((( x \wedge b) \vee c) \wedge a) \vee b) \wedge c) \vee a$ *has no fixed points in* $\mathrm{FL}(a, b, c)$. This is proved using some lemmas of Whitman [49]. We shall forego this proof and present instead a modular example. Let $L_1$ and $L_2$ be the modular lattices diagrammed in Figure 1. Let $f(x) = ((((x \wedge a) \vee b) \wedge c) \vee d) \wedge a$. Clearly if $f(x) = x$, then $x \leq a$. In $L_1$ one easily checks that $f(0) = 0$ and this is the only fixed point. In $L_2$ $f(a) = a$ and this is the only fixed point. Let $L$ be the sublattice of $L_1 \times L_2$ generated by $(a, a)$, $(b, b)$, $(c, c)$, and $(d, d)$. The unary polynomial $f$ acts on $L_1 \times L_2$ by $f(x, y) = (f(x), f(y))$ for $(x, y) \in L_1 \times L_2$. Thus $(0, a)$ is the only fixed point for $L_1 \times L_2$. The proof is completed by showing $(0, a) \notin L$.

Since $L$ is generated by $(a, a)$, $(b, b)$, $(c, c)$, and $(d,d)$, $(0, a) \in L$ if and only if there is a lattice word $w(a, b, c, d)$ such that $w^{L_1}(a, b, c, d) = 0$ and $w^{L_2}(a, b, c, d) = a$, where of course $w^{L_i}(a, b, c, d)$ is the evaluation of $w(a, b, c, d)$ in $L_i$.

In order to prove that this is not possible we need a lemma. In $\mathrm{FL}(a, b, c, d)$ define inductively $a_i$, $b_i$, $c_i$, $d_i$ by $a_0 = a$, $b_0 = b$, $c_0 = c$, $d_0 = d$ and $a_{i+1} = a_i \wedge (b_i \vee c_i) \wedge (b_i \vee d_i) \wedge (c_i \vee d_i)$ with $b_{i+1}$, $c_{i+1}$, and $d_{i+1}$ defined similarly. We claim that if $u(a, b, c, d) \in \mathrm{FL}(a, b, c, d)$ and $u^{L_2}(a, b, c, d) \geq a$

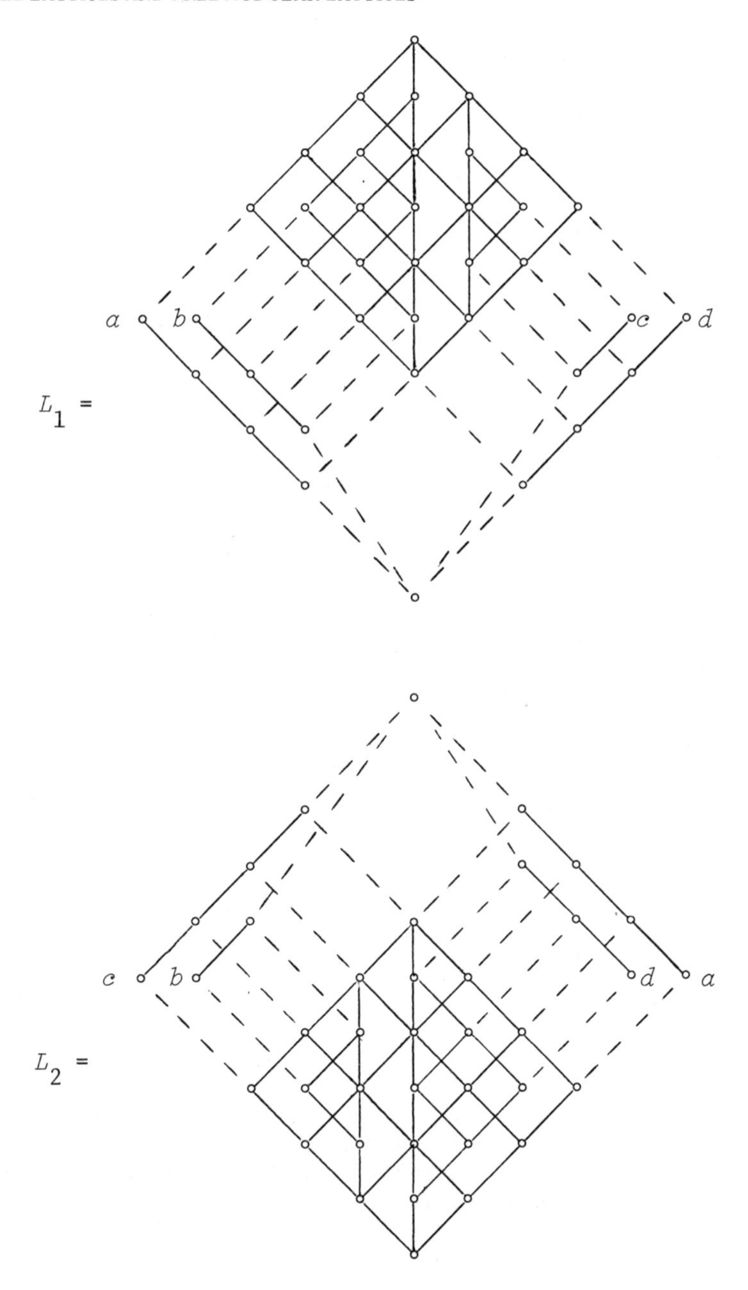


FIGURE 1

holds in $L_2$ then $u(a, b, c, d) \geq a_i$ holds in FL$(a, b, c, d)$ for some $i$ and that similar statements hold for $b$, $c$, and $d$. We prove this lemma by induction on the length of $u$. Suppose $u^{L_2} \geq a$ and suppose $u = u_1 \vee u_2$ where $u_1$ and $u_2$ have shorter length than $u$. If either $u_1^{L_2} \geq a$ or $u_2^{L_2} \geq a$ holds, it is easy to complete the proof. So we assume neither is above $a$. By examining $L_2$ we see that this implies that one of them contains $b$ and the other $c$, or one of them contains $b$ and the other $d$, or one of them contains $c$ and the other $d$. Assume the former. Then by induction $u_1 \geq b_i$ and $u_2 \geq c_j$ for some $i$ and $j$. Letting $k = \max\{i,j\}$ we have $u_1 \geq b_k$ and $u_2 \geq c_k$ and hence, $u = u_1 \vee u_2 \geq b_k \vee c_k \geq a_{k+1}$. The cases where $u$ is a generator or a meet are easier and left to the reader.

Now if $w^{L_2} = a$ in $L_2$ then $w \geq a_i$ for some $i$. But then $w^{L_1} \geq a_i^{L_1}$. It is easy to see that $a_i^{L_1} > 0$ for all $i$. Hence $w^{L_1} \neq 0$, showing that $(0, a) \notin L$.

It is possible to draw $L$. However it has a strange twist which makes its diagram awkward.

In summary, F does not contain all lattices but does contain all complete lattices and all locally finite lattices. A common generalization of these last two classes is the class of *locally complete lattices*; i.e., those lattices such that every finitely generated sublattice is a complete lattice. Clearly this class is contained in F. *Does F contain any other lattices?*

## §5 MODULAR LATTICES

Problems in the theory of modular lattices are usually more difficult than the corresponding problems for the class of all lattices or of distributive lattices. Free modular lattices are much less understood than free lattices. In this section we discuss some of the results and problems concerning free modular lattices which are related to order.

For free lattices we have Whitman's algorithm for deciding if two lattice words are equal. We now know that *there is no algorithm to decide if two lattice words are equal in the free modular lattice.* It follows that *the variety of modular lattices is not generated by its finite members.* These results depend on the "projectivity" of certain partial modular lattices. We will briefly outline the main ideas of the proofs.

The classical von Stadt coordinatization of projective geometries of dimension at least three shows how to define a ring from the geometry. Von Neumann showed how to extend these ideas to obtain a ring from any modular lattice which contains a certain partial lattice known as a 4-frame (here we shall use the term frame). A frame in a modular lattice is a set of seven elements $\{a_1, a_2, a_3, a_4, c_{12}, c_{13}, c_{14}\}$ such that
$a_1 \wedge (a_2 \vee a_3 \vee a_4) = a_1 \wedge a_2 \wedge a_3 \wedge a_4$ and symmetrically
$a_1 \vee c_{1j} = a_1 \vee a_j = a_j \vee c_{1j}$ and $a_1 \wedge c_{1j} = a_1 \wedge a_j = a_j \wedge c_{1j}$
for $j = 2, 3, 4$. This says $\{a_1, a_2, a_3, a_4\}$ generate a 16 element Boolean algebra with $a_1, a_2, a_3, a_4$ as atoms and $a_1, a_j, c_{1j}$ generate $M_3$. In a modular lattice containing a frame we let

$$R_L = \{x \in L:\ x \vee a_2 = a_1 \vee a_2,\ x \wedge a_2 = a_1 \wedge a_2\}$$

and define for $x, y \in R_L$

$$x + y = ([(x \vee c_{13}) \wedge (a_2 \vee a_3)] \vee [(y \vee a_3) \wedge (a_2 \vee c_{13})]) \wedge (a_1 \vee a_2)$$

(*)

$$xy = ([(x \vee c_{23}) \wedge (a_1 \vee a_3)] \vee [(y \vee c_{13}) \wedge (a_2 \vee a_3)]) \wedge (a_1 \vee a_2)$$

where

$$c_{23} = (c_{12} \vee c_{13}) \wedge (a_2 \vee a_3).$$

Von Neumann shows that with these operations $R_L$ is a ring with 1. To get an idea of how these definitions arise let $R$ be a ring with 1 and let $L$ be the lattice of submodules of $R^4$ as a left $R$ module. Let $a_1 = \{(x, 0, 0, 0):\ x \in R\}, \ldots, a_4 = \{(0, 0, 0, x):\ x \in R\}$, $c_{12} = \{(x, -x, 0, 0):\ x \in R\}, \ldots, c_{14} = \{(x, 0, 0, -x):\ x \in R\}$. The reader can check that $R(-1, r, 0, 0) \in R_L$ and that $R(-1, r, 0, 0) + R(-1, s, 0, 0) = R(-1, r+s, 0, 0)$ where the first "+" is defined by (*).

Frames are projective in the class of modular lattices. This means that if we map a free modular lattice, FM($X$), onto the lattice $L$ of submodules described above, we can find a frame $a_1, a_2, a_3, a_4, c_{12}, c_{13}, c_{14}$ in FM($X$) which maps to $a_1, a_2, a_3, a_4, c_{12}, c_{13}, c_{14}$ in $L$. In particular we get a ring in FM($X$).

Unfortunately it need not be $R$. In fact it is probably $\mathbb{Z}$ in most cases.

Now let $K_p$ and $K_q$ be countable fields of distinct finite characteristics $p$ and $q$. The lattice $L_p$ of subspaces of a four dimensional vector space over $K_p$ has a frame $a_1, \ldots, c_{14}$ similar to the one defined above for $L$. Also $L_q$, the lattice of subspaces of a four dimensional vector space over $K_q$, has a frame $a'_1$, $a'_2$, $a'_3$, $a'_4$, $c'_{12}$, $c'_{13}$, $c'_{14}$. In $L_q$ the lattice of all subspaces of $a'_1 \vee a'_2$ is just the lattice of subspaces of a two dimensional vector space over $K_q$. It is easy to see that this lattice is $M_\omega$, the lattice with 0, 1, and countably many atoms. Similarly the quotient sublattice $1/a_3 \vee a_4$ of $L_p$ is $M_\omega$. Form the lattice $L$ from $L_p$ and $L_q$ by identifying these two copies of $M_\omega$. This is a modular lattice by the results of R.P. Dilworth and M. Hall [15], [28].

Consider a homomorphism of FM($X$) onto $L$. We can find frames $\overline{a_1}, \ldots, \overline{c_{14}}$ and $\overline{a'_1}, \ldots, \overline{c'_{14}}$ in FM($X$) which map to the corresponding frames in $L$. These frames determine two rings $R_p$ and $R_q$, and by choosing $\overline{a_1}, \ldots, \overline{c_{14}}$ properly, $R_p$ has characteristic $p$ and $R_q$ has characteristic $q$. Furthermore the two frames can be lined up so that there is a natural bijection from $R_p$ onto $R_q$.

If there was a lattice homomorphism $g$ from FM($X$) to a finite lattice such that $g(\overline{a_1}) \neq g(\overline{a_1} \wedge \overline{a_2})$, then $g$ is one-to-one on the frames and these frames would determine two nontrivial finite rings $S_p$ and $S_q$ of characteristics $p$ and $q$ with $|S_p| = |S_q|$. But this is clearly impossible since a finite ring (with 1) of characteristic $p$ must have order $p^n$. This implies that the variety of modular lattices is not generated by its finite members. By using some ingenious skew-field constructions of A. Macintyre and P.M. Cohn, it is possible to extend these ideas to show that the word problem for the free modular lattice on five or more generators is unsolvable.

The free modular lattice on three generators is finite [14]; however, FM(4) is infinite and it is an open problem to determine whether it has a solvable word problem. In the remaining part of this section we review some of the important problems for free modular lattices and their connection with the word problem for FM(4).

As we pointed out earlier, covers in free modular lattices are particularly important in structure theory. If $u \succ v$ in FM($n$), then by Dilworth's characterization of lattice congruences [17] there is a unique largest congruence $\psi(u, v)$ not containing $(u, v)$. Thus associated with every cover, $u \succ v$, of FM($n$) is a subdirectly irreducible modular lattice, namely $M = \text{FM}(n)/\psi(u, v)$. If $x_1, \ldots, x_n$ are the generators of FM($n$) and $a_1, \ldots, a_n$ are their images in $M$, then since $(u, v) \notin \psi(u, v)$, $u(a_1,\ldots,a_n) \neq v(a_1,\ldots,a_n)$. Thus the relation $u = v$ fails in $M$ for $a_1, \ldots, a_n$. Now the modularity of FM($n$) implies that $\theta(u, v)$, the smallest congruence identifying $u$ and $v$, is an atom of the congruence lattice of FM($n$) ([7] 10.3 and 10.5). Hence $\theta(u, v) \wedge \psi(u, v) = 0$ in this congruence lattice. This says that FM($n$) is a subdirect product of $M$ and $\text{FM}(n)/\theta(u, v)$. Of course $\text{FM}(n)/\theta(u, v)$ satisfies the relation $u = v$. In fact it is not hard to show that if $L$ is any modular lattice generated by $n$ elements then either $L$ satisfies the relation $u = v$ for its generators or $L$ is a subdirect product of $M$ and a lattice $L_1$ which satisfies the relation $u = v$.

There are several important open questions about these ideas.

(1) *If* $u \succ v$ *in* FM($n$) *is* $\text{FM}(n)/\psi(u, v)$ *finite?*

(2) *Is* FM($n$) *weakly atomic?*

In the case of free lattices both of the above questions can be answered yes [37], [9]. For modular lattices we only know that at least one of them has a negative answer. As pointed out above, the elements $\overline{a_1} > \overline{a_1} \wedge \overline{a_2}$ in FM(5) become identified under any homomorphism of FM(5) into a finite lattice. If FM(5) were weakly atomic there would be elements $u$, $v$ with $a_1 \geq u \succ v \geq \overline{a_1} \wedge \overline{a_2}$. But then the natural map from FM(5) to $M = \text{FM}(5)/\psi(u, v)$ does not identify $u$ and $v$ and thus does not identify $a_1$ and $a_1 \wedge \overline{a_2}$. Hence $M$ must be infinite. That is *either weak atomicity fails for* FM(5) *or there is an infinite lattice of the form* $\text{FM}(5)/\psi(u, v)$, $u \succ v$.

A subdirectly irreducible modular lattice $L$ is called a *splitting modular lattice* if there is a lattice equation $\varepsilon$ which fails in $L$ such that every variety of modular lattices either satisfies $\varepsilon$ or contains $L$. For example, $M_3$ is a splitting modular lattice since we can take $\varepsilon$ to be the distributive law. Also every lattice of the form $\text{FM}(n)/\psi(u, v)$ for $u \succ v$ is a splitting modular lattice with $\varepsilon$: $u = v$.

(3) *Is every splitting modular lattice of this form?*

Although these concepts are not well understood, we do have several examples. In [19] infinitely many lattices of the form $FM(4)/\psi(u, v)$, for $u \succ v$, are given. In [21] it is shown that the lattice of subspaces of an n-dimensional vector space over a finite prime field is a splitting lattice if $4 \leq n < \infty$. On the other hand, $M_4$, the six element lattice of length two, is not a splitting lattice [10].

Although we do not know if the word problem for FM(4) is solvable, C. Herrmann has shown that the free modular lattice generated by two complemented pairs does have a solvable word problem [29]. In fact he lists all subdirectly irreducible modular lattices which can be generated by two complemented pairs. Most of these lattices are of the form $FM(4)/\psi(u, v)$ for $u \succ v$ in FM(4). In his proof he assumes $L$ is a subdirectly irreducible modular lattice generated by two complemented pairs but not on his list. Then $L$ must satisfy all the relations $u = v$. Herrmann uses these relations in an ingenious way to eventually show that $0 = 1$ in $L$; i.e., $L$ is trivial.

There is some hope this sort of technique can be applied to FM(4). Indeed there is some evidence that four generated subdirectly irreducible modular lattices are a much more restricted class than five generated ones. Representation theorists have considered endomorphisms of vector spaces which preserve four given subspaces. These lead to algebras of "tame representation type" whereas with five subspaces one gets algebras of "wild representation type" [3], [4], [26].

On the other hand the class of four generated subdirectly irreducible lattices is not too restrictive, as the next theorem shows.

THEOREM 5. *There are* $2^{\aleph_0}$ *projective planes generated by four points. Thus there are* $2^{\aleph_0}$ *simple, complemented modular lattices which are four-generated.*

In order to prove this result we need to recall Marshall Hall's free plane construction [27]. We view a projective plane $\pi$ as a set of points and set of lines and an incidence relation between them (which says whether or not a given point is on a given line.) The axioms are: (i) each pair of distinct points are on a unique line, (ii) for each pair of distinct lines there is a unique point which is on both, (iii) $\pi$ contains a quadrangle (four points, no three on a line). Projective planes may be viewed as modular lattices with the points as atoms, the lines as coatoms and $p \leq \ell$ if and only if $p$ is on $\ell$.

A collection $\pi_0$ of points, lines, and an incidence relation between them is a partial plane provided each pair of distinct points is on at most one line. It follows that each pair of lines has at most one point on both. If $\pi_o$ is a partial plane containing a quadrangle, it can be completed to a plane as follows. For each pair of distinct points $p_1$ and $p_2$ in $\pi_o$ with no line containing both, we add a new line $\ell$ and extend the incidence relation by saying $p_1$ and $p_2$ are on $\ell$. The result is a new partial plane $\pi_1$. $\pi_2$ is formed by a dual process from $\pi_1$. Thus we obtain a sequence $\pi_o \subseteq \pi_1 \subseteq \pi_2 \subseteq \ldots$. We call $\pi_n$ the $n$-step free extension of $\pi_o$. Then it is easy to see that $\pi = \bigcup \pi_i$ is a projective plane extending $\pi_o$.

We shall construct our $2^{\aleph_o}$ planes by modifying the above construction. In each case $\pi_o$ will consist of four points and no lines. Then $\pi_1, \pi_2, \ldots$ shall be constructed so that (i) $\pi_i \subseteq \pi_j$ if $i < j$ (this implies that both incidence and non-incidence is preserved); (ii) $\pi_n$ is a partial plane generated by $\pi_o$; (iii) if $n$ is even, $\pi_{n+1} - \pi_n$ has only lines and every pair of points in $\pi_n$ lies on a line in $\pi_{n+1}$, and dually for $n$ odd. If $x \in \pi_n - \pi_{n-1}$, we say $x$ has rank $n$ and write $r(x) = n$; note lines always have odd rank and points even. Note that these conditions imply $\pi = \bigcup \pi_n$ will be a projective plane generated by four points. Also note that these conditions imply that each line $\ell$ of $\pi$ contains only finitely many points with rank less than $r(\ell)$ and that all but at most one of these points must have rank $r(\ell) - 1$.

Let $p_1, p_2, p_3, \ldots$ be a set of prime numbers, with $p_i \geq 5$. Let $\rho_i$ be a projective plane coordinatized by the field with $p_i$ elements, $i = 1, 2, 3, \ldots$. Recall that every quadrangle in $\rho_i$ generates it. We shall construct $\pi$ with $\rho_i \subseteq \pi$, $i = 1, 2, 3, \ldots$. Let $\pi_o$ have only four points and let $\pi_6$ be the 6-step free extension of $\pi_o$. The reader can check that $\pi_6$ has 8 points with no two joined by a line. Use 4 of these points to generate a subplane isomorphic to $\rho_1$. We shall denote this subplane $\rho_1$. This construction differs from the free plane construction in that, for example, if $p, q, r \in \rho_1$ are points in $\pi_n$ with no two on a line in $\pi_n$, in $\pi_{n+1}$ we can add one line which contains all three provided that $p, q, r$ are collinear in $\rho_1$. All other lines are added freely.

In $\pi_{12}$ the other four points will have again created 8 points no two on a line. In $\pi_{18}$ four of these points will have created 8 new points, no two on a line. Now $\rho_1 \subseteq \pi_n$ for some $n$ and it is clear that there are 8 points of rank greater than $n$ with no two on a line. Four of these points are used to generate $\rho_2$. Continuing in this way we obtain a plane $\pi$ with $\rho_i \subseteq \pi$, $i = 1, 2, \ldots$. Notice that if $x \in \rho_i$ and $y \in \rho_j$, $i \neq j$, then $r(x) \neq r(y)$. Also notice that if $q_1, \ldots, q_k$ are points on $\ell$ with $r(q_i) < r(\ell)$, $i = 1, \ldots, k$ and $k > 2$ then $\ell$ and $q_1, \ldots, q_k$ are in some $\rho_i$.

Now suppose $\rho_o$ is the plane over $GF(p_o)$ where $p_o \geq 5$ is a prime with $p_o \neq p_i$, $i = 1, 2, \ldots$. We claim $\rho_o$ is not a subplane of $\pi$. Suppose $\rho_o \subseteq \pi$. Pick an element of highest rank in $\rho_o$; say it is a point $q$. In $\rho_o$ there are $p_o + 1$ lines $\ell_1, \ldots, \ell_{p_o+1}$ containing $q$ and all but at most one of these must have rank $r(q) - 1$. By the dual of the above remarks, $q$, $\ell_1, \ldots, \ell_{p_o+1}$ must lie in some $\rho_t$. Choose one of these lines, say $\ell_1$, with $r(\ell_1) = r(q) - 1$. Then every point of $\rho_o$ on $\ell_1$ has rank less than $r(\ell_1)$ or rank equal to $r(q)$. If $\ell_1$ contains exactly two points of $\rho_o$ of lower rank, then, since $p_o \geq 5$, $\ell_1$ contains $q_1, q_2 \in \rho_o$ with $q_1 \neq q \neq q_2$ and $r(q_1) = r(q_2) = r(q)$. Clearly $q_1, q_2 \in \rho_t$. If there is another line, say $\ell_2$ with this same situation, we obtain $q_3, q_4$ on $\ell_2$ with $q_3, q_4 \in \rho_t$. But $q_1, q_2, q_3, q_4$ form a quadrangle and so generate $\rho_o$. Hence $\rho_o \subseteq \rho_t$, which is impossible.

Thus for all but possibly two of the lines $\ell_1, \ldots, \ell_{p_o+1}$, all points except $q$ and possibly one other must have rank less than $r(\ell_j) = r(q) - 1$. There are more than two such points and hence $\ell_j$ and all points of $\rho_o$ of lower rank on $\ell_j$ must lie in some $\rho_i$ but $\ell_j \in \rho_t$ so the points of lower rank must also be in $\rho_t$. It is now easy to see that there is a quadrangle of $\rho_o$ in $\rho_t$, which is impossible.

Since there are $2^{\aleph_0}$ subsets of the primes, we have $2^{\aleph_0}$ nonisomorphic projective planes generated by four points.

This result improves the theorem of I. Rival and B. Sands [42] that there are $2^{\aleph_0}$ four-generated simple (nonmodular) lattices. They, however, show that every countable lattice can be embedded into a four-generated simple lattice. *Is this true in the modular case?* The author conjectures it is not.

The existence of $2^{\aleph_0}$ simple four-generated modular lattices means serious modifications of Herrmann's techniques will be required in order to solve the word problem for FM(4). On the other hand these lattices do not rule out a positive solution. In fact, since every partial plane can be extended to a plane, it is easy to show that *one can algorithmically decide whether or not two lattice terms are equal in all projective planes.*

REFERENCES


[1] M.E. Adams and D. Kelly (1977) Chain conditions in free products of lattices, *Algebra Universalis* 7, 235-244.

[2] M.E. Adams and D. Kelly (1977) Disjointness conditions in free products of lattices, *Algebra Universalis* 7, 245-258.

[3] S. Brenner (1967) Endomorphism algebras of vector spaces with distinguished sets of subspaces, *J. Algebra* 6, 100-114.

[4] S. Brenner (1974) On four subspaces of vector spaces, *J. Algebra* 29, 587-599.

[5] P.M. Cohn (1977) *Skew Field Constructions*, London Math. Soc. Lecture Note Series no. 27, Cambridge University Press, Cambridge.

[6] P. Crawley and R.A. Dean (1959) Free lattices with infinite operations, *Trans. Amer. Math. Soc.* 92, 35-47.

[7] P. Crawley and R.P. Dilworth (1973) *Algebraic Theory of Lattices*, Prentice-Hall, Englewood Cliffs, N.J.

[8] A.M. Davis (1955) A characterization of complete lattices, *Pacific J. Math.* 5, 311-319.

[9] A. Day (1977) Splitting lattices generate all lattices, *Algebra Universalis* 7, 163-169.

[10] A. Day, C. Herrmann, and R. Wille (1972) On modular lattices with four generators, *Algebra Universalis* 2, 317-323.

[11] R.A. Dean (1956) Completely free lattices generated by partially ordered sets, *Trans. Amer. Math. Soc.* 83, 238-249.

[12] R.A. Dean (1956) Component subsets of the free lattice on $n$ generators, *Proc. Amer. Math. Soc.* 7, 220-226.

[13] R. A. Dean (1961) Sublattices of a free lattice, in Lattice Theory, *Proceedings of Symposia in Pure Mathematics, II*, Amer. Math. Soc., Providence, 31-42.

[14] R. Dedekind (1900) Über die drei Moduln erzeugte Dualgruppe, *Math. Ann.* 53, 371-400.

[15] R.P. Dilworth (1941) The arithmetic theory of Birkhoff lattices, *Duke Math. J.* 8, 286-299.

[16] R.P. Dilworth (1945) Lattices with unique complements, *Trans. Amer. Math. Soc.* 57, 123-154.

[17] R.P. Dilworth (1950) The structure of relatively complemented lattices, *Ann. of Math.* 51, 348-359.

[18] D. Duffus and I. Rival (1980) A note on weak embeddings of distributive lattices, *Algebra Universalis* 10, 258-259.

[19] R. Freese (1973) Breadth two modular lattices, *Proc. Houston Lattice Theory Conf.*, Department of Mathematics, University of Houston, Houston, 409-451.

[20] R. Freese (1979) The variety of modular lattices is not generated by its finite members, *Trans. Amer. Math. Soc.* 255, 277-300.

[21] R. Freese (1979) Projective geometries as projective modular lattices, *Trans. Amer. Math. Soc.* 251, 329-342.

[22] R. Freese (1980) Free modular lattices, *Trans. Amer. Math. Soc.* 261, 81-91.

[23] R. Freese and J.B. Nation (1978) Projective lattices, *Pacific J. Math.* 75, 93-106.

[24] R. Freese and J.B. Nation (1979) Finitely presented lattices, *Proc. Amer. Math. Soc.* 77, 174-178.

[25] F. Galvin and B. Jónsson (1961) Distributive sublattices of a free lattice, *Canad. J. Math.* 13, 265-272.

[26] I.M. Gelfand and V.A. Ponomarev (1970) Problems in linear algebra and classification of quadruples of subspaces in a finite dimensional vector space, *Coll. Math. Soc. J. Bolyai* 5, Hilbert Space Operators, 164-237.

[27] M. Hall (1943) Projective planes, *Trans. Amer. Math. Soc.* 54, 229-277.

[28] M. Hall and R.P. Dilworth (1944) The imbedding problem for modular lattices, *Ann. of Math.* 45, 450-456.

[29] C. Herrmann (1976) On modular lattices generated by two complemented pairs, *Houston J. Math.* 2, 513-523.

[30] C. Herrmann, A parameter for subdirectly irreducible modular lattices with four generators, preprint.

[31] C. Herrmann, M. Kindermann, and R. Wille (1975) On modular lattices generated by 1 + 2 + 2, *Algebra Universalis* 5, 243-252.

[32] B. Jónsson (1961) Sublattices of a free lattice, *Canad. J. Math.* 13, 256-264.

[33] B. Jónsson, Varieties of lattices: some open problems, preprint.

[34] B. Jónsson and J.E. Kiefer (1962) Finite sublattices of a free lattice, *Canad. J. Math.* 14, 487-497.

[35] B. Jónsson and J.B. Nation (1977) A report on sublattices of a free lattice, *Coll. Math. Soc. J. Bolyai* 17, Contributions to Universal Algebra (Szeged), North Holland, 223-257.

[36] A. Macintyre (1973) The word problem for division rings, *J. Symbolic Logic* 38, 428-436.

[37] R. McKenzie (1972) Equational bases and nonmodular lattice varieties, *Trans. Amer. Math. Soc.* 174, 1-43.

[38] A. Mitschke and R. Wille (1976) Finite distributive lattices projective in the class of all modular lattices, *Algebra Universalis* 6, 383-393.

[39] J.B. Nation (to appear) Finite sublattices of a free lattice, *Trans. Amer. Math. Soc.*

[40] J. von Neumann (1960) *Continuous Geometry* (I. Halperin, ed.), Princeton University Press, Princeton, N.J.

[41] I. Rival, W. Ruckelshausen, and B. Sands (1981) On the ubiquity of herringbones in finitely generated lattices, *Proc. Amer. Math. Soc.* 82, 335-340.

[42] I. Rival and B. Sands (1981) How many four-generated simple lattices? Universal Algebra Semester, Stefan Banach Mathematical Centre Tracts, Warsaw.

[43] I. Rival and B. Sands (1978 and 1979) Planar sublattices of a free lattice, I and II, *Canad. J. Math.* 30, 1256-1283 and *Canad. J. Math.* 31, 17-34.

[44] I. Rival and B. Sands (to appear) Pictures in lattice theory, *Ann. Discrete Math.*

[45] I. Rival and R. Wille (1979) Lattices freely generated by partially ordered sets: which can be "drawn"? *J. Reine Angew. Math.* 310, 56-80.

[46] H. Rolf (1958) The free lattice generated by a set of chains, *Pacific J. Math.* 8, 585-595.

[47] A. Tarski (1955) A lattice-theoretical fixpoint theorem and its applications, *Pacific J. Math.* 5, 285-309.

[48] P. Whitman (1941) Free lattices, *Ann. of Math.* 42, 325-330.

[49] P. Whitman (1942) Free lattices II, *Ann. of Math.* 43, 104-115.

[50] R. Wille (1973) Über modulare Verbände die von einer endlichen halbgeordneten Menge frei erzeugt werden, *Math. Z.* 131, 241-249.

[51] R. Wille (1977) On lattices freely generated by finite partially ordered sets, *Coll. Math. Soc. J. Bolyai* 17, Contributions to Universal Algebra, Budapest, 581-593.

ADDENDUM

In §4 the proof that $(0,a) \notin L$ may be simplified as follows. It is easy to see that there are homomorphisms $\varphi_i : L_i \to M_4$, $i = 1,2$, where $M_4$ is the length two lattice with atoms $a,b,c,d$, such that $\varphi_i(a) = a, \ldots, \varphi_i(d) = d$, $i = 1,2$. It is easy to check that $M = \{(x_1,x_2) \in L_1 \times L_2 : \varphi_1(x_1) = \varphi_2(x_2)\}$ is a sublattice of $L_1 \times L_2$ and that $L \subseteq M$. Since $(0,a) \notin M$, $(0,a) \notin L$.

Some of the ideas in §4 have also been considered by G. Grätzer and D. Kelly in their study of $\mathfrak{m}$-complete lattices. They show that FL$(a,b,c)$ is not in $F$ by a proof which is similar to the one hinted at in the text.

Two questions about covers in free lattices were raised. David Kelly asked if there is an element of FL$(n)$ with no upper cover and no lower cover. Q.F. Stout asked if the statement "$w$ has an upper cover" is recursive in $w$.

# AN INTRODUCTION TO THE THEORY OF CONTINUOUS LATTICES

Michael W. Mislove
Department of Mathematics
Tulane University
New Orleans, Louisiana 70118

ABSTRACT

The theory of continuous lattices traces its beginnings to the work of Dana Scott and of Jimmie Lawson. In the late 1960's, Scott was looking for mathematical models of the $\lambda$-calculus of Church and Curry. Since this logic is type free, any such model would have to allow elements to represent *functions* as well as *sets*. This led Scott to look for cartesian closed categories, and for objects in such a category which are isomorphic to their own function space of self-maps. He found such a category and such an object by first considering certain $T_0$-spaces, which he called *injective*. A $T_0$-space $I$ is *injective* if, given any pair of $T_0$-spaces $X \subset Y$ and any continuous function $f : X \to I$, there is an extension $f' : Y \to I$ for $f$. (Here, $X \subset Y$ means $X$ is a subspace of $Y$.) An example of an injective space is the so-called *Sierpinski space* $2 = \{0,1\}$ endowed with the topology making $\{1\}$ open, but $\{0\}$ not open. In fact, the injective spaces are characterized by the fact that they are precisely the retracts of products of copies of the Sierpinski space. It turns out that the category of injective spaces and continuous maps between them is exactly the category for which Scott was searching, but to see this more clearly, we first translate this situation into an order-theoretic setting. This is accomplished by defining the *specialization order* on a topological space $X$, whereby $x \leq y$ if and only if every open set containing $x$ must also contain $y$. Said another way, $x \leq y$ iff $y \in \{x\}^-$. While this order is always reflexive and transitive, it is antisymmetric precisely when $X$ is a $T_0$-space. Now, under this order, the injective spaces correspond to objects which Scott called



*I. Rival (ed.), Ordered Sets, 379–406.*

*continuous lattices*, and the continuous functions between the injective spaces can be characterized in completely algebraic terms. We clarify this aspect below.

At about the same time that Scott was searching for his models of the λ-calculus, Jimmie Lawson was beginning his investigations into a certain class of locally compact topological semilattices. He called these semilattices *semilattices with small semilattices* because his hypothesis was that each had a neighborhood basis of subsemilattices at each point. Due to his extensive work in the area, such semilattices have come to be called *Lawson semilattices*. That these two rather disparate lines of research should merge into one theory came about as follows.

In 1974, Karl Hofmann, Al Stralka and the author wrote a monograph [HMS] exploring in detail the Pontryagin duality between compact zero-dimensional semilattices, on the one hand, and discrete semilattices on the other. From our viewpoint, the crucial result in this work is the fact that each compact zero-dimensional semilattice is actually an algebraic lattice endowed with a unique, algebraically determined topology. Moreover, each algebraic lattice becomes a compact zero-dimensional semilattice when endowed with this topology. Since a compact zero-dimensional semilattice is, in particular, a Lawson semilattice, it was only natural that Hofmann and Stralka should endeavor to extend this result to the more general class of compact Lawson semilattices. They accomplished this in [HS], where they gave a completely algebraic description of compact Lawson semilattices and the continuous semilattice homomorphisms between them. Stralka then realized that the objects which Scott defined in [S] to be continuous lattices were exactly the same as those objects which algebraically describe compact Lawson semilattices in [HS]. This led to a great deal of activity in the area, centering around a write-in seminar (known as SCS), and culminated in the writing of the comprehensive treatise [C] about the theory of continuous lattices.

In order to give an exact definition of continuous lattices, we begin by recalling the definition of an algebraic lattice. First, we call a subset $D$ of a complete lattice $L$ *directed* if, given $x,y \in D$ there is some $z \in D$ with $x,y \leq z$. An element $x$ of $L$ is then called *compact* if, for any directed set $D$, $x \leq \sup D$ implies $x \leq d$ for some $d \in D$. For example, the zero of $L$ is compact. If we denote the set of compact elements of $L$ by $K(L)$, then we say that the complete lattice $L$ is *algebraic* if $x = \sup\{y \in K(L) : y \leq x\}$ holds for each $x \in L$. Algebraic lattices arise naturally in algebra; for example, the lattice of ideals of a ring is algebraic, where the compact elements are exactly the finitely generated ideals.

Now, the unique topology referred to above relative to which each algebraic lattice becomes a compact zero-dimensional semilattice is easily described: A subbasis is the family $\{\uparrow k : k \in K(L)\} \cup \{L \setminus \uparrow x : x \in L\}$, where $\uparrow x = \{y \in L : x \leq y\}$. Since each element of this basis is a subsemilattice, it is easy to see that $L$ is a Lawson semilattice when endowed with this topology. It is also true that the topology so defined is compact and zero-dimensional. That each compact zero-dimensional semilattice is actually an algebraic lattice endowed with this topology follows from two observations: First, any such semilattice $S$ has enough continuous semilattice homomorphisms $\phi : S \to 2$ into the two-point semilattice (with the discrete topology) to separate the points, and for each such homomorphism, $\inf \phi^{-1}(1)$ is a compact element of $S$. Thus, for each $x \in S$, we have $x = \sup\{\wedge\phi^{-1}(1) : \phi(x) = 1\}$, thus realizing each element as the sup of compact elements.

The more general notion of a continuous lattice arises from a weakening of the notion of a compact element. For a complete lattice $L$, we say the element $y$ is *compact in* the element $x$ if, for any directed set $D$ with $x \leq \sup D$, there is some $d \in D$ with $y \leq d$. We also say $y$ is *way-below* $x$ in this case, and we denote this relation by $y \ll x$. Thus, the compact elements of $L$ are exactly those elements which are way-below themselves. We then define the complete lattice $L$ to be a *continuous lattice* if $x = \sup\{y \in L : y \ll x\}$ for every $x \in L$. Clearly any algebraic lattice is a continuous lattice, since a compact element is way-below any element above it. However, there are many more continuous lattices than just the algebraic lattices; for example, the unit interval (or any complete chain, for that matter) is a continuous lattice, where $y \ll x$ exactly when $y < x$ or else $y = 0$. That every continuous lattice has a unique compact Hausdorff topology relative to which it is a Lawson semilattice follows by generalizing the topology indicated above for algebraic lattices in the natural way; namely, we take as subbasis the family $\{\twoheaduparrow y : y \in L\} \cup \{L \setminus \uparrow x : x \in L\}$, where $\twoheaduparrow y = \{x \in L : y \ll x\}$. This topology is called the *Lawson topology*. While it is not true that $\twoheaduparrow y$ is a subsemilattice in general, nonetheless $L$ becomes a compact Lawson semilattice when endowed with this topology. Furthermore, the continuous semilattice homomorphisms between compact Lawson semilattices are characterized as being those functions between continuous lattices which preserve all infima and *directed* suprema.

To close this circle of ideas, we point out how injective spaces arise in this context. Indeed, it is clear that the fundamental notion for continuous lattices is the idea that directed nets should converge to their suprema. We can define the weakest topology on a continuous lattice $L$ for which this holds by letting $U \subset L$ be open iff $U = \uparrow U$ and, for every directed set $D$ in $L$, if sup $D \in U$, then $D \cap U \neq \emptyset$. This is the *Scott topology* on $L$, and sets such as $U$ above are called *Scott open sets*. Clearly the Lawson topology is a refinement of the Scott topology (well, it isn't actually clear, but it does follow from the definition of a continuous lattice). In any event, it turns out that any continuous lattice, when endowed with the Scott topology is an injective $T_0$-space, and every injective $T_0$-space so arises. Moreover, the continuous functions between injective spaces are precisely those functions which preserve directed sups! Thus, the appropriate category for Scott's construction of models of the $\lambda$-calculus is precisely the category of continuous lattices and Scott continuous functions. As it happens, this category is cartesian closed, and using fairly straightforward techniques (which are nonetheless rather clever) Scott constructs a continuous lattice which is isomorphic in a natural way to its space of Scott continuous self-maps. While he shows this can be done starting with any continuous lattice whatsoever -- and, as a result, it follows that any $T_0$-space is embeddable in an injective space which is isomorphic to its space of continuous self-maps -- his application to the mathematical theory of computing uses the two point lattice as the building block (see [S-2] for the details).

Continuous lattice theory has made considerable contributions to general topology via spectral theory, and we want to close this introduction by indicating how this has come about. If we recall Stone's basic result about Boolean algebras, it says that each Boolean algebra arises as the algebra of compact-open subsets of some compact zero-dimensional space, and, for a given Boolean algebra, the space in question is the space of maximal ideals of the algebra endowed with a suitable topology (see [St]). This spectral theory generalizes to a category of continuous lattices in the following way. If one examines Stone's theorem carefully, it becomes clear that two facts are crucial to the proof. First, that the maximal ideals are the prime elements in the lattice of all ideals of the algebra, and second that the lattice of ideals of the algebra is a distributive algebraic lattice. This led Hofmann and Lawson to consider the space Spec $L$ of all non-identity primes of an arbitrary distributive lattice $L$. Recalling that the *lower topology* on the lattice $L$ has the sets of the form $L \setminus \uparrow x$ for a subbasis, it turns out that Spec $L$, when endowed with the relative lower topology from $L$, becomes a locally quasicompact sober space (a space is *locally quasicompact* if each point has a neighborhood basis of quasicompact sets; it is *sober* if each closed subset which cannot be written as the union of two proper closed subsets has a unique dense point). Moreover, the map sending each point $x$ of $L$ to the open subset (Spec $L)\setminus \uparrow x$ of Spec $L$ is an isomorphism of the distributive continuous lattice $L$ onto the lattice $O(\text{Spec } L)$ of open subsets of Spec $L$. On the other hand, for any locally quasicompact sober space $X$, the lattice $O(X)$ of open subsets of $X$ is a distributive continuous lattice, and if $f : X \to Y$ is a continuous proper map between such spaces, then the map $f^{-1} : O(Y) \to O(X)$ preserves all sups, finite infs, and the way-below relation. The details of this duality are contained in [HL]. This theory also leads to many new and interesting questions about the theory of continuous lattices and its applications to other areas of mathematics. For example, a general theory of local compactness (in the absence of Hausdorff separation) has been developed in [HM], and this in turn has led to the laying of the foundations for a general theory of $k$-spaces (again without Hausdorff separation) in [SCS-HL].

## 0. INTRODUCTION.

In the late 1960's Dana Scott was searching for models of the λ-calculus of Church and Curry, with the goal of laying a firm mathematical foundation for the theory of computation. This logic is *type free*; i.e., as opposed to classical logic where elements are clearly identified as to type - element, set, function, etc. - a type free system allows elements to simultaneously represent functions as well as sets. Thus, any model for such a system must allow elements to operate as selfmaps on the model. For, if $x$ and $y$ are elements of the model, then $x$ can represent a function as well as an element, and so the expression $x(y)$ must make sense, as must $y(x)$. Hence, to be a λ-calculus model, an object $A$ must be isomorphic to its space of selfmaps $\mathcal{A}(A,A)$ in whatever category $\mathcal{A}$ to which $A$ belongs. In particular, the space of selfmaps $\mathcal{A}(A,A)$ must be an object of the category $\mathcal{A}$ for any object $A$ in $\mathcal{A}$; categories with this property, along with concomitant natural functorial properties for the space $\mathcal{A}(A,A)$ are called *cartesian closed*. Therefore, Scott was seeking examples of cartesian closed categories $\mathcal{A}$ and objects $A$ in $\mathcal{A}$ with $A \simeq \mathcal{A}(A,A)$.

At about the same time that Scott was looking for models of the λ-calculus, Jimmie Lawson was beginning his investigation into the structure theory of certain topological semilattices. These semilattices, which soon came to be called *Lawson semilattices*, have the defining property that each point has a neighborhood basis of subsemilattices. The canonical example of such a semilattice is the unit interval $I = [0,1]$ endowed with the usual topology and minimum multiplication. One of Lawson's principal results is that, for a large class of locally compact Lawson semilattices, there are enough continuous semilattice homomorphisms into $I$ to separate the points [L-1]; in particular, each compact Lawson semilattice is embeddable as a semilattice in a product $I^X$ for some set $X$. Since Lawson had also shown that a map between compact semilattices is a continuous semilattice homomorphism precisely when it preserves all infima and all directed suprema (i.e., suprema of directed subsets ) [L-2], and since he had also shown that each semilattice admits at most one compact topology relative to which it is a compact topological semilattice, it became clear that a completely algebraic characterization of compact Lawson semilattices was a feasible goal. The first step in this direction was obtained by Karl Hofmann, Al Stralka, and the author in their 1974 monograph [HMS], where the category of compact zero-dimensional semilattices was shown to be equivalent to the category of complete algebraic lattices and maps preserving all infima and all directed suprema. The final step was the work of Hofmann and Stralka [HS] where they showed that compact Lawson semilattices are indeed completely algebraic objects. In fact, Stralka then realized that the algebraic description of compact Lawson semilattices which they had obtained was exactly the same as the definition of Scott's *continuous lattices* given in [S-1], which was Scott's first published account of his search for $\lambda$-calculus models. This resulted in the formation of the write-in seminar SCS (for Seminar on Continuity in Semilattices), which created a great deal of activity around the topic of continuous lattices and their applications, and which eventually led to the writing of the comprehensive treatise [C] on the theory.

In this paper, we shall briefly outline both the results described above, but we shall not present all of the details of either approach. The reason for this is that both approaches are very completely treated (at least from a purely mathematical viewpoint) in the work [C], and we see no need to rehearse the presentation given therein. Moreover, we also want to present material which goes beyond the material outlined above, and, although most of this is also given in [C], we hope that our presentation here motivates the readers to follow up on those points which interest them by consulting [C].

## I. THE BASICS: COMPACT ELEMENTS AND THE WAY-BELOW RELATION.

We begin by laying down some groundrules. First, we shall be concerned almost exclusively with complete lattices; i.e., with partially ordered sets in which all subsets have an infimum and a supremum. In those instances where we want to consider more general structures, we shall explicitly state our assumptions. The notation we employ is fairly standard by now: A subset $D$ of a lattice is *directed* if $x,y \in D$ imply there is some $z \in D$ with $x,y \leq z$; the subset $F$ is *filtered* if $x,y \in F$ imply there is some $z \in F$ with $z \leq x,y$. An *upper set* is a subset $U$ with $U = \{x \in L : u \leq x \text{ for some } u \in U\} = \uparrow u$; a *lower set* is defined dually. An *ideal* is a directed lower set, and a *filter* is a filtered upper set. The zero of a lattice is denoted $0$, and the identity by $1$.

We introduce the concept of a continuous lattice as a generalization of the notion of an algebraic lattice, and so we first recall the definition of a compact element. Recall then that an element $x$ of a lattice $L$ is *compact* if, whenever a directed family $D$ of elements of $L$ satisfies $x \leq \sup D$, then there is some $d \in D$ with $x \leq d$. The set of all compact elements of $L$ is denoted $K(L)$. For example, the zero of any lattice is always a compact element; it may be the only one, as is the case of the unit interval I. On the other hand, every element of a finite lattice is compact. Algebraic lattices are those in which compact elements abound. Namely, a complete lattice L is *algebraic* if every element of L is the supremum of the compact elements which it dominates; in symbols, $x = \sup\{y \in K(L) : y \leq x\}$ for all $x \in L$. As an elementary example, consider the power set lattice $2^X$ of the set X in its natural order. Clearly any finite subset of X is a compact element of $2^X$, and equally obvious is the fact that each subset of X is the sup(= union) of its finite subsets. Hence, $2^X$ is an algebraic lattice. Another more interesting example is obtained as follows: Let R be a ring with unit, and let L denote the lattice of all two-sided ideals of R. Then clearly L is a complete lattice, and any principal ideal of R is a compact element of L. This latter follows from the fact that the sup of a directed family of ideals is just the union, and so a principal ideal is contained in the sup of a directed family of ideals precisely when the generator, and hence the principal ideal itself, is contained in one of the ideals of the family. Moreover, the finitely generated ideals of R are then precisely the compact elements of L, and it is then clear that L is algebraic, since every ideal is the union of the finitely generated ideals contained in it. Thus, L is an algebraic lattice.

A moment's thought makes it clear that there is an abundance of algebraic lattices; after all, any finite lattice is algebraic, and the product of any family of algebraic lattices is again algebraic. (For the latter, one notes that the compact elements of the product of algebraic lattices are precisely those elements which are compact in every factor, and which are 0 in almost all factors.) We point out that this leads to a characterization of algebraic lattices: Indeed, for each compact element x of the algebraic lattice L, the characteristic function f of the upper set of x is a map of L into 2, the two-point lattice, which preserves all infima (since $\uparrow x$ is closed under all infima), and also preserves directed suprema (since x is compact). Since L is algebraic, the family of all such characteristic functions actually separates the points of L, and so we can realize L as a subset of some power of 2 ($2^{K(L)}$ to be precise) which is closed under all infima and all directed suprema. We shall see a parallel to this for continuous lattices later.

The notion which is crucial to the concept of an algebraic lattice is that of a compact element, and so, to generalize the idea of an algebraic lattice, we generalize this notion. We first note that the statement that the element $x$ is compact is the same thing as saying that $x$ is compact *in* $x$. By this, we are simply emphasizing the fact that the definition of compactness is a two-step process. To wit: For any directed set $D$,

(1) If $x \leq \sup D$, then

(2) $x \leq d$ for some $d \in D$.

Thus, we could say that the element $x$ is *relatively compact in* the element $y$ if, for any directed set $D$,

(1) Whenever $y \leq \sup D$, then

(2) $x \leq d$ for some $d \in D$.

Clearly any compact element is relatively compact in itself, and, in fact, in any element which dominates it. But, to illustrate the difference between compactness and relative compactness, we consider the following class of examples. Let $X$ be a topological space, and let $\mathcal{O}(X)$ denote the lattice of all open subsets of $X$. Then $\mathcal{O}(X)$ is clearly a complete lattice in which the sup of a directed family is the union of the family. Thus, the compact elements of $\mathcal{O}(X)$ are precisely those open subsets of $X$ which are compact in the usual topological sense. Hence, $\mathcal{O}(X)$ is an algebraic lattice if and only if every open set is the union of compact open sets, which is the same thing as saying that $X$ has a basis of compact open subsets. If $X$ is Hausdorff, this means, of course, that $X$ is locally compact and totally disconnected.

As to the condition of relative compactness in this example, if $U$ and $V$ are open subsets of $X$, then $U$ is relatively compact in $V$ if and only if every open cover of $V$ admits a finite subcover of $\underline{U}$ (the terminology here is quite appropriate). Now, certainly if $\overline{U} \subset V$ and $\overline{U}$ is compact, then $U$ is relatively compact in $V$; more generally, if there is some quasicompact subset $Q$ of $X$ with $U \subset Q \subset V$, then $U$ is relatively compact in $V$ (by *quasicompact*, we simply mean that $Q$ satisfies the Heine-Borel property; when we use the term *compact*, we are also assuming that the ambient space is Hausdorff). Thus, in a connected locally compact Hausdorff space $X$, the only compact element of $\mathcal{O}(X)$ is the empty set, but there are many open sets $U \subset V$ with $U$ relatively compact in $V$.

As another example, consider any complete chain; e.g., the unit interval $I$ in its natural order. For $I$, the only compact element is $0$. But, for any complete chain $C$, if $x,y \in C$ and $x < y$, then $x$ is relatively compact in $y$. This follows from the fact that any directed subset $D$ of $C$ either lies in $\downarrow x$, in which case $\sup D \leq x$, or else $D \setminus \downarrow x \neq \emptyset$, in which case there is some $d \in D$ with $x \leq d$.

This is a convenient place to establish some more prevalent terminology than that which we have so far established. While the statement that " $x$ is relatively compact in $y$ " is very suggestive of the relation between $x$ and $y$, the more widely adopted terminology is " $x$ is *way-below* $y$ ". This is usually denoted $x << y$, and $\Downarrow y = \{x \in L : x << y\}$ is also a convenient notation, as is $\Uparrow x = \{y \in L : x << y\}$.

We also want to point out some useful properties of the way-below relation. First, if $x << y$ and $y \leq z$, then $x << z$ since any sup which dominates $z$ must also dominate $y$. Also, if $w \leq x << y$, then $w << y$. A more interesting fact is that the set $\Downarrow y$ is directed: Indeed, if $x,z << y$, then $x \vee z << y$. (This follows from the fact that, if $D \subset L$ is directed with $y \leq \sup D$, then there are $d_1, d_2 \in D$ with $x \leq d_1$ and $z \leq d_2$; but, since $D$ is directed, there is then some $d \in D$ with $d_1, d_2 \leq d$, and so $x \vee z \leq d$ as required.) Finally, as we have already noted, $x << y$ if $x$ is compact and $x \leq y$, and so $0 << y$ for all $y \in L$.

We can now define the notion of a continuous lattice; this is accomplished by simply replacing the role which compact elements play in an algebraic lattice with the concept of relative compactness. Thus, a *continuous lattice* is a complete lattice $L$ such that $y = \sup \Downarrow y$ for each $y \in L$. Clearly then, any algebraic lattice is a continuous lattice, and our comments above show that any complete chain is a continuous lattice.

For a topological space $X$, the question of when $\mathcal{O}(X)$ is a continuous lattice is somewhat more delicate. We have already seen that the open sets $U$ and $V$ satisfy $U << V$ if $U^-$ is compact and a subset of $V$. Since any open subset $V$ is the union of such subsets $U$ provided $X$ is locally compact Hausdorff, it follows that $\mathcal{O}(X)$ is a continuous lattice for any locally compact Hausdorff space $X$. We now show that a converse for this holds; namely, that a Hausdorff space $X$ satisfies $\mathcal{O}(X)$ is a continuous lattice only if $X$ is locally compact. Suppose then that $\mathcal{O}(X)$ is a continuous lattice, and let $U,V \in \mathcal{O}(X)$ with $U << V$. We first show that $U^- \subset V$ : Indeed, let $x \in X \backslash V$. Then for each $v \in V$, there are disjoint open subsets $U_v$ containing $x$, and $W_v$ containing $v$. The family of $W_v$'s is then an open cover of $V$ which admits a finite subcover for $U$, assuming that $U << V$. But, the intersection of the corresponding open sets containing $x$ is then an open set which contains $x$ and misses $U$, thus showing that $x \notin U^-$. Hence $U^- \subset V$ as claimed. We next show that $U^-$ is compact. Let $O_i$ be an open cover of $U^-$. If $x \in U^-$, then there is some index $i$ with $x \in O_i \cap V$, and since $\mathcal{O}(X)$ is continuous, there is then some open set $U_x$ containing $x$ with $U_x << O_i \cap V$. What we have just shown then implies that $U_x^- \subset O_i \cap V$. Now, the family of

$U_x$'s covers $U^-$, and so the addition of the open set $V \backslash U^-$ creates an open cover of $V$. Since $U << V$, it follows that there are finitely many of the $U_x$'s, $U_1, \ldots, U_n$ which cover $U$. Thus, $U^- \subset U_1^- \cup \ldots \cup U_n^- \subset O_1 \cup \ldots \cup O_n$, which generates a finite subcover for $U^-$ from the open cover $O_i$. It then follows that $U^-$ is compact as we claimed.

To summarize the previous paragraph, we have shown that, for a Hausdorff space $X$, the lattice $\mathcal{O}(X)$ is continuous if and only if $X$ is locally compact, and, if this is the case, then $U << V$ in $\mathcal{O}(X)$ iff $U^- \subset V$ and $U^-$ is compact. This result was first noticed by Day and Kelly [DK], and a more general result is also available. Namely, Hofmann and Lawson showed in [HL] that for a sober space $X$, the lattice $\mathcal{O}(X)$ is continuous if and only if $X$ is a locally quasicompact space. (A space is *sober* if the primes of $\mathcal{O}(X)$ are precisely those open sets of the form $X \backslash \{x\}^-$; it is *locally quasicompact* if each point has a neighborhood basis of quasicompact sets.) We shall say more about this later on when we address the topic of spectral theory for continuous lattices.

As a final introductory example, we present an analogue to the algebraic lattice of ideals of the ring $R$ with unit cited above. Recall that a *C*-algebra* is a Banach algebra $A$ with involution $* : A \to A$ such that $\|a^*a\| = \|a\|^2$ for each $a \in A$. For example, $C_0(X)$, the algebra of all continuous complex-valued functions vanishing at infinity on the locally compact space $X$ is the prototypical commutative C*-algebra. In fact, if $\Delta(A)$ denotes the space of maximal ideals of the commutative C*-algebra $A$, then $\Delta(A)$ is locally compact in the weak *-topology, and $A \simeq C_0(\Delta(A))$. Moreover, any closed ideal $B$ of $A$ is itself a commutative C*-algebra, and $B \simeq C_0(Y)$, where $Y = \{x \in \Delta(A) : f(x) \neq 0$ for some $f \in B\}$. Thus, the lattice of closed ideals $\mathrm{Id}(A)$ of the commutative C*-algebra $A$ is isomorphic to the space of closed subsets of the locally compact Hausdorff space $\Delta(A)$. Since this lattice is isomorphic under complementation to the lattice of open subsets of $\Delta(A)$, it follows that $\mathrm{Id}(A)$ is a continuous lattice.

A much more general result is known in this regard. For any C*-algebra $A$, the lattice of all closed two-sided ideals of $A$ is a continuous lattice. A very clever proof of this fact is due to Karl Hofmann, and can be found in [C], Example 1.20, p.47ff. This result can be combined with the spectral theory for distributive continuous lattices to show that the prime spectrum of any C*-algebra is a locally quasicompact sober space.

## II. TOPOLOGIES ON CONTINUOUS LATTICES

The definition of the way below relation is very suggestive of a topology. In fact, if we agree to restrict our attention to directed nets (i.e., to nets $x_i$ in the lattice $L$ such that $i \le j$ implies $x_i \le x_j$), then the statement that $y$ is way-below $x$ is precisely the same as saying that $\uparrow y$ is a neighborhood of $\uparrow x$. We are then led to the *Scott topology* on the complete lattice $L$ : A subset $U \subset L$ is *Scott open* if $U = \uparrow U$ and if $D \cap U \neq \emptyset$ for any directed subset $D$ of $L$ with $\sup D \in U$. Clearly, $L \backslash \downarrow x$ is Scott open for any $x \in L$, and so the Scott topology is always $T_0$. The continuous lattices can be characterized solely in terms of their Scott topology, but in order to elucidate that characterization, we must first establish an important property of the way below relation on a continuous lattice.

If $L$ is a lattice, then $\twoheaduparrow y \cap \twoheaduparrow z = \twoheaduparrow (y \vee z)$, as is readily verified. If $L$ is continuous, then $x = \sup \twoheaddownarrow x$ for each $x \in L$, and so $\uparrow x = \cap\{\twoheaduparrow y : y << x\}$. Thus, we can generate a $T_0$-topology on $L$ by taking the sets $\twoheaduparrow y$ for a basis. We now show that this topology is the Scott topology on $L$, provided $L$ is continuous. Indeed, if $U$ is a Scott open subset of $L$ and if $x \in U$, then, since $\twoheaddownarrow x$ is directed and $x = \sup \twoheaddownarrow x$, there is some $y << x$ with $y \in U$. Hence $x \in \twoheaduparrow y \subset U$, so that each Scott open set is open in the topology described above. Conversely, suppose that $y \in L$. To show that $\twoheaduparrow y$ is Scott open, we first establish the so-called *interpolation property* for the way below relation on L:

If $y << x$ in $L$, then there is some $z \in L$ with $y << z << x$.

To prove this, we define the set $D = \{y \in L : y << z << x$ for some $z \in L\}$. If $y_1, y_2 \in D$, then there are $z_1, z_2 \in \twoheaddownarrow x$ with $y_i << z_i << x$ for $i = 1,2$, and so $y_1 \vee y_2 << z_1 \vee z_2 << x$, whence $y_1 \vee y_2 \in D$. Thus $D$ is directed, and we let $x^* = \sup D$, noting that $x^* \le x$. If $x^* < x$, then we find some $z << x$ with $z \not\le x^*$ since $L$ is continuous. But then $z \not\le x^*$ implies there is some $y << z$ with $y \not\le x^*$. Thus $y << z << x$, which implies that $y \in D$. This means that $y \le \sup D = x^*$, a contradiction. Thus, we must have $x^* = x$, which establishes the interpolation property.

Now, returning to the question of the Scott topology on a continuous lattice $L$, we show that $\twoheaduparrow y$ is Scott open for each $y \in L$, which then completes the proof that the Scott topology has the family $\{\twoheaduparrow y : y \in L\}$ for a basis if $L$ is continuous. Now, suppose that $y \in L$ and $x \in \twoheaduparrow y$. Then $y << x$, and so the interpolation property implies there is some $z \in L$ with $y << z << x$. Now, suppose that $D \subset L$ is directed with $x \le \sup D$. Then

$z \ll x$ implies that there is some $d \in D$ with $z \leq d$, and so $y \ll z$ then implies that $d \in \Uparrow y$. Hence, $\Uparrow y \cap D \neq \emptyset$, so $\Uparrow y$ is Scott open, as claimed.

We now characterize continuous lattices in terms of their Scott topology. To start, we make the following definition: A $T_0$-space $X$ is *injective* if, given $T_0$-spaces $Y \subset Z$ and a continuous map $f : Y \to X$, there is a continuous extension $f' : Z \to X$. For example, the so-called *Sierpinski space* $2 = \{0,1\}$ with the topology making $\{1\}$ open, but $\{0\}$ not open, is readily seen to be injective. Also, any product $\Pi X_i$ of injective spaces is injective (given $f : Y \to \Pi X_i$, just define $f' : Z \to \Pi X_i$ using $f_i' : Z \to X_i$ for each index $i$, the existence of which is guaranteed by the fact that $X_i$ is injective), and any retract of an injective space is injective (if $X' \subset X \xrightarrow{r} X'$ with $X$ injective, and if $f : Y \to X'$ is given, then $r \circ f' : Z \to X'$ is the required map, where $f' : Z \to X$ is the map guaranteed by the fact that $X$ is injective). Conversely, an injective space is a retract of any space in which it is embedded, since the identity map $1 : X \to Z$ must have a continuous extension $1' : Z \to X$ if $X$ is injective. This notion of injectivity is due to Scott [S-1], and our development of the characterization of continuous lattices in terms of injective spaces closely follows his presentation.

We now show that each continuous lattice is an injective $T_0$-space in the Scott topology, and in fact that each injective $T_0$-space is some continuous lattice with its Scott topology. We first recall a result due to Cech, which states that every $T_0$-space $X$ satisfies the property that the map $x \to \chi_x : X \to 2^{\mathcal{O}(X)}$ is an embedding from $X$ into the product of $\mathcal{O}(X)$ copies of the Sierpinski space. Next, we note that the Sierpinski space is just the two-point lattice $2$ endowed with its Scott topology, and, in fact, the product $2^A$ of $A$ copies of the Sierpinski space is just the algebraic lattice consisting of the product of $A$ copies of the two-point lattice endowed with its Scott topology, for any set $A$. Finally, we note that the definition of the Scott topology means that a map $f : L \to L'$ between continuous lattices is continuous in the Scott topology if and only if $f$ preserves sups of directed sets (for details see [C], Proposition 2.1, p.112). Thus, given a continuous lattice $L$ and its Scott topology $\sigma(L)$, to show $(L,\sigma(L))$ is an injective $T_0$-space we need only show that there is some retraction $r : 2^{\sigma(L)} \to L$, since we already know that $L$ is embedded in $2^{\sigma(L)}$ and that this is an injective

$T_0$space. The definition of $r$ is somewhat delicate:

$$r(y) = \sup\{\inf U : y(U) = 1 \text{ and } U \in \sigma(L)\}.$$

Since $L$ is continuous, it follows that $r(\chi_x) = x$ for all $x \in L$, and so we shall be done if we show that $r$ preserves sups of directed sets. So, let $D \subset L$ be a directed set, and let $y = \sup D$. Since $r$ is clearly monotone, we have that $\sup r(D) \leq r(y)$. Suppose that $x \ll r(y)$, and let $U = \{z \in L : x \ll z\}$. Then $U$ is a Scott open subset of $L$, and clearly $r(y) \in U$. Thus, the definition of $r$ implies that $1 = \chi_{r(y)}(U) \leq y(U)$, and since $y = \sup D$, there is some $d \in D$ with $d(U) = 1$. But, then $r(d) = \sup\{\inf V : d(V) = 1 \text{ and } V \in \sigma(L)\} \geq \inf U \geq x$, as $U = \Uparrow x$. This shows that $x \leq \sup r(D)$ for every $x \ll r(y)$, and so $r(y) \leq \sup r(D)$. Now, since we already have $\sup r(D) \leq r(y)$, we conclude that $r(\sup D) = \sup r(D)$ for directed sets $D \subset L$, so that $r$ is Scott continuous. Thus $(L,\sigma(L))$ is an injective $T_0$-space for a continuous lattice $L$.

Conversely, suppose that $X$ is an injective $T_0$-space. We want to show that $X$ is a continuous lattice endowed with its Scott topology. First, we need to establish a partial order on $X$, and that is accomplished by setting $x \leq y$ in $X$ if $x \in \{y\}^-$. Said another way, we have $x \leq y$ iff every open set which contains $x$ must also contain $y$. This order is called the *specialization order* on the space $X$, and it is always reflexive and transitive. The specialization order is anti-symmetric precisely when the space $X$ is a $T_0$-space. If, for example, we consider the Sierpinski space, then clearly it gives rise to the two-point lattice in the specialization order. Moreover, it is readily seen that any power of the Sierpinski space gives rise to the corresponding power of the two-point lattice when endowed with the specialization order, and so the product of $A$ copies of the Sierpinski space gives the algebraic lattice $2^A$ when given the specialization order. Now, we know that any injective $T_0$-space $X$ is a retract of the product of $\mathcal{O}(X)$ copies of the Sierpinski space, and so, to complete our proof that every injective $T_0$-space is a continuous lattice in the specialization order, we first observe the following:

If $L$ is a continuous lattice and $p : L \to L$ is a Scott continuous map with $p \circ p(x) = p(x)$ for all $x \in L$, then $p(L)$ is a continuous lattice in the inherited order.

The proof of this is straightforward: Since $p$ is Scott continuous, $p$ is also monotone, and so $p$ is a *projection operator* on $L$. It is widely known (and easy to show) that $p(L)$

is a complete lattice if $L$ is one under these circumstances. So, to show $p(L)$ is continuous, we show that $p(x) = \sup\{p(y) : p(y) << p(x)$ in $p(L)\}$ for every $x \in L$. So, let $x = p(x) \in p(L)$, and let $y << x$ in $L$. We claim that $p(y) << x$ in $p(L)$. In fact, if $D \subset p(L)$ is directed with $x \leq \sup D$ (in $p(L)$), then since $p(L) \subset L$, we have that $x \leq \sup D$ in $L$. This means that there is some $d \in D$ with $y \leq d$ since $y << x$ in $L$, and so $p(y) \leq p(d) = d$. Thus $p(y) << x$ as claimed. But, since $p$ preserves directed sups, we then conclude that $x = \sup\{p(y) : y << x$ in $L\} \leq \sup\{p(y) : p(y) << x$ in $p(L)\}$, so that $p(L)$ is indeed continuous.

Now, if $X$ is an injective $T_0$-space, then $X$ is a retract of $2^{\mathcal{O}(X)}$, and $2^{\mathcal{O}(X)}$ is a continuous lattice by our remarks above. Moreover, the retract $r : 2^{\mathcal{O}(X)} \to 2^{\mathcal{O}(X)}$ is a Scott continuous projection operator since the topology on $2^{\mathcal{O}(X)}$ is the Scott topology. Thus, the image of the retract, $X$, is a continuous lattice by the result we proved above. But, since $X$ is a subspace of $2^{\mathcal{O}(X)}$ in its original topology, tbe inherited order on $X$ is just the specialization order from the topology of $X$, and so $X$ is a continuous lattice in the specialization order.

As our final observation in this regard, we note that it is clear that the specialization order on a continuous lattice endowed with the Scott topology is just the original order on the lattice. Conversely, if $X$ is an injective $T_0$-space, then we show that the Scott topology on $X$ as a continuous lattice when endowed with the specialization order is the original topology on $X$ : Indeed, since $X$ is injective, we have the sequence $X \subset 2^{\mathcal{O}(X)} \to X$, where the first arrow is an embedding, and the second a retraction. Applying the specialization order, we obtain the sequence $(X,\leq) \subset (2^{\mathcal{O}(X)},\leq) \to (X,\leq)$. Now, endowing each term with the Scott topology yields the sequence $(X,\sigma(X)) \subset (2^{\mathcal{O}(X)},\sigma(2^{\mathcal{O}(X)})) \to (X,\sigma(X))$. But the original topology on $2^{\mathcal{O}(X)}$ is the Scott topology, and so it follows that $X$ in its original topology and the space $(X,\sigma(X))$ are the same subspace of $2^{\mathcal{O}(X)}$, and so they are the same. Thus $\mathcal{O}(X) = \sigma(X)$ as claimed.

We summarize the preceding as follows:

THEOREM: (1) *Each continuous lattice* $L$ *is an injective* $T_0$*-space when endowed with the Scott topology, and the specialization order is the original order on* $L$.

(2) *Each injective* $T_0$*-space* $X$ *is a continuous lattice when given the specialization order, and in fact the topology of* $X$ *is the Scott topology.*

Scott discovered continuous lattices in his search for models of the $\lambda$-calculus. In fact, he realized that the continuous maps between injective spaces are precisely the Scott continuous maps, i.e., those maps preserving sups of directed subsets. He thus found the category of continuous lattices and Scott continuous maps to be the same as the category of injective $T_0$-spaces and continuous maps. His work in [S-1] shows that the space $[L \to M]$ of Scott continuous maps between continuous lattices $L$ and $M$ is also a continuous lattice, and he then shows how, starting with any continuous lattice $L$, a continuous lattice $L'$ can be constructed with $L' \simeq [L' \to L']$. Thus, $L'$ is a suitable candidate for a model for the $\lambda$-calculus. A consequence of Scott's construction is that the original lattice $L$ is embeddable in the lattice $L'$ under a Scott continuous map, and so any injective space can be embedded in an injective space which is isomorphic to its space of continuous self maps. This work has been generalized to the setting of certain continuous posets (which are more applicable to computation) by Niño [N], and the details of the relevant material for the continuous lattices are also contained in [C], Section IV-4.

Having outlined the role of continuous lattices in Scott's work, we now turn our attention to Lawson's work on topological semilattices. Recall that a *compact semilattice* is a compact Hausdorff space $S$ endowed with a continuous, associative and commutative idempotent multiplication. The unit interval $I$ in its natural order is the canonical example. We shall assume that each such semilattice has an identity element $1$. Now, in any such semilattice, every subset $X$ has an infimum, namely $\inf X = \lim\{\inf F : F \subset X \text{ is finite}\}$, where $\inf \emptyset = 1$, and the limit is with respect to the compact topology of the semilattice. Thus each compact semilattice is a partially ordered set with largest element in which every subset has an inf, and so each such semilattice is a complete lattice in its natural order. We shall show which compact semilattices correspond to continuous lattices.

We note that any compact semilattice satisfies the property that every directed subset converges to its sup, and every filtered subset converges to its inf. For example, if $D \subset S$ is directed, then $d \leq \sup D$ for each $d \in D$, and so $x \leq \sup D$ for each cluster point $x$ of the net $D$, since $(\lim D)(\sup D) = \lim(d \sup D) = \lim D$. Conversely, if $x$ is a cluster point of the net $D$, then since $\uparrow d$ is closed and $D \cap \uparrow d$ is residual in $D$ for each $d \in D$, it follows that $x \in \uparrow d$ for each $d \in D$, and so $\sup D \leq x$. Hence $\sup D = \lim D$ for a directed subset $D$ of $S$.

We begin our characterization of which compact semilattices are actually continuous lattices by describing the topology relative to which each continuous lattice becomes a compact semilattice. First, recall that the *lower topology* on any complete lattice $L$ has for a subbasis the family $\{L\backslash\uparrow x : x \in L\}$. We then define the *Lawson topology* on a complete lattice $L$ to be the common refinement of the Scott topology and the lower topology on $L$, and we denote this topology by $\lambda(L)$. Thus, for a continuous lattice $L$, the Lawson topology has for a subbasis the family $\{\Uparrow y : y \in L\} \cup \{L\backslash\uparrow x : x \in L\}$. We now show that the space $(L,\lambda(L))$ is a compact semilattice if $L$ is a continuous lattice:

(1) $\lambda(L)$ is Hausdorff: If $x,y \in L$ with $x \neq y$, then we can assume that $x \not\leq y$. Since $L$ is continuous, there is then some $z \in L$ with $z \ll x$ and $z \not\leq y$. The sets $\Uparrow z$ and $L\backslash\uparrow z$ are then disjoint open sets containing $x$ and $y$, respectively.

(2) $\lambda(L)$ is compact: We appeal to Alexander's Theorem, and show that any open cover from the subbasis given above has a finite subcover. Let $\{\Uparrow y_i : i \in I\} \cup \{L\backslash\uparrow x_j : j \in J\}$ be such an open cover of $L$. If $x = \sup x_j$, then $x \notin L\backslash\uparrow x_j$ for each $j \in J$ implies that there is some index $i$ with $x \in \Uparrow y_i$. Then, since $y_i \ll x = \sup x_j$, there is a finite subset $F \subset J$ with $y_i \leq \sup\{x_j : j \in F\}$. It is then routine to show that the family $\Uparrow y_i \cup \{L\backslash\uparrow x_j : j \in F\}$ covers $L$.

(3) The map $(x,y) \to xy : L \times L \to L$ is continuous: Suppose that $xy \in \Uparrow z$, for some $z \in L$. Then the interpolation property implies that there is a $w \in L$ with $z \ll w \ll xy$, and so $x,y \in \Uparrow w$. Moreover, if $r,s \in \Uparrow w$, then $w \leq r,s$ implies that $w \leq rs$, and so $z \ll rs$. Thus $(\Uparrow w)(\Uparrow w) \subset \Uparrow z$. On the other hand, if $xy \in L\backslash\uparrow z$ for some $z \in L$, then either $x \in L\backslash\uparrow z$ or else $y \in L\backslash\uparrow z$, since $\uparrow z$ is a subsemilattice. Assuming the former, we then have $x \in L\backslash\uparrow z$, and certainly $y \in \Uparrow 0$, and $(L\backslash\uparrow z)(\Uparrow 0) \subset L\backslash\uparrow z$ as $L\backslash\uparrow z$ is an ideal. These arguments show that the map is continuous.

Thus, each continuous lattice is a compact semilattice when endowed with the Lawson topology. However, not every compact semilattice arises in this fashion (although counterexamples are extremely difficult to obtain; see [C], Section VI-4). In any event, it turns out that it is relatively easy to distinguish those compact semilattices which are continuous lattices endowed with the Lawson topology. We call a topological semilattice a *Lawson semilattice* if each point has a neighborhood basis of subsemilattices. If $L$ is a continuous lattice, then it is simple to show that the Lawson topology on $L$ has this property. Indeed, for each $x \in L$, the set $L \backslash \uparrow x$ is an ideal, and hence is a subsemilattice neighborhood of each of its points. In general, the set $\twoheaduparrow y$ is not a subsemilattice, but an argument we have already used shows that each point in $\twoheaduparrow y$ has a subsemilattice neighborhood which is also a subset of $\twoheaduparrow y$. Indeed, if $x \in \twoheaduparrow y$, then we can choose a sequence $y_n$ with $y \ll y_{n+1} \ll y_n \ll x$ by utilizing the interpolation property. If we then let $F = \cup \uparrow y_n$, arguments we have already given show that $F$ is a Scott open subset of $\twoheaduparrow y$ which contains $x$ and is a subsemilattice. Thus, each continuous lattice is a compact Lawson semilattice when endowed with the Lawson topology.

Conversely, suppose that $S$ is a compact Lawson semilattice. Then we have already noted that $S$ is a complete lattice in its natural order. To show that $S$ is a continuous lattice, we choose an $x \in S$ and an open set $U$ which contains $x$. Since $S$ is a Lawson semilattice, there is then a subsemilattice neighborhood, $V$, of $x$ contained in $U$, and we may assume that $V$ is compact. If $y = \inf V$, we then have that $x \in V \subset \uparrow y$, and $y \in U$. We claim that $y \ll x$ : In fact, if $D \subset S$ is directed and $x \leq \sup D$, then the continuity of the inf operation implies that $x = \sup xD$ (since we have already shown that directed sets converge to their sup in any compact semilattice). But, since $\sup xD = \lim xD$, and since $x \in (\uparrow y)^\circ$, it follows that there is some $d \in D$ with $xd \in (\uparrow y)^\circ$. But then $y \leq xd \leq d$, and so we have shown that $y \ll x$, as claimed. Thus, inside any open set $U$ containing $x$, there is some $y$ with $y \ll x$. Since $S$ is Hausdorff, it follows that $x = \sup \twoheaddownarrow x$, and so $S$ is a continuous lattice. It remains to show that the topology on $S$ is the Lawson topology. As we have already seen, the sets $\uparrow x$ and $\downarrow x$ are both closed in any compact semilattice, and so the set $S \backslash \uparrow x$ is open in $S$ for each $x \in S$. For any $x \in S$, we have just shown that $x = \sup\{y \in S : x \in (\uparrow y)^\circ\}$, and this set is clearly directed. Thus, if $z \ll x$, then there is some $w \in S$ with $z \ll w \ll x$ by the interpolation property, and so there is some $y \in S$ with $x \in (\uparrow y)^\circ$ and $w \leq y$. Hence, we have that $x \in (\uparrow y)^\circ \subset \uparrow w \subset \twoheaduparrow z$. This then shows that each set $\twoheaduparrow y$ is open in the original topology of $S$, and we have already shown that this also holds for each set of the form $S \backslash \uparrow x$. Thus, the Lawson topology is coarser than the original topology of $S$, and since each of these topologies is compact and Hausdorff, we conclude that they are the same.

We commented above that a map between continuous lattices is Scott continuous if and only if it preserves sups of directed sets. Continuous semilattice morphisms between compact Lawson semilattices (in fact, between compact semilattices in general) are characterized by the fact that they preserve directed suprema and all infima (see [L-2]). Moreover, one of the basic representation results about compact Lawson semilattices is that each can be embedded in a product $I^A$ of $A$ copies of the unit interval for some set $A$ (see [C], Section IV-2 and Section VI-3). The original path to these results was not as smooth as that given above. Indeed, Karl Hofmann and Al Stralka gave an algebraic description of the category of compact Lawson semilattices and continuous semilattice morphisms in [HS], and it was only after Stralka came across Scott's paper [S-1] that this fairly straightforward presentation emerged. Finally, while the details given above were known to each of the authors of [C] early on, it was Jim Lea [Le] who published the first account of this equivalence between continuous lattices and compact Lawson semilattices. We close this section by summarizing our results:

THEOREM: *The category of compact Lawson semilattices and continuous semilattice maps is equivalent to the category of continuous lattices and maps preserving all infima and all directed suprema.*

## III. IRREDUCIBLES, PRIMES, AND THE LEMMA

We now turn our attention to the fine structure theory of continuous lattices from a lattice theoretic viewpoint. This will lead to a characterization of distributive continuous lattices - remarkably, one that is in purely topological terms. We begin by recalling that an element $x$ of a lattice $L$ is *irreducible* if $x = ab$ implies that either $x = a$ or $x = b$. The element $x$ is *prime* if $ab \leq x$ implies that $a \leq x$ or $b \leq x$ for all $a,b \in L$. Clearly, every prime is irreducible, and the converse holds in case $L$ is distributive. We denote the set of irreducible elements of the lattice $L$ by Irr $L$, the set of primes by PRIME $L$, and the set of non-identity primes by Spec $L$.

We say that a subset $A$ of the lattice $L$ *order generates* $L$ if $x = \inf(\uparrow x \cap A)$ for each $x \in L$. It is easy to show that any complete lattice $L$ in which PRIME $L$ order generates must be distributive and meet-continuous (i.e., $x(\sup D) = \sup xD$ for any directed subset $D$ of $L$ and any element $x$ of $L$; see, e.g., [C], Lemma 3.13, p.71).

The important fact about irreducibles in a continuous lattice is that they order generate the lattice. To prove this, we first note that any maximal element in the complement of a filter is irreducible in a complete lattice. Indeed, if $F$ is a filter in the complete lattice $L$, and if $x$ is a maximal element in $L\setminus F$, then $x = ab$ for some $a,b \in L$ implies that either $a \in L\setminus F$ or $b \in L\setminus F$ since $F$ is a filter. But $x = ab$ means $x \leq a$ and $x \leq b$, and $x$ maximal in $L\setminus F$ then implies that either $x = a$ or $x = b$. Thus $x$ is irreducible. Now, to show that the irreducibles in a continuous lattice order generate the lattice, we suppose that $L$ is a continuous lattice and that $x$ and $y$ are distinct elements of $L$. Assuming that $x \not\leq y$, we can then find an element $z \in L$ with $z \ll x$ and $z \not\leq y$. Using the now well-worn argument, we can then produce a sequence $z_i$ with $z \ll z_{i+1} \ll z_i \ll x$, and so $F = \cup \uparrow z_i$ is a Scott open filter with $x \in F$ and $y \notin F$. But the fact that $F$ is Scott open implies that $L \backslash F$ is Scott closed, which is to say that there must be a maximal element $p$ in $L \backslash F$ lying above $y$. Our argument above shows that $p$ is then irreducible, and so we have shown that $y = \inf(\uparrow y \cap \mathrm{Irr}\, L)$ for each $y \in L$.

Thus, the irreducible elements in a continuous lattice order generate, and so our comments above imply that a continuous lattice is distributive if and only if the primes order generate. Moreover, the primes in a continuous lattice have an unusual property, in that they are in some sense topologically prime. We make this precise by stating the following lemma, which is due to Gierz and Keimel [GK]:

THE LEMMA: *Let* $L$ *be a continuous lattice,* $L'$ *a complete lattice, and* $i : L \to L'$ *a map preserving sups of directed sets. If* $p$ *is a prime element of* $L'$, *and if* $K \subset L$ *is a compact set (in the Lawson topology) with* $\inf i(K) \leq p$, *then* $i(k) \leq p$ *for some* $k \in K$. *In particular, if* $L = L'$ *and* $K \subset L$ *is compact with* $\inf K \leq p$, *then* $p \in \uparrow K$.

To prove The Lemma, we suppose that $i(k) \not\leq p$ for each $k \in K$. Since $L$ is continuous and $i$ preserves directed sups, we can then find $y_k \ll k$ with $i(y_k) \not\leq p$ for each $k \in K$. Since $K$ is compact and $\Uparrow y_k$ is Scott open, there are then finitely many $y_k$'s, say $y_1, \ldots, y_n$ with $K \subset \cup \Uparrow y_j$, and so $i(K) \subset \cup \uparrow i(y_j)$. But then $\inf i(K) \geq i(y_i) \ldots i(y_n)$, and since $p$ is prime and $i(y_j) \not\leq p$ for each $j$, it follows that $i(y_1) \ldots i(y_n) \not\leq p$, and hence that $\inf i(K) \not\leq p$. The result then follows by contraposition.

The Lemma can be utilized to produce a variety of applications of continuous lattices to other areas of mathematics. As an example, we cite the following from functional analysis: Suppose $K$ is a compact convex subset of a locally convex topological vector space $V$. Then Krein-Milman applies, and so $K$ is the closed convex hull of its extreme points. Now, the space $\Gamma(K)$ of closed subsets of $K$ is a continuous lattice under reverse inclusion since $K$ is compact Hausdorff. Moreover, the map $p : \Gamma(K) \to \Gamma(K)$ which associates to each closed subset $A$ of $K$ the closed convex hull $p(A)$ of $A$ is clearly a projection operator. We show that $p$ is Scott continuous: Indeed, if $D$ is a directed family in

$\Gamma(K)$ with $K_0 = \sup D = \cap D$, then clearly $p(K_0) \subset \cap p(D)$. Now, since $V$ is locally convex, the closed convex set $p(K_0)$ has a neighborhood basis of closed convex sets, and if $C$ is any such neighborhood, then $C$ is also a neighborhood of $K_0$. Hence, there is some $K_i \in D$ with $K_i \subset C$, and so $p(K_i) \subset p(C) = C$. Thus, $\cap p(D) \subset p(K_i) \subset C$, and so $\cap p(D) \subset p(K_0)$. As we have already noted the reverse inclusion, we conclude that $p(K) = \cap p(D)$, so that $p$ is indeed Scott continuous.

Now, $p : \Gamma(K) \to \Gamma(K)$ is a Scott continuous projection operator on the continuous lattice $\Gamma(K)$, and so the result we proved in Section I shows that the image of $p$, namely the lattice of closed convex subsets of $K$, is also a continuous lattice (this result has been observed implicitly by Fuchsteinner [Fu], and explicitly by Lawson). Moreover, the primes in this lattice are clearly the singleton sets which contain extreme points of K. We can now use this information to prove a converse to the Krein-Milman Theorem. Namely, suppose that $A$ is a closed subset of the compact convex set $K$ such that the closed convex hull of $A$, $p(A)$, equals $K$. Then we show that $A$ contains the extreme points of $K$: In fact, the map $x \to \{x\} : K \to \Gamma(K)$ is an embedding of the compact set $K$ into $\Gamma(K)$. Moreover, the image of the set $A$ under this map satisfies $\inf p\{\{x\} : x \in A\} = K$. Now, the Lawson topology on $\Gamma(K)$ is the usual Vietoris topology, and so the set $\{\{x\} : x \in A\}$ is compact in $\Gamma(K)$. Finally, we have $K = p(A) = \inf p(\{\{x\} : x \in A\})$, and so, if $e$ is any extreme point of $K$, then $e \in p(A)$. The Lemma then implies that $\{e\} \in \uparrow p(\{\{x\} : x \in A\})$, or, what is the same thing, $e \in A$. This proves the desired result.

The above discussion is taken from [GK], where several other similar applications of continuous lattices via The Lemma are provided. For instance, The Lemma is used to prove Jónsson's Lemma from universal algebra, a theorem of Cunningham and Roy about extreme points in the closed unit ball of the dual of certain Banach spaces, as well as a generalization of the result of Gelfand and Kolmogoroff that the closed prime ideals in the algebra $C(X)$ of continuous complex-valued functions on the compact space $X$ are all zero sets of some point of $X$. While the use of The Lemma to prove this variety of results does not necessarily yield a simpler proof in any instance, nonetheless, it does provide a unifying framework within which to view the results.

## IV. THE SPECTRAL THEORY OF DISTRIBUTIVE CONTINUOUS LATTICES

We have already seen in Section I that, for a Hausdorff space $X$, the lattice $\mathcal{O}(X)$ of open subsets of $X$ is a continuous lattice if and only if X is locally compact. In this section, we examine this relationship between topological spaces and continuous lattices from a more formal viewpoint, namely that of spectral theory. We shall describe a duality between a category of topological spaces and continuous maps and a category of continuous lattices and appropriate morphisms.

We start by making an observation about the lattice of open sets of any topological space X. Since the sup of any subset of $\mathcal{O}(X)$ is just the union, and since the inf of any finite subset is the intersection, the lattice $\mathcal{O}(X)$ satisfies the following distributive law:

$$x \wedge (\sup A) = \sup(x \wedge A) \quad \text{for any} \quad x \in \mathcal{O}(X) \quad \text{and any} \quad A \subset \mathcal{O}(X).$$

A complete lattice L which satisfies this law is called a *complete Heyting algebra*, and so a lattice must be a complete Heyting algebra to be the lattice of open subsets of some topological space. In particular, such a lattice is distributive.

Now, if we start with a space X, and consider the lattice $\mathcal{O}(X)$, we want to be able to recover the space X from $\mathcal{O}(X)$ alone. To be sure, there is the map $\phi : X \to \mathcal{O}(X)$ by $\phi(x) = X\backslash\{x\}^-$, and so we certainly want this map to be one-to-one. This means, of course, that distinct points of X must have distinct closures, or that X is a $T_0$-space. Moreover, it is clear that $\phi(x)$ is a prime element of $\mathcal{O}(X)$ for each $x \in X$, and so the natural place to seek a copy of X in $\mathcal{O}(X)$ is from among the prime elements. However, not all primes in $\mathcal{O}(X)$ correspond to images of points of X under the map $\phi$; for example, the whole space X is certainly prime in $\mathcal{O}(X)$, but $X \neq \phi(x)$ for any $x \in X$. This particular problem is easily avoided; we simply restrict our attention to the subset Spec $\mathcal{O}(X)$ of non-identity primes of $\mathcal{O}(X)$. However, it might still happen that there are non-identity primes in $\mathcal{O}(X)$ which are not of the form $\phi(x)$ for any $x \in X$. To avoid this, we make an additional assumption on our space X which assures us that this cannot happen. We call the space X a *sober space* if each prime U in $\mathcal{O}(X)$ is of the form $U = \phi(x)$ for a *unique* $x \in X$. The assumption that x is unique for U ensures that X is also a $T_0$-space.

Thus, for a sober space X, the lattice $\mathcal{O}(X)$ is a complete Heyting algebra, and the map $\phi : X \to \text{Spec}\ \mathcal{O}(X)$ is a bijection. Moreover, Spec $\mathcal{O}(X)$ order generates the lattice $\mathcal{O}(X)$, since $U = \cap\{X\backslash\{x\}^- : x \notin U\}$ for all open subsets U of X. We adopt the terminology that the complete lattice L *has enough primes* if the primes order generate L. Thus, we can say that the lattice $\mathcal{O}(X)$ is a complete Heyting algebra with enough primes.

Next, we recover the topology of X from $\mathcal{O}(X)$. That is, we define a topology on Spec L for any complete lattice L which gives the original topology of $X$ to Spec $L$ in case $L = \mathcal{O}(X)$. For an element $x \in L$, we define the *hull of* $x$ to be the set $h(x) = \uparrow x \cap \text{Spec } L$. The *hull-kernel topology* on Spec $L$ is then defined to be the topology having the family $\{\text{Spec } L \backslash \uparrow x : x \in L\}$ as its open sets. Thus, the family of hulls of points of $L$ form the closed sets in the hull-kernel topology.

First, we show that Spec $L$ is a sober space in the hull-kernel topology for any complete Heyting algebra $L$ with enough primes. In fact, we have, for $x, y \in L$, $\text{Spec } L \backslash \uparrow x \cap \text{Spec } L \backslash \uparrow y = \text{Spec } L \backslash \uparrow (x \wedge y)$, since Spec $L$ consists of primes. Moreover, for any subset $A \subset L$, we also have $\text{Spec } L \backslash \uparrow (\sup A) = \cup \{\text{Spec } L \backslash \uparrow a : a \in A\}$. Thus, the family $\{\text{Spec } L \backslash \uparrow x : x \in L\}$ is a topology for Spec $L$. Next, if $\text{Spec } L \backslash \uparrow x$ is a prime element of this topology, then $\text{Spec } L \backslash \uparrow x \supset \text{Spec } L \backslash \uparrow y \cap \text{Spec } L \backslash \uparrow z$ implies that either $\text{Spec } L \backslash \uparrow y \subset \text{Spec } L \backslash \uparrow x$ or else $\text{Spec } L \backslash \uparrow z \subset \text{Spec } L \backslash \uparrow x$. Since $\text{Spec } L \backslash \uparrow (y \wedge z) = \text{Spec } L \backslash \uparrow y \cap \text{Spec } L \backslash \uparrow z$, we conclude that $\text{Spec } L \backslash \uparrow x$ is prime iff $\uparrow (y \wedge z) \supset \uparrow x$ implies either $\uparrow y \supset \uparrow x$ or else $\uparrow z \supset \uparrow x$, which is clearly equivalent to $y \wedge z \leq x$ implies either $y \leq x$ or $z \leq x$; i.e., $\text{Spec } L \backslash \uparrow x$ is prime iff $x$ is prime in $L$. Thus the hull-kernel topology on Spec $L$ is sober, since the closure of any point is the hull of the point. Finally, it is clear that the map $x \to \text{Spec } L \backslash \uparrow x : L \to \mathcal{O}(\text{Spec } L)$ is an isomorphism in case $L$ is a complete Heyting algebra with enough primes.

Now, we show that Spec $L$ in the hull-kernel topology is homeomorphic to the space $X$ in case $L = \mathcal{O}(X)$ and $X$ is sober. Indeed, if $U \in \mathcal{O}(X)$, then $\text{Spec } \mathcal{O}(X) \backslash \uparrow U = \{X \backslash \{x\}^- : U \not\subset X \backslash \{x\}^-\}$ $= \{X \backslash \{x\}^- : \{x\}^- \cap U \neq \emptyset\}$. But, $\{x\}^- \cap U \neq \emptyset$ implies there is some $y \in \{x\}^- \cap U$, and so $x \in U$ as $U$ is open. Thus, $\{x\}^- \cap U \neq \emptyset$ iff $x \in U$, so $\text{Spec } \mathcal{O}(X) \backslash \uparrow U = \{X \backslash \{x\}^- : x \in U\} = \{\phi(x) : x \in U\}$, where $\phi : X \to \text{Spec } \mathcal{O}(X)$ by $\phi(x) = X \backslash \{x\}^-$. Thus, for any sober space $X$, the map $\phi$ is a homeomorphism of $X$ into Spec $\mathcal{O}(X)$ in the hull-kernel topology.

Next, of course, we need to consider morphisms between the objects of the two categories. On the topological side, we clearly want continuous maps between sober spaces to be our morphisms. However, if $f : X \to Y$ is such a map, $f$ does not map open sets in $X$ to open sets in $Y$. Rather, it is the map $f^{-1} : \mathcal{O}(Y) \to \mathcal{O}(X)$ which we want. In fact, the lattice properties of the map $\mathcal{O}(f) = f^{-1}$ are easily listed: $\mathcal{O}(f)$ preserves all sups and all finite infs, since $f^{-1}$ preserves all unions and all intersections. Thus, for the morphisms between complete Heyting algebras with enough primes, we take those maps preserving all sups and all finite infs; i.e., those lattice maps which preserve all sups. Now, given the morphism $f^{-1} : \mathcal{O}(Y) \to \mathcal{O}(X)$, it is not so easy to retrieve the map $f : X \to Y$. Thus, we make the following detour:

For complete lattices $L$ and $M$, a *Galois connection* between $L$ and $M$ is a pair of maps $g : L \to M$ and $d : M \to L$ satisfying $d(g(x)) \leq x$ and $g(d(y)) \geq y$ for each $x \in L$ and each $y \in M$. The map $g$ is called the *upper adjoint to* $d$, and the map $d$ is called the *lower adjoint to* $g$. Moreover, $g$ preserves all infs, and $d$ preserves all sups. In fact, given a map $g : L \to M$ preserving all infs, the map $d : M \to L$ by $d(y) = \inf g^{-1}(\uparrow y)$ preserves all sups, and the pair $(g,d)$ is a Galois connection between $L$ and $M$; dually, given any map $d : M \to L$ preserving all sups, the map $g : L \to M$ by $g(x) = \sup d^{-1}(\downarrow x)$ preserves all infs, and the pair $(g,d)$ is a Galois connection between $L$ and $M$.

Now, we need the following result about Galois connections between complete lattices. Namely, given a Galois connection $g : L \to M$ and $d : M \to L$, we know that $g$ preserves all infs and $d$ preserves all sups. Moreover, if $d$ also preserves finite infs, then $g(\text{Spec } L) \subset \text{Spec } M$ : Indeed, if $p$ is a prime in $L$, let $a,b \in M$ with $ab \leq g(p)$. Then $d(ab) \leq d(g(p)) \leq p$, since $d$ is monotone and we have a Galois connection. But $d$ preserves finite infs, so we have $d(a)d(b) = d(ab) \leq p$, and since $p$ is prime, either $d(a) \leq p$ or $d(b) \leq p$. This implies that either $g(d(a)) \leq g(p)$ or $g(d(b)) \leq g(p)$. But, using the fact that we have a Galois connection once again, we have that either $a \leq g(d(a)) \leq g(p)$ or $b \leq g(d(b)) \leq g(p)$, and so $g(p)$ is a prime as claimed.

Thus, given complete Heyting algebras $L$ and $M$, each having enough primes, and given $f : L \to M$ preserving all sups and finite infs, the upper adjoint of $f$, $f' : M \to L$ maps $\text{Spec } M$ to $\text{Spec } L$, and so we define the map $\text{Spec } f : \text{Spec } M \to \text{Spec } L$ to be the restriction and corestriction of $f'$ to $\text{Spec } M$ and $\text{Spec } L$, respectively. We now show that $\text{Spec } f$ is continuous in the hull-kernel topologies : If $x \in L$, then since $f'$ preserves all infs, it follows that the set $f'^{-1}(\uparrow x)$ has an inf, say $y$, and since $f'^{-1}(\uparrow x)$ is clearly an upper set in $M$, it follows that $f'^{-1}(\uparrow x) = \uparrow y$. Hence, $(\text{Spec } f)^{-1}(\text{Spec } L \setminus \uparrow x) = \text{Spec } M \setminus f'^{-1}(\uparrow x) = \text{Spec } M \setminus \uparrow y$, and so $(\text{Spec } f)^{-1}(U)$ is open if $U$ is open.

We have already observed that, if $L$ is a complete Heyting algebra with enough primes, then the map $i_L : L \to \mathcal{O}(\text{Spec } L)$ by $i_L(x) = \text{Spec } L \setminus \uparrow x$ is an isomorphism. Thus, we have the following diagram which we now show is commutative:

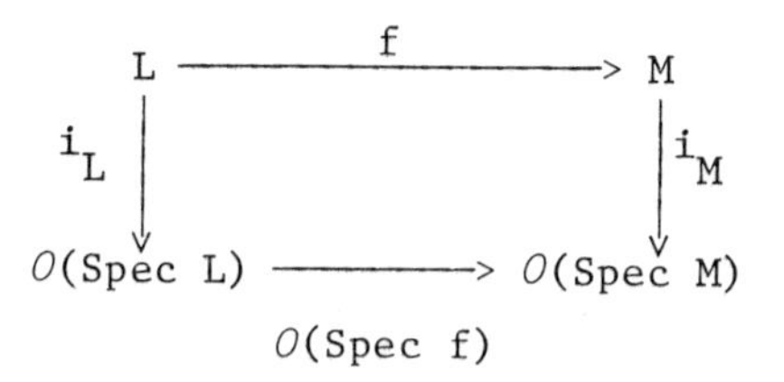


If $x \in L$, then $i_M(f(x)) = \text{Spec } M \setminus \uparrow f(x)$. On the other hand, $O(\text{Spec } f)(i_L(x)) = O(\text{Spec } f)(\text{Spec } L \setminus \uparrow x) = \text{Spec } M \setminus \uparrow y$, where $y = \inf\{z \in M : f'(z) = x\}$. But the definition of $f'$ implies that $y = \inf\{z \in M : f'(z) = x\} = f(x)$, and so we have that $O(\text{Spec } f)(i_L(x)) = \text{Spec } M \setminus \uparrow f(x) = i_M(f(x))$.

A similar argument shows that, for sober spaces $X$ and $Y$ and a continuous map $f : X \to Y$, the diagram below commutes:

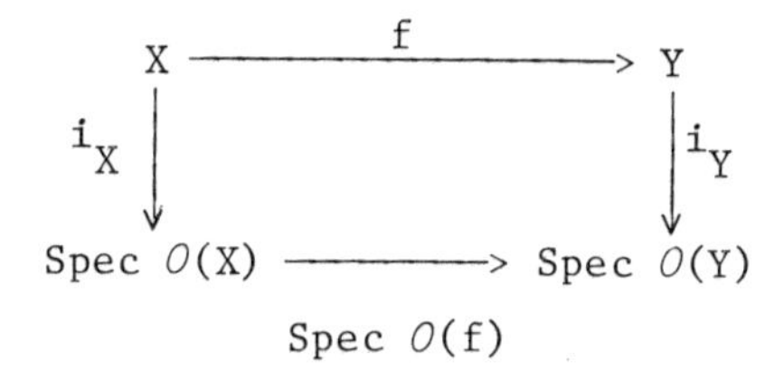


where $i_X : X \to \text{Spec } O(X)$ is given by $i_X(x) = X \setminus \{x\}^-$.

Summarizing, we have the following:

THEOREM: *The category* $SOB$ *of sober spaces and continuous maps is dual to the category* $HEYT_0$ *of complete Heyting algebras with enough primes and lattice maps preserving all sups under the functors* $O : SOB \to HEYT_0$ *and* Spec : $HEYT_0 \to SOB$.

It is now a relatively simple matter to bring continuous lattices into the picture. In fact, we have already commented that, for any locally quasicompact sober space $X$, the lattice $O(X)$ is a continuous lattice. Thus, on the topological side, we distinguish the full subcategory $LQSOB$ of the category $SOB$ of locally quasicompact sober spaces and continuous maps, and we note that $O(X)$ is a distributive continuous lattice for any such space $X$. On the other hand, we can also show that Spec $L$ is a locally quasicompact sober space in the hull-kernel topology for any continuous lattice $L$. Indeed, we first show that $L$ is a complete Heyting algebra with enough primes: We already know that Spec $L$ order generates $L$ for any distributive continuous lattice $L$, and so we only need show that finite infs distribute over all sups in L. So, let $A \subset L$ and let $x \in L$, and consider $x \wedge (\sup A)$. Since we always have $\sup x \wedge A \leq x \wedge (\sup A)$ in any complete lattice, we must show the reverse inequality. So, let $y \in L$ with $y \ll x \wedge (\sup A)$. Then, since $\sup A = \sup\{\sup F : F \subset A \text{ is finite}\}$, and this latter is directed, we can find a finite subset $F \subset A$ with $y \leq x \wedge (\sup F)$. But L is a distributive lattice, so that $x \wedge (\sup F) = \sup x \wedge F$. Now, $\sup x \wedge F \leq \sup x \wedge A$ is clear, and so we have $y \leq \sup x \wedge F \leq \sup x \wedge A$. Hence, if $y \ll x \wedge (\sup A)$, then $y \leq \sup x \wedge A$, and since L is continuous, it follows that $x \wedge (\sup A) \leq \sup x \wedge A$. Since we already have the reverse inequality, we conclude that $x \wedge (\sup A) = \sup x \wedge A$, and so L is a complete Heyting algebra with enough primes. Thus, to complete our argument, we must show that Spec L is locally quasicompact in the hull-kernel topology. To that end, suppose that $p \in \text{Spec } L$ and $x \in L$ with $p \in \text{Spec } L \backslash \uparrow x$. We must produce a quasicompact neighborhood Q of p with $Q \subset \text{Spec } L \backslash \uparrow x$. Since $p \notin \uparrow x$, and since L is continuous, we can use the interpolation property to find a sequence $x_n$ with $x_0 \ll x_{n+1} \ll x_n \ll x$ for each $n \geq 1$, and $x_0 \not\leq p$. Thus, if $Q = \text{Spec } L \setminus F$, where $F = \cup\{x_n : n \geq 1\}$, then clearly Q is a neighborhood of p in Spec L. Moreover, suppose that the family $\{\text{Spec } L \backslash \uparrow y_j : j \in J\}$ is an open cover of Q in the hull-kernel topology. Then we must have that $\uparrow(\sup y_j) \cap Q = \emptyset$, so that $\sup y_j \in F$. But F is Scott open, and so there are finitely many $y_j$'s, say $y_1, \ldots, y_m$ with $y_1 \vee \ldots \vee y_m \in F$. It then follows that the family $\{\text{Spec } L \backslash \uparrow y_j : j = 1, \ldots, m\}$ is the desired finite subcover for Q, and so Q is quasicompact.

We can summarize the results of the preceding paragraph as follows:

THEOREM: *The category LQSOB of locally quasicompact sober spaces and continuous maps is dual to the category CHEYT of continuous distributive lattices and lattice maps preserving all sups under the functors O : LQSOB → CHEYT and* Spec : *CHEYT → LQSOB*.

Actually, this is only the beginning of the role spectral theory plays in the theory of continuous lattices, for we can easily specialize the duality given above to more special lattices than just the distributive continuous lattices. For example, we can consider the subcategory of distributive algebraic lattices, and realize a duality for these objects, and as it turns out, the category of locally quasicompact sober spaces having a basis of quasicompact open sets; or, we can consider the subcategory of completely distributive lattices, and find that it corresponds to the locally quasicompact sober spaces which are *continuous posets* in the specialization order. There is one other duality that does deserve mention here, since it is rather special. Namely, suppose that we have a compact partially ordered space with closed partial order. It is rather routine to show that the family of closed lower sets of the space form a continuous lattice (the main ingredient in the proof is the result of Nachbin's that each closed lower set has a basis of neighborhoods which are themselves lower sets). The question we want to ask is which continuous lattices arise in this fashion. First, we notice that the natural embedding of the space $X$ into the lattice of closed lower sets which sends each point to its lower set, actually maps $X$ onto the primes of the lattice, and so the prime elements of the lattice are closed in this instance (in the Lawson topology). Moreover, given any distributive continuous lattice $L$ in which the primes are closed in the Lawson topology, it can be shown that $L$ is isomorphic to the lattice of closed lower sets of the compact partially ordered space which is given by the set of prime elements of $L$ endowed with the *Lawson topology*. Moreover, a completely intrinsic characterization has been found for when the primes of a distributive continuous lattice are closed in the Lawson topology: namely, it is equivalent to the property that the way below relation is *multiplicative;* i.e., if $a \ll x$ and $b \ll y$ in $L$, then $ab \ll xy$ in $L$.

There are many further topics to be addressed which have developed from the theory of continuous lattices. For example, we have alluded above to the concept of a continuous poset. Lawson [L-3] has developed a duality for these objects which supercedes the duality for compact zero-dimensional semilattices and algebraic lattices, among other dualities. Moreover, he has also shown that each completely distributive lattice is completely determined by its set of coprime elements, which is a continuous poset in the natural order. Furthermore, the concept of local quasicompactness plays a central role in the spectral theory for continuous lattices outlined above, and it turns out that a general theory for k-spaces (without separation) has been outlined by Lawson and Hofmann. Some of the details of this, as well as other relevant topics are contained in the paper [HM]. Finally, the complete details of the outline given here for continuous lattices is contained in Chapter V of [C]; this particular theory was first developed by Hofmann and Lawson in their papers [HL-1] and [HL-2].

## REFERENCES

C Gierz, G., K. H. Hofmann, K. Keimel, J. Lawson, M. Mislove, and D. Scott, A Compendium of Continuous Lattices, Springer-Verlag, Heidelberg, New York (1980), 371p.

DK Day, B. J., and G. M. Kelly, On topological quotient maps preserved by pull-backs or products, Proc. Camb. Phil. Soc. 67 (1970), 553-558.

GK Gierz, G. and K. Keimel, A lemma on primes appearing in algebra and analysis, Houston Journal of Math. 3 (1977), 207-224.

Fu Fuchsteinner, B., Lattices and Choquet's theorem, Jour. Funct. Anal. 17 (1974), 461-465.

HL-1 Hofmann, K. H. and J. Lawson, Irreducibility and generation in continuous lattices, Semigroup Forum 13 (1976/77), 307-353.

HL-2 Hofmann, K. H. and J. Lawson, The spectral theory of distributive continuous lattices, Trans. Amer. Math. Soc. 246 (1978), 285-310.

HM Hofmann, K. H. and M. Mislove, Local compactness and continuous lattices, in: Proceedings of the Bremen Workshop on Continuous Lattices, Lecture Notes in Math., Springer-Verlag.

HMS Hofmann, K. H., M. Mislove, and A. Stralka, The Pontryagin Duality of Compact 0-dimensional Semilattices and Its Applications, Lecture Notes in Mathematics 396.

L-1 Lawson, J., Vietoris mappings and embeddings of topological lattices, University of Tennessee Dissertation, 1967.

L-2 Lawson, J., Intrinsic topologies in topological lattices and semilattices, Pac. Journal of Math. 44 (1973), 593-602.

L-3 Lawson, J., The duality of continuous posets, Houston Journal of Math., to appear.

Le Lea, J. W., Jr., Continuous lattices and compact Lawson semilattices, Semigroup Forum 13 (1976/77), 387-388.

N Niño, J., Dissertation, Tulane University, 1981.

S-1 Scott, D., Continuous lattices, in: Lecture Notes in Mathematics 274, 97-136.

S-2 Scott, D., Data types as lattices, SIAM Journal of Comput. 5 (1976), 522-587.

# ORDERED SETS IN GEOMETRY

Garrett Birkhoff
Department of Mathematics
Harvard University
Cambridge, Massachusetts 02138

ABSTRACT

Partial orderings play a major role in geometry, especially in combinatorial settings. Most of these partial orderings involve lattices, and a rich profusion of lattices arise in this way. A brief review will be given of many old results, and some new ones, about lattices arising in this way.

I. Order and Incidence: Flats. As is well-known, the 'flats' (= points, lines, planes, etc.) in any projective or affine geometry form a *geometric lattice*.

In the (self-dual) projective case, this lattice is *modular*; in the affine case, it is a 'Hilbert lattice'. In $n \geq 3$ dimensions, these lattices can be 'coordinatized': for some division ring $D$, they are the lattices of *linear* subspaces of $D^{n+1}$ and of *affine* subspaces of $D^n$, respectively. (See [Art], [B-B], [Hil], [LT1, p. 60], and [V-Y].)

Analogous lattices arise in infinite-dimensional function spaces, as the lattices of (closed) *subspaces* of Hilbert space left *invariant* by some self-adjoint linear *operator*. Von Neumann [JvN] coordinatized these by a brilliant extension of Wedderburn theory to *regular* rings. These subspace lattices also describe 'the logic of quantum mechanics' [JvN, pp. 105-125].

II. Order and 'Polarity': Orthogonality. The lattices discussed in §I are associated with homogeneous and inhomogeneous *linear* equations, respectively, by an obvious 'polarity' [LT1, §32]. Any equation

(1) $$a_0x_0 + a_1x_1 + \ldots + a_nx_n = 0 ,$$



*I. Rival (ed.), Ordered Sets, 407–443.*

respectively

(2) $$a_1x_1 + a_2x_2 + \dots + a_nx_n = a_0 ,$$

defines a binary relation $a \rho x$ (meaning $\langle a,x \rangle = 0$) between *points* $x = (x_0,x_1,\dots,x_n)$ or $x = (x_1,x_2,\dots,x_n)$ and *coefficient vectors* $a = (a_0,a_1,\dots,a_n)$. The sets $S \subset D^{n+1}$ that are *closed* under the resulting polarity are just the (affine and linear) *flats* in the usual geometric sense.

When $D$ is an ordered *formally real field*, the equation $\langle a,x \rangle = 0$ defines an *orthogonality* relation $a \perp x$ in the usual sense.

Even more interesting is the analogue of (1) for *algebraic varieties* of higher degree. Simplest among these (after 'flats') are the circles, spheres, and hyperspheres, defined by equations of the form

(3) $$a_0r^2 + \sum_{k=1}^{n} a_kx_k + c = 0 ,$$

where $r^2 = \sum_{k=1}^{n} x_k^2$ . The analogues of 'lines' are here played by circles. (Since precisely one circle passes through any three points, any subset of two points is 'closed'.)

Next simplest are *conics* (and quadrics, etc.), of which just one passes through any 5 coplanar points (any subset of 4 points is 'closed'). 'Pencils' of circles, spheres, conics, quadrics, etc. can be handled very naturally in this context. [LT3, pp. 84, 124].

III. Order and Convexity: Segments. Much as replacing 'inclusion' by (self-dual) *incidence* is very natural geometrically, so it is natural to replace 'order' by *betweenness*. On a line, this leads to the notions of 'interval' and *convexity*. Hence, if $D$ is an *ordered* division ring or field, we have a much richer, though still elementary, geometric theory. [Hil]

As one, highly combinatorial aspect of this, we have the theory of *convex polyhedra*, and their lattices of facets. As another, we have lattices of convex *subsets*. As still another, we have the theory of linear inequalities (hence linear *programming*). [Gru]

Homology. The combinatorial topology of Poincaré is based on the concept of a polyhedral complex, in terms of which one can define not only 'dimension', but also homology groups and many other topological *invariants* of manifolds.

Several authors have tried to define the (global) topology of complexes in terms of invariants of such complexes. ([A-H], [Tuc], [Whi], [Rot]).

IV. Projective Metrics. Stemming from Laguerre, through Felix Klein and Hilbert, is the concept of a projective metric. In its simplest form, this defines $d(x,y)$ on any line as the *logarithm* of the cross-ratio $X(a,x,y,b)$, $a$ and $b$ being the points where the line pierces the (strictly convex) *skin* of the 'ball' $\|x\| \leq a$. If the 'ball' is an *ellipsoid*, this construction gives hyperbolic geometry. ([Hil], [Pog]).

From any strictly convex 'ball', the *geodesics* for this distance ('metric') are straight lines and form a Hilbert lattice, as they also are for the symmetric Riemann spaces of E. Cartan.

Positive Linear Operators. Finally, in any convex *cone* $C$ (of 'positive' vectors), the preceding construction defines a *quasimetric*, with respect to which any linear transformation such that $f(C) < C$ is a *contraction*.

In the special case that $C$ is the positive *orthant* of all $x$ with all $x_i \geq 0$, the preceding construction gives the Perron-Frobenius theory of positive (and non-negative) *matrices*, and its extension to positive linear operators [LT3, Ch. XVI].

REFERENCES

[Art] E. Artin (1957) *Geometric Algebra*, Interscience.

[A-H] P. Alexandroff and H. Hopf (1935) *Topologie, I.*, Springer.

[B-B] M.K. Bennett and G. Birkhoff, Algebraic foundations of geometry (unpublished manuscript).

[C-R] H. Crapo and G.-C. Rota, *Geometric Lattices*, M.I.T. Press, Second ed. in preparation.

[Gru] Branko Grünbaum (1967) *Convex Polytopes*, Interscience.

[Hil] D. Hilbert (1930) *Grundlagen der Geometrie*, 7th ed. Teubner.

[Jvn] J. von Neumann (1962) *Collected Works, Vol. IV,* Pergamon Press.

[LTh] G. Birkhoff, *Lattice Theory,* $n$-th ed. (First ed., 1940; Second ed., 1948; Third ed., 1967.), *Amer. Math. Soc.*

[Pog] A. Pogorelov (1979) *Hilbert's Fourth Problem,* Holt, Rinehart, and Winston.

[Poi] H. Poincaré (1895; 1890) Analysis Situs, *J. Ec. Polyt.* 1, 1- ; *Rendic. Palermo* 13, 285- .

[Rot] G.-C. Rota (1971) On the ... Euler characteristic, *Studies in Pure Math. Presented to R. Rado,* Academic Press, 221-233.

[Sko] L.A. Skornyakov (1964) *Complemented Modular Lattices and Regular Rings,* Oliver and Boyd.

[S-T] Ernst Snapper and R.J. Troyer (1971) *Metric Affine Geometry,* Academic Press.

[Tuc] A.W. Tucker (1933; 1936) An abstract approach to manifolds, *Annals of Math.* 34, 191-243; 37, 92-100.

[V-Y] O. Veblen and J.W. Young (1910, 1918) *Projective Geometry,* 2 Vols.

[Whi] J.H.C. Whitehead (1963) *Mathematical Works, Vol. iii,* Macmillan.

[Zas] T. Zaslavsky (1975) Partitions of space by hyperplanes, *Mem. Amer. Math. Soc.* 1, 1-102.

## 1. PROJECTIVE GEOMETRIES

My talk will attempt to provide a general overview of (partially) ordered sets or *posets* arising in geometry, and to bring out the *geometric insights* achieved by considering the context in which they arose, giving only a "broad brush" discussion. As other speakers have observed, the relevant concepts (including the hyperbolic metric of §5), are almost all definable in terms of a single basic binary relation, that of set-theoretic 'inclusion' or *incidence*.

I shall begin by paraphrasing a statement in Ore's preface to his fundamental essays "On the foundations of abstract algebra".[1] I claim that:

> "In the discussion of incidence relations in geometry, one is not primarily interested in the "points" of a space, but in the inclusion relations between certain *distinguished subsets*, like straight lines and planes in affine and projective space, circles and spheres in inversive geometry, and algebraic varieties in algebraic geometry."

In geometry, as in algebra, these distinguished subsets typically form so-called "Moore families" closed under arbitrary intersection, hence *complete lattices* in which meets are intersections. Equivalently, defining the "closure" of any set $S$ to be the intersection $\bar{S} = \cap T_\alpha$ of all the distinguished subsets $T_\alpha$ which contain $S$, the usual closure axioms such as $S \subset \bar{S} = \bar{\bar{S}}$ are satisfied: see §8.

Except in §7, I will be talking mainly about 'graded' lattices of finite length, in which height corresponds to geometric "dimension", and inclusion is referred to as "incidence". Indeed one of the main advantages of the lattice-theoretic approach to geometry is its 'dimension-free' character. This becomes increasingly useful in geometric lattices of length $n > 4$, for which ordinary visualization fails.

Among the many kinds of ordered sets arising in geometry, the complemented modular lattices associated with (abstract) $n$-dimensional *projective geometries* have the most substantial theory, and I will discuss them first.

As I showed in 1934, and Menger showed independently slightly later,[2] not only do the subspaces of any projective geometry form a complemented modular lattice of finite length, but conversely any complemented modular lattice of finite length is a product of projective geometries and Boolean algebra. More precisely, to go from classical Boolean algebra to projective geometry it suffices to replace the distributive law by the weaker modular law, which states that

$$a \wedge (b \vee c) = (a \wedge b) \vee (a \wedge c) = (a \wedge b) \vee c$$

holds whenever $a \geqq c$, and to assume that every line contains at least 3 points.

The geometric background for this result was available in Veblen and Young's two-volume "Projective Geometry" (1910, 1918). There it is meticulously proved that *incidence axioms* similar to those of Hilbert's path-breaking *Grundlagen der Geometrie* are fully equivalent to the *algebraic* concept of $PG(D;n-1)$, definable using homogeneous coordinates -- at least for $n \geqq 3$. This remarkable result, adumbrated a half-century earlier by von Staudt, made clear the key roles played by the theorems of Desargues and Pappus in *coordinatization* by what Artin later suggestively called *geometric algebra.* [3]

Thus, it was known by 1910 that, although any *finite projective plane* that satisfies the theorem of Desargues also satisfies the theorem of Pappus,[4] there exist many non-Desarguesian finite projective planes. It is also known, although tricky to prove (cf. A. Seidenberg, *Amer. Math. Monthly* 83 (1976), 192-194), that "Pappus implies Desargues".

Beginning with Marshall Hall's highly original 1943 paper,[5] and stimulated by the subsequent development of high-speed computers, a systematic attempt has been made during the past 40 years to enumerate *all* finite projective planes. This problem is easily shown to be equivalent to the search for all finite complemented modular lattices of height three. The search is analogous to the problem of finding all finite simple groups, which was initiated by L.E. Dickson in 1900 and has only recently been crowned with success. Since the *un*complemented finite modular lattices of height three are trivial to enumerate (if a point is uncomplemented, every line must contain it, and dually), it is also equivalent to the problem of enumerating all *finite modular lattices of height three.*

In treating modular lattices of height four or less, the dimension-free lattice-theoretic approach is probably *not* the most effective one; it is probably better to refer to "points", "lines", and "planes", and to resort to a case-by-case analysis.

## 2. COMPLEMENTED MODULAR LATTICES

The discovery of a simple and straightforward connection between complemented modular lattices and 'flats' in projective geometry quickly stimulated the lattice-theoretic construction by von Neumann of his *continuous-dimensional projective geometries*, or "continuous geometries". These helped to complete von Neumann's enumeration of the possible structures of rings of self-adjoint operators on Hilbert space, which in turn was an outgrowth of his classic "*Mathematische Grundlagen der Quantenmechanik* (Springer, 1932; authorized English translation by Robert T. Beyer, Princeton University Press, 1955).

### Orthomodular Lattices

In turn, this area of application stimulated considerable interest in the "logic of quantum mechanics". The logic of quantum mechanics is generally believed to be associated with the orthocomplemented *weakly modular* lattice $OC(\hbar)$ of all *closed* subspaces of (separable) Hilbert space. This may be contrasted with the complemented *distributive* lattice (Boolean algebra) $M/N$ of all measurable subsets of phase space, modulo null sets which plays the analogous role in the logic of classical mechanics.[6] It may also be contrasted with the complemented modular lattice of *all* subspaces of $\mathbb{R}^n$.

The orthocomplemented lattice $OC(\hbar)$ differs from most of the other lattices I will discuss, in that it involves an extraneous operation $x^\perp$ *not* definable in terms of inclusion. Nevertheless, because of its connection with physics, it and its analogues have been studied by many mathematicians.

Mathematical progress in this area is reviewed in Gudrun Kalmbach's *Orthomodular Lattices*, to be published in 1982 by Academic Press.[7] I will now make a small contribution to this admirable book by solving its Problem 27.

Namely, consider the modular lattice of all subspaces of $Q^4$. This lattice carries two inequivalent complementations, associated with the positive definite quadratic forms

$$\langle \underline{x}, B\underline{x} \rangle = x^2 + y^2 + z^2 + t^2 \quad \text{and} \quad \langle \underline{x}, C\underline{x} \rangle = x^2 + y^2 + z^2 + 2t^2 ,$$

respectively. These are inequivalent, because if $P^T BP = C$, then $|C| = |P|^2 \cdot |B|$, while $|P|^2 = 2$ has no rational solution.

To go more deeply into the theory of infinite-dimensional orthomodular lattices would necessarily involve measure theory, and other aspects of abstract real (and complex) analysis that are irrelevant to the purposes of this meeting. Instead, the rest of my talk will be mainly concerned with *finite-dimensional, non-modular geometric lattices*.

Before leaving the modular case, I want to re-emphasize the central role in modern *algebra* played by the concepts which von Neumann investigated so brilliantly. Namely, completely reducible representations (by matrices) of groups, rings, and linear algebras are to some extent characterized by their lattices of invariant subspaces.[8] And these can be quite general complemented *modular* lattices $M = M_1 \times M_2 \times \ldots \times M_r$ of finite length, whose *indecomposable* 'factors' $M_i$ correspond to *irreducible* representations. Von Neumann's more general "regular rings" have lattices of ideals (left-, right-, and two-sided) with a similar structure.

As a final relevant reference to what is known about modular geometric lattices, I will also cite 1932 papers of Pontrjagin and Kolmogoroff.[9] In it, they extended an old theorem of Frobenius to prove that if the 'level sets' of elements of given height have a *locally compact* topology in which the operations $\wedge$ and $\vee$ are continuous, then $D$ must be the real field $\mathbb{R}$, the complex field $\mathbb{C}$, or the division ring of real quaternions.

## 3. HISTORICAL REMARKS

In the next three sections, I will discuss the geometric lattices whose elements are the 'flats' of affine and hyperbolic $n$-space, (partially) ordered by inclusion. To appreciate the geometrical significance of these lattices, one should keep in mind some 19th century developments in the foundations of geometry, with special emphasis on the work of Cayley, von Staudt, Riemann, Helmholtz, Klein, Lie, Poincaré, and Hilbert.

Following the rediscovery by Poncelet of (real) projective geometry as a self-contained subject, and the discovery by Bolyai and Lobachevsky of (real) hyperbolic or 'non-Euclidean' geometry, the foundations of geometry became emancipated from the axioms of Euclid. These geometries were soon identified by their common properties of having constant curvature and, equivalently, a three-parameter group[10] of rigid motions in two dimensions (and an $n(n+1)$-parameter group in $n$-dimensions).

### "Projective Geometry is All Geometry"

During the second half of the century, geometrical thinking was greatly influenced by Cayley's famous dictum (endorsed by

Felix Klein), that "projective geometry is all geometry". What Cayley meant was that the *configurations of geodesics* (in Menger's terminology, the lattices of "flats") of Euclidean (affine) and hyperbolic geometry can be embedded in that of projective geometry, with preservation of intersections and unions.

Using homogeneous coordinates $(x_0, x_1, x_2)$ the first statement becomes fairly obvious: perspectivity from the origin embeds the *affine* plane $x_0 = 1$ in the projective plane; one need only delete from the latter the line (circle) at infinity. To embed the hyperbolic plane in the projective plane is only a little more difficult.

Namely, following Poincaré, one can map the hyperbolic plane conformally onto the unit disk $x^2 + y^2 < 1$, so that geodesics go into circles orthogonal to the boundary; this gives a Riemann surface with distance differential $ds$ satisfying

$$ds^2 = (dx^2+dy^2)/(1-x^2-y^2) .$$

By *Ptolemaic* projection from the south pole, one can then map this disk conformally[11] onto the northern hemisphere. Geodesics are mapped into circles perpendicular to the equator. Finally, *vertical* projection maps these back onto straight line segments interior to the unit disk.

This gives Cayley's model of the hyperbolic plane, in which geodesics are straight line segments interior to the unit disk.

Hilbert's Influence

In his influential book *Grundlagen der Geometrie*, David Hilbert extended and generalized the work of Pasch, Fano, Veronese, and other geometers by axiomatizing 'geometries' over various kinds of "number systems" (division rings and fields). In his book, he also took a position opposed to Cayley and Klein, by assigning a central position to *three-dimensional* (affine) *Euclidean* geometry, and ignoring the projective plane. Using constructions based on parallels going back to the Greeks, which he called "Streckenrechnung" but which might better be referred to as *geometric algebra*, he essentially showed that his incidence and parallel axioms gave rise to *affine geometry* in $D^3$, $D$ an arbitrary division ring.[12] Hilbert also greatly generalized hyperbolic geometry; see §6.

However, in their classic treatise cited in §1, Veblen and Young reverted to the classic projective standpoint of Cayley and Klein. As we saw there, it was probably fortunate for lattice theory that they did!

## 4. LATTICES AND AFFINE GEOMETRY

In view of the classic connections between the lattices of flats in projective, affine (Euclidean) and hyperbolic ('non-Euclidean') geometry just recalled, it is natural to try to derive analogues for *affine* and *hyperbolic* geometry of the lattice-theoretic characterization of (abstract) *projective* geometry that I described in §§1-2.

And indeed, Karl Menger made repeated efforts to do this. Thus, already in his 1935 paper (op. cit., p. 416, Law +6), he observed (in effect)[13] that the geometric lattice $AG(D^n)$ of all affine "flats" in $D^n$ enjoys the following property:

H. *If* $p > 0$, *the interval sublattice* $[p,I]$ *is modular.*

It seems appropriate to call geometric lattices satisfying Axiom H *Hilbert lattices*, for the following reason.

Every geometric lattice of height two is modular. Therefore, every geometric lattice of height three is a Hilbert lattice. Hence Property H is first non-trivial in a geometric lattice of height 4. In this case it asserts precisely that if two 'planes' $\Pi$ and $\Pi'$ with $d(\Pi) = d(\Pi') = 3$ have a *point* in common -- i.e. if $\Pi \wedge \Pi' > 0$, -- then (by modularity of $[\Pi \wedge \Pi', I]$):

(1) $d[\Pi \wedge \Pi'] = d[\Pi] + d[\Pi'] - d[\Pi \vee \Pi'] = 3 + 3 - 4 = 2$ , -- i.e., $\Pi \wedge \Pi'$

is a line. Hence Property H is equivalent to Hilbert's Incidence Axiom I7.[14]

Under the name of "incidence geometries", Hilbert lattices were studied further by S. Gorn, S. Sasaki, and O. Wyler. In particular, Sasaki characterized those "incidence geometries" (Hilbert lattices) that satisfy Hilbert's incidence and parallel axioms. However, his condition was not really lattice-theoretic.

I shall next describe a more nearly lattice-theoretic characterization of geometric lattices (of height $n + 1$) which can be represented as the *affine geometry* $AG(D^n)$ of all 'flats' in the affine space $D^n$. This characterization is a very recent result due to M.K. Bennett. Her result can be motivated as follows:

*Hyperbolic Geometry.* It is easy to see that, even when $d[I] = 4$, Axiom H does not suffice to characterize the $AG(D^n)$ among geometric lattices. Thus, it is also satisfied in the hyperbolic geometry of all affine flats having a nonvoid intersection with the interior of the real unit ball. Yet this obviously fails to satisfy any form of Euclid's parallel axiom. In particular, the three lines $\ell$, $\ell'$, $m$ in the sketch of Fig. 1a generate the (simple) 6 element sublattice of $HG(\mathbb{R}^2)$ (see §5) whose diagram is sketched in Fig. 1b. Very recently, M.K. Bennett has proved the following result.

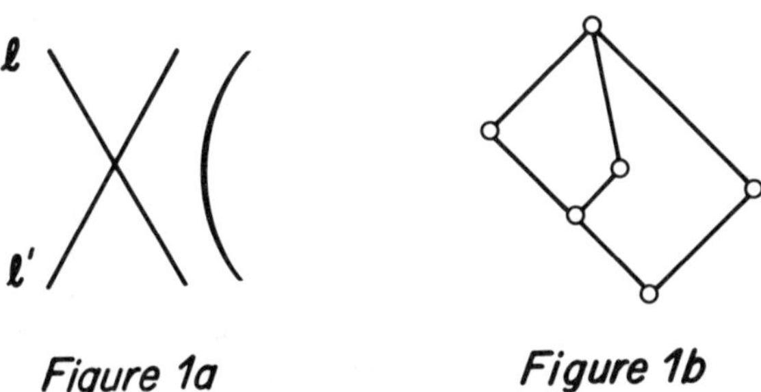


*Figure 1a* *Figure 1b*

THEOREM 1. *A Hilbert lattice of finite length $n \geqq 4$ is isomorphic to some $AG(D^n)$ if and only if it contains no sublattice of the form of Fig. 1b, and every coatom has a parallel complement.*

Her proof is reproduced in the Appendix of this paper; the question at issue concerns the existence and uniqueness of parallels. One can assume this for hyperplanes, in the following form.

A. *Given a hyperplane $h$ and a point $p$ not in $h$, there exists a unique hyperplane $p_1 \geqq p$ with $h \wedge h' = 0$.*

If one calls a Hilbert lattice satisfying Condition A a "pseudo-affine" lattice, one can show that many properties of affine geometry hold. These include the following.

THEOREM 2. *In a pseudo-affine lattice of height 4 or more, given a line $\ell$ and a point $q$ not on $\ell$, there is a unique parallel to $\ell$ through $q$ in the plane $\ell \vee q$.*

THEOREM 3. *Given $a$, $0 < a < 1$ in a pseudo-affine lattice, and a point $p$ not in $a$, there exists an $a_1 = p$ that is properly parallel to $a$.*

COROLLARY. *Every initial interval sublattice $[0,a]$ of an affine lattice is affine.*

Unfortunately, if we delete from $AG(\mathbb{R}^2)$ the "points" $P:(1,1)$ and $Q:(1,-1)$, then the "lines" $\ell : y = x$ and $\ell' : y = -x$ intersect in $O:(0,0)$, yet both are parallel to (i.e., fail to intersect) the line $m : x = 1$. Hence Axiom A does not characterize affine spaces as Hilbert lattices.

## 5. LINEAR AND CYCLIC ORDER

Most lattice-theoretic studies of affine and projective geometries have concentrated on unified theories, valid over any *division ring* $D$ or at least over any field $F$. This is probably because the greatest service of lattice theory in this area has been to provide *dimension-free* theorems. Therefore, it seems appropriate to try to make them also *coordinate-free*: independent of $D$.

However, in Poincaré's opinion,[15] the main significance of Hilbert's *Grundlagen der Geometrie* consisted precisely in the clarity with which it brought out (following earlier work of Italian geometers) the *differences* between the geometric properties of geometries having different number systems as coordinates.

Evidently, affine and projective spaces over a division ring $D$ become much more interesting geometrically when $D$ is an ordered division ring. In particular, most of the familiar basic properties of linear *order betweenness* and *convexity* hold in any such affine "space". I will next briefly recall these properties, emphasizing *closure* operators (§1), and other lattice-theoretic concepts.

If $D$ is any division ring, then the linear span $L(X)$ of a set $X \subset D$, defined as the (linear) subspace of all finite linear combinations (with variable $m$):

(5.1) $$L(X) = \{\lambda_1\underline{x}_1 + \ldots + \lambda_m\underline{x}_m\} , \quad \underline{x}_j \in D ,$$

and the affine span $A(X)$ of $X$, defined as the *flat*

(5.2) $$A(X) = \{\lambda_1\underline{x}_1 + \ldots + \lambda_m\underline{x}_m\} , \quad \Sigma\, \lambda_j = 1 ,$$

are evidently both *closure* operations in the usual sense [LT3, Chap. V, §1].[16] If $D$ is ordered, then so is the "conical" span defined by

(5.3) $$C(X) = \{\lambda_1\underline{x}_1 + \ldots + \lambda_m\underline{x}_m\} , \quad \forall\, \lambda_j \geq 0 .$$

Moreover, so is the "convex" span, defined by

(5.4) $$B(X) = \{\lambda_1\underline{x}_1 + \ldots + \lambda_m\underline{x}_m\} , \quad \forall\, \lambda_j \geq 0,\ \Sigma\,\lambda_j = 1 .$$

LEMMA. *Let $c_1(X)$ and $c_2(X)$ be two closure operators on a set $X$. Then $c(X) = c_1(X) \cap c_2(X)$ is also a closure operator on $X$.*

We omit the proof. More generally, we have the following result, not stated in [LT3, Chap. V, §1]:[16]

THEOREM 4. *The Moore families of subsets of (i.e., closure relations on) any set X themselves form a Moore family of subsets of* $2^X$.

COROLLARY. *The Moore families of subsets of any set X form a complete lattice.*

*Caution.* Obviously, $B(X) \subset A(X) \cap C(X)$ for all $X$. But for $X = \{(1,0), (1,1), (0,1)\}$, $A(X) \supset C(X)$, and the cone $A(X) \cap C(X) = C(X)$ *properly* contains the triangle $B(X)$. This is because $A(X)$ contains the origin, (0,0).

The theories of ordered division rings and convexity are very large. About them, I will make a few relevant observations.

First, Hilbert's original construction of a non-commutative ordered division ring, by extended formal power series, was a major advance in its time. Of course, any ordered division ring contains the rational field $Q$ as a *subfield*, and one can apply variants of Hilbert's method to construct a great variety of *non-commutative* polynomial extensions of any ordered fields.

As regards *ordered fields*, the great breakthrough was of course provided by the Artin-Schreier theory of *formally real* fields,[17] applied by Artin to prove that every positive definite rational function is a sum of a finite number of squares of rational functions.

L. Fuchs has reviewed the subject of *fully ordered rings and fields* in Chap. VIII of his monograph "*Partially Ordered Algebraic Systems*" (Pergamon, 1963), and it would waste your time to try to summarize this chapter here. I will only recall the connection with the Pythagorean Theorem: a field $F$ is "orderable" if and only if it is "formally real", in the sense that $x_1^2 + \ldots + x_n^2 = 0$ in it implies $x_1 = \ldots = x_n = 0$. This has implications for the theory of *orthocomplemented* modular lattices (see §2), because $S \cap S^\perp = 0$ for all $S$ (there are no "null lines") precisely when $F$ is formally real.

Likewise, E.V. Huntington has made very careful studies of the general properties of (pure) linear (alias "serial") and cyclic order in *Trans. Amer. Math. Soc.* 38 (1935), 1-9, and his readable monograph, *The Continuum and Other Types of Special Order.* I have only a few remarks to add to what he wrote there. The main point is, of course, that over any *ordered* division ring $D$, affine lines are linearly ordered and projective lines are cyclically ordered.

Firstly, Huntington's Postulates A-D for linear order are equivalent to the statement that three distinct points have only one *median*, (strictly) between the other two. His remaining Postulate 9 asserts (in effect) that of any 'tetrad' of four distinct points, two are interior and two are extreme.

Huntington defines cyclic order as an *anti*-symmetric ternary property, analogous to the binary < relation. It would be more natural from a geometric standpoint to define cyclic order in terms of the notion that any tetrad consists of two pairs of "opposite" points. It would be desirable to characterize these functions $f : (S^4)^* \to \underline{3}$ from ordered tetrads to partitions of four points which are associated with cyclic order in this way. (The asterisk signifies that the four points must be distinct.)

## 6. HYPERBOLIC GEOMETRIES AND CONVEXITY

To recapitulate §3, every (Riemannian) differentiable manifold with constant curvature is *locally* similar to either Euclidean $n$-space $\mathbb{R}^n$, to the unit $n$-sphere in $\mathbb{R}^{n+1}$, or to hyperbolic $n$-space. Hence, up to similarity, its covering surface is (metrically) *globally* similar to one of these three models. Moreover, similarity transformations carry geodesic configurations into geodesic configurations having the same incidence relations.

Furthermore, the three *lattices* in which these geodesic configurations are embedded are closely related to each other, the relation becoming especially simple if the unit $n$-sphere is replaced by projective $n$-space, $PG_n(\mathbb{R})$, so as to make exactly one geodesic pass through any two points.

Indeed, projective and Euclidean $n$-space differ (as sets of 'flats' ordered by inclusion) only by the presence or absence of a (projective) hyperplane at $\infty$. And finally, analogous geometric lattices of 'flats' can be defined in (abstract) projective and affine 'geometries' over any division ring. However, the situation is very different for hyperbolic $n$-space; only generalizations to 'hyperbolic geometries over *ordered fields*' are known.

The projective *metric* under which the algebraic straight lines are geodesics (i.e., shortest) is given by[18]

$$\rho(P,Q) = \log|\chi(O,P,Q,R)| \ , \qquad (6.1)$$

where $O$ and $R$ are the points where the extended straight line $\overrightarrow{PQ}$ pierces the "sphere at infinity". The resulting lines are isometric with $\mathbb{R}$, and their translations can be extended to global collineations of hyperbolic space (free mobility).

Inspired by Minkowski's geometry of numbers, in which an arbitrary (possibly symmetric) *convex ball* in $\mathbb{R}^n$ is allowed to define distance (thus inspiring the construction of Banach spaces and Finsler spaces), Hilbert used (6.1) to construct a much wider family of real hyperbolic geometries. These admit to be sure a much smaller group of global collineations, but all such hyperbolic spaces have analogous *locally* isomorphic categories of collineations.[19]

Hilbert's fourth problem, the subject of a recent book by Pogorelov reviewed by H. Busemann (*Bull. Amer. Math. Soc.* 4 (1981), 87-90), is concerned with the determination of *all* hyperbolic geometries having the properties noted by Hilbert. A natural question concerns the properties of analogous 'hyperbolic geometries' constructed over a general ordered division ring. Even though log $\chi$ is a transcendental function not readily defined even in fields, so that lines may well not be homogeneous, one can treat $|\chi(O,P,Q,R)|$ as an 'écart' or pseudo-distance, and define a pseudo-metric topology, congruence, etc.

## 7. POSITIVE CONVEX CONES

Similar families of problems concerns positive convex cones in $D^n$, where $D$ is an ordered division ring as always in this paper. More generally, one can consider infinite-dimensional, lattice-ordered linear metric spaces over any such $D$ -- to do this well seems like an interesting Ph.D. topic.

The most natural (convex) *positive cone* in $D^n$ is the set of $n$-vectors $\underline{x} = (x_1,\ldots,x_n)$ with all $x_i \geq 0$; this cone always defines a vector lattice. I suspect that the results of Murray Mannos, summarized in [LT2, pp. 237 and 240][16], apply to any $D$, i.e., that any finite-dimensional vector lattice over $D$ can be built up from $D$ by repeated formation of direct sums and lexicographic union.

Over any formally real (= orderable) field $F$, the *Lorentz cone* consisting of all $\underline{x} = (x_0,x_1,\ldots,x_n)$ in $F^n$ satisfying

$$(7.1) \qquad x_0^2 \geq x_1^2 + \ldots + x_n^2 , \quad \text{all } x_i \geq 0 ,$$

is also geometrically interesting, because its intersection with the hyperplane $x_0 = 1$ is the natural generalization of hyperbolic space. Over any *Pythagorean field*, in which $\sqrt{a^2+b^2}$ always exists, it is invariant under an analogue of the real Lorentz group.

For a decade beginning in 1957, I gave considerable thought to *positive linear operators* acting on such a space, concentrating on the case $F = \mathbb{R}$ of greatest 'real world' interest. I showed that the natural extension of the Laguerre-Hilbert metric (6.1) to a 'pseudo-metric' in any (closed) positive cone had the following desirable property,which I think can be extended to positive cones in any $F^n$, and even to their infinite-dimensional analogues.

THEOREM 5. *Let $C$ be a positive cone in a vector space $V$ over an ordered field $F$, and let $f : V \to V$ be a linear mapping such that $f(C) \subset C$. Then $f$ is a* contraction *in the pseudo-metric defined by* $|\chi(O,P,Q,R)|$.

With minor changes, the proof of Theorem 2 of [LT3, p. 383][16] can be adapted to this case, the point being that log $\chi$ is isotone. Of course, the conditions guaranteeing *uniform* convergence to a limiting dominant direction are much harder to generalize.[20] (It is this convergence that is most important in population theory and nuclear reactor theory.)

Clearly, my brief and unorthodox discussion of *convexity* has omitted many other interesting questions as well; I shall comment on a few of them in §9.

## 8. CLOSURE ALGEBRAS

General topology (alias point-set theory) is also a fruitful source of lattices. The Moore family of sets closed under a distributive *closure* operation gives, naturally, a *distributive* lattice of "closed" sets, and $T_0$-spaces are defined essentially in this way.

Thus [LT1, §§17, 91, 92, 104][16] the class of distributive lattices of finite length, each of which is necessarily finite, is identical with the class of lattices of closed subsets of $T_0$-spaces.

If no restriction is placed on the cardinality or dimensionality of the sets involved, we get a very interesting class of complete, completely distributive *Brouwerian* lattices more closely associated with logic than geometry proper. I call your attention to the fact that *pseudo*-complements and *regular* open sets are natural constructs in this context [LT1, §124],[16] as are

*Stone lattices* and Stone's Boolean rings without units not quite equivalent to Boolean algebras. Perhaps the deepest and most seminal ideas of all may be found in Stone's original papers (*Trans. Amer. Math. Soc.* 41 (1937), 375-481, and *Cas. Mat. Fys.* 67 (1937), 1-25). Here Stone established his classic 'crypto-isomorphism' between Boolean algebras and '*clopen*' (= closed and open) subsets of zero-dimensional (= totally disconnected (bi) compact) topological spaces.

This circle of ideas was also studied most thoroughly in two scholarly papers by McKinsey and Tarski (*Annals of Math.* 45 (1944), 141-191, and ibid, 47 (1946), 122-162). Out of these studies emerged many new theorems, and also the 'universal' algebraic concept of a 'functionally free' algebra. In a survey lecture giving general orientations, younger mathematicians should be reminded that many of the underlying ideas were extended by Jónsson and Tarski to a more general class of *Boolean algebras with operators* (*Amer. J. Math.* 73 (1951), 891-939, and 74 (1952), 127-167). This included much of the algebra of relations, and indeed helped to suggest a standard axiom system for the latter.

However, no new *examples* of interesting posets or Boolean algebras emerged from these studies, to my knowledge. In my opinion, the most interesting examples of the genre I have been discussing are: (i) the metric Boolean algebra $M/N$ of all measureable sets modulo sets of measure zero, (ii) the lattice of open sets modulo nowhere dense sets, and (iii) the free Borel algebra with $\aleph$ generators, which for $\aleph = \aleph_0$ is isomorphic with the algebra of all Borel subsets of the Cantor ternary set, or for that matter of almost any other uncountable metrizable space. See [LT1, §122 and §124-125], and [LT3, Chap. XI, §§2-3].[16]

## 9. COMBINATORIAL TOPOLOGY

Ordered sets arise in two other very important geometrical contexts: combinatorial topology and algebraic geometry. I will make a few historical remarks about the former, and then (in §10) try to suggest new areas of study associated with the latter.

I know that Lefschetz changed the name of "combinatorial" topology to "algebraic" topology, and that his terminological innovation has received widespread approval.[21] However, I wish to emphasize very specifically the *combinatorial* aspects of the subject. I am thinking very specifically of the *polyhedral complexes* originally proposed by Poincaré as a visually suggestive device for classifying manifolds according to their topological structure.

A good point of departure for what I have to say about combinatorial topology is provided by two articles by A.W. Tucker in the *Annals of Math.* in 1933 (vol. 34, pp. 191-243, and vol. 37, 92-100). Tucker's "cell spaces" are nothing else than posets, and the papers (written under the "stimulus and guidance" of Lefschetz) suggest that many topological concepts should be studied in this generality; see [LT1, p. 15].[16]

Besides Tucker and Lefschetz, Alexander was much intrigued by this lattice-theoretic approach to topology in the late 1930's. His theory of "gratings", and his concept of a "sub-base" of open sets (neighborhoods), are however more in the direction of bridging the gap between combinatorial and algebraic topology, than in that of combinatorial topology *per se*. Finally, J.H.C. Whitehead analyzed lattice-theoretic approaches to *duality* in topology in the late 1950's.[22]

By including the empty set $\emptyset$ and the manifold $M$ as 'cells', one gets a lattice; in fact a Moore family, whose natural building blocks are (convex?) polyhedra. Thus all the results of Grünbaum, Klee, and others are relevant. If one omits $\emptyset$ and $M$, then the *topological sum* becomes the cardinal sum of the general 'extended arithmetic' of posets, and a space is *connected* if and only if the poset of its cells is connected. If one includes $M$ but not $\emptyset$, then the *Cartesian product* becomes the cardinal product in the same sense [LT1, p. 15].[16] Thus an $n$-cube becomes $[(\underline{1}+\underline{1}) \oplus 1]^n$.

Recent years have seen continuing interest in characterizing the class of graded lattices of facets of convex polytopes, as well as in the problem of generalizing the Euler characteristic and homology groups to finite unions of compact convex sets in $\mathbb{R}^n$. Especially noteworthy are papers by Hadwiger, Klee, and Rota.[23] For more recent results, see A. Björner, *J. Combinatorial Theory* 30 (1980), 90-100, and the paper by Björner, Garsia, and Stanley in these *Proceedings*. It would seem timely to correlate the newer definitions and results with the earlier work of Tucker and Alexander.

## 10. ALGEBRAIC GEOMETRY

The concept of *polarity* (or *Galois connection*) unifies and suggests generalizations of several of the preceding constructions. Thus, over any division ring $D$, the lattice $PG(D^{n+1})$ of linear subspaces of any $D^{n+1}$; the lattice $AG(D^n)$ of affine 'flats' of any $D^n$, and the lattice of all regular open sets of any metrizable space are associated, respectively, with the polarities (Galois connections) defined by the relation $\emptyset(\underline{x}) = 0$ for the following sets $\Phi_i$ of functions:

(10.1) $\emptyset(\underline{x}) = x_1a_1 + \ldots + x_na_n = \langle\underline{x},\underline{a}\rangle$,

(10.2) $\emptyset(\underline{x}) = a_0 + x_1a_1 + \ldots + x_na_n$ ,

(10.3) $\emptyset(\underline{x})$ is a continuous function.

Namely, if $S \in D^n$ is given, and $S^*$ denotes the set of $\emptyset \in \Phi_i$ such that $\emptyset(\underline{x}) = 0$ for all $\underline{x} \in S$, then the *closure* $\bar{S}$ of $S$ is the set of all $\underline{y} \in D^n$ such that $\emptyset(\underline{y}) = 0$ for all $\emptyset \in S^*$. (For $D$ an ordered field, the lattice of convex sets in $D^n$ can be defined similarly by *inequalities* $\emptyset(\underline{x}) \geq 0$.)

Similar constructions give other interesting geometric lattices.

Inversive Geometry

For example, over any field $F$ in which $1 + 1 \neq 0$, consider the set $\Phi_4$ of all functions of the form

(10.4) $\emptyset(\underline{x}) = A\,\Sigma\,x_k^2 + 2\,\Sigma\,b_kx_k + c$

from $F^n$ to $F$. The polarity defined by (10.4) yields as 'closed' sets all *flats*, together with their transforms under the inversive group (circles, spheres, ... , hyperspheres), and point-pairs as well.

In the lattice $SG(F^n)$ so defined, all point-pairs are closed sets, but through any three points there passes a unique circle (or straight line); through any four points a unique sphere (or plane); and so on. Therefore, the height of the lattice is $n + 2$; moreover any interval sublattice $[a,1]$ with $a$ not 0 or an atom is modular. These "inversive lattices" were studied in 1950-51 by A.J. Hoffman and W. Prenowitz,[24] and Professor Ray-Chaudhuri has discussed them at this meeting, but I suspect there is more to be found out.

Varieties of Higher Degree

Analogous geometric lattices are associated with algebraic varieties of higher degree over any field $F$ whose characteristic exceeds the degree. Of course, not all important geometric properties hold in this generality. Thus over the rational field $Q$ ordered as usual, the point (1,0) is inside the circle

$\Gamma : x^2 + y^2 = 4$, yet the straight line $x = 1$ through this point fails to intersect $\Gamma$. However, surprisingly many geometric theorems hold over most fields.

In particular, as in [LT3, Chap. V, §8],[16] the *algebraic varieties* of $F^n$ always constitute a $\cap$-ring of sets for any field $F$, hence a distributive lattice. Moreover this lattice satisfies the descending chain condition (DCC), by [LT3, pp. 180-186]. In this distributive lattice $AV(F^n)$, the zero-dimensional varieties constitute the generalized Boolean algebra of all *finite* sets of points. (Hence if $F$ is finite, every subset of $F^n$ is an algebraic variety, and so the algebraic varieties form a Boolean algebra.) The preceding generalized Boolean algebra is a *connected component* of the graph defined by the covering relation in $AV(F^n)$, and it is a lattice ideal $N_0$ . Finite unions of algebraic curves form an analogous ideal in the quotient-lattice $AV(F^n)/N_0$ , and so on (Krull, Ore).

More interesting as ordered sets are the *geometric lattices* $L_m(F^n)$ defined by polarity from the polynomials $p(\underline{x}) = p(x_1, \ldots, x_n)$ of degree $m$ or less in $n$ variables with coefficients in $F$. The fact that these form a geometric lattice follows from the results of [LT3, Chap. IV, §3].[16] Moreover, in any such lattice $L_m(F^n)$, each 'algebraic variety' of degree $m$ or less is a meet of co-atoms: the pencil of varieties of degree $m$ containing it.

As an example, consider the geometric lattice $L_2(F^2)$, where $F$ is any field in which $1 + 1 \neq 0$. The lattice is of height 6, as is obvious since the polynomials $Ax^2 + 2Bxy + Cy^2 + Dx + Ey + F$ have six arbitrary coefficients. Its coatoms are the irreducible *conics* $q(x,y) = 0$,[25] and the pairs of straight lines $(ax+by+c)(dx+ey+f) = 0$. The elements of height 4 are the vertices of quadrangles, and configurations consisting of a line and a point. Those of rank 3 are the straight lines and the sets of vertices of triangles; those of rank 2 are the point-pairs.

The interval sublattices of the $L_2(F^2)$ are easily deduced from the preceding geometrical description; note that the set of vertices of a quadrangle is the intersection of its two pairs of opposite sides. To be equally specific about the geometrical structure of algebraic varieties in $L_2(F^3)$, $L_3(F^2)$, and so on, offers scope for considerable geometrical research. It should also provide a large class of interesting geometric lattices.

## 11. COMBINATORIAL GEOMETRIES

I have described in some detail the posets associated with *geometrical configurations* that have arisen in classical synthetic geometry, combinatorial topology, and algebraic geometry. These configurations all define posets through the inclusion relation $S \subset T$, and its symmetrization to 'incidence' (meaning that $S \subset T$, or $T \subset S$).

I shall now describe more briefly posets associated with analogous combinatorial configurations, which I will call *combinatorial geometries* because of their close relation to geometrical configurations such as finite projective planes and graphs (one-dimensional 'polyhedral' complexes).[26] I shall begin with several related families of combinatorial geometries corresponding to graded posets of height one, for which lattice-theoretic concepts are primarily of conceptual interest.

### Tactical Configurations

As Peter Dembowski has emphasized, a natural generalization of finite plane projective geometries is provided by 'tactical configurations' in the sense of E.H. Moore. [27] The simplest tactical configurations are *cubic graphs*, or systems of 'points' and 'lines' (alias 'blocks') in which every line contains 2 points, and every point is on 3 lines. These can also be viewed (adjoining 0 and 1) as those geometric lattices of height 3, whose points and lines have the properties stated above. Kirkman-Steiner *triple systems*, in which every line contains 3 points, and every point-pair is in a unique line, constitute another simple family of tactical configurations.[28]

### Incidence Geometries

Much more general are so-called "incidence structures" of 'points' and 'lines' or 'blocks', connected by 'incidence relations' or 'flags'. Lest you think that I am straying from my subject, I observe that an 'incidence structure' is just another name for a *graded poset of height one*, whose rank 0 elements are called 'points' and whose rank 1 elements are called 'blocks' or *lines*. Hence a tactical configuration is just such a poset in which every line contains the same number $r \geq 2$ of points, and every point is contained in the same number $k \geq 2$ of lines.

Since (as Dembowski emphasizes) finite projective planes are tactical configurations, it seems appropriate to call 'incidence structures' *incidence geometries*, and I shall do so. Evidently, any incidence geometry whose automorphism group is transitive on points and lines, is a tactical configuration --

note also that the dual and the complement[29] of any tactical configuration are themselves tactical configurations; self-dual tactical configurations are of especial interest.

Redundant Terminology.

Not only is an 'incidence structure' just another name for a graded poset of height one; it is also just another name for a *binary relation* between two disjoint sets, $P$ and $L$, and likewise for a *bipartite graph*, having a specified partition into two nonvoid subsets. To add to the confusion, a general (simple) graph is the same as ("crypto-isomorphic" with) a *symmetric* binary relation; as mentioned before, it is also a one-dimensional unoriented (simplicial) *complex* in the sense of Poincaré.

Clearly, by the trivial adjunction of 0 and 1, the poset defined by any graph becomes a graded, relatively complemented, *lattice* of height 3 in which each 'line' (element of rank 2) contains just 2 'points'. More generally, it is easy to prove the following results.

THEOREM 6. *The Dedekind-MacNeille completion of any graded poset of height one (alias "incidence geometry") is a lattice L having the same points, dual points, and group of automorphisms.*

Every relatively complemented lattice of finite height $n > 2$ can be obtained by the preceding construction, since its join-irreducibles are all points and its meet-irreducibles are all dual points.[30]

By definition, the incidence geometry of Theorem 6 is a tactical configuration if and only if, in the lattice $L$ mentioned there,

(i) every coatom ('line') contains $r$ points, and

(ii) every atom ('point') is in $k$ coatoms.

Less trivial lattice-theoretically is the class of 'partially balanced incomplete block designs' or *PBIB designs*.[31] These are tactical configurations in which every point-pair $\{p,q\}$ is contained in the same number $\lambda > 0$ of 'lines' (coatoms). We have, for tactical configurations:

THEOREM 7. *The lattice L of Theorem 6 is a* geometric lattice *of height three if and only if the given incidence geometry is a PBIB design with* $\lambda = 1$. *It is a* modular lattice *of height three if and only if the incidence geometry is a projective plane.*

## Hadamard Matrices

Graded posets of arbitrarily large height also arise naturally in combinatorics. Thus Hadamard matrices (Hall, op. cit., Chap. 14)[31] provide another highly symmetric class of graphs (and hence of incidence geometries). By definition, these are $m \times m$ matrices satisfying $HH^T = mI$.[32] Hence their rows (and columns) are sets of $m$ orthogonal vectors $\underline{v}_k$ (and $\underline{w}_k$) with coordinates $x_i = v_{ki} = \pm 1$. The $2m$ vectors $\pm\underline{v}_k$ clearly form the vertices of a regular convex polytope in $\mathbb{R}^m$; moreover the distance apart of non-opposite vertices is $\sqrt{m}$. Hence the polytope has $2m(2m-2)(2m-4)/6$ equilateral triangular faces. Such polytopes exist in $m > 2$ dimensions only for $m = 4n$; they have $8n$ vertices and $32n(2n-1)(2n-1)/3$ triangular faces. More generally, they define a $(m-1)$-dimensional *simplicial complex* which represents the $(m-1)$-sphere in $\mathbb{R}^m$. If opposite vertices are identified, one gets a simplicial subdivision of projective $m$-space.

## Hamming Codes[33]

Likewise, for any positive integer $s$, there is a 'Hamming code' having word-length $n = 2^s - 1$ and message-length $m = n - s$ -- hence $s$ 'check digits'. There is also an 'augmented Hamming code' having word-length $n = 2^s$ and $s + 1$ 'check digits', hence the same message length $2^s - s - 1$. As binary $(m,n)$-codes, the code words thus each select a subset of $2^m$ elements of $\underline{2}^n$. Considered over $GF(2)$, these form a linear subspace of $\underline{2}^n$ (the null space of the 'parity check matrix') having remarkable algebraic properties.

However, we will consider the code words as constituting a subset of $2^m$ vertices of the *unit cube* $\{0,1\}^n$ in $\mathbb{R}^n$. These all lie on the unit $(n-1)$-sphere with center $(1/2, \ldots, 1/2)$ and diameter $\sqrt{n}$. Moreover they span a *convex polytope* in $\mathbb{R}^n$ having $(1/2, \ldots, 1/2)$ in its interior, and each code word has $2^s - 1$ nearest neighbors. By joining each vertex (code word) with its nearest neighbors, we get a *graph* which is a tactical configuration with $r = 2$ and $k = n - 1$. Its perpendicular bisectors of the edges of this graph decompose the unit $(n-1)$-sphere into $2^m$ regular $(n-2)$-cells, each having $2^s - 1$ faces.

## 12. CONCLUDING REMARKS ON SYMMETRY

I will conclude my brief overview of posets arising in geometry by some remarks about their *symmetry groups*, which are among their most interesting properties. Most of the examples I have described, in §11 and elsewhere, have extensive groups of symmetries.

Thus the automorphisms of any Desarguesian projective geometry $PG_n(D)$ are its collineations, and correspond (via homogeneous coordinates) to the semilinear transformations $\underline{y} = A\underline{x}^{\phi}$, where $A$ is a nonsingular $(n+1) \times (n+1)$ matrix and $\phi$ is an automorphism $x_i \to x_i^{\phi}$ of $D$. To determine the automorphisms of *non*-Desarguesian finite projective (necessarily plane) geometries is an active field of current research.[34] Likewise, the automorphism group of the affine geometry $AG_n(D)$ has a normal subgroup $T$ of translations; moreover if $D = F$ is commutative, $AG_n(F)/(S \cup T)$, where $S$ is its center (all scalar matrices), is essentially isomorphic to $PG_{n-1}(F)$. The lattices of 'flats' of hyperbolic geometries have much smaller automorphism groups, being limited to their isometries (rigid motions) with respect to the metric (6.1) and its analogues.

The automorphisms and many other properties of affine and projective "designs" are described in Dembowski, op. cit. §§2.1-2.3. Those of triple systems and Hadamard designs are discussed there in §2.4.

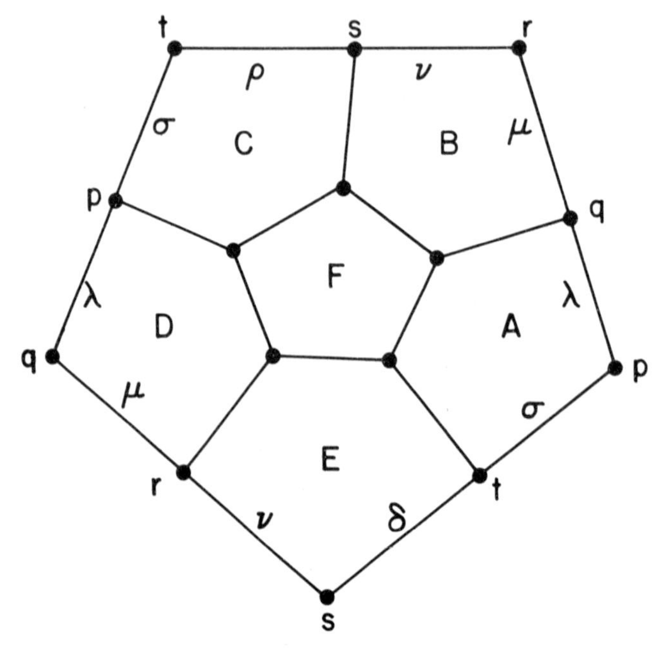


Figure 2

Likewise, the dodecahedron can be considered as a graph, as a two dimensional complex, or as a convex polytope; all these posets have the same 'icosahedral' group of symmetries, which is associated also with the dual icosahedron. This group was studied in depth by Felix Klein, if only admired by Plato; moreover the icosahedron is a simplicial complex. The faces of the solid dodecahedron $\Delta$ and dual icosahedron $\Delta'$ constitute *graded* lattices [LT1, §18][16]; the graphs of their vertices and edges are *tactical configurations;* as just remarked, all have isomorphic automorphism groups of order 20, with center of order 2 (unlike the symmetric group) and quotient-group $A_5$ .

Projectively, the dodecahedron reduces to the *hexahedron* with 6 pentagonal faces and 15 edges drawn in Fig. 2. As the lettering indicates, opposite vertices and edges are identified; the automorphism group of its poset is $A_5$ .

## Some General Theorems

The preceding examples can be viewed as special cases of the following general result, which I proved in 1946.[35]

THEOREM 8. *Any group is isomorphic to the group* Aut $L$ *of a suitable distributive lattice* $L$.

More relevant to the present survey are the following results of R. Frucht.[36]

THEOREM 9. *If* $g$ *is a finite group of order* $g$ *with* $n$ *generators, then there exists a cubic graph* $G$ *with at most* $5(n+2)g$ *nodes such that* Aut $G = g$.

Combining this result with Theorem 7, we obtain the following corollary.

COROLLARY. *Under the hypotheses of Theorem 9, there exists a lattice of height 3 having* $g$ *for its group of automorphisms.*

## Possible Extensions

Theorem 9 solves in the affirmative a problem posed by D. König in 1938: is every abstract finite group the group of all automorphisms of some graph? König's problem has the following natural extension.

PROBLEM A. Given a transitive permutation group $G$ acting on a set $S$, construct a graded poset $P$ such that the action of Aut $P$ on the points of $P$ is equivalent (by bijections) to the action of $G$ on $S$.

PROBLEM B. For given $G$ acting on $S$, find the smallest poset $P$ solving Problem A.

It is easy to show that graphs do not suffice for solving Problem A, since any graph admitting a doubly transitive group of automorphisms must have the symmetric group for its group of automorphisms (be totally disconnected or complete). In searching for solutions to Problems A and B, therefore, convex polytopes and polyhedral complexes seem like promising candidates. I shall conclude my talk with a few remarks about these candidates, emphasizing simple groups.

Regular Polytopes

Long before Klein studied the icosahedron, the regular solids had been studied in depth by Pappus and Kepler; an up-dated account of some of their discoveries has been included by Coxeter in a scholarly sequel to his classic study of polytopes generated by reflections.[37] Clearly, any *convex polytope* constitutes a polyhedral complex representing a sphere, in which each 1-cell contains exactly 0-cells; hence we may speak of the *graph* of the polytope (or other complex).

In many cases, including the regular dodecahedron discussed above, this graph determines the complex. Namely, the 2-cells of the complex correspond to the shortest 1-cycles of the graph, and are bounded by them. Likewise, the smallest 2-cycles[38] of the resulting subcomplex of 1-cells and 2-cells bound the 3-cycles of the complex, thus determining them and so on. Hence the automorphism group of the complex is the same as that of its graph in such cases; the implications for Problem B above are obvious.

Simple Lie Algebras

E. Witt[39] seems to have been the first to see the connection between Coxeter's work and the root diagrams of simple Lie algebras, a connection worked out in depth by Chevalley and greatly extended by Tits.[40] Both Coxeter and Chevalley recognized the basic role played by reflections of period two, in hyperplanes corresponding to the 'cleavage planes' of crystals.

Simple Finite Groups

Interest in these connections was greatly intensified by the recognition (also by Chevalley) of the connections between simple Lie algebras and finite simple groups, whose enumeration has recently been successfully completed. By the Feit-Thompson Theorem, every noncyclic finite simple group has an *involution* (analogous to a reflection), and the determination of all finite simple groups has been helped materially by an analysis of the centralizers of these involutions. An enormous amount of deep research has been published along the lines sketched above, some of it concerned with realizations of simple groups as symmetry groups of simplicial complexes, convex polytopes, and incidence geometries.

Much of this literature is ably summarized in two collections of survey articles,[41] rich in stimulating ideas. I hope that the study of these articles and the other references I have given, in relation to the ideas expressed in this survey, will lead to substantial progress during the next decade in solving Problems A and B stated above.

REFERENCES


1 O. Ore, *Annals of Math.* 36 (1935), 406-437 (esp. p. 406) and 37 (1936), 265-292.

2 G. Birkhoff, *Annals of Math.* 36 (1935), 743-748, (also *Bull. Amer. Math. Soc.* 40 (1934, p. 209); K. Menger, ibid. 37 (1936), 456-481.

3 I will not try to evaluate the contributions of von Staudt, Pasch, Peano, Veronese and other predecessors of Hilbert. The *idea* of geometric algebra, of course goes back to Euclid.

4 It is, of course, a corollary of Wedderburn's Theorem that every finite division ring is commutative. No synthetic proof of this result has yet been found.

5 Marshall Hall, Projective planes, *Trans. Amer. Math. Soc.* 54 (1943), 229-277. See also P. Dembowski, *Finite Geometries*, Springer, 1968.

6 G. Birkhoff and J. von Neumann, *Annals of Math.* 37 (1936), 823-843.

7 See also the review article by S. Holland in *Trends in Lattice Theory* (James C. Abbott, ed.), van Nostrand, 41-126, (1970), and my article in *Proc. Symposium Pure Math.* (R.P. Dilworth, ed.), Amer. Math. Soc. (1961).

8 Representations of finite groups and semisimple algebras, over $\mathbb{R}$ and $\mathbb{C}$, *are* completely reducible.

9 *Annals of Math.* 33 (1932), 163-176.

[10] The closely related 'free mobility' property was identified much earlier by Helmholtz, the explicit recognition of the group property had to wait for Lie. The spherical, Euclidean, and hyperbolic also have a special position in conformal geometry, through the Uniformization Theorem of Klein-Poincaré-Koebe. This says that every Riemann surface is a conformal image of precisely one of these three.

[11] Likewise, Mercator projection from the center maps the unit sphere conformally onto a developable cylinder, which can be rolled into a plane map.

[12] In recent decades, Artin and Snapper-Troyer have followed Hilbert's example, while Hartshorne has followed that of Veblen and Young.

[13] The concept of geometric lattice (alias "matroid lattice") was only identified in 1935, after the time that Menger's paper was completed.

[14] In the notation of the seventh (German) edition of Hilbert's *Grundlagen der Geometrie*. Hilbert lattices have also been studied by O. Wyler, *Duke Math. J.* 20 (1953), 601-610.

[15] See his review of Hilbert's book in *Bull. Amer. Math. Soc.* 9 (1902), 158-166.

[16] G. Birkhoff, *Lattice Theory*, Amer. Math. Soc. (1st ed., 1940; 2nd ed., 1948; 3rd ed., 1967).

[17] E. Artin and O. Schreier, *Abh. Math. Sem. Univ. Hamburg* 5 (1926), 85-99; E. Artin, ibid. 100-115. Artin mentions Hilbert's *Grundlagen* (Chap. VIII) on p. 100.

[18] Felix Klein observes, rather sadly, that the metric of (5.1) was discovered by Laguerre.

[19] In the sense of Lie groups, as envisaged by Lie, generalized to categories in the sense of Eilenberg and MacLane.

[20] Generally speaking, one can ask which results in T. Rockafellar's *Convex Analysis* (Princeton Univ. Press, 1970) can be treated by rational methods, and which depend on commutativity. Similar questions can be asked about linear programming.

[21] See the Preface to S. Lefschetz, *Algebraic Topology*, Amer. Math. Soc., 1942, and the comments by N. Steenrod in the Symposium honoring Lefschetz, *Algebraic Geometry and Topology*, Princeton Univ. Press (1957), 24-43.

22 See Publications ##78, 82, and 87 in vol. IV of his *Mathematical Works*; also #12 in vol. II.

23 H. Hadwiger, *J. Reine Angew. Math.* 194 (1955), 101-110 and *Monatsh. Math.* 64 (1960), 349-354; V. Klee, *Amer. Math. Monthly* 70 (1963), 119-127; G.-C. Rota, *Studies in Pure Mathematics in Honor of R. Rado* (M. Mirsky, ed.), Academic Press (1971), 221-233.

24 W. Prenowitz, *Canad. J. Math.* 2 (1950), 100-119; A.J. Hoffman, *Trans. Amer. Math. Soc.* 71 (1951), 216-242. See also H.S.M. Coxeter, *Abh. Math. Sem. Univ. Hamburg* 29 (1966), 217-242, and P. Dembowski, op. cit. in ftnt. 5, Chap. 6.

25 In $L_2(F^2)$, the 'closure' of a set of 5 points, no 3 collinear, is the conic through them defined by Pascal's Theorem. Likewise, in inversive geometry, the 'closure' of any set of 3 points is the circle (or straight line) through them.

26 See. O. Veblen's *Analysis Situs*, Amer. Math. Soc. 1916 (sec. ed. 1931). This well-written book explains the basic concepts of combinatorial topology in the one-and two-dimensional contexts in which they are best understood, before treating the general $n$-dimensional case.

27 *Amer. J. Math.* 18 (1896), 264-305. The bibliography of P. Dembowski, *Finite Geometries*, Erg. 24, Springer, 1968, contains many other references. Curiously, Dembowski never defines the word "geometry". In his *Geometrische Configurationen* (Hirzel, Leipzig, 1929), F. Levi calls tactical configurations "geometrical configurations".

28 See M. Hall's article in *Studies in Combinatorics* (G.-C. Rota, ed.), MAA Study in Mathematics #17,(1978), 218-254. Tactical configurations in which every point-pair is on $\lambda$ lines are called *block designs*.

29 The *complement* of a tactical configuration $[P,L,A]$ has for its incidence matrix the complement $A'$ of $A$.

30 See B. Banaschewski, *Z. Math. Logik Grundlagen Math.* 2 (1956), 117-130; J. Schmidt, *Arch. Math.* 7 (1956), 241-249. See also G. Markowsky, *Trans. Amer. Math. Soc.* 203 (1975), 185-200, and *J. Algebra* 48 (1977), 305-320, for the considerations involved.

31 See M. Hall, *Combinatorial Theory*, Blaisdell (1967), Chaps. 10, 15, 16. Dembowski calls these "designs", for brevity.

32 PBIB designs have incidence matrices satisfying $AA^T = (k-\lambda)I + \lambda J$, where $k(k-1) = \lambda(v-1)$; see Hall, op. cit., (10.1.3) and (16.4.1).

33 See for example J.H. van Lint, *Coding Theory*, Springer Lecture Notes #201, 1971, or G. Birkhoff and T.C. Bartel, *Modern Applied Algebra*, McGraw-Hill, 1970.

34 See N.L. Biggs and A.T. White, *Permutation Groups and Combinatorial Structures*, Cambridge Univ. Press, 1979, and its review in the *Bull. Amer. Math. Soc.* 5 (1981), 197-201.

35 *Rev. Un. Mat. Argentina* 11 (1946), No. 1.

36 *Canad. J. Math.* 1 (1949), 365-378, and 2 (1951), 417-419.

37 H.S.M. Coxeter, *Regular Polytopes*, MacMillan, New York (1963).

38 In a non-oriented complex, a "$k$-cycle" is a $k$-chain over $\mathbb{Z}_2$ with $\partial C = 0$.

39 *Abh. Math. Sem. Univ. Hamburg* 14 (1941), 289-337.

40 C. Chevalley, *Tohoku Math. J.* 7 (1955), 14-66; also *Amer. J. Math.* 77 (1955) 778-782. The imaginative and wide-ranging work of Tits is summarized in his monograph.

41 M.B. Powell and G. Higman (eds.), *Finite Simple Groups*, Academic Press, 1971; and M.J. Collins (ed.), *Finite Simple Groups II*, Academic Press, 1980.

APPENDIX

## A LATTICE CHARACTERIZATION OF AFFINE $n$-SPACE

M.K. Bennett

### 1. INTRODUCTION

The essential novelty of the lattice theoretic approach to geometry, as Professor Birkhoff pointed out (these *Proceedings*) is its identification of all flats (or subspaces) of a given geometry as equally basic; whereas synthetic axioms distinguish 'lines' and 'planes' from 'points', and the algebraic (analytic) approach to a given geometry usually has $n$-tuples of ring or field elements as its basic entities.

As was first shown by Birkhoff [B1] and Menger [M1], *projective geometries* have a very simple lattice theoretic characterization, as irreducible complemented modular lattices.

By an *incidence space* [G1] we mean a collection of points, lines and planes satisfying Hilbert's Axioms of Incidence ([Verknüpfung], [H1]), with suitable dimension modifications. (Hilbert was dealing only with 3-dimensional geometry.) Lattices of *flats* (point-sets which contain, along with any two points, the line they determine) of finite dimensional incidence spaces were characterized by Sasaki [S1] and Wyler [W1] as what may be called *Hilbert lattices*, i.e. geometric lattices (of finite height) in which subintervals $[a,1]$ are modular for any $a > 0$. Projective and hyperbolic geometries[1] are Hilbert lattices, as are *affine geometries*, our main concern here.

An *affine space* (as treated, for example, in [A1], [S3], etc.) is an incidence space satisfying this parallel axiom for lines:

(P) *Through any point not on a line, there is a unique line parallel to the given line,* (parallel lines *being coplanar and non-intersecting*).

Any affine space of dimension $n \geq 3$ satisfies Desargues' Theorem; and its flats may be considered as the affine subspaces (translates of linear subspaces) of the module $D^n$ for some division ring $D$ (see [A1]).

Sasaki [S2], building on the partial characterization of Menger [M1], characterized affine geometries of arbitrary (not necessarily finite) dimension as essentially Hilbert lattices (without the constraint of finite height) in which this lattice theoretic parallel axiom holds:

(P') *Let* $p,q,r$ *be atoms with* $p \wedge (q \vee r) = 0$. *Then there exists a unique* $\ell$ *where* $p < \ell < p \vee q \vee r$ *and* $\ell \wedge (q \vee r) = 0$ .

This (P') is a restatement in lattice theoretic terms of the synthetic parallel axiom (P); and, as such, is not truly lattice theoretical. The purpose of this APPENDIX is to offer a more natural lattice theoretical characterization of finite dimensional affine geometry.

## 2. FINITE DIMENSIONAL AFFINE GEOMETRIES

In the sequel, $L$ will always represent a Hilbert lattice, while 'lines' and 'planes' in $L$ are, as usual, elements of height 2 and 3 respectively. Elements $x$ and $y$ in $L$ are called *parallel* (written $x \parallel y$) when $x \wedge y = 0$ and $x \vee y$ covers both $x$ and $y$. According to this definition, parallel elements have the same dimension; in particular, two *lines* are parallel if and only if they form a non-modular pair ([M2], p. 79). Furthermore, the lattice theoretic and synthetic definitions of parallel lines agree.

Our purpose is to give conditions which, in a Hilbert lattice, are equivalent to (P), which we may break into two parts, uniqueness and existence, as follows.

(P1) *Through any point not on a line there is at most one parallel to the line.*

(P2) *Through any point not on a line there is at least one parallel to the line.*

Condition (P1) suggests precluding from being a sublattice, the six-element lattice denoted $K_6$ drawn in Fig. 1. $K_6$ is subdirectly irreducible, and the variety it generates covers that generated by the five-element non-modular lattice [M3].

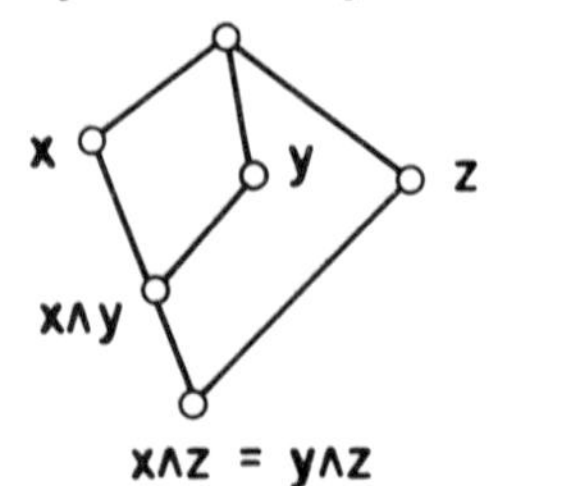


If $K_6$ is a sublattice of any Hilbert lattice, its bottom element must be zero since intersecting elements form modular pairs. In case $x \wedge y$ is an atom (point) $p$, with $x$, $y$ and $z$ lines, then $x$ and $y$ are lines through $p$ parallel to $z$. For this reason, any Hilbert lattice not containing $K_6$ as a sublattice satisfies (P1); by the same token, any hyperbolic geometry contains many copies of $K_6$ (at least one for each point not on a line).

More surprisingly, $K_6$ cannot be a sublattice *anywhere* in an affine geometry, regardless of the dimensions of the various elements.

2.1 THEOREM. *Let $L$ be an affine geometry. Then $L$ has no sublattice isomorphic to $K_6$.*

Proof: If the height of $L$ is $< 3$, $K_6$ cannot be a sublattice of $L$. If the height of $L$ is 3, then we have an affine plane, $x$ and $y$ must be intersecting lines; and $z$ must be a line parallel to both $x$ and $y$, a contradiction. If the height of $L$ is greater than 3, then we have a coordinatizing division ring $D$. If $D \neq Z_2$, then any atom under $x \vee y$ is under $x_1 \vee y_1$ where $x_1$ and $y_1$ are atoms under $x$ and $y$ respectively. In particular any atom

$z_1$ under $z$ is under some $x_1 \vee y_1$, where neither $x_1$ nor $y_1$ is under $x \wedge y$. Since $x_1 \leq (x \wedge y) \vee z$, by considering $x_1$ as an affine combination of elements from $x \wedge y$ and $z$, we must have $(x_1 \vee p) \parallel (z_1 \vee z_2)$ for some atoms $p \leq x \wedge y$, and $z_2 \leq z$. By the parallel axiom, $(y_1 \vee p) \wedge (z_1 \vee z_2) \neq 0$, and since this atom is under $y \wedge z$, we have $y \wedge z \neq 0$. But this contradicts the fact that $L$ is a Hilbert lattice, i.e. $y$ and $z$ would form a modular pair unless $y \wedge z = 0$.

In case $D = Z_2$, again by the Hilbert lattice property, the bottom element of $K_6$ must be 0. By considering possible affine combinations, given an arbitrary atom $z_1 \leq z$, there exist $x_1$, $y_1$, and $y_2$ such that $(z_1 \vee x_1) \parallel (y_1 \vee y_2)$ (or the analogous situation with the roles of $x$ and $y$ reversed). Since there are exactly two points on each line, we have $(y_1 \vee z_1) \parallel (x_1 \vee y_2)$ as well. Now using the fact that $x_1 < (x \wedge y) \vee z$, we obtain $p \leq x \wedge y$ such that $(x_1 \vee p) \parallel (z_1 \vee z_2)$ for some $z_2 \leq z$. This means that $(z_1 \vee x_1) \parallel (p \vee z_2)$; hence $z_2 \leq y_1 \vee y_2 \vee p \leq y$, a contradiction.

Parallelism of hyperplanes (coatoms) has a natural lattice theoretic interpretation since two hyperplanes are parallel if and only if they are *complementary* (their meet is 0 and their join is 1). The analogue of (P) for hyperplanes, Professor Birkhoff's Axiom (A), is valid in affine geometry.

(A) *Through any point not on a hyperplane, there exists a unique parallel to the hyperplane.*

By itself, Axiom (A) does not characterize Hilbert lattices which are affine geometries since the flats of any affine space with one point removed form a Hilbert lattice in which (A) holds. However, with $K_6$ precluded as a sublattice, a weaker form of (A) may be used to characterize finite dimensional affine geometries.

2.2 THEOREM. *Let $L$ be a Hilbert lattice of height $n > 3$ satisfying*

($\alpha$) *each coatom has a complementary coatom.*

*Then exactly one of the following alternatives holds:*

($\beta$) *$L$ is an affine geometry, or*

($\gamma$) *$L$ contains $K_6$ as a sublattice.*

Proof. By 2.1 ($\beta$) and ($\gamma$) are not both true. Assume ($\gamma$) fails. Then $L$ is the lattice of flats of an incidence space in which (P1) is valid. It remains to be shown that through any point not on a line there is at least one parallel to the line. We use a 'reverse induction' procedure to do this.

By ($\alpha$) any hyperplane has a hyperplane parallel to it. If $x \in L$ is such that $d(x) = n - 2$ (where $d(1)$, the height of $L$, is $n$), then we take $x \vee q = h$ for some atom $q$, and let $p$ be an atom such that $p \wedge h = 0$. Then $p \vee x$ is a hyperplane $h'$. Let $h'' \parallel h'$. If $h'' \parallel h$, then we would have $K_6$ as follows:

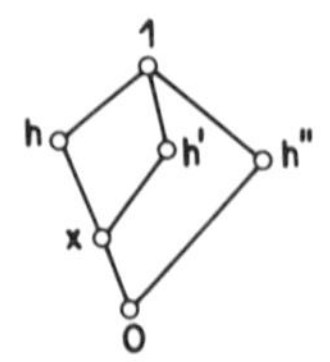


Thus $h'' \wedge h = y \neq 0$, and by (1) $d(y) = n - 2$. Furthermore $x \wedge y = 0$ since $x \leq h'$ and $y \leq h''$. But $x \vee y \leq h$, and since $x \neq y$, we must have $x \vee y = h$. Thus $x \parallel y$.

Now let $r$ be an atom such that $r \wedge x = 0$. If $r \wedge (x \vee y) = 0$, $d((r \vee x) \wedge (r \vee y)) = n - 2$. Let $z = (r \vee x) \wedge (r \vee y)$. If $t$ is an atom under $(r \vee y) \wedge x$, we have $r \leq t \vee y \leq x \vee y$, a contradiction. Hence $(r \vee y) \wedge x = 0$, so $z \wedge x = 0$. But $z \vee x = [(r \vee x) \wedge (r \vee y)] \vee x = (r \vee x) \wedge (r \vee y \vee x) = r \vee x$, which covers $x$ (and therefore also $z$). Hence $z \parallel x$.

If $r \leq x \vee y$, we can choose $r'$ such that $r' \wedge (x \vee y) = 0$, and obtain $z'$ such that $r' \leq z'$ and $z' \parallel x$, where $d(z') = n - 2$. Then $r \wedge (z' \vee x) = 0$, so we can find $z$ such that $r \leq z$, $d(z) = n - 2$, and $z \parallel x$ as above.

We therefore have:

(*) *Given $x$ of height $n - 2$, and $r$ an atom such that $r \wedge x = 0$, there exists $z$ such that $r \leq z$, $d(z) = n - 2$ and $z \parallel x$.*

Assuming that (*) holds for $d(x) = n - k$, we prove (*) for $d(x) = n - (k+1)$, by the same method as above. Let $x \vee q = h$ (not now a coatom), $p \wedge h = 0$, and $x \vee p = h'$. By (*) there is $h''$ such that $q \leq h''$ and $h'' \parallel h'$ with $d(h') = n - k$. Again using $K_6$ as above, we get $h'' \wedge h \neq 0$, and $(h'' \wedge h) \parallel x$. The same argument used above gives the existence of $z \geq r$ such that $z \parallel x$ and $d(z) = d(x)$, completing the proof of (*) for $d(x) = n - (k+1)$.

Finally we obtain (*) for $d(x) = 2$, or the 'existence part' of the parallel axiom, which was all we needed for $L$ to be the lattice of flats of a finite dimensional affine geometry.

The characterization of finite dimensional affine geometries follows immediately:

2.3 THEOREM. *$L$ is an $n$-dimensional affine geometry* ($n>3$) *if and only if*

(1) *$L$ is a Hilbert lattice of height $n > 3$,*

(2) *each coatom has a complementary coatom, and*

(3) *$L$ does not contain $K_6$ as a sublattice.*

## 3. CONCLUDING REMARKS

Since any Hilbert lattice with no sublattice $K_6$ is the lattice of flats of an incidence geometry in which (P1) holds, we shall call such a lattice a *semi-affine* geometry. Given a complemented modular lattice $M$ of finite height, and a proper order ideal $S \subseteq M$, we may form a *Wilcox lattice* ([M2], p. 12) $L = (M \setminus S) \cup \{\overline{0}\}$ whose join operation is that of $M$ and whose meets are given by $x \wedge_L y = 0$ if $x \wedge_M y \in S$ and $x \wedge_L y = x \wedge_M y$ otherwise. Any finite-dimensional affine geometry is a Wilcox lattice (where $M$ is a projective geometry and $S$ the principal ideal generated by a coatom). It follows from some results in Maeda ([M2], p. 93) that any semi-affine geometry of height $\geq 4$ is a Wilcox lattice. Conversely, it can be shown that $K_6$ cannot be a sublattice of any Wilcox lattice; hence we have:

3.1 THEOREM. *Let $L$ be a lattice of height $n \geq 4$. Then $L$ is a Wilcox lattice if and only if $L$ is a semi-affine geometry.*

More general work on embedding geometric lattices into complemented modular lattices has been done by R. Wille [*Proc. Perugia Conf.* (A. Barlotti, ed.), 1970] under the hypothesis that $[x,1]$ is modular for any $x$ of height $n - 4$.

We conclude by remarking that our characterization of affine geometries may be extended to include the infinite dimensional case as well. The main problem in doing so is that the induction argument in 2.2 will not go through; thus (2) of 2.3 cannot be used. It can be replaced, however, by either one of two stronger conditions, the first of which is:

(2') *If $a \wedge b = 0$, there exist $a_1 \geq a, b_1 \geq b$ with $a_1 \| b_1$ and $a \vee b = a_1 \vee b_1$ .*

We shall use the second replacement in our characterization theorem. Since the affine hull of a subset $X$ in a vector space $V$ consists of all *finite* affine combinations of the elements in $X$, in geometries of infinite length we need to have *compact atomisticity*, i.e. if $p$ and $\{p_i\}_{i \in I}$ are atoms with $p \le \bigvee_I p_i$, there is a finite subset $F \subseteq I$ with $p \le \bigvee_F p_i$. We are now in position to characterize affine geometries of arbitrary dimension.

3.2 THEOREM. *Let $L$ be a complete compactly atomistic lattice such that $M(x,y)$ ($x$ and $y$ form a modular pair) implies $M(y,x)$ in $L$. Then $L$ is an affine geometry if and only if*

(1) $x \wedge y \neq 0$ *implies* $M(x,y)$,

(2) $M(x,y)$ *for all* $y \in L$ *implies* $x = 0$, $x = 1$ *or* $x$ *is an atom,*

(3) $L$ *does not contain* $K_6$ *as a sublattice.*

FOOTNOTE

[1] We use 'geometry' to refer to a lattice of flats, as opposed to a synthetic 'space'.

REFERENCES

[A1] E. Artin (1957) *Geometric Algebra*, Interscience, New York.

[B1] G. Birkhoff (1935) Combinatorial relations in projective geometries, *Ann. of Math.* 36, 743-748.

[B2] G. Birkhoff (1967) *Lattice Theory*, Third ed., Amer. Math. Soc., Providence.

[G1] S. Gorn (1940) On incidence geometry, *Bull. Amer. Math. Soc.* 46, 158-167.

[H1] D. Hilbert (1902) *Foundations of Geometry*, Transl. by E.J. Townsend, Open Court, La Salle, Ill. Reprinted, 1950.

[M1] K. Menger (1936) New foundations of projective and affine geometry, *Ann. of Math.* (2), 37, 456-482.

[M2] F. Maeda and S. Maeda (1970) *Theory of Symmetric Lattices*, Springer-Verlag, New York.

[M3] R.N. McKenzie (1972) Equational bases and nonmodular lattice varieties, *Trans. Amer. Math. Soc.* (174), 1-43.

[S1] U. Sasaki (1953) Lattice theoretic characterization of geometries satisfying "Axiome der Verknüpfung", *Hiroshima Math. J. Ser. A*, 16, 417-423.

[S2] U. Sasaki (1952) Lattice theoretic characterization of affine geometry of arbitrary dimension, *Hiroshima Math. J. Ser. A*, 16, 223-238.

[S3] E. Snapper and R. Troyer (1971) *Metric Affine Geometry*, Academic Press, New York.

[W1] O. Wyler (1953) Incidence Geometry, *Duke Math. J.* 20, 601-610.

# RESTRUCTURING LATTICE THEORY: AN APPROACH BASED ON HIERARCHIES OF CONCEPTS

Rudolf Wille
Fachbereich Mathematik
Technische Hochschule Darmstadt
6100 Darmstadt
Federal Republic of Germany

ABSTRACT

Lattice theory today reflects the general status of current mathematics: there is a rich production of theoretical concepts, results, and developments, many of which are reached by elaborate mental gymnastics; on the other hand, the connections of the theory to its surroundings are getting weaker and weaker, with the result that the theory and even many of its parts become more isolated. Restructuring lattice theory is an attempt to reinvigorate connections with our general culture by interpreting the theory as concretely as possible, and in this way to promote better communication between lattice theorists and potential users of lattice theory.

The approach reported here goes back to the origin of the lattice concept in nineteenth-century attempts to formalize logic, where a fundamental step was the reduction of a concept to its "extent". We propose to make the reduction less abstract by retaining in some measure the "intent" of a concept. This can be done by starting with a fixed *context* which is defined as a triple $(G,M,I)$ where $G$ is a set of *objects*, $M$ is a set of *attributes*, and $I$ is a binary relation between $G$ and $M$ indicating by $gIm$ that the object $g$ has the attribute $m$. There is a natural Galois connection between $G$ and $M$ defined by $A' = \{m \in M \mid gIm$ for all $g \in A\}$ for $A \subseteq G$ and $B' = \{g \in G \mid gIm$ for all $m \in B\}$ for $B \subseteq M$. Now, a *concept* of the context $(G,M,I)$ is introduced as a pair $(A,B)$ with $A \subseteq G$, $B \subseteq M$, $A' = B$, and $B' = A$, where $A$ is called the *extent* and $B$ the *intent* of the concept $(A,B)$. The *hierarchy of concepts* given by the relation subconcept-superconcept is captured by the definition $(A_1,B_1) \leq (A_2,B_2) \Leftrightarrow A_1 \subseteq A_2 (\Leftrightarrow B_1 \supseteq B_2)$ for concepts $(A_1,B_1)$ and $(A_2,B_2)$ of $(G,M,I)$. Let $L(G,M,I)$ be the



*I. Rival (ed.), Ordered Sets, 445–470.*

set of all concepts of $(G,M,I)$. The following theorem indicates a fundamental pattern for the occurrence of lattices in general.

THEOREM: *Let* $(G,M,I)$ *be a context. Then* $(\mathfrak{L}(G,M,I),\leq)$ *is a complete lattice* (called the *concept lattice* of $(G,M,I)$) *in which infima and suprema can be described as follows:*

$$\bigwedge_{i\in J}(A_i,B_i) = \left(\bigcap_{i\in J}A_i, \left(\bigcap_{i\in J}A_i\right)'\right),$$

$$\bigvee_{i\in J}(A_i,B_i) = \left(\left(\bigcap_{i\in J}B_i\right)', \bigcap_{i\in J}B_i\right).$$

*Conversely, if* $L$ *is a complete lattice then* $L \cong (\mathfrak{L}(G,M,I),\leq)$ *if and only if there are mappings* $\gamma : G \to L$ *and* $\mu : M \to L$ *such that* $\gamma G$ *is supremum-dense in* $L$, $\mu M$ *is infimum-dense in* $L$, *and* $gIm$ *is equivalent to* $\gamma g \leq \mu m$ *for all* $g \in G$ *and* $m \in M$*; in particular,* $L \cong (\mathfrak{L}(L,L,\leq),\leq)$.

Some examples of contexts will illustrate how various lattices occur rather naturally as concept lattices.

(i) $(S,S,\neq)$ where $S$ is a set.
(ii) $(\mathbb{N},\mathbb{N},1)$ where $\mathbb{N}$ is the set of all natural numbers.
(iii) $(V,V^*,\perp)$ where $V$ is a finite-dimensional vector space.
(iv) $(\mathcal{V},\mathrm{Eq}(\mathcal{V}),\models)$ where $\mathcal{V}$ is a variety of algebras.
(v) $(G\times G,\mathbb{R}^G,\sim)$ where $G$ is a set of objects, $\mathbb{R}^G$ is the set of all real-valued functions on $G$, and $(g_1,g_2) \sim \alpha$ iff $\alpha g_1 = \alpha g_2$.

Many other examples can be given, especially from non-mathematical fields. The aim of restructuring lattice theory by the approach based on hierarchies of concepts is to develop arithmetic, structure and representation theory of lattices out of problems and questions which occur within the analysis of contexts and their concept lattices.

## 1. RESTRUCTURING LATTICE THEORY

Lattice theory today reflects the general status of current mathematics: there is a rich production of theoretical concepts, results, and developments, many of which are reached by elaborate mental gymnastics; on the other hand, the connections of the theory to its surroundings are getting weaker: the result is that the theory and even many of its parts become more isolated. This is not only a problem of lattice theory or of mathematics. Many sciences suffer from this effect of specialization. Scientists and philosophers are, of course, aware of the danger of this growing isolation. In [17], H. von Hentig extensively discusses the status of the humanities and sciences today. One consequence is his charge to "restructure" theoretical developments in order to integrate and rationalize origins, connections, interpretations, and applications. In particular, abstract developments should be brought back to the commonplace in perception, thinking, and action. Thus, *restructuring lattice theory* is understood as an attempt to unfold lattice-theoretical concepts, results, and methods in a continuous relationship with their surroundings (cf. Wille [40], [41]). One basic aim is to promote better communication between lattice theorists and potential users of lattice theory.

The important current monographs on lattice theory are by G. Birkhoff [3], P. Crawley and R.P. Dilworth [8], and G. Grätzer [15]. They are primarily concerned with "systematically developing a body of results at the heart of the subject" [8, p. V]. In an approach to restructure lattice theory we are first confronted with the question: *Why develop lattice theory*? G. Birkhoff in [3] refers to an earlier article "What can lattices do for you?" [4], where he attempts to rationalize the statement: "lattice theory has helped to simplify, unify and generalize mathematics, and it has suggested many interesting problems". In his first survey paper on "lattices and their applications" [2], G. Birkhoff set up a more general viewpoint for lattice theory: "lattice theory provides a proper vocabulary for discussing order, and especially systems which are in any sense hierarchies". Notwithstanding this broad point of view, it was the abstract level of accepting lattice theory as a theory of "substructures" in mathematics which brought the breakthrough in the 1930's and established lattice theory as a mathematical discipline (cf. Mehrtens [24]).

The approach to lattice theory outlined in this paper is based on an attempt to reinvigorate the general view of order. For this purpose we go back to the origin of the lattice concept in nineteenth-century attempts to formalize logic, where the study of hierarchies of concepts played a central rôle (cf. Schröder [35]). Traditional philosophy considers a *concept* to be determined by its "extent" [extension] and its "intent" [intension, comprehension]: the *extent* consists of all objects (or entities) belonging to the concept while the *intent* is the multitude of all attributes (or properties) valid for all those objects (cf. Wagner [36]). The *hierarchy* of concepts is given by the relation of "subconcept" to "superconcept", i.e. the extent of the subconcept is contained in the extent of the superconcept while, reciprocally, the intent of the superconcept is contained in the intent of the subconcept. Paradigmatic is the example: "human being" is a subconcept of the superconcept "being". Listing all objects and attributes belonging to a given concept is almost never possible. For practical reasons then, there is the proposal to limit a "context" by fixing the set of objects and the set of attributes (cf. Deutsches Institut für Normung [11], [12]). In set-theoretical language, this gives rise to lattices whose elements correspond to the concepts of the context and whose order comes from the hierarchy of concepts. The approach reported here takes these "concept lattices" as the basis and discusses how parts of arithmetic, structure and representation theory of lattices may be developed out of problems and questions which occur within the analysis of contexts and their hierarchies of concepts.

## 2. CONCEPT LATTICES

We start defining a *context* as a triple $(G,M,I)$ where $G$ and $M$ are sets, and $I$ is a binary relation between $G$ and $M$; the elements of $G$ and $M$ are called *objects* [Gegenstände] and *attributes* [Merkmale], respectively. If $gIm$ for $g \in G$ and $m \in M$ we say: the object $g$ *has* the attribute $m$. For $A \subseteq G$ and $B \subseteq M$ we define

$$A' := \{m \in M \mid gIm \text{ for all } g \in A\},$$
$$B' := \{g \in G \mid gIm \text{ for all } m \in B\}.$$

The mappings given by $A \mapsto A'$ and $B \mapsto B'$ are said to form a *Galois connection* between the power sets of $G$ and $M$, i.e. they fulfill the following basic properties (cf. Birkhoff [3; pp. 122-125]).

PROPOSITION. *For a context* $(G,M,I)$:

(1) $A_1 \subseteq A_2$ *implies* $A_1' \supseteq A_2'$ *for* $A_1, A_2 \subseteq G$,

(1') $B_1 \subseteq B_2$ *implies* $B_1' \supseteq B_2'$ *for* $B_1, B_2 \subseteq M$,

(2) $A \subseteq A''$ *and* $A' = A'''$ *for* $A \subseteq G$,

(2') $B \subseteq B''$ *and* $B' = B'''$ *for* $B \subseteq M$.

Now, a *concept* [Begriff] of the context $(G,M,I)$ may be defined as a pair $(A,B)$ where $A \subseteq G$, $B \subseteq M$, $A' = B$, and $B' = A$; $A$ and $B$ are called the *extent* and the *intent* of the concept $(A,B)$, respectively. The hierarchy of concepts is captured by the definition

$$(A_1,B_1) \leq (A_2,B_2) :\Leftrightarrow A_1 \subseteq A_2 \quad (\Leftrightarrow B_1 \supseteq B_2)$$

for concepts $(A_1,B_1)$ and $(A_2,B_2)$ of $(G,M,I)$; $(A_1,B_1)$ is called the *subconcept* of $(A_2,B_2)$, and $(A_2,B_2)$ is called the *superconcept* of $(A_1,B_1)$. To formulate the basic theorem (cf. Banaschewski [1], Schmidt [34]) which indicates a fundamental pattern for the occurrence of lattices in general, we need some further definitions. Let $\mathfrak{B}(G,M,I)$ be the set of all concepts of the context $(G,M,I)$ and let $\underline{\mathfrak{B}}(G,M,I) := (\mathfrak{B}(G,M,I),\leq)$. A subset $D$ of a complete lattice $L$ is called *infimum-dense* (*supremum-dense*) if $L = \{\bigwedge X \mid X \subseteq D\}$ $(L = \{\bigvee X \mid X \subseteq D\})$.

THEOREM. *Let $(G,M,I)$ be a context. Then $\underline{\mathfrak{B}}(G,M,I)$ is a complete lattice, called the* concept lattice *[Begriffsverband] of $(G,M,I)$ in which infima and suprema can be described as follows:*

$$\bigwedge_{j\in J} (A_j,B_j) = \Big(\bigcap_{j\in J} A_j, \big(\bigcap_{j\in J} A_j\big)'\Big),$$

$$\bigvee_{j\in J} (A_j,B_j) = \Big(\big(\bigcap_{j\in J} B_j\big)', \bigcap_{j\in J} B_j\Big).$$

*Conversely, if $L$ is a complete lattice then $L \cong \underline{\mathfrak{B}}(G,M,I)$ if and only if there are mappings $\gamma: G \to L$ and $\mu: M \to L$ such that $\gamma G$ is supremum-dense in $L$, $\mu M$ is infimum-dense in $L$, and $gIm$ is equivalent to $\gamma g \leq \mu m$ for all $g \in G$ and $m \in M$; in particular, $L \cong \underline{\mathfrak{B}}(L,L,\leq)$.*

PROOF. Obviously, the relation $\leq$ is an order on $\mathfrak{B}(G,M,I)$. Let $(A_j,B_j) \in \mathfrak{B}(G,M,I)$ for $j \in J$. By the preceding proposition, $\bigcap_{j\in J} A_j \subseteq (\bigcap_{j\in J} A_j)''$ and $(\bigcap_{j\in J} A_j)'' \subseteq A_k'' = A_k$ for each $k \in J$ wherefore $(\bigcap_{j\in J} A_j)'' = \bigcap_{j\in J} A_j$ and hence $(\bigcap_{j\in J} A_j, (\bigcap_{j\in J} A_j)') \in \mathfrak{B}(G,M,I)$. If $(A,B) \in \mathfrak{B}(G,M,I)$ such that $(A,B) \leq (A_j,B_j)$ for all $j \in J$, i.e. $A \subseteq A_j$ for all $j \in J$, then $A \subseteq \bigcap_{j\in J} A_j$ and hence $(A,B) \leq (\bigcap_{j\in J} A_j, (\bigcap_{j\in J} A_j)')$. Thus, $(\bigcap_{j\in J} A_j, (\bigcap_{j\in J} A_j)')$ is the infimum of the $(A_j,B_j)$ $(j \in J)$. The proof for the supremum goes analogously. Now, let $\varphi$ be an isomorphism from $\underline{\mathfrak{B}}(G,M,I)$ onto a complete lattice $L$. We define $\gamma g := \varphi(\{g\}'',\{g\}')$ for $g \in G$ and $\mu m := \varphi(\{m\}',\{m\}'')$ for $m \in M$. Since $A = \bigcup_{g\in A} \{g\}''$ and $B = \bigcup_{m\in B} \{m\}''$ for every $(A,B) \in \mathfrak{B}(G,M,I)$, $\gamma G$ is supremum-dense in $L$ and $\mu M$ is infimum-dense in $L$, respectively. Furthermore, $gIm \Leftrightarrow \{g\}'' \subseteq \{m\}' \Leftrightarrow \gamma g \leq \mu m$. Conversely, let $L$ be a complete lattice and let $\gamma: G \to L$ and $\mu: M \to L$ arbitrary mappings having the desired properties. We define a mapping $\varphi: \underline{\mathfrak{B}}(G,M,I) \to L$ by $\varphi(A,B) := \bigvee\gamma A$ for all $(A,B) \in \underline{\mathfrak{B}}(G,M,I)$. Obviously, $\varphi$ is order-preserving. For $x \in L$, let $\psi x := (\{g \in G | \gamma g \leq x\}, \{m \in M | x \leq \mu m\})$. Since $\gamma G$ is supremum-dense in $L$ and $\mu M$ is infimum-dense in $L$, $\bigvee\gamma\{g \in G | \gamma g \leq x\} = x = \bigwedge\mu\{m \in M | x \leq \mu m\}$. Therefore, because of $gIm \Leftrightarrow \gamma g \leq \mu m$, $\psi x$ is a concept of $(G,M,I)$ and $\varphi\psi x = x$. Clearly, $(A,B) \leq \psi\varphi(A,B)$. As $\gamma g \leq \bigvee\gamma A \leq \mu m$ always implies $gIm$, we even have $(A,B) = \psi\varphi(A,B)$. Furthermore, $\psi$ is order-preserving. Now, we can summarize: $\varphi$ is an isomorphism from $\underline{\mathfrak{B}}(G,M,I)$ onto $L$ and $\varphi^{-1} = \psi$. If we choose $G := L$, $M := L$, $I := \leq$, $\gamma$ and $\mu$ as the identity on $L$, we get $L \cong \underline{\mathfrak{B}}(L,L,\leq)$.

For a finite lattice $L$ with $J(L)$ as its set of $\vee$-irreducible elements and $M(L)$ as its set of $\wedge$-irreducible elements, the theorem yields $L \cong \underline{\mathfrak{B}}(J(L),M(L),\leq)$, and the context $(J(L),M(L),\leq)$ is smallest with the property that its concept lattice is isomorphic to $L$. In general, the proposition and the theorem makes apparent a dual situation which often gets a concrete meaning by interchanging objects and attributes. Formally, we consider $(M,G,I^{-1})$ as the *dual context* of the context $(G,M,I)$ where the mapping given by $(A,B) \mapsto (B,A)$ is an antiisomorphism from $\underline{\mathfrak{B}}(G,M,I)$ onto $\underline{\mathfrak{B}}(M,G,I^{-1})$.

## 3. EXAMPLES

Some examples may illuminate how contexts give rise to concept lattices. The first example, a context of our planets, is described by the following table (cf. [25]) in which crosses indicate the relation between the objects and the attributes; the concept lattice is represented by its Hasse diagram where the labels indicate the mappings $\gamma$ and $\mu$ of the preceding theorem.

| | size | | | distance from sun | | moon | |
|---|---|---|---|---|---|---|---|
| | small | medium | large | near | far | yes | no |
| Mercury | × | | | × | | | × |
| Venus | × | | | × | | | × |
| Earth | × | | | × | | × | |
| Mars | × | | | × | | × | |
| Jupiter | | | × | | × | × | |
| Saturn | | | × | | × | × | |
| Uranus | | × | | | × | × | |
| Neptune | | × | | | × | × | |
| Pluto | × | | | | × | × | |

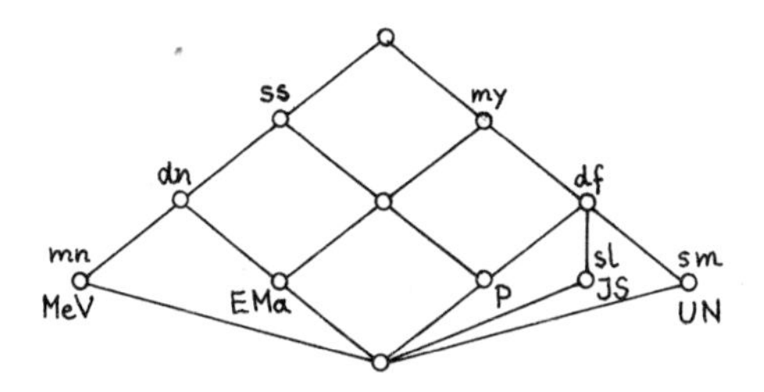


The second example is taken from the German national division of health and welfare (see Podlech [27]). The context indicates which personal information is needed by law for the different tasks of a local medical subdivision; tasks (objects) and information (attributes) are the following: 1. confirmation of requests for preventive care, 2. evaluations of regulations of insurance benefits, 3. confirmation of incapacity to work to insure success of treatment, 4. confirmation of correct diagnosis of incapacity to work, 5. work preliminary to rehabilitation, 6. verification of sickness, 7. advice to clients, 8. advice on general preventives, 9. carrying on the statistics of the medical service, a. name and address of the client ..., b. career history ..., c. kind of membership ..., d. name of responsible agency ..., e. family medical history, f. vocational education ..., g. number of certificates.

| | a | b | c | d | e | f | g |
|---|---|---|---|---|---|---|---|
| 1 | × | × | × | × | × | × | |
| 2 | × | × | × | × | × | × | |
| 3 | × | × | × | × | × | × | |
| 4 | × | × | × | × | | | |
| 5 | × | × | × | × | × | × | |
| 6 | × | | × | × | | × | |
| 7 | × | × | × | | × | × | × |
| 8 | | | | | | | |
| 9 | | | × | | | | × |

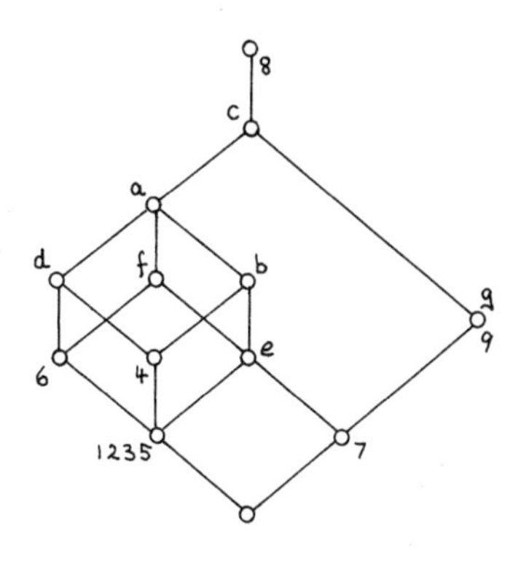


The concept lattice of a task-information-context $(G,M,I)$ gives a natural hierarchical classification of the tasks and makes apparent the dependencies between information. But for certain purposes the concept lattice of the *complementary context* $(G,\overline{M},\overline{I})$ with $\overline{M} := \{\overline{m} | m \in M\}$ and $g\overline{I}\overline{m} :\Leftrightarrow \neg(gIm)$ is more suitable, for instance, if one wants to assign the tasks to different parties such that the access to personal information is mostly limited. Both lattices are combined in the concept lattice of the *dichotomic context* $(G, M\dot{\cup}\overline{M}, I\dot{\cup}\overline{I})$, and it might be interesting to analyse how this lattice is dependent on the two smaller ones (cf. proposition in Section 4). Here we only present the Hasse diagrams with respect to the context above.

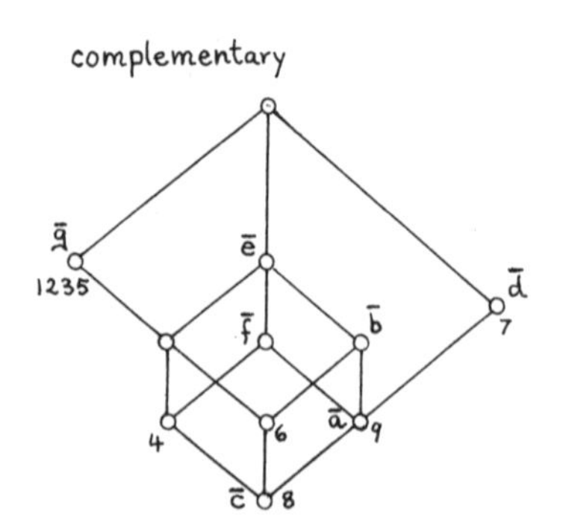


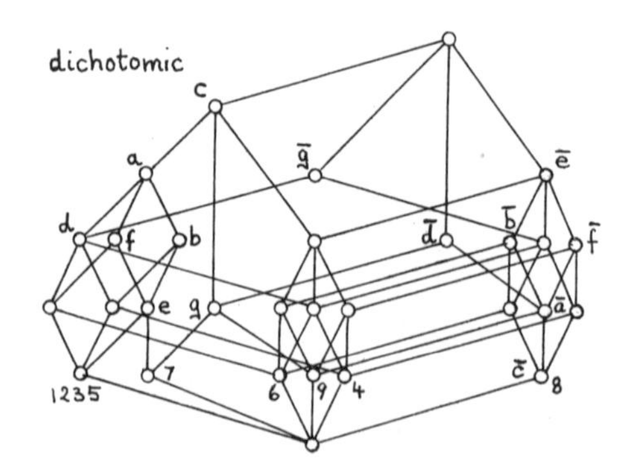


Experience has shown that concept lattices of "empirical" contexts which occur in rich multitude in many fields usually do not belong to the mostly studied classes of lattices as those of distributive, modular, semimodular, orthomodular etc. lattices. The situation becomes different if we consider contexts in mathematics (cf. Birkhoff [3]). For a set $S$, the concept lattice of $(S,S,\neq)$ is isomorphic to the Boolean lattice of all subsets of $S$ since $(A,S\setminus A)$ is a concept of $(S,S,\neq)$ for every $A \subseteq S$. If $|$ is the divisibility relation on the set $\mathbb{N}$ of all natural numbers, $\underline{\mathfrak{B}}(\mathbb{N},\mathbb{N},|)$ is distributive. Let $V$ be a finite-dimensional vector space, let $V^*$ be its dual space, and let $a \perp \varphi :\Leftrightarrow \varphi a = 0$ for $a \in V$ and $\varphi \in V^*$. Then $\underline{\mathfrak{B}}(V,V^*,\perp)$ is isomorphic to the complemented modular lattice of all linear subspaces of $V$. If $H$ is a Hilbert space and $\perp$ its orthogonality relation, then $\underline{\mathfrak{B}}(H,H,\perp)$ is isomorphic to the orthomodular lattice of all closed linear subspaces of $H$. For a variety $\mathcal{V}$ of algebras and the set $Eq(\mathcal{V})$ of all equations valid in each algebra of $\mathcal{V}$, the concept lattice of $(\mathcal{V},Eq(\mathcal{V}),\models)$ is isomorphic to the coalgebraic lattice of all subvarieties of $\mathcal{V}$. Many further examples could be given. These examples may suggest an extensive study of special classes of lattices. Here we want to continue with the general approach. We only mention one further class of examples which will be discussed more in Section 8: If $(P,\leq)$ is a (partially) ordered set then $\underline{\mathfrak{B}}(P,P,\leq)$ is up to isomorphism the Dedekind-MacNeille completion of $(P,\leq)$.

## 4. THE DETERMINATION PROBLEM

The concept lattice can be understood as a basic answer to two fundamental questions concerning a context, namely the question of an appropriate classification of the objects and the question about the dependencies between the attributes. Hence an important problem is: *How can one determine the concept lattice of a given context* $(G,M,I)$? We get the most simple answer by forming $(X'',X')$ for all $X \subseteq G$ (or $(Y',Y'')$ for all $Y \subseteq M$). Of course, this method is far too costly and should only be used to determine single concepts. More successful, in particular if no computer is involved, is the approach using the following formulas for $(A,B) \in \mathfrak{B}(G,M,I)$:

$$A = \bigcap_{m \in B} \{m\}' \quad , \qquad\qquad B = \bigcap_{g \in A} \{g\}' \quad .$$

For instance, in the context of our planets one first determines the extents $\{ss\}'$, $\{sm\}'$, $\{sl\}'$, $\{dn\}'$, $\{df\}'$, $\{my\}'$, and $\{mn\}'$; then one gets all the other extents as intersections from those. From the extents it is easy to obtain the intents. For a larger context one may use both formulas simultaneously.

For the use of computers, it would be desirable to have "good" algorithms determining the concept lattice of a concept. Of course, designing those algorithms one has to consider how the concept lattice is handled afterwards. For instance, it might be presented by its Hasse diagram (see Section 4), it might be used to establish a measurement concerning the objects (see Section 8), or it might be a tool for a description of the dependencies between the attributes. In particular, one may elaborate the classification of objects further in the line of cluster analysis (for a recent survey on cluster analysis see Bock [6]): For many contexts one would like to have suitable partitions of the set of objects whose blocks are extents, i.e. they can be described by attributes; first experiences with randomly chosen contexts have shown that there are usually few such partitions if the cardinalities of their blocks are not becoming too small.

In certain situations, the determination problem can be attacked by a structural method which idea is to construct the concept lattice of a context by the concept lattices of some suitable subcontexts. We recall that the order $R$ of the *direct product* of the two ordered sets $(P,R_1)$ and $(P,R_2)$ is defined by $(x_1,x_2)R(y_1,y_2) :\Leftrightarrow x_1R_1y_1$ and $x_2R_2y_2$. The *horizontal sum* of two bounded ordered sets $(P_1,R_1)$ and $(P_2,R_2)$ is obtained from their *cardinal sum* $(P_1\dot{\cup}P_2,R_1\dot{\cup}R_2)$ by identifying the smallest elements and the greatest elements of both ordered sets, respectively. The *vertical sum* of $(P_1,R_1)$ and $(P_2,R_2)$ is obtained from their *ordinal sum* $(P_1\dot{\cup}P_2,R_1\dot{\cup}R_2\dot{\cup}P_1\times P_2)$ by identifying the greatest element of

$(P_1,R_1)$ with the smallest element of $(P_2,R_2)$.

THEOREM. *Let* $(G_1,M_1,I_1)$ *and* $(G_2,M_2,I_2)$ *be contexts with* $G_1 \cap G_2 = \emptyset$, $M_1 \cap M_2 = \emptyset$, $G_i' = \emptyset$ *and* $M_i' = \emptyset$ *for* $i = 1,2$; *let* $L_i := \underline{\mathfrak{B}}(G_i,M_i,I_i)$ *for* $i = 1,2$. *Then*

(1) $\underline{\mathfrak{B}}(G_1 \dot\cup G_2, M_1 \dot\cup M_2, I_1 \dot\cup I_2)$ *is isomorphic to the horizontal sum of* $L_1$ *and* $L_2$,

(2) $\underline{\mathfrak{B}}(G_1 \dot\cup G_2, M_1 \dot\cup M_2, I_1 \dot\cup I_2 \dot\cup G_1 \times M_2)$ *is isomorphic to the vertical sum of* $L_1$ *and* $L_2$,

(3) $\underline{\mathfrak{B}}(G_1 \dot\cup G_2, M_1 \dot\cup M_2, I_1 \dot\cup I_2 \dot\cup G_1 \times M_2 \dot\cup G_2 \times M_1)$ *is isomorphic to the direct product of* $L_1$ *and* $L_2$.

PROOF. (1) Obviously, $(A_i,B_i)$ with $\emptyset \neq A_i \subseteq G_i$ and $\emptyset \neq B_i \subseteq M_i$ is a concept of $(G_1 \dot\cup G_2, M_1 \dot\cup M_2, I_1 \dot\cup I_2)$ if and only if it is a concept of $(G_i,M_i,I_i)$ $(i = 1,2)$. If $A \subseteq G_1 \cup G_2$ and $A \cap G_i \neq \emptyset$ for $i = 1,2$ then $A' = \emptyset$ and hence $A'' = G_1 \cup G_2$; furthermore $\emptyset'' = \emptyset$. This proves (1).
(2) $(G_1 \dot\cup G_2, M_1 \dot\cup M_2, I_1 \dot\cup I_2 \dot\cup G_1 \times M_2)$ has exactly the concepts $(\emptyset, G_1 \cup G_2)$, $(A_1, B_1 \cup M_2)$ with $A_1 \neq \emptyset$ and $(A_1,B_1) \in \underline{\mathfrak{B}}(G_1,M_1,I_1)$, and $(A_2 \cup G_1, B_2)$ with $A_2 \neq \emptyset$ and $(A_2,B_2) \in \underline{\mathfrak{B}}(G_2,M_2,I_2)$.
(3) $(G_1 \dot\cup G_2, M_1 \dot\cup M_2, I_1 \dot\cup I_2 \dot\cup G_1 \times M_2 \dot\cup G_2 \times M_1)$ has exactly the concepts $(A_1 \cup A_2, B_1 \cup B_2)$ with $(A_1,B_1) \in \underline{\mathfrak{B}}(G_1,M_1,I_1)$ and $(A_2,B_2) \in \underline{\mathfrak{B}}(G_2,M_2,I_2)$.

In cases where the preceding theorem cannot be applied, one may still reduce a given context to smaller contexts to apply the following proposition which can be easily verified using (3).

PROPOSITION. *Let* $(G,M,I)$ *be a context, let* $\{G_j \mid j \in J\}$ *be a partition of* $G$, *and let* $\{M_k \mid k \in K\}$ *be a partition of* $M$. *Then*

(4) *a* $\wedge$*-embedding of* $\underline{\mathfrak{B}}(G,M,I)$ *into the direct product of the* $\underline{\mathfrak{B}}(G_j,M,I \cap G_j \times M)$ $(j \in J)$ *is given by* $(A,B) \mapsto (A \cap G_j, (A \cap G_j)')_{j \in J}$,

(4') *a* $\vee$*-embedding of* $\underline{\mathfrak{B}}(G,M,I)$ *into the direct product of the* $\underline{\mathfrak{B}}(G,M_k,I \cap G \times M_k)$ $(k \in K)$ *is given by* $(A,B) \mapsto ((B \cap M_k)', B \cap M_k)_{k \in K}$.

One hint is to apply (4) and (4') several times to reach subcontexts whose concept lattices can be easily determined; then in the direct product of these lattices one may find the images of the concepts of the original context under the order embedding given by (4) and (4'). The concept lattice of a context $(G,M,I)$ is a chain if and only if $G$ and $M$ can be linearly ordered such that $I$ becomes an order filter in $G \times M$. Thus, in the light of the described procedure, it might be a challenging problem to find for a given context $(G,M,I)$ the smallest number of applications of (4) and (4') which yields an order embedding of $\underline{\mathfrak{B}}(G,M,I)$ into a direct product of (two-element) chains.

## 5. THE DESCRIPTION PROBLEM

Connected with the description problem is the basic problem: *How can one describe the concept lattice of a given context* $(G,M,I)$? "Good" solutions of this problem are helpful for further analysis and communication, especially with non-mathematicians. The most communicative description is given by Hasse diagrams. If we label in a Hasse diagram of $\underline{\mathfrak{B}}(G,M,I)$ the circle representing $(\{g\}'',\{g\}')$ by a name for $g$ $(g \in G)$ and the circle representing $(\{m\}',\{m\}'')$ by a name for $m$ $(m \in M)$, as in the examples in Section 3, the diagram retains the full information about $(G,M,I)$ because $gIm$ is equivalent to $(\{g\}'',\{g\}') \leq (\{m\}',\{m\}'')$ by the theorem in Section 2. Although Hasse diagrams are frequently used, it is surprising that there is no "theory of readable Hasse diagrams". Drawing a Hasse diagram is still more an art than a mechanical skill that even a computer can work out. Because of the absence of a theory, we shall only discuss the development of a Hasse diagram by an example.

As an example, we consider the context of human Rhesus blood types by Wiener and Wechsler (see Diem, Lentner [13]). Its objects are the 28 Rhesus phenotypes which have been established up to 1956, and its attributes are the following seven test serums: a. Anti-$Rh_0$, b. Anti-rh', c. Anti-rh", d. Anti-$rh'^{w}$, e. Anti-hr', f. Anti-hr", g. Anti-hr. The symbol + in the table below indicates when a blood type reacts positively with a test serum; no reaction is indicated by the symbol –. We analyse the context as a dichotomic context (cf. Section 3), i.e. we split each attribute $x$ into the two attributes $x+$ and $x-$. The concept lattice of the dichotomic context is represented on the next page by a (partial) Hasse diagram in which several line segments are left out to allow easier reading. Besides the circle representing the least element, the diagram consists of three congruent components. If one moves the component above to one of the components below parallel along the line segment joining the top circles, then the traces of the circles will just add the missing line segments between the two components. Analogously, one has to move the sub-components in each component parallel along the joining line segments; but the trace of a circle is a missing line segment only if the circle reaches another. To get more familiar with the diagram, one may do the exercise to find

| | a | b | c | d | e | f | g |
|---|---|---|---|---|---|---|---|
| 1 | – | – | – | – | + | + | + |
| 2 | – | + | – | – | + | + | + |
| 3 | – | + | – | – | – | + | – |
| 4 | – | + | – | + | + | + | + |
| 5 | – | + | – | + | – | + | – |
| 6 | – | – | + | – | + | + | + |
| 7 | – | – | + | – | + | – | – |
| 8 | – | + | + | – | + | + | – |
| 9 | – | + | + | – | + | + | + |
| 10 | – | + | + | – | – | + | – |
| 11 | – | + | + | – | + | – | – |
| 12 | – | + | + | – | – | – | – |
| 13 | – | + | + | + | + | + | – |
| 14 | – | + | + | + | – | + | – |
| 15 | + | – | – | – | + | + | + |
| 16 | + | + | – | – | + | + | + |
| 17 | + | + | – | – | – | + | – |
| 18 | + | + | – | + | + | + | + |
| 19 | + | + | – | + | – | + | – |
| 20 | + | – | + | – | + | + | + |
| 21 | + | – | + | – | + | – | – |
| 22 | + | + | + | – | + | + | – |
| 23 | + | + | + | – | + | + | + |
| 24 | + | + | + | – | – | + | – |
| 25 | + | + | + | – | + | – | – |
| 26 | + | + | + | – | – | – | – |
| 27 | + | + | + | + | + | + | – |
| 28 | + | + | + | + | – | + | – |

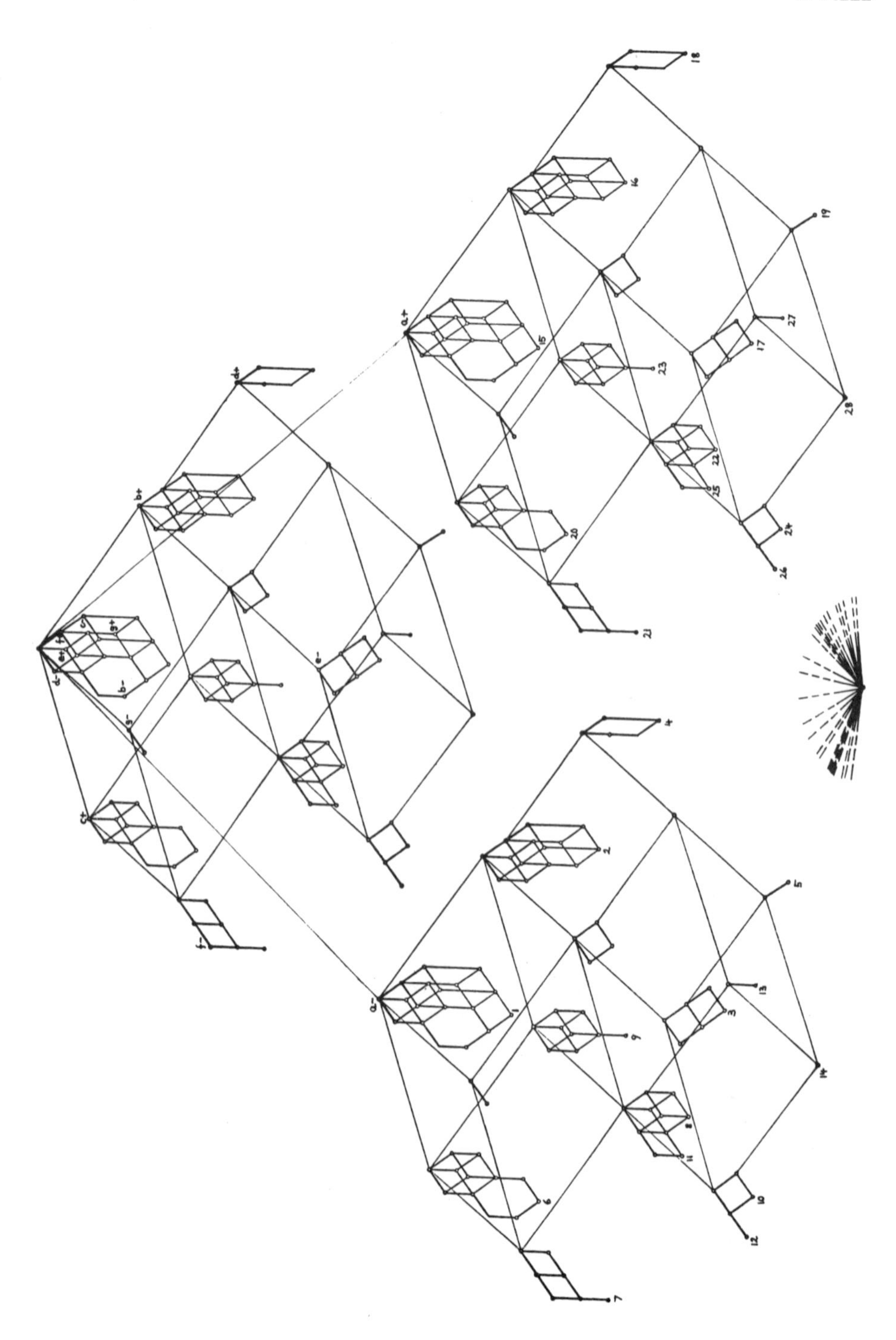
d+
b+
a+
a-
c+
f-
e-
g-
b-
1
2
3
4
5
6
7
8
9
10
11
12
13
14
15
16
17
18
19
20
21
22
23
24
25
26
27
28

out which attributes are dependent on $\{a-,c+,g+\}$, i.e. contained in $\{a-,c+,g+\}''$ (take the meet of the elements labelled by $a-$, $c+$, and $g+$, and list all labelled elements above this meet: the result will be $\{a-,c+,g+\}'' = \{a-,c+,d-,e+,f+,g+\}$). The development of the Hasse diagram representing 286 elements has followed the scheme: Search for a suitable chain $L\times L =: R_0 \supset R_1 \supset R_2 \supset \ldots \supset R_n := id_L$ of equivalence relations with convex equivalence classes on the lattice $L$ such that $[a]R_{i-1}/R_i$ has dimension $\leq 3$ for all $a \in L$ and $i = 1,\ldots,n$. In our example we have $n = 3$; $L/R_1$ and $[1]R_1/R_2$ may be represented by the following Hasse diagrams:

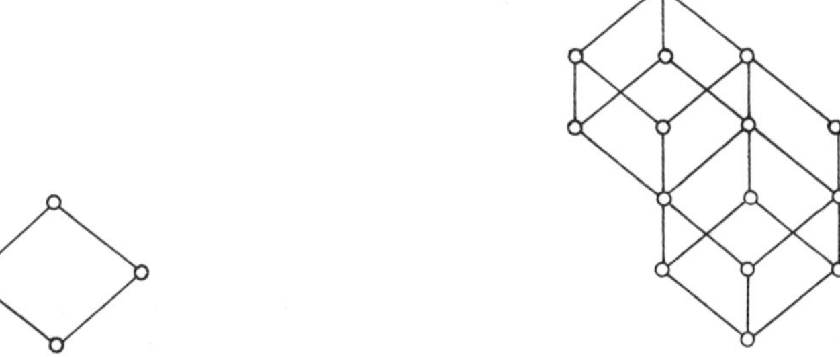

Besides the description by Hasse diagrams, there are other useful methods for describing concept lattices. For instance, a concept lattice might be completely determined by some relative sublattice (e.g. its scaffolding) which can be diagrammed (cf. Wille [39]). A rich multitude of descriptions arises from the general idea of measurement which especially leads to numerical representations; this will be discussed further in Section 7.

## 6. FROM STRUCTURES TO CONCEPT LATTICES

Data models are often not contexts as defined in Section 2, i.e. relational structures with only unary relations; they frequently use many-placed relations and operations. Nevertheless, it is also desirable to have hierarchies of concepts for arbitrary relational structures available. For this, the general scheme is to derive (unary) attributes from the given structure and to form the concept lattice of the context consisting of the elements of the structure and these derived attributes. For instance, an $n$-ary relation $R$ on a set $G$ may be resolved into unary relations defined by $R_{(g_1,\ldots,g_{i-1},g_{i+1},\ldots,g_n)}(x) :\Leftrightarrow R(g_1,\ldots,g_{i-1},x,g_{i+1},\ldots,g_n)$ where $(g_1,\ldots,g_{i-1},g_{i+1},\ldots,g_n) \in G^{n-1}$ and $i \in \{1,\ldots,n\}$. The derivation always depends on which concepts will be of interest with respect to the data. In this section we restrict our considerations to structures of "nominal character"; further structures are discussed in Section 7.

In many situations, the term "attribute" has an even more

general meaning, i.e. an attribute may allow several values instead of one. A person may have any one of several values for the attribute "profession", for instance, "artist", "bus driver", "carpenter", etc.; there are also people with no profession (no value). We capture the general case by the definition of a *many-valued context* which is a quadruple $(G,M,W,I)$ such that $G$, $M$, and $W$ are sets, and $I$ is a binary relation between $G$ and $M \times W$ where $gI(m,w_1)$ and $gI(m,w_2)$ always imply $w_1 = w_2$; the elements of $G$, $M$, and $W$ are called *objects*, *(many-valued) attributes*, and *attribute values*, respectively. If $gI(m,w)$ for $g \in G$, $m \in M$, and $w \in W$ we say: the object $g$ *has* value $w$ for the attribute $m$. If $|W| = n$ then $(G,M,W,I)$ is called an *$n$-valued context*. A one-valued context can be considered as a context as defined in Section 2. Now, in the case where the attribute values are understood as nominal data, we suggest the derivation to "one-valued" attributes by just taking $(G,M \times W,I)$ as the derived context; let us indicate this case by calling $(G,M,W,I)$ a *nominal context*. Hence, for a nominal context $(G,M,W,I)$, its concept lattice is just $\mathfrak{L}(G,M \times W,I)$. In this setting, the context of our planets in Section 3 can be understood as a nominal context with $M :=$ {size, distance from sun, moon} and $W :=$ {small, medium, large, near, far, yes, no}. The dichotomic contexts turn out to be always 2-valued nominal contexts.

It is common to assume that an $n$-valued context $(G,M,W,I)$ is *complete*, i.e. for every $g \in G$ and $m \in M$ there exists a $w \in W$ with $gI(m,w)$; complete $n$-valued contexts are just a translated version of "relational data models" (cf. Date [9]). For attacking the determination problem and the description problem, the knowledge of special properties satisfied by the concept lattice of a complete nominal context is helpful. For a characterization of the concept lattices of complete $n$-valued nominal contexts the following definition is used. An element $d$ of a complete lattice $L$ has *valence* $n$ if $n$ is the smallest cardinality of a subset $D$ of $L \setminus \{0\}$ containing $d$ which is maximal with respect to the property that $x \wedge y = 0$ for all $x,y \in D$ with $x \neq y$. Recall that a lattice is *atomistic* if each lattice element is a supremum of atoms.

THEOREM. *A complete lattice $L$ is isomorphic to a concept lattice of a complete $n$-valued nominal context if and only if $L$ is atomistic and has an infimum-dense subset of elements of valence $\leq n$.*

PROOF. We assume that $L$ has the desired properties. Let $A$ be the set of all atoms of $L$ and let $A(b) := \{a \in A \mid a \leq b\}$ for $b \in L$; furthermore, let $M$ be the set of all elements of valence $\leq n$ in $L$ and let $W$ be a set with $|W| = n$ and $1 \in W$. For each $m \in M$ there must be a subset $W_m$ of $W$ with $1 \in W_m$ and an injective map $\iota_m : W_m \to L$ such that $\iota_m 1 = m$ and $\{A(\iota_m w) \mid w \in W_m\}$ is a partition of $A$. For $a \in A$, $m \in M$, and $w \in W$, we define $aI(m,w)$ if and only if $w \in W_m$

and $a \leq \iota_m w$. Since $\{\iota_m w \mid m \in M$ and $w \in W_m\}$ is infimum-dense in $L$ and $A$ is supremum-dense in $L$, $L$ is isomorphic to $\underline{\mathfrak{B}}(A, M\times W, I)$ by the theorem of Section 2 where $(A,M,W,I)$ is a complete $n$-valued nominal context. Conversely, let $(G,M,W,I)$ be a complete $n$-valued nominal context. If $\{g_1\}'' \neq \{g_2\}''$ for $g_1, g_2 \in G$, then w.o.l.g. there exists an $(m,w) \in M\times W$ with $g_1 I(m,w)$ but not $g_2 I(m,w)$. By the completeness, there is a $\bar{w} \in W$ with $g_2 I(m,\bar{w})$. Now, $\{(m,w)\}' \cap \{(m,\bar{w})\}' = \emptyset$ implies $\{g_1\}'' \cap \{g_2\}'' = \emptyset$. Hence $\{\{g\}'' \mid g \in G\}$ is a partition of $G$. Thus, $\underline{\mathfrak{B}}(G,M\times W,I)$ is atomistic. Obviously, a concept $(\{(m,w)\}', \{(m,w)\}'')$ has valence $\leq n$ in $\underline{\mathfrak{B}}(G,M\times W,I)$ for all $(m,w) \in M\times W$. Since the set of all these concepts is infimum-dense in the concept lattice, the theorem is proved.

COROLLARY. *A finite lattice $L$ is isomorphic to a concept lattice of a finite complete $n$-valued nominal context if and only if $L$ is atomistic and every $\wedge$-irreducible element of $L$ has valence $\leq n$.*

COROLLARY. *A finite lattice $L$ is isomorphic to a concept lattice of a finite complete 2-valued nominal context if and only if $L$ is atomistic and every $\wedge$-irreducible element of $L$ has a pseudo-complement.*

The concept lattice of a nominal context $(G,M,W,I)$ only clarifies the dependencies between the "one-valued" attributes of $M\times W$ but not the dependencies between the many-valued attributes of $M$ which are in general defined as follows (since the elements of $M$ can be understood as partial maps from $G$ into $W$, we write $mg = w$ instead of $gI(m,w)$): For $m \in M$ and $B \subseteq M$, the attribute $m$ is *dependent* on $B$ if $mg_1 = mg_2$ for all $g_1, g_2 \in G$ whenever $ng_1 = ng_2$ for all $n \in B$ or equivalently, if there exists a mapping $\alpha: W^B \to B$ such that $\alpha(ng)_{n\in B} = mg$ for all objects $g$ for which $ng$ is defined for all $n \in B$ (special dependencies as monotone, linear, quadratic, etc. usually refer to restrictions on the mappings $\alpha$). The closed subsets of $M$ with respect to the defined dependence are exactly the intents of the context $(G\times G, M, \sim)$ where $(g_1,g_2) \sim m :\Leftrightarrow mg_1 = mg_2$. Here lattices of (partial) equivalence relations arise.

## 7. MEASUREMENT

Numerical descriptions of concept lattices may be viewed as a part of that area concerned with the measurements of objects. Concept lattices may even arise from many-valued contexts whose attribute values are numerical; there the attributes can be understood as some kind of measure of objects. The common theory of measurement defines a scale (measure) to be a (relatively closed) homomorphism from an empirical relational structure into a numerical relational structure (cf. Pflanzagl [26], Krantz, Luce,

Suppes, Tversky [20], Roberts [31]). In our approach, it is desirable to have a type-free notion of measure (instead of the type-fixed notion homomorphism) wherefore we separate the notions of scale and measure. A *scale* is defined as a *numerical context* $\Sigma := (G_\Sigma, M_\Sigma, I_\Sigma)$ where $G_\Sigma$ is a set of numbers or number-valued functions. Then a $\Sigma$-*measure* of a context $(G,M,I)$ is a mapping $\mu$ from $G$ into $G_\Sigma$ such that $\mu^{-1}A$ is an extent of $(G,M,I)$ for every extent $A$ of $\Sigma$; $\mu$ is called *full* if $\mu^{-1}$ induces an isomorphism from the concept lattice of $(\mu G, M_\Sigma, I_\Sigma \cap \mu G \times M_\Sigma)$ onto $\underline{\mathfrak{B}}(G,M,I)$. The description problem appears now within the *representation problem* which asks for (full) $\Sigma$-measures of a given context into an appropriate scale $\Sigma$. Paradigms for scales are the *real ordinal scale* $\Sigma_O := (\mathbb{R}, M_O, \in)$ with $M_O := \{(-\infty, r] \mid r \in \mathbb{R}\}$, the *real interval scale* $\Sigma_I := (\mathbb{R}, M_I, \in)$ with $M_I := M_O \cup \{\{r-s, r, r+s\} \mid r \in \mathbb{R},\ s \in \mathbb{R}^+\}$, the *real ratio scale* $\Sigma_R := (\mathbb{R}, M_R, \in)$ with $M_R := M_I \cup \{\{r:s, r, r\cdot s\} \mid r \in \mathbb{R},\ s \in \mathbb{R}^+\}$, and their multidimensional analogues.

Our setting of measurement provides a general scheme for deriving a context from a many-valued context $(G,M,W,I)$. If an attribute $m \in M$ is understood as a map into a scale $\Sigma$, then we propose to resolve $m$ into the set $\{m\} \times M_\Sigma$ where an object $g \in G$ has the attribute $(m,n)$ for $n \in M_\Sigma$ if $g \in m^{-1}\{n\}'$; this translation guarantees that $m$ will be a $\Sigma$-measure of the derived context. For instance, if all attributes in $M$ are maps into the real ordinal scale $\Sigma_O$, we obtain the context $(G, M \times \mathbb{R}, I_\leq)$ where $gI_\leq(m,r) :\Leftrightarrow mg \leq r$; in such a case, we call $(G,M,W,I)$ an *ordinal context*. The embedding of $\underline{\mathfrak{B}}(G, M \times \mathbb{R}, I_\leq)$ into $\mathbb{R}^M$ will be discussed in a more general frame below.

A main question in the theory of measurement is whether a $\Sigma$-measure $\mu_0$ of a context $(G,M,I)$ is *unique* in the sense that for every $\Sigma$-measure $\mu$ of $(G,M,I)$ which admits a bijection $\beta: \mu_0 G \to \mu G$ inducing an isomorphism between the corresponding concept lattices and satisfying $\beta_0 \mu_0 = \mu$, there exists a bijection $\sigma: G_\Sigma \to G_\Sigma$ which induces an automorphism of $\underline{\mathfrak{B}}(\Sigma)$ and extends $\beta$. This indicates that the representation problem also leads to a study of automorphisms of concept lattices. For instance, the automorphisms of the concept lattice of $\Sigma_O$, $\Sigma_I$, and $\Sigma_R$ are induced by the order automorphisms, the positive affine mappings, and the positive linear mappings of $\mathbb{R}$, respectively (the set of all permutations of $G_\Sigma$ inducing automorphisms on $\underline{\mathfrak{B}}(\Sigma)$ is usually considered as the *scale type* of $\Sigma$).

How lattice-theoretic developments are connected with a systematic study of the representation problem may be demonstrated in the case of order measurement. Let $(\Omega_t)_{t \in T}$ be a family of complete chains (of numbers) and let $\Omega$ be the direct product of the $\Omega_t$ $(t \in T)$; then $\Omega_\leq := (\Omega, \Omega, \leq)$ is called an *order scale of dimension* $|T|$ *and of length* $l(\Omega)$. Since $\Omega$ is a complete lattice, the mapping given by $x \mapsto ((x],[x))$ is an isomorphism from $\Omega$ onto $\mathfrak{B}(\Omega_{\leq})$ by the theorem of Section 2 (we recall that $(x] := \{y \mid y \leq x\}$ and $[x) := \{z \mid x \leq z\}$).

PROPOSITION. *For a full* $\Omega_{\leq}$*-measure* $\mu$ *of a context* $(G,M,I)$, *let* $\overline{\mu}(A,B) := \bigvee \mu A$ *for all* $(A,B) \in \mathfrak{B}(G,M,I)$. *Then the mapping given by* $\mu \mapsto \overline{\mu}$ *is a bijection from the set of all full* $\Omega_{\leq}$*-measures of* $(G,M,I)$ *onto the set of all* $\bigvee$*-embeddings of* $\mathfrak{B}(G,M,I)$ *into* $\Omega$; *in particular,* $\mu g = \overline{\mu}(\{g\}'',\{g\}')$ *for all* $g \in G$.

PROOF. If $\mu$ is a full $\Omega_{\leq}$-measure of $(G,M,I)$ then an isomorphism from $\mathfrak{B}(G,M,I)$ onto $\mathfrak{B}(\mu G,\Omega,\leq)$ is given by $(A,B) \mapsto (\mu A,(\mu A)')$; in particular, $\mu A$ is an extent of $(\mu G,\Omega,\leq)$ for each extent $A$ of $(G,M,I)$ and therefore $(\bigvee \mu A] \cap \mu G = \mu A$. Thus, $\overline{\mu}$ is injective. Let $(A_j,B_j) \in \mathfrak{B}(G,M,I)$ for $j \in J$ and let $a := \bigvee_{j \in J} \bigvee \mu A_j$. There exists a concept $(A,B)$ of $(G,M,I)$ with $\mu A = (a] \cap \mu G$. Since $a \leq \bigvee \mu C$ for every concept $(C,D)$ of $(G,M,I)$ while $\bigvee_{j \in J} (A_j,B_j) \leq (C,D)$, it follows $(A,B) = \bigvee_{j \in J} (A_j,B_j)$. Hence $\overline{\mu}$ is a $\bigvee$-embedding of $\mathfrak{B}(G,M,I)$ into $\Omega$. Since $(\mu g] \cap \mu G$ is the smallest extent of $(\mu G,\Omega,\leq)$ containing $g$ $(g \in G)$, we have $(\mu g] \cap \mu G = \mu\{g\}''$ and hence $\mu g = \overline{\mu}(\{g\}'',\{g\}')$. Now, let $\iota$ be a $\bigvee$-embedding of $\mathfrak{B}(G,M,I)$ into $\Omega$. We define $\underline{\iota} g := \iota(\{g\}'',\{g\}')$ for all $g \in G$. Let $A$ be an extent of $(G,M,I)$. Obviously, $(\bigvee \underline{\iota} A] \cap \underline{\iota} G \supseteq \underline{\iota} A$. Let $\underline{\iota} g \leq \bigvee \underline{\iota} A$. Because of $\bigvee \underline{\iota} A \leq \iota(A,A')$, it follows $\iota(\{g\}'',\{g\}') \leq \iota(A,A')$ and hence $(\{g\}'',\{g\}') \leq (A,A')$ wherefore $g \in A$ and $\underline{\iota} g \in \underline{\iota} A$. Thus, $(\bigvee \underline{\iota} A] \cap \underline{\iota} G = \underline{\iota} A$. This means that $\underline{\iota}$ induces an injective map from $\mathfrak{B}(G,M,I)$ into $\mathfrak{B}(\underline{\iota} G,\Omega,\leq)$. Let $a \in \Omega$ and let $A := \underline{\iota}^{-1}(a] = \{g \in G \mid \iota(\{g\}'',\{g\}') \leq a\}$. Then $\iota(A'',A') = \iota \bigvee \{(\{g\}'',\{g\}') \mid g \in A\} = \bigvee \{\iota(\{g\}'',\{g\}') \mid g \in A\} \leq a$ which implies $A = A''$. Therefore $\underline{\iota}$ is an $\Omega_{\leq}$-measure of $(G,M,I)$ and induces a bijective map from $\mathfrak{B}(G,M,I)$ onto $\mathfrak{B}(\underline{\iota} G,\Omega,\leq)$. Obviously, $(A,B) \leq (C,D)$ implies $\underline{\iota} A \subseteq \underline{\iota} C$. Conversely, if $\underline{\iota} A \subseteq \underline{\iota} C$ then $\iota(A,B) \leq \iota(C,D)$ and hence $(A,B) \leq (C,D)$. Now we can summarize that $\underline{\iota}$ induces an isomorphism from $\mathfrak{B}(G,M,I)$ onto $\mathfrak{B}(\underline{\iota} G,\Omega,\leq)$, i.e. $\underline{\iota}$ is a full $\Omega_{\leq}$-measure of $(G,M,I)$. Finally, $\underline{\overline{\mu}} g = \overline{\mu}(\{g\}'',\{g\}') = \bigvee \mu\{g\}'' = \mu g$ for each full $\Omega_{\leq}$-measure $\mu$ of $(G,M,I)$, and $\overline{\underline{\iota}}(A,B) = \bigvee \underline{\iota} A = \bigvee \{\iota(\{g\}'',\{g\}') \mid g \in A\} = \iota \bigvee \{(\{g\}'',\{g\}') \mid g \in A\} = \iota(A,B)$ for each $\bigvee$-embedding $\iota$ of $\mathfrak{B}(G,M,I)$ into $\Omega$. This finishes the proof of the proposition.

By the preceding proposition, full order scaling can be analysed purely lattice-theoretically. For instance, the smallest dimension of an order scale $\Omega_{\leq}$ which admits a full $\Omega_{\leq}$-measure of $(G,M,I)$ equals the $\bigvee$-*dimension* of $\mathfrak{B}(G,M,I)$ which is defined as the smallest number of complete chains whose direct product admits a $\bigvee$-embedding of the lattice. The smallest length of such an order scale is given by the $\bigvee$-*rank* of $\mathfrak{B}(G,M,I)$ which is defined as the smallest length of a direct product of complete chains which admits a $\bigvee$-embedding of the lattice. The following theorem gives a basic analysis of $\vee$-embeddings into direct products of complete chains in the case of finite lattices (cf. Lea [22], Ritzert [29]).

THEOREM. *Let $L$ be a finite lattice. If $\chi := \{C_t \mid t \in T\}$ is a partition of $M(L)$ into chains then $\hat{\chi}: L \to \times_{t\in T}(C_t \cup \{1_L\})$ defined by $\hat{\chi}a := (a_t)_{t\in T}$ with $a_t := \min\{c \in C_t \cup \{1_L\} \mid a \leq c\}$ is a $\vee$-embedding. If $\iota: L \to \Omega$ is any $\vee$-embedding in a direct product of complete chains then there exists always a partition $\chi := \{C_t \mid t \in T\}$ of $M(L)$ into chains and a $\vee$-homomorphism $\kappa$ from $\Omega$ onto $\times_{t\in T}(C_t \cup \{1_L\})$ such that $\kappa$ maps $M(\Omega) \cup \{1_\Omega\}$ onto $M(\times_{t\in T}(C_t \cup \{1_L\})) \cup \{1_{\times C_t}\}$ and $\hat{\chi} = \kappa \circ \iota$.*

PROOF. Since every element of $L$ is a meet of $\wedge$-irreducible elements, $\hat{\chi}$ is injective. Obviously, $a_t \vee b_t = (a \vee b)_t$ for all $a,b \in L$ and $t \in T$. Thus, $\hat{\chi}$ is a $\vee$-embedding of $L$ into $\times_{t\in T}(C_t \cup \{1_L\})$. Let $\Omega := \times_{s\in S} \Omega_s$ and let $\pi_s: \Omega \to \Omega_s$ the canonical projection for $s \in S$. If $\pi_s \iota a = \pi_s \iota b^s$ and $a \leq b^s$ for $a \in M(L)$, $s \in S$, and $b^s \in L$, then $\iota a = \bigwedge_{s\in S} \iota b^s$ and hence $a = \bigwedge_{s\in S} b^s$ wherefore $a = b^s$ for some $s \in S$. Thus, there exists an injective map $\sigma: M(L) \to S$ such that $\pi_{\sigma a}\iota a = \pi_{\sigma a}\iota b$ implies $a \geq b$ for all $a \in M(L)$ and $b \in L$. $M(\Omega)$ contains exactly one element $m_{\sigma a}$ with $\pi_{\sigma a} m_{\sigma a} = \pi_{\sigma a}\iota a$ for each $a \in M(L)$. If $\sigma a = \sigma b$ and $m_{\sigma a} < m_{\sigma b}$ for $a,b \in M(L)$ then $a < b$, because otherwise we would have $b < a \vee b$ but $\pi_{\sigma b}\iota b = \pi_{\sigma b}\iota(a \vee b)$. Now, let $C_{\sigma a}$ be the chain consisting of all elements $b \in M(L)$ such that $m_{\sigma a}$ and $m_{\sigma b}$ are comparable in $\Omega$; furthermore, let $T := \sigma M(L)$. Then $\chi := \{C_t \mid t \in T\}$ is a partition of $M(L)$ into chains. For $x \in \Omega$ we define $\kappa x := (x_t)_{t\in T}$ with $x_t := \min\{b \in C_t \cup \{1_L\} \mid x \leq m_{\sigma b}\}$ where $m_{\sigma 1_L} = 1_\Omega$. Again, $x_t \vee y_t \vee z_t \vee \ldots = (x \vee y \vee z \vee \ldots)_t$ for all $x,y,z,\ldots \in \Omega$ and $t \in T$; hence $\kappa$ is a $\vee$-homomorphism. Obviously, $\kappa(M(\Omega) \cup \{1_\Omega\}) = M(\times_{t\in T}(C_t \cup \{1_L\})) \cup \{1_{\times C_t}\}$ and $\kappa$ is surjective. If $\iota a \leq m_{\sigma b}$ for $a \in L$ and $b \in M(L)$ then $a \leq b$, because otherwise we would have $b < a \vee b$ but $\pi_{\sigma b}\iota b = \pi_{\sigma b}\iota(a \vee b)$. Therefore $(\iota a)_t = a_t$ for all $a \in L$ and $t \in T$. Hence $\hat{\chi} = \kappa \circ \iota$.

For the first assertion of the following corollary we used, of course, the theorem of R.P. Dilworth on the width of ordered sets [14].

COROLLARY. *For a finite lattice $L$, the $\vee$-dimension of $L$ is equal to the width of $M(L)$ and the $\vee$-rank of $L$ is equal to the cardinality of $M(L)$.*

COROLLARY. *A finite context $(G,M,I)$ admits a unique full $\Omega_{\leq}$-measure with $\Omega = \times_{t\in T} \Omega_t$ if and only if there is exactly one* *partition* $\{C_t | t \in T\}$ *of* $M(L)$ *into chains such that the length of* $\Omega_t$ *is equal to the cardinality of* $C_t$ *for all* $t \in T$; *then, in particular, the dimension and the length of the order scale* $\Omega_\leq$ *will be equal to the width and the cardinality of* $M(\underline{\mathfrak{B}}(G,M,I))$, *respectively.*

To end this section, we apply the results of our lattice-theoretical analysis to a simple example of an ordinal context. The context below is the list of the general grades on conduct (c), diligence (d), attentiveness (a), and orderliness (o) given to a class of the Ludwig-Georgs-Gymnasium (Darmstadt) for the winter term 1980/81 (the best grade is 1).

| | c | d | a | o |
|---|---|---|---|---|
| Anna | 3 | 4 | 3 | 4 |
| Berend | 3 | 4 | 3 | 4 |
| Christa | 2 | 2 | 2 | 2 |
| Dieter | 1 | 1 | 1 | 1 |
| Ernst | 2 | 2 | 2 | 2 |
| Fritz | 2 | 1 | 2 | 2 |
| Gerda | 2 | 2 | 2 | 3 |
| Horst | 2 | 2 | 2 | 3 |
| Ingolf | 2 | 3 | 3 | 2 |
| Jürgen | 2 | 2 | 3 | 2 |
| Karl | 2 | 3 | 2 | 2 |
| Linda | 2 | 1 | 2 | 2 |
| Manfred | 2 | 2 | 2 | 2 |
| Norbert | 3 | 3 | 2 | 3 |
| Olga | 1 | 1 | 2 | 2 |
| Paul | 1 | 1 | 1 | 1 |
| Quax | 2 | 2 | 2 | 2 |
| Rudolf | 3 | 4 | 3 | 3 |
| Stefan | 1 | 1 | 1 | 1 |
| Till | 1 | 1 | 1 | 1 |
| Uta | 1 | 1 | 2 | 2 |
| Volker | 2 | 2 | 3 | 2 |
| Walter | 3 | 4 | 3 | 4 |
| Xaver | 1 | 1 | 1 | 1 |
| Zora | 2 | 2 | 2 | 2 |

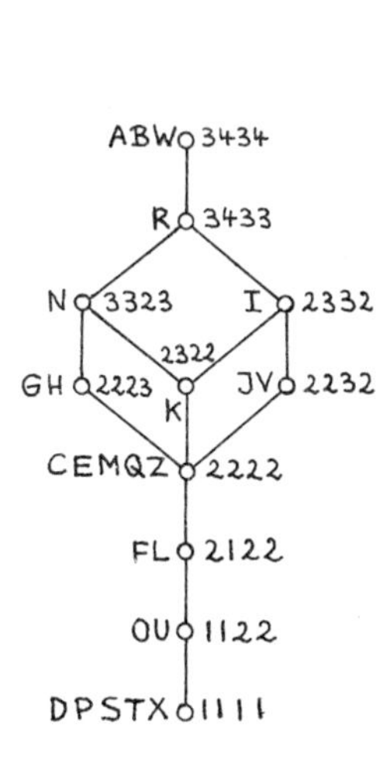


The concept lattice has eight $\wedge$-irreducible elements, namely $\gamma R$, $\gamma N$, $\gamma I$, $\gamma G$, $\gamma J$, $\gamma F$, $\gamma O$, and $\gamma D$; its width is two. Thus, the smallest dimension (length) of an order scale $\Omega_\leq$ allowing a full $\Omega_\leq$-measure is two (eight); for instance, one may take as chains $\{\gamma A, \gamma R, \gamma N, \gamma G, \gamma F, \gamma O, \gamma D\}$ and $\{\gamma A, \gamma I, \gamma J\}$. Since the concept lattice has an automorphism interchanging $\wedge$-irreducible elements, the context admits no unique full $\Omega_\leq$-measure of dimension two. Surprisingly, only the order scale built out of eight chains of length one admits a unique full $\Omega_\leq$-measure. In spite of the missing uniqueness, the conclusion seems reasonable that the given data are two-dimensional with even one main dimension. In general, it may be fruitful to explore whether the described order measurement can be used as basis for a factor analysis of ordinal data (cf. Rummel [32]).

## 8. COMPLETIONS OF PARTIAL CONCEPT LATTICES

Up to now, it was always assumed that the concept lattices are derived from known contexts. But in many situations one has only a vague knowledge of a context although many of its concepts are fairly clear. For instance, it seems to be impossible to write down a comprehensive context of musical instruments, but everyone uses concepts as violin, trumpet, guitar, string instrument, etc. and meaningful sentences such as "a violin is a string instrument". In such cases of concepts of a vague context we have a modified determination problem: *How can one extend a partial lattice of concepts to a complete concept lattice of some context?* The following theorem (see MacNeille [23], Schmidt [34]) describes the smallest solution, usually called the *Dedekind-MacNeille completion.*

THEOREM. *For an ordered set* $(P,\leq)$ *an embedding* $\iota$ *of* $(P,\leq)$ *into* $\underline{\mathfrak{B}}(P,P,\leq)$ *is defined by* $\iota x := ((x],[x))$ $(x \in P)$; *especially,* $\iota\bigwedge X = \bigwedge\iota X$ *and* $\iota\bigvee X = \bigvee\iota X$ *if* $\bigwedge X$ *and* $\bigvee X$ *exist in* $(P,\leq)$, *respectively. If* $\lambda$ *is any embedding of* $(P,\leq)$ *into a complete lattice* $L$ *then there exists an order embedding* $\kappa$ *of* $\underline{\mathfrak{B}}(P,P,\leq)$ *into* $L$ *such that* $\lambda = \kappa\circ\iota$.

PROOF. By definition, the concepts of $(P,P,\leq)$ are exactly the pairs $(A,B)$ such that $A,B \subseteq P$ and $A' = \{x \in P | x \leq y$ for all $y \in B\}$, $B' = \{y \in P | x \leq y$ for all $x \in A\}$; in particular, $((x],[x))$ with $x \in P$ are concepts of $(P,P,\leq)$ what confirms $\iota$ as an embedding. If $\bigwedge X$ exists in $(P,\leq)$, then $(\bigwedge X] = \bigcap_{x\in X} (x]$ and hence $\iota\bigwedge X = ((\bigwedge X],[\bigwedge X)) = (\bigcap_{x\in X} (x],(\bigcap_{x\in X} (x])') = \bigwedge\iota X$ by the theorem of Section 2; the assertion for $\bigvee X$ is proved dually. Now, let $\lambda$ be an embedding of $(P,\leq)$ into a complete lattice $L$. We define $\kappa(A,B) := \bigvee\lambda A$ for $(A,B) \in \underline{\mathfrak{B}}(P,P,\leq)$. Obviously, $\lambda = \kappa\circ\iota$. If $\kappa(A_1,B_1) \leq \kappa(A_2,B_2)$ then $\bigvee\lambda A_1 \leq \bigvee\lambda A_2 \leq \lambda b$ for all $b \in B_2$ and hence $b \in A_1' = B_1$ for all $b \in B_2$ because $\lambda$ is an embedding. Thus, $B_1 \supseteq B_2$ and therefore $(A_1,B_1) \leq (A_2,B_2)$; in particular, this includes the injectivity of $\kappa$. Since $\kappa$ is order-preserving, $\kappa$ is the desired order embedding.

Instead of closing a partial concept lattice to its smallest completion, one is usually more interested to unfold the concepts as far as possible to reach a rich context. Since meet and join are the fundamental operations for concepts which may produce new concepts from known ones, they form a natural tool for the desired extensions. For instance, we may start with some known concepts of musical instruments; then we try to produce new concepts by forming various meets and joins and to separate them by listing more and more objects and attributes. Such a procedure would follow the free generating process in a lattice. Unfortunately, the common set-theoretical language does not allow us to speak of a "free complete lattice"; hence, there is in this sense no largest solution of the modified determination problem available. But if we strictly bound the cardinality of subsets of which meets and joins are taken by some fixed cardinal number $\alpha$, then there exists a lattice freely $\alpha$-generated by a given partial concept lattice (see Crawley, Dean [7]); for the final completion one may use the preceding theorem. Now, these large solutions of the modified determination problem give rise to a modified description problem: *How can one describe a lattice freely $\alpha$-generated by a partial lattice?* The common answer uses lattice terms to describe the lattice elements and gives a procedure how the order of the lattice can be derived from the describing terms (see Whitman [37], [38], Dean [10], Crawley, Dean [7], Jónsson [19], Lakser [21], Grätzer, Lakser, Platt [16]).

Again, descriptions by Hasse diagrams are desirable. But it seems that already a lattice freely ($\aleph_0$-) generated by three unordered elements cannot be described by a readable Hasse diagram. Nevertheless, many lattices freely generated by a partial lattice can be "drawn". So far, only the case of finite ordered sets has been completely analysed (if one agrees that a lattice having a sublattice freely generated by three unordered elements does not have a readable Hasse diagram). Those finite ordered sets whose free completions can be "drawn" are characterized by the theorem below (see Rival, Wille [30]). It is surprising that all these free completions can be embedded into a lattice freely generated by the six-element ordered set $H$; the Hasse diagrams of $H$ and of its free completion are on the next page. The Hasse diagrams of all these free completions can be obtained by combining some of the ten diagrams given in [30]. In the case where a free completion of a partial lattice cannot be "drawn", it is still interesting to explore how far, following the generating process, the Hasse diagram can be developed.

THEOREM. *For a finite ordered set P the following conditions are equivalent*:

(*i*) *A lattice freely generated by P does not contain a sublattice freely generated by three unordered elements*;

(*ii*) *A lattice freely generated by H has a subset isomorphic to P which freely generates a sublattice*;

(*iii*) *P contains no subset having one of the following Hasse diagrams*:

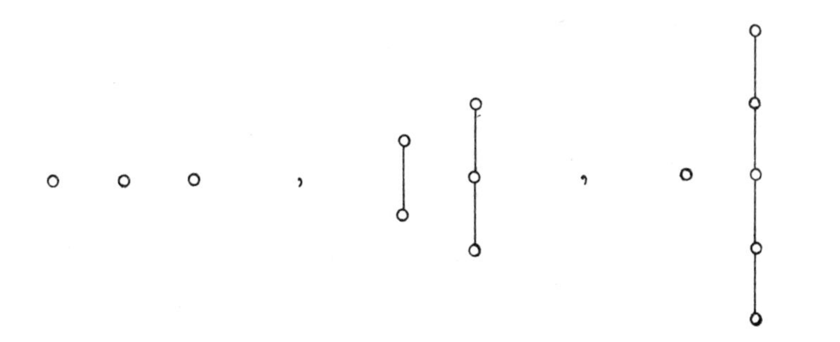

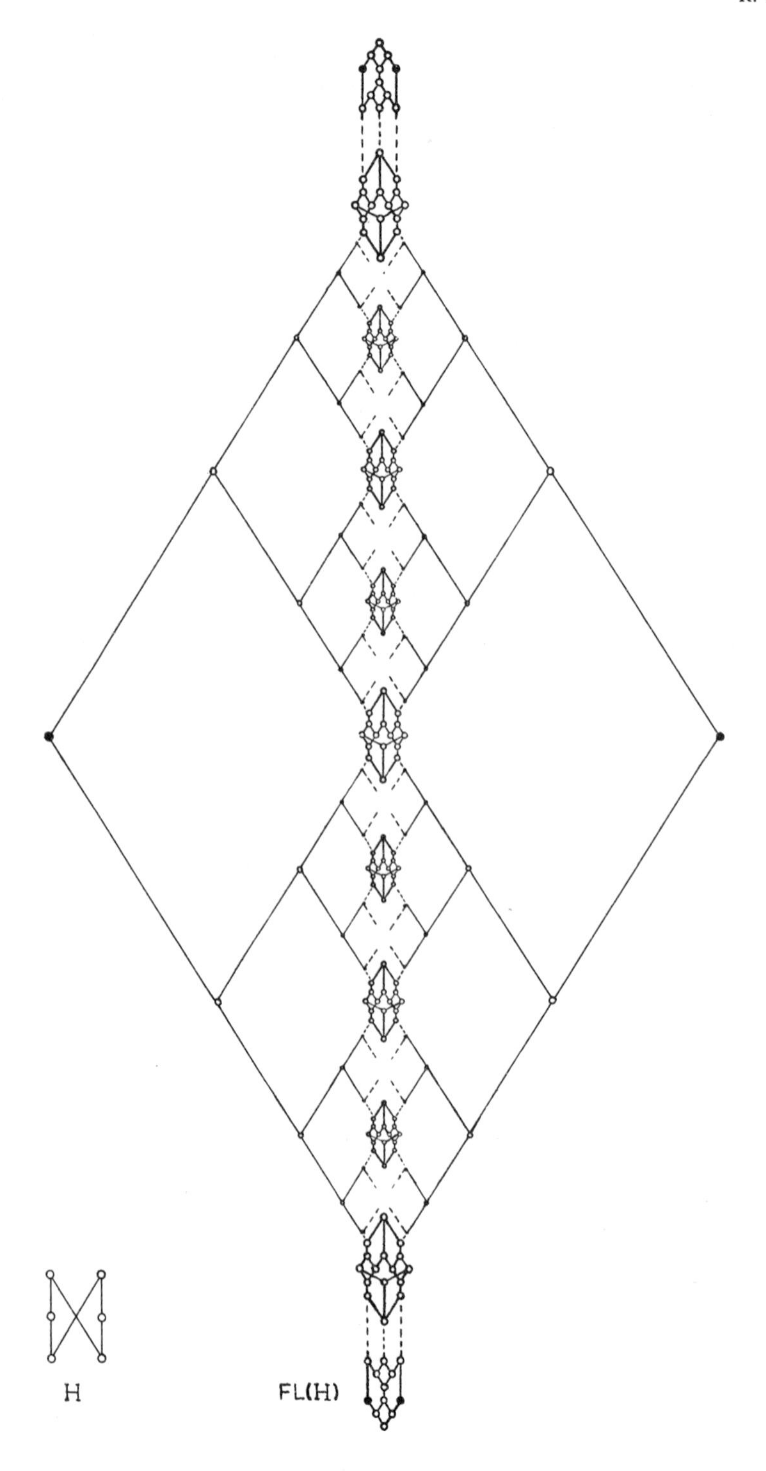
H
FL(H)

## 9. FURTHER REMARKS

This restructuring of lattice theory does not pretend to be complete in any sense. Many ties of lattice theory to its surroundings have not been mentioned. A comprehensive development of lattice theory should integrate and elaborate much more: for instance, origins as in logic (cf. Rasiowa [28]), connections as in geometry (cf. Birkhoff [5]), interpretations as in computer science (cf. Scott [33]), and applications as in quantum mechanics (cf. Hooker [18]). Especially, its significance to other mathematical disciplines which is still considered to be the main source has to be clarified. Besides the interpretation by hierarchies of concepts, other basic interpretations of lattices should be introduced; an important rôle is already played by the interpretation of lattices as closure systems. Feedback will always be given by the communication with potential users of lattice theory.

Coming back to the initial question: Why develop lattice theory? we may conclude that there is no short answer. The justification of lattice theory arises from its place in the landscape of our culture in general.

ACKNOWLEDGEMENTS. Thoughts and ideas in this paper have been matured to a great extent within the "Arbeitsgemeinschaft Mathematisierung" at the TH Darmstadt in which we have worked on the theme of restructuring mathematics now for more than three years. I would like to thank all members of the Arbeitsgemeinschaft for their stimulation, contributions, and continuous support. I would also like to thank all those who have joined discussions at the several lectures which I gave on the subject during the last two years. I am especially grateful to Ivan Rival for his encouragement and help during the writing of the final manuscript. This work was partially supported by the N.S.E.R.C. of Canada under the International Scientific Exchange Programme.

## REFERENCES

[1] B. Banaschewski (1956) Hüllensysteme und Erweiterungen von Quasi-Ordnungen, *Z. Math. Logik Grundlagen Math.* 2, 117-130.

[2] G. Birkhoff (1938) Lattices and their applications, *Bull. Amer. Math. Soc.* 44, 793-800.

[3] G. Birkhoff (1967) *Lattice Theory*, Third edition, Amer. Math. Soc., Providence, R.I.

[4] G. Birkhoff (1970) What can lattices do for you? in: *Trends in Lattice Theory* (J.C. Abbott, ed.) Van Nostrand-Reinhold, New York, 1-40.

[5] G. Birkhoff (1982) Ordered sets in geometry, in: *Symp. Ordered Sets* (I. Rival, ed.) Reidel, Dordrecht-Boston, 107.

[6] H.-H. Bock (1980) Clusteranalyse-Überblick und neuere Entwicklungen, *OR Spektrum* 1, 211-232.

[7] P. Crawley and R.A. Dean (1959) Free lattices with infinite operations, *Trans. Amer. Math. Soc.* 92, 35-47.

[8] P. Crawley and R.P. Dilworth (1973) *Algebraic Theory of Lattices*, Prentice-Hall, Englewood Cliffs, N.J.

[9] C.J. Date (1977) *An Introduction to Data Base Systems*, Second edition, Addison-Wesley, Reading, Mass.

[10] R.A. Dean (1956) Completely free lattices generated by partially ordered sets, *Trans. Amer. Math. Soc.* 83, 238-249.

[11] Deutsches Institut für Normung (1979) DIN 2330, *Begriffe und Benennungen, Allgemeine Grundsätze*, Beuth, Köln.

[12] Deutsches Institut für Normung (1980) DIN 2331, *Bergriffs-systeme und ihre Darstellung*, Beuth, Köln.

[13] K. Diem and C. Lentner (1968) Wissenschaftliche Tabellen, 7. Aufl., *J. R. Geigy AG*, Basel.

[14] R.P. Dilworth (1950) A decomposition theorem for partially ordered sets, *Ann. of Math.* (2) 51, 161-166.

[15] G. Grätzer (1978) *General Lattice Theory*, Birkhäuser, Basel-Stuttgart.

[16] G. Grätzer, H. Lakser, and C.R. Platt (1970) Free products of lattices, *Fund. Math.* 69, 233-240.

[17] H. von Hentig (1972) *Magier oder Magister? Über die Einheit der Wissenschaft im Verständigungsprozess*, Klett, Stuttgart.

[18] C.A. Hooker (ed.) (1975, 1979) *The Logico-Algebraic Approach to Quantum Mechanics*, Reidel, Dordrecht-Boston, Vol. I and Vol. II.

[19] B. Jónsson (1962) Arithmetic properties of freely α-generated lattices, *Canad. J. Math.* 14, 476-481.

[20] D.H. Krantz, R.D. Luce, P. Suppes, and A. Tversky (1971) *Foundations of Measurement*, Vol. I, Academic Press, New York.

[21] H. Lakser (1968) *Free Lattices Generated by Partially Ordered Sets*, Ph. D. Thesis, Univ. of Manitoba, Winnipeg.

[22] J.W. Lea (1972) An embedding theorem for compact semi-lattices, *Proc. Amer. Math. Soc.* 34, 325-331.

[23] H.M. MacNeille (1937) Partially ordered sets, *Trans. Amer. Math. Soc.* 42, 416-460.

[24] H. Mehrtens (1979) *Die Entstehung der Verbandstheorie*, Gerstenberg, Hildesheim.

[25] *Observer's Handbook 1981* (1980) Royal Astronomical Society Canada, Univ. Toronto Press, Toronto.

[26] J. Pflanzagl (1968) *Theory of Measurement*, Physica-Verlag, Würzburg-Wien.

[27] A. Podlech (1981) *Datenerfassung, Verarbeitung, Dokumentation und Information in den sozialärztlichen Diensten mit Hilfe der elektronischen Datenverarbeitung* (manuscript) TH Darmstadt.

[28] H. Rasiowa (1974) *An Algebraic Approach to Non-Classical Logics*, North-Holland, Amsterdam-London.

[29] W. Ritzert (1977) *Einbettung halbgeordneter Mengen in direkte Produkte von Ketten*, Dissertation, TH Darmstad.

[30] I. Rival and R. Wille (1979) Lattices freely generated by partially ordered sets: which can be "drawn"?, *J. reine angew. Math.* 310, 56-80.

[31] F.S. Roberts (1979) *Measurement Theory*, Addison-Wesley, Reading, Mass.

[32] R.J. Rummel (1970) *Applied Factor Analysis*, Northwestern Univ. Press, Evanston.

[33] D.S. Scott (1976) Data types as lattices, *SIAM J. Comput.* 5, 522-587.

[34] J. Schmidt (1956) Zur Kennzeichnung der Dedekind-MacNeilleschen Hülle einer geordneten Menge, *Arch. Math.* 7, 241-249.

[35] E. Schröder (1890, 1891, 1895) *Algebra der Logik I, II, III*, Leipzig.

[36] H. Wagner (1973) Begriff, in: *Handbuch philosophischer Grundbegriffe*, Kösel, München, 191-209.

[37] Ph. M. Whitman (1941) Free lattices, *Ann. of Math.* (2) 42, 325-330.

[38] Ph. M. Whitman (1942) Free lattices, II, *Ann. of Math.* (2) 43, 104-115.

[39] R. Wille (1977) Aspects of finite lattices, in: *Higher Combinatorics* (M. Aigner, ed.) Reidel, Dordrecht-Boston, 79-100.

[40] R. Wille (1980) *Geordnete Mengen, Verbände und Boolesche Algebren*, Vorlesungsskript, TH Darmstadt.

[41] R. Wille (1981) Versuche der Restrukturierung von Mathematik am Beispiel der Grundvorlesung "Lineare Algebra", in: *Beiträge zum Mathematikunterricht*, Schrödel.

# PART V

# ENUMERATION

# EXTREMAL PROBLEMS IN PARTIALLY ORDERED SETS

Douglas B. West
Department of Mathematics
Princeton University
Princeton, New Jersey 08544

ABSTRACT

The central problem among extremal problems in partially ordered sets (posets) is to find the largest set of mutually incomparable elements in a poset. The first results were obtained by Sperner in 1928 [Sr]: *the maximum-sized set of mutually incomparable subsets of an n-element set consists of those subsets with* $\lfloor n/2 \rfloor$ *elements, or those with* $\lceil n/2 \rceil$ *elements.* A completely different approach was taken by Dilworth in 1950 [Di]: *the size of the largest incomparable collection* (antichain) *equals the smallest number of totally ordered collections* (chains) *whose union includes all of the elements of the poset.*

Sperner's theorem is typical of attempts to characterize extremal configurations explicitly for a class of posets. Dilworth's theorem applies to all posets but gives a less "precise" result: it gives only the maximum size (although many algorithms which compute that will also find a maximum collection).

Both of these theorems have been proved in many ways and, especially in recent years, both have been generalized in many ways. We also consider several other extremal problems, some arising from the correspondence between the antichains and the order ideals of a poset.



*I. Rival (ed.), Ordered Sets, 473–521.*

## 1. Introduction

### 1.1. Scope and goals

This paper contains a history of results on extremal problems in partially ordered sets (posets). The foremost problem of this sort is that of finding the largest collection of mutually incomparable elements of the poset; we survey results on this question and on numerous other questions it has spawned. Due to the volume of results, we will never do more than hint at proofs. An excellent introduction to techniques of proof in this field can be found in the survey article by Greene and Kleitman [GK3]. The sequence of material in this paper parallels theirs.

The first result was obtained by Sperner in 1928 [Sr1]. Sperner's Theorem states that the maximum-sized incomparable collections of subsets of an $n$-set consist uniquely of those having $\lfloor n/2 \rfloor$ elements or those having $\lceil n/2 \rceil$ elements. A completely different approach was used by Dilworth in 1950 [Dil]. Dilworth's theorem asserts that the size of the largest incomparable collection in *any* poset equals the smallest number of totally ordered collections which cover the poset. Sperner's Theorem typifies attempts to characterize extremal configurations explicitly for a "well-behaved" class of partially ordered sets. Dilworth's Theorem applies more generally to all posets, but gives a less "complete" result.

We will examine a wide variety of extremal problems, grouped as follows: extensions and variations of Sperner's Theorem and Dilworth's Theorem, maximization of collection size subject to other restrictions, and optimization of other objective functions for collections of fixed size, especially minimizing intersections of order ideals. The material in Section 4 on maximizing the size of collections has been surveyed many times before, but cannot be ignored in a survey of extremal problems. In Section 4 we stick to our promise of no proofs and try to emphasize generalizations beyond subsets of a set.

For the reader new to this field, in other sections we try to explain the basic results. For the experienced reader, the main value of this paper will be the list of references. We have attempted to provide a thorough list of references. The less directly related a problem is to maximizing incomparable collections, the less complete will be its references in our list. We apologize in advance for misstatement, misattribution, or omission of results.

### 1.2 Notation and terminology

Terminology in this subject is not so chaotic as in graph theory, but there are still some non-standard usages, and $k$ is a very overworked parameter. We have tried to be consistent in choosing parameter names, conforming to the majority rather than respecting the conventions chosen by individual researchers. Our method of describing various conditions on collections in Section 4 was born of necessity. The reader may treat the following definitions as a glossary to be consulted as needed.

A *partially ordered set* or *poset* is a set together with a transitive, anti-symmetric order relation on its elements. A *lattice* is a poset in which any pair of elements $a,b$ have a unique least upper bound $a \vee b$ (called their *join)* and a unique greatest lower bound $a \wedge b$ (called their *meet)*. The *(direct) product* of two posets $P$ and $Q$ is a poset $P \times Q$ whose elements are the ordered pairs $\{(a,b) : a \in P,\ b \in Q\}$. In $P \times Q$, $(a,b) \leqslant (a',b')$ if and only if $a \leqslant a'$ in $P$ and $b \leqslant b'$ in $Q$.

A subset of a poset will be called a *collection* or *family* of elements, usually denoted $F$. A family of unrelated or incomparable elements is called an *antichain,* in contrast to a *chain,* which is a totally ordered family. A chain is also called a *linearly ordered* set. We use $C_k$ to denote a chain with $k$ elements. A $k$*-family* is a family containing no chain of $k+1$ elements. We speak of *maximal* families and *maximum* families satisfying various conditions. "Maximal" means not contained in any other; "maximum" means maximum-sized. The size of a maximum antichain is called the *width* of the poset. An element $x$ is *maximal (minimal)* in a poset $P$ if there is no $y \in P$ with $y > x$ $(y < x)$. Element $y$ *covers* element $x$ if $y > x$ and if $y \geqslant z > x$ implies $y = z$.

Antichains have also been called "Sperner families" [HH], "Sperner hypergraphs" [Be2], "amonotonic subcollections" [Ba1,2], and "clutters" [DGH], et al. Some authors have required a "Sperner" family to be a maximum antichain, in order to give meaning to the term "Sperner $k$-family". Others let it be any antichain; we use the latter convention.

Questions related to Sperner's Theorem make sense only for posets which have a well-defined *rank function,* also called *ranked* posets. A poset $P$ is *ranked* if its elements can be partitioned into subsets $P_0, P_1, \ldots, P_n$ such that an element in $P_k$ covers only elements in $P_{k-1}$, if any. The *rank* of such a poset is $n$. The rank function is defined on its elements; if $x \in P_k$, then $\text{rank}(x) = k$. Traditionally, a rank function satisfies the additional requirement that all minimal elements have rank 0 and all maximal elements have rank $n$. However, we will not need this restriction. The number of elements with rank $k$ is called the $k^{th}$ *Whitney number* (of the second kind), and denoted $N_k$. The Whitney numbers are *unimodal* if there is a rank $k$ such that $N_i \leqslant N_{i+1}$ for $i<k$ and $N_i \geqslant N_{i+1}$ for $i \geqslant k$. They are *symmetric* if $N_k = N_{n-k}$. The subsets $P_0, \ldots, P_n$ are called the *ranks* of $P$. If $F$ is a collection of elements in $P$, let $f_k = |F \cap P_k|$. The numbers $\{f_k\}$ are called the *(rank) parameters* of $F$. The collection of all elements in $P_k$ related to any element of $F$ is the *k-shadow* of $F$.

The best-studied poset is the lattice of subsets of an $n$ element set ($n$-set), ordered by inclusion, which we denote $B_n$ (Boolean algebra). Various other notations have been used. In referring to the underlying set, we use the notation $[n] = \{1, 2, \cdots, n\}$. Here the integers serve only as labels for the set elements. Similarly, when we say a maximum family is unique or "essentially unique", we mean except for obvious permutations of the indices, and perhaps also complementation. Elements of $B_n$, being subsets of set, are traditionally written as capital letters. In $B_n$, $A \wedge B = A \cap B$, $A \vee B = A \cup B$, $\text{rank}(A) = |A|$, and $N_k = \binom{n}{k}$. For notational clarity, we use $B_n^k$ to denote the $k^{th}$ rank of $B_n$, rather than using an additional subscript. Extremal results about families in $B_n$ have often been phrased in hypergraph language. This eliminates confusion between elements of $B_n$ and elements of the underlying $n$-set by calling the former "edges". However, the generalizations belong to poset theory, so we use poset terminology exclusively. We do note that there has been a resurgence of hypergraph terminology where the edges are subsets of a poset, rather than elements of $B_n$ (see [Sa1]).

Many results about $B_n$ generalize to the lattice of divisors of an integer, ordered by divisibility. If $M = \Pi p_i^{e_i}$ for $n$ distinct primes $\{p_i\}$, then the lattice of divisors of $M$ is isomorphic to a product of $n$ chains with sizes $\{e_i+1\}$. The divisor lattice is also isomorphic to multisets of an $n$-set, with multiplicities bounded by $\{e_i\}$. Then the poset can be written as $M(e_1, \ldots, e_n)$, with elements written as sequences of integers $\bar{a} = (a_1, \ldots, a_n)$, $0 \leqslant a_i \leqslant e_i$, and with $\bar{a} \leqslant \bar{b}$ if and only if $a_i \leqslant b_i$ for $1 \leqslant i \leqslant n$. Hence $M(\bar{e})$ has also been called the set of lattice points in a parallelotope, but we stick to the multiset or chain-product terminology. Note that $\bar{a} \wedge \bar{b}$ and $\bar{a} \vee \bar{b}$ are the sequences with $(\bar{a} \wedge \bar{b})_i = \min\{a_i, b_i\}$, $(\bar{a} \vee \bar{b})_i = \max\{a_i, b_i\}$. Also, $\text{rank}(\bar{a}) = \Sigma a_i$, $N_k(M(\bar{e}))$ is the coefficient of $x^k$ in $\Pi(1+x+\cdots+x^{e_i})$, and $B_n = M(1,1,\ldots,1)$. We will adopt the convention $e_1 \geqslant \cdots \geqslant e_n$ unless otherwise noted, which enables us in Section 4 to refer quickly to the largest or smallest component.

Some results about $B_n$ also generalize to the lattice of subspaces of a finite vector space, ordered by inclusion. $L_n(q)$ denotes such a lattice for a space of dimension $n$ over a $q$-element field. The rank function is dimension, the meet and join of two subspaces are their intersection and the smallest subspace containing both, and the Whitney numbers are the *Gaussian coefficients*

$$N_k = \begin{bmatrix} n \\ k \end{bmatrix}_q = \frac{(q^n-1)(q^{n-1}-1)\cdots(q^{n-k+1}-1)}{(q^k-l)(q^{k-1}-1)\cdots(q-1)}.$$

A lattice $P$ is *distributive* if all triples $a,b,c \in P$ satisfy $a \wedge (b \vee c) = (a \wedge b) \vee (a \wedge c)$ (or, equivalently, all triples satisfy $a \vee (b \wedge c) = (a \vee b) \wedge (a \vee c)$). The distributive lattices include $M(\bar{e})$ but not $L_n(q)$. We will not be concerned with these defining properties or lattice-theoretic results, but a few of the extremal results will extend to distributive lattices.

We postpone until Section 4 most definitions of subsets of posets other than chains and antichains. However, we make one exception here for a concept closely related to antichains, which will crop up occasionally throughout the paper. An *(order) ideal* is a subset $F$ of a poset such that $y \in F$ and $x < y$ implies $x \in F$. This property is also called "closed below." Ideals are in 1-1 correspondence with antichains, since any antichain "generates" an ideal and the maximal elements of any ideal form an antichain.

## 2. The Sperner property—special posets

THEOREM: *The unique maximum antichain in $B_n$ is $B_n^{\lfloor n/2 \rfloor}$ (or $B_n^{\lceil n/2 \rceil}$).*

Sperner's Theorem has been proved in many ways. These proofs strengthen the theorem by showing that various poset properties possessed by $B_n$ suffice to imply a conclusion similar to or stronger than that of Sperner's Theorem.

A ranked poset has the *Sperner property* or "is Sperner" if its largest antichain consists of the elements of the largest-sized rank. For a given $k$, it has the *$k$-Sperner property* if the largest $k$-family consists of the $k$ largest ranks. It has the *strong Sperner property* (or is "strongly Sperner") if it is $k$-Sperner for all $k$. Most of these poset properties that imply the Sperner or strong Sperner property imply further that maximum-sized families are obtained only by taking full ranks.

Chronologically, Erdös [Er1] provided the first improvement, generalizing Sperner's argument to show that $B_n$ is strongly Sperner. Proofs that $M(\bar{e})$ is Sperner appeared in [dBTK] and [Bu]. In fact, the lattices $M(\bar{e})$ and $L_n(q)$ are strongly Sperner, as noted in [Sc1] and [GK1], respectively.

Some natural posets are not Sperner at all. Dilworth and Greene [DG] provided a non-Sperner class of geometric lattices. Rota [Ro] asked whether the lattice of partitions of an $n$-set, ordered by refinement, has the Sperner property. This prompted much study of this lattice, including [Mu], [Mi1,2], [Sp1], [Gg1], and [Ca1]. Finally Canfield [Ca2] showed that for large enough $n$, the lattice is not Sperner. The smallest such value is about $10^{25}$, making it surprising that it took only a decade to settle the conjecture.

Some early proofs of Sperner's Theorem were algebraic in nature. Kleitman, Edelberg, and Lubell [KEL] noted the following: Every finite poset $P$ contains a maximum antichain invariant under every order-preserving automorphism. In $B_n$, any $k$-set is mapped to any other by some order-preserving automorphism, so such an antichain must consist of a single rank. Freese [Fe] obtained another proof of their lemma based on results of Dilworth [Di2]. Greene and Kleitman [GK1] obtained a similar result for $k$-families, thereby proving that $B_n$ and $L_n(q)$ are strongly Sperner.

## 2.1 Dilworth decompositions

A *Dilworth decomposition* of a poset $P$ is a partition of $P$ into disjoint chains such that there is an antichain which intersects each chain. (This phrasing appears in [OS] and applies also to infinite posets.) After the appearance of Dilworth's Theorem, it was recognized that $P$ is Sperner if it can be partitioned into disjoint chains that all intersect the largest rank. If, in addition, $P$ has a partition into chains such that every chain $C$ contains an element from each of the $\min\{k,|C|\}$ largest ranks, then $P$ is $k$-Sperner. When the Whitney numbers are symmetric and unimodal, as for $B_n$, $M(\bar{e})$, and $L_n(q)$, we can ask for still more. If $P$ has rank $n$, a *symmetric chain decomposition* is a partition of $P$ into chains such that every chain contains one element of each rank $j,j+1,\ldots,n-j$ for some $j$. Such a $P$ is called a *symmetric chain order*.

DeBruijn, Tengbergen, and Kruyswijk [dBTK] discovered a symmetric chain decomposition of chain products $M(\bar{e})$. They expressed it inductively. The same decomposition was rediscovered in an explicit manner by Leeb [unpublished] and by Greene and Kleitman [GK2]. Representing the elements of $B_n$ as $\{0,1\}$ vectors, the decomposition in [GK2] consists of replacing the 0's and 1's by left and right parentheses and putting elements in the same chain if they have the same "matched parenthesization" structure. This generalizes to $M(\bar{e})$. [Gg1] contains a thorough discussion of these results.

Brown [Bw] was the first to focus on level-by-level construction of chain decompositions, using Hall's matching theorem to rediscover the existence of a Dilworth decomposition of $B_n$. Aigner [Ai1] expressed the [dBTK] decomposition for $B_n$ in a lexicographic level-by-level form. Later, in [Ai2], he extended this to chain products and proved a more general theorem about symmetric decomposition. Closely related questions are examined in [Lw] and [Le]. White and Williamson [WhW] showed that there is a large class of linear orders for which a "lexical" level-by-level matching produces "good" results, such as symmetric chain decompositions. No explicit symmetric chain decomposition is known for $L_n(q)$, though Griggs [Gg2] proved that one exists.

More recently, interest in chain decomposition of other posets has grown. Metropolis and Rota [MR] provided a "somewhat" symmetric Dilworth decomposition of the lattice of faces of the $n$-cube ($B_n$ is the lattice of faces of the $n$-simplex). Their construction is analogous to that of [GK2]. They extended this in [MRSW] to more general posets.

The existence of a symmetric chain decomposition remains an open question for several other classes of posets. Stanley conjectures that the lattice $L(m,n)$ is a symmetric chain order. One way to describe this lattice is as the set of non-decreasing sequences $0\leqslant a_1,\leqslant \cdots \leqslant a_m\leqslant n$, ordered by $\bar{a}\leqslant\bar{b}$ if and only if $a_i\leqslant b_i$ for $1\leqslant i\leqslant m$. Equivalently, it is the lattice of order ideals in the product of an $m$-element chain with an $n$-element chain. Symmetric chain decompositions have been provided for $L(m,n)$ when $\min\{m,n\}\leqslant 4$ in [Ld2], [We], [Ri]. They are much more complicated than the bracketing construction for $M(\bar{e})$.

Other lattices for which this question remains unsettled include 1) the weak Bruhat order on the permutations of $n$ elements, in which $\pi$ covers $\sigma$ if $\pi$ is obtained from $\sigma$ by transposing an adjacent pair of elements to increase the number of inversions by 1, and 2) an order on the $\{0,1\}$-vectors of length $n$ in which $\bar{a}$ covers $\bar{b}$ if $\bar{a}$ is obtained from $\bar{b}$ by introducing a 1 in the first position or by moving a 1 one digit to the right. These questions are raised by Edelman and by Lindström (the latter communicated by Leeb), respectively.

## 2.2 The LYM Property

Sperner's Theorem was also proved using an elegant counting argument by Lubell [Lb], Yamamoto [Ya], and Meshalkin [Me]. [Lb] is in fact the shortest paper cited in this bibliography; it occupies less than a page. The crux of the LYM argument is that each $k$-set belongs to $k!(n-k)!$ of the $n!$ maximal chains in $B_n$, and no two elements of an antichain can appear in a single chain. If $f_k$ denotes the number of elements of rank $k$ in $F$ and the resulting inequality is divided by $n!$, this translates into $\sum_{k=0}^{n} \frac{f_k}{N_k} \leqslant 1$ for any antichain F. If every antichain in a poset $P$ satisfies this inequality, called the *LYM inequality*, then $P$ has the *LYM property*. This property clearly implies the Sperner property. Furthermore, it is easy to show that any $k$-family is the union of $k$ antichains and thus prove that the LYM property also implies the strong Sperner property.

Baker [Bk] noted that this maximal chain argument applies whenever the numbers of elements covered by (or covering) a given element depends only on its rank. Drake [Dr] characterized the maximum antichains for a poset satisfying this property; [Ks] gave only a sufficient condition.

Kleitman [Kl13] abstracted the essence of the maximal chain argument to show that the existence of a regular covering by chains implies the LYM property. A *regular covering by chains* of $P$ is a collection **C** of maximal chains, not necessarily distinct, such that each element of rank $k$ occurs in the same proportion of the chains in **C**.

Graham and Harper [GHa] introduced a concept called the *normalized matching property*. A ranked poset $P$ has this property if for any $k$ and any collection $F \subset P_k$ the inequality $\frac{|F|}{N_k} \leqslant \frac{|F^*|}{N_{k+1}}$ holds, where $F^*$ is the $(k+1)$-shadow of $F$, i.e. the collection of elements in $P_{k+1}$ that cover elements of $F$. The truth of this was noted for $B_n$ by Sperner [Sr2] and for $M(\bar{e})$ by Anderson [An2], both before [GHa] defined the concept.

Kleitman [Kl13] proved the elegant result that the LYM property, the normalized matching property, and the existence of a regular covering by chains are all equivalent. [GK3] contains a summary of the proof, along with analogues of the LYM inequality for weight functions more general than cardinality and "ordered set-systems" more general than chains. They use this approach to unify results from [EKR], [Bo], [GKK], [Er1], [Ka5], [KSp], [Sc1], [Hs1], and [Kl8], which we will describe later.

Recently, Kleitman and Saks [KSa] obtained a stronger form of the LYM inequality for $B_n$. They found a way of increasing the individual terms without letting the sum exceed 1. In particular, if $q$ is the least value such that $\sum_{k=0}^{q} \frac{f_k}{\binom{n-1}{k-1}} > 1$, then

$$\sum_{k=0}^{q} \frac{q}{k} \frac{f_k}{\binom{n}{k}} + \sum_{k=q+1}^{n} \frac{n-q}{n-k} \frac{f_k}{\binom{n}{k}} \leqslant 1.$$

Their argument can be iterated to obtain further improvements, increasingly complicated as they approach the full generality of the Kruskal-Katona theorem discussed in Section 6.

The LYM inequality was generalized in a different direction by Levine and Lubell [LL], to a set $X$ of positive real numbers. Note that the *kth* elementary symmetric function $\delta_k(X)$ reduces to $\binom{n}{k}$ when all the variables equal 1. Let $w(A)$ be the product of the real numbers in the subset $A \subset X$, and let $F$ be an antichain. They proved that $\sum_{A \in F} \frac{w(A)}{\delta_{|A|}(X)} \leq 1$.

The LYM property holds for $B_n$, for $M(\bar{e})$ (by normalized matching [An2]), and for $L_n(q)$ (by regular coverings). When $\max\{m,n\} \geq 4$, it does not hold for the lattice $L(m,n)$ mentioned earlier, as noted in [Gg4],[We]. Spencer [Sp] showed that for large enough $n$, the lattice of partitions of an $n$-set also does not have the LYM property.

Griggs [Gg2] proved that the LYM property, together with symmetric unimodal Whitney numbers, suffices to guarantee the existence of a symmetric chain decomposition. Of course, the LYM property is not a necessary condition.

### 2.3 Direct Products

Kleitman [Kl1] and Katona [Ka2] noted that Sperner's Theorem could be strengthened by forbidding only relations of a particular sort. Let the $n$-set be divided into two parts $S_1$ and $S_2$, and let $F$ be a family of subsets such that no pair is related if their difference lies entirely within one of the parts $S_i$. Then $\max |F| \leq \binom{n}{\lfloor n/2 \rfloor}$.

More generally, consider the product of any two ranked posets $P$ and $Q$. A *semiantichain* is a collection $F$ of elements in $P \times Q$ such that a relation $(a,b) < (a',b')$, among elements of $F$ implies $a < a'$ and $b < b'$. In other words, the differences can't lie entirely in one component. $P \times Q$ has the *two-part Sperner Property* if its largest semiantichain consists of the elements of its largest rank.

By essentially the same argument used originally for $B_n$, the two-part Sperner property holds for products of multiset lattices [Sc1] and for products of symmetric chain orders [Ka4]. The symmetric chain decompositions of the component orders induce a partition of the product into "symmetric rectangles." A semiantichain cannot contain more than one element from any row or column of one of these rectangles. This constrains the semiantichain so that choosing the elements of middle rank produces the largest possible family. See [Gg1] for fuller discussion.

Products of three posets don't behave as nicely. We want to exclude only relations where the difference lies entirely in one part of a three-way division of $S$. Unfortunately, even for $n=3$ these families can exceed $\binom{n}{\lfloor n/2 \rfloor}$. "Three-part Sperner theorems" consist of proving sufficient additional conditions on $F$ for the maximum family to remain "the middle rank." Katona's initial result [Ka7] was improved by Griggs and Kleitman [GgK], and later Griggs [Gg3] gave an independent condition. These theorems all apply to products of symmetric chain orders. There appear to be no results on how large the family can be when these additional conditions are dropped, even for the case $B_n$.

Some results have been obtained for products in which the two-part Sperner property does not hold. For example, consider rectangles determined by integer points in the plane, $(0,0) \leq (a,b) \leq (M,N)$. These form a poset, ordered by inclusion, which is the product of the posets consisting of intervals in the two directions. West and Kleitman [WK] showed that the largest semiantichain consists of all the squares. More generally, they showed that if both posets have partitions into chains which meet consecutive ranks starting with rank 0, then the largest semiantichain in the product consists of all ordered pairs $(a,b)$ such that $\text{rank}_P(a)=\text{rank}_Q(b)$. G.W. Peck [Pe1] generalized the simpler case of this "anti-Sperner" theorem, finding an appropriate restriction so that the maximum family in $d$ dimensions would consist of all the $d$-dimensional cubes.

Considering direct products in another way, we can ask what Sperner-type properties are preserved under products. The following classes of posets are closed under the operation of direct product (in all cases the Sperner Property is included).

a) Symmetric chain orders [Ka4]. See also [Gg1].

b) LYM orders with logarithmically concave Whitney numbers $(N_k^2 \geq N_{k-1}N_{k+1})$ [Hr2], [HK].

c) Strongly Sperner posets with symmetric unimodal Whitney numbers [PSS], [Ca3].

The two known proofs of the last result both require linear algebra. [PSS] also contains a result applicable when symmetry does not hold. Define $P$ and $Q$ to be *compatible* if there exists a rank $k$ such that for all $i$ and $j$ we have $N_i(P)<N_j(P)$ if and only if $N_{k-i}(Q)\leq N_{k-j}(Q)$. Then the product of two strongly Sperner rank-unimodal posets that are compatible has the Sperner property and unimodal Whitney numbers. Finally, Saks [Sa2] strengthened this by dropping the unimodality condition; the product of strongly Sperner compatible posets is Sperner.

We note one other early paper on direct products that appears to have gone unnoticed. Liu [Lu] considered the direct product of $P$ with a $k$-element chain $C_k$ long before the work of Saks and others (see Section 3). Let $d_k(P)$ be the size of the largest $k$-family in $P$. Liu showed $d_k(P\times C_{l+1})\leq \sum_{t=0}^{\min\{l-1,k-1\}} d_{k-1+l-2t}(P)$. With this he obtained the $k$-family results of Erdös [Er1] and Schönheim [Sc1]. He also weakened the $k$-family restriction on the left side of the inequality to obtain a slight sharpening of these results. Finally, by considering a partition of $Q$ into chains, he obtained a bound on $k$-families in the product $P\times Q$. No one has developed his approach further.

### 2.4 Applications and Miscellany

The most celebrated application of Sperner's Theorem is to a problem known as the Littlewood-Offord problem. Given a set of $n$ vectors in a metric space, each having magnitude more than one, how many of its subsets can sum to vectors lying in a single sphere of diameter 1? The maximum is bounded by $\binom{n}{\lfloor n/2\rfloor}$. This was proved for dimension 1 in [Er1] using Sperner's Theorem and for dimension 2 in [Ka2], [Kl1] by using two-part Sperner theorems. Kleitman [Kl11] supplied the proof for arbitrary dimension by mimicking the [dBTK] construction of symmetric chains. The bound $\binom{n}{\lfloor n/2\rfloor}$ is attained when the vectors are all equal. This proof also gives the analogue for the union of $k$ such spheres. Recent papers have considered bounds when the sphere has larger size or the vectors must differ by some angle—see [Kl14], [Gg6]. Jones [Jo2] discussed a probabilistic version.

Erdös also posed a related problem for dimension 1. Given a set of $n$ *distinct* real numbers, how many of its subsets can sum to a single value, or more generally to one of $k$ values? He conjectured that the maximum is achieved when the set consists of the $n$ integers of smallest absolute value. The recent proof took several stages. Stanley [Sn] used algebraic geometry to give a condition equivalent to the Sperner property. Using that, he showed that a particular poset is Sperner. This implied the analogue of Erdös' conjecture when the $n$ real numbers must be strictly positive: the extremal set consists of the $n$ smallest positive integers. G.W. Peck [Pe2] used this result to prove Erdös' conjecture.

Sperner's Theorem gives extremal results for other problems as well. In theoretical computer science, the "keys" in a data base relation on $n$ attributes form an antichain in $B_n$, so their maximum number is $\binom{n}{\lfloor n/2 \rfloor}$—see [BkD]. The "basic elements" are a subset of the keys; a subsequent paper [BHFK] found bounds (not best possible) for the size of this antichain. [Ka8] contains an "iterated Sperner theorem" and two other applications to theoretical computer science.

Baumert, et al. [BMRR] proved an interesting probabilistic result that involves $\binom{n}{\lfloor n/2 \rfloor}$. Suppose $A$ and $B$ are chosen independently from $B_n$ according to some probability distribution. Then the probability that $A$ contains $B$ is at least $\binom{n}{\lfloor n/2 \rfloor}^{-1}$. Using several new concepts, Shearer and Kleitman [SK] gave a shorter proof which generalized this result. They defined two Dilworth decompositions to be *orthogonal* if no pair of elements occurs together in a chain in both partitions. The *partition number* of a poset is the maximum number of pairwise orthogonal Dilworth decompositions. To obtain the result above, they use the fact that the partition number of $B_n$ is at least two. They conjecture it equals $\lceil \frac{n+1}{2} \rceil$. We note it is easy to show this is an upper bound.

Meshalkin [Me] proposed another generalization of Sperner families which is vaguely similar to $k$-families. Consider partitions of $[n]$ into $r$ labelled blocks. A collection of these partitions is called an *$s$-system of order $r$* if the corresponding blocks from each partition form $r$ Sperner families in $B_n$. An $s$-system of order $r$ is like an $(r-1)$-family except that sets can repeat, the antichains are the same size, and corresponding sets must be disjoint.

Meshalkin [Me] proved that the number of partitions in an $s$-system of order $r$ is at most $\max \frac{n!}{\prod_{i=1}^{r} n_i!}$, where $\sum_{i=1}^{r} n_i = n$. Hochberg and Hirsch [HH] proved that there is an essentially unique system that achieves this bound. Together these results reduce to Sperner's Theorem when $r=2$. Hochberg [Ho1,2] studied various additional restrictions on these families.

Our final application concerns the counting of antichains in $B_n$. The total number equals the size of the free distributive lattice on $n$ generators—a problem posed by Dedekind. We list it as an application here because recent bounds have been obtained by analyzing the symmetric chain decomposition mentioned in Section 2.1. Antichains are also in 1-1 correspondence with monotone Boolean functions, which are order-preserving maps from $B_n$ to $\{0,1\}$. This monotonicity places restrictions on the allowable functions. Hansel [Ha1] suggested specifying the values of such a function by considering the elements according to the chain containing them in the deBruijn-Tengbergen-Kruyswijk decomposition, beginning with the smallest chains and proceding upward in size. He showed the function can never be undetermined on more than two elements of any chain when that chain is encountered. This yields an upper bound of $3^{\binom{n}{\lfloor n/2 \rfloor}}$.

Kleitman [Kl10] and Kleitman and Markowsky [KMk] (summarized in [Kl12]) refined this argument to improve the upper bound to $2^{\binom{n}{\lfloor n/2 \rfloor}(1+(\frac{\log n}{n}))}$. The lower bound of $2^{\binom{n}{\lfloor n/2 \rfloor}(1+O(\frac{1}{n}))}$ is due to Yamamoto [Ya].

Recently Balbes [Bl] obtained recursive formulas for the numbers of antichains in $B_n$ with $m_0, \ldots, m_n$ members at ranks $0, \ldots, n$. [Bl] and [Kl10] surveyed the earlier literature. Sands [Sa] examined extremal problems in enumeration of antichains in arbitrary posets, and Rosenthal [Rs] discussed an application to searching problems. Sands proved the existence of a constant $r$ (between 4.3 and 9) such that, in any poset with only two ranks, there always exists an element which belongs to more than $1/r$ and less than $1-1/r$ of the antichains. The analogous questions for longer posets are unresolved. Note that the number $A(P)$ of antichains in $P$ can be computed recursively by $A(P)=A(P-U_x)+A(P-D_x)$ for any $x \in P$, where $U_x=\{y: y \geqslant x\}$ and $D_x=\{y: y \leqslant x\}$. See [BK] for other recursive formulas. If $x$ can be chosen suitably, then these computations can be done "quickly".

## 3. Dilworth's Theorem—general posets

THEOREM: *Let $d(P)$ be the size of the largest antichain in $P$, and let $m(P)$ be the minimum number of chains whose union includes all of $P$. Then $d(P)=m(P)$.*

Like Sperner's Theorem, Dilworth's Theorem has been proved in many ways. However, most of the early proofs were applications of known techniques and did not provide generalizations. One could argue that Sperner's Theorem had a lead time of 22 years over Dilworth's Theorem, so that we should expect generalizations of Dilworth's Theorem to have a comparable delay. To some extent, this pattern is being followed. However, alternate proofs of the basic theorem followed rapidly.

### 3.1 Miscellaneous Proofs

Dilworth [Di1] originally proved the theorem by induction on $|P|$. The proofs cited most often use the König-Hall matching theorem (max matching = min vertex covering) or the Ford-Fulkerson network theorem (max flow = min cut). These first appeared in [DH], [Fu]. Perles [Pr1] streamlined Dilworth's proof; this argument is summarized in [Pz2]. A similar proof appears in [Tv]. The idea is to construct and remove a chain that intersects all maximum antichains, invoking induction on the width of $P$. The ability to do so rests on Dilworth's later observation [Di2] that the maximum antichains of $P$ form a lattice, ordered by $F \leqslant G$ if every element of $F$ is less than some element of $G$. The lattice operations are $F \vee G = \max(F \cup G)$ and $F \wedge G = \min(F \cup G)$, where $\max(Q)$ for a collection $Q \subset P$ denotes the elements of $Q$ which are maximal in $Q$ with respect to the ordering on $P$.

As an aside, although Dilworth's Theorem always holds when there is a fixed finite value for $d(P)$, we note that it does not always hold when the antichains can be arbitrarily large. Let $\beta$ be any infinite cardinal. Perles [Pr2] and Wolk [Wo] both construct examples of a poset which contains no infinite antichain but cannot be decomposed into fewer than $\beta$ chains. However, in the countable case, Oellrich and Steffens [OS] proved that every countable poset with no infinite chains has a Dilworth decomposition. They also characterized the countable "tree"-posets which have infinite chains but still have Dilworth decompositions. Results for maximal antichains of subsets of continua ("complete amonotonic decompositions") appear in [Ba1,2].

**3.2 The Greene-Kleitman result.**

The primary extant generalization of Dilworth's Theorem is that by Greene and Kleitman. It computes the size of the largest $k$-family in a poset in terms of a dual covering problem.

Consider an arbitrary poset $P$. Let $d_k$ be the size of its largest $k$-family. (Saks calls this the $k^{th}$ *Dilworth number).* Given any chain partition $\mathbf{C}$ of a poset let $m_k(\mathbf{C}) = \sum_{C \in \mathbf{C}} \min\{k, |C|\}$. No chain contributes more than $k$ to a $k$-family, so $d_k \leqslant m_k(\mathbf{C})$. If equality holds then $\mathbf{C}$ is called a *k-saturated* partition. Dilworth decompositions are 1-saturated partitions. If $\mathbf{C}$ is $k$-saturated for all $k$, then $\mathbf{C}$ is a *completely saturated* partition. It is easy to show that not every poset has a completely saturated partition, but Greene and Kleitman [GK1] proved the following remarkable fact: For every poset $P$ and every $k$, there exists a partition $\mathbf{C}$ which is both $k$ and $(k+1)$-saturated. This ability to move from $k$ to $k+1$ has been named the $t$-phenomenon ("transistion phenomenon") by Hoffman. The Greene-Kleitman proof first extended Dilworth's algebraic result [Di2] on antichains by showing that the $k$-families of $P$ form a lattice, with the maximum $k$-families forming a sublattice. Then a delicate examination of the sequence of differences $\Delta_k = d_k - d_{k-1}$ gave the result. The $t$-phenomenon implies that the $\Delta_k$ form a decreasing sequence. Later, Fomin [Fo] independently obtained these results by a shorter, more direct combinatorial proof, and then Frank [Fk] obtained them using network flows.

Hoffman and Schwartz [HS] proved a slight generalization by transforming this extremal problem into a related "transportation" problem, where duality is known to hold. We will not discuss integer programming aspects of the dual combinatorial problems in this section, because that approach is examined by Hoffman [Hf] in this volume.

Recently, Saks [Sa1] gave a short proof of the existence of $k$-saturated partitions by examining a special type of 1-saturated partition in the product of $P$ with a $k$-element chain. He uses a special case of another result [Sa3], that $d_1(P \times Q) \leqslant \sum_k \Delta_k(P)\Delta_k(Q)$. His proof does not yield the $t$-phenomenon. In [Sa2] he gives a bound on the Dilworth numbers of a product poset in terms of the Dilworth numbers of its factors.

Saks' investigation suggests that the Greene-Kleitman result may be a special case of a more general result involving direct products. West and Tovey [WT1] discuss such a generalization, originally suggested by West and Saks. Define a *unichain* in a direct product $P \times Q$ to be the product of an element from one poset with a chain from the other. They conjecture that the size of the largest semiantichain equals the smallest number of unichains which cover the direct product; obviously it is never greater. They prove a special case which includes all previous two-part Sperner Theorems and also products of LYM orders that have completely saturated partitions. They also conjecture a generalization of the $t$-phenomenon to direct products. The conjecture reduces to the Greene-Kleitman result when either poset is a chain. In attempting to prove the conjecture, they obtained a sufficient condition [WT2] for the largest semiantichain to be an antichain; it requires a similar equality between two covering problems.

We mention one more duality result that has a product flavor, but for which no relation to the ideas above has yet been found. Chaiken, et al. [CKSS], studied rectilinearly convex regions in the plane. Consider points and rectangles in the region, with two points independent if no rectangle contains both. Then the largest number of independent points equals the smallest number of rectangles that cover the region.

### 3.3 Conjugate duality

The Greene-Kleitman Theorem proves a duality between a union of $k$ antichains and a single covering by chains. Since antichains and chains are dual objects, there may be a dual theorem. For $k=1$ this was noted by Mirsky [Mr]: A poset which has no $(m+1)$-element chain is the union of $m$ antichains. Greene [Gr3] extended this result for all $k$. Let $\hat{d}_k$ equal the size of the largest union of $k$ chains, let $\hat{\Delta}_k=\hat{d}_k-\hat{d}_{k+1}$, and let $\hat{m}_k(\mathbf{A})=\sum_{A\in\mathbf{A}}\min\{k,|A|\}$ for any antichain partition $\mathbf{A}$. Say that $\mathbf{A}$ is *k-saturated* if $\hat{d}_k=\hat{m}_k(\mathbf{A})$. Greene showed there always exist simultaneously $k$ and $(k+1)$-saturated antichain partitions. Even more striking, he proved that $\Delta$ and $\hat{\Delta}$ are conjugate partitions of $|P|$. (Partitions $\lambda$ and $\mu$ are *conjugate* if $\sum_{i=1}^{k}\lambda_i=\sum_{i=1}^{k}\min\{k,\mu_i\}$. There is a more common and intuitive graphical definition, but the similarity of this definition to $m_k$ and $\hat{m}_k$ may make Greene's result less mystifying.) The conjugate partition result and the $t$-phenomenon for both chains and antichains were also proved by Fomin [Fo] and later by Frank [Fk].

[Gr2] is an excellent early survey of the related features of chains and antichains. Examples of conjugate partitions with a combinatorial interpretation appear as early as [Gr1]. Saks [Sa3] provided a synthesis and extension of previous conjugate duality results. This has stimulated considerable interest in packing or covering problems which have a *rank sequence* associated with them – e.g., the largest union of $k$ of the collection. Results of this sort, further questions, and applications of these ideas appear in [Sa3], [AB1,2], [Ga], [Al], [AC] and [ACS]. For example, [AC] and [ACS] include duality results on relations in a poset. They define two pairs of related elements to be *independent* if the elements do not all lie in a single chain of $P$. Then the minimum number of independent collections of pairs needed to cover all the relations of $P$ equals the maximum number of relations in a single chain. However, in this case the conjugate duality does not hold.

### 3.4 Applications and Miscellany

Griggs [Gg4] defined several chain-related properties that a poset might have. For example, $P$ has property $T$ if for each value $k$ it has a collection of chains, each containing an element from each of the $k$ largest ranks, which are disjoint and together cover the *kth* largest rank. He showed that strongly Sperner posets have property $T$, and that the converse is true for posets with unimodal Whitney numbers. Shorter proofs appeared in [GSS]. Griggs also discussed the relationship between property $T$ and the existence of completely saturated partitions. It remains an open question whether posets with the LYM property always have completely saturated partitions. [Ta] gives a necessary and sufficient condition for a given maximal chain to belong to a $k$-saturated partition. The condition is $d_k(P)=d_k(\hat{P})$, where $\hat{P}$ is a modification of $P$ obtained by adding a small number of elements. This is useful, since the values $d_k$ or $\hat{d}_j$ can be computed "quickly" for any poset; Frank [Fk] provided a network flow algorithm to compute them all.

The dimension of a poset $P$– first introduced by Dushnik and Miller [DM]– is the smallest number of linear orders whose intersection equals the relations in $P$. We will not say much about dimension and other poset parameters whose computation is an extremal problem, because they are discussed in [KT], in this volume. However, we do note that the width of a poset is an upper bound on its dimension, as shown by Hiraguchi [Hu]. Dilworth [Di2] showed that the dimension of a distributive lattice equals the width of a related poset. Trotter [Tr1] considered a variant which is related to the problem of partitioning $P$ into chains with less than $k$ elements. Finally, Pretzel [Pz1] introduced a variant called the "stable dimension", which he showed equals the maximum size of a 2-family.

Applications of Dilworth's Theorem appear in [DD], [MTN] and [GS]; the last reviews solution methods. The "Dilworth number" of a graph is defined in [FHa]; it is the width of a poset related to the graph. In number theory, Riddel used Dilworth's theorem to obtain a lower bound for a problem of Moser, as mentioned in [EKs]. [Po] constructed a countable poset covered by two chains which contains all countable posets having width 2. Finally, results related to [Di2] appear in [BS] and [Fa], and results related to saturated partitions appear in [DD] and [Gg5].

## 4. Maximizing family size

In this section we consider other conditions that can be placed on families of poset elements. These do not require that $F$ be an antichain. We ask for the maximum family under various combinations of these restrictions, which may or may not include the antichain condition. First we describe the conditions, and then give the maximum families for various combinations. Later we discuss maximum families for various weakened versions of the conditions, and also some questions involving more than one family at a time.

Even more so than with the other sections, this section concentrates on listing a large number of results in a small space. It may seem tedious on first reading and may serve best as a reference tool. The reader who becomes fatigued is advised to skip to section 5.

### 4.1. Standard restrictions and their combinations

At first, these problems were all posed and studied for the poset $B_n$, so we will also state the conditions that way. They can be generalized to ranked lattices. For example, restrictions on the sizes of intersections and unions can become restrictions on the ranks of meets and joins. Some results have been generalized to $M(\bar{e})$ and fewer to $L_n(q)$; we will mention them.

Survey articles on these questions include parts of [EK2], [Ka9], [BD4], [Be2], and [BS]. [Be2] is oriented towards graphical aspects of hypergraphs. [BS] deals only with collections of $k$-sets (uniform hypergraphs). [BD4] deals with a few subsets of the restrictions. [Ka9] is the most thorough, mentioning material both of this section and of section 6. [EK2], in addition to presenting results on the maximum-sized collection (beware typographical errors), asks for the number of maximum families and the minimum size of a maximal family. Except for Dedekind's problem, only a few isolated results are known about the latter questions.

In this section, $A$ and $B$ will be arbitrary elements of some family $F \subset B_n$. We first consider combinations of the following restrictions on $F$. (The letters $I, U, C$ were first used in this way in [BD4].)

$C$: $A \not\subset B$. No pair is related; $F$ is an antichain.

$I$ (or $I_t$): $|A \cap B| \geq 1$ ($|A \cap B| \geq t$). Every pair intersects (in at least $t$ elements). Families satisfying $I_t$ have also been called *t-intersecting* families.

$U$ (or $U_s$): $|A \cup B| \leq n-1$ ($|A \cup B| \leq n-s$). This is dual to condition $I$. Most results that specify only one of them could specify the other instead, and we will usually state results using $I$.

$rI_t$ (or $rU_s$, etc.): Any $r$-tuple of elements of $F$ has the property previously specified for pairs. For example, $|A_1 \cap \cdots \cap A_r| \geq t$. The condition $rI_t$ can be read as "$r$-wise $t$-intersecting."

$I_t^*$: The use of an asterisk denotes various restrictions added to exclude families where *all* the elements contain some fixed $t$-set. However, $I_t$ must still hold.

$S_{=k}$ ($S_{\leq k}$): The size – i.e., rank – of elements must equal (must not exceed) $k$.

When a family satisfies some set of conditions $X$, we may call it an *X-family*.

To organize the results and make it easier to remember the implications between them, we provide a poset (see Fig. 1). The elements of the poset are subsets of the above conditions for which results are known about maximum families. In this poset, $X \geq Y$ if the conditions $X$ imply the conditions $Y$, in which case $\max_Y |F| \geq \max_X |F|$.

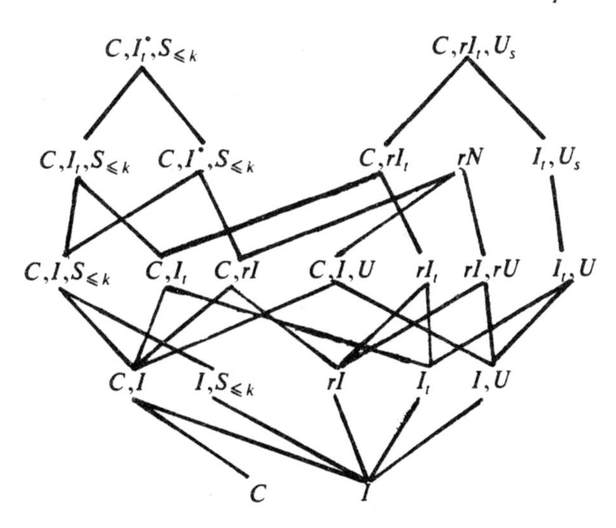


Fig. 1.

In many cases, one set of conditions can be obtained from another by restricting parameter values. We include them as separate points for several reasons:

1) history– The more specialized problem was probably solved separately and earlier.
2) The more specialized problem may be interesting or applicable in its own right.
3) The more general problem may not yet be solved for posets more general than $B_n$.

### 4.2. Maximum families

Unless otherwise specified, each result in this section gives the maximum size of a family in $B_n$, subject to the conditions which head that paragraph. For compactness in describing the results, we define additional notation. Recall that $B_n^k$ and $M_k(\overline{e})$ are the $k^{th}$ ranks of $B_n$ and $M(\overline{e})$, and $[j]=\{1,\cdots,j\}$. In addition, let $[i,j]=\{i,i+1,\cdots,j\}$, $B_n^k[j]=\{A\in B_n^k\colon [j]\subset A\}$, and $B_n^k[\hat{j}]=\{A\in B_n^k\colon [j]\not\subset A\}$. In the last definition, note that $A\cap[j]=\emptyset$ is not required. As before, when we say an extremal family is unique, we mean unique up to permutation or complementation. We proceed to the results.

$C$: Again, $\max|F|=\binom{n}{\lfloor n/2\rfloor}$, achieved uniquely by $F=B_n^{\lfloor n/2\rfloor}$.

$I$: By itself, this is trivial and noted in [EKR]: $\max|F|=2^{n-1}$, achieved by $F=\bigcup_{k=1}^{n} B_n^k[1]$. Another extremal family, writing $n$ as $2m+1$ or $2^n$, is $F_{2m+1}=\bigcup_{k=m+1}^{2m+1} B_n^k$, $F_{2m}=(\bigcup_{k=m+1}^{2m} B_n^k)\cup(B_n^m[\hat{1}])$. The latter is the configuration that generalizes . . .

$I_t$: Katona [Ka1] showed that the best approach is to force intersections by using elements with large cardinality. Suppose $t\leqslant n$ and $q=\lceil\frac{n+t}{2}\rceil$. Then

$$\max|F|=\sum_{i=q}^{n}\binom{n}{i}+\begin{cases}0 & (n+t\text{ even})\\ \binom{n-1}{q-1} & (n+t\text{ odd})\end{cases},$$

achieved by $F=\bigcup_{k=q}^{n} B_n^k$ if $n+t$ is even and by $F=(\bigcup_{k=q}^{n} B_n^k)\cup B_n^{q-1}[\hat{1}]$ if $n+t$ is odd.

There are several ways to generalize intersection to multisets. For the first, we say $I_t$ requires $\operatorname{rank}(\overline{a}\wedge\overline{b})\geqslant t$ for all $\overline{a},\overline{b}\in F$. The case $t=1$ was solved (in divisor terminology) by Erdös and Schönheim [ES] and by Woodall [unpublished]. Suppose $M(\overline{e})$ is a product of n chains. Then $\max|F|=\frac{1}{2}\sum_{A\subset[n]}\max\{w(A),w(\overline{A})\}$, where the weight of a set is the product of the corresponding $e_i$. This is achieved by letting $F$ consist of all $\overline{a}$ which are non-zero precisely on sets of indices $A$ for which $w(A)>w(\overline{A})$ (appropriately modified in the case of equality). In a subsequent paper with Herzog [EHS], they determined the minimum size of a maximal $I_t$-family of multisets. It is $e_n\prod_{i<n}(e_i+1)$, achieved by $F=\{\overline{a}:a_n\geqslant 1\}$. Gereb [Ge] gave another proof and showed that these are essentially the only minimum maximal families. Note that this definition of $I_t$ applies to all ranked lattices. A slight variation, in

which $I_t$ requires $\bar{a} \wedge \bar{b}$ to be nonzero in at least $t$ components, applies only to multisets. They coincide when $t=1$, but little is known about either when $t>1$. We will not refer to this vairation again, using instead the following definition.

What we call the second generalization of intersection to multisets is more natural for some questions, although it does not extend to more general ranked posets. Here we say that $I_t$ holds for $F \subset M(\bar{e})$ if for all $\bar{a}, \bar{b} \in F$ the vector $\bar{a}+\bar{b}$ exceeds the vector $\bar{e}$ in at least $t$ positions. In particular, viewing $M(\bar{e})$ as divisors of some integer $M$, the condition $I$ means that $ab \nmid M$. Greene and Kleitman [GK3] pointed out that under this interpretation the largest $I$-family in $M(\bar{e})$ has size $\lfloor \frac{1}{2} \mid M(\bar{e}) \mid \rfloor$. For $I_t$, we expect the analogue of Katona's [Ka1] result to hold as follows. Suppose all $e_i$ are odd, $n+t$ is odd, and $q = \lceil \frac{n+t}{2} \rceil$. Then we conjecture the maximum $I_t$-family is $F = \{\bar{a}: a_i > e_i/2$ in at least $q$ positions$\}$. If $n+t$ is even, add $\{\bar{a}: a_i > e_i/2$ in $q-1$ positions other than $a_1\}$. If the $e_i$ are not all odd, an appropriate modification must be made.

$C,I$: Let $q = \lceil \frac{n+1}{2} \rceil$. Purdy conjectured that here $\max |F| = \binom{n}{q}$, as achieved by $F = B_n^q$. Brace and Daykin [BD4] proved this for odd values of $n$, and later Schönheim [Sc3] proved it for even values of $n$.

$C,I_t$: Milner [Ml2] completely solved this. Let $q = \lfloor \frac{n+t+1}{2} \rfloor$. Then $\max |F| = \binom{n}{q}$. This is achieved uniquely by $F = B_n^q$ when $n+t$ is even; when $n+t$ is odd, $F = B_n^{q-1}[t] \cup B_n^q[\hat{t}]$ also works. Here again the multiset generalization would be interesting. Milner's result actually generalizes Sperner's Theorem, which is the degenerate case $t=0$.

$C,I,S_{\leqslant k}$: This is the Erdös-Ko-Rado theorem in its basic form, perhaps the best-known result of this section. For $k \leqslant n/2$, $\max |F| = \binom{n-1}{k-1}$, achieved by $F = B_n^k[1]$, uniquely if $k < n/2$. Some authors strengthen the condition $C,S_{\leqslant k}$ to $S_{=k}$ to speak of "uniform hypergraphs." However, in theorems about $k$-sets the restriction $S_{=k}$ can often be weakened to $C,S_{\leqslant k}$ with no increase in the maximum family.

Proofs and generalizations of the EKR theorem can be found, for example, in [Ka6], [Bo], [GKK], and [Lo]. The most direct proof is due mostly to Katona [Ka6] and proceeds as follows. Under the condition $C,I,S_{\leqslant k}$, any circular arrangement of $[n]$ contains at most $k$ members of $F$ appearing as contiguous segments. Totalling this over all circular arrangements and eliminating multiple counting proves the result. This argument was strengthened in [Bo] and in [GKK] to obtain an LYM-type inequality: when $F$ satisfies $C,I,S_{\leqslant n/2}$, $\sum_{A \in F} \binom{n-1}{|A|-1}^{-1} \leqslant 1$. Lovász [Lo] proved the $I,S_{=k}$ result using an eigenvalue method to compute the Shannon capacity of a related graph. Stanton [St] extended this to get an EKR-type theorem for "Chevalley groups." Erdös and Kleitman [EK2] asked whether the minimum size of a maximal $I,S_{=k}$-family is $\binom{2k-1}{k}$, achieved by restricting the subsets to the first 2k-1 elements.

Greene and Kleitman [GK3] generalized the EKR-bound to multisets, using the "generalized Macaulay theorem" discussed in Section 6. Let $F$ be an $I,S_{=k}$-family with $k \leqslant \Sigma e_i/2$, using the second generalization of $I$. Generally speaking, $|F|$ is maximized by having all pairs intersect in the $n$th component, but this must be modified slightly when $e_n$ is even. For arbitrary $e_n$, the maximum family is $F=\{\bar{a}\in M_k(\bar{e})\colon a_i>e_i/2 \text{ for some } i \text{ and } a_j\geqslant e_j/2 \text{ for all } j>i\}$.

Chvátal considered related questions for $B_n$. He discovered [Ch1] a weakened intersection property that can still hold in an $S_{=k}$-family with more than $\binom{n-1}{k-1}$ elements. He also proved [Ch2] a slight generalization of the EKR theorem involving a partial order he defined on $B_n^k$. If $G$ is an $I$-subfamily of an order ideal $F$ in that partial order, then $|G|\leqslant|F\cap B_n^k[1]|$.

Liggett [Lg] strengthened the EKR-theorem (in its $I,S_{=k}$ version) by proving $|F|/(|F|+|\hat{F}|)\leqslant k/n$, where $2k\leqslant n$, $F\subset B_n^k$, and $\hat{F}$ is the collection of $k$-sets disjoint from $k$-sets in $F$. He generalizes this further and gives a statistical application. In its simplest case, the application states: If $\{Y_i\}$ are independent $\{0,1\}$-random variables such that $\mathrm{Prob}(Y_i=1)=p$ with $p\geqslant\frac{1}{2}$, and $\{\alpha_i\}$ are non-negative weights summing to 1, then $\mathrm{Prob}(\Sigma\alpha_i Y_i\geqslant\frac{1}{2})\geqslant p$.

$C,I_t,S_{\leqslant k}$: This is the full Erdös-Ko-Rado [EKR] result: $\max|F|=\binom{n-t}{k-t}$ for $n>n_0(k,t)$, achieved uniquely by $F=B_n^k[t]$. For small values of n, larger families exist, but not much is known about them. [EKR] contains the first upper bound for $n_0(k,t)$, strengthened in [Hs1]. Finally, Frankl [Fr4] proved Erdös' conjecture that $n_0(k,t)<(k-t+1)t(1+\epsilon)$; furthermore, he proved that $n_0(k,t)=(k-t+1)(t+1)$ for $t\geqslant 15$. Hsieh [Hs1] proved the analogous theorem for $L_n(q)$: the maximum number of $k$-dimensional subspaces of $L_n(q)$ which pairwise intersect with dimension at least $t$ is $\begin{bmatrix}n-t\\k-t\end{bmatrix}_q$, for $n\geqslant n_0(k,t)$. He also showed $n_0(k,t)=2k+1$ if $q\geqslant 3$, $2k+2$ if $q=2$.

$C,I_t^*,S_{\leqslant k}$: We consider various ways to exclude the family $B_n^k[t]$. When $t=1$, $k\leqslant n/2$ and the members of $F$ cannot all contain any single element, Hilton and Milner [HM] showed that $\max|F|=\binom{n-1}{k-1}-\binom{n-k-1}{k-1}+1$. Letting $D=[2,k+1]$, this is achieved by $F=D\cup B_n^k[1]-\{A\in B_n^k[1]\colon A\cap D=\varnothing\}$. More generally, suppose $F$ is a $C,I,S_{\leqslant k}$-family with $k<n/2$ such that leaving out less than $s$ of its elements leaves a collection with void intersection. They showed that

$$\max|F|=\begin{cases}\binom{n-1}{k-1}-\binom{n-k}{k-1}+n-k & 2<k\leqslant s+2\\ \binom{n-1}{k-1}-\binom{n-k}{k-1}+\binom{n-k-s}{k-s-1}+s & k\leqslant 2 \text{ or } k\geqslant s+2\end{cases}$$

Frankl [Fr8] generalized their $t=1$ result in a different direction. For $n>n_0(k)$ he determined the largest $I_t,S_{=k}$-family such that $|\bigcap_{A\in F}A|<t$. Letting $D=[n-k+t,n]$, the family is (uniquely)

$$\{A\in B_n^k\colon [t]\subset A,\ A\cap D\neq\varnothing \text{ or } D\subsetneq A,\ |A\cap[t]|=t-1\} \quad \text{for } k>2t+1$$
$$\{A\in B_n^k\colon |A\cap[t+2]|\geqslant t+1\} \quad \text{for } k\leqslant 2t+1.$$

He also obtained extensive and elegant results for $I_t^*,S$-families where $I_t^*$ means $I_t$ plus the condition that no element of $[n]$ can appear in more than a fraction $c$ of the elements of $F$. These results all require $n>n_0(c,k)$. An earlier unpublished result of Erdös, Rothschild and Szemerédi states that for $c>2/3$, the maximum family is $F=\{A\in B_n^k\colon|A\cap[3]|\geqslant 2\}$. (Note: the sets with $[3]\subset A$ don't cause a problem, because for large enough $n$ the overall proportion of these is small enough that the proportion containing a single element gets arbitrarily close to 2/3.) More generally, Frankl [Fr8] showed the following for $n > n_0(k,t)$. Suppose $F$ is an $I_t, S_{=k}$-family where, for some $\epsilon$ with $0<\epsilon<1/(t+2)$, no element of $[n]$ appears in more than $(1-\epsilon)\,|F|$ of the sets. Then $\max|F| = (t+2)\binom{n-t-2}{k-t-1}+\binom{n-t-2}{k-t-2}$, achieved uniquely by $F=\{A\in B_n^k: |A\cap[t+2]|\geqslant t+1\}$. He obtained further, more delicate results of this sort. Also, for $t=1$ he verified the conjecture of Erdös, Rothschild, and Szemerédi that for $3/7<c<1/2$, and $n\geqslant n_0(c,k)$, the largest $I_t, S_{=k}$ family consists of the $k$-sets which contain at least one triple from some projective plane on 7 elements (7-3 Steiner system). Along the same lines, Füredi [Fd] recently obtained some constructive bounds on the minimum size of a maximal $I, S_{=k}$-family; these also involve projective planes.

$I, S_{h,k}$: In [Hi3], Hilton weakened the condition $S_{=k}$ to $h\leqslant |A|\leqslant k$ for $A\in F$. Adding the intersection condition has the expected effect; $\max|F| = \sum_{i=h}^{k}\binom{n-1}{i-1}$, achieved uniquely. If in addition we require $\bigcap_{A\in F} A=\varnothing$, he obtained the corresponding generalization of the [HM] result on $C, I^*, S_{\leqslant k}$-families mentioned above: $\max|F| = 1+\sum_{i=h}^{k}\left[\binom{n-1}{i-1}-\binom{n-h-1}{i-1}\right]$.

$rI_t$: The extremal $I$-family $B_n[1]$ is in fact $r$-wise intersecting for all $r\leqslant 2^{n-1}$, but this is not true for the extremal $I_t$-families. Frankl [Fr11] recently obtained the following result for the symmetric condition $rU_t$. Let $E(n,k,t,s)=\{A\in B_n: |A\cap[t+ks]|\leqslant s\}$, where $t+ks\leqslant n$. Then the maximum size of an $rU_t$-family is $\max_s |E(n,k,t,s)|$ for $k>2$, $t\leqslant k2^k/150$. For $t=1$, the largest family is always $E(n,k,1,1)$, as shown in [BD1]. See also [Fr4a].

$rI^*$: What is the maximum size of an $r$-wise intersecting family where some $(r+1)$-tuple has void intersection? [EK2] asks also for the minimum size of a maximal family.

$C, rI$: Frankl [Fr2] obtained an LYM-type inequality for $C, rI, S_{\leqslant k}$-families, generalizing the EKR theorem and the LYM-type inequality mentioned earlier for $k=n/2$. If $k\leqslant n(r-1)/r$, then $\sum_{A\in F}\binom{n-1}{|A|-1}^{-1}\leqslant 1$. He used this to determine the largest $C,3I$-family in $B_n$, for $n>n_0$. Letting $q=\lfloor\frac{n-1}{2}\rfloor$, it is unique and consists of $\binom{n-1}{q}+\epsilon$ elements, where $\epsilon=0$ for odd $n$ and $\epsilon=1$ for even $n$. See also [Go1]. Finally, Gronau [Go2] showed that for $r\geqslant 4$, the maximum $C, rI$-family has size $\binom{n-1}{q}$, achieved uniquely by $B_n^{\lfloor\frac{n-1}{2}\rfloor+1}[1]$. He also completed Frankl's results on $r=3$ to all values of n.

$I, U$: This result was conjectured by Brace and Daykin [BD4] and proved in various ways by Daykin and Lovász [DL], Schönheim [Sc2], and Seymour [Se], and later by Hilton [Hi4] and Anderson [An3]. If $F$ is an $I, U$-family in $B_n$, then $\max|F| = 2^{n-2}$. This is achieved by $F=\{A\in B_n: a\in A,\ b\notin A\}$, where $a,b$ are fixed elements of $[n]$.

Extensions to multisets are known under both generalizations of $I$. First let $I,U$ mean $\mathrm{rank}(\bar{a}\wedge\bar{b})>0$, $\mathrm{rank}(a\vee b)<\mathrm{rank}(M(\bar{e}))$ and let $f(\bar{e})$ be the Erdös-Schönheim bound on the largest $I$-family. Anderson [An3] obtained $f(\bar{e})^2/|M(\bar{e})|$ as an upper bound on the size of $I,U$-families, but it is not best possible.

For $I,U$-families, the second generalization is more appropriate. Rephrasing the definition, condition $I,U$ requires $\bar{a}+\bar{b},\bar{a}+\bar{b}-\bar{e}\notin M(\bar{e})$ for all $\bar{a},\bar{b}\in F$. In divisor terminology, $ab\nmid M$ and $M\nmid ab$. Daykin, Kleitman and West [DKW] proved that an $I,U$-family in $M(\bar{e})$ has size at most $\frac{1}{4}|M(\bar{e})|$. This result is best possible when $\bar{e}$ has at least two odd values, say $e_i$ and $e_j$, in which case the bound is achieved by $F=\{\bar{a}\in M(\bar{e})\colon a_i<e_i/2,\ a_j>e_j/2\}$. It is almost best possible when one value is odd, and its status has not been explored when all $e_i$ are even.

$I_t,U$: Katona [Ka9] conjectured and Frankl [Fr1] proved that the largest $I_t,U$-family in $B_n$ is the largest $I_t$-family in $B_n[\hat{1}]$.

$I_t,U_s$: Wang [Wa] obtained recursive bounds here. Let $f(n,t,s)$ be the maximum size of an $I_t,U_s$-family in $B_n$; he showed that $f(n,t,s)\leqslant f(n-1,t-1,s)+f(n-1,t+1,s)$. This yields the known optimum for $s=0$ and $s=1$, but for $s\geqslant 2$ the family he constructs is smaller than the solution to the recurrence. Bang, Sharp, and Winkler (and they presume also Frankl) have an interesting conjecture for the optimal family. For some fixed value of $q$, let $G$ be the maximum $I_t$-family in $B_q$, and let $H$ be the maximum $U_s$-family in $B_{n-q}$. Form $F$ by pairing each element from $G$ on $[1,q]$ with each element from $H$ on $[q+1,n]$. In other words, the first $q$ components are devoted to satisfying $I_t$, and the last $n-q$ to satisfying $U_s$. They conjecture that the largest $F$ so formed is the maximum $I_t,U_s$-family. They report that Frankl has results for $rI_t,rU_s$-families that do not hold when $r=2$.

$C,I,U$: This case is similar to $I_t,U$. The maximum family is the largest $C,I$-family in $B_n[\hat{1}]$, as shown by Brace and Daykin [BD4]. Nothing is known about extensions to multisets.

$C,rI_t$: The complementary result is easier to phrase. We want the union of any $r$ elements of $F$ to contain at most $n-t$ elements. Express $n-t$ as $qr+s$ with $0\leqslant s<r$, and let $h=\min\{q,s/2\}$. Frankl [Fr0] showed that $\max|F|=\sum_{i=0}^{h}\binom{n-s}{q-i}\binom{s}{\lfloor s/2\rfloor+i}$ for $n>n_0(r,t)$, achieved by $F=\{A\in B_n^{q+\lfloor s/2\rfloor}\colon |A\cap[n-s]|\leqslant q\}$.

$C,rI_t,U_s$: Frankl [Fr10] recently considered this condition, which implies each of the last three conditions listed above. Phrased in complementary terms, we want an $s$-intersecting family such that the union of any $r$ elements has size at most $n-t$. Frankl obtained an upper bound on $|F|$ for large $n$ and determined when equality can hold. The bound is a sum of products like that for $C,rI_t$-families, but more complicated.

$rN$: Finally, we mention a simple but very strong restriction that implies all of $C$, $rI$, and $rU$. We say that a family is *r-independent* if, for any distinct $A_1,\ldots,A_r\in F$ and every choice of $B_i=A_i$ or $B_i=\bar{A}_i$, $|B_1\cap\cdots\cap B_r|\geqslant 1$. For $r=2$, the condition is equivalent to $C,I,U$, and the question was posed and partially solved by Renyi. The answer is $\max|F|=\binom{n-1}{\lfloor n/2\rfloor-1}$, as proved in [KSp] and [Ka8]. Kleitman and Spencer [KSp] also explored the case $r>2$, obtaining

$2^{cn/r2^r} \leqslant \max |F| \leqslant 2^{dn/2^r}$, where $c$ and $d$ are constants and $n$ is sufficiently large. The lower bound came from a probabilistic counting argument. No results are known about the extensions to multisets, nor about the requirement $|\cap B_j| \geqslant t$, suggested by Katona [Ka8] for the case $r=2$.

### 4.3. Weakened restrictions

Various ways to weaken conditions $C$ and $I$ have been examined. We did not include their combinations in Fig. 1, but they would appear below the corresponding points using $I$, $I_t$, $rI_t$, or $C$. The poset of conditions discussed here appears in Fig. 2. The dotted lines are relations to previously discussed conditions. Possible replacements for condition $I$ in $B_n$ include

$rP_t$: Any $r$-tuple $A_1, \ldots, A_r \subset F$ contains a pair $A_i, A_j$ such that $|A_i \cap A_j| \geqslant t$. In particular, $2P_t = I_t$.

$J_s$: No subcollection of $2(s+1)$ distinct elements of $F$ can be partitioned into $s+1$ disjoint pairs, called $s+1$ "simultaneously disjoint pairs." In particular, $J_0 = I$.

The conditions we consider in place of the antichain condition are

$Q_1$: $F$ contains no $A,B,C$ with $A \cup B = C$ or $A \cap B = C$.
$Q_2$: $F$ contains no $A,B,C$ with $A \cup B \subset C$.
$Q_3$: $F$ contains no $A,B,C$ with $A \cup B \supset C$.
$Q_4$: $F$ contains no $A,B,C$ with $A \cup B = C$.
$Q_5$: $F$ contains no $A,B,C$ with $A \cup B = C$ with $A \cap B = \varnothing$.
$R_k$: $F$ contains no copy of $B_k$.
$C_k$: $F$ contains no chain of $k+1$ elements.
$kQ_6$: $F$ contains no $A \subset B$ with $|B-A| \geqslant k$.

The antichain condition implies all of these except $Q_3$, which is independent of it.

First we consider the weakenings of $I$.

$J_s$: Generally, this condition allows the addition of $s$ elements beyond the corresponding intersecting family, for appropriate values of $s$. Hilton [Hi2,2a] studied the substitution of $J_s$ for $I$ in the following cases.

$J_s$: $\max |F| = 2^{n-1} + s$.

$C, J_s$: $\max |F| = \binom{n}{\lceil \frac{n+1}{2} \rceil}$, regardless of $s$, unless $n$ is even and $s$ is very large.

$J_s, S_{\leqslant k}$: $\max |F| = \sum_{i=1}^{k} \binom{n-1}{i-1} + s$ if $2 \leqslant k \leqslant n/2$ and $s \leqslant n-4$.

$C, J_s, S_{\leqslant k}$: $\max |F| = \binom{n-1}{k-1} + s$ if $k \leqslant n/2$ and $s < \binom{n-k-1}{k-1}$.

$J_s, S_{h,k}$: $\max |F| = \sum_{i=h}^{k} \binom{n-1}{i-1} + s$ [Hi2a].

Hilton [Hi2a] also obtained partial results for families which do not have $s+1$ pairwise disjoint $r$-tuples, with all the elements in all the $r$-tuples being distinct. For $s=0$, this becomes condition $rP$. Clements [Cl7] determined the minimum number of simultaneously disjoint pairs in a collection of fixed size, which is a problem of the type

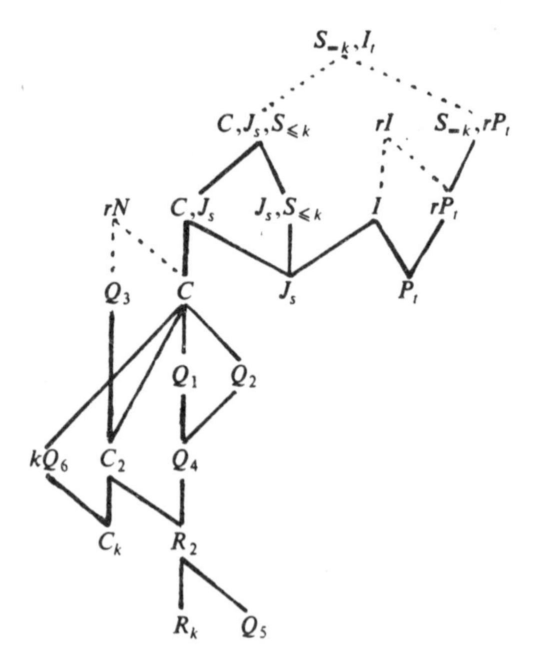


Fig. 2.

examined in Section 6.

$rP_t,S_{=k}$: Erdös [Er2] solved this for the case $t=1$, showing that then $\max|F|=\binom{n}{k}-\binom{n-r+1}{k}$, achieved by $F=\{A\in B_n^k\colon |A\cap[r-1]|\geq 1\}$. Bollobás, Daykin, and Erdös [BDE] obtained a result that includes this and the [HM] result on $C,I^*,S_{\leq k}$-families.

For general $t$, the solution is due to Hajnal and Rothschild [HR]. The case $r=2$, $t=1$ gives the EKR theorem; in fact, this is its broadest generalization not dealing with generalized weights. For an $rP_t,S_{=k}$-family in $B_n$, they showed that $\max|F|=\sum_{i=1}^{r-1}(-1)^{i+1}\binom{r-1}{i}\binom{n-it}{k-it}$ when $n>n_0(k,r,t)$. This is achieved uniquely by letting $F$ be the family of $k$-sets that contain at least one of $r-1$ disjoint $t$-sets.

Katona [Ka9] suggested a generalization not yet attempted. That is, what is the maximum family having at most $s$ $r$-tuples of $rs$ distinct $k$-sets so that the $r$-tuples each have no pair intersecting with cardinality at least $t$? This is the common generalization of the conditions examined in [Hi2a] and [HR].

For the conditions replacing the antichain condition, extremal families generally are not known. Most of the results are asymptotic.

$Q_1$ (no pair and their join or meet): Extremal families may be known here. [EK2] credits Clements with a construction achieving $|F|=\binom{n}{\lfloor n/2\rfloor}+1$ when $n$ is even. We note here that $|F|=\binom{n}{\lfloor n/2\rfloor}\frac{n+1}{n}$ is achieved for odd values of $n$ by $F=B_n^{(n+1)/2}[1]\cup B_n^{(n-1)/2}[\hat{1}]$, a fact we have not seen elsewhere. It is conjectured that both constructions are optimal.

$Q_2$ (no pair and a superset of their join): $\max|F|\sim\binom{n}{\lfloor n/2\rfloor}(1+c/n)$ [EK2].

$Q_3$ (no pair and a subset of their join): $\max|F|/2^n\to 0$ as $n\to\infty$ [EK2].

$Q_4$ (no two and their join): The best known result for $B_n$ is $\max|F|\leqslant\binom{n}{\lfloor n/2\rfloor}(1+c/\sqrt{n})$ by Kleitman [Kl15], improving his earlier result [Kl13]. He conjectured $\max|F|\leqslant\binom{n}{\lfloor n/2\rfloor}(1+c/n)$. In the divisibility ordering on the natural numbers, $Q_4$-families were considered in [ESS] and [An1], but cardinality was not the objective function maximized there.

$R_2$ (no pair plus their join and meet): For $B_n$, Erdös and Kleitman [EK1] determined $c_1 2^n/n^{1/4}\leqslant\max|F|\leqslant c_2 2^n/n^{1/4}$ for two constants with $c_1<c_2$. Anderson [An2] obtained an upper bound for the multiset case $M(\bar{e})$. Little is known about $R_k$ when $k>2$, even for $B_n$.

$Q_5$ (no disjoint pair plus their join): If $\varnothing$ were included in these families, $R_2$ would imply this condition. Hence these families can be larger. In fact, Kleitman [Kl8] showed that for $n=3q+1$, $\max|F|=\sum_{i=q+1}^{2q+1}\binom{n}{i}$. He obtained similar but not best possible results when $n\neq 3q+1$ and asked for the best possible. Frankl [Fl5] considered the generalization where no $k$ disjoint elements and their join can appear. Not surprisingly, he obtained $\max|F|=\sum_{i=q+1}^{kq+k-1}\binom{n}{i}$ for $n=(k+1)q+k-1$.

$kQ_6$ (no related pair with large difference): Erdös [Er1] proved that the largest $kQ_6$-family consists of the $k$ middle ranks. Katona [Ka5] considered the opposite restriction, forbidding $A\subset B$ with $|B-A|<k$. He found $\max|F|=\max_q\sum_i\binom{n}{q+ki}$. Both results can be proved using generalizations of the LYM inequality.

$C_k$: These are the $k$-families, discussed previously. [ABH] considers a weaker condition than $C_k$ on products of equal-length chains, somewhat reminiscent of the conditions in [Pe1] mentioned in Section 2.3.

### 4.4. Multiple families

Next, we consider multiple families satisfying the same kinds of conditions we have treated before. The simplest version is to maximize the union of $q$ families, each satisfying the given condition.

$C$: These unions are the $k$-families when $q=k$, and they have been discussed. Maximum $k$-families have not been determined for any interesting class of posets not having the $k$-Sperner property. A possible place to start would be with the class of examples given by Dilworth and Greene [DG].

$I$: Kleitman [Kl2] showed that $\max|F|=2^n-2^{n-q}$, achieved by $F=\{A\in B_n\colon |A\cap[q]|\geqslant 1\}$. The individual families can be expressed as $F_i=\{A\in B_n\colon i\in A\}$.

$I,S_{=k}$: Similar to the above. [EK2] pointed out that $\max|F|=\sum_{i=1}^{q}\binom{n-i}{k-i}$, achieved by $F=\{A\in B_n^k: |A\cap[q]|\geqslant 1\}$, from the $q$ families $F_i=\{A\in B_n^k: i\in A\}$.

There are also a number of results on pairs of related families. Let $F$ and $G$ be two families such that each element of one is related to each element of the other in some way, usually by an intersection condition. Then their sizes cannot both exceed certain bounds. These theorems generalize several of those mentioned previously, which follow by letting the two families be the same.

$I,S_{-k}$: The first such theorem was conjectured by Milner and proved by Kleitman [Kl6]. Let $F$ and $G$ be $S_{-j}$ and $S_{-k}$ families, with $j+k\leqslant n$, such that each element of $F$ intersects every element of $G$. Then $|F|\leqslant\binom{n-1}{j-1}$ or $|G|<\binom{n-1}{k-1}$, which implies the EKR theorem. A slightly stronger statement (see [Kl16]) is that $|G|\leqslant\binom{n}{k}-|F^*|$, where $F^*$ is the $(n-j)$-shadow of $F$. Hilton [Hi5] studied collections of $q$ $S_{-k}$-families where each element intersects every element in every other family. He obtained $\Sigma|F_i|\leqslant\binom{n}{k}$ for $n/k\geqslant q$ and $\Sigma|F_i|\leqslant q\binom{n-1}{k-1}$ for $n/k\leqslant q$. By taking $q$ large enough, this also yields the EKR theorem.

$I_t,S_{-k}$: Hsieh [Hs2] generalized the result of [Kl6]. Let $F$ and $G$ be $S_{-j}$ and $S_{-k}$ families such that $\mathrm{rank}(A\wedge B)\geqslant t$ when $A\in F,B\in G$. In $B_n$, $|F|\leqslant\binom{n-t}{j-t}$ or $|G|<\binom{n-t}{k-t}$, provided $n\geqslant t+(t+1)(j-t)(k-t+1)$ with $j>k$. Hsieh [Hs2] also proved the analogue for $L_n(q)$: $|F|\leqslant\begin{bmatrix}n-t\\ j-t\end{bmatrix}_q$ or $|G|\leqslant\begin{bmatrix}n-t\\ k-t\end{bmatrix}_q$, provided $n\geqslant j+k+2$ ($n\geqslant j+k+1$ if $q=3$).

$I^*,S_{-k}$: Hilton [Hi1] generalized the result of Hilton and Milner [HM] to pairs of families.

$C$: Let $F$ and $G$ be such that no element of $F$ is related to any element of $G$. Seymour [Se] proved that $|F|^{1/2}+|G|^{1/2}\leqslant 2^{n/2}$. Hilton [Hi6] added the condition that neither family can have any element along with its complement. Then $|F|+|G|\leqslant 2^{n-1}$. Daykin, Kleitman, and West [DKW] generalized the weaker form $|F\cup G|\leqslant\frac{1}{2}|P|$ to arbitrary finite lattices $P$ with polarity. A *polarity* is an order-reversing bijection whose square is the identity, such as complementation.

Daykin, et al. [DFGH] recently obtained bounds involving $q$ families such that elements from different families are incomparable. Let $\gamma_i=\sum_j(F_j)_i$ be the sum of the *ith* rank parameters. They obtained an LYM-type inequality, $\Sigma\gamma_i\binom{n}{i}^{-1}\leqslant\max\{q,n+1\}$. Furthermore, they showed that $\Sigma|F_i|\leqslant\max\{2^n,q\binom{n}{\lfloor n/2\rfloor}\}$, which yields Sperner's Theorem when q is large enough. This followed from a slightly stronger result and is reminiscent of Hilton's result [Hi5] above for $I,S_{-k}$.

$C,I_t$: Frankl [Fr3] proved that if $F$ and $G$ are two antichains such that any element of $F$ intersects any element of $G$ in at least $t$ members, then $\min\{|F|,|G|\}\leqslant\binom{n}{q}$, where $q=\lfloor\frac{n+k+t}{2}\rfloor$. [Fr3] contains another result on a pair of collections which discusses the $j$-shadow of a collection of $k$-sets. This generalizes a result in [Ka1].

$I_t$: Frankl [Fr3] mentioned results like those above for two or three collections, not necessarily antichains, proved by himself and by Ahlswede and Katona. More recently, Daykin and Frankl [DF1] considered $q$ families such that choosing an element from each and taking the intersection leaves a set of size at least $t$ (they phrased it for unions). Defining $\alpha$ by $\alpha^q = 2\alpha - 1$, $\frac{1}{2} < \alpha < 1$, they showed that $\min |F_i| < 2^n \alpha^q$. They also provided an example with $\min |F_i| > c 2^n \alpha^q / \sqrt{q}$.

*$rI$,$rU$*: Take $q$ families of distinct subsets. Choose one element each from any $r$ of them. Suppose these $r$ must always have non-void intersection and have union $\neq [n]$. Hilton [Hi6] proves $\Sigma |F_i| \leq (r-1)2^n$ if $r-1 \leq q \leq 4(r-1)$, $\Sigma |F_i| \leq q 2^{m-2}$ if $q \geq 4(r-1)$. He proves a similar bound when only the union condition is retained, but the families themselves must be intersecting.

*Infinite sets:* A result by Ehrenfeucht and Mycielski [EM], roughly put, says that for certain intersecting multiple families, there exists a finite set that intersects all pairs from different families. See also [Ka10].

*Non-uniform conditions:* Daykin and Frankl [DF3] very recently suggested the following idea. Take a graph on $q$ vertices and a condition to be applied to pairs of families. Given $q$ families, require the pair $(F_i, F_j)$ to satisfy the condition if $(i,j)$ is an edge in the graph (for certain conditions, the graph may be a directed graph). They obtained bounds for a large number of interesting examples. Needless to say, this opens up a new area of discussion.

### 4.5 Miscellaneous other restrictions

*Antichains composed of complementary pairs:* a maximal family of this sort must consist of an intersecting family and its complements. This yields $\max |F| = 2\binom{n-1}{\lfloor n/2 \rfloor - 1}$. Bollobás [Bo] proved this, obtaining an LYM-type inequality. Greene and Hilton [GHi] used a slightly weaker condition, requiring only $f_k = f_{n-k}$ for the rank parameters of $F$, and they obtained a similar inequality. Gronau [Go3] extended the idea to $M(\bar{e})$, also obtaining an LYM-type inequality. (In $M(\bar{e})$, the complement of $\bar{a}$ is $\bar{e} - \bar{a}$.)

*Antichains containing no element and its complement:* Purdy [Pu] proved that $\max |F| = \binom{n}{\lfloor n/2+1 \rfloor}$, achieved by $F = B_n^{\lfloor n/2+1 \rfloor}$. For even values of $n$ this can also be achieved by taking half of $B_n^{n/2}$. Clements and Gronau [CG] extended this to $M(\bar{e})$, using the Kruskal-Katona theorem of Section 6. The relationship between this condition and the previous condition is discussed in [GHi].

*Distinguished elements:* Lih [Lh] determined the largest antichain in which all elements must intersect a particular $k$-set. Not surprisingly, $\max |F| = \binom{n}{\lfloor n/2 \rfloor} - \binom{n-k}{\lfloor n/2 \rfloor}$. Griggs [Gg7] generalized this, but both results follow from the product theorem on LYM orders, as remarked in [HWD]. Hilton [Hi3] obtained generalizations of a number of EKR-type results by adding a distinguished-set condition.

*$I_t$*: Besides the two ways mentioned in Section 4.2, there is yet a third way to generalize $I_t$ to $M(\bar{e})$. Frankl and Füredi [FF] suggested the requirement that $\bar{a}$ equal $\bar{b}$ in at least $t$ positions, for any $\bar{a}, \bar{b} \in F$. For products of equal-length chains ($e_i = k-1$ for $1 \leq i \leq n$), they obtained an EKR-type result. They conjectured $\max |F| = k^{n-t}$ if and only if $n \leq t+1$ or $k \geq t+1$, and they proved it for $t \geq 15$. As in the EKR theorem, the extremal family is of the form $\{\bar{a} \in M(\bar{e}) : a_1 = \cdots = a_t = 1\}$.

Special cases of this conjecture had been considered before. Kleitman [Kl4] determined the extremal families when $k=2$. In $B_n$, this condition reduces to a condition on the symmetric difference: $|A \oplus B| \leq n - t$. The optimal families consist of all elements differing from a fixed element in at most $(n-t)/2$ positions, plus $\binom{n-1}{(n-t-1)/2}$ of those differing by one more if $n-t$ is odd. Berge [Be0] had proved the conjecture when $t=1$, as had Livingston [Lv]. Frankl and Deza [FD] considered the analogous problem on sets of permutations.

$S_{=k},I_{=t}$: If all pairwise intersections have equal size $t$, Then $F$ is called a *weak* $\Delta$*-system*. If the pairwise intersections are all identical, then it is called a *strong* $\Delta$*-system*. $S_{=k}$ is also required. Results about these families were surveyed by Erdös [Er3]. We mention only three that have an Erdös-Ko-Rado flavor, and several reminiscent of an "$r$-wise" condition.

One extension [HM] of the EKR theorem said that an intersecting family with enough members must be a "strong intersecting family"-everything in the collection contains a particular element of $[n]$. Deza [De] proved a slight generalization of the conjecture of Erdös and Lovász that a weak $\Delta$-system with more than $k^2-k+2$ members must be a strong $\Delta$-system. (In fact, the condition $S_{=k}$ can be dropped as long as $k=\max_{A\in F}|A|$.) Deza, Erdös and Frankl [DEF] extended this result to a situation where many cardinalities of intersection are allowed. In fact, their result generalizes the EKR theorem. If the allowed intersection cardinalities are $\{1, \ldots, k-1\}$ for $k \leqslant n/2$, then their result gives the size and composition of the largest $I,S_{=k}$-family. Deza, Erdös, and Singhi [DES] have further results in the case that exactly two values of the intersection cardinality are allowed.

Erdös and Rado took a different tack. They proved that, given $m > m_0(k,r)$ $k$-sets (no restrictions on $n$), some $r$ of them form a strong $\Delta$-system. They obtain some bounds on $m_0$. [AHS] contains the exact value of $m_0(2,r)$ and better bounds for $m_0(k,3)$. [ES] contains further improvements.

$S_{=k},rI_{\neq t}$: Specifically, suppose no $r$ of the elements of $F \subset B_n^k$ form a weak $\Delta$-system with pairwise intersection size $t$. Let $f(n,r,t,k)=\max|F|$. Briefly, the results are:

[EKR]: $f(n,2,0,k)=\binom{n-1}{k-1}$, for $n \geqslant 2k$.

[Fr7]: $f(n,2,1,k)=\binom{n-2}{k-2}$, for $k \geqslant 4, n \geqslant n_0(k)$.

[DE]: $f(n,r,1,k) < c(k,r)\binom{n-2}{k-2}$.

[DE],[Fr9]: $f(n,r,1,3) > \begin{cases} r(r-1)(n-2r) & \text{for } r \text{ odd} \\ r(r-3/2)(n-2r+1) & \text{for } r \text{ even} \end{cases}$

[Fr9]: $f(n,r,1,3) < \frac{5}{3}r(r-1)n$

$f(n,3,1,3)=6(n-6)+2$, for $n \geqslant 54$

Finally, a weakened intersection condition: suppose any subset of the family $F \subset B_n$ must have a non-void intersection unless their union covers $[n]$. Li [Li1] showed that $\max|F|=2^{n-1}+1$ and characterized the maximum and maximal families. Kleitman observed that applying the condition only to pairs (i.e. $A \cap B=\varnothing \Rightarrow A \cup B=[n]$) allows $\max|F|=2^{n-1}+\binom{n-1}{\lfloor n/2\rfloor-1}$. Li [Li2] also considered both of these questions on multisets, with the interpretation "$\bar{a}$ intersects $\bar{b}$" meaning "$\text{rank}(\bar{a}\wedge\bar{b})>0$." The results differ from those for sets. The two (achievable) upper bounds are respectively the [EHS] and [ES] bounds $e_n(\prod_{i<n}(e_i+1))$ and

$\frac{1}{2} \sum_{A \subset [n]} \max\{w(A), w(\bar{A})\}$ which we mentioned previously in connection with *I*-families.

## 5. Order Ideals and Inequalities

In this section, we consider extremal problems on intersections of (order) ideals or their dual objects, (order) filters. A few extremal results that use different objective functions but involve ideals and filters appear in the Section 6. An *(order) filter* is a collection "closed above", just as an ideal was defined to be closed below—i.e., if $x$ belongs to a filter $F$ and $y > x$, then $y \in F$. Ideals and filters are also called "down-sets" and "up-sets" or "hereditary families". In $B_n$ an ideal is also called a simplicial complex. Let $F^-$ ($F^+$) be the ideal (filter) generated by a family $F$, i.e. $F^- = \{x : x \leqslant y$ for some $y \in F\}$.

### 5.1 Kleitman's Inequality

This area of research began with the following simple but important result of Kleitman [Kl2]: *For ideals $F$ and $G$ in $B_n$, $|F|\,|G| \leqslant 2^n |F \cap G|$.* That this also holds in $M(\bar{e})$ appeared first in [An3] and [GK3], independently. Actually, the inequality was first stated in this form by Seymour [Se]; Kleitman gave an equivalent formulation.

The inequality $|A|\;|B| \leqslant |P|\;|A \cap B|$ holds for all pairs of ideals in posets $P$ still more general than $M(\bar{e})$. We call this inequality *Kleitman's inequality.* Daykin [Da4] and Jones [Jo1] independently showed that the class of posets satisfying Kleitman's inequality is closed under products. Daykin also showed in [Da4] that this class contains the distributive lattices, which in turn include $M(\bar{e})$. However, it does not contain $L_n(q)$ – see [GK3]. Jones showed in [Jo1] that in fact a stronger property involving probability distributions is preserved under products.

Before proceeding to stronger inequalities, we mention a dual formulation, which in fact was the form in which Kleitman first proved the result. For an ideal $F$ and filter $G$, $|F|\,|G| \geqslant |M(\bar{e})|\;|F \cap G|$. This inequality produced the quick proof in [An3] that the largest I,U-family in $B_n$ has size $2^{n-2}$, and also the corresponding upper bound there for I,U-families in $M(\bar{e})$. This inequality was also used in [Se] to prove the upper bound of $|F|^{\frac{1}{2}} + |G|^{\frac{1}{2}}$ mentioned in Section 4.4 for a pair of incomparable collections. Frankl and Hilton [FH] found another application, giving a lower bound on the number of elements comparable to a collection of fixed size, a problem raised in [BD4]. Their result holds both in $B_n$ and $M(\bar{e})$; if $|F| = m$, then $|F^+ \cup F^-| \geqslant 2\sqrt{m\,|M(\bar{e})|} - m$.

Daykin, Kleitman, and West [DKW] published the first strengthening of Kleitman's inequality for a class of finite lattices. For any subsets $A, B$ of a lattice, let $A \wedge B = \{a \wedge b : a \in A,\, b \in B\}$ and $A \vee B = \{a \vee b : a \in A,\, b \in B\}$. Note that if $A$ and $B$ are ideals, then $A \wedge B = A \cap B$. Daykin and Kleitman proved that in $M(\bar{e})$, $|A|\,|B| \leqslant |M(\bar{e})|\;|A \wedge B|$. In fact, for fixed sizes of $A$ and $B$, $|A \wedge B|$ is minimized when $A$ and $B$ are ideals. This is not true for all distributive lattices, but West conjectured that for distributive lattices $|A \wedge B|$ is minimized by letting $A$ and $B$ each be the intersection of a filter and an ideal.

Collecting these ideas in [DKW], they also determined the ideals $A$ and $B$ that minimize $|A \wedge B|$ over $|A|=\alpha$, $|B|=\beta$. These results hold when $M(\bar{e})$ is the product of $n$ $k$-chains. In that case, $\min |A \wedge B|$ is achieved by letting $A$ be the $\alpha$ lexicographically smallest $n$-digit vectors and letting $B$ be the $\beta$ lexicographically smallest when the components are viewed in reverse order. Intuitively, this makes sense, because then $\bar{0}$ and other small-rank elements appear as meets with much duplication. For the product of unequal chains, they conjectured that for each $\alpha,\beta$ there is some ordering of the chains (not requiring $e_1 \geqslant \cdots \geqslant e_n$) such that the optimal sets are obtained in the same way from the lexicographic orders.

Knowing the optimal families in the product of $k$-chains, they could study the accuracy of the inequality. They showed

$$0 \leqslant |M(\bar{e})| (\min |A \wedge B|) - |A| |B| \leqslant \frac{kn}{4}.$$

Except for trivial cases, the error is 0 only when $k \mid \alpha$, $k \mid \beta$, and $k^n \mid \alpha\beta$. The upper bound of $kn/4$ is very small compared to $|M(\bar{e})|$, so that actually $\min |A \wedge B| = \lceil \alpha\beta/k^n \rceil$. They found values of $\alpha$ and $\beta$ producing error close to $kn/4$, but the exact maximum error is not yet known.

Daykin improved Kleitman's inequality still further, showing that in a distributive lattice $|A| |B| \leqslant |A \wedge B| |A \vee B|$ for all collections $A,B$. Later in [Da5], he showed that among the lattices, this property characterizes the distributive lattices. He phrased the theorem in a way that includes infinite lattices also. Thus it includes a very early result of Behrend [Bh], who proved Kleitman's inequality for intersections of a certain class of ideals in the divisibility ordering of the natural numbers.

**5.2 The FKG Inequality.**

To obtain more general inequalites, we must leave the province of cardinality. To begin, we rephrase Kleitman's inequality on ideals as $\frac{|F|}{2^n} \cdot \frac{|G|}{2^n} \leqslant \frac{|F \cap G|}{2^n}$. If elements are viewed as "equally likely", this means $\mathrm{Prob}(x \in A \mid x \in B) \geqslant \mathrm{Prob}(x \in A)$. This is satisfying, if not surprising. It is reminiscent of Chebyshev's inequality for non-decreasing sequences, which we can write as $\frac{\Sigma a_i}{n} \cdot \frac{\Sigma b_i}{n} \leqslant \frac{\Sigma a_i b_i}{n}$. This says that the sequences are positively correlated.

Fortuin, Kastelyn and Ginibre [FKG], in order to unify several results in statistical mechanics, generalized Chebyshev's results to all finite distributive lattices. Let $\mu$ be a positive measure (or probability distribution) on a distributive lattice $L$, satisfying the condition $\mu(x)\mu(y) \leqslant \mu(x \vee y)\mu(x \wedge y)$ for all elements $x,y \in L$. Define the expected value of a function $f$ by $E(f) = (\sum_{x \in L} f(x)\mu(x)) / (\sum_{x \in L} \mu(x))$. Suppose that $f$ and $g$ are both non-decreasing (or both non-increasing) functions on $L$. Then $E(f)E(g) \leqslant E(fg)$. This elegant result is called the *FKG inequality*. Its combinatorial consequences are far-reaching, since it appered in a journal of physics. For example, it includes all the above-mentioned results extending Kleitman's inequality to $M(\bar{e})$ and to all distributive lattices, although most of these results appeared later. Since cardinality satisfies the condition on $\mu$, that extension follows by letting $f$ and $g$ be the characteristic functions on two ideals. Seymour and Welsh [SW] eventually remarked on this and discussed other combinatorial implications of the FKG inequality.

Holley [Hy] generalized the FKG inequality in a way that enabled him to consider two distributions instead of two increasing functions.

Ahlswede and Daykin [AD1] obtained a remarkable result that contains all of this, and did so by a simple induction proof. Let $f_i$ be four non-negative real-valued functions on a distributive lattice, and let $f_i(A) = \sum_{a \in A} f_i(a)$. If $f_1(a)f_2(b) \leqslant f_3(a \vee b)f_4(a \wedge b)$ for all $a, b \in L$, then $f_1(A)f_2(B) \leqslant f_3(A \vee B)f_4(A \wedge B)$ for all $A, B \subset L$. In [AD2], they discussed the implications of this inequality for previous results. There they also showed that all pairs of Boolean set functions that do not behave as in the above theorem are isomorphic to $(\oplus, \wedge)$, where $\oplus$ denotes symmetric difference. However, even that pair obeys the theorem when all the $f_i$ stand for cardinality. Recently, Daykin [Da7] determined the full behavior of $(\oplus, \wedge)$, proving that $f_1(A)f_2(B) \leqslant [\frac{1}{2}(1+\sqrt{2})]^n f_3(A \oplus B)f_4(A \wedge B)$ when it always holds that $f_1(a)f_2(b) \leqslant f_3(a \oplus b)f_4(a \wedge b)$. This followed from a more general theorem about products [Da7]. [Da5a],[Da6] contain intermediate results.

A continuous analogue of the FKG inequality for integrals in $\mathbf{R}^n$ was obtained by Ruzsa [Ru]. Preston [Ps] generalized Holley's inequality to the continuous case, and Batty and Bollman [BB] proved a continuous analogue of the Ahlswede-Daykin inequality. Recently, Shepp [Sh] applied the FKG inequality to obtain, among other results, an easy proof of an inequality of Graham, Yao, and Yao [GYY] on conditional probabilility in posets. Kleitman and Shearer [KSh] have a similar paper. More details of this and other matters involving the FKG inequality can be found in [Gh], in this volume. Finally, Lih [Lh2] recently used network flow arguments and a new idea about majorization of sequences to show that the lattices where the FKG and Holley results hold are precisely the distributive lattices.

### 5.3 Chvátal's conjecture

Before leaving the subject of ideals, we mention one completely different problem. Let a *principal* filter in $B_n$ be one generated by a single element of rank 1, i.e. by a singleton set. Some years ago, Chvátal [Ch2] conjectured that the largest intersecting family contained in an ideal is always the intersection of that ideal with some principal filter. In a sense, the EKR theorem and his generalizations of it are special cases of this conjecture. I.e., he proved it for ideals which are "compressed" with respect to a given element of $[n]$, say 1. "Compressed" means that substituting 1 for anything else in an element of the ideal gives another element of the ideal.

Berge [Be2] proved another special case. Daykin and Hilton [DH] have provided another proof of Berge's result and some generalizations. Wang also has partial results. Kleitman and Magnanti [KMg] proved that the intersection of any ideal with a filter generated by a 2-set could not be larger that its intersection with either of the principal filters. For this they used an extremal result on "latent subsets", which are the subsets of $B_n$ strictly contained in elements of some family $F$. More precisely, let $F^L = |F^- - F|$. They showed that if $F$ and $G$ are two families such that each member of $F$ intersects each member of $G$, then $|F^L||G^L| \geqslant |F||G|$. Kleitman [Kl16] recently discussed progress on the Chvátal conjecture. He suggested a stronger conjecture and noted that it holds for all $I$-families containing a 2-set and also in some other special cases.

## 6. Extremal problems on families of fixed size

### 6.1 The Kruskal-Katona and generalized Macaulay theorems

The seminal result in this area is due independently to Kruskal [Kr1] and to Katona [Ka3]. The question is, over all collections $G \subset B_n^k$ of size $m$, which has the smallest $(k-1)$-shadow? They showed that the optimal $G$ consists of the first $m$ elements in the anti-lexicographic (or lexicographic) ordering of $B_n^k$. The *anti-lexicographic ordering* (AL-ordering) of $B_n^k$, denoted $AL(B_n^k)$, is the ordering which has $\bar{a} < \bar{b}$ if $a_i < b_i$ in the *last* index where $\bar{a}$ and $\bar{b}$ differ – e.g. (11100,11010,10110) are the first three elements in $AL(B_5^3)$. The result makes sense, since when $G$ is chosen that way many of the $m \times k$ $(k-1)$-sets generated for the $(k-1)$-shadow are duplicates. The choice of the AL-ordering meshes with $e_1 \geqslant \cdots \geqslant e_n$ and simplifies the description in the next paragraph.

They also showed how to compute the minimum number of distinct covered sets. To explain this we need to discuss numerical aspects of the AL-ordering. Given $k$, there is a unique way to express $m = \binom{m_k}{k} + \binom{m_{k-1}}{k-1} + \cdots + \binom{m_j}{j}$ with $m_k > m_{k-1} > \cdots > m_j \geqslant j$, called the *k-nomial representation* of $m$. It can be found by letting $m_k$ be as large as possible without $\binom{m_k}{k}$ exceeding $m$, then maximizing $m_{k-1}$ on the difference, and so on. Alternatively, let the AL-ordering start at 0 and interpret $B_n^k$ as binary vectors indexed on positions 0 through $n-1$. Consider the $m^{th}$ vector in the ordering (labelled $m-1$). Suppose it has its $k^{th}$ 1 in position $m_k$. To put a 1 in that position, $\binom{m_k}{k}$ vectors have been passed by, since the first $\binom{m_k}{k}$ vectors in $AL(B_n^k)$ are those with $k$ 1's in positions 0 through $m_k-1$. Placing the $(k-1)^{th}$ vector requires further skipping, and so on, until the entire $k$-nomial representation is generated. Note that the left justified 1's require no skipping, corresponding to the fact that the $k$-nomial representation could be defined to continue down to $\binom{m_1}{1}$ with the previously unspecified values of $m_i$ being $i-1$. Kruskal [Kr1] called the first $m$ vectors of the AL-ordering a "cascade" of sets, since they consist of the $k$-subsets of $[0, m_k-1]$ plus the $k$-sets formed by adding $m_k$ to $(k-1)$-subsets of $[0, m_{k-1}-1]$, and so on. Actually, $n$ is irrelevant here, so the $k^{th}$ rank of the infinite Boolean algebra could be treated similarly.

It is now easy to obtain the number of vectors covered by the first $m$ vectors of $AL(B_n^k)$. The covered vectors form an initial segment of $AL(B_n^{k-1})$. The highest of these in $AL(B_n^{k-1})$ is obtained by dropping the *smallest-indexed* 1 from the $m^{th}$ vector in $AL(B_n^k)$. By the positional counting argument just described, the resulting vector is the $\partial_k m^{th}$ vector in $AL(B_n^{k-1})$, where $\partial_k m$ is obtained from the $k$-nomial representation of $m$ by $\partial_k m = \binom{m_k}{k-1} + \binom{m_{k-1}}{k-2} + \cdots + \binom{m_j}{j-1}$. This operation is called "lowering the denominators" in [GK3].

We have been discussing $B_n$ in terms of vectors, so it is natural to generalize to $M(\bar{e})$. Such a theorem was proved by Clements and Lindström [CL] ( by a different method of proof) before hearing of the Kruskal-Katona result. They started from the much earlier theorem of Macaulay [Ma] which handled the case $e_1 = \cdots = e_n = \infty$—i.e., the $n$-dimensional non-negative integer vectors. (Sperner [Sp2] gave another proof of Macaulay's result.) Clements and Lindström proved that the $m$-element collection in $M_k(\bar{e})$ having the smallest $(m-1)$-shadow consists of the first $m$ vectors in $\mathrm{AL}(M_k(\bar{e}))$. They call this the "generalized Macaulay theorem."

The theorem can be phrased in an alternative way that sometimes helps in proofs. For $G \subset M_k(\bar{e})$, let $\Gamma(G)$ be the $(k-1)$-shadow of G, and let $\Upsilon(G)$ be its $(k+1)$-shadow. Let $F(m,k)$ be the first $m$ vectors in $\mathrm{AL}(M_k(\bar{e}))$, and let $L(m,k)$ be the last $m$ vectors. $\{F(m,k)\}$ and $\{L(m,k)\}$ are the *compressed* sets of vectors in $M_k(\bar{e})$. Compression operators $C$ and $C^*$ can be defined for $G \subset M_k(\bar{e})$ by $C(G) = F(|G|,k)$, $C^*(G) = L(|G|,k)$. For an arbitrary $G \in M(\bar{e})$, recall that $G_k = G \cap M_k(\bar{e})$, and define $f(G) = \bigcup_k f(G_k)$ for $f \in \{\Gamma, \Upsilon, C, C^*\}$.

Now the generalized Macaulay theorem of Clements and Lindström can be expressed as follows: *If* $G \subset M_k(\bar{e})$, *then* $\Gamma(C(G)) \subset C(\Gamma(G))$, *and* $\Upsilon(C^*(G)) \subset C^*(\Upsilon(G))$. In other words, the family with the smallest shadow is a compressed family. The minimum size of the shadow is harder to compute for $M(\bar{e})$ than for $B_n$, but once again the covered vector highest in $\mathrm{AL}(M_{k-1}(\bar{e}))$ is found by subtracting 1 from the first non-zero position of the $m^{th}$ vector in $\mathrm{AL}(M_k(\bar{e}))$. This was studied in more detail in [Cl3] for computing the smallest $j$-shadow for a collection of $m$ elements of $M_k(\bar{e})$. Notation for computing the numbers by "lowering denominators" appears in [GK3].

The crux of the Clements-Lindström proof is as follows: 1) Partition $G$ into sets $G^j$ that have the $d^{th}$ component $j$, for some fixed $d$. 2) Apply the "induced" compression operators to the sets $G^j$, ignoring the presence of the $d^{th}$ component and leaving it unchanged. 3) By induction on the number of chains, the shadow of $G$ cannot increase. 4) When no change is produced by performing this operation for any position $d$, $G$ must be fully compressed, except for one special case that is easily handled separately.

Daykin [Da1] later gave a short proof of the Kruskal-Katona theorem by looking at the pair $A,B$ with minimal $|B-A|$ such that $A \in F(|G|,k)$, $B \notin F(|G|,k)$, $A \notin G$, $B \in G$. He showed that replacing $B$ by $A$ can never increase the shadow. Other proofs were given by Hansel [Ha2], Hilton [Hi6], and Daykin again [Da3].

If $G \subset M(\bar{e})$ then $C(G)$ can be considered a "canonical form" for all $G$ having the same rank parameters $g_0, g_1, \cdots$. This is especially true if $\Gamma(G) \subset G$, in which case $G$ is an order ideal. The Clements-Lindström method of proof involving iterative replacement is sometimes called the "method of canonical forms." In [DKW], the lexicographic nature of the ideals achieving the smallest meet is proved this way (see Section 5). [Wa] (see Section 4.2) also used it.

### 6.2 Applications

The generalized Macaulay theorem, or even the Kruskal-Katona theorem, is extremely versatile and has produced short proofs of many results. Several of these have been mentioned already; we cite other applications here.

Katona [Ka3] immediately used the theorem to solve Erdös' question on $(k-1)$-representation of $S_{-k}$-families. A $(k-1)$-representation finds for each $A_i \in G$ a distinct $B_i \in B_n^{k-1}$ such that $B_i \subset A_i$. The largest $m$ such that any collection of $m$ $k$-sets has a $(k-1)$-representation is $m = \sum_{i=1}^{k} \binom{2i-1}{i}$. This is best possible for $k \leqslant n/2$.

Generalizations, e.g. to multisets, appear in [Cl1] and [Cl4].

Another early application gave a necessary and sufficient condition for the existence of an antichain with specified rank parameters. Simply put, we must have $f_i+\gamma_i(F)\leqslant N_i$, where $N_i$ is the $i^{th}$ Whitney number and $\gamma_i(F)$ is the minimum shadow at rank $i$ of the members at higher ranks. The generalized Macaulay theorem gives achievable lower bounds for the $\gamma_i$. Consider $B_n$ and let $f_k$ be the first nonzero $f_i$. Then the necessary and sufficient condition, obtained independently by Clements [Cl4] and by Daykin, Godfrey, and Hilton [DGH], is

$$f_k+\partial_{k+1}(f_{k+1}+\partial_{k+2}(f_{k+2}+\cdots+\partial_n(f_n)\cdots))\leqslant\binom{n}{k}$$

Clements [Cl6] studied this further, giving an algorithm for computing the maximum size of an antichain whose rank parameters above $k$ have been fixed. In essence, this means calculating $\binom{n}{k}-\partial_{k+1}(f_{k+1}+\partial_{k+2}(f_{k+2}+\cdots+\partial_n(f_n\cdots))$. He also provided [Cl8] an algorithm for computing a number of values $\partial_k(m)$ without computing all the $k$-nomial representations. See also [Da3].

Dividing the [DGH],[Cl4] condition on antichains by $\binom{n}{k}$ leads to a strengthening of the LYM inequality, as noted in [GK3]. There is a corresponding strengthening of the LYM-type inequality for $C,I,S_{\leqslant n/2}$-families to give a characterization of their rank-parameter sequences. In particular, Greene and Kleitman [GK3] showed that there is an intersecting antichain in $B_n$ with rank parameters $(0,f_1,\cdots,f_{\lfloor n/2\rfloor})$ if and only if there is an antichain in $B_{n-1}$ with rank parameters $(f_1,\cdots,f_{\lfloor n/2\rfloor})$.

Daykin [Da2] used the Kruskal-Katona theorem to give a short proof of the EKR theorem. Greene and Kleitman [GK3] used the generalized Macaulay theorem to generalize the EKR theorem to multisets (see Section 4.2).

Kleitman and Milner [KMl] conjectured that antichains in $B_n$ could be "stuffed into the upper half". That is, if $f_0,\cdots,f_n$ are the rank parameters of an antichain, then there is another antichain with rank parameters $g_i=0$ for $i<n/2$, $g_{n/2}=f_{n/2}$ if n is even, and $g_i=f_i+f_{n-i}$ for $i>n/2$. This was proved for $B_n$ by Daykin, Godfrey, and Hilton [DGH], and for $M(\overline{e})$ by Clements [Cl9], using the same methods.

For $M(\overline{e})$, Clements studied the monotonicity properties of $|\Gamma(F(m,k))|$ in [Cl10]. In [Cl11], he studied two extremal problems over antichains $G$ with $|G|=m$. He gave procedures for calculating 1) the minimum size of the order ideal generated by $G$ and 2) the minimum sum of ranks of elements in $G$. This extended earlier results. Kleitman and Milner [KMl] had shown for $B_n$ that when $|G|\geqslant\binom{n}{k}$ with $k\leqslant n/2$, the sum of ranks in $G$ must be at least $k|G|$, with equality achieved only by $G=B_n^k$. Clements [Cl3a] had improved this result. Daykin [Da3a] extended these by obtaining the $m$-element $k$-family in $B_n$ with the minimum sum of ranks.

For the problem of minimizing the size of the generated ideal, Clements [Cl5] had given a procedure for finding the $m$-element antichain in $B_n$ that generates the smallest ideal, as an application of inequalities involving $|\Gamma(G)|$ and $|\Upsilon(G)|$. His results are algorithms only, with no formulas given for min $|G|$ for general $m$. On the other hand, earlier Kleitman [Kl3] explicitly solved the case where $m=\binom{n}{k}$. In particular, he showed that if $|G|\geqslant\binom{n}{k}$ with $k\leqslant n/2$, then $|G^-|\geqslant\sum_{i=0}^{k}\binom{n}{i}$, again with equality only for $B_n^k$. Frankl and Hilton [FH] obtained a result with a similar flavor for the union of the generated ideal and generated filter, as a corollary of their result on $|G^+ \cup G^-|$ mentioned in Section 5: If $|G| \geqslant 2^{m-2i}$, then $|G^+ \cup G^-| \geqslant 2^{m+1-i} - 2^{m-2i}$. Equality is achieved when $G = \{A : [i] \subset A$ and $i+1, \ldots, 2i \notin A\}$.

### 6.3 Maximum adjacencies or minimum boundary

We saved one of the earliest and most important applications till the end because there are several other similarly-phrased problems. Harper was the first to study the rather natural question of how to choose $m$ of the $2^n$ vertices of the $n$-cube to obtain the maximum number of adjacent pairs. This is directly related to the coding problem of how to place the labels $0, \cdots, 2^n - 1$ on the vertices to minimize the average difference between adjacent vertices. Later, in [Hr1a], Harper also determined how to label the vertices to minimize the maximum difference between adjacent vertices.

For the former problem, Harper [Hr1] showed that a "greedy" algorithm can be used: at each stage add a vertex that neighbors the most previously chosen vertices. ( [Hr1] is a source of the term "shadow"— a $k$-dimensional subcube has $n-k$ "shadows" obtained by changing one of the fixed coordinates.) Bernstein [Br] filled a gap in Harper's proof. Lindsey [Ly] generalized the result to $n$-cubes with longer sides, i.e. $M(\bar{e})$ for equal length chains. He also characterized the optimal collections.

Kruskal [Kr2] noted the analogy between this problem and that of choosing points on the $n$-simplex; Harper's problem on $B_n$ is a special case of a generalized Macaulay theorem. Finally, Clements [Cl2] solved the problem for general $M(\bar{e})$ – as usual, take the initial segment of the AL-order. Lindström [Ld1] also generalized the solution, giving a full answer to Kruskal's question. Both Clements and Lindström point out that the answer to the generalized problem is equivalent to the generalized Macaulay theorem.

Kleitman, Kreiger, and Rothschild [KKR] solved the analogous problem for Hamming distances in the product of equal-length chains. Define points to be *adjacent* if they differ in only one coordinate. For equal $e_i$, again the maximum number of adjacencies belongs to the initial segment of lexicographic ordering.

Wang and Wang generalized Harper's problem to the following "discrete isoperimetric" problem. Given $|G| = m$, find $G$ to minimize the number of *boundary points*. These are points not in the set but at distance 1 from some element of the set. They called the optimal collections of points "standard discrete spheres". In [WW1], they re-solved the generalized Harper problem, pointed out its equivalence to the Macaulay theorem, and obtained a similar result for the set of $n$-dimensional vectors without the positivity restriction. In [WW2], they extended their results to the discrete torus, obtaining a generalization of the generalized Macaulay theorem. The minimum boundary is obtained by taking the first $m$ points in the ordering where, with $e_1 \geqslant \cdots \geqslant e_n$, $\bar{a} < \bar{b}$ if $\Sigma a_i < \Sigma b_i$ or if $\Sigma a_i = \Sigma b_i$ and $\bar{a} < \bar{b}$ in the AL-ordering.

Ahlswede [Ah1] examined another question of maximizing adjacencies. Given $m$ elements of $B_n$, how should one choose them to maximize the number of intersecting pairs. Not suprisingly, the optimal collection is an order filter, obtained by taking all the elements above rank $k$ and an initial segment from $\mathrm{AL}(B_n^k)$, for the appropriate $k$. Frankl [Fr6] earlier showed that a similar construction, differing only in the manner of choosing the $k$-sets, minimizes the number of disjoint pairs among $m$ elements chosen from $B_n$.

Kleitman [Kl9] studied the minimum number of related pairs in a collection of fixed size. In $B_n$, again there is a "canonical order" of sorts. Begin by taking all the

$\lfloor n/2 \rfloor$-sets, then the $(\lfloor n/2 \rfloor+1)$-sets when those are exhausted, then the $(\lfloor n/2 \rfloor-1)$-sets, and so on until $m$ elements are chosen. Within the ranks the order is immaterial. He asks whether this ordering also minimizes the number of $k$-element chains for any $k>2$. We ask for what other strongly Sperner posets a similar result holds, though the ordering within the ranks may matter in more general cases.

Finally, Kleitman [Kl16] has a conjecture on a problem that sounds like these, but for which there is no canonical ordering such that the initial segment always maximizes the adjacencies. Consider a collection of $k$-sets in $B_n$, and define $A$ and $B$ to be "adjacent" if a single element substitution turns one into the other. We want to choose $m$ $k$-sets to contain the maximum number of adjacencies.

Let the optimal collection be $F(n,m,k)$. It is easy to see that $F(n,m,k)$ also minimizes the number of adjacencies between $F$ and $B_n^k-F$, among all $F$ with $|F|=m$, and hence also that $F(n,m,k)=B_n^k-F(n,\binom{n}{k}-m,k)$. Kleitman conjectures that $F(n,m,k)$ can be computed recursively as follows:

1) If $k>n/2$, take the complements of the sets in $F(n,m,n-k)$.

2) If $k\leqslant n/2$, $m\leqslant\binom{n-1}{k-1}$, take the sets in $F(n-1,m,k-1)$ and adjoin the element $n$ to each.

3) If $k\leqslant n/2, \frac{1}{2}\binom{n}{k}\leqslant m\leqslant\binom{n-1}{k}$, take $F(n-1,m,k)$.

4) If $k\leqslant n/2$, $m$ other than above, apply $F(n,m,k)=B_n^k-F(n,\binom{n}{k}-m,k)$.

Kleitman gives an alternate, more compact but less explicit phrasing, which we translate into our notation as follows: *For any value of $m$, there is an optimal $F$ such that either $F$ or its complement in $B_n^k$ or the complements of the elements of $F$ as subsets of $[n]$ or the complement of that set in $B_n^{n-k}$ does not contain any elements containing the last element in $[n]$.* The best statement must be somewhere between two. Tovey [Ty] also restates the conjecture and discusses applications.

Ahlswede and Katona [AK] earlier proved the case $k=2$, though Kleitman was unaware of their result at the time of his conjecture. For $k=2$ we are allowed $m$ edges in a graph on $n$ vertices and want to maximize the number of pairs of incident edges. The solution is always to choose what Ahlwede and Katona call a *quasi-complete graph* or a *quasi-star*. The former is a complete subgraph on a subset of $r$ vertices, plus $s$ edges from them to a single additional vertex, where $m=\binom{r}{2}+s$. The latter is the complement of the quasi-complete graph on $\binom{n}{2}-m$ edges. It has several vertices joined to all others, and one of the remaining vertices joined to as many more as needed. Which type should be chosen depends again on what range of values contains $m$. Kleitman's conjecture reduces to this; note that the last vertex is not used in any edge of the quasi-complete graph until $m>\binom{n}{2}-\binom{n-1}{1}$, and so forth. The methods used by Ahlswede and Katona are complicated, but perhaps they will generalize.

## 7. Miscellaneous miscellany

There are numerous other extremal problems on set-systems that we have not discussed. For the most part, generalizations of these questions have not been suggested, even to multisets. We give a brief list here.

Valency: An alternate objective function to cardinality was suggested by Brace and Daykin [BD4]. Let $\nu(F)=\min_i |\{A \in F: i \in A\}|$. They considered the problem of maximizing valency under various combinations of the conditions $I, U$, and $C$, obtaining numerous results for $B_n$. Recently, Daykin and Frankl [DF1] studied the maximum valency of an $rU$-family in $B_n$. They conjectured $\max \nu(F) \leqslant 2^{n-r-1}$ for $r \geqslant 3$. This was proved for $r \geqslant 25$ and $n \leqslant 2r$ in [Da5a]. For $r=2$, they showed $\max \nu(F) \leqslant 2^{n-2}(1-n^{-1})$ and provided an example with $\nu(f) > 2^{n-2}(1-n^{-.651})$. (Note that $rU$ can be read as "no $r$ elements of $F$ form a cover." The smoothness of that phrase may explain why some results have been phrased in terms of $U$ instead of $I$.)

Covering problems: Kneser [Kn] conjectured that the minimum number of $I, S_{-k}$-families needed to cover all of $B_n^k$ is $n-2k+1$. This was proved recently by Lovasz [Lo2] and by Bárány [Bn], using topological theorems of Borsuk and Gale. A more standard covering problem was solved by Baranyai [By]. Here $B_n^k$ is to be covered by collections of pairwise-disjoint sets; at least $\lceil \binom{n}{k} / \lfloor n/k \rfloor \rceil$ such collections are needed, and that many suffice. Section 3 of [BS] contains further discussion and references about covering problems.

Convex families: A subset $F$ of a poset is *convex* if $x, y \in F$ and $x<z<y$ imply $z \in F$. Alternatively, a convex set is the intersection of a filter and an ideal, so this is a natural place to look for generalizations of the results in Section 5. Not much has been done. Daykin and Frankl [DF2] formulated an attractive conjecture for $B_n$. Let $V$ be a convex set in $B_n$, with $w(V)$ the size of the largest antichain contained in it. They conjectured that $w(V)/|V| \geqslant w(B_n)/|B_n|$. If this is true, does it also hold for $M(\bar{e})$? In addition, letting $X$ be the collection of elements incomparable to an antichain $F$, $D$ any down-set, and $U$ any up-set, they conjectured

$$|X|\,|X|\,|X \cap D \cap U|\, w(B_n) \leqslant |B_n|\,|X \cap D|\,|X \cap U|\,(w(B_n)-|F|).$$

When $F=\varnothing$, this reduces to Kleitman's inequality. These conjectures are related to a result they proved strengthening the LYM inequality. If $P$ has the LYM property and $F \subset P$ is a $k$-family, then

$$\max\left\{\frac{k\,|F|}{d_k(P)}, \sum_{x \in F} \frac{1}{N_x}\right\} \leqslant k \frac{|F^+ \cup F^-|}{|P|},$$

where $N_x$ is the size of the rank containing $x$.

Jamison-Waldner [Ja] also looked at convex sets. He defined the *Helly number* $h(P)$ of a poset to be the minimum number such that whenever any $h$ of a collection of convex families of $P$ have non-empty intersection the intersection of all families in the collection must also be non-void. In other words, any $h(P)$-wise intersecting collection of convex families in $P$ is completely intersecting. He proved that the Helly number of a poset is the size of its largest convex 2-family.

Ramsey-type problems: The $\Delta$-system problem mentioned already falls into this category. Along a different line, [MS] and [DL] discuss the number of distinct set differences that must occur among $m$ sets. It is $m$, and they applied this to other problems we have discussed. Finally, for results on how many sets that together cover $[n]$ are needed to force a subcover of $[n]$ by $t$ of them, see [BD1,2,3,5],[GM],[EHM].

Other extremal problems: See [Lp],[Su],[EKs] for random examples. Also, various extremal poset problems have been examined in the setting of acyclic digraphs–i.e., without the requirement of transitivity. In particular, Linial [Ln2] extended the Greene-Kleitman theorem. See also [Ga],[Ln1]. Finally, Katona [Ka11,12] recently began a study of continuous versions of extremal results on $B_n$.

Concluding remark: Sperner's Theorem has a long history and has spawned an impressive volume of results, though it seems that most of the easy work has been done. The ramifications of Dilworth's Theorem, on the other hand, are now emerging in many areas. By examining the proceedings of this conference alone, one can see that the analogues and extensions of this theorem are currently of great interest to most researchers in ordered sets.

**Acknowledgement**

The author wishes to thank Deborah Stein and Patricia Suyemoto for their help in preparing the manuscript, and especially David Shmoys for his invaluable assistance in compiling the list of references and his many additional hours of help on typesetting and proofreading. Without his assistance, this paper would not exist in its present form.

## References

[AHS] H.L. Abbot, D. Hanson, N. Sauer, Intersection theorems for systems of sets. *J. Comb. Theory A* 12(1972), 381-389.

[Ah] R. Ahlswede, Simple hypergraphs with maximal number of adjacent pairs of edges. *J. Comb. Theory B* 28(1980), 164-167.

[AD1] R. Ahlswede and D.E. Daykin, An inequality for the weights of two families of sets, their unions and intersections. *Z. Wahrsch. V. Geb.* 43(1978), 183-185.

[AD2] R. Ahlswede and D.E. Daykin, Inequalities for a pair of maps $S \times S \rightarrow S$ with $S$ a finite set, *Math. Zeitschr.* 165(1979),267-289.

[AK] R. Ahlswede and G.O.H. Katona, Graphs with maximal number of adjacent pairs of edges. Acta Math. Acad. Sci. Hung. 32(1978), 97-120.

[Ai1] M. Aigner, Lexicographic matching in Boolean algebras. *J. Comb. Theory B* 14(1973), 187-194.

[Ai2] M. Aigner, Symmetrische Zerlegung von Kettenprodukten. *Monatshefte für Mathematik* 79(1975), 177-189.

[AB1] M.O. Albertson and K.A. Berman, The chromatic difference sequence of a graph. *J. Comb. Theory B* 29(1980), 1-12.

[AB2] M.O. Albertson and K.A. Berman, Critical graphs for chromatic difference sequences. *Disc. Math.* 31(1980), 225-233.

[Al] M.O. Albertson, A new paradigm for duality questions, preprint.

[AC] M.O. Albertson and K.L. Collins, Duality and perfection for edges in cliques, to appear.

[ACS] M.O. Albertson, K.L. Collins, and J.B. Shearer, Duality and perfection for $r$-clique graphs. *Proc. 12th S.E. Conf. Comb., Graph Th. and Computing.*

[ABH] B. Alspach, T.C. Brown, P. Hell, On the density of sets containing no $k$-element arithmetic progression of a certain kind. *J. London Math. Soc.* (2) 13(1976), 226-234.

[An1] I. Anderson, An application of a theorem of deBruijn, Tengbergen, and Kruyswijk. *J. Comb. Theory* 3(1967), 43-47.

[An2] I. Anderson, On the divisors of a number. *J. London Math. Soc.* 43(1968), 410-418.

[An3] I. Anderson, Intersection theorems and a lemma of Kleitman. *Disc. Math.* 16(1976), 181-185.

[Ba1] H.L. Baker, Jr., Complete amonotonic decompositions of compact continua. *Proc. Amer. Math. Soc.* 19(1968), 847-853.

[Ba2] H.L. Baker, Maximal independent collections of closed sets. *Proc. Amer. Math. Soc.* 32(1972), 605-610.

[Bk] K.A. Baker, A generalization of Sperner's lemma. *J. Comb. Theory* 6(1969), 224-225.

[BSt] K.A. Baker and A.R. Stralka, Compact, distributive lattices of finite breadth. *Pacific J. Math.* 34(1970), 311-320.

[Bl] R. Balbes, On counting Sperner families. *J. Comb. Theory A* 27(1979), 1-9.

[Bn] I. Bárány, A short proof of Kneser's conjecture. *J. Comb. Theory A* 25(1978),325-326.

[By] Z. Baranyai, The edge coloring of complete hypergraphs, I. *J. Comb. Theory B* 26(1979), 276-294.

[BB] C.J.K. Batty and H.W. Bollmann, Generalised Holley-Preston inequalities on measure spaces. *Z. Wahrsch. V. Geb.* 53(1980), 157-173.

[BMRR] D. Baumert, E.R. McEliece, E.R. Rodemich, and H. Rumsey, A probabilistic version of Sperner's theorem, to appear.

[Bh] F.A. Behrend, Generalization of an inequality of Heilbron and Rohrbach, *Bull. Amer. Math. Soc.* 54(1948), 681-684.

[BkD] A. Békéssy and J. Demetrovics, Contribution to the theory of data base relations. *Disc. Math.* 27(1979), 1-10.

[BHFK] A. Békéssy, J. Demetrovics, L. Hannák, P. Frankl, and G. Katona, On the number of maximal dependencies in a data base relation of fixed order. *Disc. Math.* 30(1980), 83-88.

[Be0] C. Berge, Nombres de coloration de l'hypergraphs $h$-parti complete. *Hypergraph Seminar* (C. Berge and D. Ray-Chaudhuri), Springer-Verlag Lect. Notes 411(1974), 13-20.

[Be1] C. Berge, A theorem related to the Chvátal conjecture. *Proc. 5th British Comb. Conf. Aberdeen (1975),* (C.St.J.A. Nash-Williams and J. Sheehan), Utilitas, (1976), 35-40.

[Be2] C. Berge, Packing problems and hypergraph theory: a survey. *Annals of Disc. Math.* 4(1979), 3-37.

[BK] J. Berman and P. Kohler, Cardinalities of finite distributiive lattices. *Mitt. Math. Sem. Giessen* (2) (1976), 103-124.

[Br] A.J. Bernstein, Maximally connected arrays on the $n$-cube. *SIAM J. Appl. Math.* 15(1967), 1485-1489.

[C16] G.F. Clements, Sperner's theorem with constraints, *Disc. Math.* 10(1974), 235-255.

[C17] G.F. Clements, Intersection theorems for sets of subsets of a finite set. *Quart. J. Math. Oxford* (2) 27(1976), 325-337.

[C18] G.F. Clements, The Kruskal-Katona method made explicit. *J. Comb. Theory A* 21(1976) 245-249.

[C19] G.F. Clements, An existence theorem for antichains. *J. Comb. Theory A* 22(1977), 368-371.

[C110] G.F. Clements, More on the generalized Macaulay theorem II, *Disc. Math.* 18(1977), 253-264.

[C111] G.F. Clements, The minimal number of basic elements in a multiset antichain. *J. Comb. Theory A* 25(1978), 153-162.

[CG] G.F. Clements and H.-D.O.F. Gronau, On maximal antichains containing no set and its complement. *Disc. Math.* 33(1981), 239-247.

[CL] G.F. Clements and B. Lindström, A generalization of a combinatorial theorem of Macaulay. *J. Comb. Theory* 7(1969), 230-238.

[DH] G.B. Dantzig and A. Hoffman, On a theorem of Dilworth, *Linear inequalities and related systems,* (H.W. Kuhn and A.W. Tucker) Annals of Math Studies, 38(1966), 207-214.

[Da1] D.E. Daykin, A simple proof of Katona's theorem. *J. Comb. Theory A* 17(1974), 252-253.

[Da2] D.E. Daykin, Erdös-Ko-Rado from Kruskal-Katona, *J. Comb. Theory A* 17(1974), 254-255.

[Da3] D.E. Daykin, An algorithm for cascades giving Katona-type inequalities. *Nanta Math.* 8(1975)2, 78-83.

[Da3a] D.E. Daykin, Antichains in the lattice of subsets of a finite set. *Nanta Math.* 8(1975)2, 84-95.

[Da4] D.E. Daykin, Poset functions commuting with the product and yielding Chebyshev type inequalities. *Problemes Comb. et Theorie des Graphs* (J.C. Bermond), C.N.R.S. Colloques, Paris (1976), 93-98.

[Da5] D.E. Daykin, A lattice is distributive iff $|A|\ |B| \leqslant |A \wedge B|\ |A \vee B|$ *Nanta Math.* 10(1977), 58-60.

[Da5a] D.E. Daykin, Minimum subcover of a finite set, *Amer. Math. Monthly* 85(1978), 766.

[Da6] D.E. Daykin, Inequalities among the subsets of a set. *Nanta Math.* 12(1979), 137-145.

[Da7] D.E. Daykin, A hierarchy of inequalities. *Studies in Applied Math.* 63(1980), 263-274.

[Da8] D.E. Daykin, Functions on a distributive lattice with a polarity *Quart. J. Math. Oxford,* to appear.

[DF1] D.E. Daykin and P. Frankl, Families of finite sets satisfying union conditions, preprint.

[DF2] D.E. Daykin and P. Frankl, Inequalities for subsets of a set and LYM posets, preprint.

[DF3] D.E. Daykin and P. Frankl, Extremal problems on families of subsets of a set generated by a graph, preprint.

[DFGH] D.E. Daykin, P. Frankl, C. Greene, and A.J.W. Hilton, A generalization of Sperner's theorem, preprint.

[DGH] D.E. Daykin, J. Godfrey, and A.J.W. Hilton, Existence theorems for Sperner families. *J. Comb. Theory A* 17(1974), 245-251.

[DH] D.E. Daykin and A.J.W. Hilton, Pairings from down-sets in distributive lattices, to appear.

[DKW] D.E. Daykin, D.J. Kleitman, and D.B. West, The number of meets between two subsets of a lattice. *J. Comb. Theory A* 26(1979), 135-156.

[DL] D.E. Daykin and L. Lovász, The number of values of a Boolean function, *J. Lond. Math. Soc.* (2) 12(1976), 225-230.

[DD] D. Deckard and L.K. Durst, Unique factorization in power series rings and semigroups. *Pacific J. Math.* 16(1966) 239+.

[De] M. Deza, Solution d'un probleme de Erdös-Lovász. *J. Comb. Theory B* 16(1974), 166-167.

[DEF] M. Deza, P. Erdös and P. Frankl, Intersection properties of systems of finite sets. *Combinatorics*, Colloq. János Bolyai, 18(1976), 251-256. Also, J. Lond. Math. Soc. (3)36(1978), 369-384.

[DES] M. Deza, P. Erdös and N.M. Singhi, Combinatorial problems on subsets and their intersections. *Studies in Foundations and Combinatorics* (G.-C. Rota) vol. 1, Academic Press, (1978), 259-265.

[Di1] R.P. Dilworth, A decomposition theorem for partially ordered sets. *Annals of Math.* 51(1950). 161-165.

[Di2] R.P. Dilworth, Some combinatorial problems on partially ordered sets. *Combinatorial Analysis* (Proc. Symp. Appl. Math.), Amer. Math. Soc. (1960), 85-90.

[DG] R.P. Dilworth and C. Greene, A counterexample to the generalization of Sperner's theorem. *J. Comb. Theory* 10(1971), 18-21.

[DoD] T.A. Dowling and W.A. Denig, Geometries associated with partially ordered sets. *Higher Combinatorics* (M. Aigner), D. Reidel Publishing Co. (1977), 103-116.

[Dr] D.A. Drake, Another note on Sperner's lemma. *Canad. Math. Bull.* 14(2)(1971), 255-256.

[DE] R. Duke and P. Erdös, Systems of finite sets having a common intersection.

[DM] B. Dushnik and E. Miller, Partially ordered sets. *Amer. J. Math.* 63(1961), 600-610.

[EM] A. Ehrenfeucht and J. Mycielski, On families of intersecting sets. *J. Comb. Theory A* 17(1974) 259-260.

[Er1] P. Erdös, On a lemma of Littlewood and Offord. *Bull. Amer. Math. Soc.* 51(1945), 898-902.

[Er2] P. Erdös, A problem of independent $r$-tuples, *Ann. Univ. Sci. Budapest* 8(1965), 93-95.

[Er3] P. Erdös, Problems and results on finite and infinite combinatorial analysis. *Combinatorics,* Colloq. János Bolyai 10(1973), 403-424.

[EHS] P. Erdös, M. Herzog, and J. Schönheim, An extremal problem on the set of noncoprime divisions of a number. *Israel J. of Math.* 8(1970), 408-412.

[EK1] P. Erdös and D.J. Kleitman, On collections of subsets containing no 4-member Boolean algebra, *Proc. Amer. Math. Soc.* 28(1971) 507-510.

[EK2] P. Erdös and D.J. Kleitman, Extremal problems among subsets of a set. *Discrete Math.* 8(1974), 281-294.

[EKR] P. Erdös, Chao Ko, and R. Rado, Intersection theorems for systems of finite sets. *Quart. J. Math.* Oxford (2) 12(1961), 313-320.

[EHM] P. Erdös, A. Hajnal, and E.C. Milner, On the complete subgraphs of graphs defined by systems of sets, *Acta Math. Acad. Sci. Hung.* 17(1966), 159-229.

[EKm] P. Erdös and J. Komlós, On a problem of Moser. *Combinatorial Theory and its Applications,* Balatonfured 1969, Colloq. János Bolyai (1970), 365-367.

[ER] P. Erdös and R. Rado, Intersection theorems for systems of sets. *J. London Math. Soc.* 35(1960), 85-90.

[ES] P. Erdös and J. Schönheim, On the set of non-pairwise coprime divisions of a number. *Combinatorial Theory and its Applications,* Balatonfured 1969, Colloq. János Bolyai (1970).

[ESz] P. Erdös and E. Szemerédi, Combinatorial properties of systems of sets. *J. of Comb. Theory A* 24(1978), 308-313.

[ESS] P. Erdös, A. Sarkosy, and E. Szermeredi, On the solvability of the equations $[a_i,a_j]=a_r$ and $(a_i',a_j')=a_r'$ in sequences of positive density. *J. Math. Anal. Appl.* 15(1966), 60-64.

[Fa] K. Fan, On Dilworth's coding theorem. *Math. Z.* 127(1972), 92-94.

[FH] S. Foldes and P.L. Hammer, Split graphs having Dilworth number two. *Canad. J. Math.* 29(1977), 666-672.

[Fo] S.V. Fomin, Finite partially ordered sets and Young tableaux. *Soviet Math. Dokl.* 19(1978), 1510-1514.

[FKG] C.M. Fortuin, P.W. Kasteleyn and J. Ginibre, Correlation inequalities on some partially ordered sets. *Comm. Math. Phys.* 22(1971), 89-103.

[Fk] A. Frank, On chain and antichain families of a partially ordered set. *J. Comb. Theory B* 29(1980), 176-184.

[Fr0] P. Frankl, A theorem on Sperner systems satisfying an additional condition. *Studia Sci. Math. Hung.* 9(1974), 425-431.

[Fr1] P. Frankl, The proof of a conjecture of G. O. Katona. *J. Comb. Theory A* 19(1975), 208-213.

[Fr2] P. Frankl, On Sperner families satisfying an additional condition. *J. Comb. Theory A* 20(1976), 1-11.

[Fr3] P. Frankl, Generalizations of theorems of Katona and Milner. *Acta Math. Acad. Sci. Hung.* 27(1976), 359-363.

[Fr4] P. Frankl, The Erdös-Ko-Rado theorem is true for $n=ckt$. *Combinatorics,* Colloq. János Bolyai 18(1976), 365-375.

[Fr4a] P. Frankl, Families of finite sets satisfying an intersection condition. *Bull. Austral. Math. Soc.* 15(1976), 73-79.

[Fr5] P. Frankl, Families of finite sets containing no $k$ disjoint sets and their union. *Per. Math. Hung.* 8(1977), 29-31.

[Fr6] P. Frankl, On the minimum number of disjoint pairs in a family of finite sets. *J. Comb. Theory A* 22(1977), 249-251.

[Fr7] P. Frankl, On families of finite sets no two of which intersect in a singleton. *Bull. Austral. Math. Soc.* 17(1977), 125-134.

[Fr8] P. Frankl, On intersecting families of finite sets. *J. Comb. Theory A* 24(1978), 146-161.

[Fr9] P. Frankl, An extremal problem for 3-graphs. *Acta Math. Acad. Sci. Hung.* 32(1978), 157-160.

[Fr10] P. Frankl, A Sperner-type theorem for families of finite sets. *Per. Math. Hung.* 9(1978), 257-267.

[Fr11] P. Frankl, Families of finite sets satisfying a union condition. *Disc. Math.* 26(1979), 111-118.

[FD] P. Frankl and M. Deza, On the maximum number of permutations with given maximal or minimal distance. *J. Comb. Theory A* 22(1977), 352-360.

[FF] P. Frankl and Z. Füredi, The Erdös-Ko-Rado theorem for integer sequences. *SIAM J. Alg. Disc. Meth.* 1(1980), 376-381.

[FHi] P. Frankl and A.J.W. Hilton, On the minimum number of sets comparable with some members of a set of finite sets. *J. Lond. Math. Soc.* (2) 15(1977), 16-18.

[Fs] R. Freese, An application of Dilworth's lattice of maximal antichains. *Disc. Math.* 7(1974), 107-109.

[Fu] D.R. Fulkerson, Note on Dilworth's decomposition theorem for partially ordered sets. *Proc. Amer. Math. Soc.* 7(1956), 701-702.

[Fd] Z. Füredi, On maximal intersecting families of finite sets. *J. Comb. Theory A* 28 (1980), 282-289.

[Ga] E.R. Gansner, Acyclic digraphs, Young Tableaux and nilpotent matrices, to appear.

[Ge] M. Gereb, On the smallest maximal set of non-pairwise coprime divisions of a number. *Math. Lapok.* 24(1973), 375-380.

[GS] I. Gertsbakh and H.I. Stern, Minimal resources for fixed and variable job schedules. *Operations Research* 26(1978), 68-85.

[Gh] R.L. Graham, Linear extensions of partial orders and the FKG inequality, in this volume.

[GHa] R.L. Graham and L.H. Harper, Some results on matching in bipartite graphs, *SIAM J. Appl. Math.* 17(1969), 1017-1022.

[GYY] R.L Graham, A.C. Yao, and F.F. Yao, Some monotonicity properties of partial orders. *SIAM J. Alg. Disc. Meth.* 1(1980), 251-258.

[Gr1] C. Greene, An extension of Schensted's theorem. *Advances in Math.* 14(1974), 254-265.

[Gr2] C. Greene, Sperner families and partitions of a partially ordered set. *Combinatorics* (Hall and J.H. Van Lint), Math. Center Tracts 56(1974), 91-106.

[Gr3] C. Greene, Some partitions associated with a partially ordered set. *J. Comb. Theory A* 20(1976),69-79.

[GHi] C. Greene and A.J.W. Hilton, Some results on Sperner families. *J. Comb. Theory A* 26(1979), 202-209.

[GKK] C. Greene, G.O.H. Katona and D.J. Kleitman, Extensions of the Erdös-Ko-Rado theorem. *Studies in Appl. Math.* 55(1976), 1-8.

[GK1] C. Greene and D.J. Kleitman, The structure of Sperner $k$-families. *J. Comb. Theory A* 20(1976), 41-68.

[GK2] C. Greene and D.J. Kleitman, Strong versions of Sperner's Theorem. *J. Comb. Theory A* 20(1976), 80-88.

[GK3] C. Greene and D.J. Kleitman, Proof techniques in the theory of finite sets. *Studies in Combinatorics,* (G.-C. Rota), MAA Studies in Math. 17(1978) 22-79.

[Gg1] J.R. Griggs, Symmetric chain orders, Sperner theorems and loop matching, PhD Thesis, M.I.T. (1977).

[Gg2] J.R. Griggs, Sufficient conditions for a symmetric chain order. *SIAM J. Appl. Math.* 32(1977), 807-809.

[Gg3] J.R. Griggs, Another three-part Sperner theorem. *Studies in Applied Math.* 57(1977), 181-184.

[Gg4] J.R. Griggs, On chains and Sperner $k$-families in ranked posets. *J. Comb. Theory A* 28(1980),156-168.

[Gg5] J.R. Griggs, Poset measure and saturated partitions, preprint.

[Gg6] J.R. Griggs, Recent results on the Littlewood-Offord problem, to appear.

[Gg7] J.R. Griggs, Collections of subsets with the Sperner property, preprint.

[GgK] J.R. Griggs and D.J. Kleitman, A three-part Sperner theorem. *Disc. Math.* 17(1977), 281-289.

[GSS] J.R. Griggs, D. Sturtevant, and M. Saks, On chains and Sperner $k$-families in ranked posets, II. *J. of Comb. Theory A* 29(1980), 391-394.

[Go1] H.-D.O.F. Gronau, On Sperner families in which no 3 sets have an empty intersection. *Acta Cybernetica* 4(1978), 213-220.

[Go2] H.-D.O.F. Gronau, On Sperner families in which no $k$ sets have an empty intersection. *J. Comb. Theory A* 28(1980), 54-63.

[Go3] H.-D.O.F. Gronau, On maximal antichains consisting of sets and their complements. *J. Comb. Theory A* 29(1980), 370-375.

[GM] R.K. Guy and E.C. Milner, Graphs defined by coverings of a set. *Acta Math. Acad. Sci. Hungar.* 19(1968), 7-21.

[HR] A. Hajnal and B. Rothschild, A generalization of the Erdös-Ko-Rado theorem on finite set systems. *J. Comb. Theory A* 15(1973), 359-362.

[Ha1] G. Hansel, Sur le nombre des fonctions Booleenes monotones de n variables. *C.R. Acad. Sci. Paris* 262(1966), 1088-1090.

[Ha2] G. Hansel, Complexe et decompositions binomiales, *J. Comb. Theory A* 12 (1972), 167-183.

[Hr1] L.H. Harper, Optimal assignments of numbers to vertices. *SIAM J. Appl. Math.* 12(1964) 131-135.

[Hr1a] L.H. Harper, Optimal numberings and isoperimetric problems on graphs. *J. Comb. Theory* 1(1966), 385-393.

[Hr2] L.H. Harper, The morphology of partially ordered sets. *J. Comb. Theory A* 17(1974), 44-58.

[Hi1] A.J.W. Hilton, An intersection theorem for two families of finite sets. *Comb. Math. and its Applications,* Proc. Conf. Oxford 1969, Academic Press (1971), 137-148.

[Hi2] A.J.W. Hilton, Simultaneously disjoint pairs of subsets of a finite set. *Quart. J. Math. Oxford* (2) 24(1973),81-95.

[Hi2a] A.J.W. Hilton, Simultaneously independent $k$-sets of edges of a hypergraph.

[Hi3] A.J.W. Hilton, Analogues of a theorem of Erdös, Ko, and Rado on a family of finite sets. *Quart. J. Math Oxford* (2) 25(1974), 19-28.

[Hi4] A.J.W. Hilton, A theorem on finite sets. *Quart. J. Math. Oxford* (2) 27(1976), 33-36.

[Hi5] A.J.W. Hilton, An intersection theorem for a collection of families of subsets of a finite set. *J. London Math. Soc.* (2) 15(1977), 369-376.

[Hi6] A.J.W. Hilton, Some intersection and union theorems for several families of finite sets. *Mathematika* 25(1978), 125-128.

[Hi7] A.J.W. Hilton, A simple proof of the Kruskal-Katona theorem and of some associated binomial inequalities. *Per. Math. Hungar.* 10(1979), 25-30.

[HM] A.J.W. Hilton and E.C. Milner, Some intersection theorems for systems of finite sets. *Quart. J. Math. Oxford* (2) 18(1967),369-384.

[Ho1] M. Hochberg, Restricted Sperner families and $s$-systems. *Disc. Math.* 3(1972),359-364.

[Ho2] M. Hochberg, Doubly incomparable $s$-systems. *Second Int. Conf. on Comb. Math.* (A. Gewirtz and L.V. Quintas), Annals of N.Y. Acad. of Sci. 319(1979), 281-283.

[HH] M. Hochberg and W.M. Hirsch, Sperner families, $s$-systems, and a theorem of Meshalkin. *Int'l. Conf. on Comb. Math.* (A. Gewirtz and L.V. Quintas), N.Y. Acad. of Sciences, (1970), 224-237.

[HS] A.J. Hoffman and D.E. Schwartz, On partitions of a partially ordered set. *J. Comb. Theory B* 23(1977), 3-13.

[Hf] A.J. Hoffman, Ordered sets and linear programming, in this volume.

[Hy] R. Holley, Remarks on the FKG inequalities. *Comm. Math. Phys.* 36(1974),227-231.

[Hs1] W.N. Hsieh, Intersection theorems for systems of finite vector spaces. *Disc. Math.* 12(1975), 1-16.

[Hs2] W.N. Hsieh, Familes of intersecting vector spaces. *J. Comb. Theory A* 18(1975), 252-261.

[HK] W.N. Hsieh and D.J. Kleitman, Normalized matching in direct products of partial orders. *Studies in Appl. Math.* 52(1973), 285-289.

[Ja] R.E. Jamison-Waldner, Partition numbers for trees and ordered sets. *Pacific J. Math.*, to appear.

[Jo1] L.K. Jones, Note sur une loi de probabilite conditionelle de Kleitman. *Disc. Math.* 15(1976), 107-108.

[Jo2] L.K. Jones, On the distribution of sums of vectors. *SIAM J. Appl. Math.* 34(1978), 1-6.

[Ka1] G.O.H. Katona, Intersection theorems for systems of finite sets. *Acta Math. Acad. Sci. Hung.* 15(1964), 329-337.

[Ka2] G.O.H. Katona, On a conjecture of Erdös and a stronger form of Sperner's theorem. *Stud. Sci. Math. Hung.* 1(1966),59-63.

[Ka3] G.O.H. Katona, A theorem of finite sets. *Theory of graphs,* Proc. Coll. Tihany, 1966. Akademiai Kiado (1966), 187-207.

[Ka4] G.O.H. Katona, A generalization of some generalizations of Sperner's theorem. *J. Comb. Theory B* 12(1972), 72-81.

[Ka5] G.O.H. Katona, Families of subset having no subset containing another one with small difference. *Nieuw Arch. Wiskd.* (3) 20(1972), 54-67.

[Ka6] G.O.H. Katona, A simple proof of the Erdös-Ko-Rado theorem. *J. Comb. Theory B* 13(1972), 183-184.

[Ka7] G.O.H. Katona, A three part Sperner theorem. *Stud. Sci. Math. Hung.* 8(1973), 379-390.

[Ka8] G.O.H. Katona, Two applications of Sperner-type theorems (for search theory and truth functions). *Per. Math. Hungar.* 4(1973) 19-23.

[Ka9] G.O.H. Katona, Extremal problems for hypergraphs. *Combinatorics* (M. Hall ahd J.H. Van Lint), Math. Center Tracts 56(1974), 13-42.

[Ka10] G.O.H. Katona, On a conjecture of Ehrenfeucht & Mycielski.

[Ka11] G.O.H. Katona, Continuous versions of some extremal hypergraph problems. *Combinatorics,* Colloq. János Bolyai 18(1976), 653-677.

[Ka12] G.O.H. Katona, Continuous versions of some extremal hypergraph problems, II. *Acta Math. Acad. Sci. Hung.* 35(1980), 67-77.

[KT] D. Kelly and W.T. Trotter Jr., Dimension theory for ordered sets, in this volume.

[Ks] M.J. Klass, Maximum antichains: a sufficient condition. *Proc. Amer. Math. Soc.* 45(1974), 28-30.

[Kl1] D.J. Kleitman, On a lemma of Littlewood and Offord on the distribution of certain sums. *Math. Zeitsch.* 90(1965),251-259.

[Kl2] D.J. Kleitman, Families of non-disjoint subsets. *J. Comb. Theory* 1(1966), 153-155.

[Kl3] D.J. Kleitman, On a combinatorial problem of Erdös. *Proc. Amer. Math. Soc.* 17(1966), 139-141.

[Kl4] D.J. Kleitman, On a combinatorial conjecture of Erdös. *J. Comb. Theory* 1(1966) 209-214.

[Kl5] D.J. Kleitman, On subsets containing a family of non-commensurable subsets of a finite set. *J. Comb. Theory* 1(1966), 297-299.

[Kl6] D.J. Kleitman, On a conjecture of Milner on $k$-graphs with non-disjoint edges. *J. Comb. Theory* 5(1968), 153-156.

[Kl7] D.J. Kleitman, Maximum number of subsets of a finite set no $k$ of which are pairwise disjoint. *J. Comb. Theory* 5(1968) 157-163.

[Kl8] D.J. Kleitman, On families of subsets of a finite set containing no two disjoint sets and their union. *J. Comb. Theory* 5(1968), 235-237.

[Kl9] D.J. Kleitman, A conjecture of Erdös-Katona on commeasurable pairs among subsets of an n-set. *Theory of Graphs,* Proc. Coll. Tihany 1966, Akademiai Kiado (1968), 215-218.

[Kl10] D.J. Kleitman, On Dedekind's problem: The number of monotone Boolean functions. *Proc. Amer. Math. Soc.* 21(1969), 677-682.

[Kl11] D.J. Kleitman, On a lemma of Littlewood and Offord on the distribution of linear combinations of vectors. *Advances in Math.* 5(1970), 155-157.

[Kl12] D.J. Kleitman, The number of Sperner families of subsets of an $n$ element set. *Infinite and finite sets,* Colloq. János Bolyai (Keszthely) 10(1973), 989-1001.

[Kl13] D.J. Kleitman, On an extremal property of antichains in partial orders: the LYM property and some of its implications and applications. *Combinatorics* (Hall and J.H. Van Lint), Math. Centre Tracts 56(1974), 77-90.

[Kl14] D.J. Kleitman, Some new results on the Littlewood-Offord problem. *J. Comb. Theory A* 20(1976), 89-113.

[Kl15] D.J. Kleitman, Extremal properties of collections of subsets containing no two sets and their union. *J. Comb. Theory A* 20(1976), 390-392.

[Kl16] D.J. Kleitman, Hypergraphic extremal properties. *Proc. Brit. Comb. Conf.* (1980).

[KEL] D.J. Kleitman, M. Edelberg and D. Lubell, Maximal sized antichains in partial orders. *Disc. Math.* 1(1971), 47-53.

[KKR] D.J. Kleitman, M.M. Krieger, and B.L. Rothschild, Configurations maximizing the number of pairs of Hamming-adjacent lattice points. *Studies in Appl. Math.* 50(1971)2.

[KMg] D.J. Kleitman and T.L. Magnanti, On the latent subsets of intersecting collections. *J. Comb. Theory A* 16(1974), 215-220.

[KMk] D.J. Kleitman and G. Markowsky, On Dedekind's problem: the number of monotone Boolean functions II. *Trans. Amer. Math. Soc.* 45(1974), 373-389.

[KMl] D.J. Kleitman and E.C. Milner, On the average size of sets in a Sperner family. *Disc. Math.* 6(1973) 141-147.

[KSa] D.J. Kleitman and M. Saks, Stronger forms of the LYM inequality, to appear.

[KSh] D.J. Kleitman and J.B. Shearer, Some monotonicity properties of partial orders. *Studies in Appl. Math*, to appear.

[KSp] D.J. Kleitman and J. Spencer, Families of $k$-independent sets. *Disc. Math.* 6(1973), 255-262.

[Kn] R. Kneser, Aufgabe 360. *Jahresbericht d. Deutschen Math. Verein* (2)58(1955).

[Kr1] J.B. Kruskal, The number of simplices in a complex. *Mathematical Optimization Techniques*, Univ. of Calif Press, Berkeley and Los Angeles (1963), 251-278.

[Kr2] J.B. Kruskal, The number of $s$-dimensional faces in a complex: An analogy between the simplex and the cube. *J. Comb. Theory* 6(1969) 86-89.

[Le] K. Leeb, Sperner theorems with choice functions. Vordnunck, to appear.

[LL] E. Levine and D. Lubell, Sperner collections on sets of real variables. *Int'l. Conf. on Comb. Math.* (A. Gewirtz and L.V. Quintas), N.Y. Acad. of Sci., (1970), 272-276.

[Lw] M. Lewin, Choice mappings of certain classes of finite sets. *Math. Z.* 124(1972), 23-36.

[Li1] S.-Y. R. Li, An extremal problem among subsets of a set. *J. Comb. Theory A* 23(1977), 341-343.

[Li2] S.-Y. R. Li, Extremal theorems on divisors of a number. *Disc. Math.* 24(1978), 37-46.

[Lg] T.M. Liggett, Extensions of the Erdös-Ko-Rado theorem and a statistical application. *J. Comb. Theory A* 23(1977), 15-21.

[Lh1] K.-W. Lih, Sperner families over a subset. *J. Comb. Theory A* 29(1980), 182-185.

[Lh2] K.-W. Lih, Majorization on finite partially ordered sets. *SIAM J. Alg. Disc. Meth.*, to appear.

[Ly] J.H. Lindsey, Assignment of numbers to vertices. *Amer. Math. Monthly* 71(1964) 508-516.

[Ld1] B. Lindström, Optimal number of faces in cubical complexes. *Arkiv. for Mat.* 8(1971) 245-257.

[Ld2] B. Lindström, A partition of L(3,n) into saturated symmetric chains. *Europ. J. Comb.* 1(1980), 61-63.

[Ln1] N. Linial, Covering digraphs by paths. *Disc. Math.* 23(1978) 257-272.

[Ln2] N. Linial, Extending the Greene-Kleitman theorem to directed graphs. *J. Comb. Theory A* 30(1981), 331-334.

[Lp] W. Lipski Jr., On strings containing all subsets as substrings. *Disc. Math.* 21(1978), 253-259.

[Lu] C.L. Liu, Sperner's theorem on maximal-sized antichains and its generalizations. *Disc. Math.* 11(1975), 133-139.

[Lv] M.L. Livingston, An ordered version of the Erdös-Ko-Rado theorem. *J. Comb. Theory A* 26(1979), 162-165.

[Lo] L. Lovász, The Shannon capacity of a graph. *IEEE Trans. Info Th.* IT-25(1979), 1-7.

[Lo2] L. Lovász, Kneser's conjecture, chromatic number, and homotopy. *J. Comb. Theory A* 25(1978),319-324.

[Lb] D. Lubell, A short proof of Sperner's lemma. *J. Comb. Theory* (1966), 299.

[Ma] F.S. Macaulay, Some properties of enumeration in the theory of modular systems. *Proc. Lond. Math. Soc.* 26(1927), 531-555.

[MS] J. Marica and J. Schönheim, Differences of sets and a problem of Graham. *Canad. Math. Bull.* 12(1969), 635-637.

[Me] L.D. Meshalkin, A generalization of Sperner's Theorem on the number of subsets of a finite set. *Teor. Verojatnosti Primener* 8(1963), 219-220 (in Russian) - *Theory Probab. Appl.* (English trans.) 8(1963), 203-204.

[MR] N. Metropolis and G.-C. Rota, Combinatorial structure of the faces of the $n$-cube. *SIAM J. Appl. Math.* 35(1978), 689-694. Also *Bull. Amer. Math. Soc.* 84(1978), 284-286.

[MRSW] N. Metropolis, G.-C. Rota, V. Strehl, and N. White, Partitions into chains of a class of partially ordered sets. *Proc. Amer. Math. Soc.* 71(1978), 193-196.

[Mi1] S. Milne, Restricted growth functions and incidence relations of the lattice of partitions of a $n$-set. *Advances in Math.* 26(1977), 290-305.

[Mi2] S. Milne, Rank row matchings of partition lattices, the maximum Stirling number and Rota's conjecture.

[Ml1] E.C. Milner, Intersection theorems for systems of sets. *Quart. J. Math. Oxford* 18(1967), 369-384.

[Ml2] E.C. Milner, A combinatorial theorem on systems of sets. *J. London Math. Soc.* 43(1968), 204-206.

[Mr] L. Mirsky, A dual of Dilworth's decomposition theorem. *Amer. Math. Monthly* 78(1971), 876-877.

[Mu] R. Mullin, On Rota's problem concerning partitions. *Aequationes Math.* 2(1969), 98-104.

[NTN] G.L. Nemhauser, L.E. Trotter Jr., and R.M. Nauss, Set partitioning and chain decomposition. *Management Sci.* 20(1974), 1413-1473.

[OS] H. Oellrich and K. Steffens, On Dilworth's decomposition theorem. *Disc. Math.* 15(1976), 301-304.

[Pe1] G.W. Peck, Maximum antichains of rectangular arrays. *J. Comb. Theory A* 27(1979), 397-400.

[Pe2] G.W. Peck, Erdös conjecture on sums of distinct numbers. *Studies in Appl. Math.* 63(1980), 87-92.

[Pr1] M.A. Perles, A proof of Dilworth's decomposition theorem for partially ordered sets. *Israel J. of Math.* 1(1963), 105-107.

[Pr2] M.A. Perles, On Dilworth's theorem in the infinite case. *Israel J. of Math.* 1(1963), 108-109.

[Po] M. Pouzet, Ensemble ordonne universel recouvert par deux chaines. *J. Comb. Theory B* 25(1978), 1-25.

[Ps] C.J. Preston, A generalization of the FKG inequalties. *Comm. Math. Phys.* 36(1974), 233-241.

[Pz1] O. Pretzel, On the dimension of partially ordered sets. *J. Comb. Theory A* 22(1977), 146-152.

[Pz2] O. Pretzel, Another proof of Dilworth's decomposition theorem. *Disc. Math.* 25(1979), 91-92.

[PSS] R.A. Proctor, M.E. Saks and D.G. Sturtevant, Product partial orders with the Sperner property. *Disc. Math.* 30(1980), 173-180.

[Pu] G. Purdy, A result on Sperner collections. *Utilitas Math.* 12(1977), 95-99.

[Ri] W. Riess, Zwei Optimierungsprobleme auf Ordnungen. *Arbeitsbereichte des institute für mathematische Maschinen und Datenverarbeitung (Informatik) II* 5(1978), 50-57.

[Rs] E. Rosenthal, Searching in posets, preprint.

[Ro] G.-C. Rota, A generalization of Sperner's theorem, research problem. *J. Comb. Theory* 2(1967),104.

[Ru] I.Z. Ruzsa, Probabilistic generalization of a number-theoretic inequality. *Amer. Math. Monthly* 83(1976), 723-725.

[Sa1] M. Saks, A short proof of the existence of $k$-saturated partitions of partially ordered sets. *Advances in Math.* 33(1979), 207-211.

[Sa2] M. Saks, Dilworth numbers, incidence maps and product partial orders. *SIAM J. Alg. Disc. Meth.* 1(1980), 211-216.

[Sa3] M. Saks, Duality properties of finite set systems. Ph.D. Thesis, M.I.T. (1980).

[Sd] W. Sands, Counting antichains in finite partially ordered sets, preprint.

[Su] N. Sauer, On the density of families of sets. *J. Comb. Theory A* 13(1972), 145-147.

[Sc1] J. Schönheim, A generalization of results of P. Erdös, G. Katona and D.J. Kleitman concerning Sperner's theorem. *J. Comb. Theory* 11(1971), 111-117.

[Sc2] J. Schönheim, On a problem of Daykin concerning intersecting families of sets. *Proc. Brit. Comb. Conf. Aberystwyth 1973,* Lond. Math. Soc. Lect. Notes 13(1974), 139-140.

[Sc3] J. Schönheim, On a problem of Purdy related to Sperner systems. *Canad. Math. Bull.* 17(1974), 135-136.

[Se] P.D. Seymour, On incomparable collections of sets. *Mathematika* 20(1973), 208-209.

[SW] P.D. Seymour and D.J.A. Welsh, Combinatorial applications of an inequality from statistical mechanics. *Math. Proc. Camb. Phil. Soc.* 77(1975), 485-495.

[SK] J. Shearer and D.J. Kleitman, Probabilities of independent choices being ordered. *Studies in Appl. Math.* 60(1979),271-276.

[Sh] L.A. Shepp, The FKG Inequality and some monotonicity properties of partial orders. *SIAM J. Alg. Disc. Meth.* 1(1980), 295-299.

[Sp] J.H. Spencer, A generalized Rota conjecture for partitions. *Studies in Appl. Math.* 53(1974), 239-241.

[Sr1] E. Sperner, Ein Satz uber Untermengen einer endlichen Menge. *Math. Z.* 27(1928), 544-548.

[Sr2] E. Sperner, Über einer kombinatorische Satz von Macaulay und seine Anwendung aug die Theorie der Polynomideale. *Abh. Math. Sem. Hamburg* 7(1930), 149-163.

[Sn] R.P. Stanley, Weyl groups, the hard Lefschetz theorem, and the Sperner property. *SIAM J. Alg. Disc. Meth.* 1(1980), 168.

[St] D. Stanton, Some Erdös-Ko-Rado theorems for Chevalley groups. *SIAM J. Alg. Disc. Meth.* 1(1980),160.

[Sk] B.S. Stechkin, Yamamoto inequality and gatherings. Transl. from *Matematicheskie Zametki* 19(1976), 155-160.

[Ta] V.S. Tanaev, Optimal decomposition of a partially ordered set into chains. *Doklady Akad. Nauk BSSR* 23(1979), 389-391. *Math. Rev.* 80m:06003.

[Ty] C.A. Tovey, Polynomial local improvement algorithms in combinatorial optimization. Ph.D. Thesis, Stanford Dept. of Op. Research (1981).

[Tr] W.T. Trotter Jr., A note on Dilworth's embedding theorem. *Proc. Amer. Math. Soc.* 52(1975), 33-39.

[Tv] H. Tverberg, On Dilworth's decomposition theorem for partially ordered sets. *J. Comb. Theory* 3(1967), 305-306.

[Wa] D.-L. Wang, On systems of finite sets with constraints in their unions and intersections. *J. Comb. Theory A* 23(1977). 344-348.

[WW1] D.-L. Wang and P. Wang, Extremal configurations on a discrete torus and a generalization of the generalized Macaulay theorem. *SIAM J. Appl. Math.* 33(1977), 55-59.

[WW2] D.-L. Wang and P. Wang, Discrete isoperimetric problems. *SIAM J. Appl. Math.* 32(1977), 860-870.

[We] D.B. West, A symmetric chain decomposition of $L(4,n)$. *Europ. J. Comb.* 1(1980), 379-383.

[WK] D.B. West and D.J. Kleitman, Skew chain orders and sets of rectangles. *Disc. Math.* 27(1979), 99-102.

[WT1] D.B. West and C.A. Tovey, Semiantichains and unichain coverings in direct products of partial orders. *SIAM J. Alg. and Disc. Meth.* 2(1981), Sept.

[WT2] D.B. West and C.A. Tovey, A sufficient condition for the two-part Sperner property.

[WhW] D.E. White and S.G. Williamson, Recursive matching algorithms and linear orders on the subset lattice. *J. Comb. Theory A* 23(1977), 117-127.

[Wo] E.S. Wolk, On decompositions of partially ordered sets. *Proc. Amer. Math. Soc.* 15(1964), 197-199.

[Ya] K. Yamamoto, Logarithmic order of free distributive lattices. *J. Math. Soc. Japan* 6(1954), 343-353.

# ENUMERATION IN CLASSES OF ORDERED STRUCTURES

Robert W. Quackenbush
Department of Mathematics and Astronomy
The University of Manitoba
Winnipeg, Canada R3T 2N2

ABSTRACT

Let $K$ be a class of finite structures. There are two obvious enumeration questions to ask of $K$: what is the set of cardinalities of members of $K$ (the *spectrum problem*) and, for each $n \geq 1$ what is the number of pairwise non-isomorphic members of $K$ of cardinality $n$ (the *fine spectrum problem*). If we restrict our attention to classes of partially ordered structures, the spectrum problem is usually trivial since the standard classes considered usually contain the class of chains, while the fine spectrum problem is usually hopeless since it seems impossible to enumerate ordered sets much more complicated than chains or antichains.

But there are interesting classes where the spectrum problem is intractable and where it is tractable but not trivial. Surely the most famous unsolved enumeration problem for classes of ordered structures is Dedekind's problem: what is the cardinality of the free distributive lattice on $n$ free generators. The answer is known for $n \leq 7$. An indication of just how difficult this problem is can be seen by noting that the answer for $n = 7$ was known in 1965 while in 1981 the answer for $n = 8$ remains unknown. For a much more tractable problem, let $3^P$ be the lattice of all isotone maps from the finite ordered set $P$ into the 3-element chain and take

$$K = \{3^P \mid P \text{ is a finite poset}\} ;$$

what is the spectrum of this $K$? Typical fine spectrum problems would be to enumerate the number of lattices of order $n$ and the number of posets of order $n$. Similarly, we would like to enumerate the number of structures in the important subclasses of these two classes of ordered structures.



*I. Rival (ed.), Ordered Sets, 523–554.*

In almost all problems of this type an exact answer is beyond hope and so we seek an approximate answer. For a spectrum problem we can be optimistic and ask for the exact answer modulo a finite number of exceptions. This is the usual approach to the spectrum problem for classes of block designs. Somewhat surprisingly, this approach works well for some classes of ordered structures. Let $M_n$ be the $(n+2)$-element lattice with $n$ atoms; then for each $n$ the spectrum of $\{(M_n)^P \mid P \text{ is a finite poset}\}$ is known modulo finitely many exceptions [MQ]. If we are pessimistic, then we only look for asymptotic results. We can ask for the asymptotic density of the spectrum. More usually, as with Dedekind's problem we know that the asymptotic density is zero and that there is at most one structure of any cardinality; in this case we ask for an asymptotic estimate for the cardinality of the $n^{\text{th}}$ largest member of the class as a function of $n$. For Dedekind's problem, the best published result [KM] gives an asymptotic formula for the logarithm of the cardinality (thus the error is multiplicative rather than additive).

For the fine spectrum problem, we seek an asymptotic formula for the number of structures of order $n$. An asymptotic estimate is known for the logarithm of the number of $n$-element posets [KR]. For the number of $n$-element lattices, only upper and lower bounds for the logarithm are known, [KW] and [KL], respectively. Fortunately, the rates of growth of these estimates are the same. Interestingly, there do not seem to be any known reasonable estimates for the number of $n$-element distributive lattices.

An entirely different aspect of enumeration is exact enumeration for small values of $n$. For the smallest values of $n$, paper and pencil usually suffice; for large values of $n$ the task is hopeless. In between we use the computer to do our counting for us. It would also be nice to be able to use the computer to test conjectures on moderately sized structures. Thus we are interested in finding good algorithms for generating all structures up to some moderate size. There seems to be relatively little in the literature on this topic related to ordered structures, although there is an extensive literature for graphs and designs. In [Da] there is an algorithm for generating posets, while in [Ky] there is one for lattices. In designing such algorithms, one seeks high efficiency on the one hand (mere number crunching is in extremely bad taste) and the ability to apply the algorithm to important subclasses on the other; unfortunately, these two goals are not necessarily compatible. An algorithmic approach to the spectrum problem occurs in trying to compute the cardinalities of free lattices in equational classes other than distributive lattices; this has been done in [BK].

An important and interesting aspect of this algorithmic approach is the isomorphism problem: how hard is it to tell whether two $n$-element structures in the class $K$ are isomorphic. The isomorphism problems for ordered sets and lattices are easily seen to be hard (i.e., polynomial-time equivalent to the isomorphism problem for graphs). The difficulty of the isomorphism problem for a class seems closely related to the growth rate of the fine spectrum of the class.

There are relatively few results and techniques in the area of enumeration of ordered structures; the basic ones will be examined in detail. That there are only meager results in this area has two advantages. The first is that there is much interesting mathematics to be done. The second may be viewed as a philosophical rewording of the first: the area is in its infancy and so this is an opportune time to set out a coherent plan of investigation.

## 1. INTRODUCTION.

Given a class K of finite structures, there are two common questions asked by enumerators: what is the set of sizes of members of K (the *spectrum* problem), and for each n ≥ 1 what is the number of pairwise non-isomorphic structures of size n in K (the *fine spectrum* problem). In this paper we will survey classes of ordered structures with respect to these two enumeration problems; on the whole they will be unresponsive.

The degree of difficulty of the spectrum problem ranges from easy (chains) to intermediate (boolean algebras) to hopelessly hard (simple complemented modular lattices of length 3). Similarly, the degree of difficulty of the fine spectrum problem ranges from easy (cyclic groups) to intermediate (abelian groups) to hopelessly hard (groups). But for a given class, the spectrum problem is usually much easier than the fine spectrum problem, for example, lattices or groups. In order to solve the fine spectrum problem, one must prove a strong structure theorem as for boolean algebras or abelian groups. For the spectrum problem, there is a common reason for it to be easy, namely that the class contains a natural subclass containing one member of size n for each n ≥ 1, for example, lattices and chains, groups and cyclic groups. There is also a common reason for the spectrum problem to be of moderate difficulty, namely that the class is sufficiently structured that straightforward combinatorial arguments give necessary conditions for the spectrum but the class is sufficiently unstructured that using direct and recursive constructions, one can construct a member of each size satisfying the necessary conditions. This is the case with Steiner triple systems and other classes of block designs; we will see that is is also the case for certain classes of lattices.

We are only interested in natural classes of structures. It is all too easy to write down a list of properties and ask to enumerate the class of all structures satisfying this list, for example, simple complemented modular lattices of length 3. It is also all too easy to take a trivial problem and make a slight change to arrive at a problem that is no longer trivial. For instance, the class of distributive lattices contains all chains and so has a complete spectrum. To modify the problem, recall that every distributive lattice can be represented as $\underline{2}^P$, the lattice of all isotone maps from the ordered set P into the 2-element chain $\underline{2}$. For an ordered set Q define K(Q) to be $\{Q^P \mid P \text{ is an ordered set}\}$; here $Q^P$ is the ordered set of all isotone maps from P to Q. If $Q \not\cong \underline{2}$, then K(Q) contains no non-trivial chains; what is the spectrum of K(Q)?

While these specific examples may appear contrived, they are actually quite natural. The classes K(Q) for Q a bounded ordered set will be discussed in the next section where it will be indicated how they arise naturally in universal algebra. As for the first class, a simple complemented modular lattice of length 3 is the lattice of flats of a projective plane; thus, determining the spectrum of the class of simple complemented modular lattices of length 3 is equivalent to determining the spectrum of the class of projective planes. Not only can two spectrum problems be equivalent, but a spectrum problem can be equivalent to a fine spectrum problem. For instance, Dedekind's problem, the problem of determining the spectrum of the class of free distributive lattices, is equivalent to the problem of determining the fine spectrum of the class of simplicial complices.

Enumeration problems for ordered sets tend to be either trivial or very hard. This is because the arithmetic of ordered structures tells very little about the cardinalities of the structures; there are no analogs of the Sylow theorems for groups. Usually we will not be able to give an exact solution to our enumeration problem but only an approximate solution. The best approximate solution is where we know the answer exactly except at a finite number of places. This is the kind of solution we should look for with spectrum problems; typically, it is the kind of solution given for classes of block designs. We will see that for certain lattices L it is also the case for the spectrum of K(L). Barring this we seek an asymptotic estimate in terms of n for the $n^{th}$ smallest member of Spec(K), the spectrum of K; this is what is sought for Dedekind's problem. Failing this we seek an asymptotic estimate for the density of Spec(K); obviously, the density of the spectrum of the class of free distributive lattices is 0.

For an approximate solution to a fine spectrum problem, we can at best hope for an asymptotic solution. Let $f_K(n)$ be the number (of isomorphism classes) of structures of size n in K. If $\lim_{n\to\infty} (g(n) - f_K(n))/g(n) = 0$, then g(n) is *asymptotic* to $f_K(n)$; that is, the percentage error tends to 0 as n tends to $\infty$. Often, all that is known is an asymptotic estimate for $\log(f_K(n))$;

usually this means that the percentage error tends to $\infty$. Sometimes we only know crude estimates for $\log(f_K(n))$; this is the case for the class of lattices where it is known that $cn^{3/2} \leq \log_2(f_K(n)) \leq dn^{3/2}$ with $\log_2(c) = 2^{1/2}/4$ and d is about 6.1.

An entirely different aspect of enumeration is exact enumeration for small values of n. With the advent of modern computers, it has become both practical and useful to use the computer to generate small models on which to test conjectures. Thus we seek efficient algorithms for generating all structures in K of small size n or less. This takes us into the theory of algorithms. However, we will not go very far into it since we are interested in efficiency for low values of n rather than asymptotic classifications of efficiency.

There is one aspect of the theoretical side of algorithms which we will look at in more detail. This is the isomorphism problem: how hard is it to tell whether two n-element structures from K are isomorphic. The canonical hard isomorphism problem is that for graphs. The isomorphism problem for a class K will be called *hard* if it is polynomial-time equivalent to the isomorphism problem for graphs. Thus it is easy to show that the isomorphism problems for ordered sets and for lattices are both hard while for chains and for boolean algebras they are trivial. It seems that the faster $f_K(n)$ grows, the more difficult the isomorphism problem becomes.

The remainder of the paper is organized as follows: In section 2 some spectrum problems and the key ideas behind them are discussed. In section 3 the same is done for some fine spectrum problems. Section 4 contains a review of some published algorithms for enumerating all lattices and all ordered sets and discusses how they might be improved. The last section is devoted to the isomorphism problem.

Finally, note well that I am using the following convention: all structures are finite unless explicitly stated otherwise. This convention has already been used throughout this introduction.

## 2. SPECTRUM PROBLEMS.

Let K be a class of ordered structures. For the spectrum problem for K to be non-trivial, it must contain neither the class of chains nor the class of anti-chains. In this section we will consider two types of such K: K(Q) as defined in section 1 and $F(V)$, the class of free lattices in some variety $V$ of lattices.

I promised a justification for considering $K(Q)$; it lies in the profound classification of precomplete algebras by I. Rosenberg [Ro]. One of the foremost problems in the theory of multiple-valued logic (or the theory of clones) is: given a set $F$ of functions on the base set $A$, when is $F$ complete (that is, when is every function on $A$ a composition of functions from $F$). Call $F$ precomplete if it is not complete but augmenting $F$ by any function $g$ not a composition from $F$ makes $F \cup \{g\}$ complete. Rosenberg has given a complete description of all precomplete $F$ in terms of 6 classes of relations. One of these classes is the class of bounded orders, that is, orders with a least element 0 and a greatest element 1. Given a bounded ordered set $P$, let $F_P$ be all isotone functions (of all arities) on $P$; then $F_P$ is precomplete. The composition of isotone functions is again isotone so $F_P$ cannot be complete. Let $g$ be a non-isotone function on $P$; it is a straightforward exercise to show that $F \cup \{g\}$ is complete.

Given any set $F$ of functions on $A$ we form the universal algebra $\langle A; F\rangle$; a natural question to ask is: what does $e(\langle A; F\rangle)$, the variety (equational class) generated by $\langle A; F\rangle$, look like. If $P = \underline{2}$, the 2-element chain, then it is well known that for $F = \{\vee, \wedge, 0, 1\}$, every isotone function on $\underline{2}$ is a composition from $F$. But $e(\langle\{0, 1\}; F\rangle) = BD$, the variety of bounded distributive lattices. In R. Quackenbush [Qu], it is shown for the bounded ordered set $Q$ that $BD$ and $e(\langle Q; F_Q\rangle)$ are equivalent categories if and only if $Q$ is a lattice. This was done by showing that every finite algebra in $e(\langle Q; F_Q\rangle)$ is isomorphic to $Q^P$ for some ordered set $P$. Thus for $Q$ a lattice, $K(Q)$ is (up to isomorphism) the finite part of $e(\langle Q; F_Q\rangle)$. As categories, $K(Q)$ and $K(\underline{2})$ are isomorphic, but underlying sets are not preserved by this isomorphism. Thus $K(Q)$ and $K(\underline{2})$ have different fine spectra and indeed, different spectra.

2a. $\mathrm{Spec}(K(M_n))$.

If $Q \neq \underline{2}$, then $K(Q)$ contains no non-trivial chain and it is not clear what $\mathrm{Spec}(K(Q))$ is. If we recall that $(Q_1 \times Q_2)^P \cong (Q_1)^P \times (Q_2)^P$, then we see that with $\underline{2}^n$ the $n^{th}$ power of 2, $\mathrm{Spec}(K(\underline{2}^n))$ is all $n^{th}$ powers of the natural numbers. Let $\underline{n}$ denote the n-element chain. What is $\mathrm{Spec}(K(\underline{3}))$? Since $\underline{3} \cong \underline{2}^{\underline{2}}$, since $((R)^{Q})^{P} \cong ((R)^{P^{Q}})$ and since

$P^{\underline{2}} = \{(a, b) \mid a, b \in P, a \leq b\}$ is the order relation determined by P, $\underline{3}^P$ is isomorphic to the order relation determining the distributive lattice $\underline{2}^P$. This is interesting, but not very helpful; instead, let us turn to R. McKenzie and R. Quackenbush [MQ] for help.

Let $\underline{3}(P) = |\underline{3}^P|$; if P is small, it is easy to compute $\underline{3}(P)$. The basic idea is to compute $\underline{3}(P)$ for small P and then use recursive constructions for bigger P, but we must get explicit formulas for $\underline{3}(P)$ when it is recursively constructed. It is clear that we must do something of this sort since otherwise we run into Dedekind's problem in the guise of computing $\underline{2}(\underline{2}^n)$, the size of $\underline{2}^{(\underline{2}^n)}$ (if we cannot compute $\underline{2}(\underline{2}^n)$, then we cannot compute $\underline{3}(\underline{2}^n)$). Two obvious recursive constructions are the disjoint union and the linear sum. Let $P \cup Q$ denote the disjoint union of P and Q; $P + Q$ denotes the linear sum of P and Q with midpoint m (that is, the disjoint union of P, Q and $\{m\}$ additionally ordered so that each element of P is less than m and m is less than each element of Q). Then it is easy to see that $\underline{2}(P \cup Q) = \underline{2}(P)\underline{2}(Q)$, $\underline{3}(P \cup Q) = \underline{3}(P)\underline{3}(Q)$, $\underline{2}(P + Q) = \underline{2}(P) + \underline{2}(P)$, $\underline{3}(P + Q) = \underline{3}(P) + \underline{3}(Q) + \underline{2}(P)\underline{2}(Q)$.

The building blocks for these recursive constructions are $\underline{1}$ and $D_1$, the 2-element antichain. Inductively, $D_{k+1}$ is the linear sum (without midpoint) of $D_k$ and $D_1$, and $E_k = \underline{k} \cup \underline{1}$. Next, for $0 \leq j \leq k$ define $P_{k,j}$ to be $\underline{3(k-j)} + D_j$ and $Q_{k,j}$ to be $\underline{2(k-j)} + E_j$; lastly, define $R_{k,j,i}$ to be $P_{k,j} + Q_{k,i}$. A straightforward but tedious computation yields the formula:

$$\underline{3}(R_{k,j,i}) = \frac{25}{2}k^2 + \frac{55}{2}k + 15 - \frac{1}{2}i^2 - \frac{1}{2}i - j. \qquad (2.1)$$

This is true for $0 \leq i,j \leq k$ so $\underline{3}(R_{k,j,i}) \in \mathrm{Spec}(K(\underline{3}))$ for all $k \geq 1$ and $0 \leq i,j \leq k$. For fixed k as i and j range from 0 to k, the values of $\underline{3}(R_{k,j,i})$ range over an interval. For $k \geq 47$, consecutive intervals overlap. This proves that the spectrum of $K(\underline{3})$ contains all but finitely many positive integers.

THEOREM 2.2 [MQ]. *Spec*$(K(\underline{3}))$ *is cofinite.*

The authors next consider the lattices of length 2: $M_n$ is the lattice of length 2 with exactly n atoms. They proceed to determine $\mathrm{Spec}(K(M_n))$; let $\underline{M}_n(P) = |(M_n)^P|$. Since

$\underline{M}_n(P \cup Q) = \underline{M}_n(P)\underline{M}_n(Q)$ and $\underline{M}_n(P + Q) = \underline{M}_n(P) + \underline{M}_n(Q) + n\underline{2}(P)\underline{2}(Q)$, it is again straightforward to compute $\underline{M}_n(R_{k,j,i})$. When this is done, the crucial term $(\frac{1}{2}i^2 + \frac{1}{2}i + j)$ appears again but this time multiplied by $(n^2 - 2n)$. For $n = 1$, this yields the result for $\underline{3}(R_{k,j,i})$. For $n = 2$, $M_2 \cong \underline{2}^2$ so that $\underline{M}_2(P) = (\underline{2}(P))^2$; note that $n^2 - 2n = 0$ for $n = 2$. For $n \geq 3$, the authors show that any congruence class modulo $(n^2 - 2n)$ having non-zero intersection with $\mathrm{Spec}(K(M_n))$ has all but finitely many of its elements in $\mathrm{Spec}(K(M_n))$. They next show that the term $n^2 - 2n$ cannot be avoided:

THEOREM 2.3 [MQ]. *$\underline{M}_n(P) \equiv \underline{2}(P) + \frac{n}{2}(\underline{2}(P)^2 - \underline{2}(P))$ modulo $(n^2-2n)$.*

PROOF. Let $\underline{M}_{n,j}(P)$ be the number of isotone maps from P into $M_n$ each of whose images contains exactly j atoms of $M_n$. Then $\underline{M}_n(P) = \sum_{j=1}^{n} \underline{M}_{n,j}(P)$. But $\underline{M}_{n,j}(P) = \binom{n}{j}\underline{M}_{j,j}(P)$, and $\underline{M}_{j,j}(P)$ is divisible by $j!$. Hence, $\underline{M}_n(P) \equiv \underline{M}_{n,0}(P) + \underline{M}_{n,1}(P) + \underline{M}_{n,2}(P)$ modulo $(n^2 - 2n)$. From here the result quickly follows.

This gives the following result.

THEOREM 2.4 [MQ]. *Let $n \geq 3$. If $m \in \mathrm{Spec}(K(M_n))$, then for some $0 \leq a < n^2 - 2n$, $m \equiv (n - 2)(\frac{a^2 - a}{2}) + a^2$ modulo $(n^2 - 2n)$. Conversely, all but finitely many such positive integers m are in $\mathrm{Spec}(K(M_n))$.*

This result, stating that certain necessary congruence conditions on the spectrum are asymptotically sufficient (that is, have only finitely many exceptions), is remarkably similar to the proof of the existence conjecture for block designs of Richard M. Wilson [Ws]. Many of these classes of block designs can be viewed as varieties of quasigroups. Algebraically, lattices and quasigroups are quite different. It would be very interesting to understand why these classes have such similar spectra.

What about $\mathrm{Spec}(K(P))$ for other P? The next easiest case is $\underline{4}$. The authors report that the separation of variables phenomenon of equation (2.1) fails and that they are unable to determine the spectrum of the resulting polynomial. A severe impediment to the solution of the spectrum problem for arbitrary P is that the degree of the polynomial obtained by their method equals the length of the longest chain in P.

2b. Dedekind's Problem.

Now let us examine the spectrum problem for free algebras in locally finite varieties of lattices (that is, varieties where all finitely generated lattices are finite). Our easiest problem is that of determining the size of the free distributive lattice on n free generators, FD(n). This is Dedekind's problem and its exact solution appears hopeless; not only is there no known closed form expression for $|FD(n)|$, but there seems to be no practical recursive algorithm for computing $|FD(n)|$. The value of $|FD(7)|$ has been known since 1965 [Ch], but $|FD(8)|$ remains undetermined. Thus we will examine asymptotic estimates for $|FD(n)|$. Then we will examine some techniques for determining the size of n-generated free lattices in small varieties for small n. Surprisingly, good use can be made of a rather sophisticated universal algebraic concept, weak independence. We will also get some understanding of why these problems are so hard: the recursive procedure for determining $|FD(n+1)|$ depends not on $|FD(n)|$ but rather on a deep understanding of FD(n) itself.

The asymptotic results for $|FD(n)|$ are those of D. Kleitman and G. Markowsky [KM]. As mentioned earlier, $|FD(n)|$ is two less than the number of antichains in $\underline{2}^n$; this is also $\underline{2}(\underline{2}^n) - 2$ (if we speak of bounded distributive lattices, the term -2 disappears). The trivial lower bound for $|FD(n)|$ is $2^{E(n)}$ where $E(n) = \binom{n}{[n/2]}$; this is because E(n) is the size of the largest antichain in $\underline{2}^n$, occurring at the middle level). The upper bound says that there are not many more elements that this:

$$\log_2 |FD(n)| \le (1 + O(\log_2(n)/n))E(n). \tag{2.5}$$

The authors obtain their upper bound by a detailed and delicate analysis and modification of the proof of G. Hansel [Ha] that $|FD(n)| \le 3^{E(n)}$. Hansel's proof is quite elegant; here we present the description of it by Kleitman and Markowsky.

THEOREM 2.6 [KM]. $|FD(n)| \le 3^{E(n)}$.

PROOF. Think of FD(n) as all (non-constant) isotone functions from $\underline{2}^n$ into $\underline{2}$. Partition $\underline{2}^n$ into E(n) disjoint chains as follows: Instead of thinking of $\underline{2}^n$ as sequences of 1's and 0's,

[KL] W. Klotz and L. Lucht, Endliche Verbände (1971) *J. Reine Angew. Math.* 247, 58-68.

[Ko] A. Korsunov (1977) Solution of Dedekind's problem on the number of monotonic boolean functions, *Dokl. Akad. Nauk SSR* 223(4) (Transl. by Amer. Math. Soc. in *Soviet Math. Dokl.* 18(2), 442-445).

[Ky] S. Kyuno (1979) An inductive algorithm to construct finite lattices, *Math. Comp.* 33, 409-421.

[Mc] B McKay (1981) Practical graph isomorphism, *Congressus Numeratum* 30, 45-88.

[MQ] R. McKenzie and R. Quackenbush (1981) The spectrum of a lattice-primal algebra, *Disc. Math.* 35,

[Qu] R. Quackenbush (1979) Pseudovarieties of finite algebras isomorphic to bounded distributive lattices, *Disc. Math.* 28, 189-192.

[Re] R. Read (1978) Every one a winner, *Ann. Disc. Math.* 2, 107-120.

[Ro] I. Rosenberg (1970) Über die functionale Vollständigkeit in den mehrwertigen Logiken, *Rozpravy Ceskoslovenske Akad. Ved Rada Met. Prirod. Ved.* 80, 3-93.

[Sa] B. Sands (1981) Counting anti-chains in finite partially ordered sets, *Disc. Math.* 35, 213-226.

[Wa] A. Waterman (1967) The free lattice with 3 generators over $N_5$, *Portugal. Math.* 26, 285-288.

[Wi] R. Wille (1976) Subdirekte Produkte vollständiger Verbände, *J. Reine Angew. Math.* 283/284, 53-70.

[Ws] R. Wilson (1972) An existence theory for pairwise balanced designs I, Composition theorems and morphisms, *J.Comb. Th.* 13, 220-245.

$i < j$ and if $C_i$, $C_j$ have the same length, then $C_i \in I_k$ and $C_j \in I_m$ with $k \le m$. Let $\tau$ be an ordering of $\{1,\ldots,J(n)\}$. Then for each $(\sigma, \tau)$ we have an ordering of $\{\sigma(C_1),\ldots,\sigma(C_{E(n)})\}$, in total, $(n!)(J(n)!)$ orderings.

We want to find a small number N such that every element of FD(n) can be constructed by some ordering $(\sigma, \tau)$ using at most two choices per chain with at most N exceptions. This will give the following bound:

$$|FD(n)| \le n!\cdot J(n)!\cdot\binom{E(n)}{N}\cdot 2^{E(n)-N}\cdot 3^N. \qquad (2.7)$$

In this inequality, there are two numbers which must be made small: J(n) and N. By an explicit construction of a partition, the authors prove that $\log_2(J(n)) \le n^{1/2}\ln(n)\log_2(n)$. The big difficulty is bounding N.

Given $(\sigma, \tau)$, the authors modify Hansel's algorithm for defining isotone functions. Suppose we know f on the first k chains and want to define it on the $k+1^{st}$ chain, $\sigma(C_j)$. We know f is undetermined at two or fewer points; if one or none, then proceed as before. Suppose f is undefined at R, S with $R \subset S$; then $\sigma^{-1}(S) - \sigma^{-1}(R) = \{i\}$ for some $1 \le i \le n$. Call the elements of $\sigma^{-1}(S)$ open or closed according to whether in $C_j$ they represent open or closed left parentheses. If the number of closed elements of $\sigma^{-1}(S)$ greater than i is at least half the number of closed elements of $\sigma^{-1}(S)$, define $f(R) = 0$ and $f(S)$ arbitrarily. Otherwise define $f(R) = 1 = f(S)$.

As the last big step, the authors show by a probabilistic argument that for every isotone f there is a $(\sigma, \tau)$ such that f can be constructed by their algorithm from $(\sigma, \tau)$ with at most $\frac{c}{n}E(n)$ exceptions (c is a constant independent of n). Thus with $N = \frac{c}{n}E(n)$ they arrive at:

THEOREM 2.8 [KM]. *$\log_2|FD(n)| \le E(n)(1 + \frac{c}{n}\log_2(n))$.*

An asymptotic formula for |FD(n)| has been announced by A. D. Korsunov [Ko], but its proof has not yet appeared in print.

2c. Computing the sizes of some small free lattices.

Let FBD(n) denote the free bounded distributive lattice on n free generators. As mentioned earlier, the difficulty with trying to determine $|FBD(n)|$ recursively is that $|FBD(n)|$ depends on the fine structure of FBD(n-1), not simply on $|FBD(n-1)|$. More explicitly,

$$FBD(n) \cong \underline{2}^{(\underline{2}^n)} \cong \underline{2}^{(\underline{2}^{n-2}\times\underline{2}^2)} \cong (\underline{2}^{\underline{2}^{n-2}})^{(\underline{2}^2)} \cong FBD(n-2)^{(\underline{2}^2)}.$$

Let a and b be the atoms of $\underline{2}^2$; an isotone map f from $\underline{2}^2$ into FBD(n-2) can be determined by mapping a and b arbitrarily and then making $f(0) \leq f(a)\wedge f(b)$ and $f(1) \geq f(a)\vee f(b)$. In an ordered set P let $(x] = \{y \mid y \leq x\}$ and $[x) = \{y \mid y \geq x\}$. Then we see that:

$$|FBD(n+2)| = \Sigma\{\,|(x\wedge y]|\cdot|[x\vee y)| \;:\; x,y \in FBD(n)\}. \qquad (2.9)$$

Except for a misprint, this is the formula used in J. Berman and P. Köhler [BK] to verify the previously determined value of $|FBD(7)|$. Since $|FBD(5)| = 7581$, this computation is feasible; since $|FBD(6)| = 7,828,354$, $|FBD(8)|$ cannot be computed from this formula.

Let us turn to the problem of computing the free lattices in some varieties slightly larger than distributive lattices; they cannot be too much larger as FL(3), the 3-generated free lattice, is infinite, as is FM(4), the 4-generated free modular lattice. As is well known, every non-trivial variety V of lattices contains D, the class of distributive lattices and if $V \neq D$, then V contains either $V(M_3)$, the variety generated by $M_3$ or $V(N_5)$, the variety generated by $N_5$. Let $FM_3(n)$ and $FN_5(n)$ be the n-generated free lattices in $V(M_3)$ and $V(N_5)$, respectively; let W be the variety generated by $\{M_3, N_5\}$ and FW(n) the n-generated free lattice in W. It is natural to try to compute the sizes of these free lattices for small n. This has been done by J. Berman and B. Wolk [BW] for n = 3, 4. Since $|FW(4)| = 824,356,943,788$, this is not an exercise in number crunching; indeed, we will see that some rather sophisticated lattice theory and universal algebra are involved. We describe the authors' techniques in what follows.

Suppose we want to compute $|FV(3)|$. There is a natural homomorphism from FV(3) onto FD(3); it partitions FV(3) into $|FD(3)| = 18$ congruence classes and each of these congruence classes is itself a lattice. Thus we can compute $|FV(3)|$ by computing the size of each of the 18 sublattices. Actually, we need only enumerate 3 lattices (and only 2 if V is self-dual). To see this, let x, y, z be the free generators of FV(3). Thus every element of FV(3) can be written as p(x,y,z) for some ternary lattice term p. The congruence class containing p is

$[p_*, p^*] = \{q(x,y,z) \in FV(3) \mid p_* \le q \le p^*\}$; here $p_*$ is the disjunctive normal form of the term p obtained by repeated use of the inequality $(x_1 \vee x_2) \wedge x_3 \ge (x_1 \wedge x_3) \vee (x_2 \wedge x_3)$ and $p^*$ is the conjunctive normal form of p. If p is a join or meet of variables, then $p_* = p^* = p$; for n = 3, there are 11 such cases giving 11 1-element classes. The congruence classes containing $(x \vee y) \wedge z$, $(x \vee z) \wedge y$ and $(y \vee z) \wedge x$ are isomorphic and similarly for their duals; if V is self-dual, then all 6 classes are either isomorphic or dually isomorphic. Finally, the upper median polynomial, $m(x,y,z) = (x \vee y) \wedge (x \vee z) \wedge (y \vee z)$, defines the last congruence class.

If we wish to compute $|FV(4)|$, then a similar analysis applies with respect to FD(4). But now $|FD(4)| = 166$ and there are only 26 1-element classes; the automorphism group of FD(4) cuts the number of other classes down to 24 and in the self-dual case this becomes 12. We will use this technique to compute $FM_3(4)$ and $FN_5(4)$.

That $|FM_3(3)| = 28$ is a standard exercise in lattice theory; in fact, $FM_3(3) = FM(3)$. $FM_3(3)$ decomposes into 11 1-element classes, 6 copies of $\underline{2}$ and one copy of $M_3$. That $|FN_5(3)| = 99$ is also well known but less easy; see A. Waterman [Wa]. Besides the 11 1-element classes, $FN_5(3)$ decomposes into 6 copies of $\underline{2}^2$ and one of $\underline{2}^6$. The authors compute $FN_5(3)$ in the following way: $FN_5(3)$ is a subdirect product of copies of $N_5$ and $\underline{2}$. Since $\underline{2}$ is a homomorphic image of $N_5$, we can represent $FN_5(3)$ as a subdirect power of $N_5$. There are 6 maps from $\{x,y,z\}$ onto the generating set of $N_5$; hence $FN_5(3) \subseteq (N_5)^6$. The authors then computed $FN_5(3)$ as a sublattice of $(N_5)^6$ via a straightforward computer program which computes subalgebras of powers of algebras.

To compute $FM_3(4)$, note that it is a subdirect product of copies of $M_3$ and $\underline{2}$. There are 84 maps from the generators of $FM_3(4)$ onto the atoms of $M_3$ and 14 onto $\underline{2}$ so $FM_3(4) \subseteq (M_3)^{84} \times \underline{2}^{14}$. However, the 6-element automorphism group of $M_3$ reduces the 84 copies of $M_3$ to 14 = 84/6 copies. Since $\underline{2} \subseteq M_3$, $FM_3(4) \subseteq (M_3)^{28}$. This is a bit too big for the authors' program, so they decompose $FM_3(4)$ via FD(4) into 166 sublattices, only 12 of which need be computed. Let $C = [p_*, p^*]$ be one of these 12 classes.

Obviously $C \subseteq (M_3)^{28}$ also. However, if $\alpha$ is a homomorphism from $FM_3(4)$ onto $\underline{2}$, then $\alpha(p_*) = \alpha(p^*)$ so that this $\alpha$ may be ignored and $C \subseteq (M_3)^{14}$; similarly, we may ignore any such $\alpha$ from $FM_3(4)$ onto $M_3$ if $\alpha(p_*) = \alpha(p^*)$. We are now ready to compute C via the authors' program.

However, we have forgotten one minor detail; we must supply the program with a generating set for C. Unfortunately, there seems to be no easy way of finding a generating set of C. The authors compute $I_\vee(C)$, the set of join irreducible elements of C. In order to do this, they start with a result of R. Wille [Wi].

LEMMA 2.10 [Wi]. *Let the finite lattice L be a subdirect product of the lattices $L_j$, $j \in J$ and let $\pi_j$ be the canonical homomorphism from L onto $L_j$. Then*

$$I_\vee(L) = \bigcup_{j \in J} \{\wedge \pi_j^{-1}(x) \mid x \in I_\vee(L_j)\}.$$

Let $\alpha$ be a homomorphism from $FM_3(4)$ onto $M_3 = \{0, a_1, a_2, a_3, 1\}$; what is $\alpha^{-1}(a_i)$? The authors answer this with another lemma; recall that m(x,y,z) is the upper median polynomial.

LEMMA 2.11 [BW]. *Let V be a modular variety; let X be a free generating set for FV(n), and let $\alpha: FV(n) \to M_3$ be onto. Let $a_i' = \bigwedge \{x \in X \mid \alpha(x) \geq a_i\}$ for $i = 1, 2, 3$. Then $\alpha^{-1}(a_i) = a_i' \wedge m(a_1', a_2', a_3')$.*

Using these lemmas, |C| can be computed; it ranges from 5 to 1647, and so $|FM_3(4)|$ is computed:

THEOREM 2.12 [BW]. *$|FM_3(4)| = 19{,}982$.*

Turning to $FN_5(4)$, the authors prove a lemma analogous to 2.10; let $N_5 = \{0, a_1, a_2, a_3, 1\}$ with $a_2 < a_3$.

LEMMA 2.13 [BW]. *Let V be a variety of lattices containing $N_5$; let X be a free generating set for FV(n), and let $\alpha: FV(n) \to N_5$ be onto. Let $a_i' = \bigwedge \{x \in X \mid \alpha(x) \geq a_i\}$ for $i = 1, 2, 3$. Then $\alpha^{-1}(a_i) = a_i' \wedge (a_1' \vee a_2')$.*

Let $C = [p_*, p^*]$; sometimes $I_\vee(C)$ is too large to permit computation of C via the authors' program (for one C, $|I_V(C)| = 36$). Fortunately, each C is a distributive lattice: let $\alpha$ be as above with $\alpha(p_*) < \alpha(p^*)$; let $\theta$ be the smallest non-trivial congruence on $N_5$, and note that $N_5/\theta$ is distributive. Since $p_* = p^*$ in a distributive lattice, $(\alpha(p_*), \alpha(p^*)) \in \theta$; thus $\alpha(p_*) = a_2$ and $\alpha(p^*) = a_3$ so that $\alpha(C) \cong \underline{2}$. It turns out that $I_V(C)$ is sufficiently simple that $|C|$ can be computed. The largest C has size $5^{12}$; for this C, $I_V(C)$ is 12 disjoint copies of $\wedge$.

THEOREM 2.14 [BW]. *$|FN_5(4)| = 540,792,672$.*

Finally, let $W = V(M_3) \vee V(N_5)$ be the smallest variety containing $V(M_3)$ (the variety generated by $M_3$) and $V(N_5)$; the authors compute $FW(4)$. Surprisingly, this becomes a trivial exercise after some universal algebra. Let V be the smallest variety containing the varieties $V_1$ and $V_2$: $V = V_1 \vee V_2$. We say $V_1$ and $V_2$ are *independent* if there is a term $t(x_1,x_2)$ such that $t(x_1,x_2) = x_i$ is an identity in $V_i$. This is equivalent to each algebra in V being a unique direct product of algebras from $V_1$ and $V_2$, and to $FV(n) = FV_1(n) \times FV_2(n)$ for $n \geq 1$. Thus if we know $|FV_i(n)|$ and if $V_1$, $V_2$ are independent, then it is trivial to compute $|FV(n)|$.

Now, $V(M_3)$ and $V(N_5)$ are not independent since $W' = V(M_3) \cap V(N_5)$ is just D, the variety of distributive lattices. However, we can apply independence to W "modulo D". If $V = V_1 \vee V_2$, we say that $V_1$ and $V_2$ are *weakly independent* if for any terms $t_1(x_1,\ldots,x_n)$ and $t_2(x_1,\ldots,x_n)$ with $t_1(x_1,\ldots,x_n) = t_2(x_1,\ldots,x_n)$ being an identity of $V' = V_1 \cap V_2$, then there is a term $t(x_1,\ldots,x_n)$ such that $t(x_1,\ldots,x_n) = t_i(x_1,\ldots,x_n)$ is an identity of $V_i$. To see what this means in terms of free algebras, let $c \in FV'(n)$ and let $c(V)$, $c(V_1)$ and $c(V_2)$ be the corresponding congruence classes in $FV(n)$, $FV_1(n)$ and $FV_2(n)$, respectively. Weak independence is equivalent to $c(V) = c(V_1) \times c(V_2)$. Since we have already computed $c(V(M_3))$ and $c(V(N_5))$ for $c \in FD(4)$, to compute $|FW(4)|$

we need only know that $V(M_3)$ and $V(N_5)$ are weakly independent. With the help of lemmas 2.10 and 2.12, the authors show that $V(M_3)$ and $V(N_5)$ are indeed weakly independent.

THEOREM 2.15 [BW]. *$|FW(4)| = 824,356,943,788$.*

## 3. FINE SPECTRUM PROBLEMS

In this section we shall estimate the number $P(n)$ of (isomorphism classes of) ordered sets of size $n$ and the number $L(n)$ of (isomorphism classes of) of lattices of order $n+2$ (for lattices, 0 and 1 come for free). What does the average ordered set or lattice look like; is it long and thin, short and fat or is it intermediate? Naturally, we expect it to be intermediate; it is surprising, then, to find that they tend to be short and fat.

To see why, let $P = \{a_1,\ldots,a_n\}$ be an ordered set such that if $a_i < a_j$, then $i < j$. Define the $n \times n$ matrix $M(P) = [P_{i,j}]$ by $P_{i,j} = 1$ if $a_i < a_j$, and $P_{i,j} = 0$ otherwise. Thus $M(P)$ is an upper triangular matrix with zero diagonal; at most $\binom{n}{2}$ entries can be 1 so there are at most $2^{C(n,\ 2)}$ such matrices (we will be encountering complicated binomial coefficients and so I will often write $C(m,\ n)$ for $\binom{m}{n}$). But not all such matrices qualify: if $P_{i,j} = P_{j,k} = 1$, then we must have $P_{i,k} = 1$ (this is just transitivity). Most 0-1 upper triangular matrices will not satisfy this condition. Is there an easily identifiable subset that does? Yes; consider those matrices satisfying: if $P_{i,j} = 1$, then $1 \le i \le [n/2]$ and $[n/2] < j \le n$. Since we never have $P_{i,j} = P_{j,k} = 1$, transitivity is trivial; the corresponding ordered sets have just two levels: $\{a_1,\ldots,a_{[n/2]}\}$ and $\{a_{[n/2]+1},\ldots,a_n\}$. If $n = 2m$, then the number of these matrices is $2^{m^2}$; the exponent is a half of $\binom{n}{2}$, the exponent of the crude upper bound. Thus transitivity favors short, fat ordered sets.

In our computations above, we have ignored the problem of isomorphism. For 2-level ordered sets, an automorphism permutes the minimal elements among themselves. Hence the number of isomorphism classes is at least $2^{m^2}/(m!)^2 = 2^{m^2+o(m^2)}$. The upper bound will also be of this last form; thus our estimates are so crude that we can ignore isomorphisms. Similar results apply for L(n) except that the bounds are cruder: the upper and lower bounds have exponents differing by a multiplicative factor.

3a. Estimating P(n).

As we have seen, $m^2+o(m^2) \le \log_2(P(2m)) \le 2m^2$; for arbitrary n we have $n^2/4 + o(n^2) \le \log_2(P(n)) \le n^2/2$. To get a better upper bound, we turn to D. Kleitman and B. Rothschild [KR], where it is shown that:

$$\log_2(P(n)) = n^2/4 + 3n/2 + O(\log_2(n)). \qquad (3.1)$$

They do this by showing that $P(n) = (1 + O(1/n))Q(n)$ where $Q(n)$ is the number of 3-level ordered sets on $\{1,\dots,n\}$ such that

(i) the middle level contains about n/2 elements and the other levels about n/4 elements each, and

(ii) each element covers about half the elements in the level immediately below it (if any), and is covered by about half the elements in the level immediately above it (if any).

The method of proof is typical of asymptotic enumeration problems: boil away a large number (in this case, 13) of negligible classes until you have distilled out your class of "average" structures (in this case, those 3-level ordered sets just described).

We will not give the proof in any detail, but merely look at a few steps which give the flavor of the arguments; we assume that $P_n^*$ is the class of ordered sets based on $\{1,\dots,n\}$. Let $E_n^*$ be the members of $P_n^*$ containing a 30-element chain, and let $E(n) = |E_n^*|$. Let $Y_n^*$ be the members of $P_n^*$ consisting of 3 levels, $L_1$, $L_2$ and $L_3$ such that if b covers $a \in L_i$, then $b \in L_{i+1}$ and such that for each $b \in L_{i+1}$, b covers a for some $a \in L_i$. Let $X_n^*$ be the members of $Y_n^*$ such that

$$n/2 - \log_2(n-1) < |L_2| < n/2 + \log_2(n-1)$$

and such that

$$n/4 - (n-1)^{1/2}\log_2(n-1) < |L_i| < n/4 + (n-1)^{1/2}\log_2(n-1)$$

for i = 1, 3. Let $X(n) = |X_n^*|$. For an element a of an ordered set, let $\gamma(a)$ be the subset of elements that either cover a or are covered by a. Finally, let $Q_n^*$ be the members of $X_n^*$ such that for $a \in L_1 \cup L_3$,

$$n/4 - (n-1)^{7/8} < |\gamma(a) \cap L_2| < n/4 + (n-1)^{7/8}$$

and such that for $a \in L_2$ and i = 1, 3,

$$n/8 - (n-1)^{7/8} < |\gamma(a) \cap L_i| < n/8 + (n-1)^{7/8}.$$

Let $Q(n) = |Q_n^*|$.

LEMMA 3.2 [KR]. *$\log_2(E(n+1)/P(n-29)) < 9n$ for n large enough.*

PROOF. To get a member of $E_{n+1}^*$, choose 30 elements from $\{1,\ldots,n\}$; this can be done in $\binom{n+1}{30}$ ways. Turn these 30 elements into a chain C; this can be done in 30! ways. Now define an ordered set S on the n-29 remaining elements; this can be done in P(n-29) ways. To connect C with S, add a new zero and a new unit to C to get C'. To each $a \in S$ choose the least element in C' above a and the greatest element in C' below a; this can be done in $\binom{32}{2}^{n-29}$ ways. Thus,

$$E(n+1) \leq \binom{n+1}{30}(30!)\binom{32}{2}^{n-29}P(n-29).$$

LEMMA 3.3 [KR]. *$\frac{1}{4}n^2+\frac{3}{2}n-3\log_2(n) < \log_2(X_n) < \frac{1}{4}n^2+\frac{3}{2}n+\log_2(n)$ for n large enough.*

PROOF. For the lower bound, count the number of 3-level ordered sets such that $|L_1| = [n/4]$, $|L_2| = [n/2]$ and $|L_3| = n-[n/4]-[n/2]$: choose $L_2$ (in C(n, [n/2]) ways); choose $L_1$ (in C(n-[n/2], [n/4]) ways); choose how to connect each $a \in L_2$ with $L_1$ (in $2^{[n/4]}-1$ ways), and finally choose how to connect each $a \in L_3$ with $L_2$ (in $2^{[n/2]}-1$ ways). For the upper bound, choose $L_2$ (in at most $2^n$ ways); choose $L_1$ (in at most $2^{n/2+\log_2(n)}$ ways), and choose how to connect $L_1 \cup L_3$ with $L_2$ (in at most $2^{n^2/4}$ ways).

LEMMA 3.4 [KR]. *There is a constant C such that*

$$E(m+1)/X(m+1) < C/13(m+1).$$

PROOF. This is actually part of the main theorem of [KR] which is an inductive proof that $P(n) = (1 + O(1/n))Q(n)$. Thus for $n < m+1$, we will assume that we know that
$P(m) < (1 + C/m)X(m)$. Choose C large enough to work for small m.

By Lemma 3.3, $X(n)/X(n+1) < 2^{-n/2+5\log_2(n)}$. Thus,

$$\frac{E(m+1)}{X(m+1)} = \frac{E(m+1)}{P(m-29)} \cdot \frac{P(m-29)}{X(m-29)} \cdot \frac{X(m-29)}{X(m-28)} \cdots \frac{X(m)}{X(m+1)}$$

$$< 9m\left(1+\frac{C}{m-29}\right)2^{-(m-29)+5\log_2(m)} \cdots 2^{-m/2+5\log_2(m)}$$

$$< \frac{C}{13(m+1)} \quad \text{for m large.}$$

Having shown that $P(m) = (1 + O(1/m))X(m)$, the authors show that $X_m^* - Q_m^*$ can be boiled off, leaving:

THEOREM 3.5 [KR]. *$P(n) = (1 + O(1/n))Q(n)$.*

Of course, we get (3.1) from $P(n) = (1 + O(1/n))X(n)$ and Lemma 3.3.

## 3b. Estimating L(n).

A lattice always has a least element 0 and a greatest element 1; we discard these elements and consider only the remaining elements. Thus, $L(n) = |L_n^*|$ where $L_n^*$ is the class of all ordered sets on $\{1,\ldots,n\}$ such that adding a new 0 and a new 1 turns the ordering into a lattice ordering. $L_n^*$ is the set of members of $P_n^*$ such that:

$$\text{if } a < c,\ a < d,\ b < c,\ b < d,\ a \neq b \text{ and } c \neq d, \text{ then there is an e such that } a < e < c \text{ and } b < e < d. \tag{3.6}$$

Clearly, not many members of $P_n^*$ satisfy this condition; indeed, the exponent of $P(n)$ is $n^2$ and we shall see that the exponent of $L(n)$ is $n^{3/2}$. For a lower bound, we want to find a subclass of $L(n)$ that is easy to count. Recall that for $P(n)$ we had a set of covering relations such that deleting any subset of them left us still in the class. In $L_n^*$ we may not delete edges arbitrarily. But there is a nice subclass of $L_n^*$ where we can delete edges arbitrarily. Let $K_n^*$ be the members of $L_n^*$ consisting of just two levels with the lower level containing $[n/2]$ elements; if in any member of $K_n^*$ we delete an edge, the new ordered set again belongs to $K_n^*$. Let $K(n) = |K_n^*|$.

This is the approach used by W. Klotz and L. Lucht [KL]. Given $S \in K_{2n}^*$, let us return to the matrix M(S) introduced at the start of this section. The authors compute the maximum number of 1's in $M(S) = [S_{i,j}]$ as follows. Condition (3.6) is equivalent to M(S) containing no 2×2 submatrix of the form $\begin{pmatrix}1&1\\1&1\end{pmatrix}$. Thus for $j < k$, there is at most one i such that $S_{i,j} = S_{i,k} = 1$. Let $t_i$ be the number of 1's in the $i^{th}$ row of M(S). Then $\sum_{i=1}^{n} \binom{t_i}{2} \le \binom{n}{2}$, so that $\sum_{i=1}^{n} (t_i-1/2)^2 \le n(n-3/4)$. But now the maximum of $\sum_{i=1}^{n} t_i$ subject to this condition occurs when $t_1 = t_2 = \ldots = t_n = t$. Hence, $t = 1/2 + (n-3/4)^{1/2}$ and the number of 1's in M(S) is at most $n/2 + n(n-3/4)^{1/2}$. Can we achieve this maximum? For p a prime, let $G_p$ be the ordered set of 1-dimensional and 2-dimensional subspaces of the 3-dimensional vector space over GF(p). Then $G_p \in K_{2r}^*$ with $r = p^2+p+1$; the number of 1's in $M(G_p)$ is $(p+1)(p^2+p+1)$ which exactly equals the computed upper bound. Since any of the $(p+1)(p^2+p+1)$ covering relations may be deleted and we still remain within $K_{2r}^*$, this tells us that

$$L(2p^2+2p+2) \ge 2^{(p+1)(p^2+p+1)},$$

$$\text{or } \log_2(L(n)) \ge (2^{1/2}/4)n^{3/2} + o(n^{3/2})$$

for $n = 2p^2+2p+2$ with p a prime.

Of course, we can replace p with a prime power or any other order for which the reader can produce a projective plane. But this does not cover all n. To do so, the authors invoke the prime number theorem; since consecutive prime numbers are not too far apart, they are able to show:

THEOREM 3.7 [KL]. *$\log_2(L(n)) \ge (2^{1/2}/4)n^{3/2} + o(n^{3/2})$.*

What about an upper bound for L(n)? In [KL] Klotz and Lucht show that $\log_2(L(n)) \le (2^{1/2}/3)n^{3/2}\log_2(n) + O(n^{3/2})$. A sharper upper bound has been found by D. Kleitman and K. Winston [KW] who show that $\log_2(L(n)) \le n^{3/2}\log_2(\alpha) + o(n^{3/2})$ where $\alpha$ is about 6.11343.

Their proof is inductive and is based on the idea of decomposing $L \in L_n^*$ into two lattices of size n/2. Recall that $I \subseteq L$ is an *order ideal* if $a \in I$, $b \in L$ and $b \le a$ imply $b \in I$; $I\cup\{1\}$ is a lattice under the ordering inherited from L (it will be a sublattice of L only when I is an ideal of L). Note that $F = L-I$ is an order filter so $F\cup\{0\}$ is also a lattice. We can always choose I of size $[n/2]+1$. All covering relations within I and within F are now accounted for. What remains is to account for covering relations from I to F; no member of I covers a member of F. Thus these coverings yield a 2-level lattice with $I-\{0\}$ the lower level and $F-\{1\}$ the upper level; that is, a member of $K_n^*$.

This analysis gives us the inequality, $L(n) \le L([n/2])L(n-[n/2])K(n)$. Suppose we can prove that $\log_\beta(K(n)) = n^{3/2} + o(n^{3/2})$ for some $\beta > 1$; then we can prove inductively that $\log_\alpha(L(n)) \le n^{3/2} + o(n^{3/2})$. Because of the term $o(n^{3/2})$, we are empowered to begin our induction. Then,

$$\begin{aligned}\log_\alpha(L(n)) &\le \log_\alpha(L([n/2])) + \log_\alpha(L(n-[n/2])) + \log_\alpha(K(n))\\ &\le [n/2]^{3/2} + (n-[n/2])^{3/2} + n^{3/2}\log_\alpha(\beta) + o(n^{3/2})\\ &< ((2^{1/2}/2) + \log_\alpha(\beta))n^{3/2} + o(n^{3/2}) = n^{3/2} + o(n^{3/2})\end{aligned}$$

provided that $(2^{1/2}/2) + \log_\alpha(\beta) = 1$; that is, $\alpha = \beta^{2+2^{1/2}}$.

The task remaining, then, is to estimate K(n). The authors do this by counting the number of ways to go from a member of $K_n^*$ to a member of $K_{n-1}^*$. Let $K \in K_n^*$ with K having $n_1$ atoms and $n_2$ co-atoms. Let $c \in K$ be a co-atom containing the fewest number $d_1$ of atoms of K. Similarly, let $a \in K$ be an atom contained in the fewest number $d_2$ of co-atoms of K. Let $K_1 = K-\{c\}$ and $K_2 = K-\{a\}$; $K_1$ and $K_2$ are in $K_{n-1}^*$. Let $w_1$ be the maximum over $K' \in K_{n-1}^*$ of the number of ways of adding a co-atom containing $d_1$ atoms to K' to get a member of $K_n^*$. Define $w_2$ with respect to $d_2$ similarly. We map $K_n^*$ to $K_{n-1}^*$ as follows: if $w_1 < w_2$, then map K to $K_1$; otherwise map K to $K_2$. Suppose that we can show that this map is at most $\beta^{\frac{3}{2}n^{1/2}+o(n^{1/2})}$-to-one; then $K(n) \leq \beta^{n^{3/2}+o(n^{3/2})}$. To see this, note that by induction,

$$K(n) \leq \beta^{(n-1)^{3/2}+o(n^{3/2})}\beta^{\frac{3}{2}n^{1/2}+o(n^{1/2})}.$$

But $(n-1)^{3/2} + \frac{3}{2}n^{1/2} = n^{3/2} + o(1)$ so that the claim follows.

Since $d_i$ can take on at most n values, our mapping is at most $(2n)\min(w_1,w_2)$-to-one. If $\min(w_1,w_2) \leq \beta^{\frac{3}{2}n^{1/2}+o(n^{1/2})}$, then since $2n = \beta^{\log_\beta(2n)} < \beta^{o(n^{1/2})}$, so is $(2n)\min(w_1,w_2)$. To prove this, the authors define a tree $T(K_1)$ over the atoms of $K_1$. A subset S of the atoms of $K_1$ is *eligible* if no two members of S are covered by a co-atom of $K_1$. The nodes of $T(K_1)$ are the eligible sets of atoms. The root is the empty set and its offspring are the atoms of $K_1$. In general, the offspring of node N will contain one more atom than N. Suppose we have just defined $N_1,\ldots,N_d$ to be the offspring of N (so that $d = d(N)$ is the out-degree of N); thus $N_i-N = \{a_i\}$ for $1 \leq i \leq d$. Let $A_i = \{a_i,\ldots,a_d\}$. For atom a, let I(a) be the set of atoms a' such that $a \vee a'$ is a co-atom. We assume that $|I(a_i)\cap A_i| \geq |I(a_j)\cap A_i|$ for $1 \leq i,j \leq d$. Now define the offspring of $N_i$ to be those eligible $N_i\cup\{a_j\}$ for $i < j$; note that $N_d$ has no offspring. It is easily seen that every eligible set appears in $T(K_1)$ exactly once. From its definition, we see that $w_1$ is the maximum over $K' \in K^*_{n-1}$ of the number of nodes in $T(K')$ of size $d_1$.

The authors now prove a lemma stating that large enough nodes have relatively few offspring. The proof of the lemma uses the ordering of offspring just defined.

LEMMA 3.8 [KW]. *Let N be a node of $T(K_1)$ satisfying $|N| > (n_1/(d_2)^2)\log_2(n_2) = s$; then $d(N) < n_1/d_2$.*

LEMMA 3.9 [KW]. *$min(w_1,w_2) < \beta^{\frac{3}{2}n^{1/2}+o(n^{1/2})}$ with $\beta$ about 1.699.*

PROOF. $w_2 < (n_1)^{d_2}$ so we are done unless $d_2 > \dfrac{n^{1/2}\log_2(\beta)}{\log_2(n_1)}$.
Let s be as in Lemma 3.8. Since $C(n_2, s) < (n_2)^{s}$, $\log_2(C(n_2, s)) = o(n^{1/2})$. Since $w_1 < C(n_2, d_1)$, if $d_1 \le s$, then $\log_2(w_1) = o(n^{1/2})$ and we are done. Thus $s < d_1$ so that Lemma 3.8 applies. Hence, $w_1 < C(n_2, s)C((n_1-1)/d_2, d_1-s)$
$< C((n_1-1)/d_2, d_1-s)\beta^{o(n^{1/2})}$. Let
$f(x) = (x^{(x^{1/2})})/(x-1)^{(x-1)/(x^{1/2})}$; then Stirling's formula tells us that $C(n_1d_2, d_1-s) \le f(n_1/d_2(d_1-s))^{(n_1d_1/d_2)^{1/2}}$. Since the maximum value of $f(x)$ for $x > 0$ is $\beta^{3/(2^{1/2})}$,

$$w_1 < \beta^{(3/(2^{1/2}))(n_1d_1/d_2)^{1/2}+o(n^{1/2})}.$$

Finally, $\min(n_1d_1/d_2, n_2d_2/d_1) \le n/2$. If $n_1d_1/d_2 \le n/2$, then we are done; otherwise we dualize.

THEOREM 3.10 [KW]. *$L(n) < \alpha^{(n^{3/2}+o(n^{3/2}))}$ with $\alpha$ about 6.113.*

3c. Estimating D(n).

Let D(n) be the number of distributive lattices of order n+2. Estimating D(n) seems to be much more difficult than estimating P(n) or L(n). Since the length of a distributive lattice equals the number of join irreducibles, few distributive lattices are short so that the method of Klotz and Lucht is useless. If we try to imitate Kleitman and Winston by slicing a distributive lattice in half, we find that we must slice it into a prime ideal and a prime filter in order to retain distributivity. This means that given a distributive lattice we want to find a prime ideal of approximately half its size. This problem is very difficult in itself and only little progress has been made on it; see B. Sands [Sa].

If we try to use the duality between distributive lattices and ordered sets, we run afoul of exponentiation. We constructed $2^{n^2/4}$ ordered sets of size n. Each has an antichain of size n/2, and so the corresponding distributive lattice has size at least $2n^{/2}$. Let $m = 2^{n/2}$; this gives us $m^{\log_2(m)}$ distributive lattices of size about m. This suggests that $\log_2(D(n))$ might look like $(\log_2(n))^2$.

It is easy to do much better than this. Consider the ordered set $\underline{n} \cup \underline{n}$; for $1 \leq i \leq n$ we choose whether to make the $i^{th}$ element of the left chain above the $i^{th}$ element of the right chain. This yields $2^n$ ordered sets of size 2n; pictorially, they are skewed ladders with some rungs missing. The corresponding distributive lattices have order $(n+1)^2$ or less. By adding long enough chains at the bottom we get $2^n$ distributive lattices of order $(n+1)^2$ so that $\log_2(D(n))$ is at least $n^{1/2}$. Notice that these lattices are all planar: a distributive lattice is planar if and only if its ordered set of join irreducibles is a union of two chains. If we try this same construction using 3 chains of length n, then we have 2n choices to make. Thus we construct $2^{2n}$ distributive lattices of order at most $(n+1)^3$; this yields a logarithmic growth rate of only $n^{1/3}$. This suggests that the lower bound of $2^{n^{1/2}}$ may not be all that bad. More surprisingly, it suggests that most distributive lattices are "almost planar" in some vague sense.

## 4. CATALOGS OF ORDERED SETS

In this section we consider the problem of finding efficient algorithms for generating the isomorphism classes of all ordered sets or all lattices on a small number of elements. The obvious approach is to generate all the structures and then weed out duplicates. For n-element lattices, start with all (n-1)-element lattices and insert new atoms in all possible ways. As a new lattice is generated, check whether it is isomorphic to a previously generated lattice. If it is, then delete it; if it isn't, then add it to the list. For ordered sets, proceed in a similar manner, adding new minimal elements.

Isomorphism testing is very time consuming and the algorithms above require an extremely large number of isomorphism tests. We will be much better off if we can find some method of generation which does not require isomorphism testing. Such a method does exist; it was discovered by R. Read [Re] for graphs, and in C. Colbourn and R. Read [CR] it was extended to other structures including ordered sets and lattices. Such algorithms are said to be *orderly*.

Let us see how we generate the n-element ordered sets in an orderly manner. Each n-element ordered set P is based on {1,...,n} and the natural ordering of {1,...,n} is an extension of the ordering on P. Thus each ordering is represented by an upper triangular n×n matrix of 0's and 1's; a 1 occurs in the (i,j) entry if and only if j covers i in P. If we order these n(n-1)/2 entries linearly (say, lexicographically) we get a 0-1 vector of length n(n-1)/2; this is the *representation* of P. Thinking of the representation as a binary number, among all the ordered sets isomorphic to P, there is one which has the largest representation; this is the *canonical representation* of P.

We wish to generate all and only canonical representations. Our procedure is to start with the totally unordered set. Having generated all canonical representations with exactly k 1's (that is, all ordered sets with exactly k covering relations), we wish to generate the canonical representation with k+1 1's. Notice that if v is a canonical representation with k+1 1's and v' is obtained from v by changing the rightmost 1 to a 0, then v' is also a canonical representation. Thus each canonical representation with k+1 1's can be generated and generated uniquely by taking some canonical representation with k 1's and changing some 0 to the right of the rightmost 1 to 1.

Let us take a canonical representation with k 1's and change some 0 to the right of the rightmost 1 to 1. Two difficulties may arise. The first is that the new vector may not be a representation; we may have created a shortcut and so no longer have a Hasse diagram. It is easy to check whether this is the case; if it is the case, do not add this vector to our list. Suppose we do have a representation; it may not be canonical. If it is not canonical, do not add it to our list; if it is, do add it to our list.

What this orderly approach has done is to replace checking whether a new representation is isomorphic to a previously generated representation with checking whether a representation is canonical. At first glance, checking whether a representation is canonical seems to involve computing the automorphism group of the ordered set. However, we are only interested in larger representations. Thus let v be a representation obtained from the canonical representation v' by changing a 0 to the right of the rightmost 1 to 1; let P be the ordered set corresponding to

v. To each 0 to the left of the rightmost 1 of v form the vector consisting of the initial segment of v to the left of this 0 followed by a 1. Now check whether this vector is the initial segment of a representation of an ordered set isomorphic to P. If v is not canonical, then the canonical representation of P will be found in this manner.

For lattices we want to use the same ideas. As before, we ignore 0 and 1. Thus an (n+2)-element lattice is represented by an n-element ordered set satisfying condition (3.6). For orderly generation to work, we need to show that each lattice with k+1 covering relations can be obtained from some lattice with k covering relations by adding a new covering relation; equivalently, we can delete one of the k+1 covering relations and obtain an ordered set satisfying (3.6). Unfortunately, this is not always possible; for $n \geq 4$, the lattice $\underline{2}^n$ has the property that deleting any covering relation yields an ordered set which is not a lattice. This is contrary to what is stated in section 7 of [CR]; this section is extremely garbled and should be ignored.

This leaves us with no good algorithm for generating lattices. The naive method mentioned at the start of this section has been used in S. Kyuno [Ky] to generate lattices of order $\leq 9$.

## 5. THE ISOMORPHISM PROBLEM

The isomorphism problem for the class K* asks how hard it is to tell whether two n-element structures in K* are isomorphic. Thus let $\alpha$ be an algorithm which given two members of K* determines whether they are isomorphic, and let $m(\alpha,n)$ be the maximum number of steps needed by $\alpha$ to determine whether two n-element structures of K* are isomorphic. How fast does $m(\alpha,n)$ grow with n? There are three standard growth rates: polynomial, subexponential and exponential. We say that $\alpha$ is a *polynomial algorithm* if there is some polynomial $p(x)$ such that $m(\alpha,n) \leq p(n)$ for n sufficiently large; we say $\alpha$ is an *exponential algorithm* if there is a polynomial $p(x)$ such that $\log_2(m(\alpha,n)) \leq p(n)$ for n sufficiently large. The subexponential algorithms are between these two classes of algorithms; typically, we would have $\log_2(m(\alpha,n)) \leq (\log_2(n))^k$ for some fixed $k > 1$. Polynomial algorithms are good, while exponential algorithms are bad.

The isomorphism problem for graphs has been extensively

investigated; see C. Colbourn [Co], D. Corneil [Cr] and B. McKay [Mc] for recent surveys on this problem. It is known that the isomorphism problem for graphs can be solved in exponential time; despite a large effort by many people, no subexponential algorithm for graph isomorphism is known. Thus the isomorphism problem for graphs (undirected, without loops) is the canonical example of a hard isomorphism problem.

How does the graph isomorphism problem compare to, say, the groupoid isomorphism problem (a groupoid is a set with a binary operation on the set)? The groupoid isomorphism problem is at least as hard as the graph isomorphism problem: let $G$ be a graph with vertex set $\{1,\ldots,n\}$; we define a groupoid $T(G) = \langle\{1,\ldots,n\}; *\rangle$ by $a*b = 1$ if $(a,b)$ is an edge of $G$ and $a*b = n$ otherwise; clearly, $G \cong G'$ if and only if $T(G) \cong T(G')$.

We expect that the groupoid isomorphism problem is not too much harder than the graph isomorphism problem. Indeed, we shall now define a transformation $T$ from groupoids to graphs such that for groupoids $A$ and $B$, $A \cong B$ if and only if $T(A) \cong T(B)$ and such that if $|A| = n$, then $|T(A)| = n^2+4n+3$. Let $A = \langle\{1,\ldots,n\}; *\rangle$. The vertex set for $T(A)$ will be

$$(A \times (A \cup \{n+1,n+2,n+3,n+4\})) \cup \{1,2,3\}.$$

The edge set is the union of the following three sets:

1) $\{((i,n+j),\ (i,n+k)) \mid 1 \le i \le n,\ 1 \le j < k \le 3\}$
2) $\{((i,n+2),\ 1),\ ((i,n+3),\ 2),\ ((i,n+3),\ 3),$
   $((i,n+3),\ (j,n+4)) \mid 1 \le i,j \le n\}$
3) $\{((i,n+1),\ (i,j)),\ ((j,n+2),\ (i,j)),$
   $((k,n+3),\ (i,j)) \mid 1 \le i,j \le n \text{ and } i*j = k\}$.

Thus the groupoid isomorphism problem is polynomial-time equivalent to the graph isomorphism problem; there is a polynomial (subexponential) algorithm for the groupoid isomorphism problem iff there is a polynomial (subexponential) algorithm for the graph isomorphism problem.

Similarly, the isomorphism problem for directed graphs with loops (i.e. sets with an arbitrary binary relation) is polynomial-time equivalent to the groupoid isomorphism problem and hence to the graph isomorphism problem. In fact, given a finite type $\tau$ (a finite set of operations and relations each of finite arity), the isomorphism problem for $K^*(\tau)$ (the class of all structures of type $\tau$) is polynomial-time equivalent to the graph isomorphism problem provided that $\tau$ contains either an operation of arity $\ge 1$ or a relation of arity $\ge 2$.

It is an easy exercise to show that the isomorphism problems for ordered sets and for lattices are both polynomial-time equivalent to the graph isomorphism problem. However, the quasigroup isomorphism problem is known to be subexponential (due to the fact that a subquasigroup has at most half the size of its quasigroup); see [Co]. What is the difficulty of the isomorphism problem for distributive lattices? Trivially, it is at worst equivalent to the graph isomorphism problem. Since distributive lattices and ordered sets are dually isomorphic categories and since ordered sets have a hard isomorphism problem, one might think that the same is true for distributive lattices. But ordered sets tend to be short and if P is a short ordered set, then $\underline{2}(P)$ will be exponential in $|P|$. Thus, there may very well be a subexponontial (or even polynomial) algorithm for testing isomorphism of distributive lattices.

There seems to be a strong connection between the degree of hardness of the isomorphism problem for a class $K^*$ and the rapidity of growth of the fine spectrum function of $K^*$, $f_{K^*}(n)$. Perhaps foolishly, I am going to conjecture that this is always the case under reasonable circumstances.

CONJECTURE. Let $J^*$ and $K^*$ be classes of structures of finite type (both operations and relations are permitted) each defined by a first order sentence and such that for some polynomial $q(x)$, $f_{J^*}(n) \leq q(f_{K^*}(n))$ for n large enough. Then there is a polynomial $p(x)$ such that for every isomorphism testing algorithm $\alpha$ for $K^*$ there is an isomorphism testing algorithm $\beta$ for $J^*$ such that $m(\beta,n) \leq p(m(\alpha,n))$ for n large enough.

Note that if we weaken the assumption to $f_{J^*}(n) \leq f_{K^*}(q(n))$, then we could take $J^*$ to be lattices and $K^*$ to be planar distributive lattices. But planar lattices have a polynomial-time isomorphism problem (since planar graphs do). The conjecture would then assert that the graph isomorphism problem is polynomial-time, and this I do not believe.

It would be very useful to have an orderly algorithm for generating interesting classes of lattices. This may not be possible since imposing a new restriction may kill the orderliness of the algorithm. It may be possible to partition lattices into disjoint classes and use a different algorithm on each. For instance, partition lattices into dismantlable and non-dismantlable lattices. An (n+1)-element dismantlable lattice can be obtained from an n-element dismantlable lattice by adding a new doubly irreducible element. Hence, we may be able to use an orderly algorithm here. For non-dismantlable lattices, we proceed as before. If n is large, only a negligible percentage of n-element lattices are dismantlable; asymptotically, we have gained nothing. But for small n, most lattices are dismantlable; practically, we may be able to gain a great deal from this approach.

## REFERENCES

[BK] J. Berman and P. Köhler (1976) Cardinalities of finite distributive lattices, *Mitt. Math. Sem. Giessen* 121, 103-124.

[BW] J. Berman and B. Wolk (1980) Free lattices in some small varieties, *Alg. Univ.* 10, 269-289.

[Ch] R. Church (1965) Enumeration by rank of the elements of the free distributive lattice with 7 generators, *Notices Amer. Math. Soc.* 11, 724.

[Co] C. Colbourn (1980) *The complexity of graph isomorphism and related problems*, Dept. of Comp. Sci., U. of Toronto, Tech. Rep. 142/80.

[CR] C. Colbourn and R. Read (1979) Orderly algorithms for generating restricted classes of graphs, *J. Graph Th.* 33, 187-95.

[Cr] D. Corneil (1979) Recent results on the graph isomorphism problem, *Congressus Numeratum* XXII(1979), 13-33.

[Da] S. K. Das (1977) A machine representation of finite $T_0$ topologies, *J. Assoc. Comp. Mach.* 24, 676-692.

[Ha] G. Hansel (1966) Sur les nombres des fonctions booléennes monotones de n variable, *C. R. Acad. Sci. Paris Sér.* A-B 262, A1088-A1090.

[KM] D. Kleitman and G. Markowsky (1975) On Dedekind's problem: the number of isotone boolean functions. II, *Trans. Amer. Math. Soc.* 213, 373-390.

[KR] D. J. Kleitman and B. L. Rothschild (1975) Asymptotic enumeration of partial orders on a finite set, *Trans. Amer. Math. Soc.* 205, 205-220.

[KW] D. J. Kleitman and K. J. Winston (1980) The asymptotic number of lattices, *Ann. Disc. Math.* 6, 243-249.

think of it as sequences of ('s and )'s, respectively. In a given sequence, parentheses are either open or closed. To label parentheses in a string open or closed, proceed as follows:

> If the leftmost unlabeled parenthesis is ), then label it open. Otherwise it is (; if there is no unlabeled ) to the right of it, then label it open. Otherwise move right to the first unlabeled ) and label it closed; move left to the first unlabeled parenthesis [necessarily (] and label it closed. Iterate this procedure.

Call two sequences equivalent if their closed parentheses are of the same shape in the same positions. Thus for $n = 4$, an equivalence class is {(()(, )()(, )())} with the second and third parentheses being closed in each case. Translating ( to 1 and ) to 0, the equivalence classes now become the chains of the decomposition. Each chain is centred about the middle of $\underline{2}^n$ and skips no levels; the length of each chain is the number of open parentheses in any member. Let $a < b < c$ be three consecutive elements of one of these chains and let d be the relative complement of b in [a, c]; then the closed parentheses of d are those common to a, b, c plus one additional pair. Hence d belongs to a chain of length 2 smaller than the chain containing a, b and c. Now let $f:\underline{2}^n \to \underline{2}$ be isotone; if f is known on all chains of length less than $n - p$ in our partition, then this relative complementation result implies that f is known at all but at most 2 (necessarily consecutive) elements of any chain of length $n - p$ in our partition. Since $|\underline{2}^{\underline{2}}| = 3$, f is determined by at most 3 choices on each chain; hence, the theorem is proved.

Clearly, not every chain requires 3 choices. Thus we would like to order the chains so as to minimize the number of chains on which 3 choices can be made. As no useful uniform bound seems to exist, the authors take advantage of two degrees of freedom. Let $\{C_1,\ldots,C_{E(n)}\}$ be Hansel's set of chains ordered so that shorter chains precede longer. The first degree of freedom is that many different orderings are possible among chains of equal length. The second is that many different sets of chains can be used: if $\sigma$ is an automorphism of $\underline{2}^n$, then $\sigma$ preserves relative complementation and so Hansel's argument works for $\{\sigma(C_1),\ldots,\sigma(C_{E(n)})\}$ also.

To order chains of equal length, let $T_i$ be the element of $C_i$ of size $E(n)$ (thinking of $\underline{2}^n$ as subsets of $\{1,\ldots,n\}$). Let $J(n)$ be the smallest integer for which there is a partition $I^* = \{I_1,\ldots,I_{J(n)}\}$ of the subsets of $\{1,\ldots,n\}$ of size $E(n)$ such that if distinct $R, S \in I_i$, then $|R \cap S| < E(n) - n^{1/2}\ln(n)$; that is, R and S are "sufficiently disjoint". Now we insist that if

ADDENDUM

As I feared, the conjecture in Section 5 was foolish. Maurice Pouzet showed me the following counterexample: Let $K$ be graphs and $K'$ labeled graphs. There are more $n$-element structures in $K'$ than $K$, but the isomorphism problem for $K'$ is clearly polynomial-time since the only possible isomorphism between two $n$-element members of $K'$ is the map preserving labels. Moreover, labeled graphs can be interpreted as structures with two binary relations, $C$ and $R$, such that $C$ is a chain and $R$ is a graph. To resurrect the conjecture, we need to add some condition which guarantees that the number of potential isomorphisms is not significantly reduced, such as requiring that there be sufficiently many $n$-element structures with large automorphism groups. But I do not know how to formulate this in a more reasonable manner, and so I leave it to the interested reader to think about this question.

# THE MÖBIUS FUNCTION OF A PARTIALLY ORDERED SET

Curtis Greene
Department of Mathematics
Haverford College
Haverford, Pennsylvania 19041

ABSTRACT

The history of the Möbius function has many threads, involving aspects of number theory, algebra, geometry, topology, and combinatorics. The subject received considerable focus from Rota's by now classic paper in which the Möbius function of a partially ordered set emerged in clear view as an important object of study. On the one hand, it can be viewed as an enumerative tool, defined implicitly by the relations

$$f(x) = \sum_{y \leq x} g(y) \quad \text{and} \quad g(x) = \sum_{y \leq x} \mu(y,x) f(y)$$

where $f$ and $g$ are arbitrary functions on a poset $P$. On the other hand, one can study $\mu$ for its own sake as a combinatorial invariant giving important and useful information about the structure of $P$.

These expository lectures will trace this historical development and recent progress in the theory from both of these points of view. Topics to receive special attention are these:

(i) the Möbius function as a geometric invariant, for geometric lattices and other ordered structures associated with geometries;
(ii) algebraic and homological methods;
(iii) a catalog of interesting families of partially ordered sets for which the Möbius function is known.



*I. Rival (ed.), Ordered Sets, 555–581.*

This paper is a brief expository account of some basic results in the theory of Möbius functions on partially ordered sets. It is not a survey, and no attempt will be made to be complete. What this paper represents is a summary, with complete proofs, of the basic results upon which the subject rests. For the most part, these are taken from Rota's pivotal paper, "On the foundations of combinatorial theory I: The theory of Möbius functions" [Ro1], and from several papers of Crapo ([Cr1], [Cr2]), in which Rota's work was extended significantly. We have endeavored to refine and condense the proofs as much as possible, although this has also led us to abandon interesting (but less efficient) lines of development. For the reader interested in learning more, there is a substantial bibliography including many recent papers of considerable importance.

If there is a "modern era" of the Möbius function, it begins in 1964 with Rota's paper [Ro1]. One could argue, of course, that the story begins earlier: the Möbius function of elementary number theory has a rich and varied history stretching back to the last century. The inclusion-exclusion principle has its roots in combinatorial antiquity. Significant steps were taken before 1964 (e.g. by Weisner, P. Hall, Dilworth, and others) toward developing and using a theory of Möbius inversion on arbitrary partially ordered sets. However Rota's paper marks the time at which the Möbius function emerged in clear view as a fundamental invariant, which unifies both enumerative and structural aspects of the theory of partially ordered sets. The title of [Ro1] is a startling prophesy, fulfilled to a great extent by the paper itself, and also by many lines of research which continue to this day.

We hope this paper will help whet the appetite of those not yet fully acquainted with this interesting and important subject.

## 1. DEFINITIONS

Let $P$ be a finite partially ordered set, and let $f$ and $g$ be functions on $P$ taking values in some ring $R$. Suppose that $f$ and $g$ are related by the formula

$$f(x) = \sum_{y \leq x} g(y) \qquad (1.1)$$

for all $x$ in $P$. By elementary reasoning, one concludes that the values of $g$ can be expressed as integral linear combinations of the values of $f$: i.e. $g(x) = f(x)$ if $x$ is a minimal element, $g(x) = f(x) - f(y) - f(z) - f(w) + 2f(u)$ if the ideal generated by $x$ is as shown in Figure 1.

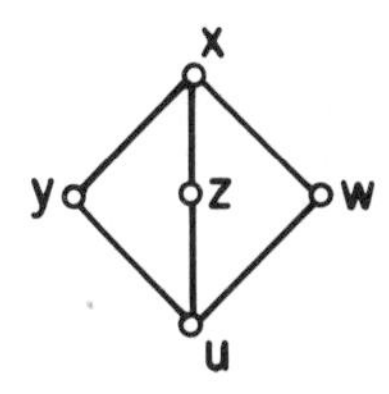


Indeed, for each $x$ in $P$ there is a formula

$$g(x) = \sum_{y \leq x} \mu_P(y,x) f(y) \tag{1.2}$$

where $\mu_P(y,x)$ is a unique integer valued function on $P \times P$, depending only on $P$ (not on $f$ or $g$), assuming nonzero values only when $y \leq x$. The function $\mu_P$ is called the *Möbius function of $P$*, and (1.2) is known as the *Möbius inversion formula*.

Formula (1.2) lends itself readily to the computation of $\mu_P$. Taking $g(x)$ to be a Kronecker delta, i.e. $g(x) = 1$ iff $x = a$, it follows from (1.1) that $f(x) = 1$ iff $x \geq a$. It follows from (1.2) that

$$\sum_{a \leq y \leq x} \mu_P(y,x) = \begin{cases} 1 & \text{if } x = a \\ 0 & \text{if } x \neq a \end{cases} \tag{1.3}$$

from which $\mu_P$ can be computed recursively by the rule

$$\mu_P(x,x) = 1 \qquad \mu_P(x,y) = - \sum_{x < z \leq y} \mu_P(z,y) . \tag{1.4}$$

Although we have considered only finite posets, it is not hard to see that the same conclusions hold when $P$ is *locally finite*, that is, when every interval $[x,y]$ in $P$ is finite. Most of the theory to be discussed in this paper extends to the locally finite case with only trivial changes. For many applications (see, for example, [DRS]) it is convenient to have the greater generality, but we will not develop the elementary theory from this point of view.

The "classical" Möbius function (see, e.g. [HW], pp. 232-258) is obtained by taking $P$ to be the (locally finite) lattice $\mathbb{Z}$ of positive integers ordered by divisibility. Clearly

$\mu_{\mathbb{Z}}(a,b)$ depends only on the quotient of $b$ and $a$, and we write $\mu_{\mathbb{Z}}(a,b) = \mu(b/a)$ in this case.

## 2. INCIDENCE FUNCTIONS

Let $F$ be a field. A function $\theta : P \times P \to F$ is said to be an *incidence function on P with values in F* if $\theta(x,y) = 0$ whenever $x \not\leq y$. Important examples of incidence functions are

| | |
|---|---|
| *The Kronecker delta:* | $\delta(x,y) = I(x,y) = 1$ iff $x = y$ |
| *The Zeta function:* | $\zeta(x,y) = 1$ iff $x \leq y$ |
| *The strict Zeta function:* | $N(x,y) = 1$ iff $x < y$ |
| *The Möbius function:* | $\mu(x,y)$, defined above. |

Our notation follows [Rol]. The set of all incidence functions on $P$ with values in $F$ is naturally endowed with the structure of an $F$-algebra, with multiplication defined by

$$\theta * \tau(x,y) = \sum_{x \leq z \leq y} \theta(x,z)\tau(z,y) \ . \tag{2.1}$$

This algebra is called the *incidence algebra of P over F*, denoted $I(P,F)$. The incidence algebra can of course be viewed as a matrix algebra, consisting of all $|P| \times |P|$ matrices over $F$ in which entries corresponding to pairs $x \not\leq y$ are required to be zero. Every matrix algebra defined in this way by "patterns of zeros" is in fact an incidence algebra.

Formula (1.3) is equivalent to the statement that $\mu$ is a right inverse of $\zeta$ in $I(P,F)$. Since $I(P,F)$ is a finite dimensional algebra, it follows immediately that $\mu$ is also a left inverse for $\zeta$, which is equivalent to

$$\mu(x,x) = 1 \qquad \mu(x,y) = - \sum_{x \leq z \leq y} \mu(x,z) \tag{2.2}$$

In other words, $\mu_P(x,y) = \mu_{P^*}(y,x)$, where $P^*$ denotes the dual of $P$, and $x \leq y$ in $P$. (This important fact is not combinatorially obvious!) The resulting dual version of the Möbius inversion formula is:

$$f(x) = \sum_{y \geq x} g(y) \iff g(x) = \sum_{y \geq x} \mu(x,y) f(y) \ . \tag{2.3}$$

A great deal of information about $P$ can be derived from the incidence algebra $I(P,F)$. Indeed, a highly nontrivial theorem of Stanley [S1] states that for any locally finite poset $P$, the structure of $I(P,F)$ as an $F$-algebra determines the order structure of $P$ as a partially ordered set.

We mention two results of combinatorial interest, which illustrate typical incidence algebra methods. The first is a theorem of P. Hall [Hal].

*For $k = 1,2,\ldots$ let $C_k$ denote the number of chains of $k$ elements whose endpoints are $x$ and $y$. Then*

$$\mu(x,y) = C_1 - C_2 + C_3 - C_4 \cdots . \tag{2.4}$$

The proof consists of expanding $\mu = \zeta^{-1} = (I+N)^{-1} = I - N + N^2 - N^3 + \ldots$, and noting that $C_{k+1} = N^k(x,y)$.

This result has an important homological interpretation, which has been the basis of an enormous amount of recent work applying algebraic and topological methods to combinatorial problems. (See, for example, [Ba1], [BB1], [BGS], and numerous related references.) Define the *order complex of $P$*, denoted $\Delta(P)$, to be the simplicial complex whose points are the elements of $P$, and whose simplices are the chains of $P$. Then

$$\chi(x,y) = 1 + \mu(x,y) \tag{2.5}$$

where $\chi(x,y)$ denotes the *Euler characteristic* of the order complex of the open interval $(x,y)$ in $P$ (i.e. the set of all $z$ such that $x < z < y$). For this reason, $\mu(x,y)$ is sometimes referred to as the *reduced Euler characteristic* of the open interval $(x,y)$.

The second result is a "reciprocity" theorem due to Stanley [S7]:

*Let $P$ be a finite partially ordered set with $0$ and $1$, in which the longest chain has $L + 1$ elements. Define $Z(P,n) = \zeta^n(0,1)$.*

(i) *$Z(P,n)$ is a polynomial in $n$ of degree $L$,*

(ii) $Z(P,-1) = \mu(0,1)$,

(iii) $Z(P,-n) = \mu^n(0,1)$ *for $n = 1,2,\ldots$ .*

All products are taken in the incidence algebra. Thus $Z(P,n) = \zeta^n(0,1) = \zeta*\zeta*\ldots*\zeta(0,1)$, which counts the number of "multichains" $0 = x_0 \leq x_1 \leq \ldots \leq x_n = 1$ of length $n$ in $P$. To prove (i) write $\zeta^n = (I+N)^n$, so that

$$\zeta^n(0,1) = \sum_{k=0}^{n} \binom{n}{k} N^k(0,1) = \sum_{k=0}^{L} \binom{n}{k} N^k(0,1) \qquad (2.6)$$

since $N^k(0,1) = 0$ for $k > L$. Since $\binom{n}{k}$ is a polynomial of degree $k$, for each $k$, it follows that $Z(P,n)$ is a polynomial of degree $L$. Replacing $n$ by $-n$ in the polynomial $\binom{n}{k}$ gives

$$Z(P,-n) = \sum_{k} \binom{-n}{k} N^k(0,1) \ . \qquad (2.7)$$

But the right hand side is just the result of expanding $\mu^n = \zeta^{-n} = (I+N)^{-n}$ by the binomial theorem. Thus $Z(P,-n) = \mu^n(0,1)$. Note that this argument depends only on the fact that $N$ is a nilpotent element of $I(P,F)$.

## 3. THE PRODUCT FORMULA

Let $P \times Q$ denote the *cartesian product* of posets $P$ and $Q$, i.e. $(a,b) \le (c,d)$ in $P \times Q$ iff $a \le c$ in $P$ and $b \le d$ in $Q$. The following easy formula yields $\mu_{P\times Q}$ in terms of $\mu_P$ and $\mu_Q$ :

$$\mu_{P\times Q}((a,b),(c,d)) = \mu_P(a,c)\mu_Q(b,d) \ . \qquad (3.1)$$

This property ("multiplicativity") lies at the heart of many applications of the classical Möbius function $\mu_{\mathbb{Z}}$ in number theory. Indeed, we can use (3.1) to compute the values of $\mu_{\mathbb{Z}}$ : if $N = P_1^{e_1} P_2^{e_2} \dots P_k^{e_k}$, the lattice of divisors of $N$ is isomorphic to a product $C_1 \times C_2 \times \dots \times C_k$, where $C_i$ denotes a chain of length $e_i + 1$. The Möbius function of a chain is easy to compute: $\mu(x,y) = -1$ if $x$ is covered by $y$, and otherwise $x < y$ implies $\mu(x,y) = 0$. It follows from (3.1) that

$$\mu_{\mathbb{Z}}(a,b) = \mu(b/a) = \begin{cases} (-1)^k & \text{if } b/a \text{ is squarefree} \\ 0 & \text{otherwise .} \end{cases} \qquad (3.2)$$

We can also think of divisors of $N$ as restricted multisets, with the multiplicities of elements (in this case primes) restricted by $(e_1,e_2,\dots,e_k)$. When the $e_i$'s are all equal to one, the multisets are sets and the lattice is a Boolean algebra. Thus

$$\mu_{B(U)}(S,T) = (-1)^{|T-S|} \tag{3.3}$$

where $S \subseteq T \subseteq U$, and $B(U)$ denotes the Boolean algebra of subsets of $U$.

The concept of "multiplicative function" on partially ordered sets other than divisors of an integer has been explored extensively in [DRS], and [Sm].

## 4. THE PRINCIPLE OF INCLUSION-EXCLUSION

The principle of inclusion-exclusion is just Möbius inversion over a Boolean algebra. Let $X$ be a set, and let $U$ denote a set of "properties" which elements in $X$ either "have" or "don't have". If $S \subseteq U$, let $f(S)$ denote the number of objects in $X$ having all the properties in $S$, and let $g(S)$ denote the number of objects in $X$ having all the properties in $S$ but no others. Then

$$f(S) = \sum_{S \subseteq T} g(T) \ , \quad \text{so that} \quad g(S) = \sum_{S \subseteq T} \mu_{B(U)}(S,T) f(T)$$

by Möbius inversion over the lattice $B(U)$ of subsets of $U$. Using (3.3) we obtain, for example,

$$g(\Phi) = \sum_{T \subseteq U} (-1)^{|T|} f(T)$$

which is the well-known "sieve" formula for the number of objects in $X$ having none of the properties in $U$. Extensions and applications of this formula abound in the combinatorial literature, and are too numerous to mention here.

## 5. THE CLOSURE AND GALOIS CONNECTION THEOREMS

In applications of the Möbius inversion formula, structural considerations frequently make it useful to group the summands in special ways. In some cases, this amounts to doing Möbius inversion over a different poset, where the relation between the two posets is given by a *closure operator* (that is, a function $x \longrightarrow \overline{x}$ satisfying (i) $\overline{x} \geq x$, (ii) $x \leq y \Rightarrow \overline{x} \leq \overline{y}$, and (iii) $\overline{\overline{x}} = \overline{x}$.) The following simple formula, due to Rota, relates the Möbius functions which arise in this way:

*Let $x \longrightarrow \overline{x}$ be a closure operator on $P$, and assume $P$ has a $0$. Let $\overline{P}$ denote the poset of closed elements in $P$ (i.e. the elements $x$ for which $\overline{x} = x$). Then for all $\overline{z} \in \overline{P}$,*

$$\sum_{\substack{x \\ \bar{x}=\bar{z}}} \mu_P(0,x) = \begin{cases} 0 & \text{if } \bar{0} > 0 \\ \mu_{\bar{P}}(\bar{0},\bar{z}) & \text{if } \bar{0} = 0 \end{cases} . \quad (5.1)$$

This formula is enormously useful as a tool for computing Möbius functions. Indeed, in most cases where the Möbius function of a class of posets is known, the derivation rests to a large extent on (5.1) or one of its close relatives. The proof is elementary:

For $\bar{z}$ in $\bar{P}$, define

$$g(\bar{z}) = \sum_{\substack{x \in P \\ \bar{x}=\bar{z}}} \mu_P(0,x) \quad \text{and} \quad f(\bar{z}) = \sum_{\substack{\bar{x} \in \bar{P} \\ \bar{y} \le \bar{z}}} g(\bar{y}) .$$

Since $f(\bar{x})$ sums $\mu_P(0,x)$ over an interval $[0,\bar{z}]$ in $P$, it follows that $f(\bar{z}) = 0$ if $\bar{z} > 0$, and $f(z) = 1$ if $\bar{z} = 0$. Solving for $g$ in terms of $f$ by Möbius inversion over $\bar{P}$ gives

$$g(\bar{z}) = \sum_{\bar{y} \le \bar{z}} \mu_{\bar{P}}(\bar{y},\bar{z})f(\bar{y}) = \begin{cases} 0 & \text{if } \bar{0} > 0 \\ \mu_{\bar{P}}(\bar{0},\bar{z}) & \text{if } \bar{0} = 0 \end{cases}$$

and the theorem is proved.

Several extremely important formulas are immediate consequences of the closure theorem.

Closures on a Boolean Algebra. *Let $S \to \bar{S}$ be a closure operator on the Boolean algebra of all subsets of the finite set $U$, such that the empty set is a closed set. Let $L$ denote the lattice of closed sets. If $\bar{T} \in L$, then*

$$\mu_L(\Phi,\bar{T}) = \sum_{\bar{S}=\bar{T}} (-1)^{|S|} . \quad (5.2)$$

In other words, $\mu(\Phi,\bar{T})$ is the difference between the number of even sets whose closure is $U$ and the number of odd sets whose closure is $U$. The proof is immediate, by application of (5.1).

Weisner's Formula. *Let $L$ be a finite lattice, and let $a$ be an element of $L$ such that $a > 0$. Then for all $b \ge a$*

$$\mu(0,b) = - \sum_{\substack{x<b \\ x \vee a=b}} \mu(0,x) . \quad (5.3)$$

To prove (5.3), define a closure operator $x \to x \vee a$, for each $x$ in $L$, and apply (5.1). (Note that $\overline{0} > 0$.)

The Galois Connection Formula. A situation related to (5.1), but slightly more general, occurs when $P$ and $Q$ are related by a *Galois connection*: that is, there exist maps $p \to \overline{p}$ from $P$ to $Q$ and $q \to \overline{q}$ from $Q$ to $P$ such that (i) both maps are order reversing, and (ii) $\overline{\overline{p}} \geq p$ and $\overline{\overline{q}} \geq q$ for all $p \in P$ and $q \in Q$. We will call a pair of maps satisfying (i) and (ii) *Galois maps*, and each is called the *adjoint* of the other. (Warning: this terminology is not standard!) We review some of the well-known facts about Galois connections (see [Bi1]):

(1) $\overline{p} \geq q$ iff $\overline{q} \geq p$;

(2) the inverse image of every set in $Q$ of the form $\{q|q \geq x\}$ is a set in $P$ of the form $\{p|p \leq y\}$;

(3) each Galois map determines its adjoint uniquely, and any map satisfying (2) has a Galois adjoint;

(4) the maps $\overline{\overline{p}} \to p$ and $q \to \overline{\overline{q}}$ are closure operators on $P$ and $Q$;

(5) the posets $\overline{P}$ and $\overline{Q}$ of closed elements are dually isomorphic (under the Galois maps);

(6) each $\overline{p}$ is $Q$-closed, and each $\overline{q}$ is $P$-closed;

(7) if $P$ and $Q$ are lattices, then $p \to \overline{p}$ is a Galois map if and only if it takes joins into meets, and in this case $q \to \overline{q}$ is determined by $\overline{q} = \sup\{p|\overline{p} \geq q\}$.

For a very clear discussion of Galois connections, as well as extensions of the present theory to arbitrary relations and order preserving maps, see the initial sections of [Ba2].

The following theorem of Rota relates $\mu_P$ and $\mu_Q$ when $P$ and $Q$ are connected by Galois maps. The statement below, as well as the idea of proof, can be found in [Cr2].

*Let $P$ and $Q$ be finite partially ordered sets related by a Galois connection, as defined above. Let $a \in P$ and $b \in Q$. Then*

$$\sum_{\substack{p \\ \overline{p}=b}} \mu_P(a,p) = \sum_{\substack{q \\ \overline{q}=a}} \mu_Q(b,q) . \qquad (5.4)$$

*The common value of both sides is zero unless $a \in \overline{P}$ and $b \in \overline{Q}$, in which case it is equal to*

$$\mu_{\overline{P}}(a,\overline{b}) = \mu_{\overline{Q}}(b,\overline{a}) . \qquad (5.5)$$

To prove the theorem, observe that the left hand side of (5.4) is zero if $\bar{\bar{a}} > a$, by the closure theorem. If $\bar{\bar{b}} > b$, it is zero because every element $\bar{p}$ must be $Q$-closed. If $a$ and $b$ are both closed, the closure theorem gives

$$\sum_{\substack{p \\ \bar{p}=b}} \mu_P(a,b) = \mu_{\bar{P}}(a,\bar{b}) \ . \tag{5.6}$$

Since $\bar{P}$ and $\bar{Q}$ are dually isomorphic, equation (5.5) is true and hence (5.4) follows by symmetry.

Although the Galois connection theorem is essentially just two closure theorems combined, it permits a somewhat wider range of applications. For example:

The "Little Cross-Cut Theorem" ([Ro1]). *Let $L$ be a finite lattice. Let $C$ be a subset of $L$ such that the order ideal generated by $C$ consists of $L - \{1\}$. Then*

$$\mu_L(0,1) = \sum_{\substack{S \subseteq C \\ \inf S = 0}} (-1)^{|S|} \ . \tag{5.7}$$

To prove (5.7), define a Galois connection between $L$ and the Boolean algebra $B(C)$ of subsets of $C$, as follows: for $x \in L$ define $\bar{x} = \{c | c \geq x\}$. For $S \subseteq C$, define $\bar{S} = \inf S$. It is easy to see that both maps are Galois, and (5.7) follows immediately from the Galois connection theorem (5.5), taking $a = 0$ in $L$, and $b = \Phi$ in $B(C)$.

An important corollary of the little cross-cut theorem is the following result of P. Hall [Ha2]:

The "Join of Atoms" Theorem. *Let $L$ be a finite lattice. If 0 is not a meet of coatoms or 1 is not a join of atoms, then* $\mu_L(0,1) = 0$.

Another important consequence is the following:

*In any finite lattice, the value of $\mu(0,1)$ depends only on the join-sublattice spanned by the atoms of $L$ (or, dually, on the meet-sublattice spanned by the coatoms of $L$).*

## 6. THE CROSS-CUT AND COMPLEMENTATION THEOREMS

In this section, we consider two important theorems: Rota's cross-cut theorem [Ro1], and Crapo's complementation theorem [Cr1].

The reader will notice a superficial relation with the results of the previous section, but the cross-cut and complementation theorems are actually much deeper. A considerable body of important research has resulted from attempts to understand, interpret, and extend these two theorems. (See for example [Fol], [Ba2], [BB1], [BB2], [BW1], [Bj2], [Ed1].)

The Cross-Cut Theorem. *Let $L$ be a finite lattice, and let $C$ be a subset of $L-\{0,1\}$ which is a cross-cut, that is, an antichain which meets every maximal chain of $L$. Then*

$$\mu_L(0,1) = \sum_{\substack{S\subseteq C \\ \sup S=1 \\ \inf S=0}} (-1)^{|S|} . \tag{6.1}$$

We give two proofs. The first, due to Edelman [Ed1], is similar in spirit to Rota's original proof. Recall that the *Zeta polynomial* $Z(P,n)$ counts the number of multichains $0 = x_0 \le x_1 \le \ldots \le x_n = 1$ in $P$.

For each $S \subseteq C$, let $L/S$ denote the sublattice $[0,\inf S] \cup [\sup S,1]$. Let $f(S)$ denote the number of $0-1$ multichains $K \subseteq L$ such that every element of $K$ is comparable with every element of $S$. Let $g(S)$ denote the number of multichains counted by $f(S)$ but not by $f(T)$ for any subset of $S$. Then clearly $f(S) = Z(L/S,n)$, and

$$g(S) = \sum_{\substack{T \\ S\subseteq T\subseteq C}} (-1)^{|T-S|} Z(L/T,n) \tag{6.2}$$

by inclusion-exclusion. Evaluating (6.2) at $S = \Phi$ and setting $n = -1$ gives

$$0 = \sum_{T\subseteq C} (-1)^{|T|} \mu_{L/T}(0,1) \tag{6.3}$$

since $Z(P,-1) = \mu(0,1)$, and $g(\Phi) = 0$ follows from the fact that $C$ is a cross-cut. But $\mu_{L/T}(0,1) = \mu_L(0,1)$ if $T = \Phi$. Otherwise $\mu_{L/T}(0,1) = 0$ unless $\inf T = 0$ and $\sup T = 1$, by the join-of-atoms theorem (or by the definition of $\mu$). In the latter case, $L/T$ is a two-element chain, and $\mu_{L/T}(0,1) = -1$. Rearranging terms in (6.3) gives (6.1).

The second proof is due to Crapo [Cr2], and makes use of *interval posets*: for any poset $P$, define the *interval poset* of

$P$, denoted $I(P)$, to be the set of intervals of $P$ (including the empty interval), ordered by set-inclusion. In other words, $[a,b] \leq [c,d]$ iff $c \leq a$ and $b \leq d$. Then

$$\mu_{I(P)}([a,b],[c,d]) = \mu_P(c,a)\mu_P(b,d)$$
$$\mu_{I(P)}(\Phi,[c,d]) = -\mu_P(c,d) \ . \tag{6.4}$$

We omit the straightforward proof. The first case is obvious, the second is somewhat surprising. For example, as a corollary we see that if $\mu_P$ is nonvanishing on $P$ then $\mu_{I(P)}$ is nonvanishing on $I(P)$.

Next we apply (6.4) to the proof of the cross-cut theorem. It is easy to see that the following maps define a Galois connection between the Boolean algebra $B(C)$ of subsets of $C$ and the dual of $I(L)$.

$$S \longrightarrow [\inf S, \sup S]$$
$$\{c \in C \mid a \leq c \leq b\} \longleftarrow [a,b] \ . \tag{6.5}$$

Taking $a = \Phi$ and $b = [0,1]$ in the Galois connection formula (5.4) we obtain

$$\sum_{\substack{I \in I(P) \\ \bar{I} = \Phi}} \mu_{I(L)}(I,[0,1]) = \sum_{\substack{S \subseteq C \\ \sup S = 1 \\ \inf S = 0}} (-1)^{|S|} . \tag{6.6}$$

The left hand side can be rewritten

$$-\mu_L(0,1) + \sum_{a \leq b \leq C} \mu(0,a)\mu(b,1) + \sum_{C \leq a \leq b} \mu(0,a)\mu(b,1) \tag{6.7}$$

which is easily seen to be equal to $-\mu_L(0,1) + \mu_L(0,1) + \mu_L(0,1) = \mu_L(0,1)$, and the proof is complete.

We turn next to Crapo's complementation theorem, which states:

*Let $L$ be a finite lattice, and let $a \in L - \{0,1\}$. Then*

$$\mu_L(0,1) = \sum_{x \leq y} \mu_L(0,x)\mu_L(y,1) \tag{6.8}$$

*where the sum ranges over all complements $x$ and $y$ of $a$.*

We prove (6.8) by a method which follows the general lines of Crapo's original argument. First let $P$ be any partially ordered set with a 0 and 1. If $C \subseteq P$ is a dual order ideal (= upward-closed subset), then

$$\mu(0,1) = \sum_{\substack{x \leq y \\ x,y \in C}} \mu(0,x)\mu(y,1) . \tag{6.9}$$

The proof is immediate: for each $x \in C$ the sum of $\mu(y,1)$ over $y \geq x$ is zero unless $x = 1$.

Next, let $L$ be a finite lattice. If $a \in L - \{0,1\}$, let $C$ denote the set of elements $x \in L$ for which $a \vee x = 1$. We will show that for each $y \in C$

$$\sum_{\substack{x \leq y \\ x \in C}} \mu(0,x) = 0 \tag{6.10}$$

unless $a \wedge y = 0$. Clearly this suffices to prove the theorem. Observe that the map $x \to (x \vee a) \wedge y$ defines a closure operator on the interval $[0,y]$, such that $\overline{0} = 0$ if and only if $a \wedge y = 0$, and $\overline{x} = y$ if and only if $x \vee a = 1$. Now (6.10) follows directly from the closure theorem.

A corollary of the complementation theorem is the following striking result:

*Let $L$ be a finite lattice in which some element $x$ fails to have a complement. Then* $\mu_L(0,1) = 0$.

Much recent work has focused on topological implications of this corollary. Suppose that $L$ is not complemented. Then $\mu(0,1) = 0 \Rightarrow \chi(\Delta(\overline{L})) = 1$, where $\overline{L}$ denotes the poset $L - \{0,1\}$, $\Delta$ denotes the order complex, and $\chi$ denotes the Euler characteristic. We conclude that if $L$ is not complemented, then $\Delta(\overline{L})$ has the Euler characteristic of a point. In fact much more is true: $\Delta(\overline{L})$ has the homology of a point [Ba2], and is even *contractible* [Bj2]. As noted in [BB1], either of these last facts implies the *order fixed point property* for $L$ :

*Let $L$ be a finite lattice which is not complemented. Then every order-preserving map from $\overline{L}$ into $\overline{L}$ has a fixed point.*

This is a deep result, which can be viewed as a direct descendent of Crapo's complementation theorem.

## 7. THE FORMULAS OF BACLAWSKI AND BJÖRNER-WALKER

We consider first a simple formula which, incredibly, seems not to have been noticed until recently ([Ba5]). If $P$ is any poset with 0 and 1, and $x \in P - \{0,1\}$, then

$$\mu_P(0,1) = \mu_{P-x}(0,1) + \mu_P(0,x)\mu_P(x,1) \ . \qquad (7.1)$$

More generally, if $S \subseteq P - \{0,1\}$, then

$$\mu_{P-S}(0,1) = \sum (-1)^k \mu_P(0,x_1)\mu_P(x_1,x_2)\ldots\mu_P(x_k,1) \qquad (7.2)$$

where the sum is over all chains $x_1 < x_2 < \ldots < x_k$ in $S$, including the empty chain. Formula (7.2) follows readily from (7.1) by induction, and (7.1) is not difficult. Instead of proving it, we will prove a more general formula due to Björner and Walker [BW] which contains (7.1) as a special case. Again let $S$ be a subset of $P - \{0,1\}$, but assume now that $S$ is convex: that is, $x < y$ in $S$ implies $[x,y] \subseteq S$. Then

$$\mu_P(0,1) = \mu_{P-S}(0,1) + \sum_{\substack{x \le y \\ x,y \in S}} \mu(0,x)\mu(y,1) \ . \qquad (7.3)$$

Taking $S$ to be a single point gives (7.1). Formula (7.3) is a striking generalization of the Crapo complementation theorem to arbitrary posets. It is nontrivial, however, to show that it even implies Crapo's theorem! Björner and Walker prove that if $P$ is a lattice, and $S$ is the set of complements of an element $a \in L$, then $\Delta(P-S)$ is contractible. This implies $\mu_{P-S}(0,1) = \chi(\Delta(P-S)) - 1 = 0$, and the complementation theorem follows.

To prove (7.3) recall that by Hall's theorem (2.4),

$$\mu(0,1) = \sum_{C \subseteq P} (-1)^n \qquad (7.4)$$

where the sum is over all chains $C$: $0 = x_0 < x_1 < \ldots < x_n = 1$ in $P$. Clearly, the chains disjoint from $S$ contribute a total of $\mu_{P-S}(0,1)$ to this sum. The sum appearing on the right in (7.3) counts the number of pairs of chains $0 = x_0 < x_1 < \ldots < x_i = x$ and $y = y_0 < y_1 < \ldots < y_j = 1$, such that $x \le y$, with sign $(-1)^{i+j}$. Each such pair can be linked together to form a chain $C$ in $P$, with $n$ equal to either $i + j$ or $i + j + 1$ depending on whether $x = y$ or $x < y$. On the other hand, if $C$ is a $0-1$ chain in $P$ and $|C \cap S| = k$, then $C$ occurs as a "join" of exactly $k$ pairs such that $x = y$ and $k - 1$ pairs such that $x < y$. (Here we use the fact that $C \cap S$ is a segment in $C$, which is true by the convexity of $S$.) Thus each $C$ with $|C \cap S| = k > 0$ is counted $k(-1)^n + (k-1)(-1)^{n-1} = (-1)^n$ times by the sum on the right in (7.3). Thus (7.3) agrees with (7.4) and the proof is complete.

## 8. THE FORMULAS OF DOUBILET, WILF, AND LINDSTRÖM

In this section we consider an inversion formula which has perhaps not received all the attention it deserves. It was used by Doubilet [Db1] as part of a systematic derivation of classical identities for symmetric functions by Möbius inversion over the lattice of partitions of a set. Let $f$ and $g$ be functions on a finite lattice $L$ satisfying

$$f(x) = \sum_{\substack{y \\ x \wedge y = 0}} g(y) . \tag{8.1}$$

Clearly such a relation is not always invertible (for example, if $L$ is a 3-element chain). However the following is true:

*Let $L$ be a finite lattice such that $\mu(0,x) \neq 0$ for all $x \in L$. Then* (8.1) *can be solved for $g$ in terms of $f$. In fact,*

$$g(x) = \sum_{y} \alpha(x,y) f(y) ,$$

*where*

$$\alpha(x,y) = \sum_{t} \frac{\mu(x,t)\mu(y,t)}{\mu(0,t)} . \tag{8.2}$$

Although (8.2) looks complicated, it is easily seen to be equivalent to a much simpler formula, which is often useful in its own right. Let $G(x) = \sum_{y \geq x} g(y)$. Then

$$\sum_{0 \leq t \leq x} \mu(0,t) G(t) = \sum_{\substack{y \\ x \wedge y = 0}} g(y) . \tag{8.3}$$

Formula (8.3) is almost obvious if one remembers the way $\mu$ acts as a sieve for upper sums. The fact that $L$ is a lattice is essential in the proof (which we omit).

To derive (8.2) from (8.3), first solve for $G$ in terms of $f$ by Möbius inversion of (8.3). Thus:

$$\mu(0,x) G(x) = \sum_{t \leq x} \mu(t,x) f(t) . \tag{8.4}$$

Then solve this for $g$ in terms of $G$, by a second Möbius inversion. Formula (8.2) follows immediately.

Formula (8.2) is closely related to an elegant theorem discovered independently by Lindström [Li1] and Wilf [Wi1]:

Let $L$ be a finite lattice (or $\wedge$-semilattice) and $f$ be a function defined on $L$. Define $F(x) = \sum_{z \leq x} f(z)$. Then

$$\det(F(x\wedge y))_{x,y} = \Pi_x f(x) \quad . \tag{8.5}$$

More generally, Lindström shows that if $f(x,t)$ is a function defined for all $x,t \in L$, and $F(x,t) = \sum_{z \leq x} f(z,t)$, then

$$\det(F(x\wedge y,x))_{x,y} = \Pi_x f(x,x) \quad . \tag{8.6}$$

Formula (8.5) generalizes a well-known theorem of Polya and Szego, which states that

$$\det(gcd(i,j))_{1\leq i,j\leq b} = \Pi_{k=1}^{n} \ \phi(k) \tag{8.7}$$

where $\phi$ is the Euler $\phi$-function.

To prove (8.6), define a matrix $M = (M(x,y))$, by setting $M(x,y) = \zeta(x,y)f(x,y)$. Clearly $M$ is triangular and $\det M = \Pi_x f(x,x)$. On the other hand,

$$M^T\zeta = \left(\sum_z f(z,x)\zeta(z,x)\zeta(z,y)\right)_{x,y}$$

$$= \left(\sum_{z\leq x\wedge y} f(z,x)\right)_{x,y} = (F(x\wedge y,x)) \quad .$$

Thus $\det(F(x\wedge y,x)) = \det M^T\zeta = \det M$. When $f(x,t) = f(x)$, that is, when $f(x,t)$ depends only on $x$, we have

$$(F(x\wedge y))_{x,y} = \zeta^T D_f \zeta \tag{8.8}$$

where $D_f = \mathrm{diag}(f(x))$. Of course (8.5) follows immediately since $\det \zeta = 1$.

The connection between the formulas of Doubilet and Lindström-Wilf is this: if we take $f(x) = \mu(0,x)$, then $F(x\wedge y) = \sum_{z\leq x\wedge y} \mu(0,z) = \delta(0,x\wedge y)$. Thus the matrix $R = (F(x\wedge y))_{x,y}$ is just an incidence matrix for the relation $x \wedge y = 0$. Using (8.8) we can interpret (8.2) as saying

$$f = Rg \iff g = R^{-1}f = \mu D_\mu^{-1} \mu^T f \quad . \tag{8.9}$$

In other words, $R$ is invertible $\Longleftrightarrow$ $D_\mu = \text{diag}(\mu(0,x))$ is invertible.

As an application of (8.5) we give a remarkably simple proof, due to Lindström (unpublished, communicated to us by Björner), of a theorem of Dowling and Wilson [DW1]:

*Let $L$ be a finite $\wedge$-semilattice such that $\mu(0,x) \neq 0$ for all $x \in L$. Then there exists a permutation $\phi : L \to L$ satisfying $x \wedge \phi(x) = 0$ for all $x \in L$.*

The proof follows immediately from the fact that $R$ is invertible (which is true by either (8.2) or (8.8)): since $\det R \neq 0$, some term in the expansion of $\det R$ must be nonzero. This term yields a permutation $\phi$ having the desired properties.

When $L$ is a finite geometric lattice, the hypotheses of this theorem are satisfied. Rota observed [Ro1] that in fact $(-1)^{r(y)-r(x)}\mu(x,y) > 0$ for all $x < y$, i.e. "alternates in sign". (This follows easily by induction from Weisner's formula (5.3).) An immediate consequence is the following important theorem, due to Dowling and Wilson [DW1]:

*Let $L$ be a finite geometric lattice of rank $n$, and let $W_i$ denote the number of elements of rank $i$ in $L$. Then for each $k \le n/2$, $W_1 + \dots + W_k \le W_{n-k} + \dots + W_{n-1}$.*

To see this, replace $\wedge$ by $\vee$ in the previous theorem, and observe that if $x \vee \phi(x) = 1$ for all $x$, then $\phi$ maps elements of rank $\le k$ injectively into elements of rank $\ge n-k$. The inequality follows immediately. The proof in [DW1] is actually not much different from the one we have given: it uses the same determinant argument, but proves the nonsingularity of $R$ directly from (8.3). (For an example of a terribly long and ugly proof (for $k = 1$) see [Gr1].)

## 9. OTHER IMPORTANT LANDMARKS

In the preceding sections, we have tried to give a rigorous treatment of the "basic" theorems, together with a sampling of other results to indicate some directions in which the theory has been applied. Our list of references gives a more complete picture of the subject, with an emphasis on topics of current interest. As a "guide" to the references, and an invitation to read further, we conclude with a section consisting of brief remarks on several other important topics.

The Möbius Algebra of a Poset

Let $P$ be a finite lattice, and $F$ a field. The *Möbius algebra* $A_\vee(P)$ of $P$ over $F$ is, by definition, the semigroup algebra of $P$ over $F$, relative to the operation $\vee$. A fundamental theorem, due to Solomon [Sol] is that $A_\vee(P)$ has basis of orthogonal idempotents, consisting of elements of the form

$$\sigma_p = \sum_{y \le p} \mu(y,p)y . \tag{9.1}$$

A dual result holds for the semigroup algebra $A_\wedge(P)$, relative to the operation $\wedge$. Solomon showed that, surprisingly, the definition of multiplication in $A_\vee(P)$ can be extended to any finite partially ordered set, whether or not it is a lattice, so that the elements defined by (9.1) form a basis of orthogonal idempotents: define

$$pq = \sum_r \lambda(p,q,r)r ,$$

where

$$\lambda(p,q,r) = \sum_{\substack{p \le t \le r \\ q \le t \le r}} \mu(t,r) \tag{9.2}$$

and extend to all of $A_\vee(P)$ by linearity. An easy proof of Solomon's theorem is given in [Gr2]. Many of the identities discussed in Sections 1-7 of this paper can be given easy algebraic proofs using the Möbius algebras [Gr2]. The trick is to keep track of what happens to the basic idempotents under appropriate maps (e.g. closures, Galois connections, etc.) and then compare coefficients. One of the main results of [Gr2], which actually contains a proof of the Galois connection formula (5.4), is the following: *if* $f : P \to Q$ *is a map between posets, then* $f$ *extends to a homomorphism* $\overline{f} : A_\vee(P) \to A_\vee(Q)$ *if and only if it forms one part of a Galois connection between* $P$ *and* $Q^*$.

The Valuation Ring of a Distributive Lattice

Let $L$ be a distributive lattice, and $F$ a field. Rota [Ro2] defined the *valuation ring* $V(L)$ of $L$ over $F$ to be the quotient ring $A_\vee(L)/M$, where $M$ is the ideal in $A_\vee(L)$ consisting of all linear combinations of elements of the form $x + y - x \vee y - x \wedge y$. (Distributivity of $L$ is necessary in order that $M$ be an ideal.) The crucial property of $V(L)$ is that all valuations on $L$ (i.e. functions $f : L \to F$ such that $f(x \vee y) + f(x \wedge y) = f(x) + f(y)$)

factor through $V(L)$, and thus can be studied by looking at linear functionals on $V(L)$. Many results in this direction can be found in [Ro2], [Ro3], and [Ge]. An important observation, due to Davis [Da] is that, when $L$ is finite, Rota's $V(L)$ is isomorphic to Solomon's $A_{\vee}(P_L)$, where $P_L$ denotes the poset of join-irreducibles of $L$. (See [Gr2] for another proof.) However the definition of $V(L)$ does not require the finiteness of $L$, so the concept of valuation ring is strictly more general than that of Möbius algebra.

Homology and Homotopy Properties of Posets

We have already seen several situations (e.g. sections 6 and 7) in which it was important to know that $\mu_P(0,1)$ could be interpreted as the reduced Euler characteristic of the order complex $\Delta(P - \{0,1\})$. In [Ro1] Rota introduced another class of simplicial complexes having reduced Euler characteristic $\mu_P(0,1)$. These can be defined whenever $P$ is a lattice. For any cross-cut $C$, define the *cross-cut complex* $K(C)$ to be the complex whose vertices are the elements of $C$ and whose simplices are the subsets $S \subseteq C$ which are nonspanning (i.e. such that $\bigvee S \neq 1$ or $\bigwedge S \neq 0$). An easy argument using the cross-cut theorem shows that $\mu_P(0,1) = \chi(K(C)) - 1$. Rota conjectured that the complexes $K(C)$ have the same homology groups, for any cross-cut $C$ [Ro1]. The Euler-Poincaré formula for $\chi$ would thus explain why formula (6.1) for $\mu$ is independent of the cross-cut $C$. Rota's conjecture was proved by Folkman [Fo1], who showed further that each complex $K(C)$ has the same homology as the order complex $\Delta(P)$. Folkman's theorem was extended by Mather [Ma] who showed that the $K(C)$'s all have the same homotopy type, and by Lakser [La1], who showed that the homotopy type is the same as that of $\Delta(P)$. Further extensions and refinements, e.g. to infinite lattices, arbitrary posets, can be found in [La1], [Q1], [Bj1], [Wa]. When $P$ is a finite semimodular lattice, it follows easily from Folkman's theorem that the homology of $\Delta(P)$ is trivial except in the top dimension ([Fo1], see also [Bj1]). Since every interval of a semimodular lattice is semimodular, this is equivalent to the statement that $P$ is *Cohen-Macaulay* (see [BGS]). Indeed, when $P$ is Cohen-Macaulay, one can always interpret $|\mu(0,1)|$ as the dimension of the top homology group of $\Delta(P)$. Other papers dealing with topological themes closely related to these are [Fa], [Ba1], [BW], and [Wa].

Distributive, Supersolvable, Semimodular, Admissible Lattices; Lexicographically Shellable Posets; Rank-Selected Möbius Functions

In an important series of papers [S3], [S4], [S6] Stanley showed that for a large class of finite (ranked) lattices $L$, the

Möbius function can be calculated by a rule of the form

$$(-1)^r\mu(0,1) = \text{number of maximal chains in } L \text{ with "descents" at every rank.} \quad (9.3)$$

Here $r$ denotes the rank of $L$ and "descent" is a concept defined appropriately in each case. More generally, if $\mu_S$ denotes the Möbius function of the poset obtained from $L$ by removing all ranks except those corresponding to a set $S$ of indices, then

$$(-1)^r\mu_S(0,1) = \text{number of maximal chains in } L \text{ with "descent set" } S. \quad (9.4)$$

Although these ideas are intimately related with the theory of Cohen-Macaulay posets (see [BGS]), the papers referred to above use nothing but elementary lattice theory. For example, if $L$ is a finite distributive lattice, and $\lambda : P_L \to \{1,2,\ldots,n\}$ is a linear extension of its poset of join-irreducibles, one can associate a permutation $\pi_1,\pi_2,\ldots,\pi_n$ with each maximal chain $0 = x_0 < x_1 < \ldots < x_n$ by defining $\pi_i = \lambda(x_i - x_{i-1})$, where $x_i - x_{i-1}$ denotes the unique join-irreducible $p \le x_i$, $p \nleq x_{i-1}$. A "descent" is an index $i$ such that $\pi_i > \pi_{i+1}$. In [S3] it is shown that (9.3) and (9.4) hold for any finite distributive lattice. In [S4] the argument is extended to the case when $L$ is "supersolvable", i.e. $L$ has a distinguished maximal chain which generates a distributive sublattice with every other chain. In [S6] a somewhat different labelling is used, and (9.3) and (9.4) are shown to hold for all semimodular lattices. A general notion of *admissible labelling* (of join-irreducibles) is introduced in [S6] which includes all of the previous cases. In [Bj2] Björner introduced the concept of *lexicographically shellable posets*, which generalize Stanley's admissible lattices in that coverings are labelled directly, without reference to join-irreducibles. Once again, one can prove that (9.3) and (9.4) hold. An important theorem, proved by Stanley in [S3] and "carried along" through each successive stage of generalization is the following. *Suppose that the "rank-selected" Möbius function* $\mu_S(0,1) \neq 0$ *for a particular set* $S$ *of ranks. Then* $\mu_T(0,1) \neq 0$ *for all* $T \subseteq S$. For a complete treatment of this subject see [BGS], where in the notation of that paper, $\mu_S(0,1) = (-1)^{|S|+1}\beta(S)$.

Geometric Lattices

As noted earlier, the Möbius function of a finite geometric lattice is nonzero and alternates in sign. The argument which proves this (Weisner's formula) also can be used to calculate $\mu$ in many important cases: e.g. $\mu(0,1) = (-1)^{n-1}(n-1)!$ for the partition lattice $\Pi_n$, $\mu(0,1) = (-1)^n q^{\binom{n}{2}}$ for the lattice of flats of an $n$-dimensional projective geometry over $GF(q)$. Although many results in this area are tied up with applications, (e.g. coloring problems, geometric enumeration (see e.g. [Z7])) there are two methods of "calculating" the Möbius function of a geometric lattice which are also of theoretical interest.

1. The Broken Circuit Theorem ([Wh], see also [Ro1], [Br1]). Suppose that the atoms of a geometric lattice $L$ have been given a linear ordering $1,2,\ldots,P$. Then $\mu_L(0,1)$ equals the number of bases (= independent spanning sets of atoms) which contain no "broken circuits". (A *broken circuit* is a minimal dependent set of atoms with its largest element removed.)

2. The Tutte-Grothendieck (Deletion-Contraction) Method (see [Br1] or [Z7]). Let $p$ be an atom of a geometric lattice $L$. Then

$$\mu_L(0,1) = \mu_{L-p}(0,1) - \mu_{L/p}(0,1) \qquad (9.5)$$

where $L - p$ denotes the join-sublattice of $L$ generated by the atoms not equal to $p$, and $L/p$ denotes the interval sublattice $[p,1]$. (This formula can be derived easily from the cross-cut theorem). It turns out that $\mu$ is only one of many "combinatorial invariants" which obey recursions of the form (9.5). In [Br1] it is shown that all functions $\phi$ defined (somewhat more generally) on *matroids* $G$ and satisfying

$$\phi(G) = \phi(G-p) + \phi(G/p), \quad \phi(G \oplus H) = \phi(G)\phi(H) \qquad (9.6)$$

for all points $p$ which are out "direct factors" of $G$ are obtained by specialization of a "universal" polynomial $T_G(x,y)$, which depends only on $G$. $T_G(x,y)$ is called the *Tutte polynomial* of $G$, and also satisfies (9.6). In this notation, if $L$ is the lattice of flats of $G$, then $|\mu_L(0,1)| = T_G(1,0)$. If $L^*$ is the lattice of flats of the matroid "orthogonal" to $G$, then $|\mu_{L^*}(0,1)| = T_G(0,1)$. The theory of "Tutte-Grothendieck invariants" (terminology due originally to Rota) is surveyed extensively in [Br1]. The idea of "deletion and contraction" is the starting point for an extensive series of investigations in enumerative geometry involving arrangements of hyperplanes [Z1], [Z3], [Z4], acyclic orientations [Gr5], [Z4], and other applications [Z5], [Z6]. (See also [OS1]).

ACKNOWLEDGEMENTS

The author has received valuable advice during the preparation of this paper from P. Edelman, K. Baclawski, and R. Stanley, whom he now wishes to thank.

REFERENCES

[Ai] M. Aigner (1980) *Combinatorial Theory*, Springer-Verlag (Berlin-Heidelberg, New York).

[Ba1] K. Baclawski (1975) Whitney numbers of geometric lattices, *Advances in Math.* 16, 125-138.

[Ba2] K. Baclawski (1977) Galois connections and the Leray spectral sequence, *Advances in Math.* 25, 191-215.

[Ba3] K. Baclawski (1979) The Möbius algebra as a Grothendieck ring, *J. Algebra* 57, 167-179.

[Ba4] K. Baclawski (1980) Cohen-Macaulay ordered sets, *J. Algebra* 63, 226-258.

[Ba5] K. Baclawski (1981) Cohen-Macaulay connectivity and geometric lattices, preprint.

[BB1] K. Baclawski and A. Björner (1979) Fixed points in partially ordered sets, *Advances in Math.* 31, 263-287.

[BB2] K. Baclawski and A. Björner (to appear) Fixed points and complements in finite lattices, *J. Combinatorial Theory (A)*.

[Bi1] G. Birkhoff (1967) *Lattice Theory*, 3rd edition. Amer. Math. Soc. Colloq. Publ. 25, Providence.

[Bj1] A. Björner (1979) On the homology of geometric lattices, preprint.

[Bj2] A. Björner (1980) Shellable and Cohen-Macaulay partially ordered sets, *Trans. Amer. Math. Soc.* 260, 159-183.

[Bj3] A. Björner (1981) Homotopy type of posets and lattice complementation, *J. Combinatorial Theory (A)* 30, 90-100.

[BGS] A. Björner, A. Garsia, R. Stanley (1982) An introduction to Cohen-Macaulay partially ordered sets, in: *Symp. Ordered Sets* (I. Rival, ed.) Reidel, Dordrecht-Boston, 583-615.

[BW] A. Björner and J. Walker (1981) A homotopy complementation formula for partially ordered sets, preprint.

[BG] E. Bender and J. Goldman (1975) On applications of Möbius inversion in combinatorial analysis, *Amer. Math. Monthly* 82, 789-803.

[Br1] T. Brylawski (1972) A decomposition for combinatorial geometries, *Trans. Amer. Math. Soc.* 171, 235-282.

[CF] P. Cartier and D. Foata (1969) Problèmes combinatoires de commutation et réarrangements, *Springer Lecture Notes* No. 85, Springer-Verlag, New York/Berlin.

[CLL] M. Content, F. Lemay, and P. Leroux (1980) Catégories de Möbius et fonctorialités: un cadre général pour l'inversion de Möbius, *J. Combinatorial Theory (A)* 28, 169-190.

[Cr1] H. Crapo (1966) The Möbius function of a lattice, *J. Combinatorial Theory* 1, 126-131.

[Cr2] H. Crapo (1968) Möbius inversion in lattices, *Arch. Math.* 19, 595-607.

[Da] R.L. Davis (1970) Order algebras, *Bull. Amer. Math. Soc.* 76, 83-87.

[Di1] R. Dilworth (1954) Proof of a conjecture on finite modular lattices, *Annals of Math. II* 60, 359-364.

[Db1] P. Doubilet (1972) On the foundations of combinatorial theory VII: Symmetric functions through the theory of distribution and occupancy, *Studies in Applied Math.* 51, 377-395.

[DRS] P. Doubilet, G.-C. Rota, and R. Stanley (1972) On the foundations of combinatorial theory VI: The idea of generating function, *Proc. Sixth Berkeley Symposium on Mathematical Statistics and Probability*, 267-318.

[Do1] T. Dowling (1973) A class of geometric lattices based on finite groups, *J. Combinatorial Theory* 14, 61-86.

[DO2] T. Dowling (1977) Complementing permutations in finite lattices, *J. Combinatorial Theory (B)* 23, 223-226.

[DW1] T. Dowling and R. Wilson (1974) The slimmest geometric lattices, *Trans. Amer. Math. Soc.* 196, 203-215.

[DW2] T. Dowling and R. Wilson (1975) Whitney number inequalities for geometric lattices, *Proc. Amer. Math. Soc.* 47, 504-512.

[Fa] F. Farmer (1979) Cellular homology for posets, *Math. Japonica* 23, 607-613.

[Fol] J. Folkman (1966) The homology groups of a lattice, *J. Math. Mech.* 15, 631-636.

[Ge] L. Geissinger (1973) Valuations on distributive lattices I, II, *Arch. Math. (Basel)* 24, 230-239, 337-345.

[Gr1] C. Greene (1970) A rank inequality for geometric lattices, *J. Combinatorial Theory* 2, 357-364.

[Gr2] C. Greene (1973) On the Möbius algebra of a partially ordered set, *Advances in Math.* 10, 177-187.

[Gr3] C. Greene (1975) An inequality for the Möbius function of geometric lattices, *Studies in Applied Math.* 54, 71-74.

[Gr4] C. Greene (1976) Weight enumeration and the geometry of linear codes, *Studies in Applied Math.* 55, 119-128.

[Gr5] C. Greene (1977) Acyclic orientations, (lecture notes) in: *Higher Combinatorics* (M. Aigner, ed.) D. Reidel, Dordrecht, 66-68.

[GZ] C. Greene and T. Zaslavsky (1980) On the interpretation of Whitney numbers through arrangements of hyperplanes, zonotopes, nonRadon partitions, and orientations of graphs, preprint.

[Ha1] P. Hall (1932) A contribution to the theory of groups of prime power order, *Proc. London Math. Soc. II*, Ser. 36, 39-95.

[Ha2] P. Hall (1936) The Eulerian functions of a group, *Quart. J. Math.* (Oxford), 134-151.

[HW] G.H. Hardy and E.M. Wright (1954) *An Introduction to the Theory of Numbers*, Oxford.

[Kl1] V. Klee (1963) The Euler characteristic in combinatorial geometry, *Amer. Math. Monthly* 70, 119-127.

[La1] H. Lakser (1971) The homology of a lattice, *Discrete Math.* 1, 187-192.

[Li1] B. Lindström (1969) Determinants on semilattices, *Proc. Amer. Math. Soc.* 20, 207-208.

[Ma] J. Mather (1966) Invariance of the homology of a lattice, *Proc. Amer. Math. Soc.* 17, 1120-1124.

[OS1] P. Orlik and L. Solomon (1980) Combinatorics and topology of complements of hyperplanes, *Inventiones Math.* 56, 167-189.

[OS2] P. Orlik and L. Solomon (1980) Unitary reflection groups and cohomology, *Inventiones Math.* 59, 77-94.

[Qu] D. Quillen (1978) Homotopy properties of nontrivial $p$-subgroups of a group, *Advances in Math.* 28, 101-128.

[Ro1] G.-C. Rota (1964) On the foundations of combinatorial theory I: Theory of Möbius functions, *Zeit. für Wahrscheinlichkeitstheorie* 2, 340-368.

[Ro2] G.-C. Rota (1971) On the combinatorics of the Euler characteristic, in: *Studies in Pure Mathematics* (L. Mirsky, ed.) Academic Press, London, 221-233.

[Ro3] G.-C. Rota (1973) The valuation ring of a lattice, *Proc. Houston Lattice Theory Conference.*

[Ro4] G.-C. Rota (1969) Baxter algebras and combinatorial identities II, *Bull. Amer. Math. Soc.* 75, 330-334.

[Ro5] G.-C. Rota (1975) *Finite Operator Calculus,* Academic Press, New York.

[Ro6] G.-C. Rota and D. Smith (1977) Enumeration under group action, *Ann. Scuola Normale Superiore Pisa,* Ser. 4, Vol. 4, 637-646.

[Sm] D.A. Smith (1967, 1969) Incidence functions as generalized arithmetic functions I, II, *Duke Math. J.* 34, 617-633, 36, 15-30.

[So1] L. Solomon (1967) The Burnside algebra of a finite group, *J. Combinatorial Theory* 2, 603-615.

[So2] L. Solomon (1979) Partially ordered sets with colors, in: *Relations between Combinatorics and Other Parts of Mathematics, Proc. Symp. in Pure Math.* 24, American Math. Soc., Providence, 309-329.

[S1] R. Stanley (1970) Incidence algebras and their automorphism groups, *Bull. Amer. Math. Soc.* 76, 1236-1239.

[S2] R. Stanley (1971) Modular elements of geometric lattices, *Algebra Universalis* 1, 214-217.

[S3] R. Stanley (1972) Ordered structures and partitions, *Mem. Amer. Math. Soc.* 119.

[S4] R. Stanley (1972) Supersolvable lattices, *Algebra Universalis* 2, 197-217.

[S5] R. Stanley (1973) Acyclic orientations of graphs, *Discrete Math.* 5, 171-178.

[S6] R. Stanley (1974) Finite lattices and Jordan-Hölder sets, *Algebra Universalis* 4, 361-371.

[S7] R. Stanley (1974) Combinatorial reciprocity theorems, *Advances in Math.* 14, 194-253.

[S8] R. Stanley (1976) Binomial posets, Möbius inversion, and permutation enumeration, *J. Combinatorial Theory* 20, 336-356.

[S9] R. Stanley (1981) Some aspects of groups acting on finite posets, preprint.

[Wa] J.W. Walker (1981) Homotopy type and Euler characteristic of partially ordered sets, preprint.

[Wel] L. Weisner (1935) Abstract theory of inversion of finite series, *Trans. Amer. Math. Soc.* 38, 474-484.

[Wh] H. Whitney (1932) A logical expansion in mathematics, *Bull. Amer. Math. Soc.* 38, 572-579.

[Wil] H. Wilf (1968) Hadamard determinants, Möbius functions, and the chromatic number of a graph, *Bull. Amer. Math. Soc.* 74, 960-964.

[Z1] T. Zaslavsky (1975) Facing up to arrangements: face count formulas for partitions of space by hyperplanes, *Mem. Amer. Math. Soc.* 154.

[Z2] T. Zaslavsky (1977) A combinatorial analysis of topological dissections, *Advances in Math.* 25, 267-285.

[Z3] T. Zaslavsky (1981) The geometry of root systems and signed graphs, *Amer. Math. Monthly* 88, 88-105.

[Z4] T. Zaslavsky (1979) Arrangements of hyperplanes; matroids and graphs, *Proc. Tenth Boca Raton Conf. on Combinatorics, Graph Theory, and Computing*, Utilitas Publ. Co. (Winnipeg), 895-911.

[Z5] T. Zaslavsky (to appear) Signed graph coloring, *Discrete Math.*

[Z6] T. Zaslavsky (1980) The slimmest arrangements of hyperplanes: I. Geometric lattices and projective arrangements; II. Basepointed geometric lattices and Euclidean arrangements, preprints.

[Z7] T. Zaslavsky (in prep.) The Möbius function and the characteristic polynomial, in: *Combinatorial Geometries* (H. Crapo, G.-C. Rota, N. White, eds.).

# AN INTRODUCTION TO COHEN-MACAULAY PARTIALLY ORDERED SETS

A. Björner
Department of Mathematics
University of Stockholm
Box 6701
S-113 85 Stockholm, Sweden

A.M. Garsia
Department of Mathematics
University of California - San Diego
La Jolla, California 92093

R.P. Stanley
Department of Mathematics
Massachusetts Institute of Technology
Cambridge, Massachusetts 02139

ABSTRACT

Combinatorics, algebra and topology come together in a most remarkable way in the theory of Cohen-Macaulay posets. These lectures will provide an introduction to the subject based on the work of Baclawski, Hochster, Reisner and the present authors (see references).

Combinatorial properties of some important posets will be surveyed. This leads in a natural way to the concept of a Cohen-Macaulay poset. This concept can be formulated in terms of a certain ring associated with the poset. On the other hand, Cohen-Macaulay posets can be defined in terms of topological properties. A fundamental theorem of Reisner gives the connection between these two definitions. Examples of Cohen-Macaulay posets include semimodular lattices (in particular, distributive and geometric lattices), supersolvable lattices, face-lattices of polytopes, and Bruhat order. This theory has given birth to the concept of lexicographically shellable posets. Their ubiquity and usefulness give them a central position in the subject.



*I. Rival (ed.), Ordered Sets, 583–615.*

The theory of Cohen-Macaulay posets has many applications both to combinatorics and to other branches of mathematics, including algebra and topology. It can be used, for instance, to obtain information about such numerical invariants of posets as number of chains and Möbius function. Cohen-Macaulay posets can be used to study rings of interest to algebraic geometry and establish the topological type of certain simplicial complexes. Contributions to representation theory can be made by considering groups acting on Cohen-Macaulay posets. Examples of these and other applications will be discussed, together with open problems.

## 0. INTRODUCTION

The theory of Cohen-Macaulay posets derives much of its appeal from its interactions with three distinct branches of mathematics - combinatorics, commutative algebra, and algebraic topology. Our aim here is to convey the flavor of this subject by discussing the basic results and some of their main applications. A comprehensive account of Cohen-Macaulay posets would be too lengthy to include here and will appear elsewhere [BGS], but it is hoped that this introduction will stimulate the reader to look further into the subject.

The combinatorial aspects of Cohen-Macaulay posets have their origins in the following natural problem. Given some set S of objects which one wants to enumerate or in some way "determine," is it possible to partition S into (finitely many) "nice" subsets $S_1,\ldots,S_r$ which can easily be handled individually? The first significant example of such an approach toward counting a set of objects may be found in MacMahon [MM] (especially vivid in Section 439), and was later systematized in [St$_1$] (and less comprehensively in [Kn]). The theory of P-partitions developed in [St$_1$] may be formulated in terms of distributive lattices, and is outlined here in Section 1. It is then natural to generalize this theory to a wider class of lattices. Successive generalizations appear in [St$_2$] and [St$_3$], finally culminating in the theory of lexicographic shellability [Bj$_1$]. Section 2 is devoted to this topic.

Section 3 is devoted to the basic topological concepts associated with posets. The prototype for this subject is the theorem of P. Hall [Ha, (2.21)][Ro, p.346], which may be interpreted as expressing the values of the Möbius function of a poset P as a reduced Euler characteristic. Since Euler characteristics may be computed from certain homology groups, it is natural to go further and look at the actual homology groups. This idea was suggested by Rota [Ro, pp. 355-6] and further considered by Mather [Mat], Folkman [Fo], and Lakser [La]. The work of Folkman on geometric lattices was made part of a general theory by Baclawski [Ba$_{2,4}$], who was the first to consider Cohen-Macaulay posets from a purely topological point of view. Subsequent topological investigations related to the Cohen-Macaulay property appear in [Ba$_6$],[Bj$_{2,3}$], [BW$_1$], [BWr], [Fa], [Mu], [Qu] and [Wr].

In Section 4 we introduce the fundamental commutative ring $R_P$ associated with a poset P, and show some connections with the preceding combinatorial concepts. In particular we give a ring-theoretic definition of a Cohen-Macaulay poset. Although Cohen-Macaulay rings date back to the pioneering work of Macaulay [Mac], it was not until the paper [$Hoc_1$] of Hochster that it became apparent that Cohen-Macaulay rings (or, at least, special classes of them) were closely related to combinatorics. This connection was made more explicit in [$St_5$] and [Re], where a ring $R_\Delta$ was attached to an arbitrary simplicial complex $\Delta$. A poset P may be regarded as a special kind of simplicial complex (see Section 3), and the rings $R_P$ were first singled out for special study in [$St_6$, Section 8]. Subsequently the poset ring, *per se*, was studied in [$Ba_7$], [BG], [$Ga_1$], [$Ga_2$], [$Hoc_2$], and [$St_8$].

In Section 5 we discuss what is perhaps the main theorem on Cohen-Macaulay posets (due to G. Reisner [Re]), which demonstrates the equivalence of the algebraic and topological approaches. Finally in Section 6 we present a brief glimpse of further results in order to illustrate the many fascinating ramifications and applications of the theory, and to bring the reader near some exciting areas of current research.

The following notation will be adhered to throughout. $\mathbb{N}$ denotes the set of nonnegative integers and $\mathbb{P}$ the positive integers. If $d \in \mathbb{N}$, then $[d] = \{1,2,\ldots,d\}$. A disjoint union is denoted by the symbol $\uplus$. All posets P will be finite throughout this paper. If y covers x in P (i.e., if $x < y$ and no $z \in P$ satisfies $x < z < y$) then we write $x \to y$. A poset P is said to be *pure* if every maximal chain has the same length $d+1$ (or cardinality $d+2$). If P is pure and has a $\hat{0}$ and $\hat{1}$ (i.e., $\hat{0} \leq x \leq \hat{1}$ for all $x \in P$), then P is said to be *graded* of rank $d+1$. If P is any poset, then $\hat{P}$ denotes P with a $\hat{0}$ and $\hat{1}$ adjoined. If P has a $\hat{0}$ and $\hat{1}$, then $\bar{P}$ denotes P with the $\hat{0}$ and $\hat{1}$ removed and is called the *proper part* of P. The set of maximal chains of P is denoted $M(P)$, and the set of all chains $C(P)$.

Assume throughout the remainder of this section that P is graded of rank $d+1$. If $x \in P$, then $r(x)$ denotes the *rank* of x, i.e., the length $\ell$ of the longest chain $\hat{0} = x_0 < x_1 < \cdots < x_\ell = x$ with top element x. In particular, $r(\hat{0}) = 0$ and $r(\hat{1}) = d+1$. If $x \leq y$ in P we write $r(x,y) = r(y) - r(x)$. The set of elements of P of rank i is denoted $P_i$, and the cardinality $|P_i|$ of this set is denoted $W_i$ and is known as a *Whitney number of the second kind*. If $c: \hat{0} = x_0 < x_1 < \cdots < x_{i+1} = \hat{1}$ is a chain of P, then the set $\{r(x_1),\ldots,r(x_i)\}$ is called the *rank set* of c.

Let $\mu$ denote the Möbius function of P [Ro]. For our purposes the most convenient definition of $\mu$ is the following. If $x \leq y$ in

P, then define $\mu(x,y) = c_0 - c_1 + c_2 - \cdots$, where $c_i$ is the number of chains $x = x_0 < x_1 < \cdots < x_i = y$ of length i between x and y. We write $\mu(P)$ for $\mu(\hat{0},\hat{1})$. The *Whitney numbers* $w_i$ *of the first kind* are defined by

$$w_i = \sum_{x \in P_i} \mu(\hat{0},x) .$$

In particular, $w_0 = 1$ and $w_{d+1} = \mu(P)$. Recall that the *characteristic polynomial* $p_P(t)$ is defined by

$$p_P(t) = \sum_{i=0}^{d+1} w_i t^{d+1-i}$$

Now suppose that S is any subset of $[d] = \{1,2,\ldots,d\}$. Define the S-*rank-selected subposet* $P_S = \{x \in P\colon x = \hat{0},\ x = \hat{1}, \text{ or } r(x) \in S\}$. Thus for instance $P_{[d]} = P$ and $P_i = \bar{P}_{\{i\}}$. Let $\alpha(S) = \alpha_P(S)$ denote the number of maximal chains of $P_S$. Equivalently, $\alpha(S)$ is the number of chains in P whose rank set is S. Note in particular that $\alpha(i)$ (short for $\alpha(\{i\})$) is equal to $W_i$. Define

$$\beta(S) = \beta_P(S) = \sum_{T \subseteq S} (-1)^{|S-T|} \alpha(T). \tag{1}$$

Thus by the Principle of Inclusion-Exclusion,

$$\alpha(S) = \sum_{T \subseteq S} \beta(T) . \tag{2}$$

Note that it follows from our definition of $\mu$ that $\beta(S) = (-1)^{|S|-1}\mu(P_S)$. In particular, for $i = 1,2,\ldots,d$,

$$(-1)^i w_i = \beta(1,2,\ldots,i-1) + \beta(1,2,\ldots,i).$$

Recall also that the *zeta polynomial* $Z(P,n)$ of P, as defined in [$St_4$] and further studied in [$Ed_1$], may be defined for graded posets by either of the two equivalent conditions

$$Z(P,n) = \sum_{S \subseteq [d]} \alpha(S) \binom{n}{|S|+1}$$

$$\sum_{n \geq 0} Z(P,n) t^n = \frac{\sum_{S \subseteq [d]} \beta(S) t^{|S|+1}}{(1-t)^{d+2}} . \tag{3}$$

Thus we see that the numbers $\beta(S)$ (or equivalently, $\alpha(S)$) are closely related to many properties of P of known interest and importance. The prominence of the numbers $\beta(S)$ will become even more apparent in the following pages, where they will count certain maximal chains in P and will appear in the Hilbert series of the poset ring, as Betti numbers of rank-selected subposets, and as group characters.

## 1. DISTRIBUTIVE LATTICES

Let P be a finite poset with $d+1$ elements. An *order ideal* (also called a *semi-ideal* or *decreasing subset*) of P is a subset I of P such that if $x \in I$ and $y < x$, then $y \in I$. Let J(P) denote the set of all order ideals of P, ordered by inclusion. Then J(P) is graded of rank $d+1$, and is in fact a distributive lattice. Conversely, any finite distributive lattice is of the form J(P) for a unique poset P [Bi, p. 59, Thm. 3]. Consider, for instance, the poset P given by

(4)

Then J(P) looks like

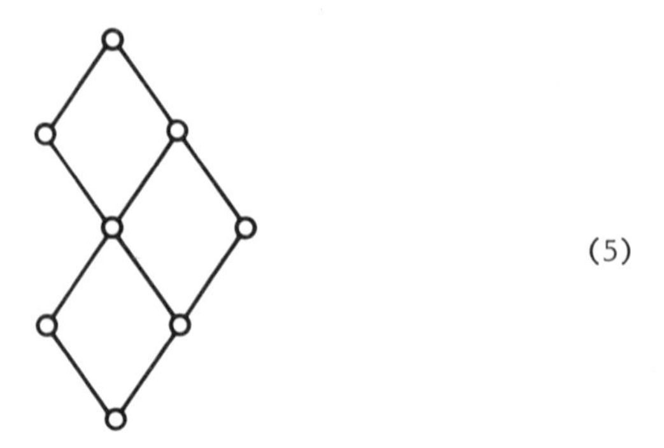

(5)

The values of $\alpha_{J(P)}(S)$ and $\beta_{J(P)}(S)$ are given by

| S | $\alpha(S)$ | $\beta(S)$ |
|---|---|---|
| $\phi$ | 1 | 1 |
| 1 | 2 | 1 |
| 2 | 2 | 1 |
| 3 | 2 | 1 |
| 12 | 3 | 0 |
| 13 | 4 | 1 |
| 23 | 3 | 0 |
| 123 | 5 | 0 |

Several empirical facts are evident. For instance, $\beta(S) \geq 0$ and several values of $\beta(S)$ are equal to 0. The fact that $\beta(1,2,3) = 0$ follows from the well-known fact that if L is a finite distributive lattice then $\mu(L) = 0$ unless L is a boolean algebra [Ro, pp. 349-50]. But the reason, say, for $\beta(1,3) = 1$ above is not so clear. What is needed is a suitable combinatorial interpretation of the numbers $\beta(S)$.

Consider the general case $L = J(P)$, where $|P| = d+1$. We first associate with every maximal chain of L a permutation of the set $[d+1]$. Fix an order-preserving bijection $\omega:P \to [d+1]$. Let $\phi = I_0 \subset I_1 \subset \cdots \subset I_{d+1} = P$ be a maximal chain m of L. Each $I_i$ is an i-element order ideal of P. Hence $I_i - I_{i-1}$ contains a unique element $x_i$ of P. Associate with m the permutation $\pi(m) = (\omega(x_1),\omega(x_2), \ldots, \omega(x_{d+1}))$ of $[d+1]$. The set $\mathcal{L}(P) = \mathcal{L}(P,\omega) := \{\omega(m):m \in \mathcal{M}(L)\}$ is called the *Jordan-Hölder set* of P. If $\pi = (a_1,\ldots,a_{d+1})$ is any permutation of $[d+1]$, then the *descent set* of $\pi$ is defined by $\mathcal{D}(\pi) = \{i:a_i > a_{i+1}\}$. It is then easy to verify that $\alpha_L(S)$ is equal to the number of permutations $\pi \in \mathcal{L}(P)$ whose descent set is contained in S.

For instance, if we define $\omega$ for the poset P of (4) by

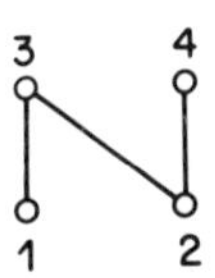


then we have the following table:

| $\pi \in \mathcal{L}(P)$ | $\mathcal{D}(\pi)$ |
|---|---|
| 1234 | $\phi$ |
| 2134 | 1 |
| 1243 | 3 |
| 2143 | 1,3 |
| 2413 | 2 |

Since, for example, four of the above descent sets are contained in $\{1,3\}$, we conclude $\alpha(1,3) = 4$.

Now let $\gamma(S) = \gamma_P(S) = |\{\pi \in \mathcal{L}(P):\mathcal{D}(\pi) = S\}|$. Clearly

$$\sum_{T \subseteq S} \gamma(T) = |\{\pi \in \mathcal{L}(P):\mathcal{D}(\pi) \subseteq S\}| = \alpha(S).$$

But the numbers $\beta_L(T)$ are uniquely defined by (2), so we conclude $\beta_L(S) = \gamma_P(S)$. Hence:

1.1 THEOREM: *Let* $L = J(P)$. *Then* $\beta_L(S)$ *is equal to the number of permutations* $\pi \in L(P)$ *whose descent set is* $S$.

This shows in particular that $\beta_L(S) \geq 0$ for a distributive lattice L, and much additional information can be gleaned from Theorem 1.1 [$St_1$]. Moreover, the numbers $\beta(S)$ play a central role in the theory of P-partitions [$St_1$], which deals with order-preserving (or order-reversing) maps of a poset P into a chain. For instance, if $\Omega(P,n)$ denotes the number of order-preserving maps $\sigma: P \to [n]$, then

$$\sum_{n\geq 0} \Omega(P,n)t^n = \Big(\sum_{\pi\in L(P)} t^{1+|\mathcal{D}(\pi)|}\Big)(1-t)^{-d-2}$$

$$= \Big(\sum_{S\subseteq[d]} \beta(S)t^{1+|S|}\Big)(1-t)^{-d-2} . \tag{6}$$

Note that it follows from (3) or otherwise that $\Omega(P,n) = Z(J(P),n)$.

At this point it is natural to ask for a generalization of the preceding theory to wider classes of graded posets Q. What structural properties of Q guarantee that $\beta_Q(S) \geq 0$? Is there a natural way of associating a permutation (or possibly a sequence) with each maximal chain of Q so that Theorem 1.1 remains valid? These and related questions will be considered in the next section.

## 2. LEXICOGRAPHICALLY SHELLABLE POSETS

Let P be a graded poset of rank $d+1$. Assume that each edge $x \to y$ in P has been labeled by an integer $\lambda(x \to y)$. Then each k-chain $x_0 \to x_1 \to \cdots \to x_k$ is naturally labeled by a k-tuple $\lambda(x_0 \to x_1 \to \cdots \to x_k) = (\lambda(x_0 \to x_1), \lambda(x_1 \to x_2), \ldots, \lambda(x_{k-1} \to x_k)) \in \mathbb{Z}^k$. We will compare such k-tuples by their *lexicographic order*: $(a_1, a_2, \ldots, a_k) <_L (b_1, b_2, \ldots, b_k)$ if and only if $a_i < b_i$ in the first coordinate where they differ.

2.1 DEFINITION: The edge-labeling $\lambda$ will be called an *EL-labeling* if for every interval $[x,y]$ in P:

(i) There is a unique chain $a_{x,y}: x = x_0 \to x_1 \to \cdots \to x_k = y$

such that $\lambda(x_0 \to x_1) \leq \lambda(x_1 \to x_2) \leq \cdots \leq \lambda(x_{k-1}, x_k)$,

(ii) for every other chain $b: x = y_0 \to y_1 \to \cdots \to y_k = y$ we have $\lambda(b) >_L \lambda(a_{x,y})$.

The poset P is said to be *edge-wise lexicographically shellable*, or *EL-shellable*, if it admits an EL-labeling.

The reader is invited to check that the following edge-labeling is an EL-labeling.

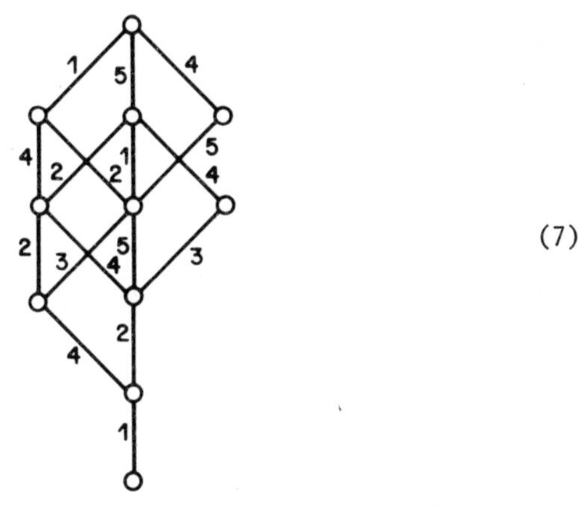


(7)

For each maximal chain $m: \hat{0} = x_0 \to x_1 \to \cdots \to x_{d+1} = \hat{1}$ we define the *descent set* $\mathcal{D}(m) = \{i \in [d] : \lambda(x_{i-1} \to x_i) > \lambda(x_i \to x_{i+1})\}$. There is the following combinatorial interpretation of the $\beta$-invariant of an EL-shellable poset P.

2.2 THEOREM. *For* $S \subseteq [d]$, $\beta(S)$ *equals the number of maximal chains* m *in* P *such that* $\mathcal{D}(m) = S$. *In particular,* $\beta(S) \geq 0$.

Every interval of P is EL-shellable under the same edge-labeling, so we deduce the following interpretation of the Möbius function of P.

2.3 COROLLARY: *When* $x < y$ *in* P, $(-1)^{r(x,y)}\mu(x,y)$ *equals the number of chains* $x = x_0 \to x_1 \to \cdots \to x_k = y$ *such that* $\lambda(x_0 \to x_1) > \lambda(x_1 \to x_2) > \cdots > \lambda(x_{k-1} \to x_k)$. *In particular,* $(-1)^{r(x,y)}\mu(x,y) \geq 0$.

2.4 EXAMPLE: Distributive lattices. Recall the discussion of Section 1. Let $L = J(P)$ be a finite distributive lattice of rank $d+1$, and let $\omega: P \to [d+1]$ be an order-preserving bijection. If $I \to I'$ is an edge in $J(P)$ then $I' - I = \{x\}$, and we can label $\lambda_\omega(I \to I') = \omega(x)$. This is an EL-labeling. The lattice (5) considered in Section 1 gets the following labeling $\lambda_\omega$ from the choice of $\omega$ considered there.

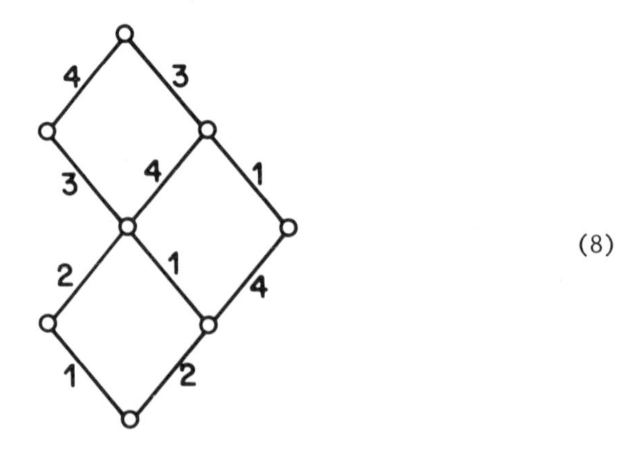


(8)

2.5 EXAMPLE: Semimodular lattices. Let L be a finite (upper) semimodular lattice. Give the join-irreducibles of L a linear order $e_1, e_2, \ldots, e_s$ which extends their partial order (i.e., $e_i < e_j$ implies $i < j$). Then

$$\lambda(x \to y) = \min \{i \mid x < x \vee e_i = y\}$$

defines an EL-labeling of L. Clearly, if L is distributive this coincides with the labeling method described in Example 2.4. The first picture below shows an ordering of the join-irreducibles of a semimodular lattice, the second picture shows the induced EL-labeling.

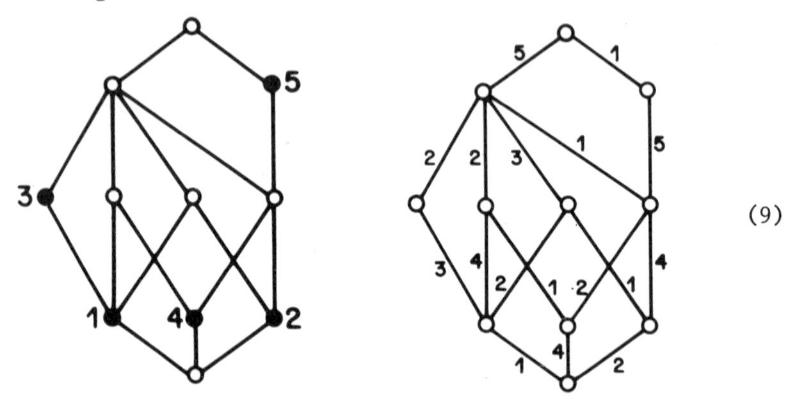


(9)

2.5 EXAMPLE: Supersolvable lattices. A finite lattice L is said to be *supersolvable* if it contains a maximal chain M which together with any other chain in L generates a distributive sublattice. Each such distributive sublattice can be labeled by the method previously described so that M receives the increasing label $(1,2,\ldots,d+1)$. It turns out that this will assign a unique label to each edge of L, and the resulting global labeling of L is an EL-labeling. If G is a supersolvable finite group then the lattice L(G) of subgroups of G is a supersolvable lattice (hence the terminology). The following picture shows an EL-labeling of the subgroup lattice of the Abelian group $\mathbb{Z}_4 \times \mathbb{Z}_4$ produced by the method described.

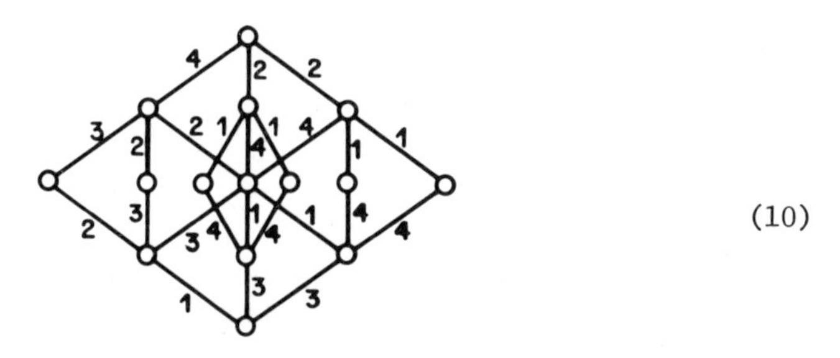

(10)

The notion of a lexicographically shellable poset as defined above was introduced in [$Bj_1$], the motivating examples being the finite distributive [$St_1$], supersolvable [$St_2$], and semimodular lattices [$St_3$]. In an EL-shellable poset the maximal chains derive their labels from the edges traversed. In some situations it is not known how to assign labels to the individual edges in a coherent manner, but it is nevertheless possible to assign labels directly to the maximal chains in such a way that the basic properties encountered above are preserved. This leads to the more general definition of lexicographic shellability considered in [$BW_1$] and [$BW_2$].

Let P be a graded poset of rank $d+1$. To each maximal chain $m$: $\hat{0} = x_0 \to x_1 \to \cdots \to x_{d+1} = \hat{1}$ we assign a label $\lambda(m) = (\lambda_1(m), \lambda_2(m), \ldots, \lambda_{d+1}(m)) \in \mathbb{Z}^{d+1}$, satisfying the following condition: If two maximal chains m and m' coincide along their first k edges, then $\lambda_i(m) = \lambda_i(m')$ for $i = 1,2,\ldots,k$. If $[x,y]$ is an interval and c a saturated chain from $\hat{0}$ to x the pair $([x,y],c)$ will be called a *rooted interval* and denoted $[x,y]_c$. Every maximal chain b in a rooted interval $[x,y]_c$ has a well-defined induced label $\lambda^c(b) \in \mathbb{Z}^{r(x,y)}$, namely $\lambda^c_i(b) = \lambda_{r(x)+i}(m)$ for $i = 1,2,\ldots,r(x,y)$ and any maximal chain m containing c and b.

2.7 DEFINITION: A labeling $\lambda$ of the maximal chains of P as above is a *CL-labeling* if for every rooted interval $[x,y]_c$ in P:

(i) there is a unique maximal chain $a = a_{x,y,c}$ in $[x,y]$ such that $\lambda^c_1(a) \le \lambda^c_2(a) \le \cdots \le \lambda^c_{r(x,y)}(a)$,

(ii) if b is any other maximal chain in $[x,y]$ then $\lambda^c(b) >_L \lambda^c(a_{x,y,c})$.

The poset P is said to be *chain-wise lexicographically shellable*, or *CL-shellable*, if it admits a CL-labeling.

This definition is illustrated in the following picture which shows a rank 3 poset and a CL-labeling of its maximal chains.

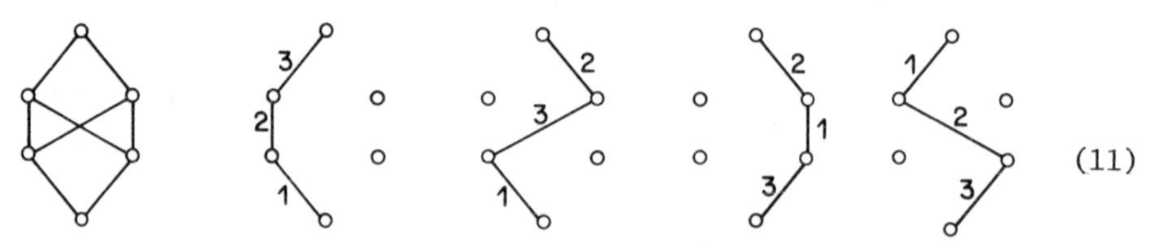


(11)

It is clear that an EL-labeling of P induces a CL-labeling, hence EL-shellable implies CL-shellable. The reason for calling these posets "lexicographically shellable" is that the lexicographic order of the labels gives a shelling order to the maximal chains. For a definition of "shelling order" and the general notion of "shellable posets" see $[Bj_1]$.

The notion of *descent set* generalizes immediately from EL-labelings to CL-labelings: For a maximal chain m in P and a CL-labeling $\lambda$, $\mathcal{D}(m) = \{i \varepsilon [d] \mid \lambda_i(m) > \lambda_{i+1}(m)\}$. Theorem 2.2 remains true for CL-shellable posets. Also, a CL-labeling of P restricts to a CL-labeling of any rooted interval $[x,y]_c$, so Corollary 2.3 suitably reformulated also remains true.

2.8 EXAMPLE: Bruhat order. Let (W,S) be a *Coxeter group* (see [Bo, Ch. 4]). For instance, W could be the symmetry group of a regular polytope or the Weyl group of a root system, and S the set of reflections across the walls of a fundamental chamber. *Bruhat order* is a partial ordering of W which appears repeatedly in the geometry and representation theory of algebraic groups and Lie algebras. For instance, let G be a connected complex semisimple algebraic group and B a Borel subgroup. The Schubert varieties $\bar{C}_w$ in the irreducible complex projective variety G/B are indexed by the corresponding Weyl group W, and Bruhat order describes the disposition of Schubert varieties: $w \le w'$ if and only if $\bar{C}_w \subseteq \bar{C}_{w'}$. We will here explicitly formulate Bruhat order only for the symmetric groups $S_n$ and refer to [De] or $[BW_1]$ for the general definition.

Let $S_n$ be the group of all permutations of [n], and let $\pi = a_1 a_2 \cdots a_n \varepsilon S_n$. The elements covering $\pi$ in Bruhat order are precisely those permutations $\pi'$ obtained from $\pi$ by interchanging a pair of elements $a_i$ and $a_j$ with $i < j$, satisfying: (i) $a_i < a_j$, and (ii) if $i < k < j$ then $a_k < a_i$ or $a_k > a_j$. It follows that the poset rank of $\pi$ equals its inversion number, i.e., the number of pairs (i,j) with $i < j$ and $a_i > a_j$. Here is the Bruhat order of $S_3$:

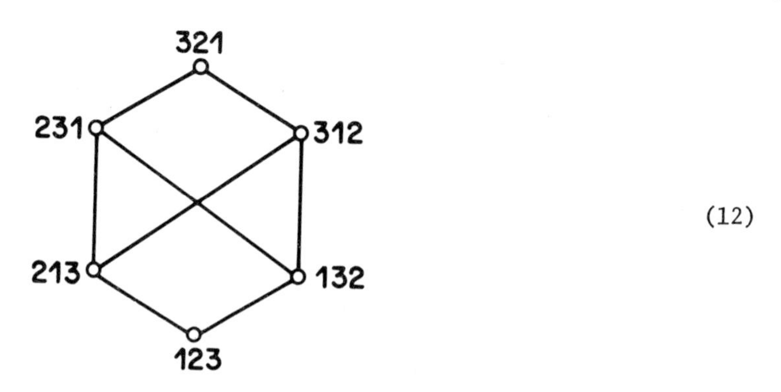


(12)

Let (W,S) be a Coxeter group (not necessarily finite). Each interval [w,w'] in the Bruhat order of W is a finite graded poset, which can be shown to be CL-shellable [$BW_1$]. For instance, the labeling algorithm applied to $S_3$ produces the labeling of maximal chains in picture (11). A slightly stronger formulation is possible. For $J \subseteq S$, let $W^J$ denote the set of minimal coset representatives modulo the parabolic subgroup $W_J$ (as defined e.g. in [Bo, p. 19]). Then every interval [w,w'] in the Bruhat order of $W^J$ is CL-shellable.

It is not known whether Bruhat order is in general EL-shellable. However, EL-labelings have been found (cf. [$Ed_2$],[Pr]) for the classical finite Weyl groups corresponding to root systems of type A, B, C and D.

2.9 EXAMPLE: Face lattices. Let $\Delta$ be a finite simplicial or polyhedral complex with all maximal faces of the same dimension. The faces of $\Delta$ ordered by inclusion form a graded lattice $L(\Delta)$ (the improper faces $\phi$ and $\Delta$ included), the *face-lattice* of $\Delta$. The face-lattices (or their duals) of the following complexes are known to be CL-shellable (cf. [$BW_2$]): (1) The boundary complex of a convex polytope, (2) the Coxeter complex of a finite Coxeter group, (3) the Tits building of a finite group with BN-pair, (4) the independent set complex of a matroid, (5) the broken circuit complex of a matroid. Face-lattices of simplices, cubes and cross-polytopes of all dimensions are known to be EL-shellable [$Bj_1$, p. 174]; picture (13) shows an EL-labeling of the face-lattice of the 2-cube. It is not known whether face-lattices of convex polytopes are in general EL-shellable.

2.10 EXAMPLE: Totally semimodular posets. A finite poset P is said to be (upper) *semimodular* if whenever two distinct elements u,v both cover $t \in P$ there exists a $z \in P$ which covers both u and v (cf. [Bi, p. 39]). P is said to be *totally semimodular* if it has $\hat{0}$ and $\hat{1}$ and all intervals [x,y] are semimodular. Totally semimodular posets are CL-shellable [$BW_2$].

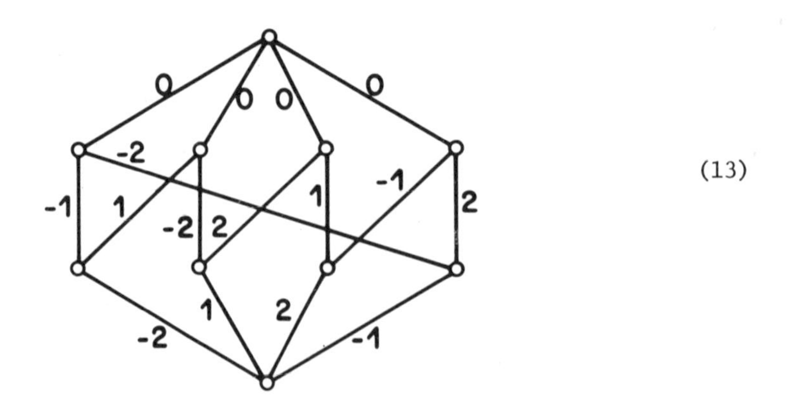


(13)

## 3. ORDER HOMOLOGY

Let P be a finite poset, and let k be a field or the ring of integers $\mathbb{Z}$. For each $i \in \mathbb{Z}$ let $C_i(P,k)$ denote the free k-module on the basis of the i-chains $x_0 < x_1 < \cdots < x_i$ of P. We consider $\phi$ to be a (-1)-chain, so $C_{-1}(P,k) \cong k$. Also, $C_i(P,k) = 0$ for $i < -1$ and $i > \text{length}(P)$. Define a map $d_i\colon C_i(P,k) \to C_{i-1}(P,k)$ by linearly extending

$$d_i(x_0 < x_1 < \cdots < x_i)$$

$$= \sum_{j=0}^{i} (-1)^j (x_0 < x_1 < \cdots < x_{j-1} < x_{j+1} < \cdots < x_i)$$

when i-chains exist, and setting $d_i = 0$ otherwise. Then $d_i d_{i+1} = 0$; that is, $\operatorname{Im} d_{i+1} \subseteq \ker d_i$. Define, for $i \in \mathbb{Z}$,

$$\tilde{H}_i(P,k) = \ker d_i / \operatorname{Im} d_{i+1} \,. \tag{14}$$

The k-modules $\tilde{H}_i(P,k)$ are the *order homology groups of P with coefficients in* k. Clearly, $\tilde{H}_i(P,k) = 0$ for $i < -1$ and $i > \ell(P)$. Also, $\tilde{H}_{-1}(P,k) = 0$ if and only if $P \neq \phi$, and $\tilde{H}_0(P,k) = 0$ if and only if P is connected (as a poset). In fact, $\tilde{H}_0(P,k)$ is a free k-module of rank one less than the number of connected components of P.

Let $c_i(P)$ denote the number of i-chains of P. For k a field we have the *Euler-Poincaré formula*

$$\sum_{i=-1}^{\ell(P)} (-1)^i c_i(P) = \sum_{i=-1}^{\ell(P)} (-1)^i \dim_k \tilde{H}_i(P,k) . \tag{15}$$

As was mentioned in the Introduction the left side of (15) equals $\mu(\hat{P})$. Hence, the Möbius function $\mu(\hat{P})$ is the *Euler characteristic* of order homology:

$$\mu(\hat{P}) = \sum_{i=-1}^{\ell(P)} (-1)^i \dim_k \tilde{H}_i(P,k) . \tag{16}$$

The numbers $\dim_{\mathbb{Q}} \tilde{H}_i(P,\mathbb{Q})$ are called the *Betti numbers* of P.

Define a simplicial complex $\Delta(P)$ in the following way: the vertices of $\Delta(P)$ are the elements of P and the i-faces of $\Delta(P)$ are the i-chains $x_0 < x_1 < \cdots < x_i$. The order homology of P defined in (14) is then the (reduced) simplicial homology of $\Delta(P)$, as familiar from combinatorial topology. In fact, the properties stated above are just a recapitulation in poset terminology of some basic facts from simplicial topology. As a finite simplicial complex, the *order complex* $\Delta(P)$ determines a compact topological space $|\Delta(P)|$, the *geometric realization* of P. Thus one may speak of topological type, homotopy type, etc., of posets. The following picture shows two posets and next to them their geometric realizations:

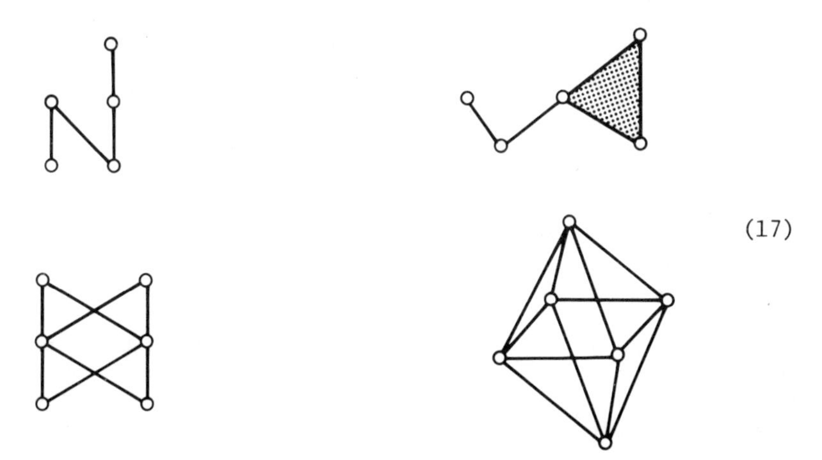

(17)

To obtain a poset with preassigned homology groups or topological type, one can take a simplicial complex $\tau$ with the desired properties and look at the poset $P_\tau$ of faces of $\tau$ ordered by inclusion. Then $|\Delta(P_\tau)|$ and $|\tau|$ are homeomorphic (the former being the barycentric subdivision of the latter). In particular, any compact triangulable space is the geometric realization of a finite poset (in fact, of the proper part $\bar{L}$ of a finite lattice L).

Suppose now that P is a lexicographically shellable poset of rank $d+1$. The order homology of $\bar{P}$ can easily be computed by standard methods of algebraic topology (the Mayer-Vietoris exact sequence): $\tilde{H}_i(P,k) = 0$ for $i \neq d-1$, and $\tilde{H}_{d-1}(P,k) \cong k^{|\mu(P)|}$. Since each interval $[x,y]$ in P is lexicographically shellable, we have found that for each open interval $(x,y)$ in P "homology vanishes" below the top dimension. This brings us to the key definition.

3.1 DEFINITION: Let P be a graded poset. We say that P is *Cohen-Macaulay over* k (CM/k, for short) if for each open interval $(x,y)$ in P: $\tilde{H}_i((x,y),k) = 0$ for $i \neq \delta(x,y)$, where $\delta(x,y) = r(x,y)-2$ is the dimension of $\Delta((x,y))$.

In particular,

3.2 THEOREM: *If* P *is lexicographically shellable, then* P *is Cohen-Macaulay over* $\mathbb{Z}$ *and all fields* k.

In a lexicographically shellable poset the Möbius function $\mu(x,y)$ has a combinatorial interpretation (Corollary 2.3) which shows that $(-1)^{r(x,y)}\mu(x,y) \geq 0$. More generally, in a Cohen-Macaulay poset by (16): $(-1)^{r(x,y)}\mu(x,y) = \dim_k H_{\delta(x,y)}((x,y),k) \geq 0$.

In a similar way the nonnegativity of the numbers $\beta(S)$ (cf. Theorem 2.2) extends to any Cohen-Macaulay poset P (cf. Theorem 5.2).

3.3 THEOREM: *For each* $S \subseteq [d]$, *the top homology group of* $\bar{P}_S$ *is free over* k *of rank* $\beta(S)$.

The definition of Cohen-Macaulayness given here applies only to *graded* posets. However, it is possible to define the notion in a similar manner for any finite poset P. It can be shown that this definition implies that P is graded ($\hat{0}$ and $\hat{1}$ may have to be added), so there is in fact no loss of generality.

The property of being Cohen-Macaulay depends on the ring k as well as the poset. The universal coefficient theorem for homology shows that CM-ness over $\mathbb{Z}$ is the strongest property and CM-ness over $\mathbb{Q}$ the weakest. More precisely, CM/$\mathbb{Z}$ implies CM/k for *all* fields k, while CM/k for *some* field k implies CM/$\mathbb{Q}$.

All graded posets of rank one and two are CM. A graded poset P of rank 3 is CM if and only if the proper part $\bar{P}$ is connected. From rank 4 on CM-ness cannot be as easily characterized. Consider, for instance, the following rank 4 poset P.

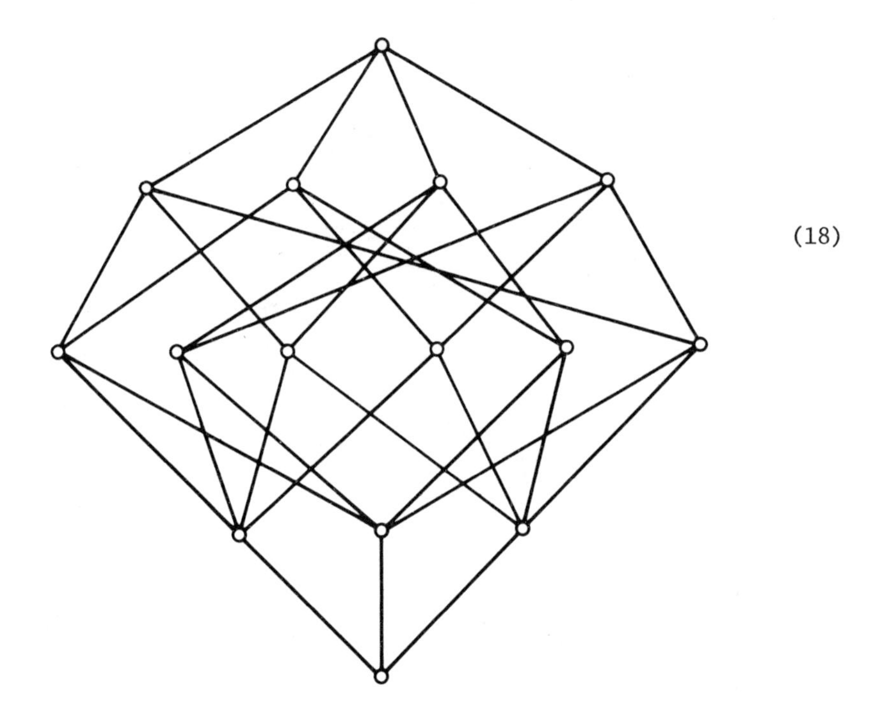

(18)

It can be seen by inspection that all rank 3 intervals are connected, so to test for CM-ness one must only compute the homology of the proper part $\bar{P}$. Now, $|\Delta(\bar{P})|$ is the real projective plane, so

$$\tilde{H}_i(\bar{P}, \mathbb{Z}) = \begin{cases} \mathbb{Z}_2, & i = 1 \\ 0, & i \neq 1 \end{cases}$$

Hence, P is not CM over $\mathbb{Z}$ or over $\mathbb{Z}_2$, but P is CM over fields of characteristic $\neq 2$.

For another example, consider the following poset.

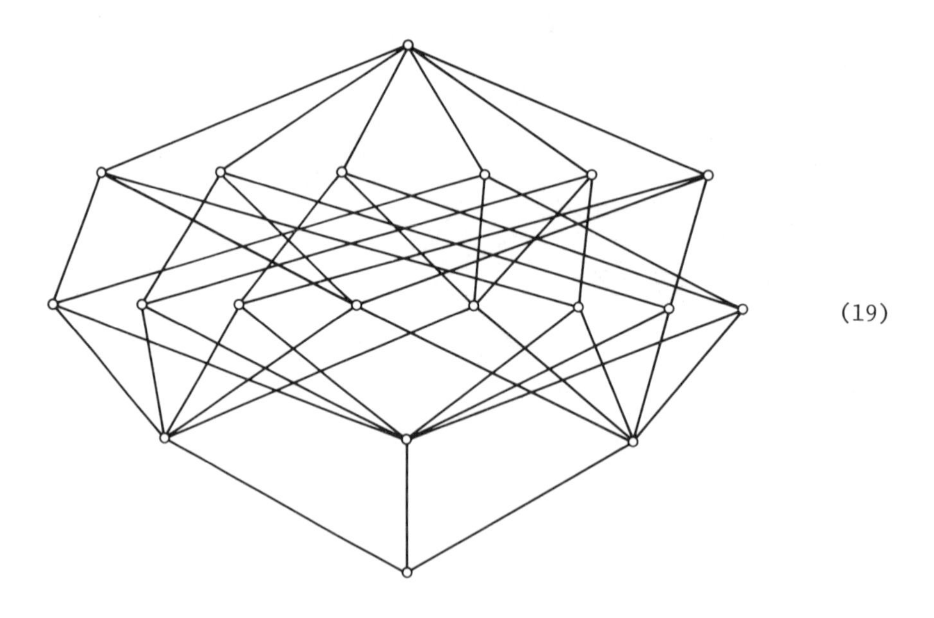

(19)

Here $\tilde{H}_i(\bar{P}, \mathbb{Z}) = 0$ for all i (the geometric realization of $\bar{P}$ is the dunce hat), so P is CM/$\mathbb{Z}$. However, P is not (lexicographically) shellable.

Let us finally mention the remarkable fact that the CM-ness of a poset is a topological property: If $P_1$ and $P_2$ are two finite graded posets and $|\Delta(\bar{P}_1)|$ is homeomorphic to $|\Delta(\bar{P}_2)|$, then $P_1$ is CM/k if and only if $P_2$ is CM/k.

## 4. THE POSET RING

With any finite poset P we will associate a commutative ring $R_P$ whose structure closely reflects the combinatorial and topological properties of P. In this way we gain new insight into the significance of Cohen-Macaulay posets. Assume that P has a $\hat{0}$ and $\hat{1}$, and let k be any field (which for our purposes can be taken to be the rational numbers $\mathbb{Q}$). (Much of the theory goes through for an arbitrary commutative ring k, but for simplicity we take k to be a field throughout this and the next section.) If $\bar{P} = \{x_1, \ldots, x_n\}$, then form the polynomial ring $R = k[x_1, \ldots, x_n]$, where the elements of $\bar{P}$ are regarded as independent indeterminates.

Let $I_P$ be the ideal of R generated by all products $x_i x_j$ where $x_i$ and $x_j$ are incomparable elements of $\bar{P}$. Set $R_P = R/I_P$. We call $R_P$ the *poset ring* corresponding to P. (In some papers it is not required that P have a $\hat{0}$ and $\hat{1}$, and our $R_P$ would be denoted $R_{\bar{P}}$.) More generally, one can define a ring $R_\Delta$ associated with an arbitrary simplicial complex $\Delta$ [St$_{5,6}$] [Re], so that the ring $R_{\Delta(\bar{P})}$ associated with the order complex of $\bar{P}$ coincides with the poset ring $R_P$. In this general setup $R_\Delta$ is called the "Stanley-Reisner ring" of $\Delta$. Here, however, we will be concerned exclusively with posets.

The ring $R_P$ is an algebra over the field k. A k-basis for $R_P$ consists of all monomials of the form $y_1^{a_1} y_2^{a_2} \cdots y_s^{a_s}$, where $a_i > 0$ and $y_1 < y_2 < \cdots < y_s$. We can identify this monomial with the multichain (= chain with repeated elements)

$$y_1^{a_1} < y_2^{a_2} < \cdots < y_s^{a_s}$$

of P, where the notation indicates that $y_i$ is repeated $a_i$ times. Thus we can regard elements of $R_P$ as linear combinations of multichains.

In general, a k-algebra A is said to be $\mathbb{N}^m$-*graded*, where $m \in \mathbb{P}$, if it is given a vector space direct sum decomposition $A = \coprod_{\alpha \in \mathbb{N}^m} A_\alpha$ satisfying $A_\alpha A_\beta \subseteq A_{\alpha+\beta}$. We then call an element $x \in A_\alpha$ *homogeneous of degree* $\alpha$, denoted $\deg x = \alpha$. If A is finitely-generated as a k-algebra, then each $A_\alpha$ is a finite-dimensional vector space over k. In this case, the *Hilbert series* of A is defined to be the formal power series

$$F(A,t) = \sum_{\alpha \in \mathbb{N}^m} (\dim_k A_\alpha) t^\alpha ,$$

where $t = (t_1, \ldots, t_m)$ and $t^\alpha = t_1^{\alpha_1} \cdots t_m^{\alpha_m}$ for $\alpha = (\alpha_1, \ldots, \alpha_m)$. For poset rings there are several possible ways to define an $\mathbb{N}^m$-grading, of which only one will concern us here. Namely, if P is graded of rank $d+1$ and $x \in P_i$, $1 \le i \le d$, then we define $\deg x$ to be the $i^{\text{th}}$ unit coordinate vector $e_i$ in $\mathbb{N}^d$. This makes $R_P$ into an $\mathbb{N}^d$-graded k-algebra where $(R_P)_\alpha$ has a k-basis consisting of all multichains containing $\alpha_i$ elements of rank i, where $\alpha = (\alpha_1, \ldots, \alpha_d)$. Let us compute the Hilbert series $F(R_P, t)$ with respect to this grading. By grouping multichains in P according to their supports (i.e., the chain of elements of P which appear at least once in the multichain), we see that

$$F(R_P,t) = \sum_{S\subseteq[d]} \alpha(S) \prod_{i\in S} \frac{t_i}{1-t_i} .$$

When put over the common denominator $\prod_{i=1}^{d}(1-t_i)$, the coefficient of $\prod_{i\in S} t_i$ in the numerator becomes $\sum_{T\subseteq S} (-1)^{|S-T|}\alpha(T) = \beta(S)$. There follows:

4.1 THEOREM: *We have*

$$F(R_P,t) = \left(\sum_{S\subseteq[d]} \beta(S) \prod_{i\in S} t_i\right) \prod_{i=1}^{d} (1-t_i)^{-1} .$$

If A is any finitely-generated $\mathbb{N}^m$-graded k-algebra and $x\in A_\alpha$, then it is easily seen that

$$F(A,t) \leq \frac{F(A/x,t)}{1-t^\alpha} \qquad (\text{termwise} \leq) ,$$

with equality if and only if x is a non-zero-divisor in A [$St_7$, Theorem 3.1]. Define the *rank-level parameters* $\theta_1,\ldots,\theta_d$ of $R_P$ by

$$\theta_i = \sum_{x\in P_i} x ,$$

so $\theta_i$ is homogeneous of degree $e_i$. The quotient ring $Q_P = R_P/(\theta_1,\ldots,\theta_d)$ inherits an $\mathbb{N}^d$-grading from $R_P$. The following result is now an immediate consequence of Theorem 4.1.

4.2 THEOREM: *Let* $\theta_1,\ldots,\theta_d$ *be the rank-level parameters of the graded poset* P *of rank* $d+1$. *Let* $Q_P = R_P/(\theta_1,\ldots,\theta_d)$. *Then*

$$F(Q_P,t) = \sum_{S\subseteq[d]} \beta(S) \prod_{i\in S} t_i \tag{20}$$

*if and only if* $\theta_i$ *is not a zero-divisor in the ring* $R_P/(\theta_1,\ldots,\theta_{i-1})$, $1\leq i\leq d$ .

We define the ring $R_P$ to be *Cohen-Macaulay* if it satisfies either of the two equivalent conditions of Theorem 4.2. Note that our definition is only applicable to *graded* posets. It is possible to define what is meant for any finitely-generated graded algebra to be Cohen-Macaulay. One can then prove that if P is any finite poset with $\hat{0}$ and $\hat{1}$ such that $R_P$ is Cohen-Macaulay, then P is indeed graded. Thus it costs us nothing to restrict our attention to graded posets from the beginning.

An important property of a Cohen-Macaulay poset ring $R_P$ is that there is a simple canonical form for its elements. First note that if $x \in P_i$ then $x\theta_i = x^2$ in $R_P$. Hence $x^2 = 0$ in $Q_P$, so $Q_P$ is spanned by all *chains* (or squarefree monomials) in P. (This also follows from (20)). Choose a k-basis $\mathcal{B}$ for $Q_P$ consisting of chains of P. We call $\mathcal{B}$ a set of *separators* for $R_P$. By (20), the number of chains $0 < y_1 < \cdots < y_s < 1$ in $\mathcal{B}$ satisfying $S = \{r(y_1),\ldots,r(y_s)\}$ is equal to $\beta(S)$. The next result is a consequence of the previous theorem.

4.3 COROLLARY: *Let $R_P$ be Cohen-Macaulay, and let $\mathcal{B}$ be a set of separators. Then every element f of $R_P$ can be written uniquely in the form*

$$f = \sum_{\eta \in \mathcal{B}} \eta \cdot p_\eta(\theta_1,\ldots,\theta_d), \tag{21}$$

*where $p_\eta$ is a polynomial in $\theta_1,\ldots,\theta_d$ (with coefficients in k).*

It is an interesting problem to decide efficiently when a poset ring is Cohen-Macaulay and to find a set of separators when this is the case. The following result appears in [$Ga_2$], together with an algorithm for testing CM-ness.

4.4 THEOREM: *Let $\mathcal{B}$ be a collection of chains of P, such that $\beta_P(S)$ elements of $\mathcal{B}$ have rank set S, for all $S \subseteq [d]$. Let $\Phi$ be the incidence matrix between $\mathcal{B}$ and the set M of maximal chains of P, i.e., if $c \in \mathcal{B}$ and $m \in M$ then*

$$\Phi_{cm} = \begin{cases} 1, & \text{if } c \subseteq m \\ 0, & \text{if } c \nsubseteq m \end{cases}.$$

*The following two conditions are equivalent:*
*(i) $R_P$ is CM and $\mathcal{B}$ is a set of separators,*
*(ii) the matrix $\Phi$ is invertible (over k).*

Now suppose P is lexicographically shellable, and let

$$m: \hat{0} = y_0 \to y_1 \to \cdots \to y_d \to y_{d+1} = \hat{1}$$

be a maximal chain of P. Define the *restriction*

$$\mathcal{R}(m) = \{y_i : \lambda(y_{i-1} \to y_i) > \lambda(y_i \to y_{i+1})\},$$

and let $\mathcal{B} = \{\mathcal{R}(m) : m \in M\}$. If one orders $M$ lexicographically and orders $\mathcal{B}$ correspondingly, then the matrix $\Phi$ of Theorem 4.4 is upper triangular with 1's on the main diagonal. Hence from Theorem 4.4 we deduce:

4.5 COROLLARY: *Let* P *be lexicographically shellable, and let* $B = \{R(m) : m \in M\}$. *Then* $R_P$ *is Cohen-Macaulay (over any field* k*), and* B *is a set of separators.*

Corollary 4.5 can be formulated for arbitrary shellable posets [$Ga_2$, Thm. 4.2], and can be generalized to arbitrary shellable simplicial complexes [KK].

4.6 EXAMPLE: Let P be the distributive lattice

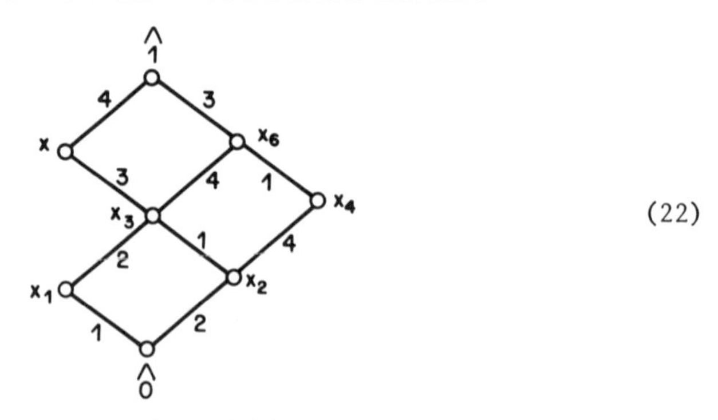


(22)

with EL-labeling as shown (cf. (8)). We then have the following table.

| maximal chain m | $R(m)$ |
|---|---|
| $\hat{0} \to x_1 \to x_3 \to x_5 \to \hat{1}$ | $\phi$ |
| $\hat{0} \to x_1 \to x_3 \to x_6 \to \hat{1}$ | $x_6$ |
| $\hat{0} \to x_2 \to x_3 \to x_5 \to \hat{1}$ | $x_2$ |
| $\hat{0} \to x_2 \to x_3 \to x_6 \to \hat{1}$ | $x_2 < x_6$ |
| $\hat{0} \to x_2 \to x_4 \to x_6 \to \hat{1}$ | $x_4$ |

It follows that every element f of $R_P$ can be written uniquely in the form

$$f = p_1 + x_6 p_2 + x_2 p_3 + x_2 x_6 p_4 + x_4 p_5 ,$$

where $p_i$ is a polynomial in the variables $\theta_1 = x_1 + x_2$, $\theta_2 = x_3 + x_4$, $\theta_3 = x_5 + x_6$.

## 5. REISNER'S THEOREM

Perhaps the most important result in the theory of Cohen-Macaulay posets is the following. While we state it for the case of a field k, the theorem and its attendant background carry over straightforwardly to more general commutative rings k, in particular to the important case $k = \mathbb{Z}$.

5.1 THEOREM: *Fix a field* k *and let* P *be a finite graded poset. Then* P *is Cohen-Macaulay (in the topological sense of Definition 3.1) if and only if the poset ring* $R_P$ *is Cohen-Macaulay.*

There is no really simple proof known of this fundamental result. The original proof of Reisner [Re] (which dealt with arbitrary simplicial complexes) used two tools from commutative algebra: (i) the theory of local cohomology first developed by Grothendieck, and (ii) the theory of the "purity of Frobenius" developed by Hochster and Roberts which enables one to compute local cohomology in characteristic p. Subsequently a proof was given by Hochster [$Hoc_2$] which replaced local cohomology by properties of the Koszul complex. Hochster's proof yields a homological criterion for $R_P$ to be Cohen-Macaulay which requires considerable topological arguments [Mu] to show is equivalent to Definition 3.1. A proof similar to Hochster's, but more elementary, was given by Baclawski and Garsia [BG, Prop. 6.2]. Hochster also gave a further proof using local cohomology but avoiding purity of Frobenius. This proof is unpublished by Hochster but reproduced in [$St_{10}$]. For readers familiar with homological algebra it is the shortest and most elegant proof of Reisner's theorem to date. Finally, at this very symposium on Ordered Sets, Stanley and Walker succeeded in finding a proof devoid of all but the simplest ring theory and using nothing from topology beyond the Mayer-Vietoris sequence.

Reisner's theorem is merely the first step in a beautiful theory connecting the combinatorics and topology of P with the algebraic properties of $R_P$. We will give the briefest glimpse of what lies beyond Reisner's theorem in Section 6e. For the conclusion of this section we will content ourselves with an illustration of the usefulness of Reisner's theorem in proving purely combinatorial statements concerning P.

Let P be Cohen-Macaulay of rank $d+1$, and let $S \subseteq [d]$. Let I be the ideal of $R_P$ generated by $\bigcup_{i \in [d]-S} P_i$. Then $R_P/I = R_{P_S}$, where $P_S$ is the S-rank-selected subposet of P. Moreover, $Q_P/\bar{I} = Q_{P_S}$, where $\bar{I}$ denotes the image of I in $Q_P$. But since

$$F(Q_P,t) = \sum_{T \subseteq [d]} \beta_P(T) \prod_{i \in T} t_i \quad ,$$

there follows

$$F(Q_P/\bar{I},t) = \sum_{T \subseteq S} \beta_P(T) \prod_{i \in T} t_i \quad .$$

Hence $R_{P_S}$ is Cohen-Macaulay, and we have proved the result alluded to in Section 3:

5.2 THEOREM: *If* P *is Cohen-Macaulay of rank* d+1, *then so is* $P_S$ *for all* $S \subseteq [d]$.

While it is possible to give a purely topological proof of this result, the details are rather messy [Mu][$Ba_4$]. The ring-theoretic approach yields the above almost trivial proof, which first appeared in [$St_8$].

## 6. FURTHER DEVELOPMENTS

The main purpose of preceding sections was to introduce the reader to the notion of a Cohen-Macaulay poset starting from scratch. In this section we will briefly mention some areas in which there is current research activity. For a fuller treatment see our exposition in [BGS].

(a) *Applications to ring theory.* An elaborate theory now exists showing how properties of a poset ring $R_P$ can be "transferred" to other rings of interest in invariant theory, representation theory and algebraic geometry. The idea to put the poset ring to such use seems to be due independently to DeConcini and Garsia. Thus DeConcini and collaborators [$DEP_2$], [DL], [DP], [Ei] have developed the notion of "Hodge algebras," while Garsia's ideas have been further developed by him and Baclawski [$Ga_1$], [$Ga_2$], [BG], [$Ba_5$], [GS] leading to the notion of "lexicographic rings." Preambles to these developments appear in the work of Hodge (see below) and of Rota and his colleagues [DRS], [DKR] on "straightening laws." This latter work in turn has its roots in the work of Alfred Young.

The interest to algebraic geometry in these developments derives from the fact that certain rings related to algebraic varieties, in particular the homogeneous coordinate rings of certain projective varieties, are closely related to the poset rings of Bruhat order and other related posets. The archetypal example, due to Hodge [Hod], [HP, p. 378], concerns the Grassmann variety $G_{rn}$ of r-planes in $\mathbb{C}^n$. Recent work of DeConcini, Lakshmibai, Musili and Seshadri [DL], [LMS] extends the range of the method to wide classes of "generalized Schubert varieties." It is a noteworthy development that the Cohen-Macaulayness of the homogeneous coordinate rings of these varieties is deduced from the Cohen-Macaulayness of the corresponding Bruhat poset rings, and hence is made to ultimately depend on the pure combinatorics of lexicographic shellability.

(b) *Group actions.* Let G be a finite group of automorphisms of a Cohen-Macaulay poset P of rank d+1. For each $S \subseteq [d]$, G permutes the chains of rank set S. Call the character of this permutation representation $\alpha_S$. Thus, $\alpha_S(g)$ is the number of chains of rank set S fixed by $g \in G$. Also, the action of G on $\bar{P}_S$ induces an action on the homology $\tilde{H}_{|S|-1}(\bar{P}_S, \mathbb{C})$. Call the character of this complex representation $\beta_S$. The following character relations can be deduced from the Hopf trace formula:

$$\alpha_S(g) = \sum_{T \subseteq S} \beta_T(g)$$
$$\beta_S(g) = \sum_{T \subseteq S} (-1)^{|S-T|} \alpha_T(g). \tag{23}$$

Note that these formulas evaluated at the identity $e \in G$ reduce to the fundamental relationship (1)-(2) between the numbers $\alpha(S)$ and $\beta(S)$ mentioned in the Introduction. The formulas (23) were anticipated by Solomon [$So_1$, p. 389] and first proved by Stanley [$St_9$].

The typical example of the situation described above is that of the symmetric group $S_n$ acting on the Boolean algebra $B_n$ of all subsets of an n-element set. Here the characters $\alpha_S$ were first considered by Frobenius and $\beta_S$ by Solomon [$So_2$, § 6]. There is an explicit formula for decomposing $\beta_S$ into irreducible characters, which is due to Solomon [$So_2$] (see also [$St_9$, Thm. 4.37]). Another example is that of the general linear group $GL_{d+1}(q)$ acting on the lattice of subspaces of a $(d+1)$-dimensional vector space over $GF(q)$. Here the characters $\alpha_S$ were first considered by Steinberg [Ste], and $\beta_{[d]}$ is now known as the Steinberg character. For other examples, see [$St_9$] and [$Bj_4$].

(c) *Alternating Möbius function.* Let us call a graded poset P of rank d+1 *alternating* if $\beta(S) = (-1)^{|S|-1}\mu(P_S) \geq 0$ for all $S \subseteq [d]$. Cohen-Macaulay posets are alternating (Theorem 3.3), but the converse is not necessarily true, as the following example shows.

(24)

A simple condition which assures that a graded poset P is alternating is the following: P is said to be *ER* (also called *partitionable*) if for each maximal chain m there is a subchain $R(m) \subseteq m$, the *restriction* of m, such that $C = \biguplus_{m \in M} [R(m), m]$. This condition says that for each chain c there is exactly one maximal chain m such that $R(m) \subseteq c \subseteq m$. The poset above is ER. Also, all lexicographically shellable posets are ER; for $R(m)$ simply take the subchain of m where the descents in the label occur (cf. Corollary 4.5). It is not known whether all Cohen-Macaulay posets are ER.

The nonnegativity of the numbers $\beta(S)$ for a Cohen-Macaulay poset has several interpretations. Thus, we have seen that $\beta(S)$ occur as Betti numbers for the homology of rank-selected subposets and determine the Hilbert series of the poset ring. The following purely combinatorial interpretation appears in [$St_8$]: If P is a Cohen-Macaulay poset of rank $d+1$ then there exists a subcomplex $\Lambda$ of the order complex $\Delta(\bar{P})$ such that for any $S \subseteq [d]$ the number of faces of $\Lambda$ having rank set S is equal to $\beta(S)$. It follows that if $\beta(S) \neq 0$ and $T \subseteq S$, then $\beta(T) \neq 0$. The proof is based on manipulations with the poset ring. This result can be viewed as giving a necessary condition that a collection of integers $\beta(S)$, $S \subseteq [d]$, are the numbers $\beta(S)$ of a Cohen-Macaulay poset. The condition is not quite sufficient for posets, but if one allows slightly more general objects (so called "Cohen-Macaulay completely balanced complexes") then the condition is both necessary and sufficient [BS].

(d) *Further topological developments*. The topological definition 3.1 of a Cohen-Macaulay poset imposes a condition on the homology of each open interval. By sharpening the attention to homotopy type we get a related notion: A graded poset P is said to be *homotopy Cohen-Macaulay* if for every open interval (x,y) in P the homotopy groups $\pi_i(|\Delta((x,y))|)$, $i < \delta(x,y)$, are trivial. Thus P is homotopy CM if and only if P is CM/$\mathbb{Z}$ and when $\delta(x,y) \geq 2$, $|\Delta((x,y))|$ is simply connected. This notion was introduced by Quillen [Qu] for the purpose of studying certain posets of subgroups of a group G. For instance, he showed that if $G = GL_n(k)$, where k is a field, and if $p \neq$ char k and k has a primitive p-th root of unity, then the poset of elementary Abelian p-subgroups of G is homotopy Cohen-Macaulay. For the full subgroup lattice there is the following characterization which follows from results of Björner, Iwasawa and Stanley [$Bj_1$, p. 167]: The subgroup lattice L(G) of a finite group G is homotopy Cohen-Macaulay if and only if G is a supersolvable group. A lexicographically shellable poset is homotopy CM and a homotopy CM poset is CM/$\mathbb{Z}$, but neither of the converse implications are true. For instance, the poset of Figure (19) is homotopy CM but not lexicographically shellable.

For a variety of topological developments related to the Cohen-Macaulay property of posets we refer the reader to [$Ba_{1-4}$] [$Bj_{2,3}$], [BWr], [Fa], [Mu], [Qu] and [Wr].

(e) *Further ring-theoretical developments:* With every finite poset P one can associate its poset ring $R_P$, and as a very general program one might ask in what way various ring-theoretic properties of $R_P$ influence the structure of P. In this paper we have been concerned with merely one instance of this, viz., the Cohen-Macaulay property. Even for the Cohen-Macaulay case one can seek to probe much further into the ring-theoretic structure, asking for explicit minimal free resolutions, description of the canonical modules, etc. ... . Let us here merely exemplify by mentioning Gorenstein posets. A poset ring $R_P$ (or poset P) is said to be *Gorenstein* (with respect to k) if the ideal generated by the rank-level parameters is *irreducible*, i.e., cannot be expressed as an intersection of two strictly larger ideals. When is a graded poset P Gorenstein? There is a characterization of this property in terms of the homology of open intervals of P which is analogous to Reisner's theorem, [$\text{Hoc}_2$, § 6], [$\text{St}_6$, § 8]. This can be formulated as follows: P is Gorenstein if and only if P is CM and the Möbius function of nuc P satisfies $\mu(x,y) = (-1)^{r(x,y)}$. Here "nuc P" denotes P minus those elements $\neq \hat{0},\hat{1}$ which are comparable to all $x \in P$. Somewhat surprisingly, this condition for Gorensteinness is equivalent to the apparently weaker condition that P is CM and the Möbius function of nuc P satisfies $\mu(x,y) = 1$ only for intervals $[x,y]$ of length 2, together with the additional requirement that $\mu(\hat{0},\hat{1}) \neq 0$.

Consider, for instance, the two distributive lattices

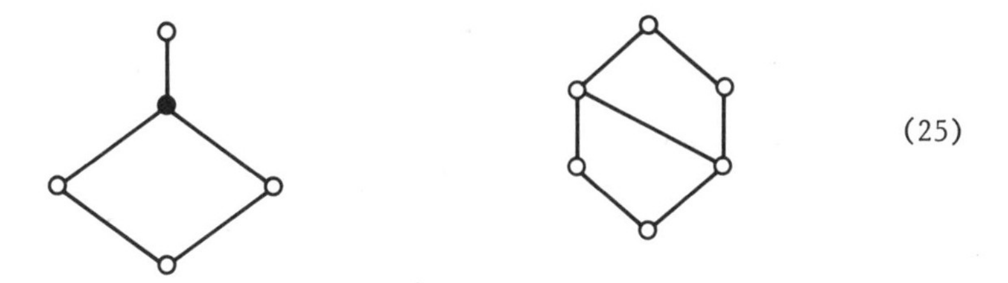

(25)

The first one is Gorenstein but the second is not (the open circles denote elements of nuc P). Nevertheless, the order complexes of their proper parts are homeomorphic. On the other hand, if $\bar{P}$ and $\bar{Q}$ are *nonacyclic* (i.e., their order homology is non-zero) and if $|\Delta(\bar{P})|$ and $|\Delta(\bar{Q})|$ are homeomorphic, then P is Gorenstein if and only if Q is Gorenstein. It follows from the above characterization of Gorenstein posets that Boolean algebras, (full) Bruhat order, and face lattices of convex polytopes (cf. Section 2) are Gorenstein.

The poset ring $R_P$ of a Gorenstein poset P satisfies a certain "self-duality" property. While we cannot enter into the details here, we can mention one simple combinatorial manifestation of this fact. Suppose P is Gorenstein of rank $d+1$ and that $P = \text{nuc}\, P$.

If S is any subset of [d], then $\beta(S) = \beta([d]-S)$. More generally [$Sta_9$, Prop. 2.2], if Q is any (induced) subposet of $\bar{P}$ then $\mu(\hat{Q}) = (-1)^d \mu(P-Q)$. This result extends to the situation of group actions as follows [$Sta_9$, Thm. 2.4]. Let P be as above, and let $G \subset \operatorname{Aut} P$. Then the $\beta$ characters of Section 6b satisfy $\beta_S = \beta_{[d]}\beta_{[d]-S}$.

Further ring-theoretic aspects of CM posets are studied in [$Ba_{5,6}$], [BG], [$DEP_{1,2}$], [$Ga_{1,2}$], [$Ho_2$], and [$St_{6,8,10}$].

REFERENCES

[$Ba_1$] K. Baclawski (1975) Whitney numbers of geometric lattices, *Advances in Math* 16, 125-138.

[$Ba_2$] K. Baclawski (1976) *Homology and combinatorics of ordered sets*, Ph.D. Thesis, Harvard University.

[$Ba_3$] K. Baclawski (1977) Galois connections and the Leray spectral sequence, *Advances in Math.* 25, 191-215.

[$Ba_4$] K. Baclawski (1980) Cohen-Macaulay ordered sets, *J. Algebra* 63, 226-258.

[$Ba_5$] K. Baclawski (1981) Rings with lexicographic straightening law, *Advances in Math.* 39, 185-213.

[$Ba_6$] K. Baclawski (to appear) Cohen-Macaulay connectivity and geometric lattices, *European J. Combinatorics*.

[$Ba_7$] K. Baclawski (to appear) Canonical modules of partially ordered sets.

[BG] K. Baclawski and A. M. Garsia (1981) Combinatorial decompositions of a class of rings, *Advances in Math.* 39, 155-184.

[Bi] G. Birkhoff (1967) *Lattice Theory (3rd. ed.)*, Amer. Math. Soc. Colloq. Publ. No. 25, Amer. Math. Soc., Providence,RI.

[$Bj_1$] A. Björner (1980) Shellable and Cohen-Macaulay partially ordered sets, *Trans. Amer. Math. Soc.* 260, 159-183.

[$Bj_2$] A. Björner (1981) Homotopy type of posets and lattice complementation, *J. Combinatorial Theory* A 30, 90-100.

[$Bj_3$] A. Björner (to appear) On the homology of geometric lattices, *Algebra Universalis*.

[Bj$_4$] A. Björner (to appear) Shellability of buildings and homology representations of finite groups with BN-pair.

[BGS] A. Björner, A. M. Garsia and R. P. Stanley (to appear) *Cohen-Macaulay partially ordered sets.*

[BS] A. Björner and R. P. Stanley (to appear) The number of faces of a Cohen-Macaulay complex.

[BW$_1$] A. Björner and M. Wachs (to appear) Bruhat order of Coxeter groups and shellability, *Advances in Math.*

[BW$_2$] A. Björner and M. Wachs (to appear) On lexicographically shellable posets.

[BWr] A. Björner and J. W. Walker (to appear) A homotopy complementation formula for partially ordered sets, *European J. Combinatorics.*

[Bo] N. Bourbaki (1968) *Groupes et algèbres de Lie*, Éléments de Mathématique, Fasc. XXXIV, Hermann, Paris.

[DEP$_1$] C. DeConcini, D. Eisenbud and C. Procesi (1980) Young diagrams and determinantal varieties, *Invent. Math.* 56, 129-165.

[DEP$_2$] C. DeConcini, D. Eisenbud and C. Procesi (to appear) Hodge algebras.

[DL] C. DeConcini and V. Lakshmibai (to appear) Arithmetic Cohen-Macaulayness and arithmetic normality for Schubert varieties, *Amer. J. Math.*

[DP] C. DeConcini and C. Procesi (to appear) Hodge algebras: a survey, in *Proceedings of the Conference on Schur Functors* 1980, Torun, Poland.

[De] V. V. Deodhar (1977) Some characterizations of Bruhat ordering on a Coxeter group and determination of the relative Möbius function, *Inv. Math.* 39, 187-198.

[DKR] J. Désarménien, J. P. S. Kung and G.-C. Rota (1978) Invariant theory, Young bitableaux, and combinatorics *Advances in Math.* 27, 63-92.

[DRS] P. Doubilet, G.-C. Rota and J. Stein (1974) On the foundations of combinatorial theory IX: Combinatorial methods in invariant theory, *Studies in Appl. Math.* 8, 185-216.

[$Ed_1$] P. H. Edelman (1980) *The zeta polynomial of a partially ordered set*, Ph.D. thesis, Massachusetts Institute of Technology.

[$Ed_2$] P. H. Edelman (1981) The Bruhat order of the symmetric group is lexicographically shellable, *Proc. Amer. Math. Soc.* 82, 355-358.

[Ei] D. Eisenbud (1980) Introduction to algebras with straightening laws, in *Ring theory and algebra III* (Proc. Third Oklahoma Conference) (B. R. McDonald, ed.), Dekker, New York, 243-268.

[Fa] F. D. Farmer (1979) Cellular homology for posets, *Math Japonica* 23, 607-613.

[Fo] J. Folkman (1966) The homology groups of a lattice, *J. Math. Mech.* 15, 631-636.

[$Ga_1$] A. M. Garsia (1979) Méthodes combinatoires dans la théorie des anneaux de Cohen-Macaulay, *C. R. Acad. Sci. Paris. Sér.* A 288, 371-374.

[$Ga_2$] A. M. Garsia (1980) Combinatorial methods in the theory of Cohen-Macaulay rings, *Advances in Math.* 38, 229-266.

[GS] A. M. Garsia and D. Stanton (to appear) Group actions on Stanley-Reisner rings and invariant theory, *Advances in Math.*

[Ha] P. Hall (1936) The Eulerian functions of a group, *Quart. J. Math.* 7, 134-151.

[$Hoc_1$] M. Hochster (1972) Rings of invariants of tori, Cohen-Macaulay rings generated by monomials, and polytopes, *Annals of Math.* 96, 318-337.

[$Hoc_2$] M. Hochster (1977) Cohen-Macaulay rings, combinatorics, and simplicial complexes, in *Ring Theory II* (Proc. Second Oklahoma Conference) (B. R. McDonald and R. Morris, ed.), Dekker, New York, 171-223.

[Hod] W. V. D. Hodge (1943) Some enumerative results in the theory of forms, *Proc. Camb. Phil. Soc.* 39, 22-30.

[HP] W. V. D. Hodge and D. Pedoe (1952) *Methods of Algebraic Geometry, Vol. II*, Cambridge Univ. Press (reprinted 1968), London.

[KK] B. Kind and P. Kleinschmidt (1979) Schälbare Cohen-Macaulay-Komplexe und ihre Parametrisierung, *Math. Z.* 167, 173-179.

[Kn] D. E. Knuth (1970) A note on solid partitions, *Math. Comp.* 24, 955-967.

[La] H. Lakser (1971) The homology of a lattice, *Discrete Math.* 1, 187-192.

[LMS] V. Lakshmibai, C. Musili and C. S. Seshadri (1979) Geometry of G/P, *Bull. Amer. Math. Soc.* 1, 432-435.

[Mac] F. S. Macaulay (1916) *The Algebraic Theory of Modular Systems*, Cambridge Tracts in Mathematics and Mathematical Physics, No. 19, Cambridge Univ. Press, London.

[MM] P. A. MacMahon (1915,1916) *Combinatory Analysis, Vols. 1-2*, Cambridge Univ. Press, London (reprinted by Chelsea, New York, 1960).

[Mat] J. Mather (1966) Invariance of the homology of a lattice, *Proc. Amer. Math. Soc.* 17, 1120-1124.

[Mu] J. Munkres (1976) Topological results in combinatorics, Preprint, MIT, Cambridge, Mass.

[Pr] R. A. Proctor (to appear) Classical Bruhat orders and lexicographic shellability.

[Qu] D. Quillen (1978) Homotopy properties of the poset of non-trivial p-subgroups of a group, *Advances in Math.* 28, 101-128.

[Re] G. Reisner (1976) Cohen-Macaulay quotients of polynomial rings, *Advances in Math.* 21, 30-49.

[Ro] G.-C. Rota (1964) On the foundations of combinatorial theory: I. Theory of Möbius functions, *Z. Wahrscheinlichkeitstheorie und Verw. Gebiete* 2, 340-368.

[$So_1$] L. Solomon (1966) The order of the finite Chevalley groups. *J. Algebra* 3, 376-393.

[$So_2$] L. Solomon (1968) A decomposition of the group algebra of a finite Coxeter group, *J. Algebra* 9, 220-239.

[$St_1$] R. P. Stanley (1972) Ordered structures and partitions, *Mem. Amer. Math. Soc.* 119.

[$St_2$] R. P. Stanley (1972) Supersolvable lattices, *Algebra Universalis* 2, 197-217.

[$St_3$] R. P. Stanley (1974) Finite lattices and Jordan-Hölder sets, *Algebra Universalis* 4, 361-371.

[$St_4$] R. P. Stanley (1974) Combinatorial reciprocity theorems, *Advances in Math.* 14, 194-253.

[$St_5$] R. P. Stanley (1975) Cohen-Macaulay rings and constructible polytopes, *Bull. Amer. Math. Soc.* 81, 133-135.

[$St_6$] R. P. Stanley (1977) Cohen-Macaulay complexes, in *Higher Combinatorics* (M. Aigner, ed.), Reidel, Dordrecht.

[$St_7$] R. P. Stanley (1978) Hilbert functions of graded algebras, *Advances in Math.* 28, 57-83.

[$St_8$] R. P. Stanley (1979) Balanced Cohen-Macaulay complexes, *Trans. Amer. Math. Soc.* 249, 139-157.

[$St_9$] R. P. Stanley (to appear) Some aspects of groups acting on finite posets, *J. Combinatorial Theory* A.

[$St_{10}$] R. P. Stanley (in preparation) Some interactions between commutative algebra and combinatorics, Report, Dept. of Math., Univ. of Stockholm, Stockholm, Sweden.

[Ste] R. Steinberg (1951) A geometric approach to the representation of the full linear group over a Galois field, *Trans. Amer. Math. Soc.* 71, 274-282.

[Wa] M. Wachs (to appear) On the relationship between shellable and Cohen-Macaulay posets.

[Wr] J. W. Walker (1981) *Topology and combinatorics of ordered sets*, Ph.D. Thesis, Massachusetts Institute of Technology.

## DIAGRAM OF IMPLICATIONS

The relationships among most of the poset properties discussed in this paper are summarized by the diagram below. While we have included all implications of which we are aware, we have not indicated which reverse implications are open or are known to be false. For instance, it is unknown whether CL-shellable implies EL-shellable, and whether shellable implies CL-shellable. On the other hand, it follows from a recent deep result of R. Edwards that spheres (i.e., the poset of faces of a triangulation of a sphere, with a $\hat{1}$ adjoined) need not be homotopy CM.

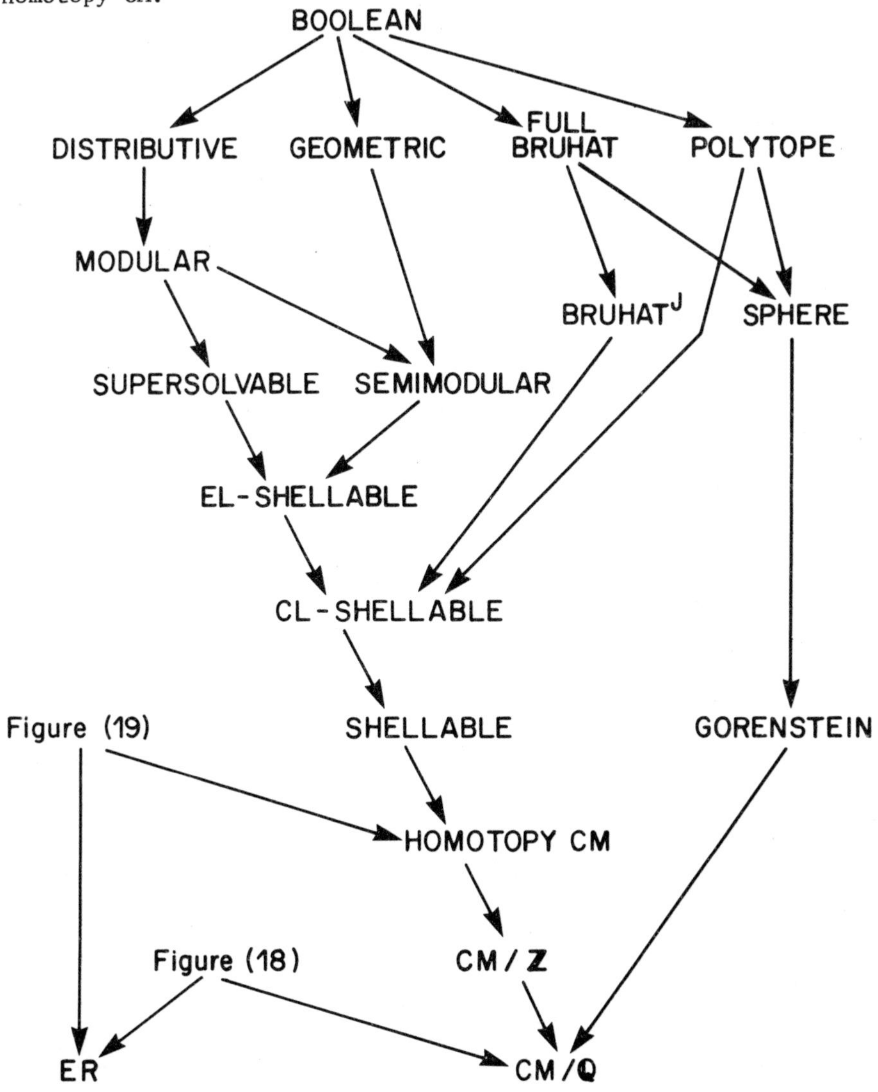

# PART VI

# APPLICATIONS OF ORDERED SETS TO COMPUTER SCIENCES

# ORDERED SETS AND LINEAR PROGRAMMING

A.J. Hoffman
Mathematical Sciences Department
Research Division
International Business Machines Corporation
Box 218
Yorktown Heights, New York 10598

## ABSTRACT

Our themes are: the use of linear programming ideas to prove theorems about ordered sets, and the use of ideas from ordered sets to prove theorems in linear programming.

In the first group, we will discuss Dilworth's theorem (for which we will present two linear programming demonstrations, one using the notion of a totally unimodular matrix [5], the other using nothing [22]) and the generalizations of it by Greene and Kleitman ([16],[17]), which we generalize still further [24]. An amusing sidelight is that their results on $k$- and $k$+1-saturation are derived in the linear programming context as properties of parametric linear programs, and are not needed (as they are in G-K) to prove the main theorems.

It is proper also to think of the max flow min cut theorem as a theorem on ordered sets, if one returns to the original Ford-Fulkerson paper[[11]. We shall present a generalization [20] which emphasizes that connection.

Analogous to the interplay between chains and antichains in Dilworth's theorem is the relation between paths and cuts in the max flow min cut theorem. This leads us to consider the cut-packing theorem, for which we also offer a linear programming proof.

We turn to the second group of topics. Here our principal tool is the concept of lattice polyhedron [25], whose defining rows are indexed by elements of a lattice. These polyhedra have the property that all their vertices are integral and also that certain "dual" problems have, among their optimal vertices, at



*I. Rival (ed.), Ordered Sets, 619–654.*

least one which is integral. We show that many theorems asserting integrality of linear programming problems are subsumed by this idea, including some of the theorems discussed in the first group, and we give details about the "most general" lattice polyhedra [19].

Finally, we use the concept to give a characterization of the lattice of cuts of the paths considered in the max flow min cut theorem [21].

## 1. Introduction.

Our themes are: the use of linear programming ideas to prove theorems about partially ordered sets, especially Dilworth's theorem [6] and related results such as its generalizations by Greene [16], and Greene and Kleitman [17]; the use of concepts from partially ordered sets to prove some combinatorial theorems in a linear programming context, focussing principally on "lattice polyhedra" [25], [21], [19]. The moral in both discussions is that linear programming provides an appropriate machine for studying problems where one asks if the maximum of one kind of thing is the minimum of some other kind of thing. The ingenuity required to construct ad hoc mechanisms to make the proof go is replaced by the less demanding task of finding the right fuel to make the l.p. machine go. Also, l.p. provides a technology for actually calculating the quantities we will be discussing.

## 2. Basic Theorems of Linear Programming.

We summarize for easy citation the few easy results of linear programming we need. (See [29] or any standard text).

(2.1) A polyhedron $Q$ has at least one vertex if and only if $Q$ contains no line.

(2.2) If a polyhedron $Q$ has at least one vertex, then $Q = K + C$, where $K$ is the convex hull of the vertices of $Q$ and $C$ is the cone of all vectors $y$ such that, for each $x \epsilon Q$ and each $\lambda \geq 0$, $x + \lambda y \epsilon Q$.

(2.3) A concave (convex) function bounded from below (above) on a polyhedron $Q$ with at least one vertex has a minimum (maximum) on $Q$, which is attained at a vertex of $Q$. (Note that a linear function is both concave and convex.)

(2.4) If $x^0 \epsilon Q$, and

(a) if $Q \equiv \{x \mid Ax \leqq b\}$, then $x^0$ is a vertex of $Q$ if and only if the submatrix of $A$ formed by $T(x^0) \equiv \{i \mid (Ax^0)_i = b_i\}$ has rank the number of columns of $A$,

(b) if $Q \equiv \{x \mid Ax \leqq b, x \geqq 0\}$, then $x^0$ is a vertex of $Q$ if and only if the submatrix of $A$ formed by rows in $T(x^0)$ and columns in $\text{Supp}(x^0) \equiv \{j \mid x_j^0 > 0\}$ has rank $|\text{Supp}(x^0)|$,

(c) if $Q \equiv \{x \mid Ax = b, x \geq 0\}$, then $x^0$ is a vertex of $Q$ if and only if the submatrix of $A$ formed by $\text{Supp}(x^0)$ has rank $|\text{Supp}(x^0)|$.

Further, in each of (a), (b), (c) the coordinates of $x^0$ (in (b) and (c), the nonzero coordinates of $x^0$) can be calculated by solving the appropriate system of equations.

(2.5) If $c$ is a vector such that $\max (c,x) \mid x \epsilon Q \equiv \{x \mid Ax \leqq b, x \geqq 0\}$ occurs at $x^0$, then $\max (c,x) \mid x \epsilon Q$, $\text{Supp}(x) \subset \text{Supp}(x^0)$ is $(c,x^0)$.

(2.6) The function $\max (c,x) \mid x \epsilon Q$ is a concave function of $b$ (of (2.5)) for the $Q$ specified in each of (2.4) (a), (b), (c).

(2.7) If $A$ is totally unimodular (i.e., every nonsingular square submatrix of $A$ of every order has determinant $\pm 1$), and if $b$ is integral, then every vertex of each $Q$ specified in (2.4) (a), (b), (c) is integral. (Use the last remark in (2.4) and Cramer's rule.)

(2.8) If $A$ has $m$ rows and $n$ columns, and has rank $m$, if $x^0$ is a vertex of $Q \equiv \{x \mid Ax = b, x \geqq 0\}$, with $|\text{Supp}(x^0)| = m$, if $B$ is the (nonsingular) submatrix of $A$ formed by columns in $\text{Supp}(x^0)$, then $(c,x^0) = \max (c,x) \mid x \epsilon Q$ if and only if

$$d' B^{-1} A_j \geqq c_j \quad \text{for all columns } A_j \text{ of } A,$$

where $d$ is the $m$ vector formed by the coordinate of $c$ corresponding to $\text{Supp}(x^0)$. (This is the well-known optimality criterion in the simplex method for solving linear programming problems.)

(2.9) If $Q$ has at least one vertex, then every vertex of $Q$ is integral if and only if for every integral vector $c$ such that $(c,x)$ is bounded from above on $Q$, $\max (c,x) \mid x \epsilon Q$ is an integer.

(Since (2.9) is not yet in textbooks, although pictorially obvious, here is a proof. Assume $Q$ given in the most general form - (2.4)(a) - and $x^0$ a vertex of $Q$, to be proved integral. By (2.4)(a), there is a nonsingular matrix (say formed by the first $n$ rows of the $m$ by $n$ matrix $A$) such that

$$\sum_j a_{ij} x_j^0 = b_i, \quad i = 1,\dots,n$$
$$\sum_j a_{ij} x_j^0 \leqq b_i, \quad i = n+1,\dots,m$$

If $0 < \lambda \epsilon R^n$, then it is clear from the definition of $Q$ and (2.4) that the linear function $\sum_j \left( \sum_{i=1}^{n} \lambda_i a_{ij} \right) x_j$ is bounded from above on $Q$ by $\sum_{i=1}^{n} \lambda_i b_i$ and attains this bound on $Q$ uniquely at $x^0$. If we can find a $\lambda$ so that, for each $j$, $\sum_i \lambda_i a_{ij}$ is an integer $d_j$, and for each $k = 1,\dots,n$ find a $\lambda^k$ so that $\sum_i \lambda_i^k a_{ij} = d_j$ for $j \neq k$ and $\sum_i \lambda_i^k a_{ik} = d_k + 1$, then it will follow from the hypothesis of (2.9) that every coordinate of $x^0$ will be an integer. To accomplish this, let $B$ be the submatrix of $A$ formed by its first $n$ rows, and let

$$f = \max_{\substack{z \mid |z_k| \leq 1 \\ \text{for all } k}} \ \max_j \ |z' B^{-1}|_j .$$

Let $e \epsilon R^n$ have all coordinates 1. If $g > f$, then using the greatest integer function [ ] for vectors, we have

$$[ge'B] = ge'B - z', \quad \text{where } 0 \leqq z < e, \text{ so}$$
$$[ge'B] = ge'B - z'B^{-1}B = (ge' - z'B^{-1})B = \lambda' B, \quad \text{with } \lambda > 0.$$

A similar argument shows that, if $e^k$ has all coordinates 0 except for the $k^{th}$, which is

1, then

$$[ge'B] + (e^k)' = ge'B + z', \quad \text{where} \quad -e \leqq z \leqq e,$$

and so

$$[ge'B] + (e^k)' = (\lambda^k)'B, \quad \text{where} \quad \lambda^k > 0).$$

(2.10) Duality Theorem. There is a version of the duality theorem for polyhedra in each of the forms (2.4) (a), (b), (c); indeed, there are versions general enough to include each of (a), (b) (c) as special cases, but we give here the specific cases we need.

Consider the pair of linear programs (one of which is called *primal,* the other *dual,* with no convention on which is which): Given

$$Q = \left\{ x \,\middle|\, \begin{matrix} Ax \leqq b \\ x \geqq 0 \end{matrix} \right\}, \quad Q^* = \left\{ y \,\middle|\, \begin{matrix} y'A \geqq c' \\ y \geqq 0 \end{matrix} \right\}$$

Problem I: $\max\ (c,x) \mid x \epsilon Q$

Problem II: $\min\ (b,y) \mid y \epsilon Q^*$.

Then

(a) if $Q \neq \phi$, and $(c,x)$ is bounded from above on $Q$, then $Q^* \neq \phi$.

(b) if $Q^* \neq \phi$, and $(b,y)$ is bounded from below on $Q^*$, then $Q \neq \phi$.

(c) if $x \epsilon Q$ and $y \epsilon Q^*$, then $(c,x) \leqq (b,y)$

(d) if $Q \neq \phi$ and $Q^* \neq \phi$, then

$$\max\ (c,x) \mid x \epsilon Q = \min\ (b,y) \mid y \epsilon Q^*.$$

(2.11) Duality Theorem. Let

$$Q = \{x \mid Ax = b, x \geqq 0\}, \quad Q^* = \{y \mid y'A \geqq c'\}.$$

Problem I: $\max\ (c,x) \mid x \epsilon Q$

Problem II: $\min\ (b,y) \mid y \epsilon Q^*$.

Then (a) - (d), cited in (2.10), hold.

## 3. Dilworth's Theorem [6]

The smallest cardinality of a set of chains whose union contains all elements of a partially ordered set $P$ is the largest cardinality of an anti-chain of $P$.

The first [5] l.p. proof of Dilworth's theorem was inspired by an l.p. formulation [4] of the following problem. Given a set $L$ of loaded journeys with specified origin and destination dates to be undertaken by tankers (this problem goes back to the days when all tankers were the same size), what is the smallest number of tankers needed to accomplish the schedule? Recognizing that $L$ is a partially ordered set - journey 1 precedes journey 2 if a tanker can unload at the destination of journey 1 and still have time to reach the loading point of journey 2 by the specified date - and that a chain in $L$ corresponds to a possible sequence of journeys for a single tanker, it follows that the solution of the tanker problem is the smallest cardinality of a set of chains whose union contains all elements of $L$.

By employing the partially ordered set aspect of the tanker problem, one is led by [3] to consider the following linear programming problem: assume a partially ordered set $P$ has $n$ elements, $P = \{1,2,\ldots n\}$, with ordering designated by "$<$". Define an $(n+1)$ by $(n+1)$ matrix $C = (c_{ij})$ by $c_{00} = 1$, $c_{0j} = 0$, $j = 1,\ldots,n$, $c_{i0} = 0, i = 1,\ldots,n$ and

$$c_{ij} = \begin{Bmatrix} 0 & \text{if} & i<j \\ -\infty & \text{if} & i<j \end{Bmatrix} \text{ for } i,j>0 \tag{3.1}$$

in particular $c_{ii} = -\infty$ for $i>0$.

Maximize $\Sigma c_{ij}x_{ij}$, where

$$(3.2) \qquad \begin{aligned} X = (x_{ij}) \geqq 0, \ \sum_{j=0}^{n} x_{0j} = \sum_{i=0}^{n} x_{i0} = n, \\ \sum_{j=0}^{n} x_{ij} = 1 \ \text{ if } \ i>0, \sum_{i=0}^{n} x_{ij} = 1 \ \text{ if } \ j>0. \end{aligned}$$

The conditions (3.2) define a polyhedron in $R^{(n+1)^2}$, every vertex of which is integral. This is a special case of the fact that, if $\sum_{i=1}^{m} a_i = \sum_{j=1}^{n} b_j$, where each $a_i$ and $b_j$ is a nonnegative integer, then every vertex of

$$(3.3) \qquad \left\{X = (x_{ij} \mid x_{ij} \geqq 0, \sum_j x_{ij} = a_i \text{ for all } i, \sum_i x_{ij} = b_j \text{ for all } j\right\}$$

is integral, a fact that is well known to linear programmers in the language: "the transportation problem has all its vertices integral". To prove this fact, one can (working with square matrices of order $\Sigma a_i = \Sigma b_j$) use that it follows from Birkhoff's theorem [2] that the set of doubly stochastic matrices is the convex hull of the permutation matrices, or one can invoke the results of §2. For the conditions (3.3) define a polyhedron of the type (2.4)(c), the relevant matrix is totally unimodular [23] and (2.7) applies.

Now it is easy to see that every vertex of (3.2) has all coordinates 0 or 1, (except for $x_{00}$) and $x_{ij} = 1 \Rightarrow c_{ij} \neq -\infty$. Any such $X$ defines a decomposition of $P$ into chains and conversely. Instead of a formal proof, consider the following Hasse diagram of a partially ordered set $P$ on nine elements.

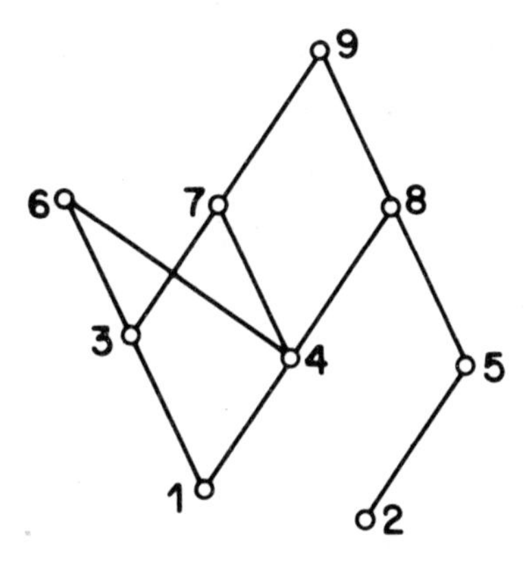


$C$ is the matrix

| | 0 | 1 | 2 | 3 | 4 | 5 | 6 | 7 | 8 | 9 |
|---|---|---|---|---|---|---|---|---|---|---|
| 0 | 1 | 0 | 0 | 0 | 0 | 0 | 0 | 0 | 0 | 0 |
| 1 | 0 | $-\infty$ | $-\infty$ | 0 | 0 | $-\infty$ | 0 | 0 | 0 | 0 |
| 2 | 0 | $-\infty$ | $-\infty$ | $-\infty$ | $-\infty$ | 0 | $-\infty$ | $-\infty$ | 0 | 0 |
| 3 | 0 | $-\infty$ | $-\infty$ | $-\infty$ | $-\infty$ | $-\infty$ | 0 | 0 | $-\infty$ | 0 |
| 4 | 0 | $-\infty$ | $-\infty$ | $-\infty$ | $-\infty$ | $-\infty$ | 0 | 0 | 0 | 0 |
| 5 | 0 | $-\infty$ | $-\infty$ | $-\infty$ | $-\infty$ | $-\infty$ | $-\infty$ | $-\infty$ | 0 | 0 |
| 6 | 0 | $-\infty$ | $-\infty$ | $-\infty$ | $-\infty$ | $-\infty$ | $-\infty$ | $-\infty$ | $-\infty$ | $-\infty$ |
| 7 | 0 | $-\infty$ | $-\infty$ | $-\infty$ | $-\infty$ | $-\infty$ | $-\infty$ | $-\infty$ | $-\infty$ | 0 |
| 8 | 0 | $-\infty$ | $-\infty$ | $-\infty$ | $-\infty$ | $-\infty$ | $-\infty$ | $-\infty$ | $-\infty$ | 0 |
| 9 | 0 | $-\infty$ | $-\infty$ | $-\infty$ | $-\infty$ | $-\infty$ | $-\infty$ | $-\infty$ | $-\infty$ | $-\infty$ |

The chains (1,3,6), (4,7), (2,5,8,9) correspond to the matrix $X$

| | 0 | 1 | 2 | 3 | 4 | 5 | 6 | 7 | 8 | 9 |
|---|---|---|---|---|---|---|---|---|---|---|
| 0 | 6 | 1 | 1 | 0 | 1 | 0 | 0 | 0 | 0 | 0 |
| 1 | 0 | 0 | 0 | 1 | 0 | 0 | 0 | 0 | 0 | 0 |
| 2 | 0 | 0 | 0 | 0 | 0 | 1 | 0 | 0 | 0 | 0 |
| 3 | 0 | 0 | 0 | 0 | 0 | 0 | 1 | 0 | 0 | 0 |
| 4 | 0 | 0 | 0 | 0 | 0 | 0 | 0 | 1 | 0 | 0 |
| 5 | 0 | 0 | 0 | 0 | 0 | 0 | 0 | 0 | 1 | 0 |
| 6 | 1 | 0 | 0 | 0 | 0 | 0 | 0 | 0 | 0 | 0 |
| 7 | 1 | 0 | 0 | 0 | 0 | 0 | 0 | 0 | 0 | 0 |
| 8 | 0 | 0 | 0 | 0 | 0 | 0 | 0 | 0 | 0 | 1 |
| 9 | 1 | 0 | 0 | 0 | 0 | 0 | 0 | 0 | 0 | 0 |

Note that the 1's in the $0^{\text{th}}$ row corresponding to $j>0$ tell us the smallest elements of the 3 chains, the 1's in the $0^{\text{th}}$ column give us the largest elements of the 3 chains. Conversely, given such an $X$, we could easily discover the chains. That $x_{00} = 6$ means that we have found 3 chains and $n-3 = 6$. From (2.3) and the above, we see that the maximum of $\Sigma\, c_{ij}x_{ij}$, subject to (3.2), is $n-p$, where $p$ is the smallest cardinality of a set of chains whose union contains all elements of $P$. By (2.11), this number is minimum $n(u_0 + v_0) + \sum_1^n u_i + \sum_1^n v_j$, $u_i + v_k \geq c_{ij}$ for all $i_{ij}$. Further, by (2.3) and (2.7), we may assume that all $u$'s and $v$'s are integral. Additional manipulation allows us to assume that $v_0 = 0, u_0 = 1, u_i = 0$ or 1 for $i>0, v_j = 0$ or $-1$ for $j>0$. Properly interpreting these numbers leads to the conclusion that the solution to the minimum problem is $n-q$, where $q$ is the largest cardinality of an anti-chain. From (2.11), $p = q$, which is Dilworth's theorem.

Here is another l.p. proof of Dilworth's theorem [22]. Let $A$ be a (0,1) matrix with rows corresponding to chains of $P$, columns corresponding to elements of $P$, with $A_{Cj} = \begin{Bmatrix} 1 & \text{if} & j \in C \\ 0 & \text{if} & j \notin C \end{Bmatrix}$. We shall prove

(3.4) every vertex of

$$Q \equiv \{x \mid Ax \leqq \vec{1}, x \geqq 0\}$$

is integral, and

(3.5) [28] if $A$ is a(0,1) matrix such that (3.4) holds, then for every nonnegative integral $w$, the programming problem minimize $(\vec{1},y) \mid y'A \geqq w', y \geqq 0$ has an optimum vector $y$ which is integral.

Before proving (3.4) and (3.5), let us see how they prove Dilworth's theorem. Consider first the programming problem

$$\max\ (\vec{1},x) \mid x \epsilon Q. \tag{3.6}$$

By (2.3), the maximum occurs for some (0,1) vector $x^0 \epsilon Q$. By the definition of $A$, $\{j \mid x_j^0 = 1\}$ is an anti-chain of $P$. On the other hand if $U$ is an anti-chain of $P$, the (0,1) vector $x$ defined by $x_j = 1$ if and only if $j \epsilon U$ is in $Q$. From these remarks, it follows that (3.6) is the largest cardinality of an anti-chain of $P$. By (2.10) and (3.5), this number equals the smallest cardinality of a set of chains whose union contains all elements of $P$.

We first prove (3.5), by induction on $\Sigma\, w_j$. If $\Sigma\, w_j = 0$, the theorem is obvious, so assume $\Sigma\, w_j = k$, some integer, and the theorem has been proved for all nonnegative integer vectors $w$ such that $\Sigma\, w_j < k$. Consider the problem dual to the one described in (3.5). By (2.3), its value is some integer, say $q$, so by (2.10) the linear programming problem of (3.5) has value $q$. Let $\bar{y}$ be an optimum solution. If not integral, assume (say) that $\bar{y}_1 = r + \theta$., where $r \geq 0$ and integral, $0<\theta<1$. Let $A'_1$ be the first row of $A$. Then $\hat{y} = (r, y_2, \ldots)$ satisfies $\hat{y}'A \geqq w' - \theta A_1' \geqq w' - A_1'$. Since $\hat{y}$ and $A$ are nonnegative, this implies, using the positive part function $(\ )_+$ for vectors, that

$$\hat{y}'A \geqq (w' - A_1')_+. \tag{3.7}$$

Now $\sum_j (w'-A_1')_{+j} < k$ unless $A_{1j} > 0 \Rightarrow w_j = 0$. But if this were so $\bar{y}_1$ has no business being positive! So we can invoke the induction hypothesis to show there exists an integral $z$ minimizing

$$\sum y_i \mid y \geqq 0, y'A \geqq (w'-A')_+.$$

Further, by (3.7) $\Sigma z_i \leqq \Sigma \hat{y}_i < q$, so $\Sigma z_i \leqq q-1$. But then the integral vector

$$\hat{z} = (z_1 + 1, z_2, \ldots) \text{ satisfies}$$

$\hat{z}'A \geqq w', \hat{z} \geqq 0, \Sigma \hat{z}_i \leqq q$. Therefore $\Sigma \hat{z}_i = q$. This proves (3.5).

We will prove (3.4) by induction on $|P|$. Let us first note that we may assume

(3.8) each element of $P$ is comparable to at least one other element of $P$, by the induction hypothesis. Second, we may assume that if $x^0$ is a vertex of $Q$,

(3.9) every $x_j^0 > 0$, else by (2.4)(c) the induction hypothesis applies. Third, we may assume

(3.10) every $x_j^0 < 1$,

otherwise (3.8) contradicts (3.9).

Let $T(x^0) = \{\text{chains } C \mid \sum_{j \epsilon c} x_j^0 = 1\}$. By (2.4)(c), $x^0$ is the unique solution to

(3.11) for all $C \epsilon T(x^0)$, $\sum_{j \epsilon c} x_j = 1$.

Let $S = \{j \epsilon P \mid j \text{ is minimal in some } C \epsilon T(x^0)\}$. Then

(3.12) $j \epsilon S$ implies $j$ minimal in $P$.

Otherwise, assume $k < j$, $j$ minimal in $C_1 \epsilon T(x^0)$. Then $x_k^0$ positive implies, using (3.9), that, on chain $C_2$ consisting of $k \cup C_1$,

$$\sum_{j \in C_2} x_j^0 = x_k^0 + \sum_{j \in C_1} x_j^0 = x_k^0 + 1 > 1,$$

violating $x^0 \in Q$. Then (3.11) and (3.12) imply that $z = (z_1, \ldots)$ defined by

$$z_j = \begin{Bmatrix} 1 & \text{if} & j \in S \\ 0 & \text{if} & j \notin S \end{Bmatrix}$$

satisfies $\sum_{j \in c} x_j = 1$ for all $C \in T(x^0)$. By (3.11) we have a contradiction. Hence every $x_j^0$ is 0 or 1, which is (3.4).

## 4. The Perfect Graph Theorem [28].

Call a graph $G$ perfect if, for every induced subgraph $H \subset G, \chi(H) = \kappa(H)$, where $\chi(H)$ is the chromatic number of $H$, $\kappa(H)$ is the size of the largest clique contained in $H$. For any graph $G$, let $\overline{G}$ to be the complementary graph (same points, the edges of $\overline{G}$ are precisely the non-edges of $G$). Berge introduced the concept of perfect graph. He conjectured [1] and Lovasz proved [28]

(4.1) if $G$ is perfect, so is $\overline{G}$.

Note that, if $G$ is a comparability graph (the underlying undirected graph of a partially ordered set), the fact that $G$ is perfect is a trifle, since cliques of $G$ correspond exactly to chains of $P$. The statement that $\overline{G}$ is perfect is Dilworth's theorem. Hence (4.1) includes Dilworth's theorem. We will prove (4.1) by a variant of the argument in [28].

Let $A$ be the incidence matrix of cliques versus points of the perfect graph $G$; i.e.,

$$A_{Kj} = \begin{Bmatrix} 1 & \text{if point } P_j \in \text{ clique } K \\ 0 & \text{otherwise} \end{Bmatrix}.$$

Then, for this matrix $A$, (3.4) and (3.5), together with (2.10), yield (4.1). The proof of (3.5) was taken from [28], and to prove (3.4) we use from [28] the duplication lemma.

(4.2) if $G$ is a perfect graph on $n$ points, $r = (r_1,\ldots,r_n)$ is a vector with nonnegative integral coordinates, and $G\langle r\rangle$ is the graph formed from $G$ in which, for each $j$, $P_j$ is replaced by a clique on $r_j$ points, then $G\langle r\rangle$ is perfect.

Let $x^0$ be a nonzero vertex of $Q$. Clearly, every coordinate of $x^0$ is rational, so we may write $x_j^0 = \frac{r_j}{s}$, where $s$ is a positive integer, and each $r_j$ is a nonnegative integer. Then it is easy to see that $\kappa(G\langle r\rangle) = s$, and $K_{\langle r\rangle}$ is a clique of $G\langle r\rangle$ of size $s$ if and only if $\sum_j A_{Kj}x_j^0 = 1$ for the corresponding clique $K$ of $c$.

By (4.2), $G\langle r\rangle$ is perfect, so the set of points colored blue in a coloring of the points of $G\langle r\rangle$ in $s$ colors contains exactly one point of every clique of $G\langle r\rangle$ of size $s$. But this means the corresponding set of points of $G$ meets every clique in $T(x^0)$ in exactly one point. By (2.4) (c), this means every $x_j^0$ is 0 or 1.

## 5. Greene-Kleitman's theorem: [17] The t-phenomenon.

Let $P$ be a partially ordered set, $t$ a nonnegative integer. Let $f(t)$ be the largest cardinality of a subset $S \subset P$ such that, for each chain $C \subset P$, $|S \cap C| \leqq t$. For any collection $\mathscr{C}$ of disjoint chains $\{C_i\}$ of $P$ whose union contains all elements, let $g(\mathscr{C},t) = \sum_i \min(t, |C_i|)$. Let $g(t) = \min_{\mathscr{C}} g(\mathscr{C},t)$. Greene and Kleitman proved

(5.1) $g(t) = f(t)$ for every nonnegative integer $t$. Their proof used a lemma interesting by itself:

(5.2) For every integer $t \geqq 0$, there exists a collection $\mathscr{C}$ of disjoint chains whose union is $P$ such that $g(t) = g(\mathscr{C},t)$ and $g(t+1) = g(\mathscr{C},t+1)$. Following [24], we will prove a generalization of (5.1) and (5.2). First, if $C$ and $D$ are chains with at least one common element $x$, define

$$(C,x,D) \equiv \{y \mid y \in C, y < x\} \cup \{x\} \cup \{y \mid y \in D, x < y\}.$$

Next, call a nonnegative integral function $r(C)$ defined on all chains of $P$ a "switch function on chains" if

$$r(C) \leqq r(D) \text{ when } C \text{ is a subchain of } D, \text{ and}$$

$$\text{if } x \in C \cap D, \text{ then } r(C) + r(D) = r(C,x,D) + r(D,x,C).$$

Let $f(r)$ be the largest cardinality of a subset $Q \subset P$ such that $|Q \cap C| \geqq r(C)$ for all chains $C$. Also, for any partitioning of $P$ into a collection of disjoint chains $\mathscr{C} = \{C_i\}$, define $g(\mathscr{C},r) = \sum_i \min(r(C_i), |C_i|)$. Define $g(r) = \min_{\mathscr{C}} g(\mathscr{C},r)$. Then

(5.3) $g(r) = f(r)$.

(5.4) For every switch function $r$ on chains, there exists a collection $\mathscr{C}$ of chains partitioning $P$ such that $g(r) = g(\mathscr{C},r)$ and $g(r+1) = g(\mathscr{C},r+1)$.

The constant function $t$ is clearly a switch function, so (5.3) and (5.4) specialize to (5.1) and (5.2). It is interesting to note that in [17], (5.2) is used to prove (5.1). In our approach (5.3) ⊃ (5.1) is proved first. The study of (5.4) ⊃ (5.2) is reduced to showing that a certain parametric linear programming problem (to be defined later) changes at integer values of the parameter $t$. This "$t$-phenomenon" has been shown to hold elsewhere [16], [18], [12]. To prove (5.3), we first observe that every switch function arises as follows: there is a nonnegative integral function $r(a)$ defined on elements of $P$, and a nonnegative integral function $r(a,b)$ defined on pairs $a$, $b$ where $a<b$. Then, if

$$C = \{a_1 < \dots < a_s\}, r(C) = r(a_1) + r(a_1,a_2) + \dots + r(a_{s-1},a_s).$$

We now define a "transportation problem": minimize $\sum_{i,j=0}^{n} c_{ij}\, x_{ij}$, where

(5.5) $\quad c_{i0} = 0$ for all $i \geqq 0, c_{0j} = r(j)$ for $j>0.$

$$c_{ii} = 1 \text{ for } i>0, c_{ij} = \begin{cases} r(i,j) & \text{if } i<j \\ \infty & \text{if } i \not< j, i \neq j \end{cases}.$$

and $X = (x_{ij})$ satisfies

(5.6) $$x_{ij} \geqq 0 \text{ for all } i \text{ and } j, \sum_j x_{0j} = \sum_i x_{i0} = n,$$

$$\sum_j x_{ij} = 1 \text{ for all } i>0, \sum_i x_{ij} = 1 \text{ for all } j>0.$$

Through divers manipulations we learn that the solution to this problem is $g(r)$, the solution to the dual problem is $f(r)$, and (2.11) shows (5.3). Similar manipulations will establish a recent result [27] on chains in acyclic directed graphs.

To prove (5.4), we first show, using (2.8), that is $A$, $b$, $c$, $d$ are integral and $\begin{bmatrix} A \\ d' \end{bmatrix}$ is totally unimodular, then if $Q \equiv \{x \mid Ax = b, c \geqq 0\}$ and for each $0 \leqq t \leqq 1$, a solution to the parametric linear programming problem min $(c + td,x) \mid x \in Q$ exists, then there is an integral $x^0 \in Q$ such that $(c,x^0) = \min_{x \in Q} (c,x)$ and $(c + d,x^0) = \min_{x \in Q} (c + d,x)$.

This lemma is applied to the current situation by letting $Q$ be the polyhedron specified by (5.6), $c$ the vector given by (5.5), $d$ the vector specified by $d_{01} = \ldots = d_{0n} = 1$, all other $d_{ij}$ are 0. The fact that the appropriate $\begin{bmatrix} A \\ d \end{bmatrix}$ is totally unimodular is in [24].

One can also imitate the second l.p. proof of Dilworth's theorem. Recall the matrix $A$ of (3.4), let $t \geqq 0$ and integral, and let $S \subset P$. Then the argument used in proving (3.4) can be easily adapted to prove

(5.7) every vector of $Q(S,t) \equiv \{x \mid Ax \geqq t\vec{1},\ x \geqq 0,\ x_j \geqq 1 \text{ for } j \in S\}$ is integral. (Note, however, that this statement does not hold if $A$ is the matrix of §4 arising from perfect graphs.)

Next we show

(5.8) For every nonnegative integral $w$, the problem: minimize $t(\vec{1},y) + \sum_{j \in S} z_j \mid y \geqq 0$, $z \geqq 0$, $(y'A + z)_j \geqq w_j$ for $j \in S$, $(y'A) \geqq w_j$, for $j \notin S$ has an optimum $(y;\ z)$ which is integral.

The proof of (5.8) uses the fact that we can apply induction on $\Sigma\, w_j$ to show that every $z_j$ can be assumed integral. The problem then becomes minimize $t(\vec{1}.y) \mid y \geqq 0$, $y'A \geqq v'$, where $v$ is $w$ reduced by the integral vector $\bar{z}$. But the optimum $y$ does not depend on $t$, so (3.5) applies. Now the case $S = P$, $w = \vec{1}$ is (5.1).

To prove (5.2) we use a device from [18]. Recall the function $f(t)$ described at the beginning of this section. Consider the problem

(5.9) maximize $\Sigma\, x_j + \alpha(f(t + 1) - f(t)) \mid Ax + \alpha\vec{1} \geqq (t + 1)\vec{1},\ 0 \leqq x \leqq \vec{1}$. We know from the foregoing that the maximum occurs at an integral $(x;\ \alpha)$, for the column $\vec{1}$ corresponds to a new element in every chain. Consider the problem

(5.10) maximize $\Sigma x_j \mid Ax \geqq (t+1-\alpha)\vec{1},\ 0 \leqq x \leqq \vec{1},\ \alpha \geqq 0,$

and call the maximum $k(\alpha)$. If $f$ is the optimum value in (5.9) then

$$f = \max_{\alpha \geqq 0} \left(k(\alpha) + (f(t+1) - f(t))\alpha\right)$$

Now, by (2.5), $k(\alpha)$ is concave, so $k(\alpha) + (f(t+1)-f(t))\alpha$ is concave. For $\alpha = 0$ and 1, the function is the same at $\alpha = 0$ and $\alpha = 1$, namely $f(t+1)$. The integrality of solutions to (5.9), combined with the last two sentences, shows that $f$ is $f(t+1)$.

By (2.11), ($c$) and ($d$), it is easy to show that, since there is an optimal solution to (5.9) in which $\alpha > 0$, then every solution to the dual problem: minimize $(t+1)(\vec{1},y) + \Sigma z_j \mid y'A + z \geqq \vec{1},\ y \geqq 0,\ z \geqq 0,\ (\vec{1},y) \geqq f(t+1) - f(t)$ satisfies the condition that

(5.11) $\quad (\vec{1},y) = f(t+1) - f(t).$

On the other hand, by (2.10), the minimum in this problem is $f(t+1)$ and is attained at integral $(\bar{y},\bar{z})$. Therefore,

(5.12) $\quad (t+1)(\vec{1},\bar{y}) = f(t+1).$

Combined with (5.11), we get

(5.13) $\quad t(\vec{1},\bar{y}) = f(t),$

Together (5.12) and (5.13) will imply (5.2) if we know $\bar{y}$ has no coordinate larger than 1, which it is obviously possible to arrange.

## 6. Greene's Theorem [16].

If, in (5.1) and (5.2) the word chain is replaced by anti-chain, then the resulting statements are true. For example, (5.1) then becomes: for every nonnegative integer t, the largest cardinality of a set $S$ containing at most t elements of each anti-chain is the smallest value of $\sum_i \min(t, |A_i|)$, where $\{A_i\}$ is a partition of $P$ into disjoint anti-chains.

Here our discussion of Greene's theorem will plagiarize concepts from "superperfect" graphs [15].

Let $B$ be the incidence matrix of anti-chains versus elements of the partially ordered set $P$, so that

$$B_{Aj} = \begin{cases} 1 & \text{if element } j \text{ is in anti-chain } A \\ 0 & \text{if otherwise} \end{cases}$$

Let $S \subset P$, and $t \geqq 0$ and integer. Define $Q(S,t) = \{x \mid x \geqq 0,\ Bx \leqq t\vec{1},\ x_j \leqq 1 \text{ if } j \in S\}$. Let $c \geqq 0$ and integer. Then

(6.1) The linear program

(6.1a) $\quad \max\ (c,x) \mid x \in Q(S,t)$

and the dual program

(6.1b) $\quad \min\ t(\vec{1},y) + \sum_{j \in S} z_j\ y \geqq 0,\ z \geqq 0,\ (y'B)_j \geqq c_j \text{ if } j \notin S,\ (y'B)_j + z_j \geqq c_j \text{ if } j \in S$

each have optimal vectors which are integral.

It is easy to see that (2.11) and (6.1) for the case $S = P$ and $c = \vec{1}$ yield's Greene's theorem. Greene also established the $t$-phenomenon, which we describe in the language

(6.2) For every integer $t \geqq 0$, there is an integer vector $(y;z)$ which is optimum in (6.1b) for $t$ and $t + 1$.

To introduce the notion of superperfect graph (which arose in joint work with Ellis Johnson on scheduling ship construction), we first define $G(w)$, which is a graph to which a vector $w = (w_1, w_2, \ldots)$ of nonnegative numbers has been associated with the points. A coloring of $G(w)$ is a mapping of points to open intervals $f(i) = I_i$, where $i$ and $j$ adjacent implies $I_i \cap I_j = \phi$, and the length of $I_i$ is $w_i$. The measure of a coloring $f$ is

$$m(f) = m(\cup I_i),$$

and the chromatic number of $G(w)$ is

$$\chi(G(w)) = \inf_f m(f).$$

Note that $w = \vec{1}$ yields $\chi(G(\vec{1})) = \chi(G)$.

The expression $\kappa(G(w))$ is

$$\kappa(G(w)) = \max_{\substack{K \text{ a clique} \\ \text{of } G}} \sum_{i \in K} w_i$$

Obviously, $\kappa(G(w)) \leqq \chi(G(w))$, and we call a graph superperfect if, for all $w$, $\kappa(G(w)) = \chi(G(w))$. (Recall from § 4 that if, for all $(0,1)w$, $\kappa(G(w)) = \chi(G(w))$, then $G$ is called perfect.)

It is easy to see that every comparability graph is superperfect, but it is also true that there are superperfect graphs that are not comparability graphs (see [15] for this information and also for the content of the next paragraph). Let $G$ be a graph, and let $B$ be the incidence matrix of independent sets versus points. (When $G$ is a comparability graph, it is the same $B$ as we have above.) Then $G$ is a superperfect graph if and only if, for every $w \geqq 0$, the linear programming problem

(6.3) minimize $(\vec{1}, y) \mid y \geqq 0, y'B \geqq w$

has an optimal vector $y$ such that the submatrix of $B$ formed by rows corresponding to Supp($y$) has the *consecutive* 1's *property* (i.e., the rows of this matrix can be permuted so that, in each column, the 1's occur consecutively). Such a matrix is totally unimodular [23]. Now consider (6.1b) and let $\bar{y};\bar{z}$ be optimal. By considering the problem

$$(6.4) \quad \min\ (\vec{1},y) \mid y \geqq 0,\ y'B \geqq c' - \bar{z}',$$

we see from (6.3) that (6.1b) has an optimum solution $(\hat{y};\hat{z})$ in which the support of $(\hat{y};\hat{z})$ corresponds to a totally unimodular matrix. Using (2.5), this assures us that (6.1b) has an optimum (y ; z) which is integral. On the other hand, this fact and the duality theorem (2.10), applied to (2.9) prove every vertex of $Q(S,t)$ is integral. Using (2.3), we infer (6.1). The proof of (6.2) is the same as the analogous proof occurring at the end of §5.

## 7. Paths and Cuts.

Analogous to the concepts of chain and anti-chain in a partially ordered set are the concepts of path and cut in a directed graph. Let $\vec{G}$ be a directed graph with two distinguished points $a$ and $b$. Let $A$ be the incidence matrix of $a-b$ directed paths in $\vec{G}$ versus edges of $G$, and let $c \geqq 0$ and integral. A vector $y$ such that

$$y'A \leqq c' y \geqq 0$$

is called a flow, and the maximum of $(\vec{1},y)$ under these conditions is called the maximum flow.

An $a-b$ cut $C$ is a collection of edges such that every $a-b$ path meets at least one edge of $C$. Then the max flow min cut theorem [11], [20] asserts that the pair of dual linear programs

$$\max\ (\vec{1},y) \mid y'A \leqq c'$$

and

$$\min\ (c,x) \mid Ax \geqq \vec{1}, x \geqq 0$$

each have optimal vectors which are integral. Note that Supp$(x^0)$, where $x^0$ is integral and optimal, corresponds to an $a-b$ cut.

A generalization implicit in [11] and explicit in [20] may be of interest to specialists in ordered sets. Let $U$ be a set, $\mathscr{P} = \{S_1, S_2, \ldots\}$ a distinguished family of subsets (called paths) each of which is simply ordered. (But it is possible for $p <_{s_1} q$ and $q <_{s_2} p$).

A set of paths is "closed with respect to switching" if whenever $p \in S_1 \cap S_2$, there exists some $S_i \in \mathscr{P}$ (denoted by $(S_1, p, S_2)$ - recall §5) such that

$$(S_1, p, S_2) \subset \{q \mid q <_{s_1} p\} \cup \{p\} \cup \{q \mid p <_{s_2} q\}.$$

A nonnegative integral function $r(S_i)$ defined on the paths is a "switch function on paths" if, whenever $p \in S_1 \cap S_2$,

$$r(S_1, p, S_2)) + r(S_2, p, S_1) \geqq r(S_1) + r(S_2).$$

With these definitions, one can prove that the pair of dual linear programs (with $c \geqq 0$ and integral)

(7.1) $Ax \geqq r, x \geqq 0$, min $(c,x)$,

(7.2) $y'A \leqq c', y \geqq 0$, max $(r,y)$

have respective optimal vectors which are integral [20].

Further

$$(7.3)\qquad Ax \geqq r, 0 \leqq x \leqq \vec{1},\ \min\ (c,x)$$

and

$$(7.4)\qquad y'A - z' \geqq c',\ y \geqq 0,\ \max\ (r,y) - (\vec{1},x)$$

have respective optimal vectors which are integral (generalizing a result of John Suurballe) and a $t$-phenomenon occurs for (7.4). These results are similar to those described in §5, and have similar proofs.

Parallel to the max flow min cut theorem is the cut packing theorem (see [26]) - I do not know its origin.

Let $B$ be the incidence matrix of cuts versus edges; $c \geqq 0$ and integral. Then the pair of dual linear programs

$$(7.5)\qquad Bx \geqq \vec{1}, x \geqq 0,\ \min\ (c,x)$$

and

$$(7.6)\qquad y'B \leqq c', y \geqq 0,\ \max\ (\vec{1},y)$$

have respective optimal vectors which are integral. Note that, in (7.5) an integral optimal $\bar{y}$ will be (0,1), and Supp $(\bar{y})$ will be edges of a shortest $a-b$ directed path, if the numbers in $c$ are taken as lengths of edges.

There are analogues of (7.1) - (7.4), but we defer them to later when we introduce the appropriate abstract setting.

## 8. Lattice Polyhedra: Definitions [26], [25], [19].

In this and the next two sections, we discuss lattice polyhedra and their role as a useful, though not ubiquitous tool, in the enterprise of proving combinatorial theorems in a linear programming environment. Since the latter frequently involves showing that optima occur at integral points, the term lattice polyhedron was concocted because both mathematical meanings of lattice were pertinent.

Following Dilworth [7], a real function $f:L \rightarrow R$ defined on a lattice $L$ is

(8.1) submodular if $f(a \vee b) + f(a \wedge b) \leq f(a) + f(b)$

(8.2) supermodular if $f(a \vee b) + f(a \wedge b) \geq f(a) + f(b)$

(8.3) modular if $f(a \vee b) + f(a \wedge b) = f(a) + f(b)$.

A function $f:L \rightarrow \{0,1,-1\}$ is consecutive if

(8.4) $a<b \Rightarrow f(a)f(b) \geq 0$, and

(8.5) $a<b<c, f(a)f(c)>0 \Rightarrow f(b) = f(a) = b(c)$.

The meaning of (8.5) is that if $f(a)$ and $f(c)$ are both $+1$ (or both $-1$), then $f(b)$ is $1$ $(-1)$.

Let $r$ be integer and submodular, $d \geq 0$ and integer. Let $A$ be a $(0,1,-1)$ matrix, with rows indexed by $L$, every column of which is consecutive and supermodular. Then

(8.6) Every vertex of $Q(r,d) \equiv \{x \mid Ax \leq r, 0 \leq x \leq d\}$ is integral (assuming $Q(r,d) \neq \phi$).

(8.7) If $Q(r,d) \neq \phi$, then for every integer $c$, the programming problem

$$y'A + z' \geq c', y \geq 0, z \geq 0, \text{ minimize } (r,y) + (d,z)$$

has an optimum vector which is integral.

From (2.9), we see that (8.6) will follow from (8.7). To prove (8.7), we first establish, for any partially ordered set $P$, a certain partial order of $R^{|P|}$. To do this, it is sufficient to construct a positive cone $C \subset R^{|P|}$ by

$$(8.8) \qquad C \equiv \{x \epsilon R^{|P|} \mid \text{every maximal element } a \epsilon \text{Supp}(x) \text{ has } x(a) > 0\},$$

and say $x \prec y$ if $y - x \epsilon C$. It is easy to see that every polytope contains maximal elements.

Let $(y^0; z^0)$ be optimal in the linear programming problem (8.7). Consider the polytope

$$(8.9) \qquad y'A \geqq c' - z^{0'}, y \geqq 0, \ (r,y) = (r,y^0), (\vec{1},y) = (\vec{1},y^0),$$

and let $\bar{y}$ be maximal with respect to the positive cone $C$ of (8.8), arising from the partially ordered set $L$. Then

$$(8.10) \qquad \text{Supp}(\bar{y}) \text{ is a chain in } L.$$

For, if (8.10) were false, assume $\bar{y}(a) > 0, \bar{y}(b) > 0$, $a$ and $b$ incomparable in $L$. Then, for small $\varepsilon > 0$, the vector $\hat{y}$ which agrees with $\bar{y}$ on all coordinates except

$$\hat{y}(a \vee b) = \bar{y}(a \vee b) + \varepsilon, \hat{y}(a \wedge b) = \bar{y}(a \wedge b) + \varepsilon,$$
$$\hat{y}(a) = \bar{y}(a) - \varepsilon, \hat{y}(b) = \bar{y}(b) - \varepsilon$$

is nonnegative and satisfies $(\vec{1},\hat{y}) = (\vec{1},y^0)$. By the supermodularity of each column of $A$,

$$\hat{y}'(A) \geqq c' - z^{0\prime}.$$

By the submodularity of $r$,

$$(r,\hat{y}) \leqq (r,y^0),$$

but they must be equal, otherwise $(y^0;z^0)$ did not solve (8.7).

So $\hat{y}$ is in the polytope (8.9), but $\hat{y}-\bar{y} \epsilon C$, a contradiction.

Now (8.10), and the fact that each column of $A$ is consecutive, shows that Supp$(\hat{y})$ corresponds to a submatrix of $A$ every column of which has all entries $+1$ or 0, or all entries $-1$ or 0, and has consecutive 1's or $-1$'s, hence totally unimodular. If we restrict the l.p. of (8.7) to Supp$(\hat{y};z^0)$, then from (2.5) we have a proof of (8.7).

Similarly, we can prove

(8.11) If $r$ is integer and supermodular, the columns of $A$ consecutive and submodular, $d \geqq 0$ and integral, and if $Q \equiv \{x \mid Ax \geqq r,\ 0 \leqq x \leqq d\} \neq \phi$, then every vertex of $Q$ is integral, and for every integer $c$, the l.p. problem

$$y'A - z' \leqq c', y \geqq 0, z \geqq 0, \text{ maximize } (y,r)-(z,d)$$

has an optimum vector which is integral.

Also,

(8.12) If $r$ is integer and submodular, the columns of $A$ modular, $e \leqq d$ integral, and if $Q = \{x \mid Ax \leqq r, e \leqq x \leqq d\} \neq \phi$, then every vertex of $Q$ is integral, and for every integer $c$, the l.p. problem

$$y'A + z' - w' = c', y \geqq 0, z \geqq 0, w \geqq 0, \text{ minimize } (r,y) + (d,z) - (e,w)$$

has an optimum vector which is integral.

We now indicate how to construct consecutive sub- or supermodular functions $f:L \to \{0,\pm 1\}$ from certain basic functions. We say that $g:L \to \{0,1\}$ is upper if $a<b$ implies $g(a) \leqq g(b)$, lower if $a<b$ implies $g(a) \geqq g(b)$. Note that any upper or lower $g$ is consecutive. The following are easy to prove

(8.13) If $g_1$ and $g_2$ are respectively upper and lower, then $f:L \to \{0,\pm 1\}$ defined by $f = g_1 + g_2 - 1$ is consecutive. Further, if both $g_1$ and $g_2$ are respectively sub-, super- or modular, so if $f$.

(8.14) If $f$ is consecutive and submodular, $f_+$ is consecutive and submodular.

Although not needed in the sequel, it is interesting to note that these constructions are fairly general. One can show

(8.15) If $g:L \to \{0,1\}$ is consecutive and nonzero, there exist $g_1$ and $g_2$, respectively upper and lower such that $g = (g_1 + g_2 - 1)_+$ and

(a) $g_1$ and $g_2$ are supermodular if $g$ is supermodular;

(b) $g_1$ and $g_2$ are submodular if $g$ is submodular and $L$ distributive;

(c) $g_1$ and $g_2$ are modular if $g$ is modular and $L$ distributive.

(8.16) If $f:L \to \{0,\pm 1\}$ is consecutive and nonzero, and

(a) if $f$ is supermodular and not nonpositive, there exist $g_1$ and $g_2$ supermodular and respectively upper and lower, with $f = g_1 + g_2 - 1$;

(b) if $f$ is submodular and not nonnegative, there exist $g_1$ and $g_2$ submodular and respectively upper and lower, with $f = g_1 + g_2 - 1$;

(c) if $f$ is modular and $L$ is distributive, there exist $g_1$ and $g_2$ modular and respectively upper and lower such that $f = g_1 + g_2 - 1$.

In addition to (8.13) and (8.14), the following construction is useful. Let $L_1$ and $L_2$ be disjoint lattices, and let $g_1$ and $g_2:L\to\{0,1\}$ be upper and lower supermodular. Define the lattice $L = L_1 \cup L_2$, where the ordering in $L_1$ and $L_2$ is hereditary, and $a<b$ for $a\in L_1, b\in L_2$. Then $f:L\to\{0,1\}$, defined to agree with the $g_i$, is consecutive and supermodular.

## 9. Lattice Polyhedra: Applications [19].

(9.1) Continuing the discussion of the preceding paragraph, let a lattice $L$ be given, and $g:L\to\{0,1\}$ upper and supermodular. Let $\bar{L}$ be the reverse of $L$ (interchanging $\wedge$ and $\vee$), so that $g:\bar{L}\to\{0,1\}$ is lower and supermodular. Then, from the preceding paragraph, letting $L_1 = L$ and $L_2 = \bar{L}$, we have the consecutive supermodular function $f:L\cup\bar{L}\to\{0,1\}$. Let $A$ be a (0,1) matrix indexed by the elements of $L$ so that each column corresponds to an upper supermodular $g$. Then, if $r_1,r_2$ are submodular integral functions on $L$, $d\geqq 0$ and integral then $Q = \{x \mid Ax \leqq \min(r_1,r_2), 0\leqq x\leqq d\}$ is a polyhedron of the type considered in (8.6) and (8.7), and provides alternative proofs of results on intersection of two polymatroids [8], [3],

(9.2) Let $L_0$ be a distributive sublattice (with minimal element $m$) $L$ a sublattice of $L_0, E\subset L_0$. For each $j\in E$, let

$$g_{1j}(a) = 1 \text{ if } j\wedge a = m, 0 \text{ otherwise}$$
$$g_{2j}(a) = 1 \text{ if } j\leqq a, 0 \text{ otherwise.}$$

Then $g_{1j}$ and $g_{2j}$ are respectively lower and upper supermodular. Therefore $\bar{g}_{1j} = 1-g_{1j}$ and $\bar{g}_{2j} = 1-g_{2j}$ are respectively upper and lower submodular. It follows from (8.13) and (8.14) that $f_j = (g_{1j} + g_{2j}-1)_+$ is consecutive and submodular. Hence, if $r$ is integral and supermodular on $L$, $A$ the matrix whose columns are specified by the $\{f_j\}$, rows by elements of $L$, then

$$Q\equiv\{x\in R_+^E \mid Ax\geqq r, x\geqq 0\}$$

is a polyhedron of the type considered in (8.11). In case $L_0$ is the lattice of all subsets of $\{0,1,...,n\}$, $L$ a sublattice of the lattice of all subsets containing 0, $E$ the two-element subsets, we have the cut-set polyhedron [26] for undirected graphs. A special case is the cut-packing theorem cited in §7.

(9.3) Let $G = (V,E)$ be a directed graph, $t(j)$ and $h(j)$ the tail and head of edge $j \in E$. Let $L$ be a lattice of node subsets $S \subset V$, such that $S \neq \phi, V$. Then if $e \leqq d$ are integral, $r$ integral and submodular on $L$,

$$A = (a_{sj}) = \left\{ \begin{array}{rl} 1 & \text{if } t(j) \in S, h(j) \notin S \\ -1 & \text{if } h(j) \in S, t(j) \notin S \\ 0 & \text{otherwise} \end{array} \right\},$$

then

$$Q = \{x \in R^E \mid e \leqq x \leqq d, Ax \leqq r\}$$

can be shown (with some effort) to have the properties described in (8.13), capturing [8] and [12].

(9.4) Let $P$ be a partially ordered set, $L$ the lattice of maximal anti-chains of $L(a \leqq b$ if for each $x \in a$, there is a $y \in b$ such that $x \leqq y)$. For each $j \in P$, let $g_{1j}(a) = 1$ if $j \leqq x$ for at least one $x \in a$, otherwise 0; $g_{2j}(a) = 1$ if $x \leqq j$ for at least one $x \in a$, otherwise 0. Then $g_{1j}$ and $g_{2j}$ are respectively upper and lower supermodular, so by (8.13), $f_j = g_{1j} + g_{2j} - 1$ is consecutive and supermodular.

Further, if for some $a$, $g_{1j}(a)$ and $g_{2j}(a)$ are both 0, then $j \in a$, a contradiction. So $f_j = g_{1j} + g_{2j} - 1$ is a $\{0,1\}$ function which is consecutive and supermodular, and (obviously) $f_j(a) = 1$ if and only if $j \in a$. Therefore, if $A$ is the matrix whose columns are the functions $f_j$, $r$ submodular and integral, $d \geqq 0$ and integral, then for

$$Q \equiv \{x \mid Ax \leqq r, 0 \leqq x \leqq d\},$$

(8.6) and (8.7) hold. If $r$ is constant, $d = \vec{1}$, we recover Greene's theorem of §6.

(9.5) Finally (see [19] for more), let $G = (V,E)$ be a directed graph, and $L$ a lattice of subsets of $V$. Let $A = (a_{sv})$ where $v \in V$ be defined by

$$a_{sv} = \begin{cases} 1 & \text{if } v \in S \text{ but has no successor in } S \\ -1 & \text{if } v \notin S \text{ but has a successor in } S \\ 0 & \text{otherwise} \end{cases},$$

Using (8.13), we see that, if $d \geqq 0$ and integral, r submodular and integral, then (8.6) and (8.7) apply to

$$Q \equiv \{x \mid Ax \leqq r, 0 \leqq x \leqq d\}.$$

From this,one can infer that, if $c: V \to R$ satisfies $\Sigma \{(c(v): v \in C\} \leqq 0$ for every circuit $C$ of $G$, then setting $r = \vec{1}, d = \infty$ in the $Q$ just defined, the problem of maximizing, over all paths $D$ in $G$, $\sum_{v \in D} c(v)$ can be formulated as

$$\max (c,x) \mid x \in Q.$$

## 10. Lattice Clutters. [21]

Assume that $A$ is a (0,1) matrix in which it is false that the columns containing the 1's in one row are a subset of the columns containing the 1's of another row. If we regard the rows as incidence vectors of a family of sets, the family is called a *clutter*.

Suppose now that the rows of $A$ are indexed by a partially ordered set $L$, and suppose further that, for each $a,b \in L$ there exist $a \vee b$ and $a \wedge b$ satisfying

(10.1) $a \wedge b = b \wedge a$; $a \wedge b \leqq a,b$; $a \leqq b \Rightarrow a \wedge b = a$; $a \vee b = b \vee a$; $a,b \leqq a \vee b$; $a \leqq b \to a \vee b = b$.

We do not assume $L$ is a lattice. But we do assume that each column of $A$ is consecutive.

Assume now that, with respect to the defined $\wedge$ and $\vee$, each column of $A$ is submodular. Then the associated family of sets will be called a *submodular* clutter. Similarly, if each column of $A$ is supermodular, then the associated family of sets will be called a *supermodular* clutter. We will describe, in a certain sense, all supermodular and submodular clutters.

To facilitate this description, we need the concepts of blocker and anti-blocker. Let $U$ be a set (we have in mind the set of columns of $A$, so that our clutters are subsets of $U$). If $\mathcal{M}$ is a clutter on $U$, the blocker of $\mathcal{M}$, denoted by $\mathcal{B}(M)$, is the set of all minimal $B \subset U$ such that, for all $M \in \mathcal{M}$, $B \cap M \neq \phi$. Similarly, $\mathcal{A}(\mathcal{M})$ - the anti-blocker of $\mathcal{M}$ - is the set of all maximal $B \subset U$ such that $|B \cap M| \leqq 1$.

For any clutter $\mathcal{M}$ on $U$, $\mathcal{B}(\mathcal{B}(\mathcal{M})) = \mathcal{B}$ [9]. Further, if $\mathcal{M}$ is a supermodular clutter on $U$, then (8.6), (which did not need that $L$ is a lattice, only (10.1)), together with [14], establishes that $\mathcal{A}(\mathcal{A}(\mathcal{M}) = \mathcal{M}$.

Our descriptions of supermodular and submodular clutters will depend on understanding their respective anti-blockers and blockers. We first consider the submodular case. Refer back to §7. Let $U$ be a set and $\mathcal{S}$ a distinguished family of paths closed with respect to switching, as described in §7. We can picture $\mathcal{S}$ as follows. In some directed graph $G$ with points from $U$ and two additional distinguished nodes $a$ and $b$, there are edges from $a$ to each initial element of a path, to $b$ from each terminal element of a path, and from $u_i$ to $u_j$ if there is a path in which $u_i$ immediately precedes $u_j$. In this picture, $S$ is some set of paths closed with respect to switching. For example, omitting $a$ and $b$, in the directed graph (all edges point up)

(10.2)

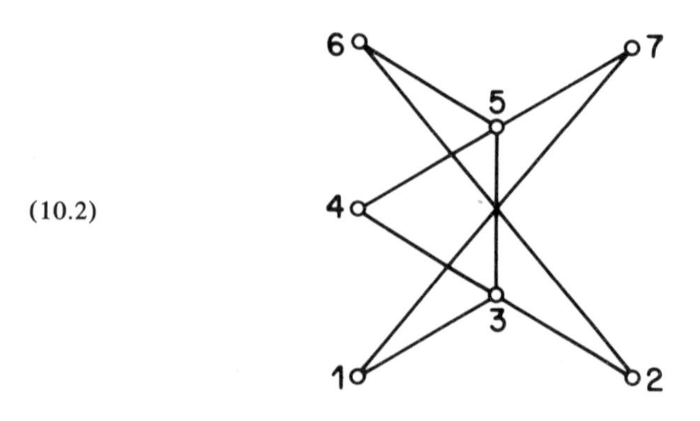


1356, 17, 26, 23457

is such a set $\mathscr{S}$. Assume also $\mathscr{S}$ is a clutter.

We now partially order $\mathscr{B}(\mathscr{S})$.

(10.3) $T_1 \leqq T_2$ if for every path $S \in \mathscr{S}$, the first element of $S$ contained in $T_2$ does not precede (in $S$) the first element of $S$ contained in $T_1$.

It turns out that $\mathscr{B}(\mathscr{S})$, under (10.3), is a lattice. We illustrate by the Hasse diagram of the blocker of (10.2)

(10.4)

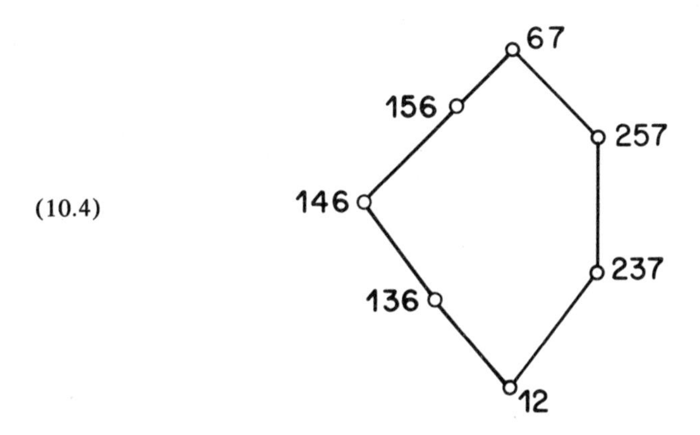


One can prove

(10.5). If $\mathscr{L}$ is a submodular clutter on $U$,$\mathscr{B}(\mathscr{L})$ its blocker, then $\mathscr{B}(\mathscr{L})$ is a clutter $\mathscr{P}$ of paths closed with respect to switching, $\mathscr{L}$ is the blocker of $\mathscr{B}(\mathscr{L})$ with the order and lattice structure of (10.3) coinciding exactly with the original $\prec$, $\vee$ and $\wedge$.

Of course, in proving (10.5), one has to use the original ordering and $\vee$ and $\wedge$ in $\mathscr{L}$ to simply order (as in §7) the elements of each set of $\mathscr{B}(\mathscr{L})$ - the paths - and show that with that ordering, the set of paths is closed with respect to switching. The situation for supermodular clutters is almost analogous. Let $P$ be a partially ordered set, $\mathscr{L}$ the set of maximal anti-chains, partially ordered as in (9.4), so that $\mathscr{L}$ becomes a lattice. As we have seen in (9.4), this is a supermodular clutter.

(10.6). If $\mathscr{L}$ is a supermodular clutter, $\mathscr{A}(\mathscr{L})$ its anti-blocker, then $\mathscr{A}(\mathscr{L})$ is the set of all maximal chains of a partially ordered set $P$, and $\mathscr{L}$ is the set of maximal anti-chains of $P$, partially ordered as in (9.4).

But the operations $\wedge$ and $\vee$ defining the original supermodular clutter do not have to be the lattice operations from the partial order. For example, if $P$ has Hasse diagram

6
3 4 5
1 2

then $\mathscr{L}$ has Hasse diagram

e
d
b c
a

where $a = \{1,2\}$, $b = \{2,3\}$, $c = \{1,5\}$, $d = \{3,4,5\}$, $e = \{3,6,5\}$. If $b \vee c$ is taken to be $e$ rather than $d$, $\mathscr{L}$ is still a supermodular clutter.

## References


[1] C. Berge, Farbung von Graphen, deren sämtliche bzw. deren ungerade Kreise starr sind, Wiss Z. Martin-Luther-Univ., Halle-Wittenburg Math. Natur, Reihe, (1961), 114-115.

[2] G. Birkhoff, Tres observaciones sobre el algebra lineal, Univ. Nac. Tucumán, Rev Ser. A, 5 (1946), 147-151.

[3] W. H. Cunningham, An unbounded matroid intersection polyhedron, Linear Algebra Appl. 16 (1977), 209-215.

[4] G. B. Dantzig and D. R. Fulkerson, Minimizing the number of tankers to meet a fixed schedule. Naval Res. Logist. Quart. 1 (1954), 217-222.

[5] G. B. Dantzig and A. J. Hoffman, Dilworth's theorem on partially ordered sets, in Linear Inequalities and Related Systems, Annals of Mathematics Study

38, H. W. Kuhn and A. W. Tucker eds., Princeton Univ. Press, Princeton, 1956, 207-214.

[6] R. P. Dilworth, A decomposition theorem for partially ordered sets, Annals of Math (2) 51, (1950), 161-166.

[7] ———, Dependence relations in a semi-modular lattice, Duke Math. J. 11 (1944), 575-587.

[8] J. Edmonds, Submodular functions, matroids and certain polyhedra, Combinatorial structures and their applications, Gordon and Breach (1970) 69-87.

[9] J. Edmonds and D. R. Fulkerson, Bottleneck extrema, J. Combinatorial Theory, 8 (1970), 299-306.

[10] J. Edmonds and R. Giles, A min-max relation for submodular functions on graphs, Studies in Integer programming, Vol. 1, Annals of Discr. Math. (1977) 185-204.

[11] L. R. Ford and D. R. Fulkerson, Maximal flow through a network, Canad. J. Math. 8 (1956), 399-404.

[12] A. Frank, Kernel system of directed graphs, Acta Sci. Math. (Szeged) 41 (1979), 63-76.

[13] ———, A weighted matroid intersection algorithm, to appear.

[14] D. R. Fulkerson, Blocking and anti-blocking pairs of polyhedra, Math. Programming 1 (1971), 168-193.

[15] M. C. Golumbic, Algorithmic graph theory and perfect graphs, Academic Press, New York, 1980.

[16] C. Greene, Some partitions associated with a partially ordered set, J. Combinatorial Theory Ser. A 20 (1976), 69-79.

[17] C. Greene and D. J. Kleitman, Strong versions of Sperner's theorem, J. Combinatorial Theory Ser. A 20 (1976), 80-88.

[18] H. Gröflin and A. J. Hoffman, On matroid intersections, Combinatorica 1 (1981), 43-47.

[19] ———, Lattice polyhedra II, to appear.

[20] A. J. Hoffman, A generalization of max flow- min cut, Math Programming 6 (1974), 352-359.

[21] ————, On lattice polyhedra III: blockers and anti-blockers of lattice clutters, Math. Programming Study 8 (1978), 197-207.

[22] ————, Linear programming and combinatorial problems, in Proceedings of a Conference on Discrete Mathematics and its Applications, M. Thompson ed., Indiana University, 1976, 65-92.

[23] A. J. Hoffman and J. B. Kruskal, Integral boundary points of convex polyhedra, in Linear inequalities and related systems, Annals of Mathematics Study 38, H. W. Kuhn and A. W. Tucker eds., Princeton University Press, Princeton (1956), 223-246.

[24] A. J. Hoffman and D. E. Schwartz, On partitions of a partially ordered set, J. Combinatorial Theory Ser. B 23 (1977), 3-13.

[25] ————, On lattice polyhedra, in Combinatorics: Proc. 5th Hung. Coll. on Comb. Keszthelev (1976), A. Hajnal and V. T. Sós eds., Bolyai J. Math. Soc., Budapest, and North Holland, Amsterdam (1978), 593-598.

[26] E. Johnson, On cut set integer polyhedra, Cahiers du Centre de Recherche Opérationnelle 17 (1975), 235-251.

[27] N. Linial, Extending the Greene-Kleitman theorem to directed graphs, J. Combinatorial Theory Ser. A 30 (1981), 331-334.

[28] L. Lovász, Normal hypergraphs and the perfect graph conjecture, Discrete Math. 2 (1972), 253-267.

[29] M. Simonnard, Linear Programming, Prentice-Hall, Englewood Cliffs, N.J., 1966.

# MACHINE SCHEDULING WITH PRECEDENCE CONSTRAINTS

E.L. Lawler
Computer Sciences Division
University of California at Berkeley
Berkeley, California 94720

J.K. Lenstra
Mathematisch Centrum
Kruislaan 413 1098 SJ Amsterdam
Postbus 4079 1009 AB Amsterdam
The Netherlands

## ABSTRACT

Machine scheduling theory is concerned with the allocation over time of scarce resources in the form of *machines* or *processors* to activities known as *jobs* or *tasks*. If the jobs can be performed in any order, they are said to be *independent*. However, this is often not the case; technological or other constraints may dictate that of two given jobs one must be performed before the other. Such *precedence constraints* of course impose a partial ordering on the jobs.

A variety of techniques has been developed for the solution of machine scheduling problems [Graham *et al.* 1979]. Such algorithmic results are often complemented by results stating that satisfactory solution techniques for related problems will probably never be found. The theory of computational complexity provides the tools to make a formal distinction between *well-solved* problems, which are solvable by an algorithm whose running time is bounded by a polynomial function of problem size, and *NP-hard* problems, for which the existence of such an algorithm is highly unlikely [Cook 1971; Karp 1972; Garey and Johnson 1979].

There appear to be only three classes of precedence constrained scheduling problems for which polynomial time algorithms exist:

I. single machine problems, in the case of min-max optimization;



*I. Rival (ed.), Ordered Sets, 655–675.*

II. certain problems with series-parallel precedence constraints;

III. certain parallel machine problems.

In this paper we shall describe these problem classes, indicate the nature of the algorithmic techniques that have been used to successfully cope with precedence constraints in each of them, and present the relevant *NP*-hardness results.

I. Single Machine Scheduling to Minimize Maximum Cost. One of the earliest results in scheduling theory is the "earliest due date rule" [Jackson 1955]. Suppose $n$ independent jobs are to be processed on a single machine. Each job $j$ ($j = 1,\ldots,n$) has a specified *processing time* $p_j$ and a *due date* $d_j$. Is it possible to sequence the jobs, from time 0 onwards, so that each is completed by its due date? The answer is yes, if and only if all jobs meet their deadlines when they are sequenced in non-decreasing due date order. Moreover, even if this order does not serve to complete all jobs on time, it will nevertheless minimize maximum *lateness*, where the lateness of a job is equal to its completion time minus its due date.

The earliest due date rule was later greatly generalized, as follows [Lawler 1973]. Suppose instead of a due date $d_j$ there is a nondecreasing *cost function* $f_j$ specified for each job $j$. If job $j$ is completed at time $t$, a cost $f_j(t)$ is incurred; thus, in the case of lateness, we have $f_j(t) = t - d_j$. It is possible to find a schedule which minimizes the maximum of the job costs, even in the presence of arbitrary precedence constraints, by the following simple rule: From among all jobs that are eligible to be sequenced last (*i.e.*, that have no successors under the precedence constraints), put that job last which will incur the smallest cost in that position. Then repeat on the set of $n - 1$ jobs remaining, and so on. The resulting algorithm can be implemented to run in $O(n^2)$ time; in the case of minimizing maximum lateness, $O(n \log n)$ time can be achieved.

This algorithm has been still further generalized to apply to the situation in which the jobs become available at arbitrary *release dates* and *preemption* is permitted (*i.e.*, the processing of any job may arbitrarily often be interrupted and resumed at a later time, without penalty) [Baker *et al.* 1980]. In contrast, the problem of finding a feasible *nonpreemptive* schedule with respect to given release dates and deadlines is *NP*-hard, even in the absence of precedence constraints [Garey and Johnson 1977; Lenstra *et al.* 1977].

II. Series-parallel Scheduling. Another classical result in scheduling theory is the "ratio rule" [Smith 1956]. Suppose again that there are $n$ independent jobs, and that in addition to a processing time $p_j$ each job $j$ has a specified *weight* $w_j$. It is possible to minimize the weighted sum of the job completion times by sequencing the jobs in nondecreasing order of the ratios $p_j/w_j$.

This result has been generalized to apply to problems with precedence constraints admitting of various types of decompositions [Sidney 1975], with precedence constraints in the form of rooted trees [Horn 1972; Adolphson and Hu 1973], and with series-parallel precedence constraints [Lawler 1978].

The series-parallel case can be solved in $O(n \log n)$ time, given a decomposition tree for the constraints, and this is optimal to within a constant factor. On the other hand, the case of arbitrary precedence contraints is *NP*-hard, even if all $p_j$ are equal or all $w_j$ are equal [Lawler 1978; Lenstra and Rinnooy Kan 1978].

The series-parallel algorithm can also be applied to a number of other scheduling problems, including one version of the two-machine flow shop problem [Sidney 1979; Monma and Sidney 1979].

III. Parallel Machine Scheduling. Suppose all jobs have a unit processing time and the problem is to schedule them on $m$ identical parallel machines so as to minimize the maximum of the job completion times. Precedence constraints can be dealt with successfully if either they are in the form of a rooted tree [Hu 1961], or there are only two machines [Fujii *et al.* 1969, 1971; Coffman and Graham 1972]. The former problem is solvable in linear time and the latter in "almost" linear time [Gabow 1980].

These techniques have been extended to polynomial time algorithms for more general problems involving release dates and/or due dates [Garey and Johnson 1976, 1977; Brucker *et al.* 1977] or arbitrary processing times and preemption [Lawler 1981].

The problem of scheduling $n$ unit time jobs on $m$ identical parallel machines subject to arbitrary precedence constraints is *NP*-hard, if $m$ is a variable specified as part of the problem instance [Ullman 1975; Lenstra and Rinnooy Kan 1978]. However, the complexity status of this problem is open if $m$ is fixed. In particular, the "three-processor problem" remains one of the most vexing open problems in scheduling theory. Many special cases have been resolved [Dolev; Warmuth], but little real progress seems to have been made.

## 1. INTRODUCTION

Machine scheduling theory is concerned with the allocation over time of scarce resources in the form of *machines* or *processors* to activities known as *jobs* or *tasks*. If the jobs can be performed in any order, they are said to be *independent*. However, this is often not the case; technological or other constraints may dictate that of two given jobs one must be performed before the other. Such *precedence constraints* of course impose a partial ordering on the job set.

Machine scheduling problems occur in many different situations. Suppose a conference organizer has to sequence $n$ lectures, some of which are based on material presented in other lectures. If each lecturer has specified the times of his arrival and departure, the organizer is faced with the task of finding a feasible schedule for $n$ jobs on a single machine subject to precedence constraints, release dates and deadlines (see Section 3). Suppose the conference participants are simultaneously arriving in $n$ airplanes, which are waiting for permission to land on a single runway. If the objective is to keep total gasoline usage as low as possible, the air controller has to minimize the weighted sum of $n$ job completion times on a single machine (see Section 4). A third example is the Chinese Cook Problem [Lawler *et al.* 1976]. Suppose the conference dinner consists of $n$ courses and is to be prepared on a stove with $m$ burners. In order to start as late as possible, the cook has to minimize maximum completion time for $n$ jobs on $m$ parallel machines; it is easily imagined that in this situation precedence constraints between the jobs are specified (see Section 5).

A variety of techniques has been developed for the solution of machine scheduling problems [Graham *et al.* 1979]. Such algorithmic results are often complemented by results stating that satisfactory solution techniques for related problems will probably never be found. The theory of *computational complexity* provides the tools to make a formal distinction between *well-solved* problems, which are solvable by an algorithm whose running time is bounded by a polynomial function of problem size, and *NP-hard* problems, for which the existence of such an algorithm is highly unlikely. In Section 2 we give a brief introduction to this theory.

There appear to be only three classes of precedence constrained scheduling problems for which polynomial-time algorithms exist: *single machine problems in the case of min-max optimization*, certain problems with *series-parallel precedence constraints*, and certain *parallel machine problems*. In Sections 3, 4 and 5 we describe these problem classes, indicate the nature of the algorithmic techniques that have been used to successfully cope with precedence constraints in each of them, and present the relevant *NP*-hardness results. In Section 6 we make some concluding remarks.

## 2. COMPUTATIONAL COMPLEXITY THEORY

For many combinatorial optimization problems, highly efficient solution techniques exist. For many other such problems, there is little hope of obtaining an optimal solution in a reasonable amount of time, and one has to choose between the use of fast heuristics, which yield only an approximate solution, or of enumerative methods, which produce an optimal solution only after an often time consuming search through the set of feasible solutions. A distinction between these problem classes can be made by means of concepts from the theory of computational complexity. An informal exposition of these concepts is given below. The reader is referred to [Karp 1975; Lenstra & Rinnooy Kan 1979] for a more extensive introduction and to [Garey & Johnson 1979] for a comprehensive treatment of the theory.

Let us define the *size* of a problem as the number of bits needed to encode its data, and the *running time* of an algorithm as the number of elementary operations (such as additions and comparisons) required for its solution.

If a problem of size $s$ can be solved by an algorithm with running time $O(p(s))$ where $p$ is a *polynomial* function, then the problem is considered to be *well solved*. The justifications of this notion are the following: *machine independence* (theoretical computing devices like Turing machines and ordinary computers are all polynomially related), *asymptotic behavior* (any polynomial-time algorithm is faster than any superpolynomial one for sufficiently large problems), *accordance with experience* (polynomial-time algorithms tend to be efficient in practice), and *susceptability to theoretical analysis* (the subject of this section). Polynomial-time algorithms exist for a wide variety of combinatorial optimization problems, such as finding shortest paths, maximum flows and maximum matchings in graphs [Lawler 1976].

Computational complexity theory deals primarily with *decision problems*, which require a yes/no answer, rather than with optimization problems. Three decision problems that will recur below are:

SATISFIABILITY
*Instance*: A conjunctive normal form expression, *i.e.*, a conjunction of clauses, each of which is a disjunction of literals $x_1,\bar{x}_1,\ldots,x_t,\bar{x}_t$, where $x_1,\ldots,x_t$ are boolean variables and $\bar{x}_1,\ldots,\bar{x}_t$ denote their complements.
*Question*: Does there exist a truth assignment to the variables such that the expression assumes the value *true*?

CLIQUE
*Instance*: A graph $G = (V,E)$ and an integer $c$.
*Question*: Does there exist a subset $C \subset V$ of cardinality $c$ such that $\{j,k\} \in E$ if $\{j,k\} \subset C$?

PARTITION
*Instance*: Positive integers $a_1,\ldots,a_t,b$ with $a_1+\ldots+a_t = 2b$.
*Question*: Does there exist a subset $S \subset \{1,\ldots,t\}$ such that $\Sigma_{j \in S}\, a_j = b$?

An instance of a decision problem is said to be *feasible* if the question can be answered affirmatively. Feasibility is usually characterized by the existence of an associated *structure* which satisfies a certain property. *E.g.*, in the case of CLIQUE the structure is a set of $c$ pairwise adjacent vertices.

We now define two important problem classes. A decision problem is in the class $P$ if, for any instance, one can determine its feasibility or infeasibility in polynomial time. It is in the class $NP$ if, for any instance, one can determine in polynomial time whether a given structure affirms its feasibility. *E.g.*, CLIQUE is a member of $NP$, since for any clique $C$ one can verify that all its vertices are pairwise adjacent in $O(c^2)$ time. Also SATISFIABILITY and PARTITION belong to $NP$.

It is clear that $P \subseteq NP$. The question if this inclusion is a proper one or if equality holds is probably the foremost open problem in theoretical computer science. However, *it is very unlikely that* $P = NP$, since $NP$ contains many notorious combinatorial problems for which, in spite of considerable research efforts, no polynomial-time algorithms have been found so far.

Cook [Cook 1971] proved that SATISFIABILITY is the most difficult problem in $NP$ by showing that every other problem in $NP$ is *reducible* to it. That is, for any instance of any problem $P \in NP$ a corresponding instance of SATISFIABILITY can be constructed in polynomial time such that solving the latter instance will solve the instance of $P$ as well. He also proved that CLIQUE is as hard as SATISFIABILITY, simply by giving a reduction of SATISFIABILITY to CLIQUE. Both problems are said to be $NP$-*complete* in the sense that (*i*) they belong to $NP$, and (*ii*) every other problem in $NP$ is reducible to them.

Karp [Karp 1972] elaborated on these ideas and identified a large number of $NP$-complete problems $P$ (including PARTITION) by verifying that $P \in NP$ and specifying a reduction of a known $NP$-complete problem to $P$. Since then, hundreds of combinatorial problems have been shown to be $NP$-complete; we refer to [Garey & Johnson 1979] for an impressive compilation of these results.

A polynomial-time algorithm for an $NP$-complete problem $P$ could be used to solve any problem in $NP$ in polynomial time by first reducing it to $P$ and then applying the algorithm for $P$, and the existence of such an algorithm would thus imply that $P = NP$. Hence, *it is very unlikely that* $P \in P$ *for any* $NP$-*complete* $P$. This observation justifies the use of heuristic or enumerative methods for its solution.

In dealing with the computational complexity of *optimization problems*, one usually reformulates the problem of finding a feasible solution of, say, minimum value as the problem of deciding whether there exists a feasible solution with value at most equal to a given threshold. If this decision problem is $NP$-complete, then the optimization problem is said to be $NP$-*hard* in the sense that the existence of a polynomial-time algorithm for its solution would imply that $P = NP$.

## 3. SINGLE MACHINE SCHEDULING TO MINIMIZE MAXIMUM COST

Suppose $n$ *jobs* are to be processed on a *single machine* which can execute at most one job at a time. Each job $j$ requires an uninterrupted *processing time* $p_j$. Moreover, *precedence constraints* between the jobs are specified in the form of a partial order $\rightarrow$; if $j \rightarrow k$, then job $j$ has to be completed before job $k$ can start. Associated with each job $j$ is a monotone nondecreasing *cost function* $f_j$; if job $j$ is completed at time $C_j$, a cost $f_j(C_j)$ is incurred. It is possible to find a schedule which minimizes the *maximum of the job completion costs* $\max_j\{f_j(C_j)\}$ by the following simple rule [Lawler 1973]. From among all jobs that are eligible to be sequenced last, *i.e.*, that have no successors under $\rightarrow$, put that job last which will incur the smallest cost in that position. Then repeat on the set of $n$-1 jobs remaining, and so on.

The correctness of this rule can be proved as follows. Let $N = \{1,\ldots,n\}$ be the set of all jobs, let $q = p_1+\ldots+p_n$, let $L \subseteq N$ be the set of jobs without successors, and for any $S \subseteq N$ let $f^*(S)$ be the maximum job completion cost in an optimal schedule for $S$. If job $l \in L$ is chosen such that

$$f_l(q) = \min_{j\in L}\{f_j(q)\},$$

then the criterion value of a schedule which is optimal subject to the condition that job $l$ is processed last is given by

$$\max\{f_l(q), f^*(N-\{l\})\}.$$

Since neither of the maximands is larger than $f^*(N)$, there exists an optimal schedule in which job $l$ is in the last position.

The algorithm can be implemented to run in $O(n^2)$ time, under the assumption that each $f_j$ can be evaluated in unit time for any value of the argument.

We next consider an extension of this problem, in which each job $j$ becomes available for processing at a specified *release date* $r_j$. In this situation it may be advantageous to relax the requirement that each job is to be processed without interruption.

Instance of PARTITION:
$t = 7$, $a_j = j$ $(j = 1,\ldots,7)$, $b = 14$.
Solution:
$S = \{1,6,7\}$.
Corresponding schedule:

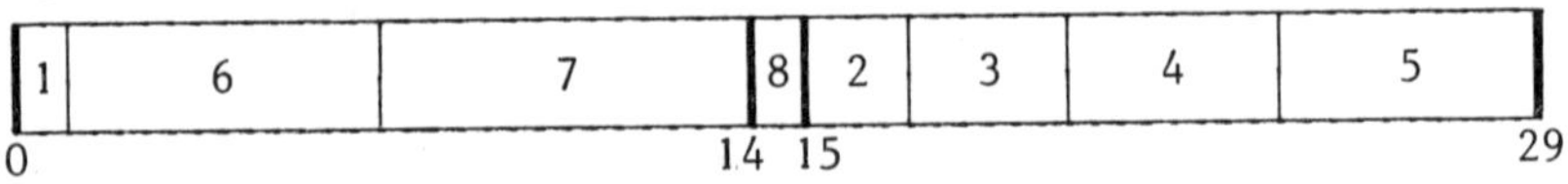

| 1 | 6 | 7 | 8 | 2 | 3 | 4 | 5 |
|---|---|---|---|---|---|---|---|

0 14 15 29

Figure 1. Illustration of the reduction from PARTITION.

The above algorithm has been generalized to the situation in which *preemption* is permitted, *i.e.*, the processing of any job may arbitrarily often be interrupted and resumed at a later time, without penalty [Baker *et al.* 1982].

In contrast, the *nonpreemptive* version of this problem is *NP*-hard. More specifically, the problem of deciding whether there exists a *feasible* schedule in which each job $j$ is processed for an uninterrupted period of length $p_j$ between a given *release date* $r_j$ and a given *deadline* $d_j$ is *NP*-complete, even in the absence of precedence constraints [Garey & Johnson 1977; Lenstra *et al.* 1977]. The proof is given below.

The feasibility problem belongs to *NP*, since any given schedule can be tested for feasibility in linear time. It suffices to prove that a known *NP*-complete problem is reducible to the feasibility problem. Consider the PARTITION problem as formulated in Section 2. Given any instance of PARTITION, defined by positive integers $a_1,\ldots,a_t,b$ with $a_1+\ldots+a_t = 2b$, we construct a corresponding instance of the feasibility problem as follows: the number of jobs is given by

$$n = t+1;$$

there are $t$ *partition jobs* $j$ $(j = 1,\ldots,n)$ with

$$r_j = 0,\quad p_j = a_j,\quad d_j = 2b+1;$$

there is one *splitting job* $n$ with

$$r_n = b,\quad p_n = 1,\quad d_n = b+1.$$

It should be obvious that PARTITION has a solution if and only if there exists a feasible schedule (*cf*. Figure 1), and hence PARTITION is reducible to the feasibility problem.

The remainder of this section deals with the special case in which the cost function of each job $j$ measures its *lateness* $f_j(C_j) = C_j-d_j$ with respect to a given *due date* $d_j$. Of course, the problem of minimizing *maximum lateness* is *NP*-hard if processing times, release dates and due dates are arbitrary, but it is possible to find an optimal schedule in polynomial time if all $p_j$ are equal, all $r_j$ are equal, or all $d_j$ are equal.

First assume that the jobs are *independent*. One of the oldest results in scheduling theory is the "earliest due date rule" [Jackson 1955], which states that the case of equal release dates is solved by sequencing the jobs in order of nondecreasing due dates; this rule can be viewed as a specialization of Lawler's algorithm. Similarly, the case of equal due dates is solved by sequencing the jobs in order of nondecreasing release dates.

The case of unit processing times and integer release and due dates is solved by an extension of Jackson's rule [Baker & Su 1974], which schedules at any time an available job with smallest due date. The case of equal processing times requires a more sophisticated approach but is still solvable in polynomial time [Simons 1978; Garey *et al.* 1981A].

Now suppose that *precedence constraints* are specified. Consider a feasible schedule in which two jobs $j$ and $k$ with $j \rightarrow k$ are completed at times $C_j$ and $C_k$, respectively; note that $C_j \leq C_k - p_k$. The feasibility of the schedule is not affected if we set

$$r_k := \max\{r_j+p_j, r_k\},$$

since we have for the starting time of job $k$ that $C_k - p_k \geq C_j \geq r_j + p_j$. The criterion value of the schedule does not change if we set

$$d_j := \min\{d_j, d_k - p_k\},$$

since we have for the lateness of job $k$ that $C_k - d_k \geq C_j - (d_k - p_k)$. Hence, we may modify release and due dates as indicated above, so that $r_j < r_k$ and $d_j < d_k$ whenever $j \rightarrow k$ [Lageweg *et al.* 1976].

The reader should have no difficulty in verifying that the algorithms for the cases that all $r_j$ are equal, all $d_j$ are equal or all $p_j = 1$ will, after modification of the $r_j$ and $d_j$ according to $\rightarrow$, automatically respect the precedence constraints. The algorithm for the case that all $p_j$ are equal has the same property. However, in the general case this modification is not sufficient and the precedence constraints have to be taken explicitly into account.

## 4. SERIES-PARALLEL SCHEDULING

Another classical result in scheduling theory is the "ratio rule" [Smith 1956]. Suppose again that $n$ *independent jobs* are to be processed on a *single machine*, and that in addition to a *processing time* $p_j$ each job $j$ has a specified *weight* $w_j$. It is possible to find a schedule which minimizes the *weighted sum of the job completion times* in $O(n \log n)$ time by sequencing the jobs in order of nondecreasing ratios $\rho_j = p_j/w_j$.

The problem becomes $NP$-hard if arbitrary *precedence constraints* are permitted, even if all $p_j$ are equal or all $w_j$ are equal [Lawler 1978A; Lenstra & Rinnooy Kan 1978]. However, a number of special cases has been dealt with successfully, including precedence constraints admitting of various types of decompositions [Sidney 1975], rooted trees [Horn 1972; Adolphson & Hu 1973], and series-parallel precedence constraints [Lawler 1978A]. The algorithm for the last case has been generalized to apply to a diverse variety of other sequencing problems [Lawler 1978B; Monma & Sidney 1979]. In the following we give a brief review of the theory of series-parallel sequencing that has resulted from this generalization.

First, let us pose a very general type of sequencing problem. Given a set of $n$ jobs and a real-valued function $f$ which assigns a value $f(\pi)$ to each permutation $\pi$ of the jobs, find a permutation $\pi^*$ such that

$$f(\pi^*) = \min_\pi \{f(\pi)\}.$$

If we know nothing of the structure of the function $f$, there is clearly nothing to be done except to evaluate $f(\pi)$ for each of the $n!$ permutations $\pi$. However, we may be able to find a transitive and complete relation $\leq$ (*i.e.* a quasi-total order) on the jobs with the property that for any two jobs $b,c$ and any permutation of the form $\alpha bc\delta$ we have

$$b \leq c \Rightarrow f(\alpha bc\delta) \leq f(\alpha cb\delta).$$

Such a relation is called a *job interchange relation*. It says that whenever $b$ and $c$ occur as adjacent jobs with $c$ before $b$, we are at least as well off to interchange their order. Hence, this relation is sometimes referred to as the "adjacent pairwise interchange property". Smith's ratio rule observes this property.

It is a simple matter to verify the following.

THEOREM 1. *If $f$ admits of a job interchange relation $\leq$, then an optimal permutation $\pi^*$ can be found by ordering the jobs according to $\leq$, with $O(n \log n)$ comparisons of jobs with respect to $\leq$.*

Now suppose that precedence constraints are specified in the form of a partial order $\to$. A permutation $\pi$ is *feasible* if $j \to k$ implies that job $j$ precedes job $k$ under $\pi$. The objective is to find a feasible permutation $\pi^*$ such that

$$f(\pi^*) = \min_{\pi \text{ feasible}} \{f(\pi)\}.$$

We need something stronger than a job interchange relation to solve this problem. A transitive and complete relation $\leq$ on subpermutations or *strings* of jobs with the property that for any two disjoint strings of jobs $\beta,\gamma$ and any permutation of the form $\alpha\beta\gamma\delta$ we have

$$\beta \leq \gamma \Rightarrow f(\alpha\beta\gamma\delta) \leq f(\alpha\gamma\beta\delta)$$

is called a *string interchange relation*. Smith's rule generalizes to such a relation in a fairly obvious way: for any string $\alpha$ we define $\rho_\alpha = \Sigma_{j\in\alpha} p_j / \Sigma_{j\in\alpha} w_j$. However, it is not true that every function $f$ which admits of a job interchange relation also has a string interchange relation.

The remainder of this section is devoted to an intuitive justification of the following result. Details can be found in [Lawler 1978B].

THEOREM 2. *If $f$ admits of a string interchange relation $\leq$ and if the precedence constraints $\rightarrow$ are series-parallel, then an optimal permutation $\pi^*$ can be found by an algorithm which requires* $O(n \log n)$ *comparisons of strings with respect to $\leq$.*

A digraph is said to be *series-parallel* if its transitive closure is *transitive series-parallel*, as given by the recursive definition below:

(1) A digraph $G = (\{j\},\emptyset)$ with a single vertex $j$ and no arcs is transitive series-parallel.

(2) Let $G_1 = (V_1,A_1)$, $G_2 = (V_2,A_2)$ be transitive series-parallel digraphs with disjoint vertex sets. Both the *series composition* $G_1 \rightarrow G_2 = (V_1 \cup V_2, A_1 \cup A_2 \cup (V_1 \times V_2))$ and the *parallel composition* $G_1 \| G_2 = (V_1 \cup V_2, A_1 \cup A_2)$ are transitive series-parallel digraphs.

(3) No digraph is transitive series-parallel unless it can be obtained by a finite number of applications of Rules (1) and (2).

A variety of interesting and useful digraphs (and their corresponding partial orders) are series-parallel. In particular, rooted trees, forests of such trees and level digraphs are series-parallel. The smallest acyclic digraph which is not series-parallel is the *Z-digraph* shown in Figure 2. An acyclic digraph is series-parallel if and only if its transitive closure does not contain the Z-digraph as an induced subgraph. It is possible to determine whether or not an arbitrary digraph $G = (V,A)$ is series-parallel in $O(|V|+|A|)$ time [Valdes *et al.* 1981].

The structure of a series-parallel digraph is displayed by a *decomposition tree* which represents one way in which the transitive closure of the digraph can be obtained by successive applications of Rules (1) and (2). A series-parallel digraph and its decomposition tree are shown in Figure 3. Each leaf of the decomposition tree is identified with a vertex of the digraph. An S-node represents the application of series composition to the subdigraphs identified with its children; the ordering of these children is important: we adopt the convention that left precedes right. A P-node represents the application of parallel composition to the subdigraphs identified with its children; the ordering of these children is unimportant. The series or parallel relationship of any pair of vertices can be determined by finding their least common ancestor in the decomposition tree.

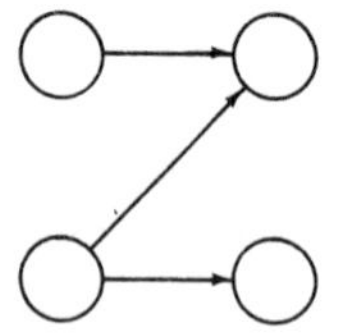

Figure 2. Z-digraph.

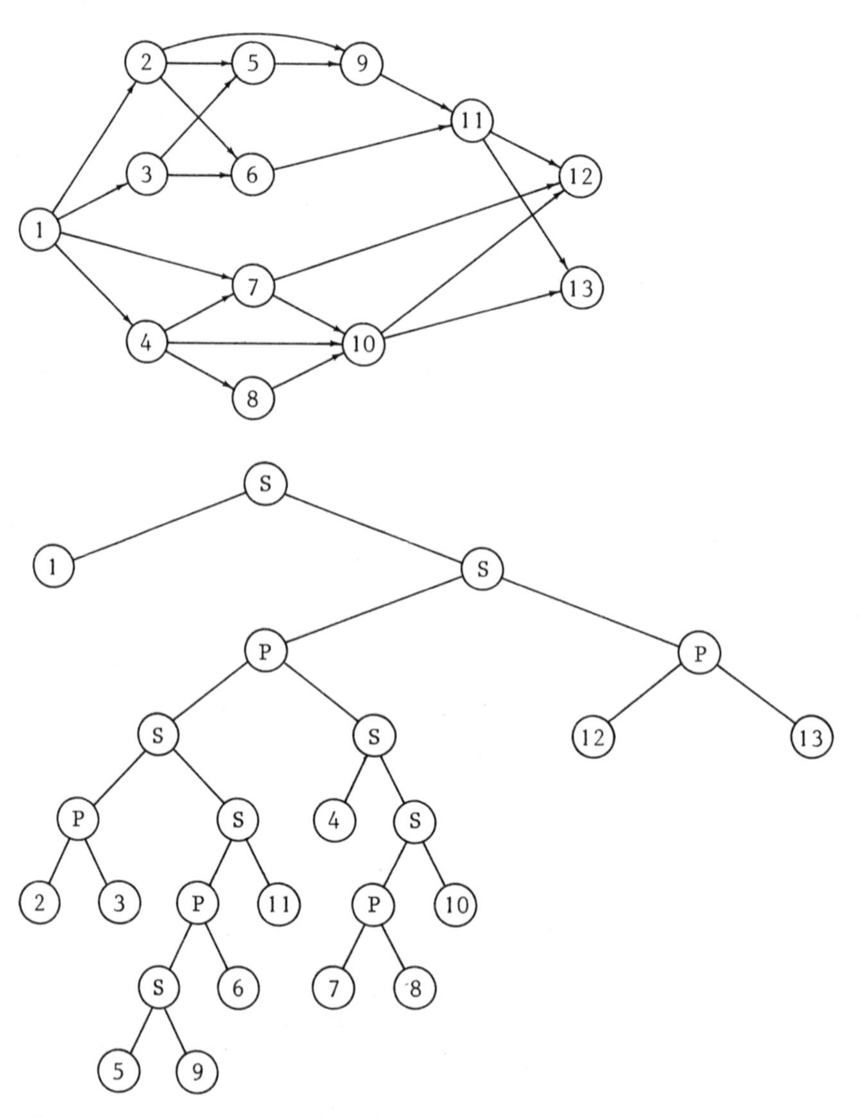


Figure 3. Series-parallel digraph and its decomposition tree.

Suppose we are to solve a sequencing problem for which a string interchange relation $\leq$ exists and a decomposition tree for the series-parallel constraints $\rightarrow$ is given. We can solve the problem by working from the bottom of the tree upward, computing a set of strings of jobs for each node of the tree from the sets of strings obtained for its children. Our objective is to obtain a set of strings at the root node such that sorting these strings according to $\leq$ yields an optimal feasible permutation.

We will accomplish our objective if the sets $S$ of strings we obtain satisfy two conditions:
(*i*) Any ordering of the strings in a set $S$ according to $\leq$ does not violate the precedence constraints $\rightarrow$.
(*ii*) At any point in the computation, let $S_1,\ldots,S_k$ be the sets of strings computed for nodes such that sets have not yet been computed for their parents. Then some ordering of the strings in $S_1 \cup \ldots \cup S_k$ yields an optimal feasible permutation.

If we order the strings computed at the root according to $\leq$, then condition (*i*) ensures that the resulting permutation is feasible and condition (*ii*) ensures that it is optimal.

For each leaf of the tree, we let $S = \{j\}$, where $j$ is the job identified with the leaf. Condition (*i*) is satisfied trivially and condition (*ii*) is clearly satisfied for the union of the leaf-sets.

Suppose $S_1$ and $S_2$ have been obtained for the children of a P-node in the tree. There are no precedence constraints between the strings in $S_1$ and the strings in $S_2$. Accordingly, conditions (*i*) and (*ii*) remain satisfied if for the P-node we let $S = S_1 \cup S_2$.

Suppose $S_1$ and $S_2$ have been obtained for the left child and the right child of an S-node, respectively. Let

$$\sigma_1 = \max_{\leq} S_1, \quad \sigma_2 = \min_{\leq} S_2.$$

If $\sigma_2 \not\leq \sigma_1$, then conditions (*i*) and (*ii*) are still satisfied if for the S-node we let $S = S_1 \cup S_2$. If $\sigma_2 \leq \sigma_1$, we assert that there exists an optimal feasible permutation in which $\sigma_1$ and $\sigma_2$ are adjacent, *i.e.*, in which $\sigma_1$ and $\sigma_2$ are replaced by their concatenation $\sigma_1\sigma_2$. (The proof of this assertion involves simple interchange arguments; see [Lawler 1978B].) This suggests the following procedure:

```
BEGIN
    σ1 := max≤ S1; σ2 := min≤ S2;
    IF σ2 ≰ σ1
    THEN S := S1∪S2
    ELSE σ := σ1σ2;
        S1 := S1-{σ1}; σ1 := max≤ S1;
        S2 := S2-{σ2}; σ2 := min≤ S2;
        WHILE σ ≤ σ1 ∨ σ2 ≤ σ
        DO   IF σ ≤ σ1
             THEN σ := σ1σ;
                 S1 := S1-{σ1}; σ1 := max≤ S1
```

```
            ELSE σ := σσ2;
                 S2 := S2-{σ2}; σ2 := min≤ S2
            FI
        OD;
        S := S1∪{σ}∪S2
    FI
END.
```

(We make here the customary assumption that $\max_{\leq} \emptyset$ and $\min_{\leq} \emptyset$ are very small and large elements, respectively.) It is not difficult to verify that conditions (*i*) and (*ii*) remain satisfied if for an S-node we compute a set of strings according to the above procedure.

The entire algorithm can be implemented so as to require $O(n \log n)$ time plus the time for the $O(n \log n)$ comparisons with respect to $\leq$.

In addition to the total weighted completion time problem, several other sequencing problems admit of a string interchange relation and hence can be solved efficiently for series-parallel precedence constraints. Among these are the problems of minimizing *total weighted discounted completion time* [Lawler & Sivazlian 1978], *expected cost of fault detection* [Garey 1973; Monma & Sidney 1979] or *minimum initial resource requirement* [Abdel-Wahab & Kameda 1978; Monma & Sidney 1979] on a single machine, and the *two-machine permutation flow shop problem with time lags* [Sidney 1979].

## 5. PARALLEL MACHINE SCHEDULING

Suppose that $n$ *unit-time jobs* are to be processed on $m$ *identical parallel machines*. Each job can be assigned to any machine, and the schedule has to respect given *precedence constraints*. The objective is to find a schedule which minimizes the *maximum of the job completion times*.

This problem is *NP*-hard in general, but it can be solved in polynomial time if either the precedence constraints are in the form of a rooted tree or if there are only two machines. Below we discuss these three basic results and mention a number of refinements and extensions.

First, let us assume that the precedence constraints are in the form of an *intree*, *i.e.*, each job has exactly one immediate successor, except for one job which has no successors and which is called the *root*. It is possible to minimize the maximum completion time in $O(n)$ time by applying an algorithm due to Hu [Hu 1961]. The *level* of a job is defined as the number of jobs in the unique path to the root. At the beginning of each time unit, as many available jobs as possible are scheduled on the $m$ machines, where highest priority is granted to the jobs with the largest levels. Thus, Hu's algorithm is a *list scheduling* algorithm, whereby at each step the available job with the highest ranking on a priority list is assigned to the first machine that becomes available. It can also be viewed as a *critical path scheduling* algorithm: the next job chosen is the one which heads the longest current chain of unexecuted jobs.

To validate Hu's algorithm, we will show that, if it yields a schedule of length $t^*$, then no feasible schedule of length $t < t^*$ exists.

Choose any $t < t^*$ and define a label for each job by subtracting its level from $t$; note that the root has label $t$ and that each other job has a label one less than its immediate successor. The algorithm gives priority to the jobs with the smallest labels. Since it yields a schedule of length larger than $t$, in some unit-time interval $s$ a job is scheduled with a label smaller than $s$. Let $s$ be the earliest such interval and let there be a job with label $l < s$ scheduled in it. We claim that there are $m$ jobs scheduled in each earlier interval $s' < s$. Suppose there is an interval $s' < s$ with fewer than $m$ jobs scheduled. If $s' = s-1$, then the only reason that the job with label $l$ was not scheduled in $s'$ could have been that an immediate predecessor of it was scheduled in $s'$; but then this predecessor would have label $l-1 < s-1$, which contradicts the definition of $s$. If $s' < s-1$, then there are fewer jobs scheduled in $s'$ than in $s'+1$, which is impossible from the structure of the intree. Hence, each interval $s' < s$ has $m$ jobs scheduled. Since each of these jobs has a label smaller than $s$, at least one job with a label smaller than $s$ must be scheduled in interval $s$, so that there is no feasible schedule of length $t < t^*$ possible. This completes the correctness proof of Hu's algorithm.

An alternative linear-time algorithm for this problem has been proposed by Davida and Linton [Davida & Linton 1976]. Assume that the precedence constraints are in the form of an *outtree*, *i.e.*, each job has at most one immediate predecessor. The *weight* of a job is defined as the total number of its successors. The jobs are now scheduled according to decreasing weights.

If the problem is to minimize the *maximum lateness* with respect to given due dates rather than the maximum completion time, then the case that the precedence constraints are in the form of an *intree* can be solved by an adaptation of Hu's algorithm, but the case of an *outtree* turns out to be $NP$-hard [Brucker *et al.* 1977]. Polynomial-time algorithms and $NP$-hardness results for the maximum completion time problem with various other special types of precedence constraints are reported in [Dolev 1981; Garey *et al.* 1981B; Warmuth 1980].

Next, let us assume that there are *arbitrary precedence constraints* but only *two machines*. It is possible to minimize the maximum completion time by a variety of algorithms.

The earliest and simplest approach is due to Fujii, Kasami and Ninomiya [Fujii *et al.* 1969, 1971]. A graph is constructed with vertices corresponding to jobs and edges $\{j,k\}$ whenever jobs $j$ and

$k$ can be executed simultaneously, *i.e.*, $j \not\rightarrow k$ and $k \not\rightarrow j$. A *maximum cardinality matching* in this graph, *i.e.* a maximum number of disjoint edges, is then used to derive an optimal schedule; if the matching contains $c$ pairs of jobs, the schedule has length $n-c$. Such a matching can be found in $O(n^3)$ time [Lawler 1976].

A completely different approach by Coffman and Graham [Coffman & Graham 1972] leads to a *list scheduling* algorithm. The jobs are labeled in the following way. Suppose labels $1,\ldots,l$ have been applied and $S$ is the subset of unlabeled jobs all of whose successors have been labeled. Then a job in $S$ is given the label $l+1$ if the labels of its immediate successors are *lexicographically minimal* with respect to all jobs in $S$. The priority list is formed by ordering the jobs according to decreasing labels. This method requires $O(n^2)$ time.

Recently, an even more efficient algorithm has been developed by Gabow [Gabow 1980]. His method uses labels, but with a number of rather sophisticated embellishments. The running time is almost linear in $n+a$, where $a$ is the number of arcs in the precedence graph.

If the problem is to find a *feasible* two-machine schedule under arbitrary precedence constraints when each job becomes available at a given integer *release date* and has to meet a given integer *deadline*, polynomial-time algorithms still exist [Garey & Johnson 1976, 1977]. These algorithms can be applied to minimize *maximum lateness* in polynomial time.

It is unlikely that the minimal common generalization of the two well-solved problems discussed above is solvable in polynomial time. More specifically, the problem of minimizing maximum completion time for $n$ unit-time jobs on $m$ identical parallel machines subject to arbitrary precedence constraints is $NP$-hard. This result is due to Ullman [Ullman 1975]; the proof to be given below is from [Lenstra & Rinnooy Kan 1978].

It suffices to prove that a known $NP$-complete problem is reducible to the scheduling problem. Consider the CLIQUE problem as formulated in Section 2. Given any instance of CLIQUE, defined by a graph $G = (V,E)$ and an integer $c$, let $v = |V|$, $e = |E|$, $d = \frac{1}{2}c(c-1)$, $c' = v-c$ and $d' = e-d$. We construct a corresponding instance of the scheduling problem as follows: the numbers of machines and jobs are given by

$$m = \max\{c, d+c', d'\}+1, \quad n = 3m;$$

there are $v$ *vertex jobs* $j$ ($j \in V$) and $e$ *edge jobs* $\{k,l\}$ ($\{k,l\} \in E$) with

$$j \rightarrow \{k,l\} \quad \text{whenever } j \in \{k,l\};$$

there are $n-v-e$ *dummy jobs* $\langle 1,i\rangle$ ($i = 1,\ldots,m-c$), $\langle 2,i'\rangle$ ($i' = 1,\ldots,m-d-c'$), $\langle 3,i''\rangle$ ($i'' = 1,\ldots,m-d'$) with

Instance of CLIQUE:

$G = (V,E)$:

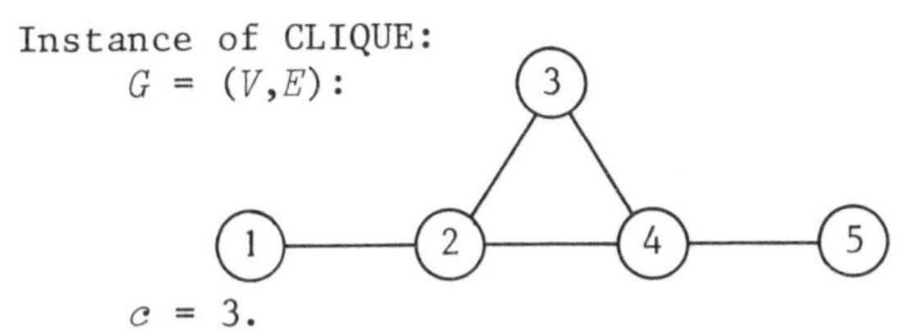

$c = 3$.

Solution:

$C = \{2,3,4\}$.

Corresponding schedule:

| | 0–1 | 1–2 | 2–3 |
|---|---|---|---|
| | 2 | {2,3} | {1,2} |
| $d'$ | 3 | {2,4} | {4,5} |
| $c$ | 4 | {3,4} | <3,1> |
| | <1,1> | 1 | <3,2> |
| $d+c'$ | <1,2> | 5 | <3,3> |
| $m$ | <1,3> | <2,1> | <3,4> |

0 1 2 3

Figure 4. Illustration of the reduction from CLIQUE.

$$\langle 1,i\rangle \to \langle 2,i'\rangle \quad \text{and} \quad \langle 2,i'\rangle \to \langle 3,i''\rangle \quad \text{for any } i,i',i''.$$

We claim that CLIQUE has a solution if and only if there exists a feasible schedule of length at most 3. The basic idea behind the reduction is the following. In any schedule of length 3, the dummy jobs create a pattern of idle machines during three unit-time intervals, available for the vertex and edge jobs. This pattern consists of $c$, $d+c'$ and $d'$ idle machines during the first, second and third interval, respectively, and it can be filled properly if and only if CLIQUE has a solution (*cf.* Figure 4).

More precisely, suppose that CLIQUE has a solution $C \subset V$. We construct a feasible schedule of length 3 by processing the $m-c$ dummy jobs $\langle 1,i\rangle$ and the $c$ clique vertex jobs $j$ ($j \in C$) in the first interval, the $m-d-c'$ dummy jobs $\langle 2,i'\rangle$, the $d$ clique edge jobs $\{k,l\}$ ($k,l \in C$) and the $c'$ remaining vertex jobs in the second interval, and the $m-d'$ dummy jobs $\langle 3,i''\rangle$ and the $d'$ remaining edge jobs in the third interval. Note that this schedule respects the precedence constraints.

Conversely, suppose that CLIQUE has no solution and assume that the dummy jobs $\langle 1,i\rangle$, $\langle 2,i'\rangle$ and $\langle 3,i''\rangle$ are processed in the first, second and third interval, respectively. Any set of $c$ vertex jobs processed in the first interval releases at most $d-1$ edge jobs for processing in the second interval and since there are only $c'$ vertex jobs left, at least one machine must remain idle during the second interval. We conclude that any feasible schedule has length at least 4.

Two observations concerning this result are in order. If $P(t)$ denotes the problem of deciding whether there exists a feasible schedule of length at most $t$, then $P(3)$ has been shown to be $NP$-complete, whereas $P(2)$ can trivially be solved in $O(n^2)$ time. The problem thus exhibits the *mystical property of* 3-*ness*, which is encountered in many other combinatorial problems. Moreover, the $NP$-completeness of $P(3)$ implies that no polynomial-time algorithm for the minimization problem can guarantee a worst-case performance bound less that 4/3, unless $P = NP$.

In the above problem formulation, the number of machines is a variable specified as part of the problem instance. The complexity status of the problem is open, however, for any fixed $m \geq 3$. In particular, the case $m = 3$ remains one of the most vexing open problems in scheduling theory.

In this section we have concentrated on problems involving *nonpreemptive* scheduling of *unit-time* jobs. There has been a parallel investigation of problems concerning *preemptive* scheduling of jobs of *arbitrary length*. Gonzalez and Johnson [Gonzalez & Johnson 1980] have developed a polynomial-time algorithm for preemptive scheduling which is analogous to that from [Davida & Linton 1976]. Further, in [Lawler 1982] algorithms are proposed which are the preemptive counterparts of those found in [Brucker *et al.* 1977; Garey & Johnson 1976, 1977]. These preemptive scheduling algorithms employ essentially the same techniques for dealing with precedence constraints as the corresponding algorithms for unit-time jobs. However, they are considerably more complex, and we shall not attempt to deal with them here.

## 6. CONCLUDING REMARKS

We have given an introductory survey of the most important polynomial-time algorithms and $NP$-hardness proofs for machine scheduling with precedence constraints. These results represent only a small fraction of the theory of deterministic sequencing and scheduling. Other models which have been studied in detail involve *non-identical parallel machines*, *multi-operation jobs*, and a wide variety of *optimality criteria*. Also, a considerable amount of work has been done on the design and analysis of *approximation algorithms*. A comprehensive survey of the entire area is given in [Graham *et al.* 1979].

Given a well-defined class $S$ of combinatorial optimization problems, it is often an easy matter to derive a *partial ordering* $\rightarrow$ on the class with the following property: if $P, Q \in S$ and $P \rightarrow Q$, then $P$ is reducible to $Q$. This implies that, if there exists a polynomial-time algorithm for $Q$, then $P$ is also well solved, and if $P$ is $NP$-hard, then the same is true for $Q$.

For an extensive class of scheduling problems, a computer program has been developed that employs such a partial ordering to determine the complexity status of all problems in the class [Lageweg *et al.* 1981A, 1981B]. More precisely, the program receives as input the known research results in the form of a listing of well-solved problems and a listing of *NP*-hard problems. It then partitions the problem class into three subclasses of well-solved, open and *NP*-hard problems, and produces as output a count of these classes as well as listings of the problems in each of the subclasses which are *minimal* or *maximal* with respect to the partial ordering. The maximal well-solved problems and the minimal *NP*-hard problems represent the strongest results that have been obtained so far. The listings of minimal and maximal open problems have proved to be very useful as guidelines for further research.

It can be shown that, in a classification scheme like this, the problem of determining the minimum number of research results which together would resolve all remaining open problems, is itself *NP*-hard.

## ACKNOWLEDGMENTS

This paper is partly based on material from [Graham *et al.* 1979; Lawler 1978B; Lenstra & Rinnooy Kan 1978]. The research by the first author was supported by NSF grant MCS78-20054.

## REFERENCES

H.M. ABDEL-WAHAB, T. KAMEDA (1978) Scheduling to minimize maximum cumulative cost subject to series-parallel precedence constraints. *Oper. Res. 26*,151-158.

D. ADOLPHSON, T.C. HU (1973) Optimal linear ordering. *SIAM J. Appl. Math.* 25,403-423.

K.R. BAKER, E.L. LAWLER, J.K. LENSTRA, A.H.G. RINNOOY KAN (1982) Preemptive scheduling of a single machine to minimize maximum cost subject to release dates and precedence constraints. *Oper. Res.*, to appear.

K.R. BAKER, Z.-S. SU (1974) Sequencing with due-dates and early start times to minimize maximum tardiness. *Naval Res. Logist. Quart. 21*,171-176.

P. BRUCKER, M.R. GAREY, D.S. JOHNSON (1977) Scheduling equal-length tasks under tree-like precedence constraints to minimize maximum lateness. *Math. Oper. Res. 2*,275-284.

E.G. COFFMAN, JR., R.L. GRAHAM (1972) Optimal scheduling for two-processor systems. *Acta Informat. 1*,200-213.

S.A. COOK (1971) The complexity of theorem-proving procedures. *Proc. 3rd Annual ACM Symp. Theory of Computing*, 151-158.

G.I. DAVIDA, D.J. LINTON (1976) A new algorithm for the scheduling of tree structured tasks. *Proc. Conf. Inform. Sci. and Syst.*,

Baltimore, MD, 543-548.

D. DOLEV (1981) Scheduling wide graphs. Unpublished manuscript.

M. FUJII, T. KASAMI, K. NINOMIYA (1969, 1971) Optimal sequencing of two equivalent processors. *SIAM J. Appl. Math.* *17*,784-789. Erratum. *20*,141.

H.N. GABOW (1980) An almost-linear algorithm for two processor scheduling. Technical Report CU-CS-169-80, Department of Computer Science, University of Colorado, Boulder.

M.R. GAREY (1973) Optimal task sequencing with precedence constraints. *Discrete Math.* *4*,37-56.

M.R. GAREY, D.S. JOHNSON (1976) Scheduling tasks with nonuniform deadlines on two processors. *J. Assoc. Comput. Mach.* *23*,461-467.

M.R. GAREY, D.S. JOHNSON (1977) Two-processor scheduling with start-times and deadlines. *SIAM J. Comput.* *6*,416-426.

M.R. GAREY, D.S. JOHNSON (1979) *Computers and Intractability: a Guide to the Theory of NP-Completeness*. Freeman, San Francisco.

M.R. GAREY, D.S. JOHNSON, B.B. SIMONS, R.E. TARJAN (1981A) Scheduling unit-time tasks with arbitrary release times and deadlines. *SIAM J. Comput.* *10*,256-269.

M.R. GAREY, D.S. JOHNSON, R.E. TARJAN, M. YANNAKAKIS (1981B) Scheduling opposing forests. Unpublished manuscript.

T. GONZALEZ, D.B. JOHNSON (1980) A new algorithm for preemptive scheduling of trees. *J. Assoc. Comput. Mach.* *27*,287-312.

R.L. GRAHAM, E.L. LAWLER, J.K. LENSTRA, A.H.G. RINNOOY KAN (1979) Optimization and approximation in deterministic sequencing and scheduling: a survey. *Ann. Discrete Math.* *5*,287-326.

W.A. HORN (1972) Single-machine job sequencing with treelike precedence ordering and linear delay penalties. *SIAM J. Appl. Math.* *23*,189-202.

T.C. HU (1961) Parallel sequencing and assembly line problems. *Oper. Res.* *9*,841-848.

J.R. JACKSON (1955) Scheduling a production line to minimize maximum tardiness. Research Report 43, Management Science Research Project, University of California, Los Angeles.

R.M. KARP (1972) Reducibility among combinatorial problems. In: R. E. MILLER, J.W. THATCHER (eds.) (1972) *Complexity of Computer Computations*, Plenum Press, New York, 85-103.

R.M. KARP (1975) On the computational complexity of combinatorial problems. *Networks* *5*,45-68.

B.J. LAGEWEG, E.L. LAWLER, J.K. LENSTRA, A.H.G. RINNOOY KAN (1981A) Computer aided complexity classification of combinatorial problems. Report BW 137, Mathematisch Centrum, Amsterdam.

B.J. LAGEWEG, E.L. LAWLER, J.K. LENSTRA, A.H.G. RINNOOY KAN (1981B) Computer aided complexity classification of deterministic scheduling problems. Report BW 138, Mathematisch Centrum, Amsterdam.

B.J. LAGEWEG, J.K. LENSTRA, A.H.G. RINNOOY KAN (1976) Minimizing maximum lateness on one machine: computational experience and some applications. *Statist. Neerlandica* *30*,25-41.

E.L. LAWLER (1973) Optimal sequencing of a single machine subject to precedence constraints. *Management Sci.* *19*,544-546.

E.L. LAWLER (1976) *Combinatorial Optimization: Networks and Matroids*, Holt, Rinehart and Winston, New York.

E.L. LAWLER (1978A) Sequencing jobs to minimize total weighted completion time subject to precedence constraints. *Ann. Discrete Math.* *2*,75-90.

E.L. LAWLER (1978B) Sequencing problems with series parallel precedence constraints. *Proc. Summer School on Combinatorial Optimization*, Urbino, Italy, July 1978, to appear.

E.L. LAWLER (1982) Preemptive scheduling of precedence-constrained jobs on parallel machines. In: M.A.H. DEMPSTER, J.K. LENSTRA, A.H.G. RINNOOY KAN (eds.) (1982) *Deterministic and Stochastic Scheduling*, Reidel, Dordrecht.

E.L. LAWLER, H.W. LENSTRA, A.H.G. RINNOOY KAN, T.J. WANSBEEK (eds.) (1976) *Een Tuyltje Boskruyd: Essays in Honor of Dr. J.-K. Lenstra*, Paris/Amsterdam/Delft/Leyden.

E.L. LAWLER, B.D. SIVAZLIAN (1978) Minimization of time varying costs in single machine sequencing. *Oper. Res.* *26*,563-569.

J.K. LENSTRA, A.H.G. RINNOOY KAN (1978) Complexity of scheduling under precedence constraints. *Oper. Res.* *26*,22-35.

J.K. LENSTRA, A.H.G. RINNOOY KAN (1979) Computational complexity of discrete optimization problems. *Ann. Discrete Math.* *4*,121-140.

J.K. LENSTRA, A.H.G. RINNOOY KAN, P. BRUCKER (1977) Complexity of machine scheduling problems. *Ann. Discrete Math.* *1*,343-362.

C.L. MONMA, J.B. SIDNEY (1979) Sequencing with series-parallel precedence constraints. *Math. Oper. Res.* *4*,215-224.

J.B. SIDNEY (1975) Decomposition algorithms for single-machine sequencing with precedence relations and deferral costs. *Oper. Res.* *23*,283-298.

J.B. SIDNEY (1979) The two-machine maximum flow time problem with series parallel precedence relations. *Oper. Res.* *27*,782-791.

B. SIMONS (1978) A fast algorithm for single processor scheduling. *Proc. 19th Annual IEEE Symp. Foundations of Computer Science*, 246-252.

W.E. SMITH (1956) Various optimizers for single-stage production. *Naval Res. Logist. Quart.* *3*,59-66.

J.D. ULLMAN (1975) *NP*-Complete scheduling problems. *J. Comput. System Sci.* *10*,384-393.

J. VALDES, R.E. TARJAN, E.L. LAWLER (1981) The recognition of series parallel digraphs. *SIAM J. Comput.*, to appear.

M.K. WARMUTH (1980) M Processor unit-execution-time scheduling reduces to M-1 weakly connected components. M.S. Thesis, Department of Computer Science, University of Colorado, Boulder.

# SOME ORDERED SETS IN COMPUTER SCIENCE

Dana S. Scott
Department of Computer Science
Carnegie-Mellon University
Pittsburgh, Pennsylvania 15213

## ABSTRACT

Strictly speaking the structures to be used are not lattices since as posets they will lack the top (or unit) element, but the adjunction of a top will make them complete lattices. The closure properties as posets, then, are closure under *inf* of any non-empty subset and *sup* of directed subsets. A family of subsets of a set closed under *intersections* of non-empty subfamilies and *unions* of directed subfamilies is a special type of poset with the closure properties where additionally every element is the directed sup (union) of the finite ("compact") elements it contains. We call such posets *finitary domains*. (With a top they are just the well known algebraic lattices.) The *continuous domains* can be defined as the continuous *retracts* of finitary domains. A mapping between domains is *continuous* if it preserves direct sups. A map of a domain into itself is a *retraction* if it is idempotent. Starting with a finitary domain, the range (= fixed-point set) of a continuous retraction -- as a poset -- is a continuous domain. Numberless characterizations of continuous domains, both topological and order-theoretic, can be found in [2]. For the most part in the lectures we shall concentrate on the finitary domains, but the continuous domains find an interest as a generalization of *interval analysis* and by the connection with spaces of *upper-semicontinuous* functions.

In [10] the presentation of the theory of finitary domains is made especially elementary by considering them as filter completions of *neighborhood systems*. Up to isomorphism every finitary domain can be represented this way, and the use of neighborhoods makes the definitions of notions and the constructions of other domains very explicit. This is helpful in discussing questions of *computability* and of *effective presentation* of domains.



*I. Rival (ed.), Ordered Sets, 677–718.*

The primary mathematical feature of the theory of domains is the fact that the category of finitary domains and continuous mappings is *cartesian closed*; that is, the category has products and function spaces and has a natural isomorphism

$$X^{(Y \times Z)} \cong (X^Y)^Z \ .$$

There are several important sub-cartesian-closed categories: the *countably based domains* (= completion of a countable neighborhood system) with continuous maps, and the effectively presented domains with *computable* maps. These categories are not like the category of sets, since every continuous self-map on a finitary domain has a fixed point; indeed, on any finitary domain $X$ there is a continuous fixed-point operator

$$\text{fix} : X^X \to X \ .$$

A notation for maps in a cartesian closed category with fixed-point operators is a rudimentary programming language: it is a *typed λ-calculus with recursion* (expressed by fixed points). Connections with recursion theory are discussed in [6].

More elaborate languages can be given a *denotational semantics* by construction of special effectively presented domains. For example a domain $D$ where

$$D \cong D \times D \cong D^D$$

can be used to give meanings to the terms of an *untyped λ-calculus*, see [1] and [11] for details and [7] for an introduction. Additional expository discussion is found in [8] and [9]. Other examples of semantical definitions, more appropriate to computer science, and of solutions to domain equations will be given. Proof systems as in [3] will also be discussed.

There are many difficult combinatorial problems about (finite) ordered sets that raise serious problems for computing both of a theoretical and a practical nature. We shall not consider such problems here; the reader can find ample examples and explanations elsewhere in this volume. Rather, we shall discuss how partially ordered sets -- even ones of the cardinality of the continuum -- can be used for the conceptual organization of a plan of how to give mathematical meaning to many constructs in higher-level computer languages. The story has been told before (cf. [11], [3], [4], and [7]), but recently I have found a way of making the details quite a bit more elementary. A fuller presentation is contained in [10], which I hope can be expanded soon into a book; the lecture notes can also be consulted for more examples and many exercises.

Even though we propose working with non-denumerable ordered sets, the constructive and computable nature of the study is kept firmly in hand by restricting attention to those posets that can be obtained as "completions" of certain countable partial semi-lattices that can be computably described (*via* recursive presentations, for example). This is true of the real numbers, but they live in a category of *totally* ordered sets; it was found that for many reasons it is more fruitful to work in a category of posets. Instead of giving the abstract definitions all at once, we begin with some examples that are meant to convince the reader of the reasonableness of the final choice of a category. The choice to be described here is not the only one, but it is a very simple one with many familiar properties; it might even be argued that it is a minimal choice. But we leave such discussions for another time.

It is often suggested that I discovered continuous lattices as a consequence of my search for models of the Church-Curry λ-calculus. But, as was explained in [2], it was exactly the other way 'round: *after* I discovered the proper category, I realized that there was enough "elbow-room" to have models for λ-calculus. I had indeed many times denied there was good mathematical content to λ-calculus, and I was searching for an *alternative*. As I argue in [8], however, the search was very 'round about and could have taken care of λ-calculus much sooner. Be that as it may, the topological and categorical import of the initial idea is now well understood (cf. [2]), and the book [1] of Barendregt ties up the models well with the formal theories (though, perhaps, the last word has not quite been said). In [9] I indicate that the so-called *categorical logic* might also throw some light on the matter, since any category whatsoever can be used as a *site*, or a basis, for the construction of more grandiose categories (called *topoi*) which have more satisfactory logical properties (more closure conditions, for instance). Whether this is a good idea remains to be seen, and we have no space to discuss it further now.

## 1. MOTIVATING EXAMPLES

The place where I actually started my investigations was from ideas in recursion theory with a very common-place structure: the partial functions from numbers to numbers. Let $\mathbb{N}$ be the set of natural numbers, and contrast

$$f : \mathbb{N} \to \mathbb{N} \quad \text{with} \quad g : \mathbb{N} \rightharpoonup \mathbb{N} \;.$$

The first property means that $f$ is a *total* (everywhere defined) function; while the second means that $g$ is only a *partial* (partially defined) function from $\mathbb{N}$ into $\mathbb{N}$ (possibly even nowhere defined). Turning now to posets we have:

EXAMPLE 1. The set $(\mathbb{N} \rightharpoonup \mathbb{N})$ of all partial number-theoretic functions is partially ordered by the relation:

$$f \subseteq g \quad \text{iff whenever } f(n) \text{ is defined, then so is } g(n) \text{ defined and } f(n) = g(n).$$

It is permissible to use the symbol "$\subseteq$" here, since -- as everyone can see -- the relation defined is the same as that of *inclusion between the graphs* of the functions regarded as sets of ordered pairs of numbers in the standard way. (Obviously in this example we could have used other sets aside from $\mathbb{N}$.) Note that the total functions are just the *maximal* elements of the poset $(\mathbb{N} \rightharpoonup \mathbb{N})$. □

Functions in $(\mathbb{N} \rightharpoonup \mathbb{N})$ are mappings between numbers, but there are as well many important *mappings between functions*. Let us call these *functionals*, or operators. For instance, if "$\circ$" denotes *composition* of functions (and it works in a natural way for partial functions), then the correspondence, say,

$$f \mapsto a \circ f \circ b \circ f \;,$$

where $a$ and $b$ are fixed functions, is a good example of a functional. Let us write

$$F(f) = a \circ f \circ b \circ f$$

for all $f \in (\mathbb{N} \rightharpoonup \mathbb{N})$. Now $(\mathbb{N} \rightharpoonup \mathbb{N})$ is a set, and $F$ is well defined on *all* arguments in this set; we can thus write in the set-theoretical sense that

$$F : (\mathbb{N} \to \mathbb{N}) \to (\mathbb{N} \to \mathbb{N}) \ .$$

But $F$ is hardly an arbitrary set-theoretical mapping; in fact, if $a,b,f$ are computable (say, partial recursive), then so is $F(f)$. The question is whether $F$ has some essential order-theoretic property that might characterize "good" functionals.

The hint about computability is not misleading, because it is rather straightforward to argue that computable functionals are "continuous", and that continuity is an easily expressible order-theoretic property. This was known for a very long time in recursion theory. We simply remark that to compute $F(f)(n)$ at an argument $n$ we need only compute $f$ at a *finite number* of well-determined arguments (namely, $f(n)$ and $f(b(f(n)))$, and then the answer comes by feeding into $a$ this last number). Now $F(f)(n)$ may not be defined -- it all depends upon a chain of events involving places where $a$, $b$, and $f$ are defined -- but when it is we can indeed find the answer. In the choice of the definition of this $F$ we generally had to compute *two* values of $f$ for each value of $F(f)$, but the reader can find for himself other functionals that demand the computation of several values (with the exact number even depending upon $n$).

Setting aside worries about computability and focusing instead just on questions of dependency, we can say that a functional $F$ is *continuous* iff for $f \in (\mathbb{N} \to \mathbb{N})$ and $n \in \mathbb{N}$, the value of $F(f)(n)$ only depends on some finite number $f(k_0),\ldots,f(k_{m-1})$ of values of $f$. (That is, if $g$ agreed with $f$ at these arguments, then $F(f)(n) = F(g)(n)$, provided that $F(f)(n)$ is defined.) This does not *look* like an order-theoretic property, but it is. We recall that in the poset $(\mathbb{N} \to \mathbb{N})$ every function $f$ is the *union* of the finite functions (functions with a finite graph) it contains. We can even write

$$f = \bigcup_{n=0}^{\infty} f_n \ ,$$

where $f_n$ is the restriction of $f$ to $\{0,1,\ldots,n-1\}$. In this representation we find that

$$f_0 \subseteq f_1 \subseteq f_2 \subseteq \cdots \subseteq f_n \subseteq f_{n+1} \subseteq \cdots \ ,$$

so that the "approximations" $f_n$ form a tower. There are many possible towers; there are lots of towers where the $f_n$ are not even finite. Continuous functionals $F$, however, always "respect" towers in the sense that

(*) $$F\left(\bigcup_{n=0}^{\infty} f_n\right) = \bigcup_{n=0}^{\infty} F(f_n) \ .$$

A small amount of thought will convince the reader that conversely, every such functional is continuous. Thus, equation (*), for all towers, is an order-theoretic definition of continuity for functionals on $(\mathbb{N} \to \mathbb{N})$. (This notion is in fact topological continuity in a suitable topology, as will be explained below.)

It is easy to find other examples of continuous functionals and to find examples of other similar domains for such functions. It is also easy to extend the definition of continuity to functionals of several variables. A nice example, which is also well known in semigroup theory is:

EXAMPLE 2. The set $\mathbb{P}$ of all partial permutations of $\mathbb{N}$ (that is, one-one functions on subsets of $\mathbb{N}$ into $\mathbb{N}$) is a subposet of $(\mathbb{N} \to \mathbb{N})$ on which composition and inverse are continuous operations. □

Regarding a set, better a subset of $\mathbb{N}$, as a partial function from $\mathbb{N}$ to a one-element set, we have similarly:

EXAMPLE 3. The set $\mathbb{S}$ of subsets of $\mathbb{N}$ is a poset under ordinary inclusion where the operations of union, $\bigcup$, and intersection, $\bigcap$, are continuous. □

Note that complementation is *not* continuous, because it follows from (*) that all continuous functionals are monotone:

(**) $$f \subseteq g \text{ always implies } F(f) \subseteq F(g) \ .$$

Clearly set complementation is not even monotone. (Property (**), by the way, does not imply (*).) If a continuous complementation operation is desired, then a different poset should be investigated: the partial functions from $\mathbb{N}$ into a *two-element* set. Just as $\mathbb{P}$ is not quite a group, this new example is not quite a Boolean algebra -- but it is none the worse for that.

Another example of a different kind which was very suggestive to me was:

EXAMPLE 4. The set $\mathbb{F}$ of all continuous functionals from $(\mathbb{N} \to \mathbb{N})$ into $(\mathbb{N} \to \mathbb{N})$ is a poset under the relation:

$$F \subseteq G \quad \text{iff} \quad F(f) \subseteq G(f) \ , \quad \text{for all } f \in (\mathbb{N} \to \mathbb{N}) \ .$$

Under this definition least upper bounds of towers exist in $\mathbb{F}$, and an operation like the passage from $a \in (\mathbb{N} \to \mathbb{N})$ to the functional

$$f \mapsto a \circ f$$

is a continuous operator from $(\mathbb{N} \to \mathbb{N})$ into $\mathbb{F}$. An example of a continuous operator going in the other direction is the *least fixed-point* operator defined for $F \in \mathbb{F}$ by

$$\mathrm{fix}(F) = \bigcup_{n=0}^{\infty} F^n(\bot) ,$$

where $\bot$ is the empty function (or least element of $(\mathbb{N} \to \mathbb{N})$, and where $F^n$ is the $n$-fold composition of $F$ with itself. Using the continuity of $F$, it can be shown by a standard argument that $f = \mathrm{fix}(F)$ is the least function in the poset $(\mathbb{N} \to \mathbb{N})$ such that $f = F(f)$. What is being stressed now, however, is not so much that this least $f$ exists, but that the whole operator fix is itself continuous. □

We are not stopping to verify all the (easy) assertions in these examples. They should be immediately plausible, however, and they should indicate that there is a starting place for a more general theory here. The question to be answered is: what is common among these examples? and what kind of poset structure should be abstracted from them?

## 2. CLOSURE SYSTEMS AND DOMAINS

Taking, as we have already done, a number-theoretic function as a set of ordered pairs, then the poset $(\mathbb{N} \to \mathbb{N})$ *is* a family of sets partially ordered by inclusion between sets. Of course, any family of sets is a poset in this way, but $(\mathbb{N} \to \mathbb{N})$ has further closure properties. For one, it is closed under taking *subsets*; BUT this will not be a property to be abstracted from examples, because this is not an order-theoretic property. What do prove to be useful properties are singled out in this definition:

DEFINITION. A *closure system* is a non-empty family of sets closed under intersections of arbitrary non-empty subfamilies and unions of towers. A poset isomorphic to a closure system I call a *domain*. □

The definition requires a few comments. In general towers have to be taken as transfinite towers of any ordinal length, but we will work with families where all the sets are contained in one countable set. In this case simple infinite chains are sufficient, and the worries about full generality are unnecessary. (Actually, even in general it is better to take the closure condition as employing unions of *directed* subfamilies and to forget about linearly ordered towers.) A closure system by my definition has a *least element* $\perp$, which is just the intersection of the whole family of sets. Often $\perp$ is the empty set $\emptyset$, but this is not necessarily so, and it is not being assumed. It is usual also to have a *greatest element* (which therefore contains all other sets as subsets), but this is definitely not being assumed. A closure system by this definition is not a directed family of sets; two sets in the system need not be included in a common set and are thus "inconsistent" with each other. We will discuss consistency more below.

It should be quite clear that $(\mathbb{N} \to \mathbb{N})$ is a closure system as is $\mathbb{P}$ and, of course $\mathbb{S}$. The question about $\mathbb{F}$ is not quite so obvious and is one of the main points of the theory. If a family of sets is defined by "closure conditions", it is a closure system. What is meant by closure conditions? There are several types. First we have:

(a) $$k \in x ,$$

where $k$ is a given constant. Then we have:

(b) $$i,j,\ldots \in x \text{ always implies } \sigma(i,j,\ldots) \in x ,$$

where $\sigma$ is a given operation under which the set $x$ is to be closed. Next we have:

(c) $$i,j,\ldots \in x \text{ always implies } \Pi(i,j,\ldots) ,$$

where $\Pi$ is a property of the $i,j,\ldots$ not mentioning the set $x$. These are the usual types of closure properties (strictly speaking (c) is more of a consistency condition), but we could also combine (b) and (c) into:

(d) $$i,j,\ldots \in x \text{ always implies either } \Pi(i,j,\ldots) \text{ or } \sigma(i,j,\ldots) \in x ,$$

which means $x$ is closed under $\sigma$ for sequences $i,j,\ldots$ *outside* $\Pi$. More generally we can take as a closure condition any universally quantified property of the set $x$ made by a finite *disjunction* of clauses:

$$\Pi(i,j,\ldots),\ \tau(i,j,\ldots) \notin x,\ \sigma(i,j,\ldots) \in x ,$$

for various properties $\Pi$, functions $\tau$ and $\sigma$, PROVIDED there is *at most one* of the third kind (that is, the positive atomic clause involving membership in $x$). Any (infinite) conjunction of closure conditions defines a closure system -- provided there is at least one set satisfying all of them. Many, many examples of closure systems can now be produced even with simple choices of clauses.

Let us return for a moment to the question of consistency. Let $\mathbb{C}$ be a closure system. We can assume for the sake of argument that $\mathbb{N}$ is the underlying set, so that $\mathbb{C}$ is a family of subsets of $\mathbb{N}$. Not every set $x \subseteq \mathbb{N}$ is consistent relative to $\mathbb{C}$, however. *$\mathbb{C}$-consistency* means simply that $x \subseteq y$ for some $y \in \mathbb{C}$; that is, $x$ can be extended to a closed set. Every $\mathbb{C}$-consistent set has a *closure* defined in the obvious way as the least closed extension:

$$\bar{x} = \bigcap \{y \in \mathbb{C} \mid x \subseteq y\} .$$

This kind of closure operation has nice properties, and we can use them to characterize closure systems in the well-known way. All that need be kept in mind over the usual simple examples is that we are not assuming that all subsets of the underlying set are consistent.

DEFINITION. A *consistency system* is a non-empty family of sets closed under taking subsets and unions of towers (or: directed unions). □

The $\mathbb{C}$-consistent sets, call this the family $\mathbb{C}_0$ for short, form a consistency system. But $\mathbb{C}_0$ is far too weak a notion to determine $\mathbb{C}$. We need the closure operation $\bar{\ }: \mathbb{C}_0 \to \mathbb{C}_0$ so as to define $\mathbb{C}$ as the "closed" elements:

$$\mathbb{C} = \{x \in \mathbb{C}_0 \mid x = \bar{x}\} .$$

For instance, $\bar{\ }$ might be something clever and $\mathbb{C}_0$ might be something boring like the power set of $\mathbb{N}$. Indeed many different closure systems can have the same consistency system. We definitely have to know the closure operator to recapture $\mathbb{C}$ from $\mathbb{C}_0$ . What are the properties? Just the familiar ones:

DEFINITION. A *closure operator* is a function $\bar{\ }$ defined on a consistency system $\mathbb{C}_0$ so that:

(i) $x \subseteq \bar{x} = \bar{\bar{x}}$ for all $x \in \mathbb{C}_0$ ;

(ii) $\bar{\ }$ commutes with unions of towers contained in $\mathbb{C}_0$ .

As a consequence, every closure operator is monotone:

(iii) $x \subseteq y$ implies $\bar{x} \subseteq \bar{y}$ for $x,y \in \mathbb{C}_0$ .

Or in words a closure operator is an inflationary, idempotent, continuous function on a consistency system. □

Condition (ii) is usually described as requiring $\overline{\phantom{x}}$ to be an *algebraic* closure operator rather than a *topological* closure, which commutes with finite unions and does not satisfy (ii).

FOLK THEOREM. *There is an (obvious) one-to-one correspondence between closure systems and closure operators.* □

The only point that might be a trifle nonstandard in our discussion is the emphasis on consistency systems, since it is more common to work in the power set of a set as the consistency system. The generalization can be seen as a natural one, however, especially when it is considered how consistency systems relate to finite sets.

PROPOSITION. *Let $\mathbb{C}_0$ be a consistency system and let $\mathbb{C}_{00}$ be the family of finite sets in $\mathbb{C}_0$. Then $\mathbb{C}_{00}$ determines $\mathbb{C}_0$, since a set $x$ is in $\mathbb{C}_0$ just in case the family of finite subsets of $x$ is included in $\mathbb{C}_{00}$. Moreover, every non-empty family of finite sets closed under taking subsets is a $\mathbb{C}_0$ for some consistency system $\mathbb{C}_0$.* □

The proof of the proposition is clear. As a direct consequence we can also see that $\mathbb{C}_0$ is a consistency system if and only if it has the property that a set belong to $\mathbb{C}_0$ just in case all its finite subsets belong to $\mathbb{C}_0$. (This assumes that $\mathbb{C}_0$ is non-empty or, equivalently, that $\emptyset \in C_0$.) In other words consistency systems are families of sets of "finite character".

The finite (consistent) sets are also interesting for the closure operator: every closed set is the union of the closures of its finite subsets. We call a set $x \in \mathbb{C}$ which is the closure of a finite subset a *finitely generated* element, or just a finite element of $\mathbb{C}$ for short. It is not difficult to characterize the finite elements of $\mathbb{C}$ by a purely order-theoretic property. Neither is it difficult to characterize exactly the type of poset the set of finite elements of a closure system is, again in purely order-theoretic terms. We allude to such a characterization in the next section. We do not stress here these order-theoretic questions, since the main purpose of the presentation is to show that there is a multitude of easily obtained closure systems (domains). What we wish to do next is to make a category out of the class of domains and to discuss construction principles in this category.

Before turning to more general matters, however, note that $\mathbb{N}$ could be regarded as (almost) a closure system -- in a straightforward and trivial sense. Instead of $\mathbb{N}$ use

$$\check{\mathbb{N}} = \{\{n\} \mid n \in \mathbb{N}\} \cup \{\emptyset\} .$$

This is a very boring closure system from the order-theoretic point of view. Note that $(\mathbb{N} \to \mathbb{N})$ can be construed (up to order-theoretic isomorphism) as the domain of continuous functions $f : \check{\mathbb{N}} \to \mathbb{N}$ where $f(\bot) = \bot$ . (These are often called *strict* functions.) In this way we can keep all our entities in the same category. A less trivial question concerns the totality of continuous functions on an arbitrary domain. We will give a complete analysis.

## 3. NEIGHBORHOOD SYSTEMS AND THE REPRESENTATION OF DOMAINS

Closure systems are actually so familiar that it would be possible to work entirely with them and to form the desired category that way. But it recently occurred to me that a suitable form of neighborhood systems is rather more elementary and suggestive, making pictures easier to draw for illustrating many notions and constructs. All I shall actually do is give another set-theoretical *representation* of closure systems, but it is one that puts the very useful finite elements more on the surface.

DEFINITION. Let $\Delta$ be a given non-empty set. A *neighborhood system* over $\Delta$ is a family $\mathcal{D}$ of subsets of $\Delta$ where

(i) $\Delta \in \mathcal{D}$; and

(ii) whenever $X, Y, Z \in \mathcal{D}$ and $Z \subseteq X \cap Y$, then $X \cap Y \in \mathcal{D}$.

A *finite* collection of neighborhoods is *consistent* (in $\mathcal{D}$) if the intersection contains another neighborhood. An *arbitrary* collection is *consistent* just in case every finite subcollection is. □

The purpose of the theory of domains is to have a suitable notion of partial element generalizing the way partial functions behave in $(\mathbb{N} \to \mathbb{N})$. A number of reasons why such a notion is to be regarded as "necessary" will be discussed in due course. The intuition behind neighborhood systems is this: think of the elements of $\Delta$ as "tokens" or rough examples of (some of) the desired partial elements. The job of a neighborhood is to collect together all of the tokens sharing the same (finite) amount of information. A token $t \in \Delta$ may not be "consistent", however, because we do not assume that

$$\{X \in \mathcal{D} \mid t \in X\}$$

is a consistent set of neighborhoods. The "message" that $t$ carries could refer to several quite different elements. For the structure of the domain the exact form of the tokens is not really important; we shall see in several ways that one domain can have many representations. The poset structure of $\mathcal{D}$ is what we regard as important. We could, of course, axiomatize the kind of semilattice structure we need; but it is so easy to represent such (partial) semilattices as neighborhood systems, the axioms are not especially needed. Surely the reader has to admit that he knows dozens of neighborhood systems made from familiar mathematical ingredients.

We have spoken about "elements" without defining precisely what we want. We remarked that the elements of the domain to be defined by $\mathcal{D}$ are not necessarily the members of the set $\Delta$. A consistent set of neighborhoods ought to have something to do with an element, and it is a (mathematical) temptation to take a *maximal* consistent set as representing an element. There is nothing wrong with that, except such elements are obviously the *total* elements of the domain defined by $\mathcal{D}$, since relative to $\mathcal{D}$ we have said as much as is consistently possible about each one. These are interesting elements, but they need not be constructively attainable, whereas partial elements are.

A straightforward way of determining an element is to give a sequence of neighborhoods that "converges" to the element. If the $X_n \in \mathcal{D}$ are chosen so that

$$X_0 \supseteq X_1 \supseteq X_2 \supseteq \cdots \supseteq X_n \supseteq X_{n+1} \supseteq \cdots ,$$

then a sense of convergence is fairly clear. But the question is: convergence to what? Perhaps that is not the right question. Suppose we have another sequence:

$$Y_0 \supseteq Y_1 \supseteq Y_2 \supseteq \cdots \supseteq Y_n \supseteq Y_{n+1} \supseteq \cdots ,$$

can we say that the two sequences are shrinking down on the "same element"? If we can, then we have an equivalence relation between neighborhoods and can identify elements simply as *being* such equivalence classes. In other words, elements are introduced as ideal elements corresponding to the notion of limit provided by the sequences of neighborhoods. Now since all we really have is the order structure on $\mathcal{D}$ (that is, the $\subseteq$-relation), we can not expect the definition to be very subtle. In fact, we shall say that the two sequences of neighborhoods are equivalent (converge to the same element(s), are equally convergent) if they go "equally far" or formally:

$$\forall n \exists m . X_n \supseteq Y_m \text{ and } \forall m \exists n . Y_m \supseteq X_n .$$

This is clearly an equivalence relation, but it is clumsy to work with equivalence classes if a canonical choice from each class can be made. The official definition not only does this but avoids having to restrict attention to countable systems $\mathcal{D}$, where simple *sequences* of neighborhoods are sufficient.

DEFINITION. The (ideal) *elements* of a neighborhood system $\mathcal{D}$ are those subfamilies $x \subseteq \mathcal{D}$ where:

(i) $\Delta \in x$;

(ii) $X, Y \in x$ always implies $X \cap Y \in x$;

(iii) whenever $X \in x$ and $X \subseteq Y \in \mathcal{D}$, then $Y \in x$.

The totality of all such elements is denoted by $|\mathcal{D}|$. □

Either as a subfamily of $\mathcal{D}$ or as a subfamily of the power set of $\Delta$, it is easy to see that the property $x \in |\mathcal{D}|$ is defined by closure conditions. We can then establish the:

PROPOSITION. *For any neighborhood system $\mathcal{D}$ the totality $|\mathcal{D}|$ is a closure system -- the domain determined by $\mathcal{D}$.* □

There is no mystery to the definition of an element: such an $x$ is just a $\mathcal{D}$-filter, and $|\mathcal{D}|$ could be called the *filter completion* of $\mathcal{D}$. Every sequence of neighborhoods, $X_n \supseteq X_{n+1}$, determines a *filter base*, and we could say that

$$x = \{Y \in \mathcal{D} \mid \exists n . X_n \subseteq Y\}$$

is the (principal) "limit" of the sequence. Clearly two sequences determine the same filter if and only if they are equivalent in the sense mentioned earlier. Such ideas have been around for a long time; what is to be added to them is the way we relate different domains in the natural category that they form.

Before discussing mappings in general, a remark on the simpler notion of *isomorphism* is appropriate. A neighborhood system $\mathcal{D}$ is a poset under inclusion; so is the corresponding domain $|\mathcal{D}|$. If $\mathcal{D}_0$ and $\mathcal{D}_1$ are two neighborhood systems, then there could be two kinds of isomorphism between them, one at the level of neighborhoods and one at the level of elements. We can prove:

PROPOSITION. *Two neighborhood systems are isomorphic (as posets) iff the corresponding domains are isomorphic. In general an isomorphism between domains makes finite elements correspond to finite elements. For a neighborhood system, the finite elements are dually isomorphic to the neighborhood system, because the finite elements are exactly the principal filters.* □

We stated earlier that tokens $t \in \Delta$ might not necessarily determine elements. Examples are easy to find showing this. Nevertheless it is always possible to represent a system in a form where the tokens are exactly the elements:

PROPOSITION. *Every neighborhood system $\mathcal{D}$ is isomorphic to the neighborhood system*

$$[\mathcal{D}] = \{[X] \mid X \in \mathcal{D}\},$$

*where for $X \in \mathcal{D}$ we have:*

$$[X] = \{x \in |\mathcal{D}| \mid X \in x\}.$$

*The underlying set for $[\mathcal{D}]$ is $|\mathcal{D}|$.* □

On the other hand, we could also equally well make the finite elements the tokens. It is more direct, for neighborhood systems, to make the neighborhoods themselves the tokens:

PROPOSITION. *Every neighborhood system $\mathcal{D}$ is isomorphic to the neighborhood system*

$$\downarrow \mathcal{D} = \{\downarrow X \mid X \in \mathcal{D}\}$$

*where for $X \in \mathcal{D}$ we have:*

$$\downarrow X = \{Y \in \mathcal{D} \mid Y \subseteq X\}.$$

*The underlying set for $\downarrow \mathcal{D}$ is $\mathcal{D}$ itself.* □

This last -- very elementary -- theorem suggests that finite elements might be used as tokens more generally, and the basic representation result shows that indeed the abstractness of closure systems is only apparent:

REPRESENTATION THEOREM. *Every closure system (as a poset) is isomorphic to the system of elements of a neighborhood system. Hence, the domains determined by neighborhood systems give us up to isomorphism all domains.* □

The proof, which we do not give in detail, is quite direct. If $\mathbb{C}$ is a closure system, let $\Delta$ be the set of consistent *finite* sets for $\mathbb{C}$. ($\Delta = \mathbb{C}_{00}$ in the earlier notation.) For $F \in \Delta$ define with the aid of the closure operator

$$\uparrow F = \{G \in \Delta \mid F \subseteq \bar{G}\} .$$

Then the system

$$C = \{\uparrow F \mid F \in \Delta\}$$

is such that $\mathbb{E}$ is isomorphic to $|C|$. It follows, therefore, that neighborhood systems give us the full range of domain structures, in what will be found to be a convenient representation.

## 4. APPROXIMABLE MAPPINGS AND THE CATEGORY OF DOMAINS

The motivating example of functionals in Section 1 is quite sufficient to suggest the definitions for the general case. Of course $(\mathbb{N} \to \mathbb{N})$ is special in that it has only countably many finite elements, in which case simple infinite towers are all that is needed to define continuity. The general domain requires the use of arbitrary directed sets of elements for that kind of definition. But if we remember that in any domain the elements are completely determined by the finite elements they contain (which always form a directed set), then the definition can be expressed in terms of neighborhoods in a much more elementary way.

DEFINITION. Let $\mathcal{D}_0$ and $\mathcal{D}_1$ be two neighborhood systems. An *approximable mapping* $f : \mathcal{D}_0 \to \mathcal{D}_1$ is a binary relation between the sets of neighborhoods such that

(i) $\Delta_0 f \Delta_1$ ;

(ii) $XfY$ and $XfY'$ always imply $Xf(Y \cap Y')$; and

(iii) $XfY$, $X' \subseteq X$, and $Y \subseteq Y'$ always imply $X'fY'$. □

The idea of this definition is that $f : \mathcal{D}_0 \to \mathcal{D}_1$ should be regarded as an input-output relation. When we say "$XfY$" we mean that if we are willing to say at least $X$ about the input, then $f$ will tell us at least $Y$ about the output -- perhaps even more. So condition (i) is the trivial requirement that *no* information about the input assures *no* information about the output. Condition (ii) is a consistency condition: if $f$ is willing, say at different times, to give two output statements for the same input information, then it must be willing to allow the *conjunction* of those output statements. Finally, condition (iii) is a monotonicity condition: given an input-output pair of neighborhoods, if either the input is strengthened or the output is weakened, then $f$ must still be willing to go along with the resulting pair. Intuitively, this should all make good sense; formally, the conditions of the definition are just those needed to prove the next result:

PROPOSITION. *Approximable mappings* $f : \mathcal{D}_0 \to \mathcal{D}_1$ *always determine continuous functions between the domains* $|\mathcal{D}_0|$ *and* $|\mathcal{D}_1|$ *by virtue of the formula:*

$$f(x) = \{Y \in \mathcal{D}_1 \mid \exists\, X \in x \,.\, XfY\} \,.$$

*Moreover, every continuous function can be obtained this way.* □

The verification of the statement is simple, for we put into (i)-(iii) the conditions required to show that the image under the relation $f$ of a filter $x \in |\mathcal{D}_0|$ is indeed a filter in $|\mathcal{D}_1|$. In the other direction, if $f : |\mathcal{D}_0| \to |\mathcal{D}_1|$ is continuous, then the neighborhood relation that determines it is just:

$$Y \in f(\uparrow X) \;,$$

where we define

$$\uparrow X = \{X' \in \mathcal{D}_0 \mid X \subseteq X'\}$$

as the principal filter corresponding to $X$. We use the same symbol for an approximable mapping $f : \mathcal{D}_0 \to \mathcal{D}_1$ and a continuous $f : |\mathcal{D}_0| \to |\mathcal{D}_1|$, because by what we have just said we have an order-preserving isomorphism between the two kinds of objects. (Recall, the order relation between continuous mappings is the pointwise order.) This observation together with the obvious fact that conditions (i)-(iii) are closure conditions proves the fundamental:

THEOREM. *The set of continuous functions between two given domains is again (isomorphic to) a domain under the pointwise partial order.* □

The result will become even clearer if we give an explicit notation to the neighborhoods of the function space.

DEFINITION. For two neighborhood systems $\mathcal{D}_0$ and $\mathcal{D}_1$, let $(\mathcal{D}_0 \to \mathcal{D}_1)$ be the neighborhood system over the set

$$\{f \mid f : \mathcal{D}_0 \to \mathcal{D}_1\}$$

consisting of all finite, non-empty intersections of sets of the form

$$[X,Y] = \{f \mid XfY\} \;,$$

where $X \in \mathcal{D}_0$ and $Y \in \mathcal{D}_1$. □

Any family of non-empty sets generates a neighborhood system by closing it under non-empty intersections. What is interesting about the construction just defined is that the neighborhood system gives back just the right set of elements to be a representation of the function space.

THEOREM. *The elements of the neighborhood system* $(\mathcal{D}_0 \to \mathcal{D}_1)$ *form a domain isomorphic to the domain of approximable mappings from* $\mathcal{D}_0$ *into* $\mathcal{D}_1$. □

As with the $[\mathcal{D}]$-construction, we find that tokens equal elements (up to isomorphism). In proving this above result we find that we can identify the finite elements of the function space by a simple formula involving finite sequences of pairs of neighborhoods. Suppose $X_i \in \mathcal{D}_0$ and $Y_i \in \mathcal{D}_1$ are given for each $i < n$. The first question is that of consistency; that is, when do we have

$$\bigcap_{i<n} [X_i, Y_i] \neq \emptyset \ ?$$

This means: when does there exist a function $f$, where for each $i < n$, $f$ maps $[X_i]$ into $[Y_i]$? The answer is: for all subsets $I \subseteq \{i \mid i < n\}$, if the set $\{X_i \mid i \in I\}$ is consistent in $\mathcal{D}_0$, then the set $\{Y_i \mid i \in I\}$ must also be consistent in $\mathcal{D}_1$. That is a very reasonable input-output consistency condition on a sequence of pairs of neighborhoods. It is easily seen to be necessary; the sufficiency is proved by showing that the consistency condition is just what is required to show that there is a minimal element $f_0$ of the neighborhood $[X_0, Y_0] \cap \ldots \cap [X_{n-1}, Y_{n-1}]$ given the formula:

$$X f_0 Y \ \text{ iff } \ \bigcap\{Y_i \mid X \subseteq X_i\} \subseteq Y \ ,$$

for arbitrary $X \in \mathcal{D}_0$ and $Y \in \mathcal{D}_1$.

We do not have time to elaborate on these observations, but they prove that if domains can be given *countable* neighborhood systems, then so can the function space. Not only this, but the whole definition is very constructive: from knowing about the poset structure of $\mathcal{D}_0$ and $\mathcal{D}_1$, we know fully about the poset structure of $(\mathcal{D}_0 \to \mathcal{D}_1)$.

It may not have been necessary to introduce two names for the same notion: approximable/continuous mappings. But the definitions look quite different, and the second is a slightly mysterious name -- until one realizes that domains can be made into topological spaces in such a way that continuous maps are exactly the topologically continuous maps. (Hint: use the sets $[X]$ for $X \in \mathcal{D}$ as a basis for a topology on $|\mathcal{D}|$.) The reason for "approximable" is that such maps can be given finite approximations, and are indeed determined by these approximations. In any neighborhood system $\mathcal{D}$, for $X \in \mathcal{D}$ and $x \in |\mathcal{D}|$ the following are equivalent:

$$X \in x \ , \ x \in [X] \ , \ \uparrow X \sqsubseteq x \ ,$$

and we can say the finite element $\uparrow X$ *approximates* $x$. (This is the same as saying that the neighborhood $X$ gives a *finite amount of information* about $x$.) Moreover, we can write:

$$x = \bigsqcup\{\uparrow X \mid X \in x\} \ ,$$

so that every element is the *union* (or *limit*) of its finite approximations; and, hence, it is determined by them. What we have shown is that the same goes for approximable mappings. The statements

$$f \in \bigcap_{i<n} [X_i, Y_i]$$

and

$$X_i f Y_i \ , \ \text{all } i < n$$

are equivalent; and we saw how to define the principal finite element of the indicated neighborhood. The mappings we are talking about are those that can be obtained through finite approximation. (Keep in mind that the precise meaning of the word "finite" is relative to the neighborhood systems under consideration.)

It should be apparent by now that the domains together with the approximable mappings form a *category*, in the technical sense of the word. Just to underline the fact, we say how to form identity and composition maps.

PROPOSITION. *For any neighborhood system* $\mathcal{D}$*, the identity map* $I_{\mathcal{D}} : \mathcal{D} \to \mathcal{D}$ *is approximable, and we have the equivalent definitions:*

(i) $I_{\mathcal{D}}(x) = x$, *for all* $x \in |\mathcal{D}|$; *and*

(ii) $X I_{\mathcal{D}} Y$ *iff* $X \subseteq Y$, *for all* $X, Y \in \mathcal{D}$. □

PROPOSITION. *For given neighborhood systems* $\mathcal{D}_0$, $\mathcal{D}_1$, *and* $\mathcal{D}_2$ *and approximable maps*

$$f : \mathcal{D}_0 \to \mathcal{D}_1 \quad \text{and} \quad g : \mathcal{D}_1 \to \mathcal{D}_2$$

*the composition* $g \circ f : \mathcal{D}_0 \to \mathcal{D}_2$ *is approximable, and we have the equivalent definitions:*

(i) $(g \circ f)(x) = g(f(x))$, *for all* $x \in |\mathcal{D}_0|$; *and*

(ii) $X(g \circ f)Z$ *iff* $\exists\, Y$ . $XfY$ *and* $YgZ$, *for all* $X \in \mathcal{D}_0$ *and* $Z \in \mathcal{D}_2$ . □

The category of domains has many good properties. We have seen how to construct function spaces; the construction of products is even easier.

PROPOSITION. *For neighborhood systems* $\mathcal{D}_0$ *and* $\mathcal{D}_1$ *over* $\Delta_0$ *and* $\Delta_1$ *assume*

$$\Delta_0 \cap \Delta_1 = \emptyset \ .$$

*Define*

$$\mathcal{D}_0 \times \mathcal{D}_1 = \{X \cup Y \mid X \in \mathcal{D}_0, \ Y \in \mathcal{D}_1\} \ .$$

*Then* $(\mathcal{D}_0 \times \mathcal{D}_1)$ *is a neighborhood system and the domain* $|\mathcal{D}_0 \times \mathcal{D}_1|$ *is order isomorphic to the usual cartesian product of* $|\mathcal{D}_0|$ *and* $|\mathcal{D}_1|$. *The pairing function on elements* $x \in |\mathcal{D}_0|$ *and* $y \in |\mathcal{D}_1|$ *can be defined by*

$$(x,y) = \{X \cup Y \mid X \in x, \ Y \in y\} \ .$$

*This is a filter in* $|\mathcal{D}_0 \times \mathcal{D}_1|$ *and the pointwise pairing of approximable maps is again approximable. The obvious projection maps are also approximable.* □

We cannot stop to verify or even mention all the details, but the category of domains not only has categorical products, but the function-space construction is categorically well behaved. There is a technical name for the well-behavedness: the category of domains and approximable maps is *cartesian closed.* This means in particular that the *evaluation map,*

$$\mathrm{eval} : (\mathcal{D}_1 \to \mathcal{D}_2) \times \mathcal{D}_1 \to \mathcal{D}_2 \ ,$$

is approximable (that is, it exists in the category). Furthermore, the systems

$$((\mathcal{D}_0 \times \mathcal{D}_1) \to \mathcal{D}_2) \quad \text{and} \quad (\mathcal{D}_0 \to (\mathcal{D}_1 \to \mathcal{D}_2))$$

are "naturally" isomorphic. This makes it possible to write down an endless number of logically defined maps in the category.

The category is much more than just a cartesian closed category, however. For instance, the *fixed-point operator,*

$$\mathsf{fix} : (\mathcal{D} \to \mathcal{D}) \to \mathcal{D} ,$$

is approximable, where

$$\mathsf{fix}(f) = f(\mathsf{fix}(f))$$

holds for all $f \in |\mathcal{D} \to \mathcal{D}|$. And there are a multitude of other equations satisfied by the fixed-point operator. In another direction, the category supports many other interesting functors besides $\times$ and $\to$. Reference [10] can be consulted for details.

## 5. DOMAIN EQUATIONS AND THE UNIVERSAL DOMAIN

In Section 1 we began with some simple examples which led us directly to the general definition of closure/neighborhood systems. In Section 4 we saw that the category of domains had some good closure properties, which of course let us multiply our examples. But the product and function-space constructs are pretty bland; what is needed is a way of obtaining more "special-purpose" examples where the domains themselves have interesting or useful structure. One way of constructing good domains is by a recursion which iterates given functors.

A convenient trick for defining quite complex neighborhood systems is to construe the neighborhoods as sets of finite sequences. A two-letter alphabet is sufficient, and we take $\Sigma = \{0,1\}$, the binary alphabet. We then let $\Sigma^*$ be the free semi-group of all binary words, where we regard $\Sigma$ itself as equal to the set of words of length 1 in the usual way. The empty word is $\Lambda$, and for $\sigma, \tau \in \Sigma^*$ we denote concatenation by $\sigma\tau$. If $X \subseteq \Sigma^*$, then

$$\sigma X = \{\sigma\tau \mid \tau \in X\} .$$

Note that if $X, Y \subseteq \Sigma^*$, then the sets $0X$ and $1Y$ are always disjoint. For forming neighborhood systems in this section we shall always take $\Delta = \Sigma^*$ and assume that the systems do not contain the empty set.

Let us continue our list of examples of special domains.

EXAMPLE 5. Let $\mathcal{B}$ be the unique family of non-empty subsets of $\Sigma^*$ defined by:

$$\mathcal{B} = \{\Sigma^*\} \cup \{0X \mid X \in \mathcal{B}\} \cup \{1X \mid X \in \mathcal{B}\} .$$

The least element in $|\mathcal{B}|$ is

$$\bot = \{\Sigma^*\} ,$$

and for $x \in |\mathcal{B}|$ and $\sigma \in \Sigma^*$ we define:

$$\sigma x = \{Y \in \mathcal{B} \mid \exists\, X \in x \,.\, \sigma X \subseteq Y\} \,. \quad \square$$

The family $\mathcal{B}$ is simply the least family of sets generated from $\Sigma^*$ by the two set-forming operations shown. That $\mathcal{B}$ is a neighborhood system is easily proved by induction using a law like:

$$\sigma X \cap \sigma Y = \sigma(X \cap Y) \;;$$

that the operation $x \mapsto \sigma x$ is approximable is also obvious since $\sigma X \subseteq Y$ is the appropriate neighborhood relation. Well, these are details; the real question is: what is the meaning of $\mathcal{B}$? The explanation is facilitated by a general definition.

DEFINITION. For any two neighborhood systems (over $\Sigma^*$) $\mathcal{D}_0$ and $\mathcal{D}_1$, let

$$\mathcal{D}_0 + \mathcal{D}_1 = \{\Sigma^*\} \cup \{0X \mid X \in \mathcal{D}_0\} \cup \{1Y \mid Y \in \mathcal{D}_1\} \,.$$

This construct is called the *sum* of the domains. □

The definition of the product could be made to look similar for systems over $\Sigma^*$:

$$\mathcal{D}_0 \times \mathcal{D}_1 = \{\{\Lambda\} \cup 0X \cup 1Y \mid X \in \mathcal{D}_0,\, Y \in \mathcal{D}_1\} \,.$$

The reason for adding in the empty string $\Lambda$ is that

$$\Sigma^* = \{\Lambda\} \cup 0\Sigma^* \cup 1\Sigma^* \,.$$

If we did not put it into the product definition, then the result $\mathcal{D}_0 \times \mathcal{D}_1$ would only be a system over $\Sigma^* - \{\Lambda\}$.

In the category of domains, both the product and the sum can be made into functors in an obvious way. The product *is* the categorical (cartesian) product. The sum, however, *is not* the categorical coproduct. Such does not exist in the category: though there is a terminal object (say, the trivial system $\{\Sigma^*\}$), there is no initial object, and no way to make disjoint sums. The uniform occurrence of $\perp$ in all domains is one of the blocks to forming disjoint sums. The elements of the above defined $\mathcal{D}_0 + \mathcal{D}_1$ could be regarded as a disjoint sum of the two posets $|\mathcal{D}_0|$ and $|\mathcal{D}_1|$ *plus* a new $\perp$-element. Though it is not the categorical coproduct, it still has many good properties and is a useful functor.

We can now detail in greater depth the idea behind the system $\mathcal{B}$. The *finite* elements of $\mathcal{B}$ are all of the form $\sigma\bot$ where $\sigma$ is a finite word (or sequence) of 0's and 1's. We find that the relationship

$$\sigma\bot \subseteq \tau\bot$$

holds if and only if $\sigma$ is an initial segment of $\tau$. Of course $\bot$ is included in everything. What do these elements approximate? Answer: *infinite* sequences over the binary alphabet; these are in fact the *total* elements of $|\mathcal{B}|$. If $\rho$ is an infinite sequence, and $\rho_n$ is the initial segment of $\rho$ of length $n$, then we can unambiguously identify $\rho$ with this element of $|\mathcal{B}|$:

$$\bigcup_{n=0}^{\infty} \rho_n \bot .$$

This is nothing more than the filter

$$\{\rho_n \Sigma^* \mid n \in \mathbb{N}\} ,$$

and it is easily proved to be maximal. So the domain determined by $\mathcal{B}$ is the domain of binary sequences, where the finite sequences are only *partial* and are extendable to the infinite *total* sequences. Moreover, $\mathcal{B}$ satisfies a *domain equation* of the form

$$\mathcal{B} \cong \mathcal{B} + \mathcal{B} .$$

Not only this, but $\mathcal{B}$ is uniquely determined up to isomorphism as the *initial solution* of the equation. (This notion can be made precise categorically by introducing the *algebras* corresponding to a functor from domains to domains; in this case the functor is

$$X \mapsto X + X .$$

An initial algebra is a "free" algebra on zero generators.) This initial algebra characterization of $\mathcal{B}$ is convenient to work with, and it gives a "recursive definition" for the domain.

EXAMPLE 6. Let $\mathcal{C}$ be the neighborhood system

$$\mathcal{C} = \mathcal{B} \cup \{\{\sigma\} \mid \sigma \in \Sigma^*\} .$$

The domain equation satisfied by $\mathcal{C}$ is:

$$\mathcal{C} \cong \{\Sigma^*\} + \mathcal{C} + \mathcal{C} .$$

We can identify the words $\sigma \in \Sigma^*$ with the principal filters $\uparrow\{\sigma\}$, which are also maximal in $|\mathcal{C}|$. □

The difference between $B$ and $C$ is that $C$ allows for two kinds of finite elements: the partial ones of the form $\sigma\bot$ , and the total ones of the form $\sigma$. Both $B$ and $C$ allow the operations $x \mapsto \sigma x$, and both allow for a more general associative concatenation operation $x,y \mapsto xy$. In $B$, the operation is trivial and satisfies the equation

$$xy = x \ .$$

In $C$ the situation is less extreme, but still the equation is satisfied for *infinite* $x$. It is an interesting question to find less one-sided domains that define a semigroup extending $\Sigma^*$, as an approximable multiplication, and allow infinite elements. There are several possibilities, but the author does not understand them very well yet.

EXAMPLE 7. Let $E$ be recursively defined by:

$$E = \{\Sigma^*\} \cup \{\{00\},\{01\}\} \cup \{10X \cup 11Y \mid X,Y \in E\} \ .$$

The domain equation satisfied by $E$ is:

$$E \cong \{\Sigma^*\} + \{\Sigma^*\} + (E \times E) \ .$$

There are two (total) distinguished elements of $E$:

$$0 = \{X \in E \mid 00 \in X\} \ , \text{ and}$$
$$1 = \{X \in E \mid 01 \in X\} \ .$$

The approximable operation $x,y \mapsto \langle x,y\rangle$ is defined by:

$$\langle x,y\rangle = \{Z \in E \mid \exists\, X \in x\ \exists\, Y \in y \ .\ 10X \cup 11Y \subseteq Z\} \ .$$

There are two projection operations $z \mapsto pz$ and $z \mapsto qz$ defined by:

$$pz = \{X \in E \mid \exists\, Z \in z \ .\ 10\Sigma^* \cap Z \subseteq 10X\} \ , \text{ and}$$
$$qz = \{Y \in E \mid \exists\, Z \in z \ .\ 11\Sigma^* \cap Z \subseteq 11Y\} \ .$$

We have satisfied in $E$ as equations:

$$p\langle x,y\rangle = x \quad \text{and} \quad q\langle x,y\rangle = y \ ,$$

but *not* the opposite:

$$\langle pz,qz\rangle = z \ ,$$

since $E$ is more than just its own cartesian square. □

What is the "meaning" of $E$? The finite elements are exactly those generated by the binary operation $x,y \mapsto \langle x,y\rangle$ from $\bot$ , 0, and 1. Those not containing $\perp$ are *total*; while any involving $\perp$ are *partial*. This can be seen from the inclusions:

$$\perp \subseteq x \text{ , and}$$

$$\langle x,y\rangle \subseteq \langle x',y'\rangle \quad \text{if} \quad x \subseteq x' \text{ and } y \subseteq y' \text{ .}$$

We could diagram such elements as *trees* with binary branching and with the ends of the branches marked with $\perp$ or 0 or 1. Besides these finite trees there are also many, many infinite trees. For example, the equation

$$a = \langle 0,\langle a,1\rangle\rangle$$

has a unique solution in $E$, and the diagram of the infinite, total tree is clear:

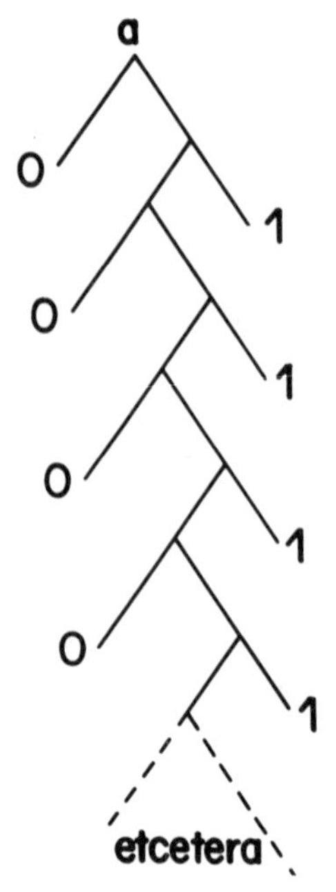


We could also think of $E$ as a domain of fully parsed *expressions* with two *atomic expressions* and one *binary mode of composition*. By leaving room for partial expressions as well as total expressions, we allow for a coherent theory of *infinite expressions*.

There is an unlimited supply of similar domains all characterized by analogous domain equations. Solutions to the domain equations can be proved to exist in many ways. What may be novel in the present exposition is that the neighborhoods (finite elements) are defined first by a recursion with a structure parallel to the desired domain equation, then the whole domain is found as the completion of the neighborhood system. This approach is particularly simple when the domain equation involves only the functions + and ×, because the required neighborhoods are so easily sorted out. For more complex equations, the method of retracts is more general and easier to apply directly, but the analysis of what domain is obtained may be far from obvious. The scope of the method of retracts is most broad when applied to the *universal domain*. Such a domain is not actually unique, but a particularly nice one can be defined recursively in a way similar to the previous examples.

EXAMPLE 8. Let $U$ be recursively defined by

$$\begin{aligned} U = \{\Sigma^*\} &\cup \{\{\Lambda\} \cup 0X \mid X \in U\} \\ &\cup \{\{\Lambda\} \cup 1Y \mid Y \in U\} \\ &\cup \{\{\Lambda\} \cup 0X \cup 1Y \mid X,Y \in U\} \ . \end{aligned}$$

It can be proved by induction that $\emptyset \notin U$, and indeed that all sets in $U$ are infinite. It must also be proved by induction that $U$ is closed under (consistent) intersections; thus, $U$ is a neighborhood system. The same kind of proof shows that it is also closed under union of two sets. □

The system $U$ is similar to those of other examples, but it is not the same as any of the ones we have yet seen. How is it different? What is its special nature? Perhaps another representation will help. Consider the *dyadic* rational numbers in the interval [0,1). Take this half-open set as $\Delta$. By a neighborhood we will mean any non-empty subset of $\Delta$ that is a finite union of half-open intervals. A picture of one would look like:

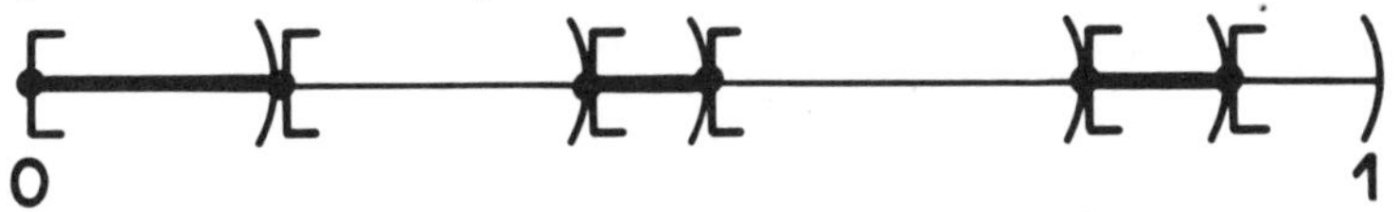


Here the dark intervals are the ones making up the desired union, and we might as well take them as separated as shown. These sets clearly form a neighborhood system $U'$. But is it isomorphic to $U$? To see that it is, consider the two shrinking transformations

$$t \mapsto t/2 \quad \text{and} \quad t \mapsto (t+1)/2 \ ,$$

which we can call

$$t \mapsto 0t \quad \text{and} \quad t \mapsto 1t$$

for short. For $X,Y \subseteq \Delta$, there are three images we can form by means of these transformations:

$$0X\ ,\ 1Y\ ,\ \text{and}\ 0X \cup 1Y\ .$$

The lemma to prove is: any non-empty union of dyadic intervals contained in $\Delta$ can be obtained from $\Delta$ by applying these three operations over and over. (The proof is by induction on the largest power of 2 needed in a denominator of a number at an endpoint of one of the intervals in the union.) With this way of thinking of the generation of $U'$ it is obvious that it is isomorphic to $U$.

The geometric representation of $U'$, however, has advantages in visualization: If we adjoin $\emptyset$ to $U'$, it becomes a *Boolean algebra*. (Note, as is clear from the picture, that the complement of a union of dyadic intervals is again such -- unless it is empty.) By taking half-open intervals, points are *excluded*: $U'$ as a Boolean algebra is *atomless*. But $U'$ is countable, and -- up to isomorphism -- there is only one countable atomless Boolean algebra, namely, the free one on $\aleph_0$-generators. This Boolean algebra is quite remarkable, for *every* countable Boolean algebra is isomorphic to a *subalgebra*! What will this mean for neighborhood systems?

Let $\mathcal{D}$ be any countable neighborhood system. Pass to the other representation of $\mathcal{D}$, say $[\mathcal{D}]$. This system $[\mathcal{D}]$ has the advantage that consistency means having a non-empty intersection. Generate a Boolean algebra of subsets of the set $|\mathcal{D}|$ from the family $[\mathcal{D}]$. This will of course be a countable Boolean algebra, because $\mathcal{D}$ was countable. So this algebra is isomorphic to a subalgebra of $U' \cup \{\emptyset\}$. But then by excluding the empty set from consideration we see that $[\mathcal{D}]$ is isomorphic to a subsystem of $U'$. All that is needed is a clear definition of *subsystem* of a neighborhood system.

DEFINITION. For two neighborhood systems $\mathcal{D}_0$ and $\mathcal{D}_1$ we write

$$\mathcal{D}_0 \triangleleft \mathcal{D}_1$$

to mean that $\mathcal{D}_0$ and $\mathcal{D}_1$ are systems over the same set $\Delta$ and that $\mathcal{D}_0 \subseteq \mathcal{D}_1$ . Moreover, we require that whenever $X,Y \in \mathcal{D}_0$ and $X \cap Y \in \mathcal{D}_1$ , then $X \cap Y \in \mathcal{D}_0$ . In other words, consistency in $\mathcal{D}_0$ is just the *restriction* of the notion to $\mathcal{D}_1$ . In case $\mathcal{D}_0$ is isomorphic to a system $\mathcal{D}_0' \triangleleft \mathcal{D}_1$ , we write this as

$$\mathcal{D}_0 \underset{\sim}{\triangleleft} \mathcal{D}_1$$

for short. □

THE EMBEDDING THEOREM. *We have* $\mathcal{D} \underset{\sim}{\lhd} U$ *for every countable neighborhood system* $\mathcal{D}$. □

This is the sense in which $U$ is "universal", but there are many such systems. For example $U + U$ is universal but *not* isomorphic to $U$. But it is interesting enough to have one universal system.

Perhaps the reader will have noticed that the relation $\lhd$ is defined by a *closure condition*? Since we have a sufficiently large, but countable $U$ already fixed, we conclude:

THE UNIVERSE THEOREM. *The system*

$$\{\mathcal{D} \mid \mathcal{D} \lhd U\}$$

*is a closure system with a countable number of finite elements: the finite subsystems of* $U$. □

The finite elements are easy to pick out, because if $X \subseteq U$ is a finite set, all we have to do is to close it under finite intersections (non-empty!) to obtain $\bar{X} \lhd U$. But $\bar{X}$ is a finite family of sets itself. Since $U$ is countable, there are only countably many finitely generated subsystems. This gives us -- as a domain -- a "universe" of all countable domains (up to isomorphism). We are used to thinking of a set of sets, but not of a ring of rings or lattice of lattices. A domain of domains is quite possible, and -- provided we can find suitable approximable operations on the "universe" -- it can be quite useful for proving the existence of a remarkable number of domains. The way to achieve this is by the *calculus of retractions*. The existence of the necessary retraction mappings is guaranteed by the next result.

PROPOSITION. (i) *If* $\mathcal{D}_0 \lhd \mathcal{D}_1$, *then there exist approximable maps*

$$i : \mathcal{D}_0 \to \mathcal{D}_1 \quad \text{and} \quad j : \mathcal{D}_1 \to \mathcal{D}_0$$

*forming a projection pair in the sense that*

$$j \circ i = I_{\mathcal{D}_0} \quad \text{and} \quad i \circ j \subseteq I_{\mathcal{D}_1} .$$

*Indeed, for* $x \in |\mathcal{D}_0|$ *and* $y \in |\mathcal{D}_1|$, *we have*

$$i(x) = \{Y \in \mathcal{D}_1 \mid \exists\, X \in x \,.\, X \subseteq Y\} \text{ and}$$

$$j(y) = y \cap \mathcal{D}_0 .$$

(ii) *Conversely, if there is a projection pair between* $\mathcal{D}_0$ *and* $\mathcal{D}_1$, *then* $\mathcal{D}_0 \trianglelefteq \mathcal{D}_1$, *because* $\mathcal{D}_0$ *is isomorphic to*

$$\{Y \in \mathcal{D}_1 \mid Y(i \circ j)Y\} \triangleleft \mathcal{D}_1 .$$

*Consequently, projection pairs determine exactly the relation* $\trianglelefteq$ *between domains.* □

Part (ii) above suggests that subsystems are also determined by mappings, and in fact we have:

THEOREM. *The following conditions are equivalent for an approximable map* $a : \mathcal{D} \to \mathcal{D}$:

(i) *There is a projection pair with* $a = i \circ j$;

(ii) $a \circ a = a \subseteq I_{\mathcal{D}}$ *and the range of* $a$ *is isomorphic to a domain;*

(iii) $a \circ a = a \subseteq I_{\mathcal{D}}$, *and whenever* $XaZ$, *then* $X \subseteq YaY \subseteq Z$, *for some* $Y \in \mathcal{D}$.

*Consequently, the maps satisfying these conditions form a domain, because there is a mapping*

$$\mathrm{sub} : (\mathcal{D} \to \mathcal{D}) \to (\mathcal{D} \to \mathcal{D})$$

*defined by*

$$X \, \mathrm{sub}(a) Z \quad \textit{iff} \quad X \subseteq YaY \subseteq Z, \text{ some } Y \in \mathcal{D}$$

*which itself satisfies these conditions on* $(\mathcal{D} \to \mathcal{D})$ *and has as its range exactly these maps on* $\mathcal{D}$. □

A mapping $a : \mathcal{D} \to \mathcal{D}$ where $a \circ a = a$ is called a *retraction*. Its range of values is the same as its fixed-point set,

$$\{x \in |\mathcal{D}| \mid x = a(x)\} .$$

As a poset, such a fixed-point set shares many of the properties of domains, but it need not be a domain in the sense we have been using the word. If $I_{\mathcal{D}} \subseteq a$, then it is a closure system (and, hence, a domain). But the case we are interested in is $a \subseteq I_{\mathcal{D}}$, and it need not be a closure system. Even when $a$ satisfies the conditions of the theorem, the fixed-point set need not be a closure system -- but it is isomorphic to one. The trick is that for *any* approximable $a$, we have

$$\mathcal{D}_a = \{Y \in \mathcal{D} \mid YaY\} \triangleleft \mathcal{D} .$$

But when $a$ satisfies the conditions, this set determines $a$, and the poset of fixed points of $a$ is isomorphic to $|\mathcal{D}_a|$ by restriction, as in the definition of $j$ above.

More than all this, however, is the additional information that, owing to the uniform way the construction was done, we find

$$\{X \mid X \triangleleft \mathcal{D}\} \underset{\sim}{\triangleleft} (\mathcal{D} \rightarrow \mathcal{D}) \ ,$$

because sub is the required retraction. Applying all this to $U$, we see

$$\{\mathcal{D} \mid \mathcal{D} \triangleleft U\} \underset{\sim}{\triangleleft} (U \rightarrow U) \underset{\sim}{\triangleleft} U \ ,$$

since $(U \rightarrow U)$ is a countable system. This may not seem very striking until the reader realizes that the embeddability of all these spaces in $U$ gives us a *notation* for functions and spaces all as *elements* of the *same* space $U$.

## 6. WHAT DOES THIS ALL HAVE TO DO WITH COMPUTER SCIENCE?

The question will have crossed the reader's mind long before this. For the answer, we have to go back to the examples of Section 1 that motivated the study in the first place. Everyone knows about partial functions $f : \mathbb{N} \rightarrow \mathbb{N}$. We could even use other sets $A$ and $B$ and consider partial functions $g : A \rightarrow B$. There are sets that can be constructively given (in the most down-to-earth manner!) and partial functions that can be effectively (recursively) defined (and computed efficiently!). Many people are content to leave the theory of computability with that, for indeed there is much to say about algorithms for functions defined on explicitly given *discrete* sets. The theory advocated here, however, without denying the importance of the discrete case, suggests that there is another level to which the analysis of computability can be applied. It is also claimed that computer science needs this extended theory.

The first step brought us to the domain (closure system) of *functions* $(\mathbb{N} \rightarrow \mathbb{N})$, and we could just as well have defined a function space $(A \rightarrow B)$ of partial functions between any two sets. These are posets, and we recognized that there are *functionals* (operators) defined on these function spaces which have simple-to-state order-theoretic properties. Moreover, these properties (continuity, monotonicity) hold of functionals that are *computably defined*.

Examples of programs that operate on functions as well as numbers are immediate: for example, take a function $f$ and iterate it $n$-times on 0, say, computing $f^n(0)$. We just suggested a functional of type

$$(\mathbb{N} \to \mathbb{N}) \to (\mathbb{N} \to \mathbb{N}) \ .$$

It is obviously a basic idea, and examples of this sort can be multiplied over and over. In other words, programs embody not only specific calculations of special functions but also *methods* for obtaining in a general way new functions from given functions. It should be self-evident that computer science needs the study of such methods, and what the author is trying to collect is the mathematical tools and ideas to pursue this study. This is where the theory of domains comes in.

Going back to the start again, we can regard $\mathbb{N}$ as a domain -- in two ways, in fact

$$\check{\mathbb{N}} = \{\{n\} \mid n \in \mathbb{N}\} \cup \{\emptyset\} \qquad \text{(closure system)}$$

$$\hat{\mathbb{N}} = \{\mathbb{N}\} \cup \{\{n\} \mid n \in \mathbb{N}\} \qquad \text{(neighborhood system)}$$

All we have done is to add $\perp$ as the "undefined" element to $\mathbb{N}$. As we have gradually shifted to neighborhood systems, we take $\hat{\mathbb{N}}$ (and the corresponding $\hat{A}$ for any set $A$) as the one to use; of course it is trivial to see that $|\hat{\mathbb{N}}|$ and $\check{\mathbb{N}}$ are isomorphic posets.

Now $(\mathbb{N} \to \mathbb{N})$ is not the same poset as $|(\hat{\mathbb{N}} \to \hat{\mathbb{N}})|$, but it is almost the same. There is an easy-to-define system

$$\mathcal{D} \lhd (\hat{\mathbb{N}} \to \hat{\mathbb{N}})$$

where $|\mathcal{D}|$ is isomorphic to $(\mathbb{N} \to \mathbb{N})$. We find for a number of reasons (e.g. category-theoretic reasons!) that it is simpler and more regular to take $(\mathcal{D}_0 \to \mathcal{D}_1)$ as the principal function-space construction.

So, starting from $\hat{\mathbb{N}}$, or $\hat{A}$ and the like, we iterate such constructs (functors) as

$$(\mathcal{D}_0 \times \mathcal{D}_1) \quad \text{and} \quad (\mathcal{D}_0 \to \mathcal{D}_1) \ ,$$

obtaining the "higher-type" domains. But more than these there are domains like $U$ that, in a certain sense, are of the highest type -- at least among countable neighborhood systems. There is clearly no reason to exclude $U$, as can be appreciated from its very simple recursive definition. And, to extract all the interesting subdomains of $U$ is the result of an elementary construction

$$\mathcal{D}_a = \{Y \in U \mid YaY\}$$

for any approximable $a : U \to U$. We thus almost have at hand a notation for definitions. What is lacking is the formal notation for functions.

Before passing to the notational questions, we have to emphasize the justification of the theory of domains from the point of view of the *theory of computability*: Not only do these higher type domains exist, but they can be *effectively given.* ("Effective" is used here in the sense of recursive function theory.) Surely the reader has seen already how elementary our examples are; that is to say, the constructions of the desired neighborhood system -- and of the desired approximable mappings -- requires very little. (Indeed with the aid of $\Sigma^*$ a large number of examples of neighborhood systems can be found with a minimum of set notation.) We can give a name to this feeling by suggesting a general procedure of "Gödel numbering" of neighborhoods. And there would be other ways of making the degree of constructivity concrete.

DEFINITION. A *computable presentation* of a neighborhood system $\mathcal{D}$ is a way of indexing the sets in $\mathcal{D}$ so we can write:

$$\mathcal{D} = \{X_n \mid n \in \mathbb{N}\} ,$$

and where we know the following two relations are recursive:

(i) $X_n \cap X_m = X_k$ ;

(ii) $\exists\, k \in \mathbb{N} \,.\, X_n \cap X_m = X_k$ .

A system with a computable presentation is *effectively given.* □

We must take care with this definition, as Kanda and Park [5] have pointed out, for presentations are not isomorphism invariants. To *give* a domain effectively, the indexing must be given together with (the Gödel numbers of) the recursive functions that decide the two relations (i) and (ii). What is being given here is just the poset structure of $\mathcal{D}$, but note we demand in (ii) that the notion of consistency of neighborhoods is effectively decidable. In view of (i), whenever $X_n$ and $X_m$ are consistent, then we can find the $k$ that indexes the intersection.

All the examples we have mentioned are effectively given. It is very easy to check that if $\mathcal{D}_0$ and $\mathcal{D}_1$ are effectively given, then so are

$$\mathcal{D}_0 \times \mathcal{D}_1 \quad \text{and} \quad \mathcal{D}_0 + \mathcal{D}_1 \ .$$

Indeed, from the explicit enumerations for the $\mathcal{D}_i$ we can construct what is needed for the compound domain. The case of

$$\mathcal{D}_0 \to \mathcal{D}_1$$

may not be so obvious, but the result is true here also. The proof rests on the facts that (1) we can test a consistency question in $(\mathcal{D}_0 \to \mathcal{D}_1)$ of the form

$$\bigcap_{i<n} [X_i, Y_i] \neq \emptyset$$

as a sequence of consistency checks in $\mathcal{D}_0$ and $\mathcal{D}_1$ (as was indicated in Section 4), and that (2) we can also test inclusions between function-space neighborhoods by means of the basic equivalence:

$$\bigcap_{i<n} [X_i, Y_i] \subseteq [X, Y] \quad \text{iff} \quad \bigcap \{Y_i \mid X \subseteq X_i\} \subseteq Y ,$$

which holds provided the first-mentioned neighborhood is non-empty. Again tests involving neighborhoods of $(\mathcal{D}_0 \to \mathcal{D}_1)$ are thrown back on (several) tests of relationships among neighborhoods of $\mathcal{D}_0$ and of $\mathcal{D}_1$. We can see in this way $(\mathcal{D}_0 \to \mathcal{D}_1)$, too, is effectively given. Therefore we have a very rich collection of such domains. What requires further comment is the way in which different effectively given domains are to be related.

Before speaking of relationships, there is a prior question about *elements*. A quite simple domain like $(\hat{\mathbb{N}} \to \hat{\mathbb{N}})$ or $U$, though effectively given, has an *uncountable* number of elements. (For countable neighborhood systems $\mathcal{D}$, the cardinality of $|\mathcal{D}|$ never goes beyond that of the continuum.) Obviously not all of these elements can be regarded as having a constructive existence. We should try to pick out those that do.

DEFINITION. Let $\mathcal{D}$ be effectively given with

$$\mathcal{D} = \{X_n \mid n \in \mathbb{N}\} .$$

An element $x \in |\mathcal{D}|$ is said to be *computable* if the set

$$\{n \in \mathbb{N} \mid X_n \in x\}$$

is recursively enumerable. □

Some justification of this definition is required, but the primary example of $(\mathbb{N} \to \mathbb{N})$ is sufficient for seeing why it is correct. Let $(\mathbb{N} \to \mathbb{N})_0$ be the set of *finite* partial functions. Then as a neighborhood system we use as neighborhoods of $(\mathbb{N} \to \mathbb{N})$ the sets

$$\{f \in (\mathbb{N} \to \mathbb{N}) \mid e \subseteq f\}$$

where $e \in (\mathbb{N} \to \mathbb{N})_0$ . Then the elements of this system correspond exactly to the elements of the set $(\mathbb{N} \to \mathbb{N})$. When is $f \in (\mathbb{N} \to \mathbb{N})$ computable? By the definition above, we first look to a standard enumeration

$$(\mathbb{N} \to \mathbb{N})_0 = \{e_n \mid n \in \mathbb{N}\}$$

to show that the domain is effectively given. Then, the meaning of the definition in this case is that the set

$$\{n \in \mathbb{N} \mid e_n \subseteq f\}$$

should be recursively enumerable. But this is just a slightly more complicated way of saying that the graph of $f$, as a subset of $(\mathbb{N} \times \mathbb{N})$, is recursively enumerable. But, as is well known, this in turn is just a way of saying that $f$ is *partial recursive*. This is the correct notion of computable element of $(\mathbb{N} \to \mathbb{N})$. Using only the *total recursive* functions in $(\mathbb{N} \to \mathbb{N})$ would be wrong, because from a definition of an enumeration there is no way of telling whether the function is total or partial. The partial recursive functions can be effectively enumerated but not the total recursive functions. The "partial" notion is the fundamental one, as experience in recursive function theory has shown.

Another way of thinking about the reason for formulating the definition in the way we have is this: the elements of a domain are in general *infinite*. Neighborhoods represent finite approximations; recall that for $X \in \mathcal{D}$ and $x \in |\mathcal{D}|$ we have:

$$X \in x \quad \text{iff} \quad \uparrow X \subseteq x \ ,$$

and "in the limit"

$$x = \bigcup\{\uparrow X \mid X \in x\} \ .$$

So when we enumerate $\{n \in \mathbb{N} \mid X_n \in x\}$ we are enumerating the finite approximations to $x$. We are expecting for $x$ to be computable that these approximations can *eventually* be listed in an effective way. We *do not* demand that we can uniformly decide whether

$$X_n \in x \ .$$

The point is that $x$ is infinite and it has to be built up "slowly", step by step. At one moment we may know that

$$X_{m_0}, X_{m_1}, \ldots, X_{m_{k-1}} \in x \ ,$$

but confronted with an arbitrary $X_n$ we cannot say -- at that moment -- that it will not turn up later. Of course if it is there we can tell by checking out

$$X_n = X_{m_i}$$

for all $i < k$. But if it is not, no quick conclusion can be reached. There are likely to be many, many interesting elements where the set

$$\{n \in \mathbb{N} \mid X_n \in x\}$$

is *recursive*, but this requirement is too strong as a definition of being a computable element. To "compute" an element we set up a process that grinds away giving out more and more information about it; but, as we cannot tell how long the process needs to reach any particular stage, we just have to wait and see what happens.

Computable elements are closely connected to computable functions. We can give a completely analogous definition of the latter concept.

DEFINITION. Let $\mathcal{D}_0$ and $\mathcal{D}_1$ be effectively given with

$$\mathcal{D}_0 = \{X_n \mid n \in \mathbb{N}\}\text{ , and}$$

$$\mathcal{D}_1 = \{Y_m \mid m \in \mathbb{N}\}\text{ .}$$

An approximable mapping $f : \mathcal{D}_0 \to \mathcal{D}_1$ is said to be *computable* if the set of pairs

$$\{(n,m) \in \mathbb{N} \times \mathbb{N} \mid X_n f Y_m\}$$

is recursively enumerable.

Again this generalizes in a natural way the notion of a partial recursive function, and from our previous discussion there ought to be a close connection between the concepts. There is.

PROPOSITION. *An approximable mapping $f : \mathcal{D}_0 \to \mathcal{D}_1$ between effectively given domains is computable as a* mapping *if and only if regarded as an element of the function space $(\mathcal{D}_0 \to \mathcal{D}_1)$ it is computable as an* element. □

The proof is based on our explicit understanding of the neighborhoods of the function space and underlines a desirable stability of the notion of computability. Further stability is provided by:

PROPOSITION. *The composition of computable mappings is computable; hence, the application of a computable mapping to a computable argument produces a computable value.* □

It remains to discuss the *scope* of the notion of computability in effectively given domains as well as the *applicability*. As we have already seen, effectively given domains are closed under the (very constructive) types of constructs we discussed earlier. Indeed, we have all the data to prove:

THEOREM. *The category of effectively given domains and computable mappings is cartesian closed.* □

This means that not only do we get the domains we want from iterating product and function-space constructs, but we get all the maps appropriate to such compound domains as computable mappings. That already provides considerable scope, but the category is much fuller than that result indicates: it is closed under many other constructs (with associated maps) including an extensive number of solutions to domain equations. This can be proved directly for each new construct as it is found, but there is another way to argue that more or less makes such results automatic. The method is based on the use of the universal domain $U$ described in the last section.

THEOREM. *Not only is the universal domain $U$ effectively given, but any effectively given domain $\mathcal{D}$ is computably isomorphic to a domain $\mathcal{D}' \lhd U$, where, if we let*

$$U = \{X_n \mid n \in \mathbb{N}\}$$

*be the given enumeration, then*

$$\{n \in \mathbb{N} \mid X_n \in \mathcal{D}'\}$$

*is recursively enumerable. Any such $\mathcal{D}' \lhd U$ is in itself effectively given, and these subdomains correspond exactly to the computable fixed points of the operator*

$$\mathsf{sub} : (U \to U) \to (U \to U) \ ,$$

*which is also a computable function.* □

The upshot of this result is that the whole theory of effectively given domains is reduced to a study of (certain) computable elements of $(U \to U)$. But the reduction can be pushed further. By the very universal nature of $U$, there are (computable!) maps

$$\mathsf{abs} : (U \to U) \to U \text{ and}$$

$$\text{rep} : U \to (U \to U)$$

(standing for "abstract" and "represent", respectively), which form a projection pair to prove

$$(U \to U) \trianglelefteq U .$$

Via rep the computable elements of $U$ therefore parameterize the computable elements of $(U \to U)$ in a many-one manner. So all the fuss is being reduced to the study of elements of $U$ alone.

The success of the reduction can be made much more visible if we turn $U$ into an "algebra" -- and it is a very rich one. Keep in mind that

$$(U \times U) \trianglelefteq U ,$$

so there are (computable!) maps

$$\text{pair} : U \times U \to U \text{ , and}$$

$$\text{fst, snd} : U \to U ,$$

where we have:

$$\text{fst}(\text{pair}(x,y)) = x \text{ ;}$$

$$\text{snd}(\text{pair}(x,y)) = y \text{ ; and}$$

$$\text{pair}(\text{fst}(z), \text{snd}(z)) \sqsubseteq z .$$

This already gives $U$ an algebraic flavor.

Next, by way of abbreviation write:

$$u(x) = \text{rep}(u)(x)$$

for any two elements $u,x \in |U|$. Since every element of $U$ represents a function, we might as well use the function-value notation to make this more transparent. But this process can be reversed (using abs). Suppose we have an expression $\sim x \sim$ . where

$$(x \mapsto \sim x \sim)$$

is an approximable mapping from $U$ into $U$. Again, by way of abbreviation write:

$$\lambda x . \sim x \sim = \text{abs}(x \mapsto \sim x \sim) .$$

The meaning of this $\lambda$-notation is that the $\lambda$-abstract on the left is the element of $U$ (chosen by abs) to represent the function defined by the expression. By way of algebra we have:

$$(\lambda x\,.\!\sim\! x\!\sim)(y) = \sim\! y\!\sim \;;\ \text{and}$$

$$\lambda x\ .\ u(x) \subseteq u\ ,$$

which are analogous to our equations and inclusions for the cartesian product $(U \times U)$.

To make this notation useful for the theory of domains, we should also recall that

$$(U + U) \underset{\sim}{\lhd} U\ .$$

We can thus introduce operations on $U$ expressing this structure. To make the equations easier to state, it is useful to remember that also:

$$T \underset{\sim}{\lhd} U\ ,$$

where $T$ is the two-element domain,

$$T = \{\{0\},\{1\},\Sigma^*\}\ .$$

We call the corresponding elements of $U$

$$\mathsf{true} \text{ and } \mathsf{false}\ .$$

They are distinguishable, and we can give them helpful rôles by using

$$\mathsf{cond} : U \times U \times U \to U\ ,$$

where cond conditionally chooses between the two elements as follows:

$$\mathsf{cond}(\bot,x,y) = \bot\ ;$$

$$\mathsf{cond}(\mathsf{true},x,y) = x\ ;$$

$$\mathsf{cond}(\mathsf{false},x,y) = y\ ;\ \text{and}$$

$$\mathsf{cond}(z,\mathsf{true},\mathsf{false}) \subseteq z\ .$$

The mapping on $z$ indicated in the last line is the projection of $U$ onto (the isomorphic copy of) $T$. We must, however, assume that the projection actually takes place by saying that the only values of $\mathsf{cond}(z,\mathsf{true},\mathsf{false})$ are true, false, and $\bot$ .

Returning to $(U + U)$, we can say in an explicit way that there is a copy of this domain inside $U$ by introducing a number of mappings

$$\text{which, inl, inr, outl, outr}$$

which satisfy a number of equations. For example:

$$\text{outl}(\text{inl}(x)) = x\ ;$$

$$\text{outr}(\text{inr}(y)) = y\ ;$$

$$\text{outl}(\text{inr}(y)) = \bot\ ;$$

$$\text{outr}(\text{inl}(x)) = \bot\ ;$$

$$\text{which}(\text{inl}(x)) = \text{true}\ ;$$

$$\text{which}(\text{inr}(y)) = \text{false}\ ;$$

$$\text{which}(z) = \text{cond}(\text{which}(z),\text{true},\text{false})\ ;$$

$$\text{cond}(\text{which}(z),\text{inl}(\text{outl}(z)),\text{inr}(\text{outr}(z))) \sqsubseteq z\ .$$

In other words (the image of) $U + U$ has two parts which the function which has tagged with true and false. (The true side is the left side and the false side is the right side.) But $U \underset{\sim}{\triangleleft} U + U$ in two ways: we can inject an element on the left by inl and on the right by inr -- and if we take them out again, nothing is lost.

We can make an essential reduction in notation if we recall the isomorphism

$$(U \to (U \to U)) \cong ((U \times U) \to U)\ .$$

Using this together with $(U \to U) \underset{\sim}{\triangleleft} U$ as embodied in the λ-notation, we can regard all the functions just introduced as *elements* of $U$. To put it another way, instead of writing:

$$\text{pair}(x,y)\ ,$$

we write:

$$\text{pair}(x)(y)\ .$$

Or more generally we can define:

$$u(x,y,\ldots,z) = u(x)(y)\ \ldots\ (z)\ .$$

We must stress that in this reduction of all the constructs to elements of $U$ we make full use of $u(x)$ as a binary operation on $U$ -- but it is the *only* binary operation we need. The structure on $U$, therefore, that we are emphasizing consists of

$$u(x) \quad \text{and} \quad \lambda x\ .\sim x\sim$$

together with these *constants*:

fix, sub ;

cond, true, false, $\bot$ ;

pair, fst, snd ; and

which, inl, inr, outl, outr .

Reductionism could actually be pushed much further if we so wanted. For example, we could just as well *define* pair by;

$$\text{pair} = \lambda x \lambda y \lambda t\ .\ \text{cond}(t,x,y)\ ,$$

but this is a bit artificial. Though we could eliminate all the last eight constants in our list above, we would just have to introduce them again at once by definition, because they are so basic.

So, however we choose to set it up, $U$ becomes an algebra in which all kinds of useful constructs can be defined with the aid of certain constants and the $\lambda$-notation. $U$ is an effectively given domain, and every operation defined in this algebra is not only *approximable* but also *computable*. (With a couple more constants, by the way, we could assure that *all* computable elements and operations on $U$ are $\lambda$-definable. But this completeness is not especially relevant to the present discussion.) The $\lambda$-notation with these constants becomes a mathematically clean *programming language* for defining a very rich collection of computable notions.

Because $U$ is a mathematically defined domain, what we have been giving is the mathematical semantics (meaning) for the definition language. It is quite another question as to how you *find* the value of some long expression written in this notation -- this is where computer science comes in properly, since it is reasonable to ask how we can use the language, how we can manipulate the expressions. Computers are needed because the calculations with the expressions are too long (*far* too long) for pencil and paper. What mathematics has done (within the theory of certain posets) is to show that the expressions *have* a value, that it makes sense to ask whether two expressions *have the same* value. The answers to these mathematical questions are neither obvious from the notation itself nor from a few rules that have been suggested for general manipulations. The theory of domains (or something similar) is needed to make the questions precise and interesting. And there can be more than one answer, since, as we have seen, there is much freedom in setting up a universal domain like $U$.

To close the circle of discussion started with our introduction of effectively given domains presented in this section, we still have to show how reduction to (elements of) $U$ is an aid to solving domain equations. The answer all has to do with the use of projections. We recall that projections $a = i \circ j$ coming from projection pairs were the right kind of retracts for thinking of the $\mathcal{D} \triangleleft U$. For short let us call them *finitary* projections. What is nice about $U$ and its algebra is:

THEOREM. *There are computable maps* $\times, +, \to$ *defined on* $U \times U$ *such that whenever* $a$ *and* $b$ *are finitary projections, then so are* $(a \times b)$, $(a + b)$ *and* $(a \to b)$ *and we have*

$$\mathcal{D}_{a\times b} \cong \mathcal{D}_a \times \mathcal{D}_b \ ;$$

$$\mathcal{D}_{a+b} \cong \mathcal{D}_a + \mathcal{D}_b \ ; \text{ and}$$

$$\mathcal{D}_{a\to b} \cong \mathcal{D}_a \to \mathcal{D}_b \ .$$

*Indeed we can define:*

$$a \times b = \lambda t \ . \ \mathsf{pair}(a(\mathsf{fst}(t)), b(\mathsf{snd}(t))) \ ;$$

$$a + b = \lambda t \ . \ \mathsf{cond}(\mathsf{which}(t), \mathsf{inl}(a(\mathsf{outl}(t))), \mathsf{inr}(b(\mathsf{outr}(t)))) \ ;$$

$$a \to b = \lambda t \ . \ b \circ t \circ a \ ;$$

*where in general*

$$f \circ g = \lambda x \ . \ f(g(x)) \ .$$

*The collection of finitary projections becomes a category when we define*

$$f : a \to b \ \text{ iff } \ f = b \circ f \circ a \ .$$

*It is equivalent to the cartesian closed category of countable neighborhood systems and approximable mappings. We also find that*

$$\mathcal{D}_a \text{ is effectively given } \ \text{iff} \ \ a \text{ is computable} \ . \quad \square$$

The consequence of this theorem is that the relevant parts of the whole theory of domains are "internalized" into $U$ and into its algebra of elements. The programming language that the algebra gives us is not only a definition language for computable functions at all types, but it is also a language for *defining the types themselves*. For example, the least fixed point of the equation

$$b = b + b \ ,$$

does in fact give a finitary projection $b$ where

$$\mathcal{D}_b \cong \mathcal{B} \ .$$

Other domain equations can be solved in a similar way using the ordinary fixed-point operator

$$\mathsf{fix} : (U \to U) \to U \ .$$

References [7] and [10] contain more details.

REFERENCES

[1] H.P. Barendregt (1981) *The Lambda Calculus: Its Syntax and Semantics*, North-Holland Publishing Co., xiv + 615 pp.

[2] G. Gierz, K.H. Hofmann, K. Keimel, J.D. Lawson, M. Mislove, and D.S. Scott (1980) *A Compendium of Continuous Lattices*, Springer-Verlag, xx + 371 pp.

[3] M.J. Gordon (1979) *The Denotational Description of Programming Languages*, Springer-Verlag, 160 pp.

[4] M.J. Gordon, A.J.R. Milner, and C.P. Wadsworth (1979) *Edinburgh LCF*, Springer-Verlag Lecture Notes in Computer Science, vol. 78, 159 pp.

[5] A. Kanda and D. Park (1979) When are two effectively given domains identical?, *Proc. of the 4th GI Theoretical Computer Science Symposium*, Springer-Verlag Lecture Notes in Computer Science, vol. 67, 170-181.

[6] D.S. Scott (1976) Data types as lattices, *SIAM J. Comput.* 5, 522-587.

[7] D.S. Scott (1977) Logic and programming languages, *Comm. ACM* 20, 634-641.

[8] D.S. Scott (1980) Lambda calculus: Some models, some philosophy, in: *The Kleene Symposium* (J. Barwise, et al., eds.) North Holland Publishing Co., 223-265.

[9] D.S. Scott (1980) Relating theories of the λ-calculus, in: *To H.B. Curry: Essays on Combinatory Logic, Lambda Calculus and Formalism* (J.P. Seldin and J.R. Hindley, eds.) Academic Press, 403-450.

[10] D.S. Scott (1981) *Lectures on a Mathematical Theory of Computation*, Oxford University PRG Technical Monograph, iv + 148 pp.

[11] J.E. Stoy (1977) *Denotational Semantics: The Scott-Strachey Approach to Programming Language Theory*, MIT Press xxx + 414 pp.

# PART VII

# APPLICATIONS OF ORDERED SETS TO SOCIAL SCIENCES

# ORDERED SETS AND SOCIAL SCIENCES

J.P. Barthélemy
École Nationale Supérieure des Telecommunications
46, Rue Barrault
75634 Paris, Cédex 13, France

Cl. Flament
Département de Psychologie
Université de Provence
29 avenue R. Schuman
13621 Aix-en-Provence, Cédex, France

B. Monjardet
Centre de Mathématique Sociale
Laboratoire Associé au C.N.R.S.
54, Bld Raspail
75270 Paris, Cédex 06, France

ABSTRACT

Problems of modeling, of analysis and of aggregation of preferences provide important examples to illustrate the connections between ordered sets and social sciences. We review these problems according to a three-level classification: order models of preferences; order and associated metric structures of sets of preferences; aggregation of preferences. We conclude with examples connected with dimensionality of partial orders and lattice fixed point problems.



*I. Rival (ed.), Ordered Sets, 721–758.*

## INTRODUCTION

In this paper we review some uses of ordered sets in the social sciences. Although relatively numerous, there is as yet no full account of these applications. In fact, it is difficult enough to describe and classify them since the mathematical criteria for a taxonomy would be quite different from those in social science. To set the stage here we have chosen a taxonomy based on a very partial description of the activities of a social scientist in which we distinguish - somewhat artificially - three levels.

At a *first level*, social scientists *construct and study some specific objects* which may be interpreted as either data or models. Of course, we consider only such objects as have a mathematical representation. Ordered sets are sometimes the best candidates for such representations: for example, modelling "nontransitive indifference" has led many authors (especially Halphen and Luce) to the notion of a semiorder; representing some hierarchical structures such as social networks and taxonomies gives rise to ordered trees, semilattices, and lattices.

At a *second level*, social scientists *tackle sets of observed data or sets of hypothesized models requiring description and comparison*. A general instance is given by the collection of individual preferences in a group of subjects. A good methodology for this level is to consider the set of all possible objects with its associated mathematical structures. Such a set can often be endowed with an ordered structure, even if the objects themselves are not represented by ordered sets. In this case, the "natural" metrics associated with the ordered structure are central tools for comparison. An important example is furnished by lattices or semilattices of binary relations such as preorders, complete preorders, equivalence relations, partial orders, weak orders, etc., and valued relations such as ultrametrics. Distributive lattices, especially products of chains and Boolean lattices, yield another very general example: they occur whenever one considers ordered $n$-tuples with entries from an ordinal scale. The scale may formalize some criterion, variable, item, or question. These lattices have the same importance for ordinal data that vector spaces over the reals have for numerical data.

At a *third level*, social scientists *summarize data or fit models to data*. If the sets of data and models are ordered structures, these problems of aggregation and fitting are problems concerning certain mappings between ordered sets. Some classical tools such as Boolean calculus, Galois connections or residuation theory, fixed point results, median theory, and lattice metrics can be profitably used. Examples of such profitable use can be found in several methods of data analysis in social science: Guttman scales, block-models, clustering, and aggregation of preferences. Also, such invariants as dimension and width of an ordered set are useful in the search for multidimensional models of ordinal data.

As already noted, the preceding classification into three levels is a somewhat artificial yet convenient way to view mathematical activity in the social sciences. Indeed, from the point of view of social scientists, the first and third levels are quite close, and, for mathematicians, there is no clear distinction among the three levels. For instance, the same type of lattice can be used to model a social hierarchy (first level) or to describe a set of possible models of preferences (second level). Nevertheless, the third level leads to the study of mappings between ordered sets and the results obtained can be consequences either of "internal" properties of the "complex" objects that we consider (first level) or of "structural" properties of the set of all objects (second level).

Problems concerning the modelling, analysis, and aggregation of preferences can nicely illustrate the connection between ordered sets and social sciences; moreover, they are well suited to our three-level classification. So, in the first three sections we present *the models of preference orders (first level), the organization of sets of preference orders (second level), and results on the aggregation of preference orders (third level)*. In the last section we give two other examples: the model of Guttman and its multidimensional extension, and a problem leading to the study of the structure of sets of fixed points. We hope these examples will demonstrate that connections between social sciences and the study of ordered sets can benefit both endeavors.

The list of about one hundred and fifty references needs explanation. An exhaustive list would require several hundred titles. We choose to omit many, but not all, significant papers which are referred to in review papers or books such as the recent works of Mirkin, 1979, Roberts, 1979, Golumbic, 1980, and Barthélemy and Monjardet, 1980. Also, we choose to include many lesser known (but not necessarily less significant) papers, for instance: French-written papers and "forgotten" papers.

# 1. MODELS OF PREFERENCE

More and more, social or decision sciences use binary relations to model individual and collective preferences. The relations used are very often asymmetric and transitive and, thus, are order relations. We first give a list of six types of orders used, more or less frequently, as models of preferences. Second, we comment on this list from an historical point of view. Finally, we point out some related or current works in this field. The discussion is restricted to order relations on a finite set although many notions and results hold more generally.

## 1.1. Preference Orders

Let $X = \{x,y,z,\ldots\}$ be a finite set whose elements are called "objects": the objects can be alternatives, candidates, motions, social states, etc. A preference relation on $X$ shall be modelled by a partial order $P$ on $X$. So, $P$ is an irreflexive, transitive binary relation on $X$. We write $(x,y) \in P$ or $xPy$ if the two objects $x$ and $y$ are related by $P$, and write $(x,y) \notin P$ or $xP^cy$ otherwise: interpret $xPy$ as a strict preference of $x$ to $y$. Also, let us write $xIy$ if and only if $(x,y) \notin P$ and $(y,x) \notin P$: interpret $xIy$ as indifference between $x$ and $y$. Also, write $xP^dy$ if and only if $yP^cx$, or, equivalently, if and only if $xPy$ or $xIy$. This relation is sometimes interpreted as a "weak" preference of $x$ to $y$. For binary relations $P$ and $Q$ on the set $X$, $PQ$ is the usual relative product, that is $xPQy$ if and only if there is some $z \in X$ such that $xPz$ and $zQy$.

For each type of partial order in the following list we give a relational definition and, if it exists, an equivalent characterization based on a real-valued representation.

M1. $P$ is a *linear order* if

(a) $P$ is a partial order and $I$ is the equality relation,

or

(b) there is a real-valued function $F$ on $X$ such that $xPy$ if and only if $F(x) > F(y)$, and $F(x) = F(y)$ implies $x = y$.

M2. $P$ is a *weak order* if

(a) $P$ is antisymmetric and $P^cP^c \subseteq P^c$,

or

(b) $P$ is a partial order and $I$ is an equivalence relation,

or

(c) there is a real-valued function $F$ on $X$ such that $xPy$ if and only if $F(x) > F(y)$.

M3. $P$ is a *semiorder* if

(a) $P$ is irreflexive, $PIP \subseteq P$, and $P^2P^d \subseteq P$,

or

(b) $P$ is a partial order and $PI \cup IP$ is a weak order,

or

(c) there is a real-valued function $F$ on $X$ such that $xPy$ if and only if $F(x) > F(y)+1$,

or

(d) there is a map of $X$ into a set of intervals of $(\mathbb{R},\geq)$ of the same length such that $xPy$ if and only if $z > t$ for all $z \in F(x)$ and $t \in F(y)$.

M4. $P$ is an *interval order* if

(a) $P$ is irreflexive and $PIP \subseteq P$,

or

(b) $P$ is irreflexive and $PI$ is a weak order,

or

(c) there is a real-valued function $F$ on $X$ and a strictly positive function $\sigma$ on $X$ such that $xPy$ if and only if $F(x) > F(y)+\sigma(y)$,

or

(d) there is a map $F$ of $X$ to a set of intervals of $(\mathbb{R},\geq)$ such that $xPy$ if and only if $z > t$ for all $z \in F(x)$ and $t \in F(y)$.

M5. $P$ is a *semitransitive order* if $P$ is irreflexive and $P^2P^d \subseteq P$.

M6. $P$ is a *partial order* if $P$ is irreflexive and transitive.

Denote by $L$, $W$, $S$, $J$, $S_t$, and $P$ the set of all linear orders, weak orders, semiorders, interval orders, semitransitive orders, and partial orders, respectively. We have the following inclusions:

$$L \subset W \subset S = J \cap S_t \subset J \cup S_t \subset P \quad .$$

There are many other possible characterizations and properties of these orders: see, for example, Fishburn, 1970, Monjardet, 1978, and Mirkin, 1979 and section 1.2 for references to the stated equivalences. Let us just note that the interval orders are exactly the irreflexive Ferrers relations (see section 4), a fact that leads to a matrix "step-type" characterization of these relations.

Remarks

1. The transitivity of the preference relation $P$ is not always a satisfactory assumption; for example in paired comparison methods or in multicriteria decision problems. In these cases, other types of relations, like tournaments and suborders must be used.

2. In order theory, the relation $I$ is generally called the incomparability relation associated with the partial order $P$. In the order models, $I$ is interpreted as an indifference relation. But, there are cases where a preference model must display both indifference and incomparability. Then, the above order models (or the below dual complete models) are not satisfactory (see 1.3).

3. A partial order $P$ is an asymmetric relation. The relation $P^d$ defined above by $xP^dy$ iff $yP^cx$ is a complete (connected) relation. The map $d$ is a duality between the set of asymmetric relations and the set of complete relations. Thus, we can associate every asymmetric model of preference, a complete model that we shall call the dual model; in this model, $xIy$ iff $xP^dy$ and $yP^dx$. The dual models of M1, M2, M3, M6 have been called complete orders (or chains, simple orders, etc.) complete preorders (or weak orders), semi-orders, and quasi-transitive relations, respectively.

1.2 Historical Comments

In economic theory, concepts and problems about preferences were first tackled by real-valued functions, the so-called utility functions; if $u$ is such a function, $u(x)$ is the utility or the value of object $x$. It was implicitly - then, later, explicitly - assumed that this utility was measurable (sums and differences of utilities are meaningful). It was Pareto and the Lausanne school (Hicks, Allen, etc.) who undertook to found the theory of value, uniquely on the ordinal nature of the valuations. Although Pareto had not a clear idea of a binary relation, it can be asserted that he was the first to model preferences by weak orders, and, especially, to show that the indifference classes are linearly ordered (Livre d'Économie Politique, 1896-97, Manuel, 1907). At the same time, logicians, for example Peirce and Schröder (Algebra und Logik der Relative, 1895), defined and studied rigorously linear and weak orders (generally called by them series and quasi-series). Arrow's book (1951) is probably the first to systematically use the "modern" relational notations and definitions of weak orders (in the complete form) to model preference in economics.

The Lausanne school ideas were strongly debated by economists, and, in particular, Armstrong (1939) pointed out that the weak order model could not account for the "well-known fact" of the indifference. He asserted that "intransitive indifference arises from the imperfect powers for discrimination of the human mind, whereby inequalities become recognizable only when of sufficient magnitude". In fact, it has been a long time since psychophysicists, like Fechner, introduced the notion of the *just noticeable difference threshold*, to precisely account for this imperfect discrimination between stimuli, measured on a real-valued scale (for example, brightness, hue, and pitch), and Poincaré noticed the related intransitivity of the indiscrimination relation. The semiorder model is a relational model that tackles such a threshold intransitivity. The name comes from Luce (1956), who stated the existence of an underlying weak order (M3b) and of a real-valued representation. Scott and Suppes (1958) proved the fundamental representation theorem (M3c). It is worth noticing that similar notions were already in the Goodman-Galanter besideness model of discrimination (1951-56), and, above all, in the Halphen "sesquitransitive" relation designed to account for the intransitivity of indifference among events ordered by subjective probability (1955); they appeared also in the Wine-Freund "decision patterns" about statistical tests on means differences (1957) and, much later, in the Fine theory of extrapolation (1970). In economics, there has recently been a lot of work using the semiorder model: choice theory (Jamison, Lau, 1973, Fishburn, 1975), decision theory (Jacquet-Lagrèze, 1975), equilibrium theory (Jamison, Lau, 1977), social choice theory (Blau, 1979, Blair and Pollak, 1979). Interval orders and semitransitive orders are two straightforward generalizations of semi-orders (compare M3a, M4a and M5a), both used to modelize intransitive indifference. The interval order model has been popularized by Fishburn (who gave it its name) and also Mirkin; both proved representation theorems M4c and M4d. But, it is worth noticing that definition M4a and interesting properties of interval orders are in two, apparently forgotten, papers of Norbert Wiener (1914-1918). Wiener studied such relations (which he called relations of complete successions), following the suggestion of B. Russel, to formalize the relation of succession among events of time and its connection with the linear order of instants of time. There, events of time are identified with intervals of the linear order of instants; their succession defines precisely an interval order. Recently, model M4 has been used in decision theory to model situations where there is some uncertainty about the numerical evaluation of decisions on criteria (see, for example, Jacquet-Lagrèze and Roy, 1981). Model M5 has been used by Chipman (1971) in a study of consumption theory without transitive indifference.

Model M6 only assumes the transitivity of strict preference, which is not quite innocuous. First, as we have already said, observed (or modelled) preference is not always transitive. Secondly, if $P$ is transitive, associated indifference $I$ must be the complement of a comparability graph (see 1.3). Model M6, or its dual form the so-called quasi-transitive relations of Sen, 1970) has been extensively used by social choice economists searching to escape the Arrow dilemma (see section 3).

Remarks

The use of asymmetric order models of preference rather than complete models has been popularized by Fishburn (1970, 1972,

1973) and has the advantage of directly setting such models in the framework of order theory. For additional material on order preference and intransitive indifference models, see, especially, Roberts (1968, 1979), Fishburn (1970a), Monjardet (1978), Mirkin (1979) and two special issues of the review "Mathématiques et Sciences Humaines" (numbers 62 and 63, 1978).

### 1.3 Other Comments and Current Problems

As we just said, the above asymmetric preference models end up in the field of order theory. And it happens, not so exceptionally, that more or less classical problems of this field are of interest to social scientists. We develop this point in section 2 for structural and metric aspects of partial orders. The dimension and width problems are other good examples (see section 4 for applications of the first in social science). Thus, it is suitable to mention some known results on the dimension of preference orders. The dimension of a weak order is obviously at most two. The dimension of a semiorder is at most three and the semiorders of dimension 3 have been characterized (Rabinovitch, 1978(a)). Some results and upper bounds on the dimension of interval orders are in Bogart, Rabinovitch and Trotter (1976) and Rabinovitch (1978b). Note that semitransitive orders contain all the bipartite orders and thus nearly all of the known maximal dimensional partially ordered sets (Bogart, Trotter, 1973).

We have already noticed that the indifference relation associated with an arbitrary preference order is not itself arbitrary. In graph theory, a *comparability graph* is an undirected graph whose edges can be asymmetrically and transitively oriented (the resulting directed graph is a partial order). These comparability graphs have several well known characterizations. Thus, indifference relations associated with arbitrary partial orders are easily characterized as complementary graphs of arbitrary comparability graphs. Now, for special classes of preference orders we obtain special classes of graphs. For instance, when $P$ is an interval order, $I$ is an interval graph, that is to say, the intersection graph of a family of intervals of a linear order. When $P$ is a semiorder, one obtains an interval graph for which each interval must have the same length. Such interval graphs have been called indifference graphs by Roberts, who carefully studied them (1968), an earlier characterization having been obtained by Flament (1962). These graphs are periodically rediscovered under varied names, for example, similarity relations (Fine, 1970, etc.), and time graphs (Maehara, 1980). Such connections between order theory and graph theory set up a fascinating subject of which Golumbic's book (1980) gives an excellent account. For earlier or related work (for example, seriation or betweenness) see also Flament (1962), Kendall (1969),

Aigner (1969), Fishburn (1971), Hubert (1974), Trotter and Moore (1976).

We end this section by mentioning some current issues. In a symmetric (or complete) preference model, we get two basic situations between two objects: preference or indifference. In several domains, and especially in multicriteria decision problems, there is at least one more basic situation: incomparability. Then, one needs more general, but yet useful preference models and the good compromise is not yet very clear. On these lines, one can mention the Fishburn generalized weak orders (1979), the Ng subsemiorder model of multidimensional choice (1977) and, especially, several contributions by the French language school on multicriteria decision making, for example Vincke (1978), Roy and Vincke (1980), Jacquet-Lagrèze and Roy (1981), Roubens and Vincke (1981). (B. Roy is currently completing a book on the subject.)

## 2. ORGANISATION OF SETS OF PREFERENCE ORDERS

Suppose subjects are asked to give their preferences among objects, by an ordered list. Several "natural" queries arise for a social scientist studying these preferences. How many different answers can be obtained? How close are two or several answers? How does one compare answers from two groups of subjects? (etc.) Now, in that case, answers can be modelled as weak orders on the set of objects. Thus, a way to tackle these problems is to consider the set of all possible weak orders with the ordered and related metric structures with which it can be endowed. In this section we shall review such structural properties for the sets of preference orders defined above.

First, we shall describe the ordinal and lattice properties of such sets and we shall give some well known examples of metrics used in order to compare preference orders. In fact, these distances are closely related to the ordinal structures of the set of considered preferences, and they appear as particular cases of "natural" metrics (geodesics, associated with valuations) defined on orders. Thus, we shall review some general results in this field. A by-product of this "abstract" approach is to improve and to extend some axiomatic characterizations of metrics classically used in social sciences, such characterizations being useful in the problem of choosing a specific metric.

### 2.1 Lattice Properties and Enumeration

We denote by $\mathcal{R}$ the set of all binary relations on $X$. With respect to set-inclusion and the operations of set-union and set-intersection, $R$ is a Boolean lattice. The subsets of partial orders defined by one of the properties M1 through M6 are not sublattices of $R$, but they can have interesting lattice properties. In fact, we shall see that several sets of preference orders belong to a special class of lattices, or semilattices, that we now define. A lattice $T$ is *meet distributive*, or *lower locally distributive*, if the interval $[x,y]$ of $T$ is a Boolean lattice whenever $x$ is the meet of the lower covers of $y$. Such lattices, generalizations of distributive lattices, have many interesting properties: they are lower semimodular, embeddable in distributive lattices with meets and rank preserved, all elements have a unique irredundant join representation by join irreducibles, and the rank, or height, of an element is the number of join irreducibles which it dominates. Moreover, these lattices can be obtained as lattices of an antiexchange closure operator on a set. (See Dilworth, 1940, Avann, 1961, Boulaye, 1968, Edelman, 1980).

We require a few more definitions. A meet distributive lattice all of whose join irreducibles are atoms is called a *meet distributive point lattice*. For an element $x$ of a set $X$ partially ordered by $\leq$, the principal order ideal defined by $x$ is the set

$$(x] = \{y \in X | y \leq x\} \quad .$$

Also, the *rank* $r(x)$, is the length of a longest chain of $X$ with supremum $x$. Finally, a meet semilattice $S$ is call *meet distributive* if for all $x \in S$, $(x]$ is a meet distributive lattice.

We have the following facts.

(a) For any partial order $P$ in $P$, $(P]$ is a meet distributive point lattice in $R$ whose atoms are defined by the ordered pairs in $P$. In fact, $(P]$ is a meet subsemilattice of $R$. For any $Q \subseteq P$, $r(Q) = |Q|$.

(b) $P$ is a meet distributive point semilattice with $n(n-1)$ atoms (where $n = |X|$) and $n!$ maximal elements. Note that $P < Q$ in $P$ if and only if $Q = P \cup \{(x,y)\}$ where $(x,y)$ satisfies $xIy$, $zPx$ implies $zPy$, and $yPz$ implies $xPz$. Also, $P$ is a meet subsemilattice of $R$.

(c) Let $L$ be a linear order. A partial order $P \subseteq L$ is called *L-normal* if $PL \cup LP \subseteq P$. Such an ordering $P$ is a semiorder and is referred to as an *L-normal semiorder*. Then, for all $L \in L$, the set of all $L$-normal semiorders is a distributive sublattice of $R$ with $n$-1 coatoms and one atom (Dean and Keller, 1968).

(d) $W$ is a meet distributive point semilattice whose atoms are the $2^n-2$ weak orders with two equivalence classes.

(e) For all $L \in L$, one can define a complemented graded lattice $L_L$ with least element $L$ and greatest element the dual linear order of $L$ (Guilbaud and Rosenstiehl, 1963). These lattices are called "permutordre" lattices.

(f) It is easy to see that $S$, $J$, and $S_t$ are not meet semilattices with respect to set-inclusion.

For additional results on this subject see Dean and Keller, 1968 (for a and c), Barbut and Monjardet, 1970 (for b and c), Frey, 1971 (e), Avann, 1972 (a), and Edelman and Klingsberg, 1981 (a).

The well-known problem of enumerating $P$ has an extensive literature. Avann, 1972, enumerates $(L]$ for $L \in L$ with $|L| \leq 6$. There are partial results on the enumeration of $J$ (Greenough and Bogart, 1979) and apparently no results on enumerating $S_t$. On the other hand, the number of all unlabelled semiorders on an $n$-set is the Catalan number $C_n = \frac{1}{n+1}\binom{2n}{n}$. This result, first proved by Wine and Freund, 1957, has been rediscovered several times. Note that Dean and Keller, 1968, obtain it since their result amounts to proving that $L$-normal semiorders are in one-to-one correspondence with unlabelled semiorders. Chandon, Lemaire, and Pouget, 1978, obtain an induction formula to enumerate all labelled semiorders of an $n$-set. Also, an induction formula for the set $W$ of all labelled weak orders is well-known. But, it is worth noting that there exists an asymptotic equivalent in the expression $O((n-1)!)$ (Barthélemy, 1980). Let us also recall that there are $2^{n-1}$ unlabelled weak orders on an $n$-set (that is, the number of compositions of $n$).

Some other related enumeration results can be found in Sharp, 1971-1972, Rogers, 1977, Moon, 1979, Rogers, Kim, and Roush, 1979.

Remarks

1. As is well-known, the semilattice $P$ of all partial orders on $X$ is dually isomorphic to the semilattice of all $T_0$-topologies on $X$ (Birkhoff, 1967). Thus, structural or enumeration results on $T_0$-topologies can be translated to results on partial orders, and conversely.

2. For each of the six partial orders, there is the problem of obtaining an "abstract" lattice theoretic characterization of the partially ordered set of all orders. For instance, it would be interesting to characterize $(L]$, for $L \in L$, among meet distributive point lattices.

Another type of question asks for a description of the sublattices of the partially ordered set of all orders (of a given sort). For instance, it is known that *every* finite lattice is isomorphic to a lattice of partial orders on a countably infinite set (Schein, 1972).

3. There exist many possible embeddings of partial orders in euclidean spaces. Taking the convex closure of such an embedding of a set of partial orders, we obtain a convex polytope. The study of such polytopes is connected with structural properties of the considered partial orders and is important, especially for optimization problems such as those in section 3.3 (see, for example, Hausmann and Korte, 1978, Young, 1978).

## 2.2 Metric Properties

In a classical paper, Kemeny (1959) defines a method to aggregate weak orders into a weak order. This method is based on the *symmetric difference distance* (s.d.d.) between weak orders and justified by giving an axiomatic characterization of this distance (see 3.3 for a detailed account of this method). The same distance has been implicitly used by Kendall (1948) when he defined the correlation coefficient $\tau$ between two linear orders. More generally, let $P$ and $Q$ be two arbitrary partial orders on $X$. The symmetric difference distance between $P$ and $Q$ is defined by:

$$\delta(P,Q) = |P \Delta Q| = |P-Q| + |Q-P| \quad . \tag{1}$$

We can make one remark connecting this distance with the ordinal properties of $\mathcal{P}$. Indeed, in the semilattice $(\mathcal{P},\subseteq,\cap)$, the rank $r(P) = |P|$, and thus we obtain the formula:

$$\delta(P,Q) = r(P) + r(Q) - 2r(P\cap Q) \quad . \tag{2}$$

Another way to define a distance in $\mathcal{P}$ is based on the quite general idea of "admissible transformations". If, for a set of "objects", there exist (reversible) transformations from one object into another, we can define a distance between two objects $o$ and $o'$ as the minimum number of transformations to obtain $o'$ from $o$. Obviously, one can often define several kinds of transformations on the same set of objects, and one must specify what are the "admissible" transformations. (See, especially, Flament, 1963, Boorman, 1970, Boorman and Arabie, 1972, for the use of such distances between "discrete structures".) Here, let an admissible transformation of a partial order $P$ be a partial order covering or covered by $P$ in $(\mathcal{P},\leq)$. Then, the associated transformation distance is:

$$\varepsilon(P,Q) = \text{length of a minimal path of the (undirected) covering graph of } (\mathcal{P},\leq) \quad . \tag{3}$$

Moreover, one can prove that formulas (2) and (3) define the same distance, that is, the s.d. distance (formula (1)). Thus, the transformation distance, given by formula (3), is easily computable from the rank-based formula (2). We give, in the following sections, broad generalizations of these remarks to arbitrary ordered sets.

### 2.2.1 The Geodesic Metric Theorem

We have just pointed out two kinds of metrics for ordered sets: the shortest path distance on one hand, and the distance $\delta$ defined by the "lattice" formula $\delta(P,S) = |P| + |S| - 2|P\cap S|$ on the other hand.

More generally, we set the problem of constructing "compatible" metrics on an ordered set. Two ways are suggested by the two examples above; they are indeed strongly related, as we shall see.

Let $(E,\leq)$ be a (finite) ordered set (in that case $\leq$ is a reflexive, antisymmetric, transitive relation on $E$) with a least element 0. Let $\varphi$ be a strictly isotone real valued function on $E$: $x < y$ implies $\varphi(x) < \varphi(y)$. We shall use the following notations:

$$\varphi^{-}\langle x,y\rangle = \operatorname*{Max}_{t\leq x,\, t\leq y} \varphi(t)$$

$$(x\cap y)_{\varphi} = \{t \mid t \leq x,\ t \leq y \text{ and } \varphi(t) = \varphi^{-}\langle x,y\rangle\}$$

$$\bar{d}_{\varphi}(x,y) = \varphi(x) + \varphi(y) - 2\overline{\varphi}\langle x,y\rangle \quad .$$

It is obvious that

$$\bar{d}_{\varphi}(x,y) = \bar{d}_{\varphi}(y,x),\ d_{\varphi}(x,y) = 0 \text{ if and only if } x = y, \text{ and}$$

$$\bar{d}_{\varphi}(x,y) = \bar{d}_{\varphi}(x,t) + \bar{d}_{\varphi}(t,y) \quad \text{for all } t \in (x\cap y)_{\varphi} \text{ and for all } t \text{ such that } x \leq t \leq y.$$

We say that $\varphi$ is a *lower valuation* when: for all $x$, $y$, $z$ in $E$, with $x \leq z$, $y \leq z$: $\varphi(x) + \varphi(y) \leq \varphi(z) + \varphi^{-}\langle x,y\rangle$.

For $x$ and $y$ in $E$, we write $x \prec y$ when $y$ covers $x$. The *covering graph* $G(E)$ of $(E,\leq)$ is the (simple undirected) graph whose vertices are the elements of $E$ and whose edges are those pairs $x$, $y$ satisfying: $x \prec y$ or $y \prec x$.

We define the $\varphi$-length of an edge $xy$ by: $\tilde{\varphi}(x,y) = |\varphi(y)-\varphi(x)|$. For any path $c$: $x = x_0\ x_1 \ldots x_p = y$, we write $\tilde{\varphi}(c) = \sum_{i=0}^{p-1} \tilde{\varphi}(x_i, x_{i+1})$, and $\delta_{\varphi}(x,y)$ for the minimum $\varphi$-length of a path between $x$ and $y$. Clearly $\delta_{\varphi}$ is a distance function on $E$. If $\varphi$ is a rank function (if $E$ is graded), we obtain the classical shortest path metric on $G(E)$. A path $c$ between $x$ and $y$ such that $\tilde{\varphi}(c) = \delta_\varphi(x,y)$ is called a *$\varphi$-geodesic* and $\delta_\varphi$ is called the *$\varphi$-geodesic metric*.

THEOREM 1 (Geodesic Metric Theorem). *Let $E$ be a finite partially ordered set with a least element and let $\varphi$ be a strictly isotone real-valued function on $E$. Then the following statements are equivalent:*

(*i*) $\delta_\varphi = d_\varphi^-$;
(*ii*) *$\varphi$ is a lower valuation;*
(*iii*) *$d_\varphi$ is a distance function.*

In particular, since $d_\varphi(x,y) = \delta_\varphi(x,t) + \delta_\varphi(t,y)$ for $t \in (x \cap y)_\varphi$, statement (i) means that there exists a $\varphi$-geodesic between $x$ and $y$ "through $(x \cap y)_\varphi$".

The dual of Theorem 1 is left to the reader. The terms used in the dual case are $(x \cup y)_\varphi$, $\varphi^+ <x,y>$, $d_\varphi^+ (x,y)$, "upper valuation", and so on. A *valuation* is a strictly isotone real-valued function which is both a lower and upper valuation, that is, which satisfies $\varphi(x) + \varphi(y) = \varphi^- <x,y> + \varphi^+ <x,y>$.

For more details on these subjects, see Monjardet, 1981.

2.2.2 Applications

We shall, now, mention two (very related) kinds of consequences of Theorem 1.

a) Metric characterizations of orders which establish a very strong connection between the structure of an ordered set and the kind of metrics we can define on it.

b) Axiomatic characterization of distance between order relations. This gives a partial answer to the question "what distance is the more convenient in order to compare preference relations, according to a given point of view?" This approach seems to go back to Kemeny (1959). It has been illustrated by Bogart (1973, 1975), Mirkin-Chernyi (1970) and Barthélemy (1979).

Rather than be encyclopedical, we shall restrict ourselves to one case (semimodular semilattices) in order to illustrate the logical sequence linking the metric characterization of semimodular semilattices (in valuation terms) to the axiomatic characterization of the s.d. distance, $\delta$, between partial orders. Many other metric characterizations of ordered sets are reviewed in Monjardet (1981) and many applications to distances are given in Barthélemy (1979).

THEOREM 2 (Characterization of semimodular ordered sets). *An ordered set is lower semimodular if and only if it admits a rank function which is a lower valuation.*

This theorem (Monjardet, 1976) admits as corollary a purely metric characterization of semimodular ordered sets.

COROLLARY 1. *Let* $((E,\leq)$ *be an ordered set. The two following statements are equivalent:*

(*i*) $(E,\leq)$ *is a lower semimodular ordered set.*
(*ii*) *There is a metric* $\delta$ *on* $E$ *satisfying* (*a*), (*b*), (*c*):
   (*a*) *For all* $x$, $y$, $t$ *with* $x \leq t \leq y$: $\delta(x,y) = \delta(x,t) + \delta(t,y)$;
   (*b*) *For all* $x$, $y$, $t$ *with* $t \in (x \cap y)_{\varphi}$: $\delta(x,y) = \delta(x,t) + \delta(t,y)$ (*with* $\varphi(t) = \delta(t,0)$);
   (*c*) $\delta(x,y) = 1$, *if* $x \prec y$.

In the case of a semilattice, a formal axiomatic characterization of $\delta$ follows.

COROLLARY 2. *Let* $L$ *be a lower semimodular meet semilattice. Then there is one, and only one, real-valued function* $\delta$ *defined on* $L\times L$ *such that:*
(1) *For all* $x,y \in L$ *with* $x \leq y$, $\delta(x,y) = \delta(y,x)$;
(2) *For all* $x,y,t \in L$ *with* $x \leq t \leq y$; $\delta(x,y) = \delta(x,t) + \delta(t,y)$;
(3) *For all* $x,y \in L$, $\delta(x,y) = \delta(x,x\wedge y) + \delta(x\wedge y,y)$;
(4) *For all* $x,y,t,z \in L$ *such that* $x \prec y$ *and* $t \prec z$, $\delta(x,y) = \delta(t,z)$;
(5) *The strictly positive minimum value of* $\delta$ *exists and is one.*

*Moreover* $\delta$ *is the shortest path distance on* $G(L)$.

Applying this corollary to the set, ordered by inclusion, of partial orders on $X$, we obtain

COROLLARY 3. *The s.d. distance* $\delta$ *on* $\mathrm{P}$ *is the only real-valued function defined on* $\mathrm{P}\times\mathrm{P}$ *such that:*
(1) $\delta(P,Q) = \delta(Q,P)$ *for all* $P,Q \in \mathrm{P}$ *with* $P \subseteq Q$;
(2) $\delta(P,Q) = \delta(P,T) + \delta(T,Q)$, *for all* $P,T,Q \in \mathrm{P}$ *such that* $P \subseteq T \subseteq Q$;
(3) $\delta(P,Q) = \delta(P,P\cap Q) + \delta(Q,P\cap Q)$, *for all* $P,Q \in \mathrm{P}$;
(4) $\delta(P,P\cup\{x,y\}) = \delta(Q,Q\cup\{z,t\})$ *for all* $P,Q \in \mathrm{P}$ *and all* $x,y,z,t \in X$ *such that* $(x,y) \notin P$, $(z,t) \notin Q$, $P \cup \{(x,y)\} \in \mathrm{P}$, $Q \cup \{(z,t)\} \in \mathrm{P}$;
(5) *The strictly positive minimum value of* $\delta$ *is one.*

This result (Barthélemy, 1979) is typical of how one can improve previous characterizations of the s.d. distance. Notice that the above axioms are independent and they do not include (but they imply) the triangle inequality, which is not necessarily an *a priori* relevant condition in social sciences.

## 3. PREFERENCE PROCEDURES

### 3.1 Aggregation Procedures

As above $X$ is a set of $n$ objects. We shall also consider a finite $v$-set $V = \{1,2,\ldots,v\}$ whose elements are called *voters* (these "voters" being either judges or experts, criteria, variables ...).

We consider two sets of partial orders on $X$: $\mathcal{D}$ (the data) and $M$ (the models). The elements of $\mathcal{D}$ are called *individual preferences* and the elements of $M$ *social preferences*. A $\mathcal{D}$-profile is a $v$-tuple $\Pi = (P_1,\ldots,P_v)$ of individual preferences ($P_i$ being the preference of the voter $i$). Thus, the set of all $\mathcal{D}$-profiles is $\mathcal{D}^v$.

A $(\mathcal{D},M,v)$-(aggregation) *procedure* is a map $F$ from $\mathcal{D}^v$ to $M$. We shall sometimes need variations on this notion of procedure. At first, it may be desirable to get, from one $\mathcal{D}$-profile, several social preferences. Thus a $(\mathcal{D},M,v)$-*multiprocedure* will be a map from $\mathcal{D}^v$ to $2^M$. Secondly, it may be desirable to add voters to the set $V$. In these cases, the set of all $\mathcal{D}$-profiles will be, instead of $\mathcal{D}^v$, the set $G(\mathcal{D}) = \bigcup_{v \in N} \mathcal{D}^v$ and a *complete* $(\mathcal{D},M)$-*procedure* will be a map from $G(\mathcal{D})$ to $M$. We can also get *complete* $(\mathcal{D},M)$-*multiprocedures* that are maps from $G(\mathcal{D})$ to $2^M$.

Clearly, not every $(\mathcal{D},M,v)$-procedure will lead to a suitable aggregation rule. Thus we shall retain as feasible procedures those that satisfy some constraints (in the very broad sense). We shall review three kinds of constraints we usually meet in the literature, in their order of historical appearance.

a) Construction rules. The way to construct a procedure is explicitly given. The two historical procedures of Borda (1784) and Condorcet (1785) and some modern procedures occurring in multicriteria analysis (Roy-Bertier, 1973) are of that kind. Historical material may be found in Guilbaud (1952) and Black (1958).

b) Axiomatic constraints. We retain only procedures satisfying some conditions. This point of view goes back to welfare economics (Arrow, 1951); it leads essentially to possibility/impossibility theorems (cf., for instance, Fishburn 1973).

c) Optimality constraints. We have at our disposal a criterion measuring *remoteness* $D(\Pi,Q)$ of a social preference $Q$ from a $\mathcal{D}$-profile $\Pi$ and we take as an aggregation of $\Pi$ a social preference minimizing this remoteness (notice that this method induces a multiprocedure). Before being advocated by Kemeny (1959), this method appeared in a statistical context (Kendall, 1948).

Clearly, the study of procedures belongs to the third level of our taxonomy. However, the methods we shall use occur also at the first or the second level: at the first level we work with the "internal" properties of social and individual preferences (for example, the Arrowian theory of aggregation). At the second one, we work with the structural properties of an ordered set of social (individual) preferences (for example, the median theory in distributive lattices). The plan we shall use to review some aggregation procedures will follow the "historical" order (a), (b), (c), that we have presented above.

### 3.2 The Pioneers: Borda and Condorcet

We begin with an example of the current "plurality rule". Consider the following voting situation: 23 voters ($v = 23$) have to choose one candidate out of three $a$, $b$, $c$ ($n = 3$). If $a$ is preferred (globally) by 9 voters, $b$ by 6 voters and $c$ by 8 voters, then $a$ is the winner. Now, assume that, among the voters choosing $b$, 4 would prefer to see $c$, as winner, rather than $a$. Then $4 + 8 = 12$ voters (a majority) would prefer to see $c$ as winner rather than $a$. This "paradox" led Borda (1784) to propose a somewhat strange solution: "the sum of rankings" method.

Assume that each voter compares the candidates pair by pair and that these comparisons lead to the following profile of linear orders:

| | | | | | |
|---|---|---|---|---|---|
| order of preference: | $a$ | $a$ | $b$ | $b$ | $c$ |
| | $b$ | $c$ | $a$ | $c$ | $b$ |
| | $c$ | $b$ | $c$ | $a$ | $a$ |
| number of voters: | 5 | 4 | 2 | 4 | 8 |

(Note that this profile is compatible with the previous ballot). Assign 2 points for a first ranked candidate, 1 for a second, 0 for a third. Then the score of $a$ is 20, the score of $b$ is 25, the score of $c$ is 24. Thus $b$ is the Borda winner. However 12 voters (hence a majority) would have preferred to see $c$ as winner, rather than $b$.

Criticizing the Borda procedure, Condorcet (1785), proposed a method based on collective paired comparisons. Let $v(x,y)$ be the number of voters choosing $y$ over $x$; then $y$ will be socially preferred to $x$ when $v(x,y) > v(y,x)$. In our example we obtain $v(a,b) = 14$, $v(a,c) = 1$, $v(b,c) = 12$. This leads to the social linear order $a < b < c$.

However, the Condorcet method leads to a drawback. In the example, we have obtained a social linear order. But, in the case of a cyclic voting situation (for instance, $v = 3$, $a < b < c$ for one voter, $c < a < b$ for another, $b < c < a$ for the third) such a social order cannot be defined. The binary relation $P$ defined by Condorcet rule $(x,y) \in P$ if and only if $v(x,y) > v(y,x)$, is not a transitive one. This is the well-known "paradox of voting", called the "effet Condorcet" in the French literature (Guilbaud, 1952). The aggregation theory of Arrow will show that we have met, indeed, a very general situation.

### 3.3. The Theory of Arrow.

Condorcet and Borda have had some successors (for instance in France, S. Laplace and in England, the Reverend Dodgson). However, in the first part of the twentieth century this kind of study became somewhat obsolete. Similar preoccupations in a different context (welfare economics) appeared in the fifties, leading to a relational formulation of the notion of a social welfare function and an impossibility theorem (Arrow, 1951). The work of Arrow has had a very numerous lineage. Here, we shall present the Arrowian theory (essentially results of Arrow, 1951, Mas-Collel and Sonnenschein, 1972, Blair and Pollak, 1979, Blau, 1979, Barthélemy, 1981) in a unified and, we hope, accomplished form (at least when individual and social preferences are orders; for other cases, see, for instance, Kelly, 1978).

3.3.1. Hypotheses on Individual and Social Preferences. We give some conditions on the sets $\mathcal{D}$ and $M$ (cf. 3.1). Concerning individual preferences, we shall *always* have a symmetry hypothesis.

$H_0$: If $P \in \mathcal{D}$ then for each permutation $\sigma$ on $X$, $P^\sigma \in \mathcal{D}$ where $P^\sigma = \{(\sigma(x),\sigma(y)) \mid (x,y) \in P\}$; if $P \in \mathcal{D}$ then $P^{-1} \in \mathcal{D}$ where $xP^{-1}y$ if and only if $yPx$.

Moreover, we consider three levels of complexity for $\mathcal{D}$.

$H_1$: For all $a,b,c \in X$ there exists $P \in \mathcal{D}$ such that $aPb$ and $bPc$.

$H_2$: For all $a,b,c \in X$ there exists $Q \in \mathcal{D}$ such that $aQb$, $aQc$, and $bI(Q)c$ where $I(Q)$ is the incomparability relation associated with $Q$ (cf 1.1).

$H_3$: For all $a,b,c \in X$ there exists $T \in \mathcal{D}$ such that $aTb$, $aI(T)c$, and $bI(T)c$.

We say that $\mathcal{D}$ is *regular* (respectively, *weakly regular*; or *quasi-regular*) if $\mathcal{D}$ satisfies $H_1$, $H_2$, and $H_3$ (respectively, $H_1$; $H_1$ and $H_2$).

It is worth noting that $\mathcal{D} = L$ if and only if $\mathcal{D}$ is weakly regular and satisfies neither $H_2$ nor $H_3$. Also, $\mathcal{D} \subseteq W$ if and only if $\mathcal{D}$ is quasi-regular and does not satisfy $H_3$. Some examples of regular sets are given by the set of all semiorders, interval orders, tree orders, lattice orders, or orders of dimension 2.

Finally, we assume that the following "extended codomain condition" is *always* satisfied.

$H_4$: $\mathcal{D} \subseteq M$.

(There is no problem in considering an individual preference as a possible social preference.)

3.3.2. Dictatorships and Oligarchies. Let $F$ be a $(\mathcal{D},M,v)$-procedure. An *F-oligarchy* is a subset $W$ of $V$ such that for each profile $\Pi = (P_1,\ldots,P_v)$

$$xP_iy \text{ for all } i \in W \text{ implies } xF(\Pi)y, \qquad \text{and}$$
$$xP_jy \text{ for some } j \in W \text{ implies } yF(\Pi)^c x \quad .$$

These conditions may be written

$$\bigcap_{i\in W} P_i \subseteq F(\Pi), \qquad \text{and}$$
$$\bigcup_{i\in W} P_i \subseteq F(\Pi)^d \quad .$$

The second condition means that each voter in $W$ is a *vetoer* (Fishburn, 1973, also, a *weak dictator*, according to Mas-Collel and Sonnenschein's terminology, 1972).

We notice that: *when D is weakly regular, any* $(\mathcal{D},M,v)$-*procedure F admits, at most, one F oligarchy*. So, when such an $F$-oligarchy exists, we say that $F$ is a $(\mathcal{D},M,v)$-*oligarchic procedure*. In the special case where this $F$-oligarchy is a singleton $\{j\}$, we say that $F$ is a $(\mathcal{D},M,v)$-dictatorship and the voter $j$ is called an *F-dictator*. Hence $j$ is an $F$-dictator if and only if: $P_j \subseteq F(\Pi)$ for each profile $\Pi = (P_1,\ldots,P_v)$. If the $F$-oligarchy $W$ (respectively, the $F$-dictator $j$) is such that $\bigcap_{i\in W} P_i = F(\Pi)$ (respectively, $P_j = F(\Pi)$), we say that $F$ is a $(\mathcal{D},M,v)$-*absolutely oligarchic procedure* (respectively, a $(\mathcal{D},M,v)$-*absolute dictatorship*).

3.3.3. Arrowian Procedures. We now state two classical "axiomatic constraints" on procedures.

For each $\mathcal{D}$-profile $\Pi = (P_1,\ldots,P_v)$, each $(\mathcal{D},M,v)$-procedure $F$, and each pair $x$, $y$ of objects, we denote by $\Pi\langle x,y\rangle$ the set of voters preferring $y$ to $x$, so $\Pi\langle x,y\rangle = \{i \mid xP_iy\}$, and by $F_{xy}(\Pi)$ the restriction of $F(\Pi)$ to the set $\{x,y\}\times\{x,y\}$.

We say that the $(\mathcal{D},M,v)$-procedure $F$ is *independent* when, for all $\Pi$, $\Pi'$ in $\mathcal{D}^v$ and each pair $x$, $y$ of objects, the equalities

$$\Pi \langle x,y\rangle = \Pi' \langle x,y\rangle, \quad \Pi \langle y,x\rangle = \Pi' \langle y,x\rangle$$

imply the equality

$$F_{xy}(\Pi) = F_{xy}(\Pi') \quad .$$

We say that $F$ is *Paretian* when $\bigcap_{i=1}^{v} P_i \subseteq F(\Pi)$, for each $\mathcal{D}$-profile $\Pi = (P_1,\ldots,P_v)$.

An *Arrowian*$(\mathcal{D},\mathcal{M},v)$*-procedure* is an independent and Paretian $(\mathcal{D},\mathcal{M},v)$-procedure.

THEOREM 3. *Let $F$ be an Arrowian $(\mathcal{D},\mathcal{M},v)$-procedure*

(1) *If $\mathcal{D}$ is quasi-regular, then $F$ is an oligarchic procedure.*
(2) *If $\mathcal{D}$ is regular, then $F$ is an absolutely oligarchic procedure.*
(3) *If $\mathcal{D}$ is weakly regular and not quasi-regular, then $F$ is an absolutely oligarchic procedure.*

The classical theorem of Arrow is established by taking $\mathcal{D} = \mathcal{M} = \mathcal{W}$ and states that an Arrowian $(\mathcal{W},\mathcal{W},v)$-procedure is a dictatorship (Arrow, 1951). Changing $\mathcal{M}$, we get: if $\mathcal{M} = \mathcal{P}$, then $F$ is an oligarchic procedure, the associated oligarchy being any subset of $V$ (Mas-Collel and Sonnenschein, 1972). If $\mathcal{M}$ is the set of semiorders, or the set of interval orders on $X$, then $F$ is a dictatorship (Blau, 1979, Blair and Pollak, 1979).

Indeed, these above mentioned results are not surprising, since the structure (a dictator or an oligarchy with $w > 1$ members) of an Arrowian procedure is strongly related to the dimension of partial orders in $\mathcal{M}$, as indicated by the following proposition in the case $\mathcal{D} = \mathcal{W}$.

PROPOSITION. *The three following statements are equivalent:*

(*i*) *there exists an arrowian $(\mathcal{W},\mathcal{M},v)$-oligarchic procedure whose associated F-oligarchy admits $w$ members ($w \leq v$);*
(*ii*) *each subset of $V$ with cardinality not greater than $w$ is an F-oligarchy for some arrowian $(\mathcal{W},\mathcal{M},v)$-procedure;*
(*iii*) *each intersection of $w$ linear orders is in $\mathcal{M}$.*

Condition (iii) asserts that each partial order of dimension not greater than $w$ is in $\mathcal{M}$.

3.3.4 From Condorcet to Arrow

The independence of a procedure $F$ means that $F$ is obtained by paired comparisons. Thus, Arrow's Theorem (or more generally our Theorem 3) provides an explanation for the paradox of voting. So, in order to escape this "paradox" one must escape Arrow's Theorem. Some attempts have been made in several directions: restricting the domain (each $\mathcal{D}$-profile is no longer allowed, cf.

Black, 1950, and Gaertner and Salles, 1981, for a review); constructing choice functions (one only considers the social winners for a profile, see Sen, 1970 and Bordes, 1981, for a review); considering other kinds of "constraints" on procedures.

### 3.4. Median Theory for Partial Orders

The work of Kemeny (1959) has been mentioned several times. We shall now give some more information about it.

The median of a statistical series is well-known. A suitable generalization of this notion can be obtained in a metric space: a *median* of a set $S$ of points is a point such that the sum of distances to the points in $S$ is a minimum. Accordingly, Kemeny defines a *median* of a $W$-profile $(P_1,\ldots,P_v)$ as a weak order $S$ such that:

$$\sum_{i=1}^{v} \delta(S,P_i) = \min_{T\in W} \sum \delta(T,P_i)$$

($\delta$ being the s.d. distance between binary relations). Hence, we obtain a $(W,W,v)$-multiprocedure $K$ (the *Kemeny procedure*), which assigns to each profile the set of its medians.

Here, we shall illustrate the Kemeny procedure by examining

(1) relationships between Kemeny and Condorcet procedures,
(2) axiomatic characterization of the Kemeny procedure,
(3) the Kemeny procedure and medians in distributive lattices.

Notice that the first two points are relevant to the first level (internal structures of preferences), the third one is relevant to the second level (organization of preferences). The first two points establish connections among several kinds of "constraints", from construction rule to optimization in (1), from optimization to axiomatics in (2). The point (2) has been developed by Young and Levenglick (1978); the points (1) and (3) have been worked out by Barbut (1961 and 1967), without reference to Kemeny.

#### 3.4.1. From Condorcet to Kemeny

Let $C$ be the set of all tournaments on $X$ (a *tournament* is a complete asymmetric relation). When $v$ is an odd number, the Condorcet rule (see 3.2) assigns to each $L$-profile $\Pi = (P_1,\ldots,P_v)$ a (unique) tournament $C(\Pi)$. It can be easily seen that $C(\Pi)$ is the (unique) solution of

$$\min_{T\in C} \sum_{i=1}^{v} \delta(T,P_i) \quad .$$

When $v$ is even, we can get several solutions by using the Condorcet rule and tie breakings. Hence, a suitable extension of the Condorcet rule, as an $(L,L,v)$-multiprocedure, is obtained by taking as social preferences associated to the $L$-profile $\Pi$, the solutions of:

$$\min_{S \in L} \sum_{i=1}^{v} \delta(S,P_i) \quad .$$

More generally, we define a $(\mathcal{D},\mathcal{M},v)$-multiprocedure by associating to each $\mathcal{D}$-profile $\Pi = (P_1,\ldots,P_v)$ the solutions of

$$\min_{S \in \mathcal{M}} \sum_{i=1}^{v} \delta(S,P_i) \quad .$$

When $\mathcal{D} = \mathcal{M} = \mathcal{W}$ this multiprocedure is nothing else than the Kemeny procedure. Strangely, Kemeny (1959) and then Kemeny and Snell (1962) do not mention the Condorcet rule. The quite trivial connection between those two procedures, very often rediscovered, was made, first, by Barbut (1967).

In the general case, changing the number $v$ of voters, we obtain a $(\mathcal{D},\mathcal{M})$-complete multiprocedure, denoted $\mathrm{Med}_{(\mathcal{D},\mathcal{M})}$ (or simply Med) and called the *median procedure*. Properties of this procedure (including combinatorial properties and effective computation possibilities) are reviewed in Barthélemy and Monjardet (1981).

3.4.2. The Young-Levenglick Theorem

First, we describe three properties of the complete multiprocedure Med.

Let $F$ be a complete $(\mathcal{D},\mathcal{M})$-multiprocedure. Suppose that a set $V$ of voters is divided into two parts $W$ and $\overline{W}$ (that is, $W \cap \overline{W} = \phi$, $W \cup \overline{W} = V$). To each $\mathcal{D}$-profile $\Pi = (P_1,\ldots,P_v)$ are associated the two restricted $\mathcal{D}$-profiles $\Pi_W = (P_i \mid i \in W)$, and $\Pi_{\overline{W}} = (P_i \mid i \in \overline{W})$. We say that $F$ is *consistent* when:

$$\text{if} \quad F(\Pi_W) \cap F(\Pi_{\overline{W}}) \neq \emptyset \quad \text{then} \quad F(\Pi) = F(\Pi_W) \cap F(\Pi_{\overline{W}}) \quad .$$

Clearly $\mathrm{Med}_{(\mathcal{D},\mathcal{M})}$ is consistent.

Another property of Med is its neutrality. A $(\mathcal{D},\mathcal{M},v)$-multiprocedure (or a complete $(\mathcal{D},\mathcal{M})$-multiprocedure) $F$ is *neutral* when for each permutation $\sigma$ on $X$ and every profile $\Pi = (P_1,\ldots,P_v)$ it follows that $F(\Pi^\sigma) = F^\sigma(\Pi)$, with $\Pi^\sigma = (P_1^\sigma,\ldots,P_v^\sigma)$ for $\Pi = (P_1,\ldots,P_v)$, and $F^\sigma(\Pi) = \{S^\sigma \mid S \in F(\Pi)\}$.

Assume now, that $\mathcal{D} = \mathcal{M} = L$ and let $F$ be an $(L,L,v)$-multiprocedure (or a complete $(L,L)$-procedure). We say that $F$ is *Condorcet* if, for all $x,y \in X$ such that $v(x,y) < v(y,x)$ for the profile $\Pi$, there is no $S \in F(\Pi)$ in which $y$ covers $x$, and, for all $x,y \in X$ such that $v(x,y) = v(y,x)$, then if $S'$ is the linear order obtained from a linear order $S$ by exchanging $x$ and $y$ in $S$, $S \in F(\Pi)$ if and only if $S' \in F(\Pi)$.

THEOREM 4. (Young and Levenglick). *The median procedure is the only complete* $(L,L)$-*multiprocedure which is consistent, neutral and Condorcet.*

This theorem provides an axiomatic characterization of median linear orders. It also provides an example of axiomatic conditions leading to a convenient procedure. Notice that the Borda rule, as a choice function, also admits such an axiomatic characterization (Young, 1974). An interpretation of the Condorcet condition in terms of "stability" may be found in Barthélemy and Monjardet (1981).

3.4.3. From Kemeny to Medians in Distributive Lattices

We have introduced the notion of a median by means of a metric. This notion admits also an algebraic approach.

In the set $R = 2^{X \times X}$ of binary relations on $X$, for each $v$-tuple $\Pi = (R_1, \dots, R_v) \in D^v$, we denote by $R_\alpha(\Pi)$, or simply $R_\alpha$, the relation defined by $xR_\alpha y$ if and only if there exists $[v+2/2]$ voters $i$ in $V$ such that $xR_i y$. Equivalently, we can write using the lattice operations of $R$,

$$R_\alpha(\Pi) = \bigcup_{\substack{W \subseteq V \\ |W| = \left[\frac{v+2}{2}\right]}} \left( \bigcap_{i \in W} R_i \right) .$$

We denote by $R_\beta(\Pi)$, or $R_\beta$, the relation defined by: $xR_\beta y$ if there exist $[v+1/2]$ voters $i$ in $V$ such that $xR_i y$. Thus

$$R_\beta(\Pi) = \bigcup_{\substack{W \subseteq V \\ |W| = \left[\frac{v+1}{2}\right]}} \left( \bigcap_{i \in W} R_i \right) .$$

Obviously $R_\alpha \subseteq R_\beta$. If $v$ is an odd number $R_\alpha = R_\beta$; in this case, if $\Pi$ is an $L$-profile, this relation is obtained by applying to $\Pi$ the Condorcet rule. When there is no "effet Condorcet", it is the median linear order of $\Pi$. More generally the *algebraic median relations* of the profile $\Pi$ will be the relations in the interval $[R_\alpha, R_\beta]$. Notice also that, using the distributivity of the lattice $(R, \cap, \cup)$ we get

$$R_\beta(\Pi) = \bigcup_{\substack{W \subseteq V \\ |W| = \left[\frac{v+2}{2}\right]}} \left( \bigcap_{i \in W} R_i \right) .$$

The way we have traced may be followed in any distributive lattice. The first noteworthy occurrence of medians in distributive lattices is in Birkhoff and Kiss (1947) in the case of three elements. This ternary case has led to many developments subsumed under the name of "median algebras". In a (finite) lattice, the notion of median has been extended by Barbut (1961) to an odd numbered family of elements. In a lattice $T$, the median interval of such a family $(e_1, \ldots, e_{2p+1})$ is the interval $[\alpha, \beta]$ with

$$\alpha = \bigvee_{\substack{|K| = p+1 \\ K \subseteq \{1, \ldots, 2p+1\}}} \left( \bigwedge_{i \in K} e_i \right), \qquad \beta = \bigwedge_{\substack{|K| = p+1 \\ K \subseteq \{1, \ldots, 2p+1\}}} \left( \bigvee_{i \in K} e_i \right) .$$

When this interval admits just one element (as is the case for the lattice $R$), we say that the family $(e_1, \ldots, e_{2p+1})$ admits one *algebraic median.*

THEOREM 5. (Barbut). *Let $T$ be a finite lattice. The following statements are equivalent:*
(*i*) *$T$ is distributive;*
(*ii*) *there is an odd number $2p+1$ such that every $(2p+1)$-family admits one algebraic median;*
(*iii*) *every odd numbered family in $T$ admits one algebraic median.*
*Moreover, if these conditions are satisfied the algebraic median of the family $(e_1, \ldots, e_{2p+1})$ is the (unique) solution of:*

$$\min_{s \in T} \sum_{i=1}^{2p+1} d_\varphi(s, e_i)$$

*where $d_\varphi$ is the $\varphi$-geodesic metric relative to some valuation $\varphi$ on the lattice $T$.*

Hence, the algebraic construction of medians is only available in distributive lattices and in this case algebraic and "metric" medians coincide. Notice that characterization (ii) can be improved by considering any Sperner family instead of the sets $K$, with $|K| = 2p+1$ (cf. Monjardet, 1980).

Moreover, the distributive case suggests the study of the

solutions of $\sum_{i=1}^{v} d_\varphi(s,e_i)$, called the *φ-medians* of $\Pi = (e_1,\ldots,e_v)$, φ being a "semi-valuation" on the ordered set $E$. At this point, only some partial results have been obtained; we mention two of these.

PROPERTY 1. *If φ is an upper valuation on a lattice $T$, then for each φ-median $m$ of $(e_1,\ldots,e_v)$, $\bigwedge_{i=1}^{v} e_i \leq m$.*

Property 1 is nothing else than a "Paretian rule".

PROPERTY 2. *If φ is a valuation on a modular lattice $T$, the φ-medians of a family in $T$ define a sublattice of $T$.*

## 4. SOME OTHER EXAMPLES

As already noted, the theory of preferences is well-suited to our three-level taxonomy of the activities of a social scientist. Many other examples of the utility of ordered sets are confined to one level without substantial extension to the others -- not because of a lack of interest, rather, because of a lack of results.

We shall present, quite briefly, an example located at level one. It is related to the social scientist's interest in dimensionality of partial orders.

If we observe a partial order on some set, we try to understand it as the result of several linear orders; the linear orders are supposed to be easier to think of in social terms. For instance, if a social dominance relation on a set of individuals is a partial order, we attempt to identify "simpler variables", themselves linearly ordered, whose product yields the dominance relation. Such variables as wealth, and academic grades may be deemed appropriate ones. In this area, a very common technique in social science gives rise to a slightly different problem of some mathematical interest.

Let $R$ be a relation on two sets $X$ and $Y$, $R \subseteq X\times Y$. It could be that $X$ is a set of individuals and $Y$ is a set of attributes or features; then $(x,y) \in R$ means that individual $x$ possesses feature $y$. Let α be the map of $X$ to the lattice $P(Y)$ of all subsets of $Y$ defined by

$$\alpha(x) = \{y \in Y : (x,y) \in R\} \quad .$$

There is a natural partial order induced on $X$: $x < x'$ if $\alpha(x) \subset \alpha(x')$. Dually, there is also a partial order induced on $Y$ by the map

$$\beta(y) = \{x \in X : (x,y) \in R\} \quad .$$

Recall that the (*order*) *dimension* dim($P$) of a partially ordered set $P$ is the minimum number of linear extensions of the order on $P$ whose intersection is that order. Now consider the dimensions of the partial orders on $X$ and $Y$. It is possible to construct, for an integer $n$, a relation $R$ such that dim($X$) = 2 and dim($Y$) = $n$. (Trotter first gave an example with $|X| = |Y| = 2n$; one can construct another example with $|X| = \left|\frac{Y}{2}\right| = n$.)

But we have:

$$\dim(X) = 1 \quad \text{if and only if} \quad \dim(Y) = 1 \quad .$$

In this case, we can present the matrix of the relation $R$, after permutation of the rows and of the columns, in this way:

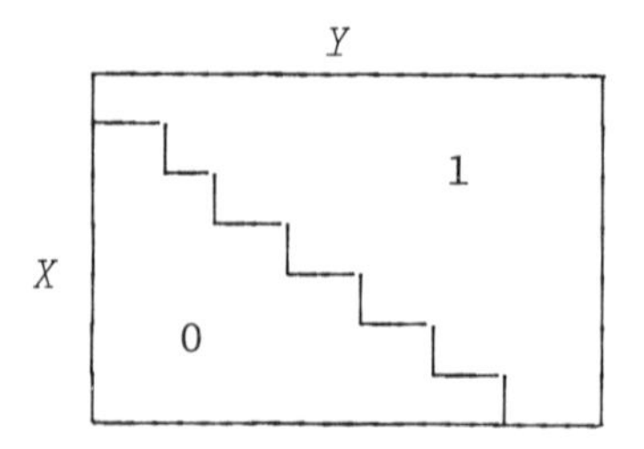


This very special case has been intensively studied for the last four decades in social sciences, especially in psychology, sociology, and archeology. Its name, for social scientists, is *Guttman's model*; some mathematicians (Riguet, 1950, Ore, 1962) name it *Ferrers relation*. It has been also called a *biorder* (Ducamp and Falmagne, 1969).

A Ferrers relation $R \subset X \times Y$ is characterized by the following axiom: for all $x,x' \in X$, $y,y' \in Y$, $(x,y) \in R$, $(x',y') \in R$, $(x,y') \notin R$ imply $(x',y) \in R$. (Note: an interval order or a semi-order $P \subset X \times X$ is a Ferrers relation, because of the axiom $PIP \subseteq P$).

This model is often too strong to allow us to analyse an empirical relation $R$. But it is possible to find a minimal family of Ferrers relations, the intersection of which is equal to the given relation $R$. Thus, it is possible to define the notion of *Ferrers dimension* of a relation $R$, denoted by $d_F(R)$ (Bouchet, 1971, Cogis, 1980, Falmagne, 1980). This Ferrers dimension can be considered as a generalization of the order dimension in the sense that, if $R \subset X \times X$ is a partial order, the Ferrers dimension of $R$ is equal to its order dimension.

Some bounds for $d_F(R)$ have been obtained. For example, with $R^c = (X \times Y) - R$, $d_F(R) \leq |R^c|$. Also, consider the graph $(R^c, \Gamma)$ where $(x,y)\Gamma(x',y')$ if $(x,y),(x',y') \in R^c$ and $(x,y'),(x',y) \in R$ (a configuration forbidden in a Ferrers relations). Let $\chi(\Gamma)$ be the chromatic number of this graph. Then $\chi(\Gamma) \leq d_F(R)$. Obviously, $d_F(R) = 1$ if and only if $\chi(\Gamma) = 1$, and $d_F(R) = |R^c|$ implies $\chi(\Gamma) = |R^c|$. Cogis proved that $d_F(R) = 2$ if and only if $\chi(F) = 2$. One might have the idea that $d_F(R) = \chi(\Gamma)$. But Trotter gave an example where $\chi(\Gamma) \leq n + 3$ and $d_F(R) \geq \frac{3n}{2}$ (with $|X| = |Y| = 6n$).

In another direction, one can consider the Galois connection associated with the relation $R$ and the Galois lattice (of Galois closed subsets). The order dimension of this lattice is $d_F(R)$. (The above results, among others, come from work of Cogis on the Ferrers dimension of digraphs.) In fact, many properties of $R$ which are of interest to social scientists, can be studied with the Galois lattice associated with $R$ (see Barbut and Monjardet, 1970, and Flament, 1979). (We note that another lattice, a Boolean one, was used by Flament, 1976, to study $R$.)

To end this (too incomplete) presentation, we mention a result concerning a classical problem in the theory of ordered sets -- the fixed point problem. The result is, in part, motivated by a social science application.

In order to obtain some combinatorial solutions to a given problem, we construct an algorithm which can be abstractly described as an isotone map $f$ on some partially ordered set in such a way that the desired solutions are exactly the fixed points of $f$. An example of such a problem is that of obtaining specific Galois closed subsets for a given relation $R$ (see Degenne et al, 1978). Let us briefly give a more specific instance. Let $X$ be a set of individuals and consider $D \subseteq X^2$ as a "social dominance" relation. Let $s$ and $p$ (for "successors" and "predecessors") be maps on the nonempty subsets of $X$ defined by

$$s(A) = \{y \in X | \text{there is some } x \in A \text{ with } xDy\} ,$$

$$p(A) = \{y \in X | \text{there is some } x \in A \text{ with } yDx\} ,$$

for all $A \subseteq X$. The subset $A$ "dominates" $s(A)$, but also shares this property with $p \circ s(A)$. To avoid ambiguity in the definition of dominant groups, we wish to find the pairs $\{A,B\}$ such that $A = p(B)$ and $B = s(A)$. Clearly, such subsets $A$ and $B$ are fixed points of the maps $p \circ s$ and $s \circ p$: that is, $A = p \circ s(A)$ and $B = s \circ p(B)$.

In general, we know the existence of trivial solutions. In fact, rather than existence results, we require theorems giving the structure of the fixed point set in order to work effectively with solutions. (From this viewpoint, we are led to some classical fixed point results such as Tarski's fixed point theorem; see also Cousot, 1977, and Flament, 1978.)

Let us state our observations concerning fixed point sets. Let $f$ be an isotone map of a complete lattice $L$ to itself. Define the maps $r$ and $e$ of $L$ to $L$ by $r(x) = x \vee f(x)$, $e(x) = x \wedge f(x)$ for all $x \in L$. Of course $r$ is increasing, $x \leq r(x)$, and $e$ is decreasing. Moreover, the map $\varphi$ given by $\varphi(x) = \sup_L(r^i(x) | i = 1,2,\ldots)$ is a closure map and $\omega$, defined by $\omega(x) = \inf_L(e^i(x) | i = 1,2,\ldots)$ is an anticlosure map. Now let $F = \{x \in L | x = f(x)\}$, the set of fixed points of $f$. Then

$$F = \varphi\circ\omega(L) = \omega\circ\varphi(L).$$

Well-known properties of $\varphi$ and $\omega$ allow that $F$ is a complete lattice, the minimum element $0_F$ of $F$ is $\varphi(0_L)$ and the maximum element $1_F$ of $F$ is $\omega(1_L)$. Moreover, for any nonempty subset $E$ of $F$,

$$\inf_F(E) = \omega(\inf_L(E)),\ \sup_F(E) = \varphi(\sup_L(E)) .$$

Of course, these kinds of results are of general order-theoretic interest aside from their usefulness for the algorithms alluded to previously.

Finally, for the sake of providing relevant references, we point out that cluster analysis -- a method of data analysis often used in social science -- is another place where order theory has been profitably applied. See Boorman, 1970, Boorman and Olivier, 1973, Degenne, Flament, and Verges, 1978, Janowitz, 1978, 1979, Mirkin, 1979, Schader, 1980, 1981, Barthélemy and Monjardet, 1981, and Leclerc, 1981.

REFERENCES

M. Aigner (1969) "Graphs and binary relations", in *The many facets of graph theory*, 1-21, New York, Springer Verlag.

W.E. Armstrong (1939) "The determinateness of the utility function", *Economics Journal*, 49, 453-467.

K.J. Arrow (1962) *Social choice and individual values*, 2nd ed., J. Wiley and Sons, New York (first edition, 1951).

S.P. Avann (1961) "Application of the join-irreducible excess function to semi-modular lattices", *Math. Annalen*, 142, 345-354.

S.P. Avann (1972) "The lattice of natural partial orders", *Aequationes Math.*, 8, 95-102.

M. Barbut (1961) "Médiane, distributivité, éloignements", Publications du Centre de Mathématique Sociale, E.P.H.E. 6ème section, Paris and *Math. Sci. Hum.* 70 (1980), 5-31.

M. Barbut (1967) "Médianes, Condorcet et Kendall", note SEMA, Paris and *Math Sci. Hum.* 69 (1980), 5-13.

M. Barbut (1967) "Echelles à distance minimum d'une partie donnée d'un treillis distributif fini, *Math. Sci. Hum.* 18, 41-46.

M. Barbut (edit. 1971) *Ordres totaux finis*, Mouton, Paris.

M. Barbut and B. Monjardet (1970) *Ordre et classification, Algèbre et Combinatoire*, 2 tomes, Paris, Hachette.

J.P. Barthélemy (1978) "Remarques sur les propriétés métriques des ensembles ordonnés", *Math. Sci. Hum.* 61, 39-60.

J.P. Barthélemy (1979) "Caractérisations axiomatiques de la distance de la différence métrique entre des relations binaires", *Math. Sci. Hum.* 67, 85-113.

J.P. Barthélemy (1980) "An asymptotic equivalent for the number of total preorders on a finite set", *Discrete Mathematics* 29, 311-313.

J.P. Barthélemy (1982) "Arrow's theorem: unusual domains and extended codomains", in P.K. Pattanaik and M. Salles (eds.) *Social choice and welfare*, North Holland, Amsterdam.

J.P. Barthélemy and B. Monjardet (1981) "The median procedure in cluster analysis and social choice theory", *Math. Soc. Sci.* 1, 3, 235-267.

G. Birkhoff (1967) *Lattice theory*, 3rd ed., Providence, Amer. Math. Soc.

G. Birkhoff and S.A. Kiss "A ternary operation in distributive lattices", *Bull. Amer. Math. Soc.* 53, 749-752.

D. Black (1958) *The theory of Committees and Elections*, Cambridge University Press, London.

D.H. Blair and R.A. Pollak (1979) "Collective rationality and dictatorship: the scope of the Arrow theorem", *J. Econ. Theory* 21, 1, 186-194.

J.K. Blau (1979) "Semiorders and collective choice", *J. Econ. Theory* 21, 1, 195-206.

K.P. Bogart (1973) "Preference structures I: Distance between transitive preference relations", *J. Math. Sociol.*, 3, 49-67.

K.P. Bogart (1973) "Preference structures II: Distances between asymmetric relations", *SIAM J. Appl. Math.*, 29, 2, 254-262.

K.P. Bogart (1982) "Some social science applications of ordered sets", *Symp. Ordered Sets* (I. Rival, ed.) Reidel, Dordrecht-Boston, 759-787.

K.P. Bogart and W.T. Trotter, Jr (1973) "Maximal dimensional partially ordered sets II. Characterization of $2n$-element posets with dimension $n$, *Discrete Mathematics*, 5, 33-43.

K.P. Bogart, I. Rabinovitch and W.T. Trotter, Jr. (1976) "A bound on the dimension of interval orders", *J. Combinatorial Theory*, A21, 319-328.

S.A. Boorman and P. Arabie (1972) "Structural measures and the method of sorting" in *Multidimensional Scaling*, vol. 1, Theory, Seminar Press, New York.

S.A. Boorman and D.C.Olivier (1973) "Metrics on spaces of finite trees", *J. of Math. Psychol.*, 10, 26-29.

J.C. Borda (1784) "Mémoire sur les élections au scrutin, *Histoire de l'Académie Royale des Sciences pour 1781*, Paris, English translation by A. de Grazia, Isis, 44 (1953).

G. Bordes (1981) "Procédures d'agrégation et fonctions de choix", in P. Batteau, E. Jacquet-Lagrèze and B. Monjardet, eds., *Analyse et agrégation des préférences*, Economica, Paris.

A. Bouchet (1971) *Étude combinatoire des ordonnés finis*, thèse d'état, Université Scientifique et Médicale de Grenoble.

G. Boulaye (1968) "Notion d'extension dans les treillis", *Revue roumaine Mathématiques pures et appliquées*, 13, 9, 1225-1231.

J.L. Chandon, J. Lemaire and J. Pouget (1978) "Dénombrement des quasi-ordres sur un ensemble fini", *Math. Sci. Hum*, 62, 61-80.

R.P. Dilworth (1940) "Lattice with unique irreducible representation", *Ann. of Math.*, 41, 2, 771-777.

J.S. Chipman (1971) "Consumption theory without transitive indifference", in *Preferences, Utility and Demand*, Harcourt Brace, New York.

O. Cogis (1980) *La dimension Ferrers des graphes orientés*, Thesis, Université P. et M. Curie, Paris.

O. Cogis (1981) "On the Ferrers dimension of a digraph", *Discrete Mathematics*, to appear.

Marquis de Condorcet (1785) *Essai sur l'application de l'analyse à la probabilité des décisions rendues à la pluralité des voix*, Paris, (reprint, Chelsea Publ. 6, New York, 1974).

C.H. Coombs (1964) *A theory of data*, Wiley, New York.

P. Cousot and R. Cousot (1977) *Constructive versions of Tarski's fixed point theorems*, Grenoble, Univ. Sci. Med.

R.A. Dean and F. Keller (1968) "Natural partial orders", *Canad. J. Math.*, 20, 535-554.

A. Degenne (1972) *Techniques ordinales en analyse des données: Statistique*, Hachette, Paris.

A. Degenne, C. Flament and P. Verges (1978) "Approche galoisienne en classification", *Statistiques et Analyse des données*, 29-34.

J.P. Doignon, A. Ducamp and J.C. Falmagne (1981) "The bidimension of a relation", Colloque international sur la théorie des graphes et la combinatoire, Luminy.

A. Ducamp and J.C. Falmagne (1969) "Composite measurement", *J. Mathematical Psychology*, 6, 359-390.

V. Duquenne (1980) *Quelques aspects algébriques du traitement de données planifiées*, third cycle thesis, Université René Descartes.

P.H. Edelman (1980) "Meet-distributive lattices and the anti-exchange closure", *Algebra Universalis*, 10, 290-299.

P.H. Edelman and P. Klingsberg (1981) "The subposet lattice and the order polynomial", discussion paper.

J.C. Falmagne (1980) "Bi-orders and bi-order dimension of a partial order", 11th European Mathematical Psychology Meeting, Liège.

J.M. Faverge (1965), *Méthodes statistiques en psychologie appliquée*, tome III, Presses Universitaires de France, Paris.

T. Fine (1970) "Extrapolation when very little is known", *Information and control*, 16, 331-360.

P.C. Fishburn (1970a) "Intransitive indifference in preference theory: a survey", *Operations Research*, 18, 2, 207-228.

P.C. Fishburn (1970b) *Utility theory for decision making*, New York, Wiley.

P.C. Fishburn (1971) "Betweenness, orders and interval graphs", *J. Pure Appl. Algebra*, 1, 2, 159-178.

P.C. Fishburn (1972) *Mathematics of decision theory*, Mouton, the Hague.

P.C. Fishburn (1973) *The theory of the social choice*, Princeton University Press, New Jersey.

P.C. Fishburn (1975) "Semiorders and choice functions", *Econometrica*, 43, 5-6, 975-977.

P.C. Fishburn (1979) "Transitivity" *Review of Economic Studies* 46, 1, 163-173.

P.C. Fishburn (1981a) "Maximum semiorders in interval orders", *SIAM J. Alg. Disc. Meth.* 2, 2.

P.B. Fishburn (1981b) "Restricted thresholds for interval orders: a class of nonfinite axiomatizability", preprint.

P.C. Fishburn and W.V. Gehrlein (1975) "A comparative analysis of methods for constructing weak orders from partial orders", *J. Math. Sociol.*, 4, 1, 93-102.

Cl. Flament (1963) *Applications of graph theory to group structure*, Prentice Hall, New York.

Cl. Flament (1962) "L'analyse de similitude", *Cahiers du Centre de Recherche Opérationnelle*, 4, 2, 63-97.

Cl. Flament (1976) *L'analyse booléenne de questionnaires*, Mouton, Paris.

Cl. Flament (1978) "Un théorème de point fixe dans les treillis", *Mathématiques et Sciences Humaines*, 61, 61-64.

Cl. Flament (1979) "Un modèle des jugements de similitude", *Math. et Sci. Hum.*, 65, 5-21.

C. Frey (1971) *Technique ordinale en analyse des données, algèbre et combinatoire*, Hachette, Paris.

W. Gaertner and M. Salles (1981) "Procédures d'agrégation avec domaines restreints" in *Analyse et agrégation des préférences* (Eds. P. Batteau, E. Jacquet-Lagrèze, B. Monjardet) Economica, Paris.

E.H. Galanter (1956) "An axiomatic and experimental study of sensory order and measure", *Psychological Review*, 63, 16-28.

M.C. Golumbic (1980) *Algorithmic graph theory and perfect graphs*, Academic Press, New York.

M. Gonzalez (1978) *L'organisation des préférences et les jugements comparatifs*, third cycle thesis, Université de Paris X.

N. Goodman (1951) *Structure of appearance*, Cambridge, Mass., Harvard University Press.

T.L. Greenough and K.P. Bogart (1979) "The representation and enumeration of interval orders", to appear.

G. Th. Guilbaud (1952) "Les théories de l'intérêt général et le problème logique de l'agrégation", *Economie appliquée*, 5, 4, 1952 reprinted in *Eléments de la théorie des jeux*, Dunod, Paris, 1968, English Transl. in *Readings in Mathematical Social Sciences*, Science Research Associates, Chicago, 262-307.

G. Th. Guilbaud (1962) "Continu expérimental et continu mathématique", notes de cours et *Math. Sci. Hum.*, 62, 11-33 (1978).

G. Th. Guilbaud and J.P. Rosenstiehl (1963) "Analyse algébrique d'un scrutin", *Math. Sci. Hum.* 4, 9-33.

E. Halphen (1953) "La notion de vraisemblance", *Publications de l'I.S.U.P.*, 4, 1, 41-92.

L. Haskins and S. Gudder (1972) "Height on posets and graphs", *Discrete Math.* 2, 357-382.

D. Hausmann and B. Korte (1978) "Adjacency on 0-1 polyhedra" *Polyhedral Combinatorics*, edited by M.L. Balinski and A.J. Hoffman, North Holland, New York, 106-127.

L. Hubert (1974) "Some applications of graph theory and related non-metric-techniques to problems of approximate seriation: the case of symmetric proximity measures", *Br. J. Math. Statist. Psychol.*, 27, 2, 133-153.

E. Jacquet-Lagrèze (1975a) "How we can use the notion of semi-orders to build outranking relations in multicriteria decision making", in *Utility subjective probability and human decision making*, ed. c. Vlek et P. Wendt, D. Reidel, et *Metra* 13, 1, (1974) 59-86.

E. Jacquet-Lagrèze (1975b) *La modélisation des préférences: pré-ordres, quasi-ordres et relations floues*, third cycle Thesis, Université Paris V.

D.T. Jamison and L.J. Lau (1973) "Semiorders and the theory of choice", *Econometrica*, 41, 901-902.

D.T. Jamison and L.J. Lau (1977) "The nature of equilibrium with semi-ordered preferences", *Econometrica*, 45, 7, 595-605.

E. Jacquet-Lagrèze and B. Roy (1981) "Aide à la décision multicritères et systèmes relationnels de préférence" in P. Batteau, E. Jacquet-Lagrèze and B. Monjardet (eds) *Analyse et agrégation des préférences*, Economica, Paris.

M.F. Janowitz (1978) "An order theoretic model for cluster analysis, *SIAM J. Appl. Math.*, 34, 1, 55-72.

M.F. Janowitz (1979) "Monotone equivariant cluster methods", *SIAM J. Applied Mathematics*, 37, 148-165.

J.S. Kelly (1978) *Arrow impossibility theorems*, New York, Academic Press.

J.G. Kemeny (1959) "Mathematics without numbers", *Daedalus* 88, 577-591.

J.G. Kemeny and J.C. Snell (1962) *Mathematical models in the social sciences*, Ginn and Co., New York.

D.G. Kendall (1969) "Incidence matrices, interval graphs, and seriation in archaeology", *Pacific J. Math.*, 28, 565-570.

M.G. Kendall (1962) *Rank correlation methods*, 3rd ed., Hafner, New York (first ed. 1948).

Y. Kergall (1974) "Étude des tresses de Guttman en algèbre à *p* valeurs", *Math. Sci. Hum.*, 46.

D.H. Krantz, R.D. Luce, P. Suppes and A. Tversky (1971) *Foundations of measurement*, vol. 1, Academic Press, New York.

B. Leclerc (1981) "Description combinatoire des ultramétriques", *Math. Sci. Hum.* 73, 5-37.

S. Leinhardt (edit. 1977) *Social networks: a developing paradigm*, Academic Press, New York.

R.D. Luce (1956) "Semiorders and a theory of utility discrimination", *Econometrica*, 24, 178-191.

R.D. Luce, R.R. Bush and E. Galanter (1963-1965) *Handbook of mathematical psychology*, 3 vol., Wiley, New York.

H. Maehara (1980) "On time graphs", *Discrete Mathematics* 32, 281-289.

J.F. Marcotorchino and P. Michaud (1979) *Optimisation en analyse ordinale des données*, Masson, Paris.

A. Mas-Collel and H. Sonnenschein (1972) "General possibility theorems for group decision", *The Review of Economic Studies*, 39, 185-192.

B.G. Mirkin (1972) "Description of some relations on the set of real-line intervals", *J. Math. Psychol.*, 9, 243-252.

B.G. Mirkin (1974) "The problems of approximation in space of relations and qualitative data analysis", *Automatika i Telemechanika* translated in *Automation and Remote control*, 35, 9, 1424-1431.

B.G. Mirkin (1979) *Group choice*, Editor P.C. Fishburn, V.H. Winston and Sons, Washington (Russian edit., 1974).

B.G. Mirkin and K.B. Cherny (1970) "On measurement of distance between partitions of a finite set of units", *Automatika i Telemechanika*, transl. in *Automation and Remote Control*, 31, 5, 786-792.

B. Monjardet (1975) "Tournois et ordres médians pour une opinion", *Math. Sci. Hum.*, 43, 55-70.

B. Monjardet (1976) "Caractérisations métriques des ensembles ordonnés semi-modulaires", *Math. Sci. Hum.*, 56, 77-87.

B. Monjardet (1978a) "An axiomatic theory of tournament aggregation", *Math. Oper. Research*, 3, 4, 334-351.

B. Monjardet (1978b) "Axiomatiques et propriétés des quasi-ordres", *Math. Sci. Hum.*, 63, 51-82.

B. Monjardet (1980) "Théorie et applications de la médiane dans les treillis distributifs finis", *Annals of Discrete Math.*, 9, 87-91.

B. Monjardet (1981) "Metrics on partially ordered sets - a survey", *Discrete Mathematics*, 35, 173-184.

J.W. Moon (1979) "Some enumeration problems for similarity relations", *Discrete Mathematics*, 26, 3, 251-260.

Y.K. Ng (1977) "Sub-semiorder. A model of multidimensional choice with preference intransitivity", *J. Math. Psychol.*, 16, 1, 51-59.

O. Ore (1962) *Theory of graphs*, Chapter 11, American Mathematical Society, Providence.

I. Rabinovitch (1978a) "The dimension of semiorders", *J. Comb. Th.* (A), 25, 50-61.

I. Rabinovitch (1978b) "An upper bound on the dimension of interval orders", *J. Comb. Th.* (A), 25, 68-71.

F. Restle (1959) "A metric and an ordering on sets", *Psychometrika*, 24-207-220.

J. Riguet (1950) "Les relations de Ferrers, *C.R. Acad. Sci. Fr.*, 231, 936-937.

F.S. Roberts (1970) "On nontransitive indifference", *J. Math. Psychol.* 7, 243-258.

F.S. Roberts (1976) *Discrete mathematical models with applications to social, biological and environmental problems*, New York, Prentice Hall.

F.S. Roberts (1978) *Graph theory and its applications to problems of society*, Philadelphia, SIAM.

F.S. Roberts (1979) *Measurement theory*, Addison Wesley.

D.G. Rogers (1977) "Similarity relations on finite ordered sets", *J. Comb. Th.* (A), 23, 88-99.

D.G. Rogers, K.H. Kim and F.W. Roush (1979) "Similarity relations and semiorders" 10th Southeastern Conference on Combinatorics, Graph theory and Computing, Baton Rouge

M. Roubens and Ph. Vincke (1980) "Représentation et ajustement des ordres d'intervalle", discussion paper, Faculté Polytechnique de Mons.

B. Roy and P. Bertier (1979) "La méthode ELECTRE II, une application au media-planning", M. Ross (Ed.) OR 72, North Holland Publishing Company.

B. Roy and Ph. Vincke (1980) "Systèmes relationnels de préférences en présence de critères multiples avec seuils", *Cahier du Centre d'Études de Recherche Opérationnelle*, 22, 1.

M. Schader (1980), "Hierarchical analysis: classification with ordinal object dissimilarities", *Metrika*, 27, 127-132.

M. Schader (1981) *Scharfe und unscharfe Klassifikation qualitativer Daten*, Verlags-gruppe Anthenäum, Hain, Scriptor, Hanstein.

B.M. Schein (1972) "A representation theorem for lattices", *Algebra Universalis*, 2, 2, 177-178.

D. Scott and P. Suppes (1958) "Foundational aspects of theories of measurement", *J. Symbolic Logic*, 23, 113-128.

A.K. Sen (1970) *Collective choice and social welfare*, Oliver and Boyd, London.

H. Sharp Jr (1971-1972) "Enumeration of transitive, step-type relations", *Acta Math. Acad. Sci. Hum.*, 22, 365-371.

S. Shye (Edit. 1978) *Theory construction and data analysis in the behavioral sciences*, Jossey-Bass, San Francisco.

W.T. Trotter Jr and J.I. Moore Jr (1976) "Characterization problems for graphs, partially ordered sets, lattices and families of sets", *Discrete Math.*, 16, 361-381.

A. Tversky (1969) "The intransitivity of preferences", *Psychol. Review*, 76, 31-48.

P. Vincke (1978) "Quasi-ordres généralisés et représentation numérique", *Math. Sci. Hum.*, 62, 35-60.

N. Wiener (1912-1914) "Contribution to the theory of relative position", *Proc. Cambridge Philos. Soc.*, 17, 441-449.

N. Wiener (1914-1916) "A new theory of measurement: a study in the logic of mathematics", *Proc. London Math. Soc.* 19, 181-205.

R.L. Wine and J.E. Freund (1957) "On the enumeration of decision patterns involving $n$ means", *Ann. Math. Statist.*, 28, 256-259.

H.P. Young (1974) "An axiomatization of Borda's rule", *J. Econ. Theory*, 9, 43-52.

H.P. Young,(1978) "On permutations and permutation polytopes", in *Polyhedral combinatorics*, editors M.L. Balinski and A.J. Hoffman, North Holland, Amsterdam.

H.P. Young and A. Levenglick (1978) "A consistent extension of Condorcet's election principle", *SIAM J. Appl. Math.*, 35, 2, 285-300.

ADDENDUM

The paper by Bogart, in this volume, contains some more detailed examples of the use of ordered sets in social science which complement the development here. Also, there is reference to work of Call and Norman on generalized semiorders, and to Margush and McMorris concerning metrics between trees or discrete structures. Concerning interval orders, we wish to point out two recent papers by Fishburn (1981a, 1981b).

ACKNOWLEDGEMENT

The authors wish to thank Dwight Duffus for his important help in the final writing of this paper.

# SOME SOCIAL SCIENCE APPLICATIONS OF ORDERED SETS

K.P. Bogart
Department of Mathematics
Dartmouth College
Hanover, New Hampshire 03755

ABSTRACT

Ordered sets arise in artifact dating in archaeology, in preference measurement in psychology, in hierarchies in sociology, in decision problems in economics and political science. In this lecture, we discuss the nature of the ordered sets that arise in each of these disciplines and the nature of the problems in ordered sets that arise. Finally we shall consider the order-theoretic aspects of two problems that arise throughout these fields: the problem of determining a consensus from a group of orderings and the problem of making statistically significant statements about orderings.



*I. Rival (ed.), Ordered Sets, 759–787.*

## 1. INTRODUCTION

The paper of Barthélemy, Flament and Monjardet [5] in this volume together with its references and those they recommend in Mirkin [47], Roberts [52], Golumbic [24], and Barthélemy and Monjardet [6], and additionally Rothenberg [56], provides a scholarly review and a rather complete literature survey of the applications of ordered sets in the social sciences. Their paper also leads the reader to much of the mathematics of ordered sets that has arisen from these studies. My goal is complementary to theirs; to present, largely by example, a tutorial for the mathematician familiar with ordered sets but not familiar with their applications in the social sciences. The treatment will necessarily be selective; I hope it will whet the appetite of the uninitiated to explore the subject laid out so thoroughly in the works above.

Section 2 of the paper is devoted to a brief survey of some kinds of orders that are studied in the social sciences. Section 3 is devoted to two apparently dissimilar problems, that of determining a consensus, that is, one ordering that somehow best represents a family of orderings, and that of finding the ordering in one family which is the closest approximation to the relation in some other family. We shall discover that *one* solution to both these problems involves essentially the same application of the idea of "social distance between orderings". This leads to some interesting algorithmic problems in ordered sets and relations: some virtually trivial, some *NP*-complete, and some completely open.

For the purposes of this paper a partial order is a transitive, irreflexive (and thus necessarily asymmetric) relation. We use $(a,b) \in P$ and $a < b$ mod $P$, or simply $a < b$ when $P$ is clear from context, interchangeably; $a \leq b$ mod $P$ has the obvious interpretation.

## 2. SOME EXAMPLES OF HOW ORDERS ARISE

In economics it is natural to assume that if given a choice between two alternatives, a person will choose one. Clearly this assumption is natural if the two alternatives are differing amounts of the same commodity. In welfare economics, the alternatives are two specific allocations of the same amount of society's wealth; it is generally presumed that a person will choose one or another. Thus the partial orderings considered in this situation are all linear orders because each pair of alternatives is compared. If the number of alternatives is countable, there is a real-valued one-to-one function $f$ (called a *utility function*) such that a person prefers $a$ to $b$ if and only if $f(a) > f(b)$. Even with a continuum of alternatives, economists traditionally assume such a function.

Of course in the real world commodities come in combinations - a vector $(c_1,c_2,\ldots,c_n)$ of amounts of $n$ different commodities is called a *commodity bundle*. Because a person might be willing to exchange a certain amount of commodity $i$ for some commodity $j$, the person might regard two commodity bundles with equal favor. In fact the *axiom of substitutability* asserts there is some amount of commodity $i$ that has equal favor with a given amount of commodity $j$. Of course, the natural componentwise partial ordering is not a linear ordering. However if we are willing to exchange bundle $A$ for bundle $B$, and bundle $B$ for bundle $C$, then we will be willing to exchange $A$ for $C$. This "transitivity of indifference" can be expressed in terms of a relation $I$ called *indifference* given by $AIB$ if neither $A < B$ nor $B < A$ in the componentwise ordering. It is an easy exercise to show that (so long as there are no more commodity bundles than real numbers) there is a utility function $f$ from the commodity bundles to the real numbers such that bundle $A$ is preferred to bundle $B$ if and only if $f(A) > f(B)$. A partial ordering whose indifference relation is transitive is called a *weak ordering*. Typically, economists assume the individual amounts $c_i$ can be any real number and require that the partial derivatives of the utility function $f$ be positive (or at least nonnegative) and frequently that they be continuous. Further it is sometimes assumed that while the amounts of individual commodities may change, the total value of all the commodities in an economy is constant. These mathematical conditions all have natural economic interpretations. The hypersurfaces of constant utility are the equivalence classes of "indifferent" commodity bundles. These hypersurfaces are the equivalence classes of weak orderings. Weak orderings and linear orderings are typically the orderings considered by economists; the interesting problems arise from the continuum of commodity values and from resolving the preferences of various economic agents. Even with weak and linear orderings, these resolutions lead to interesting order-theoretic questions, although it is safe to say that the main tools of mathematical economics lie in analysis including recently nonstandard analysis and topology.

The dividing line between welfare economics and political decision making is hard to distinguish; if a political region has a certain amount of wealth to spend on certain priorities, then it is likely to order its priorities by a linear or weak ordering as described above and spend on the highest priority projects until the wealth runs out.

Rather interesting questions of order arise from the study of committee structure - or organizational structure in general - where virtually any partial ordering might be viewed as an organization chart for a committee. Certain natural measures of power in such an organization have been developed (beginning with Harary [27], see also [34]) and they depend on order-theoretic parameters of the structure.

Frequently a committee has a linear structure denoting who will be in charge if all "higher members are absent." What is interesting here is not the structure itself, but rather how the committee decides upon it. This will be part of the topic of Section 3. A similar problem arises when a group of voters cast preferential ballots (linearly ordered lists) for members of a committee and the voting officials must produce a ranking of candidates and choose the top candidates for the committee on the basis of these individual orderings. As we see in Section 3, this problem may be approached in a similar way.

Archaeologists are interested in the evolution of characteristics of artifacts over time. If one type of artifact is always buried deeper in a site than a second type, then the scientist knows type one predates type two. However, the archaeologist cannot use the interval of burial depth to linearly or weakly order the set of artifacts in order of their evolution if a third type of artifact confuses the issue by being buried with both of them [28], [35]. An anthropologist studying the evolution of customs in a society is likely to encounter the same difficulties. Social psychologists studying stages of development in children, for example, will once again have this problem [19]. (Roberts [52] gives a number of references to such situations.) The natural question by now is what kinds of ordered sets, more general presumably than weak or linear orders, arise in these studies. In each case, the artifact, custom, or trait (henceforth called the observation) is presumably in existence over some interval of time. One observation precedes another in our partial ordering if the time interval over which it occurs entirely precedes the time interval over which the other observation exists. Thus we define an *interval order* to be a partially ordered set $(X,P)$ such that there is a mapping $f$ (not necessarily one-to-one) from $X$ to the *intervals* of the real numbers (or for that matter any other linearly ordered set you choose) such that $x < y$ mod $P$ if and only if each element of $f(x)$ is less than each element of $f(y)$. The function $f$ is called an *interval representation*. Notice that we don't bother to specify whether the intervals need be open or closed or of nontrivial length; for certain discrete representation theorems this generality is useful. (We could, however, make any of these restrictions without changing the concept we are defining.)

How can a scientist tell whether the ordered set of data collected in a given experiment is indeed an interval order? There is an elegant mathematical characterization of interval orders due to Fishburn [21] that allows us to check whether we have an interval order by checking order relationships among four-element sets of observations. So long as we find no four-element set $\{a,b,c,d\}$ such that $(a,b)$ and $(c,d)$ are the only ordered pairs of $P$ in the ordering, we have an interval order. There is now a short proof of this result developed by T.L. Greenough and the author [10]. In any interval representation of an interval ordering, there will be some number of real numbers that appear as endpoints of intervals. If the set $X$ being ordered has $n$ points, then clearly no more than $2n$ endpoints are needed; let us define the *length* of an interval ordering to be the minimum number of endpoints needed. It turns out that both the length and an interval representation of that length are obtainable from order-theoretic information about an interval ordering $P$. An outline of the development follows.

Following Rabinovitch [50] we define the *lower holdings* $H_*(x)$ of an element $x$ in a set $X$ ordered by $P$ by

$$H_*(x) = \{y | y < x\}$$

and we define the *upper holdings* $H^*(x)$ by

$$H^*(x) = \{y | y > x\} .$$

We use the suggestive symbolism $\underline{2} + \underline{2}$ to denote a partially ordered set isomorphic to the poset $\{a,b,c,d\}$ with $\{(a,b),(c,d)\}$ the partial ordering relation.

LEMMA 1. *If $(X,P)$ has no restriction isomorphic to $\underline{2} + \underline{2}$ then the distinct sets $H^*(x)$, $x \in X$ are linearly ordered by set inclusion as are the distinct sets $H_*(x)$.*

LEMMA 2. *If $(X,P)$ has no restriction isomorphic to $\underline{2} + \underline{2}$, then the number $h$ of distinct sets $H^*(x)$ equals the number of distinct sets $H_*(x)$.*

THEOREM 1. *If a finite poset $(X,P)$ has no restriction isomorphic to $\underline{2} + \underline{2}$, then it is an interval order that may be represented using $h$ distinct endpoints.*

Proof. We let $h^*(x)$ denote the number of distinct sets of upper holdings contained in the upper holdings of $x$. Formally, using vertical bars for the size of a set,

$$h^*(x) = |\{H^*(y) \mid H^*(y) \subseteq H^*(x)\}| .$$

(Intuitively $h^*$ measures how many endpoints must be above the endpoint for $x$ if $P$ is indeed an interval order.) We define $h_*(x)$ analogously. Then the function

$$f(x) = [h_*(x),\ h - h^*(x) + 1]$$

is an interval representation of $(X,P)$. (Note that $f$ may assign single point intervals to some points; in fact it always will for a linear or weak order.) To show $f$ is an interval representation we must show that $(x,z) \in P$ if and only if

$$h - h^*(x) + 1 < h_*(z) , \quad \text{or} \quad h_*(z) + h^*(x) > h + 1 .$$

Clearly the inequality holds if $(x,z) \in P$. Figure 1 shows $h$ points and their linear orderings by both the upper holdings function $H^*$ and the lower holdings function $H_*$ . Note that for $h_*(z) + h_*(x)$ to be more than $h + 1$, $H^*(z)$ must be above $H_*(x)$ in this figure.

$H_*(x_h)$ • $H^*(y_h) = \emptyset$
•
•
$H_*(x_i)$ • $H^*(y_i) = H^*(z)$
•
•
$H_*(x) = H_*(x_j)$ • $H^*(y_j)$
$H_*(x_2)$ • $H^*(y_2)$
$\emptyset = H_*(x_1)$ • $H^*(y_1)$

Figure 1.

By Lemma 1, $H^*(y_j) \supseteq H^*(z)$ and $H_*(x) \subseteq H_*(x_i)$. However $x_i$ and $y_i$ must be incomparable for otherwise they would be represented by different points on the figure. Since $y_j \neq y_i$ , $y_j$ is under some element $a$ not comparable to $z$ and similarly $x_i$ is over some element $b$ not comparable to $x$. Thus unless $(x,z) \in P$, our poset would have the restriction to $\{a,b,x,z\}$ isomorphic to $\underline{2} + \underline{2}$ . □

Interval orders of course arise naturally, though to my experience only implicitly, in the study of statistical data. If we attempt to measure a numerical observation for several subjects by giving each a series of tests with a sample numerical observation as a result, then we can assign to each subject a confidence interval where we are 95 percent (or some other percent) sure its numerical observation lies. It would be an interesting study to learn what distributions of interval orders arise from the usual, and perhaps unusual, distributions of data observed in the social sciences.

Even when we are making physical measurements we cannot expect our measuring devices to always give us exactly the same numerical answers under the same circumstances. We are accustomed to having some tolerance limits that we have to accept. These kinds of ideas arise naturally in other areas of measurement and have led to the development of an entire theory of measurement due primarily to Duncan Luce [39] and other mathematical psychologists. Here we can take either the point of view that we are using a human subject or a measuring device to measure whether one object is preferable to another, and there is some constant tolerance limit the device cannot discern. Suppose for the sake of argument there is an implicit numerical scale in which we assign a larger value to a preferable object. Then given four observations on four objects $a$, $b$, $c$, and $d$, could we simultaneously rate $a$ higher than $b$ and rate $c$ higher than $d$? Suppose $a$ gets a number higher than $b$ and $c$ gets a lower or equal number than $a$. Then the number $d$ gets is at least our tolerance level below $a$, because it is at least our tolerance level below $c$. Thus our poset of "measurable" preferences will not have a restriction isomorphic to $\underline{2} + \underline{2}$ . Thus, individuals' preference orderings (*in this model*) and orderings measurable by imperfect devices must be interval orderings.

The assumption that we are assigning numerical values with a constant tolerance limit to our observations leads to another restriction on the measured poset. We use the notation $\underline{1} + \underline{3}$ to denote a poset isomorphic to the disjoint union of a point and a three-element chain. Given four observations of objects $a$, $b$, $c$, and $d$, suppose the value of $b$ is less than that of $c$, and that $c$ is less than that of $d$. Whatever number we assign to $a$, it is either equal to that assigned to $c$, in which case observation $a$ will have the same relation to $b$ and $d$ that $c$ has or else $a$'s number is less than $c$'s, putting $a$ below $d$ by at least the tolerance limit in our ordering or, dually, $a$ will be above $b$ in our ordering. Thus if we are measuring within fixed tolerance limits, our measured poset of preferences will have no four-element subset isomorphic to $\underline{1} + \underline{3}$ . Thus in an interval representation we will not have $I_d$ above $I_c$ above $I_b$ with the interval $I_a$ overlapping in all three. Thus one interval may overlap at most two comparable intervals. Then just by moving around the endpoints of an interval representation, we may prove a result due to Scott and Suppes [57].

THEOREM 2. *A finite poset $(X,P)$ contains no four-element subset on which the restriction of $P$ is isomorphic to $\underline{2} + \underline{2}$ or $\underline{1} + \underline{3}$, if and only if it is an interval order representable by intervals of equal length.*

A proof of the Scott-Suppes Theorem by Rabinovitch [50] motivated our proof of Theorem 1.

Another way to state this theorem is that there is a number $d$ (called the "just noticeable difference") and a function $f$ from $x$ to the real numbers such that $(x,y) \in P$ if and only if $f(x) + d < f(y)$. Partial orders that may be represented in this way were first called *semiorders* by Luce [39]. The treatise by Luce et. al. [40] will provide the interested reader with a complete treatment of semiorders and the theory of measurement. Roberts' recent book provides applications and a broad view of the subject (see also [54] and [29] for applications of semiorders).

In the social science literature [52], weak orders, semiorders, and interval orders are regarded as having increasingly complex indifference relations and the intransitivity of the indifference relation (mentioned earlier in connection with weak orders) is at the core of the treatment.

The approach above, of forbidding particular subsets, has been chosen because it emphasizes the order-theoretic aspects of the subject. In fact there is a natural explanation for any finite partial ordering arising as a preference relation, so the possible indifference relations are the complements of partially orderable graphs. Thus characterizing the indifference graphs which might occur with partial orderings amounts to characterizing graphs whose complements are transitively orientable. The best known such characterizations are those of Ghouila-Houri [22] and Gilmore and Hoffman [23].

Why does every partial ordering arise as a preference relation? We might evaluate a set of objects according to several criteria, getting a linear ordering for each criterion. Then we prefer object $x$ to object $y$ if and only if $x$ rates above $y$ on each criterion. Thus our partial ordering is the intersection of the linear orders according to the various criteria. Of course from Szpilrajn's theorem [58] we know that a partial ordering is the intersection of all its linear extensions and thus can be obtained as a preference relation.

How many linear orders are required to obtain a given partial order? We say the *dimension* of a partial ordering $P$ is the smallest number of linear extensions whose intersection is $P$.

Much of the current interest in the dimension of posets was sparked by a question in a social science context: "what is the maximum dimension of a semiorder?" (Baker, Fishburn, and Roberts [3]). Rabinovitch [51] showed that this dimension was no more than 3 and characterized the three-dimensional semiorders. On the other hand, there is no upper bound on the dimension of interval orders [11].

We restrict further comments on dimension to two interesting questions. (See Kelly and Trotter [30] for a survey of dimension theory.) First, since saying an interval order has no restriction isomorphic to $\underline{1} + \underline{3}$ says it is a semiorder and has dimension no more than 3, it is natural to ask about the condition that we have no restriction isomorphic to $\underline{1} + \underline{n}$; namely what does this condition, which allows a representation using intervals of length at least one but no more than $n - 2$, say about the dimension of the poset?

For our second question we introduce still another class of posets. In our discussion of semiorders we have assumed that the "just noticeable difference" is constant. If we allow the "just noticeable difference" to vary as pairs of observations being compared vary, we can arrive at vastly more complicated orders. Consider five objects $\{a,b,c,x,y\}$. Suppose that in making comparisons between $a$, $b$ and $c$ our just noticeable difference was one and in making comparisons involving $x$ and $y$ our just noticeable difference was 2. Then the disjoint union of the two chains $a < b$ and $x < y$ could result if $a$ had an actual numerical rank of 2, $b$ of 3, $x$ of 1.5 and $y$ of 3.5. In other words we would not even have an interval order.

Call and Norman [14] have suggested the concept of an $n$-semiorder as a model for such situations. Although they do not in general require transitivity, a partial ordering is in their sense an $n$-semiorder if it contains no restriction isomorphic to the disjoint union of 2 chains of size $i_1$ and $i_2$ with $i_1 + i_2 = n + 1$. Thus a 1-semiorder does not contain 2 incomparable elements, so it is a linear order; a 2-semiorder is a weak order and a 3-semiorder is a semiorder in the usual sense. Our example in the previous paragraph is a 4-semiorder, as it contains the disjoint union of $a < b$ and $x < y$. It appears that the parameter $n$ controls just how much variation in just noticeable difference will be allowed. For example a 5-semiorder will allow all representations using just noticeable differences that vary between 1 and 2 but not all using just noticeable differences varying between 1 and 3. Call and Norman ask whether there is a Scott-Suppes type theorem for $n$-semiorders.

Of course not all families of partially ordered sets arising in the social sciences fit so nicely in the hierarchy described above. As one example, taxonomists working with the evolution of biological or social structures to form a single type as an end product find that the relation "is an ancestor of" leads to a "tree poset." $P$ is a *tree poset* if $aPb$ and $aPc$ implies either $bPc$ or $cPb$. (That is, if $a$ is an ancestor of both $b$ and $c$, then one is an ancestor of the other.) Recent contributions to this view of taxonomy have been made by McMorris [45] whose paper contains a useful bibliography. Tree posets are extremely popular as a data structure in computer science.

## 3. DISTANCE AND THE AMALGAMATION OF PREFERENCE

Some years ago a sociologist was faced with the following situation. A group of people were working together in one extended experiment, and at the end of the period each member submitted a ranking of all members of the group according to their effectiveness. The sociologist wanted a consensus ranking that gave the group's opinion of its members [31]. A central problem in welfare economics has been to amalgamate the orderings of society's members on a set of economic alternatives into society's ordering of the alternatives. A committee trying to linearly order its membership to decide who will be in charge when only part of the membership will be present is faced with the same problem. A military promotions board may be faced with ranking a number of candidates for promotion to a higher rank to be promoted as the higher-level positions become available [17], using each board member's own ranking of the candidates.

Of course we could continue this list with both hypothetical and real examples, all of which have essentially the same mathematical structure. We are given a set of linear (or perhaps weak) orderings of a set of alternatives and we ask for a consensus that reflects in the best way possible all the comparisons made by all the orderings. If we have only two alternatives, then in effect we are choosing only one, and typically we choose the one favored by the majority. Of course these are examples where we agree to make a choice if two-thirds of the orderings make that choice; for example to consider a major item not on an agenda, most deliberative bodies require a two-thirds majority in favor of considering the item.

The natural extension to more than two items is usually called *majority rule* or the "Condorcet Principle." We construct a relation $R(L_1, L_2, \ldots, L_k)$ by putting the pair $(a,b)$ in $R$ if $(a,b)$ is in a majority of the relations $L_i$. If we have an odd number of linear orderings $L_i$, then $R$ is the relation of a tournament, but need not be a linear order. The "paradox of voting" or "Condorcet effect" [26], often misnamed the "Arrow Paradox", is illustrated by the following three linear orders

$$\begin{matrix} a & b & c \\ b & c & a \\ c & a & b \end{matrix}$$

in which $R = \{(a,b),(b,c),(c,a)\}$, an intransitive tournament. Short of declaring a three-way tie, intuition gives no suggestion about resolving such situations.

At about the same time (in the eighteenth century) that Condorcet pointed this problem out, Borda independently proposed a method called the "Borda Count" by Black [7]. Borda numbered the objects being ordered in order of preference (the most preferred getting the highest number) and added the numbers across the ordering to get a score for each object. The objects are then ordered by this score and the resulting ordering will be a weak or linear ordering. Condorcet [15] pointed out an unfortunate feature of the Borda count. Consider the relations

$$\begin{matrix} a & d & b \\ d & c & a \\ c & b & c \\ b & a & d \end{matrix} .$$

The respective Borda counts are 8, 7, 7, and 8 for $a$, $b$, $c$, and $d$, so $a$ is preferred to $b$. However a majority of the orders rank $b$ above $a$, so Borda's method is certainly not going to satisfy a majority of voters. Incidentally, an important question to ask of any voting scheme is whether it resists manipulation by knowledgeable voters. Here $a$ wins the runoff for top candidate. Note that if $b$ had actually been voter one's candidate for second place, but voter one lied about his preferences, then voter one has imposed $a$ upon society.

Despite intervening work, including a large body of work in the middle of this century cited by Rothenberg [56], the real source of excitement in the field of consensus was the work of Kenneth Arrow. Arrow described axiomatically certain conditions that a "social welfare function" should satisfy. For ease of presentation, we alter somewhat his description; however, the methods and conclusions are faithful to his original work. First, we have a set $A$ of at least three alternatives. Second, we have a function $f$ called a *social welfare function* that maps $k$-tuples of the set of linear orderings of any subset $B$ of $A$ to single linear orderings of $B$. Arrow described certain natural conditions this function $f$ should satisfy in order to be reasonable.

AXIOM I. (Positive Association of Individual and Social Values) *If for a given k-tuple the function ranks a over b, then it still ranks a over b if the following kind of change is made in the k-tuple:*

(i) *each relationship between b and each $c \neq a$ is not changed in any of the orderings; and*

(ii) *any change in an order in the list involving a and c raises a's position relative to that of c.*

AXIOM II. (Independence of Irrelevant Alternatives) *If an alternative c is removed from each of the lists then the function should not change its ranking of two other alternatives a and b.*[1]

AXIOM III. (Citizen's Sovereignty) *For each pair of alternatives a and b, there is some k-tuple of each set B containing a and b for which f ranks a above b.*

AXIOM IV. (Nondictatorship) *There is no coordinate i such that for all a and b in A, f ranks a over b for all k-tuples in which ordering i ranks a over b.*

THEOREM 3. (Arrow [2]) *There is no social welfare function satisfying Axioms I to IV.*

We shall indicate a proof of Arrow's theorem shortly. However, we present an interesting sidelight illustrating the axioms. John Olguin, former Director of the Native American Program at Dartmouth, believed tribal council rule as he learned it in practice satisfied all these axioms. Briefly he described a tribal council as a subset of a tribe, self-perpetuated by any of a variety of means. The tribal council would on occasion have to produce a ranking and would discuss the situation for each $a$ and $b$ until all could agree that the supreme being wanted $a$ ranked above $b$. If they could not come to an agreement, they simply passed over the issue [48]. In fact this method satisfies all four axioms unless the tribal council contains one member. However this social welfare function provides a partial ordering, not necessarily a weak or linear ordering of the alternatives.

The proof of Arrow's theorem essentially consists of showing that tribal council rule is the only possible social welfare function satisfying the axioms and then demonstrating that the requirement that the range of the function consists of linear orderings requires that the ruling body consists of one person. The proof outlined here is essentially that found in Luce and Raiffa [41] and Roberts [52], and so details are freely omitted. What we shall actually show is that if there is a social welfare function from $k$-tuples of linear orderings to linear orderings satisfying Axioms I-III, then Axiom IV is violated. For intuition we will think of the $k$ indices that index our $k$-tuples of linear orderings as individuals. We say a set $D$ of individuals is *decisive* for alternative $x$ over alternative $y$ if whenever all members of $D$ prefer $x$ to $y$, the image of our social welfare function prefers $x$ to $y$.

LEMMA 1. *If there is one $k$-tuple of orders in which only the members of $D$ prefer $x$ to $y$ and the social welfare function prefers $x$ to $y$, then $D$ is decisive for $x$ over $y$.*

Proof. We assume there is some other $k$-tuple different from the one in the hypothesis of the theorem in which all members of $D$ prefer $x$ to $y$. We then apply a series of transformations of the type allowed by Axioms I and II to translate this $k$-tuple to the one of the hypothesis. Thus for both $k$-tuples, the social welfare function makes the same decision for $x$ and $y$. □

LEMMA 2. *The set of all individuals is decisive for each $x$ over each $y$.* □

LEMMA 3. *Let $D$ be a minimal decisive set (minimal among all pairs $(x,y)$ as well as minimal in size). Then $D$ is decisive for all pairs.*

Proof. The proof consists of examining certain sets of $k$-tuples to show there is one individual whose wishes are always followed. □

Of course Lemma 4 finishes the proof of Arrow's theorem because it contradicts Axiom IV. If we had allowed partial orders as the range of the social welfare function, our conclusion would be that $x$ is preferred to $y$ if and only if it is preferred to $y$ by all members of some fixed minimal decisive set. Thus what has been described as rule by tribal council is the only allowable social welfare function when the range consists of partial orders. The paper of Call and Norman [14] discusses the extent to which one can avoid Arrow's result by using $n$-semiorders as a range for the function.

The Arrow axioms represent a very significant weakening of the concept of majority rule. They simply abstract certain properties that (weighted) majority rule would have when it worked to produce a linear ordering. The social science literature contains a number of attacks on Arrow's axioms, especially the "independence of irrelevant alternatives." (See Rothenberg [56] for appropriate references.) On a more positive note, Goodman and Markowitz [25] propose a model for human preference, describe a different set of axioms for what a social welfare function should be, and prove that there is an essentially unique social welfare function, a "sophisticated Borda count", satisfying these axioms.

Not surprisingly, their work has also been criticized. As the Goodman-Markowitz theorem has received far less attention in the literature (in part for reasons outlined by Rothenberg [56]), it is described here in some detail.

We will again have a social welfare function that acts on $k$-tuples representing the weak or linear orderings of $k$ individuals for $m$ alternatives to produce a single weak or linear ordering of the alternatives. However, we take a different view of how we are representing the weak or linear orderings. We assume that each individual has a finite number of levels of discretion - "a change from one level to the next represents the minimum difference which is discernable to the individual" [25]. Different individuals can have different numbers of discretion levels, different individuals need not have the same difference between discretion levels, and it is not even clear that we must assume that a given individual has a constant "minimum difference which is discernable." The representation of an individual's ordering can be thought of as a column vector in which an integer in position $i$ represents the preference level the individual assigns to alternative $i$. Thus if we have $m$ alternatives and $k$ individuals, we represent our $k$-tuple of orderings by an $m \times k$ matrix $L$ of integers of preference levels.

Goodman and Markowitz [25] begin with resolutions that are not what we would consider axioms describing a social welfare function. Rather, they are warnings that if the social welfare function eventually developed has certain properties then it should not be rejected. These resolutions in part are weakenings of Arrow's axioms and in part denials of them. For a far reaching discussion of them, see Rothenberg [56].

The domain of the social welfare function will not be the matrices $L$, but rather "ranking matrices" $A$ that may be derived from them and are regarded as approximations to them. To construct column $j$ of $A$, we let $A_{ij}$ be 1 for all $i$ for which $L_{ij}$ has the highest value occurring in column $j$ of $L$. We let $A_{ij}$ be 2 for all $i$ for which $L_{ij}$ has its next highest value and so on. Since there are $m$ alternatives $A$ will have entries between 1 and $m$. $A$ contains all information about $L$ we can derive with a given set of alternatives. This is why we regard $A$ as an approximation to $L$, and as such our axioms permit us to modify $A$ in ways that do not change the nature of the information $A$ gives us about $L$. The range of the social welfare function will be weak orderings of the alternatives. Note that adding an alternative that was previously unavailable will only add another row to $L$, but could have considerable impact on the entries of $A$.

Two of the "resolutions" have an impact on how we work with the matrix $A$, so we rephrase them as Axioms 0 and 4 rather than as resolutions.

AXIOM 0. (Any state of information is possible) *The domain of the social welfare function consists of all matrices with non-negative integer entries.*[2]

AXIOM 1. (Pareto Optimality) *If no individual prefers alternative $i$ to alternative $j$ and some individual prefers alternative $j$ to alternative $i$, then $j$ is preferred to $i$ by the social welfare function.*

AXIOM 2. (Symmetry) *Permuting columns of $A$ does not change the value of the social welfare function.*

AXIOM 3. (Translation invariance) *Adding the same row vector $C$ to row $i$ and row $j$ of $A$ will not change the ordering of $i$ and $j$ by the social welfare function.*[2]

AXIOM 4. (Independence of unavailable candidates once given a state of information) *Deleting a row other than $i$ or $j$ from $A$ will not change the ordering of $i$ and $j$ by the social welfare function.*

THEOREM 4. *There is a unique social welfare function satisfying Axioms 0-4. It is given by: $i$ is preferred to $j$ if and only if the sum of the row of $A$ corresponding to $j$ is larger than the sum of the row corresponding to $i$, and otherwise the social ordering is indifferent between $i$ and $j$.*

Proof. It is of course simply a matter of checking to see that this social welfare function satisfies the axioms.

Following Goodman and Markowitz [25], we consider only two alternatives, $i$ and $j$, and assume first that

$$\sum_k A_{ik} = \sum_k A_{jk} . \qquad *$$

We wish to prove that our social welfare function will give exactly the same rank to $i$ and $j$. We construct a special vector $C$ so that when we add $C$ to both row $i$ and row $j$, one becomes a cyclic shift of the other. Thus we wish to solve the equations

$$A_{ik} + C_k = A_{j,k+1} + C_{k+1} \quad \text{for } k = 1,2,\ldots,m-1$$

and

$$A_{im} + C_m = A_{j1} + C_1 .$$

There is a solution: we assign any value we wish to $C_1$ and the equations give us the other $C$'s. The results are consistent because of equation *. Clearly we can make the entries of the revised $A$ positive with Axiom 3. Thus by Axiom 2, the social welfare function will assign equal rank to $i$ and $j$. If $A$ has more rows than just $i$ and $j$, then we apply Axiom 4 until we reach a matrix with just these two rows, so $i$ and $j$ get equal rank whenever they have equal row sums.

Now we suppose $\sum_k A_{ik} > \sum_k A_{jk}$; we will construct a vector $C$ to add to row $j$ so the added row vectors have the same row sum as row $i$. Thus "row $j$ + $C$" will be tied with row $i$; further we fix things so "row $j$ + $C$" obviously dominates row $j$. Namely, let $d = \sum_k A_{ik} - \sum_k A_{jk}$ . Note that if we had restricted the sum to those $k$ with $A_{ik} - A_{jk} > 0$, $d$ would have been larger. Thus we can pick numbers $d_1, d_2, \ldots, d_m$ so that $d_i = 0$ if $A_{ik} - A_{jk} \leq 0$, $0 \leq d_i \leq A_{ik} - A_{jk}$ if $A_{ik} - A_{jk} > 0$, and finally $\sum_i d_i = d > 0$. Thus by Axiom 1, an alternative, available or not, whose row is "row $j$ + $C$" is preferred to alternative $j$, and by the result about ties we just proved, alternative $i$ is tied with an alternative whose row is "row $j$ + $C$", so since our ordering is a weak order, alternative $i$ is preferred to alternative $j$. □

In political science there is an attempt at resolution of the paradox of voting that has interesting political content. Suppose we list the alternatives in some order and use them to label integer points on the $x$-axis, starting at 1. Given a ranking of the alternatives, we number them according to our preference among them, giving the most preferred alternative the highest number, equally preferred alternatives equal numbers and so on, we can think of our ranking defining $y$ as a function of $x$. Suppose now rather than one ranking we have several rankings, and we go through this procedure for all the rankings. We graph each of these functions. We now reorder the alternatives on the $x$-axis and repeat the process. We continue this through all orderings of the alternatives along the $x$-axis. If in one case all the graphs are unimodal - increase to a maximum value and then decrease - we say we have found an ordering of the alternatives that leads to *single peaked preference curves*.

For example, if we view the ordering of the alternatives as a political spectrum from left to right, and an individual has a certain location in the political spectrum, then alternatives farther from this location will be less desirable, so the individual's preference curve will be single peaked. The interesting theorem (paraphrased from Black [7]) is

THEOREM 5. *If there is an ordering of the alternatives such that all individuals' preference rankings are simultaneously single peaked, then the majority rule relation is a weak ordering.* □

Our next approach to a social welfare function has a varied history with rediscovery of the elementary facts occurring frequently (see [5]). The basic method is generally now associated with John Kemeny. He was asked by a sociologist exactly the question that began this section [31]. He saw the problem as one of wanting perhaps the "average" opinion of a group and realized that means and medians of finite distributions were definable in a natural way in metric spaces [32]. Thus he set about asking whether the weak orderings (he called them rankings) of a finite set had a natural metric structure that would reflect the intuitive idea that two opinions were close together or far apart. In his joint work with J.L. Snell [34] (cf. [33]) they described axiomatically a possible distance function in such a way that each axiom seems naturally to reflect the idea of measuring closeness of opinion. The axioms are: Distance is to be a metric in the usual sense. Distance is to be invariant under a permutation of the alternatives. The minimum positive distance is 1. Two other axioms require that we first introduce some definitions.

The ranking $W_2$ is *between* $W_1$ and $W_3$ if each decision made by $W_2$ is made by either $W_1$ or $W_3$ and any decision made by both $W_1$ and $W_3$ is made by $W_2$. (More briefly, as sets of ordered pairs,

$$W_1 \cap W_3 \subseteq W_2 \subseteq W_1 \cup W_3).$$

A major axiom states that if $W_2$ is between $W_1$ and $W_3$ then $d(W_1,W_2) + d(W_2,W_3) = d(W_1,W_3)$. Finally, there is an axiom whose use in their work employs the fact that the space of weak orderings was the one under consideration. They introduce the idea of a *segment* - in poset terminology, simply an open interval. The crucial axiom is that if two weak orderings agree except for a set $S$ of alternatives which is a segment of both, then the distance may be computed as if the alternatives in $S$ were the only objects being ranked.

This axiom presumes that we are really discussing a family of metrics, one on each subset of our set of alternatives. Kemeny and Snell prove

THEOREM 6. *There is a unique distance function satisfying these axioms.* □

We will not outline the Kemeny-Snell proof except to note that in their proof of uniqueness they use the fact that if $S$ is a segment, then every other alternative is either above all those in $S$ or below all those in $S$. Intervals in partial orders do not have that property.

Curiously, to show that there is a distance function satisfying the axioms, they represented the weak orderings with matrices that exhibited the antisymmetric property of orderings, namely with the weak ordering $W$ they associated the matrix $M$ with

$$M_{ij} = \begin{cases} 1 & \text{if } (i,j) \in W \\ -1 & \text{if } (j,i) \in W \\ 0 & \text{otherwise .} \end{cases}$$

They showed that if $M$ and $N$ are the matrices of weak orderings $W$ and $P$ respectively, then the distance between $W$ and $P$ is half the sum of the absolute values of the entries of the difference of $M$ and $N$. Had they used the ordinary incidence matrix, the factor of 1/2 would be unnecessary and it might have been easy to recognize that the metric they were describing was the size of the symmetric difference of $W$ and $P$ as sets of ordered pairs.

They gave a geometric interpretation of the distance for weak orderings of three objects illustrated in Figure 2. Note that the figure adds an aura of mystery to the distance by showing that some minimum distances are 2 and some are 1.

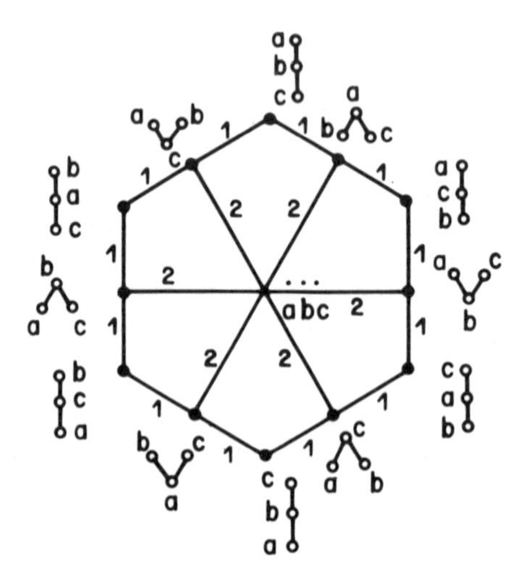


Figure 2

The authors defined the *median* of a set $S$ of weak orderings to be that weak ordering $W$ such that

$$\sum_{P \in S} d(W,P)$$

is a minimum and define the *mean* of the collection to be that ordering $V$ such that

$$\sum_{P \in S} d(V,P)^2$$

is a minimum. However they presented no efficient methods for computing either means or medians.

One can pose the consensus problem for dating archaeological artifacts from a variety of sites, for tolerance level measurements from a number of machines, and for virtually any other family of partially ordered sets. Thus, it is reasonable to ask whether the distance function can be described in the same way for arbitrary posets as for weak orders and whether this would assist in the computation of means and medians (see Bogart [8]). In fact, in many applications, preference orderings need not be transitive and are merely antisymmetric relations: the author asked the same questions for such relations in [9]. (See Bogart and Weeks [12] for a more general discussion of consensus problems for relations and signed relations including difficulties with the standard methods proposed in [52].) Such questions have been raised repeatedly; there is an extensive literature on the subject of consensus cited by Barthélemy and Monjardet [6]. Let us now see how to describe distance functions for arbitrary posets.

Since the Kemeny-Snell use of the "segment axiom" [34] was inappropriate for partial orders, it was replaced in [8] with one which says that if $P$ and $Q$ differ by the ordered pair $(a,b)$ and if $P'$ and $Q'$ differ by the same ordered pair, then the distance between $P$ and $Q$ is the same as the distance between $P'$ and $Q'$. We call this the *similarity* axiom. Because of the betweenness axiom the permutation invariance axiom may be replaced by: $d(\phi,\{(a,b)\}) = d(\phi,\{c,d\})$ for any four alternatives $a$, $b$, $c$, and $d$. In [8] the author shows that with these replacements there is a unique distance function on the partial orderings of a set satisfying the axioms and it agrees with the Kemeny-Snell distance function on weak and linear orderings. Implicit in that paper is the following theorem.

THEOREM 7. *Given any choice of the distances* $d(\phi,\{(a,b)\})$ *there is a unique metric on the partial orderings of a finite set* $S$ *satisfying the betweenness and similarity axioms.*

It has been pointed out to the author by the sociologist Levine [38], among others, that in certain social contexts it is easier for an individual to prefer $a$ to $b$ than $b$ to $a$ or easier to prefer $a$ to $b$ than to prefer $c$ to $d$ for some other $c$ and $d$. Thus Theorem 7 has relevance; its use would require value judgements or experimental work to determine the distances $d(\phi,\{(a,b)\})$.

Figure 3 illustrates how extending the Kemeny distance to partial orderings explains the strange appearances of two's as minimum distances in Figure 2. The arrows also indicate the natural partial ordering of the partial orderings of $\{a,b,c\}$.

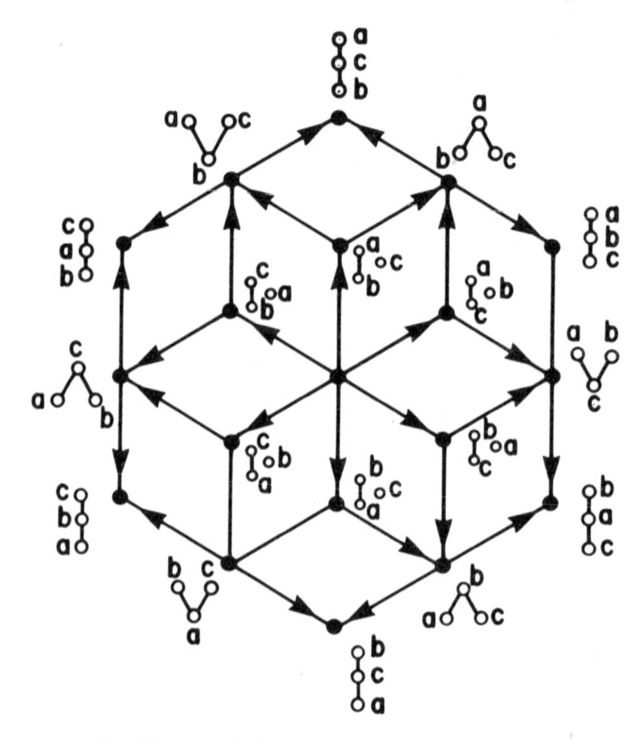

The graph of all partial orderings of a 3-element set.

Figure 3.

The reason why all distances between "adjacent orderings" is 1 is the core of the proof in [8] that the distance between partial orderings is unique; namely, given partial orderings $P$ and $Q$ with $P \subseteq Q$, it is possible to find a sequence of orderings $R_i$ such that $R_{i+1}$ has one more comparison than $R_i$ and $P = R_0 \subseteq R_1 \subseteq \ldots \subseteq R_n = Q$.

One application of distance functions is using them to compute consensus relations as means or medians. The author is one of the many independent discoverers (see [6]) of

THEOREM 8. *If the majority rule relation of an odd number of partial orderings is a partial ordering, then it is the unique median relative to the Kemeny distance* [8].

Of course the same kind of theorem holds for general asymmetric relations, for weak orderings, or any other family of interest. The natural extension of Theorem 8 to the context where the range of a social welfare function may be the set of asymmetric relations is that no matter what the domain of the median social welfare function, the majority rule relation is a median and under suitable conditions is unique [9].

There has been far less discussion of Kemeny's idea of the mean than the median as a social welfare function in the literature, perhaps because the median social welfare function is such a natural generalization of majority rule. It turns out, however, that by embedding our consensus problems in a Euclidean space, we can bring powerful geometric tools to bear. A change to a different distance function gives us such an embedding.

By restricting the idea of betweenness to a stricter notion – namely, $Q$ is *strictly between* $P$ and $R$ if and only if (1) for each $(a,b)$ in one of $P$ or $R$ but not the other, $(b,a)$ is in the other, and (2) $Q = P \cap R$ – we greatly reduce the power of the betweenness axiom. It now states that if $Q$ is strictly between $P$ and $R$, then

$$d(w_1,w_2) + d(w_2,w_3) = d(w_1,w_3) .$$

We then introduce a notion of orthogonality (the description is given in [8]); it is the natural extension saying that if $\{(a,b)\} \neq \{(c,d)\}$ then the "line" from $\phi$ to $\{(a,b)\}$ is orthogonal to the "line" from $\phi$ to $\{(c,d)\}$. Next we introduce an "orthogonality axiom" which says if a shortest sequence of orderings from $P$ to $Q$ is orthogonal to a shortest sequence of orderings from $P$ to $R$ then $d(Q,R)^2 = d(P,Q)^2 + d(P,R)^2$ (see [8] for a more precise statement). By regarding Figure 3 as a projection of a three-dimensional figure into 2-space, one can "see" this orthogonality. We then have

THEOREM 9. *There is a unique metric satisfying the strict betweenness axiom, the orthogonality axiom, the similarity axiom, the permutation invariance axiom and the minimum distance axiom.*

In fact $d(P,Q)$ is the Euclidean distance between the antisymmetric matrix representations of $P$ and $Q$ divided by $\sqrt{2}$ [8].

It is clear that the mean in the sense of Kemeny of a finite set of points in Euclidean space is simply their usual mean or average. Further we have

THEOREM 10. *If F is a finite subset of a Euclidean space and G is a subset of Euclidean space, then the points in G whose mean square distance to points in F is smallest are the points of G closest to the mean of F* [9].

This theorem gives us a method of computing means relative to our new Euclidean distance. Namely, we average the appropriate matrix representations and then search for the ordering matrix closest to the average. I am not aware, however, of an efficient search routine!

Of course in the space of asymmetric relations, no search is necessary; we find the average and then round off the entries to the nearest integer. This gives the following new social welfare function using the Euclidean distance on antisymmetric matrices.

*Euclidean Social Welfare Function.* Alternative $a$ is prefered to alternative $b$ if and only if the number of votes for $a$ over $b$ minus the number of votes for $b$ over $a$ is more than half the number of voters [9].

If this function is applied to linear orderings, then it gives in fact three-fourths majority rule! When this relation is a partial ordering, it is the mean on the space of partial orderings [9]. Thus in situations when majority rule (if it would work) would be an appropriate consensus, we would apply the Kemeny distance. In situations where significantly more than 50% majority rule might be desirable the appropriate distance function would be the Euclidean one.

Now let us return to the Kemeny metric and observe that it is the square of an Euclidean space metric; namely the Euclidean space metric on the usual incidence matrix of a relation with a one in position $(i,j)$ if $i$ and $j$ are related and a zero in that position otherwise. Thus the Kemeny median in a family $F$ of a set of relations may be computed according to Theorem 8 by averaging the incidence matrices of the relations and then finding the incidence matrix of a member of $F$ which most closely approximates this average in Euclidean distance. Again, if we allow $F$ to be the family of asymmetric relations we find this median by "rounding off". This average in Euclidean space of the incidence matrices is the weighted adjacency matrix for the weighted version of the Condorcet tournament described by Barthélemy and Monjardet [6].

This observation and Theorem 9 suggest a natural "greedy algorithm" to find the median partial ordering. Starting at one of the original partial ordering incidence matrices, check each adjacent partial ordering incidence matrix to see which is closest to the average. Repeat the process on this closest one. With little work one can see this is a polynomial time algorithm which produces an incidence matrix which is at a *local* minimum distance from the average. The same technique could be restricted to linear orderings, where linear orderings a distance two apart are considered adjacent. Jim Orlin, however, has recently shown the problem of computing the median in the family of linear orderings to be *NP*-complete [49], so the procedure is doomed to failure on the linear orderings. It is natural to suspect the same result will hold for any family of partial orderings, not just the linear ones.

The fact that we are using asymmetric transitive orderings can be captured in the following statements about the entries $x_{ij}$ of the incidence matrix: $x_{ii} = 0$, $0 \leq x_{ij} + x_{ji} \leq 1$, and $-1 \leq x_{ij} + x_{jk} - x_{ik} \leq 1$, first noted in [20] and [12]. This means we could apply integer quadratic programming to compute our median precisely. Many other methods of computing the median are described in [6]; the most efficient seem to be linear programming methods, introduced in [12] and [20] and shown to be useful for as many as 80 alternatives in [46]. (See [6] for further details and references.)

Note that we have reduced the problem of finding certain kinds of consensus orderings to finding the best approximation in a family $M(F)$ of matrices of relations to a certain rational matrix. This relates the problem to another series of approximation problems. If a psychologist has paired comparison data for an individual listening to tones and expects the subject to produce a semiorder, how do we find the best semiorder approximation? An archaeologist dating artifacts might face the same kind of problem and wish the best interval order approximation. Clearly in this context, our answer is to choose an appropriate distance function and then approximate the relation actually found with the closest member of the family of orderings of interest. The most popular distance function is likely to be the Kemeny distance function, suggesting that for ordering relations this problem too may be *NP*-complete.

The word "consensus" or the phrase "best approximation" carry with them the connotation of reliability. Just how reliable is a consensus we obtain by the median method? For just two linear orders, Kendall's tau is just a linear function of the Kemeny distance [8], [36]. Direct connections do not, however, appear with Kendall's coefficients of concordance [36]. Kemeny and Snell [34] suggest one approach to measuring reliability in general; however we now have another approach. The average of the representing matrices represents in a certain sense the true consensus; it just may not have the transitivity properties we desire. In usual statistical problems, we rely on the standard deviation or the variance as a measure of the agreement in our data. If we now knew the distribution of the variances of means of certain specially chosen points in Euclidean space representing orderings, we could then ask with what probability the variance of our consensus would occur with orderings taken at random. An empirical study would be useful; a theoretical one would be fascinating. Certainly for any applied problem, one could do a computer simulation in advance in order to compare the experimental results with random ones to determine the reliability of the consensus. Even before it became clear that these variances are variances in the usual statistical sense, Anderson suggested a study of variance as a measure of reliability [1].

Since we would expect to see a "true" consensus when the variance is relatively small, a consensus algorithm that worked quickly when the variance is small would be especially desirable. Thus in future studies of algorithms to compute consensus orderings the variance should be an essential element of the study.

Finally, although this author is not currently working in the area of axiomatics, it would be nice to have an axiomatic description of the Kemeny distance function appropriate to all classes of orderings - partial orders, trees, interval orders, semiorders, weak orders and so on. T. Margush has proposed such a uniform set of axioms and has shown there is significant redundancy in the axioms; the triangle inequality may be dropped for example [43], but more can still be done. What seems to be missing is a universal way to say that when two pairs of orderings differ in the same way, then the distance between them is the same without essentially defining the distance to be the size of the symmetric difference in that axiom.

Margush's work was motivated by his observation that the axioms given by the author for the Kemeny distance on posets, *when restricted to tree posets*, did not define a distance function. Since studying similarity and dissimilarity measures on trees has been a major source of research in numerical taxonomy (Margush's dissertation [42] provides an excellent list of references) and since Margush and McMorris were concerned with the consensus problem for tree posets as well [44], and since finally the symmetric difference metric is an appealing metric for tree posets, Margush revised the axioms for the symmetric difference distance function so that they applied to tree posets as well [43]; this led to his proposed uniform set of axioms.

In a different direction, Barthélemy [4] has shown that by studying semilattices of relations (for example the semilattice of posets) one can characterize a large class of semilattices on which a "Kemeny-type" axiomatization of distance is possible.

The author does not wish to give the impression that this concept of social distance is universally accepted or used in the social sciences. Applications of it can be found in [1] and [6]; however, its future value as a tool for social scientists in general is still in question. Anderson [1] has suggested further directions for social science applications; to my knowledge no one has followed them.

FOOTNOTES

[1]Another rephrasing of Arrow's axiom of irrelevant alternatives taken from Roberts [52] that seems less "assertive" but in fact brings the same results is: Let $A_1$ be any subset of the set of alternatives. If a $k$-tuple is modified in such a way that each individual's ranking among the elements in $A_1$ is unchanged, then the social welfare function restricted to $A_1$ is the same in both cases.

[2]I have taken the liberty of removing some conditions on $C$ and $A$ making the axioms slightly less restrictive; I don't think they are needed for an understanding of the nature of the proof. Of course the $C$ we add must still give the matrix $A$ nonnegative integer entries.

REFERENCES

[1] A. Anderson (1973) The Bogart preference structures: Applications, *J. Math. Sociology* 3, p. 67.

[2] K.J. Arrow (1951) *Social Choice and Individual Values*, John Wiley and Sons, Inc., New York.

[3] K. Baker, P. Fishburn and F. Roberts, *Partial Orders of Dimension 2, Interval Orders and Interval Graphs*, Rand Publication, P 4376.

[4] J.P. Barthélemy (1979) Caractérisations axiomatiques de la distance de la différence métrique entre des relations binaires, *Math. Sci. Humaines*, 67, 85-113.

[5] J.P. Barthélemy, Cl. Flament, and B. Monjardet (1982) Ordered sets and social sciences, in: *Symp. Ordered Sets* (I. Rival, ed.) Reidel, Dordrecht-Boston, 721-758.

[6] J.P. Barthélemy and B. Monjardet (1981) The median procedure in cluster analysis and social choice theory, *Mathematical Social Sciences*, to appear.

[7] D. Black (1958) *Theory of Committees and Elections*, University Press, Cambridge.

[8] K. Bogart (1973) Preference structures I: Distances between transitive preference relations, *J. Math. Sociology* 3, 49-67.

[9] K. Bogart (1975) Preference structures II: *SIAM J. Appl. Math.* 29, 254-262.

[10] K. Bogart and T.L. Greenough (to appear) Representation and enumeration of interval orders, *Discrete Math.*

[11] K. Bogart, I. Rabinovitch, and W.T. Trotter (1976) Bounds on the dimension of interval orders, *J. Combinatorial Theory* 21, 319-328.

[12] K. Bogart and J.R. Weeks (1979) Consensus signed digraphs, *SIAM J. Appl. Math.* 36, 1-14.

[13] V. Bowman and C. Colantoni (1973) Majority rule under transitivity constraints, *Management Sci.* 19, 1029-1041.

[14] G. Call and R.Z. Norman (1980) Generalizations of semi-orders, their representation problem and their relation to Arrow's paradox, *Congresses Numeratum* 28, p. 241.

[15] M.J.A. Condorcet (1785) *Essai sur l'application de l'analyse à la probabilité des décisions rendues à la pluralité des voix*, Paris. (Also Chelsea Publ. 6, New York, 1974.)

[16] W. Cook and R. Armstrong (1975) A committee ranking problem using ordinal utilities, manuscript.

[17] W. Cook and S. Brightwell (1978) Unifying group decisions relating to cardinal preference rankings: with an application to the operation of a military promotion board, *Cahiers du Centre d'Études de Recherche Operationelle* 20, 59-73.

[18] W. Cook and A. Snipe (1976) Committee approach to priority planning: The median ranking method, *Cahiers du Centre d'Études de Recherche Operationelle* 18.

[19] C.H. Coombs and J.E.K. Smith (1973) On the detection of structure in attitudes and developmental processes, *Psych. Rev.* 80, 337-351.

[20] J. De Cani (1969) Maximum likelihood of paired comparison ranking by linear programming, *Biometrica* 59, 537-545.

[21] P. Fishburn (1970) Intransitive indifference with unequal indifference intervals, *J. Math. Psych.* 7, 144-149.

[22] A. Ghouila-Houri (1962) Caractérisation des graphes non orientés dont on peut orienter les arêtes de manière à obtenir le graphe d'une relation d'ordre, *C.R. Acad. Sci. Paris* 254, 1370-1371.

[23] P.C. Gilmore and A.J. Hoffman (1964) A characterization of comparability graphs and of interval graphs, *Canad. J. Math.* 16, 539-540.

[24] M.C. Golumbic (1980) *Algorithmic Graph Theory and the Perfect Graph Conjecture*, Academic Press, New York.

[25] L. Goodman and H. Markowitz (1952) Social welfare functions based on individual rankings, *American Journal of Sociology* 58, 257-262.

[26] G. Guilbaud (1952) Les théories de l'intérêt général et le problème logique de l'agrégation, *Economie Appliquée* 5; English translation in *Readings in Mathematical Social Sciences*, SRA, Chicago (1966), 262-307.

[27] F. Harary (1959) A criterion for unanimity in French's theory of social power, in: *Studies in Social Power* (D. Cartwright, ed.), Institute for Social Research, Ann Arbor, Michigan, Ch. 10.

[28] F.R. Hudson, D.G. Kendall, and P. Tăutu (eds.) (1971) *Mathematics in the Archaeological and Historical Sciences*, Edinburgh Univ. Press, Edinburgh.

[29] D.T. Jamison and L.J. Lau (1973) Semiorders and the theory of choice, *Econometrica* 41, 901-912. (Note corrections in Vol. 43 (1975) 779-780.)

[30] D. Kelly and W.T. Trotter, Dimension theory for ordered sets (1982) *Symp. Ordered Sets* (I. Rival, ed.) Reidel, Dordrecht-Boston, 171-211.

[31] J. Kemeny (1969) Personal communication.

[32] J. Kemeny (1959) Generalized random variables, *Pacific J. Math.* 9, 1179-1189.

[33] J. Kemeny (1959) Mathematics without numbers, *Daedalus* 88, 577-591.

[34] J. Kemeny and J.L. Snell (1962) *Mathematical Models in the Social Sciences*, Ginn, New York.

[35] D.G. Kendall (1969) Incidence matrices, interval graphs and seriation in archaeology, *Pacific J. Math.* 28, 565-570.

[36] D.G. Kendall (1955) *Rank Correlation Methods*, Hafner Publishing Co., New York.

[37] G.B. Kolata, Overlapping genes: more than anomolies?, *Science* 176, 1187-1188.

[38] J. Levine (1974) Personal communication.

[39] R.D. Luce (1956) Semiorders and a theory of utility discrimination, *Econometrica* 24, 178-191.

[40] R.D. Luce, D.H. Kruntz, P. Suppes, and A. Tversky (1971) *Foundations of Measurement*, Academic Press, Vol. 1, Vol. 2 (to appear).

[41] D. Luce and H. Raiffa (1957) *Games and Decisions*, Wiley, New York.

[42] T. Margush (1980) *An Axiomatic Approach to Distances between Certain Discrete Structures*, Dissertation, Bowling Green State University.

[43] T. Margush (to appear) Distances between trees, *Discrete Applied Mathematics*.

[44] T. Margush and F. McMorris (1980) Consensus *n*-trees, *Bull. Math. Biology* 42.

[45] F. McMorris (1979) An algebraic problem in taxonomy, *Applied Mathematics Notes* 4, 55-61.

[46] P. Michaud and J. Marcotorchino (1979) Modèles d'optimization en analyse des données relationnelles, *Math. Sci. Humaines* 67, 7-38.

[47] B, Mirkin (1979) *Group Choice* (English translation by Y. Oliker, P. Fishburn, ed.) Halsted Press - V.A. Winston, Washington, D.C.

[48] J. Olguin (1970) Personal communication.

[49] J. Orlin (1981) Personal communication.

[50] I. Rabinovitch (1977) The Scott-Suppes theorem on semiorders, *J. Math. Psych.* 15, 209-212.

[51] I. Rabinovitch (1978) The dimension of semiorders, *J. Comb. Theory A.* 25, 50-61.

[52] F.S. Roberts (1976) *Discrete Mathematical Models*, Prentice-Hall, Englewood Cliffs, New Jersey.

[53] F.S. Roberts (1979) Indifference and Seriation, *Ann. N.Y. Acad. Sci.* 328, 177-182.

[54] F.S. Roberts (1979) *Measurement Theory with Applications to Decision Making, Utility and the Social Sciences*, Addison-Wesley, Reading, Mass.

[55] F.S. Roberts (1979) On the mobile radio frequency assignment problem and the traffic light phasing problem, *Ann. N.Y. Acad. Sci.* 319, 466-483.

[56] J. Rothenberg (1961) *The Measurement of Social Welfare*, Prentice-Hall, Englewood Cliffs, N.J.

[57] D. Scott and P. Suppes (1958) Foundational aspects of theories of measurement, *J. Symbolic Logic* 23, 113-128.

[58] E. Szpilrajn (1930) Sur l'extension de l'ordre partiel, *Fund. Math.* 16, 386-389.

# PART VIII

# PROBLEM SESSIONS

## INTRODUCTION

To those of us in Calgary who were involved in the planning of the Symposium, and here of course we mean Ivan Rival as well as the two of us, the problem session was one of the key ingredients of the meeting. We wanted to overcome what we saw as shortcomings of many problem sessions we'd attended in the past -- the dry, lecture-hall atmosphere, the single (and often too-long) session, the jumble of disconnected topics. So we soon decided that one, a problem session would happen nearly every evening, and two, each session would be restricted to a particular topic. We drew up a tentative list of topics and likely chairmen.

Not much later, a nearly ideal location for the sessions was assured, when we were able to reserve the Solarium of the Banff Centre for the entire two weeks of the meeting. This lounge-type room held about 80 people in comfortable chairs around small round tables. Accessories included a bar, which we kept open during and after each session, and a shuffleboard table. The latter attracted widespread interest, with most of us getting in our shots at least once. (One game featured the formidable team of Birkhoff and Jónsson!) Where you have mathematicians you usually have music, and evenings in the Solarium were no exception, as players such as Kirby Baker (piano), Brian Davey (guitar), Frank Farmer (lute), Robert Jamison (cello), and Oliver Pretzel (flute) entertained on several occasions. From all of us, many thanks. A fitting climax to the meeting came when, on the last evening, Flodur Voluntas inspired (and regaled) the Solarium audience with his summary of the preceding two weeks. But of course, the problem sessions were the main show, and with the addition to the Solarium of a couple of portable blackboards, they were all set to go.

There were ten sessions altogether. Those with problems to propose were asked to submit written versions to the chairman for that session. (Many did.) Also, the Banff Centre very kindly lent us a tape recorder with which, despite various problems, we recorded most of the sessions. With these aids, and the notes and memories of others, and our own frantic scribbles, we've prepared the record which follows. It has been edited with an eye to relevance to partial order. Also, some problems have been kept out because we were unable to reconstruct them sensibly. Probably, errors remain, for which we apologize in advance.

Rather than just list the problems we've included most of the background as well. Any missing information can often be obtained from the papers in this volume. We've attempted to preserve to some degree the conversational tone of the sessions



*I. Rival (ed.), Ordered Sets, 791–861.*

by, for example, including a fair number of comments from the audience. Where there's been any progress that we know of on a problem, we've indicated this in a parenthetical remark.

Thanks to all the participants at the sessions, and especially those who took notes or otherwise helped us put the problems together. And now ladies and gentlemen, abandon the shuffleboard game, get your drinks, take your seats, and let's get started!

Bill Sands
Robert Woodrow

PROBLEM SESSION 1

# ORDER TYPES

Chairman: E.C. Milner

E.C. MILNER: Before we begin this first session, let me just remind you of the notation

$$\alpha \to (\beta,\gamma)^r$$

for order types $\alpha,\beta,\gamma$. This means that whenever the $r$-element subsets of an ordering of type $\alpha$ are partitioned into two classes there must be a subset of type $\beta$ all of whose $r$-element subsets are in the first class or there must be a subset of type $\gamma$ all of whose $r$-element subsets are in the second class.

I'll now ask Professor Erdös to state one or two problems of this type.

P. ERDÖS: Well, I'll just state three or four problems.

Let $\alpha$ have no immediate predecessor (i.e. $\alpha = \beta\omega$).

*1.1 For which $\alpha$ does $\alpha \to (\alpha, \text{infinite path})^2$ ?*

Hajnal, Milner, and I (Set mappings and polarised partition relations, Combinatorial theory and its applications, *Coll. Math. Soc. J. Bolyai* 4 (1969), 327-363, esp. p. 358) showed this is true for $\alpha < \omega_1^{\omega+2}$ but our proof breaks down hopelessly for $\alpha = \omega_1^{\omega+2}$. I offer \$250 for settling the case $\alpha = \omega_1^{\omega+2}$, and \$500 for all $\alpha$.

Hajnal and I (Ordinary partition relations for ordinal numbers, *Per. Math. Hungar.* 1 (1971), 171-185) showed $\omega_1^2 \to (\omega_1\omega, 3)^2$. For \$250,

*1.2 Resolve whether $\omega_1^2 \to (\omega_1\omega, 4)^2$ .*

Nearly thirty years ago Rado and I (A partition calculus in set theory, *Bull. Amer. Math. Soc.* 62 (1956), 427-489) conjectured that $\omega^2 \to (\omega^2,n)^2$. Specker (Teilmengen von Mengen mit Relationen, *Commentarii Math. Helvetici* 37 (1957), 302-314) proved our conjecture and then showed that for $2 < k < \omega$, $\omega^k \not\to (\omega^k,3)^2$. He could not settle $\omega^\omega \to (\omega^\omega,3)^2$, and called attention to this beautiful unsolved problem. Some years later, Chang (A partition theorem for the complete graph on $\omega^\omega$, *J. Comb. Theory (A)* 12 (1972), 396-452) settled it in the affirmative, and soon after

Milner and Larson independently extended it to $\omega^{\omega} \to (\omega^{\omega}, n)^2$. In her paper (A short proof of a partition theorem for the ordinal $\omega^{\omega}$, *Ann. Math. Logic* 6 (1973/74), 120–145) Larson noted that the first unsolved problem now is $\omega^{\omega^2} \to (\omega^{\omega^2}, 3)^2$. I offer \$500 for a proof or disproof of this case, and \$1000 for settling the general question:

1.3 *for which $\alpha$ does $\omega^{\alpha} \to (\omega^{\alpha}, 3)^2$ ?*

It is probable that if $\omega^{\alpha} \to (\omega^{\alpha}, 3)^2$ holds then $\omega^{\alpha} \to (\omega^{\alpha}, n)^2$ also holds.

* * * * * * * * * *

J. GINSBURG: Let $P$ be a partial order, and let $C(P)$ be the set of all chains in $P$. For $x \in P$, let

$$a(x) = \{C \in C(P) : x \in C\}$$

and let $b(x) = C(P) \backslash a(x)$. Generate a topology on $C(P)$ by taking as subbase all sets $a(x)$ and $b(x)$ for $x \in P$. Then $C(P)$ is compact Hausdorff.

1.4 *What conditions on $P$ characterize when $C(P)$ has the countable chain condition?*

It is easy to see that $P$ cannot contain an uncountable antichain or $\kappa \times 2$ for uncountable $\kappa$. Recently, Todorčević has shown that $C(P)$ has c.c.c. for $P$ a Souslin tree.

* * * * * * * * * *

J. LARSON: This is a problem of Hagendorf. An order type $\varphi$ is *indecomposable* provided that whenever $\varphi = \varphi_1 + \varphi_2$ then $\varphi \leq \varphi_1$ or $\varphi \leq \varphi_2$, where $\leq$ is the relation of embeddability. Typical examples are $\omega$, $\omega^2$, $\omega^{\omega}$, $\omega_1$. These are also *strictly indecomposable to the right*, i.e. if $\varphi = \varphi_1 + \varphi_2$ then $\varphi \leq \varphi_2$ and $\varphi \not\leq \varphi_1$. For order types $\psi$ and $\theta$, $\psi$ is *strictly smaller* than $\theta$, written $\psi < \theta$, if $\psi \leq \theta$ but $\theta \not\leq \psi$.

1.5 *Suppose $\Phi$ is an order type satisfying*

*(i) $\Phi$ is strictly indecomposable to the right,*

*and (ii) Every $\psi < \Phi$ is embeddable in an initial segment of $\Phi$.*

*Must $\Phi$ be the order type of an ordinal?*

It is known that the answer is yes if $\Phi$ is a countable order type or if $\Phi$ is scattered (that is, $\mathbb{Q} \not\leq \Phi$). Also see Laver: On Fraïssé's order type conjecture, *Ann. of Math.* (1971), regarding $M$-types.

* * * * * * * * * *

E. HARZHEIM: Let $\alpha$ be a successor ordinal, and let $H_\alpha$ be the set of all 0-1 sequences of length $\omega_\alpha$ having a last 1. Order $H_\alpha$ lexicographically.

1.6 *Given any $f:H_\alpha \to H_\alpha$ does there exist a subset $S \subseteq H_\alpha$ of cardinality $\aleph_\alpha$ such that $f$ is monotone on $S$?*

This is true when $H_\alpha$ is replaced by $\omega_\alpha$. It is false with $H_\alpha$ replaced by the set $\mathbb{R}$ of the reals and $\aleph_\alpha$ replaced by $2^{\aleph_0}$.

J. BAUMGARTNER: Avraham has recently shown that it is consistent with $ZFC + 2^{\aleph_0}=\aleph_2$ that given any uncountable subset $S$ of $\mathbb{R}$ and $f:S \to \mathbb{R}$ then $f$ is monotone on an uncountable set.

[S. Todorčević has since obtained several results on this problem. He shows that if $2^{\aleph_\nu} \leq \aleph_\alpha$ for $\nu < \alpha$ then the answer is negative. In particular, then, the answer is negative given $GCH$. He obtains a positive result under other assumptions; for example the result for $\alpha = 1$ is positive given that $\mathbb{R}$ has the property of Avraham (mentioned above by Baumgartner).]

* * * * * * * * * *

E. COROMINAS: By a *tree* we mean a partial order $T$ for which $T_x = \{y : y < x\}$ is a chain for each $x \in T$.

Let $T$ be a tree which is a countable union of chains. Let $C(T)$ be the set of maximal chains of $T$ quasi-ordered by embeddability.

1.7 *Conjecture. The cofinality of every maximal chain of $C(T)$ is at most $\omega$.*

The conjecture is true in case $T$ is countable.

* * * * * * * * * *

M. POUZET: These problems came from discussions with Milner. For any $\kappa$, let $\mathcal{D}(\kappa)$ be the class of partial orders with no antichain of cardinality $\kappa$. Let $P$ be a partial order, and let $I(P)$ be the set of ideals of $P$ (i.e. the set of initial, up-directed subsets of $P$), ordered by inclusion.

1.8 *For each $\mu$ is there a cardinal $i(\mu)$ such that $P \in \mathcal{D}(\mu) \Rightarrow I(P) \in \mathcal{D}(i(\mu))$? In particular, can we take $i(\mu) \leq (\mu^{\underset{\smile}{\mu}})^+$?*

Bonnet (1973 in Finite and Infinite Sets) showed that $i(\omega) = \omega_1$ will do and is best possible.

Milner and Prikry have recently shown that for any $P \in \mathcal{D}(\mu)$ there is $F(\mu)$ such that $P$ can be covered by less than $F(\mu)$ ideals; moreover $F(\mu) \leq (\mu^{\underset{\smile}{\mu}})^+$. Clearly, $F(\mu) \leq i(\mu)$, so a positive answer to my first question would strengthen this result.

1.9 *What is the exact value of $F(\mu)$?*

With $GCH$ + $\exists$ $\mu$-Souslin tree, $F(\mu) = \mu^+$. Milner and Prikry showed $F(\mu) = \mu$ for $\mu$ weakly compact. I showed that for $\mu$ singular,

$$F(\mu) \leq (\sup \mu'^{\underset{\smile}{\mu'}})^+$$

where the supremum is taken over all *regular* $\mu' < \mu$, and that ($GCH$) if $\mathrm{cf}(\mu) = \omega$ then $F(\mu) = \mu$.

1.10 *Conjecture. If $V = L$ then $F(\mu) = \mu$ for $\mu$ singular or weakly compact, and $F(\mu) = \mu^+$ otherwise.*

So $\mu$ singular is the only case to resolve.

1.11 *Is it consistent with ZFC that $F(\mu) = \mu$ for all $\mu$?*

* * * * * * * * * *

I. RIVAL: Let $(U, \subseteq)$ be a nonprincipal ultrafilter on the natural numbers . Note that the set $C$ of coatoms of $U$ has no lower bounds (i.e., $\inf C \notin U$). Let $f: U \rightarrow U$ be order preserving such that $f(C)$ has no lower bound in $f(U)$ (i.e., $\inf f(C) \notin f(U)$).

1.12 *Show that $f(U)$ contains $U$ as a retract, i.e., there is $g: f(U) \rightarrow U$ and $h: U \rightarrow f(U)$, both order preserving, such that $g \circ h = \mathrm{id}_{\mu}$ .*

If this is possible then $U$ would be *irreducible*. See Duffus and Rival, A structure theory for ordered sets, *Discrete Math.* 35 (1981), 53-118.

* * * * * * * * * *

R. BONNET: Let $C$ be a chain of cardinality $\kappa \geq \omega$. $C$ is *strongly rigid* iff for every subchain $C_1$ of $C$ and every 1-1 monotone function $f: C_1 \to C$, $\{x \in C_1 : f(x) \neq x\}$ is of cardinality $< \kappa$. All such chains $C$ are of cardinality $\geq \omega_1$. An old result says that there is a strongly rigid chain $C$ of cardinality $\mathrm{cf}(2^{\aleph_0})$; in fact, $C \subset \mathbb{R}$.

1.13 (ZFC) *For which* $\kappa$ *is there a strongly rigid chain of cardinality* $\kappa$?

Chains $C_1, C_2$ of cardinality $\kappa$ are *almost disjoint* if whenever $C \leq C_1$ and $C \leq C_2$ then $|C| < \kappa$, where $\leq$ is embeddability.

1.14 *For a cardinal* $\kappa \geq \omega_1$ *find the smallest cardinal* $m = m(\kappa)$ *such that for every* $\eta < m$ *there is a family of chains* $(C_i)_{i \in I}$ *with* $|C_i| = \kappa$, $|I| = \eta$ *and the* $(C_i)$ *pairwise almost disjoint.*

It is known that if $\mathrm{cf}(2^{\aleph_0}) = \kappa$ then $m(\kappa) \geq \kappa^+$, and that $m(\aleph_1) \geq 5$ (consider $\omega_1, \omega_1^*$, an $\aleph_1$-subset of $\mathbb{R}$, a Specker chain, and a Shelah chain).

A chain $C$ is *impartible* if whenever $C \subseteq A \cup B$ for chains $A, B$ then $C \leq A$ or $C \leq B$. For example $\mathbb{Q}$ is impartible while $[0,1] \cap \mathbb{R}$ is not.

1.15 *Conjecture. There is no complete impartible chain.*

This is a conjecture of Pouzet.

* * * * * * * * * * *

R. FRAÏSSÉ: First is a problem of Sabbagh. For chains $A$ and $B$, $A \leq B$ means $A$ is embeddable in $B$. $A, B$ are *incomparable* if $A \nleq B \nleq A$. A chain $C$ is a *supremum* of $A$ and $B$ if $C \geq A$ and $C \geq B$, and $X \geq A, B \Rightarrow X \geq C$ for every chain $X$.

1.16 *Do there exist incomparable chains* $A, B$ *having a supremum?*

The answer is no when $A$ and $B$ are scattered (Hagendorf).

The following are problems of mine suggested by Hagendorf's thesis. A chain $A$ is *unsecable* if whenever $A = B + C$ then $A \leq B$ or $A \leq C$.

[This is the same notion as "indecomposable"; see Problem 1.5.]

*Right* and *left* unsecable chains are defined in the obvious way. For example, $1 + \mathbb{Q} + 1$ is unsecable, but neither right nor left unsecable.

1.17 *Let A be a chain such that $X + X < A$ for all $X < A$. Is A unsecable?*

Hagendorf showed this is true if $A$ is an ordinal.

1.18 *Suppose A is a countable chain such that $X + X < A$ for every initial interval $X < A$. Is A unsecable?*

Note that countability is required as $A = \mathbb{Q} + \omega_1$ would otherwise be a counterexample.

* * * * * * * * * *

PROBLEM SESSION 2

COMBINATORICS

Chairman: P. Hammer

A. BJÖRNER: Let $L$ be an infinite geometric lattice of rank $r$ and let $L_i = \{x \in L : \text{rank}(x) = i\}$ for each $i = 1,2,\ldots,r-1$.

2.1 *Is there a matching $f : L_1 \to L_{r-1}$, i.e., a one-to-one function from atoms to coatoms of $L$ such that $x \le f(x)$ $\forall\, x \in L_1$?*

This is known to be true if $L$ is modular or if $r = 3$ or if $|L| < \aleph_\omega$. It has been proved also for finite geometric lattices by Greene.

A more general question is:

2.2 *Does there exist a family $M$ of pairwise disjoint maximal chains in $L \backslash \{0,1\}$ such that for all $x \in L_1$ there is $m \in M$ with $x \in m$?*

I showed this is true for modular $L$, and J. Mason showed it to be true for finite $L$.

* * * * * * * * * *

R. STANLEY: Let $P$ be a finite poset and let $\sigma : P \to \mathbb{N}$ be order preserving. Set $|\sigma| = \sum_{x \in P} \sigma(x)$. For $m \ge 0$ let

$$U_m(p,q) = \sum q^{|\sigma|}$$

where the sum is over all order-preserving maps $\sigma : P \to \{0,\ldots,m\}$. This is a polynomial in $q$.

We say that $P$ is *Gaussian* if there are positive integers $h_1,\ldots,h_p$, where $p = |P|$, such that

$$U_m(p,q) = \prod_{i=1}^{p} \frac{1-q^{m+h_i}}{1-q^{h_i}}$$

for all $m \ge 0$. It turns out that chains have this property, and that $U_m(p,q)$ for a chain $P$ is a Gaussian (or $q$-binomial) coefficient. Hence the term "Gaussian".

Let $\underline{n}$ denote an $n$-element chain, and let $J(P)$ denote the lattice of order ideals of the poset $P$. Then the known connected Gaussian posets are: $\underline{r} \times \underline{s}$ for all $r,s$; $J(\underline{2}\times\underline{r})$ for all $r$; the ordinal sum $\underline{r} + \underline{2}^2 + \underline{r}$ for all $r$; $J(J(\underline{2}\times\underline{3}))$; and $J(J(J(\underline{2}\times\underline{3})))$.

2.3 *Are these all the connected Gaussian posets?*

A weaker but still open question is:

2.4 *Is every connected Gaussian poset a distributive lattice? a lattice at all?*

It is known that, for instance:

(i) $P$ is Gaussian iff every connected component of $P$ is Gaussian;

(ii) if $P$ is connected and Gaussian then every maximal chain of $P$ has the same length;

(iii) if $P$ is Gaussian then so is its dual;

(iv) no element of a Gaussian poset covers more than one minimal element.

* * * * * * * * * * 

R. JAMISON-WALDNER: Let $X$ be a partially ordered set. A subset $S \in X$ is *partitionable* provided that for some $x \in X$ there are at least two points of $S$ greater than or equal to $x$ and at least two less than or equal to $x$. (For example, any subset containing a three-element chain is partitionable.) The Helly number $h(X)$ is the supremum of the cardinalities of the nonpartitionable subsets of $X$ (see my paper "Partition numbers for ordered sets and trees", to appear in the *Pacific Journal*). Hoffman has shown that for finite $X$, $h(X)$ is the maximum cardinality of an order convex 2-family of $X$ (i.e., an order convex subset of $X$ which is the union of two antichains).

2.5 *Suppose $X$ is an infinite poset with $h(X)$ finite, and $E$ is a finite subset of $X$. Does there always exist a finite $F \subset X$ with $E \subset F$ and $h(F) = h(X)$?*

If $X$ is a semilattice, the answer is yes.

Let $X$ be a finite poset, and set $M(X)$ to be the average number of points in an order convex subset of $X$.

2.6 *For which $X$ do we have*

$$M(Y) \leq M(X)$$

*for all order convex $Y \subseteq X$?*

There are posets which fail to have this property. However the analogous property for convex subtrees of a tree is always true.

* * * * * * * * * *

R. QUACKENBUSH: Let $\pi(n)$ be the partition lattice on $n$ elements, and $D$ a 0-1 distributive sublattice. Given any atom $(i,j)$ of $\pi(n)$ there is a unique smallest element $\theta(i,j)$ of $D$ above it. We say that $\theta(i,j)$ is *principal*. Let $J(D)$ be the set of all join irreducibles of $D$. Trivially, join irreducibles are principal.

2.7 *Let $D$ be a finite distributive lattice and let $J(D) \subseteq S \subseteq D$. When is there a 0-1 embedding $\varphi: D \to \pi(n)$ for some $n$ so that $\varphi(S)$ is the set of principal elements of $\varphi(D)$?*

This is always true for $S = D$. On the other hand, $S = J(D)$ works exactly when the unit element of $D$ is join irreducible.

* * * * * * * * * *

T. TROTTER: Let $I(n)$ be the number of $n \times n$ upper triangular 0,1 matrices with no zero row or column. (For example, $I(2) = 2$ and $I(3) = 10$.) $I(n)$ also counts the number of interval orders which have interval endpoints from $\{1,2,\ldots,n\}$ for which (i) there are no duplicated holdings, and (ii) every integer is both a left and right endpoint (degenerate intervals are allowed). Greenough showed that

$$I(n) = \sum_{k=0}^{n} \binom{n}{k} S_k$$

where $S_0 = 1$ and $S_{k+1} = (2^{k+1}-1)S_k + (-1)^{k+1}$.

2.8 *Give a simple explanation of this formula.*

My other problem also arises from the theory of interval orders.

2.9 *Find the least integer $n_t$, $t$ a positive integer, so that if the 4-element subsets of $\{1,2,\ldots,n_t\}$ are colored with $t$ colors then there always exist 6 integers $i_1 < i_2 < \ldots < i_6$ so that $\{i_1,i_2,i_3,i_5\}$ and $\{i_2,i_4,i_5,i_6\}$ have the same color.*

It is known that $C_1 2^t < n_t < C_2 2^{2^t}$ for some constants $C_1, C_2$.

* * * * * * * * * *

P. HAMMER: Let $G$ be a finite graph. Introduce a preorder on the vertices of $G$ by: $i \geq j \Leftrightarrow N(j) \subseteq N(i) \cup \{i\}$. (Here, $N(j)$ denotes the set of all vertices adjacent to the vertex $j$.) Call the resulting preorder a *graphic* preorder.

2.10 *Characterize the graphic preorders.*

For example, 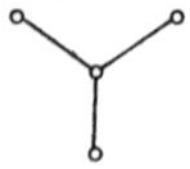 is not a graphic preorder.

The Dilworth number $\nabla(G)$ of $G$ is the width of the preorder corresponding to $G$. Let $G = \{1,2,\ldots,n\}$. It is known that $\nabla(G) = 1$ if and only if

$$\exists \text{ integers } v_1,\ldots,v_n,T \text{ such that } (i,j) \in E \text{ iff } v_i + v_j \geq T$$

and that $\nabla(G) = 2$ if and only if

$$\exists \text{ integers } v_1,\ldots,v_n,T,S \text{ such that } (i,j) \in E \text{ iff } |v_i + v_j| \geq T \text{ or } |v_i - v_j| \geq S.$$

2.11 *Characterize the graphs $G$ for which $\nabla(G) = n$, for each $n \geq 3$.*

* * * * * * * * * *

G. MCNULTY: A coloring of the points in a set of order type $\omega$ is said to be *non-repetitive* if no two adjacent intervals of the same length are colored in the same way (i.e., give the same tuple of colors).

Axel Thue proved in 1906 that $\omega$ has a non-repetitive 3-coloring.

Say that a coloring is *k-fold non-repetitive* ($k \geq 2$) if the coloring is non-repetitive on $\omega$ and the coloring restricted to $\{n:n \equiv i \bmod k\}$ is non-repetitive for each $i = 0,1,\ldots,k-1$.

2.12 *What is the smallest number of colors in a k-fold non-repetitive coloring? Generalize at will!*

A. Ehrenfeucht and I have shown that for $k = 2$, 4 colors will do and this is best possible.

Let $S_0 = \{0\} \cup \{\frac{n(n+1)}{2} : n \in \mathbb{N}\}$, and set $S_n = \{x + n : x \in S_0$ and $x > n\}$ for each $n > 0$.

2.13 *What is the smallest number of colors in a coloring that is non-repetitive on $\omega$ and on $S_n$ for all $n$?*

[These problems attracted some lively interest during the symposium and have mostly been settled. Q. Stout obtained the sharpest results, namely that 6 colors will do for all $k$ in Problem 2.12, and that 4 colors suffice and are best possible in Problem 2.13. G. McNulty showed that 4 colors work for the case $k = 3$ of Problem 2.12. H. Kierstead, R. Dilworth, and R. Freese also found partial answers.]

* * * * * * * * * *

P. WINKLER: Let $S$ be a finite set. An $I_s U_r$ *family* is a family $F$ of subsets of $S$ satisfying

$$I_s : \quad |A \cap B| \geq s \quad \forall\, A,B \in F$$

$$\text{and } U_r : \quad |S - (A \cup B)| \geq r \;\forall\, A,B \in F .$$

The following conjecture is due to Bang, Sharp, and myself, and independently to Frankl.

2.14 *Conjecture. For any finite $S$ there exists an $I_s U_r$ family $F$ of maximum size obtained by partitioning $S$ into two subsets $S_1$ and $S_2$, finding a maximum-sized family $F_1$ on $S_1$ satisfying $I_s$ and a maximum-sized family $F_2$ on $S_2$ satisfying $U_r$, and setting*

$$F = \{A \cup B : A \in F_1 \text{ and } B \in F_2\} .$$

For $r$ or $s = 1$ this was solved by Wang and Frankl.

* * * * * * * * * *

Q. STOUT: Consider the problem of locating an element $x$ in an infinite linear order by asking questions of the form "$x < a$?" or "$x \leq a$?". Corresponding to a search algorithm is a rooted binary tree (called a *search tree*), where left branches mean "yes" and right branches "no". This defines a natural lexicographic order on the leaves (i.e., terminal nodes) of the tree, and the linear order thus formed is isomorphic to the original order. The tree is strongly complete (i.e., it branches at each nonterminal node and each subtree contains a leaf). Also, every node of a search tree has finite depth, and has either a (lexicographically) maximal leaf in its left subtree or a minimal leaf in its right subtree. These properties characterize search trees.

Let $L_T(d)$ denote the number of leaves of $T$ at depth $d$.

2.15 *Let $T$ be a binary, rooted, strongly complete tree with every node at finite depth. Is there a search tree $S$ whose leaves are order isomorphic to $T$'s leaves, and such that $L_S(d) = L_T(d)$ for all $d \geq 1$?*

The answer is yes if the leaves of $T$ have the order type of an ordinal or the rationals, or if $T$ has only a finite number of infinite paths.

* * * * * * * * * *

E. HARZHEIM: A partially ordered set $P$ is $\kappa$-*universally ordered* if every poset of power $\kappa$ can be embedded in $P$. For instance, if $M$ is a set of cardinality $\aleph_\alpha$, then $(P(M),\subseteq)$ is $\aleph_\alpha$-universally ordered. More generally, suppose $|M| = \sum_{\nu<\alpha} 2^{\aleph_\nu}$ and $\aleph_\alpha$ is regular. Then I proved that if $P(M)$ is divided into $\aleph_\alpha$ subsets, at least one is $\aleph_\alpha$-universally ordered.

2.16 *Can you drop the assumption that $\aleph_\alpha$ is regular?*

There is a partial solution: all $\tau^* + \tau$ for $\tau < \omega_{\alpha+1}$ are embeddable in one class.

* * * * * * * * * *

M. POUZET: This is a problem of Rosenberg, Fraïssé, and myself. Given a relation $R$ on a set $E$, consider the restrictions $R_K$ to the $k$ element subsets $K$ of $E$, for $k = 0,1,\ldots$ . The isomorphism types form a set ordered by embedding called the *âge* of $R$ (introduced by R. Fraïssé in 1948). This ordered set is ranked, and the sequence of Whitney numbers $w_0,w_1,w_2,\ldots$ is called the *profile* of $R$. ($w_k$ is the number of isomorphism types of $k$ element sets.) I showed in 1972 and 1976 that if $E$ is infinite then the profile is increasing and if $E$ is finite then $w_0 \leq w_1 \leq \ldots \leq w_{[n/2]}$ where $n = |E|$. In what follows, assume $E$ is finite.

2.17 *Is the profile unimodal?*

2.18 *What conditions on $R$ guarantee that the profile is symmetric?*

2.19 *Under what conditions does the âge of $R$ have the Sperner property?*

It is known that the âge of $R$ has the Sperner property if $\forall\ K,K'$,

$$R_K \cong R_{K'} \Leftrightarrow R_{E\backslash K} \cong R_{E\backslash K'} .$$

* * * * * * * * * *

I. ROSENBERG: This problem arose from discussions with M. Pouzet. Let $(P,\leq)$ be a finite ordered set of width $w_p$ and height $h_p$. The set $J(P)$ of initial sections (ideals) ordered by inclusion is a sublattice of $P(P)$. For the width $w_{J(P)}$ we have the trivial but sharp inequality

$$w_P \leq w_{J(P)} \leq \binom{|P|}{\lfloor \frac{1}{2}|P| \rfloor} .$$

2.20 *What more can be said about $w_{J(P)}$? For example, give bounds for special types. When is $w_P = w_{J(P)}$?*

* * * * * * * * * *

K. BOGART: A partially ordered set $X$ is an *interval order* if there is a mapping $f$ from $X$ into intervals of $\mathbb{R}$ such that $a < b$ in $X$ iff each element of $f(a)$ is less than each element of $f(b)$.

2.21 *How many distinct labelled interval orderings does an n-element set have?*

According to Monjardet, there exists a recursion formula for semiorders (i.e., interval orders where all intervals are of equal length).

* * * * * * * * * *

PROBLEM SESSION 3

LINEAR EXTENSIONS

Chairman: K. Baker

B. SANDS: From the work of Graham, Yao, and Yao, Ivan Rival and I were led to the following problem. Let $P$ be a finite poset and let $x,y,z \in P$. Define

$$Pr(x<y) = \frac{\text{number of extensions of } P \text{ to a linear order having } x<y}{\text{number of linear extensions of } P}$$

and

$$Pr(x<y \mid x<z) = \frac{\text{number of linear extensions of } P \text{ having } x<y \text{ and } x<z}{\text{number of linear extensions of } P \text{ with } x<z}.$$

3.1 *Is* $(Pr\ x<y \mid x<z) \geq Pr(x<y)$?

Without loss $x,y,z$ may be taken as an antichain. The inequality could possibly be strict in all cases.

[Recently, L. Shepp has found a nice proof of this inequality, using the *FKG* inequality.]

* * * * * * * * * *

P. WINKLER: This problem is motivated by the notion, discussed by Fishburn, of a canonical linear extension of a partially ordered set. Suppose $P$ is a poset with $|P| = n$. Let $F$ be the family of linear extensions of $P$, i.e., 1-1 order-preserving maps of $P$ into $\{1,\ldots,n\}$. For $x \in P$, let

$$H_P(x) = \frac{1}{|F|} \sum_{f \in F} f(x)$$

denote the *average height* of $x$ in $P$. For $x,y$ incomparable in $P$, let $P \cup \{x>y\}$ denote the smallest extension of $P$ having $x > y$.

3.2 *Prove that* $H_{P\cup\{x>y\}}(x) \geq H_P(x)$ *(weak average height conjecture) and that* $H_{P\cup\{x>y\}}(x) \geq 1 + H_{P\cup\{x<y\}}(x)$ *(strong average height conjecture)..*

A positive answer to the $x,y,z$ conjecture of Rival and Sands [Problem 3.1] would yield the strong average height conjecture.

[So this problem is also solved.]

* * * * * * * * * *

R. STANLEY: Let $P$ be a partial ordering of $\{1,2,\ldots,n\}$ consistent with the usual ordering of the integers (so that $1 < 2 < \ldots < n$ would be a linear extension of $P$). Let $L(P)$ denote the set of linear extensions of $P$, regarded as permutations of $\{1,2,\ldots,n\}$. If $\pi = a_1 a_2 \ldots a_n \in L(P)$ define

$$d(\pi) = |\{i : a_i > a_{i+1}\}| \ ,$$

the number of *descents* of $\pi$. Let $M = M(P) = \max(d(\pi) : \pi \in L(P))$, and set $w_i = |\{\pi \in L(P) : d(\pi) = i\}|$ for $i = 0,1,\ldots,M(P)$. It can be shown that $M(P)$ and the $w_i$'s depend only on the isomorphism class of $P$ and not on the labelling of the elements.

About ten years ago I proved that the following are equivalent for a finite poset $P$:

(i) $w_i = w_{M-i}$ for all $0 \leq i \leq M$;

(ii) Every maximal chain of $P$ has the same length.

3.3 *Find a combinatorial proof of this theorem. More precisely, when (ii) holds describe explicitly a bijection* $f : L(P) \overset{\cong}{\to} L(P)$ *such that* $d(\pi) = M - d(f(\pi))$ *for all* $\pi \in L(P)$.

It would even be interesting to do this for the case $P \cong \underline{r} \times \underline{s}$.

P. ERDÖS: Given a partially ordered set $P$, can you get some good estimation of the largest number of descents in any linear extension?

STANLEY: Yes, it's something simple like the total number of elements of $P$ minus the number of elements in a largest chain. But there are more general questions along those lines which are harder to answer.

R. QUACKENBUSH: Do you get anything interesting if you look at the group of permutations generated by the linear extensions of $P$?

STANLEY: That I've never thought about. I imagine most of the time it'll just be the symmetric group.

* * * * * * * * * *

M. POUZET: Let $T$ be a distributive lattice. The *algebraic dimension* of $T$ is the smallest number $\kappa$ such that $T$ can be embedded as a sublattice into a product of $\kappa$ chains.

3.4 *Does algebraic dimension equal order dimension?*

This is true for distributive lattices of finite order dimension, but is unsolved in general even for Boolean algebras. If the Boolean algebra is generated by one chain the answer is yes.

My second problem is about retractions. Consider lattices which are retracts of products of chains. It is a very recent result that all countable lattices, and all distributive lattices of finite dimension, are such retracts. The *retract dimension* of $L$ is the smallest number $\kappa$ so that $L$ is a retract of the product of $\kappa$ chains. This is not the same concept as order dimension. For example, the lattice $\mathbb{Q}^2 + \mathbb{Q}^2$ has order dimension two but retract dimension three.

3.5 *Characterize the distributive lattices of finite retract dimension.*

Perhaps there is an example of a distributive lattice of order dimension two but uncountable retract dimension.

> [Pouzet now thinks that every distributive lattice of finite order dimension also has finite retract dimension. However, his problem is still open.]

* * * * * * * * * *

H. CRAPO: When an order relation $\omega$ on a set $X$ is expressed as an intersection $\omega = \bigcap_\alpha \lambda_\alpha$ of linear orders on the same set, the diagonal map

$$\Delta: (X,\omega) \to \prod_\alpha (X,\lambda_\alpha)$$

is strict monic.

3.6 *Under what conditions is $\Delta$ a coretraction?*

When this happens, $X$ will be a retract of a product of chains (see the previous problem). For $X$ finite, this is the case if and only if $X$ is a lattice; otherwise, the MacNeille completion provides a "furthest" retraction of the product toward $X$. (Of course, $X$ in general need not be complete in order for $\Delta$ to be a coretraction, for instance when $\omega$ is itself a linear order.)

*3.7* *In the general case, is there (up to isomorphism) a furthermost retraction of* $\prod_\alpha (X,\lambda_\alpha)$ *toward* $(X,\omega)$, *i.e., a "relative completion" of X within the product of chains?*

This is part of a more general question, listed as Problem 8 in my paper "Ordered sets: retracts and connections", to appear this fall in the *Journal of Pure and Applied Algebra.* The more general problem asks for a *relative completion B* within any strict monic map $C \to A$, following the work of Banaschewski and Bruns, "Categorical characterization of the MacNeille completion", *Ann. of Math.* 18 (1967), 369-377.

* * * * * * * * * *

R. FRAÏSSÉ: Let $R$ be a poset of order dimension $n$. This means that there are $n$ chains $C_1,\ldots,C_n$ on $|R|$ such that

$$R(x,y) \leftrightarrow C_1(x,y) \wedge \ldots \wedge C_n(x,y).$$

Thus there is a quantifier free formula with predicates $C_1,\ldots,C_n$ which defines $R$.

*3.8* *Can there be chains* $D_1,\ldots,D_p$ *on* $|R|$, *with* $p < n = \dim R$, *and a quantifier free formula* $\theta$ *in the p binary relations* $D_1(x,y),\ldots,D_p(x,y)$ *such that R is definable by* $\theta$?

We exclude the trivial example of an *antichain R* (of dimension 2), for which $p = 0$ since $R$ is definable in zero relations by $x \neq y$. This problem is connected to the question of the real arity of a relation, and the work of Pabion.

[During the symposium, G. McNulty and M. Pouzet independently observed that if $p = 2$ then $n = 2$ as well.]

* * * * * * * * * *

K. BOGART: The interval order $I(0,n)$ is the partially ordered set whose elements are all closed intervals with *integer* endpoints chosen from $\{0,1,\ldots,n\}$, and where one interval is less than another iff the first is *entirely* to the left of the second (so that e.g., [0,1] and [1,2] are incomparable elements of $I(0,n)$).

*3.9* *What is the dimension of* $I(0,n)$ *and how fast does it grow?*

Trotter, Rabinovitch, and I have shown that dim $I(0,10) = 3$ and dim $I(0,11) = 4$, and that

$$C_1 \log \log n \leq \dim I(0,n) \leq C_2 \log n .$$

* * * * * * * * * *

T. TROTTER: For integers $n,k$ with $0 \leq k \leq \binom{n}{2}$ let $f(n,k)$ denote the maximum number of linear extensions of a poset on $n$ points with $k$ comparable pairs.

*3.10* *Find $f(n,k)$ and determine the extremal posets.*

Kleitman and I have shown that if $0 \leq k \leq n - 1$ then $f(n,k) = \frac{n!}{k+1}$ and that the unique (up to duality) extremal poset is

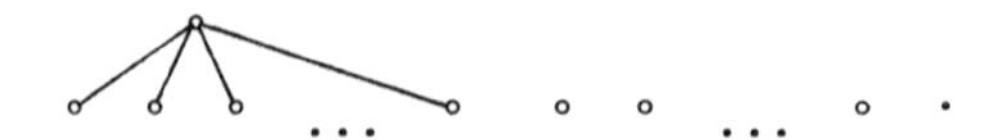

*3.11* *If $k = s(n-s)$, is $f(n,k)$ attained by the poset of height two whose comparability graph is $K_{s,n-s}$? (If so, then $f(n,k) = s!(n-s)!$.)*

* * * * * * * * * *

O. PRETZEL: Let $X$ be a poset constructed in the following way. Start with a "stack" of antichains $A_i$ $(i = 1,\ldots,t)$, $|A_i| = a_i \neq 0$. Put all elements of $A_i$ above all elements of $A_{i+1}$ for each $i$. In between $A_i$ and $A_{i+1}$ $(i = 1,\ldots,t-1)$ insert an antichain $B_i$ such that, for each pair $S,T$ of nonempty sets satisfying $S \leq A_i$ and $T \leq A_{i+1}$, $B_i$ contains a unique element below all the elements of $S$ and above all those of $T$ and incomparable to all other elements of $A_i$ and $A_{i+1}$. Thus $B_i$ has $(2^{a_i}-1)(2^{a_{i+1}}-1)$ elements. Taking the transitive closure, an element $x$ of $B_i$ lies above an element $y$ of $B_{i+1}$ if there exists $z \in A_{i+1}$ with $x > z > y$, and otherwise they are incomparable.

*3.12* *What is the dimension of $X$ in terms of $a_1,\ldots,a_t$?*

The width of $X$ is the maximum size of the $B_i$'s and so the dimension is bounded by $(2^a-1)^2$ where $a = \max a_i$. If $t = 3$ and

$a_1 = 1$, $a_2 = n$, $a_3 = 1$, and $n \geq 3$, then

$$\dim X = \binom{n}{\frac{n}{2}} .$$

If $t = 4$ and $a_1 = 1$, $a_2 = m$, $a_3 = n$, $a_4 = 1$ where $m \leq n$ and $m + n \geq 4$, then

$$\dim X = \max\left\{ \binom{n}{\frac{n}{2}}, \left\lceil \frac{2}{3}\left( \binom{m}{\frac{m}{2}} + \binom{n}{\frac{n}{2}} \right) \right\rceil \right\} .$$

* * * * * * * * * *

PROBLEM SESSION 4

## SCHEDULING AND SORTING

Chairman: R.L. Graham

A. GARSIA: This problem arose in the study of Cohen-Macaulay posets. Given an $n \times n$ matrix of zeros and ones, a *matching* is a selection of $n$ ones, one from each row and column. If the rows of the matrix are identified with $n$ boys and the columns with $n$ girls, and a one in the $i,j$ position means that boy $i$ and girl $j$ are compatible, then a matching can be thought of as a marrying off of the $n$ boys and $n$ girls into compatible pairs. Alternatively, you have a length one poset -- a bipartite graph -- with $n$ minimal elements (the boys) and $n$ maximal elements (the girls) where a minimal is less than a maximal if the corresponding boy and girl are compatible. A matching is then a 1-factor of the bipartite graph.

If a matching for a given $n \times n$ matrix exists, then there is an algorithm, the *alternating path* algorithm, which will find a matching in time $O(nE)$, where $E$ is the number of ones in the matrix (or edges in the graph).

If a matching contains compatible boy-girl pairs $(a,b)$ and $(c,d)$, and if $(a,d)$ and $(c,b)$ are also compatible pairs, then a new matching could be created by replacing the pairs $(a,b)$ and $(c,d)$ by the pairs $(a,d)$ and $(c,b)$. A *stable* matching is one that avoids this possibility; that is, for all pairs $(a,b)$ and $(c,d)$ in the matching, at least one of the pairs $(a,d)$ and $(c,b)$ is incompatible. In graph terms, this means that if $(a,b)$ and $(c,d)$ are edges in the matching then the four-cycle

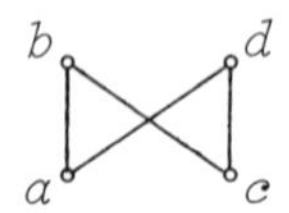


is not a subgraph of the bipartite graph.

4.1 *Is there an algorithm which will find a stable matching (if there is one) of a given $n \times n$ matrix in time $O(nET)$, where $T$ is the total number of matchings?*

4.2 *Is there a Philip Hall-type condition that will guarantee the existence of a stable matching?*

W. PULLEYBLANK: I found the following problem to be $NP$-complete: given a bipartite graph and a positive integer $k$, does there exist a matching containing $k$ or more pairs which avoids containing any alternating cycle (in the same sense that your problem requires avoiding an alternating four-cycle)? Maybe to some extent this gives you an answer. Of course, four-cycles might be special.

[Pulleyblank has since observed that his same construction shows that the problem "Given a bipartite graph and an integer $k$, does there exist a matching containing at least $k$ pairs and no alternating four-cycle?" is *NP*-complete.]

* * * * * * * * * *

D. WEST: This is the "pancake problem", an old problem posed under the name Harry Dweighter in the *Monthly*. You are given a stack of $n$ different-sized pancakes in some order, and wish to put them in decreasing order of sizes, where the only operation allowed is turning over a top portion of the stack (the computer scientists call this "sorting by prefix reversal").

4.3 *In the worst case, how many flips might be required (as a function of $n$)?*

Let this function be $f(n)$. Trivial upper and lower bounds are $n < f(n) < 2n$. The only nontrivial results are by Gates and Papadimitriou: $\frac{17}{16} n < f(n) < \frac{5}{3} n$.

GRAHAM: It's not known even to be asymptotic to a constant times $n$, is it?

WEST: No. It could oscillate, but that's unlikely.

GRAHAM: The problem's really due to Jacob Goodman, who submitted it under a pseudonym to the *Monthly*. He actually came up with it when he was sorting towels from the dryer, he said.

* * * * * * * * * *

M. AIGNER: Suppose there is a linearly ordered set of $n$ elements, and initially you know nothing about the order. Let $U_t(n)$ denote the number of comparisons that you must make to determine the $t$ top elements as an unordered set (you do not need to know the ordering among these $t$). It is known that $U_1(n) = n - 1$ and $U_2(n) = n - 2 + [\log_2(n-1)]$. Also, $U_t(n) = U_{n-t}(n)$ by symmetry.

4.4 *Is the sequence $(U_t(n) : t = 1,\ldots,n)$ unimodal?*

That is, does the problem of determining the $t$ top elements become steadily harder right up to $t = \frac{n}{2}$ ? This is certainly reasonable.

Letting $V_t(n)$ be the number of comparisons needed to select the $t$th element in the linear ordering, we have a similar problem:

4.5 *Is the sequence* $(V_t(n) : t = 1,\ldots,n)$ *unimodal?*

Here it's not nearly as likely that the answer is yes. There is no reason to believe that to select the 5th element is more difficult than to select the 4th element, for example.

* * * * * * * * * *

O. COGIS: Given any finite poset $P$ and any linear extension $L$ of $P$, define $\sigma(P,L)$ to be the minimum number of pairs $(x,y)$ of elements of $P$ such that $y$ covers $x$ in $L$ but not in $P$. That is, $\sigma(P,L)$ is the number of "jumps" between incomparables in $L$. Let $\sigma(P) = \min\{\sigma(P,L) \mid L \text{ is a linear extension of } P\}$.

4.6 *Characterize those posets $P$ for which $\sigma(P)$ can be computed by the following greedy algorithm: find a maximal chain $C$ in $P$ such that no element of $P - C$ is less than any element of $C$; delete $C$ and count* 1; *and repeat the process on $P - C$, continuing until only a chain remains of $P$.*

Recently Pulleyblank has shown that the problem of finding $\sigma(P)$ is *NP*-complete. The greedy algorithm works if $P$ is *series-parallel* (a concept introduced by Lawler and Tarjan); equivalently, if $P$ does not contain

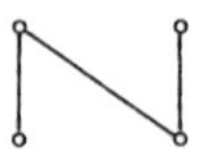

as an induced subposet. There are posets which are not series-parallel but for which the greedy algorithm works. Also, for any poset $P$ there is at least one linear extension giving $\sigma(P)$ which can be constructed by means of the greedy algorithm. So whenever the greedy algorithm fails, it's because at one stage you went the wrong way.

* * * * * * * * * *

W. POGUNTKE: This is a continuation of the previous problem. It is clear that $\sigma(P) \geq w(P) - 1$ for all posets $P$, where $w(P)$ is the width of $P$. G. Gierz and I observed that $\sigma(P) \geq d(P) - 1 \geq w(P) - 1$ where $d(P)$, the *defect* of $P$, is defined as the dimension of the kernel of the incidence matrix of $P$ over $GF(2)$. That is, let $P = \{x_1, x_2, \ldots, x_n\}$ and define an $n \times n$ matrix $A_P$ by $A_P(i,j)$ equals 1 if $x_i < x_j$ and 0 otherwise; then $d(P)$ is the nullity of the linear tranformation associated with $A_P$ over $GF(2)$.

Let $\mathcal{D} = \{P : \sigma(P) = d(P) - 1\}$.

4.7 *Give a good combinatorial characterization of* $\mathcal{D}$.

Duffus, Rival, and Winkler have shown that every poset containing no crown satisfies $\sigma(P) = w(P) - 1$, and so is in $\mathcal{D}$. But one can also show that crowns themselves are in $\mathcal{D}$, and that $\mathcal{D}$ is closed under the series and parallel constructions (so that in particular all series-parallel posets are in $\mathcal{D}$).

M. POUZET: Are you saying that $\mathcal{D}$ is the smallest class of posets closed under these operations?

POGUNTKE: No. That would be possible, but I don't know. That would be one idea, to show that by starting with crowns and the one-element poset and using the series and parallel constructions, you would get all of $\mathcal{D}$. Incidentally here is a small poset not in $\mathcal{D}$:

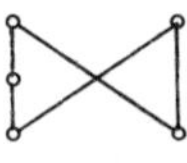

* * * * * * * * * *

W. PULLEYBLANK: This problem has to do with the *weighted* case of the previous two problems. Given a poset $P$ where each covering pair (i.e., edge) is assigned a nonnegative integral weight, we wish to find a linear extension $L$ of $P$ for which the sum of the weights of all covering pairs in $L$ is as large as possible. Let $C^*$ denote this largest possible sum of weights.

If all edges of $P$ are assigned weight 1, the problem reduces to the problem of calculating $\sigma(P)$ (see 4.6), which in the series-parallel case has been solved.

In constructing a series-parallel poset, edges are created only by the series construction and never by the parallel construction. The set of edges created by a series construction will be called a *series cut* of the poset. So the edges of every series-parallel poset can be partitioned into series cuts.

A *bifan* $B$ of $P$ consists of any element $v$ of $P$ together with all the edges incident with it. If $u$ is a lower cover and $w$ an upper cover of $v$, and if there is an element $z \neq v$ such that $u$ and $w$ are also incident with $z$, then we call $B$ a *paralleled bifan*.

A family of series cuts and paralleled bifans will be said to *cover* $P$ if each edge of $P$ occurs in the family at least as many times as the weight assigned to it.

4.8 *Conjecture. If $P$ is a weighted series-parallel poset whose covering graph is bipartite and in which there are no* 1-*edge series cuts, then $C^*$ is the minimum cardinality of a family of series cuts and paralleled bifans covering $P$.*

I can prove that the conjecture is true if, in addition, every cycle of the covering graph of $P$ contains an even number of edges from every series cut. However, the corresponding conjecture for arbitrary series-parallel posets is false.

Let me mention one other problem for the unweighted case. The problem of finding $\sigma(P)$ is *NP*-complete, even for posets of length one. Also, to any poset $P$ there corresponds a length one poset $P'$ so that finding $\sigma(P')$ would yield $\sigma(P)$. This is an existence statement; I don't know of any reasonable way to get $P'$ from $P$.

4.9 *Can you find a nice way of reducing the problem of finding $\sigma(P)$ for arbitrary $P$ to the length one case?*

M. POUZET: If $\sigma(P) = r$ and $\sigma(P-\{x\}) < r$ for all $x \in P$, must $P$ have length one?

PULLEYBLANK: I don't know.

POUZET: Because you said you reduced the general case to the length one case.

PULLEYBLANK: Right. But I don't know what the reduction looks like.

POUZET: But the reduction does not increase the size of $P$?

PULLEYBLANK: Oh, the whole thing blows up. It's a polynomial reduction, but where I had $n$ nodes before I'll probably have $n^{34}$ nodes afterward, or some such figure.

* * * * * * * * * *

M. POUZET: The jump number $\sigma(P)$ of a poset $P$ has been defined earlier (see Problem 4.6).

4.10 *If $\sigma(P) = r$ and $\sigma(P-\{x\}) < r$ for all $x \in P$, is $|P| \leq 2r$?*

If true, this bound is best possible (consider crowns). I know that this bound is correct for $r \leq 3$.

* * * * * * * * * *

J.K. LENSTRA: I will talk about a scheduling problem which actually occurred in practice. Suppose you have $m$ cottages to be rented out to $n$ clients. Client $i$ needs a cottage during the time interval $(a_i,b_i)$. Can the clients be packed into the cottages? This is a special case of the tanker problem Alan Hoffman talked about, and can be solved in $O(n \log n)$ time. It turns out that the minimum number of cottages needed equals the maximum overlap of the clients' time intervals.

Now suppose that the cottages are only available for certain time intervals, say cottage $j$ is available for the interval $(A_j,B_j)$, and you again have $n$ clients representable by intervals $(a_i,b_i)$. Furthermore we insist that each client be allowed to retain the same cottage throughout his time interval (i.e., no pre-emption). The *interval scheduling problem* is then to pack the intervals $(a_i,b_i)$ into the intervals $(A_j,B_j)$.

4.11 *Determine the complexity of the interval scheduling problem for fixed m and for arbitrary m.*

We know that the case $m = 2$ is solvable in $O(n \log n)$ time. We also know that the general case is solvable in $O\left(mn \prod_{i=1}^{n} (b_i-a_i)\right)$ time, which is pseudopolynomial for fixed $m$, but the complexity for even $m = 3$ is still unsolved.

Here is a scheduling problem which is in the Garey and Johnson catalogue. You have $m$ parallel identical machines and $n$ unit time jobs. Among the jobs you have a precedence relation which must be satisfied. The problem is to find a schedule of minimum length for processing the jobs. For $m = 2$ this can be solved in almost linear time -- the most efficient algorithm is due to Gabow. For arbitrary $m$ this problem is *NP*-hard, as proved by J. Ullman.

4.12 *What is the complexity of this problem for fixed m? In particular, what is the complexity for m = 3?*

* * * * * * * * * *

PROBLEM SESSION 5

## GRAPHS AND ENUMERATION

Chairman: B. Sands

M. ERNÉ: If there is a poset on $n$ points with exactly $k$ antichains, then of course there is a graph (namely the incomparability graph of the poset) on $n$ vertices with exactly $k$ simplices (i.e., complete subgraphs). My problem is the converse:

*5.1* *If there is a graph with n vertices and exactly k simplices, is there necessarily a poset on n points with exactly k antichains?*

The answer is yes for $n \leq 9$, and likely also for $n = 10$. Note that for both posets and graphs on $n$ elements, $n + 1 \leq k \leq 2^n$, but in this range, there are large gaps in the possible values for $k$.

Let $P(n)$ be the set of (labelled) posets with $n$ elements $\{1,2,\ldots,n\}$, and let $a(P)$ be the number of antichains of the poset $P$.

Set

$$\tilde{a}(n) = \frac{\Sigma(a(P):P\in P(n))}{|P(n)|},$$

the average number of antichains in an $n$-element poset. I can prove that

$$\log \tilde{a}(n) = \frac{n}{2} + O(\log n).$$

*5.2* *Is $\tilde{a}(n)$ asymptotically equal to $2^{n/2}\cdot c$?*

Let $C(n)$ be the number of partial orders contained in a fixed $n$-element linear order. Let $P^*(n)$ be the number of labelled posets on an $n$-element set, and $P(n)$ be the number of isomorphism classes of $n$-element posets.

Then $P(n) \leq C(n) \leq P^*(n) \leq n!P(n)$. One can prove that the sequence of $C(n)$'s is log concave, i.e., $C(n+1)^2 \leq C(n)C(n+2)$.

*5.3* *Are the sequences $\{P(n)\}$ and $\{P^*(n)\}$ log concave?*

I can show with difficulty that $2P^*(n+2)P^*(n) \geq P^*(n+1)^2$.

* * * * * * * * * * *

W. PULLEYBLANK: This problem arises in network reliability.

5.4 *Given a poset $(S,\leq)$, find a minimum cardinality set of intervals that partition $S$.*

It is clear that the minimum number is at least the maximum of the cardinalities of the set of maximal elements and of the set of minimal elements of $S$. M. Ball and G. Nemhauser proved that if $S$ is the family of independent subsets of a matroid ordered by inclusion then equality is attained.

5.5 *Is there always an optimum partition containing the interval $[x,y]$ for some $x$ minimal and $y$ maximal?*

A. BJÖRNER: With regard to your first question and your comment following, Ball and Provin have recently shown that the situation you described for matroids is more generally true for any shellable complex.

* * * * * * * * * *

M. AIGNER: This is a problem from Birkhoff's book. Let $P$ be a poset and $H$ its Hasse diagram. If $G$ is obtained from $H$ by removing the orientation of edges then $G$ is called a *Hasse graph*.

5.6 *Characterize the Hasse graphs!*

Obviously $G$ cannot contain a triangle. The smallest non-orientable graph without a triangle is the Mycielski (or Grötzsch) graph, with 11 vertices. Maybe a more reasonable question is:

5.7 *Find a decent algorithm to decide whether a graph $G$ is a Hasse graph.*

The problem is clearly $NP$, but is it $NP$-complete?

* * * * * * * * * *

P. ERDÖS: Kleitman, Rothschild and I proved that the number of (labelled) graphs on $n$ vertices without a triangle is asymptotic to the number of bipartite graphs on $n$ vertices.

5.8 *Is the number of (labelled) graphs which do not contain a $K_r$ asymptotic to the number of r-partite graphs?*

We proved that the logarithms are asymptotically equal. All this appeared in the Proceedings of the Rome Conference in 1973.

Rothschild and I conjecture the following.

5.9 *Conjecture. The number $f(n)$ of graphs on $n$ vertices which contain no triangle and which are maximal with respect to this property satisfies*

$$f(n) = 2^{(1+o(1))\frac{n^2}{8}}$$

One more question and then I'm through. Let $G(2n)$ (a graph with $2n$ vertices) have no triangle. We showed that the number of maximal independent sets of $G(2n)$ is at most $2^n$ with equality if $G(2n)$ consists of $n$ independent edges.

5.10 *Is it true that if all vertices of $G(10n)$ have valency (or degree) $\geq 3$ then the number of maximal independent sets of $G(10n)$ is at most $15^n$?*

Equality is attained if $G(10n)$ is the union of $n$ disjoint Petersen graphs. I put this conjecture with some trepidation. It may be complete nonsense.

* * * * * * * * * *

R. QUACKENBUSH: Given a class of finite structures, how difficult is it to test for isomorphism between two given $n$-element members of the class? The conjecture I will make is that under suitable circumstances this only depends on the number of $n$-element structures in the class. If $L$ is a class of finite structures, let $f_L(n)$ denote the number of $n$-element structures in $L$. If $\gamma$ is an isomorphism testing algorithm for the class $L$, we let $m_L(\gamma,n)$ denote the maximum number of steps needed by $\gamma$ to determine whether two $n$-element structures in $L$ are isomorphic.

5.11 *Conjecture. Let $J$ and $K$ be classes of structures of finite type, where each class is definable by some first order sentence, and suppose there is a polynomial $q(x)$ such that $f_J(n) \leq q(f_K(n))$ for all $n$. Then there is a polynomial $p(x)$ such that for every isomorphism testing algorithm $\alpha$ for $K$ there is an isomorphism testing algorithm $\beta$ for $J$ satisfying $m_J(\beta,n) \leq p(m_K(\alpha,n))$ for large $n$.*

M. POUZET: Can you at least prove the conjecture for classes which are axiomatized by universal sentences? This would be a test, since it is possible to characterize such classes whose number of isomorphism types is polynomially bounded.

[Pouzet has shown that, if the graph isomorphism problem is not polynomial, then the conjecture is false.]

* * * * * * * * * * *

I. RIVAL: This comes from work of B. Sands and myself. Ramsey's theorem says that $K_\omega$ or its complement are embeddable in any infinite graph $G$. We were interested in *how* the embedding can occur. Motivated by this, we showed there was an infinite $H \subset G$ so that for every $x \in G$ either (1) $x$ is unrelated to all vertices in $H$ or (2) $x$ is related to exactly one vertex in $H$ or (3) $x$ is related to infinitely many vertices in $H$. The general question concerns what this says about ordered sets, in particular via the comparability graph. We proved that if $P$ is an ordered set of finite width then there is an infinite chain $H$ in $P$ with every element of $P$ comparable to no elements of $H$ or to infinitely many elements of $H$.

*5.12* *If the length of $P$ is finite, must there be an infinite antichain $H$ such that every element of $P$ is comparable to no vertices of $H$, one vertex of $H$, or infinitely many vertices of $H$?*

Several people, including Eric Milner, have observed that the answer is yes if $P$ is bipartite.

A. HOFFMAN: Going back to your original graph theorem, is there a finite version of it?

B. SANDS: No. Take the graph with vertices $\{a_i : 1 \le i \le n\}$ and $\{b_i : 1 \le i \le n\}$, join $a_i$ to $b_j$ if $i < j$, and then connect all the $a_i$'s together as well. Now if you look at it you'll find that no matter which subset you pick you can always find a vertex of the graph which is adjacent to the wrong number of elements of the subset.

* * * * * * * * * * *

B. SANDS: This is sort of a favorite of mine at the moment.

*5.13* *Is there a number $r > 2$ so that every finite poset $P$ contains an element which is in at least $\frac{1}{r}$ and at most $1 - \frac{1}{r}$ of the order ideals of $P$?*

Eric Rosenthal communicated this form of the problem to me. It's equivalent to the following problem, which is the way I first heard of it.

5.13' *Is there a number $r > 2$ so that every finite distributive lattice $L$ contains a prime ideal $I$ satisfying*

$$\frac{1}{r} \leq \frac{|I|}{|L|} \leq 1 - \frac{1}{r} ?$$

This problem is due to C. Colbourn and I. Rival, who wanted to use it to work on the complexity of the isomorphism problem for finite distributive lattices. I can prove that, for $P$ of length one, $r = 8.807$ works.

P. ERDÖS: What about length two?

SANDS: I don't know anything about length two.

ERDÖS: You can't formulate it graph-theoretically?

SANDS: Not naturally. The result for length one of course can be put in graph-theoretic language, because a poset of length one is the same as a bipartite graph. However the problem in general seems to be not a graph theory problem but a genuine poset problem.

D. WEST: Incidentally, Rosenthal's motivation, and Mike Saks', who has been looking at the same problem, is to study the complexity of locating a particular order ideal of a poset by searching for its members.

R. JAMISON-WALDNER: Just a comment. If you replace "order ideal" by "order convex set", the answer to the problem is no.

P. WINKLER: Do you have a lower bound for the length one case?

SANDS: Yes, you can't choose $r$ to be less than 4.38.

[The bounds on $r$ for the length one case have since been improved by Erdös.]

* * * * * * * * * * *

M. POUZET: This problem is related to the reconstruction problem. Suppose that $C$ and $C'$ are two linear orders (i.e., permutations) of an $n$-element set $E$. For each $x \in E$ consider the (unique) isomorphism $\varphi_x : C|_{E-\{x\}} \to C'|_{E-\{x\}}$. The general problem is whether the (unique) isomorphism $\varphi : C \to C'$ can be reconstructed from the family $\Phi = \{\varphi_x : x \in X\}$. In particular:

5.14 *For sufficiently large $n$ is there for each pair $\{x,y\}$ a sequence $\varphi_1,\dots,\varphi_{2k+1}$ from $\Phi$ such that*

$$\varphi_{2k+1}\circ\varphi_{2k}^{-1}\circ\dots\circ\varphi_2^{-1}\circ\varphi_1\Big|\{x,y\} = \varphi\Big|\{x,y\}\ .$$

This is false for $n \leq 4$.

* * * * * * * * * *

N. ROBERTSON: This question is related to the conjecture that the class of (finite) graphs is well quasiordered under the contraction-deletion ordering of minors, where $A$ is a *minor* of $B$ if $A$ can be obtained from $B$ by contraction and deletion of edges. A particular case of the conjecture arises in graph connectivity. We say that a graph is $(h,k)$-*connected* when for any edge disjoint decomposition $G = G_1 \cup G_2$ there are at least $h$ vertices in common if both $G_1$ and $G_2$ contain circuits, and at least $k$ vertices in common if both are nonempty. To avoid degenerate situations it is usual to add that the cycle rank $r_p(G) \geq h$ and the tree number $r_b(G) \geq k$ (this forces the circuit girth $\geq h$ and edge connectivity $\geq k$).

5.15 *Conjecture. For each $h$ and $k$, there are only finitely many minimal (in the minor partial ordering) $(h,k)$-connected graphs.*

* * * * * * * * * *

PROBLEM SESSION 6

## SOCIAL SCIENCE AND OPERATIONS RESEARCH

Chairman: J.K. Lenstra

P. HAMMER: Consider the problem of maximizing

$$c_1x_1 + \ldots + c_nx_n \qquad (1)$$

where $c_1 > c_2 > \ldots > c_n > 0$ are given reals, and where the 0-1 variables $x_j$ are subject to the constraint

$$f(x_1,\ldots,x_n) < 0 \qquad (2)$$

where $f$ is a nondecreasing Boolean function (i.e. for any $i$,

$$f(x_1,\ldots,x_{i-1},1,x_{i+1},\ldots,x_n) \geq f(x_1,\ldots,x_{i-1},0,x_{i+1},\ldots,x_n)) .$$

For any nondecreasing Boolean function $g(x_1,\ldots,x_n)$, define a preorder relation between the variables, by putting $x_i \succeq_g x_j$ iff

$$g(x_1,\ldots,x_{i-1},1,x_{i+1},\ldots,x_{j-1},0,x_{j+1},\ldots,x_n) \geq g(x_1,\ldots,x_{i-1},0,x_{i+1},\ldots,x_{j-1},1,x_{j+1},\ldots,x_n) .$$

Consider the family $F$ of all nondecreasing Boolean functions $\varphi$ satisfying

$$\varphi(x_1,\ldots,x_n) \leq f(x_1,\ldots,x_n)$$

$$x_1 \succeq_\varphi x_2 \succeq_\varphi \ldots \succeq_\varphi x_n .$$

This family has a unique maximal element $\varphi^*$. It has been shown that the problems of maximizing (1) under the constraint (2) or under

$$\varphi^*(x_1,\ldots,x_n) = 0 \qquad (2')$$

have the same optimal solution. Since maximizing (1) under (2') can be done in polynomial time, it would be valuable to:

6.1 *Find classes of problems for which $\varphi^*$ can be found in polynomial time.*

* * * * * * * * * * *

T. TROTTER: The following problems are related to topics discussed by C. Flament in his lecture. They involve the *k-dimension* of a partially ordered set $P$, i.e., the smallest number of $k$-element chains whose product contains a copy of $P$.

6.2 *Find approximate algorithms for determining the k-dimension of a partial order, and analyze their efficiency (for example, greedy algorithms).*

Fix $k$. A poset $P$ is *irreducible* if the $k$-dimension of $P - \{x\}$ is less than the $k$-dimension of $P$ for every $x \in P$.

6.3 *For each k,t determine the number of irreducible posets which have k-dimension equal to t. For fixed k, is this number linear or polynomial in t?*

For each $k$ and $t$, the number of such posets is easily seen to be finite.

6.4 *Find general embedding theorems. For instance, for k fixed, is every finite poset embeddable in an irreducible of larger k-dimension?*

* * * * * * * * * *

J.-P. BARTHÉLEMY: Let $\mathcal{R}$ be a set of binary relations on an $n$-element set $X$, and let $\delta$ denote the symmetric difference distance on $\mathcal{R}$ (i.e., $\delta(r,s) = |r \Delta s|$ for $r,s \in \mathcal{R}$). If $\mathcal{R}$ is the set of linear orders on $X$ and $r_0 \in \mathcal{R}$, we know that asymptotically the distribution of $\delta(-,r_0)$ is normal. More generally, given a set $\mathcal{R}$ of binary relations on a set $X$ (for example, all weak orders, semiorders, interval orders, or partial orders on $X$), and given a relation $r_0 \in \mathcal{R}$ and a positive integer $k$, we can ask:

6.5 *What is the number of relations r in* $\mathcal{R}$ *so that* $\delta(r,r_0) = k$?

Let $\pi = (L_1,\ldots,L_v)$ be a family of linear orders and let

$$F(\pi) = \min_R \sum_{i=1}^{v} \delta(L_i,R).$$

Any $R$ giving this minimum is called a *median* of $\pi$. [BFM, §3.4.1]

6.6 *Find conditions on* $\pi$ *assuring the uniqueness of its median.*

A quite trivial condition is that the Condorcet relation [BFM, §3.2] associated with $\pi$ is linear and unique.

Let $R$ be a set of binary relations on an $n$-element set $X$. For $\pi = (r_1,\ldots,r_v) \in R^v$ and $s \in R$ let $\Delta(s,\pi) = \sum_{i=1}^{v} \delta(r_i,s)$. Define

$$F(R,v,n) = \operatorname*{Max}_{\pi \in R^v} \operatorname*{Min}_{s \in R} \Delta(\pi,s)$$

(the *maximum dispersal* of profiles $\pi$). It is known that for $R = L$ (linear orders),

$$F(L,2p,n) = p\binom{n}{2}$$

and

$$p\binom{n}{2} \le F(L,2p+1,n) \le \binom{n}{2}\frac{2p+1}{2} - \frac{1}{2}\left[\frac{n}{2}\right].$$

6.7 *Find $F(R,v,n)$ if $R$ is the set of weak orders (semiorders, interval orders, partial orders) on $X$.*

My final question concerns the search for an axiomatic characterization of median orders. This was done for linear orders by Young and Levenglick in 1978.

6.8 *Extend the Young-Levenglick Theorem to other kinds of order relations.*

A related question would be:

6.9 *Find an order-theoretic proof of the Young-Levenglick Theorem.*

* * * * * * * * * *

B. MONJARDET: Order the partial orders on a set $X$ by containment. Given a fixed linear order on $X$, the subset of all partial orders contained in it is a *meet-distributive* lattice (i.e., every element is a unique irredundant join of join-irreducibles). The entire set of partial orders on $X$ forms a meet-distributive semilattice.

6.10 *Characterize the lattice (or semilattice) of all partial orders on $X$ among meet-distributive lattices (or semilattices).*

Given a poset $P$, one can define a geodesic path metric $d$ on the covering graph $G(P)$ of $P$ [BFM, §2.2.1]. Then given $x_1,\ldots,x_v$ in $P$, a *median* of $(x_1,\ldots,x_v)$ is any element $x$ minimizing $\sum_{i=1}^{v} d(x_i,x)$.

6.11 *What can be said about uniqueness of medians in meet-distributive lattices?*

In distributive lattices, the median is unique when $v$ is odd, but not always when $v$ is even.

* * * * * * * * * *

K. BOGART: Given a family of $k$ posets, we wish to compute their median (with respect to the symmetric difference distance) as a poset. Call this the *median problem.*

6.12 *Is the median problem NP-complete?*

A poset $P = \{x_1, x_2, \ldots, x_n\}$ can be considered as an $n \times n$ incidence matrix (or an $n^2$-vector) $M$, where $M(i,j) = 1$ if $x_i \leq x_j$ and 0 otherwise. Two posets are *adjacent* if their incidence matrices differ in exactly one entry.

Given posets $P_1, \ldots, P_r$, here are two heuristic algorithms for computing their median:

(a) Starting at the Borda order [Bogart, §3], move to the adjacent partial (weak) order which in Euclidean distance on incidence matrices is closest to the average of the $P_i$'s.

(b) For each $P_i$, follow a chain of orders each of which is adjacent to the previous one and is closer to the average of the incidence matrices.

6.13 *How well do these algorithms do?*

The next question involves just noticeable differences (*jnd*'s) [Bogart, §3].

6.14 *Which ordered sets arise if you allow jnd's to vary from pair to pair of observations?*

Finally, suppose you have a point in the unit cube (of incidence matrices) and a neighborhood $N$ around it.

6.15 *What is the distribution of families of $n$ posets whose average lies in $N$? Are they normally distributed?*

* * * * * * * * * *

R. WILLE: This is a problem on similarity measures, and comes from work of S. Geist and myself. Let $M$ be a set of attributes and define an order on $P(M)^2$ by: if $A,B,C,D \subseteq M$ then $(A,B) \leq (C,D)$ if $A \cap B \subseteq C \cap D$, $A \cap \bar{B} \supseteq C \cap \bar{D}$, $\bar{A} \cap B \supseteq \bar{C} \cap \bar{D}$, and $\bar{A} \cap \bar{B} \subseteq \bar{C} \cap \bar{D}$. A similarity measure $\mu : P(M)^2 \to \mathbb{R}_0^+$ on $P(M)^2$ is *additive* if it is additive on each Boolean subinterval $[(A,\bar{A}),(A,A)]$. A subset A of $P(M)^2$ is *independent* if every order-preserving map $\alpha : A \to \mathbb{R}_0^+$ can be extended to an additive similarity measure on $P(M)^2$. Then we have the result that the independent subsets of $P(M)^2$ form a matroid of rank $\frac{(n+1)(n+2)}{2} - 1$, where $|M| = n$.

*6.16* *Represent this matroid in the real vector space of dimension* $\frac{(n+1)(n+2)}{2} - 1$.

* * * * * * * * * *

PROBLEM SESSION 7

## RECURSION AND GAME THEORY

Chairman: R. Woodrow

J. ROSENSTEIN: A linear ordering $\langle \mathbb{N}, R\rangle$ of the natural numbers is *recursive* if $R$ is a recursive binary relation. A subset $B \subseteq \mathbb{N}$ is said to be *arithmetical* if there is a natural number $k$ and a recursive predicate $P(x_1, x_2, \ldots, x_k, n)$ such that

$$n \in B \leftrightarrow (Q_1 x_1)(Q_2 x_2)\ldots(Q_k x_k)P(x_1, \ldots, x_k, n)$$

where each $Q_i$ is $\exists$ or $\forall$. More specifically, $B$ is a $\Sigma_k$-predicate if (in the representation above) the quantifiers alternate with $Q_1$ being $\exists$, and $B$ is $\pi_k$ if the quantifiers alternate beginning with $\forall$.

The Kleene-Church Arithmetical Hierarchy, outlined above, provides a measure of the complexity of a set. Recursive sets, which are $\Sigma_0$ (or $\Pi_0$) sets, are the simplest sets in this hierarchy.

In my book *Linear Orderings* (Academic Press, 1981) I construct a recursive linear ordering which is not scattered yet contains no recursive (or even $\Sigma_1$) subset of order type $\eta$ (of the rational numbers). Subsequently Lerman and I constructed a recursive linear ordering of type $\zeta\cdot\eta$ (where $\zeta$ is the order type of the integers) which has no $\Sigma_2$ subset of type $\eta$. On the other hand every recursive linear ordering of type $\zeta\cdot\eta$ has a $\Pi_2$ subset of order type $\eta$.

7.1 *Conjecture. There is a recursive linear ordering of type $\zeta^{\omega}\cdot\eta$ which contains no arithmetical subset of order type $\eta$.*

In making this conjecture, I am assuming that the following cannot be improved upon: every recursive linear order of type $\zeta^{n}\cdot\omega$ has a $\Pi_{2n}$ subset of order type $\eta$.

An order type $\tau$ is recursive if there is a recursive linear ordering of type $\tau$.

7.2 *Is there some way of characterizing the recursive order types?*

Results obtained by Lerman and myself (see Linear Orderings) provide such characterizations for various classes of order types, but the nature of those results do not give much hope for a positive answer to the question above.

K. LEEB: Are orderings known which belong to the same classical type but not the same recursive type?

ROSENSTEIN: I'm sure there are.

G. MCNULTY: $\omega$.

ROSENSTEIN: Sure. In fact there are recursive orderings of (classical) type $\omega$ which are not recursively isomorphic.

* * * * * * * * * * *

G. KALMBACH: These are problems from my book on orthomodular lattices.

7.3 *Is the word problem for free orthomodular lattices solvable?*

7.4 *Is the word problem for finitely presented orthomodular lattices unsolvable?*

Here is a suggestion. To an orthomodular lattice $L$ associate a semigroup $S(L)$, via a Galois connection (with a natural involution on the orthomodular lattice). Try to embed a semi-group with unsolvable word problem in $S(L)$. The embedding of $L$ in $S(L)$ is given by

$$a \mapsto \chi_a(x) = a' \vee (a \wedge x') .$$

7.5 *Is the variety of all orthomodular lattices generated by its finite members (or its members of finite height)?*

K. Baker showed that the subvariety generated by those of finite height is finitely based.

* * * * * * * * * * 

J. JONES: Here is an old problem which I believe is still unsolved. Consider all functions constructed from 0,1,2,... by $+$, $\cdot$, and exponentiation. For such functions $f$ and $g$, put $f \prec g$ iff $(\exists\, x)(\forall\, z>x)(f(z) < g(z))$.

7.6 *Is $\prec$ a recursive order?*

Ehrenfeucht proved that $\prec$ is a well ordering. H. Levitz studied the order types of certain subsets.

M. Rabin [1957] considered two person games of the type:

player I picks $x_1$ (a real number)

player II picks $x_2$

player I picks $x_3$

and so on.

Player I wins iff $f(x_1,\ldots,x_n) = 0$, where $f$ is a fixed recursive function. By an old theorem (on determinancy) of von Neumann and Zermelo, there must be a winning strategy for I or II. But Rabin constructed a recursive function $f(x_1,x_2,x_3)$ such that neither player had a computable winning strategy.

Suppose $f(x_1,\ldots,x_n)$ is a polynomial function. I proved (Some undecidable determined games, *Int. Journal of Game Theory*, to appear) that

(i) If $n \leq 4$ there must exist a computable winning strategy for player I or for player II.

(ii) If $n \geq 6$ there need not exist a computable winning strategy for either player.

7.7 *What about games of length $n = 5$?*

* * * * * * * * * * *

T. TROTTER: Here are some problems everyone will understand!

Hal Kierstead has recently shown that every recursive poset of width $n$ can be recursively partitioned into $(5^n-1)/4$ chains. What is the situation for comparability graphs of posets? Let the *independence number* of a graph be the size of the largest independent set.

7.8 *Does there exist a function $f:\mathbb{N} \to \mathbb{N}$ so that every recursive comparability graph $G$ of independence number $n$ can be recursively partitioned into $f(n)$ complete subgraphs?*

This question can be put in terms of a two-person game in which player I and player II alternate, player I enumerating a vertex of a comparability graph $G$ of independence number $n$ (together with its edges to the subgraph already enumerated), and player II assigning the vertex to one of $f(n)$ complete subgraphs.

Player II wins just in case the game proceeds through infinitely many stages, until $G$ is enumerated. The question asks for a recursive winning strategy for player II.

Call a graph $G$ a *Chvatal graph* if there is an orientation of the edges of $G$ so that the graph

is not an induced subgraph of $G$. Chvatal has proved that every such graph is perfect. Of course, every comparability graph is a Chvatal graph (by transitivity of partial order).

7.9 *Repeat problem 7.8 for the class of Chvatal graphs oriented as above.*

The chromatic number of any forest is two. However it is known that player I has a winning strategy for the corresponding game in which player II must color the vertices (of the forest constructed by player I) using any fixed finite number of colors. How fast does player II lose? Dually,

7.10 *What is the minimum number of colors required by II for graphs of $n$ vertices?*

* * * * * * * * * *

T. SCHNEIDER: The first question I think is intractable, the second easy.

7.11 *Classify all countably infinite (elementarily) homogeneous partial orders.*

Here the mappings are *elementary* mappings; see the paper of Schmerl (*Algebra Universalis*, 1979) for the characterization of homogeneous partial orders where mappings are simply embeddings.

7.12 *Classify all countably infinite partial orders whose first order theories are stable.*

Refer to Shelah's *Classification Theory* (North Holland).

* * * * * * * * * *

H. KIERSTEAD: Let $G = (V,E)$ be a graph. For $x \in V$ let $\delta(x)$ denote the degree of $x$, and let $\Delta(G)$ denote the maximum degree of a vertex in $G$. For $W \subset V$, $G|_W$ is the subgraph of $G$ induced by $W$. Lovasz has proved that if $n_1, n_2 \in \mathbb{N}$ with $n_1 + n_2 - 1 = \Delta(G)$ then $V$ can be partitioned into $V_1, V_2$ with $\Delta(G|_{V_1}) \le n_1$ and $\Delta(G|_{V_2}) \le n_2$.

Suppose that $\Delta(G)$ is finite and that $G$ is *highly recursive*, i.e., $G$ is recursive and $\delta(x)$ is a recursive function.

7.13 *If $G$ is highly recursive can the conclusion of the theorem of Lovasz be strengthened to assert that $V_1$ and $V_2$ are also recursive?*

* * * * * * * * * *

G. MCNULTY: Let $K$ be a class of similar structures and let $L$ be a class of similar structures on a larger type. The *K-L expansion game* is played between two players, the $K$-player and the $L$-player, the following way: At each turn the $K$-player, who has constructed a finite structure $A$ in $K$, adds one point to the structure to obtain $A' \in K$. In response, the $L$-player who has already expanded $A$ to $B \in K$, extends $B$ to $B' \in L$ so that $B'$ expands $A'$. The $L$-player wins if the play goes through infinitely many stages. The $K$-player wins otherwise.

7.14 *Find conditions on $K$ and on $L$ so that if the $K$-player has a winning strategy for the $K$-$L$ expansion game then there is a recursive structure in $K$ with no recursive expansion in $L$.*

For example, this is the case if $K$ = all ordered sets of width $\omega$, and $L$ = all ordered sets of width $\omega$, each furnished with a decomposition into $4(\omega-1)$ chains (Kierstead).

* * * * * * * * * *

M. POUZET: My problems are about axiomatizability and decidability for some theories of ordered sets.

A theorem of Schmerl asserts that $\aleph_0$-categorical theories of ordered sets of width $n$ are decidable. There are few examples of such theories. I proved (*J. Combinatorial Theory*, 1978) that no such $\aleph_0$-categorical theory can have the same universal part as the theory of ordered sets of width at most two.

7.15 *For $n \geq 3$, is there a countable ordered set of width $n$ which contains (up to isomorphism) every countable ordered set of width $\leq n$?*

In my 1978 paper, I proved that the answer is yes for $n = 2$. Note that the amalgamation property for posets of width $\leq n$ fails for each $n \geq 2$.

In many instances we have a class of structures $C$ defined by a set of universal sentences, and using a quantifier free formula $\theta$

we define a new class $\mathcal{D} = \{\langle A, \theta_A \rangle : \langle A, R_\alpha \rangle_{\alpha \in I} \in C\}$. As an example, starting with the class $C$ of ordered sets, and using the formula $F(x,y) = x \leq y$ or $y \leq x$ we obtain the class $\mathcal{D}$ of comparability graphs. While $C$ is axiomatizable by a finite number of universal sentences, $\mathcal{D}$ is not, since one must kill all odd cycles.

*7.16* *Characterize the classes C such that any class obtained from C by some quantifier free formula is finitely axiomatizable.*

The first non-trivial example is the class $C$ of chains. This result is due to C. Frasnay (1965) and uses a "recollement lemma" about permutation groups that was rediscovered by Hodges, Lachlan, and Shelah (1977).

*7.17* *Obtain the theorem of Frasnay by an algebraic argument involving linear algebra and an incidence matrix.*

I gave a general condition in 1972 containing the Frasnay result. All known examples of theories satisfying the condition are decidable. Is this true in general? More precisely,

*7.18* *Let $T$ be a universal theory in a language without function symbols and with finitely many predicate symbols. Assume that for any integer $m$ the class $C_m(T) = \{(R, U_1, \ldots, U_m) : R$ is a finite model of $T$, $U_1, \ldots, U_m$ are unary relations$\}$ has no infinite antichain. Then is $T$ decidable?*

Examples of such $T$ are the theory of chains (Higman) and the theory of ordered trees (Kruskal). I have proved that any such $T$ is finitely axiomatizable and in addition there is an $\aleph_0$-categorical theory with the same universal part as $T$. To obtain finiteness results, concepts from the study of infinite ordered sets, particularly chain conditions and the notion of well quasi-order, play a dominant rôle.

* * * * * * * * * *

J. LARSON: It is a theorem of Galvin and McKenzie, and Bonnet and Pouzet (1969) that every scattered poset can be extended to a scattered linear order. Is there a recursive version of this theorem? Let us define the *constructive scattered linear order types* to be those obtained by closure of $\{0,1\}$ under recursive well-ordered sums and recursive converse well-ordered sums.

7.19 *Can we put a condition on a poset to guarantee that it has an extension to a constructive linear order type?*

* * * * * * * * * * *

J. ROSENSTEIN (revisited): I'll just state one more problem.

7.20 *Let $T$ be a complete theory of linear orderings which has a recursive model and a prime model. Must $T$ have a recursive prime model?*

* * * * * * * * * * *

PROBLEM SESSION 8

## ORDER-PRESERVING MAPS

Chairman: I. Rival

D. DUFFUS: Given ordered sets $A$ and $B$, $A^B$ denotes the set of all order-preserving maps from $B$ to $A$, ordered pointwise (this idea was introduced by Birkhoff in the thirties). The notation "$A^B$" suggests certain natural cancellation properties. For instance, it is known that if $A^B \cong A^C$ for finite ordered sets $A$, $B$, and $C$ where $A$ is not a totally unordered set, then $B \cong C$. If $A$ is totally unordered, then $A^B \cong A^C$ only implies that $B$ and $C$ have the same number of connected components.

8.1 *Show that* $A^C \cong B^C \Rightarrow A \cong B$ *for all finite ordered sets* $A$, $B$, *and* $C$.

This is known if both $A$ and $B$ are bounded, or if $C$ is a chain. Birkhoff observed that it cannot hold for arbitrary (infinite) ordered sets.

Another question is not really about cancellation, but instead asks whether the ordered set of order-preserving maps from an ordered set to itself determines the ordered set.

8.2 *Show that* $A^A \cong B^B \Rightarrow A \cong B$ *for all finite ordered sets* $A$ *and* $B$.

Wille and I proved this statement in the finite *connected* case, and it should be true arbitrarily in the finite case.

* * * * * * * * * *

H. BAUER: My problem is a special form of one of the previous problems.

8.3 *Show that* $U^{\bowtie} \cong V^{\bowtie}$ *implies* $U \cong V$ .

The poset $\bowtie$ is a "bad" poset for cancellation and other problems, so I think that if we are able to solve this special situation then we will be able to solve the whole cancellation problem.

* * * * * * * * * *

K. KEIMEL: For connected partially ordered sets $V, W, Y, Z$, one has the following refinement property of Hashimoto: $V \times W \cong Y \times Z \Rightarrow$

$\exists\, A,B,C,D$ such that $V \cong A \times B$, $W \cong C \times D$, $Y \cong A \times C$, and $Z \cong B \times D$.

8.4 *To what extent does Hashimoto's refinement property remain true for Priestley spaces?*

By a *Priestley space* we mean a compact Hausdorff space together with a partial order which has a closed graph in the product and which has the property: given any two distinct elements one can find a clopen lower end (i.e., order ideal) which separates the two elements. We would like some *topological* connnectivity condition imposed rather than the poset connectivity condition above. One might require that the Priestley space cannot be partitioned into two disjoint (with respect to order) clopen sets, but this restriction is probably too weak.

* * * * * * * * * *

J. LAWSON: A poset can be topologized by letting the open sets be all upper ends of the poset. These spaces are sometimes called Alexandroff discrete spaces. It turns out that for posets $P$ and $Q$, a map $f:P \to Q$ will be order-preserving iff it is continuous when $P$ and $Q$ are considered as topological spaces. Now you can ask such questions as:

8.5 *Is there an order-theoretic characterization for two order-preserving maps $f,g:P \to Q$ to be homotopic?*

There is a nice characterization if both $P$ and $Q$ are finite: $f$ and $g$ are homotopic if and only if there is a fence from $f$ to $g$ in $Q^P$, that is, if and only if $f$ and $g$ are in the same connected component of $Q^P$. A related question is:

8.6 *Find conditions on a topology on $Q^P$ such that two functions from $P$ to $Q$ are homotopic if and only if they are in the same path component of $Q^P$ with respect to this topology.*

I'm not sure how much is known about the next problem. Suppose $P$ and $Q$ are finite (topologized) posets and $f:P \to Q$ is order-preserving (i.e., continuous). Consider the simplicial complexes associated with $P$ and $Q$, where the chains are the simplices, and their realizations $|P|$ and $|Q|$. Look at the realization $|f|:|P| \to |Q|$ of $f$ as a continuous function.

8.7 *Is there an algorithm which, given $f$, determines whether $|f|:|P| \to |Q|$ is null-homotopic?*

* * * * * * * * * *

A. BJÖRNER: A poset $P$ has the *fixed point property* if for every order-preserving map $f:P \to P$ there is $x \in P$ such that $x = f(x)$.

*8.8* *Conjecture. Let $P$ be a poset of finite length having an element $s$ such that*

*(i) for all $x \in P$, $s \vee x$ or $s \wedge x$ exists in $P$,*

*(ii) if $x < y$ and $s \wedge x$ does not exist but $s \wedge y$ does exist, then $(s \wedge y) \vee x$ exists.*

*Then $P$ has the fixed point property.*

For example, if $L$ is a lattice of finite length and $s$ is any element of $L - \{0,1\}$, then this conjecture would imply that $L - (\{0,1\} \cup \{\text{all complements of } s\})$ has the fixed point property. In particular if $L$ is a lattice of finite length which contains a noncomplemented element, then this conjecture implies that $L - \{0,1\}$ has the fixed point property. The conjecture has been proven for *finite* $P$ by J. Walker and myself. Our proof is topological, however; there should be a direct proof, and I'm convinced that such a proof would work also in the finite length case. If one of the conditions in (i) holds uniformly for *all* $x$, then there is a non-topological proof, in fact $P$ turns out to be an infinite dismantlable poset.

I. RIVAL: Do you have an example of a non-dismantlable poset satisfying the conditions in your conjecture?

BJÖRNER: Yes, you can take the lattice of subspaces of a four-dimensional vector space, and take an element $s$ in the middle level; then all the complements of $s$ will be on that same level. So when you remove 0,1, and all complements of $s$, you're left with a non-dismantlable poset.

* * * * * * * * * *

C. FLAMENT: I mentioned in my lecture on Sunday the following result: If $L$ is a complete lattice and $f:L \to L$ is order-preserving, then there exists a closure operator $\varphi$ (i.e., a retraction $\varphi:L \to L$ satisfying $x \leq \varphi(x)$ for all $x \in L$) and an anticlosure operator $\omega$ (i.e., a retraction $\omega$ satisfying $x \geq \omega(x)$) such that

$$\{x \in L \mid x = f(x)\} = \varphi\circ\omega(L) = \omega\circ\varphi(L) = \varphi(L) \cap \omega(L).$$

It is then possible to prove that a nonempty subset $R$ of a complete lattice $L$ is a retract of $L$ if and only if there exists a closure operator $\varphi$ and an anticlosure operator $\omega$ on $L$ such that they commute, and such that

$$R = \varphi\circ\omega(L) = \omega\circ\varphi(L) = \varphi(L) \cap \omega(L) .$$

8.9 *Which other posets have this characterization of their retracts?*

* * * * * * * * * *

H. CRAPO: The first two problems, due to Colin Davidson, arose in information retrieval and indexing of documents. They concern the abridgement of a thesaurus, in which the terms of a trade or area of knowledge form the elements of an ordered set, the order being determined by relations such as "is a type of" and "is a part of". For indexing of documents, a rather complete thesaurus is required; for most routine enquiries, a much reduced thesaurus will suffice. The idea is to find an algorithm which will produce a minimal retract of the thesaurus, subject to its containing certain key terms.

8.10 *Given a subset $A$ of a finite ordered set $C$, are all minimal supersets of $A$ which are retracts of $C$ isomorphic?*

8.11 *Find an algorithm for producing the minimal retraction of $C$ fixing $A$.*

The next problem is unrelated to the previous problems. A *connection* between ordered sets $A$ and $B$ is a pair of order-preserving maps $p:A \to B$ and $q:B \to A$ satisfying $pqp = p$ and $qpq = q$. We write $A \underset{p}{\overset{q}{\leftrightarrows}} B$ to denote a connection. Suppose $A \leftrightarrows B$ and $B \leftrightarrows C$ are connections; they define two order-preserving maps $f:B \to A \to B$ and $g:B \to C \to B$ of $B$ to itself. Two different situations are known where the composition $A \leftrightarrows C$ is a connection, namely

(i) where one of $f$ and $g$ is a closure, and the other is a coclosure;

(ii) where $f$ and $g$ commute.

8.12 *Find necessary and sufficient conditions for the composition of connections to be a connection.*

The background to this question is in my paper "Ordered sets, retracts, and connections" in the *Journal of Pure and Applied Algebra*, this year.

B. DAVEY: A remark on your first question: if $C$ is a lattice, then the answer is yes, because any minimal retract will be in fact the Dedekind-MacNeille completion of $A$.

CRAPO: Exactly. In fact Banaschewski and Bruns characterized the MacNeille completion in that way.

[Problem 8.10 has been answered in the affirmative by R. Wille. Problem 8.11 is still open, but the following "greedy" algorithm suggested during the session does *not* work: namely, choose a linear ordering $\omega$ of $C$ and send each element $x$ of $C$ to the $\omega$-least element of $C$ which is a possible image of $x$ under a retraction of $C$ fixing $A$.]

* * * * * * * * * *

M. POUZET: Let $T$ be a lattice and let $\mathrm{Sub}(T)$ denote the set of nonempty sublattices of $T$. For $S, S' \in \mathrm{Sub}(T)$ we put $S \le S'$ if for all $x \in S$, $x' \in S'$ we have $x \wedge x' \in S$ and $x \vee x' \in S'$. Note that "$\le$" is not necessarily transitive.

8.13 *Find some useful characterization of those lattices $T$ such that there is a map $\varphi: \mathrm{Sub}(T) \to T$ satisfying*

*(i) for all $S \le S'$ in $\mathrm{Sub}(T)$, $\varphi(S) \le \varphi(S')$ in $T$*

*(ii) $\varphi(S) \in S$ for all $S \in \mathrm{Sub}(T)$.*

Examples of such lattices are chains (a result of Duffus and Rival) and, more generally, finite products of chains. If $T$ is a lattice with this property, then any homomorphic image of a sublattice $S$ of $T$ is a retract of $S$; in fact $S$ will have the so-called "selection property", one of the two properties we need to prove that certain lattices are retracts of products of chains. Using this trick, I was able a few days ago to show that distributive lattices of finite dimension are retracts of products of chains. My other question is to find a class of lattices, more general than finite products of chains, with the above property.

8.14 *Let $T$ be a Boolean algebra generated by a chain. Does $T$ satisfy the property in the above question?*

* * * * * * * * * *

R. WILLE: This problem comes from joint work with my student Peter Nevermann, and concerns the classification scheme for ordered sets developed by D. Duffus and I. Rival. If $S$ is a subset of an ordered set $P$, $S_*(S^*)$ will denote the set of all elements of $P$ which are less than or equal to (greater than or equal to) all elements of $S$. Let $P$ be an ordered set, and let $\Xi(P)$ denote the set of all pairs $(D,U)$ such that $D$ is a down-set and $U$ is an up-set of $P$, $D \subseteq U_*$, and $D^* \cap U_* \ne \emptyset$. For $(D_1,U_1),(D_2,U_2) \in \Xi(P)$, define $(D_1,U_1) \le (D_2,U_2)$ iff $D_1 \subseteq D_2$ and $U_1 \supseteq U_2$; then $\le$ is an order on $\Xi(P)$. $P$ has the *strong selection* *property* (*SSP*) if there is an order-preserving map $\sigma:\Xi(P) \to P$ with $\sigma(D,U) \in D^* \cap U_*$ for all $(D,U) \in \Xi(P)$. This property together with the so-called $K$-gap property characterize exactly those ordered sets in the variety generated by any class $K$ of ordered sets with the *SSP*. Previously Ivan Rival and I had used a weaker form, the *selection property*, to handle the case when $K$ is the class of all chains.

*8.15 Which (finite) ordered sets have the SSP?*

For example, every complete (meet- or join-) semilattice, and every bounded ordered set of length $\leq 3$ has the *SSP*. On the other hand, a four-element fence does not have the *SSP*.

* * * * * * * * * *

R. QUACKENBUSH: If $P$ is a finite ordered set, $\underset{\sim}{4}^P$ denotes as usual the set of order-preserving maps of $P$ into the four-element chain.

*8.16 What is* $\{|\underset{\sim}{4}^P|: P$ *is a finite ordered set*$\}$?

As I mentioned in my lecture, there is a nice classification for these cardinalities when the posets $P$ are restricted to lattices of length two, but not in general. If you manage to solve this one, I have an ascending chain of increasingly difficult problems for you to solve.

B. JÓNSSON: Since $\underset{\sim}{4} = \underset{\sim}{2}^3$ and $\{\underset{\sim}{2}^P | P \text{ finite}\}$ is all finite distributive lattices, $\{\underset{\sim}{4}^P | P \text{ finite}\}$ is just all third powers of finite distributive lattices. Is it easier to do it that way?

QUACKENBUSH: Well, for the three-element chain you get an even easier description, namely $(\underset{\sim}{2}^P)^2$, which is just the order relation of a finite distributive lattice; a very nice description which apparently is totally useless.

* * * * * * * * * *

P. EDELMAN: Let $P$ be a finite poset whose elements are labelled with the integers $1,2,\ldots,|P|$ in a way compatible with the order relation on $P$. Define the order polynomial $\Omega(P;n)$ of $P$ to be the number of order-preserving maps from $P$ into $\underset{\sim}{n}$. It's known that

$$\sum_{n=0}^{\infty} \Omega(P;n)x^n = \frac{q(x)}{(1-x)^{|P|+1}}$$

where $q(x)$ is a polynomial. It turns out that $q(x) = \sum_{i=1}^{|P|} w_i x^i$ where $w_i$ is the number of linear extensions of $P$ with $i - 1$ descents. Richard Stanley earlier asked a problem [3.3] concerning these coefficients $w_i$. What I want to find out is:

8.17 *Is the sequence* $\{w_1,\ldots,w_{|P|}\}$ *unimodal?*

You might try to tackle this problem this way:

8.18 *Does* $q(x)$ *have all real negative roots?*

If it does then $\{w_i\}_{i=1}^{|P|}$ is not only unimodal but in fact log concave. I know that the answer to the first question is yes for $|P| \leq 6$. Also, when $P$ is a $k$-element antichain, the order polynomial is $n^k$, and $q(x)$ is the Eulerian polynomial $A(k;x)$, which does have all real negative roots.

* * * * * * * * * *

R. NOWAKOWSKI: Let $P$ be a finite partially ordered set. For all $a,b \in P$, define the distance $d(a,b)$ between $a$ and $b$ as the length of the smallest fence joining $a$ and $b$. Let $f \in P^P$ and define the *distortion* of $f$ by

$$\mathrm{dist}(f) = \min_{a\in P}\{d(a,f(a))\}$$

and the *Distortion* of $P$ by

$$\mathrm{Dist}(P) = \max_{f\in P^P}\{\mathrm{dist}(f)\}\ .$$

8.19 *Conjecture.* $\mathrm{Dist}(P\times Q) = \max(\mathrm{Dist}(P),\mathrm{Dist}(Q))$ *for all* $P,Q$.

This is true if both $P$ and $Q$ are dismantlable.

B. DAVEY: It seems to me that it's going to be pretty hard to prove, because it would imply that the product of two partially ordered sets with the fixed point property would have the fixed point property.

* * * * * * * * * *

M. ERNÉ: Consider the following two properties on a poset $P$:

(a) There is a strictly monotone map from $P$ into the chain of integers;

(b) $P$ is locally chain-bounded, i.e., for each comparable pair $x \leq y$ of points of $P$, the length of any chain between $x$ and $y$ is bounded above by some integer $n(x,y)$.

Of course, (a) implies (b).

8.20 *When does (b) imply (a)? In particular, is this true for lattices?*

One can show that (b) implies (a) for all *countable* posets, and there is an uncountable poset (which is not a lattice) for which this implication fails.

* * * * * * * * * *

H. CRAPO: If $L$ is a lattice, Retr($L$) will denote the set of all retractions of $L$ into itself, ordered pointwise.

8.21 *If L is a (complete) lattice, is* Retr($L$) *also a (complete) lattice?*

It's true for finite lattices.

* * * * * * * * * *

PROBLEM SESSION 9

## LATTICES

Chairman: R. Wille

D. DAYKIN: In the first three problems, which appear in two papers of Frankl and myself, let $S = \{1,2,\ldots,n\}$ and let $L$ be the Boolean lattice of subsets of $S$.

9.1 *Conjecture. If $V$ is a convex subset of $L$ then there is an antichain $A$ in $V$ with*

$$\frac{1}{2^n}\binom{n}{\left[\frac{1}{2}n\right]} \leq |A|/|V| .$$

That is, every convex subset of $L$ is at least as wide, in proportion to its size, as $L$ is.

The second question is a more complicated version of the first one.

9.2 *If $V,U,D \subset L$ are such that $V$ is convex, $U$ is an up-set, and $D$ is a down-set, is there an antichain $A \subset V$ with*

$$|V|^2\cdot|V\cap U\cap D|\cdot\binom{n}{\left[\frac{1}{2}n\right]} \leq 2^n\cdot|V\cap U|\cdot|V\cap D|\cdot|A| \ ?$$

This would be very powerful if true; it would include Seymour's inequality and Kleitman's inequality, and others, as special cases.

9.3 *Conjecture. Suppose $F \subset L$ is such that for all $i \in S$ more than $2^{n-4}$ members of $F$ contain $i$. Then there exist $X,Y,Z \in F$ with $X \cup Y \cup Z = S$.*

The conjecture is false if we allow some elements of $S$ to be contained in exactly $2^{n-4}$ members of $F$; just take $F$ to be all members of $L$ containing at most one of the elements 1,2,3,4.

The last problem, due to me, is unpublished.

9.4 *Which finite lattices or posets $P$ have the property that, for any subset $A$ of $P$,*

$$\frac{|A|}{|P|} \leq \frac{\text{no. of maximal chains of } P \text{ meeting } A}{\text{total no. of maximal chains of } P} \ ?$$

That is, the probability that an arbitrary point of $P$ lies in $A$ is not greater than the probability that an arbitrary maximal chain of $P$ has nonempty intersection with $A$. This property fails for certain width two nonmodular lattices, but holds for Boolean lattices (a result of Frankl, Greene, Hilton, and myself).

* * * * * * * * * * * *

G. KALMBACH: I want to pose two problems on orthomodular lattices. The first is due to G. Bruns, the second to S.S. Holland.

9.5 *Give an example of a bounded lattice which admits two orthocomplementations in such a way that the resulting ortholattices are orthomodular and are not isomorphic.*

9.6 *Characterize, completely lattice theoretically, the lattices of closed subspaces of infinite-dimensional Hilbert spaces.*

There are three examples: the lattices of closed subspaces of Hilbert spaces over the reals, the complexes, or the quaternions. They are all complete, irreducible, atomic orthomodular lattices which satisfy the exchange axiom.

* * * * * * * * * * *

R. QUACKENBUSH: I want to mention an old problem of Marshall Hall.

9.7 *Can every finite partial projective plane be embedded in a projective plane of finite order?*

As it stands, this doesn't have very much to do with lattice theory, so let me rephrase it so it has absolutely nothing to do with lattice theory:

9.7' *Equivalently, can a finite set of partial orthogonal latin squares be extended to a complete set of (perhaps larger) mutually orthogonal latin squares?*

C.C. Lindner has shown that $n$ finite mutually orthogonal partial latin squares can be extended to $n$ mutually orthogonal latin squares of (large) finite order. On the other hand, in 1973 Rudolf Wille pointed out that to answer this question in the negative, it would suffice to show that the variety of lattices generated by the incidence lattices of projective planes is not generated by its finite members.

* * * * * * * * * * *

I. RIVAL: A result of Dilworth says that for any finite modular lattice $L$, $V_k(L) = W_k(L)$ for all $k = 0,1,2,\ldots$, where $V_k(L)$ is the number of elements of $L$ covering precisely $k$ elements, and $W_k(L)$ is the number of elements of $L$ covered by precisely $k$ elements. In particular the number of join-irreducibles of $L$ equals the number of meet-irreducibles of $L$, since $|J(L)| = V_1(L) = W_1(L) = |M(L)|$.

9.8 *Conjecture. For each finite modular lattice $L$, there is a one-to-one map $f : J(L) \cup \{0\} \to M(L) \cup \{1\}$ such that $f(x) \geq x$ for all $x$.*

Such a map is called a *matching*. It is known that this conjecture is false if the elements 0 and 1 are not included, and that the corresponding conjecture for the sets $V_k(L), W_k(L)$ for larger $k$ is false. On the other hand the conjecture is true and easy to prove for finite distributive lattices. Much deeper is a collection of results of D. Duffus, to appear in the *Journal of Combinatorial Theory*, in which he verifies the conjecture under certain conditions.

* * * * * * * * * *

E. GRACZYNSKA-WRONSKA: Let $L$ be a finite lattice and $\theta$ a congruence on $L$. If $x \in L$ then $[x]\theta$ denotes the congruence class of $x$, considered as a sublattice of $L$.

9.9 *Find bounds for* $\dim(L)$ *in terms of* $\dim(L/\theta)$ *and* $\{\dim([x]\theta) : x \in L\}$ .

Bounds are known for the special cases when $L$ is an ordinal sum of lattices or when $L$ is a direct product of lattices (and $\theta$ is the obvious congruence).

There is a related problem for posets. Let $I$ be a poset, and let $(P_i : i \in I)$ be a family of pairwise disjoint posets. For all $i \leq j$ in $I$ let $h_{ij} : P_i \to P_j$ be an order-preserving mapping such that (a) for all $i$, $h_{ii}$ is the identity, and (b) if $i \leq j \leq k$ in $I$, then $h_{jk} \circ h_{ij} = h_{ik}$ . Define a poset $P$ with underlying set $\bigcup_{i \in I} P_i$ and such that $x \leq y$ in $P$ if and only if there exists $i \leq j$ in $I$ such that $x \in P_i$ , $y \in P_j$ , and $h_{ij}(x) \leq y$ in $P_j$ .

9.10 *Find bounds for* $\dim(P)$ *in terms of* $\dim(I)$ *and* $\{\dim(P_i) : i \in I\}$.

Again, ordinal sums (of bounded posets) and direct products of posets are special cases for which bounds are known.

[For Problem 9.10, B. Sands and P. Winkler found an upper bound in the case of $I$ finite.]

* * * * * * * * * *

B. DAVEY: If $L$ is projective in $\underset{\sim}{D}$, the category of bounded distributive lattices, then it is a retract of $F\underset{\sim}{D}(\kappa)$, the free bounded distributive lattice in $\kappa$ generators, for some $\kappa$. Applying the duality to this, the dual $D(L)$ of $L$ is a retract of the dual $D(F\underset{\sim}{D}(\kappa)) \cong \underset{\sim}{2}^{\kappa}$, and hence is a complete lattice. In the case that $L$ is finite we conclude that $L$ is projective in $\underset{\sim}{D}$ iff $D(L)$ (which in this case is just the subset of join-irreducibles of $L$) is a lattice.

9.11 *For arbitrary lattices $L$, find topological and order-theoretic conditions on $D(L)$ which, with the fact that $D(L)$ is a complete lattice, characterize the fact that $L$ is projective in $\underset{\sim}{D}$.*

My second question is sort of a general, waffling-type question. The duality between finite distributive lattices and finite ordered sets has often been used to answer algebraic questions concerning lattices via considerations of ordered sets.

9.12 *Give a single example of a question concerning ordered sets which can be settled by transferring it to a question about distributive lattices and then using algebraic techniques.*

For example, can we use algebraic techniques in lattices to study ordered sets with the fixed point property, or cancellation in exponents of ordered sets? I can make the translation to lattices, but it doesn't seem to help.

* * * * * * * * * *

M. POUZET: I will make some remarks about scattered lattices and scattered spaces. Let $E$ be a partial order and $I(E)$ the set of initial segments of $E$, ordered by inclusion. It is easy to show that $I(E)$ is order scattered (i.e., the rational chain is not embeddable into $I(E)$) iff it is topologically scattered (i.e., every subset of $I(E)$ has an isolated point, where the topology on $I(E)$ is the Stone topology).

*9.13* *Could this result be proved using the Birkhoff-Priestley duality, and is it possible to extend it? For example, does the same result hold for complete, completely distributive lattices?*

* * * * * * * * * * *

D. SCHWEIGERT: A lattice is order-polynomial complete if every order-preserving function of $L$ is a polynomial function.

*9.14* *Is there an infinite order-polynomial complete lattice?*

The conjecture is no.

* * * * * * * * * * *

W. POGUNTKE: This problem arose from results of B. Sands and myself, and concerns how a finitely generated lattice can "become infinite".

*9.15* *Let $L$ be a finitely generated infinite lattice. Is there always a finite antichain in $L$ that generates an infinite sublattice?*

If $L$ has width three, we showed that the answer is yes; in fact, essentially the same infinite sublattice always results.

* * * * * * * * * * *

H. BANDELT: For a lattice $L$, let $\Delta(L)$ denote the lattice of all diagonal (= reflexive) sublattices of the square $L \times L$.

*9.16* *Characterize the lattices $\Delta(L)$.*

It is known that in general $\Delta(L)$ is itself a square, is algebraic, pseudocomplemented, and 0-modular (i.e., no $N_5$ in $\Delta(L)$ contains the bottom element of $\Delta(L)$).

If $L$ is distributive then $\Delta(L)$ is distributive, but in general $\Delta(L)$ behaves rather badly.

*9.17* *Let $V$ be any nontrivial variety of lattices other than the variety of distributive lattices. Does the family of lattices $\{\Delta(L) : L \in V\}$ generate the variety of all lattices?*

For any such $V$, there will be a *nonmodular* $\Delta(L)$.

* * * * * * * * * * *

M. ERNÉ: A poset is *algebraic* if it is up-complete (i.e., arbitrary directed joins exist) and if every element is a join of compact elements. This is an obvious generalization of the concept of an algebraic lattice. It turns out that a poset is algebraic if and only if it is isomorphic to the poset of order ideals of a poset.

An old result of Birkhoff and Frink says that every algebraic lattice is weakly atomic, i.e., between every pair of elements there is a covering pair.

*9.18* *Is every algebraic poset weakly atomic?*

The problem is that while every interval of an algebraic lattice is algebraic, a fact used in the proof of the lattice case, this fails to be true for posets.

The second problem is very old, but I have not heard of any solution. A *T-lattice* is a lattice isomorphic to the lattice of open sets of some topological space; equivalently, every element of the lattice is a meet of meet-primes.

*9.19* *Let $L$ be a T-lattice and let $\varphi: L \to L'$ be a complete homomorphism onto a complete lattice $L'$. Is $L'$ a T-lattice?*

If the space associated with $L$ is locally compact or $T_1$, the answer is yes.

* * * * * * * * * *

F. YAQUB: It is known that any Post algebra of order $n$ is isomorphic to the coproduct of a Boolean algebra and an $n$-element chain (written: $B*C_n$). If you replace $C_n$ by an arbitrary chain $C$ you get a *generalized Post algebra*, and if you replace $C_n$ by an arbitrary bounded distributive lattice you get a *Post L-algebra*. Any of these classes can be realized by an abstract algebra whose set of constants is just the right-hand factor.

*9.20* *What is the largest class $K$ of finite distributive lattices such that $B*L = B*L' \Rightarrow L = L'$ for all Boolean algebras $B$ and for all $L \in K$?*

Every finite chain is in $K$; in fact Balbes and Dwinger (1971) have shown that an arbitrary chain $C$ satisfies $B*C = B*C' \Rightarrow C = C'$ (for all Boolean algebras $B$) if and only if $C$ is *rigid* (i.e., has no nontrivial automorphisms).

* * * * * * * * * *

K. BOGART: This problem has to do with information theory. The general problem is:

9.21 *What can you say about finite lattices whose Möbius function* $\mu(x,y)$ *is odd whenever* $x \leq y$?

By the Crapo complementation theorem, such lattices must be complemented; but this is all we know. The application to information theory is as follows:

9.22 *Does there exist a sequence* $(L_n)$ *of such lattices with* $|L_n| \to \infty$, *and a real number* $r > 0$ *such that for all* $n$ *there is an order ideal* $I_n$ *with* $|I_n|/|L_n| \geq r$ *and an* $\varepsilon > 0$ *so that for all* $x \in I_n$ *the principal filter* $F_x$ *generated by* $x$ *satisfies* $|F_x|/|I_n| \geq \varepsilon$?

The chances are that this strong a result will not be possible.

* * * * * * * * * *

PROBLEM SESSION 10

MISCELLANEOUS

Chairman: M. Pouzet

G. KALMBACH: Consider hypergraphs $H$ in which every line $l$ contains a finite number $k \geq 3$ of points $A(l)$. Define an *n-cycle* ($n \geq 3$) to be an $n$-tuple $(l_i)_{i=0}^{n-1}$ of lines together with nonempty subsets $A_i, B_i \subseteq A(l_i)$ which satisfy $A_i \cap B_i = \emptyset$ and $B_{i-1} = A_i$ for all $i$ (indices mod $n$). An *astroid* is a 4-cycle $(l_i)_{i=0}^{3}$ such that the distinguished subsets $A_i, B_i \subseteq A(l_i)$ satisfy $A_i \cup B_i = A(l_i)$ for all $i$. A line $l$ is *central* for a 3-cycle $(l_0, l_1, l_2)$ if $A_0 \cup A_1 \cup A_2 \subset A(l)$. An astroid $(C_i)_{i=0}^{3}$ is *central* for a 4-cycle $(l_i)_{i=0}^{3}$ (with distinguished subsets $A_i \subseteq A(l_i)$) if $A_i \subseteq A(C_{i-1}) \cap A(C_i)$ for all $i$ (indices mod 4). See my book *Orthomodular Lattices*, Section 4.

M. Dichtl (1981) has proved:

(i) $H$ represents an orthomodular poset iff every 3-cycle in $H$ has a central line;

(ii) $H$ represents an orthomodular lattice iff every 4-cycle has a central astroid.

*10.1* *Determine estimates for the minimal number of points $\sigma_n$ in a hypergraph $H_k^n$ satisfying (i) or (ii) which contains exactly n lines, each line containing exactly k points.*

For $k = 3$ and $n = 10$, $\sigma_n = 15$ (from the Petersen graph).

* * * * * * * * * * *

I. ROSENBERG: Let $(E, \leq)$ be a finite ordered set with 0 and 1. An operation $f$ on $E$ ($f: E^n \to E$) is *monotonic* if $f(x_1, \ldots, x_n) \leq f(y_1, \ldots, y_n)$ whenever $x_i \leq y_i$ for all $i$. The set of all monotonic operations forms a *clone* (i.e., a set closed under composition which contains all projections $e_i^n(x_1, \ldots, x_n) = x_i$ ; this is a multiplace analogue of the monoid of isotone self maps of $E$). A proper subclone $N$ of a clone $M$ is *maximal* in $M$ if $N \cup \{f\}$ generates $M$ (via composition) for each $f \in M \setminus N$.

10.2 *Describe all clones maximal in M and thus give an efficient characterization of systems of generators of M.*

* * * * * * * * * * *

D. Schweigert: Let $(P,\leq)$ be an infinite ordered set such that every pair $a,b \in P$ has an upper and a lower bound. Let $M$ be the clone of all order preserving operations of $P$, and let $K$ be the set of constant functions.

10.3 *Is there a finite subset F of M such that K ∪ F generates M?*

* * * * * * * * * * *

R. WILLE: Here are some problems in mathematical music theory.

A *tone system* is a pair $(T,h)$ where $T$ is a set and $h:T \to \mathbb{R}^+$ is injective; $h$ is said to be the *pitch*. The tones $t$ and $t'$ are *octaves* (written $t\omega t'$) if $h(t) = 2^z h(t')$ for some $z \in \mathbb{Z}$. A finite $K \subset T$ is a *chord*. We extend the octave relation to chords by:

$$K\Omega K' \Leftrightarrow \begin{array}{l}(\forall\ t\in K)(\exists\ t'\in K')(t\omega t') \text{ and} \\ (\forall\ t'\in K')(\exists\ t\in K)(t\omega t')\ .\end{array}$$

The translation relation $\Phi$ for chords is defined by:

$$K\Phi K' \Leftrightarrow (\exists\ r\in\mathbb{R}^+)(h(K)=r\cdot h(K'))\ .$$

Then $\Omega$ and $\Phi$ are equivalence relations on the set $P_{\text{fin}}(T)$ of finite subsets of $T$. Let $\Omega \vee \Phi$ denote the join of $\Omega$ and $\Phi$, and call $\mathcal{H} = P_{\text{fin}}(T)/\Omega \vee \Phi$ the set of *harmonic forms*. This is partially ordered by

$$H \leq H' :\Leftrightarrow (\exists\ K\in H)(\exists\ K'\in H')(K\subseteq K')$$

for all $H,H' \in \mathcal{H}$.

10.4 *Describe the ordered set* $(\mathcal{H},\leq)$. *What is its order dimension? Solve the question in particular for* $h(T) = \{2^{z/12} : z \in \mathbb{Z}\}$ *and for* $h(T) = \{2^x \cdot 3^y \cdot 5^z \cdot \ldots \cdot p^w : x,y,z,\ldots,w \in \mathbb{Z}\}$.

Now let $h(T) = \mathbb{Q}^+$. For each finite $K \subseteq T$ such that $h(K)$ consists only of odd integers, define the *consonance degree* $k(K)$ of $K$ by

$$k(K) = \frac{1}{2}\left(\max\left\{\frac{h(u)}{(h(u),h(v))} : u,v \in K\right\} - 1\right)\ .$$

10.5 *Are* $\{1,3,5,\ldots,2n+1\}$ *and* $\left\{\frac{v}{2n+1},\ldots,\frac{v}{3},\frac{v}{1}\right\}$ *with* $v = \mathrm{lcm}\{1,3,5,\ldots,2n+1\}$ *the maximal subsets of consonance degree* $\leq n$ *(up to equivalence with respect to* $\Omega \vee \Phi$*)?*

This is a generalization of Graham's $100 question.

* * * * * * * * * *

A. HOFFMAN: Consider a finite (pre)metric space (i.e. $d(x,y) = 0$ is allowed even if $x \neq y$) on $n$ points as a vector in $\mathbb{R}^{\binom{n}{2}}$. The set of all finite metric spaces on $n$ points is a polyhedral cone in $\mathbb{R}^{\binom{n}{2}}$, since it is the intersection of the half-spaces specified by the many triangle inequalities, each of which is homogeneous and linear.

10.6 *What are the extreme rays of this cone?*

Among the extreme rays are all vectors specified by: let $S$ and $T$ be non-empty subsets partitioning $\{1,\ldots,n\}$, let the distance between each pair of points in the same set be 0, and let the distance between pairs of points in opposite sets be 1.

10.7 *What are the faces of the subcone generated by these rays?*

Let $P$ be a countable poset with 0 such that every infinite chain has order type $\omega$. A function $f: P \to \mathbb{R}^+$ is a *Ramsey function* on $P$ if

(i) $f$ is order preserving with $\sup_{x \in P} f(x) = \infty$

and

(ii) there is a finite set $C = \{C_1,\ldots,C_t\}$ of chains in $P$ such that (putting $C_i = \{a_{i0} < a_{i1} < a_{i2} < \ldots\}$)
(a) $\lim_{j\to\infty} f(a_{ij}) = \infty$ for all $i$, and (b) for all $S \subset P$,
$\sup_{x \in S} f(x) = \infty$ if and only if $\sup_{x \in S} \max_{1 \leq i \leq t} \{\max j : a_{ij} < x\} = \infty$.

As an example consider the poset of finite graphs under the relation of "induced subgraph". $|V(G)|$ is a Ramsey function with $C$ consisting of two chains: the complete graphs and the independent sets. It is a fact that if $C$ and $C'$ both witness that $f$ is Ramsey, then there is a bijection between the families of chains, so $C$ is essentially unique.

10.8 *Give an equivalent definition for "f is Ramsey" that removes the reference to a family C of chains.*

* * * * * * * * * *

M. POUZET: Let $K_n$ be the complete graph on $n$ vertices and let $C$ be a colouring of the edges with $k$ colours $C_1,\dots,C_k$. By a *representation* of $\langle K_n,C\rangle$ in a Euclidean space $V$ I mean an injective map $f$ from $K_n$ into $V$ such that two edges $\{x,y\}$ and $\{x',y'\}$ have the same colour if and only if the corresponding segments $[f(x),f(y)]$ and $[f(x'),f(y')]$ have the same length. For example, if all edges have the same colour, then for $n = 3$ the image is an equilateral triangle and for $n = 4$ a regular tetrahedron.

10.9 *Describe all the* $(K_n,C)$ *which cannot be represented in the real line but such that all induced coloured subgraphs can be represented.*

For example, with $\leq 2$ colours there are only three:

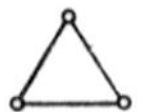
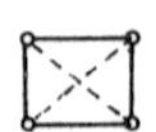

This notion of representability and the corresponding notion of the *dimension* of such a colouring (i.e., the smallest dimension of a Euclidean space in which the coloured graph can be represented) was known by E. Specker who conjectured that if $\dim(K_n,C) = n - 1$ then $C$ is monochromatic. It occurs in work of mine on the Ulam reconstruction problem. Using a result of Tutte it can be seen that the dimension is reconstructible. This notion has also been considered by Rosenfeld.

* * * * * * * * * *

M. ERNÉ: A complete lattice $L$ is *meet-continuous* iff

$$(d) \qquad x \wedge \bigvee D = \bigvee (x \wedge D)$$

for all $x \in L$ and all directed sets $D$. Equivalently $(d)$ holds for all chains $D$.

A complete lattice $L$ is *continuous* iff

$$(D) \qquad \bigwedge\{\bigvee D_i : i \in I\} = \bigvee\{\bigwedge\psi(I) : \psi \in \prod_{i\in I} D_i\}$$

holds for all families of directed sets $D_i$.

10.10 *Does it suffice that (D) hold for all families of chains $D_i$ ?*

If the *chain-below* relation

$$x <_c< y \Leftrightarrow x \in \bigcap\{\downarrow C : y \leq \bigvee C,\ C \text{ a chain}\}$$

has the interpolation property ($x <_c< y \Rightarrow \exists\, z : x <_c< z <_c< y$) then the answer is yes. However I conjecture that interpolation doesn't always follow from the chain-below relation.

* * * * * * * * * *

J. MILLER: Let $X$ be an infinite set, $\leq$ a preorder on $X$. Define a preorder $\leq'$ on $X$ by:

$$a \leq' b \text{ iff } (\forall\, x \in X)(x<a \Rightarrow x<b) \text{ and } (\forall\, y \in X)(y>b \Rightarrow y>a) .$$

(Call $\leq$ the *given* or *determining* order, $\leq'$ the *associated* order.) This construction is useful because $\leq$ can be used to put an open-interval topology $U$ on $X$ having subsets $(a<)$ and $(<a)$ for $a \in X$ as subbase. Under suitable conditions that often arise, $\leq$ is a partial order, $\leq'$ is a lattice order, and $(X, \leq', U)$ becomes a topological lattice.

If $\leq$ is a preorder, we let $\leq'' = (\leq')'$, $\leq''' = (\leq'')'$, etc.

10.11 *For $\leq$ a preorder, how many of the orders $\leq, \leq', \leq'', \leq''', \ldots$ can be distinct?*

It is known that if $\leq$ is an order-dense partial order then $\leq''$ is a partial order; and $a < b \Rightarrow a \leq'' b \Rightarrow a <' b$, and $\leq^{(2n)} = \leq''$, $\leq^{(2n+1)} = \leq'$. If also $\leq'$ is a partial order then $\leq', \leq''$ coincide.

10.12 *Given a lattice $(X, \preceq)$ and a determining order $\leq$ for $\preceq$ find weakest conditions on $\leq, \preceq$ sufficient to ensure that the topology $U$ of $\leq$ is intrinsic (i.e., every lattice isomorphism $f : X \to X$ is a homeomorphism).*

I only know a complicated set of conditions.

M. ERNÉ: For your first problem, do you have any simple example where $\leq'$ and $\leq''$ are different?

MILLER: No, I don't believe I have.

* * * * * * * * * *

J. ROSENSTEIN: Given a linear ordering $A$ of type $\tau$ we say that

(i) $A$ is *groupable* if there is an ordered group whose order type is $\tau$;

(ii) $A$ is *fieldable* if there is an ordered field of type $\tau$;

(iii) $A$ is *transitive* if for every $a,b \in A$ there is an automorphism $f$ of $A$ with $f(a) = b$;

(iv) $A$ is *doubly transitive* if for every $a < b$ and $c < d$ there is an automorphism $f$ of $A$ for which $f(a) = c$ and $f(b) = d$;

(v) $A$ is *uniform* if $A^{<a} \simeq A^{<b}$ for every $a,b \in A$;

(vi) $A$ is *exponentiable* if $A^{\alpha}$ can be defined for all ordinals $\alpha$.

The following implications are known:

$$\begin{array}{ccccc} A \text{ fieldable} & \Rightarrow & A \text{ doubly transitive} & & \\ \Downarrow & & \Downarrow & & \\ A \text{ groupable} & \Rightarrow & A \text{ transitive} & \Rightarrow & A \text{ uniform} \\ & & \Downarrow & & \\ & & A \text{ exponentiable} & & \end{array}$$

There are two questions, to be considered both for sub-orderings $A$ of $\mathbb{R}$ and for arbitrary $A$.

*10.13 Characterize those linear orderings which have each of these properties.*

*10.14 Determine in which cases, if any, the implications are reversible.*

For $A$ countable, $A$ is doubly transitive iff $A$ has the order type $\eta$ of the rational numbers, and $A$ is groupable iff $A$ has order type $\zeta^{\alpha}$ or $\zeta^{\alpha}\cdot\eta$, with $\alpha$ an ordinal and $\zeta$ the order type of the integers. More generally, $A$ is groupable iff $A$ has order type $\zeta^{\alpha}\cdot\delta$ where $\alpha$ is an ordinal and $\delta$ is a groupable dense order type. An example of a linear ordering which is groupable but not fieldable is the linear ordering of the irrational numbers. See Chapters 8 and 9 of *Linear Orderings*.

* * * * * * * * * *

M. JAMBU-GIRAUDET: For a dense chain $T$ a *cut* is a pair $(A,B)$ of non-empty subsets of $T$ with $A < B$, $A$ without a last point, $B$ without a first point, and either $T = A \cup B$ or $T = A \cup \{r\} \cup B$ for some element $r$ of $T$ between $A$ and $B$. The *character* of the cut $(A,B)$ is the pair $(\alpha,\beta)$ of ordinals where $\omega_{\alpha}$ is the cofinality of $A$ and $\omega_{\beta}$ is the co-initiality of $B$. Let $C(T)$ denote the set of characters of all cuts of a given dense chain $T$.

Take $X$ a bounded set of ordinals such that $\omega_x$ is regular for $x \in X$. Clearly, for any subset $C$ of $X^2$ there is a dense chain $T$ such that $C(T) \cap X^2 = C$; take for instance $T = \bigoplus_{(\alpha,\beta)\in C} (S\cdot(\omega_\alpha+\omega_\beta^*))$ where $C$ is lexicographically ordered and $S$ is an $\omega_\gamma$ saturated dense chain for some $\gamma >$ all $x \in X$. A chain $T$ is 2-*homogeneous* if any two of its closed bounded intervals are isomorphic. (See *Ordered Permutation Groups* by Andrew Glass, to appear in Cambridge University Press.)

10.15 *For which subsets $C$ of $X^2$ is there a 2-homogeneous chain $T$ such that $C(T) \cap X^2 = C$? In particular, are there uncountably many such $C$'s when $X = \omega$?*

Let $L$ be a sublattice and a subgroup of the set of all automorphisms of a chain $T$ such that

(i) for any $r_1 < r_2$ and $s_1 < s_2$ in $T$ there is some $f$ in $L$ with $f(r_1) = s_1$ and $f(r_2) = s_2$ ($L$ acts 2-transitively on $T$), and

(ii) for any $f$ in $L$ there are $t_1$ and $t_2$ in $T$ such that $f(s) = s$ for all $s < t_1$ and for all $s > t_2$ in $T$.

An element $f$ of $L$ is an *l-prime* if, for any nonidentity $f_1$ and $f_2$ in $L$ such that $f_1 \wedge f_2$ is the identity map on $T$, $f_1 \vee f_2$ is different from $f$.

10.16 *Is it possible that $L$ contains no l-prime?*

If there is such an $L$, there is a countable one for which $T$ is the rational line.

* * * * * * * * * *

R. FRAÏSSÉ: Given relations $A$ of base $E$ and $A'$ of base $E'$, a *local isomorphism* from $A$ towards $A'$ is an isomorphism of some restriction of $A$ (to a subset of $E$) onto some restriction of $A'$. A *local automorphism* of $A$ is a local isomorphism from $A$ towards itself. A subset $D$ of the base $E$ is called an *interval* (mod $A$) when every local automorphism $f$ of $A|_D$ can be extended to $E \backslash D$, i.e., the union $f \cup$ (Identity on $E\backslash D$) is a local automorphism of $A$. When $A$ is a chain, it is easy to check that $D$ is an interval (mod $A$) iff $D$ is an interval (i.e., $x,y \in D$ and $x < z < y \Rightarrow z \in D$). Note that any intersection of intervals is an interval and that the empty set, the base $E$ and any singleton are intervals.

Chains and other simple relations are *compactible*, i.e., if $D$ is an interval then $E\backslash D$ is a finite union of intervals. Moreover, for a chain, $E\backslash D$ is always a finite union of *disjoint* intervals.

10.17 *Give an example of a compactible relation with an interval $D$ such that $E\backslash D$ is* not *a finite union of disjoint intervals.*

* * * * * * * * * *

J. BARTHÉLEMY: Let $(T,\wedge)$ be a finite meet-semilattice, and let $d$ be a distance function on $T$. Let $E \subseteq T$. For $\pi = (x_1,\ldots,x_\nu) \in E^\nu$, define Med($\pi$) to be the set of solutions of

$$\operatorname*{Min}_{y \in E} \sum_{i=1}^{\nu} d(x_i,y) .$$

We say that $d$ is *paretian* on $E$ if for all $\pi$, $\bigwedge x_i \leq m$ for all $m \in$ Med($\pi$).

10.18 *Characterize the triples $(T,E,d)$ such that $d$ is paretian on $E$.*

If $T$ is a lattice, and $d$ is defined by an upper valuation $v$ (i.e., $d(x,y) = 2v(x\vee y) - v(x) - v(y)$, where $v(x) + v(y) \geq v(x\wedge y) + v(x\vee y)$), then $d$ is paretian on $T$.

A particular case arises when $R$ is the lattice of all binary relations on a set $S$, and $d$ is symmetric difference. Then $d$ is paretian on each of the following subsets of $S$: all linear orders, all weak orders, all equivalence relations.

10.19 *On what other sets of binary relations is $d$ paretian?*

* * * * * * * * * *

R. BONNET: For a distributive lattice $L$ we denote by $I_p(L)$ the space of all prime ideals of $L$; this is a compact totally disconnected space (as a subspace of $2^{|L|}$, with the pointwise topology.

Let $L$ be a scattered (= superatomic) Boolean algebra (or more generally a scattered distributive lattice).

10.20 *Is there a well-founded distributive lattice* A *such that the spaces $I_p$(A) and $I_p$(L) are homeomorphic?*

(Here well-founded means that $A$ has no infinite strictly decreasing sequence.) The answer is yes if $L$ is a Boolean algebra generated by a chain, in particular if $L$ is countable.

* * * * * * * * * *

B. DAVEY: Let $P$ be an $n$-element ordered set ($n > 3$?) and let $F = \{P_1,\ldots,P_n\}$ be its set of (unlabelled) one-element-deleted ordered subsets.

10.21 *Can we reconstruct $P$ from the family $F$?*

It's easy to prove that posets with 0 can be reconstructed.

Since the more familiar *graph* reconstruction conjecture is so difficult to settle, it is natural to ask:

10.22 *Is ordered set reconstruction equivalent to graph reconstruction?*

Rumour has it that this is known (at least partially), but I'd like someone to confirm it.

10.23 *Which properties of $P$ can be reconstructed from $F$? In particular, can order dimension be reconstructed?*

For instance, both the width and the length of $P$ can be determined from $F$.

* * * * * * * * * *

G. MCNULTY: A finite interval order $P$ has *count* $n$ provided $n$ is the smallest number of distinct lengths of intervals required to represent $P$ by intervals of real numbers. Hence the count one interval orders are exactly the semiorders. Suppose $P$ is a count 2 interval order. The *spectrum* of $P$ is the set $\{r > 1 : P$ can be represented by intervals of lengths 1 and $r\}$. I have two questions, the first due to Trotter.

10.24 *Is the spectrum always an interval?*

10.25 *What intervals of real numbers can be spectra of count* 2 *interval orders?*

* * * * * * * * * *

[The problem sessions ended as they began, with problems by Paul Erdös.]

P. ERDÖS: I will talk about some problems which arose during this meeting.

Daykin and I considered the following problem. First some background. An old conjecture of mine is as follows. Let $S$ be a set with $2n$ elements, and let $F = \{A_1,\dots,A_m\}$ be a family of subsets of $S$. Form a graph $G_{2n}(F)$ whose vertices are the $A$'s and such that two vertices are joined if the corresponding sets are comparable, i.e., one contains the other. My conjecture states that if $m = (2-\varepsilon)2^n$ and $n > n_0(\varepsilon)$ then $G_{2n}(F)$ has fewer than $2^{2n}$ edges. It is easy to see that for $\varepsilon = 0$ this is false. Daykin and P. Frankl proved that if $G_n(F)$ has $(1-o(1))\binom{m}{2}$ edges (i.e., apart from $o(m^2)$ edges it is complete) then $m^{1/n} \to 1$. Daykin and I conjecture the following:

10.26 *Conjecture. If $G_n(F)$ has $\alpha m^2$ edges (for some $\alpha > 0$, $\alpha$ independent of $n$) then $m < (\sqrt{2}+o(1))^n$.*

Perhaps there exists $C = C(\alpha)$ so that, for sufficiently large $n$, $m < C2^{n/2}$. If true this conjecture is easily seen to be best possible. Also, perhaps to every $\varepsilon > 0$ there is an $n$ so that if $G_n(F)$ has more than $m^{2-n}$ edges then $m < (\sqrt{2}+\varepsilon)^n$.

The next problem is due to Erné and myself. Let $G(n)$ be a graph of $n$ vertices. Denote by $f(G(n))$ the number of complete graphs contained in $G(n)$. Denote by $F(n)$ the number of possible values of $f(G(n))$ over all graphs $G(n)$. $F(n) = 0(2^n)$ is easy and we almost certainly can prove $\lim_{n\to\infty} (F(n))^{1/n} = 2$.

10.27 *Determine $F(n)$ explicitly and also determine the set of possible values of $f(G(n))$.*

Similar questions can be asked about cliques and other types of subgraphs.

Trotter and I considered the following question. Let $S$ be a set with $|S| = n$. Consider a family $(A_k)$ of subsets of $S$ where no $A_k$ contains any other and such that, for every $t$, if there is an $A_k$ with $|A_k| = t$ then there are exactly $r$ different $A$'s of size $t$. If $r = 1$ and $n > 3$, one can always give $n - 2$ sets $A_k$

and one can not give $n - 1$. If $r > 1$ and $n >$ some number $n_0 = n_0(r)$, one can always give $r(n-3)$ sets $A_k$ but one cannot give $r(n-2)$ such sets. We have no satisfactory estimate of $n_0(r)$.

*10.28* *Give estimates for* $n_0(r)$.

Let $1 \leq a_1 < a_2 < \ldots < a_k \leq n$. Assume that

$$(1) \qquad (a_j - a_i) \nmid a_j \text{ whenever } a_j - a_i \geq t \, .$$

Let $F(n;t)$ be the maximum length $k$ of any sequence satisfying (1).

*10.29* *Is it true that* $\lim_{n\to\infty} \frac{1}{n} F(n;t) = \frac{1}{2}$ *for every t?*

$F(n;1) = \left[\frac{n+1}{2}\right]$ is trivial; the odd numbers satisfy (1), and if (1) is satisfied then $a_{i+1} - a_i > 1$. Also, $F(n;2) > \frac{n}{2} + c \log n$ holds; the $a$'s are the odd integers and odd powers of 2. I cannot prove $F(n;2) < (1+\varepsilon)\frac{n}{2}$, but it is quite possible that due to old age and worse I overlook a trivial point.

* * * * * * * * * *

# PART IX

# A BIBLIOGRAPHY

## INTRODUCTION

This is an edited list of bibliographic details of articles concerned solely with *ordered sets*. While this reference list benefits from the often extensive bibliographies of the twenty-three expository articles of this volume the list itself is not a cumulative record.

The project of compiling this list began several "student-generations" ago [D. Duffus, R. Nowakowski, M. Waugh, A. Dewar]. I envisioned a bibliography devoted to *order-theoretic* topics alone. The apparent narrowness of this approach was to be compensated by the realization that the theory of ordered sets lies at the confluence of several branches of mathematics including *set theory*, *lattice theory*, *combinatorial theory*, and even aspects of modern *operations research*. And this bibliography was to reflect this fact. On the other hand, I aimed to steer clear of those topics which are well-represented and well-documented in the literature, notably *set theory* and *lattice theory*. Excellent bibliographies on these topics are available: in the recent book by J. Rosenstein [*Linear orderings*, Academic Press, 1981] and in the book by G. Grätzer [*General lattice theory*, Birkhauser, 1978].

It is difficult to evaluate articles at the best of times and with a list of this size and scope it is almost impossible. Some entries are included solely because "order" appears in the title or because they deal with specifically order-theoretic matters, no matter how superficial. There may even be entries here which have never had a citation outside of a reviewing journal. I regret this.

It is unlikely that this list contains many obvious typographical errors; it certainly does contain inaccuracies. The most obvious inaccuracies are the missing or incomplete *Mathematical Review* numbers.

The size alone of such a reference list may project an undeserved authority or convey a false sense of security. The careful scholar will endeavour to verify references whenever possible. There must be many examples of incorrect bibliographic details perpetuated unwittingly by such lists because few would challenge them by checking the original source.

I am grateful to the participants of the Symposium who spent considerable time during the Symposium itself verifying details of the entries, suggesting additions, and proposing deletions.

Ivan Rival



*I. Rival (ed.), Ordered Sets, 865–966.*

Abian, Smbat; Brown, Arthur B. (1961) "A theorem on partially ordered sets, with applications to fixed point theorems." Canad. J. Math. 13, 78-82. MR 23 #A817.

Abramson, N. (1960) "A partial ordering for binary channels." IRE Trans. IT-6, 529-539. MR 25 #4948.

Adamek, J. (1976) "The algebraic structure of products and sums in a category." Lattice Theory (Proc. Colloq, Szeged, 1974), pp. 13-27. Colloq. Math. Soc. Jaos Bolyai, Vol. 14, North-Holland, Amsterdam. MR 8439.

Adams, D.H. (1970) "A note on a paper by P.D. Finch." J. Austral. Math. Soc. 11, 63-64. MR 41 #5.

Adams, M.E. (1975) "The poset of prime ideals of a distributive lattice." Algebra Universalis 5, 141-142. MR 52 #2986.

Adnadevic, Dusan (1961) "Dimension of some partially well ordered sets with applications." Bull. Soc. Math. Phy. Serbie 13, 49-105 und 225-261, engl. Zusammenfassung 105 une 262 [Serbo-kroatisch].

Adnadevic, Dusan (1964) "On the dimension of the product of partially ordered sets." (Serbo-Croatian. English summary). Mat. Vesnik (N.S.) 1 (16), 9-12. MR 34 #7413.

Adnadevic, Dusan (1966) "On the representations of finite partially ordered sets." (Serbo- Croatian. English summary). Mat. Vesnik 3 (18), 17-21. MR 35 #1510.

Adolphson, D.; Hu, T.C. (1973) "Optimal linear ordering." SIAM J. Appl. Math. 25, 403-423.

Ahlswede, R.; Daykin, D.E. (1978) "An inequality for the weights of two families of sets, their unions and intersections." Z. Wahrscheinlichkeitstheorie verw. Gebiete, 43, 183-185.

Ahlswede, R.; Daykin, D.E. (1979) "Inequalities for a pair of maps $S \times S \rightarrow S$ with S a finite set." Math. Zeit, 165, 267-289.

Aigner, M. (1973) "Lexicographic matching in Boolean Algebras." J. Comb. Theory 14, 187-194.

Aigner, M. (1975) "Symmetrische Zerlegung von Kettenprodukten." Monatshefte fur Mathematik 79, 177-189.

Aigner, M.; Prins, G. (1971) "Uniquely partially orderable graphs." J. London Math. Soc. (2) 3, 260-266. MR 41866.

Abdel-Wahab, H.M.; Kameda, T. (1978) "Scheduling to minimize maximum cumulative cost subject to series-parallel precedence constraints." Operations Res. 26, no. 1, 141-158. MR 58 #4278.

Abian, Alexander (1968) "On definitions of cuts and completion of partially ordered sets." Z. Math Logik Grundlagen Math. 14, 299-302. MR 38 #4367.

Abian, Alexander (1969) "A fixed point theorem for nonincreasing mappings." Boll. Un. Mat. Ital. (4) 2, 200-201. MR 39 #5427.

Abian, Alexander (1970) "On the similarity of partially ordered sets." Amer. Math. Monthly 77, 1092-1094. MR 42 #5864.

Abian, Alexander (1971/72) "On the isomorphism of partially ordered sets." Algebra Universalis 1, 236-237. MR 46 #103.

Abian, Alexander (1972) "Fixed point theorems of the mappings of partially ordered sets." Rend. Circ. Mat. Palermo (2) 20 (1971), 139-142. MR 48 #2006.

Abian, Alexander (1974) "Linear extensions of partial orders." Nordisk Mat. Tidskr. 22, 21-22. MR 50 #6930.

Abian, Alexander (1974) "On generic sets and their products." Bull. Math. Soc. Sci. Math. R.S. Roumanie (N.S.) 16 (64) (1972), no. 4, 367-370. MR 50 #4411.

Abian, Alexander (1977) "Non-existence of ordered sets with denumerable many dense subsets." Bull. Math. Soc. Sci. Math. R.S. Roumanie (N.S.) 20 (68) (1976), no. 1-2, 3-4. MR 185.

Abian, Alexander; Deever, David (1966) "A short proof of a stronger version of Popruzenko's theorem." Fund. Math. 59, 71-72. MR 33 #5524.

Abian, Alexander; Deever, David (1966) "On representation of partially ordered sets." Math. Z. 92, 353-355. MR 33 #2581.

Abian, Alexander; Deever, David (1966) "Representation of partially and simply ordered sets by terminating sequences." Boll. Un. Mat. Ital, III. Ser. 21, 371-376.

Abian, Alexander; Deever, David (1967) "On representation of partially ordered sets." Math. Ann. 169, 328-330. MR 34 #4169.

Abian, Alexander; Deever, David (1967) "On the minimal length of sequences representing simply ordered sets." Z. Math. Logik Grundlagen Math. 13, 21-23. MR 36 #4978.

Aigner, Martin (1969) "Graphs and partial orderings." Monatsh. Math. 73, 385-396. MR 41 #1561.

Aigner, Martin; Prins, Geert (1972) "Segment-preserving maps of partial orders." Trans. Amer. Math. Soc. 166, 351-360. MR 45 #122.

Aizenstadt, A. Ja. [Aizenstat, A. Ja.] (1962) "On the semi-simplicity of semigroups of endomorphisms of ordered sets." (Russian). Dokl. Akad. Nauk SSSR 142 (1962), 9-11.Sov. Math. Dokl. 3, 1-3 (English). MR 27 #59.

Aizenstat, A. Ja. (1962) "On homomorphisms of semigroups of endomorphisms of ordered sets." (Russian). Leningrad. Gos. Ped. Inst. Ucen. Zap. 238, 38-48. MR 29 #3567.

Aizenstat, A. Ja. (1967) "Subgroups of the semigroups of endomorphisms of ordered sets." (Russian). Leningrad. Gos. Ped. Inst. Ucen. Zap. 302, 50-57. MR 39 #4054.

Aizenstat, A. Ja. (1968) "Regular semigroups of endomorphisms of ordered sets." (Russian). Leningrad. Gos. Ped. Inst. Ucen. Zap. 387, 3-11. MR 38 #1034.

Ajtai, M. (1973) "On a class of finite lattices." Period Math. Hungar. 4, 217-220. MR 48906.

Ajtai, M. (1973) "On a class of finite lattices." Periodica Math. Hungar. 4, 217-220.

Alekseev, V.B. (1974) "Use of symmetry in finding the width of a partially ordered set." (Russian). Diskret. Analiz Vyp. 26 Grafy i Testy, 20-35, 84. MR 58 #10629.

Alo, Richard A. (1977) "Some order and topological considerations for Riesz spaces and l-groups." Symposia Mathematica, Vol. XXI (INDAM, Rome, 1975), pp. 529-546. Academic Press, London. MR 58 #7019.

Amit, R.; Shelah, S. (1976) "The complete finitely axiomatized theories of order are dense." Israel J. Math. 23, no. 3-4, 200-208. MR 58 #5162.

Anderson, Brian D.O. (1973) "Algebraic properties of minimal degree spectral factors." (French, German and Russian summaries). Automatica-J.IFAC 9, no. 4, 491-500. MR 55 #5279.

Anderson, Frank W. (1961) "Function lattices." Proc. Sympos. Pure Math, Vol. II, pp. 198-202. American Mathematical Society, Providence, R.I. MR 25 #5392.

Anderson, I. (1968) "On the divisors of a number." J. London Math. Soc. 43, 410-418.

Anderson, W.N, Jr.; Duffin, R.J. (1969) "Series and parallel addition of matrices." J. Math. Anal. Appl. 26, 576-594. MR 39 #3904.

Aoki, Masahiko; Chipman, John S.; Fishburn, Peter C. (1971) "A selected bibliography of works relating to the theory of preferences, utility, and demand." Preferences, utility and demand pp. 437-492, Harcourt Brace Jovanovich, N.Y. MR 7768.

Appert, Antoine (1947) "Ecart partiellement ordonne et uniformite." C.R. Acad. Sci, Paris 224, 442-444. MR 8 #333.

Appleson, Robert R.; Lovasz, L. (1975) "A characterization of cancellable k-ary structures." Period. Math. Hungar. 6, 17-19. MR 51 #10195.

Arditti, J.C. (1978) "Partially ordered sets and their comparability graphs, their dimension and their adjacency." Problemes combinatoires et theorie des graphes, Orsay 1976, Colloq. int. CNRS No. 260, 5-8.

Arditti, J.C.; Cori, R. (1977) "Chaines maximales de preordres." Aequationes Math. 15, no. 1, 49-54. MR 2880.

Arditti, J.C.; Jung, H.A. (1980) "The dimension of finite and infinite comparability graphs." J. London Math. Soc. (2) 21, 31-38.

Arhangelskii, V.L. (1974) "Bisemicategories of pseudoarc partial automorphisms of interior ordered sets." (Russian). Ordered sets and lattices, No. 2, pp. 3-13. Izdat. Saratov. Univ, Saratov. MR 58 #401.

Aronszajn, N. (1952) "Characterization of types of order satisfying aplha0+aplha1=aplha1+aplha0." Fund. Math. 31, 65-96.

Arrow, Kenneth J. (1951) "Social choice and individual values." Cowles Commission Monograph No. 12. John Wiley and Sons, Inc, New York, N.Y.; Chapman and Hall, Ltd, London, xi+99 pp. MR 12 #624.

Arrow, Kenneth J. (1952) "The determination of many-commodity preferences scales by two-commodity comparisons." Metroecon. 4, 105-115. MR 18 #366.

Ash, C.J. (1977) "Dense, uniform and densely subuniform chains." J. Austral. Math. Soc. Ser. A23, no. 1, 1-8. MR 3159.

Assous, Roland (1977) "Une caracterisation du omega-n-meilleur ordre." (English summary). C.R. Acad. Sci. Paris Ser. A-B 285, no. 9, A597-A599. MR 5298.

Attardi, Giuseppe (1974) "Caratterizzazione dei reticoli continui perla teoria della computazione di Dana Scott." (English summary). Calcolo 11, 33-46. MR 51 #84.

Aumann, Robert J. (1964) "Utility theory without the completeness axiom: a correction." Econometrica 32, 210-212. MR 30 #4585.

Avann, S.P. (1961) "Application of the join-irreducible excess function to semi-modular lattices." Math. Annalen, 142, 345-354.

Avann, S.P. (1972) "The lattice of natural partial orders." Aequationes Math. 8 95-102. MR 47 #81.

Avraham, U.; Shelah, S. (1981) "Martin's Axiom does not imply that every two aleph-1-dense sets of reals are isomorphic." Israel J. Math. 38, 161-176.

Babuskin, A.I.; Basta, A.L.; Belov, I.S. (1977) "Construction of a schedule for the problem of counter-routes." (Russian, English summary insert). Kibernetika (Kiev) no. 4, 130-135. MR 58 #4279.

Baclawski, K. (1977) "Galois connections and the LeRay spectral sequence." Advances Math. 25, 191-215.

Baclawski, K.; Bjorner, A. (1979) "Fixed points in partially ordered sets." Advances Math. 31, 263-287.

Baer, R. (1946) "Polarities on finite projective planes." Bull. Amer. Math. Soc. 52, 77-93.

Baer, R.M. (1957) "Certain homomorphisms onto chains." Arch. Math. 8, 93-95. MR 19 #1034.

Baer, R.M.; Osterby, O. (1969) "Algorithms over partially ordered sets." Nordisk Tidskr. Informationsbehandling (BIT)9, 97-118. MR 40 #2575.

Baer, Robert M. (1959) "On closure operators." Arch. Math. 10, 261-266. MR 21 #5585.

Baker, K.A.; Fishburn, P.C.; Roberts, F.S. (1972/73) "Partial orders of dimension 2." Networks 2, 11-28. MR 46 #104.

Baker, K.R.; Lawler, E.L.; Lenstra, J.K.; Rinnooy Kan, A.M.G. (1982) "Preemptive scheduling of a single machine to minimize maximum cost subject to release dates and precedence constraints." Oper. Res.

Baker, Kirby A. (1969) "A generalization of Sperner's lemma." J. Combinatorial Theory 6, 224-225. MR 38 #4368.

Baker, Kirby A. (1969) "A Krein-Milman theorem for partially ordered sets." Amer. Math. Monthly 76, 282-283. MR 39 #4636.

Baker, Kirby A.; Fishburn, Peter C.; Roberts, Fred S. (1970) "A new characterization of partial orders of dimension two." International Conference on Combinatorial Mathematics. (1970). Ann. New York Acad. Sci. 175, 23-24. MR 42 #140.

Balbes, R. (1979) "On counting Sperner families." J. of Comb. Theory A 27, 1-9.

Balbes, Raymond (1971) "On the partially ordered set of prime ideals of a distributive lattice." Canad. J. Math. 23, 866-874. MR 45 #128.

Balinski, M.L. (1978) "Complementarity and fixed point problems." Ed. by M.L. Balinski and R.W. Cottle. Math. Programming Stud. No. 7. North-Holland Publishing Co, Amsterdam, 1978. pp. i-viii+1-184. MR 5120.

Ballinger, B.T. (1972) "Completions of ordered sets." Canad. Math. Bull. 15, 469-472. MR 47 #8362.

Baltasar, R. Salinas (1969) "Fixpunkte in geordneten Mengen." Publ. Sem. Mat. Zaragoza 10, 163-172 (Spanish).

Banaschewski, B. (1956) "Hullensysteme und Erweiterungen von Quasi-Ordnungen," Z. Math. Logik Grundlagen Math. 2, 117-130.

Banaschewski, B.; Bruns, G. (1967) "Categorical characterization of the MacNeille completion." Arch. Math. (Basel) 18, 369-377. MR 36 #5036.

Banaschewski, B.; Hoffmann, R.-E. (editors) (1979) "Continuous lattices: Proceedings." Lect. Notes Math. 871, Springer, 413 pp.

Banaschewski, Bernhard (1953) "Uber die Konstruktion wohlgeordneter Mengen." Math. Nachr. 10, 239-245. MR 15 #690.

Banaschewski, Bernhard (1960) "On principles of inductive definition." Z. Math. Logik Grundlagen Math. 6, 248-257. MR 25 #3868.

Banaschewski, Bernhard. (1953) "Uber den Satz von Zorn." Math. Nachr. 10, 181-186. MR 15 #690.

Barbalat, I. (1952) "Les theoremes de Zorn et Kneser dans la theorie des ensembles ordonnes." Acad. Repub. Pop. Romane. Bul. Sti. Sect. Sti. Mat. Fiz. 4, 751-762. (Romanian. Russian and French summaries). MR 15 #609.

Barbut, M.; Monjardet, B. (1970) "Ordre et classification, Algebre et Combinatoire." 2 tomes, Paris, Hachette. MR 54 #7333-4.

Barry, D.L.; Robinson, D.F. (1977) "A heuristic for scheduling jobs on several identical machines." New Zealand Oper. Res. 5, no. 1, 45-53. MR 7927.

Barthelemy, J.P. (1978) "Remarques sur les proprietes metriques des ensembles ordonnes." Math. Sci. Hum. 61, 39-60.

Barthelemy, J.P. (1979) "Caracterisations axiomatiques de la distance de la difference metrique entre des relations binaires." Math. Sci. Hum. 67, 85-113.

Barthelemy, J.P. (1980) "An asymptotic equivalent for the number of total preorders on a finite set." Discrete Mathematics, 29, 311-313.

Barthelemy, J.P.; Monjardet, B. (1981) "The median procedure in cluster analysis and social choice theory." Math. Soc. Science, 1, 225-267.

Barthelemy, Jean-Pierre (1978) "Caracterisation metrique des ensembles ordonnes arborescents." (English summary). C.R. Acad. Sci. Paris Ser. A-B 286, no. 4, A189-A190. MR 57 #12274.

Barthelemy, Jean-Pierre (1978) "Le graphe de couverture d'un ensemble ordonne. Exemple des preordres totaux et des partitions d'un ensemble fini." (English summary). C.R. Acad. Sci. Paris Ser. A-B 286, no. 20, A859-A861. MR 58 #21791.

Bartol, W. (1972) "On the existence of machine homomorphisms. II." (Russian summary). Bull. Acad. Polon. Sci. Ser. Sci. Math. Astronom. Phys. 20, 773-777. MR 47 #3125.

Bauer, H.; Keimel, K.; Kohler, R. (1981) "Verfeinerungs und Kurzungssatze fur Producte geordneter topologischer Raume und fur Funktionen (-halb-) Verbande." Lect. Notes Math. 871, Springer, 20-44.

Baumgartner, J. (1973) "All aleph-1-dense sets of reals can be isomorphic." Fund. Math. 79, 101-106. MR 47 #6483.

Baumgartner, J. (1976) "A new class of order types." Annals Math. Logic 9, 187-222.

Bean, Phillip W. (1974) "Helly and Radon-type theorems in interval convexity spaces." Pac. J. Math. 51, 363-368. MR 50 #11034.

Bednarek, A.R. (1964) "On chain-equivalent orders." Nieuw Arch. Wisk (3) 12, 137-138. MR 30 #4696.

Belding, W. Russell (1978) "Bases for the positive cone of a partially ordered module." Trans. Amer. Math. Soc. 235, 305-313. MR 57 #12336.

Belov, I.S.; Stolin, Ja N. (1974) "An algorithm for the single-route scheduling problem." (Russian). Mathematical economics and functional analysis (Russian), pp. 248-257. Izdat. "Nauka", Moscow. MR 17780.

Benado, M. (1961) "On the general theory of partially ordered sets." (Russian. Czech and French summaries). Acta Fac. Nat. Univ. Comenian. 5, 397-429. MR 27 #61.

Benado, Michael S. (1969) "Ensembles ordonnes, fonctions reelles, espaces diametriques." Publ. Inst. Math. (Beograd) (N.S.) 9 (23), 143-152. MR 40 #63.

Benado, Michael [Benado, Mihail] (1964) "Remarques sur la theorie des multitreillis. VI." Mat.-Fzy. Casopis Sloven. Akad. Vied 14. 163-207. MR 31 #904.

Benado, Michael [Benado, Mihail] (1972/73) "Ensembles ordonnes et fonctions monotones." Univ. Lisboa Revista Fac. Ci. A (2) 14, 165-194. MR 48 #8313.

Benado, Mihail (1951) "La notion de normalite et les theoremes de decomposition de l'algebre." Acad. Repub. Pop. Romane. Stud. Cerc. Mat. 1 (1950), 282-317. (Romanian, Russian and French summaries). MR 16 #212.

Benado, Mihail (1952) "Les ensembles partiellement ordonnes et le theoreme de raffinement de O. Schreier." Acad. Repub. Pop. Romane. Bul. Sti. Sect. Sti. Mat. Fiz 4, 585-591. (Romanian, Russian and French summaries). MR 15.

Benado, Mihail (1953) "Sur une generalisation de la notion de structure." Acad. Repub. Pop. Romane. Bul. Sti. Sect. Sti. Mat. Fiz. 5, 41:48. (Romanian. Russian and French summaries). MR 16 #668.

Benado, Mihail (1955) "Les ensembles partiellement ordonnes et le theoreme de raffinement de Schreier. II. Theorie des multistructures." Czechoslovak Math. J. 5 (80), 308-344. (Russian summary). MR 17 #937.

Benado, Mihail (1956) "Bemerkungen zu einer Arbeit von Oystein Ore." Rev. Math. Pures Appl. 1, no. 2, 5-12. MR 18 #275.

Benado, Mihail (1958) "Sur la fonction de Mobius." C.R. Acad. Sci. Paris 246, 863-865. MR 20 #6370,#6371.

Benado, Mihail (1959) "Bemerkungen zur Theorie der Vielverbande." Math. Nachr. 20, 1-16. MR 22 #19.

Benado, Mihail (1960) "Bemerkungen zur Theorie der Vielverbande. IV. uber die Mobius'sche Funktion." Proc. Cambridge Philos. Soc. 56, 291-317. MR 22 #10929.

Benado, Mihail (1960) "La theorie des multitreillis et son role en algebre et en geometrie." Publ. Sci. Univ. Alger Ser. A7, 41-58. MR 27 #2447.

Benado, Mihail (1960) "Sur la theorie generale des ensembles partiellement ordonnes." Publ. Sci. Univ. Alger Ser. A 7, 5-39. MR 26 #2370.

Benado, Mihail (1961) "Sur la theorie generale des ensembles partiellement ordonnes, I,II,III,IV." Proc. Japan Acad. 36, 590-594, 595-597,636-638 (1960); 37,83-84. MR 24 #A3089,#A3090.

Benado, Mihail (1961) "Sur la theorie generale des ensembles partiellement ordonnes." Publ. Sci. Univ. Alger Ser. A 8, 90pp. MR 27 #5701.

Benado, Mikail (1958) "Sur la theorie generale des ensembles partiellement ordonnes." C.R. Acad. Sci. Paris 247, 2265-2268. MR 20 #5745.

Benes, V.E. (1967) "Combinatorial solution to the problem of optimal routing in progressive gradings." Bell System Tech. J. 46, 865-881. MR 36 #3604.

Benes, Vaclav E. (1967) "On comparing connecting networks." J. Combinatorial Theory 2, 507-513. MR 36 #6204.

Beran, L. (1974) "On a construction of amalgamation I." Acta Univ. Carolinae--Math. et Phys. 14 (1973), no. 2, 31-39. MR 50 #9717.

Bird, Elliott H. (1973) "Automorphism groups of partial orders." Bull. Amer. Math. Soc. 79, 1011-1015. MR 47 #6820.

Bird, Elliott; Greene, Curtis; Kleitman, Daniel J. (1975) "Automorphisms of lexicographic products." Discrete Math. 11, 191-198.

Birkhoff, G. (1937) "Extended arithmetic." Duke Math. J. 3, 311-316.

Birkhoff, G. (1937) "Rings of sets." Duke Math. J. 3, 443-454.

Birkhoff, Garrett (1942) "Generalized arithmetic." Duke Math. J. 9, 283-302. [MF 6869]. MR 4 #20.

Birkhoff, Garrett (1952) "Some problems of lattice theory." Proc. Internat. Congr. Math. (Cambridge, Mass. Aug. 30 - Sept. 6, 1950) 2, 4-7.

Bittner, Leonhard. (1972) "Abschatzungen bei Variationsmethoden mit Hilfe von Dualitatssatzen. III." Tagungsbericht zur ersten Tagung der WK Analysis (1970).Beitrage Anal. Heft 4, 145-157. MR 49 #4569.

Bjorner, A. (1980) "Shellable and Cohen-Macaulay partially ordered sets." Trans. Amer. Math. Soc. 260, 159-183.

Bjorner, A. (1981) "Homotopy type of posets and lattice complementation." J. Combin. Theory 30, 90-100.

Bjorner, A.; Rival, I. (1980) "A note on fixed points in semimodular lattices." Discrete Math. 29, 245-250.

Blair, R.L.; Tomber, M.L. (1960) "The axiom of choice for finite sets." Proc. Amer. Math. Soc. 11, 222-226. MR 22 #10926.

Blau, J.K. (1979) "Semiorders and collective choice." J. Econ. Theory 21, 1, 195-206.

Blau, Julian H. (1957) "The existence of social welfare functions." Econometrica 25, 302-313. MR 22 #10790.

Blazewicz, Jacek (1978) "Complexity of computer scheduling algorithms under resource constraints." Fond. Control Engrg. 3, no. 2, 51-57. MR 58 #20375.

Blazewicz, Jacek; Cellary, Wojciech; Slowinski, Roma; Weglarz, Jan (1976) "Deterministic problems of scheduling tasks on parallel processors. I.Sets of independent tasks." (Polish. English and Russian summaries). Podstawy Sterowania 6, no. 2, 155-177. MR 5 .

Bleicher, M.N.; Schneider, H. (1966) "Completions of partially ordered sets and universal algebras." Acta Math. Acad. Sci. Hungar. 17, 271-301. MR 34.

Blin, J.M.; Whinston, Andrew B. (1974/75) "Discriminant functions and majority voting." Management Sci. 21, no. 5, 557-566. MR 53 #12648.

Blin, Jean-Marie; Satterthwaite, Mark A. (1977) "On preferences, beliefs, and manipulation within voting situations." Econometrica 45, no. 4, 881-888. MR 7770.

Blyth, T.S. (1975/76) "The structure of certain ordered regular semigroups." Proc. Roy. Soc. Edinburgh Sect. A 75, no. 3, 235-257. MR 5385.

Blyth, T.S. (1977/78) "Perfect Dubreil-Jacotin semigroups." Proc. Roy. Soc. Endinburgh Sect. A 78, no. 1-2, 101-104. MR 57 #9627.

Bock, R. Darrell. (1956) "The selection of judges for preference testing." Psychometrika 21, 349-366. MR 18 #548.

Bod, Peter (1976) "On closed sets having a least element." Optimization and Operations Research (Proc. Conf, Oberwolfach, 1975), pp. 23-34. Lecture Notes in Econom. Math. Systems, Vol 117, Springer, Berlin. MR 14706.

Bod, Peter. (1972) "On indifferent programming problems." (Hungarian, English and Russian summaries). Szigma-Mat.-Kozgazdasagi Folyoirat 5, 137-145. MR 48 #13192.

Bogart, K.; Rabinovich, I.; Trotter, W. (1976) "A bound on the dimension of interval orders." J. Comb. Theory, 21, 319-328.

Bogart, K.P. (1973) "Preference structures II: Distances between asymmetric relations." SIAM J. Appl. Math., 29, 2, 254-262.

Bogart, K.P.; Rabinovitch, I.; Trotter, W.T. (1976) "A bound on the dimension of interval orders." J. Combinatorial Theory (Ser. A) 21, 319-328. MR 55059.

Bogart, K.P.; Rabinovitch, I.; Trotter, W.T. Jr. (1976) "A bound on the dimension of interval orders." J. Combinatorial Theory, A21, 319-328.

Bogart, Kenneth P. (1970) "Decomposing partial orderings into chains." J. Combinatorial Theory 9, 97-99. MR 41 #110.

Bogart, Kenneth P. (1973) "Maximal dimensional partially ordered sets. I. Hiraguchi's theorem." Discrete Math. 5, 21-31. MR 47 #6562.

Bogart, Kenneth P. (1973) "Preference structures. I. Distances between transitive preference relations." Mathematical Sociology 3, 49-67. MR 52 #16728a,b.

Bogart, Kenneth P.; Trotter, William T, Jr. (1973) "Maximal dimensional partially ordered sets. II: Characterization of 2n-element posets with dimension n." Discrete Math. 5, 33-43. MR 47 #6563.

Bollobas, B. (1973) "Sperner systems consisting of pairs of complementary subsets." J. Comb. Theory A, 363-366.

Bonnet, R. (1971) "Categorie et alpha-categorie des ensembles ordonnes." Publ. Dep. Math. (Lyon) 8, fasc. 1, 69-90. MR 48 #186.

Bonnet, R. (1975) "On the cardinality of the set of initial intervals of a partially ordered set." Infinite and Finite Sets (Colloq, Keszthely, 1973; Vol. I, pp. 189-198. Colloq. Math. Soc. Janos Bolyai, Vol. 10, North-Holland, Amsterdam. MR 51 #5421.

Bonnet, Robert; Pouzet, Maurice (1969) "Extension et stratification d'ensembles disperses." C.R. Acad. Sci. Paris Ser. A-B 268, A1512-A1515. MR 39 #4055.

Borohovic, R.E.; Zenin, F.B.; Zenina, Ja.A. (1977) "An algorithm for the compilation of a schedule under the conditions of an automatic control system." (Russian). Data Transmission in Automatic Control Systems, pp. 71-81,99. Izdat. "Stiinca", Kishinev. MR .

Borsuk, K. (1954) "A theorem on fixed points." Bull. Acad. Polon. Sci. Cl. III. 2, 17-20. MR 16 #275.

Botto Mura, Roberta [Mura, Roberta]; Rhemtulla, Akbar (1977) "Orderable groups." Lecture Notes in Pure and Applied Math, Vol. 27. Marcel Dekker, Inc, N.Y.-Basel, iv + 169pp. MR 58 #10652.

Bouchet, A. (1968) "An algebraic method for minimizing the number of states in an incomplete sequential machine." IEEE Trans. Computers C-17, 795-798. MR 38 #4214.

Bowman, V.J.; Colantoni, C.S. (1972) "The extended Condorcet condition: a necessary and sufficient condition for the transitivity of majority decision. collection of articles on social choice." J. Mathematical Sociology 2, 267-283. MR 54 #4572.

Brace, A.; Daykin, D.E. (1972) "Sperner type theorems for finite sets." Proceedings, Combinatorics Conf, Oxford, 18-37.

Brock, William A. (1970) "An axiomatic basis for the Ramsey-Weizsacker overtaking criterion." Econometrica 38, 927-929. MR 46 #6792.

Brouwer, A.E.; Schrijver, A. (1974) "On the period of an operator, defined on antichains." Mathematisch Centrum Afdeling Zuivere Wiskunde ZW 24/74. Mathemaatisch Centrum. Amsterdam. i + 13 pp. MR 50 #1990.

Brouwer, L.-E.-J. (1950) "Remarques sur la notion d'ordre." C.R. Acad. Sci. Paris 230, 263-265. MR 11 #305.

Brouwer, L.-E.-J. (1950) "Sur la possibilite d'ordonner le continu." C.R. Acad. Sci. Paris 230, 349-350. MR 11 #305.

Brown, T.A.; Strauch, R.E. (1965) "Dynamic programming in multiplicative lattices." J. Math. Anal. Appl. 12, 364-370. MR 32 #2248.

Brown, T.C. (1975) "A proof of Sperner's lemma via Hall's theorem." Math. Proc. Camb. Phil. Soc. 78, 387.

Brown, W.G. (1965) "Historical note on a recurrent combinatorial problem." Amer. Math. Monthly, ***, 973-976.

Brucker, P. (1970) "Verbande stetiger Funktionen und kettenwertige Homomorphismen." Math. Ann. 187, 77-84. MR 41 #5267.

Brucker, Peter; Garey, M.R.; Johnson, D.S. (1977) "Scheduling equal-length tasks under treelike precedence constraints to minimize maximum lateness." Math. Oper. Res. 2, no. 3, 275-284. MR 17781.

Bruns, G. (1962) "Darstellungen und Erweiterungen geordneter Mengen I." J. Reine und Angew Math. 209, 167-200.

Bruns, G. (1962) "Darstellungen und Erweiterungen geordneter Mengen II." J. Reine und Angew Math. 210, 1-23.

Bruns, Gunter (1962) "Darstellungen und Erweiterungen geordneter Mengen. I. II." J. Reine Angew. Math. 209, 167-200. 210, 1-23.

Bruns, Gunter (1967) "A lemma on directed sets and chains." Arch Math. (Basel)18, 561-563. MR 36 #3683.

Buchi, J. Richard (1948) "Die Boolesche Partialordnung und die Paarung von Gefugen." Portugaliae Math. 7, 119-180.

Buchi, J. Richard (1966) "Algebraic theory of feedback in discrete systems, I." Automata Theory, Academic Press, pp. 70-101. MR 34 #7293.

Bulavskii, V.A.; Jakovleva, M.A. (1977) "Algorithms with an ordered basis for a two-component linear programming problem." (Russian). Optimizacija No. 19 (36), 29-47, 154. MR 58 #32912.

Bumby, R.; Cooper, E.; Latch, D. (1975) "Interactive computation of homology of finite partially ordered sets." SIAM J. Comput. 4, no. 3, 321-325. MR 52 #13520.

Burros, Raymond H. (1976) "Complementary relations in the theory of preference." Theory and Decision 7, no.3, 181-190. MR 55 #9829.

Burscheid, Hans-Joachim von (1972) "Uber die Breite des endlichen Kardinalen Produktes von endlichen Ketten." Math. Nach. 52, 283-295.

Butler, David A. (1977) "An importance ranking for system components based upon cuts." Operations Res. 25, no. 5, 874-879. MR 2462.

Butler, Kim Ki Hang (1975) "New representations of posets." Infinite and Finite Sets (Colloq, Keszthely, 1973), I, 241-250. Colloq. Math. Soc. Janos Bolyai, Vol. 10, North-Holland, Amsterdam. MR 51 #5422.

Butler, Kim Ki Hang; Markowsky, George (1974) "The number of partially ordered sets. II." J. Korean Math. Soc. 11, 7-17. MR 50 #1991.

Butler, Kim Ki-hang (1972) "The number of partially ordered sets." J. Combinatorial Theory Ser. B 13, 276-289. MR 46 #7045.

Cameron, Neil; Miller, J.B. (1975) "Topology and axioms of interpolation in partially ordered spaces." J. Reine Angew. Math. 278/279, 1-13. MR 11862.

Canfield, E.R. "On a problem of Rota." Advances in Math.

Canfield, E.R. "On the location of the maximum Stirling numbers of the second kind." Studies in Appl. Math.

Carruth, J.H. (1968) "A note on partially ordered compacta." Pac. J. Math. 24, 229-231.

Carruth, Philip W. (1951) "Sums and products of ordered systems." Proc. Amer. Math. Soc. 2, 896-900. MR 13 #425.

Carruth, Philip W. (1952) "Products of ordered systems." Proc. Amer. Math. Soc. 3, 983-987. MR 14 #529.

Cassidy, R.G.; Field, C.A.; Sutherland, W.R.S. (1973) "Inverse programming: theory and examples." Math. Programming 4, 297-308. MR 48 #5649.

Cebotarev, A.S.; Safranskii, V.V. (1976) "The problem of scheduling a complex of performances that is represented by a resource network model of special form." The Programmed Method of Control, No.3, pp. 19-28,173. Akad. Nauk SSR Vycisl.Cetr, Moscow. MR 58 .

Cebotareva [Chebotareva], L.K. (1967) "On the entry problem for free and almost free lattices." Sibir Mat. Zurn. 8, 399-405 (1967)(Russian), engl. Ubersetzung in Siber. Math. J. 8, 290-295.

Cerny, Martin. (1973) "Preference relations and order-preserving functions." (Czech summary). Ekonom.-Mat. Obzor 9, 275-291. MR 48 #3460.

Chacron, J. (1971) "Nouvelles correspondances de Galois." Bull. Soc. Math. Belgique 23, 167-178.

Chajda, Ivan; Niederle, Josef (1977) "Ideals of weakly associative lattices and pseudo-ordered sets." Arch. Math. (Brno) 13 no. 4, 181-186. MR 3015.

Chandon, J.L.; Lemaire, J.; Pouget, J. (1978) "Denombrement des quasi-ordres sur un ensemble fini." Math. Sci. Hum., 62, 61-80.

Chang, C.C. (1961) "Cardinal and ordinal multiplication of relational types." Proc. Symposia in Pure Math., Amer. Math. Soc., 2, 123-128.

Chang, C.C. (1961) "Ordinal factorization of finite relations." Trans. Amer. Math. Soc. 101, 259-293.

Chang, C.C.; Jonsson, B.; Tarski, A. (1964) "Refinement properties for relational structures." Fund. Math. 55, 249-281.

Chang, C.C.; Morel, Anne C. (1960) "Some cancellation properties for ordinal products of relations." Duke Math. J. 27; 171-182.

Chen, C.C.; Koh, S.H. (1972) "On the embeddability of partially ordered sets into distributive lattices." Nanta Math. 5, no. 2, 1-7. MR 48 #187.

Chen, N.F.; Liu, C.L (1975) "On a class of scheduling algorithms for multiprocessors computing systems." Parallel Processing, pp. 1-16. Lecture Notes in Comput. Sci, Vol. 24, Springer, Berlin. MR 14639.

Chipman, John S. (1960) "The foundations of utility." Econometrica 28, 193-224. MR 22 #9284.

Chipman, John S. (1971) "Consumption theory without transitive indifference." Preferences, Utility and Demand, pp. 224-253. Harcourt Brace Jovanovich, N.Y. MR 14569.

Chipman, John S. (1971) "On the lexicographic representation of preference orderings." Preferences, Utility and Demand, pp. 276-288. Harcourt Brace Jovanovich, N.Y., 1971. MR 58 #4199.

Chittenden, E.W. (1947) "On the number of paths in a finite partially ordered set." Amer. Math. Monthly 54, 404-405. MR 9 #3.

Choe, T.H.; Hong, U.H. (1976) "Extensions of completely regular ordered spaces." Pac. J. Math. 64.

Choe, Tae Ho (1958) "An isolated point in a partly ordered set with interval topology." Kyungpook Math. J. 1, 57-59. MR 21 #3350.

Choe, Tae Ho (1959) "The topologies of partially ordered set with finite width." Kyungpook Math. J. 2, 17-22. MR 21 #4928.

Choe, Tae Ho (1962) "On a continuous mapping between partially ordered sets with some topology." Kyungpook Math. J. 5, 9-13. MR 26 #52.

Choudbury, A.C. (1960) "On the isotone mapping of a lattice." Bull. Calcutta Math. Soc. 52, 39-44. MR 24 #A62.

Chung, F.R.K.; Fishburn, P.C.; Graham, R.L. (1980) "On unimodality for linear extensions of partial orders." SIAM Jour. Alg. Disc. Meth., 1, 405-410.

Chvalina, J. (1973) "On the number of general topologies of a finite set." (Russian summary). Scripta Fac. Sci. Natur. UJEP Brunensis--Math. 3, 7-22. MR 51 #181.

Chvatal, V.; Hammer, P.L. (1977) "Aggregation of inequalities in integer programming." Annals of Discrete Math., 1, 145-162.

Ciobanu, G. (1972) "A multi-project programming model with resources levelling." (French and Russian summaries). Econom. Comp. Econom. Cybernet. Stud. Res. no.3, 61-68. MR 49 #4553.

Coffman, E.G, Jr. (1975) "Deterministic sequencing: complexity and optimal algorithms." Computing Systems 1975 (Proc. Fourth Texas, Conf, Univ. Texas, Austin, Tex), Paper no. 5B-1, 12pp. Univ. Texas, Austin, Tex. MR 4795.

Coffman, E.G. Jr.; Graham, R.L. (1972) "Optimal scheduling for two-processor systems." Acta Informat. 1, 200-213.

Cohn, P.M. (1957) "Groups of order automorphisms of ordered sets." Mathematika 4, 41-50. MR 19 #941.

Collection (1975) "Collection of papers on the theory of ordered sets and general algebra." Slovenski Pedagogicke Nakladatelstvo, Bratislava, 64 pp. MR 49 #10492.

Collyer, P.J.; Sullivan, R.P. (1975) "Closure operators." J. Austral. Math. Soc. 19, 321-336. MR 51 #12629.

Colmerauer, Alain (1970) "Total precedence relations." J. Assoc. Comput. Mach. 17, 14-30. MR 43 #1775.

Cornwall, Richard R. (1970) "The use of prices to characterize the core of an economy." J. Econom. Theory 1 (1969), 353-373. MR 53 #7375.

Costovici, Gh. (1975) "Automorphisms of order with the same value at some point." (Romanian summary). Bul. Inst. Politehn. Iasi (N.S.) 21 (25), no. 3-4, sect. I, 39-40. MR 52 #13527.

Crapo, Henry (1978) "Lexicographic partial order." Trans. Amer. Math. Soc. 243, 37-51. MR 58 #10630.

Crapo, Henry (1981) "Ordered sets: Retracts and connections." J. Pure and Applied Algebra 22.

Crapo, Henry (1981) "Unities and negation: On the representation of finite lattices." J. Pure and Applied Algebra 22.

Cuesta Dutari, Norherto (1959) "Matematica del orden." [Mathematics of order.] Revista de la Real Academia de Ciencias Exactas, Fisicas y Naturales, Tomos Lii y Liii, Madrid, 513 pp. MR 21 #2601,#2602a,#2602b.

Cuesta, N. (1954) "Ordinal algebra." Rev. Acad. Ci. Madrid 48, 103-145. (Spanish). MR 16 #1091.

Cuesta, N. (1954) "Ordinal arrangement." Rev. Mat. Hisp. Amer. (4) 14,237-268. (Spanish). MR 16 #1091.

Cuesta, N. (1958) "On an article of Mac Neille." Rev. Mat. Hisp.-Amer. (4) 18, 3-9. (Spanish). MR 21 #2600.

Culik, Karel (1959) "Uber die Homomorphismen der teilweise geordneten Mengen und Verbande." Czechoslovak Math. J. 9 (84), 496-518. (Russian summary). MR 22 #10927.

Cvetkovic, D.M.; Lackovic, I.B.; Simic, S.K. (1976) "Graph equations, graph inequalities and a fixed point theorem." Publ. Inst. Math. (Beograd) (N.S.) 20 (34), 59-66. MR 178.

Czajsner, Jerzy (1969) "Some equivalence relations in relational systems." Prace Mat. 13, 81-89. MR 40 #5509.

D'Errico, Alfonso (1976) "Un modello delle preferenze nelle operazioni decisionali multi-criteriali." (English summary). Boll. Un. Mat. Ital. B (5) 13, no.1, 202-215. MR 55 #9830.

Da, Taen-Yu; DeMarr, Ralph (1978) "Positive derivations on partially ordered linear algebra with an order unit." Proc. Amer. Math. Soc. 72, no. 1, 21-16. MR 58 #10658.

Dacic, R. (1971) "A counterexample to a Kurepa's conjecture." Mat. Vesnik 8 (23), 351. MR 46 #5185.

Dai, Taen Yu; DeMarr, Ralph (1977) "Isotone functions on partially ordered linear algebras with a multiplicative diagonal map." Proc. Amer. Math. Soc. 65, no. 1, 11-15. MR 58 #5443.

Dantzig, G.B.; and Hoffman, A.J. (1956) "Dilworth's theorem on partially ordered sets. Linear inequalities and Related Systems, pp. 207-214." Annals of Mathematics Studies, no. 38. Princeton University Press, Princeton, N.Y. MR 18 #555.

Dantzig, G.B.; Fulkerson, D.R. (1954) "Minimizing the number of tankers to meet a fixed schedule." Naval Res. Logist. Quart. 1, 217-222.

Das, S.K. (1977) "A machine representation of finite T0 topologies." J. Assoc. Comp. Mach. 24, 676-692.

Dauns, John (1975) "Ordered domains." Symposia Mathematica, Vol. XXI (INDAM, Rome) pp. 565-587. Academic Press, London, 1977. MR 58 #5444.

Davey, B.A.; Duffus, D.; Quackenbush, R.W.; Rival, I (1978) "Exponents of finite simple lattices." J. London Math. Soc. (2)17, no. 2, 203-221. MR 57 #12308.

Davey, Brian A. (1973) "A note on representable posets." Algebra Universalis 3, 345-347. MR 50 #2006.

Davida, G.I.; Linton, D.J. (1976) "A new algorithm for the scheduling of tree structured tasks." Proc. Conf. Inform. Sci. and Syst., Baltimore, MD, 543-548.

Davies, Roy O.; Hayes, Allan; Rousseau, George (1971) "Complete lattices and the generalized Cantor theorem." Proc. Amer. Math. Soc. 27, 253-258. MR 42 #2990.

Davio, M; Bioul, G. (1970) "Representation of lattice functions." Phillips Res. Rep. 25, 370-388. MR 50 #12461.

Davio, M; Thayse, A. (1973) "Representation of fuzzy functions." Philps Res. Rep.28, 93-106. MR 48 #10666.

Davis, A.C. (1955) "A characterization of complete lattices." Pacific J. Math. 5, 311-319.

Davis, Gary (1975) "Automorphic compactifications and the fixed point lattice of a totally-ordered set." Bull. Austral. Math. Soc. 12, 101-109. MR 51 #1762.

Day, M.M. (1945) "Arithmetic of ordered systems." Trans. Amer. Math. Soc. 58, 1-43.

Daykin, D.E. (1976) "Poset functions commuting with the product and yielding Chebyshev type inequalities." Problemes Comb. et Theorie des Graphs, J.C. Bermond, ed., C.N.R.S. Colloques, Paris, 93-98.

Daykin, D.E. (1977) "A lattice is distributive iff $|A||B| \leq |A \vee B||A \wedge B|$." Nanta Math., 10, 58-60.

Daykin, D.E.; Godfrey, J.; Hilton, A.J.W. (1974) "Existence theorems for Sperner families." J. Comb. Theory A 17, 245-251.

Daykin, D.E.; Kleitman, D.J.; West, D.B. (1979) "The number of meets between two subsets of a lattice." J. Comb. Theory A 26, 135-156.

Daykin, D.E.; Lovasz, L. "The number of values of a boolean function." J. Lond. Math 12, 255.

Daykin, David E. (1964) "Generalised Mobius inversion formulae." Quart. J. Math. Oxford Ser (2) 15, 349-354.

Daykin, David E. (1975) "Antichains in the lattice of subsets of a finite set." Nanta Math. 8 no. 2, 84-94.

De Barros, Constantino M. (1968) "Filters in partially ordered sets." Portugal, Math. 27, 87-98. MR 42.

Dean, R.A. (1956) "Completely free lattices generated by partially ordered sets." Trans. Amer. Math. Soc. 83, 238-249.

Dean, R.A.; Keller, Gordon (1968) "Natural partial orders." Canad. J. Math. 20, 535-554. MR 37 #1279.

Deb, Rajat (1977) "On Schwartz's rule." J. Econom. Theory 16, no. 1, 103-110. MR 17706.

Debreu, Gerard (1969) "Neighboring economic agents." (French summary). La Decision,2: Agregation et Dynamique des Ordres de Preference (Actes Colloq. Internat, Aix-en-Provence, 1967), pp. 85-90. Editions du Centre Nat. Recherche Sci, Paris. MR 43 #4431.

Debreu, Gerard (1972) "Smooth preferences." Econometrica 40, 603-615. MR 48 #13193.

deBruijn, N.; Tengbergen, C.; Kruyswijk, D. (1951) "On the set of divisors of a number." Nieuw Arch. Wiskunde 23, 191-193.

Dedekind, R. (1872) "Stetigkeit und Irrationalzahlen." Vieweg, 7 ed. 1969 Braunschweig.

Defrenne, Anne (1978) "The timetabling problem: a survey." Cahiers Centre Etudes Rech. Oper. 20, no. 2, 163-169. MR 58 #20376.

Deheuvels, R. (1962) "Homologie des ensembles ordonnes et des espaces topologiques." Bull. Soc. Math. France 90, 261-321.

Demarr, Ralph [DeMarr, Ralph E.] (1964) "Common fixed points for isotone mappings." Colloq. Math. 13, 45-48. MR 32 #5556.

Derderian, J.C. (1969) "Galois connections and pair algebras." Canad. J. Math. 21, 498-501. MR 39 #100.

Derderian, Jean-Claude (1967) "Residuated mappings." Pacific J. Math. 20, 35-43. MR 35 #1511.

Derderian, Jean-Claude (1969) "A completion for partially ordered sets." J. London Math. Soc. 44, 360. MR 38 #3184.

Deschamps, J.P. (1973) "Partially symmetric switching functions." Philips Res. Rep 28, 245-264. MR 48 #13489.

Deshpande, J.V. (1968) "On continuity of a partial order." Proc. Amer. Math. Soc. 19, 383-386. MR 38 #4369.

Devadze, H.M. (1968) "Generating sets of the semigroup of endomorphisms of a finite linearly ordered set." (Russian). Leningrad. Gos. Ped. Inst. Ucen. Zap. 387, 101-111. MR 38 #1195.

Devide, V. (1963/64) "On monotonous mappings of complete lattices." Fund. Math. 53, 147-154. MR 28 #2982.

Devide, Vladimir (1963) "A note on order relations." Publ. Math. Debrecen 10, 155-156. MR 30 #35.

Dhall, Sudarshan K.; Liu, C.L. (1978) "On a real-time scheduling problem." Operations Res. 26, no. 1, 127-140. MR 5053.

Dhar, Deepak (1978) "Entropy and phase transitions in partially ordered sets." J. Math. Phys. 19, no. 8 1711-1713. MR 58 #20062.

Diego, A.; de Pousa, A.G.; Suardiaz, E. (1977) "Notes on extremal semitensions." JMath. Japon. 22no. 1,1-12. MR 15452.

Dieudonne, Jean. (1941) "Sur la theorie de la divisibilite." Bull. Soc. Math. France 69, 133-144. [MF 13242]. MR 7 #110.

Dilworth, R.P. (1950) "A decomposition theorem for partially ordered sets." Ann. of Math. (2) 51, 161-166. MR 11 #309.

Dilworth, R.P. (1960) "Some combinatorial problems on partially ordered sets." Proc. Sympos. Appl. Math, Vol. 10, pp. 85-90. American Mathematical Society, Providence, R.I. MR 22 #6737.

Dilworth, R.P.; Greene, C. (1971) "A counterexample to the generalization of Sperner's theorem." J. Comb. Theory, 10, 18-21.

Doignon, J.P.; Ducamp, A.; Falmagne, J.C. (1981) "The bidimension of a relation." Colloque international sur la theorie des graphes et la combinatoire, Luminy.

Doignon, J.P.; Falmagne, J.C. (1974) "Difference measurement and simple scalability with restricted solvability." J. Mathematical Psychology 11, 473-499. MR 51 #5071.

Dokas, Lambros (1963) "Sur certains completes des ensembles ordonnes munis d'operations: completes de Dedekind et de Kurepa des ensembles partiellement ordonnes." C.R.Acad. Sci, Paris 256, 2504-2506.

Dokas, Lambros (1964) "Sur certains completes des ensembles ordonnes munis d'operations: rectifications a mes notes precedentes." C.R. Acad. Sci. Paris 258, 30-33. MR 28 #3948.

Dokas, Lambros (1965) "Sur certains completes des ensembles ordonnes munis d'operations." C.R. Acad. Sci. Paris 261,, 9-12. MR 32 #5554.

Dokas, Lambros (1969) "Prolongement des applications des ensembles partiellement ordonnes dans un espace topologique a certains completes de ces ensembles." C.R. Acad. Sci. Paris Se. A-B 269, A616-A619. MR 41 #3327.

Domosmickaja, N.E. (1972) "Reductions of a generalized valuation of elements for partially ordered semigroups of a certain type to valuation of classes." (Russian). Perm. Politehn, Inst. Sb. Naucn. Trudov 110, 144-153. MR 57 #9628.

Dowling, T.A.; Denig, W.A. (1977) "Geometries associated with partially ordered sets." M. Aigner (ed.), Higher Combinatorics, D. Reidel Publishing Co, Dordrecht-Honnald, 103-116.

Drake, David A. (1971) "Another note on Sperner's lemma." Canad. Math. Bull. 14, 255-256. MR 46 #8916.

Drozd, Ju. a. (1974) "Coxeter transformations and representations of partially ordered sets." (Russian). Funkcional. Anal. i Prilozen. 8, vyp. 3, 34-42. MR 50 #4412.

Dubreil-Jacotin, L.; Croisot, R. (1952) "Equivalences regulieres dans un ensemble ordonne." Bull. Soc. Math. France 80, 11-35. MR 14 #529.

Ducamp, A. (1967) "Sur la dimension d'un ordre partiel." (English summary). Theory of Graphs (Internat. Sympos, Rome, 1966), pp. 103-112. Gordon and Breach, New York; Dunod, Paris. MR 36 #3684.

Duffus, D.; Jonsson, B.; Rival, I (1978) "Structure results for function lattices." Canad. J. Math. 30, no. 2, 392-400. MR 57 #12309.

Duffus, D.; Poguntke, W.; Rival, I. (1980) "Retracts and the fixed point problem for finite partially ordered sets." Canad. Math. Bull. 23, 231-236.

Duffus, D.; Rival, I. (1976) "Crowns in dismantlable partially ordered sets. Combinatorics." Colloq. Math. Janos Bolyai 18, 271-292.

Duffus, D.; Rival, I. (1979) "Retracts of partially ordered sets." J. Australian Math. Soc. (Ser. A) 27, 495-506.

Duffus, D.; Rival, I. (1980) "A note on weak embeddings of distributive lattices." Algebra Universalis, 10, 258-259.

Duffus, D.; Rival, I. (1981) "A structure theory for ordered sets." Discrete Math. 35, 53-118.

Duffus, D.; Rival, I.; Simonovits, M. (1980) "Spanning retracts of a partially ordered set." Discrete Math. 32, 1-7.

Duffus, D.; Wille, R. (1978) "Automorphism groups of function lattices." Proceedings of a Meeting on Universal Algebra, Estergom.

Duffus, D.; Wille, R. (1979) "A theorem on partially ordered sets of order-preserving mappings." Proc. Amer. Math. Soc. 76, 14-16.

Duffus, Dwight; Rival, Ivan (1978) "A logarithmic property for exponents of partially ordered sets." Canad. J. Math. 30, no. 4, 979-807. MR 58 #16432.

Duric, Milan (1973) "On a fixobject property for lcb-semigroupoids." Mat. Vesnik 10 (25), 19-20. MR 52 #107.

Dushnik, Ben (1950) "Concerning a certain set of arrangements." Proc. Amer. Math. Soc. 1, 788-796. MR 12 #470.

Dushnik, Ben and Miller, E.W. (1941) "Partially ordered sets." Amer. J. Math. 63, 600-610. [MF 4684]. MR 3.

Dwinger, Ph (1954) "On the closure operators of a complete lattice." Nederl. Akad. Wetensch. Proc. Ser. A 57-Indag. Math. 16, 560-563. MR 16 #668.

Dwinger, Ph. (1955) "The closure operators of the cardinal and ordinal sums and products of partially ordered sets and closed lattices." Nederl. Akad. Wetensch. Proc. Ser. A. 58-Indag. Math. 17, 341-351. MR 17 #7.

Eastman, W.L.; Even, S.; Isaacs, I.M. (1964/65) "Bounds for the optimal scheduling of n jobs on m processors." Management Sci. 11 268-279. MR 30 #3760.

Ecker, K. (1975) "Kritischer vergleich von algorithmen fur ein Scheduling-problem." Funfte Jahrestagung der Gesellschaft fur Informatik (Dortmund, 1975), pp. 703-714. Lecture Notes in Computer Science, Vol. 34, Springer, Berlin. MR 52 #7513.

Ehrenfeucht, A. (1959) "Decidability of the theory of one linear ordering relation." Notices Amer. Math. Soc. 6, 268-269.

Eichler, Jan; Novotny, Miroslav (1962) "Uber funktionale geordneter Mengen." (Czech and Russian summaries). Spisy Pridod. Fak. Univ. Brno 133-146. MR 27 #3565.

Eilon, Samuel; Chowdhury, I.G. (1976/77) "Minimising waiting time variance in the single machine problem." Management Sci. 23, no. 6, 567-575. MR 2323.

Eisenbud, David (1969) "Groups of order automorphisms of certain homogeneous ordered sets." Michigan Math. J. 16, 59-63. MR 39 #1373.

Ellis, David. (1952) "On immediate inclusion in partially ordered sets and the construction of homology groups for metric lattices." Acta Sci. Math. Szeged 14, 169-173. MR 14 #307.

Erdos, P.; Hajnal, A. (1962) "On a classification of denumerable order types and an application to the the partition calculus." Fund. Math. 51, 117-129. MR 25 #5000.

Erdos, P.; Rado, R. (1953) "A problem on ordered sets." J. London Math. Soc. 28, 426-438.

Erne, M. (1980) "Separation axioms for interval topologies." Proc. Amer. Math. Soc. 79 (2), 185-190.

Erne, M. (1980) "Topologies on products of partially ordered sets I, II." Alg. Universalis, 11 (3), I:295-311, II:312-319.

Erne, M. (1981) "A completion-invariant extension of the concept of continuous lattices." Continuous Lattices, Proceedings Bremen 1979, Springer Lect. Notes No. 871, 45-60.

Erne, M. (1981) "Completion-invariant extensions of the concept of continuous lattices." Lect. Notes Math. 871, Springer, 45-60.

Erne, M. (1981) "Isomorphismen und Identifikationen in der Ordnungstheorie." Math. Sem. Ber. 28, 74-91.

Erne, M. (1981) "Scott convergence and Scott topology in partially ordered sets II." Continuous Lattices, Proceedings Bremen 1979, Springer Lect. Notes no. 871, 61-96.

Erne, M. "Topologies on products of partially ordered sets, I, II, III." Algebra Universalis.

Erscev, A.E. (1971) "Semilattice endomorphisms of an ordered set." (Russian). Leningrad. Gos. Ped. Inst. Ucen. Zap. 404, 186-190. MR 5370.

Erschler, J; Roubellat, F.; Vernhes, J.P. (1976) "Finding some essential characteristics of the feasible solutions for a scheduling problem." Operations Res. 24, no. 4,774-783. MR 58 #20377.

Evans, Elliott (1977) "The Boolean ring universal over a meet semilattice." J. Austral. Math. Soc. Ser. A 23, no. 4, 402-415. MR 57 #16155.

Everett, C.J, and Ulam, S (1945) "On ordered groups." Trans. Amer. Math. Soc. 57, 208-216. [MF 12130]. MR 7 #4.

Evseev, A.E. (1964) "Lattice properties of a semigroup of endomorphisms of an ordered set." (Russian). Mat. Sb. (N.S.) 65 (107), 153-171. MR 31 #262.

Fan, Ky (1972) "On Dilworth's coding theorem." Math. Z. 127, 92-94.

Farcas, Gh. (1974) "Sur les treillis obliques de type Jordan." Revue Roumaine Math. Pur. Appl. 19, 311-314.

Farmer, F.D. (1975) "Homology of products and joins of reflexive relations." Discrete Math. 11, 23-27.

Farmer, F.D. (1975) "Homology of reflexive relations." Math. Japan 20, 303-310.

Farmer, F.D. (1977) "Characteristic for reflexive relations." Aequationes Math. 15, 195-199.

Farmer, F.D. (1979) "Cellular homology for posets." Math. Japonica 23, 607-613.

Feldman, J (1973) "Poles, intermediaires et centres dans un groupe d'opinions. (English summary) Opinions et scrutins: analyse mathematique." Math. Sci. Humaines No. 43, 39-54. MR 51 #5068.

Felscher, Walter (1958) "Beziehungen zwischen verbandsahnlichen Algebren und geregelten Mengen." Math. Ann. 135, 369-387. MR 21 #10.

Felscher, Walter (1961) "Jordan-Holder-Satze und modular geordnete Mengen." Math. Z. 75,83-114.

Fernandez, E.B.; Lang, T. (1976) "Scheduling as a graph transformation." IBM J. Res. Develop. 20, no. 6,551-559. MR 58 #20378.

Fiala, Frantisek; Novak, Vitezslav (1966) "On isotone and homomorphic mappings." Arch. Math. (Brno)2, 27-32. MR 33 #396.

Filippov, N.D. (1968) "Comparability graphs of partially ordered sets of certain types." (Russian). Ural. Gos. Univ. Mat. Zap. 6, tetrad' 3, 86-108. MR 42.

Filippov, N.D. (1968) "Sigma-isomorphisms of partially ordered sets." (Russian). Ural. Gos. Univ. Mat. Zap. 6, tetrad'' 3, 71-85. MR 42.

Filus, Lidia (1977) "Combinatorial lemma related to the search of fixed points." (Russian summary). Bull. Acad. Polon. Sci. Ser. Sci. Math. Astronom. Phys. 25, no. 7, 615-616. MR 126.

Finch, P.D. (1969) "Sasaki projections on orthocomplemented posets." Bull. Austral. Math. Soc. 1, 319-324. MR 41 #5.

Finch, P.D. (1970) "On orthomodular posets." J. Austral. Math. Soc. 11, 57-62. MR 41 #5.

Finkbeiner, Daniel T. (1951) "A general dependence relation for lattices." Proc. Amer. Math. Soc. 2 756-759. MR 13 #201.

Fishburn, P. (1970) "Intransitive indifference with unequal indifference intervals." J. Math. Psychol. 7, 144-149.

Fishburn, P.C. (1970) "Intransitive indifference in preference theory: a survey." Operations Research, 18, 2, 207-228.

Fishburn, P.C. (1970) "Intransitive indifference with unequal indifference intervals." J. Math. Psychol. 7, 144-149.

Fishburn, P.C. (1971) "Betweenness, orders and interval graphs." J. Pure Appl. Algebra, 1, 2, 159-178.

Fishburn, Peter C. (1966) "Stationary value mechanisms and expected utility theory." J. Mathematical Pschology 3, 434-457. MR 34 #1076.

Fishburn, Peter C. (1969/70) "Preferences, summation, and social welfare functions." Management Sci.16, 179-186. MR 42 #5657.

Fishburn, Peter C. (1970) "Arrow's impossibility theorem: concise proof and infinite voters." J. Econom. Theory 2, 103-106. MR 53 #10177.

Fishburn, Peter C. (1970) "Conditions for simple majority decision functions with intransitive individual indifference." J. Econom. Theory 2, 354-367. MR 7772.

Fishburn, Peter C. (1972) "Even-chance lotteries in social choice theory." Theory and Decision 3, 18-40. MR 48 #13315.

Fishburn, Peter C. (1972) "Interdependent preferences on finite sets." J. Mathematical Pschology 9, 225-236. MR 51 #7657.

Fishburn, Peter C. (1972) "Minimal preferences on lotteries." J. Mathematical Psychology 9, 294-305. MR 49 #2133.

Fishburn, Peter C. (1974) "Choice functions on finite sets." Internat. Econom. Rev. 15, 719-749. MR 52 #9981.

Fishburn, Peter C. (1974) "On the family of linear extensions of a partial order." J. Combinatorial Theory Ser. B 17, 240-243. MR 51 #3007.

Fishburn, Peter C. (1975) "Semiorders and choice functions." Econometrica 43, no. 5-6,975-977. MR 55 #14082.

Fishburn, Peter C. (1976) "On linear extension majority graphs of partial orders." J. Combinatorial Theory Ser. B21, no. 1, 65-70. MR 57 #9592.

Fishburn, Peter C. (1976) "Representable choice functions." Econometrica 44, no. 5, 1033-1043. MR 10758.

Fishburn, Peter C. (1977) "Models of individual preference and choice." Mathematical methods of the social sciences, II. Syntese 36, no. 3, 287-314. MR 58 #4202.

Fishburn, Peter C.; Gehrlein, William V. (1974) "Alternative methods of constructing strict weak orders from interval orders." Psychometrika 39, 501-516. MR 55 #5243.

Fishburn, Peter C.; Gehrlein, William V. (1975) "A comprehensive analysis of methods for constructing weak orders from partial orders." J. Mathematical Sociology 4, no. 1, 93-102. MR 53 #10367.

Fishburn, Peter C.; LaValle, Irving H. (1974) "The effectiveness of ordinal dominance in decision analysis." Operations Res.22, no. 1, 177-180. MR 53 #7521.

Fisher, James L. (1977) "Application-oriented Algebra. an Introduction to Discrete Mathematics." IEP-A Dun-Donnelley Publisher, N.Y, xviii+362pp. MR 15111.

Fisher, Marshall L. (1976/77) "A dual algorithm for the one-machine scheduling problem." Math. Programming 11, no. 3, 229-251. MR 58 #15112.

Fisher, Marshall L.; Jaikumar, Ramchandran (1978) "An algorithm for the space-shuttle scheduling problem." Operations Res. 26, no. 1, 166-182. MR 58 #4282.

Flachsmeyer, Jurgen (1978) "Dedekind-MacNeille extensions of Boolean algebras and of vector lattices of continuous functions and their structure spaces." General Topology and Appl. 8, no. 1,73-84. MR 58 #10646.

Flament, Cl. (1976) "L'analyse booleene de questionnaires." Mouton, Paris.

Flament, Cl. (1978) "Un theoreme de point fixe dans les treillis." Mathematiques et Sciences Humaines, 61, 61-64.

Flerov, Ju, A. (1971) "Multilevel dynamic games." (Russian). Studies in the theory of adaptive systems (Russian), pp.111-152.Vycisl. Centr Akad. Nauk SSSR, Moscow. MR 52 #10060.

Fofanova, T.S. (1968) "The isotone mapping of free lattices." (Russian). Mat. Zametki 4, 405-416. MR 41 #3330.

Fofanova, T.S. (1970) "Retracts of a lattice." (Russian). Mat. Zametki 7 (1970), 687-692. Math. Notes 7,415-418(English). MR 42 #1724.

Foldes, S.; Hammer, P.L. (1977) "Split graphs having Dilworth number two." Can. J. Math., 29, 666-672.

Foldes, S.; Hammer, P.L. (1978) "The Dilworth number of a graph." Annals of Discrete Math., 2, 211-219.

Folkman, J. (1966) "The homology groups of a lattice." J. Math. and Mech. 15, 631-636.

Fomin, S.V. (1978) "Finite partially ordered sets and Young tableaux." Soviet Math Dokl. 19, 1510-1514.

Ford, L.R, Jr.; Fulkerson, D.R. (1962) "Flows in networks." Princeton University Press, Princeton, N.J.1962. xii+194 pp. MR 28 #2917.

Ford, L.R.; Fulkerson, D.R. (1956) "Maximal flow through a network." Canad. J. Math. 8, 399-404.

Fortuin, C.M.; Kastelyn, P.W.; Ginibre, J. (1971) "Correlation inequalities on some partially ordered sets." Comm. Math. Phys., 22, 89-103. MR 46 #8607.

Fortunatov, V.A. (1974) "Semidirect products of a semilattice and a group." (Russian). Theory of Semigroups and its Applications, No. 3 (Russian), pp. 129-139,154. Izdat. Saratov. Univ, Saratov. MR 3288.

Foster, B.A.; Ryan, D.M. (1976) "An integer programming approach to the vehicle scheduling problem." Operational Res. Quart. 27, no. 2, part 1, 367-384. MR 58 #9200.

Fraisse, Roland (1953) "Sur certaines relations qui generalisent l'ordre des nombres rationnels." C. R. Acad. Sci. Paris 237, 540-542. MR 15 #192.

Fraisse, Roland (1953) "Sur l'extension aux relations de quelques proprietes connues des ordres." C.R. Acad. Sci. Paris 237, 508-510. MR 15 #192.

Frank, A. (1980) "On chain and antichain families of a partially ordered set." J. Comb. Theory B 29, 176-184.

Frasnay, Claude (1962) "Relations invariantes dans un ensemble totalement ordonne." C.R. Acad. Sci. Paris 255, 2878-2879. MR 27 #57.

Frasnay, Claude (1963) "Groupes finis de permutations et familles d'ordres totaux; application a la theorie des relations." C.R. Acad. Sci. Paris 257 2944-2947. MR 28 #1142.

Freese, Ralph (1974) "An application of Dilworth's lattice of maximal anti-chains." Discrete Math. 7, 107-109. MR 48 #10910.

Fridman, F.A. (1974) "Ordered choice method for integer programming." (Russian). Ekonom. i Mat. Metody 10, no. 5, 964-967. MR 9057.

Frink, O. (1954) "Ideals in partially ordered sets." Amer. Math. Monthly 6, 223-234.

Frink, Orrin. (1952) "A proof of the maximal chain theorem." Amer. J. Math. 74, 676-678. MR 14 #238.

Frink, Orrin. (1954) "Ideals in partially ordered sets." Amer. Math. Monthly 61, 223-234. MR 15 #848.

Fris, Ivan (1968) "Grammars with partial ordering of the rules." Information and Control 12, 415-425. MR 39 #5270.

Fris, Ivan (1969) "Correction to: "Grammars with partial ordering of the rules"." Information and Control 15, 452-453. MR 40 #5357.

Froda, Alexandru (1956) "Sur les reunions ordonnees d'ensembles." Acad. R.P. Romine. Stud. Cerc. Mat. 7 7-35. MR 18 #274.

Froda-Schechter, Micheline (1965) "Preordres et equivalences dans l'ensemble des familles d'un ensemble." Arch. Math. (Brno) 1, 39-56. MR 34 #2471.

Frucht W, Roberto (1964) "On problem 6 of Birkhoff." (Spanish). Scietia (Valparaiso) No. 124, 5-21. MR 34 #103.

Frucht, R. (1950) "On the construction of partially ordered systems with a given group of automorphisms." Amer. J. Math. 72, 195-199.

Frucht, Robert. (1950) "Lattices with a given abstract group of automorphisms." Canadian J. Math. 2, 417-419. MR 12 #473.

Frucht, Roberto (1948) "Uber die Konstruktion von teilweise geordneten Systemen mit gegebener Automorphismengruppe." Rev. Un. Mat. Argentina 13, 12-18 (Spanish).

Fuchs (1969) "Partially ordered algebraic systems." Pergamon Press.

Fuchs, Eduard (1965) "Isomorphismus der Kardinalpotenzen." Arch. Math. (Brno) 1, 83-93. MR 33 #5525.

Fuchs, L. (1963) "Partially ordered algebraic systems." Pergamon Press, Oxford-London-New York-Paris; Addison-Wesley Pub. Co, Inc, Reading, Mass.- Palo Alto, Calif.-London, ix + 229 pp. MR 30 #2090.

Fujii, M. [Fujii, Mamoru]; Kasami, T.; Ninomiya, K. (1969) "Optimal sequencing of two equivalent processors." SIAM J. Appl. Math. 17, 784-789. MR 40 #6945.

Fulkerson, D.R. (1956) "Note on Dilworth's decomposition theorem for partially ordered sets." Proc. Amer. Math Soc. 7, 701-702. MR 17 #1176.

Funayama, Nenosuke (1944) "On the completion by cuts of distributive lattices." Proc. Imp. Acad. Tokyo 20, 1-2. [MF 14864]. MR 7 #236.

Funayama, Nenosuke (1950) "On directed systems." I. Bull. Yamagata Univ. (Nat. Sci.) 1, 1-4. (Japanese summary). MR 16 #910.

Funayama, Nenosuke (1956) "Imbedding partly ordered sets into infinitely distributive complete lattices." Tohoku Math. J. (2) 8, 54-62. MR 18 #186.

Gabovic, E. Ja. (1976) "Totally ordered semigroups and their applications." (Russian). Usephi Mat. Nauk 31, no. 1(187), 137-201. MR 58 #21886.

Gabow, H.N. (1980) "An almost-linear algorithm for two processor scheduling." Technical Report C11-CS-169-80, Dept. of Computer Science, Univ. of Colorado, Boulder.

Gallai, T. (1967) "Transitiv orientierbare Graphen." Acta Math. Acad. Sci. Hungar. 18, 25-66. MR 35026.

Galvin, F.; Shelah, S. (1973) "Some counterexamples in the partition calculus." J. Comb. Th. A 15, 167-174. MR 48 #8240.

Gapp, William; Mankekar, P.S. Mitten, L.G. (1964/65) "Sequencing operations to minimize in-process inventory costs." Management Sci. 11, 476-484. MR 30 #3753.

Garey, M.R. (1973) "Optimal task sequencing with precedence constraints." Discrete Math. 4, 37-50.

Garey, M.R.; Graham R.L.; Johnson, D.S.; Yao, Audrew Chi Chih (1976) "Resource constrained scheduling as generalized bin packing." J. Combinatorial Theory Ser. A 21, no 3, 257-298. MR 55 #7359.

Garey, M.R.; Graham, R.L. (1975) "Bounds for multiprocessor scheduling with resource constraints." SIAM J. Comput. 4, 187-200. MR 51 #14935.

Garey, M.R.; Graham, R.L.; Johnson, D.S. (1978) "Performance guarantees for scheduling algorithms." Operations Res. 26, no. 1, 3-21. MR 2522.

Garey, M.R.; Johnson, D.S. (1976) "Scheduling tasks with nonuniform deadlines on two processors." J. Assoc. Comput. Mach. 23, 461-467.

Garey, M.R.; Johnson, D.S. (1977) "Two-processor scheduling with start-times and deadlines." SIAM J. Comput. 6, no. 3, 416-426. MR 58 #25930.

Garey, M.R.; Johnson, D.S.; Simons, B.B.; Tarjan, R.E. (1981) "Scheduling unit-time tasks with arbitrary release times and deadlines." SIAM J. Comput. 10, 256-269.

Gaul, W.; Heinecke, A. (1975) "Einige Aspekte in der Zuordnungstheorie." (English summary). Z. Operations Res. Ser. A-B 19, no. 3, A89-A99. MR 55 #12115.

Geceg, F. [Gecseg, F.] (1967) "On families of automata mappings." (Russian). Acta Sci. Math. (Szeged)28, 39-54. MR 36 #2443.

Gecseg, F. (1966) "On R-products of automata. I." Studia Sci. Math. Hungar. 1, 437-441. MR 36 #2440.

Gecseg, F. (1966) "On R-products of automata. II." Studia Sci. Math. Hungar. 1, 443-447. MR 36 #2441.

Gecseg, F. (1967) "On R-products of automata.III." Studia Sci, Math. Hungar.2, 163-166. MR 36 #2442.

Gehrlein, William V.; Fishburn, Peter C. (1977) "An analysis of simple counting methods for ordering incomplete ordinal data." Theory and Decision 8, no. 3, 209-227. MR 57 #12305.

Gemignani, Giuseppe (1958) "Su une teoria assiomatica della dipendenza." Rend. Mat. e Appl. (5) 17, 405-414. MR 21 #2610.

Geoffroy, Cynthia D.; Scheiblich, H.E. (1974) "Chain conditions on posets." Glasgow Math. J. 15, 79-81. MR 50 #12826.

Georgescu, George; Lungulescu, Bogdan (1969) "Sur les proprietes topologiques des structures ordonnees." Rev. Romaine Math. Pures Appl. 14, 1453-1456. MR 41 #1614.

Gertsbakh, I.; Stern, H.I. (1978) "Minimal resources for fixed and variable job schedules." Operations Research 26, 68-85.

Gewirtz, A. [Gerwirtz, Allan]; Quintas, Louis V. (1971) "Antitone fixed point maps." Recent Trends in Graph Theory (Proc. Conf, N.Y. 1970), pp. 117-136. Lecture Notes in Mathematics, Vol. 186. Springer, Berlin. MR 44 #121.

Ghika, Al. (1950) "Uniform inductive ordered sets." Acad. Repub. Pop. Romane. Bul. Sti. Ser. Mat. Fiz. Chim. 2, 119-124. (Romanian. Russian and French summaries). MR 13 #542.

Ghouila-Houri, A. (1962) "Characterization of nonoriented graphs whose edges can be oriented in such a way as to obtain the graph of an order relation." C. R. Acad. Sci, Paris, 1370-1371.

Gierz, G.; Hofmann, K.H.; Keimel, K.; Lawson, J.; Mislove, M.; Scott, D. (1980) "A compendium of continuous lattices." Springer-Verlag, 372 pp.

Giffler, B.; Thompson, G.L. (1960) "Algorithms for solving production-scheduling problems." Operations Res. 8, 487-503. MR 27 #4555.

Gilmore, P.C.; Hoffman, A.J. (1964) "A characterization of comparability graphs and of interval graphs." Canad. J. Math. 16, 539-548.

Ginsburg, S. (1953) "Some remarks on order tyes and decompositions of sets." Trans. Amer. Math. Soc. 74, 514-535.

Ginsburg, S. (1954) "Further results on order types and decompositions of sets." Trans. Amer. Math. Soc. 77, 122-150.

Ginsburg, S. (1955) "Order types and similarity transformations." Trans. Amer. Math. Soc. 79, 341-361.

Ginsburg, Seymour. (1954) "On the lambda-dimension and the A-dimension of partially ordered sets." Amer. J. Math. 76, 590-598. MR 15 #943.

Ginsburg, Seymour; Isbell, J.R. (1965) "The category of cofinal types. I." Trans. Amer. Math. Soc. 116, 386-393. MR 34 #1199.

Ginzburg, Abraham, Yoeli, Michael (1965) "Products of automata and the problem of covering." Trans. Amer. Math. Soc. 116, 253-266. MR 34 #1112.

Give'on, Yehoshafat (1964) "Lattice matrices." Information and Control 7, 477-484. MR 31 #6735.

Gladstien, K. (1973) "A characterization of complete trellises of finite length." Algebra Universalis 3, 341-344.

Gleason, A.M.; Dilworth, R.P. (1962) "A generalized Cantor theorem." Proc. Amer. Math. Soc. 13, 704-705. MR 26 #2365.

Glebov, N.I.; Perepelica, V.A. (1970) "The lower and upper estimates for a certain problem in scheduling theory." (Russian). Studies in Cybernetics (Russian), pp. 11-17. Izdat. "Sovet. Radio", Moscow. MR 52 #7514.

Gleyzal, A. "Order types and structure of orders." Trans. Amer. Math. Soc. 48, 451-466.

Gluckaufova, Dagmar (1969) "A certain axiomatic approach to utility theory leading to the so-called indirect utility function." (Czech. English summary). Ekonom.-Mat. Obzor 5, 423-442. MR 41 #3080.

Gluhov, M.M. (1962) "An algorithm for the construction of the free extension of a latticoid." (Russian). Malekess. Gos. Ped. Inst. Ucen. Zap. 2, cast' 2, 21-33. MR 40 #2577.

Gluhov, M.M. (1962) "Modular lattices and latticoids." (Russian). Melekess. Gos. Ped. Inst. Ucen. Zap. 2, cast' 2, 35-43. MR 40 #2578.

Gluskin, L.M. (1961) "Semigroups of isotone transformations." Uspehi Mat. Nauk 16, 157-162.

Goberstein, S.M. (1976) "Fundamental order relations on right inverse semigroups." (Russian). Modern Algebra, No. 5 pp. 20-39. Leningrad Gos. Ped. Inst. im. Gercena, Leningrad. MR 57 #16158.

Goguen, J.A. (1969) "Categories of V-sets." Bull. Amer. Math. Soc. 75, 622-624. MR 39 #5428.

Goldberg, Henry M. (1977) "Analysis of the earliest due date scheduling rule in queueing systems." Math. Oper. Res. 2, no. 2, 145-154. MR 58 #25931.

Goldie, A.W. (1952) "On direct decompositions. I." Proc. Cambridge Philos. Soc. 48, 1-22. MR 14 #9.

Goldie, A.W. (1952) "On direct decompositions. II." Proc. Cambridge Philos. Soc. 48, 23-34. MR 14 #9.

Golumbic, M.C. (1977) "Comparability graphs and a new matroid." J. Combinatorial Theory (Ser. B) 22, 68-90. MR 512575.

Golumbic, M.C. (1977) "The complexity of comparability graph recognition and coloring." Computing 18, 199-203.

Golumbic, M.C. (1980) "Algorithmic Graph Theory and Perfect Graphs." Academic Press, New York.

Gomes, Alvercio M. (1952) "Completion by cuts of a distributive lattice." Rev. Cientifica 3, no. 3-4, 56-68. MR 16 #559.

Gonzales, M.J, Jr.; Soh, J.W. (1976) "Periodic job scheduling in a distributed processor system." IEEE Trans. Aerospace and Electron. Systems AES-12. no. 5, 530-536. MR 58 #25932.

Gonzalez, Teofilo; Sahni, Sartaj (1978) "Preemptive scheduling of uniform processor systems." J. Assoc. Comput. Mach. 25, no. 1, 92-101. MR 5056.

Gorbatov, V.A.; Makarenkov, S.V. (1975) "Forbidden figures in the synthesis of many-output Boolean structures." IZV. Akad. Nauk SSSR Tehn. Kibernet. 1975, no. 2, 139-144 (Russian); translated as Engrg. Cybernetics 13, no. 1, 110-116. MR 52 #16877.

Gordon, Henry G. (1976) "Complete degrees of finite-state transformability." Information and Control 32, no. 2, 169-187. MR 54 #4862.

Gordon, V.S.; Safranskii, Ja.M. (1978) "Optimal ordering under serial-parallel precedence constraints." (Russian. English summary). Dokl. Akad. Nauk BSSR 22, no. 3, 244-247,286. MR 58 #15114.

Gouyon, Rene (1963) "Sur une "cloture" des ensembles ordonnes." C.R. Acad. Sci. Paris 256, 3410-3412. MR 27 #2445.

Grabowski, Jan (1973) "Anwendung topologischer Methoden in der Theorie de ND-Automaten." (English and Russian summaries). Elektron. Informationsverarbeit. Kybernetik 9, 615-633. MR 50 #9451.

Grabowski, Jozef (1978) "Formulation and solution of the sequencing problem with parállel machines." Optimization Techniques, Park 2, pp.400-410. Lecture Notes in Control and In Sci, Vol. 7, Springer, Berlin. MR 58 #20380.

Grabowski, Jozef (1978) "Formulation and solution of the sequencing problem with the use of parallel machines." (Polish. Russian and English summaries). Arch. Automat. Telemech. 23, no. 1-2, 91-113. MR 58 #20379.

Graham, R.L. (1969) "Bounds on multiprocessing timing anomalies." SIAM J. Appl. Math. 17, 416-429. MR 40 #2461.

Graham, R.L.; Harper, L.H. (1969) "Some results on matching bipartite graphs." SIAM J. Appl. Math. 17, 1017-1022.

Graham, R.L.; Lawler, E.L.; Lenstra, J.K.; Rinnooy Kan, A.H.G. (1979) "Optimization and approximation in deterministic sequencing and scheduling: a survey." Ann. Discrete Math. 5, 287-326.

Graham, R.L.; Yao, A.C.; Yao, F.F. (1980) "Some monotonicity properties of partial orders." SIAM Jour. Alg. Disc. Meth., 1, 251-258.

Grandmont, Jean-Michel (1978) "Intermediate preferences and the majority rule." Econometrica 46, no. 2, 317-330. MR 58 #4204.

Greene, C. (1974) "Sperner families and partitions of a partially ordered set." Hall and Van Lint, Combinatorics, Math. Center Tracts 56, 91-106.

Greene, C. (1976) "Some partitions associated with a partially ordered set." J. Combinatorial Theory Ser. A 20, 69-79.

Greene, C.; Hilton, A.J.W. (1979) "Some results on Sperner families." J. Comb. Theory A 26, 202-209.

Greene, C.; Kleitman, D.J. (1976) "Strong versions of Sperner's theorem." J. Combinatorial Theory Ser. A 20, 80-88.

Greene, C.; Kleitman, D.J. (1978) "Proof techniques in the theory of finite sets." MAA Studies in Math.

Greene, Curtis (1971) "On the Mobius algebra of a partially ordered set." Mobius Algebras (Proc. Conf., Univ. Waterloo, Waterloo, Ont.), pp. 3-38. Univ. Waterloo, Waterloo, Ont., 1971. MR 50 #1994.

Greene, Curtis (1973) "On the Mobius algebra of a partially ordered set." Advances in Math. 10, 177-187. MR 47 #4886.

Greene, Curtis; Kleitman, Daniel J. (1976) "The structure of Sperner k-families." J. Combinatorial Theory Ser. A 20, no. 1, 41-68. MR 53 #2695.

Greenough, H. Paul (1974) "An identity in the incidence algebra of a poset." Proc. West Virginia Acad. Sci. 46, no. 2, 182-184. MR 53 #7869.

Griggs, J.R. (1977) "Another three part Sperner theorem." Studies in Applied Math. 57, 181-184.

Griggs, J.R. (1980) "Note on chains and Sperner k-families in ranked posets, II." J. of Comb. Theory A 29, 391-394.

Griggs, J.R. (1980) "On chains and Sperner k-families in ranked posets." J. Comb. Theory A 28, 156-168.

Griggs, J.R.; Kleitman, D.J. (1977) "A three part Sperner theorem." Disc. Math. 17, 281-289.

Griggs, Jerrold R. (1977) "Sufficient conditions for a symmetric chain order." SIAM J. Appl. Math. 32, no. 4, 807-809. MR 146.

Grillet, P.A. (1969) "Maximal chains and antichains." Fund. Math. 65, 157-167. MR 39 #5429.

Grimonprez, Georges; Losfeld, Joseph; Van Dorpe, Jean-Claude. (1975) "Perte et gain d'information sur une sequence d'elements comparables d'un trellis et metrique associee." (English summary). C.R. Acad. Sci. Paris Ser. A-B 280, Aii, A957-A960. MR 51 #9964 .

Grimonprez, Georges; Van Dorpe, Jean-Claude (1975) "Perte et gain d'information sur une sequence d'elements comparables d'un treillis." (applications). (English summary). C.R. Acad. Sci. Paris Ser. A-B 280, Aii, A1381-A1387. MR 53 #5164.

Gronau, H.-D.O.F. (1980) "On maximal antichains consisting of sets and their complements." J. Comb. Theory A 29, 370-375.

Gudder, S.; Haskins, L. (1974) "The center of a poset." Pacific J. Math. 52, 85-89. MR 50 #1993.

Guilbaud, G. Th. (1968) "Les partitions et la notation simpliciale." Math. Sci. Humaines No. 22, 27-31. MR 39 #4059.

Gurevich, Y. (1981) "Modest Theory of short chains, I." J. Symbolic Logic 44.

Gurevich, Y.; Shelah, S. (1981) "Modest Theory of short chains, II." J. Symbolic Logic 44, 491-502. MR 81 #3038b.

Gurevich, Yuri (1977) "Monadic theory of order and topology. I." Israel J. Math. 27, no. 3-4, 299-319. MR 5259.

Gysin, R. (1977) "Dimension transitiv orientierbarer graphen." Acta Math. Acad. Sci. Hungar. 29, no. 3-4, 313-316. MR 58 #5393.

Hagendorf, Jean Guillame (1972) "Extensions immediates de chaines et de relations." C.R.Acad. Sci, Paris, Ser. A 274, 607-609.

Hagendorf, Jean Guillaume (1976) "Sur distributivite du plongement entre chaines." (English summary). C.R. Acad. Sci. Paris Ser. A-B 282, no. 24, Ai,A1391-A1394. MR 57 #576.

Hall, P. (1936) "Eulerian functions in groups." Quart. J. Math. (Oxford), 134-151.

Hammer, P.L.; Johnson, E.L.; Peled, U.N. (1974) "Regular 0-1 programs." Cahiers du Centre d'Etudes de Recherche Operationnelle, 16, 267-276.

Hammer, P.L.; Peled, U.N.; Pollatschek, M.A. (1979) "An algorithm to dualize a regular switching function." IEEE Transactions on Computers, C-28, 238-243.

Hammer, Peter L.; Nguyen, Sang (1977) "APOSS. A partial order in the solution space of bivalent programs." Modern Trends in Cybernetics and Systems, Vol. I, pp. 869-883. Springer, Berlin. MR 57 #15430.

Hammer, Peter L.; Rosenberg, Ivo G. (1974) "Linear decomposition of a positive group-Boolean function." Numerische Methoden bei Optimierungsaufgaben, Band 2, pp. 51-62. Internat. Schriftenreihe Numer. Math,Band 23, Birkhauser, Basel. MR 15521.

Hammond, Peter J. (1976) "Endogenous tastes and stable long-run choice." J. Econom. Theory 13, no. 2, 329-340. MR 55 #4079.

Hansel, Georges (1966) "Sur le nombre des fonctions booleennes monotones de n variables." C.R. Acad. Sci. Paris Ser. A-B 262, A1088-A1090. MR 36 #7439.

Hansell, R.W. (1967) "Monotone subnets in partially ordered sets." Proc. Amer. Math. Soc. 18, 854-858. MR 35 #6594.

Hansson, Bengt (1969) "Voting and group decision functions." Synthese 20, 526-537. MR 45 #4808.

Hare, William R. (1964) "Generalized order convexity." Re. Ci (Lima) 66, 93-99. MR 32 #1135.

Harper, L.H. (1974) "The morphology of partially ordered sets." J. Combinatorial Theory Ser. A 17, 44-58. MR 51 #3008.

Hartley, Roger (1976) "Aspects of partial decisionmaking--kernels of quasi-ordered sets." Econometrica 44, no. 3, 605-608. MR 54 #9622.

Hartmanis, J.; Stearns, R.D. (1964) "Pair algebra and its applications to automata theory." Information and Control 7, 485-507. MR 30 #1012.

Harzheim, Egbert (1967) "Einbettung totalgeordneter Mengen in lexikographische Produkte." Math. Ann. 170, 245-252.

Harzheim, Egbert (1968) "Uber universalgeordnete Mengen." Math. Nachr. 36, 195-213. MR 39 #101.

Harzheim, Egbert (1970) "Ein Endlichkeitssatz uber die Dimension teilweise geordneter Mengen." Math. Nachr. 46, 183-188. MR 43 #113.

Hashimoto, J.; Nakayama, T. (1950) "On a problem of G. Birkhoff." Proc. Amer. Math. Soc. 1, 141-142.

Hashimoto, Junji (1948) "On the product decomposition of partially ordered sets." Math. Japonicae 1, 120-123. MR 11 #5.

Hashimoto, Junji. (1951) "On direct product decomposition of partially ordered sets." Ann. of Math. (2) 54, 315-318. MR 13 #201.

Haskins, L; Gudder, S. (1971) "Semimodular posets and the Jordan-Dedekind chain condition." Proc. Amer. Math. Soc. 28, 395-396. MR 43 #1892.

Haskins, L; Gudder, S. (1972) "Height on posets and graphs." Discrete Math. 2, no. 4, 357-382. MR 46 #5186.

Haskins, L; Gudder, S.; Greechie, R. (1974/75) "Perspectivity in semimodular orthomodular posets." J. London Math. Soc. (2) 9, 495-500. MR 50 #12827.

Hattori, Mitsuhiro; Noguchi, Hiroshi (1966) "Synthesis of asynchronous circuits." J. Math. Soc. Japan 18, 405-423. MR 34 #8870.

Hausdorff, F. (1906-1907) "Untersuchungen uber Ordnungstypen, I-III." Ber. Math. - Phys. Kl. Kg. Sachs Ges Wiss. 58, 106-169.

Hausdorff, F. (1906-1907) "Untersuchungen uber Ordnungstypen, IV-V." Ber. Math. - Phys. Kl. Kg. Sachs Ges Wiss. 59, 84-159.

Havel, Vaclav (1955) "Remark on a generalization of the direct product of partially ordered sets." Mat.-Fzy. Casopis. Slovensk. Akad. Vied 5, 3-10. (Czech. Russian summary). MR 16 #1007.

Hay, Louise; Manaster, Alfred B.; Rosenstein, Joseph, G. (1977) "Concerning partial recursive similarity transformations of linearly ordered sets." Pacific J. Math. 71, no. 1,57-70. MR 2806.

Held, Michael; Karp, Richard M. (1962) "A dynamic programming approach to sequencing problems." J. Soc. Indust. Appl. Math. 10, 196-210. MR 25 #2925.

Helgert, Hermann Josef (1967) "A partial ordering of discrete, memoryless channels." IEEE Trans, Information Theory IT-13, 360-365. MR 35 #6485.

Hell, Pavol (1976) "Graph retractions." (Italian summary). Colloquio Inter. sulle Teorie Combinatorie (Roma,1973), Tomo II, pp. 263-268. Atti dei Convegni Lincei, No.17, Accad. Naz. Lincei, Rome. MR 58 #27586.

Henn, R. (1977/78) "Antwortzeitgesteuerte Prozessor-Zuteilung unter strengen Zeitbedingungen." (English summary). Computing 19, no. 3, 209-220. MR 9038.

Heppes, A.; Malyusz, K.; Stahl, J. (1977) "On the solution of a sorting problem." (Hungarian. English summary). Alkalmaz. Mat. Lapok 2(1976), no. 3-4, 233-235. MR 57 #18798.

Hermes, Hans (1963) "La teoria de reticulos y su aplicacion a la logica matematica." [The theory of lattices and its application to mathematical logic]. Publicaciones del Instituto de Matematicas "Jorge Juan", Madrid, 57 pp. MR 28 #5016.

Hewitt, Edwin (1943) "A problem of set-theoretic topology." Duke Math. J. 10, 309-333 [MF 8471]. MR 5 #46.

Hickman, J.L. (1977) "Quasi-minimal posets and lattices." J. Reine Angew. Math. 296, 10-13. MR 11783.

Hickman, J.L. (1977) "Rigidity in order-types." J. Austral. Math. Soc. Ser. A 24, no. 2, 203-215. MR 58 #397.

Hickman, John L. (1977) "Regressive order-types." Notre Dame J. Formal Logic 18, no. 1, 169-174. MR 2879.

Higman, G. (1952) "Ordering by divisibility in abstract algebras." Proc. London Math. Soc. (3) 2, 326-336.

Hillman, Abraham P (1955) "On the number of realizations of a Hasse diagram by finite sets." Proc. Amer. Math. Soc. 6, 542-548. MR 17 #575.

Hiraguchi, T. (1955) "On the dimension of orders." Sci. Rep. Kanazawa Univ. 4, 1-20. MR 17, p.1045.

Hiraguchi, Toshio (1951) "On the dimension of partially ordered sets." Sci. Rep. Kanazawa Univ. 1, 77-94. MR 17, p. 19.

Hiraguchi, Tosio (1956) "On the lambda-dimension of the product of orders." Sci. Rep. Kanazawa Univ. 5, 1-5. MR 20 #1638.

Hodges, Willfrid; Lachlan, A.H.; Shelah, Saharon (1977) "Possible orderings of an indiscernible sequence." Bull. London Math. Soc. 9, No. 2, 212-215. MR 57 #16085.

Hodgson, Thom J. (1976/77) "A note on single machine sequencing with random processing times." Management Sci. 23, no. 10, 1144-1146. MR 58 #32903.

Hoffman, A.J.; Schwartz, D.E. (1977) "On partitions of a partially ordered set." J. Combinatorial Theory Ser. B23, no. 1,3-13. MR 57 #12244.

Hoffmann, R.-E. (1979) "Continuous posets and adjoint sequences." Semigroup Forum 18, 173-188.

Hoffmann, R.-E. (1979) "Essentially complete To-spaces." Manus. Math. 27, 401-432.

Hoffmann, R.-E. (1979) "Sobrification of partially ordered sets." Semigroup Forum 17, 123-138.

Hofmann, K.H.; Keimel, K. (1972) "A general character theory for partially ordered sets and lattices." Mem. Amer. Math. Soc. 122, 121 pp.

Hofmann, Karl Heinrich; Keimel Klaus (1972) "A general character theory for partially ordered sets and lattices." Memoirs of the American Mathematical Society, no. 122. American Math. Soc, Providence, R.I, iv + 121 pp. MR 49 #4885.

Hoft, Hartmut; Hoft, Margret (1976) "Some fixed point theorems for partially ordered sets." Canad. J. Math., 992-997.

Hoheisel, Guido; Schmidt, Jurgen (1953) "Uber die Konstruktion einer gewissen totalen Ordnung in Baumen." Arch. Math. 4, 261-266. MR 15 #204.

Hohn, Franz (1968) "Algebraic foundations for automata theory." Applied Automata Theory, pp. 1-34. Academic Press, New York. MR 40 #1204.

Holley, R. (1974) "Remarks on the FKG inequalities." Comm. Math. Phys., 36, 227-231.

Holman, Eric W. (1969) "Strong and weak extensive measurement." J. Mathematical Psychology 6, 286-293. MR 39 #3843.

Horn, W.A. (1972) "Single-machine job sequencing with treelike precedence ordering and linear delay penalties." SIAM J. Appl. Math. 23, 189-202.

Hoyer, Robert W.; Mayer, Lawrence S. (1975) "Social preference orderings under majority rule." Econometrica 43, no. 4, 803-806. MR 2209.

Hsieh, W.N.; Kleitman, D.J. (1973) "Normalized matching in direct products of partial orders." Studies in Appl, Math. 52, 285-289. MR 53 #7870.

Hsu, N.C. (1966) "Elementary proof of Hu's theorem on isotone mappings." Proc. Amer. Math. Soc. 17, 111-114. MR 32 #5553.

Hu, T.C. (1961) "Parallel sequencing and assembly line problems." Operations Res. 9, 841-848. MR B1661.

Hwang, F.K.; Lagarias, J.C. (1978) "Minimum range sequences of all k-subsets of a set." Discrete Math. 19 (1977), no. 3, 257-264. MR 5770.

Iazeolla, G. (1976) "A graph model for scheduling processes in systems with parallel computations." Calcolo 13, no. 3, 321-349. MR 58 #20381.

Ibaraki, Toshihide; Muroga, Saburo (1971) "Synthesis of networks with a minimum number of negative gates." IEEE Trans. Computers C-20, 49-58. MR 45 #1672.

Inada, Ken-ichi (1964) "A note on the simple majority decision rule." Econometrica 32, 525-531. MR 31 #1119.

Inada, Ken-ichi. (1955) "Alternative incompatible conditions for a social welfare function." Econometrica 23, 396-399. MR 17 #171.

Inagaki, Takeshi (1952) "Sur deux theoremes concernant un ensemble partiellement ordonne." Math. J. Okayama Univ. 1, 167-176. MR 14 #27.

Ion, Ion D. (1962) "Sur les treillis de dimension finie." Comun. Acad. Republ. Popul. Romine. 12, 1107-1110, russ. und franzos. Zusammenfassung 1110 (Rumanisch).

Isbell, J.R. (1957) "On a theorem of Richardson." Proc. Amer. Math. Soc. 8, 928-929. MR 19 #720.

Isbell, J.R. (1965) "The category of cofinal types. II." Trans. Amer. Math. Soc. 116,394-416. MR 34 #1200.

Isbell, J.R.; Rubin, Herman. (1956) "Limit-preserving embeddings of partially ordered sets in directed sets." Proc. Amer. Math. Soc. 7, 812-813. MR 18 #495.

Ivanescu, Petru L. (1964) "On the minimal decomposition of finite partially ordered sets in chains." Re. Romaine Math. Pures Appl. 9, 897-903. MR 33 #2582.

Ivanescu, Petru L.; Rudeanu, Sergiu (1967) "Boolean methods in operations research." (Czech summary). Ekonom.-Mat. Obzoe 3, 422-445. MR 36 #7416.

Jacak, Witold; Tchon, Krzysztof (1975) "Events and systems." (Russian summary). Systems Sci. 1 (1975), no. 2, 77-88 1976. MR 54 #7011.

Jacquet-Lagreze, E. (1973) "Le probleme de l'agregation des preferences: une classe de procedures a seuil." (English summary). Opinions et scrutins: analyse mathematique. Math. Sci. Humaines. No. 43, 29-37. MR 51 #5069.

Jacquet-Lagreze, E.; Roy, B. (1981) "Aide a la decision multicriteres et systemes relationnels de preference." P. Batteau, E. Jacquet-Lagreze and B. Monjardet (eds), Analyse et agregation des preferences, Economica, Paris.

Jacquet-Lagreze, Eric (1974) "How we can use the notion of semiorders to build outranking relations in multicriteria decision making." (French, German, Italian and Spanish summaries). Metra 13, 59-86, 95-97. MR 57 #11662.

Jaeger, Arno (1977) "A plea for preordinators." Mathematica. Economics and Game Theory, pp. 605-615. Lecture Notes in Econ. and Math. Systems, Vol. 141, Springer, Berlin. MR 15504.

Jaffray, Jean-Yves (1975) "Semicontinuous extension of a partial order." J. Math. Econom. 2, no. 3, 395-406. MR 52 #16535.

Jakubik, J. (1963) "Die Jordan-Dedekindsche Bedingung im direkten Produkt von geordneten Mengen." Acta Sci. Math. (Szeged) 24, 20-23. MR 27 #1389.

Jakubik, J. (1972) "Weak product decomposition of partially ordered sets." Colloq. Math. 25, 177-190. MR 48 #8316.

Jakubik, J.; Novotny, M.; Novak, J.; Sekanina, m.; Sekanına, A.; Katrinak, T.; Fiala, F.; Kratochvil, P.; Novak, F.; Dorec, I.; Skula, L.; Smarda, B. (1966) "Summer session on ordered sets at Cikhaj." Casopis Pest. Mat. 91, 92-103. MR 32 #7458.

Jakubik, Jan (1956) "Sur les axiomes des multistructures." Czechoslovak Math. J. 6(81), 426-430. (Russian. French summary). MR 20 #639.

Jamison, Dean T.; Lau, Lawrence J. (1975) "Semiorders and the theory of choice." Econometrica 41 (1973), 901-912. Econometrica 43, no 5-6, 979-980. MR 55 #14081.

Jamison, Dean T.; Lau, Lawrence J. (1977) "The nature of equilibrium with semiordered preferences." Econometrica 45, no. 7, 1595-1605. MR 58 #15050.

Jamison, Robert (1979) "A convexity characterization of ordered sets." Proc. tenth S.E. Conf. Comb., Graph Theory, and Computing, Utilitas Publ., 529-540.

Jamison, Robert E. (1977/78) "Tietze's convexity theorem for semilattices and lattices." Semigroup Forum 15, no. 4, 357-373. MR 58 #2142.

Jankowska-Zorychta, Zofia (1973) "Models for scheduling jobs on machines." (Polish, Russian and English summaries). Przeglad Statyst. 20, 11-25. MR 47 #1446.

Janowitz, M.F. (1964) "On the antitone mappings of a poset." Proc. Amer. Math. Soc. 15, 529-533. MR 29 #43.

Janowitz, M.F. (1965) "Baer semigroups." Duke Math. J. 32, 85-95. MR 31 #263.

Janowitz, M.F. (1967) "Residuated closure operators." Portugal Math. 26, 221-252. MR 40 #2576.

Janowitz, M.F. (1977) "A triple construction for SM-semilattices." Algebra Universalis 7, no. 3, 389-402. MR 191.

Janowitz, M.F. (1978) "An order theoretic model for cluster analysis." SIAM J. Appl. Math. 34, no. 1, 55-72. MR 57 #16161.

Janowitz, M.F. (1979) "Monotone equivariant cluster methods." SIAM J. Applied Mathematics, 37, 148-165.

Jansen, Siegfried (1975) "Representation d'un ensemble ordonne dans un sous-treillis d'un treillis complet et completement distributif." (English summary). C.R. Acad. Sci. Paris Ser. A-B 281, no. 4, Ai, A127-A128. MR 52 #10514.

Jansen, Siegfried L. (1978) "Subdirect representation of partially ordered sets." J. London Math. Soc.(1)17, no. 2, 195-202. MR 58 #5423.

Jensen, G.A. (1969) "Homomorphisms of some function lattices." Math. Ann. 182, 127-133. MR 40 #3288.

Jensen, Paul, A. (1971) "Optimum network partitioning." Operations Res. 19, 916-932. MR 45 #8413.

Jezek, J. (1972) "Algebraicity of endomorphisms of some relational structures." Acta Univ. Carolinae-Math. et Phys. 13, no. 2, 43-52. MR 52 #10544.

Johnson, C.S, Jr. (1969) "A lattice whose residuated maps do not form a lattice." J. Natur. Sci. and Math. 9, 283-284. MR 42 #141.

Johnson, C.S, Jr. (1971) "On certain poset and semilattice homomorphisms." Pacific J. Math. 39, 703-715. MR 46 #3380.

Johnston, John B. (1956) "Universal infinite partially ordered sets." Proc. Amer. Math. Soc, 507-514.

Johnstone, P.T. (1978) "A note on complete semilattices." Algebra Universalis 8, no. 2, 260-261. MR 3017.

Jongh, D.H.J. de; Parikh, Rohit (1977) "Well-partial orderings and hierarchies." Nederl. Akad. Wetensch. Proc. Ser. A 80=Indag. Math. 39, no. 3, 195-207. MR 5371.

Jongh, D.H.J. de; Troelstra, A.S. (1966) "On the connection of partially ordered sets with some pseudo-Boolean algebras." Nederl. Akad. Wetensch. Proc. Ser. A69=Indag. Math. 28, 317-329. MR 33.

Jongh, de; Parikh, R.J. (1977) "Well-partial orderings and hierarchies." Nederl. Akad. Wetensch. Proc. Ser. A 80; Indag. Math. 39 No. 8, 195-207. MR 56 #5371.

Jonsson, B. (1966) "The unique factorization problem for finite relational structures." Colloq. Math. 14, 1-32. MR 33 #87.

Jonsson, Bjarni, and Tarski, Alfred. (1947) "Direct decompositions of finite algebraic systems." Notre Dame Mathematical Lectures, no. 5. Univ. of Notre Dame, Notre Dame, Ind. v + 64pp. MR 8 #560.

Jung, H.A. (1968) "Zu einem Satz von E.S. Wolk uber die Vergleichbarkeitsgraphen von ordnungstheoretischen Baumen." Fund. Math. 63, 217-219. MR 38 #3167.

Jung, H.A. (1978) "On a class of posets and the corresponding comparability graphs." J. Combinatorial Theory Ser. B 24, no. 2, 125-133. MR 58 #10614.

Kalai, Ehud; Schmeidler, David (1977) "An admissible set occurring in various bargaining situations." J. Econom. Theory 14, no. 2, 402-411. MR 55 #12159.

Kalinin, V.V. (1976) "Orthomodular partially ordered sets with dimension." (Russian). Algebra i Logika 15, no. 5, 535-557,605. MR 58 #16442.

Kalmbach, G. (1976) "Extension of topological homology theories to partially ordered sets." J. Reine und Angew. Math. 280, 134-156.

Kaluza, Theodor (1977) "Existenz streng monotoner Folgen in geordneten Mengen." Abh. Braunschweig. Wiss. Ges. 28, 55-57. MR 58 #27666.

Kamii, Masahi; Narushima, Hiroshi (1973) "Further results on order maps." Proc. Fac. Sci. Tokai Univ. 8, no. 1-2, 1-7. MR 47 #8363.

Kampe deFeriet, J. (1973) "Mesure de l'information par un ensemble d'observateurs independants." Trans. of the Sixth Prague Conf. on Information Theory (Technical Univ. Prague, 1971) pp 315-329. Academia, Prague. MR 50 #6643.

Kaneko, Mamoru (1975) "Necessary and sufficient conditions for transitivity in voting theory." J. Econom. Theory 11, no. 3, 385-393. MR 53 #10178.

Kannai, Yakar (1963) "Existence of a utility in infinite dimensional partially ordered spaces." Israel J. Math.1, 229-234. MR 30 #951.

Kantarija, G.V. (1976) "Optimal two-level compromise-coordinated choice." (Russian, Georgian and English summaries). Sakharth. SSR Meen. Akad. Moambe 81, no. 3, 581-584. MR 54 #12254.

Kao, Edward P.C. (1978) "A preference order dynamic program for a stochastic traveling salesman problem." Oper. Res. 26, no. 6, 1033-1045. MR 58 #20473.

Karamata, J. (1963) "Contribution a une theorie generale de la croissance des fonctions." Bull. Acad. Serbe Sci. Arts Cl. Sci. Math. Natur. Sci. Math. 31, no. 4, 29-30. MR 29 #6280.

Kas, Peter (1977) "On problems of preemptive scheduling interpreted by network flow problems." (Hungarian. English summary). Alkalmaz, Mat. Lapok 3, no. 1-2, 131-138. MR 58 #20383.

Katona, G.O.H. (1973) "A three part Sperner theorem." Stud. Sci. Math. Hung. 8, 379-390.

Katrinak, T. (1968) "Bemerkung uber pseudokomplementare halbgeordnete Mengen." Acta Fac. Rerum Natur. Univ. Comeian. Math. Publ. 19, 181-185. MR 41 #5257.

Keeney, Ralph L.; Raiffa, Howard (1976) "Decisions with multiple objectives: Preferences and Value Tradeoffs." John Wiley and Sons, N.Y.-London-Sydney, xxviii+569 pp. MR 7779.

Kelly, D. (1977) "The 3-irreducible partially ordered sets." Canad. J. Math. 29, 367-383.

Kelly, D. (1977) "The 3-irreducible partially ordered sets." Canad. J. Math. 29, 367-383. MR 5205.

Kelly, D. (1980) "Modular lattices of dimension two." Algebra Universalis 11, 101-104.

Kelly, D. (1981) "On the dimension of partially ordered sets." Discrete Math. 35, 135-156.

Kelly, D.; Rival, I. (1974) "Crowns, fences, and dismantlable lattices." Canad. J. Math. 26,1257-1271.

Kelly, David; Rival, Ivan (1975) "Certain partially ordered sets of dimension three." J. Combinatorial Theory Ser. A 18, 239-242. MR 50 #12828.

Kelly, David; Rival, Ivan (1975) "Planar lattices." Canad. J. Math. 27,,636-665.

Kelly, J.S. (1978) "Arrow impossibility theorems." New York, Academic Press.

Kerntopf, Pawel "Podstawowe pojecia matematyczne w teorii automatow: relacje. systemy algebraiczne. struktury." (Polish). [fundamental concepts in the theory of automata:] Panstwowe Wydawnictwo Naukowe, Warsaw. 1967. MR 37 #1197.

Kierstead, H. (1981) "An effective version of Dilworth's Theorem." Trans. Amer. Math. Soc.

Kirkwood, Craig W. (1977) "Superiority conditions in decision problems with multiple objectives." IEEE Trans. Systems Man Cybernet. SMC-7, no. 7, 542-544. MR 17709.

Kiseleva, T.G. (1967) "Partially ordered sets endowed with a uniform structure." (Russian. English summary). Vestnik Leningrad. Univ. 22, no. 13, 51-57. MR 35 #7303.

Kislicyn, S.S. (1967) "Partially ordered sets and k-decomposability." (Russian). Sibirsk. Mat. Z. 8, 1197-1201. MR 36 #2513.

Kislicyn, S.S. (1968) "Finite partially ordered sets and their corresponding permutation sets." (Russian). Mat. Zametki 4, 511-518. MR 39 #5430.

Kjellberg, Goran (1969) "Logical and other kinds of independence." Proc. Internat. Sympos. Switching Theory 1957, I, 117-124. Harvard Univ. Press, Cambridge, Mass. MR 22 #5515.

Klarner, David A. (1969) "The number of graded partially ordered sets." J. Combinat. Theory 6, 12-19.

Klarner, David A. (1970) "The number of classes of isomorphic graded partially ordered sets." J. Combinatorial Theory 9, 412-419. MR 42 #2989.

Klass, Michael J. (1974) "Maximum antichains: a sufficient condition." Proc. Amer. Math. Soc. 45, 28-30. MR 49 #7190.

Klein, Fritz (1938) "Ein weiterer beitrag zum problem der zerlegung der elemente gewisser harmonischer verbande." Deutsche Math. 3, 46-64.

Klein, Fritz (1939) "Theorie der Gewebe." Math. Z. 45, 107-126.

Klein-Barmen, Fritz (1950) "Zur Axiomatik der ausgeglichenen Gewebe." Math. Z. 53,70-75. MR 12 #154.

Kleindorfer, George B. (1971) "Bounding distributions for a stochastic acyclic network." Operations Res. 19, 1586-1601. MR 45 #9807.

Kleiner, M.M. (1972) "Faithful representations of partially ordered sets of finite type." (Russian). Zap. Naucn. Sem. Leningrad. Otdel. Mat. Inst. Steklov. (LOMI) 28, 42-59. MR 48 #10912.

Kleiner, M.M. (1972) "Partially ordered sets of finite type. (Russian) Investigations on the theory of representations." Zap. Naucn. Sem. Leningrad. Otdel. Mat. Inst. Steklov. (LOMI) 28, 32-41. MR 48 #10911.

Kleitman, D.; Edelberg, M.; Lubell, D. (1971/72) "Maximal sized antichains in partial orders." Discrete Math. 1, no. 1, 47-53. MR 45 #5034.

Kleitman, D.; Markowsky, G. (1975) "On Dedekind's problem. The number of isotone Boolean functions.II." Trans. Amer. Math. Soc. 213, 373-390. MR 52 #2995.

Kleitman, D.J. (1966) "Families of non-disjoint subsets." J. Comb. Theory 1, 153-155.

Kleitman, D.J. (1969) "On Dedekind's problem: The number of monotone Boolean functions." Proc. Amer. Math. Soc. 21, 677-682.

Kleitman, D.J. (1973) "The number of Sperner families of subsets of an n element set." Infinite and finite sets, Colloq. Math. Soc. Janos Bolyai, Keszthely, Hungary 10.

Kleitman, D.J. (1974) "On an extremal property of antichains in partial orders. The LYM property and some of its implications and applications." Proc. Adv. Study Inst Combin, Breukelen. pp. 77-90. Math. Centre Tr. 56,Math. Centrum, Amsterdam. MR 50 #12829.

Kleitman, D.J. (1974) "On an extremal property of antichains in partial order. The LYM property and some of its implications and applications." Hall and Van Lint, Combinatorics, Math. Center Tracts 56.

Kleitman, D.J.; Rothschild, B.L. (1975) "Asymptotic enumeration of partial orders on a finite set." Trans. Amer. Math. Soc. 205, 205-220. MR 51 #5326.

Kleitman, D.J.; Winston, K.J. (1980) "The asymptotic number of lattices." Ann. Disc. Math. 6, 243-249.

Knaster, B. (1928) "Un theoreme sur les functions d'ensembles." Ann. Soc. Polon. Math. 6, 133-144.

Kobayashi, K. [Kobayashi, Kojiro] (1969) "Classification of formal languages by functional binary transductions." Information and Control 15, 95-109. MR 40 #2473.

Koch, R.J. (1959) "Arcs in partially ordered spaces." Pac. J. Math. 9, 723-728.

Koch, R.J. (1960) "Weak cutpoint ordering in hereditarily unicoherent continua." Proc. Amer. Math. Soc. 11, 679-681.

Koch, R.J. (1965) "Connected chains in quasi-ordered spaces." Fund. Math. 56, 245-249.

Kogalovskii, S.R. (1964) "On linearly complete ordered sets." (Russian). Uspehi Mat. Nauk 19, no. 2 (116), 147-150. MR 28 #5013.

Kolibiar, M. (1958) "Bemerkung uber die Ketten in teilweise geordneten Mengen." Acta Fac. Nat. Univ. Comenian. 3, 17-22. (Slovak and Russian summaries). MR 21 #1945.

Kolibiar, M. (1961) "Bemerkungen uber Translationen der Verbande." (Czech and Russian summaries). Acta Fac. Nat. Univ. Comenian. 5, 455-458. MR 24 #A2543.

Kolibiar, M. (1962) "Bemerkungen uber Intervalltopologie in halbgeordneten Mengen, General topology and its applications to modern analysis and algebra." Prague, 252-253.

Kolibiar, M. (1962) "Bemerkungen uber Intervalltopologie in halbgeordneten Mengen. General Topology and its Relations to Modern Analysis and Algebra." Proc. Sympos., Acad. Press, Publ. House Czech. Acad. Sci., Prague, 1962, 252-253.

Kolodner, I.I. (1968) "On completeness of partially ordered sets and fixpoint theorems for isotone mappings." Amer. Math. Monthly 75, 48-49. MR 37 #110.

Komatu, Atuo (1943) "On a characterisation of join homomorphic transformation-lattice." Proc. Imp. Acad. Tokyo 19, 119-124. [MF 14801]. MR 1 #235.

Komm, Horace. (1948) "On the dimension of partially ordered sets." Amer. J. Math. 70, 507-520. MR 10 #22.

Konurbaeva, B.M. (1971) "Positional games in which the set of positions is partially ordered." (Russian. Kazakh summary). Izv. Akad. Nauk Kazah. SSR Ser. Fiz.-Mat. no. 3, 44-46. MR 45 #6456.

Kopecek, Oldrich (1967) "Allgemeine kardinaloperationen." Arch. Math. (Brno) 3, 35-44. MR 36 #5037.

Kopecek, Oldrich (1968) "Die arithmetischen Operationen fur geordnete Mengen." Arch. Math. (Brno) 4, 157-174. MR 42 #4450.

Koppelberg, Sabine (1977) "Groups cannot be Souslin ordered." Arch. Math. (Basel) 29, no. 3, 315-317. MR 5741.

Korec, I. (1966) "Construction of all partially ordered sets homomorphically mappable onto a given set." (Russian. Slovak and German summaries). Acta Fac. Natur. Univ. Comenian. 10, fasc. 5, 23-36. MR 34 #104.

Kornilova, L.E. (1969) "A complex network model. I. the scheduling and diagram of the work of plant subdivisions." (Russian). Application of Mathematical Methods (Proc. Seminar, Kiev, 1969), No. 3 (Russian).

Korobkov, V.K. (1963) "On the number of monotone functions of a Boolean algebra." (Russian). Diskret. Analiz. No. 1, 24-27. MR 31 #2181.

Koutsky, Karel. (1947) "Sur les lattices topologiques." C.R. Acad. Sci. Paris 225, 659-661. MR 9 #172.

Kovalenko, I.N.; Shpak, V.D. [Spak, V.D.] (1972) "Probablistic characteristics of complex systems with hierarchical control." Izv. Akad. Nauk SSR Tehn. Kibernet. 1972, no. 6, 30-34 (Russian); translated as Engrg. Cybernetics 10, no. 6, 964-969. MR 52 #1011 .

Kowalski, Oldrich; Pondelicek, Bedrich (1966) "On the characters of chains." (Czech. Russian and English summaries). Casopis Pest. Mat. 91, 1-3. MR 32 #7459.

Kozlov, M.K.; Shafranskiy, V.V. [Safranskii, V.V.] (1977) "Time scheduling of operation complexes with limited intensities of execution of component operations." Izv. Akad. Nauk SSSR Tehn. Kibernet. no. 5, 38-43 (Russian). MR 58 #25934.

Kozmidiadi, V.A.; Cernjavskii, V.S. (1962) "On the ordering of a set of automata." (Russian). Problems in Theory of Mathematical Machines, Collection II (Russian), pp. 34-51. Fizmatgiz, Moscow. MR 33 #8237.

Kral, Jaroslav (1969) "On multiple grammars." (Czech summary). Kybernetika (Prague) 5, 60-85. MR 40 #4034.

Krasner, Marc; Docas, Lambros [Dokas, Lambros] (1967) "Expression de certaines notions classiques en termes d'un complete des ensembles partiellement ordonnes defini par M. Docas." C.R. Acad. Sci. Paris Ser. A-B 265, A453-A456. MR 36 #5038.

Kreweras, G. (1967) "Representation polyedrique des preordres complets finis et application a l'agregation des preferences." Las Decision, 2: Agregation et Dynamique des Ordres de Preference (Actes Colloq. Internat., Aix-en-Prov.

Krishnan, V.S. (1944) "Partially ordered sets and projective geometry." Math. Student 12, 7-14. [MF 11847]. MR 6 #143.

Krishnan, V.S. (1945) "The theory of homomorphisms and congruences for partially ordered sets." Proc. Indian Acad. Sci, Sect. A. 22, 1-19. [MF 13397]. MR 7 #109.

Krishnan, V.S. (1947) "Extensions of partially ordered sets. I. General theory." J. Indian Math. Soc. (N.S.) 11, 49-68. MR 10 #95.

Krishnan, V.S. (1948) "Extensions of partially ordered sets. II Constructions." J. Indian Math. Soc. (N.S.) 12, 89-100. MR 10 #587.

Krishnan, V.S. (1950) "Les algebres partiellement ordonnees et leurs extensions." Bull. Soc. Math. France 78, 235-263. MR 13 #201.

Krishnan, V.S. (1951) "Les algebres partiellement ordonnees et leurs extensions. II." Bull. Soc. Math. France 79, 85-120. MR 14 #838.

Krishnan, V.S. (1965) "Algebra and order." Math. Student 33, 27. MR 32 #5550.

Krishnan, Viakalathur S. (1950) "L'extension d'une (<.) algebre a une (sigma*.) algebre." C.R. Acad. Sci. Paris 230, 1447-1448. MR 11 #575,#638.

Krishnan, Viakalathur S. (1953) "Closure operations on c-structures." Nederl. Akad. Wetensch. Proc. Ser. A. 56=Indagationes Math. 15, 317-329. MR 15 #675.

Kristof, Walter (1968) "Structural properties and measurement theory of certain sets admitting a concatenation operation." British J. Math. Statist. Psychology 21, 201-229. MR 39 #5168.

Krivenkov, Ju. P. (1969) "A certain form of optimality conditions." (Russian). Dokl. Akad. Nauk SSSR 188, 278-281. MR 41 #6600.

Kruskal, J.B. (1960) "Well-quasi-ordering, the Tree Theorem, and Vazsonyi's conjecture." Trans. Amer. Math. Soc. 95, 210-225. MR 22 #256.

Kruskal, J.B. (1972) "The theory of well quasi ordering: a frequently discovered concept." J. Comb. Th., A, 13, 297-305. MR 46 #5184.

Kuhn, Harold W. (1977) "Pathfix: an algorithm for computing traffic equilibria." Proceedings of the 1977 IEEE Conf. on Decision and Control, Vol. 1, pp. 831-834. Inst. Electrical Electron. Engrs, N.Y. MR 58 #20358.

Kurepa Georges (1950) "Ensembles partiellement ordonnees et ensembles partiellement bien ordonnes." Acad. Serbe Sci. Publ. Inst. Math. 3, 119-125. MR 12 #683.

Kurepa, D. (1937) "Ensembles lineaires et une classe de tableaux ramifies." Publ. Math. Univ. Belgrade 6, 129-160.

Kurepa, Dj. (1977) "Monotone mappings and cellularity of ordered sets." General Topology and its Relations to Modern Analysis and Algebra, IV, Pt. B, pp. 247-253. Soc. Czechoslovak Mathematicians and Physicists, Prague. MR 58 #16433.

Kurepa, Djuro (1976) "On the system of all maximal chains (antichains) of a graph." Publ. Inst. Math. (Beograd) (N.S.) 20 (34), 161-167. MR 58 #21872.

Kurepa, Djuro [Durepa, Duro] (1964) "Remark on recent two results of Dilworth and Gleason." Publ. Inst. Math. (Beograd) (N.S.) 4 (18), 107-108. MR 31 #101.

Kurepa, Duro (1937) "Transformations monotones des ensembles partiellement ordonnes." Comptes Rendus Paris, 205, 1033-1035.

Kurepa, Duro (1940,1941) "Transformations monotones des ensembles partiellement ordonnes." Revista da Ciencias Lima 42, 827-846; 43, 483-500.

Kurepa, Duro (1942-1943) "A propos d'une generalisation de la notion d'ensembles bien ordonnes." Acta Math. 75, 139-150.

Kurepa, Duro (1948) "Sur les ensembles ordonnes denombrables." Glasnik Mat.-Fiz. Astr. Ser. II, 3, 145-151.

Kurepa, Duro (1950) "La condition de Souslin et une propriete caracteristique des nombres reels." Comptes Rendus Acad. Sci. Paris 231, 1113-1114.

Kurepa, Duro (1950) "The problem of measure and monotonic mappings of partially ordered sets." Rad Jugoslav. Akad. Znan. Umjet. Odjel Mat. Fiz. Tehn. Nauke 277, 229-237. (Serbo-Croatian). MR 14 #960.

Kurepa, Duro (1951) "Teoriza skupova." [the theory of sets]. Skolska Knjiga, Zagreb, xix + 443 pp. MR 12 #683.

Kurepa, Duro (1952) "Le probleme de la measure et les transformations monotones des ensembles partiellement ordonnes." Bull. Internat. Acad. Yougoslave Sci. Beaux-Arts (N.S.) 6, 91-96.

Kurepa, Duro (1953) "On reflexive symmetric relations and graphs." Dissertationes Slov. Acad. Sci. Ljubljana IV, 65-92.

Kurepa, Duro (1954) "Sur les fonctions reelles dans la famille des ensembles bien ordonnes de nombres rationnels." Bull. Internat. Acad. Yougoslave, Cl. Sci. Math. Phys. Techn. (N.S.) 12, 35-42.

Kurepa, Duro (1957) "Partitive sets and ordered chains." Rad. Jugoslav. Acad. Znan. Umjet. Odjel. Mat. Fiz. Tehn. Nauke 6 (302), 197-235.

Kurepa, Duro (1957) "Partitive sets and ordered chains." Rad. Jogoslav. Akad. Znan. Umjet. Odjel. Mat. Fiz. Tehn. Nauke 6 (302), 197-235. (Serbo-Croatian summary). MR 20 #3798.

Kurepa, Duro (1959) "On the cardinal number of ordered sets and of symmetrical structures in dependence on the cardinal numbers of its chains and antichains." Glasnik Mat.-Fiz. Astronom. Ser. II, 14, 183-203. MR 23 #A2329.

Kurepa, Duro (1963) "Star number and antistar number of ordered sets and graphs." (Serbo-Croatian summary). Glasnik Mat.-Fiz. Astronom. Ser. II Drustvo Mat. Fiz. Hvatski 18, 27-37. MR 30 #1945.

Kurepa, Duro (1964) "Fixpoints of monotone mappings of ordered sets." (Serbo-Croatian summary). Glasnik Mat.-Fiz. Astronom. Ser. II Drustvo Mat. Fiz. Hrvatske 19, 167-173. MR 31 #5819.

Kurepa, Duro (1964) "Monotone mappings between some kinds of ordered sets." (Serbo-Croatian summary). Glasnik Mat.-Fiz Astronom. Ser. II Drustvo Mat. Fiz Hrvatske 19, 175-186. MR 31 #5820.

Kurepa, Duro (1975) "Fixpoints of decreasing mappings of ordered sets." Publ. Inst. Math. (N.S.) 18 (32), 111-116. MR 51.

Kurepa, G. (1935) "Ensembles ordonnes et ramifies." Publ. Math. Univ. Belgrade 4, 1-138.

Kurepa, G. (1952) "On a characteristic property of finite sets." Pacific J. Math. 2, 323-326. MR 14 #255.

Kurepa, G. (1955) "Generalisation de l'operation de Suslin, de celle d'Alexandroff et de la formule de de Morgan." Bull. Internat. Acad. Yougoslave, Cl. Sci. Math. Phys. Tech. 5, 97-107. MR 20 #4498.

Kurepa, G. (1955) "Still about induction principles." Rad Jogoslzv. Akad. Znan. Umjet. Odjel Mat. Fiz. Tehn. Nauke 302, 77-86. (Serbo-Croatian). English summary). MR 18 #270.

Kurepa, G. [Kurepa, Duro] (1961) "On rank-decreasing functions." Essays on the Foundations of Mathematics, pp. 248-258. Magnes Press, Hebrew Univ, Jerusalem. MR 29 #3382.

Kurepa, George (1958) "On regressing functions." Z. Math. Logik Grundlagen Math, 4, 148-156. MR 20 #4499.

Kurepa, Georges "Problematique des ensembles partiellement ordonnes." Premier Congres des Mathematiciens et Physiciens de la R.P.F.Y, 1949. Vol. II, Comm. et Exposes Sci, pp. 7-22. Naucna Knjiga, Belgrade, (Serbo-Croatian. French summary). MR 13 #542.

Kurepa, Georges (1939) "Sur la puissance des ensembles partiellement ordonnes." Sprawozd. Towarz. Nauk. Warszaw. Mat.-Fix. 32, no. 1/3, 61-67. (Polish summary). MR 21 #2603.

Kurepa, Gjuro (1958) "On two problems concerning ordered sets." Glasnik Mat.-Fiz. Astr. Drustvo Mat. Fiz. Hrvatske. Ser. II 13, 229-234. (Serbo-Croatian summary). MR 21 #3351.

Kurepa, Gjuro (1959) "Sur la puissance des ensembles partiellement ordonnes." (Serbo-Croatian summary). Glasnik Mat.-Fiz. Astronom. Drustvo Mat. Fiz. Hrvatski Ser. II 14, 205-211. MR 23 #A2341.

Kurisu, Tadahi (1977) "Three-machine scheduling problem with precedence constraints." J. Operations Res. Soc. Japan 20, no. 3, 231-242. MR 10907.

Kyuno, S. (1979) "An inductive algorithm to construct finite lattices." Math. Comp. 33, 409-421.

Labkovskii, V.A. (1977) "A decision making model." (Russian). Adaptive Control Systems (Russian) pp. 13-18. Akad. Nauk Ukrain. SSR Inst. Kibernet, Kiev. MR 58 #25864.

Ladegaillerie, Y. (1973) "Prefermeture sur un ensemble ordonne." (Loose English summary). Rev. Francaise Automat. Informat. Recherche Operationnelle 7, Ser. R-1, 35-43. MR 51 #294.

Ladzianska, Zuzana (1965) "Characterization of directed partially ordered set by means of betweenness relation." Mat. Fyz. Casopis, Ṡlovensk. Akad. Vied 15, 162167, English summary 167 (Russian).

Laemmel, Arthur E. (1963) "Application of lattice-ordered semigroups to codes and finite-state transducers." Proc. Sympos. Math. Theory of Automata (New York, 1962), pp. 241-256. Polytechnic Press of Polytec.

Lageweg, B.J.; Lenstra, J.K.; Rinnooy Kan, A.H.G. (1976) "Minimizing maximum lateness on one machine: computational experience and some applications." Statist. Neerlandica 30, 25-41.

Lake, John (1975) "Characterising the largest, countable partial ordering." Z. Math. Logik Grundiagen Math. 21, no. 4, 353-354. MR 52 #2888.

Lal, R.N. (1975) "Uniform posets." Indian J. Pure Appl. Math. 6, no. 12, 1526-1533. MR 58 #10631.

Lam, Shui; Sethi, Ravi (1977) "Worst case analysis of two scheduling algorithms." SIAM J. Comput. 6, no. 3, 518-536. MR 58 #15115.

Lambek, Joachim (1968) "A fixpoint theorem for complete categories." Math. Z. 103, 151-161. MR 37 #270.

Lapscher, F. (1974) "Proprietes de "longueur de chaine" dans l'ensemble des fermetures superieures sur un ensemble ordonne." Discrete Math. 7, 129-140. MR 48 #8317.

Larson, Jean A. (1978) "A solution for scattered order types of a problem of Hagendorf." Pacific J. Math. 74, no. 2,373-379. MR 58 #5422.

Larson, Roland E.; Andima, Susan J. (1975) "The lattice of topologies: a survey." Rocky Mountain J. Math. 5, 177-198. MR 52 #9143.

Laudal, O. (1965) "Sur la theorie des limites projectives et inductives." Ann. Sci. Ecole Norm. Sup. 82, 241-296.

Laudal, O. (1968) "Projective systems on trees and valuation theory." Canadian J. Math. 20, 984-1000.

Lauria, Francesco, E. (1977) "Fixpoint: an order independent property." (Italian summary). Rend. Accad. Sci. Fis. Mat. Napoli(4)43(1976), 153-165. MR 57 #9614.

Lausmaa, T. (1971) "Synthesis of logical schemes in lattice logic." (Russian, Estonian and English summaries). Eesti NSV Tead. Akad. Toimetised Fuus.-Mat. 20, 315-322. MR 46 #10505.

Laver, R. (1971) "On Fraisse's order type conjecture." Ann. Math. 93, 89-111.

Laver, R. (1973) "An order type decomposition theorem." Ann. of Math. (2) 98, 96-119. MR 47 #8361.

Lawler, E.L. (1973) "Optimal sequencing of a single machine subject to precedence constraints." Management Sci. 19, 544-546.

Lawler, E.L. (1976) "Combinatorial Optimization: Networks and Matroids." Holt, Rinehart and Winston, New York.

Lawler, E.L. (1978) "Sequencing jobs to minimize total weighted completion time subject to precedence constraints." Ann. Discrete Math. 2, 75-90.

Lawler, E.L.; Sivazlian, B.D. (1978) "Minimization of time varying costs in single machine sequencing." Oper. Res. 26, 563-569.

Lawson, J. (1980) "The duality of continuous posets." Houston Jour. Math.

Lawson, Jimmie D. (1977) "Compact semilattices which must have a basis of subsemilattices." J. London Math. Soc. (2)16, no. 2, 367-371. MR 518.

Leclerc, B. (1981) "Description combinatoire des ultrametriques." Math. Sci. Hum. 73, 5-37.

Leclerc, B.; Monjardet, B. (1973) "Orders "C.A.C."." Fund. Math. 79, 1, 11-22. MR 47 #8364.

Leclerc, Bruno (1976) "Arbres et dimension des ordres." Discrete Math. 14, no. 1, 69-76. MR 52 #7979.

Lecointe, Pierre (1968) "Etude d'une certaine classe de treillis en vue de la construction de plans d'experiences et de codes correcteurs d'erreurs." C.R. Acad. Sci. Paris Ser. A-B 266, A1206-A1209. MR 38 #6887.

Lectures (1971) "Lectures on the theory of ordered sets and general algebra." Slovenoki Pedagogicke Nakladatel'stvo, Bratislava, 98 pp. MR 51 #10171.

Leifman, L.J. (1978) "Estimation of resource requirements in optimization problems of network planning." European J. Oper. Res. 2, no. 4,265-280. MR 58 #25936.

Leininger, C.W. (1966) "On chains of related sets." Bull. Math. Biophys. 28, 103-106. MR 33 #3964.

Lenstra, J.K.; Rinnooy Kan, A.H.G. (1978) "Complexity of scheduling under precedence constraints." Operations Res. 26, no. 1,22-35. MR 2526.

Lenstra, J.K.; Rinnooy Kan, A.H.G. (1979) "Computational complexity of discrete optimization problems." Ann. Discrete Math. 4, 121-140.

Lenstra, J.K.; Rinnooy Kan, A.H.G. (1980) "Complexity results for scheduling chains on a single machine." European J. Oper Res.

Lenstra, J.K.; Rinnooy Kan, A.H.G.; Brucker, P. (1977) "Complexity of machine scheduling problems." Ann. Discrete Math. 1, 343-362.

Lenzi, Domenico (1976) "Alcune osservazioni sugli insiemi parzialmete ordinati semi-modulari di lunghezza finita." (English summary). Atti dei Convegni Lincei, No. 17, Accad. Naz. Lincei, Rome. MR 57 #12306.

Lerman, M. (1981) "On recursive linear orderings." Logic year 1979-80, Lerman, Schmerl, Soare, eds., Lecture Notes in Mathematics 859, Springer-Verlag, Berlin 1981.

Leticevskii, A.A. [Leticevskii, O.A.] (1965) "On the minimization of finite automata." (Russian). Kibernetika (Kiev) no. 1, 20-23. MR 34 #4063.

Levicev, A.V. (1977) "On a connected preorder." (Russian). Dokl. Akad. Nauk SSSR 235, no. 6, 1256-1259. MR 58 #12939.

Levine, E.; Lubell, D. (1970) "Sperner collections on sets of real variables." International Conference on Combinatorial Mathematics. MR 42 #2949.

Lew, Art (1975) "Optimal resource allocation and scheduling among parallel processes." Parallel Processing, pp. 187-202. Lecture Notes in Comput. Sci, Vol. 24, Springer, Berlin. MR 57 #15394.

Lew, John S. (1975) "Preorder relations and pseudoconvex metrics." Amer. J. Math. 97, 344-363. MR 52 #2650.

Li, Hon F. (1977) "Scheduling trees in parallel/pipelined processing environments." IEEE Trans. Computers C-26, no. 11,1101-1112. MR 58 #25937.

Li, S.-Y.R. (1978) "Extremal theorems on divisors of a number." Disc. Math. 24, 37-46.

Lindenbaum, Adolphe (1931) "Sur les ensembles ordonnes." C.R. Acad. Sci, Paris 192, 1511-1514.

Lindstrom, B. (1980) "A partition of L(3,n) into saturated symmetric chains." Europ. J. Comb. 1, 61-63.

Linial, N. (1978) "Covering digraphs by paths." Disc. Math. 23, 257-272.

Linial, N. "Extending the Greene-Kleitman theorem to directed graphs." J. Combinatorial Theory Ser. A (30), 331-334.

Liouville, Bernard (1978) "Le probleme du voyageur de commerce dans un produit cartesien de deux graphes." (English summary). RAIRO Rech. Oper. 12, no. 3, 263-275, iv. MR 58 #15118.

Lisa, Jarmila (1973) "Cardinal sums and direct products in Galois connections." Comment. Math. Univ. Carolinae 14, 325-338. MR 48 #8319.

Liu, C.L. (1969) "Lattice functions, pair algebras, and finite-state machines." J. Assoc. Comput. Mach. 16, 442-454. MR 39 #2539.

Liu, C.L. (1975) "Sperner's theorem on maximal-sized antichains and its generalization." Discrete Math. 11, 133-139. MR 50 #9718.

Liu, Jane W.S.; Liu, C.L. (1974) "Bounds on scheduling algorithms for heterogeneous computing systems." Information Processing 74, pp. 349-353. North-Holland, Amsterdam. MR 14646.

Liu, Jane W.S.; Liu, C.L. (1978) "Performance analysis of multiprocessor systems containing functionally dedicated processors." Acta Inform. 10, no. 1,95-104. MR 58 #25938.

Livsic, E.M.; Rublineckii, V.I. (1972) "The optimal splitting of an ordered set into intervals." (Russian). Numerical Math. and Computer Tech, No. III (Russian), pp. 86-89. Akad. Nauk Ukrain. SSR Fix.-Tehn. Inst. Nizkih Temperatur, Kharkov. MR 17767.

Ljapin, E.S. (1967) "Semigroups of directed transformations of ordered sets." (Russian). Mat. Sb. (N.S.) 74 (116), 39-46. MR 36 #82.

Ljapin, E.S. (1968) "Defining relations of the semigroups of all directed transformations of a finite ordered set." (Russian). Mat. Zametki 3, 657-662. MR 37 #6226.

Ljapin, E.S. (1968) "Types of elements and zeros of semigroups of directed transformations." (Russian). Leningrad. Gos. Ped. Inst. Ucen. Zap. 387, 171-180. MR 38 #5679.

Ljapin, E.S. (1970) "Directed endomorphisms of ordered sets." (Russian). Sibirsk. Mat. Z. 11, 217-221. MR 42 #142.

Loonstra, F. (1977) "Classes of partially ordered groups." Symposia Math, Vol. XXI, INDAM, Rome, 1975), pp. 335-348. Academic Press, London. MR 58 #430.

Lorenzen, Paul (1958) "Uber den kettensatz der gengenlehre." Arch. Math. 9, 1-6. MR 20 #5143.

Losfeld, Joseph (1974) "Information generalisee et relation d'order." Lecture Notes in Mathematics, Vol 398, Springer, Berlin pp. 49-61. MR 51 #9965.

Lovasz, L. (1967) "Operations with structures." Acta Math. Acad. Sci. Hungar. 18, 321-328.

Lovasz, L. (1971) "On the cancellation law among finite relational structures." Period. Math. Hungar. 1, no. 2, 145-156. MR 44 #1619.

Lovasz, L. (1972) "Direct product in locally finite categories." Acta. Sci. Math. (Szeged) 33, 319-322. MR 48 #6207.

Lovasz, L. (1972) "Normal hypergraphs and the perfect graph conjecture." Discrete Math. 2, 253-267.

Lovasz, L.; Nesetril, J.; Pultr, A. (1980) "On a product dimension of graphs." J. Combinatorial Theory (Ser. B) 20. 48-67.

Lubell, D. (1966) "A short proof of Sperner's lemma." J. Combinatorial Theory 1, 299. MR 33 #2558.

Lubell, David (1974) "Combinatorial orthogonality in abstract configurations." Discrete Math. 8, 201-203. MR 48 #8318.

Lubell, David (1974) "Orthogonal structures on finite partial orders." J. Combinatorial Theory Ser. A 17, 210-218. MR 49 #8905.

Lubell, David (1975) "Local matching in the function space of a partial order." J. Combinatorial Theory Ser. A 19, no. 2, 154-159. MR 51 #12553.

Luce, R. Duncan (1956) "Semiorders and a theory of utility discrimination." Econometrica 24, 178-191. MR 17 #1222.

Luce, R. Duncan (1966) "Two extensions of conjoint measurement." J. Mathematical Psychology 3, 348-370. MR 34 #2333.

Lukoskov, V.V. (1969) "On the organization of a hierarchy of control systems." (Russian). Algorithmization of Industrial Processes (Proc. Seminar, Kiev, 1968), No. 2 (Russian), pp. 93-104. Akad. Nauk Ukrain, SSR, Kiev. MR 39 #6631.

MacLane, Saunders (1943) "A conjecture of Ore on chains in partially ordered sets." Bull. Amer. Math. Soc. 49, 567-568. [MF 8857]. MR 5 #88.

MacNeille, H.M. (1936) "Extensions of partially ordered sets." Proc. Nat. Acad. Sci. U.S.A. 22, 45-50.

MacNeille, H.M. (1937) "Partially ordered sets." Trans. Amer. Math. Soc. 42, 416-460.

Maggu, Parkash Lal; Alam, Mohd. Shamim; Ikram, Mohd. (1974/75) "On certain type of sequencing problem with n jobs and m machines." Pure Appl. Math. Sci. 1, no. 2, 55-59. MR 5060a.

Maggu, Parkash Lal; Ikram, Mohd.; Alam, Mohd. Shamim (1974/75) "On certain theorems in sequencing theory." Pure Appl. Math. Sci. 1, no. 2, 60-62. MR 5060b.

Maksimenkov, A.V. (1978) "Computer resource scheduling with minimum disproportions." (Russian). Avtomat. i Vycisl. Tehn. (Riga) no. 4, 69-75,94. MR 58 #25939.

Maksimenkov, A.V.; Scers, A.L. (1978) "Scheduling the filling of standing order assignments." (Russian). Programmirovanie no. 1, 77-83. MR 58 #25940.

Malhara, H. (1980) "On time graphs." Discrete Mathematics, 32, 281-289.

Malinowski, Krzysztof (1975) "The properties of parametric optimization by primal methods in multilevel optimization problems." (Russian summary). Systems Sci. 1, no. 1, 29-53. MR 53 #12595.

Maluszynski, Jan (1970) "The lattice of subgrammars of context free grammars." (Polish, Russian and English summaries). Algorytmy 7, no. 12, 21-27. MR 43 #8472.

Mamikonjan, S.G. (1971) "The directed transformations of ordered sets." (Russian). Leningrad. Gos. Ped. Inst. Ucen. Zap. 464, cast 2, 117-122. MR 45 #136.

Manaster, A.; Remmel, J. (1981) "Partial Orderings of fixed finite dimension: model companions and density." J. Symbolic Logic.

Manaster, A.; Rosenstein, J. (1980) "Two dimensional partial orderings: undecidability." J. Symbolic Logic, 45, 133-143.

Manaster, A.; Rosenstein, J. (1980) "Two dimensional partial orderings; recursive model theory." J. Symbolic Logic, 45, 121-132.

Manolache, Nicolae; Popescu, Dorin (1970) "Deux extensions des ensembles preordonnes." Rev. Roumaine Math. Pures Appl. 15, 569-572. MR 42 #7737.

Marcus, Solomon (1965) "Sur une description axiomatique des liens syntaxiques." Z. Math. Logik Grundlogen Math. 11, 291-296. MR 32 #3977.

Marinescu, Gh. (1944) "Structures et nombres caracteristiques." Bull. Math. Soc. Roumaine Sci. 46, 113-119. [MF 16514]. MR 7 #549.

Markowsky, G. (1976) "Chain complete posets and directed sets with applications." Alg. Universalis 6, 53-68.

Markowsky, G. (1977) "Categories of chain complete posets." Theor. Comp. Sci. 4, 125-135.

Markowsky, G. (1981) "A motivation and generalization of Scott's notion of a continuous lattice." Lect. Notes Math. 871, Springer, 298-307.

Markowsky, G. (1981) "Propaedeutic to chain-complete posets with basis." Lect. Notes Math. 871, Springer, 308-314.

Markowsky, G.; Rosen, B. (1976) "Bases for chain complete posets." IBM J. Res. and Devel. 20, 138-147.

Markowsky, G.; Rosen, B.K. (1976) "Bases for chain-complete posets." IBM J. Res. Develop. 20, no. 2, 138-147. MR 52 #13523.

Markowsky, George (1975) "The factorization and representation of lattices." Trans. Amer. Math. Soc. 203, 185-200.

Markowsky, George (1977) "Categories of chain-complete posets." Theoret. Comput. Sci. 4, no. 2, 125-135. MR 11859.

Markowsky, George; Rosen, Barry K. (1975) "Bases for chain-complete posets." 16th Ann. Symp. on Found. of Computer Sci. (Berkeley, Ca, 1975) pp. 34-47. IEEE Computer Soc., Long Beach, Calif. MR 8440.

Markwald, W. (1954) "Zur Theorie der konstruktiven Wohlordnungen." Math. Annalen 127, 135-149.

Marley, A.A.J. (1967) "Abstract one-parameter families of commutative learning operators." J. Mathematical Pschology 4, 414-429. MR 35 #7707.

Martinez, Jorge (1972) "Unique factorization in partially ordered sets." Proc. Amer. Math. Soc. 33, 213-220. MR 45 #1806.

Mas-Colell, Andreu (1977) "On the continuous representation of preorders." Internat. Econom. Rev. 18, no. 2, 509-513. MR 58 #20287.

Mather, J. (1966) "Invariance of the homology of a lattice." Proc. Amer. Math. Soc. 17, 1120-1124.

Mathews, J.C.; Anderson, R.F. (1967) "A comparison of two modes of order convergence." Proc. Amer. Math. Soc. 18, 100-104. MR 34 #3524.

Matsushima, Y. (1960) "Hausdorff interval topology on a partially ordered set." Proc. Amer. Math. Soc. 11, 233-235.

Matsuyama, Keisuke; Kijima, Kyoichi (1973) "A study of the utility function." (English summary). Rep. Univ. Electro-Commun. 24, 303-317. MR 49 #12050.

Matusima, Yataro (1952) "On some problems of Birkhoff." Proc. Japan Acad. 28, 19-24.

Maurer, S.B.; Rabinovitch, I.; Trotter, W.T. (1980) "A generalization of Turan's theorem." Discrete Math. 32, 167-189.

Maurer, S.B.; Rabinovitch, I.; Trotter, W.T. (1980) "Large minimal realizers of a partial order II." Discrete Math. 31, 297-314.

Maurer, S.B.; Rabinovitch, I.; Trotter, W.T. (1980) "Partially ordered sets with equal rank and dimension." Congressus Numerantium 29, 627-637.

Maurer, Stephen B.; Rabinovitch, I. (1977) "Large minimal realizers of a partial order." Proc. Amer. Math. Sco. 66, no. 2, 211-216. MR 8441.

Maxson, Carlton J. (1974) "Semigroups of order-preserving partial endomorphisms on trees. I." Colloq. Math. 32, 25-37. MR 50 #13343.

May, Kenneth O. (1952) "A set of independent necessary and sufficient conditions for simple majority decision." Econometrica 20, 680-684. MR 14 #392.

May, Kenneth O. (1953) "A note on the complete independence of the conditions for simple majority decision." Econometrica 21, 172-173. MR 14 #778.

May, Kenneth O. (1954) "Intransitivity, utility, and the aggregation of preference patterns." Econometrica 22, 1-13. MR 15 #888.

Mayer-Kalkschmidt, J.; Steiner, E. (1964) "Some theorems in set theory and applications in the ideal theory of partially ordered sets." Duke Math. J. 31, 287-289. MR 28 #3940.

Mazzocca, Francesco (1976) "Problemi estremali per partizioni di un insieme preordinato finito." (English summary). Atti Accad. Naz. Lincei Rend. Cl. Sci. Fis. Mat. Natur.(8)60, no. 3, 195-197. MR 186.

McCartan, S.D. (1966) "On subsets of a partially ordered set." Proc. Cambridge Philos. Soc. 62, 583-595. MR 34 #105.

McCleary, S. (1973) "The lattice-ordered group of automorphisms of an alpha-set." Pac. J. Math. 49, 417-424.

McCord, M. (1966) "Singular homology groups and homotopy groups of finite spaces." Duke Math. J. 33, 465-474.

McGarvey, David C. (1953) "A theorem on the construction of voting paradoxes." Econometrica 21, 608-610. MR 15 #976.

McKenzie, R. (1971) "Cardinal multiplication of structures with a reflexive relation." Fund. Math. 70, 59-101.

Megiddo, Nimrod (1977) "Mixtures of order matrices and generalized order matrices." Discrete Math. 19, no. 2, 177-181. MR 15463.

Mendelsohn, Eric (1971) "An elementary characterization of the category of partly ordered sets." Math.Z. 122, 111-116. MR 45 #5057.

Mendelson, Elliott (1958) "On a class of universal ordered sets." Proc. Amer. Math. Soc. 9, 712-713. MR 20 #3075.

Menger, K. (1936) "New foundations of projective and affine geometry." Ann. of Math. 37, 456-482.

Meshalkin, L.D. (1963) "A generalization of Sperner's Theorem on the number of subsets of a finite set." Teor. Verojatnost i Primener 8, 219-220 [in Russian] - Theory Probab. Appl. [English trans.] 8:203-204.

Metcalf, F.T.; Payne, T.H. (1972) "On the existence of fixed points in a totally ordered set." Proc. Amer. Math. Soc. 31, 441-444. MR 44 #3931.

Metev, Bojan (1976) "Some metric relation spaces of relations, and their application to the ordering problem." (Bulgarian. Russian and English summaries). Problemi Tehn. Kibernet. 3, 6H5-10. MR 58 #5439.

Metropolis, N.; Rota, Gian-Carlo; Strehl, Volker; White, Neil (1978) "Partitions into chains of a class of partially ordered sets." Proc. Amer. Math. Sco. 71, no. 2,193-196. MR 58 #27667.

Meyer, Karlhorst; Nieger, Magnus (1977) "Hullenoperationen, galoisverbindungen und polaritaten." (Italian summary). Boll. Un. Mat. Ital. A(5)14, no. 2,343-350. MR 3011.

Meyer, P.J. Gustav (1974) "On the structure of orthomodular posets." Discrete Math. 9, 119-146. MR 51 #290.

Michael, E. (1960) "A class of partially ordered sets." Amer. Math. Monthly 67, 448-449. MR 22 #25.

Mierlea, V. (1971) "An algorithm for scheduling n works on m machines." (French and Russian summaries). Econom. Comp. Econom. Cybernet. Stud. Res. no. 4, 17-28. MR 45 #8247.

Milgram, A.N. (1940) "Partially ordered sets and topology." Rep. Math. Colloquium (2) 2, 3-9. [MF 5768]. MR 3 #313.

Miller, Gary L. (1977) "Graph isomorphism, general remarks." Conf. Record of the Ninth Ann. ACM Symp. on Theory of Computing (Boulder, Colo,1977), pp. 143-150. Assoc. Comput. Mach, N.Y. MR 5816.

Miller, J.B. (1979) "Introduction to a theory of coups." Algebra Universalis 9, 346-370.

Miller, J.B. (1981) "Representation of Boolean hypolattices." Bull. Austral. Math. Soc. 24, 389-404.

Miller, John Boris (1975) "Simultaneous lattice and topological completion of topological posets." Compositio Math. 30, 63-80. MR 51 #10174.

Mills, Janet E. (1978) "A class of uniform bands." Semigroup Forum 16, no. 1, 53-66. MR 58 #6013.

Mirkin, B.G. (1972) "Description of some relations on the set of real-line intervals." J. Math. Psychol., 9, 243-252.

Mirkin, B.G. (1973) "The principles of the compatibility of relations." (Russian, English summary insert). Kibernetika (Kiev) no. 2, 74-79. MR 51 #4938.

Mirsky, L. (1971) "A dual of Dilworth's decomposition theorem." Amer. Math. Monthly 78, 876-877. MR 44 #5252.

Mobius (1971) "Mobius algebras." Proceedings of a Conference, Univ. of Waterloo, Waterloo, Ont, 1971. Ed. by Henry Crapo and Geffrey Roulet. Univ. of Waterloo, Waterloo, Ont. 192pp. MR 49 #4875.

Mogilevskii, M.G. (1974) "Order relations on a symmetric inverse semigroup." (Russian). Theory of Semigroups and its Applications, No. 3 (Russian), pp. 63-71, 153. Izdat. Saratov. Univ, Saratov. MR 201.

Monjardet, B. (1969) "Remarques sur une classe de procedures de vote et les "theoremes de possibilite"." La Decision, 2:(Acts Colloq. Internat, Aix-en-Provece, 1967), pp. 177-184. Editions du Centre Nat. Recherche Sci, Paris. MR 41 #6580.

Monjardet, B. (1973) "Hypergraphs and Sperner's Theorem." Discrete Mathematics, 5, 287-289. MR 48 #5870.

Monjardet, B. (1975) "Elements ipsoduaux du treillis distributif libre et familles de Sperner ipsotransversales." (English summary). J. Combinatorial Theory Ser. A 19, no. 2, 160-176. MR 52 #2996.

Monjardet, B. (1975) "Systemes de blocage." Cahiers du CERO, 17, 293-294. MR 53 #147.

Monjardet, B. (1975) "Tournois et ordres medians pour une opinion." Math. Sci. Hum. 43, 55-70.

Monjardet, B. (1977) "Caracterisations metriques des ensembles ordonnes semi-modulaires." (English summary). Math. Sci. Humaines No. 56, 77-87,133. MR 2893.

Monjardet, B. (1978) "An axiomatic theory of tournament aggregation." Math. Oper. Research, 3, 4, 334-351.

Monjardet, B. (1978) "Axiomatiques et proprietes des quasi-ordres." Math. Sci. Hum., 63, 51-82.

Monjardet, B. (1978) "Caracterisations et propriete de Sperner des treillis distributifs planaires finis." Colloque du CNRS, Problemes combinatoires et theorie des graphes. MR 80g #6016.

Monjardet, B. (1980) "Theorie de la mediane dans les treillis distributifs finis et applications." Annals of Discrete Math. 9, 87-91.

Monjardet, B. (1981) "Metrics on partially ordered sets--a survey." Discrete Mathematics, 35.

Monk, J.D.; Meisters, G.H. (1973) "Construction of the reals via ultrapowers." Rocky Mountain J. Math. 3, 141-158. MR 50 #292.

Monma, C.L.; Sidney, J.B. (1979) "Sequencing with series-parallel precedence constraints." Math. Oper. Res. 4, 215-224.

Monteiro, Antonio; Ribeiro, Hugo (1945) "La notion de fonction continue." Summa Brasil. Math. 1, 1-8. (French. Portuguese summary) [MF 15865]. MR 7 #452.

Moreira Gomes, Alvercio. (1950) "Decomposition of partially ordered systems." Revista Cientifica 1, no. 2, 12-18. (English. Portuguese summary). MR 13 #7.

Morgado, Jose (1960) "Some results on closure operators of partially ordered sets." Portugal. Math. 19, 101-139. MR 24 #A3094.

Morgado, Jose (1961) "On the closure operators of the ordinal sum of partially ordered sets." Nederl. Akad. Wetensch. Proc. Ser. A 64=Indag. Math. 23, 546-550. MR 25 #2978.

Morgado, Jose (1962) "A characterization of the closure operators by means of one axiom." Portugal. Math. 21, 155-156. MR 26 #6080.

Morgado, Jose (1962) "Note on the system of closure operators of the ordinal sum of partially ordered sets." An. Acad. Brasil Ci. 34, 1-9. MR 25 #2374.

Morgado, Jose (1963) "On the closure operators of the cardinal product of partially ordered sets." Nederl. Akad. Wetensch. Proc. Ser. A 66=Indag. Math. 25, 65-75. MR 26 #4943.

Morgado, Jose (1965) "A single axiom for closure operators of partially ordered sets." Gaz. Mat. (Lisboa) 26, 57-58. MR 38 #84.

Morgado, Jose (1965) "Note on the system of closure operators of the ordinal product of partially ordered sets." An. Fac. Ci. Univ. Porto 48, 13-20. MR 37 #3968.

Morgado, Jose (1965) "Note on the system of closure operators of the ordinal product of partially ordered sets." Portugal. Math. 24, 189-195. MR 35 #1512.

Morgado, Jose (1973) "Note on quasi-orders, partial orders and orders." Portugal. Math. 32, 1-7. MR 47 #8367.

Morkeljunas, A.I. [Morkeliunas, A.] (1970) "Social choice and equilibrium." (Russian, Lithuanian and English summaries). Litovsk. Mat. Sb. 10, 309-325. MR 44 #3709.

Mosesjan, K.M. (1972) "Basable and strongly basable graphs." (Russian. Armenian summary). Akad. Nauk Armjan. SSR Dokl. 55, 83-86. MR 48 #5904.

Mosesjan, K.M. (1972) "Certain theorems on strongly basable graphs." (Russian. Armenian summary). Akad. Nauk Armjan. SSR Dokl. 54, 241-245. MR 48 #5903.

Mosesjan, K.M. (1972) "Strongly basable graphs." (Russian. Armenian summary). Akad. Nauk Armjan. SSR Dokl. 54, 134-138. MR 48 #5902.

Mosesjan, K.M. (1973) "Saturated graphs." (Russian. Armenian summary). Akad. Nauk Armjan. SSR Dokl. 56, 257-262. MR 49 #2472.

Mullin, R. (1968) "On Rota's problem concerning partitions." Aequationes Math. 2, 98-104. MR 38 #57.

Munch-Andersen, Bo; Zahle, Torben U (1977) "Scheduling according to job priority with prevention of deadlock and permanent blocking." Acta Informat. 8no. 2, 153-175. MR 4800.

Muntz, R.R.; Coffman, E.G, Jr. (1970) "Preemptive scheduling of real-time tasks on multiprocessor systems." J. Assoc. Comput. Mach. 17, 324-338. MR 43 #8353.

Muntz, R.R.; Coffman, E.G. (1968) "Preemptive scheduling of computation graphs." Proc. Sixth Annual Allerton Conf. on Circuit and System Theory (Monticello, Ill. 1968), pp. 515-522. Univ. of Illinois, Urbana, Ill. MR 41 #8019.

Nabeshima, Ichiro (1973) "The order of n items processed on m machines.III." J. Operations Res. Soc.Japan 16, 163-185. MR 10911.

Nachbin, L. (1965) "Topology and order." Van Nostrand.

Nakamura, Yatsuka (1970) "Entropy and semivaluations on semilattices." Kodai Math. Sem. Rep. 22, 443-468. MR 43 #7257.

Nakano, Hidegoro (1942) "Uber Erweiterungen von allgemein teilweisegeordneten Moduln. I." Proc. Imp. Acad. Tokyo 18, 626-630. MR [MF 14789]8 #387.

Nakano, Hidegoro (1944) "Uber Einfuhrung der teilweisen Ordnung im reellen hilbertschen Raum." Proc. Phys.-Math. Soc. Japan (3) 26, 1-8. [MF 15180]. MR 7 #453.

Nakayama, Tadasi (1948) "Remark on direct product decompositions of a partially ordered system." Math. Japonicae 1, 49-50. MR 10 #279.

Narens, Louis (1974) "Minimal conditions for additive conjoint measurement and qualitative probability." J. Mathematical Pschology 11, 404-430. MR 50 #15979.

Narens, Louis (1976) "Utility-uncertainty trade-off structures." J. Mathematical Pschology 13, no. 3, 296-322. MR 54 #6976.

Narushima, Hiroshi (1971) "Order preserving maps and classification of partially ordered sets." Proc. Fac. Sci. Tokai Univ. 6, no. 1-2, 27-45. MR 45.

Nash-Williams, C. St. J.A. (1963) "On well quasi ordering finite trees." Proc. Cambridge Philos. Soc. 59, 833-835. MR 27 #3564.

Nash-Williams, C. St. J.A. (1964) "On well quasi ordering lower set of finite trees." Proc. Camb. Philos. Soc. 60, 369-384. MR 29 #3399.

Nash-Williams, C. St. J.A. (1964) "On well quasi ordering trees." Theory of graphs and its applications, Proc. Sympos. Smolenice Pub. House Czechoslovak Acad. Sc., Prague, 83-84.

Nash-Williams, C. St. J.A. (1965) "On well quasi ordering infinite trees." Proc. Cambridge Philos. Soc. 61, 697-720. MR 31 #90.

Nash-Williams, C. St. J.A. (1965) "On well quasi ordering transfinite sequences." Proc. Camb. Philos. Soc. 61, 33-39. MR 30 #3850.

Nash-Williams, C. St. J.A. (1967) "On well quasi ordering trees." A Seminar on Graph Theory, Holt Rinehart and Winston, N.Y., 79-82. MR 35 #5354.

Nash-Williams, C. St. J.A. (1968) "On better quasi ordering transfinite sequences." Proc. Camb. Philos. Soc. 64, 273-290. MR 36 #5001.

Nash-Williams, C. St. J.A.; Jenkyns, J.A. (1968) "Counterexamples in the theory of well quasi ordered sets." Proof techniques in graph theory, Proc. Second Ann Arbor Graph Th. Conf., Ann Arbor, Mich., Acad. Press, N.Y., 87-91. MR 40 #7156.

Nash-Williams, C. St. J.A.; Jenkyns, J.A. (1969) "Possible directions in graph theory." Combinatorial Math and its applications, Proc. Conf. Oxford, Acad. Press, London, 191-200. MR 44 #1593.

Nazarova, L.A. (1974) "Partially ordered sets with an infinite number of indecomposable representations." (Russian). Carleton Math. Lecture Notes, No. 9, Carleton Univ, Ottawa, Ont. MR 51 #612.

Nazarova, L.A. (1975) "Partially ordered sets of infinite type." (Russian). Izv. Akad. Nauk SSR Ser. Mat. 39, no. 5, 963-991, 1219. MR 53 #10664.

Neggers, Joseph (1978) "Representations of finite partially ordered sets." J. Combin. Inform. System Sci. 3, no. 3, 113-133. MR 58 #27668.

Nemeth, E. (1977) "A note on a mini-max problem in scheduling." Proceedings of the Sixth Manitoba Conf. on Numerical Math. pp. 343-349. Congress. Numer, XVIII. Utilitas Math, Winnipeg, Man. MR 58 #32906.

Nemhauser, G.L.; Trotter, L.E. Jr.; Nauss, R.M. (1974) "Set partitioning and chain decomposition." Management Sci. 20, 1413-1473.

Nepomiastchy, Pierre (1978) "Resolution d'un probleme d'ordonnancement a resources variables." (English summary). RAIRO Rech. Oper. 12, no. 3, 249-261, iv. MR 58 #25941.

Neumann, K. [Neumann, Klaus] (1977) "Evaluation of general decision activity networks." First Symp. on Operations Research (Univ. Heidelberg, Heidelberg, 1976), Teil 1, pp. 268-275. Operations Research Verfahren, XXV. Hain, Meisenheim. MR 58 #32907.

Ng, Yew Kwang (1977) "Sub-semiorder: a model of multidimensional choice with preference intransitivity." J. Mathematical Psychology 16, no. 1,51-59. MR 57 #15305.

Nieminen, J. (1975) "Minimum decomposition of partially ordered sets into chains." (German summary). Z. Operations Res. Ser. A-B19,A63-A67. MR 2881.

Nieminen, Juhani (1977) "Join-semilattices and simple graphic algebras." Math. Nachr. 77, 87-91. MR 5874.

Nikitin, A.V. (1969) "An application of dynamic programming to the solution of certain ordering problems." (Russian). Trudy Moskov. Orden. Lenin. Energet. Inst. Vyp. 68, 165-169. MR 52 #10054.

Norinaga, Kakutaro, and Nishigori, Noboru (1952) "On axiom of betweenness." J. Sci. Hiroshima Univ. Ser. A. 16, 177-181, 399-408. MR 15 #389.

Novak, F. (1964) "On the lexicographic dimension of linearly ordered sets." Fund. Math. 56, 9-20. MR 34 #7412.

Novak, V.; Novotny, M. (1974) "Abstrakte dimension von strukturen." Z. Math. Logik Grundiagen Math. 20, 207-220. MR 53 #2765.

Novak, Vitezclav (1965) "On the lexicographic product of ordered sets." (Russian summary). Czechoslovak Math. J. 15 (90), 270-282. MR 31 #93.

Novak, Vitezclav (1969) "Eine Bemerkung uber die alpha-Dimension geordneter Mengen." Arch. Math. (Brno) 5, 187-190. MR 44 #122.

Novak, Vitezclav (1969) "On the well dimension of ordered sets." Czechoslovak Math. J. 19(94), 1-16. MR 39 #2665.

Novak, Vitezclav (1969) "Uber eine Eigenschaft der Dedekind-MacNeilleschen Hulle." Math. Ann. 179, 337-342. MR 39 #1364.

Novak, Vitezclav (1973) "Uber Erweiterungen geordneter Mengen." Arch. Math. (Brno) 9, 141-146. MR 50 #6934.

Novak, Vitezclav; Schmidt. J. [Schmidt, Jurgen] (1969) "Direct decompositions of partially ordered sets with zero." J. Reine Angew. Math. 239/240, 97-108. MR 41 #1595.

Novak, Vitezslav (1961) "The dimension of lexicographic sums of partially ordered sets." (Czech. Russian and English summaries). Casopis Pest. Mat. 86, 385-391. MR 24 #A1847.

Novak, Vitezslav (1962) "A note on a problem of T. Hiraguchi." (Czech. Russian and English summaries). Spisy Prirod. Fak. Univ. Brno 147-149. MR 29 #4710.

Novak, Vitezslav (1963) "On the pseudodimension of ordered sets." (Russian summary). Czechoslovak Math J. 13(88), 587-598. MR 31 #4742.

Novotny, M. (1960) "Uber gewisse Eigenschaften von Kardinal operationen." Spisy Prirod Fac. Univ. Brno, 465-484.

Novotny, Miroslav (1953) "On representation of partially ordered sets by sequences of zeros and ones." Casopis Pest. Mat. 78, 61-64. (Czech). MR 18 #23.

Novotny, Miroslav (1955) "Bemerkung uber die darstellung teilweise geordneter Mengen." Publ. Fac. Sci. Univ. Masaryk 451-458. (Czech and Russian summaries). MR 18 #636.

Novotny, Miroslav (1959) "Uber isotone funktionale geordneter Mengen." Z. Math. Logik Grandlagen Math. 6, 9-28. MR 21 #4117.

Novotny, Miroslav (1960) "Isotone funktionale geordneter Mengen." Z. Math. Logik Grundlagen Math. 6, 109-133. MR 24 #A61.

Nozaki, Akihiro (1966) "Formal description of grammar and translation. II." Sci. Papers College Gen. Ed. Univ. Tokyo 16, 43-48. MR 34 #1120.

Oellrich, H.; Steffens, K. (1976) "On Dilworth's decomposition theorem." Disc. Math. 15, 301-304.

Ohkuma, Tadashi (1952) "A note on the ordinal power and the lexicographic product of partially ordered sets." Koda. Math. Sem. Rep. 1952, 19-22. MR 13 #828.

Ohkuma, Tadashi (1958) "Duality in mathematical structure." Proc. Japan Acad. 34, 6-10. MR 20 #2393.

Okamoto, Sigeru (1970) "Partially ordered sets and semi-simplicial complexes. I. General theory." Math. Japan 15, 69-70. MR 45 #1152a.

Okamoto, Sigeru (1971) "Partially ordered sets and semi-simplicial complexes. II. Regular complexes." Bull. Fac. Sci. Ibaraki Univ. Ser. A No. 3, 13-20. MR 45 #1152b.

Okamoto, Sigeru (1972) "Partially ordered sets and semi-simplicial complexes. III. Fundamental groups." Bull. Fac. Sci. Ibaraki Univ. Ser. A No. 4, 13-16. MR 45 #9314.

Ordered (1971) "Ordered sets and lattices. No. 1 (Russian) A collection of articles edited by V.N. Saliza." Izdat Saratov. Univ. Saratov, 90 pp. MR 50 #9716.

Ore, Oystein (1942) "Theory of equivalence relations." Duke Math. J. 9, 573-627. [MF 7337]. MR 4 #128.

Ore, Oystein (1943) "Chains in partially ordered sets." Bull. Amer. Math. Soc. 49, 558-566 [MF 8856]. MR 5 #88.

Otrasevskaja, V.V. [Otrasevs'ka, V.V.] (1977) "Representations of one-parameter partially ordered sets." (Russian). Matrix Problems (Russian), pp. 144-149. Akad. Nauk Ukrain. SSR Inst. Mat, Kiev. MR 58 #27669.

Owings, James, C, Jf. (1969) "A canonical linear order for the maximal chains of a tree." Proc. Amer. Math. Soc. 20, 280-283. MR 39 #1363.

Packard, Dennis J.; Heiner, Ronald A. (1977) "Ranking functions and independence conditions." J. Econom. Theory 16, no. 1, 84-102. MR 14561.

Padmavally, K. (1956) "On a problem of G. Kurepa." Proc. Nat. Inst. Sci. India. Part A. 21 (1955), 368-372. MR 18 #373.

Page, E.S. (1961) "An approach to the scheduling of jobs on machines." J. Roy. Statist. Soc. Ser. B 23, 484, 492. MR 26 #5977.

Pallaschke, Diethard (1977) "Infinite-dimensional vonNeumann models." Mathematica Economics and Game Theory, pp.313-321. Lecture Notes in Econ. and Math. Systems, Vol. 141, Springer, Berlin. MR 2258.

Panwalkar, S.S.; Iskander, Wafik (1977) "A survey of scheduling rules." Operations Res. 25, no. 1,45-61. MR 2330.

Partrat, Christian (1977) "Fosses d'homomorphismes d'ensembles totalement preordonnes." (English summary). C.R. Acad. Sci. Paris Ser. A-B 284, no. 17, A1081-A1084. MR 58 #20288.

Pasenkov, V.V. (1962) "Some remarks on completions of partially ordered sets." (Russian). Uspehi Mat. Nauk 17, no. 6 (108), 155-162. MR 27 #60.

Pasini, Antonio (1975) "Some fixed point theorems of the mappings of partially ordered sets." Rend. Sem. Mat. Univ. Padova 51 (1974), 167-177. MR 52 #2972.

Pattanaik, Prasanta K. (1970) "Sufficient conditions for the existence of a choice set under majority voting." Econometrica 38, 165-170. MR 42 #1542.

Peck, G.W. (1979) "Note, Maximum antichains of rectangular arrays." J. Comb. Theory A 27, 397-400.

Pelczar, A. (1961) "On the invariant points of a transformation." Ann. Polon. Math. 11, 199-101. MR 25 #1114.

Pelczar, A. (1965) "On invariant points of monotone transformations in partially ordered spaces." Ann. Polon. Math. 17, 49-53.

Pelczar, Andrzej (1968) "On some equations in partially ordered spaces." Zeszyty Nauk. Univ. Jagiello. Prace Mat. No. 12, 43-45. MR 36 #5047.

Pelczar, Andrzej (1971) "Remarks on commuting mappings in partially ordered spaces." Zeszyly Nauk. Uniw. Jagiello. Prace Mat. Zeszyt 15, 131-133. MR 48 #5929.

Perepelica, V.A.; Sevast'janov, S.V. (1976) "A scheduling problem on a network." (Russian). Upravljaemye Sistemy No. 15 Diskret. Ekstremal. Zadaci, 48-67, 81. MR 58 #9201.

Perfect, Hazel (1974) "Addendum to a theorem of O. Pretzel." J. Math. Anal. Appl. 46, 90-92. MR 49 #156.

Perles, Micha A. (1963) "A proof of Dilworth's decomposition theorem for partially ordered sets." Israel J. Math. 1, 105-107. MR 29 #5758.

Perles, Micha A. (1963) "On Dilworth's theorem in the infinite case." Israel J. Math. 1, 108-109. MR 29 #5759.

Pesotan, H. (1974) "Dense extensions of a partially ordered set." J. Reine Angew. Math. 266, 163-182. MR 48 #10913.

Petifrere, Marianne (1976) "Concepts de metajeu permettant l'abolition de la scission entre theories non cooperative et cooperative des jeux. Colloque sur la Theorie des Jeux." (Bruxelles, 1975). Cahiers Centre Etudes Recherche Oper. 18, no. 1-2, 221. MR 5 .

Pichat, Etienne (1970) "Decompositions d'une fonction de plusieurs variables." Seminaire P. Dubreil, M.-L. Dubreil-Jacotin, L. Lesieur et C. Pisot: 1969/70, Algebre et Theorie des Nombres, Fasc. 1, Exp. 5, 20 pp. Secretariat mathematique, Paris. MR 47 #473 .

Pinus, A.G. (1971) "A remark on the paper by Ju. M. Vazenin: "Elemetary properties of transformation semigroups of ordered sets"." (Russian). Algebra i Logika 10, 327-328. MR 44 #1804.

Pinus, A.G. (1972) "The imbedding of totally ordered sets." (Russian). Mat. Zametki 11, 83-88. MR 45 #3256.

Podinovskii, V.V. (1977) "Multicriterial problems with importance-ordered criteria." Avtomat. i Telemeh. 1976, no. 11, 118-127 (Russian. English summary); trans. as Automat. Remote Control 37 (1976), no. 11, part 2, 1728-1736. MR 2215.

Poguntke, W. (1975) "Zerlegung von S-Verbanden." Math. Zeit.

Poguntke, W. (1981) "On finitely generated simple complemented lattices." Can. Math. Bull.

Poguntke, Werner; Rival, Ivan (1974) "Finite sublattices generated by order-isomorphic subsets." Arch. der Math. 25, 225-230.

Polakova, Nadezda (1963) "Note on characteristics of ordered sets." (Czech. Russian and English summaries). Casopis Pest. Mat. 88, 387-390. MR 31 #4741.

Pop, Petru (1975) "On L. Nemeti's model for shop scheduling." (Romanian. French summary). Rev. Anal. Numer. Teoria Aproximatiei 3 (1974), no. 2, 177-186. MR 58 #15119.

Popadic, Milan S. (1951) "On the number of chains in a kind of ordered finite sets." Fac. Philos. Univ. Skopje. Sect. Sci. Nat. Annuaire 4, no. 4, 10 pp. (Serbo-Croatian. English summary). MR 14 #528.

Popadic, Milan S. (1954) "On ordered sets with finite chains." Fac. Philos. Univ. Skopje. Sect. Sci. Nat. Annuaire 5 (1952), no. 1, 8 pp. (Serbo-Croatian. English summary). MR 15 #690.

Popadic, Milan S. (1970) "On the number of antichains of finite power sets." Mat. Vesnik 7 (22), 199-203. MR 41.

Popruzenko, J. (1963) "Sur une propriete des ensembles partiellement ordonnes." Fund. Math. 53, 13-19. MR 28 #2059.

Potepun, A.V. (1971) "Certain forms of convergence with respect to order and of topologies in partially ordered sets." (Russian). Sibirsk. Mat. Z. 12, 819-827. MR 44 #6572.

Pouzet, Maurice (1969) "Generalisation d'une construction de Ben Dushnik-E.W. Miller." C.R. Acad. Sci. Paris Ser. A-B269, A877-A879. MR 40 #7159.

Pouzet, Maurice (1976) "Les chaines completes sont simplifiables a droite pour la multiplication ordinale." Portugal. Math. 35, no. 3-4, 127-135. MR 5841.

Pouzet, Maurice (1978) "Ensemble ordonne universel recouvert par deux chaines." J. Combin. Theory Ser. B 25, no. 1, 1-25. MR 58 #16434.

Prabhakara Rao, K.B. (1978) "Extensions of strict partial orders in N-groups." J. Austral Math. Soc. Ser. A 25, no. 2, 241-249. MR 58 #439.

Prakash, Prem [Prem Prakash] (1977) "On the consistency of a gambler with time preference." J. Econom. Theory 15, no. 1, 92-98. MR 55 #12156.

Prangisvili, I.V. "Iterative and homogeneous planar lattices and the graphs that correspond to them." (Russian). (Proc. First All-Union Conf. Comput. Systems, Novo-sibirsk, 1967) (Russian), pp. 3-19. Izdat. "Nauka" Sibirsk. Otdel. Novosibirsk, 1968. MR 46 .

Prenowitz, Walter (1946) "Partially ordered fields and geometries." Amer. Math. Monthly 53, 439-449. MR 8 #249.

Presic, Marica D (1970) "On certain formulas for equivalence and order relations." Mat. Vesnik 7(22), 510-513. MR 43 #7337.

Pretzel, O. (1969) "A Mayer-Vietoris sequence in the homology of partially ordered sets." J. London Math. Soc. 44, 408-412.

Pretzel, O. (1977) "On the dimension of partially ordered sets." J. Comb. Theory A 22, 146-152.

Pretzel, O. (1979) "Note, Another proof of Dilworth's decomposition theorem." Disc. Math. 25, 91-92.

Pretzel, O. (1980) "A construction in the theory of partially ordered sets." Discrete Math. 32, 59-68.

Pretzel, Oliver (1967) "A representation theorem for partial orders." J. London Math. Soc. 42, 507-508. MR 35 #6588.

Proctor, R.A.; Saks, M.E.; Sturtevant, D.G. (1980) "Product partial orders with the Sperner property." Disc. Math. 30, 173-180.

Quillen, D. (1978) "Homotopy properties of the poset of nontrivial p-subgroups of a group." Advances in Math. 28, 101-128.

R.-Salinas, Baltasar (1969) "Fixed points in ordered sets." (Spanish). Consejo Sup. Investigacion. Ci. Fac. Ci. Zaragoza, Zaragoza, pp. 163-172. MR 40 #4174.

Rabin, Michael (1954) "A theorem on partially ordered sets." veon Lematematika 7, 26-29. (Hebrew. English summary). MR 15 #389.

Rabinovitch, I (1978) "The dimension of semiorders." J. Combin. Theory Ser. A 25, no. 1, 50-61. MR 58 #16435.

Rabinovitch, I. (1978) "An upper bound on the "dimension of interval orders"." J. Combin. Theory Ser. A 25, no. 1, 68-71. MR 58 #5424.

Rabinovitch, I.; Rival, I. (1979) "The rank of a distributive lattice." Disc. Math. 25, 275-279.

Rabinovitch, Issie (1978) "Posets having a unique decomposition into the minimum number of antichains." Aequationes Math. 17, no. 1, 41-43. MR 57 #9615.

Radenski, Atanas A. (1977) "Existence of a fixed point for an isotone mapping of a complete pseudolattice into itself." (Russian. English summary). Annuaire Univ. Sofia Fac. Math. Mec. 68(1973/74), 263-266. MR 58 #27677.

Rader, Trout (1978) "Induced preferences on trades when preferences may be intransitive and incomplete." Econometrica 46, no. 1, 137-146. MR 58 #15016.

Rado, R. (1949) "Covering theorems for ordered sets." Proc. London Math. Soc. (2) 50, 509-535. MR 10 #688.

Rado, R. (1954) "Partial well-ordering of sets of vectors." Mathematika 1, 89-95. MR 16 #576.

Raghavachari, M.; Mote, V.L. (1969/70) "Generalization of Dilworth's theorem on minimal chain decomposition." Management Sci. 16, 508-511. MR 46 #3068.

Rainich, G.Y. (1962) "Notes on foundations. II On Galois connections." Notre Dame J. Formal Logic 3, 61-63. MR 26 #6076.

Rajaraman, M.K. (1977) "A parallel sequencing algorithm for minimizing total cost." Naval Res. Logist. Quart. 24, no. 3, 473-481. MR 58 #4284.

Ramalho, Margarita; Almeida Costa, A. (1970/71) "Problemes de perspectivite dans las ensembles partiellement ordonnes et dans les treillis." Univ. Lisboa Revista Fac. Ci. A (2) 13, 229-240. MR 46 #8917.

Redfield, R.H. (1974) "Ordering uniform completions of partially ordered sets." Canad. J. Math. 26, 644-664. MR 50 #14693.

Remmel, J.B. (1978) "Realizing partial orderings by classes of co-simple sets." Pacific J. Math. 76, no. 1,169-184. MR 58 #10379.

Restle, F. (1959) "A metric and an ordering on sets." Psychometrika, 24, 207-220.

Reusch, Bernd (1969) "Lineare automaten." B.I.-Hochschulskripten, 708. Bibliographisches Institut, Mannheim-Vienna-Zurich, ii+149 pp. MR 41 #5124.

Richardson, M. (1946) "On weakly ordered systems." Bull. Amer. Math. Soc. 52, 113-116. [MF 15450]. MR 7 #235.

Richter, Bernd (1974) "Kombinatorik auf geordneten Mengen." J. Reine Angew. Math. 266, 76-87. MR 49 #7153.

Richter, Marcel K. (1975) "Rational choice and polynomial measurement models." J. Mathematical Pschology 12, 99-113. MR 50 #15980.

Riguet, J. (1950) "Les relations de Ferrers." C.R. Acad. Sci. Fr., 231, 936-937.

Riguet, Jacques (1948) "Produit tensorial d'ensembles ordonnes." C.R. Acad. Sci. Paris 227, 1007-1008. MR 10 #279.

Riguet, Jacques (1948) "Produit tensoriel de lattices." C.R. Acad. Sci. Paris 226, 40-41. MR 9 #265.

Riismaa, T. (1977) "An approach to the optimization of the structure of a hierarchical system." (Russian). Models and Methods for the Analysis of Economic Target-oriented Systems (Russian), 51-58. Izdat. "Nauka" Sibirsk. Otdel, Novosibirsk. MR 58 #15125.

Ritter, K. (1970) "Optimization theory in linear spaces III. mathematical programming in partially ordered Banach spaces." Math. Ann. 184, 133-154. MR 41 #3115.

Rival, I. (1974) "Lattices with doubly irreducible elements." Canad. Math. Bull. 17, 91-95. MR 512837.

Rival, I. (1976) "A fixed point theorem for finite partially ordered sets." J. Combinatorial Theory (A) 21,309-318.

Rival, I. (1976) "Combinatorial inequalities for semimodular lattices of breadth two." Algebra Universalis 6, 303-311. MR 5217.

Rival, I. (1980) "The problem of fixed pints in ordered sets." Annals of Discrete Math. 8, 283-292.

Rival, I.; Sands, B. (1975) "Weak embeddings and embeddings of finite distributive lattices." Archiv. Math. 26, 346-352.

Rival, I.; Sands, B. (1978) "Planar sublattices of a free lattice I." Canad. J. Math. 30, 1256-1283.

Rival, I.; Sands, B. (1979) "Planar sublattices of a free lattice II." Canad. J. Math. 31, 17-34.

Rival, I.; Wille, R. (1979) "Lattices generated by partially ordered sets: which can be "drawn"?" J. Reine Angew Math. 310, 56-80.

Roberts, F.S. (1976) "Discrete mathematical models with applications to social, biological and environmental problems." New York, Prentice Hall.

Roberts, Fred S. (1970) "On nontransitive indifference." J. Mathematical Psychology 7, 243-258. MR 41 #3133.

Robinson, G.B.; and Wolk, E.S. (1957) "The imbedding operators on a partially ordered set." Proc. Amer. Math. Soc. 8, 551-559. MR 19 #115.

Rogers, D.G. (1977) "Similarity relations on finite ordered sets." J. Combinatorial Theory Ser. A 23, no. 1, 88-98. MR 15505.

Rogers, D.G. (1978) "Erratum: "Similarity relations on finite ordered sets"." (J. Combinatorial Theory Ser. A 23(1977), no. 1, 88-98). J. Combinatorial Theory Ser. A 25, no. 1, 95-96. MR 58 #5425.

Rogers, D.G.; Kim, K.H.; Roush, F.W. (1979) "Similarity relations and semiorders." 10th Southeastern Conference on Combinatorics, Graph theory and Computing Baton Rouge.

Rogers, Laurel A. (1977) "Ulam's theorem for partially ordered structures related to simply presented Abelian p-groups." Trans. Amer. Math. Soc. 227, 333-343. MR 503.

Rohleder, H. (1966) "Das Syntheseproblem der Schaltalgebra und seine mathematische Formulierung." Z. Math. Logik Grundlagen Math. 12, 13-22. MR 34 #8887.

Rolf, H (1958) "The free lattice generated by a set of chains." Pacific J. Math. 8, 585-595.

Rosenberg, I.G. (1967) "Maximale Gegenketten im Verband 3exp(m)." Arch. Math. (Brno) 3, 185-190. MR 39 #1367.

Rosenberg, I.G. (1970) "Double lexicographic products." Arch. Math. (Brno) 6, 221-228. MR 46 #8934.

Rosenberg, I.G. (1974) "A semilattice on the set of permutations on an infinite set." Math. Nachr. 60, 191-199.

Rosenberg, Ivo (1966) "Isotone und homomorphe Abbildungen von halbgeordneten Halbgruppoiden." (Russian summary) Spisy Prirod Fak. Univ. Brno, 411-426. MR 36 #2542.

Rosenberg, Ivo (1967) "Die transitive Hulle des lexikographischen Produktes." (Russian and English summaries). Mat. Casopis Sloven. Akad. Vied 17, 92-107. MR 37 #1281.

Rosenberg, Ivo (1970) "The refinement of two isomorphic generalized lexicographic products." Arch. Math. (Brno)6 229-235. MR 47 #106.

Rosenstein, J. (1969) "Aleph-0 categoricity of linear orderings." Fund. Math. 44, 1-5. MR 39 #3982.

Rosenstein, J. (1981) "Linear orderings." Academic Press, New York.

Rosicky, Jiri (1973) "On a characterization of the lattice of m-ideals of an ordered set." Arch Math (Brno)8(1972), 137-142. MR 49 #157.

Roskies, Ralph (1965) "A measurement axiomatization for an essentially multiplicative representation of two factors." J. Mathematical Psychology 2, 266-276. MR 31 #6687.

Rota, G.-C. (1967) "A generalization of Sperner's theorem, research problem." J. Comb. Theory 2, 104.

Rota, Gian-Carlo (1964) "On the foundations of combinatorial theory I: Theory of Mobius functions." Wahrscheinlichkeitstheorie verw. Gebiete 2, 340-368.

Roth, Alvin, E. (1977) "A fixed point approach to stability in cooperative games." Fixed points: Algorithms and Applications, pp. 165-180. Academic Press, New York. MR 2610.

Roth, J. Paul; Wagner, E.G. (1959) "Algebraic topologic methods for the synthesis of switching systems. III minimization of nonsingular Boolean trees." IBM J. Res. Develop. 3 326-344. MR 22 #5513.

Rotman, B. (1968) "On the comparison of order types." Acta Math. Acad. Sci. Hungar. 19, 311-327. MR 38 #86.

Rowicki, Andrzej (1975/76) "A note on optimal scheduling for two-processor systems." Information Processing Lett. 4, no. 2, 27-30. MR 58 #20389.

Rowicki, Andrzej (1977) "A note on optimal preemptive scheduling for two-processor systems." Information Processing Lett. 6, no. 1, 25-28. MR 14649.

Roy, B. (1978) "Electre III: un algorithme de classements fonde sur une representation floue des preferences en presence de criteres multiples." Cahiers Centre Etudes Rech. Oper. 20, no. 1,3-24. MR 58 #15017.

Roy, B.; Vincke, Ph. (1980) "Systems relationels de preferences en presence de criteres multiples avec seuils." Cahier du Centre d'Etudes de Recherche Operationnelle, 22, 1.

Rozen, V.V. (1966) "Application of the general theory of algebraic operations to the theory of ordered sets." (Russian). Certain Appl. Theory Binary Relations (Russian), pp. 3-9. Izdat. Saratov. Univ, Saratov. MR 33 #3970.

Rozen, V.V. (1966) "Isomorphisms of lattices associated with ordered sets." (Russian). Interuniv. Sci. Sympos. General Algebra (Russian), pp. 114-116. Tatu. Gos. Univ, Tartu. MR 33 #7275.

Rozen, V.V. (1971) "The partial operations min and inf in ordered sets." (Russian). Ordered Sets and Lattices, No. 1 (Russian) pp. 74-85. Izdat. Saratov Univ, Saratov. MR 51 #5425.

Rozen, V.V. (1973) "Partial operations in ordered sets." Izdat. Saratov. Univ, Saratov, 123 pp. MR 50 #4414.

Rozen, V.V. (1976) "Mixed extensions of games with ordered outcomes." (Russian). Z. Vycisl. Mat. i Mat. Fiz. 16, no. 6, 1436-1450, 1626. MR 57 #18864.

Rozen, V.V. (1977) "Axiomatization of the operations of infimum and supremum on the basis of the concept of a Galois connection." (Russian). Ordered sets and Lattices, No. 4 (Russian), pp. 98-104, 134. Izdat. Saratov. Univ, Saratov. MR 58 #402.

Rozen, V.V. (1978) "Games with ordered outcomes." Izv. Akad. Nauk SSSR Tehn. Kibernet. 1977, no. 5, 31-37 (Russian); translated as Engrg. Cybernet. 15(1977), no. 5, 18-24. MR 58 #15262.

Rubin, M. (1974) "Theories of linear orderings." Israel J. Math. 17, 392-443. MR 50 #1871.

Rudeanu, Sergiu (1963) "Axiomele laticilor si ale algebrelor Booleene." (Romanian). [Axioms for Lattices and Boolean algebras] Editura Academiei Republicii Populare Romine, Bucharest, 159pp. MR 36 #81.

Ruschitzka, Manfred (1978) "An analytical treatment of policy function schedulers." Oper. Res. 26, no. 5, 845-863. MR 58 #20390.

Ruttmann, Gottfried T. (1974) "Closure operators and projections on involution posets." J. Austral. Math. Soc. 18, 453-457.

Sabidussi, G. (1960) "Graph multiplication." Math. Z. 72, 446-457.

Sabidussi, G. (1975) "Subdirect representations of graphs." Vol. III, pp. 1199-1226. Colloq. Math. Soc. Janos Bolyai, Vol. 10, North-Holland, Amsterdam. MR 51 #10168.

Sahbazjan, K.V.; Lebedinskaja, N.B. (1977) "Minimization of the total penalty in the problem of the parallel ordering of n independent tasks." (Russian). Dokl. Akad. Nauk SSSR 237, no. 4, 790-792. MR 58 #15120.

Sahbazjan, K.V.; Lebedinskaja, N.B. (1977) "Optimal schedulings with gaps for independent tasks in a queueing system with N servers." (Russian). Zap. Naucn. Sem. Leningrad. Otdel. Mat. Inst. Steklov. (LOMI) 70, 205-231,294. MR 58 #25942.

Sakai, Shozo (1965) "Systems of postulates for a ternary relation characterizing a partial order." Fac. Sci. Shizuoka Univ. 1, no. 1, 1-4. MR 33 #7276.

Saks, M. (1980) "Dilworth numbers, incidence maps and product partial orders." SIAM J. Alg. Disc. Meth. 1, 211.

Sales, Valles, Francisco de A. (1970) "Galois mappings in ordered sets and in lattices." (Spanish). Collect. Math. 21, 19-39. MR 42 #5869.

Salles, Maurice (1975) "A general possibility theorem for group decision rules with Pareto-transitivity." J. Econom. Theory 11, no. 1, 110-118. MR 54 #4580.

Salveson, M.E. (1955) "The assembly line balancing problem." Proceedings of the Second Symposium in Linear Programming, Washington, D.C, 1955, pp. 55-101. National Bureau of Standards, Washington, D.C. MR 17 #507.

Sands, B. (1980) "Generating sets for lattices of dimension two." Discrete Math. 29, 287-292.

Sands, B. (1981) "Counting antichains in finite partially ordered sets." Discrete Math. 35, 213-228.

Sankoff, David; Sellers, Peter H (1973) "Shortcuts, diversions, and maximal chains in partially ordered sets." Discrete Math. 4, 287-293. MR 47 #1690.

Sanz Aguado, Adela (1977) "Results on a decision problem in which a partial ordering of utilities is known." (Spanish). Trabajos Estadist. Investigacion Ope. 28, no. 2-3, 99-117. MR 58 #25865.

Saposnik, Rubin (1975) "On transitivity of the social preference relation under simple majority rule." J. Econom. Theory 10, no. 1, 1-7. MR 53 #10180.

Sasaki, Yoshio (1967) "The involutive symmetry and the number of the finite partially ordered sets. I." Mem. Ehime Univ. Sect. II Ser. A5, 31-37. MR 39 #4056a.

Sasaki, Yoshio (1968) "The involutive symmetry and the number of the finite partially ordered sets. II." Rep. Univ. Electro-Commun. No. 25,59-67. MR 39 #40566.

Sato, Ryuzo (1976) "Self-dual preferences." Econometrica 44, no. 5, 1017-1032. MR 58 #15018.

Satterthwaite, Mark Allen (1975) "Strategy-proofness and Arrow's conditions: existence and correspondence theorems for voting procedures and social welfare functions." J. Econom. Theory 10, no. 2, 187-217. MR 54 #2159.

Sayeki, Yutaka (1972) "Allocation of importance: an axiom system." J. Mathematical Psychology 9, 55-65. MR 47 #10096.

Schader, M. (1980) "Hierarchical analysis: classification with ordinal object dissimilarities." Metrika, 27, 127-132.

Schelp, R.H. (1972) "A partial semigroup approach to partially ordered sets." Proc. London Math. Soc. (3) 24, 46-58. MR 44 #6578.

Schmerl, J. (1977) "On aleph-0 categoricity and the theory of trees." Fund. Math. 94, 121-128.

Schmerl, J. (1980) "Decidability and aleph-0-categoricity of theories of partially ordered sets." J. Symbolic Logic 45, 585-611.

Schmerl, J.H. (1979) "Countable homogeneous partially ordered sets." Algebra Universalis 9, 317-321.

Schmidt, E.T. (1979) "Remark on generalized function lattices." Acta Math. Acad. Sci. Hungar. 34, 337-339.

Schmidt, J. (1973) "Each join completion of a partially ordered set is the solution of a universal problem." J. Aust. Math. Soc. 16.

Schmidt, Jurgen (1953) "Uber die Minimalbedingung." Arch. Math. 4, 172-181. MR 15 #280.

Schmidt, Jurgen (1954) "Abgeschlossenheits- und Homomorphiebegriffe in der Ordnungstheorie." Wiss. Z. Humboldt-Univ. Berlin. Math.-Nat. Reihe 3, 223-225. MR 16 #343.

Schmidt, Jurgen (1955) "Lexikographische Operationen." Z. Math. Logic Grundlagen Math. 1, 127-171. MR 17 #135.

Schmidt, Jurgen (1956) "Zur Kennzeichnung der Dedekind-MacNeilleschen Hulle einer geordneten Hulle." Arch. Math. 7, 241-249. MR 18 #868.

Schmidt, Jurgen (1957) "Die transfiniten Operationen der Ordnungstheorie." Math. Ann. 133, 439-449. MR 19 #937.

Schmidt, Jurgen (1972) "Universal and internal properties of some extensions of partially ordered sets." J. Reine Angew. Math. 253, 28-41. MR 46 #105.

Schmidt, Jurgen (1974) "Each join-completion of a partially ordered set is the solution of a universal problem." J. Austral. Math. Soc. 17, 406-413. MR 50 #4415.

Schmidt, Jurgen (1978) "Binomial pairs, semi-Brouwerian and Brouwerian semilattices." Notre Dame J. Formal Logic 19, no. 3,421-434. MR 58 #10639.

Schrijver, A. (1977) "A note on David Lubell's article: "Local matching in the function space of a partial order"." (J. Combinatorial Theory Ser. A 19 (1975), 154-159). J. Combinatorial Theory Ser. A23, no. 3, 359-362. MR 15506.

Schutzenberger, M.P. (1972) "Promotion des morphismes d'ensembles ordonnes." Discrete Math. 2, 73-94.

Schutzenberger, Marcel-Paul (1948) "Sur certaines applications remarquables des treillis dans eux-memes." C.R. Acad. Sci. Paris 227, 1008-1010. MR 10 #279.

Schwebel, John C.; McCormick, Bruce H. (1970) "Consistent properties of composite formation under a binary relation." Information Sci. 2, 179-209. MR 43 #1756.

Schweizer, Urs (1976) "Convexification of preference orderings." Operations Research Verfahren, Bank XXII, pp. 192-199. Hain, Meisenheim am Glan. MR 14566.

Scott, Dana (1971) "The lattice of flow diagrams." Symposium on Semantics of Algorithmic Languages, pp. 311-366. Lecture Notes in Mathematics, Vol. 188. Springer, Berlin. MR 43 #4577.

Sedmak, Victor (1951) "Quelques applications des ensembles ordonnes." Bull. Soc. Math. Phys. Serbie 6, 12-39, 131-153. MR 18 #186.

Sekanina, A.; Sekanina, M. (1966) "Topologies compatible with ordering." Arch. Math. (Brno) 2, 113-126. MR 35 #100.

Sekanina, M. (1968) "Categories of ordered sets." Arch. Math. (Brno) 4, 25-59. MR 41 #8322a.

Sekanina, M. (1968) "Errata to: Categories of ordered sets." Arch. Math. (Brno) 4, 131. MR 41 #8322b.

Sekanina, M. (1971) "Embedding of the category of ordered sets in the category of semigroups." Acta Fac. Rer. Natur. Univ. Comenian, Math. Mimor. Cislo, 77-79.

Sekanina, Milan (1965) "On the power of ordered sets." Arch. Math. (Brno) 1, 75-82. MR 34 #7415.

Sekanina, Milan (1969/70) "Embedding of the category of partially ordered sets into the category of topological spaces." Fund. Math. 66, 95-98. MR 40 #5505.

Sekanina, Milan (1971) "On orderings of the system of subsets of ordered sets." Fund. Math. 70, no. 3, 231-243. MR 44 #2673.

Selivanov, Ju. A. (1976) "Extensions of partial orders in neorings and neofields." (Russian). Quasigroups and combinatorics. Mat. Issld. Vyp. 43, 146-156,209. MR 58 #5458.

Sethi, Ravi (1978) "On the complexity of mean flow time scheduling." Math. Oper. Res. 2 (1977), no. 4, 320-330. MR 9043.

Sevrin, L. N.; Filippov, N.D. (1966) "Partially ordered sets and their comparison graphs." (Russian). Interuniv. Sci. Sympos. General Algebra (Russian), pp. 193-194. Tartu. Gos. Univ, Tartu. MR 33 #7277.

Sevrin, L.N.; Filippov, N.D. (1970) "Partially ordered sets and their comparability graphs." (Russian). Sibirsk. Mat. Z. 11, 648-667. MR 42 #4451.

Seymour, P.D. (1973) "On incomparable collections of sets." Mathematika, 20, 208-209.

Seymour, P.D.; Welsh, D.J.A. (1975) "Combinatorial applications of an inequality from statistical mechanics." Math. Proc. Cambridge Philos. Soc. 77, 485-495. MR 51 #12554.

Shannon, Claude E. (1958) "A note on a partial ordering for communication channels." Information and Control 1, 390-397. MR 20 #6952.

Sharp, H. Jr. (1971-1972) "Enumeration of transitive, step-type relations." Acta Math. Acad. Sci. Hum., 22, 365-371.

Shearer, J.; Kleitman, D.J. (1979) "Probabilities of independent choices being ordered." Studies in Appl. Math. 60, 271-276.

Shelah, S. (1971) "The number of non-isomorphic models of an unstable first-order theory." Israel J. Math. 9, 473-487. MR 43 #4652.

Shelah, S. (1972) "On models with power-like orderings." J. Symb. Logic 37, 247-267.

Shelah, S. (1975) "The monadic theory of order." Annals of Mathematics 102, 379-419. MR 58 #10390.

Shelah, S. (1976) "Decomposing uncountable squares to countably many chains." J. Comb. Theory 21, 110-114.

Shepp, L.A. (1980) "The FKG inequality and some monotonicity properties of partial orders." SIAM Jour. Alg. Disc. Meth., 1, 295-299.

Sherman, B.F. (1974) "Cauchy completion of partially ordered groups." J. Austral. Math. Soc. 18, no. 2, 222-229. MR 57 #12333.

Shimada, Ryosaku (1972) "Simplification of many-valued logic circuits." Systems-Computers-Controls 2 (1971), no. 1, 58-59. MR 46 #3225.

Shkurba, V.V. [Skurba, V.V.] (1969) "Ordering problems." Kibernetika (Kiev) 1967, no. 2, 63-65 (Russian); translated as Cybernetics 3 (1967), no. 2, 49-51. MR 42 #2793.

Shmerl, J (1980) "Decidability and aleph-0-categoricity of theories of partially ordered sets." J. Symbolic Logic 45, 585-611.

Shmuely, Zahava (1975) "Fixed points of antitone mappings." Proc. Amer. Math. Soc. 52, 503-505. MR 51 #10182.

Shmuely, Zahava (1976) "On bounded po-semigroups." Proc. Amer. Math. Soc. 58, 37-43. MR 15524.

Sholander, Marlow (1952) "Trees, lattices, order, and betweenness." Proc. Amer. Math. Soc. 3, 369-381. MR 14 #9.

Sholander, Marlow (1954) "Medians lattices and trees." Proc. Amer. Math. Soc. 5, 808-812.

Shyr, H.J.; Thierrin, G. (1974) "Hypercodes." Information and Control 24, 45-54. MR 49 #10443.

Shyr, H.J.; Thierrin, G. (1974) "Ordered automata and associated languages." Tamkang J. Math. 5, no. 1, 9-20. MR 51 #2810.

Sidney, J.B. (1979) "The two-machine maximum flow time problem with series parallel precedence relations." Oper. Res. 27, 782-791.

Sidney, Jeffrey B. (1975) "Decomposition algorithms for single-machine sequencing with precedence relations and deferral costs." Operations Res. 23, no. 2, 283-298. MR 4801.

Sierpinski, W. (1950) "Sur les types d'ordre des ensembles lineaires." Fund. Math. 37, 253-264.

Sierpinski, W. (1953) "Sur un probleme de la theorie des relations." Ann. Scuola Norm Sup. Pisa (2) 2, 285-287.

Sierpinski, W. (1954) "Un theoreme concernant les fonctions continues dans les ensembles ordonnes." Ann. Soc. Polon. Math. 24 (1951), no. 2, 175-180. MR 15 #690.

Sik, Frantisek (1958) "Automorphismen geordneter Mengen." Casopis Pest. Mat. 83, 1-22. (Czech and Russian summaries). MR 20 #2290.

Sik, Frantisek (1961) "Uber die algebraische Charakterisierung der Gruppen von reellen Funktionen." Ann. Mat. Pura Appl. (4) 54, 295-299. MR 24 #A1848.

Sikorski, Roman (1963) "On dense subsets of Boolean algebras." Colloq. Math. 10, 189-192. MR 27 #4773.

Simbireva, E.P. (1972) "Fundamental trends in the development of the theory of partially ordered groups." (Russian). Moskov. Oblast. Ped. Inst. Ucen. Zap. 322 Metodika Mat, Element. Mat. Vyss. Algebra, Prikl. Mat. Vyp. 9, 211-238. MR 58 #16454.

Simovici, Dan A. (1976) "On the theory of reduction of semilatticial automata." An. Sti. Univ. "Al. I. Cuza" Iasi Sect. I a Mat. (N.S.) 22, no. 1, 107-110. MR 55 #12363.

Sisson, Roger L. (1959) "Methods of sequencing in job shops--a review." Operations Res. 7, 10-29. MR 20 #6321.

Skolem, Th. (1936) "Uber gewisse "Verbande" oder "lattices"." Avh. Norske Vid. Akad. Oslo, 1-16.

Skula, Ladislav (1969) "System of layers of an ordered set." Czechoslovak Math. J. 19 (94), 27-41. MR 38 #4370.

Skula, Ladislav (1977) "Eine Bemerkung zum kardinalprodukt." Czechoslovak Math. J. 27(102), no. 4,515-522. MR 15507.

Skula, Ladislaw (1971) "Kurzung unter den typen der teilweise geordneten Mengen." Acta Fac. Rerum. Natur. Univ. Comenian. Math. Minoriadne Cislo, 81-86. MR 52 #207.

Slowinski, Roman (1978) "A node ordering heuristic for network scheduling under multiple resource constraints." Found. Control Engrg. 3, no. 1,19-27. MR 58 #9202.

Slowinski, Roman; Weglarz, Jan (1978) "Solving the general project scheduling problem with multiple constrained resources by mathematical programming." Found. Control Engrg. 3, no. 1,29-39. MR 58 #20391.

Smith, David A. (1968) "On constructing a distributive lattice from a partially ordered set with DCC." Comment. Math. Univ. Carolinae 9, 515-525. MR 39 #4057.

Smith, Tony E. (1974) "On the existence of most-preferred alternatives." Internat. Econom. Rev. 15, 184-194. MR 57 #18759.

Smithson, R.E. (1971) "Fixed points of order preserving multifunctions." Proc. Amer. Math. Soc. 28, 304-310. MR 43 #114.

Smithson, R.E. (1973) "Fixed points in partially ordered sets." Pacific J. Math. 45, 363-367. MR 47 #4871.

Sokol, Ju. (1977) "The 27th problem of Birkhoff." (Russian). Ordered sets and Lattices, No. 4 (Russian), pp. 113-123,135. Izdat. Saratov Univ, Saratov. MR 58 #403.

Solnecnyi, E.M. (1976) "The theorems of Fatou and Lebesgue for sequences and systems of finite subsets in lattices." (Russian). Izv. Vyss. Ucebn. Zave. Matematika, no. 10(173),74-78. MR 577.

Spector, C. (1955) "Recursive well orderings." J. Symbolic Logic 206, 151-163.

Speed, T.P. (1972) "On the order of prime ideals." Algebra Universalis 2, 85-87. MR 46 #5188.

Speed, T.P. (1972) "Profinite posets." Bull. Austral. Math. Soc. 6, 177-183.

Spencer, J. (1971) "Minimal scrambling sets of simple orders." Acta Math. Acad. Sci. Hungar. 22, 349-353.

Spencer, J.H. (1974) "A generalized Rota conjecture for partitions." Studies in Appl. Math. 53, 239-241.

Sperner, E. (1928) "Ein Satz uber Untermengen einer endlichen Menge." Math. Z. 27, 544-548.

Sprindzuk, V.G. (1963) "Zum Dilworthschen Satz uber teilweise geordnete Mengen." Doklady Akad. Nauk BSSR 7, 803-804 (Russian).

Stadler, W. (1976) "Sufficient conditions for preference optimality." Differential game issue. J. Optimization Theory Appl. 18, no. 1, 119-140. MR 2446.

Stanley, Richard P. (1970) "Structure of incidence algebras and their automorphism groups." Bull. Amer. Math. Soc. 76, 1236-1239. MR 41 #8319.

Stanley, Richard P. (1973) "A Brylawski decomposition for finite ordered sets." Discrete Math. 4, 77-82. MR 46 #8918.

Stanley, Richard P. (1974) "Enumeration of posets generated by disjoint unions and ordinal sums." Proc. Amer. Math. Soc. 45, 295-299. MR 50 #4416.

Starke, Peter H. (1972) "Uber die Minimisierung von Automaten mit halbeordnetem Ausgabealphabet." (Russian, English and French summaries). Wiss. Z. Humboldt-Univ. Berlin Math.-Natur. Reihe 21, 531-533. MR 48 #10719.

Stearns, R.E.; Hartmanis, J. (1961) "On the state assignment problem for sequential machines. II." IRE Trans. EC-10, 593-603. MR 25 #5822.

Stearns, Richard. (1959) "The voting problem." Amer. Math. Monthly 66, 761-763. MR 21 #7799.

Steckin, B.S. (1977) "Binary functions on ordered sets." (inversion theorems). (Russian). Trudy Mat. Inst. Steklov. 143, 178-187,211. MR 58 #21660.

Stein, James D, Jr. (1970) "Hereditary and recurrent properties in lattices." J. Combinat. Theory 9, 103-107.

Stone, A.H. (1968) "On partitioning ordered sets into cofinal subsets." Mathematika 15, 217-222. MR 38 #5674.

Stone, Harold S. (1977) "Multiprocessor scheduling with the aid of network flow algorithms." IEEE Trans. Software Engrg. SE-3, no. 1, 85-93. MR 58 #4286.

Stone, M.G. (1979) "Note "On automorphisms of posets"." Discrete Math.

Stong, R.E. (1966) "Finite topological spaces." Trans. Amer. Math. Soc. 123,325-340.

Stratigopoulos, Demetrios; Stabakis, Jean (1977) "Sur les ensembles ordonnes." (f*-coupes). (English summary). C.R. Acad. Sci. Paris Ser. A-B 285, no. 3, A81-A84. MR 128.

Sturm, Teo (1967) "The isotonic extension of isotonic mappings." (Czech. Russian and German summaries). Casopis Pest. Mat. 92, 318-331. MR 36 #3685.

Sturm, Teo (1973) "A categorical conception of the theory of kernels of isotonic mappings." Fourth Sci. Conf., Coll. Transp. Eng. Zilina, Sect. 1; Mathematics, Physics, Cybernetics pp. 117-120. Vysoka Skola Dopravna v Ziline, Zilina. MR 52.

Sturm, Teo (1975) "Congruences in ordered sets." (Czech. English and Russian summaries). Kniznice Odborn. Ved. Spisu Vysoke. Uceni Tech. v Brne B-56, 131-135. MR 53 #5384.

Su, Zaw Sing; Sevcik, Kenneth C. (1978) "A combinatorial approach to dynamic scheduling problems." Oper. Res. 26, no. 5, 836-844. MR 58 #20392.

Summer (1969) "Summer session on the theory of ordered sets and general algebra." (Cikhaj, 1969). Faculty of Science, University of J.E. Purkyne, Brno, 126 pp. MR 42 #9.

Suszko, Roman (1977) "On filters and closure systems." Polish Acad. Sci. Inst. Philos. Sociol. Bull. Sect. Logic 6, no. 4, 151-155. MR 57 #16149.

Suzumura, Kotaro (1977) "Houthakker's axiom in the theory of rational choice." J. Econom. Theory 14, no. 2, 284-290. MR 55 #14093.

Swevast'janov, S.V. (1975) "The asymptotic approach to certain problems of scheduling theory." (Russian). Optimal processes (Proc. Third All-Union Conf. Problems of Theoretical Cybernetics, Novosibirsk, 1974). Upravljaemye Systemy Vyp. 14, 40-51,75. MR 58 .

Szasz, G. (1961) "Translationen der Verbande." (Czech and Russian summaries). Acta. Fac. Nat. Univ. Comenian. 5 449-453. MR 24 #A2542.

Szmielew, Wanda (1977) "Oriented and nonoriented linear orders." (Russian summary). Bull. Acad. Polon. Sci. Ser. Sci. Math. Astronom. Phys. 25, no. 7, 659-665. MR 3010a,3010b.

Szpilrajn, E. (1930) "Sur l'extension de l'ordre partiel." Fund. Math. 16, 386-389.

Szware, W. (1962) "The transportation problem." (Polish, Russian and English summaries). Zastos. Mat. 6, 149-187. MR 33 #2414.

Takahashi, Hiromitsu; Ozaki, Hiroshi (1972) "Generation of acyclic graphs from an undirected graph." Systems-Computers-Controls 2 (1971), no. 2, 34-40. MR 55 #2393.

Tamura, T. (1964) "Operations on binary relations and their applications." Bull. Amer. Math. Soc. 70, 113-120. MR 28 #3108.

Tamura, Takayuki (1956) "On a special semilattice with a minimal condition." J. Gakugei Tokushima Univ. Nat. Sci. Math. 7, 9-17. MR 19 #243.

Tanaev, V.S.; Skurba, V.V. (1975) "Introduction to scheduling theory." (Russian). Library of Mathematical Economics. Izdat. "Nauka", Moscow, 256pp. MR 58 #32904.

Tarski, A. (1955) "A lattice theoretical fixpoint theorem and its applications." Pacific J. Math. 5, 285-309.

Taskovic, M.R. (1980) "Partially ordered sets and some fixed point theorems." Publ. de L'Inst. Math. 27 (41), 241-247.

Teh, H.H. (1962) "A note on Hasse diagrams." Bull. Math. Soc. Nanyang Univ. 15-20. MR 27 #71.

Teh, H.H. (1966/67) "On a class of distributive lattice orders." Nanta Math. 1, 19-22. MR 37 #111.

Thrall, R.M. (1954) "Decision processes." edited by R.M. Thrall, C.H. Coombs and R.L. Davis. John Wiley and Sons, Inc, New York; Chapman and Hall, Ltd, London, viii+332 pp. MR 16 #605.

Tjutjukin, V.K. (1976) "Optimal two-operation scheduling in the case of parallel-serial production motion." (Russian). Primenen. Mat. Ekonom. No. 11, 60-69,140. MR 58 #4289.

Todorov, Kalco Z. (1972) "Relations that define an order for semigroups of transformations of a set with a finite chain." (Russian. English summary). Annuaire Univ. Sofia Fac. Math. 65 (1970/71), 225-234. (errata insert). MR 48 #198.

Tokarz, Marek (1975) "Functions definable in Sugihara algebras and their fragments. I." Studia Logica 34, no. 4,295-304. MR 97a.

Tokarz, Marek (1976) "Functions definable in Sugihara algebras and their fragments. II." Studia Logica 35, no. 3, 279-283. MR 97b.

Tomescu, Ioan (1972) "Ordered algebraic structures in the theory of graphs." (Romanian, English summary). Stud. Cere. Mat. 24, 469-476. MR 4.

Trebel, Michel (1971) "Incompatibilite du produit direct et du produit contracte d'ensembles ordonnes finis possedant un element nul et un element universel." C.R. Acad. Sci, Paris, Ser. A 273, 760-762. MR 06001.

Trotter, W.T. (1976) "A forbidden subposet characterization of an order-dimension inequality." Math. Systems Theory 10, 91-96. MR 57856.

Trotter, W.T. (1978) "Combinatorial problems in dimension theory for partially ordered sets." Problemes combinatoires et theorie des graphes, Colloq. int. CNRS No. 260, 403-406.

Trotter, W.T. (1978) "Some combinatorial problems for permutations." Congressus Numerantium 19, 619-632.

Trotter, W.T. (1981) "Stacks and splits of partially ordered sets." Discrete Math. 35, 229-256.

Trotter, W.T.; Bogart, K.P. (1976) "Maximal dimensional partially ordered sets III. A characterization of Hiraguchi's inequality for interval dimension." Discrete Math. 15, 389-400.

Trotter, W.T.; Bogart, K.P. (1976) "On the complexity of posets." Discrete Math. 16, 71-82. MR 52553.

Trotter, W.T.; Moore, J.I. (1977) "The dimension of planar posets." J. Combinatorial Theory (Ser. B) 22, 54-67. MR 57857.

Trotter, W.T.; Moore, J.I.; Sumner, D.P. (1976) "The dimension of a comparability graph." Proc. Amer. Math. Soc. 60, 35-38. MR 55062.

Trotter, William T. Jr. (1974) "Dimension of the crown Snk." Discrete Math. 8, 85-103. MR 49 #158.

Trotter, William T. Jr. (1974) "Irreducible posets with large height exist." J. Combinatorial Theory Ser. A 17, 337-344. MR 50 #6935.

Trotter, William T. Jr. (1975) "A note on Dilworth's embedding theorem." Proc. Amer. Math. Soc. 52, 33-39. MR 51.

Trotter, William T. Jr. (1975) "Embedding finite posets in cubes." Discrete Math. 12, 165-172. MR 51 #5426.

Trotter, William T. Jr. (1975) "Inequalities in dimension theory for posets." Proc. Amer. Math. Soc. 47, 311-316. MR 51 #5427.

Trotter, William T. Jr. (1976) "A generalization of Hiraguchi's inequality for posets." J. Combinatorial Theory Ser. A 20, no. 1, 114-123. MR 52 #10515.

Trotter, William T. Jr.; Moore, John I., Jr. (1976) "Characterization problems for graphs, partially ordered sets, lattices, and families of sets." Discrete Math. 16, no. 4, 361-381. MR 8437.

Trotter, William T. Jr.; Moore, John I., Jr. (1976) "Some theorems on graphs and posets." Discrete Math. 15, 79-84.

Tunaer, V.S. (1979) "Optimal decomposition of a partially ordered set into chains." Doklady Akad. Nauk BSSR 23, 389-391.

Tunnicliffe, W.R. (1974) "A criterion for the existence of an extension of a function on a partially ordered set." J. London Math. Soc. (2) 8, 352-354. MR 51 #5428.

Tunnicliffe, W.R. (1974) "The completion of a partially ordered set with respect to a polarization." Proc. London Math. Soc. (3) 28, 13-27. MR 48 #10914.

Tverberg, Helge (1967) "On Dilworth's decomposition theorem for partially ordered sets." J. Combinatorial Theory 3, 305-306. MR 35 #5366.

Tversky, A. (1969) "The intransitivity of preferences." Psychol. Review, 76, 31-48.

Tversky, Amos (1967) "A general theory of polynomial conjoint measurement." J. Mathematical Psychology 4, 1-20. MR 34 #5531.

Ullman, J.D. (1975) "NP-complete scheduling problems." J. Comput. System Sci. 10, 384-393.

Uryson, P. (1923,1924) "Un theoreme sur la puissance des ensembles ordonnes." Fund. Math. 5, 14-19; Fund. Math. 6, 278.

Vagner, V.V. (1978) "Introrestrictive factored and bifactored ordered sets." [Russian] Izv. Vyss. Ucebn. Zaved. Matematika no. 6 (193), 36-50. MR 80i #06008.

Vagner, V.V. (1978) "Introrestrictive factored sets and projective systems of sets." [Russian] Izv. Vyss. Ucebn. Zaved. Matematika no. 12, 18-32. MR 80d #06005.

Vaidyanathaswamy, R. (1944) "Partially ordered sets." Math. Student 12, 1-6 [MF11846], 7-14 [MF11847], 14-24 [MF11848]. MR 6 #143.

Vaidyanathaswamy, R. (1947) "Treatise on set topology. Part I." Indian Mathematical Soc, Madras, vi + 306 pp. MR 367.

Valdes, J.; Tarjan, R.E.; Lawler, E.L. (1981) "The recognition of series parallel digraphs." SIAM J. Comput.

Varecza, Arpad; Varecza, Laszlo (1974) "The number of chains in a finite Boolean algebra." (Hungarian. English summary). Acta Acad. Paedagog. Civitate Pecs Ser. 6 Math. Phys. Chem. Tech. 18, 13-19. MR 5309.

Varouchakis, Nicolas Chr. (1973) "Sur le complete de Kurepa, generalise par L. Dokas." C.R. Acad. Sci., Paris, Ser. A 276, 891-892. MR 06002.

Varouchakis, Nicolas Chr. (1974) "Sur le complete de Kurepa d'un ensemble partiellement ordonne. espaces metroides et ultrametroides." Math. Balkanica 4, 649-655. MR 51 #10175.

Vaught, R. (1961) "Denumerable models of complete theories." Infinitistic Methods, Pergamon, London.

Vazenin, Ju. M (1970) "Ordered sets and their inf-endomorphisms." (Russian). Mat. Zametki 7, 339-347. MR 41 #6729.

Vazenin, Ju. m. (1969/70) "Semigroups of inf-endomorphisms of ordered sets." (Russian). Ural. Gos. Univ. Mat. Zap. 7, tetrad' 1, 23-34. MR 43 #6345.

Vazenin, Ju. M. (1970) "Elementary properties of semigroups of transformations of ordered sets." (Russian). Algebra i Logika 9, 281-301. MR 44 #344.

Venkatanarasimhan, P.V. (1970) "Semi-ideals in posets." Math. Ann. 185, 338-348. MR 41 #3328.

Venkatanarasimhan, P.V. (1971) "Pseudo-complements in posets." Proc. Amer. Math. Soc. 28, 9-17. MR 42 #7568.

Verbeek, L.A.M. (1967) "Congruence separation of subsets of a monoid with applications to automata." Math Systems Theory 1, 315-324. MR 36 #6235.

Vesely, Vitezslav; Kopecek, Ivan (1976) "Monomorphisms in the categories of connected algebraic systems with strong homomorphisms." Arch. Math. (Brno) 12, no. 3,153-168. MR 2882.

Vilkas, E. (1973) "Compatibility of coalition orderings." (Russian, English summary). Advances in Game Theory (Proc. Second All-Union Conf. on Game Theory, Vilnius, 1971) (Russian), pp. 83-92. Izdat. "Mintis", Vilnius. MR 51 #2662.

Vincke, P. (1978) "Quasi-ordres generalises et representation numerique." Math Sci. Hum., 62, 35-60.

Vinokurov, V.G. (1963) "On the topological properties of partially ordered sets." (Russian). Uspehi Mat. Nauk 18, no. 5 (113), 151-155. MR 28 #42.

Vol'pert, A.Ja. (1969) "The limit of a multi-valued function that is defined on a partially ordered set." (Russian). Kazan. Gos. Univ. Ucen. Zap. 129, no. 3, 76-88. MR 7502.

Vorob'ev, N.N. (1967) "The problem of endomorphisms of complete semi-lattices of sets." (Russian). Theory of Games: Positional Games (Russian), pp. 183-186. Izdat. "Nauka", Moscow. MR 36 #7421.

Votjakov, A.A. (1972) "Integer programming, comparison of truncations." (Russian). Ekonom. i Mat. Metody 8, 107-116. MR 45 #3072.

Waligorski, Stanislaw (1969) "Algebraic theory of automata." (Polish, English and Russian summaries). Algorytmy 6, no. 11, 131 pp. MR 41 #3190.

Wallace, A.D. (1954) "Partial order and indecomposability." Proc. Amer. Math. Soc. 5, 780-781. MR 16 #275.

Wallman, H. (1938) "Lattices and topological spaces." Ann. of Math. 39, 112-126.

Ward, A.S. (1955) "On relations between certain intrinsic topologies in certain partially ordered sets." Proc. Camb. Phil. Soc. 51, 254-261.

Ward, L.E. (1954) "Partially ordered topological spaces." Proc. Amer. Math. Soc. 5, 144-161.

Ward, L.E. (1965) "Concerning Koch's Theorem on the existence of arcs." Pac. J. Math. 15, 347-355.

Ward, L.E. (1965) "On a conjecture of R.J. Koch." Pac. J. Math. 15, 1429-1433.

Ward, Martin (1970) "Totally unordered subsets of a partially well ordered set." Bull. Austral. Math. Soc. 2, 179-182. MR 41 #3329.

Watnick, R. (1981) "Constructive and recursive scattered order types." Logic Year 1979-80, M. Lerman, J. Schmerl, and R. Soare, eds., Lecture Notes in Mathematics 859, Springer-Verlag, Berlin.

Weck, S. (1981) "Scott convergence and Scott topology in partially ordered sets, I." Lect. Notes Math. 871, Springer, 372-383.

Weglarz, Jan (1976) "Time-optimal control of resource allocation in a complex of operations framework." IEEE Trans. Systems Man Cybernet. SMC-6, no. 11, 783-788. MR 4803.

Wehrung, Donald A. (1978) "Interactive identification and optimization using a binary preference relation." Operations Res. 26, no. 2,322-332. MR 58 #20497.

Wesley, Eugene (1976) "Borel preference orders in markets with a continuum of traders." J. Math. Econom. 3, no. 2, 155-165. MR 55 #11955.

Wesolowski, Tadeusz (1978) "A certain characterization of the supremum in a partially ordered set." (Polish. English summary). Zeszyty Nauk. Wyz. Szkoly Ped. w Opolu Mat. 20 Algebra, Zastos. Mat, 77-80. MR 58 #27494.

West, D.B. (1980) "A symmetric chain decomposition of L(4,n)." Europ. J. Combinatorics 1, 379-383.

West, D.B.; Kleitman, D.J. (1979) "Skew chain orders and sets of rectangles." Disc. Math. 27, 99-102.

White, D.E.; Williamson, S.G. (1977) "Recursive matching algorithms and linear orders on the subset lattice." J. Comb. Theory A 23, 117-127.

Wiegandt, Richard (1959) "On the general theory of Mobius inversion formula and Mobius product." Acta Sci. Math. Szeged 20, 164-180. MR 21 #7176.

Wiener, N. (1912-1914) "Contribution to the theory of relative position." Proc. Cambridge Philos. Soc., 17, 441-449.

Wilf, Herbert S. (1977) "A unified setting for sequencing, ranking, and selection algorithms for combinatorial objects." Advances in Math. 24, no. 3,281-291. MR 149.

Wille, R. (1973) "Uber modulare Verbande die von einer endlichen halbgeordneten Menge frei erzeugt werden." Math. Z., 131, 241-249.

Wille, R. (1974) "On modular lattices of order dimension two." Proc. Amer. Soc. 43, 287-292. MR 48327.

Wille, R. (1977) "On lattices freely generated by finite partially ordered sets." Contributions to universal algebra, pp.581-593. Colloq. Math. Soc. Janos Bolyai, Vol.17, North-Holland, Amsterdam. MR 58 #10642.

Wille, R. (1980) "Cancellation and refinement properties for function lattices." Houston J. Math. 6, 431-437.

Wille, Rudolf (1975) "Note on the order dimension of partially ordered sets." Albegra Universalis 5, no. 3, 443-444. MR 52 #13536.

Williams, Mary B. (1970) "Deducing the consequences of evolution: a mathematical model." J. Theoret. Biol. 29, no. 3, 343-385. MR 43 #4494.

Williamson, S.G. (1976) "Ranking algorithms for lists of partitions." SIAM J. Comput. 5, no. 4, 602-617. MR 5311.

Williamson, S.G. (1977) "On the ordering, ranking, and random generation of basic combinatorial sets." Combinatoire et Representation du Groupe Symetrique, pp. 309-339. Lecture Notes in Math, Vol. 579, Springer, Berlin. MR 58 #10477.

Wolk, E.S. (1956) "Some representation theorems for partially ordered sets." Proc. Amer. Math. Soc. 7, 589-594. MR 18 #635.

Wolk, E.S. (1957) "Dedekind completeness and a fixed-point theorem." Canad. J. Math. 9, 400-405. MR 19 #243.

Wolk, E.S. (1958) "Order-compatible topologies on a partially ordered set." Proc. Amer. Math. Soc. 9, 514-529. MR 20 #3079.

Wolk, E.S. (1962) "The comparability graph of a tree." Proc. Amer. Math. Soc. 13, 789-795.

Wolk, E.S. (1964) "On decompositions of partially ordered sets." Proc. Amer. Math. Soc. 15, 197-199. MR 28 #2981.

Wolk, E.S. (1965) "A note on the comparability graph of a tree." Proc. Amer. Math. Soc. 16, 17-20.

Wolk, E.S. (1967) "Partially well ordered sets and partial ordinals." Fund. Math. 60, 175-186. MR 35 #93.

Wolk, E.S. (1968) "The topology of a partially well ordered set." Fund. Math. 62, 255-264. MR 37 #2637.

Wolk, E.S. (1969) "A characterization of partial order." (Loose Russian summary). Bull. Acad. Polon. Sci. Ser. Sci. Math. Astronom. Phys. 17, 207-208. MR 40 #5506.

Wong, James S.W. (1967) "Common fixed points of commuting monotone mappings." Canad. J. Math. 19, 617-620. MR 35 #1513.

Wright, J.B.; Wagner, E.G.; Thatcher, J.W. (1977) "A uniform approach to inductive posets and inductive closure." Mathematical Foundations of Computer Science.pp.192-212, Lecture Notes in Comput. Sci, Vol. 53, Springer, Berlin. MR 58 #10632.

Wright, J.B.; Wagner, E.G.; Thatcher, J.W. (1978) "A uniform approach to inductive posets and inductive closure." Theoret. Comput. Sci. 7, no. 1,57-77. MR 58 #404.

Wyler, O. (1981) "Dedekind complete posets and Scott topologies." Lect. Notes Math. 871, Springer, 384-389.

Yamamoto, K. (1954) "Logarithmic order of free distributive lattices." J. Math. Soc. Japan 6, 343-353.

Yeh, Raymond T. (1968) "Generalized pair algebra with applications to automata theory." J. Assoc. Comput. Mach. 15, 304-316. MR 38 #5532.

Yoeli, M. (1961) "The cascade decomposition of sequential machines." IRE Trans. EC-10, 587-592. MR 24 #B873.

Zak, Yu. A. [Zak, Ju.O.] (1978) "Certain properties of scheduling problems." Avtomat. i Telemeh. 1978, no. 1,123-132 (Russian. English summary); trans: Automat. Remote Control 39, no. 1, pt. 2, 99-107. MR 58 #20395.

Zaks, I.B. (1974) "Completion of ordered sets." (Russian). Ordered sets and Lattices, No. 2 (Russian), pp. 28-31. Izdat. Saratov Univ, Saratov. MR 58 #405.

Zapletal, Josef (1973) "Distinguishing subsets in semilattices." Arch Math. (Brno) 9, 73-82. MR 51 #767.

Zavadskii, A.G. (1977) "Differentiation with respect to a pair of places." (Russian). Matrix Problems (Russian), pp.115-121. Akad. Nauk Ukrain. SSR Inst. Mat, Kiev. MR 58 #21874a.

Zavadskii, A.G.; Nazarova, L.A. (1977) "Partially ordered sets of same type." (Russian). Matrix Problems (Russian), pp. 122-143. Akad. Nauk Ukrain. SSR Inst. Mat, Kiev. MR 58 #21874b.

Zeeman, E.C. (1962) "The topology of the brain and visual perception." Topology of 3-manifolds and related topics (Proc. The Univ. of Georgia Institute, 1961), pp. 240-256. Prentice-Hall, Englewood Cliffs, N.J. MR 25 #3796.

Zeigler, Bernard P. (1974/75) "Maximally independent sets, minimal generating sets, and irreducibles." Math. Systems Theory 8, no. 2, 176-181. MR 52 #208.

Zelinka, Bohdan (1970) "Les graphes de Hasse des modeles analytique de langages." Bull. Math. Soc. Sci. Math. R.S. Roumanie (N.S.) 13 (61) (1969), no. 3, 389-392. MR 43 #7274.

Zinder, Ja. A.; Podcasova, T.P. (1976) "Minimization of ordered criteria in a single-stage deterministic service system." (Russian). Automated Control Systems and Data Processing (Russ) pp. 20-25. Akad. Nauk Ukrain. SSR Inst. Kibernet, Kiev. MR 58 #4266.

Zizovic, M.R. (1974) "Some topologies on a poset." Math. Balkanica 4, 743-744. MR 52 #220.

Zukovin, V.E.; Kaloev, M.A. (1975) "Multicriterion scaling of qualitative characteristics in systems analysis." (Russian, Georgian and English summaries). Sakharth. SSR Mecn. Akad. Moambe 79, no. 1, 53-56. MR 53 #15316.

Zvjagina, R.A. (1971) "A general method of solution of linear programming problems of block structure." (Russian). Optimizacija Vyp. 1 (18), 22-40. MR 45 #6420@w.

Zvjagina, R.A. (1971) "Linear programming problems with matrices of arbitrary block structure." (Russian). Dokl. Akad. Nauk SSSR 196, 755-758. MR 45 #6419.

Zvjagina, R.A. (1971) "The construction of hierarchic orders with prescribed conditions on the congruence." (Russian). Optimizacija Vyp. 1 (18), 41-54. MR 45 #6409.

Zybina, L.D. (1965) "Aggregates of relations determining the ordering of partial endomorphisms." (Russian). Leningrad. Gos. Ped. Inst. Ucen. Zap. 274, 122-142. MR 34 #110.

Zybina, L.D. (1965) "On relations determining the order of certain endomorphisms of an ordered set." (Russian). Leningrad. Gos. Ped. Inst. Ucen. Zap. 274, 106-121. MR 33 #5526.

Zybina, L.D. (1971) "The orderedness of certain endomorphisms of an ordered set." (Russian). Leningrad. Gos. Ped. Inst. Ucen. Zap. 464, cast 2, 29-39. MR 45 #6705.

Zybina, L.D. (1975) "Relations that determine the ordering in certain ordered semi-groups of idempotents." (Russian). Modern Algebra, No. 3 (Russian), pp. 58-69. Leningrad. Gos. Ped. Inst, Leningrad. MR 208.

Zybina, L.D. (1976) "Semigroups that admit some special stable orderings." (Russian). Modern Algebra, No. 5 (Russian), pp. 57-61. Leningrad. Gos. Ped. Inst. im. Gercena, Leningrad. MR 58 #16453.

Zybina, L.D. (1977) "Orderable semigroups that admit maximal two-element chains." (Russian). Modern Algebra, No. 6 (Russian), pp. 108-113. Leningrad. Gos. Ped. Inst, Leningrad. MR 58 #21890.